CW00705590

Kingston upon Hull City Libraries
WITHDRAWN
FROM STOCK
FOR SALE

ENCYCLOPEDIA OF POLYMER SCIENCE AND TECHNOLOGY

VOLUME 14

Thermogravimetric Analysis
to
Wire and Cable Coverings

Interscience Publishers
a division of John Wiley & Sons, Inc.
New York • London • Sydney • Toronto

ENCYCLOPEDIA OF POLYMER SCIENCE AND TECHNOLOGY

Plastics, Resins, Rubbers, Fibers

VOLUME 14

Thermogravimetric Analysis
to
Wire and Cable Coverings

Library of Congress Catalog Card Number: 64-22188

Printed in the United States of America

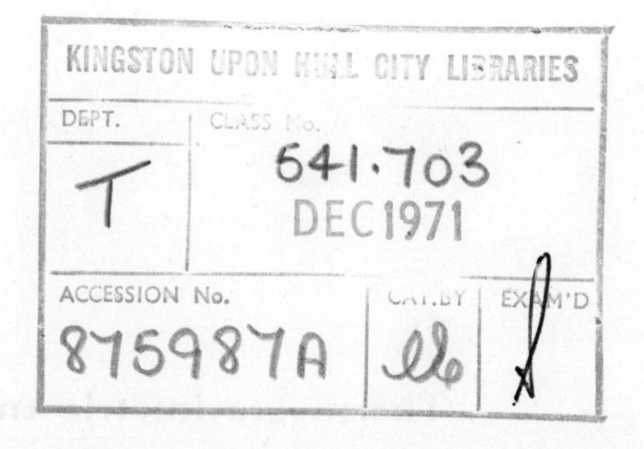

CONTENTS

EDITORIAL STAFF FOR VOLUME 14

Executive Editor: NORBERT M. BIKALES
Staff Editors: GLORIA O. SCHETTY, Editorial Supervisor • JANET PERLMAN • MICHALINA BICKFORD • JOSEPH M. RICCIARDI

CONTRIBUTORS TO VOLUME 14

Kent H. Alverson, *Spaulding Fibre Company, Inc.,* Vulcanized Fiber

J. C. Ambelang, *The Goodyear Tire and Rubber Company,* Tires

T. Anyos, *Stanford Research Institute,* Vinylarene Polymers

Thomas A. Augurt, *Lederle Laboratories, American Cyanamid Company,* Vinylamine Polymers

A. L. Barney, *E. I. du Pont de Nemours & Co., Inc.,* Elastomers under Vinylidene Fluoride Polymers

J. L. Benton, *BP Chemicals International Ltd.,* Polymerization and Properties under Vinyl Chloride Polymers

Norbert M. Bikales, *Consultant,* Polymers under Vinyl Ether Polymers

Bernard R. Bluestein, *Witco Chemical Corporation,* Waxes

C. A. Brighton, *BP Chemicals International Ltd.,* Introduction, Polymerization, Copolymers, Properties, Compounding, Fabrication, Applications, Chlorinated Poly(vinyl Chloride) under Vinyl Chloride Polymers

Joseph E. Clark, *National Bureau of Standards,* Weathering

Fredric S. Cohen, *Polaroid Corporation,* Vinyl Fluoride Polymers

A. B. Conciatori, *Celanese Research Company,* Vinylidene Cyanide Polymers

William H. Daly, *Louisiana State University,* Vinylcycloalkane Polymers; Vinyl Ketone Polymers

J. E. Dohany, *Pennwalt Corporation,* Plastics under Vinylidene Fluoride Polymers

A. A. Dukert, *Pennwalt Corporation,* Plastics under Vinylidene Fluoride Polymers

J. P. Dux, *American Viscose Division, FMC Corporation,* Fibers under Vinyl Chloride Polymers

F. G. Edwards, *The Dow Chemical Company,* Vinylidene Chloride Polymers

John D. Ferry, *University of Wisconsin,* Viscoelasticity

J. K. Gillham, *Princeton University,* Torsional Braid Analysis

G. Hardy, *Research Institute for the Plastics Industry (Hungary),* Vinylene Carbonate Polymers

J. Heller, *Pharmetrics, Inc.,* Vinylarene Polymers

J. B. Kinsinger, *Michigan State University,* Viscometry

John Kozacki, *Cadillac Plastic and Chemical Company,* Tubing

Paul Kraft, *Stauffer Chemical Company,* Vinyl Fluoride Polymers

Norman Kudisch, *Witco Chemical Corporation,* Waxes

G. R. Lappin, *Tennessee Eastman Company,* Ultraviolet-Radiation Absorbers

David W. Levi, *Picatinny Arsenal,* Thermogravimetric Analysis

Martin K. Lindemann, *Chas. S. Tanner Co.,* Vinyl Alcohol Polymers

David C. Lini, *ARCO Chemical Company,* Vinyl Bromide Polymers

Donald H. Lorenz, *GAF Corporation,* *N*-Vinyl Amide Polymers; Monomers under Vinyl Ether Polymers

G. C. Marks, *BP Chemicals International Ltd.,* Properties under Vinyl Chloride Polymers

Lloyd W. Myers, *Western Electric Company,* Wire and Cable Coverings

Seizo Okamura, *Kyoto University,* Vinylcarbazole Polymers; Vinylpyridine Polymers

Harold E. Parker, *Spaulding Fibre Company, Inc.,* Vulcanized Fiber

Richard B. Peterson, *Ren Plastics, Inc.,* Tooling with Plastics

George E. Pickering, *Arthur D. Little, Inc.,* Toys

Richard A. Preibisch, *Spaulding Fibre Company, Inc.,* Vulcanized Fiber

S. S. Preston III, *Pennwalt Corporation,* Plastics under Vinylidene Fluoride Polymers

Kermit C. Ramey, *ARCO Chemical Company,* Vinyl Bromide Polymers

Leo Reich, *Picatinny Arsenal,* Thermogravimetric analysis

T. A. Riehl, *The Goodyear Tire and Rubber Company,* Tires

Jeffrey R. Sherry, *Branson Sonic Power Company,* Ultrasonic Fabrication

R. W. Stackman, *Celanese Research Company,* Vinylidene Cyanide Polymers

Shigeo Tazuke, *Kyoto University,* Vinylcarbazole Polymers; Vinylpyridine Polymers

D. C. Thompson, *E. I. du Pont de Nemours & Co., Inc.,* Elastomers under Vinylidene Fluoride Polymers

L. E. Trapasso, *Celanese Research Company,* Vinylidene Cyanide Polymers

R. A. Wessling, *The Dow Chemical Company,* Vinylidene Chloride Polymers

Robley C. Williams, Jr., *Yale University,* Ultracentrifugation

James R. Wolfe, Jr., *E. I. du Pont de Nemours & Co., Inc.,* Vulcanization

David A. Yphantis, *University of Connecticut,* Ultracentrifugation

ABBREVIATIONS AND SYMBOLS

a	exponent in empirical relationship between intrinsic viscosity and molecular weight
A	ampere(s)
A	anion (eg, HA)
A_1, A_2, A_3	coefficients in virial expansion of osmotic pressure as power series in concentration
Å	angstrom unit(s)
AATCC	American Association of Textile Chemists and Colorists
abs	absolute
ac	alternating current
ac-	alicyclic (eg, *ac*-derivatives of tetrahydronaphthalene)
ACS	American Chemical Society
addn	addition
AIChE	American Institute of Chemical Engineers
AIP	American Institute of Physics
alc	alcohol(ic)
alk	alkaline (not alkali)
Alk	alkyl
-alt-	alternating, as in alternating copolymer
anhyd	anhydrous
ANSI	American National Standards Institute, Inc.
approx	approximate(ly)
aq	aqueous
ar-	aromatic (eg, *ar*-vinylaniline)
Ar	aryl
as-	asymmetric(al) (eg, *as*-trichlorobenzene)
ASA	American Standards Association
ASME	American Society of Mechanical Engineers
ASTM	American Society for Testing and Materials
atm	atmosphere(s), atmospheric
at. no.	atomic number
at. wt.	atomic weight
av	average
bbl	barrel(s)
Bé	Baumé
Bhn	Brinell hardness number
bp (as in bp_{11})	boiling point
Btu	British thermal unit(s)
C	catalyst; Celsius (centigrade); coulomb(s)
C-	denoting attachment to carbon (eg, *C*-acetylindole)
C_m	thermodynamic constant of Flory-Huggins dilute-solution theory
C_M	chain-transfer constant for monomer
C_P	chain-transfer constant for polymer
C_S	chain-transfer constant for solvent
ca	circa, approximately
cal	calorie(s)
calcd	calculated
cfm	cubic foot (feet) per minute
Ci	Curie(s)
CI	Colour Index (number)
cm	centimeter(s)
-co-	copolymerized with
coeff	coefficient
compd, cpd	compound (noun)
concd	concentrated
concn	concentration
cond	conductivity
const	constant
cor	corrected

cP	centipoise(s)
cpd, compd	compound (noun)
cps	cycles per second
crit	critical
cryst	crystalline
cSt	centistokes
cu	cubic
d, ρ	density (conveniently, specific gravity)
d	differential operator
d-	*dextro*-, dextrorotatory
D-	denoting configurational relationship (as to *dextro*-glyceraldehyde)
D	Debye(s)
dc	direct current
dec, decomp	decompose(s)
den	denier
den/fil	denier per filament
deriv	derivative
diam	diameter
dielec	dielectric (adj.)
dil	dilute
distd	distilled
dl	deciliter
dl-, DL	racemic
DOT	Department of Transportation
DP	degree of polymerization
$\overline{DP}$	average degree of polymerization
DS	degree of substitution
DTA	differential thermal analysis
dyn	dyne(s)
e	base of natural logarithms; electron; polarity factor in Alfrey-Price equation
E	Young's modulus of elasticity
ed.	edited, edition, editor
eg	for example
elec	electric(al)
emf	electromotive force
equil	equilibrium(s)
equiv	equivalent
esp	especially

ESR	electron-spin resonance
est(d)	estimate(d)
estn	estimation
esu	electrostatic unit(s)
eu	entropy unit(s)
eV	electron volt(s)
expt(l)	experiment(al)
f	frictional coefficient
f_1, f_2	mole fractions in monomer feed
F	Fahrenheit; farad(s)
F	faraday constant
F_1, F_2	mole fractions in copolymer
Fed, fedl	federal (eg, Fed Spec)
fl oz	fluid ounce(s)
fob	free on board
fp	freezing point
ft	foot (feet)
ft-lb	foot-pound(s)
g	gram(s)
g	gravitational acceleration
G	gauss
G	Gibbs free energy
gal	gallon(s)
g/den	gram(s) per denier
gem-	geminal (attached to the same atom)
g-mol	gram-molecular (as in g-mol wt)
g-mole	gram-mole
H	parameter relating turbidity–concentration ratio to molecular weight
hp	horsepower
hr	hour(s)
hyd	hydrated, hydrous
Hz	Hertz; cycles/sec
i, insol	insoluble
^{i}(eg, Pr^{i})	iso (eg, isopropyl)
i-	inactive (eg, *i*-methionine); iso (eg, *i*-propyl)
I	initiator
I_0	intensity of incident light
I_{abs}	intensity of absorbed light
ICC	Interstate Commerce Commission
ID	inner diameter
ie	that is

in.	inch(es)
insol, i	insoluble
IR	infrared
IUPAC	International Union of Pure and Applied Chemistry
J	joule(s)
J	elastic compliance
k	reaction rate constant
K	Kelvin
K	dissociation constant; constant in intrinsic viscosity–concentration relationship
kc	kilocycle(s)
kcal	kilogram-calorie(s)
keV	kilo electron volt(s)
kg	kilogram(s)
kV	kilovolt(s)
l	liter(s)
l-	*levo*-, levorotatory
L-	denoting configurational relationship (as to *levo*-glyceraldehyde)
lb	pound(s)
LD_{50}	dose lethal to 50% of the animals tested
liq	liquid
ln	logarithm (natural)
log	logarithm (common)
m	meter(s)
m-	meta (eg, *m*-xylene)
M	metal
M_1,M_2	monomers
M	molar (as applied to concentration; not molal); molecular weight
M_c	molecular weight per crosslink unit
$\bar{M}_n$	number-average molecular weight
$\bar{M}_v$	viscosity-average molecular weight
$\bar{M}_w$	weight-average molecular weight
$\bar{M}_z$	*z*-average molecular weight
max	maximum
Mc	megacycle
MCA	Manufacturing Chemists' Association
mcal	millicalorie(s)
meq	milliequivalent(s)
MeV	million electron volt(s)
mg	milligram(s)
min	minimum; minute(s)
misc	miscellaneous
mixt	mixture
ml	milliliter(s)
MLD	minimum lethal dose
mm	millimeter(s)
mM	millimole(s)
mo(s)	month(s)
mol	molecule, molecular
mol wt, *M*, mw	molecular weight
mp	melting point
mV	millivolt(s)
mw, *M*, mol wt	molecular weight
mμ	millimicron(s)
^{n}(eg, Bu^n), *n*-	normal (eg, normal butyl)
n (as n_D^{20})	index of refraction (for 20°C and sodium light)
n-,^{n}	normal (eg, *n*-butyl)
n	number of mers in polymer (as, $\leftarrow CH_2CH_2 \rightarrow_n$)
N	normal (as applied to concentration)
N-	denoting attachment to nitrogen (eg, *N*-methylaniline)
neg	negative (adj.)
NEMA	National Electrical Manufacturers' Association
NF	*National Formulary* (American Pharmaceutical Association)
NMR	nuclear magnetic resonance
no.	number
o-	ortho (eg, *o*-xylene)
O-	denoting attachment to oxygen (eg, *O*-acetylhydroxylamine)
OD	outer diameter

Oe	oersted(s)
owf	on weight of fiber
owg	on weight of goods
oz	ounce(s)
p-	para (eg, *p*-xylene)
P	poise(s)
P	reactivity of radical in Alfrey-Price equation
$P(\theta)$	angular light-scattering function
phr	parts per hundred of rubber or resin
pos	positive (adj.)
ppm	parts per million
ppt(d)	precipitate(d)
pptn	precipitation
prepd	prepared
prepn	preparation
psi	pound(s) per square inch
psia	pound(s) per square inch absolute
psig	pound(s) per square inch gage
pt(s)	part(s)
Q	reactivity of monomer in Alfrey-Price equation
qual	qualitative
quant	quantitative
qv	which see (quod vide)
r	roentgen(s)
r_1, r_2	monomer reactivity ratios
R	univalent hydrocarbon radical (or hydrogen); Rankine
$R_{cell}OH$	cellulose
$R_{st}OH$	starch
R_i	rate of initiation
R_p	rate of propagation; rate of polymerization
R_t	rate of termination
R_θ	Rayleigh scattering ratio at angle θ
rep	roentgen(s) equivalent physical
resp	respectively
rh	relative humidity
rms	root mean square
rpm	revolutions per minute
rps	revolutions per second
s, sol	soluble
^{s}(eg, Bu^s), *sec*-	secondary (eg, secondary butyl)
δ	sedimentation constant
s-, *sym*-	symmetrical (eg, *s*-dichloroethylene)
S-	denoting attachment to sulfur (eg, *S*-methylcysteine)
satd	saturated
SCF	standard cubic foot (feet) (760 mmHg, 60°F)
sec	second(s)
sec-,^{s}	secondary (eg, *sec*-butyl)
SFs	Saybolt Furol second(s)
sl s, sl sol	slightly soluble
sol, s	soluble
soln	solution
sp	specific
SPE	Society of Plastics Engineers
spec	specification
sp gr	specific gravity
SPI	Society of the Plastics Industry
sq	square
St	stokes
STP	standard temperature and pressure (760 mmHg, 0°C)
SUs	Saybolt Universal second(s)
sym-, *s*-	symmetrical (eg, *sym*-dichloroethylene)
t, temp	temperature
^{t}(eg, Bu^t), *t*-, *tert*-	tertiary (eg, tertiary butyl)
t	time, efflux time
t-, *tert*-,^{t}	tertiary (eg, *t*-butyl)
T_g	glass-transition temperature
T_m	crystalline melting point
TAPPI	Technical Association of the Pulp and Paper Industry
tech	technical
temp, t	temperature

tert-, *t-*,[t]	tertiary (eg, *tert*-butyl)
TGA	thermogravimetric analysis
theoret	theoretical
torr	mmHg
USP	(*The*) *United States Pharmacopeia* (Mack Publishing Co., Easton, Pa.)
UV	ultraviolet
V	volt(s)
v-, *vic-*	vicinal (attached to adjacent atoms)
vol	volume(s) (not volatile)
v s, v sol	very soluble
vs	versus
W	watt(s)
x	number of structural units in polymer molecule
$\bar{x}_n$	number-average degree of polymerization
$\bar{x}_w$	weight-average degree of polymerization
yd	yard(s)
yr	year(s)
z	dissymmetry of scattered light
α	first in a series
Γ	coefficient in virial expansion of osmotic pressure
η	viscosity
$[\eta]$	intrinsic viscosity
η_{inh}	inherent viscosity
η_r	relative viscosity
η_{sp}	specific viscosity
Θ	temperature at which polymer–solvent interactions are zero
π	osmotic pressure
ρ,d	density
τ	turbidity; relaxation
ω	last in a series
Ω	ohm(s)
Ω-cm	ohm-centimeter(s)

Quantities

Some standard abbreviations (prefixes) for very small and very large quantities are as follows:

deci (10^{-1})	d
centi (10^{-2})	c
milli (10^{-3})	m
micro (10^{-6})	μ
nano (10^{-9})	n
pico (10^{-12})	p
femto (10^{-15})	f
atto (10^{-18})	a
deka (10^{1})	dk
hecto (10^{2})	h
kilo (10^{3})	k
mega (10^{6})	M
giga (10^{9})	G (or B)
tera (10^{12})	T

T continued

THERMOGRAVIMETRIC ANALYSIS

The term "thermobalance" seems to have been introduced by Honda in 1915 to describe an instrument that he constructed which continuously measured weight changes of a substance at gradually varying temperatures (1,2). In 1926, Saito (3) described an improved balance, and in succeeding years various other Japanese workers made further improvements which allowed hundreds of pyrolysis curves to be obtained (2). From 1923 to 1938, French workers, eg, Guichard and his students, developed thermobalance instrumentation and studied the scope and limitations of the method (1,2). In 1943, Chevenard developed a thermobalance, which bears his name, that was both rugged and sensitive enough to be employed in an industrial laboratory. In 1946, Duval and his co-workers began studies of the pyrolysis of analytical precipitates employing the Chevenard thermobalance, and in the following four years they recorded the pyrolysis curves of several hundred inorganic precipitates.

Most of the early work using the thermobalance involved inorganic materials and did not involve estimations of kinetic parameters in pyrolysis, such as frequency factor, reaction order, and activation energy. One of the early attempts to estimate these parameters for the pyrolysis of organic compounds such as coal and polystyrene was made in 1951 by Van Krevelen et al. (4). However, the methods proposed were tedious and involved many assumptions. In 1958, Freeman and Carroll (5) developed a widely used method for the determination of reaction kinetics using the thermobalance. During the past few years, much progress has been made in developing more suitable methods for the determination of kinetic parameters, and some of these methods will be discussed in this article, along with the application of TGA to specific materials.

Primarily, dynamic thermogravimetric analysis (TGA) and some associated complementary methods in polymer degradation studies will be discussed. We will define TGA as a continuous process that involves the measurement of sample weight as the reaction temperature is changed by means of a programmed rate of heating. Isothermal and essentially isothermal methods will not be covered. However, results obtained by these methods will be compared, where possible, with results obtained by TGA.

In the discussion that follows, a more or less systematic presentation of methods involved in TGA based upon their "exactness" and versatility is given. This is followed by a description of qualitative and arbitrary quantitative estimations of thermal stabilities of polymers based upon their thermograms. Finally, the application of TGA methods to studies of the behavior of some polymeric materials undergoing pyrolysis is covered. For a fuller treatment see Ref. 6.

Other articles in the Encyclopedia relevant to thermogravimetric analysis include DEGRADATION; DEPOLYMERIZATION; and DIFFERENTIAL THERMAL ANALYSIS.

Instrumentation

Figures 1 and 2 show block schematic diagrams of two of the many commercially available thermobalances, ie, the Thermo-Grav and the Chevenard recording thermobalances. (For fuller coverage, see Refs. 7–9.) The Thermo-Grav is manufactured by the American Instrument Co., Inc., Silver Spring, Md.; the Chevenard is manufactured by the Société A.D.A.M.E.L., Paris, France.

To use the Thermo-Grav balance, a preweighed sample is placed in a crucible holder suspended from precision springs. The armature of a linear variable differential transformer (LVDT) which is mounted on the suspension rod actuates the coil as the sample changes weight during heating. The corresponding electrical signal is demodulated, amplified, and used as the input for the vertical Y axis of the recorder. The horizontal X axis of the recorder is used to indicate temperature (up to 1000°C) or time for either TGA or isothermal experiments, respectively. In TGA, a chromel–alumel thermocouple near the crucible records the sample temperature. Various nominal furnace heating rates may be chosen by means of a selector switch, and a periodic electrical signal results in a spike on the TGA thermogram. The exact heating rate can be estimated from these spikes. The weighing column may be evacuated for runs in vacuo, or various gases may be introduced into the system during a run at atmospheric pressure. A major disadvantage of this balance is that high-boiling volatile material may deposit on portions of the weighing system.

The Chevenard balance tends to avoid such deposits by an arrangement whereby the weighing system is situated below the sample to be pyrolyzed. This balance is also a deflection-type instrument containing a crucible which is supported by a quartz rod with an end ring. A balance beam is also suspended from the rod, and, by means of a tungsten filament, beam movements actuate an LVDT which emits electrical signals corresponding to weight changes. As in the Thermo-Grav, the electrical signals are demodulated and used to actuate a strip chart recorder. A two-channel recorder plots furnace thermocouple temperature (up to 900°C) and sample weight versus time. A wide range of heating rates may be employed and, as in the case of the Thermo-Grav, oil dashpots are used to damp out parasitic oscillations. The vacuum model can be used for controlled-atmosphere operation as well as for vacuum operation.

It may also be mentioned here that thermoanalyzers are available which can simultaneously record temperature and sample weight loss along with derivative thermogravimetric (DTG) (rate of weight loss versus temperature or time) and differential thermal analysis (DTA) curves (Fig. 3) (10).

As the reader may surmise, there are many experimental variables that may influence the determination of various parameters obtained from pyrolyses of polymers using TGA techniques. Thus, activation energy associated with a pyrolysis may be affected by the shape of the crucible; the location and arrangement of temperature measuring devices; the size, shape, and physical characteristics of the sample (eg, particle size); the furnace arrangement; and the surrounding sample atmosphere.

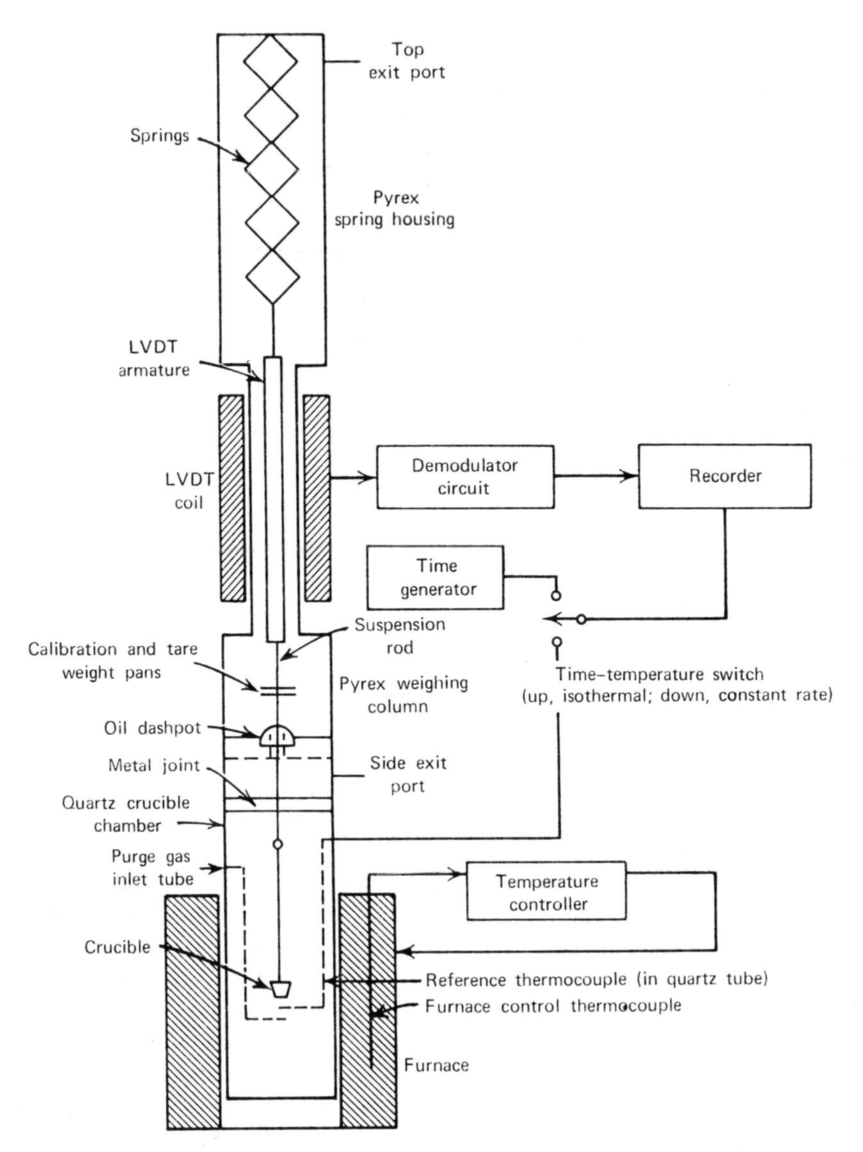

Fig. 1. Schematic diagram of Thermo-Grav recording thermobalance.

Nevertheless, valid and consistent results can be obtained by observing such precautions as use of a sample small enough to ensure temperature uniformity during decomposition as well as direct sample-temperature measurement; use of samples with a uniform sample size and which are uniformly packed in the crucible; and adjustment of the gas flow, pressure, and sample shape to reduce the effects of effluent gases from the sample.

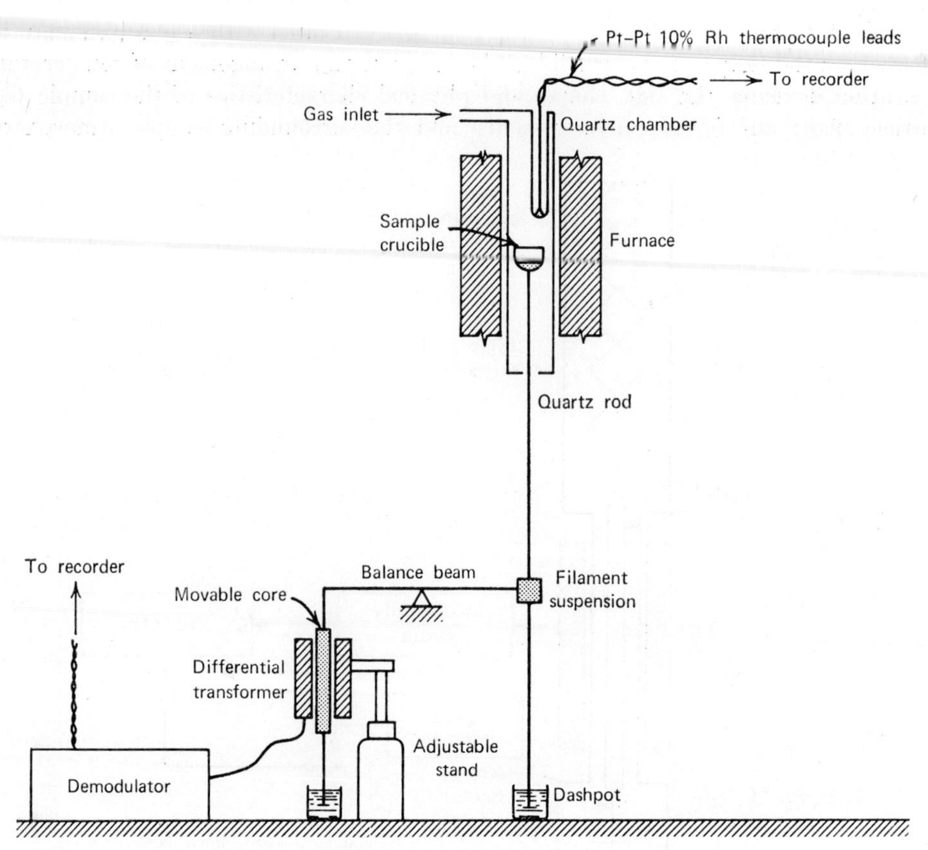

Fig. 2. Schematic diagram of Chevenard recording thermobalance.

Determination of Kinetic Parameters

In many polymer pyrolyses the TGA curve follows a relatively simple sigmoidal path. Thus the sample weight decreases slowly as reaction begins, then decreases rapidly over a comparatively narrow temperature range, and finally levels off as the reactant becomes spent. The shape of the curve depends primarily upon the kinetic parameters involved, ie, upon reaction order (n), frequency factor (A), and activation energy (E). The values of these parameters can be of major importance in the elucidation of mechanisms involved in polymer degradation (11,12) and in the estimation of thermal stability (13). However, it should be realized that the expressions utilized to evaluate these parameters are generally valid for fluid systems but are of questionable validity in solid-state reactions. Therefore, too much significance should not be given to the values of these parameters without substantiating evidence.

Thermogravimetric curves may be more complex than described above. Thus, if a material degrades by a multistep mechanism which involves rate-controlling steps of similar order, and if the activation energies of the rate-controlling steps are of a similar magnitude, a relatively simple trace may be obtained which provides an overall activation energy for the sample degradation. However, if the values of E of the rate-controlling steps differ sufficiently, the TGA trace may involve two or more sigmoids, and if the reaction orders for the various rate-controlling steps have values greater than zero, two or more inflection points may be observed. The separate sigmoidal traces may be individually analyzed for values of E, n, and A by methods similar to those employed for TGA curves that possess one sigmoid. However, if values of E for the rate-controlling steps are not sufficiently different, then one reaction may overlap another and analyses of the TGA curves may be difficult, if not impossible. TGA studies give values of overall kinetic parameters which, in themselves, may often shed little light on the mechanism involved in any particular pyrolysis. It is therefore often necessary to complement TGA studies with differential thermal analysis and with chromatographic, infrared, and mass spectrographic methods.

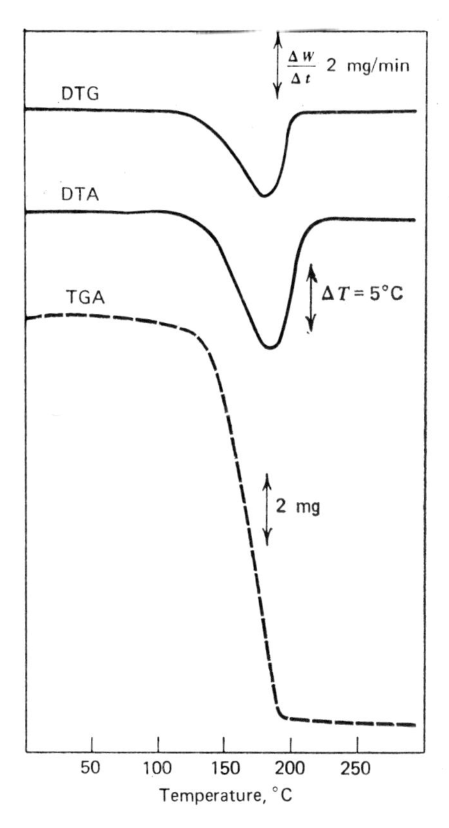

Fig. 3. Simultaneous TGA/DTG/DTA of the dehydration of calcium oxalate monohydrate (10). Heating rate is 10°C/min, sample weight 93.3 mg.

There are several advantages to using TGA methods rather than isothermal methods in the determination of kinetic parameters (14,15). These include the following:

1. Considerably fewer data are required. The temperature dependence of the volatilization rate may be determined over various temperature ranges from the results of a single experiment, whereas several separate experiments are required for each temperature range if isothermal methods are employed.

2. The continuous recording of weight loss versus temperature ensures that no features of the pyrolysis kinetics are overlooked.

3. A single sample is used for the entire TGA trace, thereby avoiding a possible source of variation in the estimation of kinetic parameters.

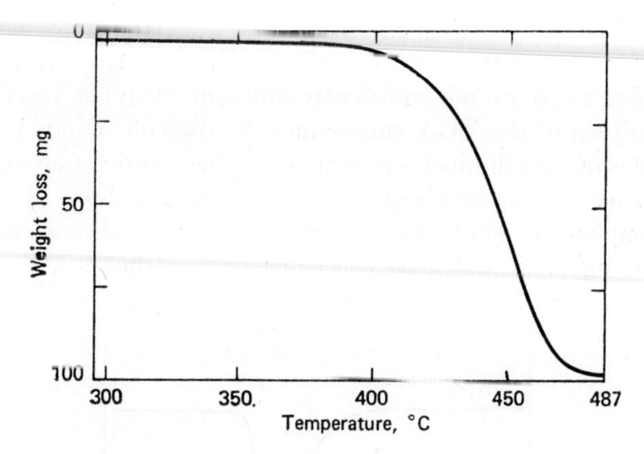

Fig. 4. Thermogravimetric curve for the degradation of polyethylene at 1 mm Hg pressure (16). Sample weight is 100 mg, heating rate 5°C/min.

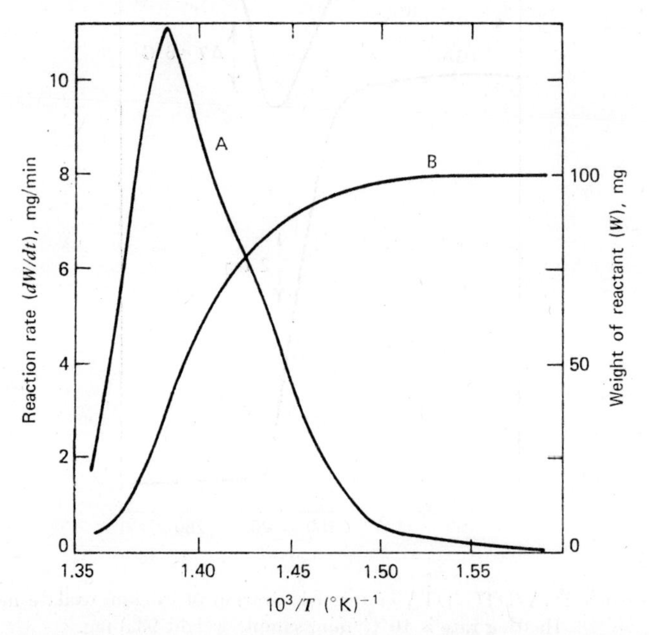

Fig. 5. Graph of the first derivative of the thermogravimetric curve (dW/dt) (curve A) and the weight of the reactant (curve B) as a function of reciprocal absolute temperature for the degradation of polyethylene in vacuum (16).

4. In the isothermal method, a sample may undergo premature reaction and this may make the subsequent kinetic data difficult, if not impossible, to analyze properly.

A major disadvantage is that precise temperature control for kinetic experiments is much more difficult with TGA than with isothermal methods.

Several so-called exact methods have been proposed for estimating kinetic parameters from TGA curves. These methods are based upon the assumptions that thermal and diffusion barriers are negligible and that the Arrhenius equation is valid. The former is reasonable since, in general, small quantities of powdered samples are employed in TGA studies. In the following discussion, the first four methods described

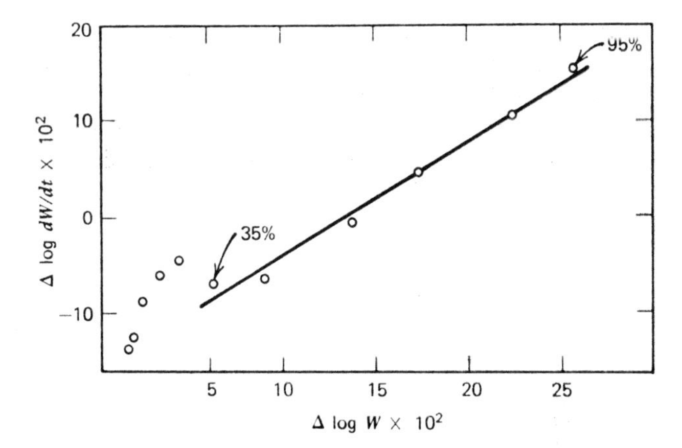

Fig. 6. Kinetics of the thermal degradation of polyethylene in vacuum (14). Data are taken from Figure 3 (16).

allow the estimation of E and n (and A). Succeeding methods described involve the determination of only a single kinetic parameter. Finally, methods are presented that involve additional assumptions and/or approximations.

Method of Freeman and Carroll (5,16). For reactions which involve a decomposition of the type

$$\text{A (solid)} \rightarrow \text{B (solid)} + \text{C (gas)}$$

a general rate expression may be written as equation 1, where W is weight of active

$$R_T = -dW/dT = (A/RH)(e^{-E/RT}W^n) \tag{1}$$

material remaining for a particular reaction and RH is the rate of heating. When equation 1 is applied at two different temperatures and the resulting expressions are subtracted from one another (RH is constant), equation 2 is obtained where R_t =

$$\left.\begin{array}{c}\Delta \log R_T \\ \text{or} \\ \Delta \log R_t\end{array}\right\} = n\,\Delta \log W - (E/2.303R)\,\Delta(1/T) \tag{2}$$

$RH(R_T)$. From equation 2 it can be seen that values of E and n may be calculated from a single TGA curve. Thus, $\Delta \log R_t$ should be linear with $\Delta \log W$ when $\Delta(1/T)$ is held constant. The slope of the resulting linear curve will give a value for n, whereas the intercept will give a value for E.

Figure 4 shows a primary thermogram for a sample of polyethylene. In Figure 5 the curves obtained using data from Figure 4 are shown. Values of R_t and W at equally spaced intervals of $1/T$ may thus be obtained. From such values, equation 2 may be plotted as shown in Figure 6. Above 35% reaction it was found that n was essentially unity and that E had a value of 67 ± 5 kcal/mole. These values appear to agree favorably with values reported by Madorsky (17), who used isothermal procedures. Other dependencies were found for the polyethylene degradation in vacuum below 35% conversion and are indicated in Figure 7.

These results obtained for polyethylene degradation illustrate advantages for procedures such as the method of Freeman and Carroll. Comparatively little data are required and the kinetics can be studied continuously over an entire temperature range. This can be particularly important in cases of polymer pyrolyses where the kinetic parameters change with conversion.

An important disadvantage of this method lies in the need for obtaining slopes from steep portions of the primary thermogram. As a result, there is often sufficient

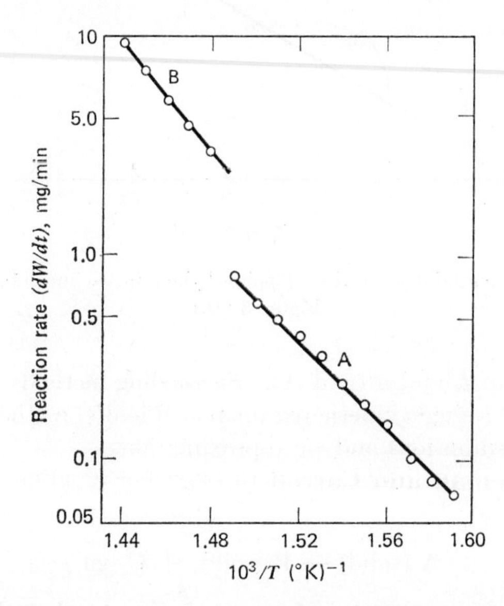

Fig. 7. Temperature dependence of the low-temperature thermal degradation of polyethylene in vacuum: curve A, up to 3% degradation; curve B, from 3 to 15% degradation (16).

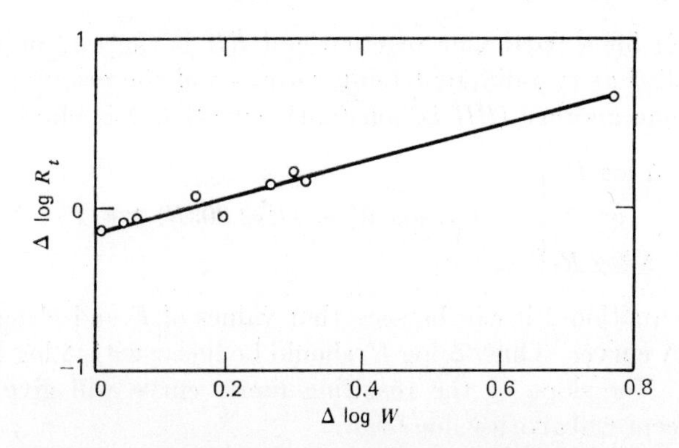

Fig. 8. $\Delta \log R_t$ plotted against $\Delta \log W$ for the thermal degradation of polytetrafluoroethylene (Teflon, Du Pont) in vacuum (11).

scatter in the plot to make accurate evaluation of kinetic parameters difficult. The case of the thermal degradation of polytetrafluoroethylene in vacuum (Fig. 8) is an example of this. There is often enough scatter for small changes in the slope of the line to produce relatively large changes in E and n. It was recently found (18) that the method of Freeman and Carroll could be applied to the second step of the thermal decomposition of thorium 8-quinolinol chelate but that a linear relationship could not be obtained by this method for the first step of the degradation (18).

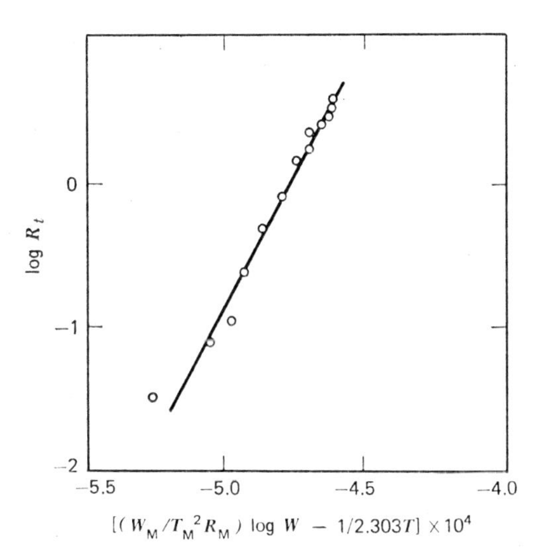

Fig. 9. Log R_t plotted against $(W_M/T_M^2R_M) \log W - (1/2.303T) \times 10^4$ for polytetrafluoroethylene at a heating rate of 6°C/min (24).

Methods Involving Maximization of Rate (19,20). A method for evaluating kinetic parameters that emphasizes the position of the inflection point on the primary thermogram has been reported by Reich and co-workers (19) and by Fuoss and co-workers (20). If equation 1 is differentiated with respect to T and the result is set equal to zero, equation 3 is obtained after rearranging (19,20). In this equation

$$n = (E/R)(W_M/R_M T_M^2) \tag{3}$$

W_M, R_M, and T_M are the weight of active material remaining, the slope, and the temperature, respectively, at the inflection point on the primary thermogram. If equation 1 is converted into a logarithmic expression and equation 3 is substituted into this expression, equation 4 is obtained (19).

$$\log R_t = \log A + (E/R)[(W_M/R_M T_M^2) \log W - (1/2.303T)] \tag{4}$$

In a plot of log R_t from equation 4, it can be readily seen that the slope of the resulting linear relation will give the value of E and that the intercept will give the value of A (Fig. 9). After the value of E has been obtained, the value of n may be obtained from equation 3. Values of kinetic parameters obtained by this method for the pyrolysis of polytetrafluoroethylene in vacuum were found to agree well with values in the literature.

Fuoss and co-workers (20) presumed the reaction order in order to estimate E from equation 3. Apparently, they were unaware of equation 4. The value of E was estimated from a single slope at the inflection point of the primary thermogram. This procedure may be difficult to apply to samples whose TGA curves become very steep. Also, if the value of n is not integral, as is usually assumed, but fractional, the value of E may be in considerable error. Nevertheless, Fuoss and co-workers were able to obtain values for E and A by their procedure which appear to be in good agreement with values in the literature (Table 1).

Equation 4 resembles equation 2 in various respects, and it might be anticipated that both methods would have similar advantages and disadvantages. It may appear

Table 1 Reaction Rate Constants for Various Polymers (20)

Polymer	T, °K	W_{initial}, mg	dW/dT	E, kcal/mole Observed	E, kcal/mole Reported	A
polytetrafluoroethylene[a]	848	78.9	−3.70	67	67	2.4×10^{16}
				69	80	
					80.5	
acrylic resin[b]	643	71.3	−4.84	56	32–55	2.0×10^{18}
polystyrene	667	83.7	−7.35	77	58	5.0×10^{24}
				74	60 ± 5	
					45	
$CaC_2O_4 \rightarrow CaCO_3 + CO$	758	16.0	−0.957	68	74	8.2×10^{18}

[a] Teflon, Du Pont.
[b] Lucite, Du Pont.

in theory that the former expression is limited in use since it can be applied only to reaction orders with values larger than zero. However, in practice, the value of n would not be expected to be exactly zero even under the most stringent experimental techniques employed. Therefore, this method should have wider applicability to pyrolysis kinetics than might appear on first inspection.

Method of Multiple Heating Rates (21–23). If the value of the constant heating rate is changed from run to run, other conditions remaining the same, different TGA curves will be obtained (Fig. 10). Equations can be readily obtained from equation 1.

$$\ln R_t = \ln A - E/RT + n \ln W \tag{5}$$

When W is held constant, a plot of $\ln R_t$ versus $1/T$, employing data from the different TGA curves, should give a linear relation whose slope will give the value of E and whose intercept the value of A. It may be advisable to carry out a series of such plots at various (constant) values of W, as shown in Figure 11. In this manner, average

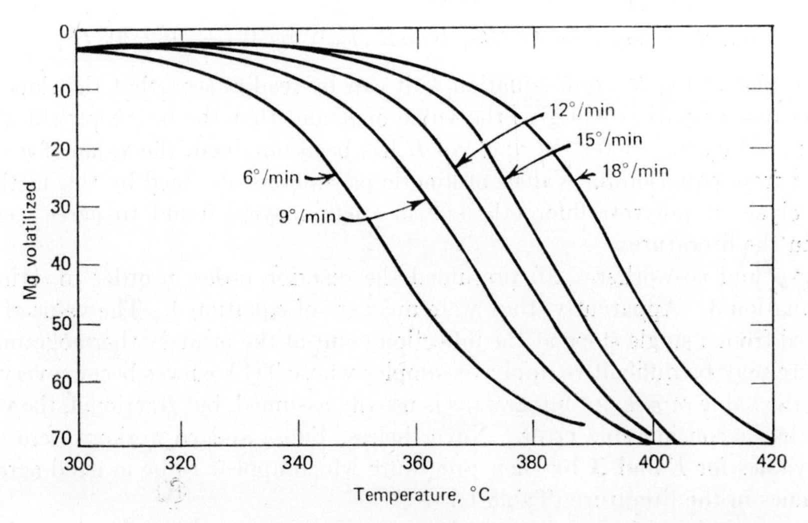

Fig. 10. Primary thermograms for m-phenylenediamine-cured halogenated epoxides at various heating rates (11).

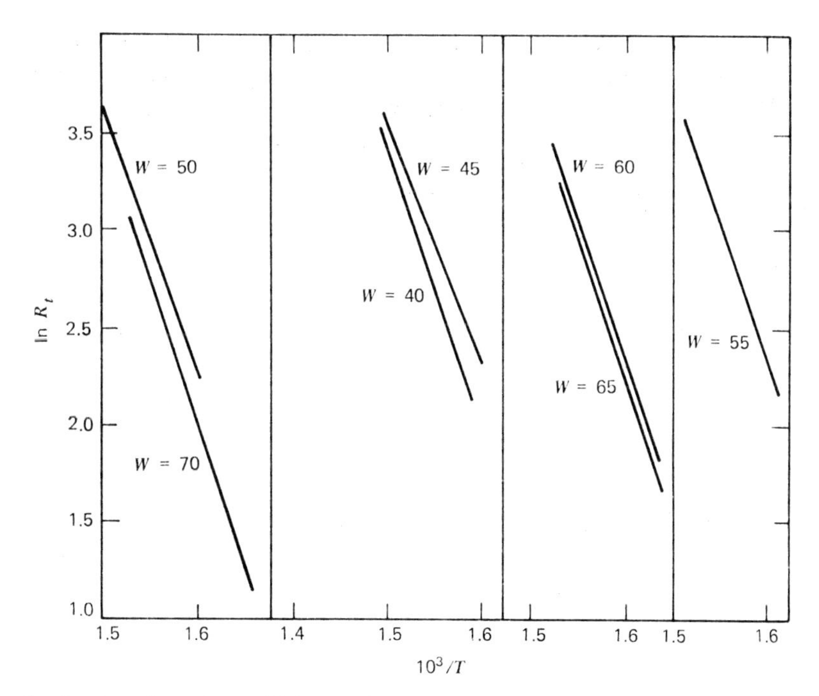

Fig. 11. Ln R_t plotted against $1/T$ for *m*-phenylenediamine-cured halogenated epoxide at the indicated values of W (11).

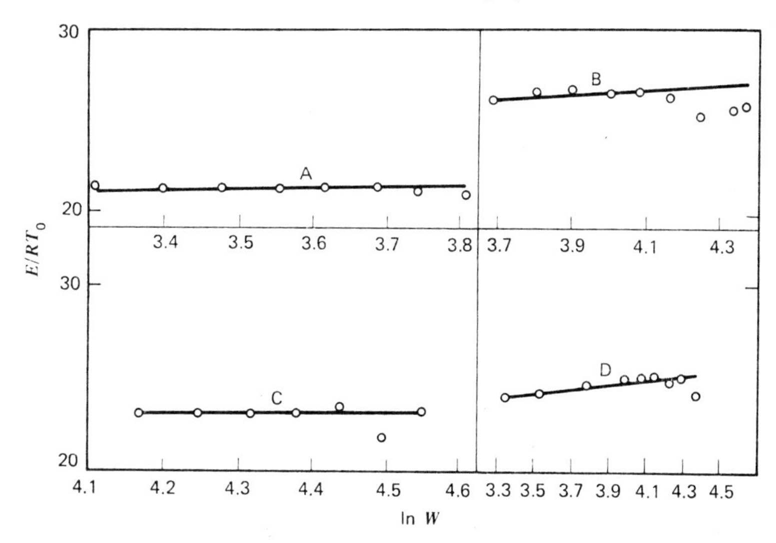

Fig. 12. Determination of reaction order for epoxy resins (11). Key: curve A, uncured epoxy resin based on epoxide of the formula

$$\overset{\displaystyle O}{\overbrace{CH_2-CH}}CH_2O-C_6H_4-\underset{\displaystyle CF_3}{\overset{\displaystyle CF_3}{C}}-C_6H_4-OCH_2\overset{\displaystyle O}{\overbrace{CH-CH_2}}$$

(6F); curve B, 6F cured with *m*-phenylenediamine; curve C, uncured Epon 820 (Shell Chemical Co.); curve D, Epon 820 cured with *m*-phenylenediamine.

values of E and A may be obtained over a range of conversion. The series of curves obtained can indicate the conversion at which the pyrolysis kinetics begin to vary.

In order to evaluate n, we may employ equation 6, which applies at $\ln R_t = 0$.

$$E/RT_0 = \ln A + n \ln W \tag{6}$$

In this case, a plot of E/RT_0 versus $\ln W$ should give a linear relation whose slope will be n (see Fig. 12).

Anderson (22,23) has employed a variation of the above method. He carried out TGA experiments at three different rates of heating and subsequently solved three simultaneous expressions, each of the form of equation 5, for values of E, n, and A by means of a computer.

The above method has been tried using a number of constant values of W and several rates of heating, and it appears to give satisfactory values for overall activation energy. Although more data are required than in other procedures, this very fact tends to enhance confidence in the results obtained. Changes in kinetics (and mechanism) may also be detected rather satisfactorily by this method. However, scatter in the plot for estimating n may make the determination of this quantity somewhat less precise than the evaluation of activation energy.

Method of Variable Heating Rate for a Single Thermogram (24). A method involving a different type of thermogram has recently been proposed (24). In one variation of this method the polymer sample is heated at a given rate, eg, 6°C/min, and after about 30% decomposition, the heat input is raised so that the heating rate approaches a higher value (15°C/min in Figure 13). In another variation (see Figure 14), the heating cycle is reversed. An initial high heating rate of about

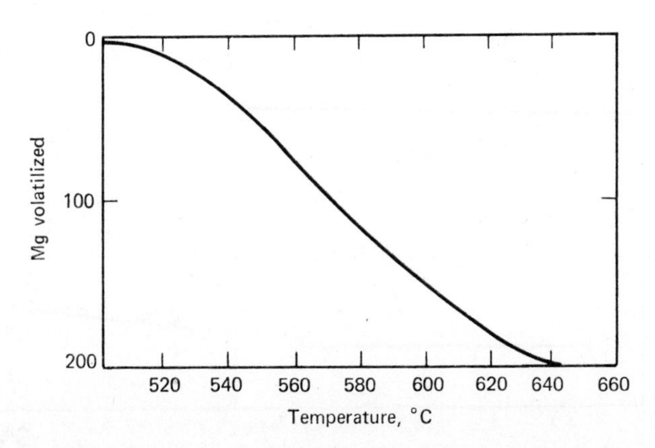

Fig. 13. Variable heating rate thermogram for polytetrafluoroethylene (24).

16°C/min was employed in obtaining this curve, and after about 30% decomposition the heat input was drastically reduced so that decomposition occurred while the material was actually cooling.

For either of the cases described above, equations 7 and 7a may be written. When R_T is held constant

$$\Delta \log (RH)/\Delta(1/T) = n[\Delta \log W/\Delta(1/T)] - (E/2.303R) \tag{7}$$

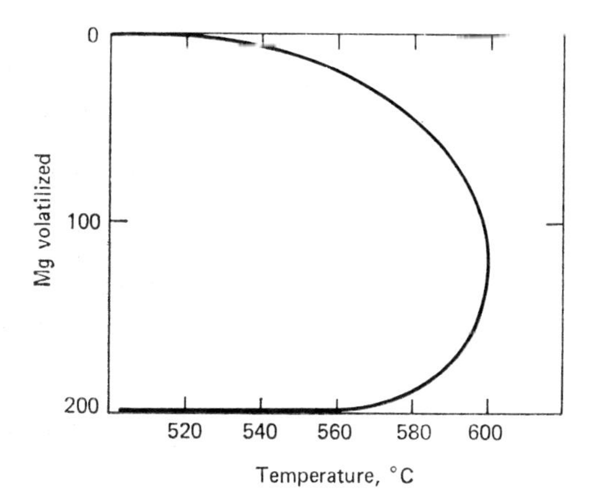

Fig. 14. Variable heating rate thermogram for polytetrafluoroethylene (24).

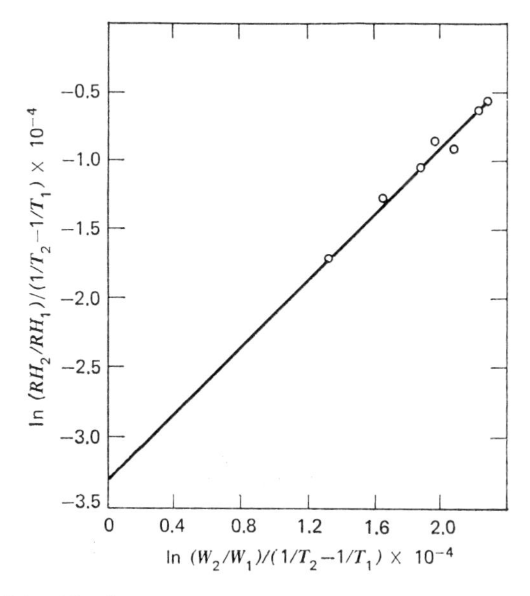

Fig. 15. Plot for obtaining kinetic parameters for polytetrafluoroethylene by equation 7 (24). Data are from Figure 13.

and when temperature is held constant

$$n = (\Delta \log R_t)/(\Delta \log W) \qquad (7a)$$

Equations 7 and 7a are readily obtained from equation 1.

In using equation 7, data from thermograms such as those shown in Figures 13 and 14 were employed. The method consisted of obtaining pairs of values of RH, T, and W from sections of these curves which gave equal values of $R_T(dW/dT)$. Then, appropriate plots, as illustrated in Figures 15 and 16, gave lines whose slopes yielded values for n and whose intercepts gave E. The value of n was checked independently by constructing isotherms on the original thermogram (Figure 14) and by substituting values into equation 7a. Values of kinetic parameters obtained by this method for the pyrolysis of polytetrafluoroethylene (Teflon) in vacuum were in satisfactory agreement with literature values. Thus, n varied from 0.89 to 1.16 and E ranged from 66 to 74 kcal/mole.

Equations 7 and 7a appear to offer some advantages. Only one thermogram need be obtained and the method is not particularly laborious. In addition, the value of n may be readily checked from a thermogram such as that shown in Figure 14. The method can also indicate whether the kinetic order has changed at various conversions. However, a thermogram such as that in Figure 13 is not as useful as that shown in Figure 14. The one in Figure 13 does not allow an independent check of n and also requires the use of primary data that may be close enough to the extremities of the thermogram to be subject to a greater possibility of error. It should be noted that when a thermogram is obtained such as that in Figure 14, experimental conditions should be such that the resulting curve is unsymmetrical but shows distinct curvature

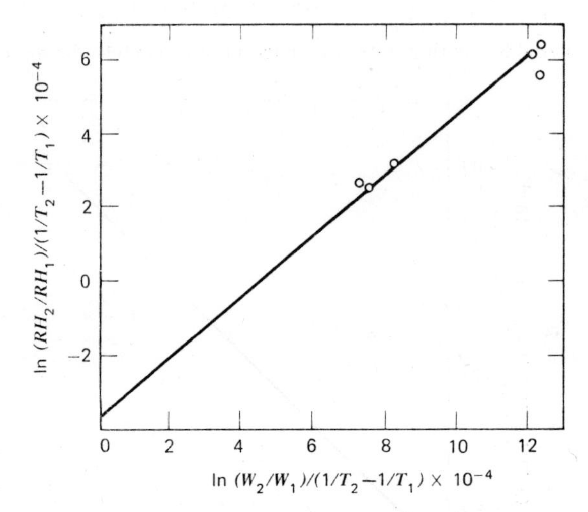

Fig. 16. Plot for obtaining kinetic parameters for polytetrafluoroethylene by equation 7 (24). Data are from Figure 14.

on both sides. Otherwise, it will be difficult, if not impossible, to use equation 7. This condition can be rather troublesome from the experimental point of view.

Methods Involving Approximate Integration of the Rate Equation. Various such methods are discussed here.

Methods of Reich (25,26). If equation 1 is expanded in an asymptotic series (27,28), and it is assumed that $(2RT/E) \ll 1$ (which is usually valid in polymer pyrolyses), equation 1 becomes equation 8.

$$-\int_{W_0}^{W} dW/W^n = [A/(RH)](RT^2/E)e^{-E/RT} \tag{8}$$

Using two thermograms for the same material having different heating rates, we may write equations 9 or 9a for any particular active residual weight, W.

$$[A/(RH)_1](RT_1^2/E)e^{-E/RT_1} = [A/(RH)_2](RT_2^2/E)e^{-E/RT_2} = \text{constant} \tag{9}$$

or

$$\ln\left[\frac{(RH)_2}{(RH)_1}\left(\frac{T_1}{T_2}\right)^2\right] = (E/R)\left(\frac{1}{T_1} - \frac{1}{T_2}\right) \tag{9a}$$

When more than two thermograms are employed, it is more convenient to express equation 9 as equation 9b.

$$\ln\,[(RH)/T^2] = -E/RT + \ln\left[\frac{AR}{E(\text{constant})}\right] \qquad (9b)$$

When various specific values of W are chosen and $\ln\,[(RH)/T^2]$ is plotted versus $1/T$, a series of nearly parallel lines should be obtained whose slopes will give values of E.

Some advantages of the above method are: (a) no prior knowledge of reaction order or kinetic process is required; (b) only primary data are used from the thermogram; (c) the method consumes relatively little time; (d) no curve fitting or laborious plots are necessary; (e) values of E may be obtained at various conversions thereby determining whether any change in mechanism is occurring as the conversion changes. Some disadvantages of the method are: (a) the reaction order is indeterminate; (b) at least two thermograms are required; (c) it is sensitive to changes in temperature.

In order to illustrate the applicability of the method, results obtained by equation 9 were compared with those obtained by theoretically exact methods for an uncured and a cured epoxy resin. The results are listed in Table 2. From this table, it can be seen that the values of E obtained by equation 9 and standard methods are in good agreement irrespective of reaction order. However, the limitations of the method presented must not be overlooked.

Table 2. Application of Equation 9 (25)

RH, °C/min	Reaction order (standard methods)	E, kcal/mole		W/W_0	
		Eq. 9	Standard method	This method	Standard methods
		Uncured epoxy resin			
10.5, 14.0	0	17	17	0.50	0.60–0.95
11.0, 14.0		15		0.70	
14.0, 19.5		15		0.70	
		Cured epoxy resin			
6.0, 12.0	1	26	28	0.50	0.50–0.95
8.5, 12.0		26		0.50	

Reich (26) has presented another method which involves a combination of equations 1 and 9. When these expressions are combined, eliminating the term $A/(RH)$, and generalizing by allowing for an inactive residue, equations 10a and 10b are obtained

$$E/R = S/W_c \ln\,(W_{0,c}/W_0) \qquad \text{for } n = 1 \qquad (10a)$$

$$E/R = S(1 - n)/W_c^n(W_{0,c}^{1-n} - W_c^{1-n}) \qquad \text{for } n \neq 1 \qquad (10b)$$

where $S = dW_c/d(1/T)$

$W_{0,c} = W_0 - W_R$

$W_c = W - W_R$

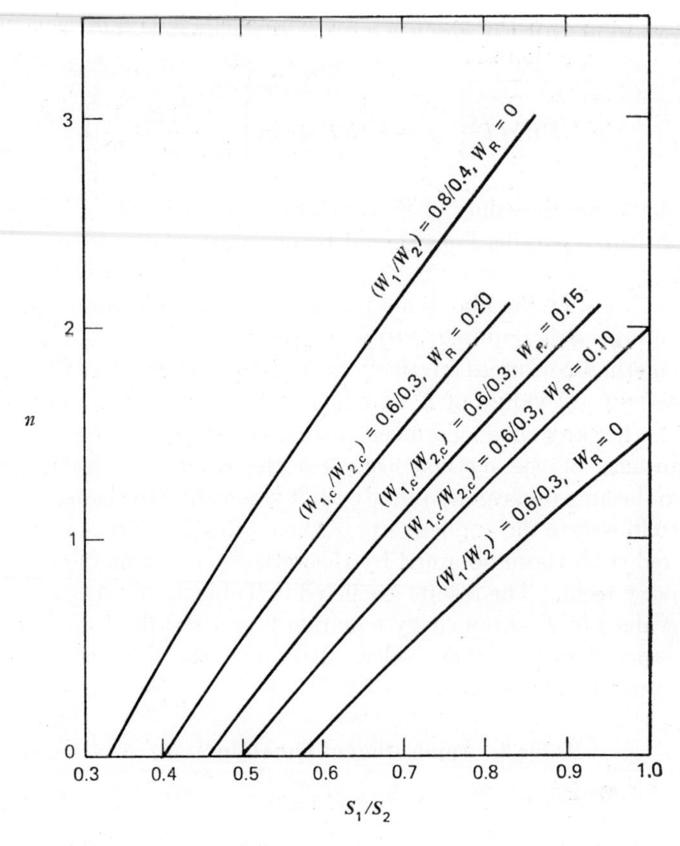

Fig. 17. Theoretical plots of reaction order, n, against S_1/S_2 for various values of W_1/W_2 or $W_{1,c}/W_{2,c}$ and W_R (26).

W is the weight fraction of material remaining at time t and W_R is the weight fraction of inactive material remaining after a pyrolysis. From equations 10a and 10b the following expressions may be obtained, respectively:

$$S_1/S_2 = (W_{1,c}/W_{2,c}) \log (W_{0,c}/W_{1,c})/\log (W_{0,c}/W_{2,c}) \qquad \text{for } n = 1 \qquad (10c)$$

and

$$S_1/S_2 = (W_{1,c}/W_{2,c})^n \left(\frac{1 - (W_{1,c}/W_{0,c})^{1-n}}{1 - (W_{2,c}/W_{0,c})^{1-n}}\right) \qquad \text{for } n \neq 1 \qquad (10d)$$

Prior to determining the overall activation energy of a pyrolysis, it is necessary to estimate n. From equations 10c and 10d it can be seen that for a particular, arbitrarily selected ratio of $(W_{1,c}/W_{2,c})$ and value of W_R the corresponding ratio S_1/S_2 may be calculated for various values of n. Such calculated values were used to construct the theoretical curves shown in Figure 17. In this figure, various values of (W_1/W_2) or $(W_{1,c}/W_{2,c})$ and of W_R are given, from which the ratio S_1/S_2 could be calculated. The curves in Figure 17 can be used to estimate n from experimental TGA curves. After the value of n has been determined, it may be substituted into equations 17a and 17b to obtain the value of E. After E and n have been estimated, the value of A may then be obtained by means of an expression such as equation 1.

When this method was applied to TGA curves for polytetrafluoroethylene and polyethylene the kinetic parameters obtained were in satisfactory agreement with values given in the literature.

It may be of interest to note here that equations 10c and 10d may be employed to obtain a simple expression involving n. Thus, at values of W equal to 0.8 and 0.1, values of S_1/S_2 (designated as $S_{0.8}/S_{0.1}$) were calculated for various values of n, with $W_R = 0$. From these values the following approximate expression may be written for values of n from about ¼ to 2

$$(S_{0.8}/S_{0.1}) \propto D(T_{0.8}/T_{0.1})^2 \approx 0.8n \tag{10e}$$

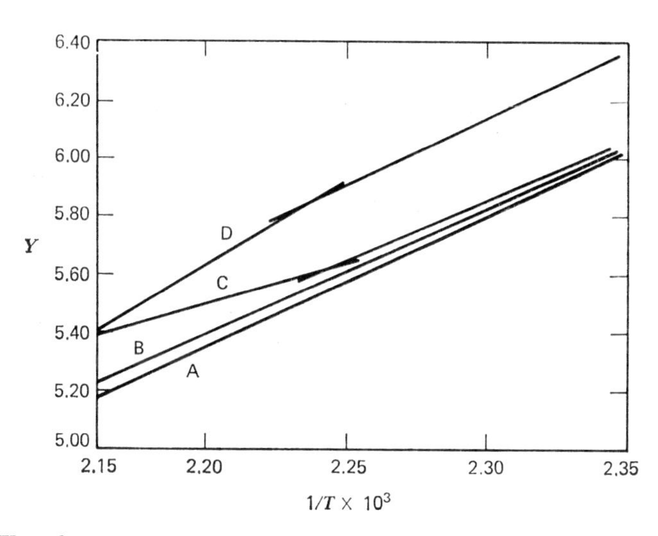

Fig. 18. Water loss from calcium oxalate monohydrate. $Y = -\log[(1 - W^{1-n})]/[T^2(1-n)$ for n = 0, 1/2, and 2/3, and $-\log(-\log W/T^2)$ for n = 1. Key: curve A, n = 2/3; curve B n = 1/2; curve C, n = 0; and curve D, n = 1 (29).

where $D \equiv (dW/dT)_{0.8}/(dW/dT)_{0.1}$. Since the value of $(T_{0.8}/T_{0.1})$ is generally a little less than unity, equation 10e may be written as equation 10f. Equation 10f

$$D \approx n \tag{10f}$$

may be used when it is desired to obtain a rapid, approximate estimation of n from a primary thermogram which possesses a value of $W_R = 0$. From experimental thermograms obtained in the authors' laboratory for degradation of polytetrafluoroethylene in vacuum at different heating rates, values of D were determined which were consistently near unity, as anticipated.

Method of Coats and Redfern (29). Coats and Redfern (29) developed a method for estimating E by use of an integrated form of the rate equation which was similar to equation 8. Thus, when $(2RT/E) \ll 1$ and the logarithm of both sides of equation 8 is taken, equation 11 is obtained, where I denotes the term involving the integral.

$$\ln\left[\frac{\int_W^{W_0} dW/W^n}{T^2}\right] = \ln\frac{I}{T^2} = \ln\frac{AR}{(RH)E} - \frac{E}{RT} \tag{11}$$

The value of I may be readily evaluated once n is known or assumed. In the latter case, when a plot of ln (I/T^2) against $1/T$ gives a linear relation, then it is considered that the correct value of n was assumed. When $(2RT/E)$ is not neglected, the logarithmic term on the righthand side of equation 11 becomes ln $[AR/(RH)E][1 - (2RT/E)]$. Coats and Redfern indicate that for most values of E and for the temperature range over which decomposition reactions generally occur, the preceding logarithmic term is essentially constant.

Figure 18 illustrates the trial and error procedure employed to study the loss of water from calcium oxalate monohydrate. From this figure it can be seen that the results best fit a ⅔ order assumption, and the corresponding value of $E = 21.7$ kcal/mole. These results are in close agreement with other reported values.

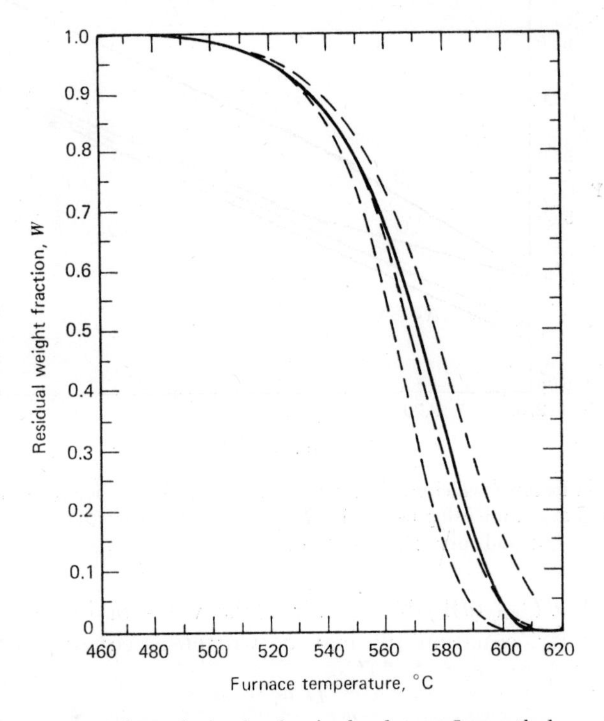

Fig. 19. Thermogravimetric analysis of pulverized polytetrafluoroethylene at 180°C/hr in dry nitrogen (27). Solid curve is experimental; lower, middle, and upper dashed curves are calculated with $E = 80, 70$, and 60 kcal/mole, respectively.

Method of Doyle (27). Doyle (27) attempted to evaluate E and A as constants of the equation of the thermogram rather than of the rate equation. He obtained equation 12 for the equation of the thermogram. In equation 12 U, having values x,

$$g(h) = \frac{EA}{(RH)R}\left(\frac{e^{-x}}{x} - \int_x^\infty \frac{e^{-U}}{U}\,dU\right) \tag{12}$$

replaces E/RT. Equation 12 may also be written as equation 13, where $p(x)$ is the term in parentheses in equation 12.

$$g(h) = \frac{EA}{(RH)R}\,p(x) \tag{13}$$

Common logarithmic values of $p(x)$ are tabulated (27), as well as first differences in log $p(x)$ for use in interpolating. Before E can be evaluated, it is necessary to determine x_a, the value of x at T_a (subscript a denotes point functions); E can then be found by using equation 14.

$$E = RT_a x_a \tag{14}$$

Methods are illustrated by Doyle for estimating x_a and hence E by this equation for the thermogram. In each case determination of E (and A) requires successive approximations or curve fitting. Figure 19 shows an experimental curve and several calculated curves for polytetrafluoroethylene using trial values of E.

Although this method appears to be workable, it is a very laborious one. Also, considering the approximations involved in the derivation, eg, reaction order must be presumed, it does not seem likely that it will find wide use in studies of polymer degradation.

Method of Flynn and Wall (30). Equation 13a can be obtained from equations 1 and 13.

$$\int_0^W \frac{-dW}{W^n} \equiv g(h) = \frac{EA}{(RH)R}\, p(E/RT) \tag{13a}$$

In logarithmic form, equation 13a becomes equation 13b.

$$\log g(h) = \log (EA/R) - \log (RH) + \log p(E/RT) \tag{13b}$$

For values of $E/RT \geqslant 20$, log p (E/RT) may be closely approximated by $-2.315 - 0.457E/RT$, and equation 13b becomes equation 13c.

$$\log g(h) \approx \log (EA/R) - \log (RH) - 2.315 - 0.457E/RT \tag{13c}$$

Upon differentiating equation 13c at constant degree of conversion equation 13d

$$-d \log (RH)/d(1/T) \approx 0.457E/R \tag{13d}$$

is obtained. From equation 13d it can be seen that from the slope of a plot of log (RH) against $1/T$, the activation energy may be calculated. This procedure may be repeated at various degrees of conversion, thereby testing the constancy of E with respect to conversion and temperature.

It may also be noted here that Ozawa (31) also developed expressions similar to equation 13d and set up appropriate theoretical master curves of conversion against log $(AE/(RH)R)p(E/RT)$ for simple nth order reactions. By means of such master curves, various kinetic parameters could be estimated from the value of the displacement of an experimental curve along the abscissa that was required to overlap an appropriate master curve.

Miscellaneous Methods. In addition to the foregoing, several other attempts have been made to estimate kinetic parameters based on simplifying assumptions and approximations. Some of these are briefly outlined below.

Newkirk (32) and Smith (15) assumed that thermal decompositions may be fitted by first-order kinetics with respect to weight loss. Calculations were made of first-order rate coefficients at each of a large number of temperatures on the continuous weight-loss curve. This was followed by construction of the appropriate Arrhenius plot, which was then examined for linearity. Linearity or near linearity over a wide

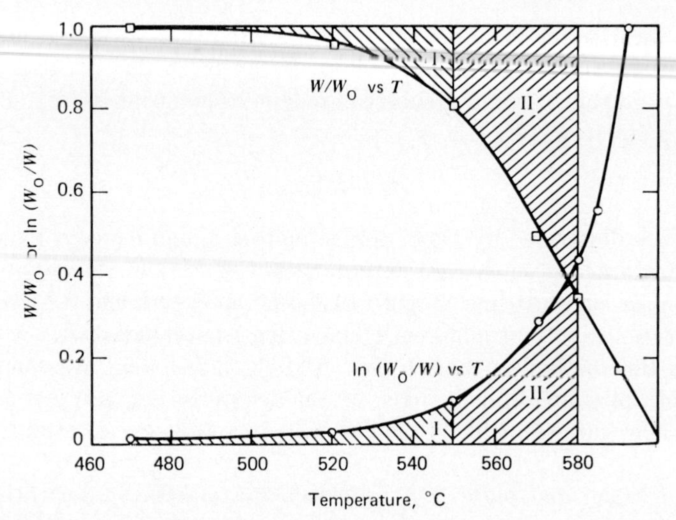

Fig. 20. Zero- and first-order rate plots (13). Areas I and II are discussed in the text.

range of conversion was considered as evidence for a first-order kinetic form of the rate-determining reaction.

Moiseev (33) integrated the rate equation. However, for data treatment he used the rate equation as such, employing graphical differentiation to obtain rates. By selecting values of n, he was able to find the best fit to the experimental data by means of the criterion of linearity.

In conjunction with developing arbitrary thermal stability indexes for various polymers undergoing pyrolysis, Reich and Levi (13) employed an approximate, rapid method for obtaining E. This method involved the measurement of areas, A_{I} and A_{II}, as shown in Figure 20. Two arbitrary temperatures were selected on the TGA curve and E was evaluated from expression 14a. The reaction order was arbitrarily assumed

$$E = \frac{\log\left(\dfrac{A_{(\mathrm{I}+\mathrm{II})}}{A_{\mathrm{I}}}\right) 4.6\, T_1 T_2}{T_2 - T_1} \tag{14a}$$

to be zero in order to calculate overall values of E from equation 14. This was justified from the fact that for many polymeric degradations it is often difficult to distinguish clearly between zero- and first-order kinetics, up to about 50% conversion in constant-temperature experiments. (It is interesting to note that Smith (15) has indicated that an incorrect assumption of the value of n may not necessarily alter the value of E very much.)

Because of an error in the derivation of equation 14, which does not affect the validity of this expression (25,34), the authors feel that an outline of the derivation should be given. Assuming that for relatively low conversions zero-order kinetics apply, we may write, from equation 1

$$\int_1^W -\frac{dW}{T^2} = \left(\frac{A}{RH}\right) \int_0^T \frac{e^{-E/RT}\, dT}{T^2} \tag{15}$$

or

$$(1/T_{\mathrm{av}})^2(1 - W) = [AR/(RH)E]e^{-E/RT} \tag{16}$$

where

$$(1/T_{\mathrm{av}})^2 = \int dW/T^2 / \int dW \approx \text{constant}$$

Equation 16 may further be converted into expression 17

$$\left(\frac{1}{T_{\text{av}}}\right)^2 \left(\frac{1}{T'_{\text{av}}}\right)^2 \int_0^T (1 - W)dT = [A/(RH)] \left(\frac{R}{E}\right)^2 e^{-E/RT} \quad (17)$$

where

$$\left(\frac{1}{T'_{\text{av}}}\right)^2 = \int \frac{(1 - W)}{T^2} dT / \int (1 - W)\, dT \approx \text{constant}$$

The constancy of the average temperatures, as defined above, was checked by means of a TGA curve for the degradation of a cured epoxy resin in vacuum. Up to about 40% conversion the following values were obtained, employing 10% conversion intervals:

$$(1/T_{\text{av}})^2 = (2.2 \pm 0.05) \times 10^{-6}$$

and

$$(1/T'_{\text{av}})^2 = (2.1 \pm 0.02) \times 10^{-6}\ (°\text{K})^{-2}$$

In equation 17 the integral represents the area of the upper portion of the TGA curve bounded by the zero-reaction horizontal line ($W/W_0 = 1$) and extending from temperature T to the temperature at which the curve and the zero-reaction horizontal line meet (see Fig. 20). If this area is denoted by A_{II}, equation 17 may be transformed, for two different temperatures T_1 and T_2 into equation 14. Similar considerations may be applied to first-order kinetics. In this respect, it may be mentioned that this method gave values of E for polytetrafluoroethylene of 68 and 69 kcal/mole when zero- and first-order expressions, respectively, were used. This method will be referred to again in the next section.

Estimation of Thermal Stability from TGA Curves

Many factors can affect the thermal stability of polymers, including melting or softening points, bond strengths, activation energies, crosslinking, and the presence of low-molecular-weight volatile material (or impurities), "weak" links, and groups which are readily affected by heat, etc (see also DEGRADATION; DEPOLYMERIZATION; HEAT-RESISTANT POLYMERS). In order properly to assess the thermal stability of polymers, many of the preceding factors must be simultaneously considered. Since this is virtually impossible, various investigators have devised arbitrary qualitative and quantitative methods. Some of these methods, which do not involve TGA curves, are briefly described in this section.

Madorsky (35) suggested that for polymers which volatilize completely below 600°C, the relative thermal stability may be estimated by heating the polymers under equivalent conditions and comparing the amount of vaporization that occurs in a vacuum pyrolysis for ½ hr at various temperatures. Another procedure involves the comparison of temperatures (T_R) at which 50% of the weight of a given polymer is vaporized under standard pyrolysis conditions. Wright (36) also presented a method which involves a comparison of temperatures at which polymers undergo a 25% weight loss after 2 hr at an elevated temperature. Two disadvantages of this method are: (*1*) The actual percentage weight loss may not be significant unless it can be related to the degradation process, eg, a dimethylsilicone polymer can be completely degraded to silica with only a 20% weight loss. (*2*) Certain polymers, eg, poly(vinyl chloride), may crosslink during pyrolysis and this may decrease the rate of volatiliza-

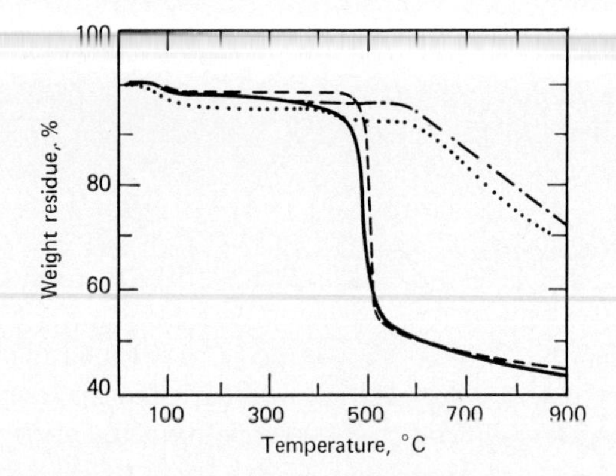

Fig. 21. Thermographic analysis of four polybenzimidazoles (41). T = 150°C/hr. Key:

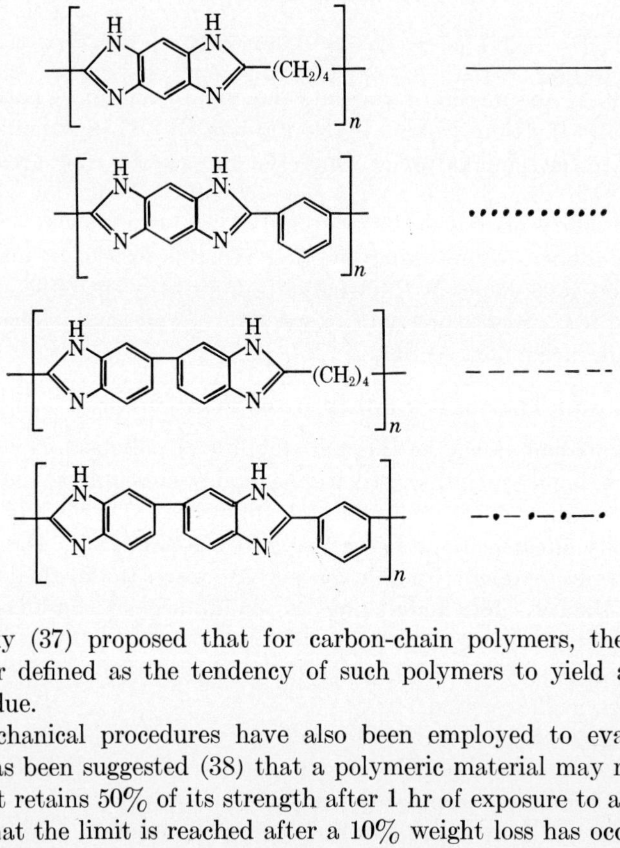

tion. Madorsky (37) proposed that for carbon-chain polymers, thermal stability might be better defined as the tendency of such polymers to yield a more or less carbonized residue.

Thermomechanical procedures have also been employed to evaluate thermal stability. It has been suggested (38) that a polymeric material may retain its "usefulness" when it retains 50% of its strength after 1 hr of exposure to a specified temperature and that the limit is reached after a 10% weight loss has occurred. It has also been specified that secondary effects such as cracking and crazing do not occur. However, this limit depends upon many assumptions and may vary widely for similar materials. Jurkov (39) studied the heat resistance of polymers by measuring changes in their moduli of elasticity caused by increasing temperature. Others (40) suggested that the use of moduli of elasticity to determine heat resistance is more significant than deformation measurements of polymers under linear compression, which have been carried out by various workers. However, in selecting the thermomechanical procedures to be used, the choice of the method of measurement, the accuracy of the measurements, as well as the complexity and availability of the apparatus must be considered. See also MECHANICAL PROPERTIES.

The following sections discuss methods for estimating thermal stability of polymers which are based essentially on TGA measurements. Some of these methods are essentially qualitative in character, others are semiquantitative in nature, and a few rest on a firm theoretical basis.

Qualitative Methods. It is often possible, merely by visual observation, to ascertain the relative thermal stabilities of various polymers. Thus, Marvel (41) has used TGA thermograms to compare stabilities of four polybenzimidazoles (qv under POLYIMIDAZOLES) in nitrogen (Fig. 21). From Figure 21 it may be seen that when an alkyl group was replaced by a phenyl, the initial decomposition temperature rose from about 400 to 500°C, the percentage residue at 900°C increased from about 45 to about 70%, and the decomposition rate decreased considerably. Many similar qualitative analyses have been reported but only a few will be described here.

Recently, Jeffreys (42) discussed the thermal stability of various phenolic resins in air. Results obtained from a static test were compared with thermobalance results (Table 3). In order to establish a criterion for evaluating the resin decomposition,

Table 3. Test Results for Evaluating Thermal Stability of Phenolic Resins (42)

Type	Resin	Static test, loss at 300°C after 1 hr, %	Thermobalance results			
			10% DT, °C	50% DT, °C	Weight-loss curve maxima	
					Temperature, °C	Value, mg/min
1	phenol–formaldehyde	7.0	430	830	415	$\geqslant 0.7$
					530	$\geqslant 1.4$
2	*m*-cresol–formaldehyde	9.5	395	650	430	$\geqslant 3.0$
3	*m*-isopropylphenol–formaldehyde	12.7	360	610	450	$\geqslant 2.4$
4	*m-tert*-butylphenol–formaldehyde	26.4	335	490	355	$\geqslant 3.2$
5	*p-tert*-octylphenol–formaldehyde	7.15[a]	360	490	470	$\geqslant 3.5$
6	*p*-dodecylphenol–formaldehyde	19.4[a]	345	460	440	$\geqslant 5.4$
7	cardanol–formaldehyde	19.6	340	485	445	$\geqslant 5.4$
8	*p*-octadecylphenol–formaldehyde	22.05[a]	340	480	470	$\geqslant 4.5$
9	phenol–benzaldehyde	11.4	360	740	broad peak at about 550°C (1.2 mg/min)	
10	phenol–furfural	14.2	320	640	broad peak at about 410°C (1.4 mg/min)	
11	*m*-isopropylphenol–furfural	35.0[a]	290	520	300	$\geqslant 1.6$
					410	$\geqslant 1.7$
12	*m-tert*-butylphenol–furfural	27.0[a]	270	440	330	$\geqslant 2.6$
13	*p-tert*-octylphenol–furfural	36.0[a]	315	490	390	$\geqslant 2.0$
					485	$\geqslant 3.0$
14	*p*-dodecylphenol–furfural	28.75[a]	350	510	475 (broad)	$\geqslant 2.9$
15	cardanol–furfural	22.2	340	485	445	$\geqslant 5.0$
16	phenol–formaldehyde (stearic acid modified)	13.0	375	710	455	$\geqslant 2.0$
17	phenol–formaldehyde (oleic acid modified)	13.1	390	665	455	$\geqslant 1.8$
18	phenol–formaldehyde (linseed fatty acids modified)	11.9	380	655	455	$\geqslant 2.4$

[a] Resins in which skin effect was observed or in which caking effect occurred.

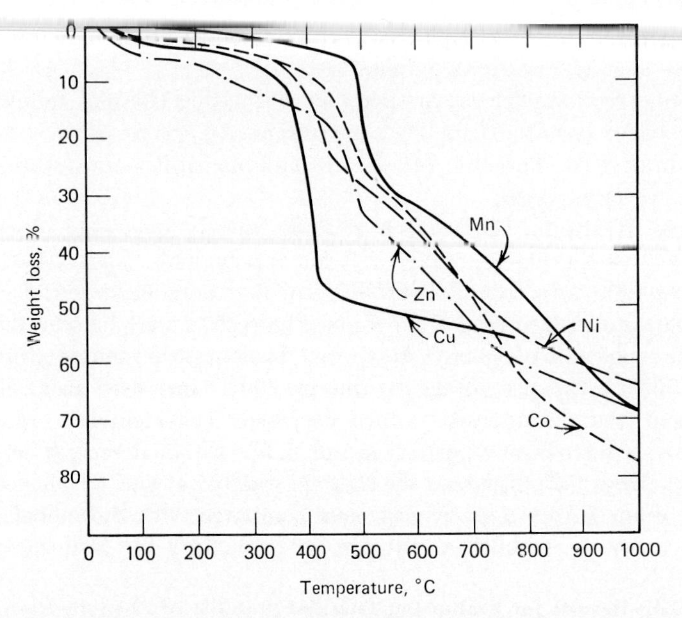

Fig. 22. Thermograms of Mn(II), Co(II), Ni(II), and Zn(II) 5,5′-[methylene bis(*p*-phenylenenitrilomethylidyne)]di-8-quinolinol coordination polymers (44). Heating rate in vacuum, 2.5°C/min.

the temperatures at which 10% decomposition (10% DT) and 50% decomposition (50% DT) had occurred were noted. Temperatures were also recorded at which there occurred maximum rates of decomposition. From Table 3 it can be seen that, based upon resin types *1*, *2*, *3*, *4*, and *7*, the thermal stability of the resins decreases with increasing molecular weight of the meta-substituted phenol, ie, stability decreases in the order, phenol > *m*-cresol > *m*-isopropylphenol > cardanol > *m*-*tert*-butylphenol. The anomalous position of the *m*-*tert*-butylphenol indicates that branching of the side

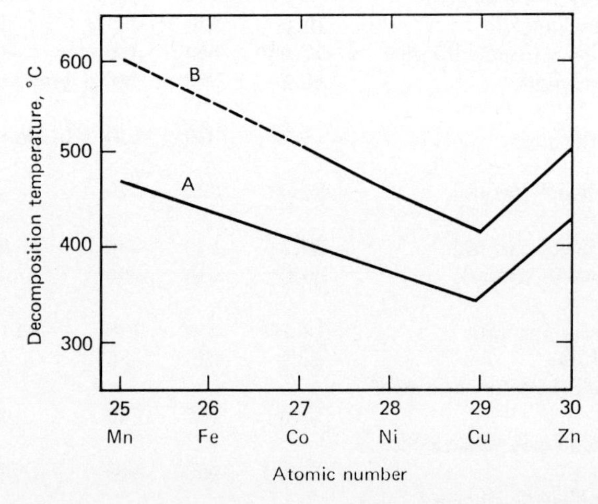

Fig. 23. Plot of decomposition temperatures of polymers against atomic number of coordinated metal. Curve A, 5,5′-[methylenebis(*p*-phenylenenitrilomethylidyne)]di-8-quinolinol polymers; curve B, bis(8-hydroxy-5-quinolyl)methane polymers (44).

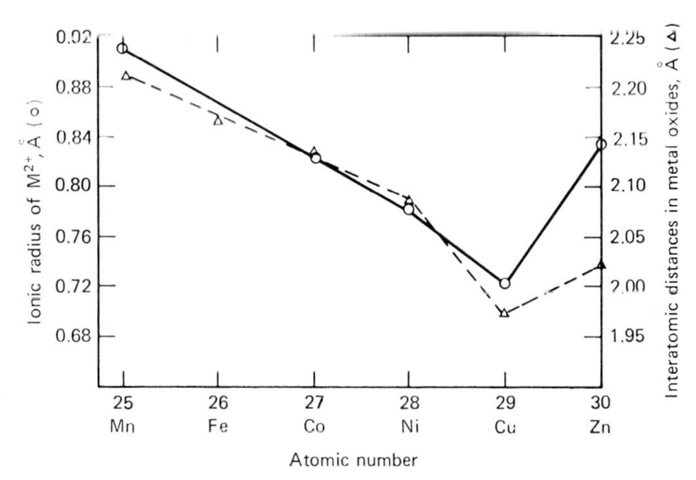

Fig. 24. Plot of ionic radius (O) and interatomic distances in metal oxides (Δ) against atomic number of metal (44).

chain has a significant effect, especially when branching occurs from the α-alkyl carbon atom which is attached to the phenolic nucleus. Other generalizations have been made by Jeffreys.

Recently, Horowitz and Perros (43,44) discussed the thermal stability of coordination polymers (qv). The thermal stability of the Mn(II), Co(II), Ni(II), and Zn(II) 5,5′-[methylenebis(*p*-phenylenenitrilomethylidyne)]di-8-quinolinol coordination polymers was studied under vacuum by means of a thermobalance. The data obtained are shown in Figure 22. After an initial weight loss, a rather sharp break occurred in each of the thermograms, indicating the onset of a decomposition process involving a rapid loss of volatile fragments. As a measure of thermal stability, decomposition temperatures of the polymers were determined. This involved drawing straight lines through the experimental points on the expanded thermogram before and after the initial sharp break in the curve that occurred at moderately low weight losses. The intersection of these two lines was then projected on the temperature axis and this temperature was arbitrarily defined as the decomposition temperature. These temperatures are plotted for each run as a function of the atomic number of the metal contained in the backbone of the polymers in Figure 23, curve A. Curve B shows a similar relationship obtained for coordination polymers of bis(8-hydroxy-5-quinolyl)-methane polymers. The following conclusions were deduced from curves A and B: the substitution of a large organic bridge for the CH_2 group in the 5,5′ position of the ligand lowers the apparent decomposition temperature of the polymers by 80 to 100°C; the decomposition mechanism, near the temperature at which accelerated weight loss occurs, appears to involve metal–ligand scissions in both polymer systems and is dependent on the nature of the metal which unites the ligands in the polymer. For the latter case, it is interesting to note that when the metal–oxygen interatomic distances for the divalent metal oxides are plotted against the atomic numbers of the metals, a relationship as shown in Figure 24 is obtained. The relationship between ionic radii of the metals and their atomic numbers is also shown in this figure. The shape of both of these plots bears a strong resemblance to those shown in Figure 23 (curves A and B) where the decomposition temperature has been plotted as a function of the atomic numbers of the metals in the polymers.

Semiquantitative Methods. Doyle (15,16) proposed indexes of thermal stability in terms of decomposition temperatures. Since the values of these indexes depended upon many procedural details, eg, type of atmospheric gas used and rate of heating, they were referred to as "procedural decomposition temperatures." Two types of such temperatures have been defined for TGA traces in inert atmosphere. One of these was called the "differential procedural decomposition temperature" (dpdt), and the other the "integral procedural decomposition temperature" (ipdt). The former was devised as a means of defining the location of bends in the TGA curves. However, since many materials decompose in steps, the dpdt is neither a unique empirical end point nor a consistently unambiguous index of incipient degradation. In cases of gradual decomposition, it may be very imprecise or even completely unavailable. For these reasons, we will not consider dpdt further.

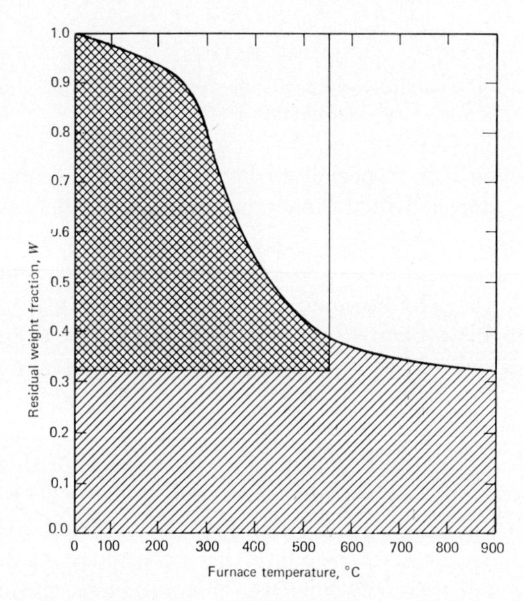

Fig. 25. Thermogram areas, A^* (crosshatched) and K^* (doubly crosshatched) (46).

The second type of index, ipdt, was devised as a way of summing up the whole shape of the "normalized" data curve. As such, it is readily available and is highly reproducible. In order to place materials on an equal procedural basis, the total experimentally accessible temperature range from 25 to 900°C was used. In Figure 25 the area of all the crosshatched region divided by the area of the total rectangular plotting region was denoted as A^*. From the value of A^*, the value of T^* may be obtained from expression 18. In this expression, it was assumed that all materials volatilize completely below 900°C and do so at a single temperature.

$$T^* = 875A^* + 25 \tag{18}$$

The value of T^* serves as a crude measure of refractoriness; however, many materials which possess high refractory weight fractions up to 900°C begin to decompose at much lower temperatures. Thus, a second area ratio, K^*, was defined as the doubly crosshatched region in Figure 25. The boundaries of the regions involved are established by the temperature T^*, by the residual weight fraction at the arbitrary temperature of 900°C, and by the horizontal zero-loss line. The ipdt may now be calculated from equation 19.

$$\text{ipdt} = 875A^*K^* + 25 \tag{19}$$

Table 4 shows values of ipdt for several polymers. Values of ipdt have been used to compare thermal stabilities of a homologous series of polymers containing azulene (47).

Table 4. Integral Procedural Decomposition Temperatures of Some Polymers (46)

Polymer	ipdt, °C
polystyrene	395
epoxy resin cured with maleic anhydride	405
poly(methyl methacrylate)[a]	345
nylon-6,6	419
polytetrafluoroethylene[b]	555
vinylidene fluoride–hexafluoropropylene copolymer[c]	460

[a] Plexiglas, Rohm & Haas Co.
[b] Teflon, Du Pont.
[c] Viton A, Du Pont.

Other thermal stability indexes have been reported (13). As mentioned previously, these indexes were devised in conjunction with equation 14. From this equation, overall activation energies for about thirty polymers were rapidly estimated, and attempts were made to correlate these values with various arbitrary parameters.

The energies obtained were correlated with the so-called 10% tangential temperature (TTN). The tangential temperature (TN) was obtained by drawing a tangent at the break in the TGA curve prior to the maximum rate portion. The temperature at the point of intersection of the tangent and the zero-reaction line is the TN temperature. The TTN is then obtained by multiplying the TN by a ratio of areas. The latter is obtained by observing where the TGA curve intersects the 10% reaction line. The total pertinent rectangular area is bounded by the zero-reaction line and extends from 25°C to the temperature obtained at the intersection of the 10% reaction line and the TGA curve. The portion of the area under the curve divided by this total area is the ratio of areas which when multiplied by TN will afford the TTN.

Another parameter referred to as the sigma temperature, ΣT, was also used. This parameter denotes the temperature corresponding to the intersection of a horizontal line with the TGA curve. This horizontal line is drawn at the ordinate whose value is determined by adding unity to the ordinate value of W/W_0 at 900°C and dividing the sum by two.

Since the decomposition of the polymer itself, and not secondary effects of the reaction, was of primary interest it was reasoned that the TN would give some measure of the onset of appreciable decomposition. This in conjunction with a measure of initial stages of volatilization, which leads to TTN, might thus be expected to represent some measure of thermal stability. For the purpose of including some effects of later stages of reaction, the ipdt was also considered. In Figure 26, values of overall energies, E_0, are plotted against two parameters, ipdt and TTN. From this figure, it may be observed that values of ipdt are relatively insensitive to changes in E_0. However, the TTN values showed a surprisingly good linear correlation with E_0 for values above 155°C. The linear correlation coefficient for the latter parameter was

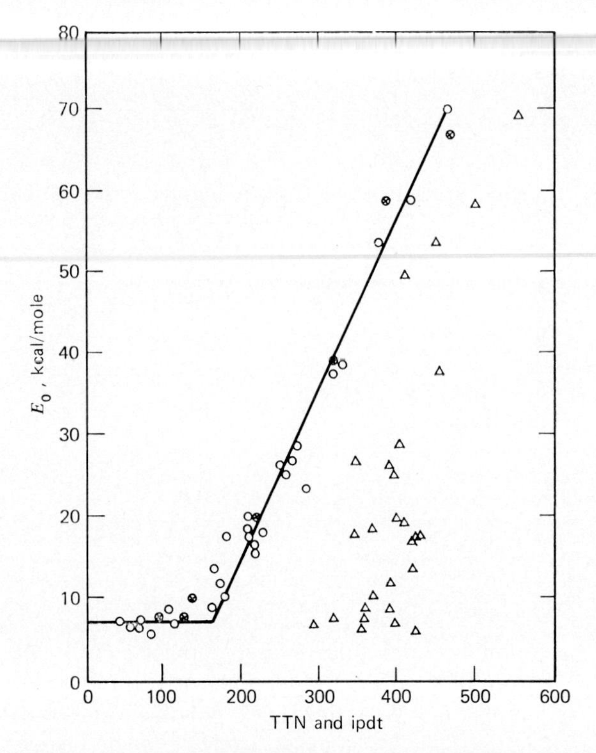

Fig. 26. E_0 plotted against the parameter TTN (heating rate 3°C/min (○); 2.5°C/min (⊗), and the parameter ipdt (△) (13).

estimated to be 0.98 for values of E_0 obtained at 3°C/min. Below 155°C, the values of E_0 leveled off at an average value of 7 for the range of TTN values investigated. When E_0 was plotted against the sum (TTN + idpt), no improvement in the linearity was observed. Smith (13) has also compared values of ipdt with overall activation energies for several polymeric systems. He also found that values of E_0 did not correlate well with values of ipdt. Figure 27 shows a plot of E_0 against a parameter which was obtained by dividing TTN by the difference between ΣT and the value of TTN. If the points represented by the dotted circles were neglected, the correlation coefficient had a value of 0.98. In contrast to other plots, there was no sharp break in the line below certain parameter values.

The values of E_0 obtained from the TGA curves were strongly dependent upon the initial portions of the curves as well as upon later portions. Therefore, a TTN parameter rather than the TN should give a better correlation with E_0, as has been observed. Since the ipdt places much emphasis on the late portions of the TGA curve, it would be anticipated that the ipdt would be relatively insensitive to changes in E_0.

A correlation between E_0 and the thermal stability of a polymer is probably always quite complex. It is presumed that E_0 gives an overall picture of the pyrolysis energetics. It probably involves energies associated with diffusion processes, bond ruptures, volatilization, etc. Prior to designating any literal interpretation to the magnitude of E_0, it should be realized that knowledge of the processes involved in the degradation must be available. Any correlation presented in terms of E_0 is at best semiquantitative and should not be used for the determination of precise activation

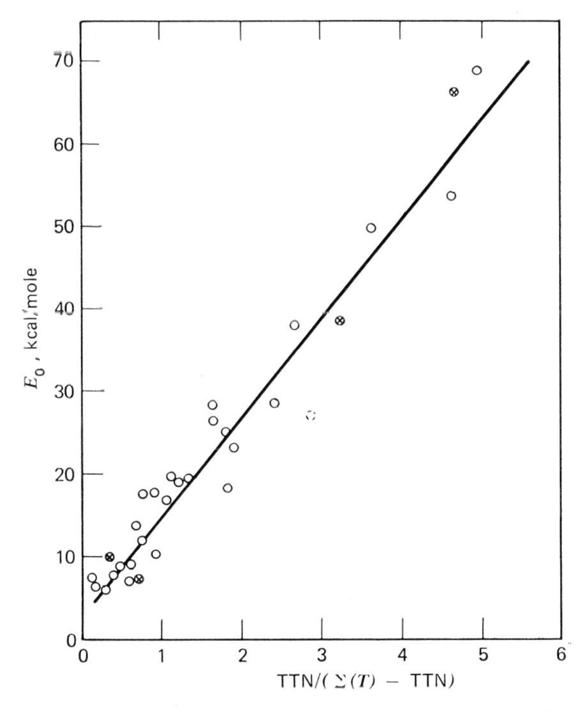

Fig. 27. E_0 plotted against the parameter TTN/Σ(T) − TTN; heating rate 3°C/min (O); 2.5°C/min (⊗).

energies at the expense of the more elaborate methods. However, such rapid correlations may be of considerable value if the thermal stability of a rather large number of experimental polymers must be compared.

Quantitative Methods. The isothermal "life" of a polymer may be quantitatively correlated with TGA curves. Thus, Doyle (48) has obtained equation 20,

$$\log t_i \approx -E/2.3RT + \text{constant} \tag{20}$$

based upon various simplifying assumptions. In this equation t_i is isothermal aging time, and T is absolute thermogravimetric analysis temperature corresponding to the equivalent aging time t_i. Based upon isothermal experiments and TGA curves for polytetrafluoroethylene in a nitrogen atmosphere, a value of E of 67 kcal/mole was obtained.

An expression similar to equation 20 may be obtained by the use of equation 8. When this expression is used in conjunction with the isothermal rate (eq. 21) expression equation 22 is obtained, for any value of n.

$$\int \frac{dW}{W^n} = kt_i \tag{21}$$

$$\log (t_i/T^2) = -E/2.3RT + \log \frac{AR}{(RH)Ek} \tag{22}$$

From equation 22 it can be seen that from a single TGA curve and from a single isothermal degradation experiment the value of E may be readily obtained, irrespective of reaction order. Equation 22 has been plotted for the degradation of polytetra-

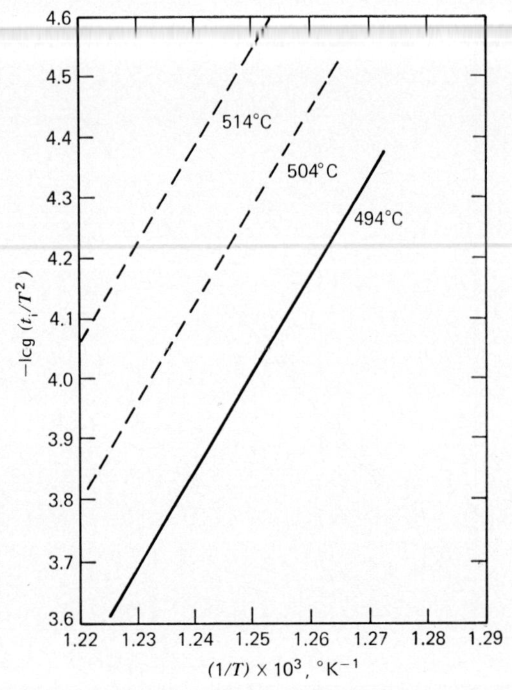

Fig. 28. Log (t_i/T^2) plotted against $1/T$ for degradation of polytetrafluoroethylene.

fluoroethylene in vacuum, as shown in Figure 28, using an isothermal temperature of 494°C. From the slope of the linear relation obtained, a value of E of 74 kcal/mole was obtained. Once the value of E has been obtained, the isothermal stability of the material in question may be quantitatively estimated for other temperatures (other than 494°C). Thus, plots for other temperatures should involve a series of parallel lines whose vertical separation, Δ, may be ascertained from equation 23, where, T_i' and

$$\Delta = (E/2.3R)(1/T_i' - 1/T_i) \qquad (23)$$

T_i are the particular isothermal temperatures involved. In this manner, the dotted lines in Figure 28 were constructed for various temperatures, based upon the line obtained at 494°C.

Thermal Degradation Behavior of Some Polymers by TGA Methods

In this section some studies of polymer degradation which are relatively thorough and which involve primarily TGA methods are discussed. Auxiliary work carried out simultaneously with these methods is also mentioned. Where possible, results from isothermal studies of polymer degradation are compared with TGA results.

Styrenated Polyester. Thermal properties of a styrenated polyester (Laminac 4116, American Cyanamid Co.) were studied by Anderson and Freeman (49). The polymer was synthesized by condensation of a glycol and two dicarboxylic acids, one of which was unsaturated. Crosslinking with styrene, used both as solvent and copolymer, was effected by the use of free-radical initiators (see also POLYESTERS, UNSATURATED). TGA, DTA, infrared, and mass spectrometer techniques were used to study the thermal degradation of this polymer both in air and in argon. Based upon infrared analysis, the unit basic structure of the polyester was taken to be (**1**).

$$
\begin{array}{l}
\quad\quad\quad \sim\!\!\sim\!\text{—CH—}\!\sim\!\!\sim \\
\quad\quad\quad\quad\;\; | \\
\quad\quad\quad\quad\; \text{CH}_2 \\
\quad\quad\quad\quad\;\; | \\
\quad\; \text{C}_6\text{H}_5\text{—CH} \quad\quad\quad\quad \text{CH}_3 \\
\quad\quad\quad\quad\;\; | \quad\quad\quad\quad\quad\quad | \\
\text{H(OOCCHCHCOOCHCH}_2\text{OOCC}_6\text{H}_4\text{COO)}_n\text{H} \\
\quad\quad\quad\quad\quad\;\; | \\
\quad\quad \text{C}_6\text{H}_5\text{—CH} \\
\quad\quad\quad\quad\quad\;\; | \\
\quad\quad\quad\quad\quad \text{CH}_2 \\
\quad\quad\quad\quad\quad\;\; | \\
\quad\quad\quad\quad\quad \text{CH}_2 \\
\quad\quad\quad\quad\quad\;\; | \\
\quad\quad\quad \sim\!\!\sim\!\text{—CH—}\!\sim\!\!\sim
\end{array}
$$

(**1**)

Figure 29 shows TGA-derived curves for the thermal degradation of polyester (**1**) in both air and argon. From this figure it can be seen that for both types of atmospheres, the polymer began to degrade at about 200°C. However, in argon, the reaction was complete at about 450°C whereas, in the case of air, there was another reaction stage from about 450 to 550°C. In air, four reaction stages appeared, ie, from 200 to 260°C; from 260 to 360°C; from 360 to 450°C; and from 450 to 550°C. There also appeared to be extensive overlap between the reactions involved in the second and third stages. In argon, degradation occurred in two stages with extensive overlap. The first stage occurred between 200 and 365°C and the second between 365 and 450°C. Figure 30 shows a comparison between the thermal degradation of polyester

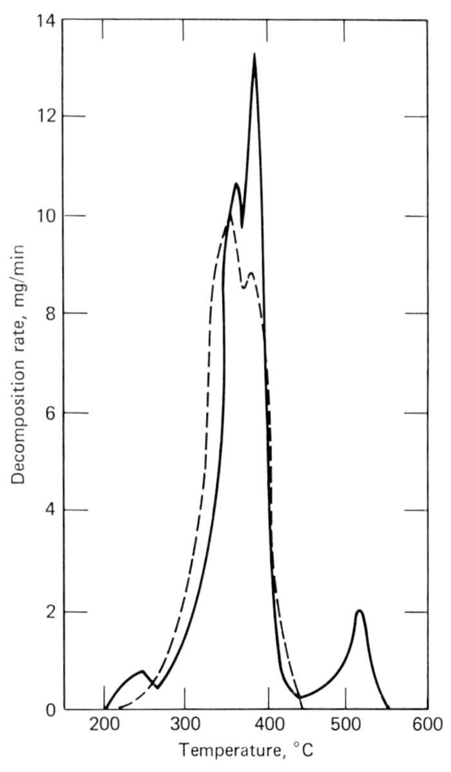

Fig. 29. Derivative thermogravimetric curves for the thermal degradation of a styrenated polyester (Laminac 4116, American Cyanamid Co.) (49). Solid curve, in air; dashed curve, in argon (49).

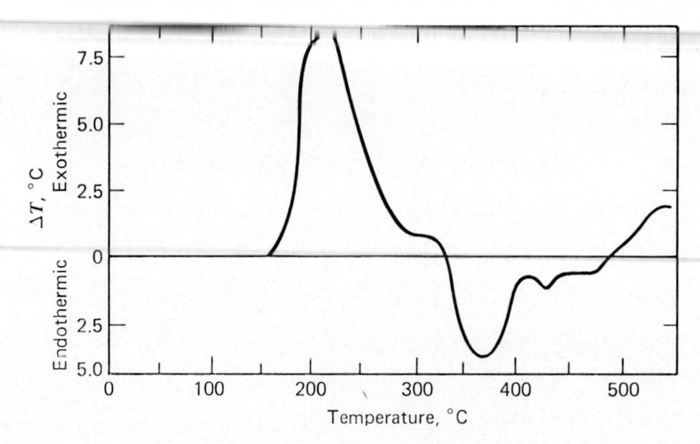

Fig. 30. Differences in temperature differential from DTA between a styrenated polyester (Laminac 4116) heated in air and in argon as a function of sample temperature ($T_{air} - T_A$) (49).

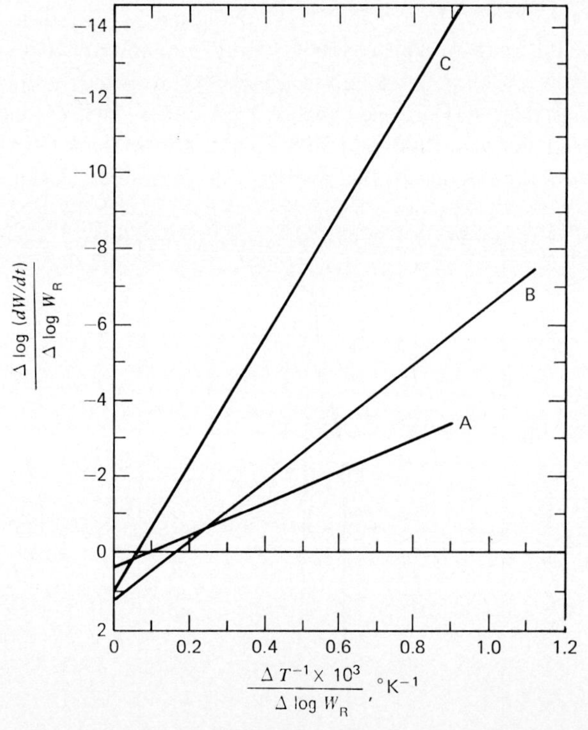

Fig. 31. Kinetics of the thermal degradation of a styrenated polyester (Laminac 4116) in air (49). Key: curve A, initial stage of reaction; curve B, second and third stages of reaction; curve C, fourth stage of reaction.

(1) in air and in argon by means of a difference plot of the DTA curves obtained in these atmospheres (the curve in argon was subtracted from that in air). Two regions of relative exotherm are apparent, from 150 to 290°C and from 470 to 550°C.

Also, degradation in air over the temperature range of 290 to 413°C is more endothermal than decomposition in argon. Upon heating the polyester in air from room temperature to about 500°C, the following noncondensable compounds were

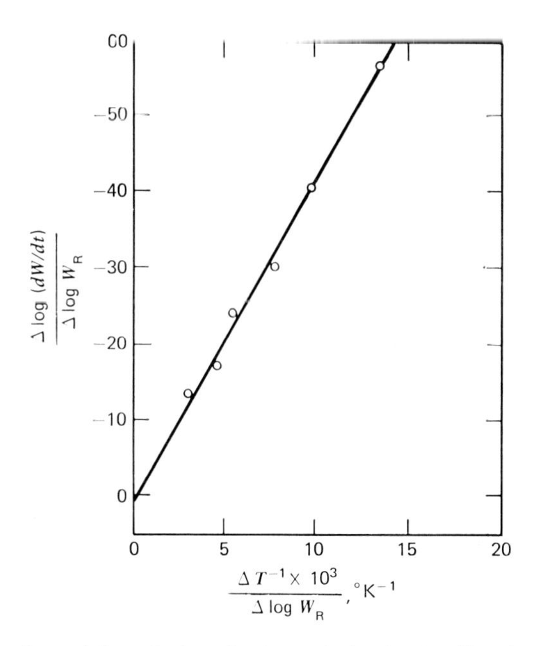

Fig. 32. Kinetics of the thermal degradation of a styrenated polyester (Laminac 4116) in argon (49).

obtained: carbon dioxide, hydrogen, methane, and propylene. The condensable compounds consisted of benzaldehyde and unsaturated hydroxy esters (between 200 and 300°C); phthalic anhydride and an oily liquid containing hydroxy esters (300–400°C); and a mixture of phthalic acid, phthalic anhydride, and a liquid containing low-molecular-weight esters of propylene glycol (400–500°C).

The degradation kinetics were determined by the method of Freeman and Carroll (5). Figure 31 gives plots for the degradation of a styrenated polyester in air. Three linear relationships were obtained which represented the initial stage of the reaction, the combination of second and third stages, and the fourth (and last) stage of the reaction. The corresponding values of E, n, and A are listed in Table 5. Figure 32 shows a similar graph for the reaction of a styrenated polyester in argon. The linear relationship obtained represented the combined first and second stages of reaction. Values of E, n, and A in this atmosphere are also given in Table 5.

Table 5. Kinetic Parameters for Thermal Degradation of a Styrenated Polyester[a] (49)

Stage of reaction	Temp range, °C	Order of reaction	Energy of activation, kcal/mole	First-order frequency factor, sec^{-1}
		In air		
1	200–260	0.4	19	1.3×10^5
2, 3	260–450	1.2	35	2.7×10^9
4	450–550	1.0	79	4.2×10^{19}
		In argon		
1, 2	200–450	1.0	20	4.8×10^3

[a] Laminac 4116, American Cyanamid Co.

On the basis of the various results obtained, several schemes were postulated (49). Thus, for the degradation behavior in air it was postulated that the initial exothermal reaction stage involved the formation of an unstable hydroperoxide intermediate, for example, as shown in equation 24.

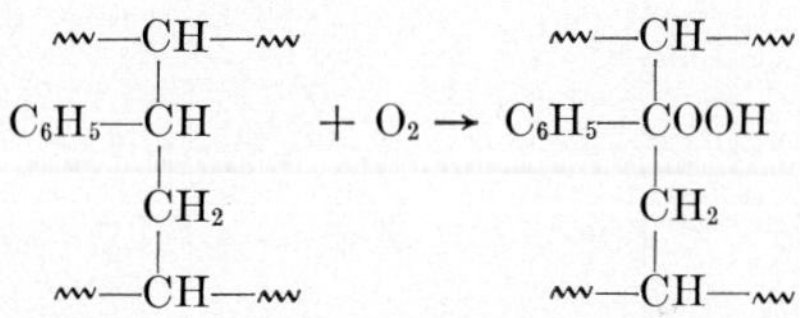

$$\begin{array}{ccc} \sim\!\sim\text{—CH—}\sim\!\sim & & \sim\!\sim\text{—CH—}\sim\!\sim \\ | & & | \\ C_6H_5\text{—CH} & + O_2 \rightarrow & C_6H_5\text{—COOH} \\ | & & | \\ CH_2 & & CH_2 \\ | & & | \\ \sim\!\sim\text{—CH—}\sim\!\sim & & \sim\!\sim\text{—CH—}\sim\!\sim \end{array} \tag{24}$$

Radical formation (equation 25) and subsequent cleavage (equations 26 and 27) could lead to products (**2**) and (**3**), respectively.

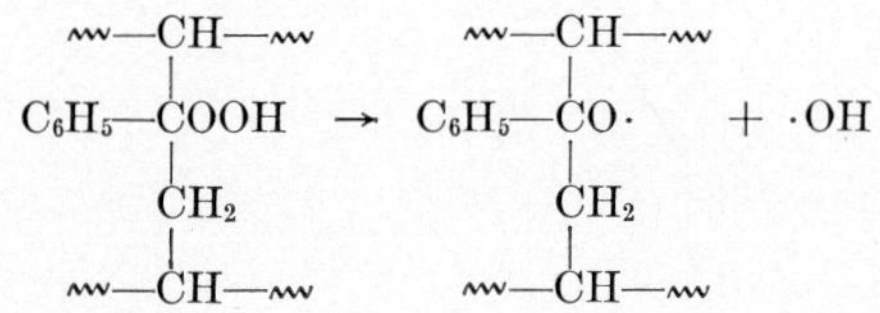

$$\begin{array}{cccc} \sim\!\sim\text{—CH—}\sim\!\sim & & \sim\!\sim\text{—CH—}\sim\!\sim & \\ | & & | & \\ C_6H_5\text{—COOH} & \rightarrow & C_6H_5\text{—CO}\cdot & + \cdot OH \\ | & & | & \\ CH_2 & & CH_2 & \\ | & & | & \\ \sim\!\sim\text{—CH—}\sim\!\sim & & \sim\!\sim\text{—CH—}\sim\!\sim & \end{array} \tag{25}$$

$$\begin{array}{ccccc} \sim\!\sim\text{—CH—}\sim\!\sim & & \text{O} & & \\ | & & /\!/ & & \\ C_6H_5\text{—CO}\cdot & + \cdot OH \rightarrow & C_6H_5\text{—C} & + & \sim\!\sim\text{—CH—} \\ | & & | & & | \\ CH_2 & & CH_2 & & OH \\ | & & | & & \\ \sim\!\sim\text{—CH—}\sim\!\sim & & \sim\!\sim\text{—CH—}\sim\!\sim & & \\ & & & & (\mathbf{2}) \end{array} \tag{26}$$

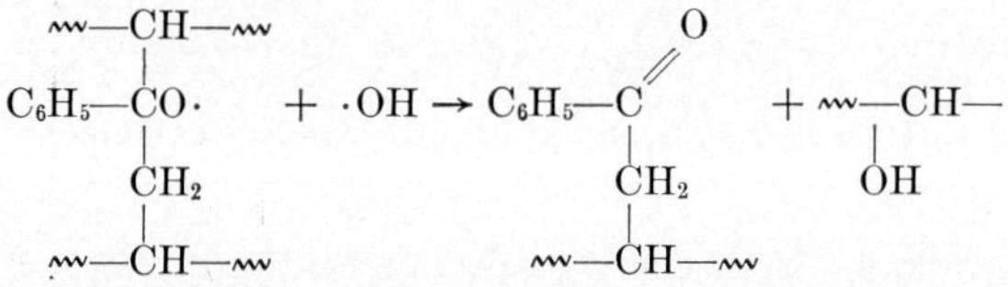

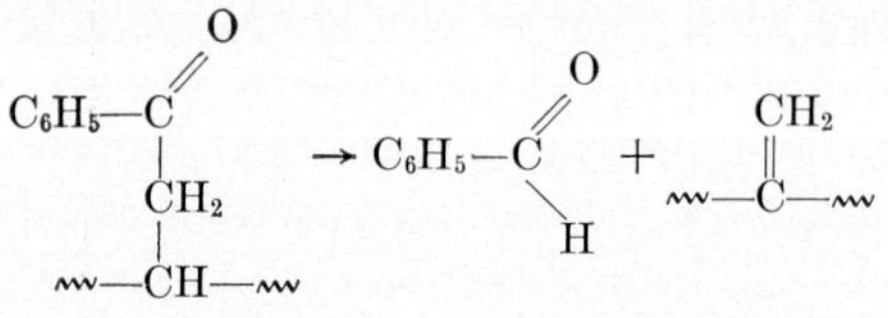

$$\begin{array}{ccccc} \text{O} & & & & \\ /\!/ & & \text{O} & & \\ C_6H_5\text{—C} & & /\!/ & & CH_2 \\ | & \rightarrow & C_6H_5\text{—C} & + & \| \\ CH_2 & & \backslash & & \sim\!\sim\text{—C—}\sim\!\sim \\ | & & \text{H} & & \\ \sim\!\sim\text{—CH—}\sim\!\sim & & & & \\ & & & & (\mathbf{3}) \end{array} \tag{27}$$

The value of E of 19 kcal/mole for the initial degradation phase in air falls in the range of values reported for hydroperoxide formation.

Since the second and third reaction stages in air were endothermal, bond rupture was indicated. Values of E, n, and A for these stages were of the order expected for the rupture of chemical bonds (50–52). Based upon the gaseous products obtained from these stages, the following free-radical mechanism was suggested (eqs. 28–30). Cleavage occurs at the carboxyl oxygen atoms to form phthalic anhydride (eq. 28).

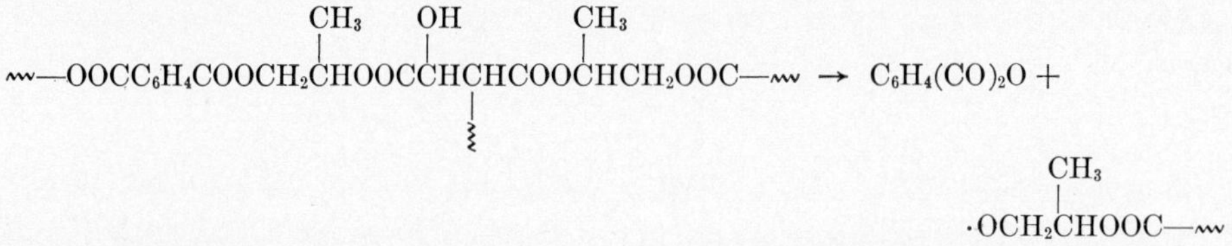

$$\sim\!\sim\text{—OOCC}_6\text{H}_4\text{COOCH}_2\overset{\overset{\displaystyle CH_3}{|}}{\text{C}}\text{HOOC}\underset{\underset{\displaystyle \wr}{|}}{\overset{\overset{\displaystyle OH}{|}}{\text{C}}}\text{HCHCOO}\overset{\overset{\displaystyle CH_3}{|}}{\text{C}}\text{HCH}_2\text{OOC—}\sim\!\sim \;\rightarrow\; C_6H_4(CO)_2O + \cdot\text{OCH}_2\overset{\overset{\displaystyle CH_3}{|}}{\text{C}}\text{HOOC—}\sim\!\sim \tag{28}$$

The hydroxy ester may then be formed by reaction with hydrogen (eq. 29).

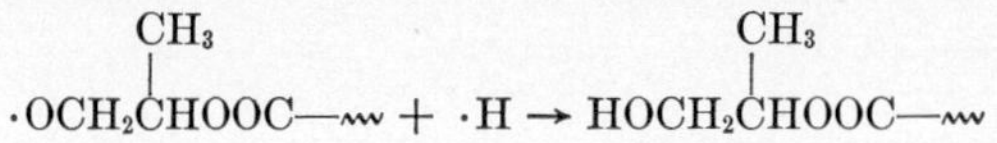

$$\cdot\text{OCH}_2\overset{\overset{\displaystyle CH_3}{|}}{\text{C}}\text{HOOC—}\sim\!\sim + \cdot H \rightarrow \text{HOCH}_2\overset{\overset{\displaystyle CH_3}{|}}{\text{C}}\text{HOOC—}\sim\!\sim \tag{29}$$

Decarboxylation of the polyester (eq 30) should occur readily and accounts for the formation of propylene and carbon dioxide.

$$\sim\!\!\sim\!\!\sim\text{—COOCH}_2\underset{}{\overset{\text{CH}_3}{\overset{|}{\text{C}}}}\text{HOOC—}\sim\!\!\sim\!\!\sim \rightarrow 2\,\text{CO}_2 + \text{CH}_2\text{=CHCH}_3 \quad (30)$$

In the fourth degradation stage (450–550°C) the exothermal trend and the high value of E of 79 kcal/mole indicate that oxidation of carbon, formed in the third reaction stage, occurs. This is supported by reported values of E for the reaction between carbon and oxygen (80 kcal/mole). Furthermore, a residue of carbon is found after degradation is complete in argon.

Polytetrafluorethylene. The isothermal decomposition of polytetrafluoroethylene (Teflon) in vacuum was reported by Madorsky and co-workers (53) to involve first-order kinetics. However, Wall and Michaelsen (54) later indicated, by the use of isothermal methods in a nitrogen atmosphere, that first-order kinetics obtain above about 510°C, whereas below 480°C the polymer degrades by a zero-order law. Subsequently, Madorsky and Straus (55) carried out another series of isothermal experiments in vacuum at lower temperatures than they had previously employed (below 485°C). They concluded that the degradation of polytetrafluoroethylene by heat was a first-order reaction through the temperature range of 425 to 513°C, and obtained a value of E of 80.5 kcal/mole. The preceding discordance of reaction order obtained by isothermal methods was the topic of several succeeding studies that employed TGA methods. See also TETRAFLUOROETHYLENE POLYMERS.

Anderson (56) studied the pyrolysis of polytetrafluoroethylene in vacuum, by means of TGA techniques, over the temperature range of 450 to 550°C. Employing the method of Freeman and Carroll (5) of estimating kinetic parameters, he found that, based upon eleven replicate TGA experiments, the values of E and n were 74.8 ± 3.9 kcal/mole and 1.02 ± 0.07, respectively. Others (11,19) have also used the method of Freeman and Carroll with the maximization method and isothermal techniques and have found values of E and n of a similar order of magnitude, even for temperatures below 480°C. Anderson (22,57) also employed thermogravimetric cycling experiments to determine whether the kinetics and mechanism are consistent as the polymer is degraded. Cycling experiments were carried out between 25°C and a fixed higher temperature; the ratio W_R/W_i for each cycle should follow a curve expressed by equation 31, where W_R is the weight of polymer remaining after cycle M,

$$(W_R/W_i) = (1 - W/W_i)^M \quad (31)$$

W_i is initial weight, and W is weight loss during the first cycle. From Figure 33 it can be seen that good agreement is obtained for polytetrafluoroethylene between theoretical and experimental cycling values. From this figure it can also be observed that the agreement between theoretical and experimental cycling values is unsatisfactory for a copolymer of tetrafluoroethylene and hexafluoropropylene. TGA curves indicate that there were two stages in the decomposition of this copolymer. Values of E varied from about 35 to 60 kcal/mole as the conversion rose from 10 to 30%. The cycling experiments (up to 500°C, which involved the initial stage) also indicate that the initial decomposition stage involves chain segments of lowered stability (see DEGRADATION). Wall and Straus (58) have reported that a similar copolymer produced certain volatile products during decomposition which were more abundant in the early pyrolysis stages. When the cycling experiments were carried up

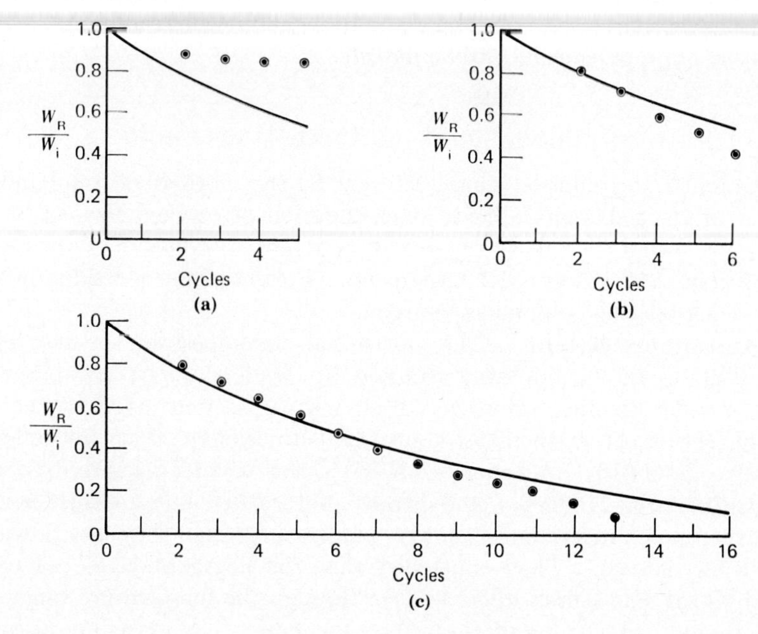

Fig. 33. Thermogravimetric cycling curves: W_i, initial weight at 25°C, W_R, residual weight at cycling temperature. The solid line represents the theoretical curve (57). Key: (**a**) fluorinated ethylene–propylene copolymer at 500°C; (**b**) fluorinated ethylene–propylene copolymer at 540°C; (**c**) polytetrafluoroethylene at 540°C.

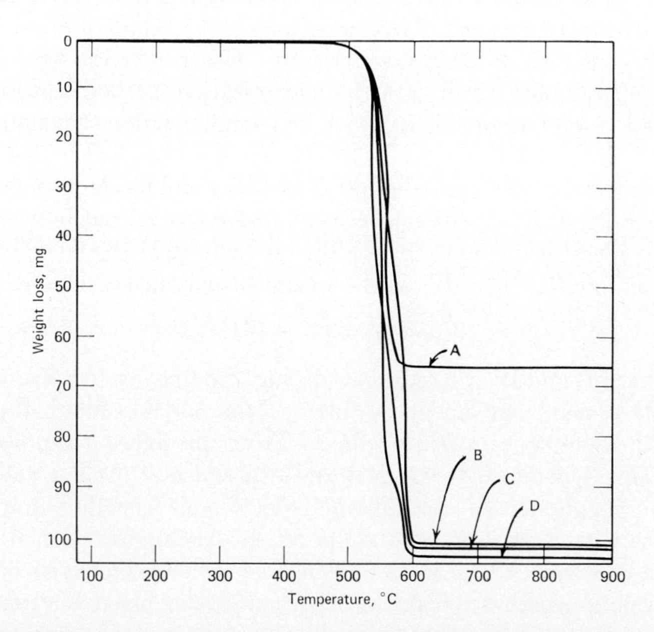

Fig. 34. Thermogravimetric analysis curves of polytetrafluoroethylene and polytetrafluoroethylene–silica mixtures in air (59). Curve A, 50% polytetrafluoroethylene, 50% silicon dioxide (100.0 mg sample); curve B, 90% polytetrafluoroethylene, 10% silicon dioxide (101.0 mg sample); curve C, polytetrafluoroethylene (102.0 mg sample); curve D, 75% polytetrafluoroethylene, 25% silicon dioxide (104.0 mg sample).

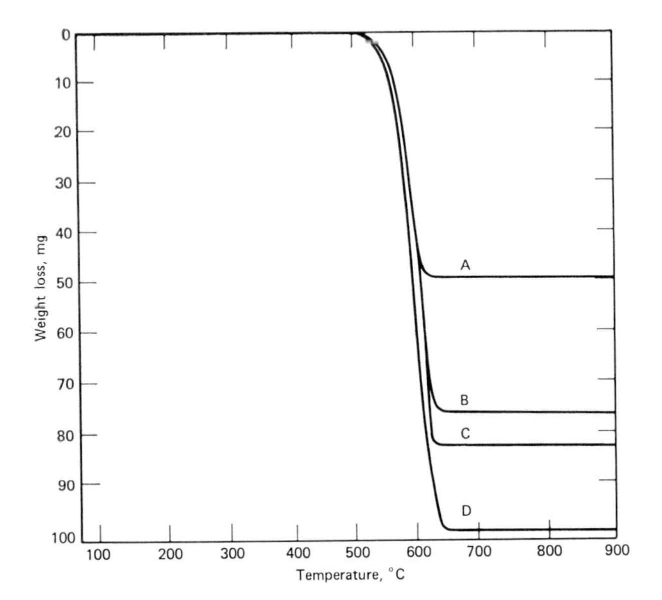

Fig. 35. Thermogravimetric analysis curves of polytetrafluoroethylene and polytetrafluoroethylene–silica mixtures in helium (59). Curve A, 50% polytetrafluoroethylene, 50% silicon dioxide (100.0 mg sample); curve B, 75% polytetrafluoroethylene, 25% silicon dioxide (103.0 mg sample); curve C, 90% polytetrafluoroethylene, 10% silicon dioxide (92.0 mg sample); curve D, polytetrafluoroethylene (100.0 mg sample).

to 540°C (which involved the second stage), the corresponding results indicate that the pyrolysis products consist of other volatile products during the later degradation stages. Thus, it has been found that at 10% conversion, the light volatiles contained 85% of C_3F_6 and 9% of C_2F_4, whereas at 92% decomposition, the volatiles contained 19% of C_3F_6 and 68% C_2F_4.

Light and co-workers (59) recently reported TGA methods for the analysis of silica-filled polytetrafluoroethylene. Figures 34 and 35 present TGA curves in air and helium, respectively, for pure powdered polytetrafluoroethylene and three mixtures of polytetrafluoroethylene with colloidal silica. In helium, the decomposition temperature range for all the samples studied was from about 500 to 650°C. At the completion of decomposition, all of the polytetrafluoroethylene volatilized, leaving a percentage of silica residue which agreed well with the theoretical value (assuming no interaction between polytetrafluoroethylene and silica). However, in air, the decomposition temperature range was from about 500 to 600°C, and based upon the amount of silica remaining it was apparent that interaction had occurred between silica and oxidation products of polytetrafluoroethylene. In order to account for the results obtained in air, the following stoichiometric equation (eq. 32) was proposed.

$$-(C_2F_4)-_n + SiO_2 + 2\,O_2 \rightarrow 2\,COF_2 + SiF_4 + 2\,CO_2 + (n-2)\,C_2F_4 + C_2F_4 \text{ (recombination products)} \quad (32)$$

From equation 32 it may be calculated that polytetrafluoroethylene–silica mixtures containing up to 23% of the silica will be completely volatilized (see Fig. 34). For a mixture containing 50% silica it can be calculated that there should be a total weight loss of 65%. An actual loss of 66% was observed (see Fig. 34). The stoichiometry of the reaction, the need for air to be present, and the identification of pyrolysis products by infrared analysis suggested several steps in the decomposition mechanism. The first step was considered to involve a simple thermal decomposition (unzipping) of the polytetrafluoroethylene which may occur in either inert or oxygen atmosphere (eq. 33).

$$\text{-(}C_2F_4\text{)-}_n\ (s) \rightarrow C_2F_4\ (g) + [C_3F_6\ (g) + \text{cyclo-}C_4F_8\ (g)] \quad (33)$$

In an inert atmosphere, no further reaction would be anticipated. However, in air, the following was postulated

$$C_2F_4 + O_2 \rightarrow 2\ COF_2 \quad (34)$$

and

$$2\ COF_2 + SiO_2 \rightarrow SiF_4 + 2\ CO_2 \quad (35)$$

It should be realized that lesser amounts of many other products have been identified, by mass spectrometry (60), from the thermal oxidation of polytetrafluoroethylene. Also, the value of E of 79 kcal/mole obtained from the isothermal oxidation of polytetrafluoroethylene (61) indicates that the attack of the oxygen occurs mainly on the gaseous pyrolysis products of the polytetrafluoroethylene and not on the polytetrafluoroethylene per se.

Several investigators (62,63) have proposed mechanisms to account for the degradation behavior of polytetrafluoroethylene in the absence of air. Thus, from TGA, isothermal, molecular-weight, etc, techniques, the following experimental observations must be accounted for: the pyrolysis product is mainly monomer; the pyrolysis rate is independent of molecular weight; the pyrolysis follows first-order kinetics over the entire conversion range; and the molecular weight decreases during vacuum pyrolysis. The following plausible scheme involves random initiation followed by depropagation and bimolecular termination (by disproportionation):

$$\sim\!\sim \xrightarrow{k_i} \sim\!\sim\cdot + \cdot\!\sim\!\sim$$

$$\sim\!\sim\cdot \xrightarrow{k_d} \sim\!\sim\cdot + C_2F_4$$

$$\sim\!\sim\cdot + \cdot\!\sim\!\sim \xrightarrow{k_t} \sim\!\sim CF{=}CF_2 + CF_3{-}CF_2{-}\!\sim\!\sim$$

From the above scheme, we may write equations 36–39, in which R_i = initiation

$$R_i = 2\,k_i\,\bar{N}_n[P] \quad (36)$$

$$d[M]/dt = k_d[R\cdot]V = -(1/m)dW/dt \quad (37)$$

$$-d[R\cdot]/dt = k_t[R\cdot]^2 \quad (38)$$

$$[P] = W/Vm\bar{N}_n = \rho/m\bar{N}_n \quad (39)$$

rate; $\bar{N}_n$ = number-average degree of polymerization; [R] and [P] denote the concentration of radicals and polymer molecules, respectively; [M] is the number of moles of monomer; V = volume of polymer; and m denotes the monomer molecular weight. Furthermore, we may write equation 40 for the rate of change in the number of polymer molecules (N).

$$dN/V\,dt = -k_i\bar{N}_n[P] + k_t[R\cdot]^2 \quad (40)$$

From the definition of number-average molecular weight, $\bar{M}_n = W/N$, equations 41 and 42 are obtained.

$$d\bar{M}_n/dt = (\bar{M}_n/W)\,dW/dt - (\bar{M}_n^2/W)\,dN/dt \quad (41)$$

$$d\bar{M}_n/dt = -(\bar{M}_n/W)\,m\,k_d[R\cdot]V + (\bar{M}_n^2/W)k_i\,\bar{N}_n[P]V - (\bar{M}_n^2/W)\,k_t[R\cdot]^2V \quad (42)$$

From steady-state considerations,

$$k_t[\text{R}\cdot]^2 = 2\,k_i\bar{N}_n[\text{P}]$$

or

$$[\text{R}\cdot] = 2^{1/2}(k_i\,\bar{N}_n[\text{P}]/k_t)^{1/2} \tag{43}$$

The kinetic chain length (L) may be defined by equation 44.

$$L = k_d/k_t[\text{R}\cdot] = (k_d/k_t)(k_t^{1/2}/2k_i\bar{N}_n[\text{P}]^{1/2}) = k_d(m/2\rho k_ik_t)^{1/2} \tag{44}$$

Inserting the expression for L (eq. 44) into the expression for $d\bar{M}_n/dt$ (eq. 41) gives equation 45.

$$d\bar{M}_n/dt = -\bar{M}_nk_i\left(\frac{\bar{M}_n}{m} + 2L\right) \tag{45}$$

When $L \ll \bar{M}_n/m$,

$$\frac{1}{\bar{M}_n} - \frac{1}{M_{n,0}} = \frac{k_i}{m}t \tag{46}$$

Furthermore, from equation 37 and equations 39 and 43, the following equation can be obtained:

$$-dW/dt = 2^{1/2}k_d\left(\frac{mk_i}{\rho k_t}\right)^{1/2} W \tag{47}$$

From the above expression, equation 48 can be obtained.

$$k(\text{exptl}) = 2k_iL \tag{48}$$

Finally, when the expression for t from equation 47 is substituted into equation 46, equation 49 is obtained.

$$\frac{\bar{M}_{n,0} - \bar{M}_n}{\bar{M}_n\bar{M}_{n,0}} = (1/2mL)\ln\frac{W_0}{W} \tag{49}$$

From equations 44, 47, and 48 it can be seen that the rate of weight loss of the polytetrafluoroethylene should be first-order in W and should be independent of molecular weight, as observed. Also, from equation 49, the value of $\bar{M}_n$ should decrease with conversion, as observed. It should be noted here that if end groups influenced the pyrolysis rate and terminal initiation was the predominant initiation step, then the pyrolysis rate would be a function of the molecular weight. This is not observed experimentally.

From equation 47 the following approximate expression may be written for the activation energies for the various steps in the mechanism (assuming that $E_t \approx 0$)

$$E(\text{exptl}) \approx E_d + \tfrac{1}{2}E_i \tag{50}$$

Since the observed activation energy possesses a value of about 80 kcal/mole and assuming that the E_i = 74 kcal/mole (calculated from thermodynamic data for the bond energy of the C–C bond in polytetrafluoroethylene (64)), the value of E_d should be about 43 kcal/mole. It is interesting to note that the activation enthalpy for depropagation of polytetrafluoroethylene at 480°C was estimated to be 44 kcal/mole (63).

Bibliography

1. C. Duval, *Inorganic Thermogravimetric Analysis*, 1st ed., Elsevier Publishing Co., New York, 1953.
2. A. E. Newkirk and E. L. Simons, *Rept. No. 63-RL-3498C*, General Electric Co., Nov. 1963.
3. H. Saito, *Bull. Imper. Acad. Japan* **2,** 58 (1926).
4. D. W. Van Krevelen, C. Van Heerden, and F. J. Huntjens, *Fuel* **30,** 253 (1951).
5. E. S. Freeman and B. Carroll, *J. Phys. Chem.* **62,** 394 (1958).
6. L. Reich and D. W. Levi, *Macromol. Rev.* **1,** pp. 173 ff. (1967).
7. W. W. Wendlandt, *Thermal Methods of Analysis*, Interscience Publishers, a division of John Wiley & Sons, Inc., New York, 1964.
8. P. D. Garn, *Thermoanalytical Methods of Investigation*, Academic Press, Inc., New York, 1965.
9. P. E. Slade, Jr., and L. T. Jenkins, eds., *Techniques and Methods of Polymer Evaluation*, Marcel Dekker, Inc., New York, Vol. 1, 1966.
10. *Tech. Bull. T-102,* Mettler Instrument Corp., Princeton, N.J.
11. D. W. Levi, L. Reich, and H. T. Lee, *Polymer Eng. Sci.* **5,** 135 (1965).
12. H. L. Friedman, *U.S. Dept. Commerce, Office Tech. Serv., PB Rept.* **145,** 182 (1959).
13. L. Reich and D. W. Levi, *Makromol. Chem.* **66,** 102 (1963).
14. A. W. Coats and J. P. Redfern, *Analyst* **88,** 906 (1963).
15. D. A. Smith, *Rubber Chem. Technol.* **37,** 937 (1964).
16. D. A. Anderson and E. S. Freeman, *J. Polymer Sci.* **54,** 253 (1961).
17. S. L. Madorsky, *J. Polymer Sci.* **9,** 133 (1952).
18. J. H. Van Tassel and W. W. Wendlandt, *J. Am. Chem. Soc.* **81,** 813 (1959).
19. L. Reich, H. T. Lee, and D. W. Levi, *J. Polymer Sci.* [B] **1,** 535 (1963).
20. R. M. Fuoss, I. O. Salyer, and H. S. Wilson, *J. Polymer Sci.* [A] **2,** 3147 (1964), errata p. 5027.
21. H. L. Friedman, *Paper, Am. Chem. Soc. Meet., Div. Polymer Chem., New York, Sept. 1963; Polymer Preprints* **4** (2), p. 662; in *Thermal Analysis of High Polymers*, B. Ke, ed., Interscience Publishers, a division of John Wiley & Sons, Inc., New York, 1964, p. 183.
22. H. C. Anderson, *Paper, Am. Chem. Soc. Meet., Div. Polymer Chem., New York, Sept. 1963; Polymer Preprints* **4** (2), p. 655.
23. H. C. Anderson, *J. Polymer Sci.* [B] **2,** 115 (1964).
24. L. Reich, H. T. Lee, and D. W. Levi, *J. Appl. Polymer Sci.* **9,** 351 (1965).
25. L. Reich, *J. Polymer Sci.* [B] **2,** 621 (1964).
26. L. Reich, *J. Appl. Polymer Sci.* **9,** 3033 (1965).
27. C. D. Doyle, *J. Appl. Polymer Sci.* **5,** 285 (1961).
28. L. A. Dudina and N. S. Yenikolopyan, *Vysokomolekul. Soedin.* **5,** 986 (1963).
29. A. W. Coats and J. P. Redfern, *Nature* **201,** 68 (1964).
30. J. H. Flynn and L. A. Wall, *J. Polymer Sci.* [B] **4,** 323 (1966).
31. T. Ozawa, *Bull. Chem. Soc. Japan* **38,** 1881 (1965).
32. A. E. Newkirk, *Anal. Chem.* **32,** 1558 (1960).
33. B. M. Moiseev, *Russ. J. Phys. Chem.* **37,** 357 (1963).
34. C. D. Doyle, *Makromol. Chem.* **80,** 220 (1964).
35. S. L. Madorsky and S. Straus, *J. Res. Natl. Bur. Std.* **55,** 223 (1955).
36. W. W. Wright, in *Thermal Degradation of Polymers,* Monograph No. 13, Society of Chemical Industry, London, 1961, p. 248.
37. S. L. Madorsky and S. Straus, p. 60, Ref. 36.
38. F. R. Eirich and H. F. Mark, p. 43, Ref. 36.
39. C. N. Jurkov, *Dokl. Akad. Nauk SSSR* **47,** 493 (1946).
40. I. F. Kanavetz and L. G. Batalov, *SPE Trans.* **1,** 63 (1961).
41. C. S. Marvel, *SPE J.* **20,** 220 (1964).
42. K. D. Jeffreys, *Brit. Plastics* **36,** 188 (1963).
43. E. Horowitz and T. P. Perros, *J. Inorg. Nucl. Chem.* **26,** 139 (1964).
44. E. Horowitz and T. P. Perros, *J. Res. Natl. Bur. Std.* **69A,** 53 (1965).
45. C. D. Doyle, *WADD Tech. Rept. 60-283,* June 1960.
46. C. D. Doyle, *Anal. Chem.* **33, 77** (1961).
47. S. S. Stivala, G. R. Sacco, and L. Reich, *J. Polymer Sci.* [B] **2,** 943 (1964).
48. C. D. Doyle, *J. Appl. Polymer Sci.* **6,** 639 (1962).
49. D. A. Anderson and E. S. Freeman, *J. Appl. Polymer Sci.* **1,** 192 (1959).

50. I. Marshall and A. Todd, *Trans. Faraday Soc.* **49,** 67 (1953).
51. S. Straus and L. A. Wall, *J. Res. Natl. Bur. Std.* **60,** 39 (1958).
52. G. Blyholder and H. Eyring, *J. Phys. Chem.* **61,** 682 (1957).
53. S. L. Madorsky, V. E. Hart, S. Straus, and V. A. Sedlak, *J. Res. Natl. Bur. Std.* **51,** 327 (1953).
54. L. A. Wall and J. D. Michaelsen, *J. Res. Natl. Bur. Std.* **56,** 27 (1956).
55. S. L. Madorsky and S. Straus, *J. Res. Natl. Bur. Std.* **64A,** 513 (1960).
56. H. C. Anderson, *Makromol. Chem.* **51,** 233 (1962).
57. H. C. Anderson, in *Thermal Analysis of High Polymers*, B. Ke, ed., Interscience Publishers, a division of John Wiley & Sons, Inc., New York, 1964, p. 175.
58. L. A. Wall and S. Straus, *J. Res. Natl. Bur. Std.* **65A,** 227 (1961).
59. T. S. Light, L. F. Fitzpatrick, and J. P. Phaneuf, *Anal. Chem.* **37,** 79 (1965).
60. R. E. Kupel, M. Nolan, R. G. Keenan, M. Hite, and L. D. Scheel, *Anal. Chem.* **36,** 386 (1964).
61. J. M. Cox, B. A. Wright, and W. W. Wright, *J. Appl. Polymer Sci.* **8,** 2951 (1964).
62. H. L. Friedman, *Aerophys. Res. Memo 37*, Tech. Info. Series R59SD385, General Electric Co., June 19, 1959.
63. J. C. Siegle, L. T. Muus, T.-P. Lin, and H. A. Larsen, *J. Polymer Sci.* [A] **2,** 391 (1964).
64. C. R. Patrick, *Tetrahedron* **4,** 26 (1958).

Leo Reich and David W. Levi
Picatinny Arsenal

THERMOMECHANICAL PROPERTIES.

See MECHANICAL PROPERTIES; TORSIONAL BRAID ANALYSIS

THERMOPLASTIC POLYMER

A thermoplastic polymer is one that is capable of being repeatedly softened when heated and hardened when cooled (1). See also POLYMER; PLASTIC; THERMOSETTING POLYMER.

Bibliography

1. "Nomenclature Relating to Plastics," *ASTM D 883-69*, American Society for Testing and Materials, Philadelphia, Pa., 1969.

THERMOSETTING POLYMER

A thermosetting polymer is one that is capable of being changed into a substantially infusible or insoluble product when cured by the application of heat or by chemical means (1). The cured polymer may be termed thermoset. See also CURING; POLYMER; THERMOPLASTIC POLYMER.

Bibliography

1. "Nomenclature Relating to Plastics," *ASTM D 883-69*, American Society for Testing and Materials, Philadelphia, Pa., 1969.

THETA PARAMETER.

See SOLUTION PROPERTIES

THIAZOLE POLYMERS.

See POLYBITHIAZOLES; POLYBENZOTHIAZOLES

THIN-LAYER CHROMATOGRAPHY.

See CHROMATOGRAPHY

THIOKOL POLYMERS. See POLYSULFIDE POLYMERS

THIOUREA RESINS. See AMINO RESINS

THIXOTROPY. See COLLOIDS; MELT VISCOSITY

THREE-DIMENSIONAL NETWORK. See CROSSLINKING

TILE. See FLOORING MATERIALS.

TIRE CORD. See TIRES

TIRES

The first rubber tires were solid rubber carriage tires exhibited in London in 1851. They were patented by R. W. Thomson in 1867, and by 1870 were widely used on bicycles as well as carriages. Solid tires were used on urban delivery trucks well into the 1920s, although the requirements of speed and comfort put pneumatic tires on passenger cars in the early years of the century. At that time the automobile industry was in its early period of growth and carriage tires were evolving into automobile tires. Since solid tires were usable to a maximum of 18 mph, long-distance trucking became practical only when pneumatic truck tires were developed. Solid tires are no longer on the roads but are still used as "industrials," eg, on lift trucks for indoor materials handling. The use of polymers in automobiles in parts other than tires is discussed under AUTOMOTIVE APPLICATIONS. See also ELASTOMERS, SYNTHETIC; RUBBER, NATURAL.

Historical Background

Although Thomson patented a pneumatic tire in 1845, his design was not used commercially, and J. B. Dunlop of Dublin is usually regarded as the first inventor; his patent of 1888 described a layer of fabric sandwiched between two layers of rubber. In the following year, the Pneumatic Tyre Company, later becoming the Dunlop Rubber Company, was founded. In the same year both W. E. Bartlett and T. B. Jeffrey patented a detachable pneumatic tire of the "clincher" type, a design that prevailed for the next twenty years. These tires were of a capital omega cross section with a rubber bead which was "clinched" by the rim; Dunlop's original tires had been cemented to the rim. In 1905 the straight-sided tire with a wire bead was introduced by the Goodyear Tire & Rubber Company but two decades passed before it completely replaced the "clincher."

The history of tire design and composition is largely an account of the development of the pneumatic tire, which has become a composite of a number of parts, each with different functions, materials, and properties, as shown in Figure 1 and Table 1. In 1904 Continental Caoutchouc Co. made the first pneumatic tire with a flat rather than rounded tread surface. Two years later nonskid buttons, or lugs, were added by American tire makers to the previously smooth tread surface to give increased traction. From this beginning have developed the familiar tread patterns of highway tires and the special tread designs for traction in snow, for agricultural machinery, for road building, and for mining equipment.

The early pneumatic tires were inflated to relatively high pressures to sustain the loads; the ride was relatively harsh by present standards. The "balloon" tire first appeared in 1923 with an increased cross-sectional area, enabling it to carry the same

Table 1. Properties and Composition of Tire Components

	Tread			Body		
	Cap	Undertread	Sidewalls	Plies	Inner liner	Bead insulation
			Paramount properties			
Requirements	resistance to wear, cut growth, groove cracking, and skid	low hysteresis; resistance to heat	resistance to weather and flex cracking	adhesion to cord and tread; low hysteresis; resistance to heat	low air permeability; adhesion to plies; resistance to flex fatigue	stiffness; adhesion to bead wire and plies
			Examples of composition			
Polymers	styrene–butadiene rubber	natural rubber	styrene–butadiene rubber	natural rubber	chlorinated butyl rubber	styrene–butadiene rubber
	polybutadiene	polyisoprene	neoprene	polyisoprene		natural rubber
	natural rubber	polybutadiene	ethylene–propylene terpolymer	polybutadiene	natural rubber	
	polyisoprene	styrene–butadiene rubber		styrene–butadiene rubber	styrene–butadiene rubber	
Pigmentation						
reinforcing	SAF[a]	HAF–LS[a]	GPF[a]	GPF[a]	GPF[a]	SRF[a]
	HAF[a]		FEF[a]	SRF[a]	FT[a]	
	HS–HAF[a]		SiO_2	SiO_2	MT[a]	
	ISAF[a]					
	SiO_2					
color			TiO_2			
			ZnO			
			colorants			

[a] Types of carbon black. See the article on CARBON for detailed descriptions.

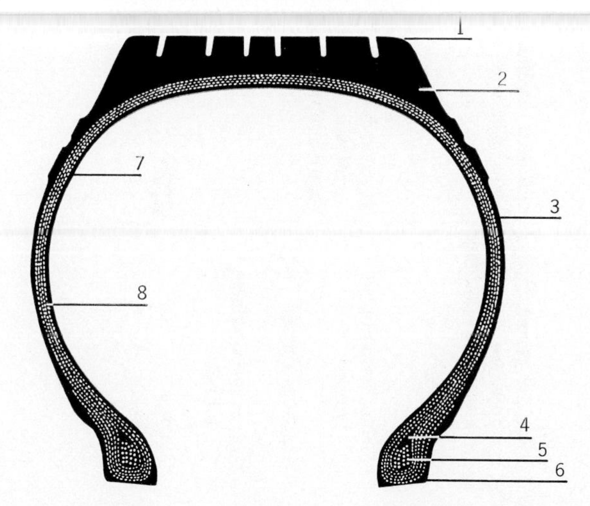

Fig. 1. Components of a tubeless tire: 1, tread cap; 2, tread base; 3, sidewall; 4, bead insulation; 5, bead wire; 6, chafer; 7, inner liner; 8, cord.

load at 32–26 psi inflation that earlier tires carried at 40–45 psi. The lower pressure gave the balloon tires greater cushioning and increased traction. Further development in this direction has reduced the designed inflation to 24 psi for the more comfortable ride expected today. (See also Fig. 2.)

Pneumatic tires in widest use through the 1940s required an air container or *inner tube* of rubber inside the casing of rubber reinforced with textile. The tubeless tire introduced in 1947 had an *inner liner,* a layer of rubber cured inside the casing to contain the air, and a *chafer* around the bead contoured to form an airtight seal with the rim. In 1954 tubeless tires were adopted for new cars (original equipment) although tubes are still extensively used, particularly in truck, agricultural, and bicycle tires.

In 1892 J. F. Palmer reinforced bicycle tires by spirally winding a single cord in place of canvas, the square-woven textile originally proposed by Dunlop. But as long ago as 1887, J. Mosely patented a fabric construction with 98% warp and in 1888 supplied it to Dunlop for experimental bicycle tires. This textile promised an economic advantage over the single-cord process but the tire was not patented until 1908. Automobile tires with cord construction were first applied in 1900 and were produced in the 1900s. The last square-woven ply fabric was used in 1925. The insulation of parallel cords by rubber avoids the fiber-to-fiber friction of square-woven fabric with the resultant heat generation and abrasion between strands. As automobiles operated at ever higher speeds, cord tires became increasingly advantageous and more than repaid their higher initial cost.

The plies of rubber-coated textile are assembled in three basic constructions, bias, radial, and bias-belted (see Fig. 3 and Table 2). Bias tires have an even number of plies with cords at an angle of 30–38° from the tread center line. Passenger-car bias tires commonly have two or four plies, with six for heavy-duty service. Over-the-road truck tires are at present built with six to twelve plies, although the larger earthmover types may contain thirty or more.

The radial-ply tire, in which one or two plies are set at an angle of 90° from the center line and a *breaker* or *belt* of rubber-coated wire or textile is added under the tread, was introduced in France in 1946. This construction gives a different tread–road

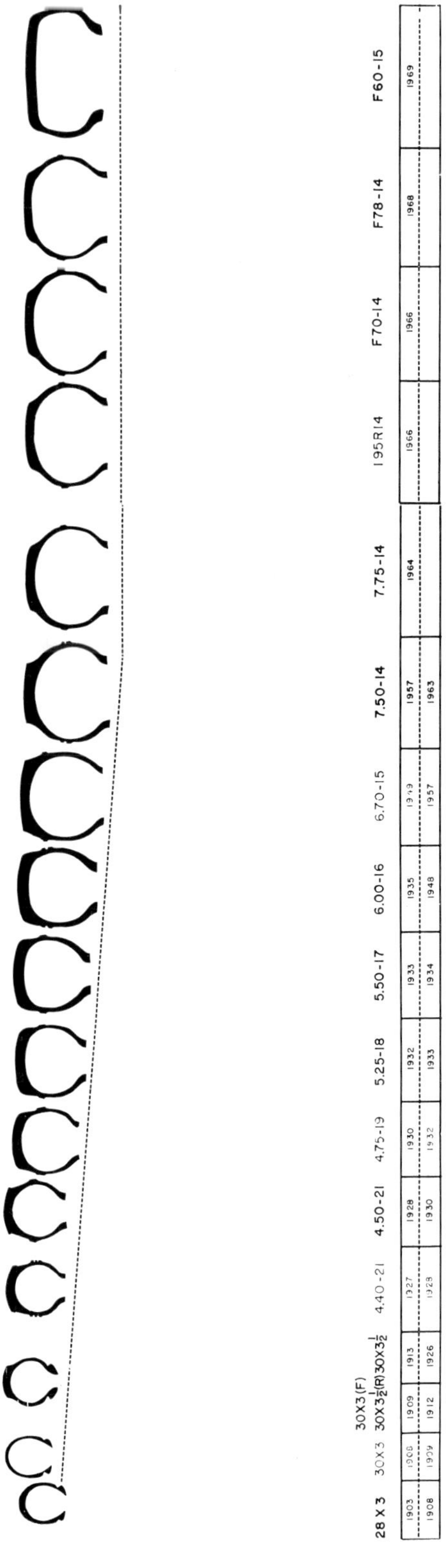

28 X 3	30X3	30X3(F) 30X3½(R)	30X3½	4.40-21	4.50-21	4.75-19	5.25-18	5.50-17	6.00-16	6.70-15	7.50-14	7.75-14	195R14	F70-14	F78-14	F60-15
1903	1908	1909	1913	1927	1928	1930	1932	1933	1935	1949	1957	1964	1966	1966	1968	1969
1908	1909	1912	1926	1928	1930	1932	1933	1934	1948	1957	1963					

Fig. 2. Evolution of automobile tire sizes and profiles from early clincher designs.

Table 2. Dimensions of Equivalent-Sized Tires in Different Basic Constructions (14)

Construction	Designation	Test rim width, in.	Minimum size factor,[a] in.	New tire section width,[b] in.
bias ply	8.25–15	6	35.57	8.20
bias-belted	G78–15	5½	35.36	8.05
radial ply	20.5R15	6	35.20	8.10

[a] Sum of outside diameter and section width.
[b] Maximum FMVSS-109 value 7% above values shown.

interaction, resulting in decreased rate of wear. The sidewall is thin and very flexible; the handling and ride qualities are noticeably different from those of a bias-ply tire. This construction has been highly successful for truck tires, especially in Europe, and is used for wire-reinforced tires. The riding and steering qualities of radial tires require different suspension systems and have not up to the present found widespread use on automobiles in North America except on sports cars, although they are increasingly popular on most European cars.

The bias-belted tire is a more recent development. This tire has much of the treadwear and traction advantage of the radial tire, but the shift from bias to bias-belted tires requires less radical change in vehicle suspension systems and in tire-building machines. Since these features make it more attractive to both automobile and tire manufacturers, it is becoming the dominant tire in the American market.

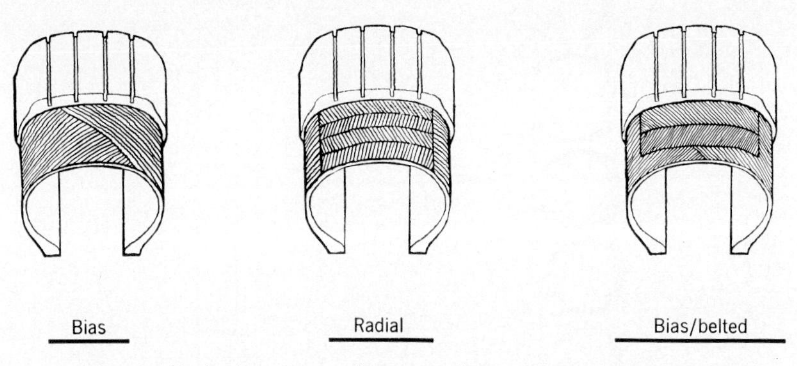

Fig. 3. Basic tire constructions according to ply cord structure.

The strength of a tire casing is determined largely by the properties of the cord used in the body or carcass. Cotton, preferably the long staple Egyptian variety, was almost exclusively used until the 1940s. In 1931 the first tires were produced from rayon; the higher tenacity, which made possible thinner, cooler-running tires, brought it into general use with styrene–butadiene rubber in truck tires during World War II. Later, rayon also replaced cotton in passenger tires but recently, in turn, has been giving way to nylon and polyester fibers. The strength–weight relation of nylon along with its balance of other properties makes it particularly well adapted to heavy-duty truck, earthmover, and airplane tires, where it is dominant today.

Polyester tire cord came on the market around 1963 and gained rapidly in the automobile-tire field, particularly at the expense of rayon. Developments in glass fiber in the 1960s made such fibers suitable for breakers or belts for belted tires of either radial- or bias-ply construction. Steel cable is favored for belts in radial-ply tires in

Table 3. Performance of Fibers in Tire Cord (1)

	Rayon	Nylon	Polyester	Glass
gage, in.	0.037	0.031	0.030	0.021
tenacity, g/den	4.0	7.5	7.5	8.0
cord strength, lb				
2%/sec	61	67	67	75
5000%/sec	71	71	78	95
elongation at break, %	13.0	19.0	17.0	4.0
initial modulus rating[a]	100	60	100	1000
moisture regain, %	11.0	3.5	0.3	0.1
dimensional stability				
thermal shrinkage, %	0.9	6.0	3.0	0.1
growth, %	2.0	8.0	3.0	0.1
wet strength, %	60	90	99+	99+
flex fatigue rating[a]	100	150	165	10
heat generation rating[a]	100	53	107	1000
heat resistance rating[a]	100	150	210	1000

[a] Ratings based on 100; higher numbers indicate superiority.

Europe. Comparison of properties of tire cords from different fibers is presented in Table 3.

Natural rubber was the base of nearly all commercial rubber compounds until World War II (see RUBBER, NATURAL). ("Methyl rubber" was a synthetic which had limited use under blockade conditions in Germany during World War I.) Styrene–butadiene copolymer was the first satisfactory synthetic tire rubber (see also BUTADIENE POLYMERS). Developed in Germany, and later in the United States as World War II became imminent, it was used successfully in all except the heaviest truck, airplane, and off-the-road tires. By the time natural rubber was reintroduced from Indonesia and Malaysia, styrene–butadiene rubber had demonstrated a qualitative superiority in many applications, which assured its continuing use.

The first of the anionic solution elastomers, polybutadiene, was introduced in volume in 1961 (see BUTADIENE POLYMERS). It showed treadwear advantages over styrene–butadiene and natural rubber, and found its place to the extent of 25–35% of the rubber hydrocarbon in tread compounds. Synthetic polyisoprene (see ISOPRENE POLYMERS) was the second polymer of this type to become available; it is currently used as a replacement for natural rubber. Solution-polymerized styrene–butadiene rubber is the latest of the solution polymers; it can be expected to replace part of the older emulsion-polymerized in tires. Butyl rubber (see BUTYLENE POLYMERS) experienced a brief life in premium casings in the period 1959–1962; its superior riding qualities were unable to outweigh its deficiency in treadwear. Because of its low air permeability, butyl rubber has remained the preferred polymer for tubes since its introduction in 1943.

Of the special-purpose rubbers, neoprene (see 2-CHLOROBUTADIENE POLYMERS) and chlorinated butyl rubber are used in the parts of a tire exposed to severe weathering conditions, eg, as a veneer for sidewalls and for white compounds. Ethylene–propylene terpolymer (see Ethylene-1-Olefin Copolymers under ETHYLENE POLYMERS) is also highly resistant to ozone and weathering attacks, but limited compatibility with general-purpose rubbers has thus far largely restricted its use in tires to colored compounds and sidewall veneers. Tires made entirely of ethylene–propylene terpolymer are technically possible, but have shown no overall advantage over tires made from presently available rubbers (2).

The extraordinary reinforcing qualities of carbon black were discovered in 1904 by S. C. Mote of the India Rubber Gutta Percha and Telegraph Works, Silvertown, England. Tread life was increased by a factor of four or more. (Extensive discussion of carbon blacks can be found in the articles on CARBON; FILLERS; ELASTOMERS, SYNTHETIC.) Channel black came into general use between 1910 and the end of World War I. Furnace blacks made from oil rather than gas were first marketed in 1922 but highly reinforcing furnace blacks, in a class with channel black, did not appear until 1947 when HAF (high abrasion furnace) black was introduced. Since that time furnace blacks have displaced channel blacks from tires in all but a few special applications.

Textiles for Tires

The manufacture of tires begins with the preparation of the fabric, namely, spinning, twisting, and application of adhesives before coating with rubber. The spinning, drawing, and twisting of textile tire cord are done by essentially the same processes as are used for textile fibers described in this Encyclopedia under MAN-MADE FIBERS, MANUFACTURE, POLYAMIDE FIBERS, and POLYESTER FIBERS; see also TEXTILE PROCESSING. Special processes for rayon tire cord are reviewed by Takayama and Matsui in Ref. 3; see also RAYON. The finish applied to the yarn is developed for tire peformance as well as mill processing; the twisting is adapted to the type of tire cord and the type of tire. The yarn is first twisted in the *ply twisting stage;* two or more plies are twisted together in the *cable twisting stage.*

Examples of cords currently used in tire plies are as follows:

Rayon	*Nylon*	*Polyester*
1100/2	840/2	1000/2
1650/2		
1650/3	1260/2	1000/3
2200/3	1260/3	1300/3

A glass-fiber cord used in automobile tire belts is 330/1. (The numbers indicate the construction of the cord: thus, 1000/3 means 1000-denier yarn twisted to make the plies and three plies twisted together to make the cables or cords. *Twist,* eg, 10/10, indicates the number of turns per inch in plies and cables, respectively.) Fatigue resistance tends to increase with twist whereas cord strength declines. Performance of the various fibers used in tire cord is indicated in Table 3.

To facilitate processing with adhesives, as well as calendering (qv), the cables are woven 23–35 ends/in., with a minimum number of staple fiber *pick threads, fill threads,* or *weft.* The above *end count* is used in fabric for automobile tires; high end counts give greater plunger strength to the tire carcass at the possible expense of separation resistance. Creel calendering of cord, without the intervening weaving, has limited use in preparation of pickless fabric. This is the usual practice for wire carcasses.

Wire for tire bodies is made from high tensile steel, with ultimate tensile stresses ranging from 350,000 to 390,000 psi. The wires are twisted into cables: for example, three 0.0058-in. diameter wires are twisted together, and seven of these are then twisted with two or four turns per inch to make the cable. The terminology of wire cord is different: the above cord is described as $7 \times 3 \times 0.0058$. Wire for tire belts or break-

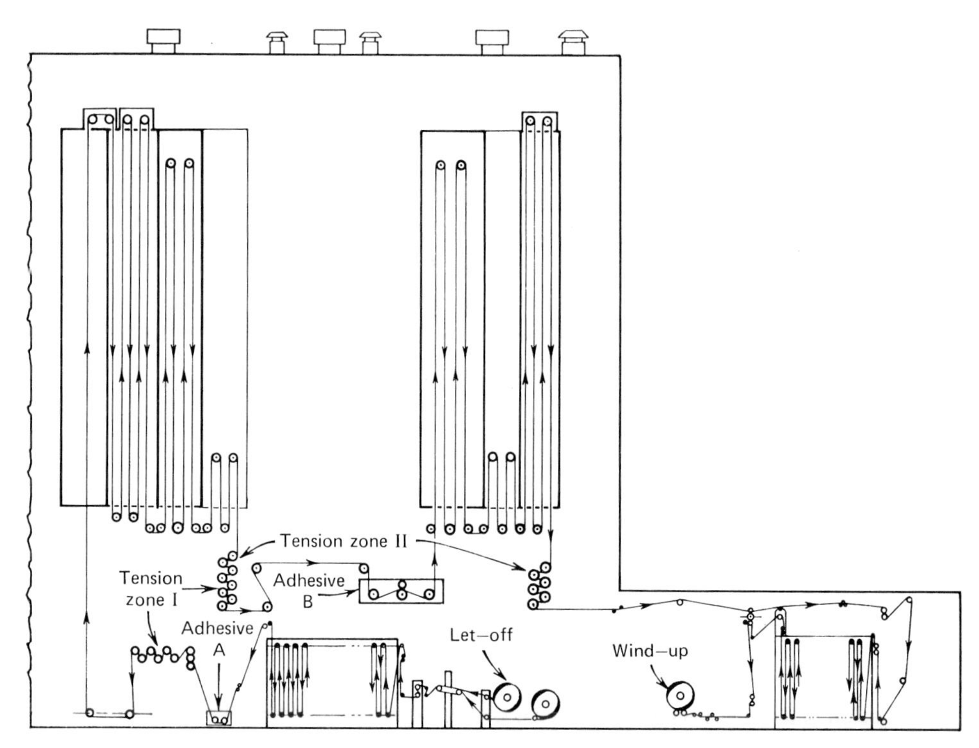

Fig. 4. Two-cycle tire-cord dipping unit. Tension is controlled by pull rolls at either end of each tension zone. The ovens are divided into compartments at 150–350°F for drying, and at 400–500°F for heat treatment.

ers is designed for greater stiffness rather than for flexibility. A breaker cable can have three strands of 0.0080-in. diameter surrounded by six single wires of 0.0150-in. diameter.

Beads are made from high-carbon steel wire insulated with a rubber compound. A typical bead wire falls within the following specification:

	%
carbon	0.65–0.75
manganese	0.60–0.90
phosphorus	0.04 maximum
sulfur	0.05 maximum

Diameter: 0.037-in. for automobile, truck, etc., tires
0.050-in. for wire ply tires
0.032-in. for bicycle tires

To achieve the necessary adhesion to rubber, bead wire is bronze-plated by passing it through a solution of copper and tin salts to deposit 0.5–1.0 g/kg of wire.

Processing of Ply Fabric. To form an adequate cured bond to rubber, textile tire cord must be treated with adhesive before coating with the carcass compound. By far the most widely used adhesives are aqueous systems containing rubber latex, resorcinol, and formaldehyde, which are allowed to react partially before application (see also Adhesives under ADHESION AND BONDING; PHENOLIC RESINS).

Since the physical properties of the cord are affected by water, the dipped fabric must be dried under tension. The application of heat also brings about further reaction of the resorcinol–formaldehyde resol with itself, with the dispersed rubber of the latex, and possibly with the fiber. Figure 4 is a diagram of a cord processing unit which subjects the cord to two cycles of dipping and drying under tension. The actual installation containing this unit occupies a seven-story building.

Although resorcinol–formaldehyde–latex dips develop adequate adhesion of rayon and nylon to rubber, polyester fiber requires additional treatment, for example, a pretreatment with a reaction product of an isocyanurate, resorcinol, and formaldehyde dispersed in a resorcinol–formaldehyde–latex dip.

Processing of glass cord is different from the foregoing in that individual glass fibers are coated with a resorcinol–formaldehyde–latex dip before twisting. In addition to bonding the glass to rubber, the dip insulates the fibers and prevents inter-fiber abrasion (see also FIBERS, INORGANIC). Wire cord does not undergo an aqueous- or solvent-dipping operation; adhesion to rubber is secured by electroplating with brass or by galvanizing.

Compounding and Processing of Rubber for Tires

General principles of the compounding of rubber are discussed in the article on RUBBER COMPOUNDING AND PROCESSING. The rubbers and other components going into each part of a tire are selected to optimize the required properties of that component, as outlined in Table 1. Rubber is prepared for fabrication into tire components by blending the elastomer with reinforcing agents, softeners, curing agents, and antidegradants in an internal mixer of the Banbury type. From the mixer the compound may be dropped onto a mill and sheeted, then cut off in slabs or, in large-scale plants, pelletized into approximately $\frac{1}{2} \times \frac{3}{4}$-in. cylinders.

When required for forming into components, the sheets or pellets of compounded rubber are first subjected to a short mastication on a "warm-up" mill, where mechanical working and heat increase its plasticity. The rubber is taken off the mill as a continuous ribbon directly to the calender, extruder, or bead insulator. Warming and plasticating the compounded rubber can also be accomplished in a cold-feed extruder; the rubber enters in a continuous sheet and emerges as a ribbon for the calender or tread tuber or contoured for tread rubber or sidewalls. See also RUBBER COMPOUNDING AND PROCESSING.

The processed fabric is coated with rubber compound in a calendering operation (see CALENDERING). The tread is usually formed by extrusion or "tubing" (Fig. 5). If a tread base or sidewall of different composition is to be included in the tire, it may be extruded simultaneously with the tread cap through a duplex unit, as depicted in Figure 6.

Fabrication of an automobile tire bead begins with insulation of the wire by the rubber compound: multiple strands of wire are fed into the insulator and emerge as a ribbon; several turns of the ribbon make up the bead with the wires forming a rectangular cross section. The ends are taped together with a strip of frictioned fabric or the entire ring may be spirally wrapped with tire cord. Beads for heavy-duty tires are spirally wrapped with square-woven fabric tape; the largest tires require an additional envelope of coated cord or square-woven fabric called a "flipper." For heavy-duty airplane and earthmover tires, eight or nine strands may be included in the ribbon and two or even three beads employed, each within a wrapper and flipper.

Tire Building. The tire is built from the inside out on a steel or inflated rubber drum. The coated cord fabric, precut to required width and specified angle, is delivered to the building machines. The inner liner of calendered rubber is wrapped around the drum, followed by the plies with the cords at the specified angle in alternating directions. The beads are affixed at the two ends of the cylinder and the ends of the plies turned back to wrap around the bead; the chafer, a strip of rubber-coated square-woven fabric, may be wrapped around the bead. The tire at this stage is pictured in Figure 7. The belt, or breaker, if used, is applied on top of the last ply, then the tread and sidewalls

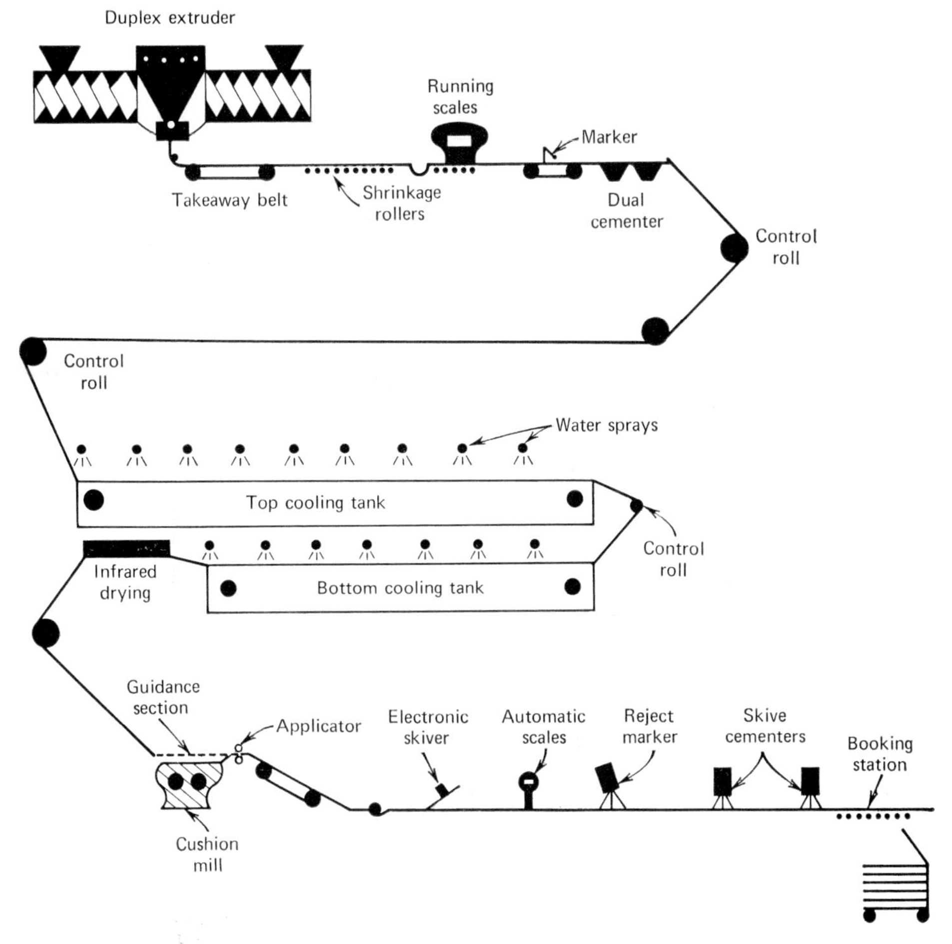

Fig. 5. Tread extrusion unit.

(Fig. 8). The drum is collapsed, permitting the "green" tire to be removed as an elongated cylinder.

The building of radial-ply tires requires an additional step, namely, partial shaping to bring the green tire closer to its ultimate dimensions before application of the belt, which has a very low extensibility. Equipment requirements are discussed in Ref. 4.

Large-sized tires are assembled by sliding bands, or preformed sleeves of coated fabric, over the drum, as illustrated in Figure 9. Instead of being extruded, the tread may be laminated, that is, built up to a programmed contour by winding a strip of

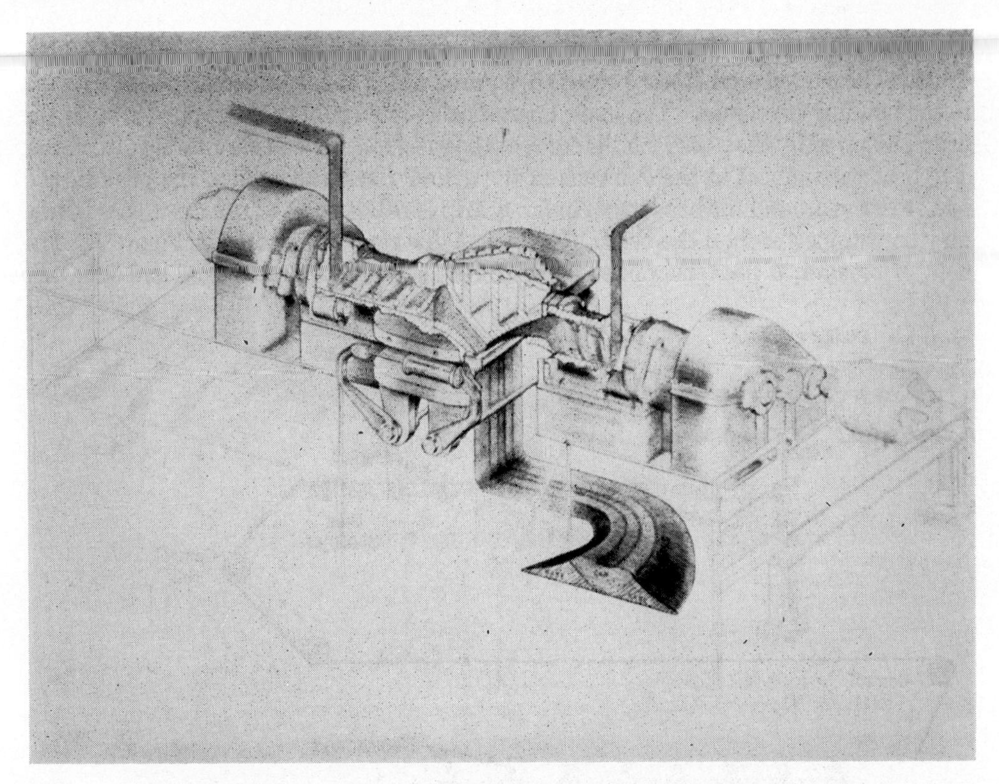

Fig. 6. Duplex tread extrusion showing ribbons of cap and base compounds entering barrel and extruding simultaneously.

Fig. 7. Building machine showing tire with inner liner, plies, and beads in place.

calendered tread compound around the tire carcass after its assembly on the drum (Fig. 10).

Curing. Curing entails the application of specified temperatures and pressures over a period of time to effect vulcanization of the rubber compounds in the tire. The rubber molecules become crosslinked in a series of reactions with sulfur, accelerators, and activators. The process is usually combined with shaping as a preliminary step in which the cylindrical green tire is compressed and forced against the mold in which it is

Fig. 8. Building machine showing "green" tire complete with chafer, tread, and sidewalls.

Fig. 9. Building machine and band ready to be pulled over steel drum.

Fig. 10. Tire with tread being laminated by strip from roll in foreground.

Fig. 11. Automatic curing presses: molds open, bladders collapsed.

to be cured. The cured tire has the final form and physical properties required for service. See also RUBBER CHEMICALS; VULCANIZATION.

In the automatic curing presses most widely used in modern plants, a continuous temperature is maintained by steam in the shell around the outer mold of steel or aluminum. Heat and pressure are supplied in cycles of steam, air, or superheated water

Fig. 12. Molds open, green tires loaded.

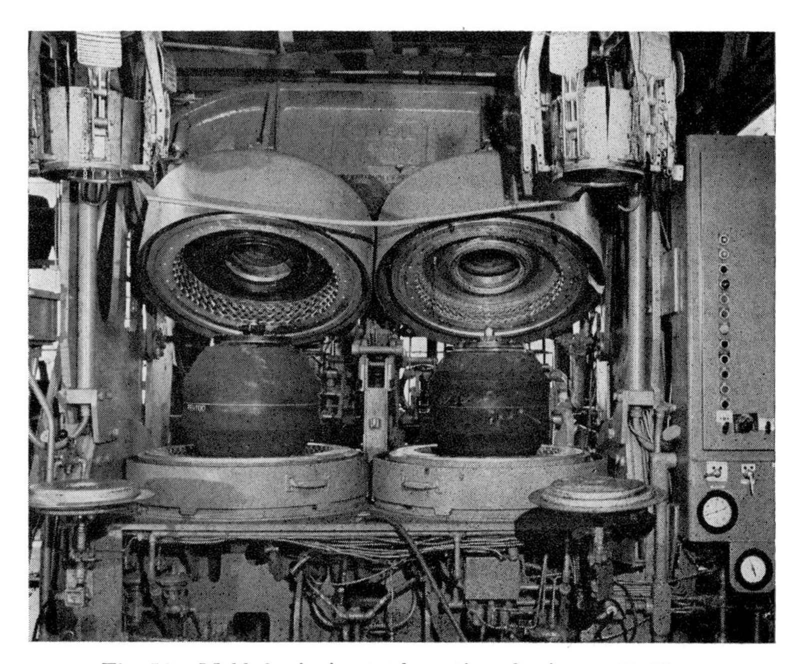

Fig. 13. Molds beginning to close: tires shaping on bladders.

through an attached "bladder" of heat-resistant rubber, depicted in Figure 11. The cylindrical green tire is placed over the bladder (Fig. 12); as the press closes, steam enters the bladder and the tire is expanded by the downward pressure of the mold plus the outward thrust of the bladder (Fig. 13). When the press is completely closed (Fig. 14) the tread and sidewall rubber is forced into the pattern of the mold before

Fig. 14. Molds closed: cure in progress.

Fig. 15. Molds opening: cure complete.

Fig. 16. Molds open: tires lifted off bladders.

vulcanization begins. At the end of the curing cycle, as set by the timer, the bladder may be cooled by internal water spray before it is collapsed by release of pressure, and the mold is opened (Figs. 15 and 16).

Testing

Both laboratory and road tests have been developed by tire manufacturers to gage performance from the time the tire leaves the factory throughout its service life, including mileage to wear-out of the tread and endurance of the carcass. Laboratory tests are designed to evaluate a specific property of a tire rather than the performance under service conditions in which the interaction of a number of properties comes into play.

Testing on a highway, though more expensive, gives results more convincing than testing in a laboratory. Road tests may be conducted on public highways, but the trend is to private roads or proving grounds where control of traffic and speeds permit greater flexibility of conditions and limitation of variables. A commonly used location is the southwestern United States, where climate and availability of land are favorable.

Laboratory Testing

Endurance. The tire is inflated and held under load against a flat-faced wheel driven at a constant specified speed. An example is the endurance test in the Federal Motor Vehicle Safety Standards (Table 4). The load is increased for specific mileages. Failure consists in cord fatigue, separations of tread, between plies, of bead, or chunking of the tread. A variation of this test utilizes a constant load with "step speeds," periods of successively higher speeds, an example of which is the Laboratory High-Speed Endurance Test (Table 4).

Heat Generation. The tire is run against a wheel as in the indoor endurance test. The load can be varied and the machine modified with braking capabilities; the drive may be on either the tire or the flywheel. The temperature in any part of the tire can be checked at intervals by inserting a needle pyrometer, or may be recorded continuously during operation by telemetry from a thermistor embedded in the tire.

Dynamic Weather Resistance. The machine is similar to the wheel used for endurance tests but is run at lower speeds and in the open air so that the effect of weathering on a tire in simulated road service can be observed. Alternatively, a stream of ozonized air can be directed against the sidewall, or the entire flywheel and tire assembly can be enclosed in an ozone box. An example of extensive use of the former is described in Ref. 5.

Table 4. Examples of Some Tests for Tires[a]

Test	Inflation, psi	Load range, lb	Test equipment	Speed, mph	Time, hr	Procedure
laboratory, endurance[b]	24	1380 (4 hr) 1500 (6 hr) 1620 (24 hr)	67.23-in. flywheel	50	34 (at 100 ± 5°F)	
laboratory, high-speed[b]	30	1380	67.23-in. flywheel	50 75 80 85	2 0.5 0.5 0.5	cool to 100°F and adjust inflation

Test	Inflation, psi	Test equipment	Rate, in./min	Procedure	Minimum requirements
carcass strength, plunger energy	24	0.75-in. hemispherical plunger	2	breaking energy at five positions in centerline W = energy, in.-lb F = force, lb P = penetration, in. $W = F \times P/2$	rayon: 1650 in.-lb nylon, polyester: 2600 in.-lb
bead unseating force	24	anvil on outer sidewall	2	bead unseating force at four positions around tire circumference	2500 lb

[a] Examples from Federal Motor Vehicle Safety Standards, 8.25-15 Bias Ply (14).
[b] Tires must complete these tests with no evidence of separations, chunking, or broken cords.

Bruise Resistance. The best-known test for bruise resistance is the static plunger test, in which a ¾-in. diameter steel plunger is forced into the tire, which is then inflated with a tube until break. The plunger moves at 2 in./min (10 ft/hr). The energy required to break is calculated from the load and deflection. This test is included in Federal Safety Standards as Carcass Strength (Table 4).

Recent attention has been directed to measurement of tire strength at higher speeds, approaching actual road conditions. By use of a high-speed plunger the rate of penetration can be increased to 30 mph on currently available hydraulically driven instruments. Rates of penetration over 50 mph can be reached with projectiles.

Whereas in all the previously described tests the tire is at rest, the flywheel plunger penetrates the tire while it is in motion. The tire rotates against a flywheel as in the endurance tests, but after a warm-up period, a plunger protrudes from the surface of the flywheel at a height controlled by a pneumatic piston. The rotation of the tire is continued at 50 mph or faster until tire failure.

Road bruise tests have been designed on the same principle; a plunger of variable height projects from the road surface. The tire on a vehicle is driven over the plunger at increasing heights until failure.

Highway Testing

Treadwear. The rate of wear of a tire is a stochastic value influenced by a wide range of variables, eg, surface conditions of the road, topography, weather, vehicle, wheel position, speed, traffic conditions, and the driver's responses (6–8). For this reason values of rate of wear cannot be considered meaningful unless the conditions are specified in minute detail. Designed rotation schedules and statistical analysis are requisites for quantitative comparisons and reliable interpretation of the data (9,10). The low degree of replicability of absolute wear rates long ago led to running control tires concurrently with experimental tires and calculating a relative wear rate. However, the relative wear is also dependent on conditions and the constancy of control tires has to be assumed.

In spite of the qualifications and the high cost of road tests, they are still regarded as indispensable since no laboratory abrasion tests have shown satisfactory correlation with wear under service conditions (9). Figure 17 shows a section of a 20-mile road constructed for this purpose; the banked curves give a slower rate of wear than the level course since the side force on a tire turning on a level road is one of the causes of rapid wear (7,8) (see also ABRASION RESISTANCE). Automobiles are usually run at 60–70 mph, 24 hours a day; the depth of the tread grooves is checked periodically with a depth gage. The wear rate is expressed in 0.001 in. loss/1000 miles, or the tread life in miles/0.001 in. loss.

Fig. 17. Test track for treadwear with level curves (fast wear) and banked curves (slow wear).

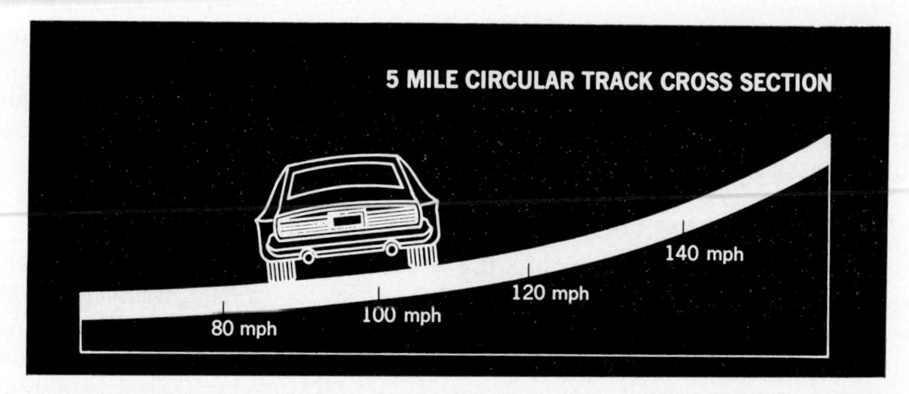

Fig. 18. Cross section of high-speed five-mile circular test track.

Skid Resistance. Skid tests have not yet reached the degree of standardization of the preceding tests for treadwear. Stopping distance is one index of skid resistance, an example of which was published by the Federal Highway Administration (11). The vehicle is accelerated to 60 mph, the brakes applied just short of locking, and the distance measured to stop. The test in this form is used more for evaluation of vehicles than of tires. The National Safety Council conducts annual tests under winter driving conditions. Stopping distance and braking distance (brakes locked) on ice from 20 mph are among the measurements carried out (12).

The instrumented two-wheel trailer is a device to make measurements of the frictional force and is specified by ASTM Standard E 274-65T (13). A pair of tires is mounted on the trailer wheels. After a 5-min warm-up period, the speed is accelerated to the specified level (up to 70 mph) as recorded by a tachometer. The trailer brakes are then locked and the wheel torque measured by a load cell as the speed decreases. The coefficient of skid resistance, SN, is calculated as follows:

$$SN = \frac{F}{W - (HF/L)} \times 100$$

where F = frictional force, lb
W = trailer weight, lb
H = hitch height, in.
L = trailer wheelbase, in.

The trailer is provided with nozzles to spray water on the road surface ahead of the wheels.

Endurance. Tires may be tested for highway endurance by running at normal speeds and loads to high mileages, eg, 20,000 miles for passenger tires, and up to 40,000 miles or more for truck tires. Groove cracking, separations, and fatigue are the types of failure encountered.

High-speed endurance may be run on specially constructed roadways, such as a circular track of 5 miles circumference. The track is banked, as indicated in Figure 18, to balance the centrifugal force at speeds up to 140 mph. After a warm-up period the vehicle is run either at constant or at increasing speeds to tire failure or for a standard duration. Prolonged high-speed operation results in groove cracking, chunking, and separation of tread from carcass.

Federal Safety Standards

The United States Department of Transportation, since January 1, 1968, has set standards of tire safety that include laboratory tests of strength and performance. The specific conditions and standards are included in Refs. 14–16. An example of the requirements for one size and type of tire is given in Table 4. The laboratory tests required include the following:

(*1*) *Endurance.* This test is described above.

(*2*) *High-speed performance.* The tire is driven against the same wheel used for the endurance test, but at speeds above 75 mph. The same types of failure are observed.

(*3*) *Carcass strength.* The strength of a tire carcass is expressed in terms of breaking energy by the static plunger test described above under Bruise Resistance.

(*4*) *Resistance to bead unseating.* The tire is mounted horizontally and pressure applied by an anvil to the sidewall until the bead is unseated from the rim. The apparatus is pictured in the references given above.

Additionally, the Department of Transportation specifies treadwear indicators, permanent sidewall labeling, and dimensions. Examples of the dimensional requirements are given in Table 2.

Nondestructive Tests

Nondestructive tests are applied to assure structural uniformity of the tire and detect heterogeneity within a tire component. Of the tests thus far developed, force variation (17,18) has been adapted to final inspection of passenger tires with limits specified by the automobile manufacturers. In one type of testing machine the tire is driven under load against a drum; the lateral and radial forces are measured by transducers connected with the axle of the drum.

Other nondestructive tests (19) have made use of instrumental techniques to locate irregularities within the tire structure by x rays or ultrasonic vibrations. Infrared radiography and laser beam holography identify surface variations that may result from internal irregularities.

Economic Aspects

The production figures at five-year intervals given in Table 5 reflect world events as well as technical developments. The drop in automobile casings and increase in

Table 5. Production of Tire Casings and Tubes in the United States[a]

	Casings[b]			
Year	Automobile	Truck and bus	Agricultural	Tubes[c]
1938	35,789	5,117		37,850
1943	7,620	12,803	914	15,010
1948	66,738	14,576	5,160	70,030
1953	81,431	14,690	3,756	74,430
1958	83,691	12,911	3,651	41,260
1963	122,184	16,889	3,820	39,660
1968	177,526	25,526	5,741	47,349
1969	180,481	27,211	5,951	44,250

[a] Thousands of units.
[b] Refs. 20 and 21.
[c] Refs. 21 and 22.

Table 6. 1967–1968 Tire-Casing Production by Country—Western Europe and the Americas[a] (23)

	Automobile		Truck and bus		Agricultural	
Country	1967	1968	1967	1968	1967	1968
Brazil	3,647		1,845			
Canada	13,998		1,849		489	
France	25,485	25,791	3,393	3,391		
Germany (West)	21,990	25,259	2,033	2,048		
Italy		16,135		1,964		
Spain	3,374		842		486	
United Kingdom	21,270	24,774	3,780	4,610	806	860
United States	142,027	177,526	28,839	25,526	5,554	5,741

[a] Thousands of units.

Table 7. Tire Production, Eastern Europe, 1967–1968[a] (24)

Country	1967	1968
Bulgaria	365	333
Czechoslovakia	1,908	2,162
Germany (East)	2,553	
Hungary	698	591
Poland	1,776[b]	2,295[b]
Rumania	1,661	1,837
USSR	20,664[c]	31,820[c]
Yugoslavia	1,578	1,780

[a] Thousands of units.
[b] Includes tires and tubes for animal-drawn vehicles.
[c] Includes off-the-road tires.

Table 8. Trends in Automobile Tire Construction[a] (25)

Construction	1960	1965	1968	1969[b]	1974[b]
bias ply	105	146	154	134	20
radial ply			3	4	25
bias-belted			15	40	175
total	105	146	172	178	220

[a] United States shipments, millions of units.
[b] Estimated.

truck and bus tires between 1938 and 1943 reflects wartime needs and cutbacks in civilian goods. The increase in agricultural tires between 1943 and 1948 resulted from a new development in tire applications, as well as an increasing mechanization of the farm. The drastic drop in tube production after 1953 marks the introduction of the tubeless tire as original equipment on the 1954 model year of automobiles.

Country-to-country comparisons are given in Table 6 for North America and Western Europe, and in Table 7 for Eastern Europe.

The present major trends in tire construction are to the bias-belted construction as the standard automobile tire in the United States, to polyester as the body cord, and to glass-fiber belts (see Table 8). A possible turn in the trend, to wider application of radial-ply tires, could mean a slower decline in rayon usage and even introduction of wire for belts as in Europe. Wire-belted–bias-ply tires are another alternative (26).

Table 9. Trends in Tire Cord Usage[a] (25)

Fiber	1960	1965	1968	1969[b]	1974[b]
rayon	221	192	145	120	25
nylon	126	235	310	305	300
polyester		5	90	125	275
glass			8	20	90

[a] In United States, millions of pounds.
[b] Estimated.

Tires of this construction have been announced by the major manufacturers and produced on a limited scale. Widespread use of wire in the United States on either radial or bias plies must await expanded production facilities for the wire itself. Nylon can be expected to hold its position in heavy-duty tires while losing ground in the replacement market for automobile tires (see Table 9).

Fabricless pneumatic tires from cast liquid elastomer were made on an experimental basis in 1959 and have been the subject of numerous patents. Recent activity (27) has heightened interest in this construction, although its future depends on the properties attainable in cast rubber (28). Solid industrial tires cast from polyurethan elastomers have been manufactured since the 1950s.

The evolution of pneumatic tires has accompanied the development and diversification of motor vehicles. First automobiles, then trucks, airplanes, farming implements, and mining and earthmoving equipment have required new designs, fabrics, and rubber compositions. Within each class, vehicles designed for greater power and speed under extreme environmental conditions have presented a continuing challenge to the technologist to meet with tires that perform effectively and safely.

Bibliography

1. F. J. Kovac, *Tire Reinforcing Systems,* Goodyear Tire & Rubber Co., Akron, Ohio, 1968.
2. K. Satake, T. Sone, and M. Hamada, *J. Inst. Rubber Ind.* **4,** 21 (1970).
3. T. Takayama and J. Matsui, *Rubber Chem. Technol.* **42,** 159 (1969).
4. W. O. Murtland, *Rubber World* **162,** 53 (June 1970).
5. T. Sweeney and E. R. Thornley, *J. Inst. Rubber Ind.* **1,** 326 (1967).
6. S. Davison, M. A. Deisz, D. J. Meier, and R. J. Reynolds, *Rubber Chem. Technol.* **38,** 457 (1965).
7. H. R. Knips, *Abnützung der Reifenlauffläche,* Gummiwerke Fulda GmbH, Fulda, Germany, 1966.
8. Gummiwerke Fulda GmbH, *Einsatzbedingungen and Lebensdauer Gummibereifung* **45**(3), 21; (4), 30; (5), 32; (6), 32 (1969).
9. G. G. Richey, J. Mandel, and R. D. Stiehler, *Rubber World* **143,** 84 (Nov. 1960).
10. S. Spinner and F. W. Barton, "Some Problems in Measuring Tread Wear of Tires," *Natl. Bur. Std. (U.S.) Tech. Note 486* (Aug. 1969).
11. Subchapter A, Motor Vehicle Safety Regulations, Part 375.2 Consumer Information, *Federal Register* **34**(99), 8112 (May 23, 1969).
12. G. P. Hajela, *Resume of Tests on Winter Surfaces, 1939–1966,* National Safety Council Committee on Winter Driving Hazards, University of Wisconsin, Madison, Wis., 1968.
13. "Tentative Method for Skid Resistance of Pavements Using a Two Wheel Trailer," ASTM E 274-65T, *1969 Book of ASTM Standards, PT11,* American Society for Testing and Materials, Philadelphia, Pa., 1969.
14. *Motor Vehicle Safety Standard No. 109, New Pneumatic Tires—Passenger Cars, Effective Jan. 1, 1968,* U.S. Dept. of Transportation, Federal Highway Administration, National Highway Safety Bureau.
15. *SAE J918b,* in *SAE Handbook,* Society of Automotive Engineers, Inc., New York, 1970, p. 867.

16. *1969 Year Book*, The Tire and Rim Association, Akron, Ohio.
17. C. Hofelt, Jr., H. D. Tarpinian, and C. Z. Draves, Jr., "Measuring Tire Uniformity," Paper No. 650522, *Soc. Automotive Engineers Trans.* (1966).
18. *SAE J332*, in *SAE Handbook*, Society of Automotive Engineers, Inc., New York, 1970, p. 874.
19. R. F. Wolf, *Rubber Age* **102,** 58 (April 1970).
20. *Rubber Statist. Bull.* **23,** 41 (1969); **24,** 41 (1970).
21. *Rubber Industry Facts*, Rubber Manufacturers Association, 1969, 1970.
22. *Chemical Economics Handbook*, Stanford Research Institute, Menlo Park, Cal., 1968.
23. *Rubber Statistical News Sheet*, Secretariat of the International Rubber Study Group, London, Feb. 1968–Jan. 1970.
24. *Rubber World* **160,** 65 (May 1969).
25. E. V. Anderson, *Chem. Eng. News* **47,** 39 (July 14, 1969).
26. R. F. Wolf, *Rubber Age* **101,** 51 (April 1969).
27. *Autoproducts* **2** (April 1970).
28. A. Pickett, *Polymer Age* **1,** 109 (May 1970).

General Bibliography

History

P. Schidrowitz and T. R. Dawson, *History of the Rubber Industry*, W. Heffer & Sons, Cambridge, England, 1952.

Compounding

G. G. Winspear, ed., *The Vanderbilt Rubber Handbook*, R. T. Vanderbilt Co., Inc., New York, 1968, p. 451.

Current Trends in Automobile Tires

C. Law, *Rubber World* **159,** 33 (March 1969).
R. F. Wolf, *Rubber Age* **101,** 51 (April 1969).
E. L. Carpenter, *Chem. Eng. News* **48,** 31 (April 27, 1970).
R. F. Wolf, *Rubber Age* **102,** 72 (May 1970).
W. O. Murtland, *Rubber World* **162,** 53 (June 1970).
A. Pickett, *Polymer Age* **1,** 109 (May 1970).

J. C. Ambelang and T. A. Riehl
The Goodyear Tire and Rubber Company

TOBACCO, RECONSTITUTED. See FOOD AND DRUG APPLICATIONS

TOLUENE–FORMALDEHYDE RESINS. See HYDROCARBON–FORMALDEHYDE RESINS

TOLUENESULFONAMIDE–ALDEHYDE RESINS. See AMINO RESINS

TOLYLENE DIISOCYANATE. See POLYURETHANS

TOOLING WITH PLASTICS

Tooling with plastics is not a new concept in this rapidly changing technological age. Plastic tools were successfully constructed and used prior to the early 1940's. It seems strange, then, that after more than thirty years, tooling with plastics is not widely known or accepted by industry.

Some industries, notably automotive and aircraft, have utilized plastic tools to great advantage. In these industries plastics are looked on not as substitutes for clay, wood, plaster, aluminum, steel, etc, but rather as a group of products that have their own engineering properties and fabrication techniques which make them the logical choice for certain applications. Plastics have advantages and limitations, and their selection by the tool engineer should be made on a cost-performance basis. A general understanding of the types of materials, the methods of construction, and the major applications is important to ascertain the place plastics occupy in the tooling field. From this understanding the advantages and disadvantages of plastics compared to other materials become apparent and the selection of tooling materials can be made. See also CASTING; LAMINATES; REINFORCED PLASTICS.

Materials

Plastic tools can be constructed from various classes of plastics. For rigid tool construction, the most common types are the epoxy, polyester, and phenolic resins. Flexible tooling applications may utilize modified versions of these, or use silicone, polysulfide, and polyurethan elastomers.

Phenolic Resins. Phenolic resins were among the first commercial materials used in plastic tooling. Introduced in the late 1930's, the phenolic resins have been largely superseded by other classes of thermosetting resins. However, in some areas of plastic tooling they are still used because of their low cost and mass-casting qualities. The disadvantages of phenolic resins include brittleness, the need for overcuring, and the corrosive nature of the acidic curing agent. The dimensional stability of the tool is limited by the migration of the by-product water produced when the phenolic resin cures. See also PHENOLIC RESINS.

Polyester Resins. Polyester resins, after they became available in the early 1940's, were used in a variety of plastic tooling applications, such as the construction of trimming, drilling, and checking fixtures. Polyester resins are still used today in tool fabrication where broad tolerances are acceptable and advantage can be gained through their comparatively low cost. Probably the largest application for polyester tools is in the construction of molds for boats and similar polyester shells. The high degree of shrinkage during cure is not critical for these applications. Also, although mold life is limited, a high gloss can be maintained on the mold surface. See also POLYESTERS, UNSATURATED.

Epoxy Resins. Since the introduction of epoxy resins to plastic tooling in the late 1940's and early 1950's, these materials have become the mainstay of this industry. They combine high dimensional stability with low shrinkage during cure, thus making possible the accurate and stable reproduction of a model at room temperature. Through the use of glass cloth, lightweight laminated tools with high mechanical strength and excellent wear characteristics can be readily produced. Since the epoxy resins adhere tightly to inserts, the introduction of metal inserts at high-wear areas is readily accomplished. Likewise, bonding of back-up structures to surface skin panels is readily accomplished without distortion. The negligibly low shrinkage of the epoxy materials makes them applicable not only in laminations, but also as surface coats and mass castings. Their flow characteristics can be controlled so that their application does not require highly skilled labor. They cure without the formation of any by-products and, without pressure, form structures that are completely resistant to ordinary atmospheric conditions and to corrosion by chemicals likely to be encountered in tooling processes. Their versatility makes possible the formulation of tooling materials ranging from tough and rubbery compositions to materials that remain rigid and hard at temperatures over about 200°C.

Plastic formulators specializing in tooling products utilize a wide variety of modifiers, fillers, and pigments as well as a wide selection of hardening agents in conjunction with the epoxy resin to offer a broad range of laminating, surface coating, casting, and paste compositions especially suited for the particular tool to be constructed. The chemistry and compounding of epoxy resins are discussed in detail in the article EPOXY RESINS.

In tools, the use of viscosity modifiers or diluents that do not react and become part of the resin molecule should be avoided, since otherwise they can migrate and cause warpage and shrinkage in the completed tool. Most epoxy tooling materials also contain fillers (qv). The primary functions of these materials are to control flow in application, reduce shrinkage, improve wear resistance, modify strength properties, modify heat of reaction, control thermal conductivity and coefficient of thermal expansion, add color, and reduce cost. Epoxy resins for tooling with plastics are available as laminating, surface coating, casting, and paste materials. They are excellent adhesives and the proper use of release and/or parting agents is necessary when the material is to be removed from any surface with which it comes in contact. See also RELEASE AGENTS.

Laminating Systems. Epoxy laminating systems perhaps rank as the most prominent tooling resins. Tools laminated with these systems form the lightest, strongest, and most durable tools in relation to other methods of plastic tool building. A typical laminating resin consists of an easily applied formulated epoxy system that exhibits good cloth penetration and coverage. The system should have a minimum tendency to drain from contoured surfaces of glass-fiber cloth, to prevent the formation of resin-starved areas.

An extension of the laminating compounds are the spray-up formulations, which are suitable for essentially the same types of applications. Proper amounts of resin and hardening agents are machine-metered and mixed before being sprayed onto a suitable surface. Simultaneously, glass roving is chopped into predetermined lengths and flocked onto the surface. The laminate is then compacted by use of rollers.

Surface Coatings. Another form of epoxy resin for tooling is the surface coat applied as a semi-pasty composition and used in conjunction with laminating systems. Structurally weaker than the laminates, the coatings serve only to obtain an accurate and detailed reproduction of a surface to be duplicated.

Castings. Epoxy casting systems represent perhaps the largest number of individual products of the resin formulators. Epoxy resins are compounded with a wide variety of fillers to produce the desired characteristics with respect to machinability, malleability, hardness, heat conductivity, abrasion resistance, and weight. The hardening agent must be selected to balance the rate of cure with the depth of the poured casting since too rapid a cure can also produce damaging heat. This consideration may limit the size of castings that can be safely poured at one time using any given formulation. On the other hand, a slowly curing formulation, designed for large masses, requires inordinately long cure times to gel thin sections. Certain casting compounds are also developed specifically for applications requiring a measure of heat resistance. Thus, resin formulators must offer a large number of casting systems to meet the various requirements.

Pastes. The paste systems constitute another type of formulated epoxy resin. These also exist in many forms for a diversity of end uses. One form, for example, gels to a solid that is easily worked with woodworking tools. In combination with laminating systems, this material offers a convenient way to develop master models.

All of the epoxy tooling applications that will be discussed in subsequent sections utilize one or more of these four epoxy systems. Although in many cases the plastic is used in conjunction with wood, plaster, or metal, the finished product is usually classed as a plastic tool.

Silicones. Room-temperature-vulcanizing silicone rubber is used for tooling applications. Generally used as flexible molds for prototype plastic cast parts or as foaming molds for rigid polyurethan furniture, they have the advantage of being self-releasing. Application is limited by the relatively high cost and by inhibition of cure when in contact with a number of dissimilar materials. See also SILICONES.

Polyurethans. At present, polyurethan elastomers are being investigated for possible tooling applications. These formulations are characterized by extreme toughness and resistance to abrasion. They are suited for die pads, press brake form pads, foundry patterns, coreboxes, and flexible molds. Handling is sometimes difficult owing to their short working life and extreme sensitivity to moisture. See also POLYURETHANS.

Tool Fabrication

Plastic tools are fabricated by four main methods of construction: they are laminated, surface cast, mass cast, or made from paste materials. In deciding which method would be best for a specific job, such factors as listed in Table 1 should be considered.

Laminated Tools. In making laminated tools, a plastic surface coating is applied to a model. When the surface coat is tackfree, a "glass–paste" mixture is used to fill all sharp corners and detail that might cause voids or bubbles under the first layer of glass cloth. The glass–paste mixture may be made of a laminating plastic mixed with chopped glass fibers or cotton flock. Alternate layers of laminating resin

Table 1. Desirability Factors for Plastics in Tools[a]

	Laminate	Metal core, surface cast	Mass cast	Paste
dimensional stability	*1*	*2–3*	*4*	*2–3*
shrinkage (during cure)	*1*	*2–3*	*4*	*2–3*
least weight	*1*	*4*	*3*	*2*
labor cost	*4*	*3*	*1*	*2*
material cost (weight basis)	*4*	*3*	*1*	*2*
strength	*1*	*2*	*3*	*4*
toughness	*1*	*2*	*4*	*3*

[a] *1*, most desirable; *2*, satisfactory; *3*, fair; *4*, least desirable.

and glass cloth are applied until the desired thickness is reached, although more than eighteen layers (3/8 in. thick) should not be applied at any one time since the heat generated by the resin can cause excessive shrinkage or warpage of the finished tool. If a framework is to be added to the laminated tool facing, this should be done immediately after laminating, before the plastic tool is removed from the model. As a general rule, at least 6 hr should elapse after the last liquid plastic has been applied, before the tool is removed from the model. The framework is best made of the same materials as the tool facing, otherwise the performance of the tool could be affected by temperature changes, moisture, aging, or the tool's weight.

Surface-Cast Tools. Three different methods have proved successful in surface casting. In the *pour method*, the core (obtained, for example, by pouring molten metal into a sand mold) is suspended over the die model and sealed around the edges. Resin is then poured through sprues or through an open end. In the *squash method*, liquid or paste plastic is placed on the die model. The core is then placed on the material and allowed to settle to a predetermined position. In the *pressure pot method*, the core is suspended over the die model and sealed around the edges. Liquid is forced into place at low points and allowed to vent at high points.

After selection of the method to be used, the core and the die model are set up. It is usually advisable to try the core in the mold before the resin is added, especially when the squash method is being considered. Entrapping of air by the casting resin as it enters the mold must be avoided by drilling vent holes through the core or by elevating one end. The core must always be on top of the die model since this will keep any small air bubbles away from the working surface of the tool. When pouring into the mold, a steady stream of material must be maintained until the complete casting is finished, otherwise air may become trapped.

Mass-Cast Tools. Mass casting is comparatively simple. When the mold has been prepared, the required amount of material is poured into the cavity, allowed to cure, and then removed from the mold. When an unusually large mass is being poured (more than 6 in. in thickness), it is often desirable to add a filler such as sand and gravel, aluminum grain, glass balls, volcanic ash, pumice, cork, or ground nut shells (see FILLERS). This prevents a high exotherm and greatly reduces the amount of shrinkage during cure. Although the simplicity of this method would seem to offer the greatest saving in time and materials, its major disadvantage is the low strength of most mass-cast tools. High shrinkage due to heat generated in curing is another disadvantage, although this is being ameliorated by newly developed formulations.

Tools from Paste. The principal advantage of paste plastic is that it does not flow. It can, therefore, be applied to vertical surfaces, does not have to be cast to form a level surface, and does not have to be retained in position with special equipment.

One of the important uses of paste plastics has been in the making of splined masters, particularly in the aircraft industry. A splined master is a master model that is constructed of templates which define the finished surface at various locations along the tool face. A plastic paste is then "splined" or swept in between the templates to complete the contour between the points established by the templates. In constructing a splined master model, accuracy is of prime importance. The base structure is the foundation of the model and must be accurate, stable, and self-supporting. Templates used in this type of tool construction are held in a common plane, usually perpendicular to the base, and on a reference line established on the base plane. A series of templates

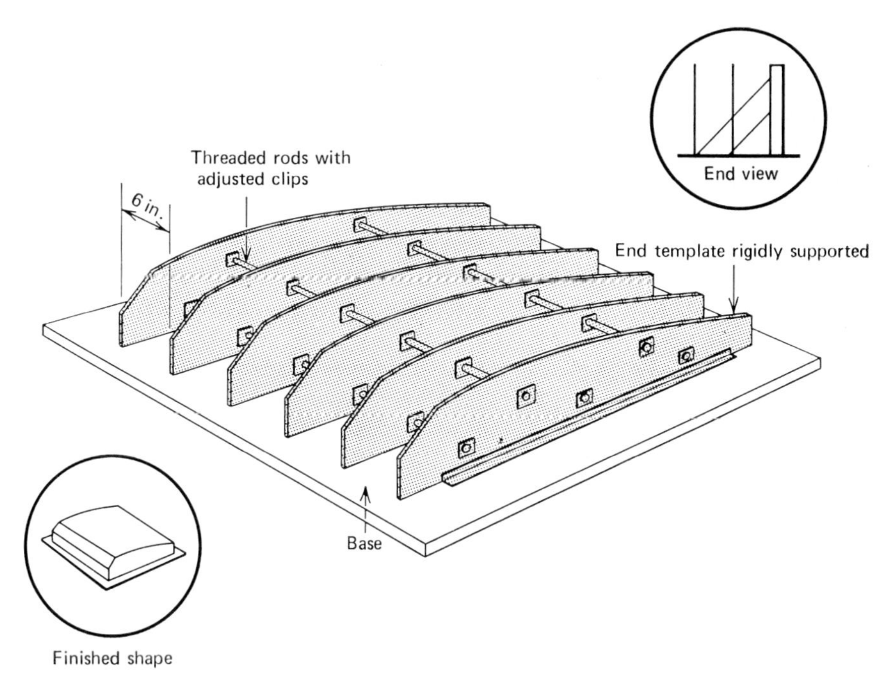

Fig. 1. Template set-up for splined master used in aircraft application.

make up the desired three-dimensional surface (see Fig. 1). Once the template set-up is completed, strips of galvanized wire cloth are cut to fit the spaces between the templates. Rods are slid into predrilled holes in the templates and the wire cloth pulled up behind the rods and attached with "hog rings" or wire ties. A base coat of nonflowing paste plastic is applied to cover the rods and wire cloth. When this coat hardens, a second base coat is applied, filling each space up to the template edges. A slab of hard rubber is used to clean the template edges and remove enough plastic so that the cured surface of the second application will be approximately 1/32 in. below the template edges. When this coat hardens, a finish coat is applied in the same general manner as before, using steel splining tools for removing excess material. Three or four such coats may be required to complete the master. By having the splining tool touching three or more adjacent templates, exact dimensions of compound arcs and curves between templates are maintained.

Carvable materials are used extensively for making engineering changes and to repair master models. The construction of extremely stable models, by building them oversized of epoxy paste and then machining them to the final dimensions, is gaining wide acceptance. See also MACHINING.

Applications

Plastics are used in tooling to best advantage when complex contours and fine detail are encountered. Their ability to be cured in the desired shape accounts for significant savings when compared to costs of machining metal. When a multiple number of tools is required, it is a simple procedure to build plastic replicas. It is also often less expensive to build more than one plastic tool than to invest in one metal set-up. Savings can also be made in lead time since plastic tools can be constructed quickly and are easy to repair or modify. They are lighter than metals, which is advantageous when the structures must be moved.

Fixtures. A field in which epoxy resins are extensively used is in the construction of the various fixtures used for trimming, drilling, routing, assembly, and checking of manufactured objects. Their value comes from the ease with which plastic tools, conforming closely to the contour of the object being manufactured, can be built.

Where possible, fixtures are constructed using the formulated epoxy resins in cast or paste form. Casting resins and pastes are the fastest and usually most economical

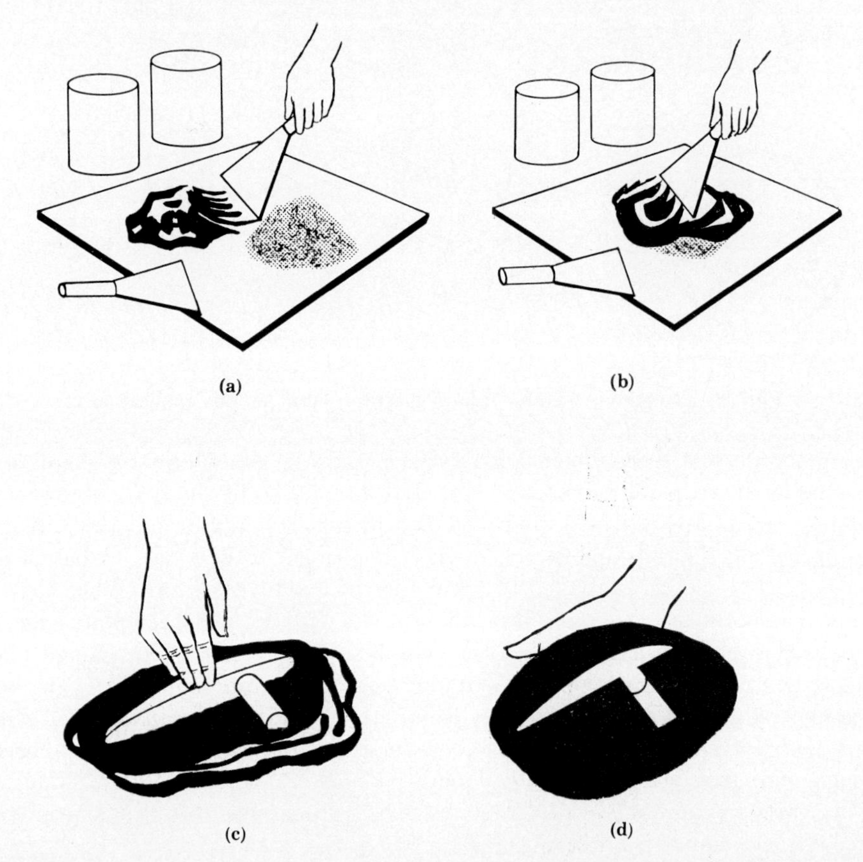

Fig. 2. Construction of a checking fixture: (**a**) the proper proportions of resin and hardener are measured (two putty knives are preferred); (**b**) the resin and hardener are thoroughly mixed to form a paste; (**c**) the part is forced into the paste; (**d**) the form is removed after the paste has set up, leaving the final impression.

compounds for plastic tool construction. Many checking fixtures (used for quality control to maintain shape and size accuracy in manufacture) and holding fixtures (used to hold a part during drilling, tapping, machining, etc) are made simply by forcing (or "squashing") a part, which has been prepared with release agent, into an epoxy paste. When the paste sets up, the part is removed and an impression of the part remains in the plastic (Fig. 2). Care must be taken that the part not be embedded too deeply in the plastic, otherwise removal after the paste has hardened becomes very difficult. A checking fixture for tubing is shown in Figure 3.

Fig. 3. All-epoxy checking fixture for tubing.

When fabricating drilling fixtures, drill bushings can be positioned on the original mold and cast in place, thus becoming an integral part of the tool. If maximum accuracy is needed, oversized holes can be drilled into a cured plastic fixture and the drill bushings subsequently potted into their exact locations (see EMBEDDING).

If the fixtures are large or have to be lightweight or extremely accurate, construction by epoxy–glass fiber cloth lamination may be indicated.

Foundry Applications. The most successful foundry applications of plastics are in patterns, coreboxes, and corebox inserts. Three common techniques are used in the construction of plastic pattern equipment, ie, laminating, surface casting, and mass casting.

Laminating, although probably the oldest technique, is now seldom used. The laminating process is very time consuming and, if an intricate shape is involved, it is difficult to prevent voids between the laminate and the surface coat. These are not discovered until the pattern or corebox is used for the first time.

Surface casting is the most successful and probably the most widely used method of pattern and corebox construction. This method consists of casting an epoxy resin surface, usually $\frac{1}{4}$–$\frac{1}{2}$ in. thick, on a core or back-up. These cores can be made quickly and inexpensively either by roughing out a shape from wood or by making a rough metal casting. Although wood makes the most inexpensive core, it is decidedly inferior to metal castings and is used only when a minimum number of castings are to be made. Equipment made by the surface casting method, using a cast metal core, has been used to produce runs of over eighty thousand castings.

Mass casting of plastic foundry equipment has had limited success. The major limitation stems from the inability of specific casting formulations to cover satisfactorily the range of sizes and thicknesses needed. When cast in thick sections, the curing produces an excessive amount of heat, which results in shrinkage and warpage. However, mass castings can be used for smaller parts of fairly uniform thickness.

Epoxy resins can be used for repair and modification of plastic or iron pattern foundry equipment. In the case of repair, epoxy pastes are troweled on, allowed to

Fig. 4. Plastic draw die mounted on die set.

cure, and then filed to the desired finished shape. When a modification or engineering change is required, an existing pattern is built up to the new dimensions using clay. After appropriate sealing and parting agents are applied, a plaster (or preferably an epoxy-faced plaster) mold is constructed over the built-up pattern. After this mold has been cured, it is removed and the clay cleaned off the pattern. The mold is then replaced over the pattern and epoxy casting material poured to fill the void. Release agent is applied to the mold, but *not* to the pattern. The mold can, therefore, be removed and used for modifying additional patterns; on the other hand, the casting material will adhere to the pattern, thereby creating the new surface.

Another advantage of plastic coreboxes which should be mentioned is the ease with which vents and screens can be installed, relocated, replaced, or removed. It is a simple matter to fill any unwanted holes with the same type of resin used in the original construction or with an epoxy–paste repair material.

Dies. Dies represent yet another area in which plastics are used to advantage. Their assets include the following: (*1*) They cost as little as one half of a comparable steel die. (*2*) They can be in use within days, compared to a minimum of several weeks required for manufacture of a steel die. (*3*) Many plastic dies are light enough to be moved by hand. (*4*) They do not corrode. (*5*) Repair is fairly simple and moderate design changes can be made easily. (*6*) They do not damage prefinished and moderately polished surfaces, but are not recommended to draw highly polished steel or aluminum because the plastic will "pick up" foreign particles with resulting imperfections in the sheet. In addition to their use for production, plastic dies are also used to check out steel designs and to aid the design of the most economical blank forms for feedstock systems.

There are three considerations that must be weighed to determine whether a plastic die is suited for a given job.

Metal Thickness. Steel in thicknesses up to 0.045 in., aluminum up to 0.060 in., and brass up to 0.030 in. can be successfully formed with plastic dies.

Die Contour. Parts with corner radii of $^{3}/_{16}$ in. or greater are suitable for forming with plastic dies. Sharp corners and fine details usually cannot be produced without wrinkles. The more complex the die cavity, the more likely it can be made more economically of plastic than of steel. Straight-line panels, flat surfaces, and concentric forms are generally more easily machined in steel than built up using plastic. Since plastic dies do not withstand side loading well, parts should be fairly symmetrical. Draw-off and flanging operations are not recommended.

Economy. Plastic dies are more economical than steel for production runs from a few parts up to as many as 80,000 parts, depending on metal thickness and configuration.

It should also be remembered that plastic tooling generally cannot be used on automated, high-speed presses because of the high temperatures generated. Plastic dies can be damaged more easily than steel dies. Also, the workpiece must be placed in position more carefully than is customary when dealing with metal dies. The most successful plastic dies are usually made in shops having personnel experienced in making both plastic and steel dies. It takes a thorough knowledge of metal as well as plastic to judge what method to use in specific cases. A plastic draw die is shown in Figure 4.

Testing. Closely associated with die construction are methods of checking the accuracy of the die cavities. This is particularly true of die-casting dies and injection molds. In the past, test materials used for this function were plaster of Paris, solder, dental waxes, and low-melting metal alloys. When poured into intricate, thin-walled die cavities, such as for transmission housings and grille work, these materials did not fill the entire cavity. Deep cores and thin walls restricted the flow of material and created air pockets; the materials would chill before filling the entire cavity. A method of die checking was developed whereby a resilient epoxy compound is pressure cast into the "shot hole" of a closed and clamped die. The die cavity is previously prepared with a release agent. A wooden plug is shaped to fit the shot hole. A small hole is drilled through this plug and the epoxy resin is forced through it under approximately 30–90 psi pressure. Complete filling is indicated by resin flowing out of vents or between the die halves at the opposite side of the injection mold. After it has hardened, the plastic is removed and inspected for wall thickness and excess flashing. Errors in die construction can thus be corrected at this stage, before actual use in production.

Molds. A rapidly growing field of use for epoxy tools is for vacuum-forming molds (see THERMOFORMING). Formerly the construction of a plastic mold for this operation used epoxy resins that had to be cured at elevated temperatures to develop the heat resistance needed in this process. However, with the development of so-called room-temperature-curing high-heat systems, it is now possible to construct extremely large molds, thus obviating the need for an oven large enough to cure the assembly. The gradual heat build-up in the vacuum-forming operation completes the cure and enables the tool to resist the required temperature of approximately 180°C. A large mold is shown in Figure 5.

Epoxy vacuum-forming molds can be made by first painting a surface coat onto a model. This is followed by three to four layers of a glass-fiber cloth laminate.

Fig. 5. Plastic mold (center) used to construct large vacuum-formed parts for house trailers.

Fig. 6. Injection molds constructed of epoxy resin used in the production of prototypes.

Aluminum spheres coated with epoxy resin are then troweled in place to a depth of ¾ in. to 1 in. It is important here to have a dry slurry with only enough plastic to act as a binder for the spheres. This produces a porous section through which a vacuum may be drawn. Attachment of a base plate and suitable fittings to a vacuum pump completes the porous vacuum chamber. Small holes are now drilled at numerous points through the surface coat and laminated section. Under vacuum, these holes allow the heated thermoplastic sheet to drape uniformly over the mold contour. This plastic–metal construction performs the dual function of supplying both rigidity and adequate heat transfer.

Other areas of forming or processing plastics that utilize plastic tooling include molds for injection molding, matched molds for forming reinforced plastics, and molds for both flexible and rigid polyurethan foams. Injection molds are generally limited to prototypes or very short production runs. Limited production is also the general rule for both matched-molded and blow-molded parts. Foamed parts are amenable to larger production runs in plastic tools; this low-cost tooling is used because of the large number of replicate molds required. In all these cases, high-temperature-resistant tooling plastics are recommended. A limiting factor to production runs is the great insulating factor of plastic tools (see THERMAL PROPERTIES). It is, therefore, extremely difficult to remove the heat at the mold surface during production runs. See also MOLDS.

Prototypes. As used here, the word "prototype" includes a model, pattern, tool, or skin of any form or object. Normally it is three dimensional (see, eg, Fig. 6). Prototype models based on epoxy resins, in the thickness range of metal sheet, can be used to position a part in the correct plane for metal drawing; they can also establish the base lines or tip position of the punch and die. A visual inspection of a prototype of this nature would bring to light critical drawing areas where trouble might occur. This in turn might suggest an engineering change, such as an increase in the radius or the addition of a little more draft to a sidewall. Plastic prototypes can also be used as field-test samples, and in market evaluation tests before management commits itself to more extensive, permanent tooling. Prototypes are often used by plant lay-out engineers to insure proper clearance of parts moving along assembly lines.

Future Outlook

In the foreseeable future epoxy resins can be expected to remain the principal material for producing rigid plastic tools. Resin formulators are directing their efforts toward developing improved compounds that are more versatile, "foolproof," and easier to use under the great variety of industrial conditions encountered. Where size and quantity warrant the expenditure, plastic handling can be semiautomated. Spray-up equipment already exists for use in producing laminated forms. Similar equipment can be used to meter, mix, and apply casting resins, thereby reducing human error. Paste compositions are being developed that can be handled in a similar fashion. As a result, many phases of tool construction will be faster.

Bibliography

Fundamentals of Tool Design, American Society of Tool and Manufacturing Engineers, Prentice-Hall, Englewood Cliffs, N.J., 1962.
The Techniques of Using Epoxy Plastic Tooling Materials, Ren Plastics, Inc., Lansing, Mich., 1968.
Vacuum Form Mold Construction, Ren Plastics, Inc., Lansing, Mich.
"Tooling with Plastics," in *Plastics Engineering Handbook of the Society of the Plastics Industry, Inc.,* 3rd ed., Reinhold Publishing Corp., New York, 1960, pp. 206–208.
H. Lee and K. Neville, *Handbook of Epoxy Resins,* McGraw-Hill Book Co., New York, 1967, Chap. 18.
I. Skeist, *Epoxy Resins,* Reinhold Publishing Corp., New York, 1958, Chap. 7.

Richard B. Peterson
Ren Plastics, Inc.

TORSIONAL BRAID ANALYSIS

The mechanical spectroscopy of polymers deals with the ability of polymeric materials to store and to dissipate mechanical energy on deformation. Dynamic mechanical measurements provide a convenient means for resolving the applied mechanical energy into stored and dissipated components. For example, the application of a sinusoidal stress (or strain) to a viscoelastic material results in a phase-displaced sinusoidal strain (or stress), with an in-phase stress-to-strain ratio dependent on the storage capacity, and with a phase relationship dependent on the ratio of energy dissipated to energy stored in the deformed material. Since the mechanical properties of polymers are dependent on both the time interval and temperature of deformation, mechanical parameters are often presented using a three-dimensional rectangular coordinate system with two of the axes corresponding to time interval and temperature. The interval of time in a dynamic mechanical experiment is the period of the cyclic stress (or strain) which is equal to the reciprocal of the frequency when the latter is expressed in Hz.

Interpretation of mechanical spectra is currently being attempted in terms of the capabilities for sub- and supramolecular motions which are activated by the applied external mechanical deformations and stresses (1–3).

The various dynamic mechanical methods of investigation of the stress–strain relationships for viscoelastic materials are discussed in Refs. 4–6. Forced vibration, forced-resonance-vibration, and free-resonance-vibration methods are used at low frequencies (<5000 Hz), whereas pulse methods are used at high frequencies (10^3–10^7 Hz). In the forced vibration methods, the frequency of the excitation stress or strain is varied while the response is measured as a function of the frequency. In free-vibration methods the specimen is initially excited and its displacement with time is then monitored at assigned temperatures. Pulse methods involve measuring the time of transit and the attenuation of short pulses of high-frequency waves through the specimen.

The torsional pendulum is an example of the free-resonance-vibration dynamic mechanical method. The present article deals with the torsional braid apparatus and its applications to the semimicro mechanical characterization of polymers. Aspects of the conventional torsional pendulum are also discussed, since the torsional braid apparatus is an extension of it. See also CHARACTERIZATION; MECHANICAL PROPERTIES; TESTING; VISCOELASTICITY.

Torsional Pendulum (4,7)

One of the most versatile instruments for low-frequency dynamic mechanical studies, especially for coverage of a wide temperature range, is the torsional pendulum. The instrument is designed to operate in the frequency range from about 0.1 to 10 Hz with the sample enclosed in a chamber that can be cooled, heated, and evacuated. Instruments of this type have been described and used by many investigators. A number of commercial variants are on the market and ASTM designation D 2236 describes the recommended procedures (7). Although the methods for recording the decaying torsional oscillations of a specimen vary, all support the specimen vertically between a fixed clamp and a second clamp that is rigidly attached to an inertial member. Schematic diagrams of the elements of the two methods that are used in the design of torsional pendulums for characterizing solid polymers are shown in Figure 1. The tension on the specimen is recommended to be less than 100 psi. When the upper clamp is attached to the inertial member, the tension on the specimen can be controlled by use of counterweights which are attached so as to have a known or negligible torsional effect. The sample is set into free torsional oscillations by an initial torsional displacement of no more than 2.5 degree/cm of specimen length for samples of the recommended dimensions. The subsequent displacement–time curve, frequently automatically recorded, provides data that, together with numerical and geometrical factors, are used to obtain the material parameters which are related directly to the storage and loss of energy on mechanical deformation.

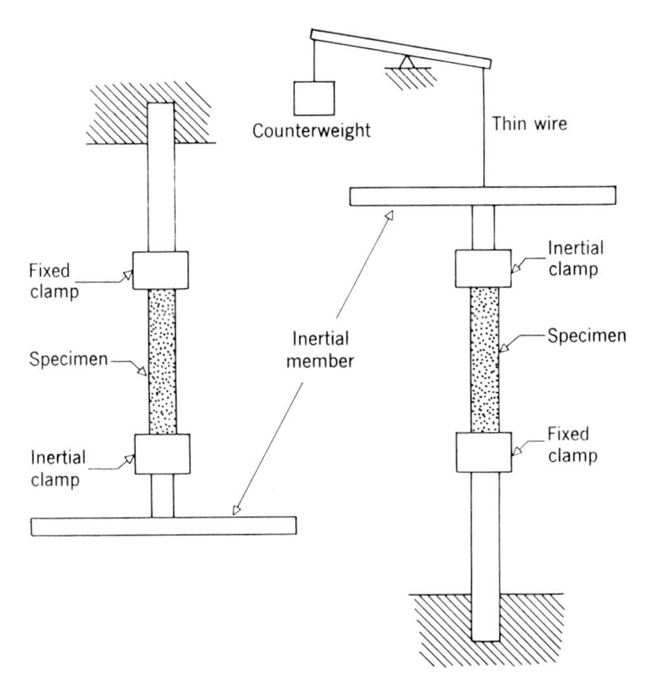

Fig. 1. Schematic diagram of torsional pendulum elements.

Theory and Reduction of Data (4–7). Two parameters, the elastic part G' of of the complex shear modulus, and the logarithmic decrement Δ, can be calculated from the damped sine waves of a homogeneous specimen undergoing free torsional oscillations. G' is the ratio of the stress in phase with the strain to the strain, and is linearly related to the energy stored in deforming the specimen (maximum energy stored per cycle $= 1/2 \times G' \times$ peak deformation squared). The parameter Δ is approximately equal to $\pi \tan \delta$ (where δ is the phase angle between the stress and strain) and relates linearly to the ratio of energy dissipated to energy stored on mechanical deformation (energy dissipated per cycle $= \pi \times G'' \times$ peak deformation squared; ratio of energy dissipated to maximum stored per cycle $= 2\pi\ G''/G' = 2\pi \tan \delta \approx 2\Delta$). For a cylindrical rod,

$$G' = \frac{8\,\pi I L}{r^4} \cdot \frac{1}{P^2} \qquad \text{dyne/sq cm}$$

whereas for specimens of rectangular cross section

$$G' = \frac{64\,\pi^2 IL}{\mu b t^3} \cdot \frac{1}{P^2} \qquad \text{dyne/sq cm}$$

where I = moment of inertia of the inertial member in g-cm^2
P = period of oscillation in sec
L = length of specimen in cm
b = width of specimen in cm
t = thickness of specimen in cm
r = radius of specimen in cm
μ = shape factor for rectangular cross sections (see Ref. 7 for numerical values)

The logarithmic decrement is calculated from successive amplitudes (A_i, A_{i+1} in degrees of rotation) of the decaying wave

$$\log_e \left(\frac{A_1}{A_2}\right) = \log_e \left(\frac{A_2}{A_3}\right) \ldots\ldots = \log_e \left(\frac{A_i}{A_{i+1}}\right)$$

In making measurements with small specimens and with specimens of irregular geometry (as in torsional braid analysis), difficulties are encountered in measuring dimensions accurately. Therefore in work discussed herein the elastic part of the complex modulus is replaced by the defining expression

$$\text{relative rigidity} = \frac{1}{P^2}$$

P is obtained by dividing the lapsed time for a conveniently large number of oscillations by that number.

A measure of the logarithmic decrement can be obtained rapidly for each member of a series of waves, by simply counting the number (n) of oscillations between two fixed but arbitrary boundary amplitudes (eg, $A_i/A_{i+n} = 20$). $1/n$ is directly proportional to Δ and is termed the *mechanical damping index* ($\Delta = 1/n \log_e [A_i/A_{i+n}]$).

Instrumentation (8,9). The author's torsional pendulum has features that permit the examination of small samples either unsupported (torsional pendulum) or supported (torsional braid analysis) through the temperature range −190°C to +700°C. A schematic description is given in Figure 2.

An analog of the decaying oscillations is obtained using a non-drag optical transducer which also serves as the inertial mass (a disk). A linear relationship between the light transmitted through the disk from a steady source to a linearly responding vacuum phototube and the displacement angle provides a differential output which is independent of the neutral position of the mechanical oscillations. Use of the transducer minimizes problems of alignment and problems associated with any change in the neutral position of the specimen. For example, with composite specimens (as in torsional braid analysis) the neutral position of the specimen changes with temperature. The optical wedge can be made by suitably vacuum coating a metallic film onto a glass disk.

An alternative and low-cost approach to a non-drag differential linear optical transducer is to use a polarizer as the inertial member and another polarizer positioned in the path of the light beam with the ideal neutral position of the mechanical oscillations

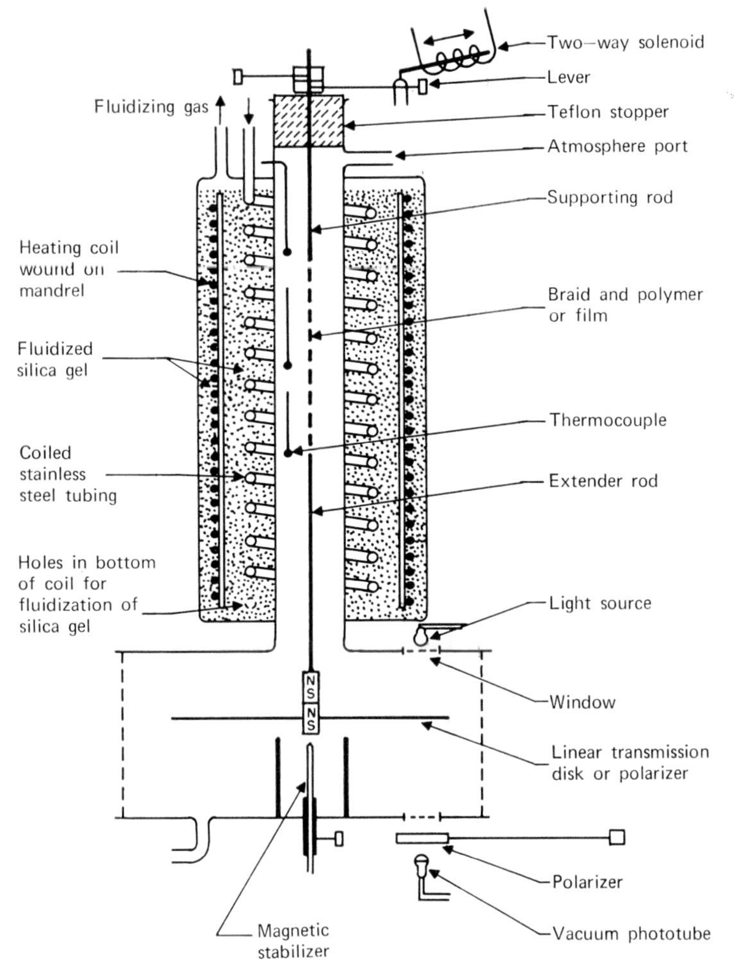

Fig. 2. Torsional pendulum and torsional braid apparatus of Gillham (schematic) (cooling coil not shown).

corresponding to the polarizers, being set 45° from the crossed and parallel positions. It can be shown that for ±15° from this position the cosine-squared function of the displacement angle from the 45° value can be approximated by a straight line to within 1% error relative to the 45° value. This angular range (45 ± 15°) for useful light transmission corresponds to one half of the total differential transmission of the pair of polarizers. The polarizer system has the advantages of the natural basis for the light attenuation (Malus' law) which permits its use at low strain levels, of low mass of the plastic polarizers which is important when dealing with small unsupported specimens (torsional pendulum), and of the fact that any nonrotational (translational) motion does not affect the transmission of light by the pair of polarizers. However, when the oscillations of the sample move outside the linear range, adjustments are made (when the system is not oscillating) in the angular position of the stationary polarizer. Typical strip-chart records of damped waves are illustrated in Figure 3.

The reaction tube surrounding the sample is embedded in a heat-transfer medium of fluidizing dry silica gel which enables the sample to be examined throughout the temperature range -190 to $+700$°C with a variation of temperature along the sample of ± 1°C in both increasing and decreasing modes of temperature change. The latter facility permits the detection of thermal hysteresis in the mechanical behavior of polymers. Hysteresis effects can arise from crystallization ⇄ fusion, dry atmosphere ⇄ water vapor, annealing ⇄ cracking, and from chemical reactions.

The pendulum is activated by turning the rod which supports both the specimen and the inertial mass. This is performed by displacement of one of the upper levers

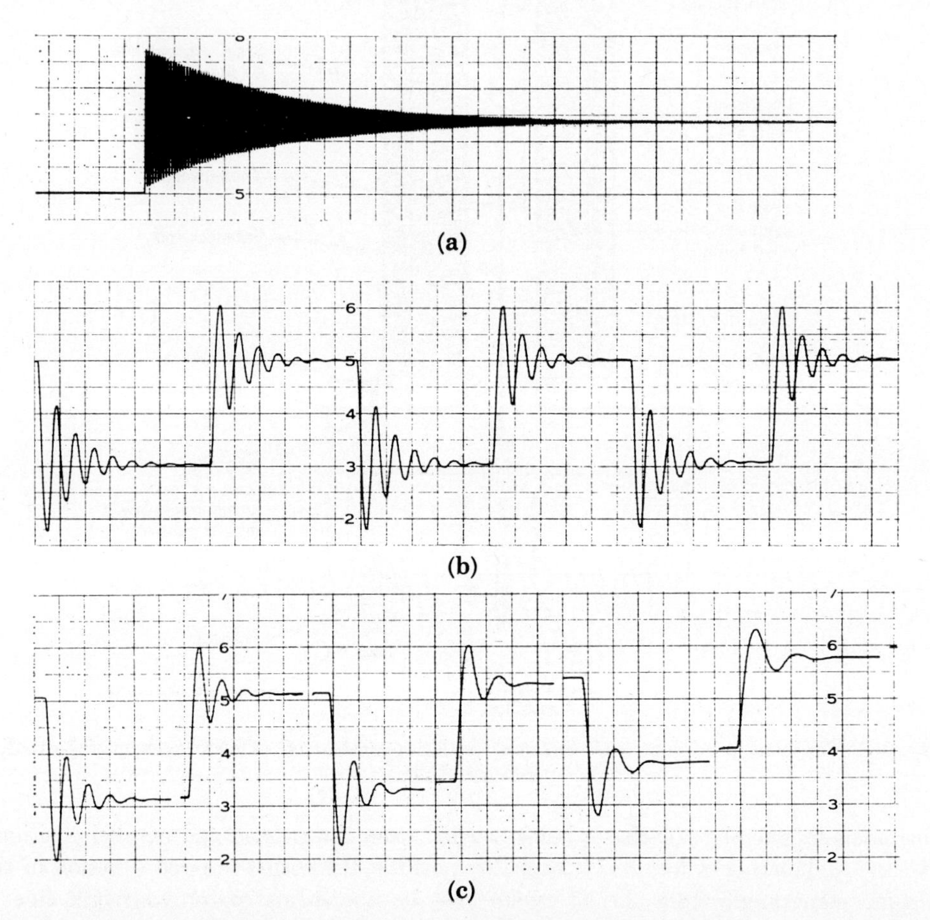

Fig. 3. Characteristic output of instrument: (**a**) glassy state; (**b**) approach to the transition; (**c**) transition region (note the drift of the neutral position).

through a small angle using a solenoid, which performs a reverse displacement a predetermined time later.

The specimen is mounted in a manner such that it can be demounted without damage. This permits the shipping of specimens for exposure to various environments (eg, radiation), so that the effect of those environments on mechanical properties can be studied. The environmental atmosphere of the specimen in the apparatus can be controlled for investigations in high vacuum, in inert, and in degradative gases.

The results of a torsional pendulum investigation of an unsupported polyimide film are shown in Figure 4 (9).

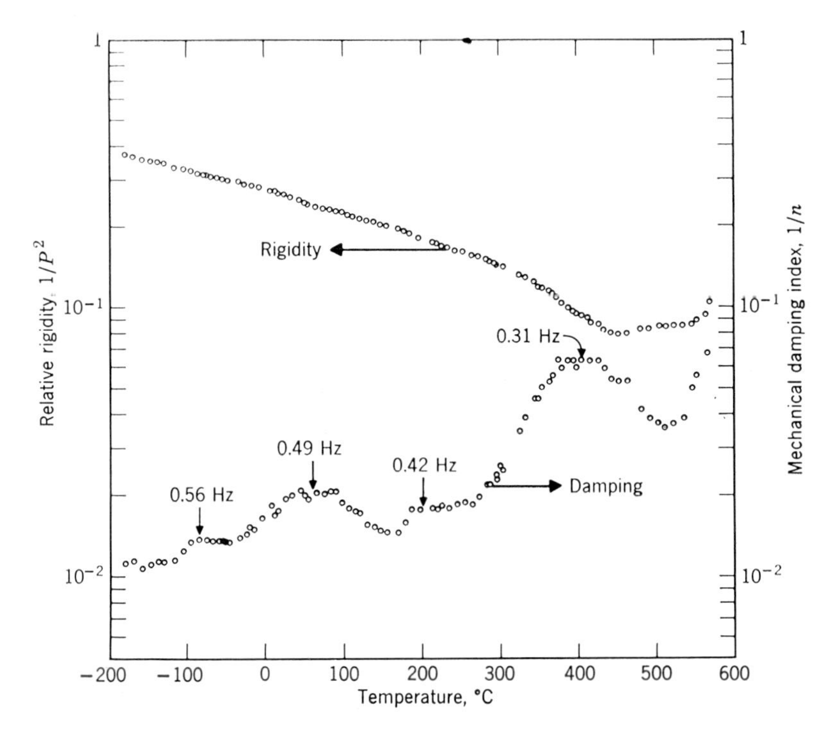

Fig. 4. Torsional pendulum analysis of a Kapton (Du Pont) polyimide film at 65 psi tensile stress. N_2 atmosphere; $\Delta T/\Delta t$ = 3°C/min.

Torsional Braid Analysis

Torsional braid analysis (8,10,11), an extension of the torsional pendulum method, was first reported by Lewis and Gillham in 1962. A specimen is prepared by impregnation of a *multifilament braid* substrate with a solution of the material that is to be tested, followed by thermal removal of the solvent. Less than 100 mg of polymer per experiment suffices. In contrast to the use of the conventional torsional pendulum, this approach permits investigation of materials that cannot support their own weight. It is therefore suitable for studies of such processes as resin curing and environmental degradation and, since thermoplastic samples may be "melted," the method allows the in situ study of the effect of prehistory on thermomechanical behavior. Thermomechanical "fingerprints" of polymer transitions and transformations through the spectrum of mechanical states may be obtained by temperature programming.

The approach is particularly useful for investigating new polymer systems, since the principal criterion for making a specimen is simple solubility of the polymer in a removable solvent. Other mechanical techniques require a more sophisticated specimen, fabrication of which demands more explicit knowledge and a larger amount of material than is often available for new polymers.

There are several reasons for using a multifilamented support rather than a single filament of rod. A substrate consisting of several thousand filaments can support a relatively large amount of material in consequence of capillary action and of the large surface-to-volume ratio of the substrate; the particular geometry of the two-phase composite minimizes the contribution of the high-modulus component to the torsional properties; and of necessity stresses are relayed through the material matrix which is forced to respond even when adhesion between the separate phases is imperfect. A braid is employed in an attempt to balance any twists in the component yarns. Most of the work performed to date has employed a loose (~3 turns/inch) braid made from 6 heat-cleaned glass yarns which form a substrate containing about 3600 single filaments. The length of the sample is from 6 to 8 inches and is prepared using polymer solutions with concentration from 5 to 100 percent. High-silica glass, high-modulus graphite, amorphous carbon, and quartz braids can be used above 600°C.

In torsional braid analysis, the relative rigidity parameter, $1/P^2$, where P is the period of oscillation, is used as a measure of the elastic part of the complex shear modulus. This expression implicitly assumes that the contributions of dimensional changes to the value of the relative rigidity are dominated by changes in the modulus of the polymer. That this is generally true follows by noting that the apparent change in modulus which results solely from a dimensional radial shrinkage of 25% in a homogeneous specimen (ie, a volume shrinkage of 44%) corresponds to a reduction in the relative rigidity parameter by a factor of 3.2.

$$\text{relative rigidity} = \frac{P_2^2}{P_1^2} \simeq \frac{r_1^4}{r_2^4} = \frac{4^4}{3^4} \approx 3.2$$

In comparison, the relative rigidity changes by a much larger factor in passing through the glass-to-rubber region of most amorphous polymers. When large percentages of polymer disappear, as in degradation, caution in interpretation is, however, necessary and the question raised is whether or not decreases in the relative rigidity are more than can be accounted for by dimensional changes. As the high-modulus inert substrate does in fact contribute to the modulus of the composite, the expected consequence of dimensional changes which accompany loss of part of the matrix on the relative rigidity parameter would be a decrease by less than the factor calculated from weight-loss measurements.

Changes in the relative rigidity and damping index are interpreted as far as is possible in terms of changes in the polymer. As a first approximation, the composite specimen is considered to behave in a manner analogous to the deformation of a stiff and weak spring in series, with the weak spring reflecting the characteristics of the viscoelastic matrix. Major and secondary relaxations are readily revealed, as are the effects of chemical reactions. Although most of the changes in the mechanical be-

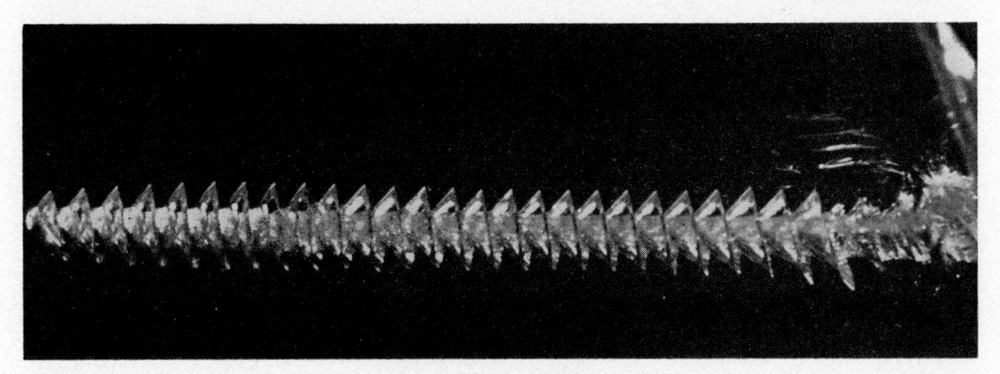

Fig. 5. Helical fracture around a multifilamented glass strand embedded in a block of brittle cross-linked polyester resin.

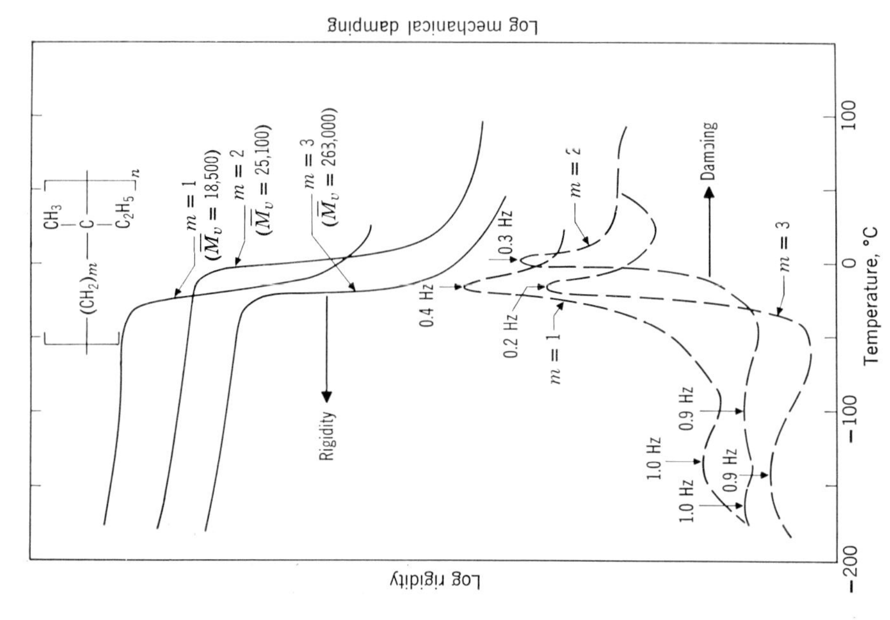

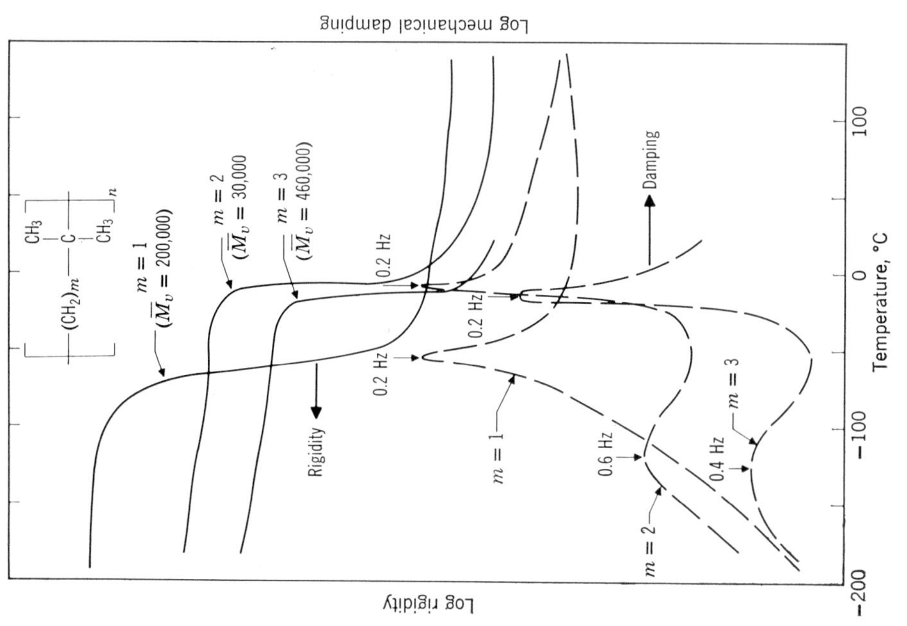

Fig. 6. Thermomechanical spectra of amorphous polyolefins deposited on glass braid (deposited from 5–10% solutions in *n*-heptane). N_2 atmosphere; $\Delta T/\Delta t = \pm 1°C/min$.

havior can be attributed to the polymer, changes which are the consequence of the composite nature of the sample are to be anticipated and form the basis of current investigations. Complications can arise from fracture of the polymer, adhesive failure, and polymer–substrate and polymer–water interactions. As an example of this current work, Figure 5 shows a helical crack which propagated around a multifilamented inclusion in a block of brittle polyester resin (12).

Thermomechanical Spectra. Thermomechanical spectra are presented for several classes of polymers: amorphous polyolefins, which relate to elastomers; semicrystalline cellulose triacetate, which relates to plastics and fibers; and a largely inorganic thermosetting polysiloxane, which relates to structural composite materials. Also discussed are the thermomechanical spectra of low-molecular-weight and model compounds and of an experimental high-temperature polybenzimidazole system.

Unless stated to the contrary, the experiments described herein were performed in a slow stream of dry nitrogen gas with changes in temperature of 1°C per minute and at frequencies of less than 1 Hz.

Amorphous Polyolefins (13). The mechanical spectra of the amorphous polyolefins shown in Figure 6 display the glass transition of each, and also the presence of secondary transitions in the glassy state. (For purposes of clarity the curves are displaced vertically without altering their shape.) It is the presence, location, and absence of secondary relaxations which confer the major and subtle differences in mechanical response between different polymeric glasses relative to their glass transitions. The presence of discrete damping maxima reveals the onset of discrete localized solid-state motions. In general, the higher the temperature at which a damping peak occurs (at a given frequency) the less localized are the contributing motions. Samples of the polyolefins referred to in Figure 6 are dimensionally unstable at room temperature and, therefore, are particularly suitable for analysis by this method.

The temperatures of transitions in nonpolar amorphous polymers are determined by the inherent flexibility of the individual molecules and by geometrical intermolecular interactions. The thermomechanical spectra of amorphous polyolefins with repeat structure $\text{-(}(CH_2)_mC(CH_3)_2\text{)-}$ and $\text{-(}(CH_2)_mC(CH_3)(C_2H_5)\text{)-}$, where $m = 1, 2$, and 3, show that the primary (T_g) and secondary transition temperatures increase in going from the first to the second member of both series and then decrease through the third eventually (by extrapolation) to the low-temperature amorphous transitions of amorphous polyethylene ($m = \infty$). Since the order of the inherent flexibility of individual nonpolar polymer molecules increases with increasing values of m, the maximum temperatures for the transitions (at $m = 2$) must arise from a dominance of geometrical intermolecular factors over the flexibility factor. Geometrical interlocking of parts of adjacent molecules is considered to be at a maximum for the second member of each series where an examination of models reveals the presence of a rather specific snug fit which is looser for higher members. Since the secondary transitions display the same temperature dependence on the values of m, geometrical interlocking is considered to be extensive. The dominant role of intermolecular geometrical interactions on the nature of transitions in the amorphous phase of solid polymers appears to be a neglected principle of polymer science.

Cellulose Triacetate (8,14). The thermomechanical spectrum for the cellulose triacetate is presented in Figure 7 together with the corresponding results for thermogravimetric (TGA) and differential thermal analyses (DTA). The glass transition in the vicinity of 190°C is accompanied by a drastic decrease in rigidity, a prominent maximum in damping, and an endothermic shift observable by DTA. The subsequent increase in rigidity at temperatures above 200°C is attributed to crystallization and is accompanied by an exothermic maximum (DTA). The melting transition (T_m) at 290°C is accompanied by an abrupt decrease in rigidity, a maximum in damping, and an endothermic maximum (DTA). The subsequent increase in rigidity, decrease in damping, exotherm (DTA), and weight loss (TGA) are attributed to crosslinking and/or chain-stiffening processes.

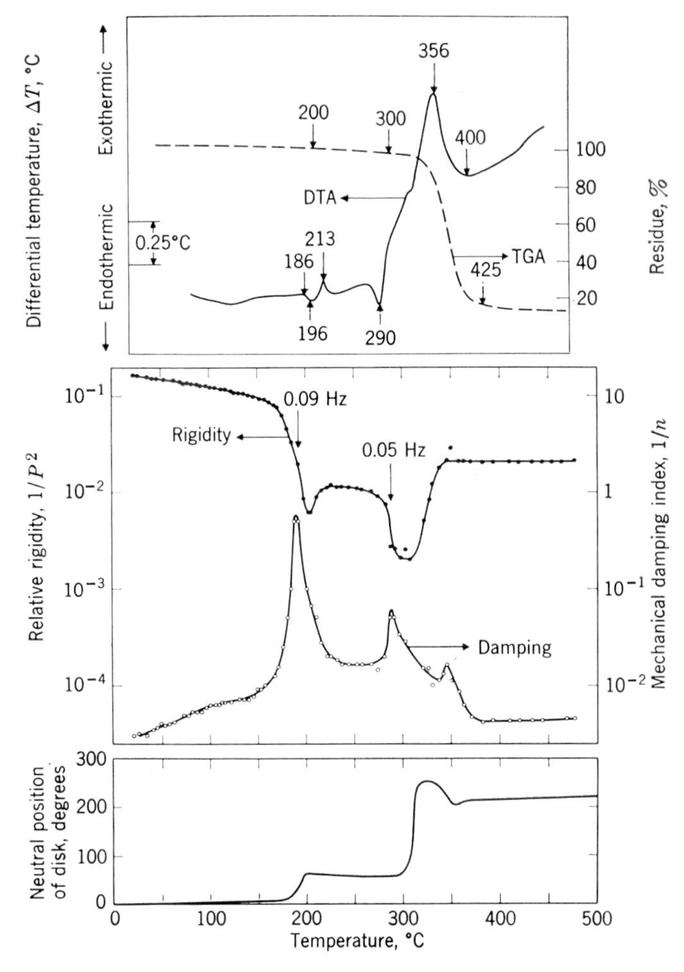

Fig. 7. Cellulose triacetate deposited on glass braid from 7% solution in methylene chloride. (**a**) Thermogravimetric and differential thermal analysis; (**b**) torsional braid analysis; (**c**) neutral position of the inertial mass as a function of temperature. N_2 atmosphere; $\Delta T/\Delta t = 2°C/min$.

The lowest diagram on Figure 7 shows the drift of the neutral position of the inertial mass as a function of temperature for a composite specimen of the cellulose triacetate and glass braid. The specimen was not oscillated. The motion is a consequence of the internal stresses that develop in the composite sample. The sense of the drift correlates with the expansion or the contraction of the matrix. It is seen that drifts which correspond to the glass and melting transitions (which involve volume

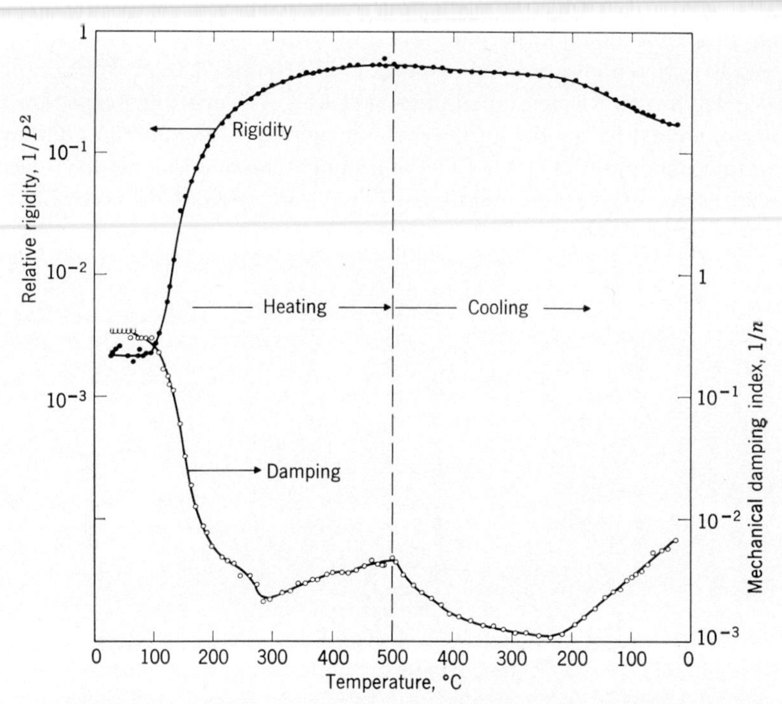

Fig. 8. Thermomechanical spectra of a thermosetting polysiloxane. N_2 atmosphere; $\Delta T/\Delta t = \pm 1°C/min$.

expansion of the matrix) are in the opposite sense to the processes corresponding to crystallization and crosslinking (which involve volume contraction of the matrix).

Thermosetting Polysiloxane (8). A composite specimen of glass braid and polysiloxane prepolymer, prepared from a commercially available alcoholic solution ("Glass Resin"-type 650, Owens-Illinois, Inc.), was heated to 500°C at 1°C min^{-1} and then immediately cooled to 25°C at the same rate. The results, presented in Figure 8, show that the rigidity increases to 500°C whereas on cooling it *decreases*. The resulting specimen was highly crazed, suggesting that the treatment of the experiment produces a highly crosslinked resin which, in the composite structure, cannot accommodate the stresses introduced thermally and by the polymerization processes. This represents an example of "overcure" in a thermosetting composite system. One might predict from the behavior of the mechanical damping in Figure 8 that resin-glass fiber composites would not be structurally useful above 280°C.

Phthalates (11). A composite specimen of di-*n*-butyl phthalate and glass braid shows two damping maxima and concomitant changes in modulus (Fig. 9). The main glass-to-liquid transition is at about −85°C: another relaxation process is made apparent by the damping maximum at about −170°C. Diisobutyl phthalate gives rise to a very similar spectrum. On the other hand, the mechanical spectrum of dicyclohexyl phthalate displays a major loss maximum (T_g) at −35°C, a sharp loss peak at −95°C, and a suggestion of a loss maximum peaking below −190°C.

In the current attempt to develop the mechanical spectroscopy of macromolecules by correlating loss maxima with the motions of specific parts of polymer molecules, systematic studies of model and small compounds by torsional braid analysis should facilitate the interpretation of dynamic mechanical spectra of polymers.

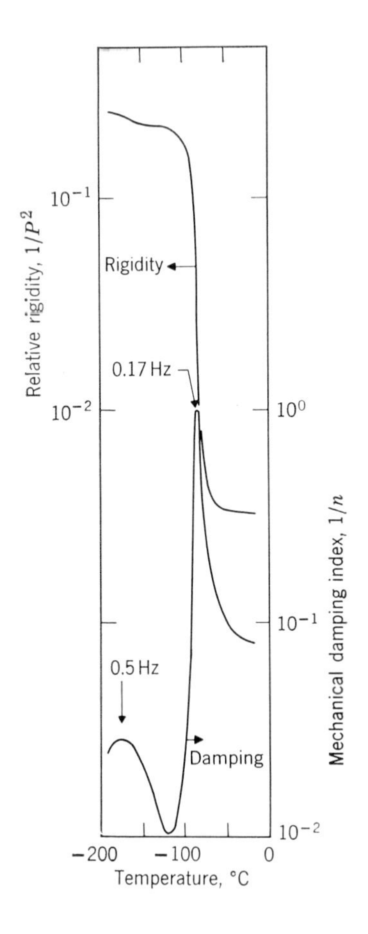

Fig. 9. Thermomechanical spectra of di-*n*-butyl phthalate.

Other relaxation techniques (eg, nuclear magnetic resonance and dielectric) will aid such investigations.

Polybenzimidazoles (11). Polybenzimidazole prepolymer, formed by partial reaction of 3,3′-diaminobenzidine and isophthalamide (see Polybenzimidazoles under POLYIMIDAZOLES), shows five distinct regions of behavior (Fig. 10).

1. At low temperatures and below 200°C, the prepolymer is a glassy material with low mechanical loss and a modulus which changes little with temperature.

2. The transition region shows a drastic drop in modulus above 220°C to a minimum at 280°C and a prominent damping maximum at 260°C. Using the latter as an index, the glass-transition temperature of the prepolymer may be designated as 260°C.

3. Above 290°C, the modulus increases while the damping passes through a maximum during the polymerization process which leads presumably to the polybenzimidazole.

4. The modulus drops above 390°C (through a damping maximum at 420°C) as most likely the result of a further transition (the glass transition of the newly formed polymer). This conclusion may be reached from isothermal behavior at 380°C.

5. Above 460°C, the modulus increases in consequence of further chemical reaction while the damping decreases to values characteristic of the glassy state. The product of this last reaction at 500°C could be structurally useful to this temperature.

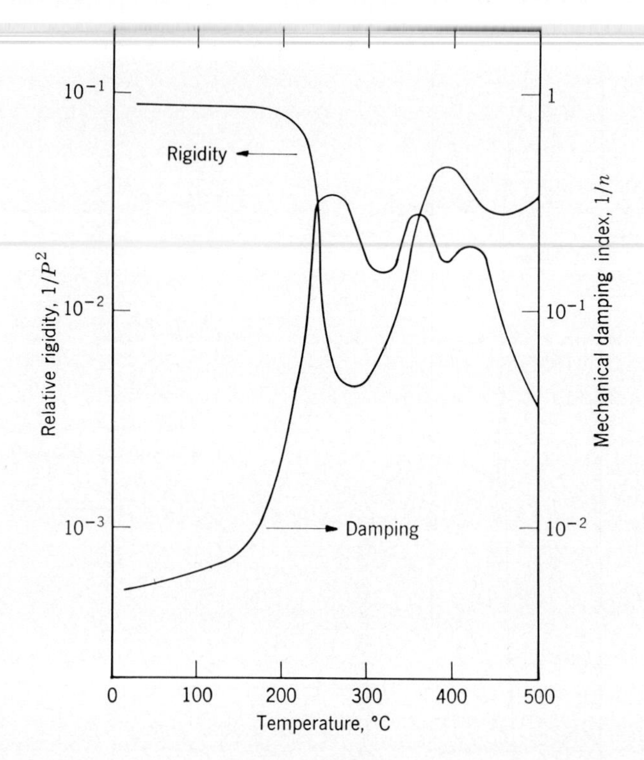

Fig. 10. Thermomechanical spectra of a polybenzimidazole prepolymer (Air Force Material Laboratory sample AF-R-151).

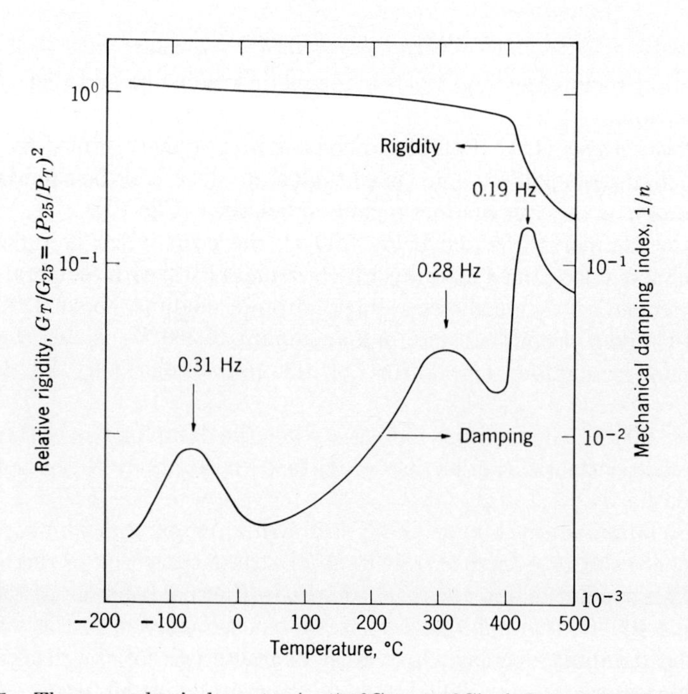

Fig. 11. Thermomechanical spectra (−170°C to 500°C) of the polybenzimidazole.

Three distinct transition regions are apparent in the thermomechanical behavior (Fig. 11) of the polybenzimidazole (prepared by isothermal treatment of the prepolymer at 380°C for 24 hours), which may be designated by the temperatures of the damping maxima: −70, 310, and 430°C. Distinct changes in the modulus are associated with each of these loss maxima.

From the results one would predict not only that the polymer would be structurally useful to above 380°C but that resistance to impact above −130°C would be good. The low-temperature damping maximum is indicative of the presence of a relaxation mechanism which could contribute to mechanical behavior in the same way that the low-temperature dispersion region of polycarbonate resins confers impact properties.

The main glass transition can be shifted progressively from 430°C to above 500°C by thermal treatment at 500°C for 4 hr.

Excellent thermomechanical properties are reported for glass cloth-reinforced laminates made from the prepolymer. For example, the data of Table 1 are for a particular laminate which had been heated in nitrogen at 510°C for 6 hr. Its thermomechanical properties were better than those of laminates prepared by curing schedules at 400°C for 6 hr. This is readily understood in terms of the torsional braid analysis.

Table 1. Thermomechanical Behavior of a Glass Cloth-Reinforced Laminate of the Polybenzimidazole

Test temperature, °F	Flexural strength, psi	Flexural modulus, psi
70	117,000	5.0×10^6
700 (after 1 hr at 700°F)	91,000	4.5×10^6

Several interpretations for the influence of thermal history on the glass-transition temperature of the polybenzimidazole are plausible. Among these are: completion of the ring-closing reactions; increase of molecular weight; crosslinking; and removal of plasticizing agents by volatilization.

The nature of the relaxation processes which are made apparent by damping maxima at −70°C and 310°C is of interest as it is difficult to visualize mobile segments in the pure structure. Explanations might be found in: the structure not being that represented by the formula for the polybenzimidazole; current tendencies to identify damping maxima with specific molecular motions being oversimplifications; specific interactions between polymer and substrate; and the presence of difficult to remove foreign entities (eg, water).

The fact that the primary (430°C) transition is affected by the thermal treatment at 500°C to a much greater extent than the relaxations which occur in the glassy state is evidence that the only significant structural changes introduced are those which affect the longer range motions associated with the primary transition. Shorter range motions, for example those associated with the relaxation at 310°C, are scarcely affected by the thermal treatment.

It is evident that the type of development represented by the polybenzimidazole system could be facilitated greatly by the torsional braid analysis method. Thermal transitions and transformations of the solid-state reactions can be monitored in terms of mechanical parameters which have both a theoretical basis (mechanical spectroscopy) and a practical relevance.

Since the composite specimen of polymer and glass fibers itself may be considered as being an element of desired gross structures (eg, glass cloth-reinforced laminates), predictions are direct. Little material is required in the experiments since the testing is nondestructive; the method is proving increasingly useful in the characterization of polymers (15).

Bibliography

1. R. F. Boyer, ed., "Transitions and Relaxations in Polymers," *J. Polymer Sci.*, Polymer Symposium No. 14 (1966).
2. R. F. Boyer, *Polymer Eng. Sci.* **8** (3), 16 (1968).
3. A. Eisenberg and M. C. Shen, *Rubber Chem. Technol.* **43** (1), 156 (1970).
4. L. E. Nielsen, *Mechanical Properties of Polymers*, Reinhold Publishing Corp., New York, 1962.
5. N. G. McCrum, B. E. Read, and G. Williams, *Anelastic and Dielectric Effects in Polymeric Solids*, John Wiley & Sons, Inc., New York, 1967.
6. J. D. Ferry, *Viscoelastic Properties of Polymers*, John Wiley & Sons, Inc., New York, 1970.
7. ASTM Standard D 2236, *Dynamic Mechanical Properties of Plastics by Means of a Torsional Pendulum*, American Society for Testing and Materials, Philadelphia, 1969.
8. J. K. Gillham in R. F. Schwenker, Jr., ed., "Thermoanalysis of Fibers and Fiber-Forming Polymers," *J. Appl. Polymer Sci.*, Applied Polymer Symposia No. 2, p. 45 (1966).
9. J. K. Gillham and M. Roller, *Polymer Eng. Sci.*, in press.
10. A. F. Lewis and J. K. Gillham, *J. Appl. Polymer Sci.* **6**, 422 (1962).
11. J. K. Gillham, *Polymer Eng. Sci.* **7** (4), 225 (1967).
12. J. K. Gillham, P. N. Reitz, and M. J. Doyle, *Polymer Eng. Sci.* **8** (3), 227 (1968).
13. J. R. Martin and J. K. Gillham, *Paper, Ann. Tech. Meet., Soc. Plastics Eng., Washington, D.C., May 1971.*
14. J. K. Gillham and R. F. Schwenker, Jr., in Ref. 8, p. 59.
15. "First Symposium on Torsional Braid Analysis," J. K. Gillham, Chairman, *Ann. Tech. Meet., Soc. Plastics Eng., Washington, D.C., May 1971.*

J. K. Gillham
Princeton University

TOXICITY. See BIOCIDES; FOOD AND DRUG APPLICATIONS

TOYS

The first use of plastics in toys occurred with the beginning of the plastics industry itself, in 1868 when John Wesley Hyatt prepared cellulose nitrate for use in billiard balls (see CELLULOSE ESTERS, INORGANIC). In the century since Hyatt's discovery, every conceivable polymeric material and processing technique has found application in some phase of the toy industry. The principal determining factors for the use of plastics in toys are easy moldability, attractive color, and of course, low price. In turn, plastics have contributed greatly to the principal sales factors dominating the toy industry, ie, realism, hygiene, and safety.

Besides the wide use of plastics in conventional toys (dolls, model vehicles, etc) an important effect of the advent of plastics has been to facilitate the use of toys as a part of the educational process. Many of the current educational toys were first available only as school equipment designed and built of conventional materials, such as wood, steel, paper, glass, etc. The production techniques of the plastics industry assisted in bringing mass-produced versions into the hands of individual children both in schools and in their own homes. Through the use of inexpensive plastic lenses and molded parts, educational equipment such as telescopes, microscopes, laboratory balances, surveyors' transits, and the like have been miniaturized and brought to a price level where they serve as "toys" for the individual child.

Several of the basic plastics processing techniques have themselves been simulated into toy replicas of industrial equipment. For example, toys featuring such functions as the molding of a poly(vinyl chloride) liquid plastisol, thermoforming of sheet plastic, and the extrusion of simple shapes are on the market.

Materials

The principal polymers used in toys are polyethylene, polystyrene, and poly(vinyl chloride). In the past, the toy industry earned an unenviable reputation for poor quality through the use of off-grade and reprocessed materials from these polymers. This trend was particularly strong in the years of short supply and high material prices immediately following World War II. The practice still exists among the less sophisticated segments of the toy industry. The better toy manufacturers, however, have learned to utilize these plastics, as well as many more recently developed materials, to upgrade plastic toys.

Impact-resistant polystyrene (see STYRENE POLYMERS) has replaced general-purpose grades for use in molding toys such as trucks and model cars. High-density polyethylene has replaced low-density polyethylene where ridigity is a functional requirement (see ETHYLENE POLYMERS). Improved plasticizers in poly(vinyl chloride) plastisols (see VINYL CHLORIDE POLYMERS) have resulted in dolls that have a more realistic feel and that do not smell bad when they are new or turn hard and crack with age. Nylon, acrylonitrile–butadiene–styrene, and acetal resins have found wide acceptance for such working parts as gears, cams, bushings, drives, clutches, and levers. Polypropylene has won a reputation as a low-cost engineering plastic in such applications as wheel hubs and molded hinges. Both high-density and low-density polyethylenes are widely used in three-dimensional hollow parts through blow molding and rotational molding processes (see MOLDING). Polyurethan and vinyl foams are used as body stuffing and as sound- and vibration-damping material for mechanical drive mechanisms. Expanded polystyrene moldings are used to simulate bones in animal skeletons; they are also used for toy furniture. Acrylic rods transmit colored patterns.

A typical example of use of several plastics in one toy is a doll for which the head is molded of poly(vinyl chloride), the body is impact-resistant polystyrene, and the hands and wrists are made of ethylene–vinyl acetate copolymer and acetal, respectively. Hair may be made of modacrylic fiber or saran. Another example illustrating the application of properties peculiar to plastics consists of a number of small figures of animals of radiation-crosslinked polyethylene. After molding, the figures are pressed under high pressure into small squares. When these squares are heated in a toy oven, a child can watch the "plastic memory" of the parts turn them back into their original shape (see MECHANICAL PROPERTIES). With another toy, the child draws a picture on a sheet of biaxially oriented polystyrene film, the shape is cut out and placed in an oven where it shrinks to approximately one quarter of its original size and several times its original thickness (see BIAXIAL ORIENTATION). Both of these toys include a small oven as part of the set. Transparent polycarbonate is used for the hoods of these ovens to allow a view of the process in action. These examples, chosen from toys currently enjoying large sales, illustrate the often ingenious ways in which toy designers utilize the properties of polymers.

Processing

A wide variety of processing techniques are used in the toy industry. Injection molding has long been the major processing method, although in recent years several other processes have come into prominence, notably blow molding, rotational molding, and thermoforming. Various other processing techniques are also used to produce special effects. These are shown in Table 1.

Table 1. Processing Techniques in the Toy Industry

Technique	Typical polymer	Typical application
injection molding	polystyrene	multiple interlocking parts
blow molding	polyethylene	wheeled vehicles
rotational molding	poly(vinyl chloride) plastisol	dolls
thermoforming	polystyrene	playing boards
calendering	poly(vinyl chloride)	inflatable toys
sheet extrusion	poly(vinyl chloride)	inflatable toys
coated fabrics	poly(vinyl chloride)	inflatable toys
foaming	polystyrene	snow, surf boards, bones
biaxial orientation	polystyrene film	shrink film
lay-up	glass-reinforced polyester	sleds
dipping	poly(vinyl chloride) plastisol	doll's boots
cold stamping	acrylonitrile–butadiene–styrene copolymer	snow shovels
embossing	polycarbonate	three-dimensional pictures
metallizing	transparent plastic sheet	mirrors

Injection Molding. Injection molding is the most widely used plastic processing technique in the toy industry because it allows versatility in product design and material selection and lends itself to close tolerance control and high-speed automatic production (see also Injection Molding under MOLDING). With the injection-molding process it is possible to produce essentially finished objects in the molding machine itself; only cutting or breaking of sprues and runners by hand or in simple fixtures is required after ejection from the molds.

Large numbers of injection-molded assembly kits for model cars, boats, airplanes, dolls, and other toys are produced as multiple interlocking parts for both factory and "do-it-yourself" assembly. In many designs, the entire toy consisting of numerous parts is produced at one shot in a single multicavity mold. There is a growing tendency to ship an entire injection-molded shot with the sprue cut off, but with all parts still attached to the runner systems. This method of shipping greatly facilitates inventory control since a missing part is more readily visible on inspection. The cost of extra material in the runner system is offset by the savings in inspection, rejection, replacement and, where applicable, assembly costs.

High costs as well as long delivery times for injection-molding tooling are obstacles to wider use of the process. Standard-sized cavities are frequently used to permit the reuse of mold bases.

Blow Molding. Blow molding has grown to be second only to injection molding in terms of volume of plastics processed for toys. The ability to produce hollow parts is the principal advantage. Since the process can produce details on external surfaces only, it cannot be used for the production of most of the designs for three-dimensional intermeshing parts, as is injection molding. Because only a female mold is used, tooling costs are usually lower (see also Blow Molding under MOLDING); toys or parts are, therefore, frequently designed to be hollow.

One of the most successful areas of application of the blow-molding process in the toy industry has been in the production of large toys, particularly wheeled vehicles that the child sits in or on. Doll parts, "piggy" banks, and vehicle wheels are other examples of blow-molded objects.

Rotational Molding. The principal products manufactured by rotational molding have been small parts, such as balls or doll parts, from liquid poly(vinyl chloride) plastisols. The process permits particularly lifelike details for heads, arms, and legs. Like blow molding, the process lends itself particularly to the production of hollow products. See Rotational Molding under MOLDING.

Most of the small poly(vinyl chloride) parts have been produced on batch-type machines (also called box ovens). These ovens require manual insertion and removal of individual molds with each cycle. Newer designs of machines in which molds are permanently mounted on moving carriages and moved alternately in and out of the oven and cooling areas have led to the production of much larger toys. Continuous rotary machines, utilizing multiple spindles and separate heating and cooling chambers, permit the use of other plastics, particularly the polyolefins.

Thermoforming (qv). Among the major uses for thermoforming in toy manufacture are three-dimensional playing boards and packaging. Thermoforming is considerably less expensive than injection or blow molding.

Fabrication of Inflatable Toys. A large portion of the market for toys and playthings consists of inflatable plastic items, such as beach balls, stand-up figures, swimming pools, and floats. The principal construction material is plasticized poly(vinyl chloride) sheet, which is printed, die cut, and heat sealed. Much of the poly(vinyl chloride) sheet for inflatable articles is still produced by calendering, although flat-sheet extrusion has proven to be competitive, particularly for "in-plant" installations. The most widely used method of printing is the silk-screen method (see DECORATING). Cutting to shape is usually accomplished by the use of steel rule dies in clicker presses. Sealing is carried out by high-frequency techniques described under DIELECTRIC HEATING. Tooling costs and start-up times for inflatable items are both extremely low, accounting in part for their popularity with toy manufacturers. Unfortunately, the longevity of these products has left much to be desired.

George E. Pickering
Arthur D. Little, Inc.

TRACER METHODS. See ISOTOPIC LABELING

TRANSESTERIFICATION. See POLYESTERS

TRANSFER. See CHAIN TRANSFER

TRANSFER MOLDING. See MOLDING

TRANSITION TEMPERATURES. See MECHANICAL PROPERTIES

TRANSMITTANCE OF LIGHT. See OPTICAL PROPERTIES

TRIACETATE FIBERS. See CELLULOSE ESTERS, ORGANIC

TRIALLYL CYANURATE POLYMERS. See ALLYL POLYMERS

TRIAZINES. See AMINO RESINS; ALLYL POLYMERS; FLUORINE-CONTAINING POLYMERS

TRIAZOLE POLYMERS. See HEAT-RESISTANT POLYMERS; POLYAMINOTRIAZOLES

TRIMETHYLENE OXIDE POLYMERS. See OXETANE POLYMERS

TRIMETHYLOLALKANES. See ALCOHOLS, MONO- AND POLYHYDRIC

TRIOXANE POLYMERS. See ALDEHYDE POLYMERS

TROMMSDORFF-NORRISH EFFECT. See FREE-RADICAL POLYMERIZATION

TUBING

This article describes the industrial uses of plastic tubing. See also PIPE.

Vinyl Plastic. By volume, vinyl tubing is probably one of the most important materials now used and, because of a favorable price-performance ratio, will continue to be widely used. The regular grade of vinyl tubing, which meets Food and Drug Administration specifications for use in a food environment, is used in applications ranging from pipeline system for harvesting maple sap to hose on beverage processing equipment. See also FOOD AND DRUG APPLICATIONS. Vinyl tubing is used in laboratory equipment, for surgical and hospital applications, and in the chemical, textile, automotive, aircraft, plumbing-heating, fuel, lubricant, material-handling, and utility industries.

Vinyl tubing exhibits resistance to both weak and strong acids and alkalies, has slight or no odor, can be formulated as self-extinguishing, can be used at operating temperatures of from −68°F to 180°F, and has good tensile strength (up to 3100 psi). An almost limitless number of sizes is available from a 1/16 in. ID–3/16 in. OD to a 4 in. ID–5 in. OD. Special sizes from 0.005 in. to 6 in. ID can be produced. Vinyl tubing can be supplied clear or colored to specifications. It can be braided for extra strength and can be produced in flexible, semirigid, or rigid grades. A mechanical grade of vinyl is also available, which is less expensive. See also VINYL CHLORIDE POLYMERS.

Nylon. Nylon is another important plastic used in tubing. It is used extensively to replace reinforced rubber hose, and copper, steel, and aluminum tubings. The basic advantages of nylon tubing include flexibility, comparatively low cost, high-pressure ratings, ease of installation, adaptability to standard fittings, fatigue resistance, solvent and chemical resistance, relatively wide operating temperature range, and resistance to fungus. Since it is translucent, it permits visual inspection of the tubing's contents. Nylon also meets specifications for FDA approval.

Because of its ability to withstand pressures of 2500 psi and its installation economy (resulting from elimination of the need to prebend or form the tubing), nylon may be used as a substitute for neoprene hose in hydraulic lines and for copper tubing in fuel, oil, lubrication, air, and pneumatic lines. Beverage and syrup lines are being manufactured from nylon tubing; these impart no taste, are resistant to fungus, and are unaffected by carbonic acid.

Nylon tubing is also used in farm machinery and equipment; construction, mining, and materials handling; metalworking machinery and equipment; other industrial machinery (food, textiles, paper, and printing); transportation equipment (automotive, aircraft, aerospace); and instruments. Mechanical applications include bearings, rollers, and machined parts. Of increasing interest is the nylon-11 formulation used for gasoline lines in the automotive industry. See also POLYAMIDE PLASTICS.

Polytetrafluoroethylene. Polytetrafluoroethylene tubing displays high resistance to heat and cold; the operating temperature range is from −450°F to +500°F. Other outstanding qualities include resistance to bacterial growth, high dielectric strength, outstanding chemical resistance (attacked only by molten alkali metals), excellent weatherability, very low coefficient of friction, and a nonsticking surface. The material will not support combustion and meets federal specifications for use in contact with foods.

Thin polytetrafluoroethylene tubing, called "spaghetti tubing," ranges from 0.015 in. ID to 0.375 in. ID. Extruded tubing ranges from 0.118 in. to 2 in. ID and up to 3 in. OD. Molded cylinders range from ½ in. to 18 in. ID and 1½ in. to 25 in. OD. Typical applications for spaghetti tubing include catheters, suture covers, needle covers, replacement veins and arteries (see MEDICAL APPLICATIONS); wire covering, terminal insulation, wire bundling in electrical industries (see ELECTRICAL APPLICATIONS); transfer lines in food processing equipment; transfer lines for solvents and corrosives in the chemical processing industry (see CHEMICALLY RESISTANT POLYMERS); and special reinforced tubing for hydraulic and steam lines.

Applications for extruded polytetrafluoroethylene tubing in the chemical processing field include fill tubes for chemical reactors, machined steam spargers, gaskets, "O" rings and expansion joints, pipe lining and covering in areas of erosion and corrosion, transfer tubes, and tower packing. Mechanical applications include machined bushings and bearings, back-up rings, ball valve seats, pump packings and seals, "O" rings, and hydraulic piston rings. Typical applications for parts machined from large molded tubes include envelope gaskets, flat gaskets, bushings, bearings, ball valve seats, and hydraulic seats. See also TETRAFLUOROETHYLENE POLYMERS.

Acrylic Plastic. Cast acrylic tubing is used in some of the more exotic applications including modern sculpture, design, and art (see FINE ARTS). The exceptional optical properties of cast acrylic tubing are matched only by the finest glass. It also possesses a good balance of mechanical properties.

Sizes of cast acrylic tubing range from 1½ in. to 18 in. OD. Larger diameters can be produced on a custom basis. Cast acrylic tubing is being widely acclaimed as a design medium for displays, animated art objects, furniture, and lighting diffusers. The tubing is used in a variety of translucent and transparent colors. Industrial applications include transparent housings for filtering elements in air- and liquid-filtration equipment, reservoirs for coolants or lubricants, testing and laboratory equipment, etc.

Extruded acrylic tubing is used in similar applications, but where optical and mechanical requirements are not as stringent. See also ACRYLIC ESTER POLYMERS; FILMS AND SHEETING; OPTICAL PROPERTIES.

Polyethylene. Another high-volume material in tubing is polyethylene. Applications for polyethylene include low-pressure transmission lines for beverages, oil, and other fluids and air. The limited operating-temperature range (−10°F to 120°F) restricts the potential uses. See also ETHYLENE POLYMERS.

Polyurethane. Flexible polyurethan tubing performs well in environments from −80°F to +225°F. It has a high resistance to abrasion, tear, impact, weathering, and to chemicals such as oil and gasoline. It is considered tasteless and odorless. Standard size ranges from 1/16 in. to 1 in. ID with various wall thicknesses. Tubes of larger diameter are also available.

Other Plastics. High-pressure thermoset laminated tubing is normally machined into bearings, housings, cores, bushings, handles, nipples, and similar articles. Reinforcements include asbestos, paper, canvas, and linen. The National Electrical Manufacturers' Association (NEMA) has classified the various formulations of laminates and standardized their designations; NEMA C, for example, is a canvas-reinforced phenolic resin. Tubes from Grade C are used for bearings. Graphitized Grade C tubing works well in bushing applications where self-lubrication is desirable. Other typical applications of various grades of laminated tubing are: piston bearings (molybdenum disulfide-filled NEMA C); housings for automobile starting cables (NEMA C); cores for magnetic tape (NEMA G-5, glass-reinforced melamine resin); textile loom bushing (NEMA LE, graphitized linen-reinforced phenolic resin); and curved printed circuits (NEMA G-10, copper-clad glass-reinforced epoxy resin). See also LAMINATES; REINFORCED PLASTICS.

Polycarbonate tubing is used for sight glasses, protective sleeving for lamps, packaging, and displays. Its primary advantage is its high impact strength. See also POLYCARBONATES.

Shrinkable tubing of vinyl, polyethylene, and tetrafluoroethylene copolymer (FEP) has gained wide acceptance as electrical insulation for wire and cables (see WIRE AND CABLE COVERINGS). Material selection is based on the degree of insulation required, the anticipated other environmental conditions and cost. Polyethylene is usually least expensive; vinyl normally falls in the medium-price range; and FEP is most costly. FEP shrinkable tubing is also used where a nonsticking surface is desirable, such as roll covers for the textile, paper, and rubber industries.

New materials include tubing that has an inner core of polyethylene, vinyl, nylon, TFE, or similar material and an outer covering of epoxy resin reinforced with continuous glass roving. This type of tube is expected to meet the needs for applications requiring high flexural strength as well as chemical and corrosion resistance.

John Kozacki
Cadillac Plastic and Chemical Company

TUNG OIL. See DRYING OILS

TURBIDIMETRIC TITRATION. See FRACTIONATION

TYNDALL EFFECT. See SCATTERING

U

ULTRACENTRIFUGATION

Sedimentation in a centrifugal field is one of the most versatile methods of obtaining information about the size and shape of macromolecules in solution. This article is intended as a general survey of the application of ultracentrifugal methods to the characterization of synthetic high polymers. More fundamental theoretical treatments and detailed descriptions of the techniques of analytical ultracentrifugation may be found in a number of excellent books and review articles on the subject (1–8). See also CHARACTERIZATION.

Macromolecules of biological origin, especially the proteins, are often monodisperse in size and molecular weight (see MONODISPERSE POLYMERS). Measurement of a given average property of a protein solution thus serves to specify that property for each molecule in the solution. In contrast, most synthetic high polymers are polydisperse. It is therefore important to determine distributions of properties of these solutes as well as their average properties. The measurement of molecular-weight distributions has been the major objective of much of the ultracentrifugal analysis of synthetic high polymers. See also MOLECULAR-WEIGHT DETERMINATION.

A number of difficulties are associated with investigations of synthetic polymers. Solutions of these molecules in many solvents are highly nonideal. Often, conditions (Θ conditions) can be found in which the effects of nonideality are minimized. High temperatures or difficult solvents, however, are frequently required to obtain such Θ conditions, and these requirements bring technical difficulties with them. Another problem results from the relatively high compressibilities of the solutes or solvents employed. Most theoretical treatments of sedimentation behavior assume incompressible systems. The investigator must be careful that the effects of differential compressibility of solvent and solute are taken into account in the interpretation of results or else are minimized by thoughtful experimental design. See also SOLUTION PROPERTIES.

Equipment

The analytical ultracentrifuge is a device that permits observation of the concentration distribution within a solution while it is being subjected to a centrifugal field. The solution is contained within a cell which is fitted into a rotor, and the rotor is then driven at speeds up to 60,000 or 68,000 rpm by a geared electric motor.

Cells. The choice of a cell depends on almost all of the conditions of an experiment: on the type of experiment, whether velocity or equilibrium sedimentation; on the temperature; on the solvent; on the optical system to be used; and on the solute concentration.

One type of ultracentrifuge cell is shown in Figure 1, and illustrates the fundamental design of most cells now in use. Two optically flat windows of quartz or

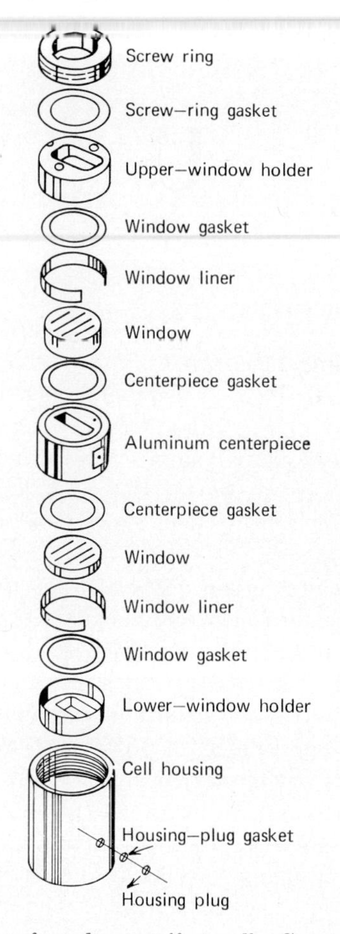

Fig. 1. Schematic illustration of an ultracentrifuge cell. Courtesy Beckman Instruments, Inc.

sapphire are placed against a centerpiece and are held in position by an outer casing of aluminum alloy, threaded at one end for a retainer ring. The solution is introduced into the sector-shaped channel in the centerpiece through a small hole. The walls of the channel delimit a sector of a cylinder defined by the axis of rotation, so that solute molecules will always sediment (and diffuse) in a radial direction. Tight seals are obtained between windows and centerpiece and also at the filling hole by the use of gaskets.

Numerous variations on the design of the centerpiece have been built, and many are available commercially. Double-sector centerpieces, with two sector-shaped channels side by side, are the most useful of these. These permit simultaneous observation of a solution and its solvent, and thus make unnecessary a separate experiment to measure redistribution of components in, or compression of, the solvent. A double-sector centerpiece is, of course, necessary with the Rayleigh interference optical system. Another useful group of designs are the "synthetic boundary" centerpieces, which facilitate the layering of solvent over solution. The concentration difference across the boundary thus formed yields directly the concentration of the solute in units of refractive-index increment. Recent centerpiece designs include those with several pairs of channels (9,10), which permit simultaneous equilibrium experi-

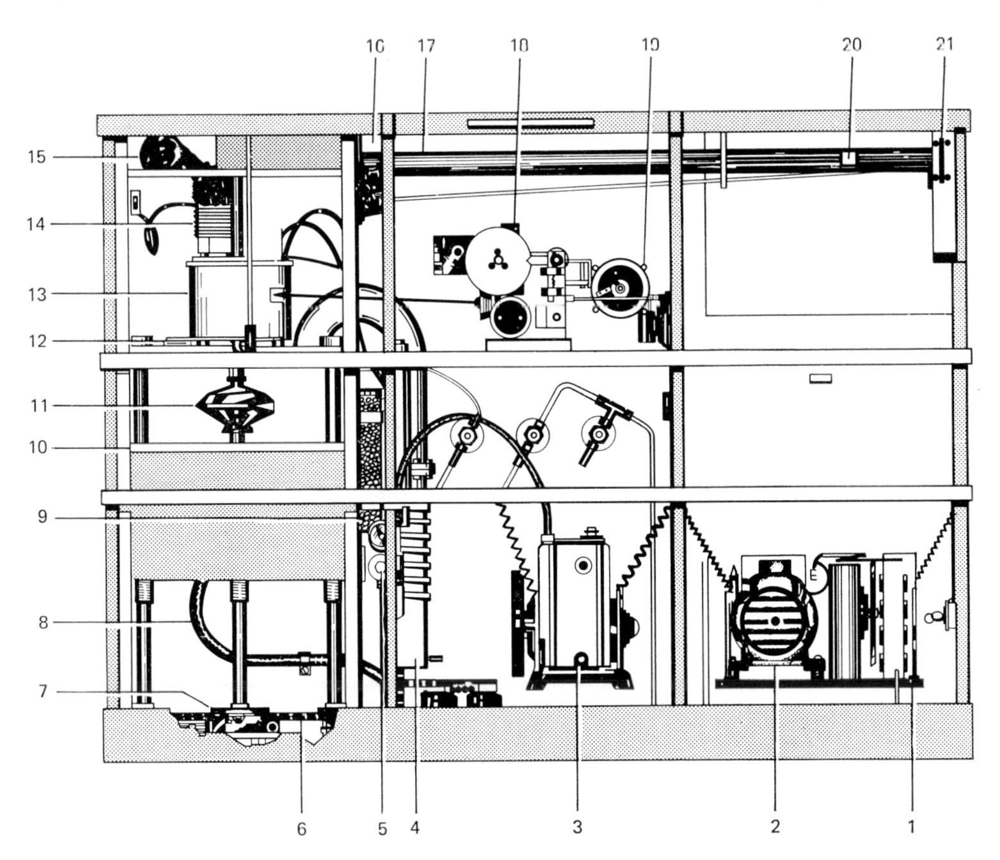

Fig. 2. Schematic illustration of an analytical ultracentrifuge. Courtesy Beckman Instruments, Inc. Key: 1, refrigeration condenser; 2, refrigeration compressor; 3, mechanical vacuum pump; 4, diffusion pump; 5, Evapotrol; 6, chamber lift mechanism; 7, schlieren interference light source; 8, capillary tube for refrigeration; 9, Drierite; 10, rotor chamber; 11, rotor; 12, drive oil gage; 13, rotor drive; 14, drive motor; 15, blower for drive motor; 16, plateshift mechanism; 17, optical tube; 18, differential gearbox; 19, synchronous motor; 20, viewer; 21, plate holder slot.

ments, "rapid equilibration" types, which approximate the equilibrium distribution by forming layers of decreasing solute concentration at the beginning of an experiment, and "band-forming" types, which pour a thin layer of solution on top of a column of solvent, so that the solute will sediment as a band.

Centerpieces made of aluminum-filled or carbon-filled epoxy plastic or of Kel-F fluorocarbon are available. Such plastic centerpieces are light, strong, easy to make, and require no gaskets between window and centerpiece. Some suffer the disadvantage of incomplete resistance to many organic solvents and all become weak at high temperatures.

Different optical path lengths may also be obtained, the most common being 1.5, 3, 6, 12, 18, and 30 mm. The length of the optical path can thus be chosen to give a convenient record on the photographic plate or scan for a wide range of solute concentrations.

Window holders also are of several types. Those for double-sector cells have two slits, one for each sector. Wide slits are employed in conjunction with the schlieren optical system, and narrow slits with the interference optical system.

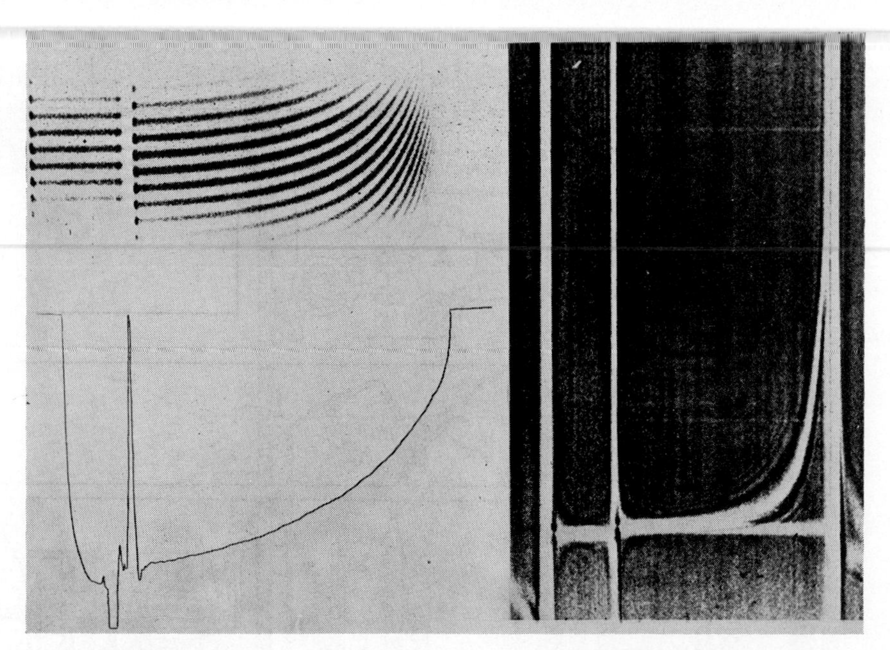

Fig. 3. Typical patterns from equilibrium experiments. Centrifugal force from left to right. Key: (**a**) (right), schlieren pattern from double-sector cell. The curved line in the center of the frame has a displacement above the horizontal straight line that is proportional to the concentration gradient. (**b**) (upper left), Rayleigh interference pattern. The vertical displacement of any one of the fringes is proportional to relative concentration. (**c**) (lower left), scanner trace. The vertical displacement of the line is proportional to absorbance. The large peak on the left of the trace is due to light reflection from the solution meniscus.

Rotors. Several types of rotors are available, differing in mass and in the number of cells each will accommodate. Four- and six-place rotors make possible the simultaneous observation of many cells, although such rotors are restricted to speeds less than those which can be attained with the light two-place rotors. The heavier rotors, two-place and multi-place, will accommodate cells with centerpieces of 30 mm optical path. By virtue of their mass, they are relatively stable at low speeds but cannot be run at the highest speeds.

Rotors are now made of either aluminum or titanium. Titanium rotors can be operated at speeds approximately 20% greater than their aluminum counterparts.

Drive and Control Systems. Figure 2 is a schematic drawing of a commercially available ultracentrifuge. The rotor hangs on a flexible shaft from the water-cooled drive and gearbox unit, and runs inside an armored vacuum chamber which can be raised and lowered. The speed of the rotor is held constant in older machines by comparison with the rotation of a synchronous motor, and in newer machines by a photoelectric counter whose output is used to regulate the current through the drive motor. The temperature of the rotor is sensed by a thermistor mounted on its base, and is regulated by refrigeration coils around the edge of the vacuum chamber and by a heater on the bottom of the chamber. Ordinarily, close control of temperature can be obtained in the range -10 to $+40$°C, and with special equipment temperatures as high as 130°C can be reached.

Optical Systems. Light sources for the optical systems are mounted beneath the vacuum chamber. Light from either of the sources passes through a collimating lens in the bottom of the vacuum chamber and strikes the rotor, which occludes the light beam except when the cell is in the optical path. The beam then passes through a condensing lens in the top of the chamber, is bent through 90° by a mirror above the drive, passes through the remainder of the optical system inside a horizontal tube, and strikes either a photographic plate or the photomultiplier of the absorption scanner at the upper right-hand corner of the instrument.

Three optical systems are in common use for observations of solute redistribution. The first two of these systems employ the refractive-index increment as a measure of solute concentration; the third utilizes light absorption. The schlieren optical system is the most commonly employed (11–14). This system displays the refractive-index gradient as a function of radial position and thus, since refractive index is closely proportional to concentration, provides values of the first derivative of the concentration with radius throughout the solutions (Fig. **3a**). In recent years, a variant of the classical Rayleigh interference optical system, more sensitive and precise than the schlieren system, has come into wide use (15). It produces an array of parallel fringes whose vertical displacement above an arbitrary level is proportional to concentration (Fig. **3b**). The interference system permits measurement only of differences in solute concentration; absolute concentration has to be obtained by the application of mass conservation or inferred by other means. The third optical system is a light-absorption scanning device (16,17) which produces a record of optical transmission or absorbance as a function of radius (Fig. **3c**). This scanner is quite useful when the solute has an optical absorption band in the visible or ultraviolet range, and the solvent does not absorb light significantly at the same wavelength. It has two primary advantages over the the schlieren and interference systems: it records a quantity (absorbance) that is proportional to absolute concentration, and it can be combined with electronic gating circuits and multicell rotors to permit simultaneous observation of many samples. Its primary disadvantage is that under most experimental conditions the scanner lacks the precision of the Rayleigh interference system.

An alternative to the scanner is the older photographic absorption system, which yields an exposure on film whose density is related to the light absorbed by the solution in the cell. The photographic record can then be scanned with a densitometer to produce a graphical record of film density as a function of radius. With careful controls and the appropriate conversion factors such records give useful estimates of solution optical density as a function of radius. This system is sensitive but clumsy for quantitative work.

In addition to the commercially available centrifuges, a few others have been built recently and deserve consideration. In particular, at least two machines are in use in which the rotor is suspended in a magnetic field and has no direct contact with the driving motor once it reaches operating speed (18,19). These machines are especially stable at low speeds, and the "free-floating" rotor should obviate some of the heat-transfer problems associated with conventional drive shaft and bearing supports.

General Techniques

Despite the great variety of techniques employed with the ultracentrifuge, there are certain common operations that can be discussed in a general way.

In sedimentation equilibrium experiments, the time required to reach equilibrium varies from a few hours to a few days. It depends on the length of the solution column, the viscosity of the solution, and the diffusion coefficient of the solute. Photographs should be taken at several times during the run. These records should then be compared to determine when equilibrium has been attained.

Data consisting of concentration (zero arbitrary) as a function of radius, or of the first derivative of concentration as a function of radius, or of true concentration as a function of radius are then read from the photographic plate or scan (see Figure 3). If weight-average molecular weights are to be obtained from refractometric records, and unless a high-speed technique is used (see below), it is necessary to find the initial concentration of the solution by a separate synthetic boundary experiment or with a differential refractometer.

Sedimentation velocity experiments are usually performed at relatively high rotor speeds for, at most, a few hours. Photographs or scans are taken automatically at known time intervals, to follow the progress of the boundary or boundaries that are formed. The experimental records obtained are a series of pictures of the concentration or its gradient as a function of radius at a sequence of times. These records can be treated in a number of ways, as described below.

It is now common practice in many laboratories to make use of the digital computer to perform much of the labor of data reduction. A number of appropriate programs have been written, and reference to them will be found in much of the literature discussed below.

Sedimentation Equilibrium

Molecular-Weight Averages. The estimation of molecular weights and their distributions from the results of sedimentation velocity experiments is commonly found in the literature. The determination of these properties by sedimentation equilibrium is less often reported but is, in principle, much less equivocal since it is soundly based on classical thermodynamics. The reason for this situation lies in the relative difficulties of interpretation of results. Recently there has been much progress toward overcoming both interpretational and experimental difficulties associated with sedimentation equilibrium. At present this approach offers much promise of becoming the method of choice.

In order to understand the principles involved, let us consider first the case of a single solute in an incompressible solvent. A sufficient requirement for equilibrium is that the total potential, chemical plus gravitational, be independent of radius. It is easily shown (1) that this condition leads to the relationship given by equation 1. Here,

$$M\omega^2 r(1 - \bar{v}\rho) = (\partial\mu/\partial c)_{P,T}\,(dc/dr) \tag{1}$$

ω is the angular speed of the rotor in radians/sec, r is the radius, ρ is the density of the solution, M is the molecular weight of the solute, $\bar{v}$ its partial specific volume, c its concentration, and μ its chemical potential. The chemical potential of the solute can be expressed, by the use of an activity coefficient, y, by equation 2.

$$\mu = \mu^0 + RT \ln (yc) \tag{2}$$

Then equation 1 becomes equation 3.

$$\frac{1}{rc}\frac{dc}{dr} = \frac{\omega^2 M(1 - \bar{v}\rho)}{RT[1 + c(\partial \ln y/\partial c)_{P,T}]} \tag{3}$$

The nonideality term, $(\partial \ln y/\partial c)_{P,T}$, can be measured, in principle, in a separate experiment. However, it is more common to assess its magnitude as a part of the sedimentation experiment. For the single solute being considered here, the term $(\partial \ln y/\partial c)_{P,T}$ may be considered, at sufficiently low concentration, to be essentially constant. One then obtains as a limiting relation:

$$\frac{1}{M_{\text{app}}} = \frac{1}{M} + Bc \tag{4}$$

where M_{app} is an "apparent" molecular weight, defined by equation 5, and B is a constant to be determined in the course of the experiment.

$$M_{\text{app}} = \frac{RT}{\omega^2(1 - \bar{v}\rho)} \frac{1}{rc} \frac{dc}{dr} \tag{5}$$

The more frequently encountered case of a polydisperse solute may be treated in a similar manner if certain assumptions are made about its nature. In this case, it is necessary to measure average molecular weights. In a solute of n components, the number-average, weight-average, and z-average molecular weights, $\bar{M}_n$, $\bar{M}_w$, $\bar{M}_z$, respectively, are defined as follows:

$$\bar{M}_n = \sum_{i=1}^{n} c_i / \sum_{i=1}^{n} (c_i/M_i) \tag{6}$$

$$\bar{M}_w = \sum_{i=1}^{n} c_i M_i / \sum_{i=1}^{n} c_i \tag{7}$$

$$\bar{M}_z = \sum_{i=1}^{n} c_i M_i^2 / \sum_{i=1}^{n} c_i M_i \tag{8}$$

Here, and in the text that follows, c_i is the concentration of the ith solute component (in weight/volume), M_i is its molecular weight, $\bar{v}_i$ is its partial specific volume, and μ_i is its chemical potential. In order to obtain these averages, equations similar to equation 1 are written.

$$M_i \omega^2 r(1 - \bar{v}\rho) = \sum_{k=1}^{n} (\partial \mu_i/\partial c_k)_{P,c_{j\neq k}} \frac{dc_k}{dr} \tag{9}$$

The effect of the presence of the other solutes upon the chemical potential of the ith solute is reflected in each of the equations. Another equation equivalent to equation 2 can be written for each solute. When each such equation is used to obtain the quantity

$$(\partial \mu_i/\partial c_k)_{P,c_{j\neq k}}$$

then one finds equation 10 for each solute (20).

$$c_i M_i = \frac{RT}{\omega^2(1 - \bar{v}_i\rho)} \left[\frac{dc_i}{dr} + c_i \sum_{k=1}^{n} (\partial \ln y_i/\partial c_k)_{P,c_{j\neq k}} \frac{dc_k}{dr} \right] \tag{10}$$

In the simple case of an ideal solution, the $(\partial \ln y_i/\partial c_k)$ vanish, and equation 11 results.

$$c_i M_i = \frac{RT}{\omega^2(1 - \bar{v}_i\rho)} \frac{dc_i}{dr} \tag{11}$$

In the case of a mixture of homologous macromolecules it is a good approximation to assume that the $\bar{v}_i$ are independent of molecular weight and equal to the measured (weight-average) $\bar{v}$ of the mixture. Then at any radius, point-average values of the molecular weights may be measured (eqs. 12 and 13).

$$M_w(r) = \frac{RT}{\omega^2(1 - \bar{v}\rho)} \frac{1}{rc} \frac{dc}{dr} \tag{12}$$

$$M_z(r) = \frac{RT}{\omega^2(1 - \bar{v}\rho)r} \left[\frac{d^2c}{dr^2} \Big/ \frac{dc}{dr} - \frac{1}{r^2} \right] \tag{13}$$

The number-average molecular weight at any radius may also be measured, provided the rotor speed is high enough that the concentration at the meniscus becomes vanishingly small with respect to the initial concentration, as in equation 14,

$$M_n(r) \cong c \Big/ \int_a^r rc\, dr \tag{14}$$

where a is the radius of the solution meniscus (9). Each of these point averages constitutes, in this case of an ideal solute with all $\bar{v}_i$ equal to each other, the true number-, weight-, or z-average molecular weight of the molecular species present at the radius where the measurement is performed. These averages will vary with radius because of the fractionating effect of the centrifugal field, and will also vary with rotor speed at a given radius. These variations may be exploited to obtain information about the molecular-weight distribution of the solute.

It is often desirable to know the *overall* value of a molecular-weight average of a solute, ie, the value which would be obtained if no centrifugal fractionation had occurred. There are several ways of doing this. One way is to invoke conservation of mass within the cell to obtain equations 15 and 16, in which a and b signify the top and

$$\overline{M}_w = \frac{RT}{\omega^2(1 - \bar{v}\rho)} \frac{c(b) - c(a)}{c_0} \frac{2}{b^2 - a^2} \tag{15}$$

$$\overline{M}_z = \frac{RT}{\omega^2(1 - \bar{v}\rho)} \frac{\left(\frac{1}{r}\frac{dc}{dr}\right)_b - \left(\frac{1}{r}\frac{dc}{dr}\right)_a}{c(b) - c(a)} \tag{16}$$

bottom, respectively, of the solution column, and c_0 is the initial concentration. The latter quantity must be obtained in a separate experiment. This method requires that the gradient at the base of the cell (or at the meniscus in the case of flotation equilibrium) be small enough so that the concentration and its first derivative can be measured there. This requirement must, of course, be balanced with the requirement that $[c(b) - c(a)]$ be large enough to be measured accurately.

Another way of obtaining overall molecular-weight averages rests on the fact that certain relations apply at the base of the cell when the concentration and its first derivative at the solution meniscus are very small (9). When the rotor speed is high enough so that this situation prevails, then

$$\overline{M}_w = M_n(r) \qquad \text{(at base)} \tag{17}$$

$$\overline{M}_z = M_w(r) \qquad \text{(at base)} \tag{18}$$

This method is useful when polydispersity is not too great. It places rather stringent requirements on the visibility of the solution column.

If the terms $(\partial \ln y_i/\partial c_k)$ in equation 10 are not negligible, further difficulties arise in the evaluation of true molecular weights. An approximation due to Fujita (21,22) appears to be adequate for use as a limiting relation at low solute concentration, and at low degrees of centrifugal fractionation. It is applicable to overall molecular-weight averages, as defined by equations 15 and 16. The $(\ln y_i)$ of equation 1 are expanded as a Taylor series (eq. 19).

$$\ln y_i = \sum_{k=1}^{n} (\partial \ln y_i/\partial c_k)_{c_j=0} c_k + \text{terms of order } (c_j c_k) \tag{19}$$

The $(\partial \ln y_i/\partial c_k)$ are denoted by $M_i B_{ik}$, and equation 10 becomes equation 20.

$$c_i M_i = \frac{RT}{\omega^2 r(1 - \bar{v}\rho)} \left[\frac{dc_i}{dr} + c_i M_i \sum_{k=1}^{n} B_{ik} \frac{dc_k}{dr} \right] \tag{20}$$

From this, Fujita has shown that equation 21 holds.

$$\frac{1}{(M_w)_{\text{app}}} = \frac{1}{\overline{M}_w} + B''c_0 + \text{terms of order } (c_0^2) \tag{21}$$

$(M_w)_{\text{app}}$ is given (see eq. 15) by equation 22.

$$(M_w)_{\text{app}} = \frac{RT}{\omega^2(1 - \bar{v}\rho)} \frac{c(b) - c(a)}{c_0} \frac{2}{b^2 - a^2} \tag{22}$$

The second virial coefficient so defined, B'', is related to the light-scattering second virial coefficient, B_{LS}, by equation 23, in which $\lambda = (1 - \bar{v}\rho)(b^2 - a^2)\omega^2/2RT$.

$$B'' = [1 + (\lambda^2 \overline{M}_z^2/12)] B_{\text{LS}} \tag{23}$$

Thus, from the results of sedimentation equilibrium experiments performed at low concentration one may obtain the true weight-average molecular weight, $\overline{M}_w$, and the limiting value of the second virial coefficient.

The measurement by Fujita and co-workers (23) of $\overline{M}_w$, $\overline{M}_z$, and B'' of polystyrene in cyclohexane provides an example of the use of the above relationships. Figure 4 shows a graph of $1/(M_w)_{\text{app}}$ plotted against c_0 obtained for four temperatures near the Θ temperature. This experiment shows that the method works, to a first approximation, even at the moderately high concentrations of several tenths of a gram per 100 ml. Nevertheless, it should be noted that the plots are not as linear as could be desired, and that approximations are necessarily involved in obtaining the final results.

A recently developed technique for obtaining weight-average molecular weights and second virial coefficients of nonideal polydisperse systems has been described by Osterhoudt and Williams (24), and by Albright and Williams (25). It makes use of a series of equilibrium experiments at different rotor speeds. Equation 21 can be restated (1) as equation 24, where the degree of approximation made by neglecting higher

$$\frac{c(b) - c(a)}{c_0} = \lambda(\overline{M}_w - B_{\text{LS}} \overline{M}_w^2 c_0 + \ldots) \tag{24}$$

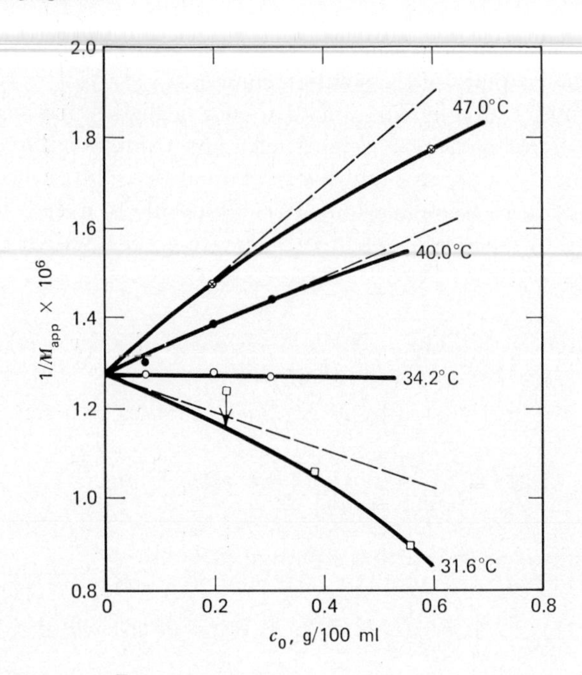

Fig. 4. Measurement of $\bar{M}_w$ and second virial coefficient, B'', of polystyrene in cyclohexane at temperatures near the Θ temperature (23). Solid lines, experimental results; dashed lines, simple nonideality. Courtesy American Chemical Society.

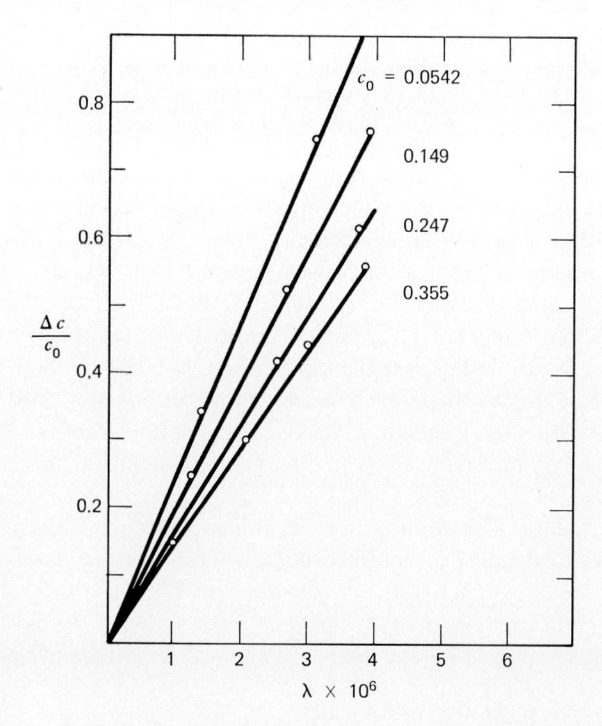

Fig. 5. Plot of $[c(b) - c(a)]/c_0$ against λ for a polystyrene (S-108; mw 267,000)–toluene system at 20°C (25). Initial concentrations are indicated in the figure. Courtesy American Chemical Society.

terms is about the same as that in equation 21. The quantity $\bar{M}_w$ may now be regarded as the limiting slope of a plot of $[c(b) - c(a)]/c_o$ against λ. If the limiting slopes of a series of such plots are plotted against c_0, a straight line will be observed at sufficiently low concentration. The intercept of this line at zero concentration is $\bar{M}_w$, and its slope is $\bar{M}_w^2 B_{LS}$. In practice, a series of experiments is performed at different rotor speeds, chosen so that λ is less than $1/\bar{M}_w$ (24). Solution columns of length 1.5 to 3 mm may be used to reduce the time required to reach equilibrium.

Albright and Williams (25) have tested this scheme by obtaining $\bar{M}_w$ and B_{LS} in two polystyrene–toluene systems and in a polyisobutylene–cyclohexane system. Figures 5 and 6 show some of their results for a polystyrene of $\bar{M}_w = 267{,}000$ in toluene.

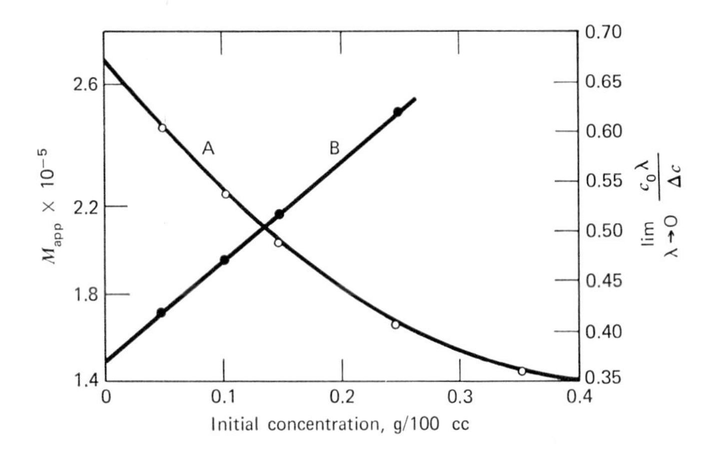

Fig. 6. Extrapolation to zero initial concentration of apparent molecular weights (25). Each point represents the slope of one of the lines in Figure 5. Curve A: plot according to equation 24; curve B: plot according to equation 21. Courtesy American Chemical Society.

In Figure 6, the extrapolation to $c_0 = 0$ of the limiting slopes obtained from the data of Figure 5 has been carried out, both by equation 24 (curve A) and by equation 21 (curve B). The values obtained for $\bar{M}_w$ were found to agree to within about 2%.

The use of data obtained at the ends of the solution column (meniscus and base) suffers from the difficulty that the location of these points on the optical image is somewhat uncertain. A method which uses points in the solution column other than the end points has been demonstrated by Scholte (26). As in the methods described above, a variety of rotor speeds is employed. From an equation of Fujita (7), the concentration gradient of a solute displaying *ideal* solution behavior is given by equation 25, where $\xi = (b^2 - r^2)(b^2 - a^2)$, and $f(M)$ is the differential molecular-weight-

$$\frac{dc}{d\xi} = -c_0 \int_0^\infty \frac{\lambda^2 M^2 \exp(-\xi\lambda M)}{1 - \exp(-\lambda M)} f(M)\, dM \tag{25}$$

distribution function. By manipulation of this equation, Fujita showed that

$$\bar{M}_w = \lim_{\lambda\to 0}\left[-\frac{1}{c_0}\left(\frac{dc}{d\xi}\right)_{\xi=\frac{1}{2}}\right] = \lim_{\lambda\to 0} q_{\frac{1}{2}}(\lambda) \tag{26}$$

It can also be shown that

$$\bar{M}_n = 16.83 \Big/ \int_0^\infty \lambda q_{1/2}(\lambda)\, d\lambda \tag{27}$$

$$\bar{M}_n = 5.31 \Big/ \int_0^\infty \lambda q_{3/4}(\lambda)\, d\lambda \tag{28}$$

$$\bar{M}_n = 2.408 \Big/ \int_0^\infty \lambda q_1(\lambda)\, d\lambda \tag{29}$$

$$\bar{M}_w\bar{M}_z/4 = \lim_{\lambda \to 0} \left\{ \frac{1}{2\lambda} \left[q_{1/4}(\lambda) - q_{3/4}(\lambda) \right] \right\} \tag{30}$$

$$\bar{M}_w\bar{M}_z\bar{M}_{z+1} = 96 \left\{ \frac{d}{d(\lambda^2)} \left[\frac{q_{1/4}(\lambda) + q_{3/4}(\lambda)}{2} \right] \right\}_{\lambda = 0} \tag{31}$$

where $q_{1/2}$, $q_{1/4}$, and $q_{3/4}$ indicate the values of $-1/c_0(dc/d\xi)$ obtained at $\xi = 1/2, 1/4$, and $3/4$, respectively. In order to determine $\bar{M}_n$, the quantity $\lambda q(\lambda)$ must be measured over the entire range in which it has significant magnitude. In practice, it seems to be possible to do this with many samples. Figure 7 shows the values of $\lambda q(\lambda)$ obtained for a sample of polyethylene in biphenyl at 123.2°C, the Θ temperature in this solvent. The extrapolations to $\lambda = 0$ seem to be fairly straightforward in the case of $\bar{M}_w$, and somewhat more difficult for $\bar{M}_z$ and $\bar{M}_{z+1}$. Scholte asserts that the overall precision of the method is about ±5% in the case of $\bar{M}_w$, about ±10% in $\bar{M}_n$ and $\bar{M}_z$, and ±25% in $\bar{M}_{z+1}$. It should be possible to adapt the method to nonideal solutions, but this has not yet been done.

Note that both the method of Albright and Williams (25) and that of Scholte (26) involve extrapolation to zero rotor speed. This procedure has the advantage that effects due to the compressibility of the solvent are eliminated, and the disadvantage that precision is lost at the lower speeds. See also DIFFUSION.

Molecular-Weight Distribution. The determination of the molecular-weight distribution of a heterogeneous polymer is a formidable task at present. Several

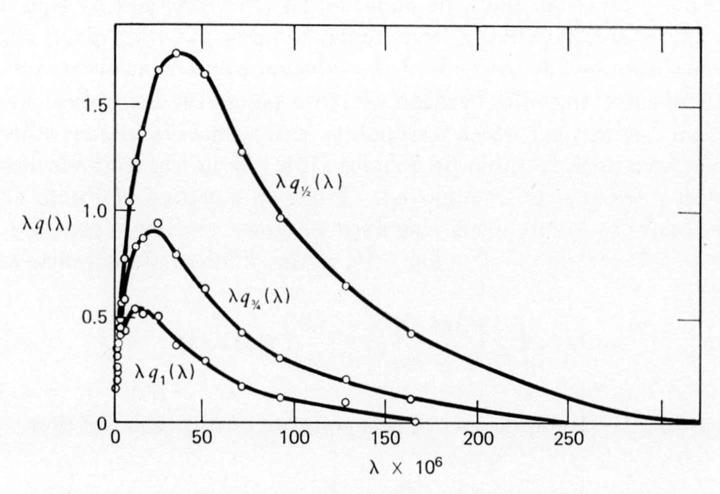

Fig. 7. Plot of $\lambda q(\lambda)$ against λ for a polyethylene–biphenyl system at its Θ temperature, showing determination of $\bar{M}_n$ at three different values of ξ (26).

procedures have been adopted to characterize such distributions from data obtained at sedimentation equilibrium. See also FRACTIONATION.

If one has knowledge, a priori, of the form of a distribution, one can estimate the width of the distribution and the location of its maximum from a knowledge of several molecular-weight averages. One such treatment was carried out by Wales and coworkers (27), in which $\bar{M}_n$, $\bar{M}_w$, and $\bar{M}_z$, and $\bar{M}_{z+1}$ were used to fit a very flexible assumed form of distribution based on Laguerre polynomials.

More recently, effort has been directed at finding experimental procedures that will allow molecular-weight distributions of arbitrary shape to be determined. Donnelly (28) has reported a method using Laplace transforms which has yielded moderate success. So far only unimodal distributions can be handled by this method and it seems to be of limited utility at present.

A different approach has been taken by Scholte (29). In his method, the rotor speed is varied, and data from several different radii in the solution column are used at each of the speeds. The starting point is equation 25:

$$\frac{dc}{d\xi} = -c_0 \int_0^\infty \frac{\lambda^2 M^2 \exp(-\xi\lambda M)}{1 - \exp(-\lambda M)} f(M)\, dM$$

The function $f(M)$ is represented as the sum of a limited number of delta functions at a succession of values of M, each denoted by M_i. The integral in equation 25 is then represented as a sum (eq. 32), where $U(\lambda,\xi) = -1/c_0(dc/d\xi)$, and f_i is the weight

$$U(\lambda,\xi) = \sum_i f_i \frac{\lambda^2 M_i^2 \exp(-\xi\lambda M_i)}{1 - \exp(-\lambda M_i)} \tag{32}$$

fraction of molecules having molecular weight M_i. A sequence of M_i is then chosen which must be broad enough to represent the entire distribution, and spaced at intervals which are determined by the precision of the available data. Then, for any value (denoted by subscript j) of $U(\lambda,\xi)$ and any trial set of f_i, one may write equation 33.

$$U(\lambda,\xi)_j = \sum_i f_i K_{ij} + \delta_j \tag{33}$$

where

$$K_{ij} = \frac{\lambda_j^2 M_i^2 \exp(-\lambda_j \xi_j M_i)}{1 - \exp(-\lambda_j M_j)}$$

A linear programming method is used to select the set of f_i which makes

$$\sum_j |\delta_j|$$

a minimum for all j. By taking a sufficiently large number of values of $U(\lambda,\xi)$, a satisfactory degree of precision in f_i may be obtained.

Results for a sample of polystyrene in cyclohexane at 35°C are shown in Figure 8. The values of $\bar{M}_n$, $\bar{M}_w$, $\bar{M}_z$, and $\bar{M}_{z+1}$ calculated from the distribution agree well with values obtained from ordinary sedimentation methods. At present, this scheme for obtaining molecular-weight distributions is restricted to ideal systems. Its extension to nonideal solutions appears to offer considerable computational difficulty, but is not to be considered impossible.

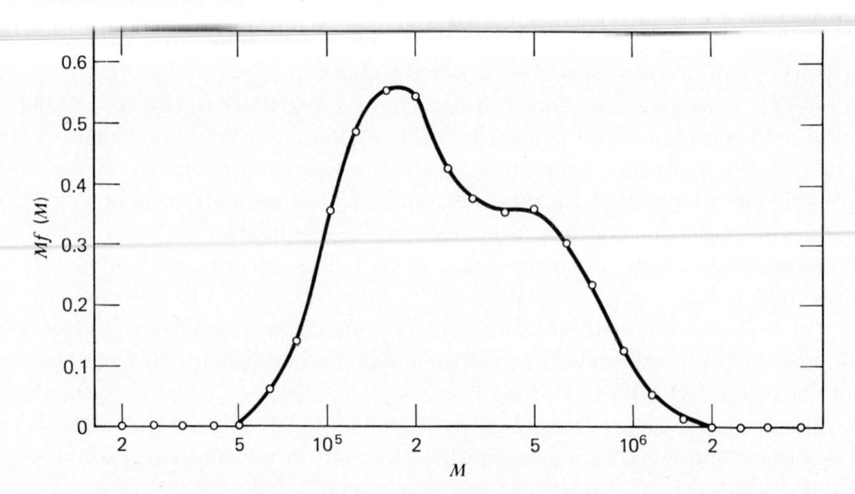

Fig. 8. Molecular-weight distribution of a polystyrene sample, obtained by the method of Scholte (29). Measurements performed in cyclohexane at 35°C. The distribution was calculated from 44 values of U(λ,ξ).

Sedimentation Velocity

Velocity sedimentation yields results less rigorously interpretable in principle than those of equilibrium sedimentation. In practice, however, it has been easier to use and has therefore been more widely employed. The primary measurement made by this technique is of the distribution of sedimentation coefficients, $g(s)$. This distribution, once determined, can be transformed by a suitable relation to a distribution of molecular weights, $f(M)$. The procedures commonly in use seem to be based on an outline proposed by Williams (30) and involve a number of uncertainties as well as much data manipulation. Large corrections must be introduced for the effects of concentration, pressure, and diffusion.

Sedimentation Coefficients. A homogeneous macromolecular solute may be described in a given solvent by a sedimentation coefficient, s, which is defined as the rate of movement of the solute, in the radial direction, per unit field. As shown in Figure 9, the movement of solute creates a more or less diffuse boundary between a centrifugal region where the concentration gradient is essentially zero (the plateau), and a centripetal region where the solute concentration is negligibly small. The sedimentation velocity of any solute may be obtained from the rate of movement of a point in this boundary whose radius, r_b, is given (20) by equations 34, 35, or 36, where

$$r_b{}^2 = r_p{}^2 - \frac{2}{c_p}\int_{r_a}^{r_b} cr\, dr \tag{34}$$

$$r_b{}^2 = \int_{r_a}^{r_p} r^2\, dc \Big/ \int_{r_a}^{r_b} dc \tag{35}$$

$$r_b{}^2 = \int_{r_a}^{r_p} r^2 \frac{dc}{dr}\, dr \Big/ \int_{r_a}^{r_p} \frac{dc}{dr}\, dr \tag{36}$$

r_p is the radius of a point in the plateau, c_p is the concentration at that point, and r_a is the radius of the meniscus. (The different forms of the expression for r_b are equivalent, but useful for different optical systems.)

The sedimentation coefficient, s, is then given by equation 37, where t is the time

$$s = \frac{1}{\omega^2 r_b}\frac{dr_b}{dt} \tag{37}$$

in seconds. The quantity 10^{-13} sec is known as the *svedberg unit*, named after Svedberg, the originator of ultracentrifugation. When s is measured by equations 34–36 for a heterogeneous solute, the value obtained will be the weight-average sedimentation coefficient.

In the case of a rapidly sedimenting homogeneous ideal solute, the boundary is quite nearly symmetrical, and r_b corresponds closely to the radius at which the concentration is one-half the value in the plateau, as well as the radius at which (dc/dr) is a maximum. For this reason it has been common practice to approximate r_b by either of these points, depending on the optical system employed. In the case of the skewed

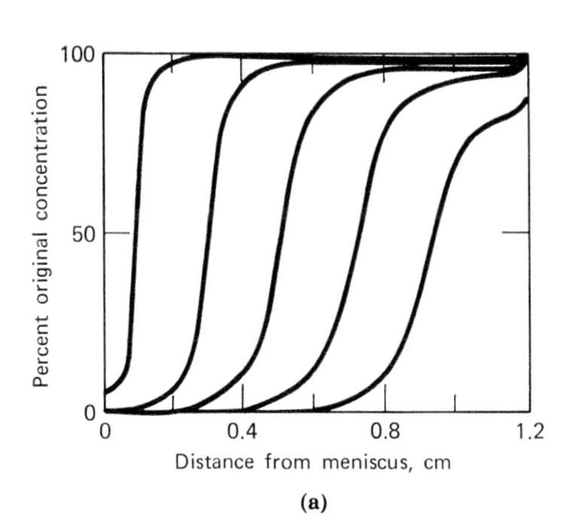

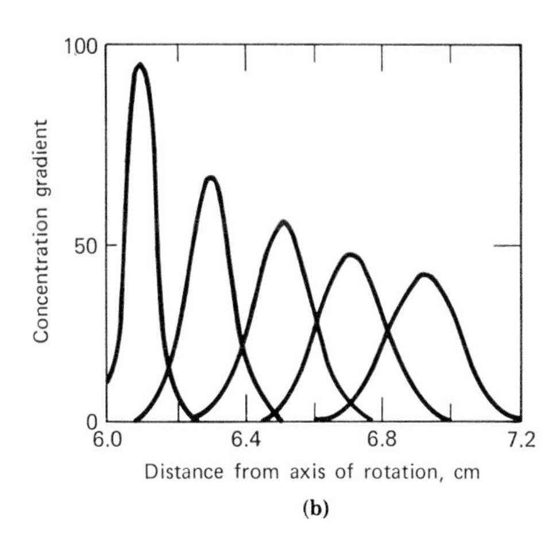

Fig. 9. Schematic diagram of the concentration (**a**) and concentration gradient (**b**) as a function of distance in the centrifuge cell during a sedimentation velocity experiment (6). Courtesy Academic Press, Inc.

boundaries produced by heterogeneous solutes, this approximation is not generally valid, and errors can be introduced by its injudicious application (31). However, only trivial errors are introduced when the skewness of the boundary originates from ordinary nonideality (32).

For a discussion of the relationship of the sedimentation coefficient to molecular weight, see DIFFUSION, Vol. 5, pp. 70–72. Measurements of this type are complicated by the difficulty of obtaining useful diffusion-coefficient averages for polydisperse solutes.

Distributions of Sedimentation Coefficient. The distribution of sedimentation coefficients may be considered to be continuous for most synthetic polymers. Measurement of this distribution may be obtained by examination of the manner in which the boundary spreads during sedimentation. In order to see how this is done, let us consider first the case of a hypothetical heterogeneous solute and make three assumptions about it. First, spreading of the sedimenting boundary depends only on differences in s, with no contribution from diffusion. Second, s is independent of c

for each component. Third, the solvent and solute are incompressible. For such a solute, $g(s)$ could be obtained (33) from the simple relation given by equation 38,

$$g(s) = \frac{1}{c_0}(dc/dr)(r/r_m^2)(\omega^2 rt) \tag{38}$$

where c_0 is the initial concentration. For real solutes, all three of the assumptions just made are violated in some degree. The strategy in what follows is to correct for the behavior of the real solute, so that the final results are independent of diffusion, concentration, and pressure.

Correction for diffusion can be introduced by making use of the fact that sedimentation proceeds in direct proportion to the time, t, whereas diffusion is proportional to $t^{1/2}$. Therefore, if apparent curves of $g(s)$ are obtained by equation 38 at a number of values of time, and appropriately extrapolated to infinite time, the effects of diffusion will be removed. A basis for extrapolation has been provided by Gosting (34), who showed that a plot of apparent values of $g(s)$ against $1/t$ is approximately linear when the condition $\omega^2 st \ll 1$ is met. A set of calculations, illustrative of the extrapolation procedure for a hypothetical Gaussian $g(s)$, is provided by Baldwin (35). A detailed example of an experimental use of the extrapolation is provided by Kotaka and Donkai (19). Extrapolations against functions of the time other than $1/t$ are occasionally used (36,37), although $1/t$ appears to be adequate for most solutions of large molecules.

Concentration dependence of the sedimentation coefficients provides a second level of complexity in the process of obtaining the true $g(s)$. If, as is usually the case for large polymers (even under Θ conditions), material at high concentration sediments more slowly than material at low concentration, a "self-sharpening" of the boundary occurs, causing the material to appear less heterogeneous than it is. In order to correct for self-sharpening, an extrapolation to zero concentration of the observed curves of $g(s)$ against s can be carried out. Examples of the application of this procedure will be found in the work of Williams and co-workers (38) on gelatins, in that of Williams and Saunders (39) on dextrans, and in many of the other references in this section. In general, there seems to be no critically adequate theory to guide the extrapolation. The usual procedure (39) is to generate "diffusion-corrected" values of $g(s)$ for several initial concentrations. For fixed values of s, these values of $g(s)$ are extrapolated linearly to infinite dilution. This procedure works well where concentration dependence is not large.

The importance of the effect of pressure on the sedimentation of materials in compressible solvents has only recently been realized. A theoretical treatment of the matter is given by Fujita (40). According to this treatment, the value of the sedimentation coefficient at 1 atmosphere, $s°$, can be related to the observed value of s by equation 39, where m is a coefficient containing the dependence on pressure of the frictional coefficient and specific volume of the solute, and of the density of the solvent.

$$s° = s/\{1 - m[(r/r_a)^2 - 1]\} \tag{39}$$

A method based on Fujita's relation has been used by Wales and Rehfeld (41), and by Blair and Williams (31) to correct observed $g(s)$ results. The quantity m was estimated by measuring the pressure dependence of the overall sedimentation coefficient of a narrow fraction of the solute. Billick (37) has used a similar method of pressure correction in the study of a sample of polystyrene in cyclohexane. Some of his results are shown in Figure 10, where the importance of the pressure correction is clearly seen.

The method discussed above can be applied to yield fairly precise and reliable results, provided the corrections for diffusion, concentration, and pressure are carefully performed (19,31,37,41,42). Certain limitations exist as to the size range of molecules that can be studied (35). The labor involved in performing the calculations required can be greatly reduced by the use of a digital computer (19,41).

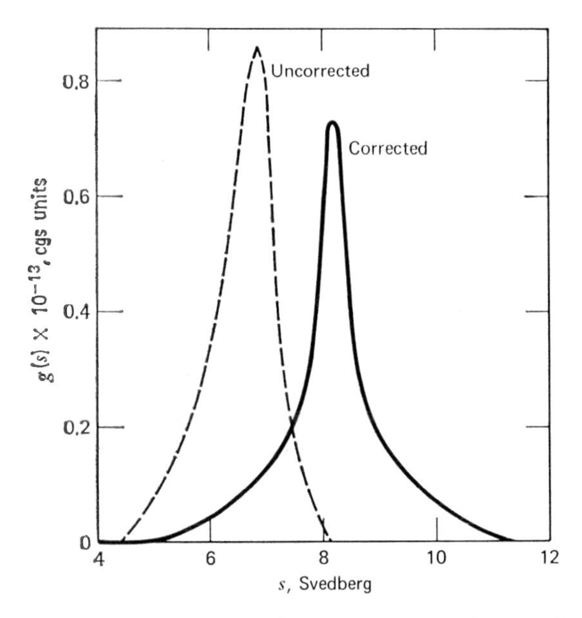

Fig. 10. Distribution of sedimentation coefficients, at $t = \infty$ and $c = 0$, for polystyrene in cyclohexane (36); shown corrected and uncorrected for pressure.

Estimation of Molecular-Weight Distributions. For homologous linear polymers, if excluded volume is not considered, it is possible to write an empirically and theoretically supported relation between sedimentation coefficient and molecular weight in the form

$$s = KM^a \tag{40}$$

where both K and a are constant under a given set of conditions. At the Θ temperature, the exponent a is approximately $\frac{1}{2}$. This relationship has been used as the basis of a number of similar methods of obtaining a molecular-weight distribution, $f(M)$, from the distribution of sedimentation coefficients. If several different molecular-weight fractions of the solute in question are available, then one may measure $\bar{M}_w$ and $\bar{s}_w$ for each of them, construct a curve of log $\bar{M}_w$ against log $\bar{s}_w$, and obtain from it the values of K and a. The desired plot of $f(M)$ against M then follows directly from the plot of $g(s)$ against s.

Examples of such work will be found in Refs. 19, 31, and 41–45. Where comparisons have been made with the results of precipitation chromatography (41), gel permeation chromatography (19), and solution precipitation fractionation (43,44), the agreement has been good. See also FRACTIONATION.

Sedimentation in a Density Gradient

If a mixed solvent composed of materials of different densities is subjected to prolonged sedimentation, an equilibrium density gradient is established by redistribution of the components within it. If a macromolecular solute is also present, this solute will redistribute to form a band in the gradient which is centered about the effective buoyant density of the macromolecule in the solvent. In the ideal case, this band is Gaussian in shape and its width is inversely proportional to the square root of the molecular weight of the macromolecule. The basic theory and methods for this technique were worked out by Meselson, Stahl, and Vinograd (46) and by Hearst and Vinograd (47). The technique has been much used in biochemical investigations, where aqueous solutions of various salts are used to form the density gradients.

A theoretical framework for density-gradient sedimentation of synthetic polymers and copolymers has been constructed by Hermans and Ende (48,49). In principle, it should be possible by this method to measure the molecular-weight distribution of a mixture of homologous polymers as well as distributions in composition of copolymers whose components vary in density. The method suffers from a number of practical difficulties. It is often difficult to obtain Θ conditions in the solvents necessary to establish the density gradient. Preferential adsorption of one component of the mixed solvent can give misleading results. Some experimental work has been done which indicates the usefulness of the method for purposes of determining the composition and the amount of contaminating homopolymers in a graft copolymer (50), and for estimating the compositional heterogeneity of a styrene–iodostyrene copolymer (51). The method has a number of theoretical advantages over other methods, and may prove quite useful if the practical difficulties can be overcome.

Glossary of Symbols

B = second virial coefficient of solute
B_{LS} = second virial coefficient of solute as used in light-scattering literature
B'' = second virial coefficient, as used by Fujita (21,22)
c_0 = initial concentration
c_p = concentration within the "plateau" region
$f(M)$ = differential molecular-weight-distribution function
$g(s)$ = differential distribution of sedimentation coefficient
M = molecular weight of a homogeneous solute
M_{app} = apparent molecular weight of a homogeneous solute
M_i = molecular weight of the ith component of a heterogeneous solute
$\overline{M}_n$ = number-average molecular weight
$\overline{M}_w$ = weight-average molecular weight
$\overline{M}_z$ = z-average molecular weight
r_a = radius of the meniscus
r_b = radius of the equivalent boundary
r_p = any radius within the "plateau" region
s = sedimentation coefficient of a homogeneous solute
$\bar{s}_w$ = weight-average sedimentation coefficient of a heterogeneous solute
$\bar{v}$ = partial specific volume
y = activity coefficient of solute
λ = derived quantity related to rotor speed and length of solution column
ξ = derived quantity related to length of solution column
ω = rotor speed

Bibliography

1. J. W. Williams, K. E. Van Holde, R. L. Baldwin, and H. Fujita, *Chem. Rev.* **58,** 715 (1958).
2. J. E. Blair, in J. Mitchell, Jr., and F. W. Billmeyer, Jr., eds., *Analysis and Fractionation of Polymers,* Interscience Publishers, a division of John Wiley & Sons, Inc., New York, 1965, p. 287.
3. J. J. Hermans and H. A. Ende, in H. F. Mark and E. H. Immergut, eds., *Newer Methods of Polymer Characterization,* Interscience Publishers, a division of John Wiley & Sons, Inc., New York, 1964.
4. E. T. Adams, Jr., in *Characterization of Macromolecular Structure* (Proceedings of Conference, April 5–7, Warrentown, Virginia), National Academy of Sciences, Washington, D.C., 1968.
5. R. L. Baldwin and K. E. Van Holde, *Fortschr. Hochpolymer.-Forsch.* **1,** 451 (1960).
6. H. K. Schachman, *Ultracentrifugation in Biochemistry,* Academic Press, Inc., New York, 1962.
7. H. Fujita, *Mathematical Theory of Sedimentation Analysis,* Academic Press, Inc., New York, 1962.
8. J. W. Williams, ed., *Ultracentrifugal Analysis in Theory and Experiment,* Academic Press, Inc., New York, 1963.
9. D. A. Yphantis, *Biochem.* **3,** 297 (1964).
10. D. A. Yphantis, *Ann. N.Y. Acad. Sci.* **88,** 586 (1960).
11. J. St. L. Philpot, *Nature* **141,** 283 (1938).
12. H. Svensson, *Kolloid-Z.* **87,** 181 (1939).
13. H. Svensson, *Kolloid-Z.* **90,** 141 (1940).
14. R. Trautman and V. W. Burns, *Biochem. Biophys. Acta* **14,** 26 (1954).
15. E. G. Richards and H. K. Schachman, *J. Phys. Chem.* **63,** 1578 (1959).
16. H. K. Schachman and S. J. Edelstein, *Biochem.* **5,** 2681 (1966).
17. S. J. Edelstein and H. K. Schachman, *J. Biol. Chem.* **242,** 306 (1967).
18. J. W. Beams, R. D. Boyle, and P. E. Hexner, *J. Polymer Sci.* **57,** 161 (1962).
19. T. Kotaka and N. Donkai, *J. Polymer Sci.* [A-2] **6,** 1457 (1968).
20. R. J. Goldberg, *J. Phys. Chem.* **57,** 194 (1953).
21. H. Fujita, *J. Phys. Chem.* **63,** 1326 (1959).
22. H. Fujita, *J. Chem. Phys.* **32,** 1739 (1960).
23. H. Fujita, A. M. Linklater, and J. W. Williams, *J. Am. Chem. Soc.* **82,** 379 (1960).
24. H. W. Osterhoudt and J. W. Williams, *J. Phys. Chem.* **69,** 1050 (1965).
25. D. A. Albright and J. W. Williams, *J. Phys. Chem.* **71,** 2780 (1967).
26. T. G. Scholte, *J. Polymer Sci.* [A-2] **6,** 91 (1968).
27. M. Wales, F. T. Adler, and K. E. Van Holde, *J. Phys. Chem.* **55,** 145 (1951).
28. T. H. Donnelly, *J. Phys. Chem.* **70,** 1862 (1966).
29. T. G. Scholte, *J. Polymer Sci.* [A-2] **6,** 111 (1968).
30. J. W. Williams, *J. Polymer Sci.* **12,** 351 (1954).
31. J. E. Blair and J. W. Williams, *J. Phys. Chem.* **68,** 161 (1964).
32. M. Dishon, G. H. Weiss, and D. A. Yphantis, *Biopolymers* **5,** 697 (1967).
33. R. Signer and H. Gross, *Helv. Chim. Acta* **17,** 726 (1934).
34. L. J. Gosting, *J. Am. Chem. Soc.* **74,** 1548 (1952).
35. R. L. Baldwin, *J. Phys. Chem.* **58,** 1081 (1954).
36. J. W. Williams, R. L. Baldwin, W. M. Saunders, and P. G. Squire, *J. Am. Chem. Soc.* **74,** 1542 (1952).
37. I. H. Billick, *J. Polymer Sci.* **62,** 167 (1962).
38. J. W. Williams, W. M. Saunders, and J. S. Cicirelli, *J. Phys. Chem.* **58,** 774 (1954).
39. J. W. Williams and W. M. Saunders, *J. Phys. Chem.* **58,** 854 (1954).
40. H. Fujita, *J. Am. Chem. Soc.* **78,** 3598 (1956).
41. M. Wales and S. J. Rehfeld, *J. Polymer Sci.* **62,** 179 (1962).
42. T. Homma, K. Kawahara, H. Fujita, and M. Heda, *Makromol. Chem.* **67,** 132 (1963).
43. H. W. McCormick, *J. Polymer Sci.* [A] **1,** 103 (1963).
44. H. J. Cantow, *Makromol. Chem.* **30,** 169 (1959).
45. M. Kalfus and J. Mitus, *J. Polymer Sci.* [A-1] **4,** 953 (1966).
46. M. Meselson, F. W. Stahl, and J. Vinograd, *Proc. Natl. Acad. Sci.* **43,** 581 (1957).
47. J. E. Hearst and J. Vinograd, *Proc. Natl. Acad. Sci.* **47,** 999 (1961).

48. J. J. Hermans and H. A. Ende, *J. Polymer Sci.* [C] **1**, 161 (1963).
49. J. J. Hermans and H. A. Ende, *J. Polymer Sci.* [C] **4**, 519 (1963).
50. H. A. Ende and V. Stannett, *J. Polymer Sci.* [A] **2**, 4047 (1964).
51. H. A Ende and J. J. Hermans, *J. Polymer Sci.* [A] **2**, 4053 (1964).

Robley C. Williams, Jr.
Yale University
David A. Yphantis
University of Connecticut

ULTRAFILTERS. See MEMBRANES

ULTRASONIC DEGRADATION. See DEGRADATION

ULTRASONIC FABRICATION

The first practical uses of ultrasonic energy were in World Wars I and II, in the underwater detection of submarines. The first application to plastics came in the 1950s when films were joined ultrasonically. But it was not until 1963, when ultrasonic energy was first applied to rigid thermoplastics, that the full potential of ultrasonic assembly operations began to be appreciated. Since that time, advances in equipment, joint design, and techniques have made ultrasonic assembly an important method of joining thermoplastics in the automotive, electronics, furniture, toy, and appliance industries.

There are five methods of ultrasonic assembly: plastics welding of injection-molded parts, metal inserting, staking, spot welding, and sewing. In each application, ultrasonic vibrations above the audible range generate localized heat by vibrating one surface against another. Sufficient frictional heat is released, usually within a fraction of a second, to cause most thermoplastic materials to melt, flow, and fuse. Ultrasonic assembly is cleaner, faster, and more economical than conventional bonding methods. It eliminates the application of heat, solvents, or adhesives, and does not require any curing time. Therefore, it permits higher production rates. See also ADHESION AND BONDING; SOUND ABSORPTION.

Equipment

The first step in producing ultrasonic energy is to change 60-Hz electric current to 20 kHz. This high-frequency current is then fed into a converter which changes the electrical energy into 20-kHz vibratory energy. The vibratory energy is transferred to the mechanical impedance transformer, which in industry is commonly called a "horn" because it is a tuned resonant section. Ultrasonic assembly equipment also includes a stand and a timer, or programmer.

The Power Supply. The purpose of the power supply is to modify the line current to appropriate frequency levels. Existing systems have power outputs at various levels up to 35,500 inch-pounds per second at the tip of the horn (see the section on Converters for a discussion of rating systems). All of the power-supply functions operate from a 117/220 volts AC, 50/60 Hz current, depending on the power required for the equipment and the local power source. Power supplies vary in output and may have vacuum-tube or, more often, solid-state circuitry. Manual controls cover frequency fine tuning and output power levels, while appropriate circuitry provides automatic frequency control.

The Converter. The converter or transducer transforms 20 kHz electrical energy, as received from the power supply, into mechanical energy vibrating at 20 kHz per second. There are two methods of conversion: electrostrictive and magnetostrictive. *Electrostrictive conversion* uses a piezoelectric element of lead zirconate titanate, which expands and contracts when excited electrically. Piezoelectric transducers operate with efficiencies in excess of 90%, depending upon the design.

In *magnetostrictive conversion*, the transducer core changes its length under the influence of an alternating magnetic flux field. The core is usually made of nickel alloy. Exciting coils are wound around the core to produce the magnetic flux. The magnetostrictive transducer is limited to less than 50% efficiency, owing to resistance and hysteresis losses.

Ultrasonic energy is rated either in watt output delivered by the power supply or in inch-pounds per second delivered at the tip of the horn. Because of differences in a transducer's conversion efficiency, wattage ratings may not be applied with the same meaning to all systems. Therefore, values in inch-pounds per second measured at the output end of the horn more nearly reflect the capabilities of the equipment, since all losses in the converter have been accounted for.

The Horn. After the electrical energy has been converted into mechanical energy, it is transmitted through the horn, or mechanical impedance transformer. The horn's function is to achieve the proper amplitude and the required energy transfer for the operation. Horns are, therefore, shaped to fit the part closely. The horn is tuned to resonate at the system's frequency, and is usually one-half wavelength long. Materials used in horn construction must have high strength-to-weight ratios in addition to good acoustical properties. Most horns are made of titanium, although some have been made of Monel alloy, beryllium, or aluminum.

The mass and shape of the horn as well as the physical properties of the material of construction influence the length at which it will resonate at the required frequency. More complex shapes and larger sizes are being made possible because of better understanding of acoustics and more sophisticated test equipment.

The Stand. The stand supports the converter–horn assembly in the proper relationship to the workpiece, and controls operating pressure, contact rate, and triggering. Pneumatic systems, because of their adaptability and ease of adjustment over a broad range, are used most frequently. High production rates are obtainable through semi- and fully automated systems.

In a typical ultrasonic assembly stand, all control functions are incorporated within the unit. Holding, or clamping, pressure is developed by a pneumatic system within the stand and is exerted by the horn to hold the part being ultrasonically fabricated in position. The regulating system used also filters the incoming air, introduces lubrication, and monitors the operating pressure. Sensing of contact with the part and adjustment of the air flow for motion control complete the features of a well-designed stand.

A pneumo-mechanical sensing of contact with the part activates the delivery of ultrasonic energy, which is then timed by a solid-state electronic programmer. If the unit is incorporated into an automatic materials-handling system, limit switches can sense cycle completion and activate the advancement of a new set of components to be assembled.

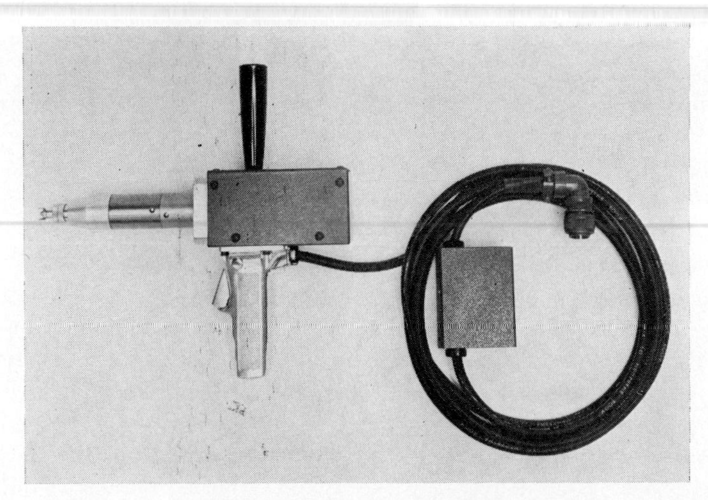

Fig. 1. Pistol-grip hand tool.

Stands are bench-mounted units and, therefore, not portable. A pistol-grip hand tool, shown in Figure 1, is totally portable and capable of spot welding, staking, and inserting. It has broadened the scope of ultrasonic assembly to include parts of very large size with hard-to-reach joints. The operator controls the pressure of the tool against the plastic manually and the time of ultrasonic-energy flow with a trigger switch in the grip.

The Programmer. The programmer for ultrasonic assembly systems controls several functions, some of which operate simultaneously, others sequentially. A foot- or hand-operated switch activates the pneumatic system to apply the horn to the part. Contact with the part signals the programmer to deliver energy for a preset period of time, adjustable over a range of 0.1–6.0 seconds. After the welding cycle is completed, a holding phase of 0.05–3.0 seconds permits the plastic to cool while pressure on the assembly is maintained. This allows the material to acquire sufficient strength so that the newly formed joint will not be separated. A final function of the programmer is to cause the horn to be retracted from the workpiece so that a new cycle may be started.

Methods

There are five basic methods of ultrasonic assembly: plastics welding, metal insertion, staking, spot welding, and sewing.

Welding. Ultrasonic welding can join injection-molded materials without the use of solvents, heat, or adhesives. The face of the ultrasonic horn is brought into contact with one of two plastic parts being welded. The horn applies a predetermined degree of pressure against the assembly, and mechanical vibrations at ultrasonic frequencies are transmitted from the horn into the plastic for a controlled period of time. Utilizing the acoustical transmitting property of the plastic, the vibrations travel until they meet the joining surfaces and are released in the form of heat.

Joint design is one of the most important requirements for a good ultrasonic weld, and the key to good joint design is the proper use of an "energy director" (see Fig. 2). An energy director is a triangular section incorporated into a joint design that defines and controls the area at which energy is dissipated. The high concentration of energy focused by the energy director results in almost immediate melting and a uniform flow in the joint area. The size of the energy director is directly related to that of the section to be welded. The base of the triangular energy director is 20% of the section width, and the height is 10% of the section. The angle at the peak is 90°, which facilitates modification of an existing mold to incorporate the energy-director design, and also permits satisfactory molding of the desired profile. This modified joint permits rapid welding and achieves maximum strength.

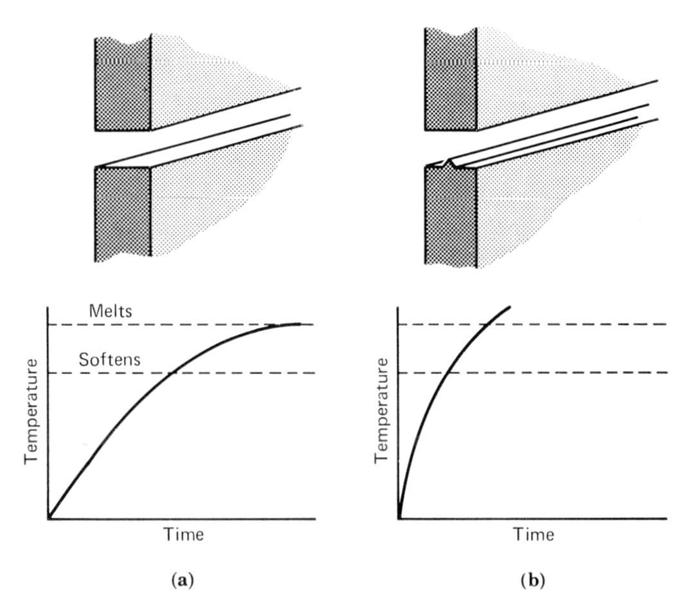

Fig. 2. Time–temperature relationships in ultrasonic welding of a thermoplastic: (**a**) for a butt joint; (**b**) for a joint incorporating an "energy director."

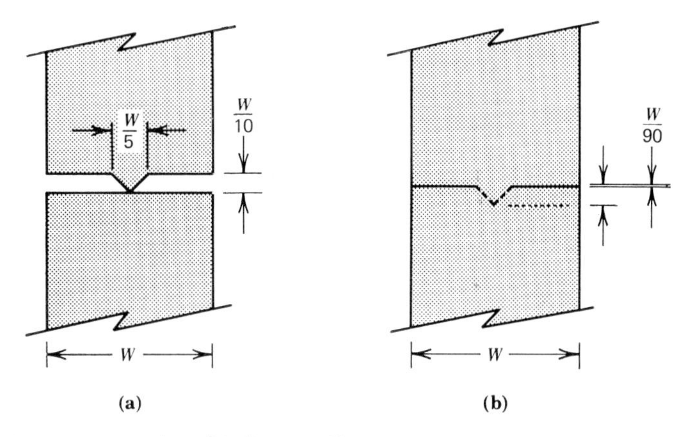

Fig. 3. Design of a butt joint with "energy director" (**a**) before welding, (**b**) after welding.

Figure 3 shows a simple butt joint modified with an energy director of the desired proportions before and after welding. The material from the energy director dissipates throughout the joint area, forming a layer of the size indicated.

Figure 4 illustrates a step joint used where a weld bead on the side would be objectionable. This joint has excellent self-aligning capability and is usually much stronger than a butt joint, since material flows into the clearance necessary for a slip fit and establishes a seal that provides strength in shear as well as in tension.

The part to be welded must be held rigidly. Fixturing nests are provided in many shapes to support the part and locate it under the horn. The nest is usually machined of aluminum or steel. Nests for oddly shaped parts may be molded in epoxy resin or a

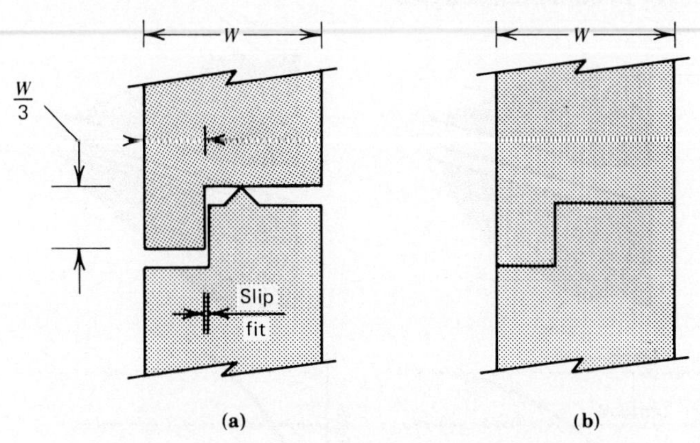

Fig. 4. Design of a step joint for ultrasonic assembly (**a**) before welding, (**b**) after welding.

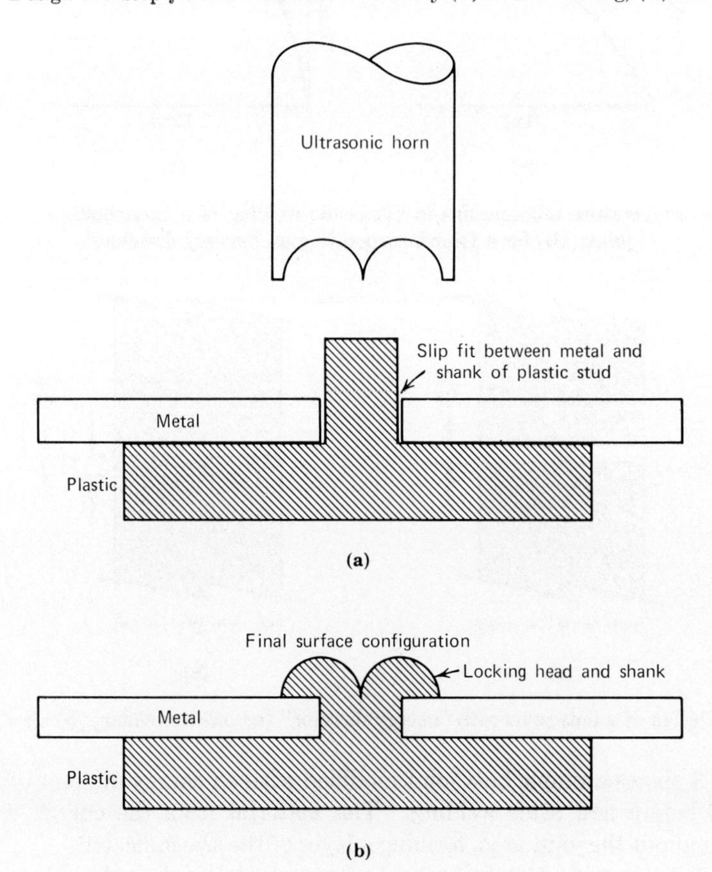

Fig. 5. Staking with ultrasonics (**a**) before contact of horn with plastic, (**b**) after completion of the operation.

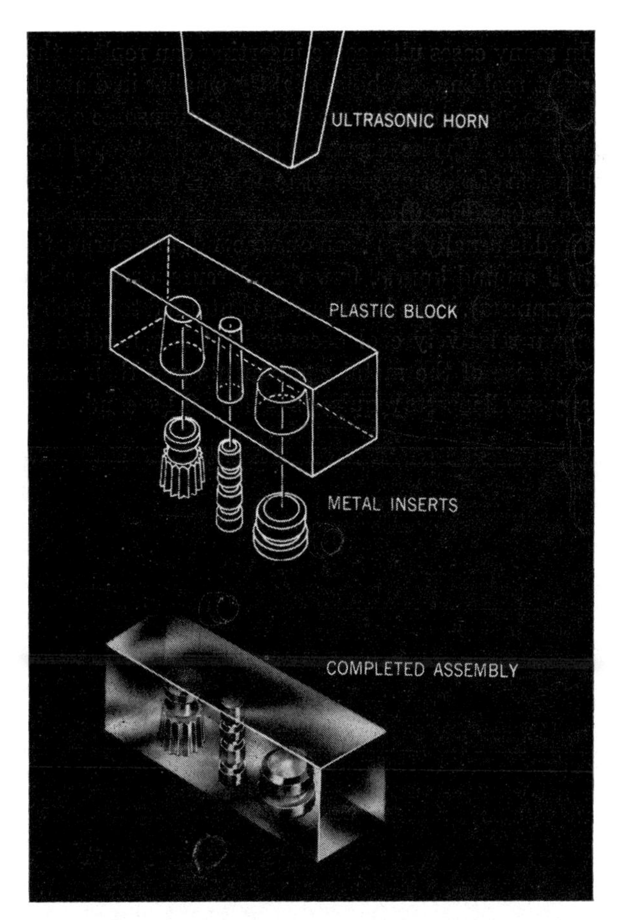

Fig. 6. Ultrasonic inserting of metal into plastic.

similar material (see TOOLING WITH PLASTICS). The nest provides accurate placement and support of the part under the horn. Thin, resilient liners are sometimes used in the nest. This cushioning improves energy flow and reduces marring of the nested part.

The success of welding and staking of plastic parts or inserting metal into plastics depends upon the proper amplitude of the horn tip. Since it may be impossible to design the correct amplitude into the horn initially because of its shape, booster horns may be necessary to increase or decrease the amplitude and produce, thereby, the proper degree of melting or flow. The nature of the plastic, its shape, and the type of work to be performed all determine the optimum horn amplitude.

Staking. Most ultrasonic staking applications involve the assembly of metal and plastic. As with ultrasonic welding, localized heat is created through the application of high-frequency vibrations. In staking, minimal pressure is used so that the energy will be released where the horn comes in contact with the plastic. The specially contoured tip in the horn meets the stud and reshapes it (see Fig. 5). Since frequency, pressure, and time are always the same during each cycle, frictional heat generated between the horn and stud is consistent and much faster than heat applied with a hot iron. The plastic is then permitted to cool for a fraction of a second before horn pressure is released.

Inserting. In many cases ultrasonic inserting can replace the conventional and costly process of insert molding. A hole, slightly smaller in diameter than the insert it is to receive, guides the insert into the plastic while ultrasonic energy and pressure are being applied by the horn. The heat created by the vibration of the metal insert is sufficient to melt a thin film of plastic around the hole momentarily, permitting the inserts to be driven into place (see Fig. 6).

The time required is usually less than one second, but during this brief contact the plastic reshapes itself around knurls, flutes, undercuts, or threads to encapsulate the inserts (see also EMBEDDING). The horn can be contoured to match the contact surface. Parts normally remain relatively cool because heat is generated only at the plastic–metal interface. The size of the metal components and their number determine the exposure time; many small inserts require as little as 0.1 second.

Fig. 7. Ultrasonic spot welding of two sheets of clear thermoplastic.

Ultrasonic insertion is advantageous because it can be carried out as a postmolding operation, and tolerances of the inserts are not critical.

Spot Welding. The new technique of spot welding thermoplastics with ultrasonic energy requires a specially shaped tool. Vibrating ultrasonically, the point of the tip penetrates the first sheet and displaces molten plastic. This material is shaped by a radial cavity in the tip and forms a neat, raised weld on the surface. Simultaneously, energy is driven through the contacted sheet and released at the interface. This produces frictional heat between the sheets to be welded. The molten plastic displaced when the tip comes in contact with the second sheet flows between the two heated sheets and forms a permanent bond. Spot welds may be made with a pistol-grip hand tool or a pneumatic tool, both of which are portable, or with a bench-model ultrasonic assembly equipment (see Fig. 7).

Spot welding requires out-of-phase vibrations between the horn and plastic surfaces. Light initial pressure is used for the out-of-phase activity at the contact area. In manual operation, proper depth is reached when the rate of penetration suddenly increases; this indicates that the tip has passed through the heated interface area. Energy is immediately turned off while pressure is retained for a fraction of a second.

A compression collar on the ultrasonic assembly equipment assures uniform welds. An adjustable stop is incorporated within the design which also controls depth of penetration.

Sewing. In ultrasonic sewing as in other methods of ultrasonic joining, ultrasonic energy is generated by a solid-state power supply, converted to mechanical vibrations, and applied to a tool called a horn. Thermoplastic materials are fed between the tip of the horn and a specially designed anvil called a stitching wheel. The vibrating horn creates frictional heat within the fabric where the fibers are compressed between the horn and wheel, causing the material to melt and fuse in the pattern of the wheel.

Since the bond is dependent upon melting, ultrasonic sewing is limited to synthetic thermoplastic materials such as nylons, polyesters, polypropylenes, modified acrylics, some vinyls, and polyurethans. Blended materials with up to 35% natural fiber content can also be sewn ultrasonically. Ultrasonics can sew knitted, woven, nonwoven, or film materials.

Ultrasonic sewing equipment can be used to seam, hem, tack, baste, pleat, slit, and buttonhole for such applications as bagging, bandages, blankets, clothing, disposables, draperies, film, filters, packaging, sails, strapping, upholstery, and wi ndo shades. Numerous stitching designs are available; see Figure 8.

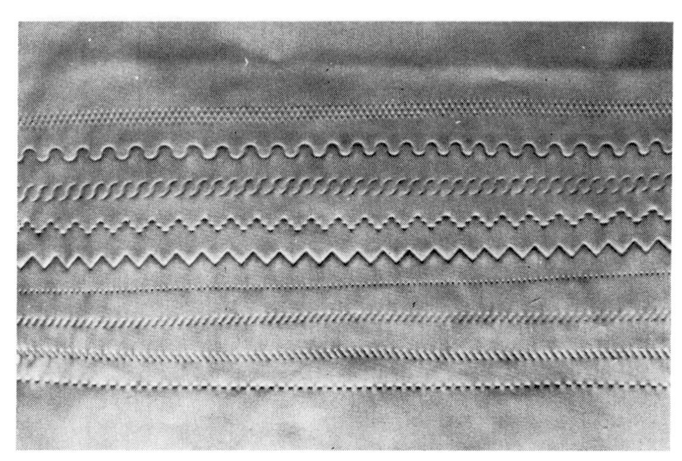

Fig. 8. Samples of stitching designs available with ultrasonic sewing.

Staking and Welding Characteristics of Plastics

Most commonly used injection-molded materials can be ultrasonically welded. Weldability depends on melting temperature, modulus of elasticity, impact resistance, coefficient of friction, and thermal conductivity. Generally, the more rigid the plastic, the easier it is to weld. Low-modulus materials, such as polypropylene and polyethylene, can often be welded provided the horn can be positioned close to the joint area. In staking, the opposite is usually true. The softer the plastic, the easier it is to stake. However, good results can be achieved with most plastics when the right combination of amplitude and force is used.

Different plastics must be chemically compatible and have similar melt temperatures to be weldable to each other. For example, polyethylene cannot be welded to polystyrene because the polymers are not chemically compatible and the melt tem-

Table 1 Suitability of Plastics in Ultrasonic Fabrication[a]

Plastics	Staking and inserting	Welding[b] Near-field	Far-field	Remarks
polystyrene				
unfilled	E	E	E	produces strong, smooth joints
rubber-modified	E	E	G–P	welding characteristics depend on degree of impact resistance
glass-filled	E	E	E	weldable with filler content up to 30%
SAN[c]	E	E	E	particularly good as glass filled compounds
ABS[d]	E	E	G	can be bonded to other polymers, such as SAN, polystyrene, and acrylics
polycarbonates	E	E	E	high melting temperature requires high energy
nylons	E	G	F	levels; oven-dried or "as-molded" parts
polysulfones	E	G	G–F	perform best, owing to hygroscopic nature of the materials
acetals	E	G	G	require high energy and long ultrasonic exposure because of low coefficient of friction
acrylic	E	E	G	weldable to ABS and SAN; applications include dials, radio cases, and meter housings; in sheet form, joints must be machined
poly(phenylene oxide)	G	G	G–F	high melting temperatures require high energy
Noryl[e]	E	G	E–G	levels
phenoxy	E	G	G–F	
polypropylene	E	G–P	F–P	horn design for welding is particularly critical;
polyethylene	E	G–P	F–P	filled compounds usually better, but need individual testing
cellulose esters	G–F	P	P	weldability varies with formulation and con-
vinyl chloride polymers	E–F	F–P	F–P	figuration; usually perform well in staking and inserting; decomposition of some formulations may occur

[a] Key: E = excellent; G = good; F = fair; P = poor.

[b] Near-field welding refers to joint ¼ inch or less from area of horn contact; far-field welding to joint more than ¼ inch from contact area.

[c] Styrene–acrylonitrile copolymers.

[d] Acrylonitrile–butadiene–styrene terpolymer.

[e] Polystyrene-modified poly(phenylene oxide), General Electric Co.

peratures are greatly different; the polyethylene would be completely melted before the polystyrene would have reached its melt temperature.

Additives or fillers can detract or add to weldability. Glass fillers tend to increase the stiffness of the plastic, thereby increasing its acoustical response and weldability. Colorants and lubricants occasionally alter such physical properties as the coefficient of friction and melting point, and may reduce weldability. Asbestos- and talc-filled materials may weld poorly because the filler absorbs the molten plastic during the welding process. A summary of the suitability of various plastics for ultrasonic fabrication is given in Table 1.

Bibliography

J. R. Frederick, *Ultrasonic Engineering*, John Wiley & Sons, Inc., New York, 1965.

Jeffrey R. Sherry
Branson Sonic Power Company

ULTRAVIOLET-RADIATION ABSORBERS

Ultraviolet-radiation absorbers are additives used to prevent the photodegradation of polymeric materials by sunlight or ultraviolet-rich artificial light. Most of these additives act as filters to prevent ultraviolet radiation from penetrating the polymer; these are correctly called ultraviolet absorbers. Some may provide protection by other mechanisms, and the broader term, *ultraviolet-radiation stabilizer*, will be used in this discussion to include all photodegradation inhibitors, regardless of the mechanism by which they act. These materials are also commonly called ultraviolet inhibitors, UVI's, light-screening agents, or light stabilizers. Although carbon black and other pigments are also used as ultraviolet-radiation stabilizers, only relatively colorless organic or organometallic compounds will be considered in this discussion (see also CARBON; PIGMENTS). The major classes of commercial ultraviolet-radiation stabilizers include derivatives of 2-hydroxybenzophenone, 2-(2*H*-benzotriazol-2-yl)-phenols, phenyl esters, substituted cinnamic acids, and nickel chelates. The stabilizer used and its concentration depend on the polymer to be stabilized and its intended use. These additives are generally used in concentrations of about 0.1–3.0 wt % and are often used in conjunction with antioxidants (qv), colorants, or other additives. The total sales of ultraviolet-radiation stabilizers in 1970 was about 2.5 million lb at prices ranging from about $2 to over $5/lb. See also STABILIZATION; WEATHERING.

Ultraviolet Radiation and Polymer Degradation

Before the function of ultraviolet stabilizers in polymer stabilization can be discussed adequately, it is necessary to review the nature of sunlight and the fundamental photochemical processes involved in the outdoor degradation of polymers.

Solar Ultraviolet Radiation. The sun emits radiation of all wavelengths from x rays to the far infrared, and much of its energy output lies in the ultraviolet region where most organic molecules absorb light (150 to 400 nm). Fortunately, the earth's atmosphere is almost completely opaque to light with wavelengths below about 285 to 300 nm. Ultraviolet light between the atmospheric cutoff of about 290 nm and the lower edge of the visible spectrum, 400 nm, is called the solar ultraviolet. Figure 1 shows the energy distribution for this spectral region.

Photochemical processes occurring in the upper atmosphere account for the absorption of light below the atmospheric cutoff. Light with wavelengths below about 150 nm is absorbed by the photochemical ionization of nitrogen and oxygen atoms at an altitude above 50 km; light from 150–290 nm is absorbed by the processes involved in the formation and dissociation of ozone at an altitude of 20 to 30 km. Only a small amount of ozone (equivalent to a 2-cm layer at standard conditions) is present, and the concentration varies with season and latitude. It is highest in late winter (northern hemisphere) and lowest in late summer. The ozone layer is thinnest at the equator where it shows little seasonal variation (1). The relative annual flux of solar ultraviolet radiation at Kingsport, Tennessee, is shown in Figure 2 (2). Outdoor exposure during the summer months is five to ten times more severe than during the winter months. Various constituents of the lower atmosphere absorb ultraviolet light to some extent; hence the solar ultraviolet intensity increases with altitude and is about two and a half times greater at 10,000 ft than at sea level. Dust, smoke, sulfur dioxide, and other atmospheric contaminants cause large local variations in the solar ultraviolet intensity. For all these reasons the rate of degradation in outdoor weathering is greatly dependent on the location of exposure. Barker (1) has suggested that short-term variation in ozone concentration may reduce the atmospheric cutoff to 270 nm or less and that this short ultraviolet light may be an important factor in the photodegradation of polymers. This suggestion has not yet been verified by experiment.

Photodegradation of Polymers. When light is absorbed by a molecule, the energy of the absorbed photon is transferred to the absorbing molecule. If the amount of energy absorbed is greater than the bond energies of the structural units of the

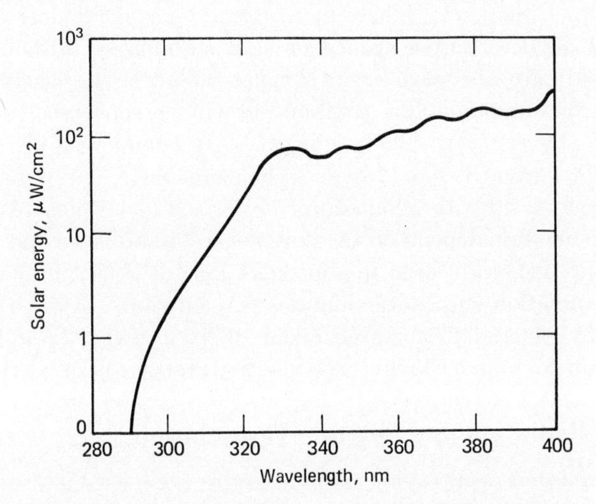

Fig. 1. Energy distribution for solar ultraviolet radiation.

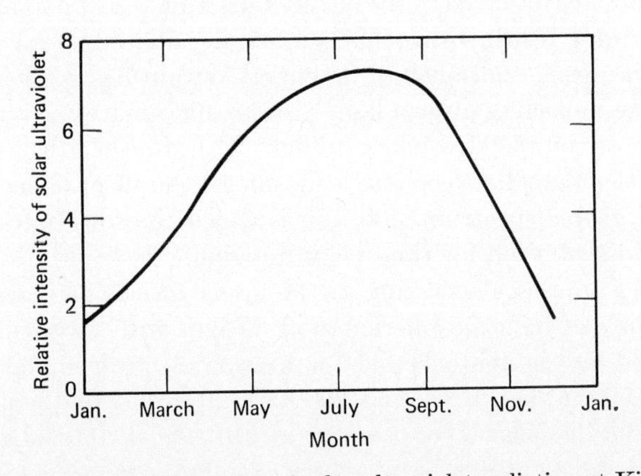

Fig. 2. Seasonal variation in intensity of solar ultraviolet radiation at Kingsport, Tenn.

molecule, bonds may be broken. The energy of a photon is inversely proportional to its wavelength. For the solar ultraviolet region, the energies involved fall in the range of 70–100 kcal/einstein (1 einstein = 1 "mole" of photons or 6×10^{23} photons). Table 1 gives the energy corresponding to various wavelengths in the solar ultraviolet and the bond energies of bonds found in organic polymers; the bond may be broken in cases where the corresponding wavelength is absorbed by the system. It can be seen that absorption of solar ultraviolet light provides more than enough energy to break polymer chains.

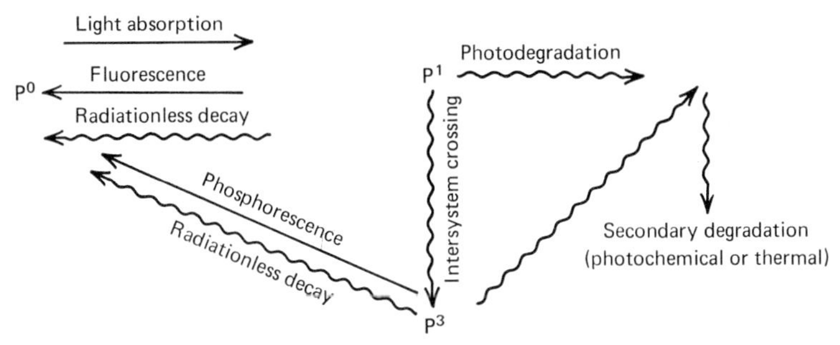

Fig. 3. Absorption of a photon by a polymer. The symbols P^0, P^1, and P^3 represent the unexcited polymer molecule, its excited singlet, and its excited triplet, respectively.

When some light-absorbing group in a polymer absorbs a photon, that group is raised to a higher energy level (excited state) which is usually a singlet (no unpaired electron spins). The singlet may then undergo a process called intersystem crossing to give a triplet (unpaired electron spins) excited state. From these excited states several alternative events may occur. These are shown in Figure 3, where the symbols P^0, P^1, and P^3 represent the unexcited polymer molecule, its excited singlet, and its excited triplet, respectively; the solid arrows indicate processes in which light is absorbed or emitted, and the wavy arrows indicate processes in which light is not absorbed or emitted. The polymer molecule can dispose of the excitation energy in several harmless ways; eg, by reemitting light by fluorescence or phosphorescence or by releasing energy as heat in radiationless decay processes. Only a small fraction of excited molecules will undergo degradation. The ratio of the number of molecules undergoing degradation to the number of quanta absorbed is called the quantum efficiency (Φ) of degradation. For most polymers Φ is from 10^{-3} to 10^{-5}; that is, only one molecule degrades for every 1,000 to 100,000 that absorb light. Despite the low value of Φ the solar ultraviolet flux of about 10^{21} photons/cm^2/yr is sufficient to break a significant fraction of the bonds for most polymers during one year of exposure.

The amount of solar ultraviolet absorbed by a polymer, the quantum efficiency of its degradation, and the nature of the chemical reactions that occur during degradation all depend on the molecular structure of the polymer and, often, on substances present in the plastic composition. Some commercial polymers are made up of structural units that have absorption maxima in the solar ultraviolet region; these include polysulfones, terephthalate polyesters, and some polyurethans. Other polymers, such as polystyrene, aliphatic polyesters, cellulose esters, poly(methyl methacrylate), and polyamides, contain groups that have an absorption peak at a somewhat shorter wavelength than 290 nm but have a significant absorption tail extending into the solar ultra-

Table 1. Energy Content of Light in Solar Ultraviolet and Bond Strengths in Organic Compounds

Wavelength, nm	Energy, kcal/einstein	Bond type	Bond energy, kcal/mole
290	100	C–H	85–100
300	95	C–C	75–80
		C–O	75–80
350	81	C–Cl	70–80
400	71	C–N	60–65

Table 2. Photodegradation Parameters for Some Commercial Polymers

Polymer	λ_{max}, nm	Φ at 254 nm	Most harmful λ, nm
poly(tetrafluoroethylene)[a]	<200	$<1 \times 10^{-5}$	
polyethylene	<200	4×10^{-2}	254
polypropylene	<200	$\sim 1 \times 10^{-1}$	
poly(vinyl chloride) (unplasticized)	~200	$\sim 1 \times 10^{-4}$	
cellulose acetate	<250	$\sim 1 \times 10^{-3}$	254
cellulose	<250	$\sim 1 \times 10^{-3}$	254
poly(methyl methacrylate)	214	2×10^{-4}	254
polycaprolactam		6×10^{-4}	254
polystyrene	260, 210	$\sim 1 \times 10^{-3}$	254
polycarbonate[b]	260	2×10^{-4}	
poly(ethylene terephthalate)	290, 240	$>1 \times 10^{-4}$	280–360
poly(aryl sulfone)	320		

[a] Teflon (E. I. du Pont de Nemours & Co.).
[b] Lexan (General Electric Co.).

violet. Polyolefins, poly(vinyl chloride), and a few other vinyl polymers have structures which should not result in any absorption in the 290–400 nm region; however, because of structural defects, such as carbonyl groups introduced by oxidation during processing or the presence of ultraviolet-absorbing catalyst residues or other impurities, even these polymers have significant solar ultraviolet absorption. Table 2 lists the absorption peaks (λ_{max}) and the approximate quantum efficiency for degradation (Φ) for a number of commercial polymers.

The values of Φ given for most polymers cited in Table 2 are not the quantum efficiencies for specific photochemical reactions, but are based on significant observable degradation in physical properties. These approximate quantum efficiencies demonstrate the wide variation in photochemical stability of various polymers, but do not reveal the nature of the degradative processes. More information concerning quantum efficiencies for photochemical reactions of specific polymers may be found under RADIATION-INDUCED REACTIONS.

The wavelength of maximum degradation, taken from a review by Rabek (3), is also given in Table 2 for a few polymers, but unfortunately, the reported measurements were usually made at only two wavelengths, 254 nm and 310 nm; light with a wavelength of 254 nm was the most harmful in almost every case. A complete action spectrum, that is, a plot of degradation rate as a function of wavelength, has been determined for only a few polymers, but would be of much greater value (39).

The chemical consequences of photodegradation include radical-induced chain scission or crosslinking; oxidation which introduces carbonyl, carboxyl, or peroxide groups into the polymer and which may also break the polymer chain; elimination reactions which introduce unsaturation; hydrolysis of ester or amide groups; and, probably, many other reactions. Detailed kinetic schemes for chain scission and crosslinking have been established for some polymers; this work was reviewed by Jellinek (4). Rigorous determination of polymer degradation mechanisms is made very difficult by the problems of product analysis in such complex systems. Some partial mechanisms for photodegradation are discussed in the article DEGRADATION. Secondary, nonphotochemical degradation reactions often follow the primary photochemical reaction. Autoxidation is the most common secondary reaction and affects almost all polymers containing aliphatic carbon–hydrogen bonds. Antioxidants (qv) greatly enhance the effectiveness of ultraviolet absorbers in the stabilization of such polymers. A more detailed discussion of polymer photochemistry and photodegradation can be found in the section on Photochemistry in the article RADIATION-INDUCED REACTIONS.

Stabilization by Additives

Additives may retard photodegradation of polymers in two ways. First, the additive itself may absorb most of the ultraviolet light and thereby leave little to be absorbed by the polymer. Such additives are properly called ultraviolet-radiation absorbers or *screening agents*. Alternatively, an additive, which itself absorbs little or no ultraviolet light, may interact with a photoexcited polymer in a way that transfers the excitation energy to the additive. If this energy transfer can occur before any other reaction of the excited polymer molecule takes place, degradation of the polymer will be prevented. Stabilizers that act in this way are usually called *excited-state quenchers*. In either case, the stabilizer must be able to dispose of its excitation energy in some harmless way before it undergoes some irreversible photochemical change which would destroy its effectiveness. Most commercial ultraviolet stabilizers act as screening agents, but quenching may also play a part in their function. The nickel chelates may act predominantly as quenchers, but they, too, absorb a considerable portion of the incident ultraviolet light. For simplicity, these two mechanisms for stabilization will be discussed separately.

Ultraviolet Absorbers. Blocking ultraviolet light before it can reach the polymer can greatly extend the outdoor life of a polymer. For example, polyethylene containing well-dispersed carbon black has survived an exposure of nearly twenty years; without carbon black, the polymer is extensively degraded after only one year. Black, opaque plastics are not acceptable for most outdoor uses; hence, soluble, relatively colorless, organic compounds which absorb light in the 290 to 400-nm region have been developed.

Many organic compounds absorb light in the desired region but few act as stabilizers. Some have little or no effect when added to polymers; many actually increase the rate of degradation. This sensitizing, or tendering, effect usually occurs through one of the following reactions, where A = the additive, A* = its photoexcited state, A· = a radical derived from A, P = the polymer, P* = a photoexcited state of the polymer, and P· = a radical derived from the polymer.

$$\mathrm{A} \xrightarrow{h\nu} \mathrm{A}^* \qquad (1a)$$

$$\mathrm{A}^* + \mathrm{P} \rightarrow \mathrm{A} + \mathrm{P}^* \qquad (1b)$$

$$\mathrm{P}^* \rightarrow \text{degradation} \qquad (1c)$$

$$\mathrm{A} \xrightarrow{h\nu} \mathrm{A}^* \qquad (2a)$$

$$\mathrm{A}^* + \mathrm{P} \rightarrow \mathrm{A}\cdot + \mathrm{P}\cdot \qquad (2b)$$

$$\mathrm{P}\cdot + \mathrm{O}_2 \rightarrow \text{autoxidative degradation} \qquad (2c)$$

To be an effective stabilizer, a compound must be able to dispose of its excitation energy without interacting with the polymer in harmful ways and without undergoing any photochemical reaction which would destroy its effectiveness. Energy is most harmlessly disposed of as heat; hence, effective ultraviolet stabilizers must have structures that provide for a rapid cascade back to the ground state through thermally excited levels of the ground state. For some compounds, this may involve a photoisomerization to some higher energy structure that reverts rapidly and efficiently back to the original structure with loss of heat energy. The quantum efficiency for return to the unexcited ground state for the stabilizer must be greater than 0.99999; that is, less than one molecule of stabilizer can be destroyed for every 100,000 that are excited, or the stabilizer will itself be destroyed before it can extend the life of the polymer to a significant degree.

To achieve maximum effectiveness and minimum color in a polymer, an ultraviolet absorber should have the highest possible absorbance from 290 to 400 nm but no absorbance above 400 nm. Cicchetti (5) has shown that, for a compound having zero absorbance below 280 nm, zero absorbance above 400 nm, and a constant absorbance between these wavelengths, the maximum possible molar extinction coefficient (ϵ) for this 290 to 400-nm region would be 20,000. If used at a level of 3 wt %, such a hypothetical, perfect ultraviolet absorber would absorb more than 99.999% of the incident ultraviolet light if it were incorporated in polypropylene; but it would absorb only about 80% of the incident light if it were incorporated in poly(ethylene terephthalate), which itself absorbs significantly in the 290 to 400-nm region. Although no real ultraviolet absorber could behave as Cicchetti's hypothetical model, several commercial ultraviolet stabilizers, discussed later, do show molar extinction coefficients of 10,000 to 15,000 through most of the region from 290 to 350 nm.

The patent literature discloses a large number of classes of compounds which are claimed to meet the requirements for an effective ultraviolet-radiation absorber; however, only four fundamentally different classes have achieved commercial significance. A few others have been distributed as experimental materials. Only the commercially significant classes will be discussed here.

Derivatives of 2-Hydroxybenzophenone. This class comprises derivatives of 2,4-dihydroxybenzophenone (**1a**), 2,2′,4-trihydroxybenzophenone (**2a**), and 2,2′,4,4′-tetrahydroxybenzophenone (**3a**). The most important members of this group are the

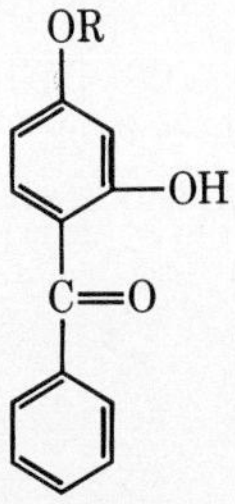

(**1a**) R = H
(**1b**) R = CH_3 to $C_{12}H_{25}$

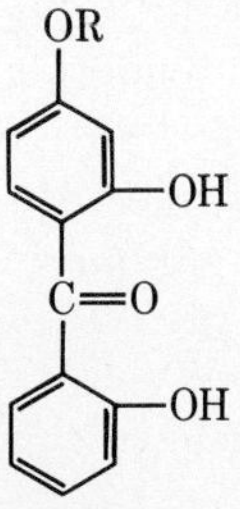

(**2a**) R = H
(**2b**) R = CH_3 to $C_{12}H_{25}$

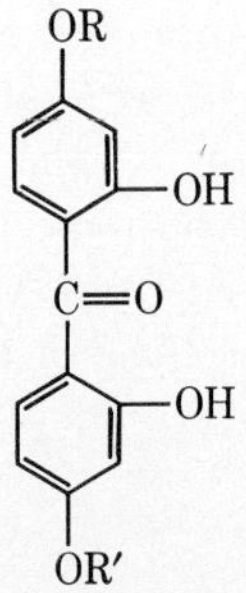

(**3a**) R, R′ = H
(**3b**) R, R′ = CH_3 to $C_{12}H_{25}$

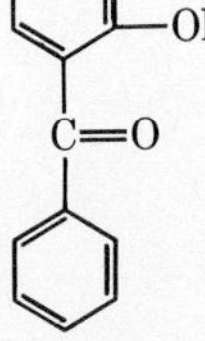

(**4**)

4-alkoxy-2-hydroxybenzophenones (**1b**). Ultraviolet-absorption curves for typical 2-hydroxybenzophenones are shown in Figure 4. The importance of the 4-substituent can be seen when the spectra of the substituted ultraviolet stabilizers are compared with that of the unsubstituted 2-hydroxybenzophenone (**4**), which has very little absorption in the solar ultraviolet region. The presence of the 4-substituent shifts the major absorption peak from about 258 nm up to the lower edge of the solar ultraviolet at about 280 nm and greatly enhances the total ultraviolet absorption over the entire

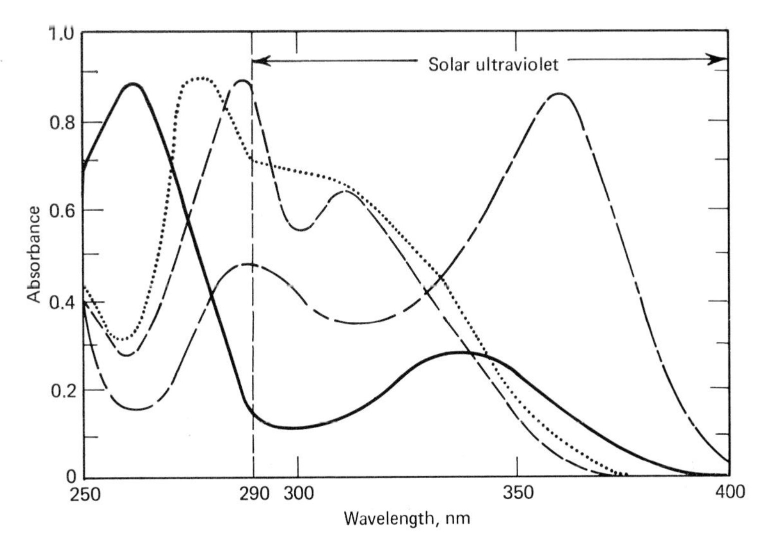

Fig. 4. Ultraviolet-absorption spectra of 2-hydroxybenzophenones (6×10^{-4} *M* in CH_2Cl_2, 0.1-cm path length). Key: solid line, 2-hydroxybenzophenone; . . . 2,4-dihydroxybenzophenone; - - - 2-hydroxy-4-methoxybenzophenone; — - — - 2,2′-dihydroxy-4,4′-dimethoxybenzophenone.

critical region. Introduction of additional electron-donating groups into the second ring of the benzophenone, as in (**2b**) and (**3b**), causes a large increase in absorption in the longer wavelength ultraviolet region and a greater total absorption throughout the critical region, but this advantage is accompanied by a disadvantageous increase in visible yellow color. Such physical properties of the stabilizer as its volatility or its compatibility with a particular polymer are also affected by the nature of the 4-substituent; this factor will be discussed later. Other substituents, such as a sulfonic acid group or a polymerizable functional group, may also be used to modify the physical properties of the compound without significant change in its ultraviolet screening effect.

Photochemistry. The photochemistry of the 2-hydroxybenzophenones has been more extensively studied than that of the other classes of ultraviolet absorbers. It is known that (**1a**) is rapidly converted to a "photoenol" (**5**) by absorption of light, and that (**5**) reverts to (**1**), with loss of energy as heat, with almost 100% efficiency (6).

RO— (benzene ring) —C(=O···H–O)— C_6H_5 $\underset{-\text{heat}}{\overset{h\nu}{\rightleftharpoons}}$ RO— (ring) =C(–O–H···O)— C_6H_5

(**1**) (**5**)

The existence of the intramolecular hydrogen bond in both (**1**) and (**5**) accounts for the rapid and efficient phototautomerism. The only important structural difference between (**1**) and (**5**) is the distribution of electrons; hence, interchange of the hydrogen is very rapid. The importance of this hydrogen bond was demonstrated by Chaudet and co-workers (7) who found that the effectiveness of a number of 2-hydroxybenzophenones as stabilizers for polyethylene increased as the strength of this bond, as measured by a nuclear magnetic resonance method, increased. More recently, Lamola and Sharp (8) found that if this intramolecular hydrogen bonding was decreased by

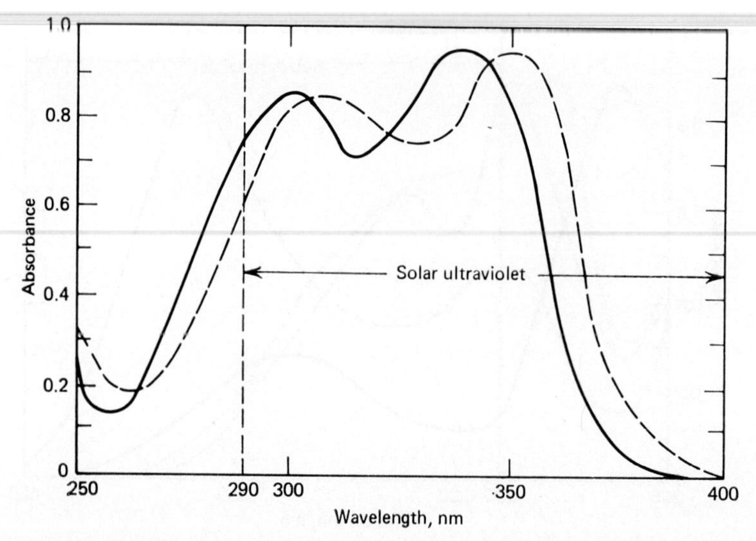

Fig. 5. Ultraviolet-absorption spectra of 2-(2*H*-benzotriazol-2-yl)phenols (6×10^{-4} *M* in CH_2Cl_2, 0.1-cm path length). Key: solid line, 2-(2*H*-benzotriazol-2-yl)*p*-cresol; dashed line, 2,4-di-*tert*-buty-6-(5-chloro-2*H*-benzotriazol-2-yl)phenol.

hydrogen bonding between (**1**) and a solvent such as ethanol, (**1**) *sensitized*, rather than retarded, certain photochemical reactions.

2-(2H-Benzotriazol-2-yl)phenols. This class comprises the Tinuvin series of compounds which have structure (**6**) where R = H or alkyl, R′ = alkyl, and X = H or Cl.

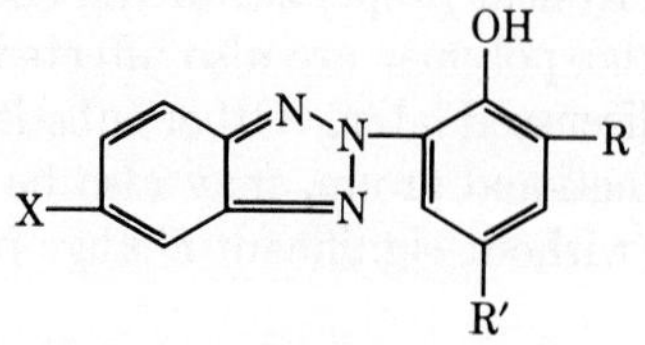

(**6**)

The ultraviolet-absorption curves for two members of this series are shown in Figure 5. When X = Cl the absorption is shifted to slightly longer wavelengths. R and R′ are mainly of importance in modifying solubility and volatility. The 2-(2*H*-benzotriazol-2-yl)phenols have somewhat higher ultraviolet absorbance than the 2-hydroxybenzophenones and have a steeper long-wavelength cutoff. These two factors make them slightly better screening agents than the benzophenones, and they have less color.

No report on the photochemistry of (**6**) has been published, but, because it possesses an intramolecular hydrogen bond and a structure formally related to the 2-hydroxybenzophenones, a phototautomerism may also occur with (**6**). The existence of a zwitterionic "photoenol" such as (**7**) is purely speculative.

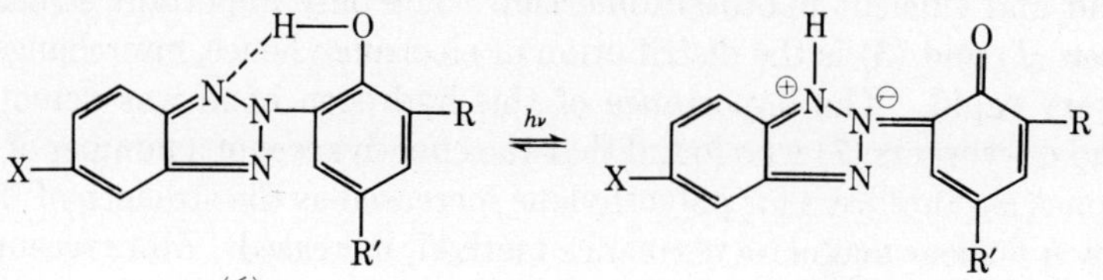

(**6**) $\overset{h\nu}{\rightleftharpoons}$ (**7**)

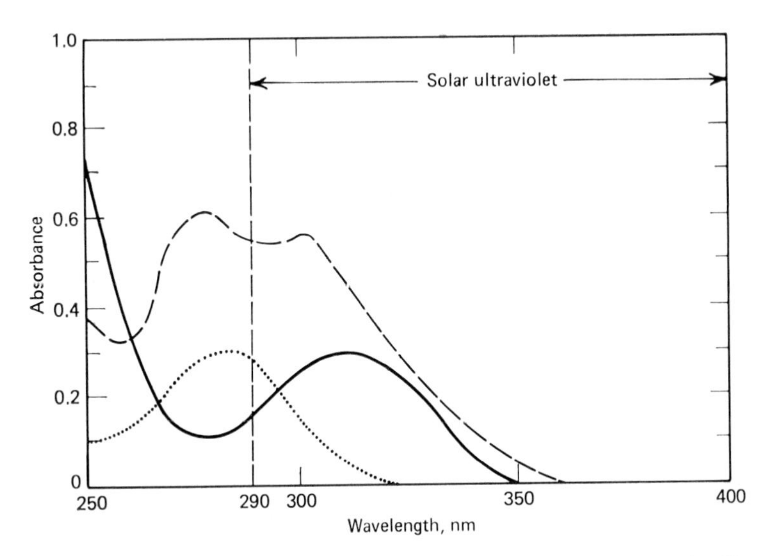

Fig. 6. Ultraviolet-absorption spectra of phenyl esters (6×10^{-4} M in CH_2Cl_2, 0.1-cm path length) Key: solid line, phenyl salicylate; dotted line, resorcinol monobenzoate; dashed line, resorcinol monobenzoate after irradiation with ultraviolet.

Phenyl Esters. The most important members of this group are resorcinol monobenzoate (**8**) and phenyl salicylate (**9a**), but substituted aryl salicylates (**9b**) and diaryl terephthalates (**10**) or isophthalates (**11**) are, or have been, sold as ultraviolet absorbers.

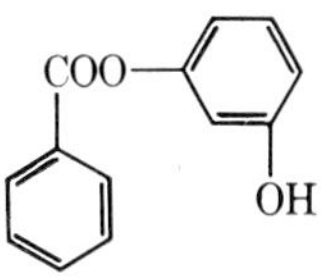

COO— OH (**8**)

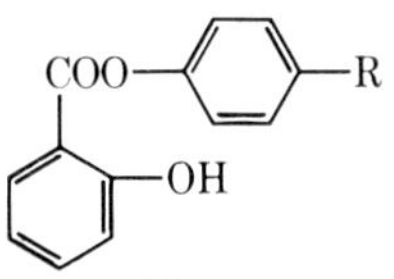

COO— —R —OH (**9a**) R = H (**9b**) R = C_4H_9 to C_8H_{17}

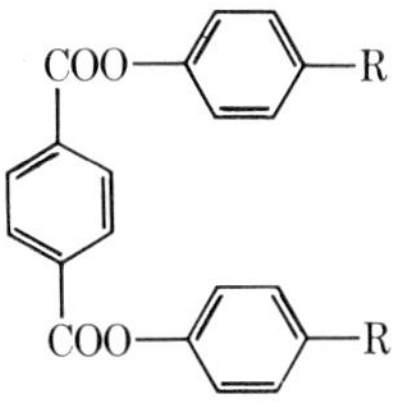

COO— —R COO— —R (**10**) R = C_8H_{17}

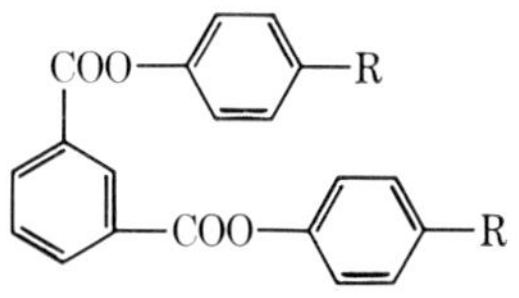

COO— —R —COO— —R (**11**) R = C_8H_{17}

These compounds have one feature in common—very low absorption in the solar ultraviolet region (Figure 6). However, after exposure to sunlight for a time, these compounds show an increase in absorption in the 290 to 400-nm region, and, after sufficient exposure, their spectra resemble those of 2-hydroxybenzophenones. These compounds owe their effectiveness to a light-catalyzed rearrangement that converts them to 2-hydroxybenzophenones (9). The products of this photo-Fries rearrangement are

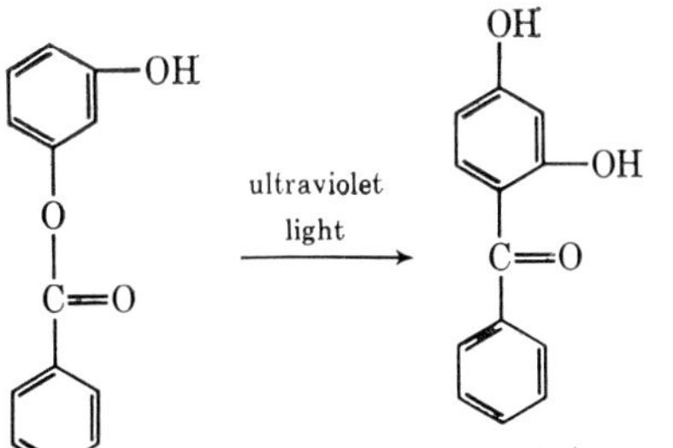

—OH O C=O (**8**) $\xrightarrow{\text{ultraviolet light}}$ OH —OH C=O (**1a**) (3)

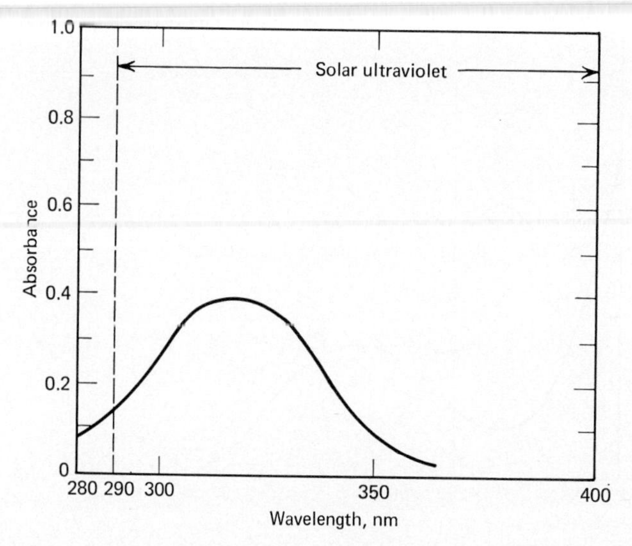

Fig. 7. Ultraviolet-absorption spectrum of ethyl 2-cyano-3,3-diphenylacrylate (6×10^{-4} *M* in CH_2Cl_2, 0.1-cm path length).

the actual stabilizers; resorcinol monobenzoate, for example, is converted to 2,4-dihydroxybenzophenone (**1a**), a very effective ultraviolet absorber (eq. 3, p. 133). Unfortunately, other products are also formed, and the conversion to effective stabilizers is only 50 to 70% efficient. In some polymers, the by-products may lead to undesirable discoloration after long exposure to light.

Substituted Cinnamic Acid Derivatives. These compounds have the general structure (**12**) where R = H or methoxy, X and Y = carboxylic acid ester or nitrile, and

$$R{-}C_6H_4{-}C(Z){=}C(X)(Y)$$

(**12**)

Z = H, alkyl, or aryl. There is no generally accepted class name for this group and such terms as substituted acrylonitrile, substituted ethylene, benzylidenemalonic acid derivative, or other uninformative names are often used. These compounds typically show a relatively low absorption maximum in the 310 to 320-nm region (Figure 7) but very little absorption above or below this region; hence, they are relatively ineffective ultraviolet filters. Nothing is known about their photochemistry except that they are quite stable to ultraviolet light. Unlike the other types of ultraviolet absorbers, the effectiveness of this class of stabilizer is greatly dependent on the nature of the polymer to which they are added. None has yet shown significant effectiveness in polyolefins, but they are moderately effective in polystyrene and poly(vinyl chloride). The effect of the substrate on stabilizer effectiveness is probably the result of the low absorption of (**12**) in the 280 to 300-nm region, where polyolefins are particularly sensitive. The most important advantage of this class is almost total absence of visible color.

Excited-State Quenchers. Photochemical reactions can be retarded by quenchers if the excitation energy of the potentially reactive molecule (donor) can be transferred to the quencher before any other photochemical event can occur. Energy can be transferred to the quencher from either the excited singlet state or excited triplet state.

Transfer from the singlet state can occur between molecules as much as 50 to 100 Å apart if there is sufficient overlap between the emission spectrum of the donor and the absorption spectrum of the quencher (resonance energy transfer). Transfer from either the singlet or triplet state can occur through collision, or near collision, of the donor and quencher molecules (exchange energy transfer) (10). Resonance energy transfer places severe structural limitations on the quencher–polymer combination, because of the spectrum-overlap requirement, and would not be expected to be a generally useful approach to polymer protection. Exchange energy transfer requires only that the quencher excited state be lower in energy than the donor excited state and that efficient transfer of energy occur at each collision; hence, the structural requirements for the quencher are much less confining than those necessary for resonance energy transfer.

Whether quenching can occur depends on the probability of collision between donor and quencher before some other photochemical fate befalls the donor; thus, quenching becomes more probable as the lifetime of the donor excited state becomes longer. Singlet states live only from 10^{-10} to 10^{-6} sec before undergoing some photochemical event; hence, the probability of quenching a singlet excited state is small. Excited state triplets live much longer, 10^{-6} to 10^{-3} sec; so quenching a triplet is much more probable than quenching a singlet. From this discussion of quenching it is evident that polymer degradation might be prevented by the following series of reactions: Q^0 and Q^3 represent the ground state and triplet excited state of the quencher, and P^0, P^1, and P^3 represent the ground state, singlet excited state, and triplet excited state of the polymer.

$$P^0 \xrightarrow{h\nu} P^1 \rightarrow P^3 \tag{4a}$$

$$P^3 + Q^0 \rightarrow Q^3 + P^0 \tag{4b}$$

$$Q^3 \rightarrow Q^0 + \text{heat} \tag{4c}$$

In the scheme given by equations 4a–4c it is assumed that Q absorbs no light and undergoes no irreversible photoreactions and that the polymer does not degrade via P^1, but only through P^3. The concentration of Q must be high enough for every polymer triplet to collide with a molecule of Q during the lifetime of P^3. From the photochemistry of simple molecules, it is known that the concentration of the quencher must approach that of the donor molecules before quenching approaches 100% efficiency. Information as to whether degradation occurs from the singlet, triplet, or both is not available for any commercial polymer.

In spite of the low probability for effective polymer stabilization, this quenching mechanism does offer a potentially useful new approach to the problem of preventing ultraviolet degradation and is the subject of a number of recent publications. Guillet and co-workers (11,12) have been able to reduce the rate of photodegradation of poly(phenyl vinyl ketone) by about 90% and the rate of degradation of a carbon monoxide–ethylene copolymer by about 50% through the use of 1,3-cyclooctadiene, a well-known triplet quencher. However, these quenching results, which were obtained by the irradiation of dilute polymer solutions, required a quencher concentration more than five times as great as the polymer concentration. These results are of considerable theoretical interest but do not represent a practical system for polymer stabilization. Other workers (13) found that low concentrations of 2-(2*H*-benzotriazol-2-yl)-*p*-cresol or, less effectively, 2-hydroxy-4-methoxybenzophenone quenched carbonyl triplets formed in irradiated polypropylene and concluded that these compounds act entirely as quenchers. However, the concentration required to quench completely the carbonyl triplets was well below the concentration required to protect the polymer against ultraviolet radiation. At useful concentrations in polyolefins, these stabilizers absorb more than 99% of the incident ultraviolet; hence, they act chiefly as screening agents even though quenching may play some part in their function.

Several nickel-containing compounds are effective stabilizers and are generally believed to act as quenchers. Organometallic complexes containing paramagnetic metal ions are effective quenchers of both singlet (14) and triplet (15) excited states in solution photochemistry. The nature of the organic ligand has a large effect on the quenching efficiency of these chelates, and it has been shown that chelates having square planar structures were most effective (16). The ketone-sensitized photo-oxidation of cumene is retarded by nickel chelates such as (**13**) and the mechanism has been shown to involve quenching of the photoexcited states of the sensitizer (40).

Briggs and McKellar (16,17) have found that the effectiveness of a series of experimental nickel chelates as ultraviolet stabilizers for polypropylene parallels their effectiveness as quenchers for anthracene triplets. However, these compounds also absorb strongly in the solar ultraviolet, and the relative importance of their screening and quenching effects remains to be determined. Ultraviolet-absorption curves for three commercial nickel-containing ultraviolet stabilizers—Ferro AM-101 (**13**), Cyasorb 1084 (**14**), and nickel dibutyldithiocarbamate (**15**)—are shown in Figure 8.

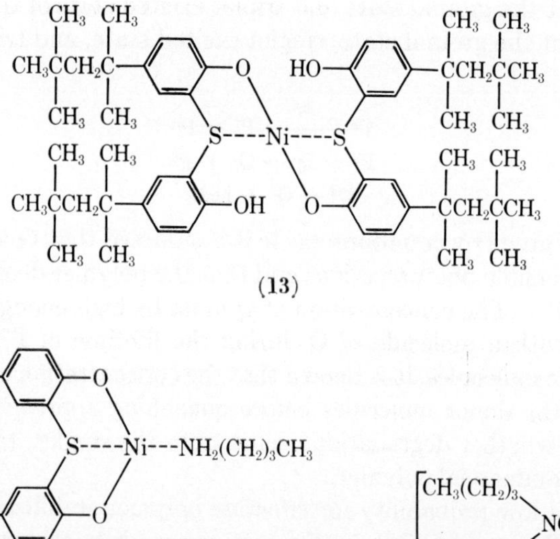

(13)

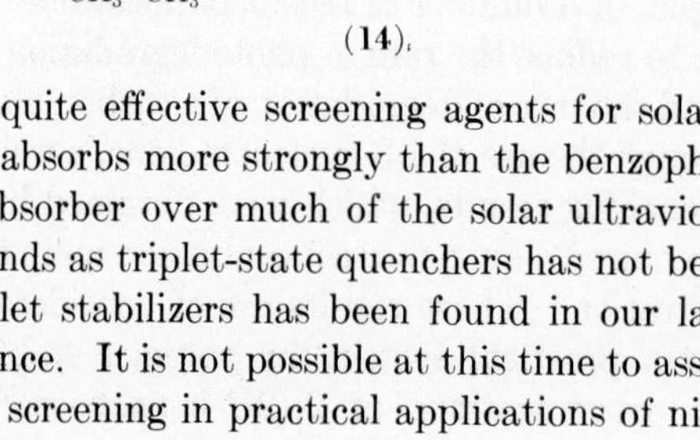

(14)

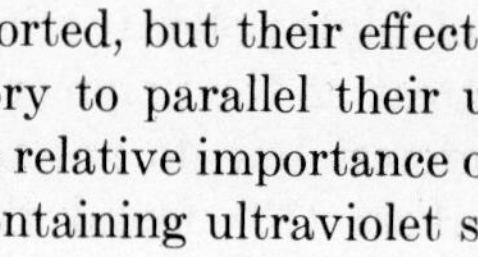

(15)

All are quite effective screening agents for solar ultraviolet; compound (**15**), in particular, absorbs more strongly than the benzophenone or benzotriazole types of ultraviolet absorber over much of the solar ultraviolet. The effectiveness of these three compounds as triplet-state quenchers has not been reported, but their effectiveness as ultraviolet stabilizers has been found in our laboratory to parallel their ultraviolet absorbance. It is not possible at this time to assess the relative importance of quenching and screening in practical applications of nickel-containing ultraviolet stabilizers. Both mechanisms probably operate and one or the other may predominate in different polymer–chelate combinations.

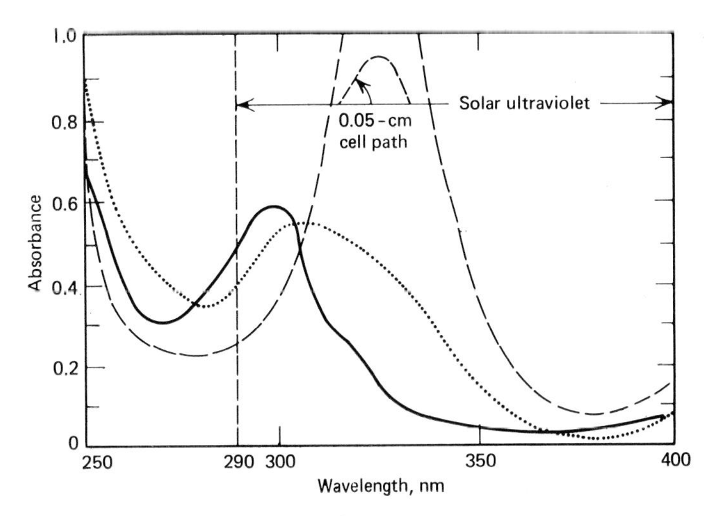

Fig. 8. Ultraviolet-absorption spectra of nickel-containing ultraviolet stabilizers (6×10^{-4} *M* in CH_2Cl_2, 0.1-cm path length). Key: solid line, Ferro AM-101; dotted line, Cyasorb 1084; dashed line, nickel dibutyldithiocarbamate.

Commercial Aspects

The use of an ultraviolet-radiation absorber in a polymer was first reported in 1945 by Meyer and Gearhart (30) who found that the outdoor life of cellulose acetate was greatly prolonged if the composition contained phenyl salicylate (salol). Resorcinol monobenzoate, a much more effective and somewhat more expensive compound than salol, was introduced in 1951 (31), but salol continued to be the only important commercial stabilizer for several years. In 1958 nearly 500,000 lb of salol was used, mostly in cellulosics, and the total market for all other ultraviolet stabilizers was only about 200,000 lb (32). Salol is still used extensively in cellulose esters and some coating formulations but is being displaced by more effective stabilizers even in these uses.

Table 3. Use of Ultraviolet-Radiation Absorbers in Various Polymers, 1958 to 1970

	Approximate use, thousands of pounds				
Polymer	1958[a]	1964[b]	1966[b]	1968[c]	1970[c]
cellulosics[d]	500	110	140	140	140
unsaturated polyesters	50	120	160	200	260
polystyrene	20	100	130	150	180
polyolefins	10	200	600	600	700
poly(vinyl chloride)	50	180	240	300	400
lacquers	10	60	160	200	400
other	10	40	200	200	240
total	*650*	*810*	*1630*	*1790*	*2320*

[a] Ref. 32.
[b] Ref. 33.
[c] Estimated use.
[d] Use for 1958 includes 360,000 lb of salol; subsequent years do not include salol.

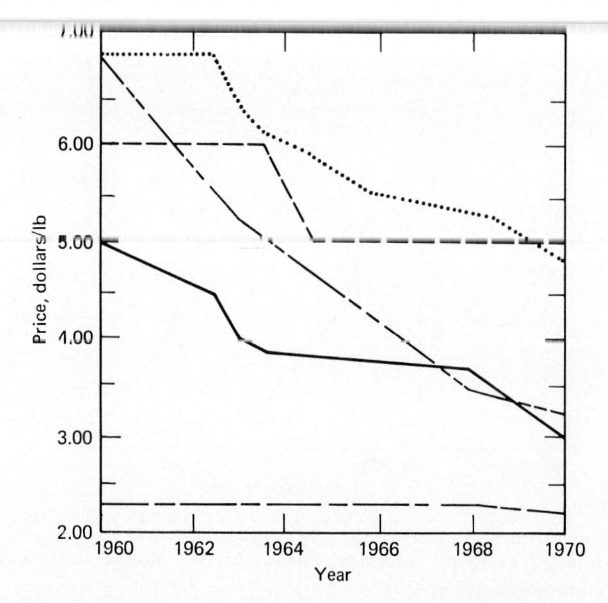

Fig. 9. Average price per pound for various types of ultraviolet absorbers, 1960–1970. Key:—— 2,4-dihydroxybenzophenone derivatives, ... 2-(2*H*-benzotriazol-2-yl)phenols; – – – substituted ethylenes; —-—- phenyl esters; —- - -— nickel chelates.

2,4-Dihydroxybenzophenone appeared on the market in 1953, followed shortly by 2-hydroxy-4-methoxybenzophenone and other derivatives. Tinuvin P, the original benzotriazole-type ultraviolet absorber, became available in 1956. The first nickel chelate, Ferro AM-101, was first sold about a year later as an additive to promote the dyeing of polypropylene, and the discovery of its effectiveness as an ultraviolet stabilizer soon followed. The sales of these new, more effective, and more costly stabilizers grew slowly until the use of polyolefins and poly(vinyl chloride) in outdoor applications blossomed in the late 1950s and early 1960s. Salol and resorcinol monobenzoate did not give the desired degree of protection in these polymers, and stabilizers of the newer types were needed. The effect of the growth of these two types of polymer on the sales of ultraviolet absorbers can be seen in Table 3.

An area of potential growth is the use of ultraviolet-radiation absorbers in automotive finishes, where they are added now to only a few types of lacquers. If, however, the use of these absorbers becomes widespread, this market could become second only to their use in polyolefins. From their introduction in the 1950s until 1968, the newer types of ultraviolet-radiation absorbers remained relatively high-priced specialty chemicals with most of the market held by a very few manufacturers. In 1968, a number of new suppliers entered the benzophenone-type stabilizer market, and a considerable weakening of the price structure occurred. Figure 9 shows the price history for various types of ultraviolet stabilizers for the past decade.

Current Status. About forty different ultraviolet stabilizers are currently available as commercial or development products, but only a few of these enjoy any significant use. Frequently, no clear distinction is made between stabilizers that are available in quantity and those which are small-scale experimental materials. Trade names are often given to experimental stabilizers that never achieve commercial production and, if no sales develop, these compounds quietly disappear from the market—

usually with no public announcement of their withdrawal. New suppliers enter the market with stabilizers similar to or the same as older commercial products but with new trade names. Hence, it is difficult to compile a list of currently available commercial ultraviolet stabilizers. Table 4 lists those stabilizers for which commercial status has been claimed and for which information about structure or type of compound has been given. Some of the compounds listed may no longer be available or may be available only in experimental quantities. No data on the sales of individual stabilizers or types of stabilizers are published. The 2-hydroxybenzophenones are the most widely used; 2,4-dihydroxybenzophenone, 2-hydroxy-4-methoxybenzophenone, 2-hydroxy-4-(octyloxy)benzophenone, and 4-(dodecyloxy)-2-hydroxybenzophenone are the most important members of this group. Resorcinol monobenzoate, used almost exclusively in cellulose esters, probably holds the second rank. The use of the benzotriazoles, particularly in polystyrene, polyesters, and other applications where low color is important, is growing rapidly as their price becomes more competitive. The nickel compounds are an important factor in the stabilization of polypropylene fiber because, in addition to ultraviolet protection, they provide sites for dyeing the fibers with metal-chelating dyes. The substituted ethylenes are used to some extent in polystyrene and poly(vinyl chloride) but are, at present, of minor commercial importance.

Selection of Stabilizers. The choice of an ultraviolet absorber for a particular polymer and use is governed largely by economic factors and by technical considerations not related to ultraviolet absorbance. Some of the most important technical considerations are:

1. High solubility of the stabilizer in the polymer.

2. Low rate of stabilizer loss from the polymer through exudation, volatilization, leaching, or other processes.

3. Absence of chemical reactivity of the stabilizer with the polymer, other additives that may be present, or catalyst residues and other contaminants in the polymer.

4. Low initial color and little or no color change of the stabilized polymer on exposure to light or heat.

5. Low toxicity of the stabilizer.

6. Ease of compounding the stabilizer with the polymer.

7. Lowest possible cost consistent with desired performance for the composition.

These factors are weighed according to their relative importance for a particular use. If a polyethylene formulation is to be used for garbage cans, color and toxicity are of minor importance, and high effectiveness and permanence are of major importance; however, in polystyrene diffusers for lighting fixtures, low initial color and freedom from discoloration on exposure are paramount considerations. Both the stabilizer used and its concentration will vary considerably in accord with the intended use of a particular polymer. The importance of some of these factors can be illustrated by the discussion of the stabilization of some specific polymers. Because the exact stabilizer system used in a particular formulation is usually not disclosed by the polymer producer, only general principles can be given.

Stabilization of Polyolefins. Polyolefins, particularly polypropylene, are subject to autoxidation in the absence of light so that it is necessary to use an effective antioxidant system as well as an ultraviolet absorber to obtain long outdoor life. The ultraviolet absorbance of a polyolefin is low; hence, ultraviolet-radiation absorbers are quite effective in preventing photodegradation.

Table 4. Commercial Ultraviolet Stabilizers[a]

Chemical name or description	Trade name and/or trademark	Supplier[b]
	2-Hydroxybenzophenones	
2,4-dihydrobenzophenone	Advastab 48	4
	Rylex H	6
	Unistat 12	16
	Uvinul 400	8
2-hydroxy-4-methoxybenzophenone	Advastab 45	4
	Cyasorb UV-9	1
	Uvinul M-40	8
4-(heptyloxy)-2-hydroxybenzophenone	Unistat 247	16
2-hydroxy-4-(octyloxy)benzophenone	Carstab 700	3
	Cyasorb UV-531	1
	Picco UV 299	14
4-(dodecyloxy)-2-hydroxybenzophenone	Eastman Inhibitor DOBP	5
	Rylex D	6
2-hydroxy-4-(2-hydroxyethoxy)benzophenone	Eastman Inhibitor HHBP	5
4-alkoxy-2-hydroxybenzophenone	Advastab 46	4
	Uvinul 410	8
2-hydroxy-4-methoxy-5-methylbenzophenone	Unistat 2211	16
2-hydroxy-3-(3-hydroxy-4-benzoylphenoxy)propyl methacrylate	Permasorb MA	13
5-benzoyl-4-hydroxy-2-methoxybenzenesulfonic acid	Cyasorb UV 284	1
	Uvinul MS-40	8
2-(2-hydroxy-4-methoxybenzoyl)benzoic acid	Cyasorb UV 207	1
2,2′-dihydroxy-4-methoxybenzophenone	Advastab 47	4
	Cyasorb UV-24	1
4-butoxy-2,2′-dihydroxybenzophenone	Cyasorb UV-287	1
2,2′-dihydroxy-4-(octyloxy)benzophenone	Cyasorb UV-314	1
2,2′, 4,4′-tetrahydroxybenzophenone	Uvinul D-50	8
2,2′-dihydr6xy-4,4′-dimethoxybenzophenone	Uvinul D-49	8
tetrasubstituted benzophenone mixture	Uvinul 490	8
not reported[c]	Mark 202A	2
	Permyl B100	7
	2-(2H-Benzotriazol-2-yl)phenols	
2-(2*H*-benzotriazol-2-yl)-*p*-cresol	Tinuvin P	9
2-*tert*-butyl-6-(5-chloro-2*H*-benzotriazol-2-yl)-*p*-cresol	Tinuvin 326	9
2,4-di-*tert*-butyl-6-(5-chloro-2*H*-benzotriazol-2-yl)phenol	Tinuvin 327	9
2-(2*H*-benzotriazol-2-yl)-4,6-di-*tert*-pentylphenol	Tinuvin 328	9
	Phenyl esters	
phenyl salicylate	salol	15
p-(1,1,3,3-tetramethylbutyl)phenyl salicylate	Eastman Inhibitor OPS	5
resorcinol monobenzoate	Eastman Inhibitor RMB	5
strontium salicylate	Sunkem SRS	15
carboxyphenyl salicylate	Sunkem CPS	15
bis(*p*-nonylphenyl) terephthalate	Stabilizer BX-721	7
bis[*p*-(1,1,3,3-tetramethylbutyl)phenyl] isophthalate	Santoscreen	11
	Substituted ethylenes	
dimethyl *p*-methoxybenzylidenemalonate	Cyasorb UV-1988	1
2-ethylhexyl 2-cyano-3,3-diphenylacrylate	Uvinul N-539	8
butyl 2-cyano-*p*-methoxy-3-methylcinnamate	UV-317	12
not reported[c]	Cyasorb UV-3100	1
	UV-Absorber 318	12
	UV-Absorber 317	12

Table 4 (*continued*)

Chemical name or description	Trade name and/or trademark	Supplier[b]
Nickel compounds		
bis[2,2′-thiobis-4-(1,1,3,3-tetramethylbutyl)phenolato]nickel	Ferro AM-101	7
[2,2′-thiobis[4-(1,1,3,3-tetramethylbutyl)phenol]ato(2-)]-(butylamine)nickel	Cyasorb UV-1084	1
nickel dibutyldithiocarbamate	NBC	6
not reported[c]	Irgastab 2002	9
nickel chelate of *p*-methylacetophenone oxime	Hegopex A	10

[a] This list may contain stabilizers that are no longer available or are available only as experimental materials. Not included are a few stabilizers introduced, usually as experimental materials, without identification as to type of compound.

[b] Suppliers code:

1. American Cyanamid Co.
2. Argus Chemical Corp.
3. Carlisle Chemical Works, Inc.
4. Advance Division, Carlisle Chemical Works, Inc.
5. Eastman Chemical Products, Inc.
6. E. I. du Pont de Nemours & Co., Inc.
7. Ferro Corp.
8. GAF Corp.
9. Geigy Chemical Corp.
10. Imperial Chemical Industries, Ltd.
11. Monsanto Co.
12. Naftone, Inc.
13. National Starch and Chemical Corp.
14. Pennsylvania Industrial Chemical Corp.
15. Sun Chemical Corp.
16. Ward, Blenkinsop & Co., Ltd.

[c] Exact structure unknown, but is of type indicated.

"Compatibility" of the additive with the polymer is a problem. This rather ambiguous term includes such properties as solubility of the additive in the polymer, the rate of diffusion of the additive through the polymer, and the rate of loss of the additive from the polymer. The most obvious manifestation of poor compatibility is the appearance of exuded additive on the polymer surface within a few days after compounding. Compatibility is a sensitive function of molecular structure and is not entirely predictable. In low-density polyethylene, only those 4-alkoxy-2-hydroxybenzophenones having alkyl groups in the C_{10}–C_{16} range resist exudation; if the alkyl group falls on either side of this range rapid exudation occurs (18). Attempts have been made to relate the compatibility of an additive with polypropylene to the structure of the additive by measuring the rates of diffusion of the additives (19,20), or by measuring the effect of the additive on crystallization temperature (19,20) or surface tension (21) of the molten polymer. The correlation of such measurements with the observed loss of stabilizer from the polymer is only fair. Some mobility of the additive in the polymer is desirable, particularly in relatively thick samples. Diffusion of the additive from the interior of the sample to the surface region replenishes the stabilizer in the place where it is most needed.

2-Hydroxy-4-(octyloxy)benzophenone and 4-(dodecyloxy)-2-hydroxybenzophenone are the ultraviolet stabilizers most widely used in polyolefins. These additives, which are generally added to the polymer feed just before extrusion of pellets, may be used at concentrations of 0.1–3.0 wt %, or more, as required for the desired weathering life. At the higher stabilizer levels, outdoor life can be extended by a factor of ten or more so that a useful outdoor life of at least ten years can be obtained with polyethylene and about three years or more with polypropylene. 2,4-Di-*tert*-butyl-6-(5-chloro-2*H*-benzotriazol-2-yl)phenol (Tinuvin 327) is also effective in polyolefins, but its relatively high cost has, so far, prevented significant use in these polymers.

Nickel-containing stabilizers, particularly bis[2,2′-thiobis-4-(1,1,3,3-tetramethylbutyl)phenolato]nickel (Ferro AM-101), are widely used in polypropylene fiber, where they show better effectiveness relative to the benzophenone types than they do in other applications. It has been suggested (22) that quenching becomes a more effective stabilization process than screening in thin films and fibers; however, better stabilization may be due to a lower loss rate for the nonvolatile nickel compounds than for the more volatile organic stabilizers. The presence of the nickel compounds allows the fiber to be dyed with metal-chelating dyes; this is a very important advantage in a polymer that has no inherent affinity for dyes. The pale green color of fiber containing the nickel compounds is masked by the dye. Nickel dibutyldithiocarbamate is a very effective stabilizer for polyolefins but can be used only where its pronounced olive-green color is not objectionable. Combinations of a hydroxybenzophenone with a nickel-containing stabilizer give excellent effectiveness but at the expense of a noticeable green color in the polymer. Nickel-containing stabilizers are thermally unstable in the presence of sulfur compounds. The sulfur may be present either in the ligand, as in bis[2,2′-thiobis-4-(1,1,3,3-tetramethylbutyl)phenolato]nickel or in the almost universally used antioxidant dilauryl 3,3′-thiodipropionate, and it may remove the nickel from the stabilizer as black nickel sulfide during extrusion of the polymer unless the temperature and residence time in the extruder are carefully controlled. Of the ester-type stabilizers now available, only *p*-(1,1,3,3-tetramethylbutyl)phenyl salicylate is effective in polyolefins. It is particularly useful in moldings with thick cross sections and has had some commercial use in such applications. The substituted ethylene type of ultraviolet absorber has shown little utility in polyolefins.

There are many uses for polyolefins containing ultraviolet-radiation absorbers. A few of them are: garbage cans and other storage containers made from high-density polyethylene or polypropylene, polyethylene bags for fertilizer or other products, polypropylene rope and twine, polypropylene outdoor carpet, outdoor advertising displays made from polyethylene or polypropylene, crates for beverage bottles, polypropylene webbing or moldings used in outdoor furniture, polyethylene film for greenhouse covering, outdoor toys, and polypropylene steering wheels and other interior trim for automobiles. See also ETHYLENE POLYMERS; OLEFIN FIBERS; PROPYLENE POLYMERS.

Stabilization of Poly(vinyl Chloride). Both rigid and flexible poly(vinyl chloride) formulations usually contain organometallic compounds added to provide heat stability. The nature of the heat stabilizer used greatly affects the outdoor performance of the polymer. Salts of barium, cadmium, lead, or zinc frequently provide considerable light stability, as well as thermal stability, to the polymer, and for many uses no additional ultraviolet absorber is needed. Organotin compounds have little effect on the photodegradation of poly(vinyl chloride), except for the sulfur-containing organotin compounds which may accelerate outdoor weathering in this polymer. If stability greater than that provided by the heat stabilizer is required, a considerable improvement in outdoor life can be obtained by adding from 0.1 to 1.0 wt % of a hydroxybenzophenone- or benzotriazole-type ultraviolet-radiation absorber. Poly-(vinyl chloride) appears to present no serious compatibility problem, and the commonly used stabilizers are 2-hydroxy-4-methoxybenzophenone or 2-(2*H*-benzotriazol-2-yl)-*p*-cresol (Tinuvin P). However, less volatile additives such as 4-(dodecyloxy)-2-hydroxybenzophenone or 2,4-di-*tert*-butyl-6-(5-chloro-2*H*-benzotriazol-2-yl)phenol (Tinuvin 327) give a longer outdoor life because of their somewhat slower loss from the polymer. The fact that poly(vinyl chloride) garden hose or swimming pool liners may carry a ten to fifteen year guarantee illustrates the effectiveness of ultraviolet absorbers in this polymer. Nickel-containing stabilizers and the ester-type absorbers are relatively ineffective in poly(vinyl chloride). The substituted ethylenes are claimed to be effective but it is not known whether they are used commercially in poly(vinyl chloride).

Poly(vinyl chloride) is widely used in applications where ultraviolet protection is required. Rigid poly(vinyl chloride) is used in guttering, window frames, wall panels, and other construction uses. Flexible poly(vinyl chloride) is used in swimming pool liners, garden hose, outdoor furniture, inflatable buildings, automobile upholstery, floor coverings, and many other applications. See also STABILIZATION; VINYL CHLORIDE POLYMERS.

Stabilization of Polystyrene. Polystyrene has poor inherent outdoor stability, and ultraviolet stabilizers do not greatly improve its performance; hence, it is little used in outdoor applications. However, polystyrene is used in fluorescent light diffusers, ceiling panels, and other indoor uses where it is exposed to relatively low levels of ultraviolet radiation which will cause the polymer to yellow. Yellowing can be prevented by the use of an ultraviolet-radiation absorber. Because good color is the major requirement for polystyrene, 2-(2*H*-benzotriazol-2-yl)-*p*-cresol (Tinuvin P) or one of the substituted ethylene-type compounds is frequently used. If a slight initial yellow color can be tolerated, 2,4-dihydroxybenzophenone or 2-hydroxy-4-methoxybenzophenone gives good protection against further yellowing. Any ultraviolet absorber tends to quench the inherent bluish fluorescence of polystyrene and cause the polymer to appear more yellow than the unstabilized polymer. Addition of an optical brightener to the polymer restores its original appearance (see BRIGHTENERS, OPTICAL). Antioxidants are also frequently added to polystyrene containing an ultraviolet-radiation absorber. If ultraviolet radiation below about 320 nm is kept low (ordinary soft glass transmits very little light below this wavelength), well-stabilized polystyrene will resist yellowing for several years, but these compositions would last for only a few months outdoors. See also STYRENE POLYMERS.

Stabilization of Unsaturated Polyesters. Glass-reinforced polyester–styrene compositions have fairly good resistance to sunlight if the proper basic resin system is used. However, yellowing and, later, surface erosion will occur unless an ultraviolet-radiation absorber is used. Compatibility or stabilizer loss does not appear to be a problem with these compositions. The composition usually has an inherent yellowish cast so that slight additional color imparted by the stabilizer is not important. Either 2,4-dihydroxybenzophenone or 2-hydroxy-4-methoxybenzophenone is commonly used as the stabilizer, but a benzotriazole may be used if lower color is important enough to offset the higher cost. The ultraviolet-radiation absorber may be dissolved in the polyester–styrene mixture so that it is uniformly distributed throughout the polymer blend; a concentration of 0.25–0.75 wt % is usually used. In applications of these compositions where a final gel coat is used, as in reinforced polyester boats or automobile bodies, better performance with less stabilizer can be obtained when the ultraviolet stabilizer is added only in the gel coat at a concentration of 0.5–2.0 wt %. The higher concentration of stabilizer in the relatively thin gel coat affords longer protection to the surface and prevents ultraviolet radiation from reaching the main body of the polymer. See also POLYESTERS, UNSATURATED.

Polymeric and Polymerizable Absorbers

In most polymers, longer outdoor life is obtained with relatively high-molecular-weight ultraviolet-radiation absorbers than with their lower-molecular-weight homologs. Slow loss of the additive probably occurs through exudation, evaporation, and leaching out by rain even when no evidence of gross incompatibility can be observed. Failure occurs when the absorber concentration falls below the required level for protection. Additive loss might be reduced considerably if the ultraviolet absorber were itself polymeric and, therefore, relatively immune to exudation, evaporation, or leaching. Alternatively, a polymerizable ultraviolet absorber could be built into the polymeric material to be protected so that any loss of stabilizer would be unlikely. Several patents claiming unusual effectiveness for such built-in or polymeric absorbers have appeared recently. Copolymers of ethylene with (**16**) (23), copolymers of (**17**) (24) or (**18**) (25) with various vinyl monomers, and polystyrene containing the homopolymer of (**19**) (26) are all claimed to be superior to compositions containing conventional monomeric ultraviolet absorbers. Polyesters or polyurethans containing (**20**) as

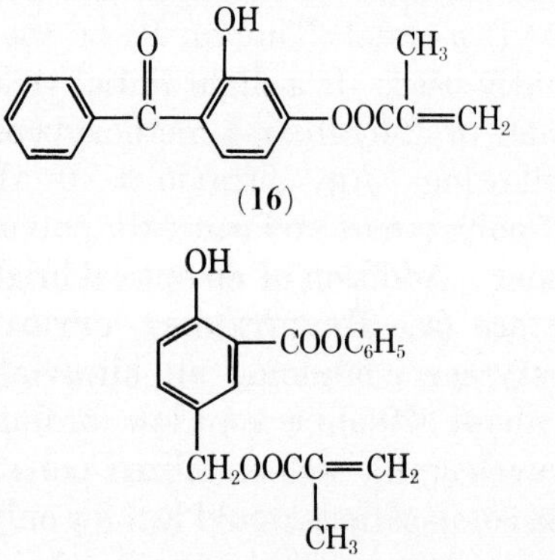

(**16**)

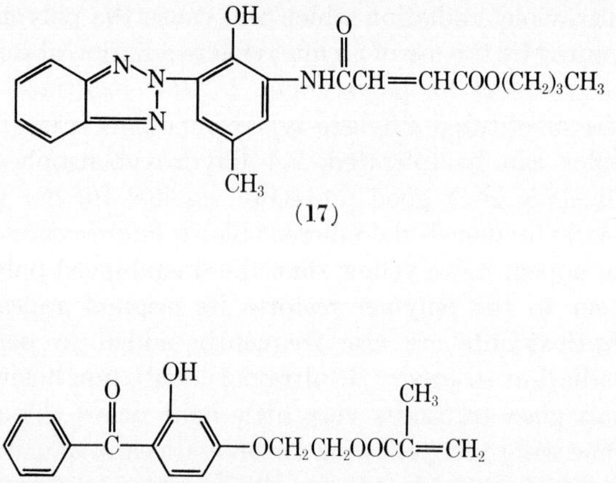

(**17**)

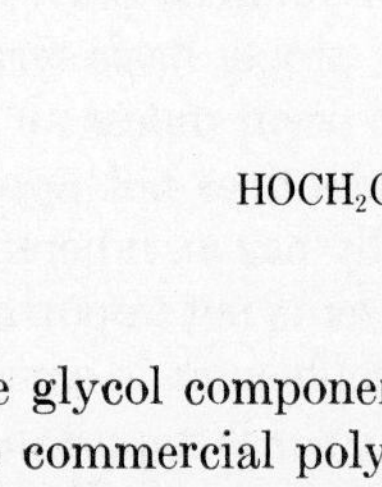

(**18**)

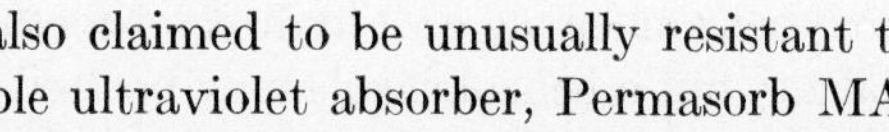

(**19**)

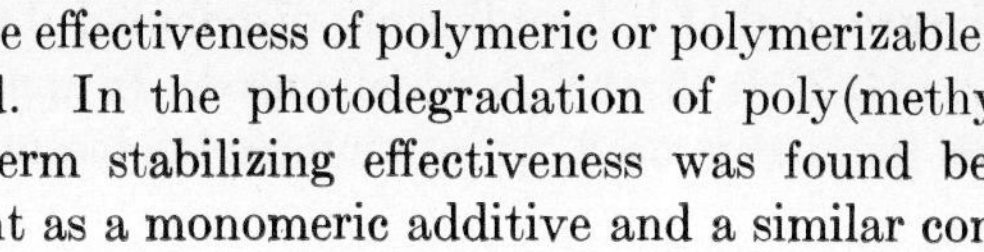

(**20**)

part of the glycol component are also claimed to be unusually resistant to sunlight (27). One commercial polymerizable ultraviolet absorber, Permasorb MA, is available.

Little data on the effectiveness of polymeric or polymerizable ultraviolet absorbers have been published. In the photodegradation of poly(methyl methacrylate), no difference in short-term stabilizing effectiveness was found between a 2-hydroxybenzophenone present as a monomeric additive and a similar compound incorporated in the polymer chain (28). In our laboratories we have found that several polymeric 2-hydroxybenzophenone-type ultraviolet-radiation absorbers were less effective in polypropylene than 4-(dodecyloxy)-2-hydroxybenzophenone, but certain copolymers of (**19**) with other vinyl monomers were more effective stabilizers for acrylic lacquers than monomeric absorbers analogous to (**19**) (29). Mutual solubility, or compatibility, factors may be as important for polymeric ultraviolet absorbers as for lower-molecular-weight additives; hence, for the potential advantages of these polymeric stabilizers to be fully utilized, it may be necessary to design a specific polymeric stabilizer for a particular polymer. When this done, polymeric ultraviolet-radiation absorbers may find a place as stabilizers for fibers, films, or coating materials.

Testing Procedures

For each specific application the correct stabilizer and its optimum concentration must be determined by tests of the various candidates over a range of concentrations. Theoretical considerations are not yet of much help in this choice. Testing under conditions that approximate the intended service environment requires months or years to obtain results; hence, test conditions that greatly accelerate degradation must be used. Both indoor tests under artificial illumination and outdoor tests can be accelerated by factors of ten or more. The means used for acceleration may change degradation pathways as well as rates. If acceleration involves a temperature increase, thermal oxidation may assume a much greater importance than it has at normal temperatures. Under intense illumination, photochemical processes may be different from those that would be experienced in natural sunlight (33). If illumination is made continuous, dark reactions which might occur under normal day–night cycles will be eliminated, and the polymer may appear to be either more or less stable because of this. Results of accelerated tests must, therefore, be used with caution in the prediction of outdoor life in an actual product use, and the larger the acceleration factor, the less reliable are the predictions of outdoor life. A test can usually be designed to give results that correlate reasonably well with the actual outdoor life for a particular polymer used in a particular application; however, the same test may be completely misleading when used to predict the outdoor life for another polymer or another use. If other additives such as pigments, plasticizers, dyes, antioxidants, mold lubricants, and antistatic agents are to be used in the finished product, these additives should be included in the test formulations, because any additive may exert a marked effect on the stability of the composition. Polyolefins and, to a lesser degree, polystyrene and cellulose esters must be stabilized against autoxidation with an effective antioxidant before satisfactory stabilization with an ultraviolet stabilizer can be obtained. Because of the wide variety of test procedures in use, this discussion will be limited to some of the basic principles. See also WEATHERING.

Outdoor Testing. Outdoor testing of samples containing ultraviolet stabilizers is usually carried out at locations having little cloudy weather, relatively high average temperatures, and that are as near to the equator as is practicable so as to maximize the solar ultraviolet flux received by the sample. Large outdoor test facilities are located near Miami, Florida, and Phoenix, Arizona. Summer temperatures and total ultraviolet flux are higher at the Phoenix site, whereas the humidity is much higher at the Miami site. Polymers react differently to these variations; some degrade faster in Florida, but most degrade faster in Arizona. Common practice is to expose samples at both locations. Samples are usually exposed on racks facing south and inclined at an angle chosen to maximize the radiation received by the surface. If a sun-following equatorial mounting is used instead of stationary racks, the total daily ultraviolet exposure is increased by a factor of about two. Further acceleration can be obtained by concentrating light on the sample with mirrors or lenses. One such device, designed by Caryl and Helmick (34), is the EMMA exposure device which combines an equatorial mount with aluminum mirrors to give a tenfold increase in ultraviolet flux received by the samples over that received by samples exposed on fixed racks.

Any of these exposure units can be fitted with water sprays to simulate rain, or samples can be mounted under glass to better approximate use conditions which might be encountered in automobile interiors and the like. For some polymers, particularly polyolefins, degradation is faster if the sample is stressed. Molded bars may be held in a sharply bent U-shape; films or fibers may be mounted under tension. To produce reliable results, the sample holder must place equal stress on all samples. Stress relaxation may occur during exposure; if sample composition affects the degree of relaxation, misleading estimates of stability will be obtained. Better correlation of test results is obtained if the duration of exposure is measured by total ultraviolet light incident on the sample rather than the total time of exposure; hence, exposure is usually reported as total langleys (g-cal/cm^2) of ultraviolet light received by the sample. The average exposure received during a summer day in Phoenix by a sample mounted on a fixed rack facing south at 45° inclination is about 590 langleys. When degradation is plotted against total langleys received, the various methods of exposing samples outdoors give fairly good correlation for most polymers. In addition to the outdoor exposure of samples in Florida and Arizona, outdoor tests are usually carried out in an industrialized area to determine the effects of air pollutants on the sample. This is particularly important in testing paints or other materials where discoloration is a serious problem.

Testing with Artificial Light Sources. Laboratory equipment fitted with artificial light sources can greatly accelerate the testing of samples containing ultraviolet stabilizers. Not only can the ultraviolet intensity be made much greater than the intensity of sunlight, but this full intensity can be maintained continuously. Because tests may be carried out in the laboratory rather than at some remote outdoor facility, artificial light exposure is particularly useful for the rapid screening of many different samples. Suitable light sources include carbon arcs, xenon arcs, mercury arcs, sunlamps, and fluorescent black lamps. None of these gives an energy distribution spectrum identical to solar ultraviolet, particularly in the short ultraviolet region. Carbon arcs have intense radiation peaks in the 390 to 400-nm region; mercury arcs and black lamps have peaks at 254, 310, and 360 nm. Xenon arcs most closely simulate sunlight, but, by the use of proper filters, any of these sources can provide a fair approximation of the solar ultraviolet spectrum (36). Combinations of sources, such as a carbon arc with fluorescent sunlamps, may also be used to improve the simulation of sunlight. The apparatus may be as simple as a sunlamp directed at the sample or as elaborate as a device in which temperature, humidity, and light intensity can be controlled. The sample may be sprayed with water, exposed to atmospheric pollutants, and subjected to any desired combination of light and dark periods. Commercial exposure devices are available from such manufacturers as Atlas Electrical Devices Co. (Fade-Ometer, Weather-Ometer, Xenon Arc-Weather-Ometer), Quartzlampen G.m.b.H. (Xenotest), American Ultraviolet Co. (fluorescent black light units), and De LaRue Fregestor, S.A. (fluorescent black light units). Acceleration factors of 100 or more can be obtained with specially designed laboratory devices.

Very highly accelerated tests may be useful for rapid screening of a large number of samples but should not be relied on for reliable prediction of outdoor life. Differences between the spectral energy distribution of the artificial light source and that of sunlight may result in changes in the photochemistry of the polymer so that the test results become misleading. High light intensity may result in high sample temperature and consequent exaggeration of thermal degradation effects. The results of testing under artificial light are often supplemented with outdoor testing of the most promising compositions before a final choice of stabilizer is made.

Early Detection of Degradation. Both indoor and outdoor tests depend on some obvious change in the sample as a measure of degradation; cracking, surface erosion, color change, loss of tensile strength, or embrittlement is frequently used as the criterion of failure. Many days, or even many months, of exposure may be required before such changes become evident. If some means could be found for detecting signs of eventual failure after only very short exposure, evaluation of ultraviolet absorbers could be carried out in a reasonable time under actual use conditions. This problem is being examined by many laboratories. It has been reported that degradation can be detected in stabilized polyethylene after a few days of outdoor exposure by the use of multiple reflectance infrared spectroscopy (37), but whether these results will accurately reflect actual outdoor life is not yet known. Contact-angle measurement can also be used to detect early degradation (38).

Bibliography

1. R. E. Barker, Jr., *Photochem. Photobiol.* **7,** 275 (1968).
2. G. C. Newland and J. W. Tamblyn, *Appl. Polym. Symp.* **4,** 119–129 (1967).
3. J. F. Rabek, *Photochem. Photobiol.* **7,** 5 (1968).
4. H. H. G. Jellinek, *Appl. Polymer Symp.* **4,** 41–60 (1967).
5. O. Cicchetti, *Advan. Polymer Sci.* **7,** 70 (1970).
6. J. G. Calvert and J. N. Pitts, Jr., *Photochemistry,* John Wiley & Sons, Inc., New York, 1967, p. 534.
7. J. H. Chaudet, G. C. Newland, H. W. Patton, and J. W. Tamblyn, *SPE Trans.* **1,** 26 (1961).
8. A. A. Lamola and L. J. Sharp, *J. Phys. Chem.* **70,** 2634 (1966).
9. D. Bellůs and P. Hrdlovič, *Chem. Rev.* **67,** 599 (1967).
10. R. O. Kan, *Organic Photochemistry,* McGraw-Hill Book Co., New York, 1966, p. 15.
11. M. Heskins and J. E. Guillet, *Marcromol.* **1,** 97 (1968).
12. F. J. Golemba and J. E. Guillet, *SPE J.* **26,** 88 (1970).
13. A. P. Pivovarov, Yu. A. Ershov, and A. F. Lukonikov, *Soviet Plastics (Engl. Transl.)* **10,** 11 (1967).
14. A. J. Fry, R. S. H. Liu, and G. S. Hammond, *J. Am. Chem. Soc.* **88,** 4781 (1966).
15. R. P. Foss, D. O. Cowan, and G. S. Hammond, *J. Phys. Chem.* **68,** 3747 (1964).
16. P. J. Briggs and J. F. McKellar, *Chem. Ind. (London)* **1967,** 622.
17. P. J. Briggs and J. F. McKellar, *J. Appl. Polymer Sci.* **12,** 1825 (1968).
18. G. R. Lappin and J. W. Tamblyn (to Eastman Kodak Co.), U.S. Pat. 2,861,053 (1958).
19. M. Dubini, O. Cicchetti, G. P. Vicario, and E. Bua, *European Polymer J.* **3,** 473 (1967).
20. O. Cicchetti, M. Dubini, P. Parrini, G. P. Vicario, and E. Bua, *European Polymer J.* **4,** 419 (1968).
21. A. Marcincin, A. Pikler, and K. Ondrejmiska, *Plast. Hmoty Kauc.* **4,** 360 (1967).
22. J. E. Bonkowski, *Textile Res. J.* **39,** 243 (1969).
23. S. Tocker (to E. I. du Pont de Nemours and Co., Inc.), Belg. Pat. 629,109 (1963).
24. H. Heller, J. Rody, and E. Keller (to J. R. Geigy, S.A.), U.S. Pat. 3,399,173 (1968).
25. J. Fertig and M. Skoultchi (to National Starch and Chemical Corp.), U.S. Pat. 3,186,968 (1965).
26. R. L. Horton and H. G. Brooks (to American Cyanamid Co.), U.S. Pat. 3,365,421 (1968).
27. R. A. Coleman (to American Cyanamid Co.), U.S. Pat. 3,391,110 (1968).

28. I. Lukac, P. Hrdlovič, Z. Manasek, and D. Bellůs, *European Polymer J. Suppl.* **1969**, 523.
29. G. R. Lappin and W. V. McConnell, U.S. Patent Office, Defensive Publication 732,020 (1969).
30. L. W. A. Meyer and W. M. Gearhart, *Ind. Eng. Chem.* **37**, 232 (1945).
31. W. M. Gearhart, R. O. Hill, Jr., and M. H. Broyles (to Eastman Kodak Co.), U.S. Pat. 2,571,-703 (1951).
32. *Oil Paint Drug Reptr.* **175** (20), 43 (1959).
33. S. B. Miller, G. R. Lappin, and C. E. Tholstrup, *Mod. Plastics, Encyclopedia Issue,* Sept. 1969, p. 300.
34. D. J. Carlsson and D. M. Wiles, *Macromol.* **2,** 597 (1969).
35. C. R. Caryl and W. E. Helmick, U.S. Pat. 2,945,417 (1960).
36. L. R. Koller, *Ultraviolet Radiation,* John Wiley & Sons, Inc., New York, 1965, pp. 20–89.
37. M. G. Chen and W. L. Hawkins, *Preprints, Am. Chem. Soc., Div. Polymer Chem.* **9** (2), 1938v (1966).
38. Fox et al., *Adv. Chem.* **87,** 72 (1968).
39. R. C. Hirt and N. Z. Searle, *Appl. Polymer Symp.* **4,** 61–83 (1967).
40. J. C. W. Chien and W. P. Connor, *J. Am. Chem. Soc.* **90,** 1001 (1968).

General References

Polymer Photochemistry

J. E. Guillet, J. Dhanraj, F. J. Golemba, and G. Hartley, *Stabilization of Polymers and Stabilizer Processes,* No. 85 in *Advances in Chemistry* Series, American Chemical Society, Washington, D.C., 1968, pp. 272–286.

Degradation and Stabilization Mechanisms in Polyolefins

O. Cicchetti, *Advan. Polymer Sci.* **7,** 70 (1970).
G. Scott, *Atmospheric Oxidation and Antioxidants,* Elsevier Publishing Co., New York, 1965, pp. 170–222.

Test Methods, Polymer Degradation and Stabilization Mechanisms, Practical Stabilization Problems

M. R. Kamal, ed., *Weatherability of Plastic Materials,* Interscience Publishers, Inc., a division of John Wiley & Sons, Inc., New York, 1967.

Ultraviolet Sources, Solar Radiation

R. E. Barker, Jr., *Photochem. Photobiol.* **7,** 275 (1968).
L. R. Keller, *Ultraviolet Radiation,* John Wiley & Sons, Inc., New York, 1965, pp. 1–139.

Protection of Polymers Against Light

H. J. Heller, *European Polymer J. Suppl.,* pp. 105–132 (1969).

G. R. Lappin
Tennessee Eastman Company

ULTRAVIOLET SPECTROSCOPY. See ULTRAVIOLET-RADIATION ABSORBERS

UNPERTURBED CHAIN. See COLLOIDS

UNSATURATED ACIDS. See ACIDS AND DERIVATIVES, ALIPHATIC

UNZIPPING. See DEGRADATION; DEPOLYMERIZATION

UREA–ALDEHYDE RESINS. See AMINO RESINS

UREAS, POLYMERIC. See POLYUREAS

URETHAN POLYMERS. See POLYURETHANS

V

VACUUM BAG MOLDING. See BAG MOLDING

VACUUM FORMING. See THERMOFORMING

VACUUM METALLIZING. See METALLIZING

VAPOR BARRIERS. See BARRIERS, VAPOR

VARNISH. See SURFACE COATINGS; DRYING OILS

VEGETABLE FIBERS. See FIBERS, VEGETABLE

VINAL FIBERS. See VINYL ALCOHOL POLYMERS

VINYL ACETAL POLYMERS. See VINYL ALCOHOL POLYMERS

VINYL ACETATE–ETHYLENE COPOLYMERS. See Polar Copolymers under ETHYLENE POLYMERS

VINYL ACETATE POLYMERS. See VINYL ESTER POLYMERS, Vol. 15

VINYL ACETATE–VINYL CHLORIDE COPOLYMERS. See VINYL CHLORIDE POLYMERS

VINYL ALCOHOL POLYMERS

POLY(VINYL ALCOHOL)

When a poly(vinyl ester), such as poly(vinyl acetate), is hydrolyzed, poly(vinyl alcohol) is obtained. This is, at present, the only industrial method of preparing this polymer. The monomer vinyl alcohol evidently does not exist in the free state, although traces have been detected indirectly (17). Whenever it is formed, vinyl alcohol rearranges to give its tautomer, acetaldehyde (eq. 1).

$$CH_2{=}CHOH \rightarrow CH_3CHO \qquad (1)$$

There is evidence that acetaldehyde contains a small amount of vinyl alcohol, especially in the presence of enolizing agents (18); however, under ordinary conditions the amount of alcohol is below the limits of analytical detection (19). On the other hand, metal derivatives of vinyl alcohol have been synthesized. For example, sodium vinylate was made from mercury bisacetaldehyde in liquid ammonia (20). Substituted vinyl alcohols such as mesityl-2-phenylvinyl alcohol are stable (21).

Manufacture

Poly(vinyl alcohol) was first made by W. O. Herrmann and W. Haehnel in 1924 (22). They hydrolyzed poly(vinyl acetate) in ethanol with potassium hydroxide and obtained a slightly yellow powder, which they identified as poly(vinyl alcohol) by analysis and by subsequent reaction with an acid anhydride to give a polymeric vinyl ester. The structure of poly(vinyl alcohol) as a poly(1,3-glycol) was elucidated by Staudinger, Frey, and Starck (23). Concurrently, Herrmann and Haehnel described the laboratory preparation by alkaline or acid hydrolysis as well as some reactions of poly(vinyl alcohol). They also suggested applications such as its use as a protective colloid to prepare stable gold or ferric hydroxide suspensions (24).

Effect of Poly(vinyl Acetate) Structure. The physical and chemical properties of poly(vinyl alcohol) are to a large extent dependent on the poly(vinyl acetate) from which the poly(vinyl alcohol) is made (see also VINYL ESTER POLYMERS); they are also influenced by the method of hydrolysis (25). By varying the polymerization conditions of vinyl acetate, such as the temperature of polymerization, the degree of conversion of monomer to polymer, the presence of solvents and modifiers during the polymerization, and the amount and kind of initiator, completely different vinyl alcohol polymers can be obtained.

When vinyl acetate is polymerized to a high degree of conversion, as is done in a suspension polymerization, the poly(vinyl acetate) is highly branched and the degree of polymerization high (Fig. 1). The average degree of polymerization of the resulting poly(vinyl alcohol), however, remains relatively constant regardless of the conversion of vinyl acetate to poly(vinyl acetate), as can be seen in Figure 1. However, high-conversion poly(vinyl acetate) yields a poly(vinyl alcohol) with a wide molecular-weight distribution, because during the hydrolysis many branches, both long and short, are separated from the main chain (26a). The presence of aldehydes in the vinyl acetate monomer, whether they are present as impurities or added as modifiers, results in aldehyde end groups (27) in the poly(vinyl alcohol); these can react with the hydroxyl group of the poly(vinyl alcohol) and lead to crosslinking and, therefore, insolubility of the polymer. The use of an initiator such as oleyl peroxide, for example, as the initiator of the vinyl acetate polymerization introduces long alkyl chains as end groups in the resulting poly(vinyl alcohol). This changes the colloidal properties of aqueous solutions of poly(vinyl alcohol) markedly (28).

The frequency of head-to-head and tail-to-tail polymerization of vinyl acetate is temperature-dependent (29). On hydrolysis these groupings lead to 1,2-glycol structures rather than 1,3-glycols (eq. 2).

$$\sim CH_2-\underset{OAc}{CH}-CH_2-\underset{OAc}{CH}-\underset{OAc}{CH}-CH_2-CH_2-\underset{OAc}{CH}-CH_2-\underset{OAc}{CH}\sim \longrightarrow$$

$$\sim CH_2-\underset{OH}{CH}-CH_2-\underset{OH}{CH}-\underset{OH}{CH}-CH_2-CH_2-\underset{OH}{CH}-CH_2-\underset{OH}{CH}\sim \qquad (2)$$

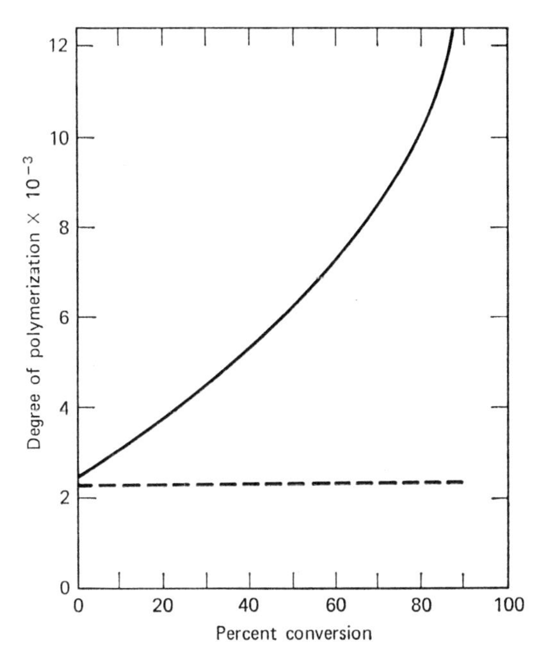

Fig. 1. Change of degree of polymerization of poly(vinyl acetate) (solid line) and of poly(vinyl alcohol) (dashed line) with percent conversion to polymer (26).

As discussed subsequently in this article, mechanical properties as related to the degree of crystallinity and chemical properties of the poly(vinyl alcohol) are somewhat dependent on the number of 1,2-glycol units.

Production Methods. There are two generally important methods for the production of poly(vinyl alcohol). The first is *hydrolysis* (eq. 3), in which a stoichiometric quantity of base and the presence of water are necessary; the second is *ester interchange* (eq. 4), in which only a catalytic amount of base and the presence of an

Direct hydrolysis

$$\sim\!\!\sim CH_2-\underset{OAc}{\underset{|}{CH}}-CH_2-\underset{OAc}{\underset{|}{CH}}-CH_2-\underset{OAc}{\underset{|}{CH}}-\sim\!\!\sim \xrightarrow[H_2O]{NaOH} \sim\!\!\sim CH_2-\underset{OH}{\underset{|}{CH}}-CH_2-\underset{OH}{\underset{|}{CH}}-CH_2-\underset{OH}{\underset{|}{CH}}-\sim\!\!\sim + NaOAc \quad (3)$$

Ester interchange

$$\sim\!\!\sim CH_2-\underset{OAc}{\underset{|}{CH}}-CH_2-\underset{OAc}{\underset{|}{CH}}-CH_2-\underset{OAc}{\underset{|}{CH}}-\sim\!\!\sim + \xrightarrow[CH_3OH]{\text{catalytic } NaOH} \sim\!\!\sim CH_2-\underset{OH}{\underset{|}{CH}}-CH_2-\underset{OH}{\underset{|}{CH}}-CH_2-\underset{OH}{\underset{|}{CH}}-\sim\!\!\sim + CH_3OAc \quad (4)$$

alcohol are required. The latter process has been largely used industrially since its discovery in 1932 by Herrmann, Haehnel, and Berg (30,31).

Poly(vinyl alcohol) is produced by both batch and continuous processes. Ester interchange is conducted with either an alkaline or an acidic catalyst with methanol or ethanol present. The alkaline interchange process is rather sensitive to the presence of water, whereas the acidic process is not.

During the ester interchange reaction two phase changes occur: the first of these is the conversion of the polymer solution into a gel, which may be soft or firm depending on the concentration of the poly(vinyl acetate) solution; the second phase change

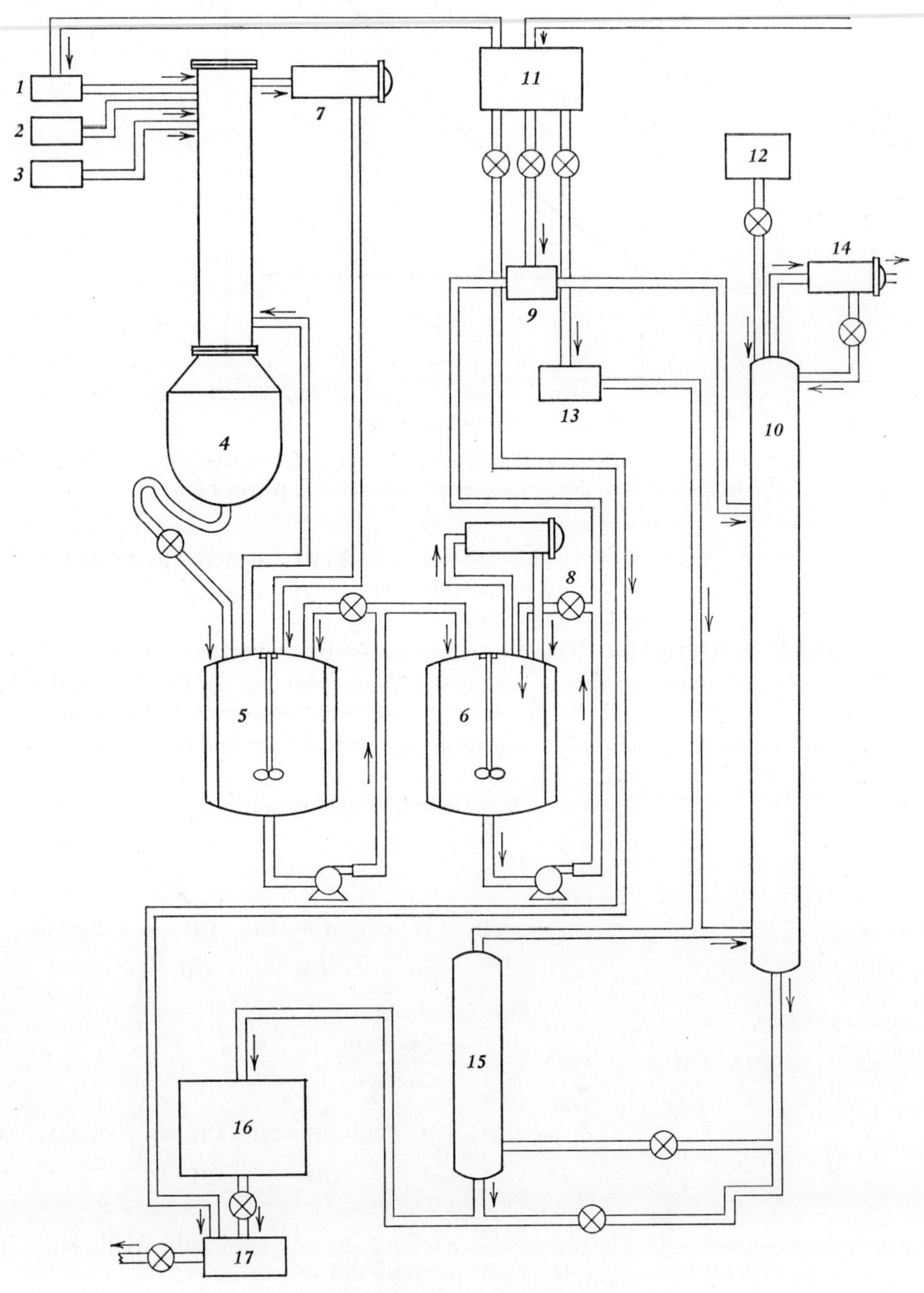

Fig. 2 (*caption on next page*)

occurs when this gel shrinks and solvent is exuded. The solvent, which is a mixture of methanol and methyl acetate, can be removed by distillation; the second-stage gel is filtered and dried after it has been washed by more solvent.

Continuous Process. A continuous process using a *screw conveyor* (32–34) for the manufacture of poly(vinyl alcohol) has been described. In one such process (32,33) (see Figs. 2 and 3) methanol, vinyl acetate, and initiator (α,α'-azobisisobutyronitrile) are fed from reservoirs *1*, *2*, and *3*, respectively, to a preheater *4* and from there to a polymerization vessel *5* equipped with reflux condenser *7*. After 25% conversion of monomer to polymer has been reached, the mixture is fed to a polymerization vessel *6* equipped with reflux condenser *8*, where the conversion reaches 50%. The total polymerization time is about 3½ hours. The mixture consisting of methanol, vinyl acetate, and poly(vinyl acetate) is then fed to the mixing chamber *9* where more methanol is added from reservoir *11*. This dilute solution is then passed into the rectification column *10*. Water is added here from reservoir *12*, to aid in the separation of vinyl acetate monomer, together with methanol vapors introduced at the bottom of column *10* from methanol vaporizers *13* and *15*. Vinyl acetate monomer together with methanol is condensed at *14* and recycled to the polymerization after analysis. The poly(vinyl acetate) "paste" is withdrawn from column *10* and stored in storage vessel *16* from where it is pumped to the dilution vessel *17* together with more methanol. The paste is then introduced to the mixing vessel *18* together with sodium hydroxide solution from vessel *19*. The mixture is then fed to the saponifier *20* which has a screw for both mixing and conveying. After the saponification step, the gelatinous mass is crushed and ground in the crusher *21* and grinder *22*, respectively, and then slurried in vessel *23*. The slurry is pumped to filter *24* where the filter cake is washed with methanol from reservoir *25*. The solids of the washed filter cake are then concentrated in the screw press *26*, ground in grinder *27*, and dried in a rotary drum dryer *28*, after which it is stored for bagging in bin *30*.

A similar process has been described in an article that also treats the kinetics of polymerization (34). It is apparent that changes in the polymerization conditions, such as changes in the vinyl acetate–methanol ratio or the degree of conversion, result in different molecular weights of poly(vinyl alcohol). The advantages of this process are the short reaction time, and the hard poly(vinyl alcohol) particles obtained, which can readily be dissolved in water.

Fig. 2. Vinyl acetate polymerization system (32). Legend:

1. methanol reservoir
2. vinyl acetate reservoir
3. initiator reservoir
4. preheater
5. polymerization kettle *1*
6. polymerization kettle *2*
7. reflux condenser for kettle *1*
8. reflux condenser for kettle *2*
9. mixing chamber
10. rectification column
11. methanol reservoir
12. water reservoir
13. methanol vaporizer
14. vinyl acetate–methanol condenser
15. methanol vaporizer
16. poly(vinyl acetate) paste storage vessel
17. paste dilution vessel
18. mixing vessel
19. sodium hydroxide storage vessel
20. saponifier
21. crusher
22. grinder
23. slurry tank
24. vacuum filter
25. methanol reservoir
26. screw press
27. grinder
28. rotary drum dryer
29. methanol condenser
30. poly(vinyl alcohol) storage

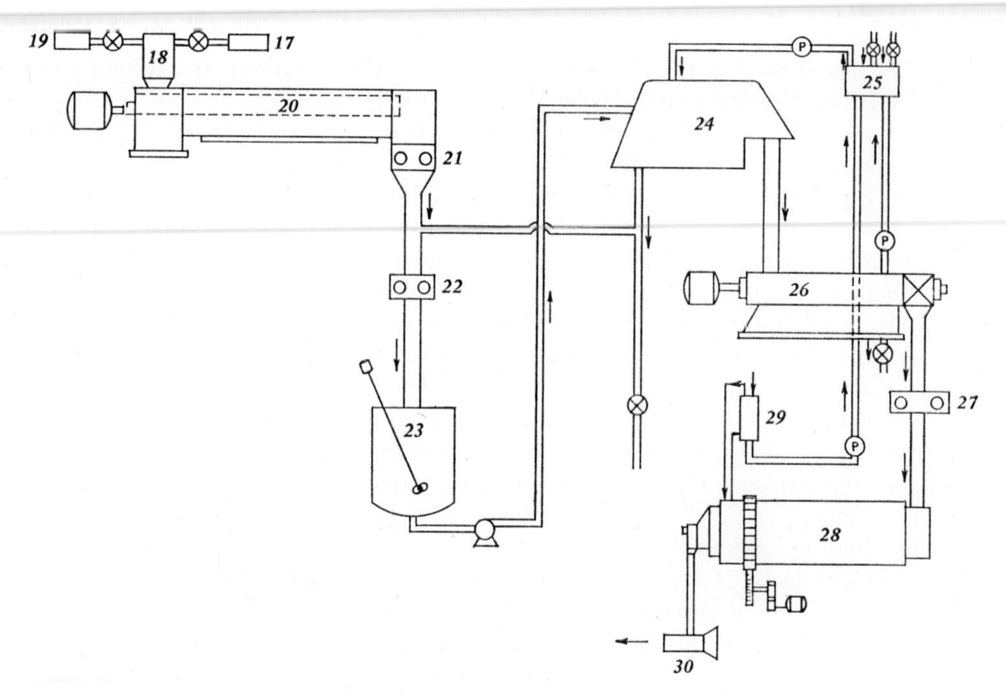

Fig. 3. Poly(vinyl acetate) saponification system (32). Legend is given in caption for Figure 2.

The continuous process may also use a *cascade system* (35–37). Such a system of continuous ester interchange is described as follows (35): a 30% solution of poly(vinyl acetate) in methanol is mixed with a solution of sodium methylate and fed into a rapidly stirred reaction vessel. The mixture is kept at reflux temperature and a certain amount of methanol and methyl acetate distilled off. This shifts the equilibrium toward the formation of poly(vinyl alcohol). The reaction is completed in a second and third reactor. The product precipitates without forming a gel, which is difficult to process. The particle size of the poly(vinyl alcohol) in this process tends to be very fine, resulting in a very dusty end product. This process has been modified to obtain larger particle sizes, which will not form dust (36,37). The resulting slurry is filtered, and the product washed and dried in a similar fashion to the process described previously.

A continuous process using a *belt saponifier* (38,39) is said to be operated by the Shawinigan Chemicals Company. Poly(vinyl acetate) beads are dissolved in methanol or the poly(vinyl acetate) is made similar to the continuous Kurashiki polymerization process (see Fig. 2). A 30% solution of poly(vinyl acetate) in methanol is fed to a mixing chamber together with a solution of sodium hydroxide (see Fig. 4). After thorough mixing, the solution is discharged onto a moving belt where it remains until the onset of syneresis. The product is then neutralized, ground, washed, and dried in the usual manner. The product has a high bulk density and can easily be dissolved in water. Poly(vinyl alcohols) of varying degrees of hydrolysis can be made by this process. This process was evolved from early work by Blaikie and Crozier (40).

Batch Process Using a Kneader (41,42). This process is the oldest and is still in operation in Europe at Wacker-Chemie. The partially hydrolyzed poly(vinyl alcohol) produced by this method is most useful as the protective colloid for vinyl acetate

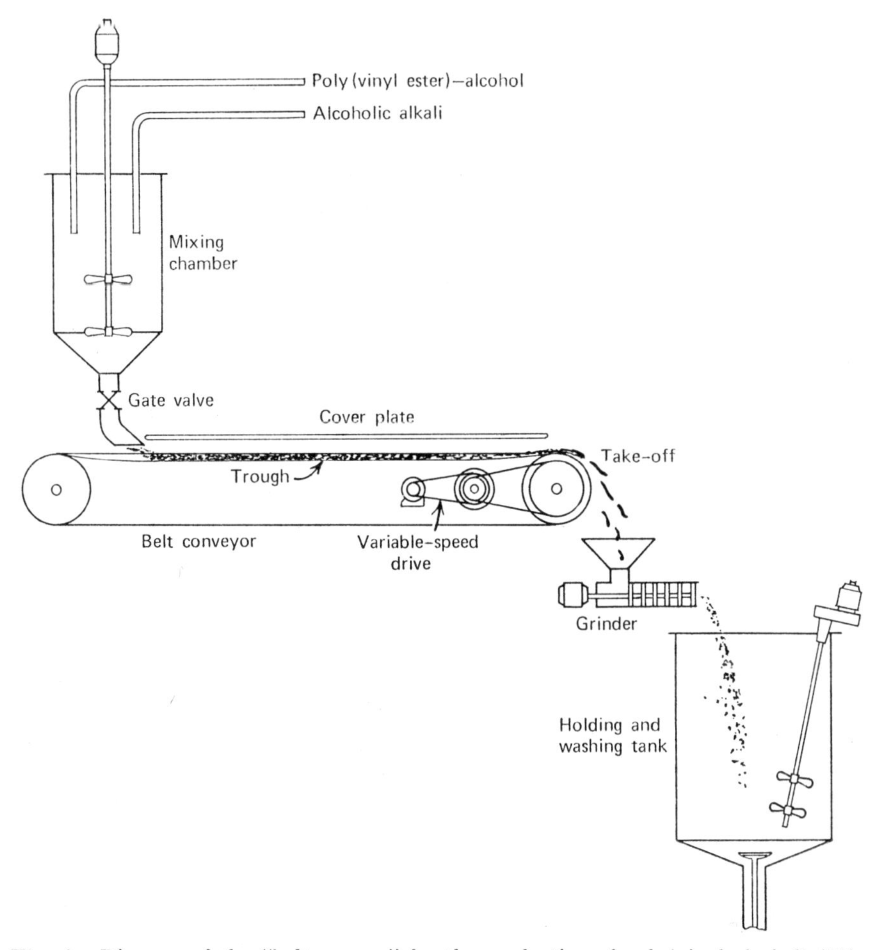

Fig. 4. Diagram of the "belt process" for the production of poly(vinyl alcohol) (38).

emulsion polymers, because of its superior emulsifying properties. A schematic diagram of the kneader process is shown in Figure 5 (42).

The poly(vinyl acetate) is charged together with methanol to the kneader *1*, where it is dissolved. Then a solution of sodium methylate in methanol is added. The quantity of sodium methylate used depends upon the degree of hydrolysis desired. The mixture is kneaded at room temperature for four hours, after which the methyl acetate and methanol are distilled off. The poly(vinyl alcohol) is obtained fairly dry.

Acidolysis. Poly(vinyl alcohol) can also be made by an acidolysis process. This can involve both an ester interchange, when working in an anhydrous system (43), and a hydrolysis when a poly(vinyl acetate) emulsion (44) is used as the starting material. Usually sulfuric acid or any other strong mineral acid is employed as the catalyst. This process is used industrially only to a very limited extent.

Other Methods. A number of other methods for the preparation of poly(vinyl alcohol) have been published in the literature. These employ starting materials other than poly(vinyl acetate). Often the poly(vinyl alcohols) made by these methods have specific properties, such as tacticity, better cold-water solubility, which may also be

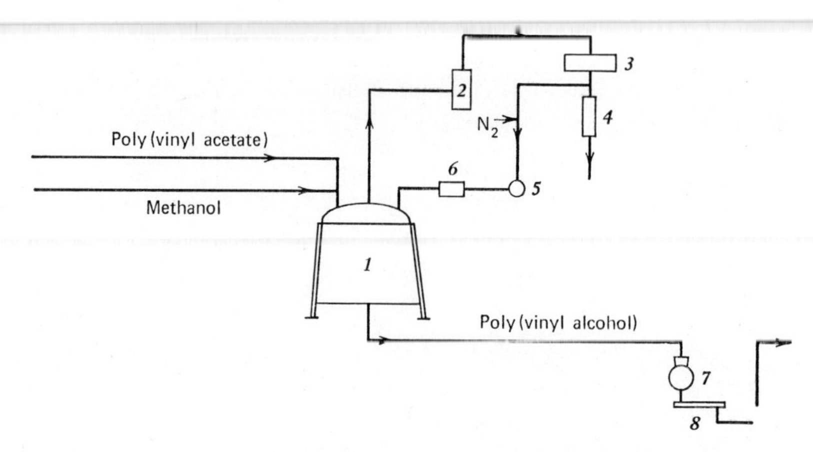

Fig. 5. Kneader process for the production of poly(vinyl alcohol) (42). Legend:

1. kneader
2. dust separator
3. condenser
4. cooler
5. blower
6. heater
7. mill
8. sieve

related to a higher or lower content of 1,2-glycol groupings, or high viscosity, or they are copolymers of vinyl alcohol with other monomers.

Stereoregular poly(vinyl alcohol) can be prepared by hydrolyzing a stereoregular poly(vinyl trifluoroacetate) (45), poly(vinyl monochloroacetate) (46), or poly(vinyl formate) (47). Syndiotactic and isotactic poly(vinyl alcohols) have also been prepared starting with poly(vinyl trimethyl silyl ether) (48), poly(vinyl *t*-butyl ether) (49,50), or poly(vinyl benzyl ether) (51).

A recent patent describes the preparation of poly(vinyl alcohol) by saponification of poly(divinyl carbonate) (52) which is very high in 1,3-glycol content. The structure of poly(vinyl alcohol) derived from poly(divinyl carbonate) has been studied (53). Much effort has been expended in trying to prepare poly(vinyl alcohol) by polymerizing acetaldehyde or metal vinylates. Although a number of papers and patents have been published on this subject (54–57), only polymers of low molecular weight and questionable structures have been described.

A poly(vinyl alcohol) with a completely 1,2-glycol structure can be made by hydrolyzing poly(vinylene carbonate) (58).

The synthesis of poly(vinyl alcohol) has been reviewed recently with emphasis on the resultant structure (59).

Physical Properties

The physical, as well as chemical, properties of poly(vinyl alcohol) depend to a greater extent on the method of preparation than is ordinarily the case with other polymers. By suitable choice not only of the polymerization conditions of the parent poly(vinyl acetate) but also of the conditions of hydrolysis, the physical and chemical properties of poly(vinyl alcohol) can be widely varied.

Commercial poly(vinyl alcohols) are white to yellow in color, and are powdery or granular. There are a number of grades available, which can be divided into two main types: the so-called fully hydrolyzed and the partially hydrolyzed poly(vinyl alcohols). The fully hydrolyzed poly(vinyl alcohols), as the name implies, have less than 1.5

Table 1. Commercially Available Poly(vinyl Alcohols)

Elvanol (Du Pont)	Vinol (Airco)	Gelvatol (Shawinigan)	Poval (Kurashiki)	Polyviol (Wacker)	Moviol (Hoechst)	Goshenol (Nippon Goshei)	Lemol (Borden)	Viscosity (4% solution), cP	Hydrolysis, mole %	Degree of polymerization
70–05		1–30	PVA–105	M 05/20	N30–98	NL–05	5–98	5–6	99	500–600
71–30	125	1–60	PVA–117	W 28/10	N75–99	NM–14	24–98	25–31	99	1700–1800
72–60	165	1–90	PVA–124		N90–98	NH–26	60–98	55–67	99	2400–2500
71–24	325	3–60	PVA–120	W 28/20	N80–99	NH–17	30–98	28–32	98	1700–1800
	350	3–90		W 48/20		NH–22	51–98GF	55–65	98	2400–2500
51–03			PVA–204	V 03/140	N30–88	GL–03		3–4	87–89	350–450
51–05	205K	20–30	PVA–205	M 05/140	N50–88	GL–05	5–88	5–6	87–89	500–600
52–22	523	20–60	PVA–217	W 25/140	N75–88	GH–17	22–88	19–23	87–89	1700–1800
50–42	540	20–90	PVA–224	W 40/140	N85–88	GH–20	42–88	39–47	87–89	2400–2500
			PVA–217EE					22–28	87–89	1700–1800
			PVA–224E					45–55	87–89	2400–2500
46–22			PVA–417	W 25/240			30–77	16–21	79–83	1700–1800
			PVA–420			KH–17		28–34	79–83	2000–2100
		0–60G	PVA–H		N90–99V			26–32	99.8	1700–1800

mole % acetate groups left in the molecule, whereas the partially hydrolyzed grades still contain as much as 20 mole % residual acetate groups. In both of these categories a number of polymers of different molecular-weight ranges are available. Recently, some copolymers of poly(vinyl alcohol) have appeared on the market, mainly imported from Japan.

Table 1 lists a number of commercially available poly(vinyl alcohols) arranged by their degrees of polymerization and their viscosities in 4% aqueous solution. The degrees of hydrolysis are also noted. The solution viscosity of poly(vinyl alcohol) in water is dependent on the average degree of polymerization (or molecular weight), as

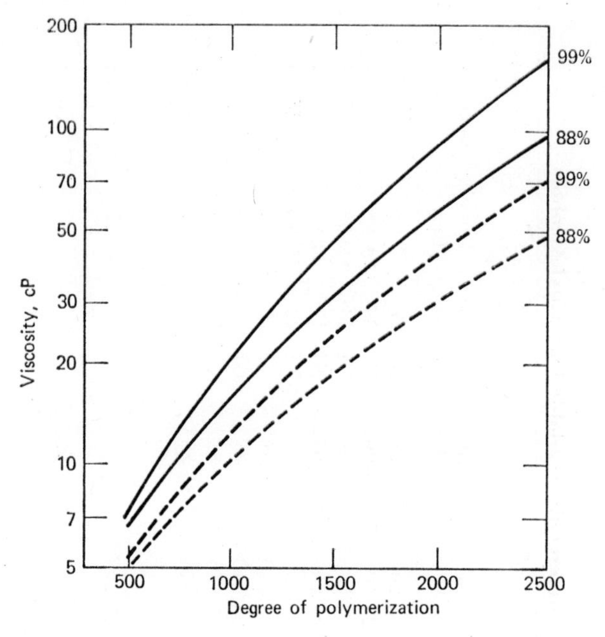

Fig. 6. Dependence of the solution viscosity of poly(vinyl alcohol) on the degree of polymerization (measured with Brookfield viscometer) (15). Percent of hydrolysis is as indicated. Solid line, 5% solution; dashed line, 4% solution.

shown in Figure 6. At the same degree of polymerization, the partially hydrolyzed poly(vinyl alcohols) are less viscous than are the fully hydrolyzed polymers.

Solid Poly(vinyl Alcohol). Because the name poly(vinyl alcohol) does not denote a definite compound, it is difficult to assign absolute physical properties to the solid poly(vinyl alcohol). The physical properties of each grade of poly(vinyl alcohol) are dependent on the degree of hydrolysis, the water content, and the molecular weight.

Table 2 lists some representative properties of a fully hydrolyzed poly(vinyl alcohol). The properties of poly(vinyl alcohol) film are discussed in the section on Films.

Melting Point. The melting point of poly(vinyl alcohol) is difficult to measure because there is no sharp melting point but rather a range from about 220 to 240°C. Melting of the crystalline regions occurs at about 220°C (189a). Measurement of the melting points of poly(vinyl alcohol) plasticized with glycerol and of partially hydrolyzed poly(vinyl alcohol) indicates that the melting point of pure poly(vinyl

Table 2. Physical Properties of Fully Hydrolyzed Poly(vinyl Alcohol)

Property	Data	Remarks
form	granular powder	
color	white to light straw	
specific gravity	1.19–1.31	
bulk density, lb/ft^3	30–40	
tensile strength, lb/in.^{2}	up to 22,000	at 50% relative humidity
refractive index	1.49–1.53	
clongation, %		
unplasticized film	up to 300	at 50% relative humidity
plasticized film	up to 600	
thermal coefficient of linear expansion, 0–50°C, plasticized	1×10^{-4}/°C	an average value; value depends upon grade and plasticizer content
specific heat, cal/g/°C	0.4	
hardness, I.C.I. automatic Sward hardness rocker, %	27–57	compared to glass, at 50% relative humidity
abrasion resistance, unplasticized	good to excellent	proportional to molecular weight
glass-transition temperature, °C	85	
heat-sealing temperature, °C	165–210	unplasticized, dry
	110–150	unplasticized, 50% relative humidity
melting point (decomposition), °C	228	
heat stability	slow degradation over 100°C; rapid degradation and decomposition over 200°C	the resin becomes progressively more water resistant as temperature is increased
compression-molding temperature, °C	120–150	plasticized
effect of light	negligible	degradation occurs under ultraviolet radiation
effect of strong acids	dissolves and/or decomposes	
effect of strong alkalies	softens or dissolves	
effect of weak acids	softens or dissolves	
effect of weak alkalies	softens or dissolves	
effect of organic solvents	negligible	
burning rate	slow (like paper)	
mold resistance	good	
gas impermeabiity	very high for most gases, low for oxygen	depends on moisture

alcohol) is 228°C, and that it decreases linearly to about 175°C for poly(vinyl alcohol)–glycerol mixture 50–50 by volume (190). For a partially hydrolyzed poly(vinyl alcohol) with about 7 mole % residual acetyl groups the melting point was found to be 170°C.

Glass-Transition Temperature. The glass-transition temperature of poly(vinyl alcohol) has been determined as 85°C. However, many studies found a transition also at 70°C (191). This is probably due to different heating rates and other experimental problems. Figure 7 (193) shows a plot of the glass-transition temperature of water containing poly(vinyl alcohols). Heat treatment of poly(vinyl alcohol) changes T_g from 73.5°C for the undrawn fiber to 87°C for the drawn fiber heat treated at 150°C (192).

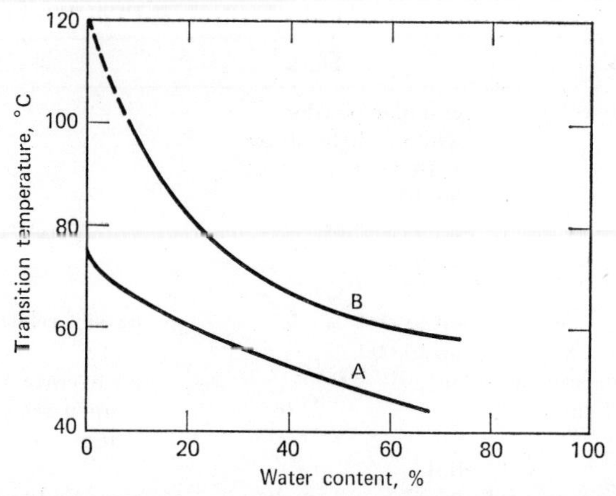

Fig. 7. Relationships between transition temperatures of poly(vinyl alcohol)–water system and water content (193). Key: curve A, glass-transition temperature, T_g; curve B, another transition temperature, T_g'.

Other Properties. The latent heat of fusion of poly(vinyl alcohol) has been determined as 1.64 ± 0.10 kcal/mole (190). The entropy of fusion is 3.3 cal/deg-mole (194). The heat capacity of poly(vinyl alcohol) is 0.00105 kcal/deg-g (195). The dielectric constant of dry poly(vinyl alcohol) at room temperature is about 40.

Mechanical Properties. The mechanical properties of poly(vinyl alcohol) are discussed in the sections on Fibers and on Films.

Table 3. Some Properties of Glycols Used as Plasticizers for Poly(vinyl Alcohol) (182)

Glycol	Structure	Melting point, °C	Boiling point, °C	Dissolving temperature of poly(vinyl alcohol), °C	Cloud point of poly(vinyl alcohol) soln,[a] °C
ethylene glycol	HO—$(CH_2)_2$—OH	−12	197	140	110
trimethylene glycol	HO—$(CH_2)_3$—OH	liquid	214	160	130
tetramethylene glycol	HO—$(CH_2)_4$—OH	195	235	<200	150
pentamethylene glycol	HO—$(CH_2)_5$—OH	liquid	239	190	175
hexamethylene glycol	HO—$(CH_2)_6$—OH	42	250	240	190
propylene glycol	CH_3—CH(OH)—CH_2—OH	liquid	187	190	150
glycerol	CH_2(OH)—CH(OH)—CH_2(OH)	19	290	160	120
2,3-butanediol	CH_3—CH(OH)—CH(OH)—CH_3	34.4	184	240	175
1,3-butanediol	CH_2(OH)—CH_2—CH(OH)—CH_3	liquid	204	185	170
diethylene glycol	HO—$(CH_2CH_2O)_2$—H	−10.5	245	210	160
triethylene glycol	HO—$(CH_2CH_2O)_3$—H	liquid	278	210	185

[a] 1% solution of poly(vinyl alcohol) in glycol.

Table 4. Mechanical Properties of Plasticized Poly(vinyl Alcohol) Films (185)[a]

Plasticizer	20 PHR[b]			40 PHR[b]			Supplier[c]
	Elongation, %	Stress at rupture, psi	Appearance	Elongation, %	Stress at rupture, psi	Appearance	
none	85	6400	clear				
glycerol	170	2000	clear			tacky, limp	
2,3-butylene glycol	260	3050	clear	280	2400	clear	Celanese Corporation of America
							Antara Chemicals
sorbitol	200	4000	clear	250	2860	clear, slightly greasy	Atlas Powder Co., Merck Chemical Division
GS-15	190	3280	clear	170	1600	clear	Nopco Chemical Co.
methylolated cyclic ethyleneurea	180	4800	clear	225	2000	clear	Monsanto Chemical Co. (Resloom E-50)
2,2-dimethyl-1,3-butanediol	250	4050	clear	280	2600	clear	Union Carbide Chemicals
Hyprin GP-25	180	2000	clear	290	1400	clear, slight tack	Dow Chemical Co.
Hyprose SP80	115	4700	clear	250	3900	clear	Dow Chemical Co.
diglycerol	220	3000	clear	235	1350	clear, tacky	Colgate-Palmolive
Stysolac AW	110	3000	hazy	150	2000	hazy, cratered	F. H. Paul & Stein Bros. (New York)
Vircol-189	200	4500	slightly hazy	incompatible			Mobil Chemical Co.

[a] Tested on Instron machine. Cross-head speed 5 in./min after 7 days conditioning at 50% relative humidity and 23°C.

[b] PHR is parts of plasticizer per 100 parts of Gelvatol 20–30.

[c] Of plasticizer.

Effect of Plasticizers. Glycols are the most common plasticizers used for poly(vinyl alcohol) (Table 3). Table 4 lists a number of other commercial plasticizers together with the mechanical properties of the plasticized poly(vinyl alcohol) films. In general, the more hydrophobic the plasticizer the less it is compatible with poly(vinyl alcohol). For example, Vircol 189, an ethoxylated phosphoric acid butyl ester, is compatible with partially hydrolyzed poly(vinyl alcohol) up to 20 parts per 100 parts of resin; however, it is not at all compatible with fully hydrolyzed poly(vinyl alcohol).

Other plasticizers which have been proposed from time to time include ethyl acid phthalate (183), tris(tetrahydrofurfuryl) phosphate (184), methylolated cyclic ethylene urea (Reslom E 50, Monsanto Co.) (185), monophenylether of poly(oxyethylene) (Pycal 94, Atlas Chemical Corp.) (186,187), and tris(chloroethyl) phosphate (188).

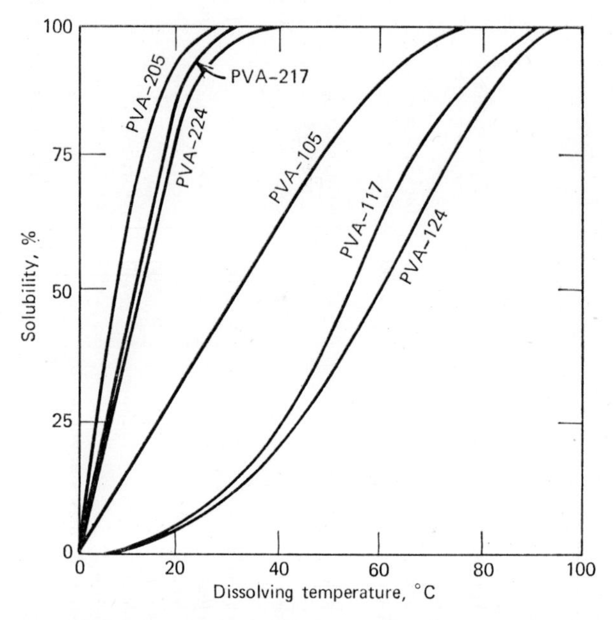

Fig. 8. Solubility of various grades of poly(vinyl alcohol) (Poval) in water (15). Solubilities are in percent of dissolved poly(vinyl alcohol) after 30 min with agitation at 2000 rpm when 4 parts of poly(vinyl alcohol) are added to 96 parts of water. Hydrolysis and degrees of polymerization are listed in Table 1.

Poly(vinyl alcohol) can also be internally plasticized, mainly by copolymerization of the parent vinyl acetate with ethylene, vinyl butyl ether, or other nonhydrolyzable comonomer and subsequent hydrolysis of the poly(vinyl acetate). Usually the water solubility of internally plasticized poly(vinyl alcohol) decreases rapidly.

Poly(vinyl alcohol) has also been treated with ethylene oxide (see under Chemical Reactions), which appeared to result in internal plasticization (189b). However, the main effect of this treatment seems to have been the binding of water to the poly(vinyl alcohol), where the bound water then is the effective plasticizer. See also PLASTICIZERS.

Solutions. The *solubility* of poly(vinyl alcohol) in water depends on the degree of hydrolysis and molecular weight, as shown in Figure 8. Fully hydrolyzed poly(vinyl alcohols) are soluble only in hot to boiling water, whereas partially hydrolyzed grades (88%) are soluble at room temperature. Poly(vinyl alcohol) with a degree of hydrolysis of 80% is soluble only at water temperatures of 10–40°C. Above 40°C the solution first becomes cloudy (the so-called cloud point) and then the poly(vinyl alcohol) precipitates.

The many hydroxyl groups of poly(vinyl alcohol) contribute to strong hydrogen bonding both intra- and intermolecularly, which greatly hinders its solubility in water. Residual acetate groups in the partially hydrolyzed poly(vinyl alcohols) weaken these hydrogen bonds so that the partially hydrolyzed poly(vinyl alcohols) are more soluble. Being hydrophobic, the acetate groups contribute to a negative heat of solution. With increasing amounts of acetate groups, the solubility in water of the partially hydrolyzed polymer decreases. Poly(vinyl alcohols) with 30 mole % acetate groups (50 weight %) are soluble only in a water–alcohol mixture.

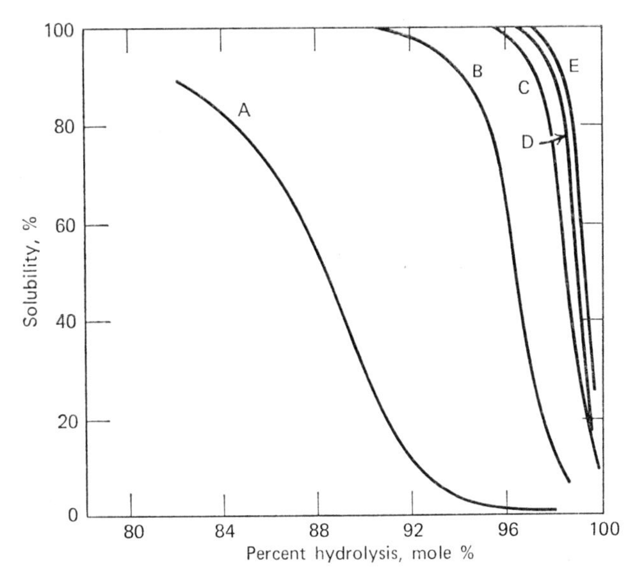

Fig. 9. Solubility in water of heat-treated poly(vinyl alcohol) (Poval) (15). Degree of polymerization, 1750; dissolving conditions, 40°C, 30 min. Key: curve A, heat treatment for 60 min at 180°C; curve B, for 10 min at 180°C; curve C, for 60 min at 100°C; curve D, for 10 min at 100°C; curve E, untreated.

Heat treatment of the poly(vinyl alcohol) increases the crystallinity and decreases the solubility. Heat treatment is, of course, less effective for partially hydrolyzed grades of poly(vinyl alcohol) (Fig. 9). Even 2 mole % of acetate groups interferes with the crystallization process during heat treatment. This is important in certain applications such as textile warp sizing where the polymer must be removed after a drying cycle.

Figures 10 and 11 show the effects of concentration and temperature on the *viscosity* of poly(vinyl alcohol) solutions. Aqueous solutions of poly(vinyl alcohol) with a high degree of hydrolysis and high molecular-weight increase in viscosity on standing, and may even gel (60,61). This increase in viscosity is temperature-dependent, the lower the temperature the faster being the increase (Fig. 12). It is, of course, also concentration-dependent. A partially hydrolyzed poly(vinyl alcohol) is completely viscosity-stable. Certain inorganic salts, urea, or the lower aliphatic alcohols, when added to solutions of fully hydrolyzed poly(vinyl alcohols), act as viscosity stabilizers.

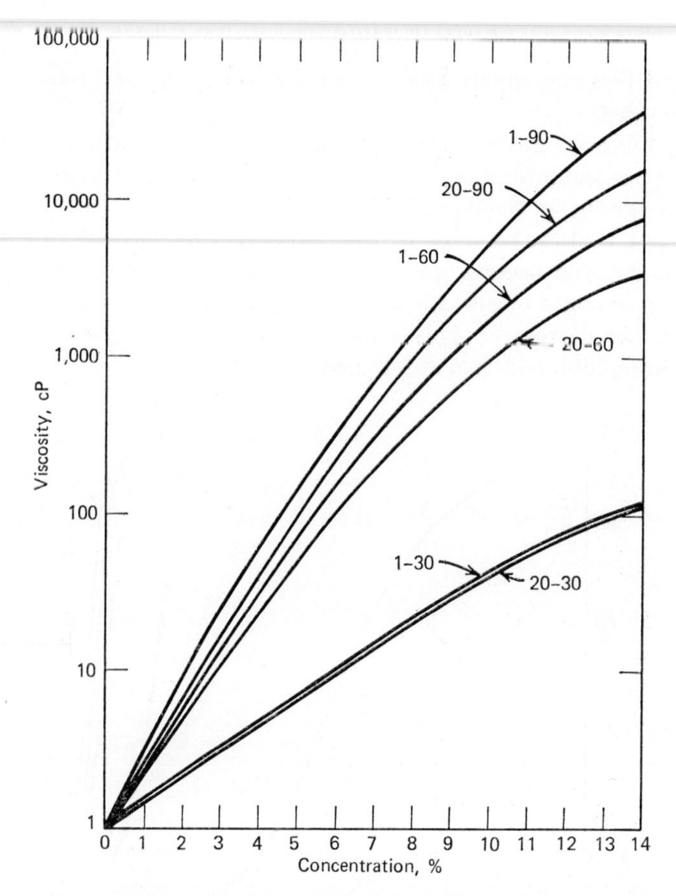

Fig. 10. Viscosity–concentration curves for various grades of poly(vinyl alcohol) (Gelvatol) at 20°C. Degrees of polymerization and percent hydrolysis are listed in Table 1.

Poly(vinyl alcohol) solutions are non-Newtonian; the shear effect increases with an increase in molecular weight, as would be expected. The effect is minimized with partially hydrolyzed poly(vinyl alcohols).

Most inorganic salts, especially sulfates and phosphates, act as *precipitants* when added to poly(vinyl alcohol) solutions (62). Table 5 lists the minimum salt concentration causing a 5% solution of a medium-viscosity, fully hydrolyzed poly(vinyl alcohol) to precipitate. Metal salts apparently affect the dissolved poly(vinyl alcohol) by increasing or decreasing intramolecular and intermolecular hydrogen bonds between hydroxyl groups (63). This was confirmed in the case of copper sulfate, in which case a complex between copper ion and poly(vinyl alcohol) is formed (64). Herrmann and Haehnel postulated very early that poly(vinyl alcohol) also forms a molecular compound with sodium hydroxide (65). This was studied by intrinsic viscosity measurements later, but could not be confirmed (66). Some detergents, notably anionics, form complexes with poly(vinyl alcohol) in solution (67) which behave like polyelectrolytes (qv). Boric acid and borax cause thickening by complex formation (see the section on Chemical Properties).

The *surface tension* of aqueous solutions of poly(vinyl alcohol) varies depending on concentration, temperature, and, most important, the degree and kind of hydrolysis.

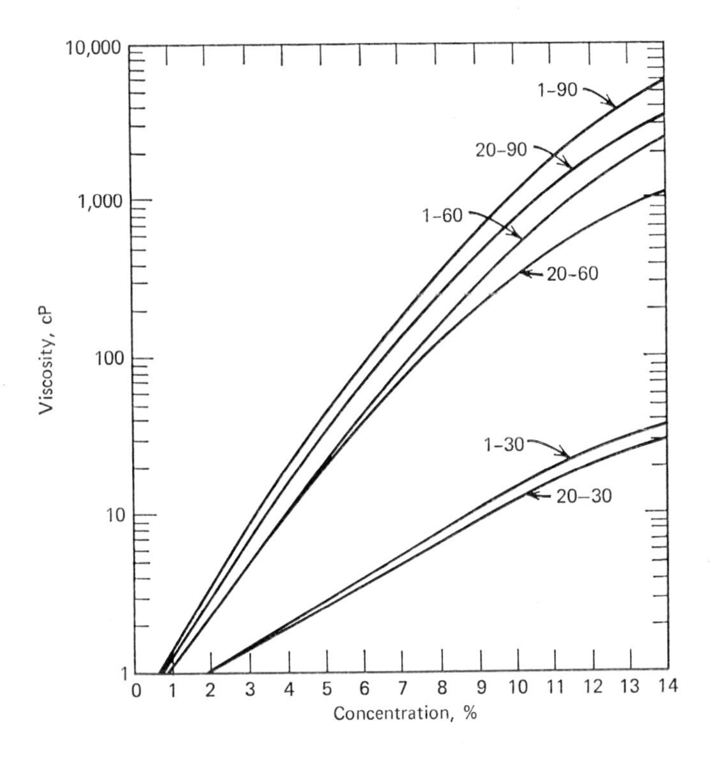

Fig. 11. Viscosity–concentration curves for various grades of poly(vinyl alcohol) (Gelvatol) at 60°C. Degrees of polymerization and percent hydrolysis are listed in Table 1.

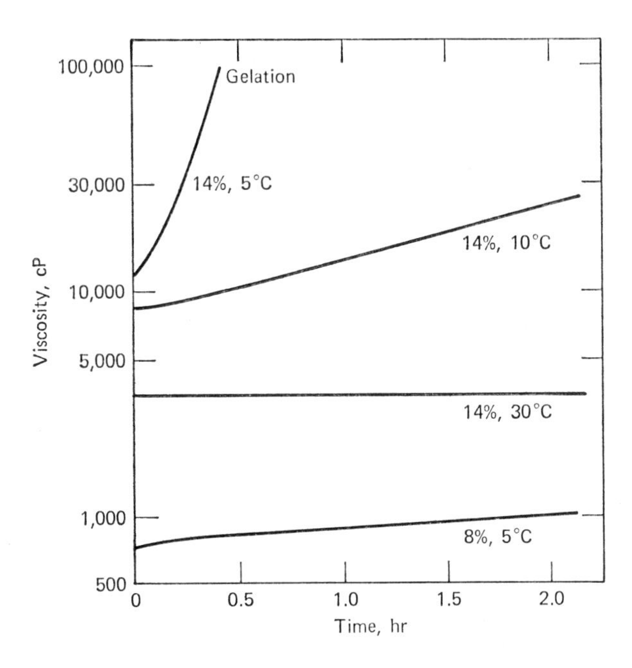

Fig. 12. Viscosity change of poly(vinyl alcohol) solution (Poval 117) with time (15).

Table 5. Minimum Salt Concentration Causing a 5% Poly(vinyl Alcohol) Solution (Poval PVA 117) to Precipitate (15)

Salts	Minimum concentration for salting out		Salting-out effect, $1/N$
	N	g/liter	
$(NH_4)_2SO_4$	1.0	66	1.00
Na_2SO_4	0.7	50	1.43
K_2SO_4	0.7	61	1.43
$ZnSO_4$	1.4	113	0.71
$CuSO_4$	1.4	112	0.71
$FeSO_4$	1.4	105	0.71
$MgSO_4$	1.0	60	1.00
$Al_2(SO_4)_3$	1.0	57	1.00
$KAl(SO_4)_2$	0.9	58	1.11
H_2SO_4			0
NH_4NO_3	6.1	490	0.16
$NaNO_3$	3.6	324	0.28
KNO_3	2.6	264	0.38
$Al(NO_3)_3$	3.6	255	0.28
HNO_3			0
NH_4Cl			0
NaCl	3.1	210	0.32
KCl	2.6	194	0.38
$MgCl_2$			0
$CaCl_2$			0
HCl			0
Na_3PO_4	1.4	77	0.71
K_2CrO_4	1.4	136	0.71
potassium citrate	0.8	38	1.25
H_3BO_3	0.8	165	1.25

It was found, for example, that the distribution of acetyl groups along the polymer chain influences the surface tension of the polymer solution. Perfectly regular distribution of acetyl groups in the polymer resulted in solutions with higher surface tension than those of polymers in which blocks of acetyl groups were present (68,69).

Figures 13 and 14 show the surface tension–concentration relationship for various poly(vinyl alcohols) of the same degree of polymerization, but of different degrees of hy-

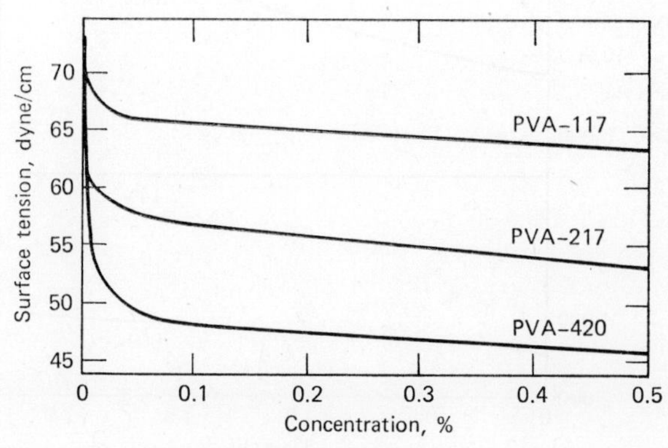

Fig. 13. Surface tension of poly(vinyl alcohol) solutions (Poval) (15) determined with du Nouy tensiometer at 20°C. Degrees of polymerization and hydrolysis are listed in Table 1.

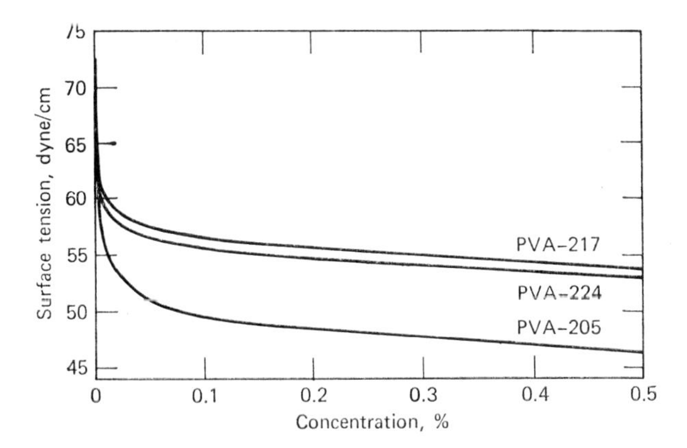

Fig. 14. Surface tension of partially hydrolyzed poly(vinyl alcohol) solutions (Poval) (15). Degrees of polymerization and hydrolysis are listed in Table 1.

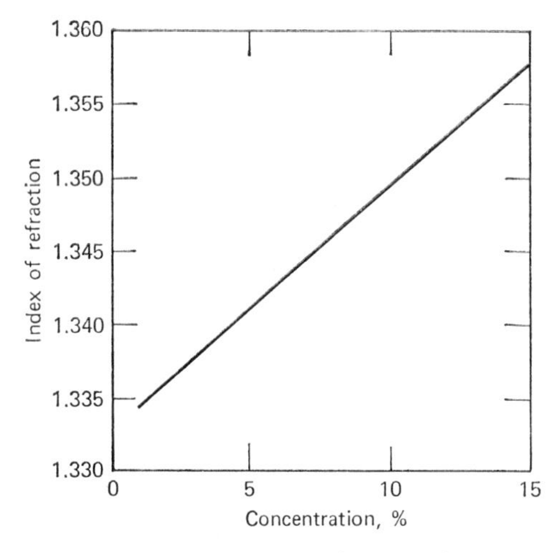

Fig. 15. Index of refraction of poly(vinyl alcohol) solution (Poval PVA 117) at 20°C (15) determined with an Abbe refractometer.

drolysis at 20°C. An increase in temperature to 40°C lowers the surface tension by about 5 dyne/cm. Figure 14 shows that solutions of low-molecular-weight poly(vinyl alcohols) have lower surface tensions than those of high molecular weight, all other factors being equal. It was found that the surface tension at 25°C of a freshly made solution of partially hydrolyzed poly(vinyl alcohol) decreases to an equilibrium value of about 50 dyne/cm after 3 hours (70).

The surface tension of solutions of poly(vinyl alcohol) is also affected by the degree of crystallinity. This has led to a proposal to foam fractionate poly(vinyl alcohol) to obtain fractions with different degrees of crystallinity (71).

The *refractive index* of poly(vinyl alcohol) solutions is highly dependent on the concentration and temperature (69). At low poly(vinyl alcohol) concentrations the index of refraction is also dependent on the degree of polymerization. Figure 15 shows the concentration dependence of high-molecular-weight, fully hydrolyzed poly(vinyl alcohol) at 20°C. Figure 16 shows the concentration dependence of the *specific gravity* for the same polymer. The density of aqueous solutions of fully hy-

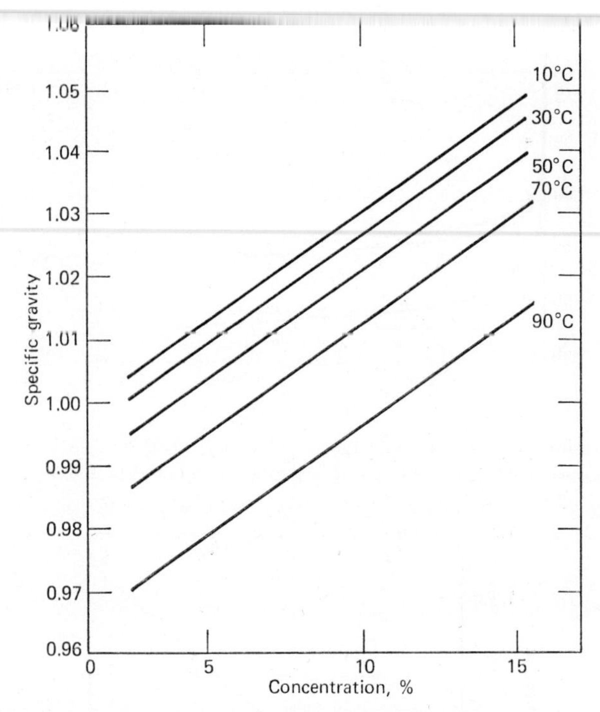

Fig. 16. Specific gravity of a poly(vinyl alcohol) solution (Poval PVA 117) (15) determined with a hydrometer.

drolyzed poly(vinyl alcohols) has been related both to molecular weight and to concentration (72,73).

Dilute Solutions. Although poly(vinyl alcohol) is a polyglycol, water is not necessarily the best solvent or even a good solvent for this polymer, since the large association of water molecules themselves to some degree prevent interaction between poly(vinyl alcohol) and water. Therefore, a small amount of acetone or *n*-propanol improves water solubility greatly (74–76). An aqueous phenol solution (77) or dimethyl sulfoxide (78) is also a good solvent for poly(vinyl alcohol), as is diethylenetriamine (78a). Poly(vinyl alcohol) in dilute solutions tends to form aggregates which can be broken up by addition of urea (79,80).

The measured heat of dilution of aqueous solutions of poly(vinyl alcohol) is far smaller than that estimated by assuming that all hydroxyl groups of the polymer were solvated (81).

Intrinsic Viscosity. Many viscosity studies of poly(vinyl alcohol) have been reported in the literature. Staudinger (82) was the first to publish intrinsic viscosity–molecular weight data. His data were used by Sakurada (83) for his recalculation of an intrinsic viscosity–molecular weight relationship that is still used by the Japanese industry today, although the values for the molecular weight are somewhat low when thus calculated. A review of most of the published intrinsic viscosity–molecular weight relationships can be found in Refs. 84 and 85. The most reliable relationship, given by equation 5, by Elias after review of existing data (86), is valid in water at

$$[\eta] = 7.31 \times 10^{-4} \bar{M}_w^{0.616} \quad \text{in dl/g} \tag{5}$$

25°C. Investigation of the shear dependence of the intrinsic viscosity of poly(vinyl alcohol) showed that the intrinsic viscosity of poly(vinyl alcohol) depends on the rate of shear at 30°C if the $\overline{DP}$ of the polymer is higher than 1700 and on the rate of shear in the range investigated (75–1000 sec^{-1}) (87).

The intrinsic viscosity of poly(vinyl alcohol) decreases with temperature (88,89). The relationship between the intrinsic viscosity and molecular weight for partially hydrolyzed poly(vinyl alcohols) has been studied (90) and is tabulated in Table 6.

Table 6. Intrinsic Viscosity–Molecular Weight Relationship as a Function of Hydrolysis (90)

Degree of hydrolysis, %	$[\eta]$–M_v relationship[a]
86.8	$[\eta] = 8.0 \times 10^{-4} \overline{M}_v^{0.58}$
93.5	$[\eta] = 7.4 \times 10^{-4} \overline{M}_v^{0.60}$
96.4	$[\eta] = 6.9 \times 10^{-4} \overline{M}_v^{0.61}$
100.0	$[\eta] = 5.95 \times 10^{-4} \overline{M}_v^{0.63}$

[a] $[\eta]$ in dl/g.

Chemical Properties

As a polyglycol poly(vinyl alcohol) undergoes chemical reactions typical of compounds with secondary hydroxyl groups; a review of the chemical reactions of poly(vinyl alcohol) has been published in Refs. 91 and 92.

Esterification. Boric acid forms a cyclic ester (**1**) with poly(vinyl alcohol), which is soluble in methanol (93–96). In an alkaline medium, a more complex form of the ester (**2**) is formed, which results in solutions of very high viscosity or insoluble gels (eq. 6). This could be due to polyelectrolyte behavior or to actual crosslinking of

H$_2$C–CH(OH)–H$_2$C–CH(OH)–H$_2$C (poly(vinyl alcohol)) + (HO)$_2$B–OH (boric acid) ⇌ H$_2$C–CH(–O–)–H$_2$C–CH(–O–)–H$_2$C with both O bonded to B–OH (**1**)

monodiol complex
(solution)

⇅ + base

H$_2$C–CH(OH)–H$_2$C–CH(OH)–H$_2$C (poly(vinyl alcohol)) + $Na_2B_4O_7{\cdot}10H_2O$ (borax) ⇌ Na$^{\oplus}$ [two chain segments H$_2$C–CH–O / H$_2$C–CH–O and O–CH–CH$_2$ / O–CH–CH$_2$ bound through four O to B$^{\ominus}$] (**2**) (6)

didiol complex
(gel)

poly(vinyl alcohol) (97,98). The presence of the borate groups results in a completely noncrystalline glassy poly(vinyl alcohol) derivative (99). Alkyl-substituted boric acids also form poly(vinyl alcohol) esters easily (100,101). Similar complexes are also formed when poly(vinyl alcohol) is treated with titanium lactate (102), titanyl sulfate (103), or vanadyl compounds (104).

Poly(vinyl nitrate) has been prepared and studied for use as an explosive and rocket fuel (105). Poly(vinyl alcohol) and sulfur trioxide react to give sodium poly(vinyl sulfate) after neutralization (106–111).

Poly(vinyl sulfonates) have been prepared by reacting poly(vinyl alcohol) and an alkanesulfonyl chloride in pyridine (112,113).

Poly(vinyl phosphates) have been obtained by the reaction of poly(vinyl alcohol) with phosphorus pentoxide (114), or with phosphoric acid in the presence of urea (115–117). Poly(vinyl arsenates) were obtained in a similar fashion (116).

The number of organic esters that can be prepared from poly(vinyl alcohol) are seemingly infinite (91,118). Only a few examples are given here.

In general, acid anhydrides, acid chlorides, and sometimes also the carboxylic acids react with poly(vinyl alcohol) to give the corresponding esters. Transesterifications are also possible. The literature on these reactions is very extensive (91,92,118). There are a few noteworthy special esterification reactions. Diketene adds to poly(vinyl alcohol) and forms the acetoacetic ester (119,120). Poly(vinyl acetoacetate) is also obtained by a transesterification reaction between poly(vinyl alcohol) and ethylacetoacetate (121).

Poly(vinyl cinnamate) has been prepared by the Schotten-Baumann reaction with cinnamoyl chloride (122,131). The reaction of a chloroformate ester with poly(vinyl alcohol) leads to a poly(vinyl carbonate) (eq. 7) (123).

$$\sim\!\!\sim CH_2{-}\underset{\displaystyle OH}{\underset{|}{CH}}{-}\sim\!\!\sim \;+\; ClCOOR \xrightarrow{\text{base}} \sim\!\!\sim CH_2{-}\underset{\displaystyle \underset{\displaystyle O}{\overset{\|}{OCOR}}}{\underset{|}{CH}}{-}\sim\!\!\sim \tag{7}$$

Polymeric drying oils were obtained from the reaction of poly(vinyl alcohol) and linseed acids (124,125). Poly(acrylic acid) or poly(methacrylic acid) reacts with poly(vinyl alcohol) to form insoluble gels (126,127). Urea and poly(vinyl alcohol) react to give a polymeric carbamate ester (eq. 8) (128–130).

$$\sim\!\!\sim CH_2{-}\underset{\displaystyle OH}{\underset{|}{CH}}{-}\sim\!\!\sim \;+\; H_2N\underset{\displaystyle O}{\underset{\|}{C}}NH_2 \longrightarrow \sim\!\!\sim CH_2{-}\underset{\displaystyle \underset{\displaystyle \underset{\displaystyle NH_2}{\underset{|}{C{=}O}}}{\underset{|}{O}}}{\underset{|}{CH}}{-}\sim\!\!\sim \;+\; NH_3 \tag{8}$$

Thiourea and poly(vinyl alcohol) react in concentrated hydrochloric acid to give an isothiuronium salt (eq. 9) (132,133). This can be hydrolyzed readily to the poly(vinyl mercaptan), which crosslinks the polymer on exposure to air.

$$\sim\!\!\sim CH_2{-}\underset{\displaystyle OH}{\underset{|}{CH}}{-}\sim\!\!\sim \;+\; H_2N\underset{\displaystyle S}{\underset{\|}{C}}H_2 \;+\; HCl \longrightarrow \sim\!\!\sim CH_2{-}\underset{\displaystyle \underset{\displaystyle H_2N{-}C{-}NH_2}{\underset{\|}{S^+Cl^-}}}{\underset{|}{CH}}{-}\sim\!\!\sim \tag{9}$$

Etherification. Poly(vinyl alcohol) can form ethers rather easily. It forms internal ethers by eliminating water. Often this reaction, which is catalyzed by strong mineral acids and alkali, leads to insolubilization.

The most commercially important ether is formed by the reaction of alkylene oxides with poly(vinyl alcohol). Ethylene oxide, for example, adds to poly(vinyl alcohol) under normal ethoxylation conditions (134,135). The products thus formed are sold for applications where cold-water solubility is important (see the section on Films). Poly(vinyl alcohol) undergoes the Michaels addition reaction with compounds with activated double bonds. Monomers that add easily include acrylonitrile (136–138), acrolein (139,140), methyl vinyl ketone (139–141), and acrylamide (142,143). Cyanoethylated poly(vinyl alcohol) has become commercially interesting in electrical applications (144). Divinyl sulfone has been used to crosslink poly(vinyl alcohol) (145).

Poly(vinyl alcohol) reacts with sodium monochloroacetate to give a glycolic acid ether (146).

$$\sim\!CH_2\!-\!\underset{\displaystyle OH}{\underset{|}{CH}}\!-\!\sim + ClCH_2COONa \longrightarrow \sim\!CH_2\!-\!\underset{\displaystyle O\!-\!CH_2\!-\!COOH}{\underset{|}{CH}}\!-\!\sim \quad (10)$$

Alkyl and aryl chlorides and bromides such as triphenylcarbinyl chloride, benzyl chloride, and allyl bromide form ethers with poly(vinyl alcohol) (147).

An especially interesting vinyl ether of poly(vinyl alcohol) is obtained by the reaction of acetylene and poly(vinyl alcohol) (146).

$$\sim\!CH_2\!-\!\underset{\displaystyle OH}{\underset{|}{CH}}\!-\!\sim + HC\equiv CH \longrightarrow \sim\!CH_2\!-\!\underset{\displaystyle O\!-\!CH\!=\!CH_2}{\underset{|}{CH}}\!-\!\sim \quad (11)$$

Miscellaneous Reactions. Poly(vinyl alcohol) has been chlorinated to give copolymers of vinyl chloride and poly(vinyl alcohol) (148,149). Poly(vinyl xanthate) can be prepared similarly to cellulose xanthate (eq. 12) (150).

$$\sim\!CH_2\!-\!\underset{\displaystyle OH}{\underset{|}{CH}}\!-\!\sim + CS_2 + \xrightarrow{\text{base}} \sim\!CH_2\!-\!\underset{\displaystyle S\!=\!C\!-\!SNa}{\underset{|}{CH}}\!-\!\sim \quad (12)$$

Poly(vinyl alcohol) reacts with benzonitrile to give substituted poly(vinyl amide) (eq. 13) (151).

$$\sim\!CH_2\!-\!\underset{\displaystyle OH}{\underset{|}{CH}}\!-\!\sim + C_6H_5CN \longrightarrow \sim\!CH_2\!-\!\underset{\displaystyle NHC(\!=\!O)C_6H_5}{\underset{|}{CH}}\!-\!\sim \quad (13)$$

Phenylisocyanate and poly(vinyl alcohol) give a phenylurethan of poly(vinyl alcohol) (152).

Poly(vinyl alcohol) forms disulfide crosslinks with S_2Cl_2 (153). Propane sulfone can sulfopropylate poly(vinyl alcohol) (eq. 14) (154).

$$\sim\!CH_2\!-\!\underset{\displaystyle OH}{\underset{|}{CH}}\!-\!\sim + \underset{\text{(cyclic: } H_2C\!-\!SO_2\!-\!O)}{H_2C\!-\!CH_2} \longrightarrow \sim\!CH_2\!-\!\underset{\displaystyle O(CH_2)_3SO_3H}{\underset{|}{CH}}\!-\!\sim \quad (14)$$

Poly(vinyl alcohol) forms a copper complex (**3**) in neutral or slightly basic solution. The complex is insoluble at this pH but can be dissolved in ammonia (155).

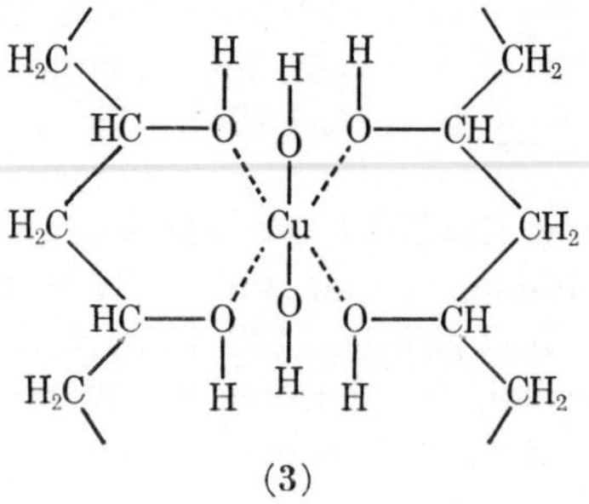

```
  H2C     H  H  H      CH2
     HC—O  O  O—CH
  H2C       Cu        CH2
     HC—O  O  O—CH
  H2C     H  H  H      CH2
```

(3)

Poly(vinyl alcohol) also forms a molecular compound with sodium hydroxide (156). Iodine reacts with poly(vinyl alcohol) to give a complex with a characteristic blue color similar to that which is formed by iodine and amylose (157). When boric acid is present it is thought that a helical inclusion compound exists in solution having the structure shown in Figure 17. Iodine gives a red-colored complex with partially hydrolyzed poly(vinyl alcohol). However, it was found that this complex formed only with partially hydrolyzed poly(vinyl alcohol) that is formed by alkaline hydroylsis. No red-colored complex formed when reacetylated poly(vinyl alcohol) was used (158). It is thought that only blocks of acetyl groups favor the formation of the iodine complex and that isolated acetyl groups such as are present in reacetylated poly(vinyl alcohol) do not form the complex. The intensity of the blue color of the poly(vinyl alcohol)–iodine complex is related to the stereoregularity of the poly(vinyl alcohol) (159). Certain organic dyes form reversible gels with poly(vinyl alcohol) solutions. For example, an addition of 2% Congo Red (based on the weight of poly(vinyl alcohol)) to a 5% poly(vinyl alcohol) solution causes a red gel to form which melts sharply at about 40°C (160). Other dyes and organic compounds that also form a temperature-reversible complex include direct azo dyes, resorcinol, catechol, and gallic acid (161).

Reaction with Aldehydes and Ketones. Industrially, the reaction products of poly(vinyl alcohol) with aldehydes are most important. Poly(vinyl butyral) and poly(vinyl formal) are the principal products, the former being used as interlayer for safety glass and as a paint resin, and the latter as a synthetic fiber (162). It is also possible to introduce other functional groups into the poly(vinyl alcohol) molecule by reaction with aldehydes having functional groups.

The following reactions (eqs. 15–17) take place between poly(vinyl alcohol) and an aldehyde:

Intramolecular acetalization of the 1,3-glycol group

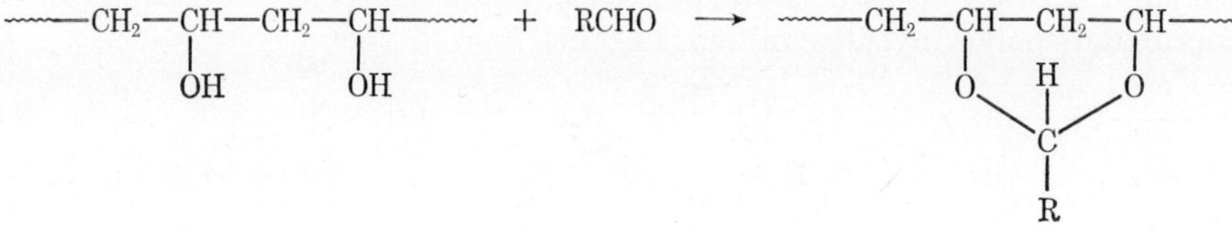

$$\sim CH_2{-}CH(OH){-}CH_2{-}CH(OH)\sim + RCHO \longrightarrow \sim CH_2{-}CH{-}CH_2{-}CH\sim \;(\text{O}{-}CHR{-}\text{O ring}) \qquad (15)$$

Intermolecular acetalization

$$\begin{array}{c} \sim\!CH_2\!-\!CH\!-\!CH_2\!-\!CH\!\sim \\ \quad | \qquad\quad | \\ \quad OH \qquad OH \\ \quad OH \qquad OH \\ \quad | \qquad\quad | \\ \sim\!CH_2\!-\!CH\!-\!CH_2\!-\!CH\!\sim \end{array} + RCHO \longrightarrow \begin{array}{c} \sim\!CH_2\!-\!CH\!-\!CH_2\!-\!CH\!\sim \\ | \\ O \\ | \\ H\!-\!C\!-\!R \\ | \\ O \\ | \\ \sim\!CH_2\!-\!CH\!-\!CH_2\!-\!CH\!\sim \end{array} \qquad (16)$$

Intramolecular acetalization of 1,2-glycol group

$$\begin{array}{c} \sim\!CH_2\!-\!CH\!-\!CH\!-\!CH_2\!\sim \\ | \quad\; | \\ OH \;\; OH \end{array} + RCHO \longrightarrow \begin{array}{c} \sim\!CH_2\!-\!CH\!-\!CH\!\sim \\ | \;\; H \;\; | \\ O \diagdown | \diagup O \\ C \\ | \\ R \end{array} \qquad (17)$$

Poly(vinyl alcohol) can be crosslinked by the reaction with polyacrolein (163) and with low-molecular-weight dialdehydes such as glutaraldehyde or glyoxal.

Ketones are more sluggish in their reaction with poly(vinyl alcohol) than are aldehydes. The rate of reaction of cyclohexanone with poly(vinyl alcohol) is, however, comparable to that of aldehydes (164).

Degradation. When poly(vinyl alcohol) powder is heated in vacuo, water is split off between 100 and 250°C and a rust brown powder is formed. Thermogravimetric and differential thermal analyses have shown that at 160°C water is split off at the maximum rate (165). Acetaldehyde, crotonaldehyde, benzaldehyde, and acetophenone have been isolated in the decomposition products (166–168). At very high temperatures (500–800°C) carbon monoxide and aromatic hydrocarbons are also formed.

When poly(vinyl alcohol) is heated to above 250°C in the presence of oxygen, induced decomposition with self-ignition may take place (169). Crosslinking of the poly(vinyl alcohol) under these conditions has also been observed (170).

The oxidation of poly(vinyl alcohol) has been studied extensively. Ammonium bichromate oxidizes poly(vinyl alcohol) in the presence of ultraviolet radiation. Depending upon the degree of oxidation, crosslinking or degradation takes place (171). Poly(vinyl alcohol) is selectively oxidized by periodic acid. Ceric ions also oxidize poly(vinyl alcohol); in the process, free radicals are formed and can lead to graft copolymers (172). Under acidic conditions, hydrogen peroxide oxidatively degrades poly(vinyl alcohol) (173). Oxidation and degradation of poly(vinyl alcohol) are also caused by air (under alkaline conditions), *t*-butyl hypochlorite, uranyl acetate, and chromic acid.

Crosslinking. Crosslinking insolubilizes and improves the water resistance and the mechanical properties of poly(vinyl alcohol). As a rule, bifunctional compounds that react with hydroxyl groups can accomplish this. Compounds that have been tested as crosslinking agents for poly(vinyl alcohol) include: dimethylolurea, trimethylolmelamine, glyoxal, glutaraldehyde, oxalic acid, diepoxides, polyacrolein, dialdehyde starch, divinyl sulfone, diisocyanates, dihydroxy diphenyl sulfone, and various organometallic (Ti, Zr) compounds.

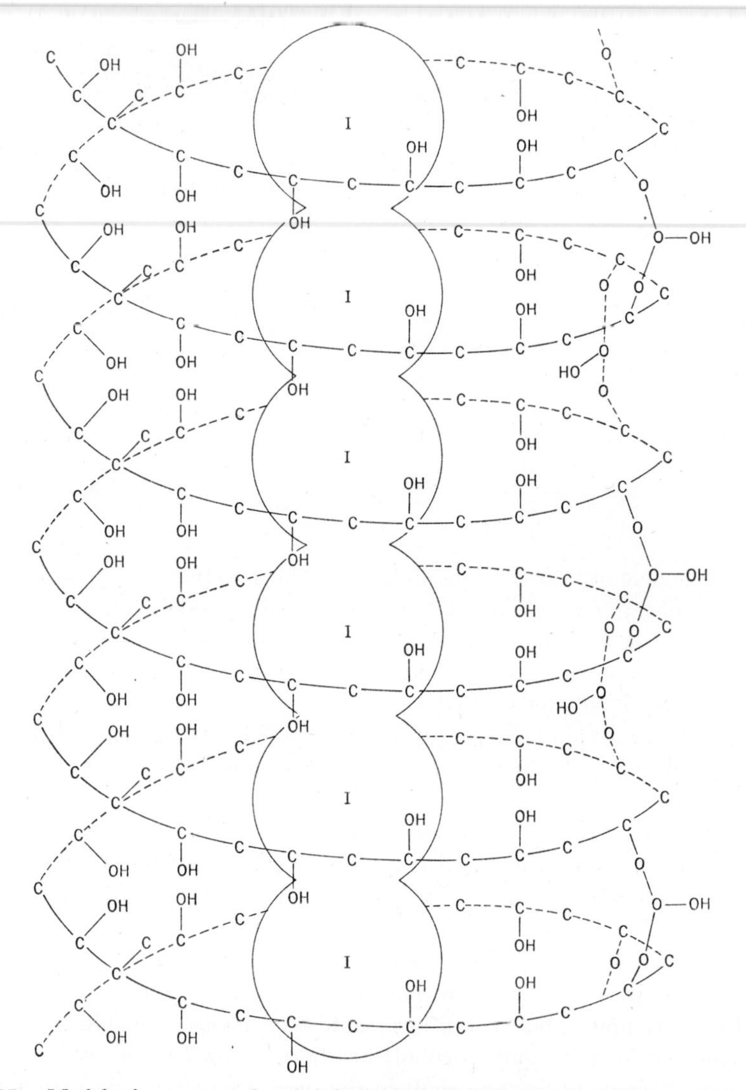

Fig. 17. Model of a proposed poly(vinyl alcohol)–iodine–boric acid complex (157).

Poly(vinyl alcohol) cannot be crosslinked by irradiation with beta rays. However, crosslinked poly(vinyl alcohol) can be obtained when the parent poly(vinyl acetate) is crosslinked by irradiation and subsequently hydrolyzed (174). A paper coating containing poly(vinyl alcohol) as the binder was insolubilized by a low-energy electrical discharge (175). After adding appropriate sensitizers, such as diazonium compounds or sodium benzoate, poly(vinyl alcohol) can be crosslinked by ultraviolet radiation (176,177).

Irradiation Effects. The effect of radiation on poly(vinyl alcohol) is highly dependent on the exact experimental conditions. Chain scission has been reported to be the most significant result of the irradiation of the solid polymer in either air or vacuum (174,178).

Heat treatment is reported to aid gelation of poly(vinyl alcohol) (179). Poly(vinyl alcohol) film swollen with water showed some crosslinking after irradiation with gamma rays, in contrast to dry films which showed degradation (180). If poly(vinyl alcohol) is irradiated in the presence of allyl methacrylate, crosslinking and insolubilization occur (181).

Toxicity. It appears that poly(vinyl alcohol) is not a very toxic material, although the U.S. Food and Drug Administration (FDA) has not cleared poly(vinyl alcohol) to be taken internally. Feeding studies of rats with poly(vinyl alcohol) revealed that low-molecular-weight poly(vinyl alcohol) was absorbed by both liver and kidney more than the high-molecular-weight poly(vinyl alcohol) (196,203). However, there are contradictory findings in the literature (197).

Poly(vinyl alcohol) has been cleared by the FDA as a component of coatings and adhesives coming into contact with fatty foods (198). It is also used as a component of cosmetics, bacteriostatic agents, and other externally applied medicines (199,200). The injection of poly(vinyl alcohol) gels into the body have been studied; however, some doubt exists about possible harmful effects of both poly(vinyl alcohol) (201) and poly(vinyl formal) (202).

Structure

Poly(vinyl alcohol) is a poly-1,3-glycol; the structure is shown in Figure 18. Poly(vinyl alcohol), on oxidation and hydrolysis, yielded acetic acid and acetone (204). Oxidation of poly(vinyl alcohol) with periodic acid gives a measure of the number of 1,2-glycol groups (205). The amount of 1,2-glycol units derived from head-to-head polymerizations is usually below 1–2%, and depends on the polymerization temperature of the vinyl acetate (see Fig. 19), but not on the polymerization method or catalyst

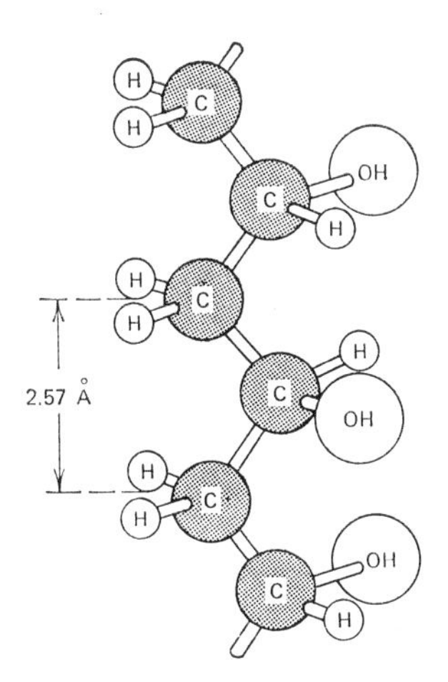

Fig. 18. Structure of poly(vinyl alcohol) showing syndiotactic placement of three consecutive hydroxyl groups.

employed (206). Other poly(vinyl esters) when hydrolyzed give different amounts of 1,2-glycol units, owing to the different activation energies of head-to-head polymerization (208). The number of 1,2-glycol structures affects some properties of poly(vinyl alcohol), such as the swelling of films in water. Low polymerization temperatures of the parent vinyl acetate result in a poly(vinyl alcohol) with a low degree of swelling (207,209). (See also p. 150 for a discussion of the effects of the structure of the parent poly(vinyl acetate).

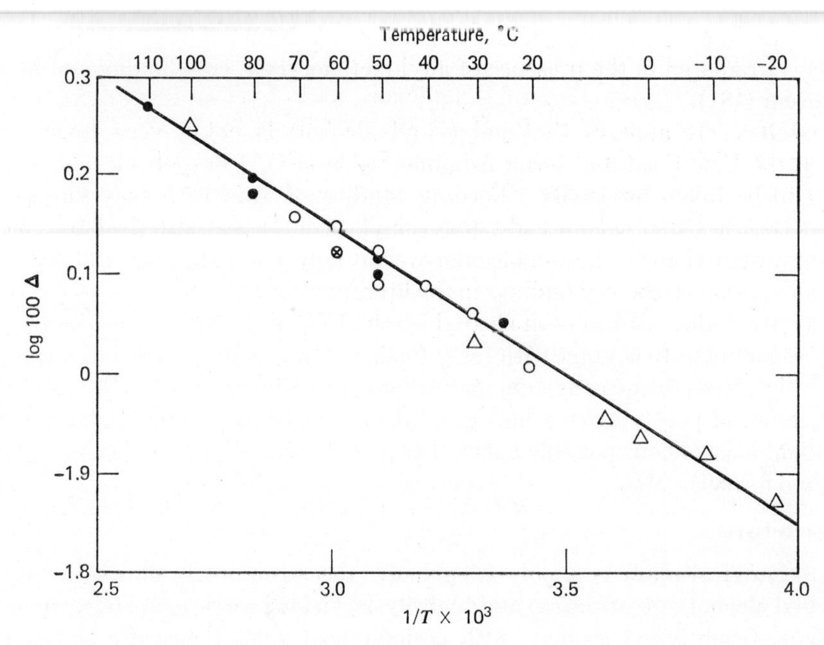

Fig. 19. Relationship between log mole percent of 1,2-glycol structure and reciprocal of absolute temperature of polymerization. Key: (●) Flory's data (205); (○) emulsion polymerization; (△) bulk polymerization; (⊗) solution polymerization (207).

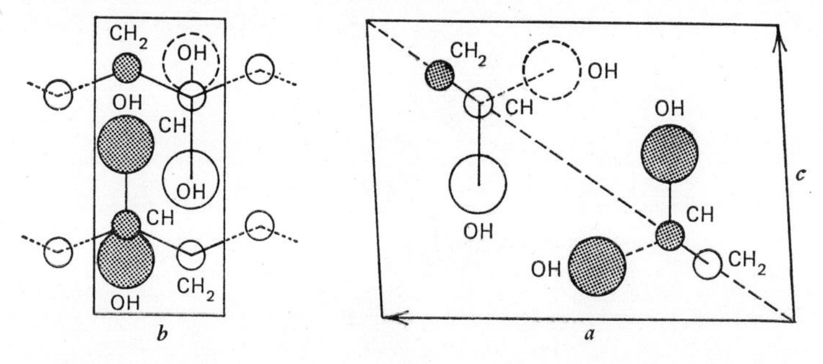

Fig. 20. Model of crystalline structure of poly(vinyl alcohol) (219).

During polymerization chain-transfer reactions with monomer, polymer, solvents, and active impurities occur more frequently than initiation and termination reactions (210). Therefore, end groups such as carboxyl and carbonyl are more frequent than end groups derived from the initiator. The formation of end groups has been studied in detail (211–213).

Carbonyl groups were also found in the main chain when benzoyl peroxide was used as the initiator. In the case of α,α'-azobisisobutyronitrile no such groups were found (214–216).

Branching reactions occur frequently during the polymerization of vinyl acetate. The main site of the branching reaction is apparently the acetyl group (217); however, some doubt has recently been cast on this theory (218).

The stereostructure of poly(vinyl alcohol) has been studied extensively. It is generally accepted that normal poly(vinyl alcohol) is atactic (219). As previously

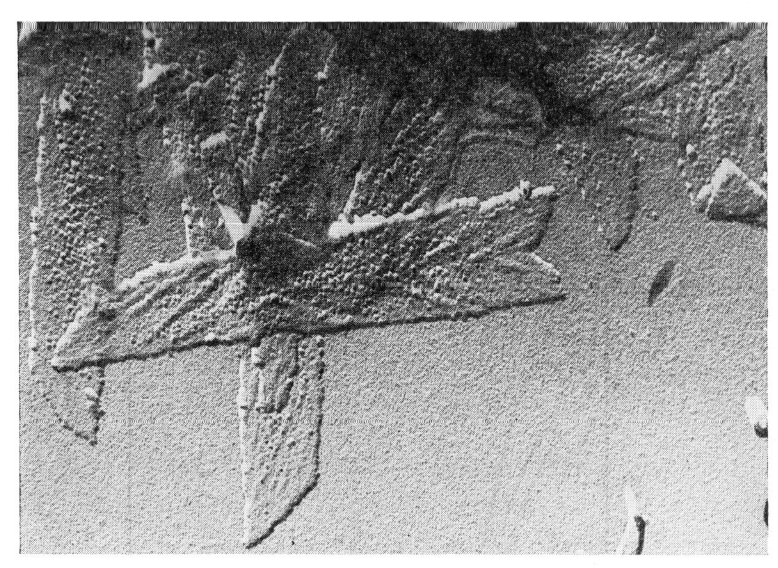

Fig. 21. Poly(vinyl alcohol) platelets crystallized from 0.01% triethylene glycol solution by cooling from 200°C (227). Courtesy Dr. V. F. Holland, Chemstrand Research Center, Inc.

mentioned, the extent of swelling of poly(vinyl alcohol) film in water depends on its 1,2-glycol content. However, stereoregularity of poly(vinyl alcohol) also plays an important role. Many studies have been made relating the polymerization conditions of vinyl esters to the stereoregularity of the derived poly(vinyl alcohol) (220–223). No effective catalyst has yet been found to yield stereoregular poly(vinyl acetate). Stereoregular poly(vinyl alcohol), however, has been prepared and its properties investigated (224,225).

The crystalline structure of poly(vinyl alcohol) is well represented by the model shown in Figure 20 (219). The following lattice constants were determined by x-ray diffraction using an x-ray counter-type diffractometer (226): $a = 7.81$ Å; $b = 2.52$ Å; $c = 5.49$ Å; $\beta = 92°10'$.

Single crystals of poly(vinyl alcohol) have been obtained; Figure 21 shows a poly(vinyl alcohol) platelet crystallized from triethylene glycol (227).

Spectra

Ultraviolet Absorption (228). Most commercial poly(vinyl alcohols) absorb strongly in the 200–400 mμ region of the ultraviolet spectrum. A low-molecular-weight poly(vinyl alcohol) has a higher absorption than does a high-molecular-weight sample (229). The absorption is due to carbonyl groups in the polymer. Acetaldehyde and oxygen, which are often present during polymerization, are responsible for the presence of carbonyl groups (see p. 150). Figure 22 shows the ultraviolet spectrum of poly(vinyl alcohol) derived from poly(vinyl acetate) polymerized in the presence of acetaldehyde (230).

Infrared Absorption (231). The infrared spectrum of poly(vinyl alcohol) can be obtained easily from films which are cast directly on a silver chloride disk. Figures 23 and 24 show the infrared spectrum of fully hydrolyzed and of partially hydrolyzed poly(vinyl alcohol), respectively. It can be seen that the carbonyl band at 1739 cm^{-1}, which is associated with the acetyl group, has almost disappeared in the case of the fully hydrolyzed poly(vinyl alcohol) but is prominent in the case of partially hydrolyzed poly(vinyl alcohol).

Table 7 gives a list of the absorption bands for a fully hydrolyzed poly(vinyl alcohol) (232). The band assignments have been reviewed in detail by Krimm (233). The intensity of the 1141-cm^{-1} band has been associated with the degree of crystallinity of poly(vinyl alcohol) (234). The bands at 916 and 850 are characteristic of syn-

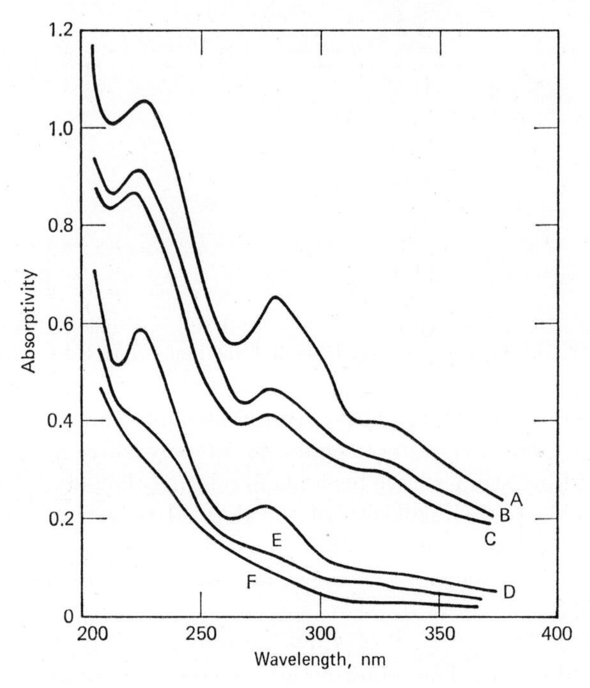

Fig. 22. Ultraviolet-absorption spectra of poly(vinyl alcohol) showing the influence of aldehyde content in vinyl acetate monomer used to prepare percent poly(vinyl acetate) (230). Aldehyde content: curve A, 0.48%; curve B, 0.19%; curve C, 0.097%; curve D, 0.02%; curve E, 0.003%; curve F, 0.001%.

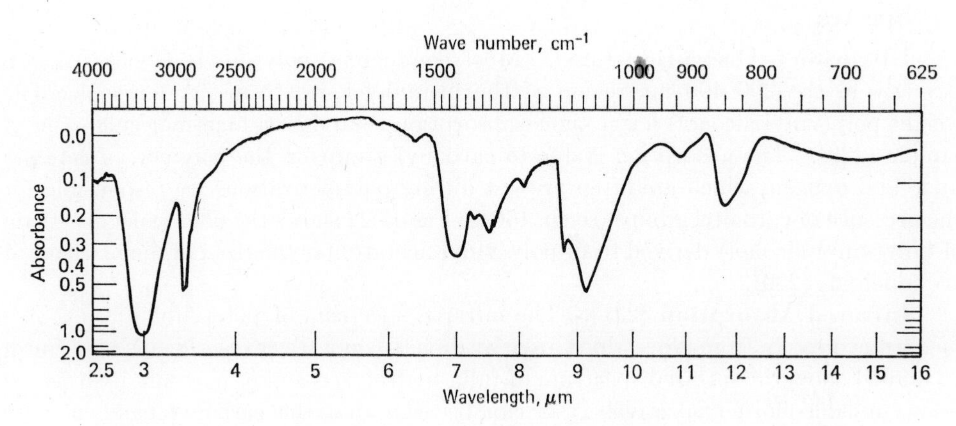

Fig. 23. Infrared-absorption spectrum of fully hydrolyzed poly(vinyl alcohol).

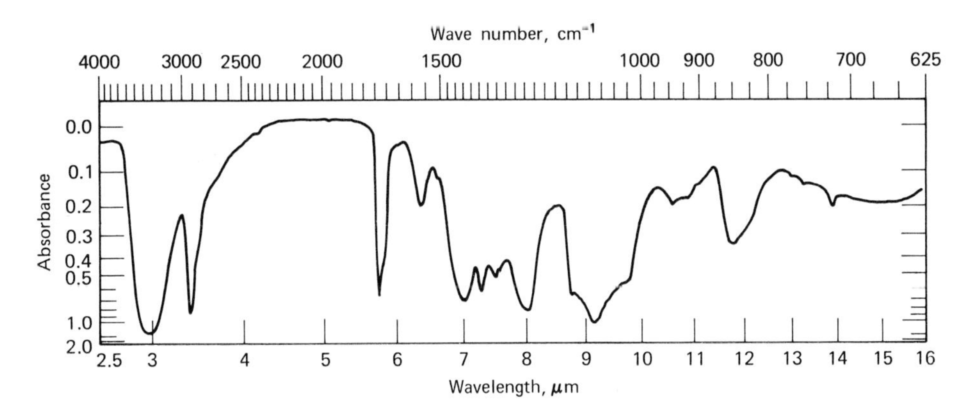

Fig. 24. Infrared-absorption spectrum of partially hydrolyzed poly(vinyl alcohol).

diotactic and of isotactic sequences, respectively (235). A discussion of the influence of stereoregularity on the infrared spectrum of poly(vinyl alcohol) is given in Ref. 239.

Nuclear Magnetic Resonance. An NMR spectrum of poly(vinyl alcohol) is shown in Figure 25. The spectrum has been obtained in phenol (236,237). The magnetic resonance spectrum of the methylene protons of poly(vinyl alcohol) is sensitive to the tacticity of the polymer (236,238). However, many of the experimental difficulties in measuring the NMR spectrum of poly(vinyl alcohol) with better resolution have been overcome (240).

Table 7. Band Assignments for Infrared Spectrum of Poly(vinyl Alcohol) (232)

Frequency, cm^{-1}	Intensity	Assignment
3340	very strong	O–H stretching
2942	strong	C–H stretching
2910	strong	C–H stretching
2840	shoulder	C–H stretching
1446	strong	O–H and C–H bending
1430	strong	CH_2 bending
1376	weak	CH_2 wagging
1326	medium	C–H and O–H bending
1320	weak	C–H bending
1235	weak	C–H wagging
1215	very weak	
1144	medium	C–C and C–O stretching
1096	strong	C–O stretching and O–H bending
1087	shoulder	
1040	shoulder	
916	medium	skeletal
890	very weak	
850	medium	skeletal
825	shoulder	CH_2 rocking
640	medium, very broad	O–H twisting
610	weak	
480	weak	
410	weak	

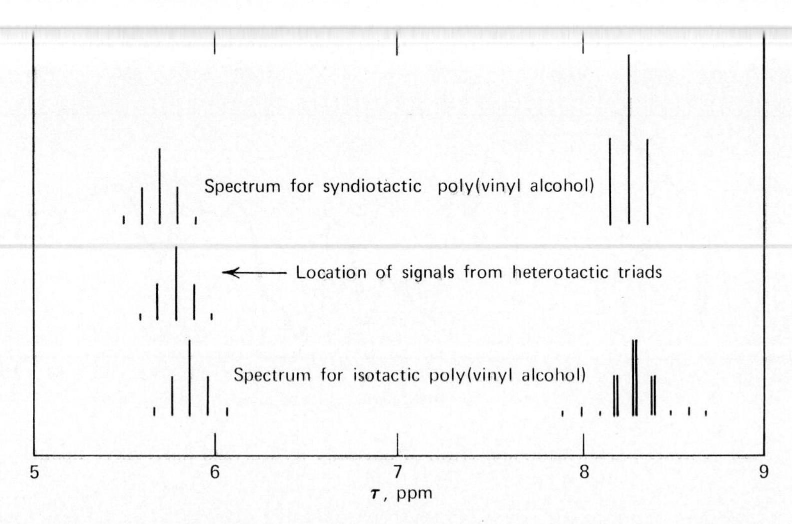

Fig. 25. Composition of the proton magnetic resonance spectrum of poly(vinyl alcohol) (236).

Vinyl Alcohol Copolymers (241)

Since vinyl acetate can be copolymerized with many other monomers (see the article VINYL ESTER POLYMERS), vinyl alcohol copolymers are readily available by the hydrolysis of vinyl acetate copolymers. They can, of course, also be prepared by chemical reactions with poly(vinyl alcohol) (see the section on Chemical Properties).

Vinyl alcohol–ethylene copolymers have been described in many patents (242, 243). Recently, commercial products have been marketed by Du Pont under the name Elvon. A patent issued to Kurashiki Rayon Co. describes a vinyl alcohol–ethylene copolymer for film manufacture (244).

Very high-viscosity vinyl alcohol copolymers have been described with octadecene or vinyl octadecyl ether (245,246) and were at one time commercially available (247). Cold-water-soluble poly(vinyl alcohol) film has been described as made from a copolymer of vinyl alcohol and methyl butenol (248). Another commercially important vinyl alcohol copolymer is Vinylite VAGH (Union Carbide), which consists of 91% vinyl chloride, 6% vinyl alcohol, and 3% vinyl acetate.

Copolymers of vinyl alcohol with acrylamide, methyl methacrylate, crotonic acid, vinylene carbonate, vinyl ketone, and vinyl Versatate, to mention only a few, have been described (241). Commercial exploitation of copolymers of vinyl alcohol has been slow, primarily because of their high cost which is partly due to the difficulties of commercial production of these copolymers.

Applications

In 1969 about 45 million pounds of poly(vinyl alcohol) was used in the United States. The principal applications were for warp sizing in the textile industry, as an emulsifier in vinyl acetate emulsion polymerization systems, as a component in aqueous adhesives, for the production of poly(vinyl acetals), eg, poly(vinyl butyral) for safety glass, and in smaller amounts as the binder for phosphorescent pigments and dyes in television tubes and in other optical applications such as polarizing lenses.

In Japan, where most of the poly(vinyl alcohol) is produced (135 million pounds in 1967), it is used as the raw material for a synthetic fiber and for film. The applications of poly(vinyl alcohol) are surveyed in the General References listed in the Bibliography.

Textile Applications. Most of the poly(vinyl alcohol) sold to the textile industry, about 20 million pounds in 1969, is used as a warp size (249–252). It contributes high weaving efficiency at low amounts of "add-on" when compared to starch, and the biological oxygen demand of poly(vinyl alcohol) is very low, which is important where stream pollution problems exist. An analytical method has been devised to determine the poly(vinyl alcohol) content of the fiber or fabric (253). Desizing has to be accomplished at fairly high temperatures if the poly(vinyl alcohol) used is fully hydrolyzed. Special poly(vinyl alcohols) dissolving at low temperatures have been developed to improve the desizing operation. See also SIZING.

Poly(vinyl alcohol) is also used as a finishing resin to impart stiffness to a fabric. It is then often insolubilized with urea–formaldehyde resins or other *N*-methylol compounds (254).

The excellent adhesive properties of poly(vinyl alcohol) solutions are used in screen printing operations, where it is an important ingredient in temporary adhesives that hold the fabric in place during the printing process (255).

There are a number of specialty applications for poly(vinyl alcohol) in the textile industry, such as binders for nonwoven fabrics, screen preparation in screen printing, and as a detergent additive to prevent soil redeposition. See also TEXTILE RESINS.

Paper Coating. Poly(vinyl alcohol) has been proposed as a pigment binder in clay coatings to replace other water-soluble binders such as starch, casein, or soy protein (256–259). Although economically competitive with casein, the use of poly-(vinyl alcohol) as a pigment binder has been retarded by its lack of water resistance. Various insolubilizing agents have been proposed and many of these work under specific conditions (see the section on Crosslinking). However, none has been found to accomplish this under the widely varying conditions of the paper-coating industry.

Fully hydrolyzed medium-viscosity poly(vinyl alcohol) is most suited to be used as a pigment binder (257). The pick strength, ie, the resistance of the coating to be "picked off" the paper during the printing process, and the water resistance increase with increasing molecular weight of the poly(vinyl alcohol). Rheological properties of poly(vinyl alcohol) containing coating colors generally present no problem on high-speed coaters. The high adhesive and cohesive strength of poly(vinyl alcohol) coatings is probably due to the many hydroxyl groups. Poly(vinyl alcohol) forms fairly stable complexes with clay (260). Figure 26 shows a comparison of pick strength of various paper-coating binders; poly(vinyl alcohol) is by far the most efficient binder (257). Because of the low binder–pigment ratio, poly(vinyl alcohol) coatings have greater brightness, opacity, and gloss than coatings with other binders.

Poly(vinyl alcohol) is also used as a binder in size press coatings, as a surface size with (261,262) and without borax added (263). It imparts a high degree of oil and grease resistance, ink holdout, and solvent resistance at very low coat weights.

Poly(vinyl alcohol) fibers have been added to pulp to improve the internal properties of paper (264). See also PAPER; PAPER ADDITIVES AND RESINS.

Adhesives. Poly(vinyl alcohol) solutions in water are excellent adhesives for gluing paper to paper and paper to wood. Poly(vinyl alcohol) has replaced starches and dextrins to a certain extent in the manufacture of gummed labels and envelopes. To improve its adhesive properties it is often compounded with other materials; for example, the wet tack is increased considerably by borating poly(vinyl alcohol) (265.

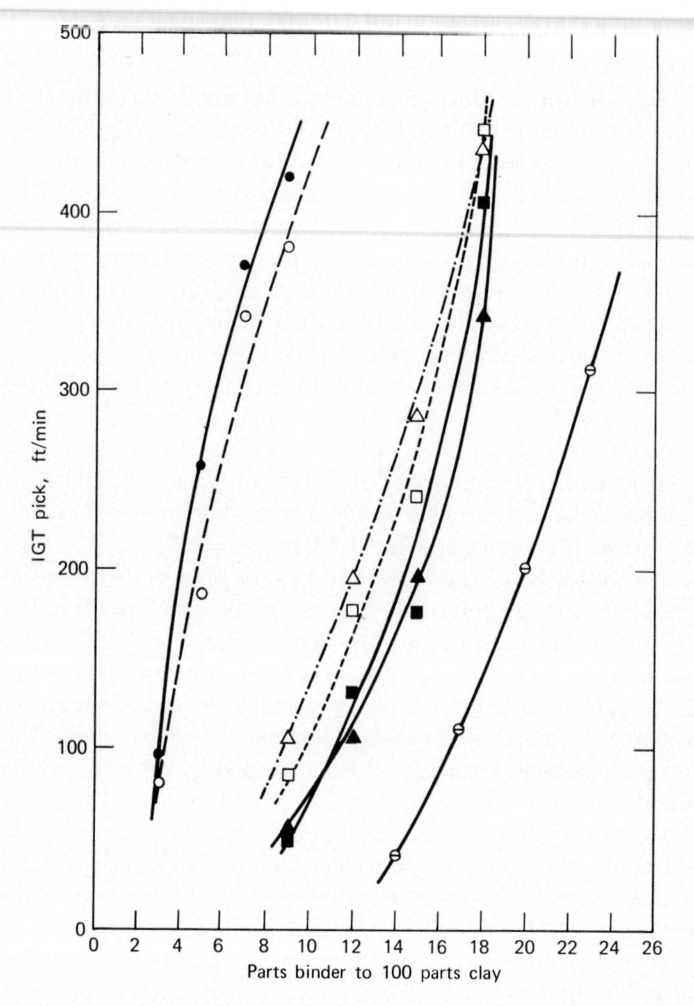

Fig. 26. Comparison of the adhesive strengths of various binders (257). Pigment is 100 parts predispersed clay. Raw stock: label, 45 lb (25 × 38 − 500). Coat weight: coated-one-side 15 lb/ream. IGT: no. 5 ink, "A" spring, 35 kg. Drying: 5 min, 194°F. Calender: 3 passes at 500 p.l.i. and 130°F. Key: (—●—) poly(vinyl alcohol) (Vinol 125); (- -○- -) ⅔ poly(vinyl alcohol)–⅓ latex (butadiene–styrene) solids; (—▲—) casein; (—·—△—·—) ⅔ casein–⅓ latex (butadiene–styrene) solids; (—■—) soy protein; (- -□- -) ⅔ soy protein–⅓ latex (butadiene–styrene) solids; (—⊖—) starch.

266). Other polymers such as starch or carboxymethylcellulose and clays can be added to the poly(vinyl alcohol) to increase the water resistance of the final bond (267).

Poly(vinyl alcohol) together with poly(vinyl acetate) constitutes the so-called "white glue" usually used for gluing wood to wood in the furniture industry and as an all-purpose household glue. See also Adhesive Compositions under ADHESION AND BONDING.

Coatings. Poly(vinyl alcohol) has been proposed for coatings for all kinds of substrates. Most of the disclosures in patents and publications are not practiced mainly because poly(vinyl alcohol) is too water-sensitive and often too expensive. On the other hand, some specialty coatings based on poly(vinyl alcohol) are made and applied where, for example, the solvent resistance of poly(vinyl alcohol) is desired (268). A good survey of suggested coatings applications is given in Ref. 269a.

Use in Light-Sensitive Systems. Poly(vinyl alcohol) is easily sensitized to light by dichromates (171,374–376). Extensive use is made of this property for sensitizing lithographic plates, silk screens, screens for rotary screen printing, and photoresists (171,377–379). Ref. 171 gives an excellent survey of the reactions leading to insolubilization (or hardening) of poly(vinyl alcohol) with dichromates under the influence of ultraviolet light.

Some of the applications of poly(vinyl alcohol), in Germany, before and during World War II, are described in detail in Refs. 286 and 287; most of these have been discontinued.

Use as Colloid Stabilizer. Poly(vinyl alcohol) solutions are surface-active and therefore can stabilize various kinds of hydrosols (288). The surface tension of poly(vinyl alcohol) solutions is dependent both on the degree of hydrolysis and on the molecular weight (289) (see Figs. 13 and 14). Usually, the poly(vinyl alcohols) having a degree of hydrolysis of 87% or 80% are used as surface-active agents. The main use of poly(vinyl alcohol) as a colloid stabilizer is in polymer dispersions (290). A large amount of the poly(vinyl acetate) emulsion sold is stabilized solely by poly(vinyl alcohol). Normally about 5% of the colloid is used on the weight of the polymer. The emulsion polymerization of vinyl acetate in the presence of poly(vinyl alcohol) has been repeatedly studied (291,292).

Suspension polymers are also often stabilized by poly(vinyl alcohol) (293–295), as well as carbon black dispersions (296) and hydrosols of metals such as gold, silver (297), or palladium and platinum (298).

Miscellaneous. Poly(vinyl alcohol) is an ingredient in plant sprays that contain insecticidal or fungicidal agents. It causes the agent to adhere better to the plant and survive a number of rains. Poly(vinyl alcohol) is also used as the binder for phosphorescent pigments in the manufacture of television picture tubes (269b,270,282). Other miscellaneous applications that have been proposed for poly(vinyl alcohol) include: quench solution for hot steel (271); antifogging coating for gas mask lenses (272); semipermeable membranes (273), eg, for desalting water by reverse osmosis; separators in batteries (274); carrier for metal catalysts in hydrogenation reactions (275); improvement of the green strength of ceramics (276); manufacture of polarizing lenses for sunglasses; photographic and nuclear film applications (278,279); molded and extruded products (280,281); soil-suspending agents in detergent formulations (283,284); soil conditioner; additives to photographic emulsions (380); and plastic component in printing plates (285).

Fibers

In 1930, a water-soluble fibrous material made of poly(vinyl alcohol) was described by Herrmann, Baum, and Haehnel (299,300). This fiber was later produced by Wacker-Chemie in Germany and sold under the name Synthofil for surgical applications (301,302).

In 1939, poly(vinyl alcohol) fibers that were stable in hot water and, therefore, were useful as a general textile fiber were developed by Sakurada, Lee, and Kawakami at Kyoto University in Japan by heat treating the wet-spun fiber (303). At almost the same time Yazawa and Meguro at the Kanegafuchi Spinning Company achieved the same goals by acetalization of the poly(vinyl alcohol) fiber (304). In 1950, the Kura-

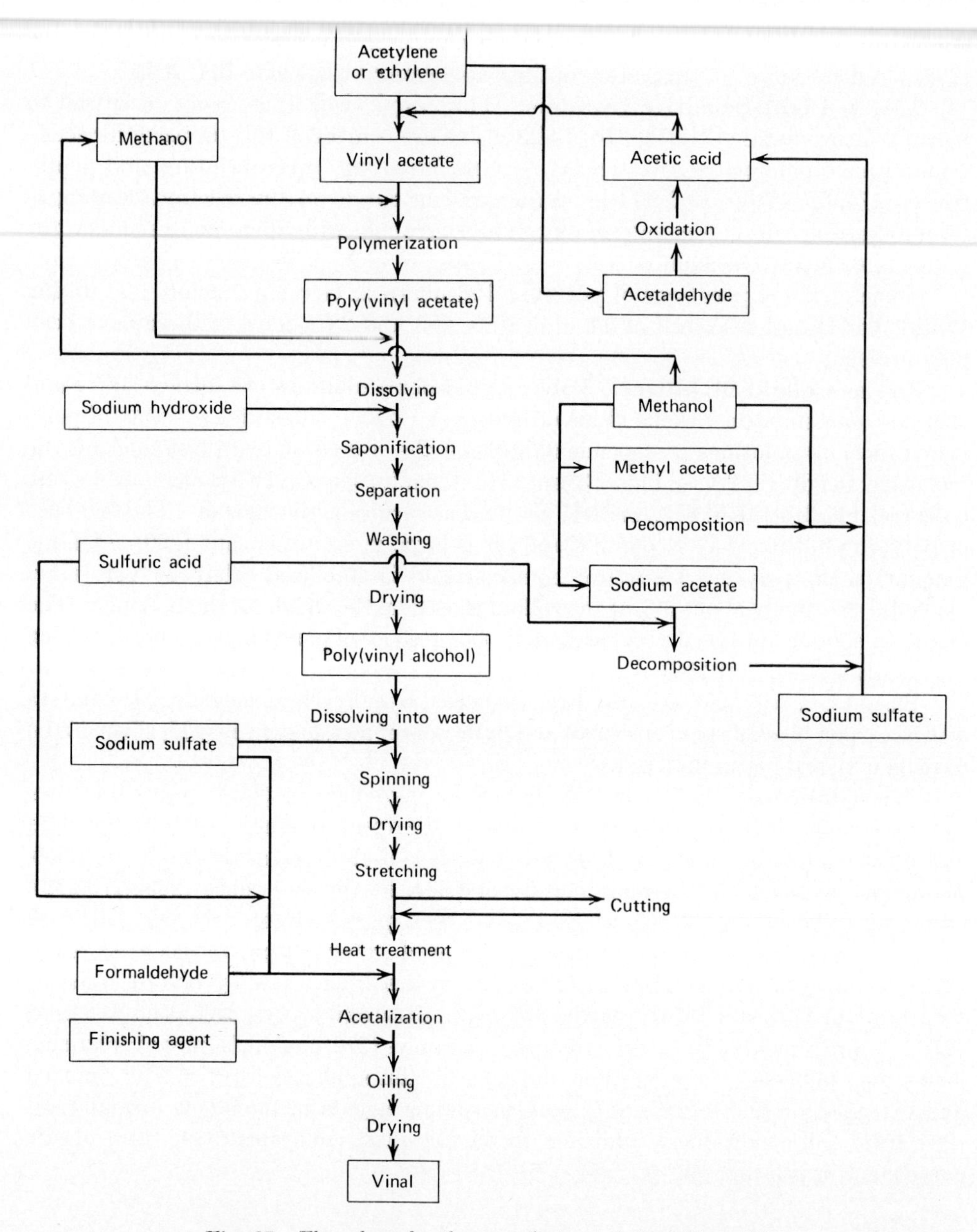

Fig. 27. Flow sheet for the manufacture of vinal fibers (7).

shiki Rayon Company began the first industrial production of poly(vinyl alcohol) fibers. In 1970, about 200 tons per day of poly(vinyl alcohol) fiber were produced in Japan by three manufacturers. There is also interest in poly(vinyl alcohol) fibers in mainland China and in the Soviet Union, as well as in various other countries (305). The fiber is called Kuralon or vinylon in Japan, vinal in the United States, and vinol in the Soviet Union. At present, poly(vinyl alcohol) fiber is not produced commercially in the United States or in Western Europe. A few million pounds of poly(vinyl alcohol) fibers are important into the United States for specialty applications.

Poly(vinyl alcohol) fiber is the most hydrophilic of the synthetic fibers, resembling cotton (qv) in this respect. Water-soluble poly(vinyl alcohol) fibers are used as a binder in paper and in making lace.

The early work on poly(vinyl alcohol) fiber in Japan is reviewed in Ref. 306; other reviews are presented in Refs. 7, 9, and 367–369. See also FIBERS.

Manufacture. A schematic presentation of the steps involved in the manufacture of poly(vinyl alcohol) fibers is given in Figure 27. For the preparation of poly(vinyl acetate), see the article VINYL ESTER POLYMERS in Vol. 15.

To achieve the desired fiber properties, the degree of polymerization of the poly(vinyl alcohol) must be about 1000–2000 and the residual acetyl content should be kept below 0.5 mole % (307,308). The degree of crystallinity and, therefore, the hot-water resistance and elongation of the fiber are affected by the number of residual acetyl groups. In order to decrease the amount of acetyl groups it is often necessary to resaponify the poly(vinyl alcohol) before dissolving it. Usually, the polymer is washed extensively with water before dissolving, because large amounts of sodium acetate in the poly(vinyl alcohol) cause discoloration during the heat-treatment step. Water-soluble poly(vinyl alcohol) fiber is made from poly(vinyl alcohol) having a relatively high percentage of residual acetyl groups.

In order to improve the quality of the fiber, the relation between the polymerization conditions used in preparing the parent poly(vinyl acetate) and the properties of the fiber obtained have been studied in detail. Since poly(vinyl alcohol) derived from poly(vinyl acetate) polymerized at low temperature has a high degree of crystallinity the hot-water resistance of the fiber is improved by lowering the polymerization temperature of the parent poly(vinyl acetate). Mechanical properties, such as tensile strength and degree of elasticity, of both wet-spun (309) and dry-spun (310) fibers are not changed by the polymerization temperature.

The spinning solution is prepared by dissolving the purified poly(vinyl alcohol) in hot water at a concentration of 15%. In case matted or dope-dyed fibers are being produced, the pigment, eg, titanium dioxide, is evenly mixed into the solution. The solution is then filtered and defoamed.

The viscosity of this solution depends on the concentration and temperature of the solution, and on the degree of polymerization and the molecular-weight distribution of the poly(vinyl alcohol). The latter are determined by the polymerization process of the parent poly(vinyl acetate). For concentrated polymer solutions the plot of log η_r versus 5 log C_2 + 3.4 log M_w has a linear region, in which equation 18 holds true (311),

$$\log \eta_r = 5 \log C_2 + 3.4 \log M_w - 9.43 \qquad (18)$$

$$11.2 < \log C_2^5 M_w{}^{3.4} < 13.2$$

where η_r is the relative viscosity of the aqueous solution of poly(vinyl alcohol) determined by a falling ball method, and where C_2 is the polymer concentration in grams per 100 ml of solution.

The slope of the viscosity–concentration plot changes abruptly at the critical concentration, C_{crit}, which is proportional to the limiting viscosity number $[\eta]$ according to equation 19 (312).

$$C_{crit} = 2.5\ (1/[\eta]) \qquad (19)$$

At the critical concentration the flow mechanism of poly(vinyl alcohol) solutions apparently changes; below the critical concentration the flow unit is the individual

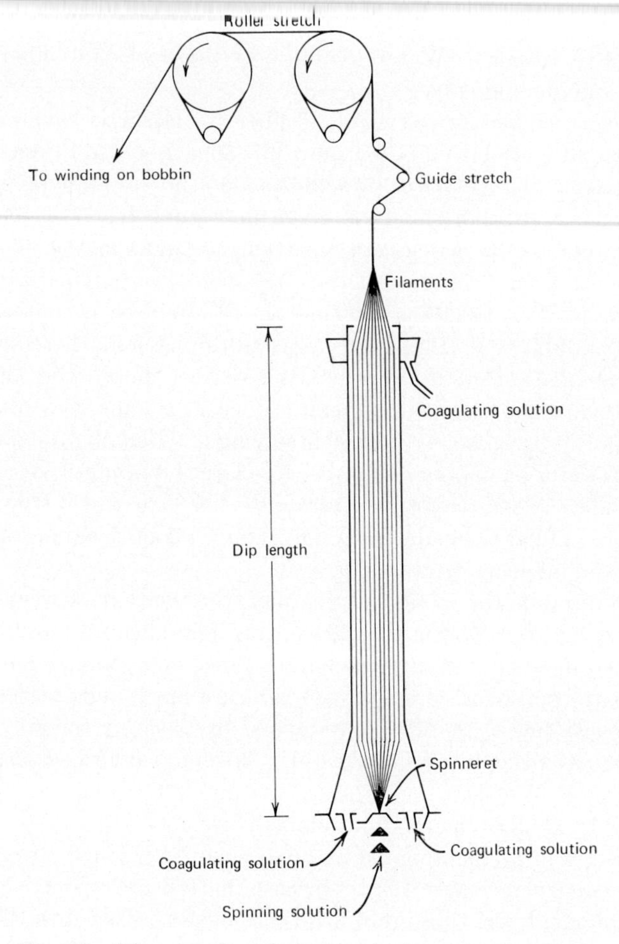

Fig. 28. Diagram of vertical wet-spinning apparatus (315).

molecule, whereas above it the solution can be considered as a fused material. Shear effects and molecular interactions are very large. Compared with other polymers C_{crit} of poly(vinyl alcohol) is very small. Considerable data have been collected to evaluate the mechanical response of concentrated poly(vinyl alcohol) solutions under large deformations (313).

Spinning. Both a wet-spinning and a dry-spinning process are used commercially to produce poly(vinyl alcohol) fibers.

In the wet-spinning process the hot (70°C), concentrated poly(vinyl alcohol) solution is transferred to a spinning machine and spun through spinnerets into an aqueous salt solution at 40 to 60°C. The fiber is formed by a dehydration and coagulation process.

The coagulation power of various salts is listed in Table 8 (314). Ammonium sulfate and sodium sulfate have the greatest coagulation power, and are often used as the coagulating agent. The pH of the coagulation bath must be regulated to prevent color formation during heat treatment of the fiber. Color formation is also prevented by adding amphoteric compounds such as zinc sulfate or magnesium sulfate to the coagulation bath. Figure 28 shows a diagram of a wet-spinning apparatus (315), in which the fibrous material and the coagulating solution go up inside a spinning tube. Either a vertical apparatus, as shown, or a horizontal apparatus (316) can be used. For stretch spinning of the fiber, a rotor guide or roller is used, and in some cases, a second coagulation bath can be used. See MAN-MADE FIBERS, MANUFACTURE.

Usually the wet-spun fibers have an even, dense outer skin and an uneven inner core. These heterogeneities lead to poor mechanical and dyeing properties (see also pp. 191, 193). To produce a homogeneous cross section, a swelling treatment is given the spun fiber by either washing with water (317) or adjusting the spinning solution or the coagulation solution (318). A transparent hollow fiber can be obtained (319) from a poly(vinyl alcohol) spinning solution to which boric acid has been added, and by spinning it through a basic coagulation bath solution at low temperature.

Table 8. Coagulation Power[a] of Various Aqueous Salt Solutions (314)

Salt	Maximum coagulation power	Salt	Maximum coagulation power
$(NH_4)_2SO_4$	8.30	$Al(NO_3)_3$	1.98
Na_2SO_4	6.95	KNO_3	1.38
K_2SO_4	2.04	HNO_3	0
$ZnSO_4$	4.65	NH_4Cl	0
$CuSO_4$	2.18	NaCl	1.50
$FeSO_4$	2.76	KCl	1.63
$MgSO_4$	6.28	$MgCl_2$	0
$Al_2(SO_4)_3$	6.73	$CaCl_2$	0
$KAl(SO_4)_2$	1.41	HCl	0
H_2SO_4	0	Na_3PO_4	2.60
NH_4NO_3	1.86	K_2CrO_4	4.08
$NaNO_3$	2.06	H_3BO_3	3.98

[a] Coagulation power is the product of the saturation concentration (in *N*) and the minimum concentration (in *N*) of the salt solution at which an addition of two drops of a 5% poly(vinyl alcohol) solution can make the mixture cloudy.

The dyeing properties of the fiber can be improved by cospinning poly(vinyl alcohol) with amino acetalized (treated with an amino aldehyde) poly(vinyl alcohol) (320) or with casein, bean protein (321), starch (322), carboxymethylcellulose, alginic soda, or gum arabic (323).

Poly(vinyl alcohol) spun together with polyacrylamide (324) or acrylic acid–acrylamide copolymers (325) gives a fiber resistant to hot water. The properties of poly(vinyl alcohol) fiber can also be improved by an emulsion blend-spinning method (326) in which a hydrophobic monomer is polymerized in the poly(vinyl alcohol) water solution before the spinning process. Vinyl chloride, for example, is used as the hydrophobic monomer in Japan to produce a poly(vinyl alcohol) fiber with the trade name of Kodelan.

The wet-spinning method yields 1–5 denier fibers in large tow bundles at exit speeds up to 100 meters per minute.

The dry-spinning method was developed later for the production of filament yarns. A concentrated solution (30–40%) is extruded out to the atmosphere, the temperature and humidity of which are controlled. Figure 29 shows a diagram of the dry-spinning apparatus (363). It is important that the spinning solution be foamless (364); it may

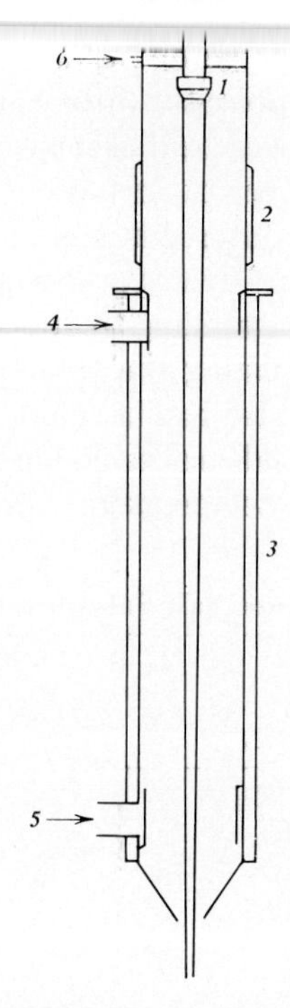

Fig. 29. Schematic drawing of spinning machine for dry spinning poly(vinyl alcohol) aqueous solution (317). Legend: *1*, spinneret; *2*, heater; *3*, steam jacket; *4*, entrance of dry air; *5*, exit of dry air; *6*, entrance of air (temperature- and humidity-controlled).

contain some additives to increase spinnability or improve color (365). Spinning speeds of 300–500 meters per minute can be achieved by this method.

Two other spinning methods have been developed for poly(vinyl alcohol), gel spinning and phase-separation spinning; however, these are not now important commercially (366).

Heat Treatment. In order to stabilize the fiber and give it a measure of water resistance, the fiber is heat treated by a drawing process (370) in hot air, or in a hot salt solution or a fused metal bath (371–373). The temperature for the hot drawing is 160–220°C and its duration is from 15 seconds to 5 minutes; the heat treatment is at 220–230°C. The crystallinity of the fiber increases markedly during this process (327). Table 9 shows the degree of crystallinity of poly(vinyl alcohol) fibers as a function of heat treatment.

In the heat treatment and drawing process, the temperature has more bearing upon swelling, density, and degree of crystallinity than the time of treatment. But

Table 9. Crystallinity of Poly(vinyl Alcohol) Fibers After Heat Treatment (327)

Hot air temp, °C	Time of heating	Crystallinity, %
air-dried	air-dried	30
80	60 min	50
120	100 sec	48
120	5 min	49
160	100 sec	53
200	100 sec	64
225	100 sec	68

both temperature and time affect the hot-water resistance and tenacity. Figure 30 shows a diagram of the spinning and heat-stretching process (328).

Acetalization. The hot-water (90°C) resistance of the heat-treated fiber is good; however, for resistance to boiling water, the fiber must be partly converted to an acetal. The acetalization is normally carried out with formaldehyde using sulfuric acid as catalyst in a 50–70°C bath containing Na_2SO_4. The degree of acetalization (about 30 mole % is required) is controlled by the composition and temperature of the bath, and the time of treatment. The acetalization reaction proceeds mainly in the amorphous region of the fiber. There may also occur some crosslinking of the poly(vinyl alcohol) (see p. 173).

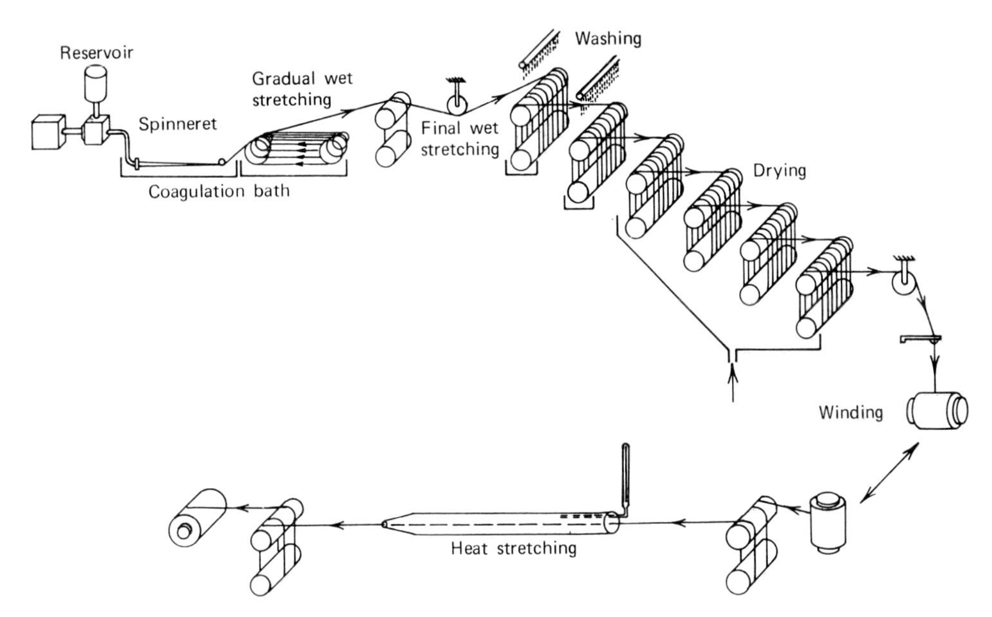

Fig. 30. Flow sheet showing successive stages in the preparation of high-strength, high-modulus poly(vinyl alcohol) fibers. Starting from the top of the drawing, the process is as follows: A hot poly(vinyl alcohol) solution is spun into a salt coagulating bath; the fibers are wet stretched in several stages by means of stepped stretching rollers, washed, dried, and heat stretched (328).

Benzaldehyde or aminoaldehyde which improve the elastic and the dyeing properties of the fiber, respectively, have also been used for acetalization. The commercial products are washed with water after acetalization, then are oil treated and dried.

Properties. The properties of poly(vinyl alcohol) fibers depend on many factors, such as properties of the poly(vinyl alcohol) used for spinning, acetalization, and spinning conditions. Good summaries of the properties and uses of poly(vinyl alcohol) fibers can be found in the literature (305,329,330).

The mechanical properties of the fiber are listed in Table 10. The tenacity of poly(vinyl alcohol) fiber is quite high and comparable to that of nylon. Typical stress–strain curves are shown in Figure 31 (332–335). The temperature dependence of the tenacity is compared with that of other fibers in Figure 32 (334).

The abrasion resistance of poly(vinyl alcohol) fiber is excellent. The moisture regain is similar to that of cotton, but decreases with heat stretching and with acetalization (Fig. 33) (336). Poly(vinyl alcohol) fibers have good dimensional stability, excel-

Table 10. Mechanical Properties of Poly(vinyl Alcohol) Fiber (331)

	Staple and tow		Filament	
	Regular-tenacity	High-tenacity	Regular-tenacity[a]	High-tenacity[b]
breaking tenacity, g/den				
standard	3.8–6.2	6.8–8.0	3.0–4.0	6.0–8.5
wet	3.2–5.0	5.3–6.4	2.1–3.2	5.0–7.6
standard loop	3.0–5.2	5.3–5.6	4.5–6.0	7.0–13.0
standard knot	2.4–4.0	4.7–5.1	2.2–3.0	2.7–4.6
ratio of standard to wet tenacity, %	72–85	78–85	70–80	75–90
breaking elongation, %				
standard	15–26	13–16	17–22	9–22
wet	16–27	14–17	17–25	10–26
elastic recovery, % (3% elongation)	70–85	72–85	70–90	70–90
initial modulus				
g/den	25–70	70–105	60–90	70–180
kg/mm²	300–800	800–1200	700–950	800–2000
specific gravity		1.26–1.30		
commercial regain, %		5.0		
water absorbency, %				
at 20°C, 65% rh	........4.5–5.0.......		3.5–4.5	3.0–5.0
20°C, 20% rh	.....................1.2–1.8.....................			
20°C, 95% rh	.....................10.0–12.0.....................			

[a] Regular-tenacity filament includes filament prepared by blend-spinning of poly(vinyl alcohol) with its derivatives—for example, with amino acetalized poly(vinyl alcohol).

[b] High-tenacity filament includes heat-treated poly(vinyl alcohol) and poly(vinyl formal) filaments.

lent light stability and rot and mildew resistance, and high durability against salts and sea water.

Irradiation with gamma rays or ultraviolet radiation seems to damage poly(vinyl alcohol) fibers less than it does other fibers (337). Poly(vinyl alcohol) fibers have a slightly higher electrical resistance than do cellulose fibers. A linear relationship exists between the mass specific resistance R_s and the moisture content of the fiber (equation 20, where n and k are constants characteristic of the fiber). Table 11 gives an

$$\log R_s = \log k - n \log M \tag{20}$$

approximate comparison of the values of n and k for various fibers (338). The chemical resistance of poly(vinyl alcohol) fibers is very good. The fiber swells in *m*-cresol or phenol and dissolves in 80% formic acid. It also dissolves in chloral hydrate. Table 12

Table 11. Comparison of Electrical Resistance of Different Fibers (338)

Sample	n	$\log k$
poly(vinyl fluoride) fiber	8.0	16.2
nylon-6 (Amilan)	7.5	15.7
cotton	11.2	16.5
cotton	10.7	16.7
viscose	12.0	21.0
ramie	11.7	18.5
wool	14.7	26.6

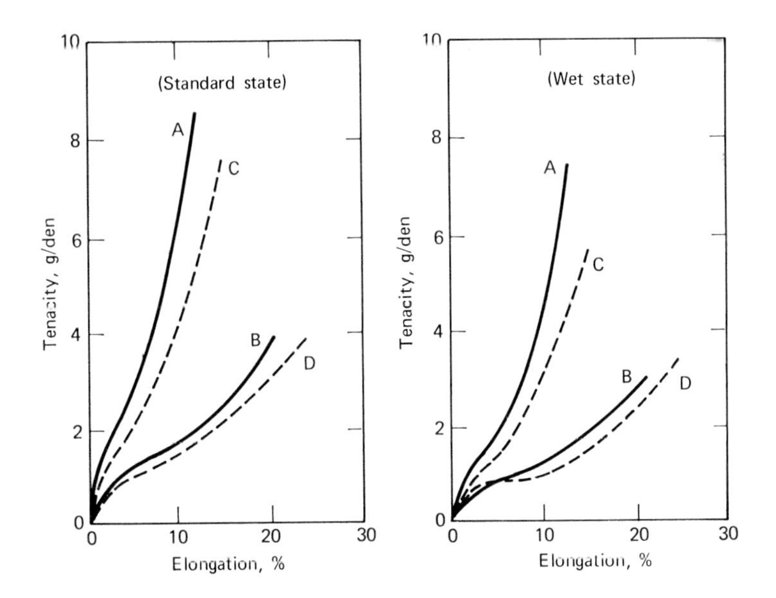

Fig. 31. Stress-strain properties of poly(vinyl alcohol) fiber (vinylon) (334). Key: curve A, high-tenacity filament; curve B, regular-tenacity filament; curve C, high-tenacity staple; curve D, regular-tenacity staple.

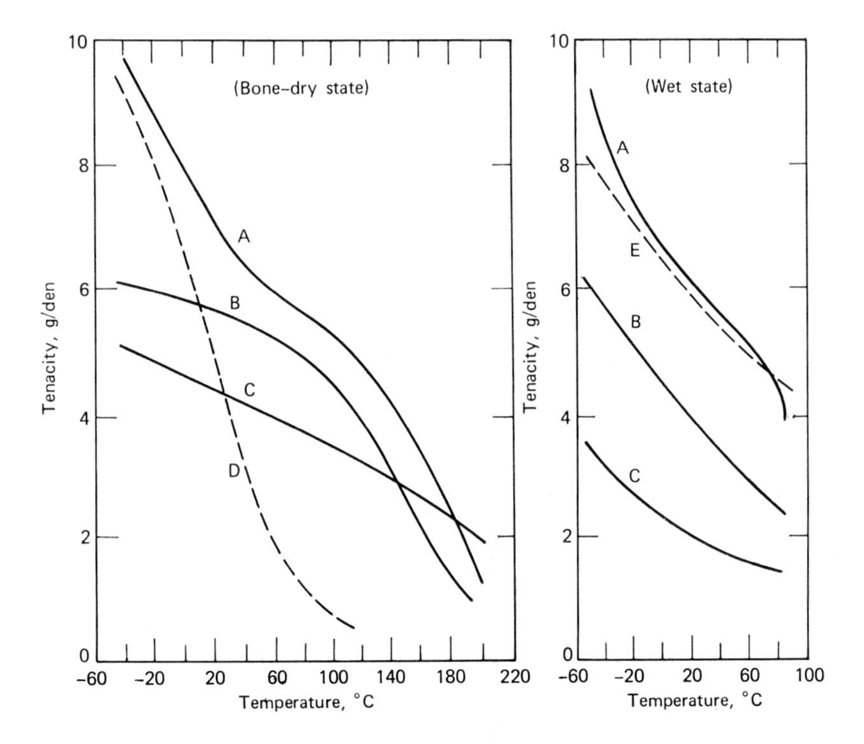

Fig. 32. Dependence of temperature on tenacity (334). Key: curve A, poly(vinyl alcohol) filament yarn; curve B, poly(vinyl formal) spun yarn; curve C, high-tenacity rayon yarn; curve D, polyethylene filament yarn; curve E, poly(ethylene terephthalate) filament yarn.

Table 12. Chemical Resistance of Poly(vinyl Alcohol) Filament Yarn and Poly(vinyl Formal) Yarn (7)

Reagent	Concn, %	Temp, °C	Period, hr	Tenacity retained, % Poly(vinyl alcohol) filament yarn	Poly(vinyl formal) spun yarn
sulfuric acid	1	20	1000	100	101
	10	20	1000	97	104
	70	20	0.1	S[a]	B[b]
hydrochloric acid	1	20	1000	99	107
	10	20	1000	101	83
	37	20	0.1	S[a]	B[a]
	1	70	100	98	102
	10	70	10	93	81
nitric acid	1	20	10	99	106
	10	20	10	94	104
	70	20	0.1	S[a]	S[a]
hydrofluoric acid	10	20	10	99	100
phosphoric acid	10	20	10	99	103
formic acid	40	20	10	98	81
acetic acid	5	100	10	93	84
	100	20	10	101	95
oxalic acid	5	100	10	76	85
sodium hydroxide	1	20	10	99	107
	10	20	10	99	100
	40	20	10	99	106
	10	100	10	102	105
hydrogen peroxide					
pH 7	0.4	20	10	97	104
	0.4	70	10	47	102
pH 11	0.2	70	10	101	100
pH 10	3	70	10	99	94
sodium hypochlorite					
pH 10	0.01	20	10	99	105
pH 11	0.4	20	10	75	95
sodium chlorite					
pH 4	0.07	100	10	95	86
pH 4	0.7	20	10	98	104
peracetic acid					
pH 4	2	20	10	95	102
pH 4	2	100	10	16	87
sodium hydrosulfite	1	70	10	99	96
"Sulfoxite" S,[c] concd	1	100	10	99	98
sodium bisulfite, pH 4	1	100	10	101	101
soap	1	100	10	97	96
sodium carbonate	1	100	10	93	97
sodium chloride	3	100	10	95	100
copper sulfate	3	100	10	95	101
zinc chloride	3	100	10	93	101
acetone	100	20	1000	99	89
amyl alcohol	100	20	1000	99	99
benzene	100	20	1000	101	98
carbon disulfide	100	20	1000	100	107
carbon tetrachloride	100	20	1000	98	100

Table 12 (*continued*)

				Tenacity retained, %	
	Concn, %	Temp, °C	Period, hr	Poly(vinyl alcohol) filament yarn	Poly(vinyl formal) supn yarn
chloroform	100	20	1000	101	103
ether	100	20	1000	100	95
ethyl alcohol	100	20	1000	99	102
ethyl acetate	100	20	1000	100	95
methyl alcohol	100	20	1000	100	95
tetrachloroethane	100	20	1000	98	101
Stoddard solvent	100	70	10	101	99
perchloroethylene	100	100	10	99	102
formaldehyde in water	10	20	1000	92	100
cottonseed oil	100	20	1000	97	108
lard	100	20	1000	98	102
ammonia in water	28	20	1000	98	94
phenol in water	5	20	10	92	61
glycerol	100	100	10	100	99
mineral oil	100	100	10	96	106

[a] S, dissolved.
[b] B, brittle.
[c] Du Pont's trademark for a reducing agent.

shows the influence of chemical agents on the tenacity (7). The softening point of wet poly(vinyl alcohol) fiber is about 120°C (339).

Dyeing and Finishing. For the dyeing of poly(vinyl alcohol) fiber, vat dyes, sulfur dyes, and neutral metal complex dyes are usually employed, and direct azoic and disperse dyes can also be used in particular cases (340,341) (see also DYEING; DYES). In general, the same dyestuffs used for cotton can also be used for poly(vinyl alcohol) fiber, although, because of the reduced number of hydroxyl groups compared with cotton, dark shades are not as easily obtained with substantive dyes. The dye affinity can be increased by acetalizing the poly(vinyl alcohol) fiber with aminoacetaldehyde or other moieties containing aldehyde groups. The microscopic structure of poly(vinyl alcohol) fiber is closely related to dyeing properties (342a). As shown in Figure 34, poly(vinyl alcohol) staple fiber made by the wet-spinning process consists of a film and an inner nucleus. The cross section is bean shaped. Figure 35 shows the cross section of poly(vinyl alcohol) filament fiber.

The wet processing of poly(vinyl alcohol) fibers follows conventional lines with minor exceptions. Fabrics are singed and desized normally. The initial whiteness of poly(vinyl alcohol) fiber is excellent. Bleaching is carried out by hypochlorites or hydrogen peroxide. The drying must be done carefully since the wet softening point of the fiber is much lower than the dry softening point. Therefore the rate of moisture removal is important (342b).

In a study of preparation of poly(vinyl alcohol) fiber from stereoregular poly(vinyl alcohol) obtained by hydrolyzing poly(vinyl trifluoroacetate), it was found that only minimal acetalization was necessary to produce water-resistant fibers of very high tenacity (343).

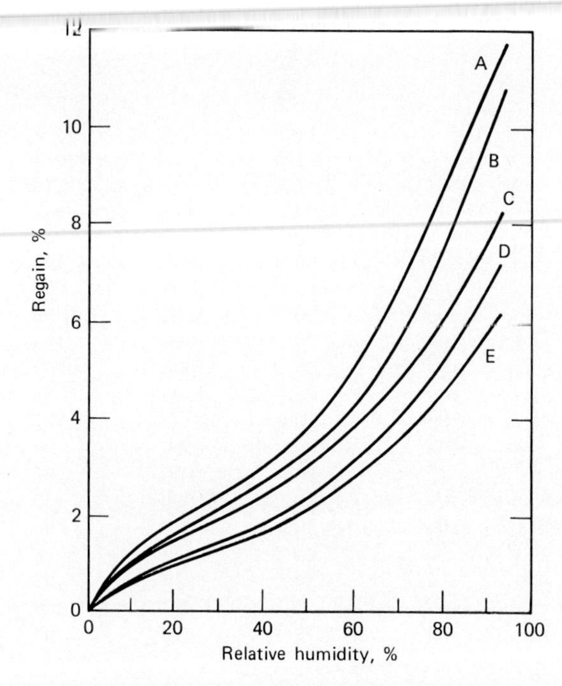

Fig. 33. Sorption isotherm of poly(vinyl formal) fiber at 40°C (336). Degree of formalization (mole %): curve A, 9.6; curve B, 20.2; curve C, 32.2; curve D, 59.6; curve E, 69.8.

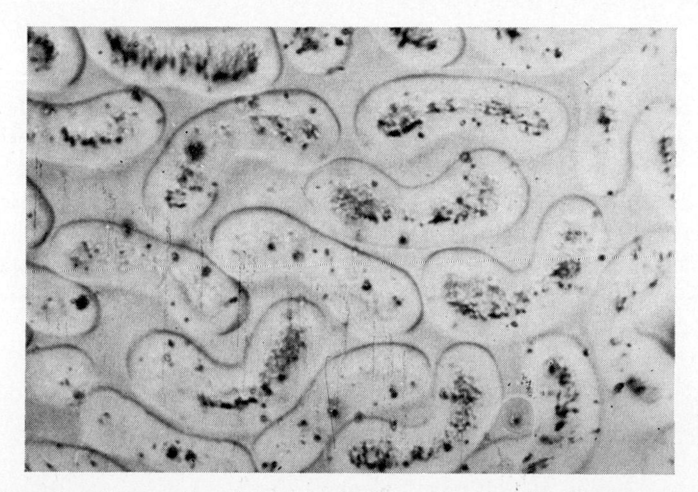

Fig. 34. Cross section of vinal staple fiber. Courtesy Air Reduction Co.

Applications. Poly(vinyl alcohol) fibers are used both in the industrial and apparel or home fibers market. Fish nets, ropes, specialty tire cord, belts, hoses, canvas, filter cloth, agricultural netting, scrim fabrics, and reinforcement for polyester plastics illustrate their industrial applications (344). Work clothes, school uniforms, raincoats, and all kinds of apparel are made from poly(vinyl alcohol) fiber. In Japan, water-soluble poly(vinyl alcohol) fiber is also used as a binder in nonwoven fabrics and for specialty papers (345–348).

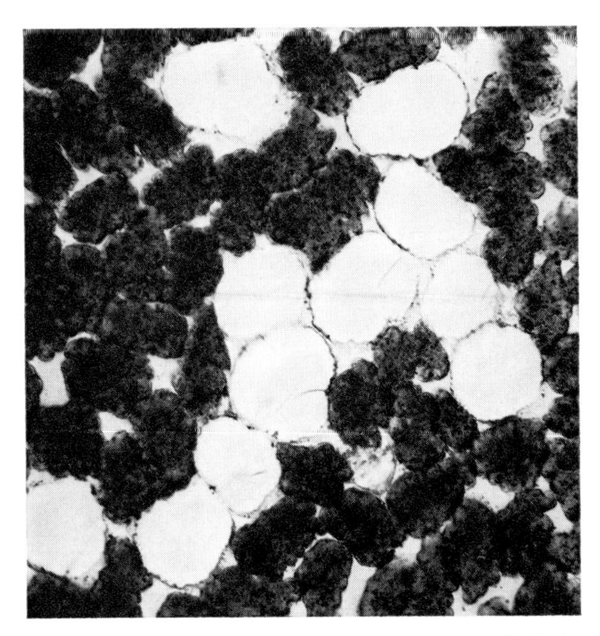

Fig. 35. Cross section of vinal filament fiber. Courtesy Air Reduction Co.

Films (349)

In Japan, a large amount of poly(vinyl alcohol) film is sold as packaging material and as a mold-release agent. The film has good transparency, luster, antielectrostatic properties, and toughness; it can be printed easily, has good gas barrier properties, and has good resistance to organic chemicals. The disadvantage of the film is its poor water resistance. There are some applications for which it is desirable to have a film soluble in cold or hot water, eg, for detergent packaging. Heat treatment of the film increases its water resistance (350) by increasing the degree of crystallinity (351).

For ordinary packaging and mold-release film a fully hydrolyzed poly(vinyl alcohol) having a degree of polymerization of at least 1000 is used. Partially hydrolyzed or chemically modified poly(vinyl alcohol) is used for such applications as film soluble in cold water.

Manufacture. The poly(vinyl alcohol) film is produced by either casting or extrusion.

In the casting process, a 10–20% poly(vinyl alcohol) solution containing plasticizers, such as glycols or glycerol, and other additives is poured onto a drying drum or belt (352). The water is evaporated and the dry film is then heat treated at 120°C. If desired, an antiblocking powder such as colloidal silica is applied to the film. The film thickness is from 20 to 100 μ. When extruding poly(vinyl alcohol) film, the polymer must be plasticized with water because the melting point of dry poly(vinyl alcohol) is close to the decomposition temperature. The Kurashiki Rayon Company has developed a specially designed extruder capable of handling poly(vinyl alcohol) pellets (353). The casting process is preferred for cold-water-soluble poly(vinyl alcohol) films, whereas the extrusion process is used to make water-resistant films.

Properties. The properties of poly(vinyl alcohol) film depend on the grade of the poly(vinyl alcohol) used to make the film. Properties of typical commercial film are listed in Table 13 (see also FILMS AND SHEETING). Since the polymer is relatively hydrophilic no antistatic agents have to be used for the film. The moisture permeability of poly(vinyl alcohol) film is of the same order as that of cellophane film. The oil and solvent resistance of the film is also very good.

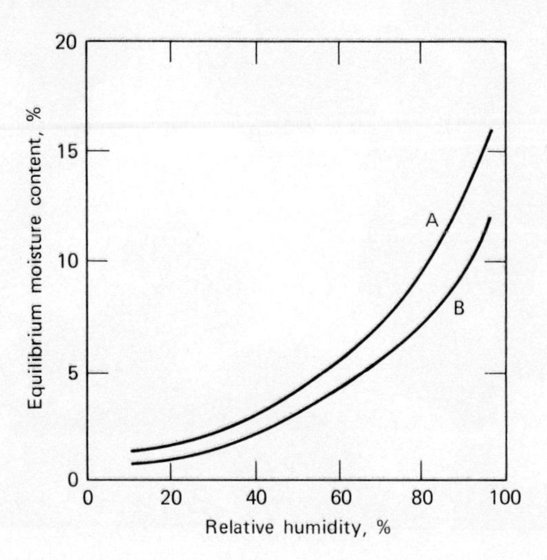

Fig. 36. Relationship between equilibrium moisture content of poly(vinyl alcohol) film and relative humidity (354). Key: curve A, plasticized poly(vinyl alcohol) film; curve B, unplasticized poly(vinyl alcohol) film.

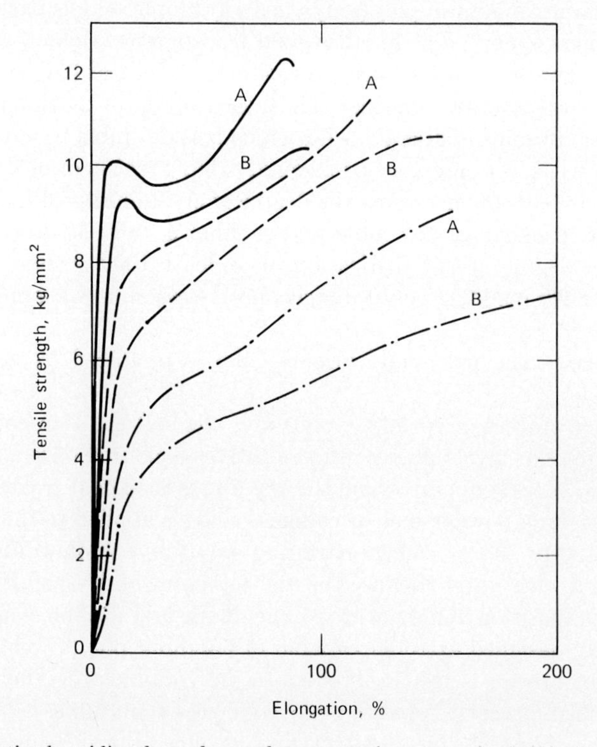

Fig. 37. Relative humidity dependence of stress–strain curves of poly(vinyl alcohol) films (354). Rate of elongation, 1000%/minute. Curves A refer to unplasticized film, curves B to plasticized film. Relative humidity (at 20°C): (——) 40%; (---) 65%; (—-—-—) 90%.

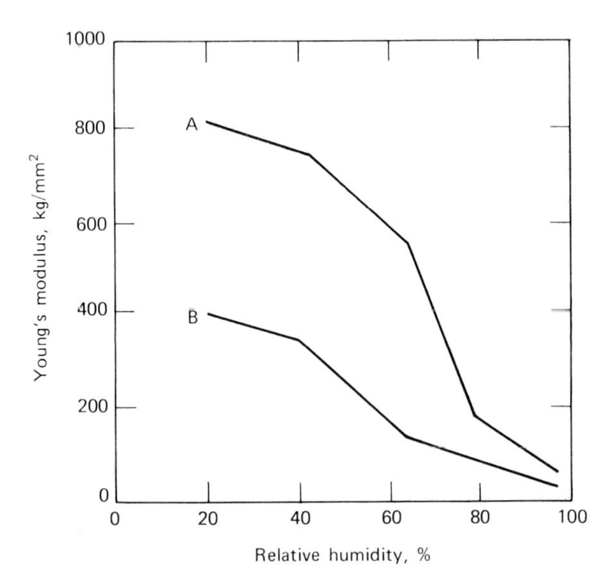

Fig. 38. Young's modulus versus relative humidity (at 20°C) (354) for poly(vinyl alcohol) film unplasticized (curve A) and plasticized (curve B).

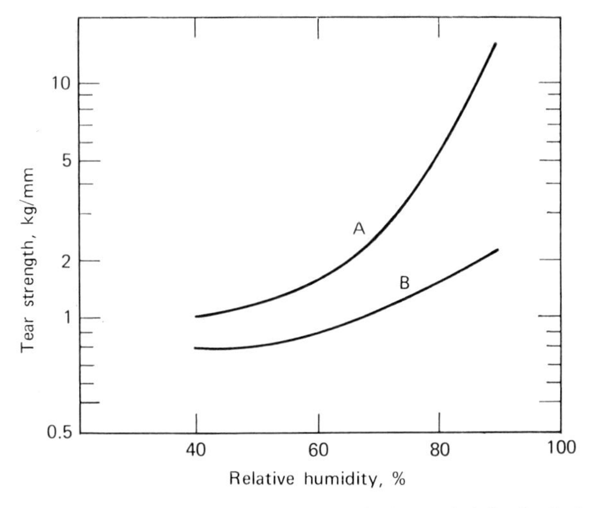

Fig. 39. Tear strength versus relative humidity (354) for poly(vinyl alcohol) film plasticized (curve A) and unplasticized (curve B). Tear strength was determined by the Elmendorff method.

The film can be heat sealed directly, or by impulse or high-frequency dielectric sealing. The heat-sealing temperature is a function of the water content of the film and varies from 203°C at 12% moisture to 230°C at 0–14% moisture. The printability of the film is excellent. As mentioned above, the film is quite hygroscopic. The equilibrium water content is shown in Figure 36 (354). It can be seen that the plasticized film has a slightly higher water content. The mechanical properties of poly(vinyl alcohol) film are again very much a function of the water content of the film. Figures 37–39 show the stress–strain curves, Young's modulus, and Elmendorff tear strengths as functions of humidity for different poly(vinyl alcohol) films and other films. Poly(vinyl alcohol) film also has high tensile and tear strength.

Table 13. Properties of Commercial Poly(vinyl Alcohol) for Packaging Films (353)

clarity,[a] % light transmitted	60–66
gloss,[b] % light reflected	81.5
moisture-vapor transmission rate,[c] g/m/24 hr (30 thickness)	1500–2000
tear strength,[d] kg/mm (Elmendorff)	
unplasticized film	2.4–7.2
plasticized film	15–85
tensile strength,[d] kg/mm²	
unplasticized film	6.2–10.5
plasticized film	4.5–6.5
elongation,[d] %	
unplasticized film	130–180
plasticized film	150–400

[a] Photometer (white-light source).
[b] Glossmeter incident light (60° angle).
[c] Relative humidities of faces of the films are 0–90% ± 2%; temperature 40 ± 1°C.
[d] Tested at 50% relative humidity; 72°F.

The gas-barrier properties of poly(vinyl alcohol) film are better than those of poly(vinylidene chloride) film. They decrease, however, with increasing humidity. For example, Figure 40 shows the oxygen permeability of poly(vinyl alcohol) film and the glass-transition temperature as functions of relative humidity. In order to improve the oxygen permeability of poly(vinyl alcohol) film at high relatively humidity, the film has been coated with poly(vinylidene chloride) (355). The gas transmissions of oxygen, carbon dioxide, and nitrogen are 0.01, 0.02, and <0.003 g/m²/24 hr-atm, respectively. The flavor-retention properties of poly(vinyl alcohol) film are also excellent (356a).

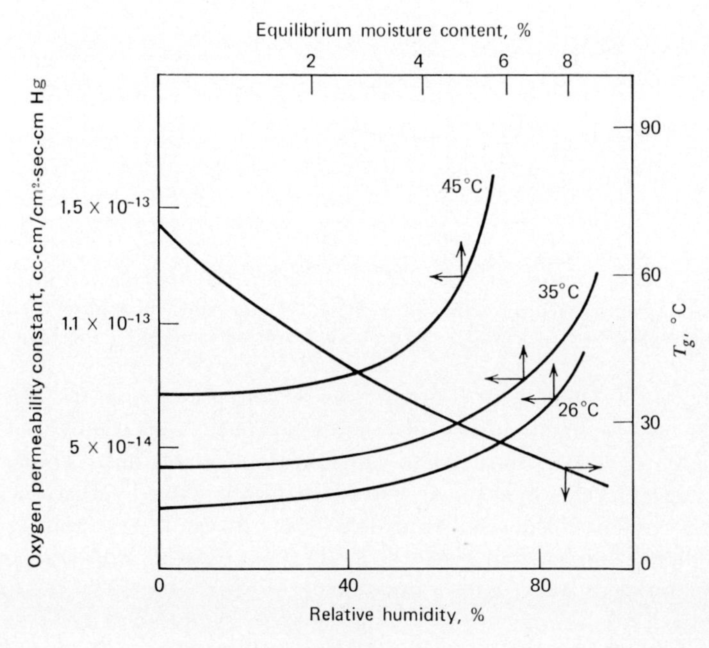

Fig. 40. Relationship between oxygen permeability constant and relative humidity for poly(vinyl alcohol) film (354).

Poly(vinyl alcohol) film can further be improved by crosslinking with glyoxal, urea–formaldehyde, or melamine–formaldehyde resins, or other crosslinking agents. Plasticizers for poly(vinyl alcohol) are discussed on p. 162.

Applications. Poly(vinyl alcohol) film has relatively poor water resistance and should therefore be used only where this property does not have an adverse effect. It can be used for general packaging applications, mainly for textiles. Specialty poly-(vinyl alcohol) film is used as cold-water-soluble film, usually plasticized by phosphate esters (356b) or glycerol (357) and glycerol derivatives (358). A good summary describing the cold-water-soluble poly(vinyl alcohol) film is given in Ref. 359.

Vinyl alcohol copolymers have also been suggested for cold-water-soluble applications (360–362). The following applications for the water-soluble films have been suggested: washing bags for hospital use; unit packages for washing agents and other chemicals; remoistenable adhesives for wallpaper. Because poly(vinyl alcohol) film has very high gas-barrier properties (especially oxygen impermeability) and flavor retention properties, it can be used in food packaging. Moisture permeability is a disadvantage in poly(vinyl alcohol) film; however, this disadvantage can be overcome by lamination with other plastic films with good water resistance and humidity resistance. It is important to prevent oxidation of foods, to ensure their shelf life, and to prevent the escape of flavor. Composite films mainly made of poly(vinyl alcohol) have good flavor retention, oxygen impermeability, and oil resistance, which is important for the packaging of oily foods.

Poly(vinyl alcohol) film is also used as a mold-release film in polyester vacuum casting. A light-polarizing film can be obtained when an iodine-containing poly(vinyl alcohol) film is stretched. These films are used in sunglasses and in other optical applications.

Bibliography

General References

1. *Properties and Applications of Poly(vinyl Alcohol)*, Society of the Chemical Industry (London) Monograph No. 30, London, 1968.
   A symposium on preparation, properties, and uses of poly(vinyl alcohol).
2. G. P. Colgan and P. Plante, "Poly(vinyl Alcohol)" in *Paper Coating Additives*, TAPPI Monograph No. 25, Technical Association of the Pulp and Paper Industry, New York, 1963, pp. 117–136.
   Discusses the use of poly(vinyl alcohol) in paper coating.
3. F. Kainer, *Polyvinylalkohole*, F. Enke, Stuttgart, 1949, in German.
   A large number of use patents are reviewed; no emphasis on scientific aspects of poly(vinyl alcohol).
4. T. Kominami, "Poly(vinyl Alcohol)," in I. Sakurada, ed., *Vinyl Acetate Resins*, Kobunshi Kagaku Publications, Kyoto, 1964, p. 233, in Japanese.
5. F. O. W. Meyer, "The Application of Poly(vinyl Alcohol) in Medicine," *Pharmazie* **4**, 264 (1949).
   Discusses the application of poly(vinyl alcohol) in medicine mainly in Germany.
6. S. A. Miller, *Acetylene*, Vol. II, Academic Press, Inc., New York, 1966, pp. 289–304.
   General review of poly(vinyl alcohol).
7. T. Osugi, "PVA Fibers," in H. Mark, S. M. Atlas, and E. Cernia, eds., *Man-Made Fibers, Science and Technology*, Vol. 3, Interscience Publishers, Inc., a division of John Wiley & Sons, Inc., New York, 1968, p. 245.
   This review deals with the application of poly(vinyl alcohol) as a fiber.
8. J. G. Pritchard, *Poly(vinyl Alcohol), Basic Properties and Uses*, Gordon and Breach, Inc., New York, 1970.
   Emphasis is on scientific aspects of poly(vinyl alcohol).

9a. I. Sakurada, ed., *Poly(vinyl Alcohol), First Osaka Symposium*, Kobunshi Gakkai, Tokyo, 1955.
9b. I. Sakurada, ed., *Poly(vinyl Alcohol), Second Osaka Symposium*, Kobunshi Gakkai, Tokyo, 1958.
9c. I. Sakurada, ed., *Poly(vinyl Alcohol), Third Osaka Symposium*, Kobunshi Gakkai, Tokyo, 1963.
Refs. 9b and 9c deal extensively with the chemistry and technology of poly(vinyl alcohol). They are written in Japanese.
10. P. Schneider, in Houben-Weyl *Methoden der Organischen Chemie*, E. Müller, ed., Vol. 14, Part 2, Georg Thieme Verlag, Stuttgart, 1963.
Gives methods for the preparation of poly(vinyl alcohol), pp. 697–703; reviews the reactions of poly(vinyl alcohol), pp. 716–730.
11. K. Toyoshima, "Development and Utilization of PVA," *Kobunshi Kako* **18,** 182–581 (1969).
This is mainly a review of the uses of poly(vinyl alcohol) in textile and paper applications.
12. S. N. Ushakov, *Poly(vinyl Alcohol) and its Derivatives* (in Russian), Izdatel. Akad. Nauk SSSR, Leningrad, 1960.
13. H. Warson, "Poly(vinyl Alcohol)," in S. A. Miller, ed., *Ethylene*, E. Benn Ltd., London, 1969, pp. 1019–1051.
Extensive review of poly(vinyl alcohol).
14. *Elvanol, Poly(vinyl Alcohol)*, E. I. du Pont de Nemours & Co., Inc., Wilmington, Del., July 1968.
Stresses all American applications for poly(vinyl alcohol).
15. *Poval, Poly(vinyl Alcohol)*, Kurashiki Rayon Co., Ltd., Osaka, 1966.
Very extensive booklet describing properties and applications of poly(vinyl alcohol) (no references); stresses applications in all fields.
16. *Vinol, Poly(vinyl Alcohol)*, Air Reduction Chemical and Carbide Co., New York, 1963.
Emphasizes paper and textile applications.

References

17. J. M. Hay and D. Lyon, *Nature* **216,** 790 (1967).
18. A. M. Sladkov and G. S. Petrov, *Zh. Obshch. Khim.* **24,** 450 (1954).
19. A. Gero, *J. Org. Chem.* **19,** 469 (1954).
20. A. N. Nesmeyanov, I. F. Lutsenko, and R. M. Khomutov, *Dokl. Akad. Nauk SSSR* **120,** 1049 (1958).
21. R. C. Fuson, R. E. Foster, W. J. Shenk, Jr., and E. W. Maynert, *J. Am. Chem. Soc.* **67,** 1937 (1945).
22. W. O. Herrmann and W. Haehnel (to Consortium für Elektrochemische Industrie), U.S. Pat. 1,672,156 (1928); also Ger. Pat. 450,286 (1927).
23. H. Staudinger, K. Frey, and W. Starck, *Ber.* **60,** 1782 (1927).
24. W. O. Herrmann and W. Haehnel, *Ber.* **60,** 1658 (1927).
25. I. Sakurada, "Chemistry of Poly(vinyl Alcohol)," in Ref. 9a, pp. 1–12.
26. R. Inone and I. Sakurada, *Kobunshi Kagaku* **7,** 211 (1950).
26a. M. K. Lindemann, unpublished work.
27. C. S. Marvel and G. E. Inskeep, *J. Am. Chem. Soc.* **65,** 1710 (1943).
28. A. Voss and W. Heuer (to Farbwerke Hoechst A.G.), U.S. Pat. 2,860,124 (1958); also Ger. Pat. 946,848 (1956).
29. H. E. Harris and J. G. Pritchard, *J. Polymer Sci.* [A] **2,** 3673 (1964).
30. W. O. Herrmann, W. Haehnel, and H. Berg (to Chemische Forschungs Gesellschaft m.b.H.), U.S. Pat. 2,109,883 (1938); also Ger. Pat. 642,531 (1932).
31. H. Berg (to Chemische Forschungs Gesellschaft m.b.H.), U.S. Pat. 2,227,997 (1941).
32. T. Kominami (to Kurashiki Rayon Co., Ltd.), U.S. Pat. 3,278,505 (1966).
33. E. Hackel, in Ref. 1, pp. 1–17.
34. H. Shohata, in Ref. 1, pp. 18–45.
35. J. E. Bristol and W. B. Tanner (to E. I. du Pont de Nemours & Co., Inc.), U.S. Pat. 2,734,048 (1956).
36. W. B. Tanner (to E. I. du Pont de Nemours & Co., Inc.), U.S. Pat. 3,296,236 (1967).
37. J. E. Bristol (to E. I. du Pont de Nemours & Co., Inc.), U.S. Pat. 3,487,060 (1969).
38. G. P. Waugh and W. O. Kenyon (to Eastman Kodak Co.), U.S. Pat. 2,642,419 (1953).

39. L. M. Germain (to Shawinigan Chemicals Ltd.), U.S. Pat. 2,643,994 (1953).
40. K. G. Blaikie and R. N. Crozier, *Ind. Eng. Chem.* **28,** 1155 (1936).
41. R. D. Dunlop, *Fiat Final Rept. No.* **1110** (1947).
42. A. Hill and D. K. Hale, *Bios Rept. No.* **1418** (1946); also *PB Rept. No.* **81539**.
43. W. O. Herrmann, W. Haehnel, and H. Berg (to Chemische Forschungs Gesellschaft m.b.H.), Ger. Pat. 642,531 (1937).
44. W. Starck (to Farbwerke Hoechst A.G.), Ger. Pat. 874,664 (1953).
45. G. H. McCain, in *Macromolecular Synthesis,* Vol. 1, C. G. Overberger, ed., John Wiley & Sons, Inc., New York, 1963.
46. C. A. Neros and N. V. Seeger (to Diamond Alkali Co.), U.S. Pat. 3,141,003 (1964).
47. K. Fujii, T. Mochizuki, S. Imoto, J. Ukida, and M. Matsumoto, *J. Polymer Sci.* [A] **2,** 2327 (1964).
48. S. Murahashi, S. Nozakura, and M. Sumi, *J. Polymer Sci.* [B] **3,** 245 (1965).
49. S. Okamura, T. Kodama, and T. Higashimura, *Makromol. Chem.* **53,** 180 (1962).
50. G. Ohbayashi, S. Nozakura, and S. Murahashi, *Bull. Chem. Soc.* (*Japan*) **42,** 2729 (1969).
51. S. Murahashi, H. Yuki, T. Sano, U. Yonemura, H. Tadokoro, and Y. Chatani, *J. Polymer Sci.* **62,** S-77 (1962).
52. Air Reduction Co., Brit. Pat. 1,129,230 (1968).
53. K. Kikukawa, S. Nozakura, and S. Murahashi, *Kobunshi Kagaku* **25,** 19 (1968).
54. S. R. Sandler and E. C. Leonard (to The Borden Co.), U.S. Pat. 3,422,072 (1969).
55. Consortium für Elektrochemische Industrie G.m.b.H., Fr. Pat. 1,365,127 (1964).
56. T. Imoto and T. Matsubara, *J. Polymer Sci.* [A] **2,** 4573 (1964).
57. Consortium für Elektrochemische Industrie G.m.b.H., Fr. Pat. 1,361,830 (1964).
58. N. D. Field and J. R. Schaefgen, *J. Polymer Sci.* **58,** 533 (1962).
59. S. Murahashi, *Pure Appl. Chem.* **15,** 435 (1967).
60. K. Shibatani, *Polymer J.* (*Japan*) **1,** 348 (1970).
61. H. Neukom, *Helv. Chim. Acta* **32,** 1233 (1949).
62. T. Osugi and K. Hishimoto, *Kobunshi Kagaku* **5,** 193 (1948).
63. H. Buc, *Ann. Chim.* (*Paris*) **8,** 431 (1963).
64. S. Saito and H. Okuyama, *Kolloid Z.* **139,** 150 (1954).
65. W. O. Herrmann and W. Haehnel, *Ber.* **60,** 1658 (1927).
66. H. Maeda, T. Kawai, and S. Sekii, *J. Polymer Sci.* **35,** 288 (1959).
67. S. Saito, *J. Colloid Interface Sci.* **24,** 227 (1967).
68. S. Hayashi, C. Nakano, and T. Motoyama, *Kobunshi Kagaku* **21,** 300 (1964); **20,** 303 (1963).
69. M. Matsumoto and Y. Ohyanagi, *J. Polymer Sci.* **31,** 225 (1958).
70. H. L. Frisch and S. Al-Madfai, *J. Am. Chem. Soc.* **80,** 3561 (1958).
71. M. Matsumoto and K. Imai (to Kurashiki Rayon Co.), U.S. Pat. 3,234,195 (1966).
72. K. Nakanishi and M. Kurata, *Bull. Chem. Soc.* (*Japan*) **33,** 152 (1960).
73. K. Nakanishi and M. Kurata, *Busseiron Kenkyu No.* **101,** 105 (1956); through *Chem. Abstr.* **51,** 6282h (1957).
74. D. W. Levi, P. C. Scherer, W. L. Hunter, and K. Washimi, *J. Polymer Sci.* **28,** 481 (1958).
75. D. W. Levi, P. C. Scherer, K. Washimi, and W. L. Hunter, *J. Appl. Polymer Sci.* **1,** 127 (1959).
76. E. Wolfram and M. Nagy, *Kolloid Z. Z. Polymere* **227,** 86 (1968).
77. M. Matsumoto and K. Imai, *J. Polymer Sci.* **24,** 125 (1957).
78. R. Naito, *Kobunshi Kagaku* **15,** 597 (1958).
78a. H. C. Haas and A. S. Makas, *J. Polymer Sci.* **46,** 528 (1960).
79. K. A. Stacey and P. Alexander, *Ric. Sci.* **25A,** 889 (1954).
80. H. Maeda, T. Kawai, and S. Seki, *J. Polymer Sci.* **35,** 288 (1959).
81. A. Amaya and R. Fujishiro, *Bull. Chem. Soc.* (*Japan*) **29,** 361, 830 (1956).
82. H. Staudinger and H. Warth, *J. Prakt. Chem.* **155,** 261 (1940).
83. I. Sakurada and T. Chiba, *Kogyo Kagaku Zasshi* **47,** 135 (1944).
84. M. Matsumoto and K. Imai, *J. Polymer Sci.* **24,** 125 (1957).
85. A. Nakajima, in Ref. 9a.
86. H. G. Elias, *Makromol. Chem.* **54,** 78 (1962).
87. M. Matsumoto and Y. Oyanagi, *Kobunshi Kagaku* **17,** 191 (1960).
88. T. Kuroiwa, *Kobunshi Kagaku* **9,** 253 (1952).
89. T. Takagi and T. Isemura, *Bull. Chem. Soc.* (*Japan*) **33,** 437 (1960).

90. A. Beresniewicz, *J. Polymer Sci.* **39**, 63 (1959).
91. S. Noma, "Chemical Reactions of Poly(vinyl Alcohol)," in Ref. 9a, pp. 81–103.
92. F. Kainer, in Ref. 3, pp. 55–92.
93. H. Deuel and H. Neukom, *Makromol. Chem.* **3,** 13 (1949).
94. S. Saito, H. Okuyama, H. Kishimoto, and Y. Fujiyama, *Kolloid Z.* **144,** 41 (1955).
95. H. Thiele and H. Lamp, *Kolloid Z.* **173,** 63 (1960).
96. E. P. Irany, *Ind. Eng. Chem.* **35,** 90 (1943).
97. K. Bolewski and B. Rychly, *Kolloid Z. Z. Polymere* **228,** 48 (1968).
98. J. Mrazek, *Chem. Prumysl* **7**, 567 (1957).
99. N. Okada and I. Sakurada, *Kobunshi Kagaku* **9,** 13 (1952); *Bull. Inst. Chem. Res. Kyoto Univ.* **26,** 94 (1951).
100. S. Kato, Y. Tsuzuki, and S. Kitajima, *Bull. Chem. Soc. (Japan)* **34,** 1107 (1961).
101. S. N. Ushakov and P. Tudorin, *Vysokomolekul. Soedin.* **6,** 934 (1964).
102. J. H. Haslam, *Adv. Chem.* **23,** 272 (1959).
103. F. K. Signaigo (to E. I. du Pont de Nemours & Co., Inc.), U.S. Pat. 2,518,193 (1950).
104. J. D. Crisp (to E. I. du Pont de Nemours & Co., Inc.), U.S. Pat. 3,518,242 (1970).
105. W. Diepold, *Explosivstoffe* **17,** 2 (1970).
106. R. V. Jones (to Phillips Petroleum Co.), U.S. Pat. 2,623,037 (1949).
107. W. Heuer and W. Starck (to I. G. Farbenindustrie A.G.), Ger. Pat. 745,683 (1938).
108. W. H. Sharkey (to E. I. du Pont de Nemours & Co., Inc.), U.S. Pat. 2,395,347 (1942).
109. A. Takahashi, M. Nagasawa, and I. Kgawa, *Kogyo Kagaku Zasshi* **61,** 1614 (1958).
110. R. Asami and W. Tokura, *Kogyo Kagaku Zasshi* **62,** 1593 (1959).
111. I. M. Fingauz, A. F. Vorobeva, G. A. Shirikova, and M. P. Dokuchaeva, *J. Polymer Sci.* **56,** 245 (1962).
112. D. D. Reynolds and W. O. Kenyon, *J. Am. Chem. Soc.* **72,** 1584 (1950).
113. J. L. R. Williams and D. G. Borden, *Makromol. Chem.* **73,** 203 (1964).
114. R. E. Ferrel, H. S. Olcott, and H. Fraenkel-Conrat, *J. Am. Chem. Soc.* **70,** 2101 (1948).
115. G. C. Daul, J. D. Reid, and R. M. Reinhardt, *Ind. Eng. Chem.* **46,** 1042 (1954).
116. K. Ashida, *Kobunshi Kagaku* **10,** 117 (1953).
117. Wacker-Chemie, Brit. Pat. 995,489 (1965).
118. P. Schneider, in Ref. 10.
119. H. Staudinger and M. Haeberle, *Makromol. Chem.* **9,** 52 (1952).
120. G. D. Jones (to General Aniline and Film Corp.), U.S. Pat. 2,536,980 (1951).
121. C. J. Berninger, R. C. Degeise, L. G. Donaruma, A. G. Scott, and E. A. Tomic, *J. Appl. Polymer Sci.* **7,** 1797 (1963).
122. M. Tsuda, *Makromol. Chem.* **72,** 174 (1964).
123. I. E. Muskat and F. Strain (to Columbia-Southern Chemical Co.), U.S. Pat. 2,592,058 (1958).
124. A. J. Seavell, *J. Oil Colourists Chem. Assoc.* **39,** 99 (1956).
125. A. E. J. Rheineck, *J. Am. Oil Chem. Soc.* **28,** 456 (1951).
126. E. F. Izard (to E. I. du Pont de Nemours & Co., Inc.), U.S. Pat. 2,169,250 (1939).
127. I. S. Okhrimenko and E. B. Dyakonova, *Vysokomolekul. Soedin.* **6,** 1891 (1964).
128. A. M. Paquin, *Z. Naturforsch.* **1,** 518 (1946).
129. K. Matsubayashi and M. Matsumoto (to Kurashiki Rayon Co.), U.S. Pat. 3,193,534 (1965).
130. I. Sakurada, A. Nakajima, and K. Shibatani, *J. Polymer Sci.* [A] **2,** 3545 (1964).
131. L. M. Minsk (to Eastman Kodak Co.), U.S. Pat. 2,725,372 (1955).
132. S. Imoto and T. Igashira (to Kurashiki Rayon Co.), U.S. Pat. 3,148,142 (1964).
133. J. Cerny and O. Wichterle, *J. Polymer Sci.* **30,** 501 (1958).
134. J. C. Lukman (to The Borden Co.), U.S. Pat. 3,125,556 (1964).
135. S. G. Cohen, H. C. Haas, and H. Slotnick, *J. Polymer Sci.* **11,** 193 (1953).
136. R. W. Roth, L. J. Patella, and B. L. Williams, *J. Appl. Polymer Sci.* **9,** 1083 (1965).
137. A. De Pauw, *Ind. Chim. Belge* **20,** Spec. No. 363 (1955).
138. L. Alexandru, M. Opris, and A. Ciocanel, *J. Polymer Sci.* **59,** 129 (1962).
139. K. Billig (to I. G. Farbenindustrie A.G.), Ger. Pat. 738,869 (1936).
140. K. Billig, *PB Rept. No.* **32987** (1937).
141. M. Tsunooka, N. Nakajo, M. Tanaka, and N. Murata, *Kobunshi Kagaku* **23,** 451 (1966).
142. M. K. Lindemann (to Air Reduction Co.), U.S. Pat. 3,505,303 (1970).
143. H. Ito, *Kogyo Kagaku Zasshi* **63,** 338 (1960). S. N. Ushakov and S. I. Kirillova, *Zh. Priklad. Khim.* **22,** 1094 (1949).

144. L. C. Flowers, *J. Electrochem. Soc.* **111,** 1239 (1964).
145. G. C. Tesoro, U.S. Pat. 3,031,435 (1960).
146. M. Hida, *Kogyo Kagaku Zasshi* **55,** 275 (1952).
147. J. G. Pritchard, in Ref. 8, p. 92.
148. H. J. Hagemeyer (to Eastman Kodak Co.), U.S. Pat. 2,484,502 (1949).
149. S. Okamura, K. Hayashi, M. Hirata, and R. Kuroda, *Large Radiation Sources in Ind., Proc. Conf. Warsaw, 1959,* **II,** 341 (1960).
150. B. G. Ranby, *Makromol. Chem.* **42,** 68 (1960).
151. M. Anavi and A. Zilkha, *Eur. Polymer J.* **5,** 21 (1969).
152. F. Mashio, T. Nakano, and Y. Kimura, *Kogyo Kagaku Zasshi* **57,** 365 (1954).
153. J. G. Pritchard, in Ref. 8, pp. 98–99.
154. E. J. Goethals and G. Natus, *Makromol. Chem.* **116,** 152 (1968).
155. S. Saito and H. Okuyama, *Kolloid Z.* **139,** 150 (1954).
156. M. Nagano and Y. Yoshioka, *Kobunshi Kagaku* **9,** 19 (1952).
157. M. M. Zwick, *J. Appl. Polymer Sci.* **9,** 2393 (1965).
158. S. Hayashi, C. Nakano, and T. Motoyama, *Kobunshi Kagaku* **20,** 303 (1963).
159. K. Imai and M. Matsumoto, *J. Polymer Sci.* **55,** 335 (1961).
160. C. Dittmar and W. J. Priest, *J. Polymer Sci.* **18,** 275 (1955).
161. G. Centola, D. Borruso, and G. Prati, *Gazz. Chim. Ital.* **85,** 1468 (1955).
162. N. Platzer, *Mod. Plastics* **28** (10), 142 (1951).
163. R. C. Schulz, *Kolloid Z.* **182,** 99 (1962).
164. P. Schneider, in Ref. 10, p. 723.
165. B. Kaesche-Krischer, *Chem. Ing. Tech.* **37,** 944 (1965).
166. T. Yamaguchi and M. Amagasa, *Kobunshi Kagaku* **18,** 645 (1961).
167. K. Ettre and P. F. Varadi, *Anal. Chem.* **35,** 69 (1963).
168. Y. Tsuchiya and K. Sumi, *J. Polymer Sci.* [A-1] **7,** 3151 (1969).
169. B. Kaesche-Krischer and H. J. Heinrich, *Chem. Ing. Tech.* **32,** 740 (1960).
170. B. Duncalf and A. S. Dunn, *Soc. Chem. Ind.* (*London*) Monograph No. 26, Gordon & Breach Publishers, New York, 1967, p. 162.
171. K. Schlaepfer, *Schweiz. Arch. Angew. Wiss. Techn.* **31,** 154 (1965).
172. G. Mino, S. Kaizerman, and F. Rasmussen, *J. Polymer Sci.* **39,** 523 (1959).
173. M. Shiraishi and M. Matsumoto, *Kobunshi Kagaku* **19,** 722 (1962).
174. A. A. Miller (to General Electric Co.), U.S. Pat. 2,897,127 (1959).
175. G. J. Arquette and L. Reich (to Air Reduction Co.), U.S. Pat. 3,061,458 (1962).
176. T. Tsunoda and T. Yamaoka, *J. Appl. Polymer Sci.* **8,** 1379 (1964).
177. T. Takakura, G. Takayama, and J. Ukida, *J. Appl. Polymer Sci.* **9,** 3217 (1965).
178. I. Sakurada and S. Matsuzawa, *Kobunshi Kagaku* **17,** 687, 693 (1960).
179. M. Matsumoto and A. Danno, *Large Radiation Sources in Industry,* International Atomic Energy Commission, Vienna, 1960, p. 331.
180. I. Sakurada, A. Nakajima, and H. Aoki, *Mem. Fac. Eng. Kyoto Univ.* **21,** 84, 94 (1959).
181. B. S. Bernstein, G. Odian, G. Orban, and S. Tirelli, *J. Polymer Sci.* [A] **3,** 3405 (1965).
182. K. Toyoshima, in Ref. 11, p. 506.
183. I. Pockel (to Cambridge Ind. Co.), U.S. Pat. 2,963,461 (1960).
184. *Kroniflex THFP,* Bulletin, FMC Corp., 1964.
185. *Plasticizers for Poly(vinyl Alcohol),* Bulletin No. G-101, Shawinigan Resins Corp., March 1960.
186. C. F. Epes, L. R. Corazzi, and A. J. Marsh (to Reynolds Metals Co.), U.S. Pat. 3,257,348 (1966).
187. L. J. Monaghan and U. A. Aliberti (to E. I. du Pont de Nemours & Co., Inc.), U. S. Pat. 3,365,413 (1966).
188. J. G. Martins (to Shawinigan Resins Corp.), U.S. Pat. 3,192,177 (1965).
189a. T. Mochizuki, *Nippon Kagaku Zasshi* **80,** 1203 (1959).
189b. P. L. Gordon and J. L. Diedrich (to The Borden Co.), U.S. Pat. 3,156,663 (1964).
190. R. K. Tubbs, *J. Polymer Sci.* [A] **3,** 4181 (1965).
191. H. Ito, I. Sekiguchi, and M. Negishi, *Kogyo Kagaku Zasshi* **59,** 834 (1956).
192. T. Osugi, in Ref. 7, pp. 257–260.
193. Y. Sone and I. Sakurada, *Kobunshi Kagaku* **14,** 574 (1957).
194. J. G. Pritchard, in Ref. 8, pp. 50–51.
195. R. W. Warfield and R. Brown, *Kolloid Z.* **185,** 63 (1962).

196. H. K. Inskip and J. H. Peterson, *Mod. Packaging* **30** (1), 137 (1957).
197. R. Lefaux, *Chemie und Toxikologie der Kunststoffe*, Krausskopf Verlag, Mainz, 1966, pp. 292–295.
198. *Federal Register*, Nov. 28, 1963; 28 F.R. 12666. *Federal Register*, Oct. 18, 1963; 28 F.R. 11192. *Federal Register*, March 1, 1960; 25 F.R. 1773.
199. V. Vitez, *Riechstoffe, Aromen, Koerperflege Mittel* **14,** 217 (1964).
200. J. B. Ward and G. J. Sperandio, *J. Soc. Cosmetic Chem.* **15,** 327 (1964).
201. W. C. Hueper, *A.M.A. Arch. Pathol.* **67,** 589 (1959).
202. V. Pardo and A. P. Shapiro, *Lab. Invest.* **15,** 617 (1966).
203. S. L. Danishevskii, *Toksikol. Vysokomolekul. Mater. Khim. Syrya Ikh. Sin. Gos. Nauch. Issled. Inst. Polim. Plast. Mass* **1966,** 94.
204. C. S. Marvel and C. Z. Denoon, *J. Am. Chem. Soc.* **60,** 1045 (1938).
205. P. J. Flory and F. S. Leutner, *J. Polymer Sci.* **3,** 880 (1948).
206. H. E. Harris and J. G. Pritchard, *J. Polymer Sci.* [A] **2,** 3673 (1964).
207. S Okamura and T. Motoyama, *Bull. Res. Inst. Chem. Fibers, Kyoto Univ.* **14,** 23 (1957); through *Chem. Abstr.* **52,** 13311d (1958).
208. I. Rosen, G. H. McCain, A. L. Endrey, and C. L. Sturm, *J. Polymer Sci.* [A] **1,** 951 (1963).
209. J. Ukida and R. Naito, *Kogyo Kagaku Zasshi* **58,** 717 (1955).
210. M. K. Lindemann, in G. E. Ham, ed., *Vinyl Polymerization*, Vol. I, Part I, Marcel Dekker Inc., New York, 1967, pp. 216–249.
211. I. Sakurada and O. Yoshizaki, *Kobunshi Kagaku* **14,** 284 (1957).
212. T. Eguchi and M. Matsumoto, *Kobunshi Kagaku* **15,** 83 (1958).
213. M. Shiraishi, *Kobunshi Kagaku* **19,** 676 (1962).
214. J. T. Clarke and E. R. Blout, *J. Polymer Sci.* **1,** 419 (1946).
215. M. Shiraishi and M. Matsumoto, *Kobunshi Kagaku* **16,** 344 (1959).
216. M. Shiraishi and K. Imai, *Kobunshi Kagaku* **16,** 637 (1959).
217. S. Imoto, J. Ukioa, and T. Kominami, *Kobunshi Kagaku* **14,** 101 (1957).
218. H. N. Friedlander, H. E. Harris, and J. G. Pritchard, *J. Polymer Sci.* [A] **4,** 649 (1966).
219. C. W. Bunn, *Nature* **161,** 929 (1948).
220. W. Cooper, "Stereochemistry of Free Radical Polymerizations," in A. D. Ketley, ed., *The Stereochemistry of Macromolecules*, Vol. 3, Chap. 3, Marcel Dekker Inc., New York, 1967, p. 235.
221. S. Murahashi and S. Nozakura, *Kogyo Kagaku Zasshi* **70,** 1869 (1967).
222. K. Fujii, T. Mochizuki, S. Imoto, J. Ukida, and M. Matsumoto, *J. Polymer Sci.* [A] **2,** 2327 (1964).
223. J. F. Kenney and G. W. Willcockson, *J. Polymer Sci.* [A-1] **4,** 679 (1966).
224. T. Osugi, in Ref. 7, p. 254.
225. J. G. Pritchard, in Ref. 8, p. 17.
226. T. Mochizuki, *Nippon Kagaku Zasshi* **81,** 15 (1960).
227. J. F. Kenney and V. F. Holland, *J. Polymer Sci.* [A-1] **4,** 699 (1966).
228. D. Moroso, A. Cella, and E. Peccatori, *Chim. Ind.* (*Milan*) **48** (2), 120 (1966).
229. H. C. Haas, H. Husek, and L. D. Taylor, *J. Polymer Sci.* [A] **1,** 1215 (1963).
230. T. Yamaguchi, *Kobunshi Kagaku* **18,** 320 (1961).
231. J. G. Pritchard, in Ref. 8, pp. 31–35.
232. C. Y. Liang and F. G. Pearson, *J. Polymer Sci.* **35,** 303 (1959).
233. S. Krimm, *Fortschr. Hochpolymer.-Forsch.* **2,** 51 (1960).
234. A. Nagai, in Ref. 9a, pp. 245–256.
235. J. F. Kenney and G. W. Willcockson, *J. Polymer Sci.* [A-1] **4,** 690 (1966).
236. J. G. Pritchard, in Ref. 8, pp. 17–24.
237. J. Bargon, K. H. Hellwege, and U. Johnson, *Makromol. Chem.* **85,** 291 (1965).
238. A. Danno and N. Hayakawa, *Bull. Chem. Soc.* (*Japan*) **35,** 1749 (1962).
239. U. N. Nikimin and B. Z. Volchek, *Russ. Chem. Rev.* **37,** 225 (1968).
240. W. C. Tincher, *Makromol. Chem.* **85,** 46 (1965).
241. H. Warson, in Ref. 13, pp. 46–76.
242. K. R. Roland (to E. I. du Pont de Nemours & Co., Inc.), U.S. Pat. 2,386,347 (1945); U.S. Pat. 2,399,653 (1946).
243. L. Plambeck (to E. I. du Pont de Nemours & Co., Inc.), U.S. Pat. 2,467,774 (1949).
244. T. Chiba, K. Hiramatsu, and K. Hirano (to Kurashiki Rayon Co.), U.S. Pat. 3,419,654 (1968).

245. H. W. Brant and W. R. Corthwaite (to E. I. du Pont de Nemours & Co., Inc.), U.S. Pat. 2,668,809 (1954).
246. Farbwerke Hoechst, Brit. Pat. 857,147 (1960).
247. *EP Series, Elvanol Poly(vinyl alcohol)*, Product Bulletin A-31 369, E. I. du Pont de Nemours & Co., Inc., Wilmington, Del., June 1963.
248. M. K. Lindemann (to Air Reduction Co.), U.S. Pat. 3,441,547 (1969).
249. E. Abrams, C. W. Rongeux, and J. N. Coker, *Textile Res. J.* **26,** 875 (1956).
250. C. R. Williams and D. D. Donermayer, *Am. Dyestuff Reptr.* **57,** 440 (1968).
251. *Warp Sizing with Du Pont Elvanol*, Product Bulletin, E. I. de Pont de Nemours & Co., Inc., Wilmington, Del., 1967.
252. C. R. Blumenstein, *Textile Ind.* **130** (7), 63 (1966).
253. W. T. Brown, E. S. Olsen, and H. J. Keegan, *Am. Dyestuff Reptr.* **56,** 703 (1967).
254. F. M. Ford (to J. Bancroft & Sons), U.S. Pat. 2,876,136 (1959).
255. Kurashiki Rayon Co., Ref. 15, p. 80.
256. G. P. Colgan and J. J. Latimer, *Tappi* **47** (7), 146A (1964).
257. G. P. Colgan and J. J. Latimer, *Tappi* **44,** 818 (1961).
258. R. H. Beeman and B. A. Beardwood, *Tappi* **46,** 135 (1963).
259. Ref. 2, p. 121.
260. Ref. 11, p. 383.
261. G. P. Colgan and C. L. Gary, *Paper Trade J.* **148,** Oct. 26 and Nov. 23, 1964.
262. *Elvanol PVA for High Gloss Printing and Greaseproof Papers*, Bulletin No. V-9-748 (1948) and V-9-158 (1958), E. I. du Pont de Nemours & Co., Wilmington, Del.
263. B. A. Beardwood and E. P. Czerwin, *Tappi* **43,** 944 (1960).
264. *Japan Pulp and Paper* **4** (2), 68 (1966).
265. R. L. Hawkins (to Air Reduction Co., Inc.), U.S. Pat. 3,135,648 (1964).
266. T. G. Kane (to E. I. du Pont de Nemours & Co., Inc.), U.S. Pat. 3,320,200 (1967).
267. R. L. Hawkins, U.S. Pat. 2,764,568 (1956).
268. K. F. Plitt, *Paint Varnish Prod.* **49** (9), 27 (1959).
269a. Ref. 3, pp. 177–183.
269b. F. Eckart, *Ann. Physik* **11,** 169 (1952).
270. G. W. Saltonstall, *Abstr. Meet. Electrochem. Soc., May 1955, Electronics Div.*, p. 124.
271. P. E. Cary, E. O. Magnus, and W. E. Herring, *Metal Progr.* **73** (3), 79 (1958).
272. W. S. Brown (to U.S. Navy Department), U.S. Pat. 2,889,298 (1959).
273. N. Kuwahara, M. Kaneko, and K. Kubo, *Makromol. Chem.* **110,** 294 (1967).
274. S. Yamamoto (to Matsushita Electric Industrial Co., Ltd.), Ger. Pat. 1,953,674 (1970).
275. D. V. Sokolskii, O. A. Tyurenkova, V. A. Dashevskii, and G. P. Yastrebova, *Zh. Fiz. Khim.* **40,** 2243 (1966).
276. E. I. du Pont de Nemours & Co., Inc., Ref. 14, p. 23.
277. E. H. Land and H. G. Rogers (to Polaroid Corp.), U.S. Pat. 2,173,304 (1939).
278. K. Tanaka, J. Demers, and P. Demers, *Can. J. Phys.* **48,** 1553 (1970).
279. T. Tsunoda, *Enka Biniiru To Porima* **8** (7), 21 (1968).
280. E. Schnabel (to Resistoflex Corp.), U.S. Pat. 2,160,371 (1939).
281. E. Schnabel (to Resistoflex Corp.), U.S. Pat. 2,177,612 (1939).
282. C. J. Prazak and K. M. Seo (to National Video Corp.), U.S. Pat. 3,472,672 (1969).
283. J. W. Hensley, *J. Am. Oil Chem. Soc.* **42,** 993 (1965).
284. J. T. Inamorato (to Colgate-Palmolive Co.), Ger. Pat. 1,145,735 (1963).
285. B. R. Whitear (to Ilford Ltd.), Ger. Pat. 1,911,497 (1969).
286. G. M. Kline, *Mod. Plastics* **23** (5), 165 (1946).
287. N. Platzer, *Mod. Plastics* **28** (7), 95 (1951).
288. G. F. Biehn and M. L. Ernsberger, *Ind. Eng. Chem.* **40,** 1449 (1948).
289. S. Hayashi, C. Nakano, and T. Motoyama, *Kobunshi Kagaku* **21,** 300 (1964).
290. J. N. Coker, *Ind. Eng. Chem.* **49,** 382 (1957).
291. M. K. Lindemann, "The Mechanism of Vinyl Acetate Polymerization," in G. E. Ham, ed., *Vinyl Polymerization*, Vol. I, Part I, Marcel Dekker Inc., New York, 1967, p. 288.
292. G. E. J. Reynolds and E. U. Gulberkian, in Ref. 1, p. 131.
293. F. H. Winslow and W. Matregek, *Ind. Eng. Chem.* **43,** 1108 (1951).
294. H. Wenning, *Kunststoffe-Plastics* **5,** 328 (1958).
295. Ref. 291, p. 290.

296. G. A. Johnson and K. E. Lewis, *Brit. Polymer J.* 1, [illegible] ([illegible]).
297. E. Fujii and M. Suguira, *Kogyo Kagaku Zasshi* **65,** 1609 (1962).
298. D. V. Sokolskii, O. A. Tyurenkova, V. A. Dashevskii, and G. P. Yastrebova, *Zh. Fiz. Khim.* **40,** 2243 (1966).
299. W. O. Herrmann, E. Baum, and W. Haehnel (to Chemische Forschungsgesellschaft), U.S. Pat. 2,072,302 (1937); see also Can. Pat. 326,531 (1932); Brit. Pat. 386,161 (1933).
300. W. O. Herrmann and W. Haehnel (to Chemische Forschungsgesellschaft), Ger. Pat. 685,048 (1939).
301. W. O. Herrmann, *Vom Ringen mit den Molekulen*, Econ-Verlag, Düsseldorf, 1963, pp. 111–124.
302. Ref. 3, pp. 250–255.
303. I. Sakurada, S. Lee, and H. Kawakami (to Nihon Kagaku Seni Kenkyusho), Japan. Pat. 147,958 (1942).
304. M. Yazawa, S. Meguro, M. Yazima, and T. Ozawa (to Kanegafuchi Spinning Co.), Japan. Pat. 153,812 (1942).
305. T. Rosner, ed., "Poly(vinyl Alcohol) Fibers," *Szczecin. Tow. Nauk., Wydz. Nauk Mat. Tech.* **5,** 198 (1966).
306. I. Sakurada, *Kolloid Z.* **139,** 155 (1954).
307. T. Ozawa and T. Matsunaza, *Kobunshi Kagaku* **7,** 248 (1950).
308. T. Ozawa and T. Akuto, *Kobunshi Kagaku* **9,** 75 (1952).
309. H. Kawakami, N. Mari, K. Kawashima, and A. Yamauchi, *Sen-i Gakkaishi* **17,** 1015 (1961).
310. M. Ishii and H. Suyama, *Sen-i Gakkaishi* **18,** 22, 27, 32 (1962).
311. Y. Oyanagi and M. Matsumoto, *J. Colloid Sci.* **17,** 426 (1962).
312. R. Naito, J. Ukida, and T. Kominami, *Kobunshi Kagaku* **14,** 117 (1957).
313. H. Nakayasu, S. Kawai, and H. Harima, *Rept. Progr. Polymer Phys. (Japan)* **8,** 205 (1965).
314. E. Nagai, *Kasen Koen Shuu (Bull. Res. Inst. Chem. Fibers, Kyoto Univ.)* **5,** 5 (1940).
315. T. Tomonari, T. Akahoshi, M. Nagai, and T. Osugi (to Kurashiki Rayon Co., Ltd.), Japan. Pat. 188,756 (1951).
316. M. M. Zwick and C. Van Bochove, *Textile Res. J.* **34,** 417 (1964).
317. T. Osugi, "Spinning of PVA: Some Problems in Wet Spinning," in I. Sakurada, ed., *Poly(vinyl Alcohol)*, Kobunshi Gakkai, Tokyo, 1955, p. 121.
318. K. Hirabayashi, N. Mori, and K. Fukumi, *Sen-i Gakkaishi* **10,** 386 (1954).
319. M. Arakawa, *Sen-i Gakkaishi* **16,** 849, 946 (1960); **17,** 317, 321 (1961).
320. T. Osugi, K. Tanabe, K. Oono, and T. Suda (to Kurashiki Rayon Co., Ltd.), Japan. Pat. 228,357 (1956).
321. F. Mashio, K. Yamaoka, H. Kawakami, S. Mizoguchi, and O. Matsuura (to Nichibo Co., Ltd.), Japan. Pat. 241,861 (1958).
322. K. Tanabe, T. Suda, S. Miyazaki, and T. Osugi (to Kurashiki Rayon Co., Ltd.), U.S. Pat. 3,044,974 (1962).
323. T. Suda and S. Miyazaki (to Kurashiki Rayon Co., Ltd.), U.S. Pat. 3,091,509 (1960).
324. H. Kawakami and K. Kawashima, *Sen-i Gakkaishi* **19,** 627 (1963).
325. H. Kawakami and K. Kawashima, *Sen-i Gakkaishi* **19,** 630 (1963).
326. S. Okamura and T. Yamashita, *Kasen Koen Shuu (Bull. Res. Inst. Chem. Fibers, Kyoto Univ.)* **12,** 60 (1955).
327. I. Sakurada and K. Fuchino, *Riken Iho (Rept. Sci. Res. Inst. Tokyo)* **21,** 1077 (1942).
328. M. M. Zwick, *Chem. Ind. (London)* **1964,** 953.
329. R. D. Wells and H. M. Morgan, *Textile Res. J.* **30,** 668 (1960).
330. *Kuralon, Poly(vinyl Alcohol) Fiber,* Technical Service Manual, Book 1–6, Kurashiki Rayon Co., Ltd., Osaka (1964).
331. *Nippon Kagaku Sen-i Kyokai* (Chemical Fibers Association, Japan), *Kasen Geppo* (Japan Chem. Fibers Monthly) **18,** 10 (1965).
332. *The Engineering Properties of Fibers,* Arthur D. Little, Inc., distributed as U.S. Dept. of Commerce, *PB Rept.* **170391,** (1966).
333. R. C. Laible and H. M. Morgan, *J. Polymer Sci.* **54,** 53 (1961).
334. H. Yabe, in Ref. 7, p. 283.
335. J. C. Smith, P. J. Shouse, J. M. Blandford, and K. M. Towne, *Textile Res. J.* **31,** 721 (1961).
336. S. Seki and T. Yano, in Ref. 9a, pp. 279–294.
337. N. Kishi, *Sen-i Gakkaishi* **14,** 948 (1958).
338. J. W. S. Hearle, *J. Textile Inst.* **44,** T117 (1953).

339. J. Majewska and B. Urbanowicz, *Faserforsch. u. Textiltechn.* **17,** 169 (1966).
340. W. H. Hindle, *Am. Dyestuff Reptr.* **49,** 463 (1960).
341. K. Tanabe, in Ref. 9b, pp. 103–113.
342a. A. Nasuno and K. Tanabe, *Sen-i Gakkaishi* **16,** 235, 241 (1960); **17,** 263 (1961); **18,** 1004, 1011 (1962); **19,** 756, 840, 900, 907 (1963).
342b. W. H. Hindle, *Mod. Textiles* **10** (11), 55 (1959).
343. W. B. Black and P. R. Cox, *Preprints, 34th Meet. Textile Res. Inst., New York, April, 1964,* pp. 53–76.
344. Ref. 330, pp. 51–78.
345. S. Okamura, M. Inagaki, and S. Hirose, *Jushi Kako* **6,** 509 (1957).
346. H. Mita and T. Saeki, *J. Japan. Tech. Assoc. Pulp and Paper Ind.* **12,** 44 (1958).
347. A. Moritamura, U. Maeda, and K. Tanabe, *J. Japan. Tech. Assoc. Pulp and Paper Ind.* **18,** 85 (1964).
348. Y. Yoshioka and T. Ashikaga, *Sen-i to Kogyo* **1,** 214 (1968).
349. F. Kainer, in Ref. 3, pp. 94–102.
350. S. Imoto, *Kogyo Kagaku Zasshi* **64,** 1671 (1961).
351. H. Tadokoro, S. Seki, and I. Nitta, *Bull. Chem. Soc. (Japan)* **28,** 559 (1955).
352. E. M. Kratz, U.S. Pat. 2,316,173 (1943); U.S. Pat. 2,346,764 (1944); U.S. Pat. 2,421,073 (1947).
353. Kurashiki Rayon Co., Ltd., Ref. 15, pp. 86–94.
354. K. Toyoshima, in Ref. 1, pp. 177–184.
355. C. H. Heiberger and D. S. Dixler (to Cumberland Chemical Corp.), U.S. Pat. 3,274,020 (1966).
356a. K. Toyoshima, in Ref. 11, pp. 506–581.
356b. M. K. Lindemann (to Air Reduction Co., Inc.), U.S. Pat. 3,157,611 (1964).
357. J. A. Robertson (to E. I. du Pont de Nemours & Co., Inc.), U.S. Pat. 2,948,697 (1960).
358. T. S. Bianco and E. M. Kratz (to Monosol Co.), U.S. Pat. 3,374,195 (1968).
359. R. Kreinhoefner and H. Reip, *Fette Seifen Anstrichm.* **63,** 855 (1961).
360. M. K. Lindemann (to Air Reduction Co., Inc.), U.S. Pat. 3,505,303 (1970).
361. T. Osugi, M. Matsumoto, and M. Maeda (to Kurashiki Rayon Co.), Can. Pat. 563,243 (1958).
362. J. N. Milne (to E. I. du Pont de Nemours & Co., Inc.), U.S. Pat. 3,106,543 (1963).
363. K. Kawai, S. Miyazaki, K. Tanabe, and T. Osugi (to Kurashiki Rayon Co., Ltd.), Japan. Pat. 232,163 (1956).
364. K. Kawai, H. Nakayasu, and H. Kurashige (to Kurashiki Rayon Co., Ltd.), Japan. Pat. 404,872 (1962).
365. S. Nakajo, E. Morita, S. Kodama, and K. Imai (to Kurashiki Rayon Co., Ltd.), U.S. Pat. 3,066,999 (1959).
366. M. M. Zwick, J. A. Duiser, and C. Van Bochove, in Ref. 1, pp. 189–207.
367. A. Alexander, *Man-Made Fiber Processing,* Noyes Development Corp., Park Ridge, N.J., 1966, pp. 31–60.
368. A. E. Akopyan, "Synthetic Fibers Based on Poly(vinyl Alcohol)," *Erevan. Armyanck. Gos. Izd.,* 1961.
369. A. I. Meos and L. A. Volf, *The Production of Fibers from Poly(vinyl Alcohol),* Leningrad, Lenizdat. (1965), 54 pp. (in Russian).
370. L. Alexandru, V. Ciobanu, I. Agachi, and T. Rizescu, *Faserforsch. u. Textiltechn.* **16,** 73 (1965).
371. T. Ogawa, in Ref. 9a, pp. 159–184.
372. E. T. Cline (to E. I. du Pont de Nemours & Co., Inc.), U.S. Pat. 2,610,360 (1952).
373. N. Mori, in Ref. 9a, pp. 147–158.
374. J. Kosar, *Light-Sensitive Systems,* John Wiley & Sons, Inc., New York, 1965, pp. 66–67.
375. B. Duncalf and A. S. Dunn, *J. Appl. Polymer Sci.* **8,** 1763 (1964).
376. W. C. Toland and E. Bassist, U.S. Pat. 2,302,816 (1942); U.S. Pat. 2,302,817 (1942).
377. A. G. Leiga and R. A. Walder, *Postprint, Reg. Tech. Conf., Soc. Plastics Eng., Photopolymers, Nov. 6, 1967,* p. 13.
378. M. Sasaki and S. Kikuchi, *Kogyo Kagaku Zasshi* **70,** 2107 (1967).
379. Ref. 3, pp. 190–199.
380. Ref. 3, pp. 199–203.

Martin K. Lindemann
Chas. S. Tanner Co.

POLY(VINYL ACETALS)

The name poly(vinyl acetal) is generically applied to a group of resins that are derived from the reaction of poly(vinyl alcohol) with aldehydes; it is also specifically applied to the reaction product of poly(vinyl alcohol) and acetaldehyde. The reaction of poly(vinyl alcohol) with ketones gives poly(vinyl ketals). Poly(vinyl acetals) were first prepared by Herrmann and Haehnel in 1924 who reacted poly(vinyl alcohol) with benzaldehyde (11). Commercial methods of poly(vinyl acetal) manufacture were developed in subsequent years by Shawinigan of Canada (12), Wacker-Chemie (13), and I. G. Farben (BASF) (14).

Only the poly(vinyl formals) and poly(vinyl butyrals) are commercially important today, the former as resin base for wire coatings, adhesives, and as poly(vinyl alcohol) fibers, the latter in metal-paint applications (wash primers), wood sealers, adhesives, and as safety-glass interlayers.

For a discussion of vinal fibers, see the section of this article entitled Poly(vinyl Alcohol).

Preparation

The acetalization of poly(vinyl alcohol) consists of condensing poly(vinyl alcohol) with an aldehyde in the presence of a strong acid. The mechanical and chemical properties of the resulting poly(vinyl acetal) will depend on the grade of poly(vinyl alcohol) that is used, ie, its degree of hydrolysis and molecular weight, as well as on the amount and kind of aldehyde. For example, fully and partially hydrolyzed poly(vinyl alcohols) have been used as well as copolymers of vinyl alcohol which are listed in Table 1. Besides formaldehyde and butyraldehyde, a great number of other aldehydes have been used to acetalize poly(vinyl alcohol); acetaldehyde, for example, was used in Germany during World War II to produce poly(vinyl acetal); poly(vinyl acetal) is still sold in small quantities in Canada, France, and Germany.

Table 1. Acetalized Vinyl Alcohol Copolymers

Comonomer	Reference
vinyl chloride	15
acrylonitrile	16
2-chloroallyl acetate	17
esters of vinylphosphonic acids	18
ethylene	19
isobutylene	20
poly(alkylene oxide) (graft copolymer)	21
acrylates	22
diallyl phthalate	23
maleic anhydride	24
divinyl ethers	25
vinylene glycol	26

Poly(vinyl alcohol) has been acetalized by the following aldehydes:

Aliphatic aldehydes (90): formaldehyde, acetaldehyde (9), propionaldehyde, butyraldehyde, nonanal (31), heptanal (31), palmitic aldehyde (28), and stearaldehyde (28).

Aliphatic dialdehydes: glyoxal (32).

Aliphatic unsaturated aldehydes: crotonaldehyde (29) and acrolein (30).

Derivatives of aliphatic aldehydes: halogen and acetaldehydes (33), amino acetaldehydes (34), acetalyl sulfide (34), alkoxyacetaldehydes (35), and aliphatic aldehyde sulfonic acids (36).

Aromatic aldehydes: benzaldehyde (11), *p*-tolualdehyde (37), 2-naphthaldehyde (37), and 9-anthracenealdehyde (38).

Substituted aromatic aldehydes: salicylaldehyde (39), *o*-chlorobenzaldehyde (40), *p*-dimethylaminobenzaldehyde (34), aminobenzaldehyde (42), *p*-vinyl benzaldehyde (43), and *o*-benzaldehyde sulfonic acid (44).

Other aldehydes: furfural (45) and glyoxylic acid (46).

Ketones: cyclohexanol (47,48), acetone (49), acetophenone (49), benzyl methyl ketone (49), methyl ethyl ketone (50), ketals (51).

More literature references to aldehydes reported to be useful for acetalizing poly(vinyl alcohol) may be found in Refs. 3, 4, 7, and 8.

The acetalization reaction is catalyzed by strong mineral acids. Generally, hydrochloric acid is preferred since dark-colored products are often obtained when using other acids, for example, sulfuric acid.

Generally, three different reactions can occur (10):

(a) Intramolecular acetalization of the 1,3-glycol group

$$\sim CH_2-\underset{OH}{CH}-CH_2-\underset{OH}{CH}\sim + RCHO \longrightarrow \sim CH_2-\underset{O}{CH}-CH_2-\underset{O}{CH}\sim \quad (O\text{–}CHR\text{–}O \text{ ring}) \qquad (1)$$

(b) Intermolecular acetalization

$$2\ \sim CH_2-\underset{OH}{CH}-CH_2-\underset{OH}{CH}\sim + RCHO \longrightarrow \begin{array}{c} \sim CH_2-\underset{OH}{CH}-CH_2-\underset{O}{CH}\sim \\ HCR \\ \sim CH_2-\overset{OH}{CH}-CH_2-\overset{O}{CH}\sim \end{array} \qquad (2)$$

(c) Intramolecular acetalization of the 1,2-glycol group

$$\sim CH_2-\underset{OH}{CH}-\underset{OH}{CH}-CH_2\sim + RCHO \longrightarrow \sim CH_2-\underset{O}{CH}-\underset{O}{CH}-CH_2\sim \quad (O\text{–}CHR\text{–}O \text{ ring}) \qquad (3)$$

Poly(vinyl alcohol) reacts with an aldehyde primarily to form six-membered rings between adjacent, intramolecular hydroxyl groups; to a lesser extent intermolecular acetals are also formed (60). The latter reaction takes place more frequently when formaldehyde is used; therefore this reaction causes crosslinking and, eventually, insolubility (10). A third possibility is the formation of five-membered-ring acetals when 1,2-glycol structures in the poly(vinyl alcohol) are acetalized.

Flory (52) postulated that the highest degree of acetalization possible is only 86.46 mole % for 1,3-glycol structures and only 81.60 mole % for 1,2-glycol structures assuming that only adjacent intramolecular hydroxyl groups are involved in the reaction and that the reaction is not reversible. Later Flory (53) revised his estimate to a possible 100% acetalization by taking reversibility of the reaction into account. It is a fact that up to now the degree of acetalization has not been reported in the literature to have exceeded 86 mole % (40), although some patents claim 100 mole % acetalization (54,55). Commercially available poly(vinyl acetals) have a degree of acetalization of 70–85 mole %.

There are four different acetalization processes:

(*1*) Precipitation method: An aqueous solution of poly(vinyl alcohol) is reacted with the aldehyde until the homogeneous-phase reaction changes to a heterogeneous reaction when the acetal precipitates. This occurs at about 30% acetalization when using acetaldehyde.

(*2*) Dissolution method: Poly(vinyl alcohol) powder is suspended in a suitable nonsolvent which dissolves the aldehyde and the final product. The reaction starts out heterogeneously and is completed homogeneously.

(*3*) Homogeneous reaction method: The reaction starts in a water solution of poly(vinyl alcohol). A solvent for the acetal which is miscible with water is added continuously to prevent precipitation.

(*4*) Heterogeneous reaction method: Poly(vinyl alcohol) in film or fiber form is reacted with the aldehyde.

(*5*) Direct conversion of poly(vinyl acetate) to poly(vinyl acetal): This is also a homogeneous method. The poly(vinyl acetate) is dissolved in a suitable solvent and is hydrolyzed by strong mineral acid and acetalized at the same time.

The properties of the acetal depend a great deal upon the method of preparation, even if the degree of acetalization is the same. The homogeneous reaction method (method *3*) is preferred if high degrees of acetalization are desired. Intermolecular acetal formation is minimized and the distribution of acetal groups is uniform using this method.

Industrially, method *3* is not used because it is too expensive. The precipitation method (method *1*) is preferred there because the acetal can be recovered and washed easily. The dissolution method (method *2*) and the direct conversion of poly(vinyl acetate) (method *5*) are preferred where the acetal is used in solution form. Method *4* is used for specific purposes such as the acetalization of poly(vinyl alcohol) fibers or film.

The distribution of acetal groups affects the physical and chemical properties of the poly(vinyl acetal). For example, Sakurada found that the x-ray diagram of a poly(vinyl alcohol) fiber did not change when the fiber was formalized to the extent of 30 mole % (56–58). Since the formalization occurs only in the amorphous areas of the fiber (59), the crystallinity of the fiber does not change. If, however, the polymer had been dissolved, formalized, and then spun, the fiber would essentially be noncrystalline (see the discussion of vinal fibers in the section on Poly(vinyl Alcohol) in this article).

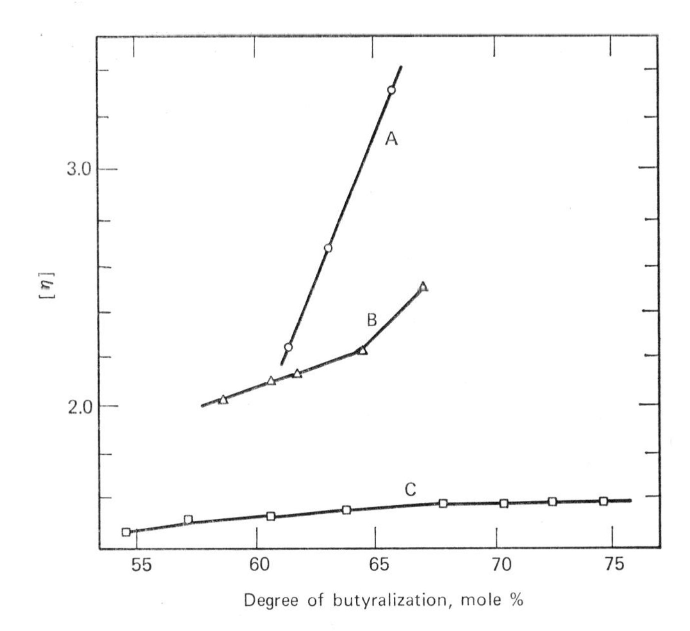

Fig. 1. Intrinsic viscosity [η], in dl/g, in tetrahydrofuran (THF) at 30°C of poly(vinyl butyral) prepared by various methods (60). (A) Precipitation method, aldehyde–poly(vinyl alcohol) molar ratio 1.2; 4% HCl. (B) Precipitation method, aldehyde–poly(vinyl alcohol) molar ratio 0.6; 5% HCl. (c) Solubilization method in isopropanol–water aldehyde–poly(vinyl alcohol) molar ratio 0.6; 1% HCl.

As mentioned above, crosslinking can occur during the acetalization process. This is shown in Figure 1 where the intrinsic viscosity of poly(vinyl butyral) increases sharply with the degree of butyralization when the precipitation method of preparation is used. However, when the solubilization method is used the intrinsic viscosity remains essentially constant. It can also be seen in Figure 1 that intermolecular acetalization is favored by high aldehyde concentrations (61).

The kinetics of acetalization of poly(vinyl alcohol) have been studied repeatedly. Generally, it is thought that the following mechanism applies to the formalization of poly(vinyl alcohol) (62, 63):

$$CH_2O + H^{\oplus} \rightleftharpoons \overset{\oplus}{C}H_2OH$$

$$\sim\!\!\underset{\underset{OH}{|}}{CH}\!-\!CH_2\!-\!\underset{\underset{OH}{|}}{CH}\!-\!CH_2\!\sim + \overset{\oplus}{C}H_2OH \underset{\text{very slow}}{\overset{\text{slow}}{\rightleftharpoons}} \sim\!\!\underset{\underset{OCH_2^{\oplus}}{|}}{CH}\!-\!CH_2\!-\!\underset{\underset{OH}{|}}{CH}\!-\!CH_2\!\sim + H_2O \qquad (4)$$

$$\sim\!\!\underset{\underset{OCH_2^{\oplus}}{|}}{CH}\!-\!CH_2\!-\!\underset{\underset{OH}{|}}{CH}\!-\!CH_2\!\sim \underset{\text{very slow}}{\overset{\text{fast}}{\rightleftharpoons}} \sim\!\!\underset{\underset{O}{|}}{CH}\!-\!CH_2\!-\!\underset{\underset{O}{|}}{CH}\!-\!CH_2\!\sim\ (\text{O}\diagdown CH_2 \diagup \text{O ring}) + H^{\oplus}$$

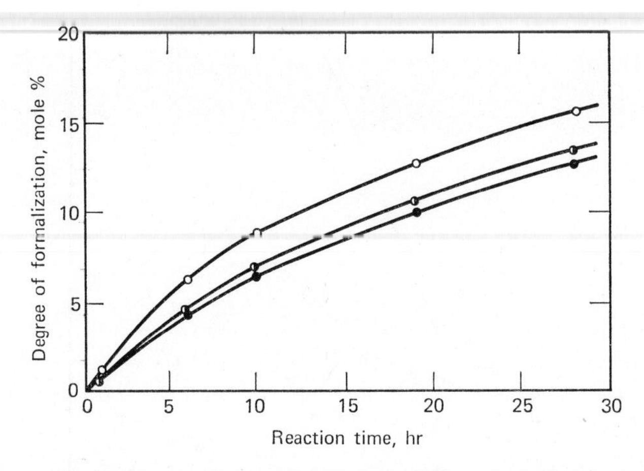

Fig. 2. Effect of tacticity of poly(vinyl alcohol) on formalization rate (64).

Sample	*Isotactic*	*Heterotactic*	*Syndiotactic*
	Tacticity, %		
○	56	32	12
◑	23	47	30
●	17	46	37

This mechanism was elucidated further (64) by studying the rate of formalization and its relationship to the tacticity of the poly(vinyl alcohol). It was shown that, in the initial period of reaction, the reaction rate of predominantly isotactic poly(vinyl alcohol) is higher than that of predominantly syndiotactic poly(vinyl alcohol) (see Figure 2). The isotactic areas in the molecular chains are more readily formalized. However, under the same reaction conditions, the ultimate degree of formalization is the same as seen in Table 2. The rate of the reverse reaction, ie, deacetalization, of predominantly isotactic poly(vinyl acetal) is smaller than that of predominantly

Table 2. Configuration of Poly(vinyl Formal) Obtained by Formalization of Three Kinds of Poly(vinyl Alcohol) Samples of Different Tacticities (64)

Original poly(vinyl alcohol) sample, type[a]	Degree of formalization, mole %		
	Total	*cis*-Formal	*trans*-Formal
○	84	70	14
◑	87	59	28
●	87	50	37

[a] See caption of Figure 2 for description of samples.

syndiotactic poly(vinyl acetal) (65). It was also determined by Matsumoto and coworkers (66) that formalization takes place selectively in the isotactic areas as can be seen in Table 3. There the mole percentage of *cis*-formal and *trans*-formal are shown as a function of total mole % formal. Only after a considerable amount of isotactic poly(vinyl alcohol) is formalized (producing the cis configuration, eq. 5) are the syndiotactic areas formalized (producing the trans configuration, eq. 6) (where R is hydrogen when formaldehyde is used).

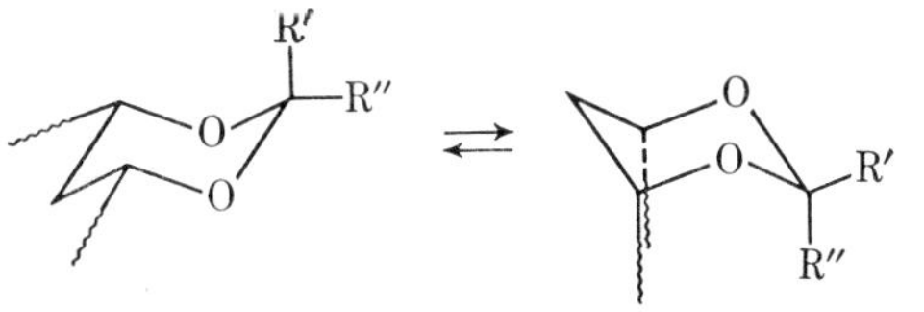

(5)

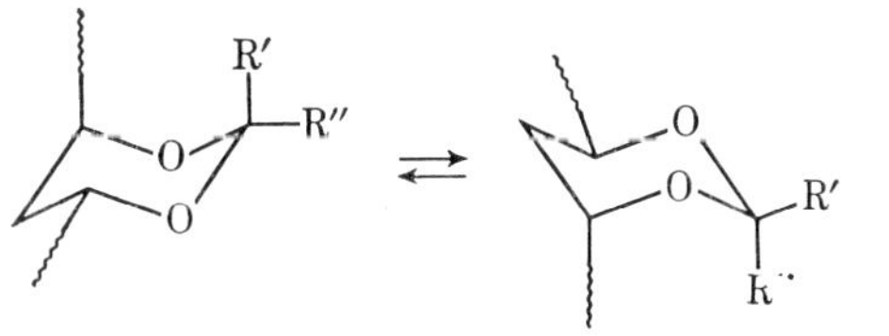

(6)

Table 3. Formalization of Commercial Poly(vinyl Alcohol) in a Homogeneous Medium (66)

	Reaction conditions		Degree of formalization		
Sample no.	Dioxane, %	Catalyst	*cis*-Formal, mole %	*trans*-Formal, mole %	Total mole %
1	50	3 *N* H_2SO_4	26.8	0.7	27.5
3	50	3 *N* H_2SO_4	40.1	2.3	42.4
9	80	3 *N* H_2SO_4	47.8	9.2	57.0
15	87	3 *N* H_2SO_4	50.1	17.0	67.1
12	100	0.14 *N* HCl	56.2	26.8	83.0

It is possible that one of the causes for the difficulty involved in obtaining highly acetalized grades of poly(vinyl alcohol) when using commercially available poly(vinyl alcohol) lies in its stereostructure.

Commercial Manufacture. There are a number of different processes practiced commercially for the manufacture of poly(vinyl acetals). The specific process selected depends on the particular acetal which is being made as well as other factors, such as availability of raw materials. There are two basic processes used in the United States. The first one employs an aqueous solution of poly(vinyl alcohol) whereas the the other one starts with poly(vinyl acetate) which is acid hydrolyzed and then subsequently acetalized.

Poly(vinyl Formal) (68). Poly(vinyl formals) usually contain a larger amount of nonhydrolyzed acetate groups than do the poly(vinyl butyrals). Figure 3 shows the general flowsheet for the manufacture of poly(vinyl formal) (67). Poly(vinyl acetate) of the desired molecular weight is dissolved in a mixture of acetic acid, water, and formaldehyde solution. Formaldehyde has to be present in excess. After this solution is pumped into the reaction kettle and heated, sulfuric acid is added. The reaction temperature is held to about 70–90°C. The reaction is completed in about 6 hr; the sulfuric acid is then neutralized after which the poly(vinyl formal) is precipitated in water, washed, and dried. Sometimes ammonium peroxydisulfate is added during the formalization step in order to avoid a yellow product (69).

Poly(vinyl Butyral). It is important for the production of poly(vinyl butyral) that the poly(vinyl acetate) be completely hydrolyzed. Therefore, the butyraldehyde is added only after the hydrolysis step is completed. There are two different processes used for the manufacture of poly(vinyl butyral).

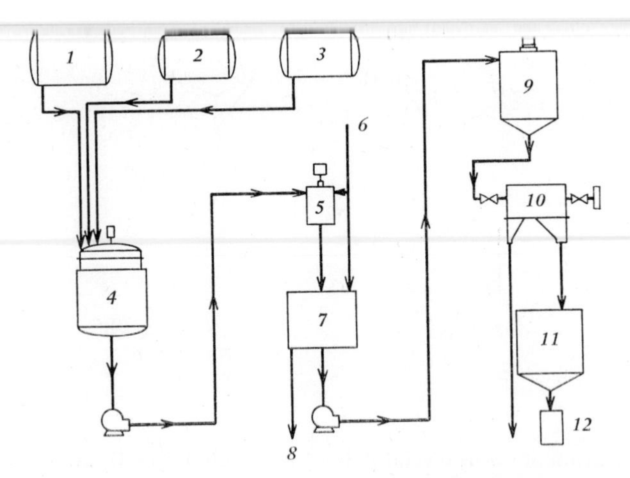

Fig. 3. Flowsheet for the manufacture of poly(vinyl formal) from poly(vinyl acetate) (67).

1. Poly(vinyl acetate) solution
2. Sulfuric acid tank
3. Sodium acetate tank
4. Reactor for hydrolysis and acetalization
5. Precipitator
6. Water tank
7. Wash tank
8. Liquor recovery system
9. Hold-up tank
10. Centrifuge
11. Dryer
12. Finished product

Acetalization of Poly(vinyl Alcohol) in Water Solution (Two-Step Process) (76). This process is used by Du Pont (70,71). A reactor is charged with a 10% solution of fully hydrolyzed poly(vinyl alcohol), strong mineral acid such as sulfuric acid (sometimes mixed with *p*-toluenesulfonic acid), and butyraldehyde. The ratio of poly(vinyl alcohol) to acid and to aldehyde is about 100:8:57. The mixture is heated for 1½ hr at 90°C. The poly(vinyl butyral) precipitates and is washed after neutralization with caustic and dried.

Acetalization in Solvent (One-Step Process) (75). This process is being practiced by the Monsanto Co. (formerly Shawinigan Resins Corp.). Figure 4 shows the flowsheet for this process (72). Vinyl acetate is polymerized in solution or suspension after which it is dissolved in methanol and hydrolyzed by strong mineral acid. The poly(vinyl alcohol) is then suspended in a mixture of ethanol and ethyl acetate; butyraldehyde and mineral acid are added to effect acetalization. The poly(vinyl butyral) solution is neutralized first, then the resin is precipitated with water, washed, and dried.

Poly(vinyl Acetal). A two-step process was used for the manufacture of poly(vinyl acetal) in Germany during World War II (73). Figure 5 shows the flowsheet for this process. In Canada, poly(vinyl acetal) was made by a one-step process similar to the one used for poly(vinyl formal) (74).

A great number of procedures for the manufacture of poly(vinyl acetals), usually modifications of the above, are disclosed in the patent literature which is extensively surveyed up to 1958 in Refs. 3 and 4.

Preparation of Poly(vinyl Acetal) Dispersions. Poly(vinyl acetal) dispersions cannot be prepared directly. A number of different processes have been disclosed for their manufacture.

Poly(vinyl butyral) powder is kneaded together with emulsifier and water at about 100°C (77). More water is added until an inversion occurs. The aqueous emul-

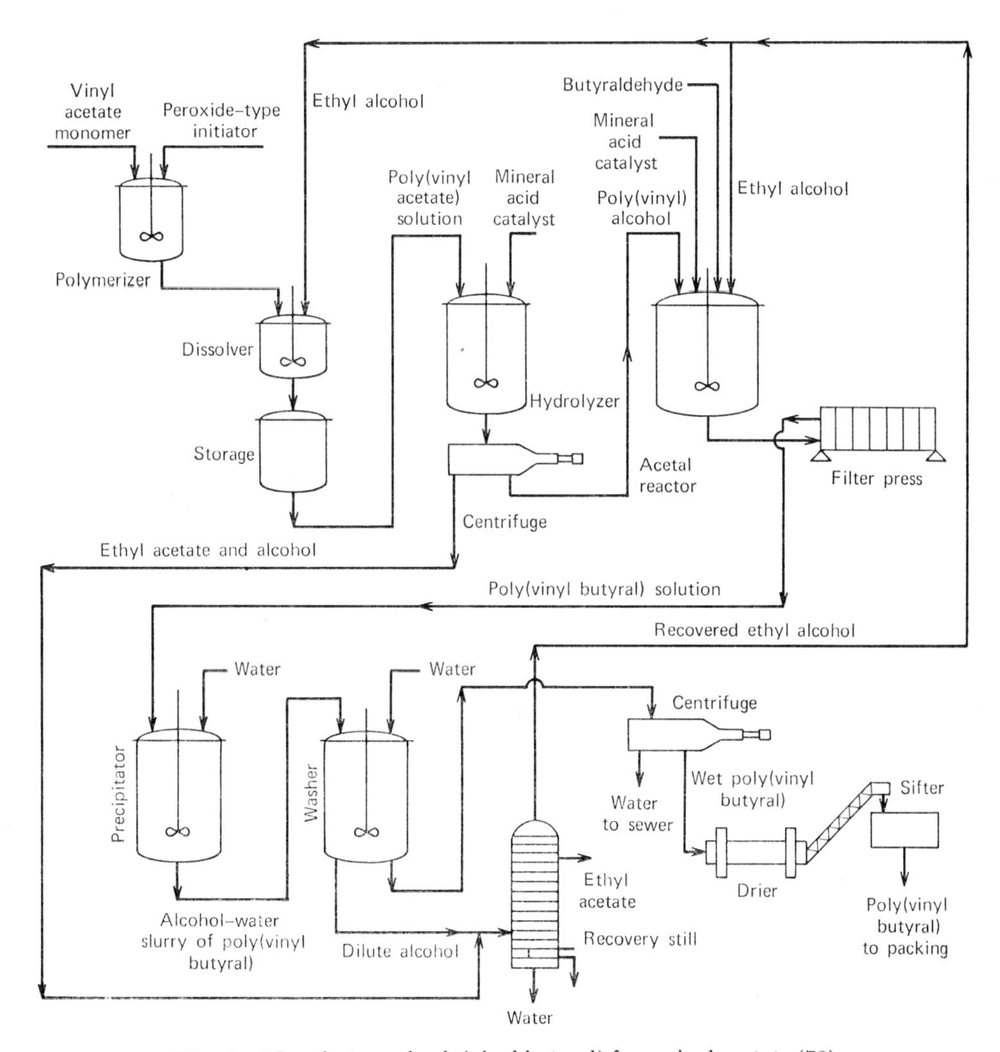

Fig. 4. Manufacture of poly(vinyl butyral) from vinyl acetate (72).

sion can then be further diluted with water. A solvent that gels poly(vinyl butyral) can be added in small amount to aid in the emulsification of the resin (78).

An aqueous poly(vinyl alcohol) solution can be acetalized in the presence of emulsifiers. The precipitating acetal is kept in suspension (79).

Another method to produce dispersions involves the dissolution of the poly(vinyl acetal) in a solvent that is not miscible with water which is then emulsified. The solvent is distilled off and a large-particle-size dispersion remains (80).

The poly(vinyl acetal) dispersions can be postplasticized easily or plasticizer can be added initially.

Poly(vinyl acetals) have been synthesized by the polymerization of divinyl acetals and *o*-vinyl acetals. A number of divinyl acetals were synthesized and polymerized (81,82). Depending on the conditions of polymerization, soluble poly(vinyl acetals) were obtained. It is interesting that by polymerizing divinyl butyral an almost 100%-substituted poly(vinyl butyral) has been made (81). This has not been possible by

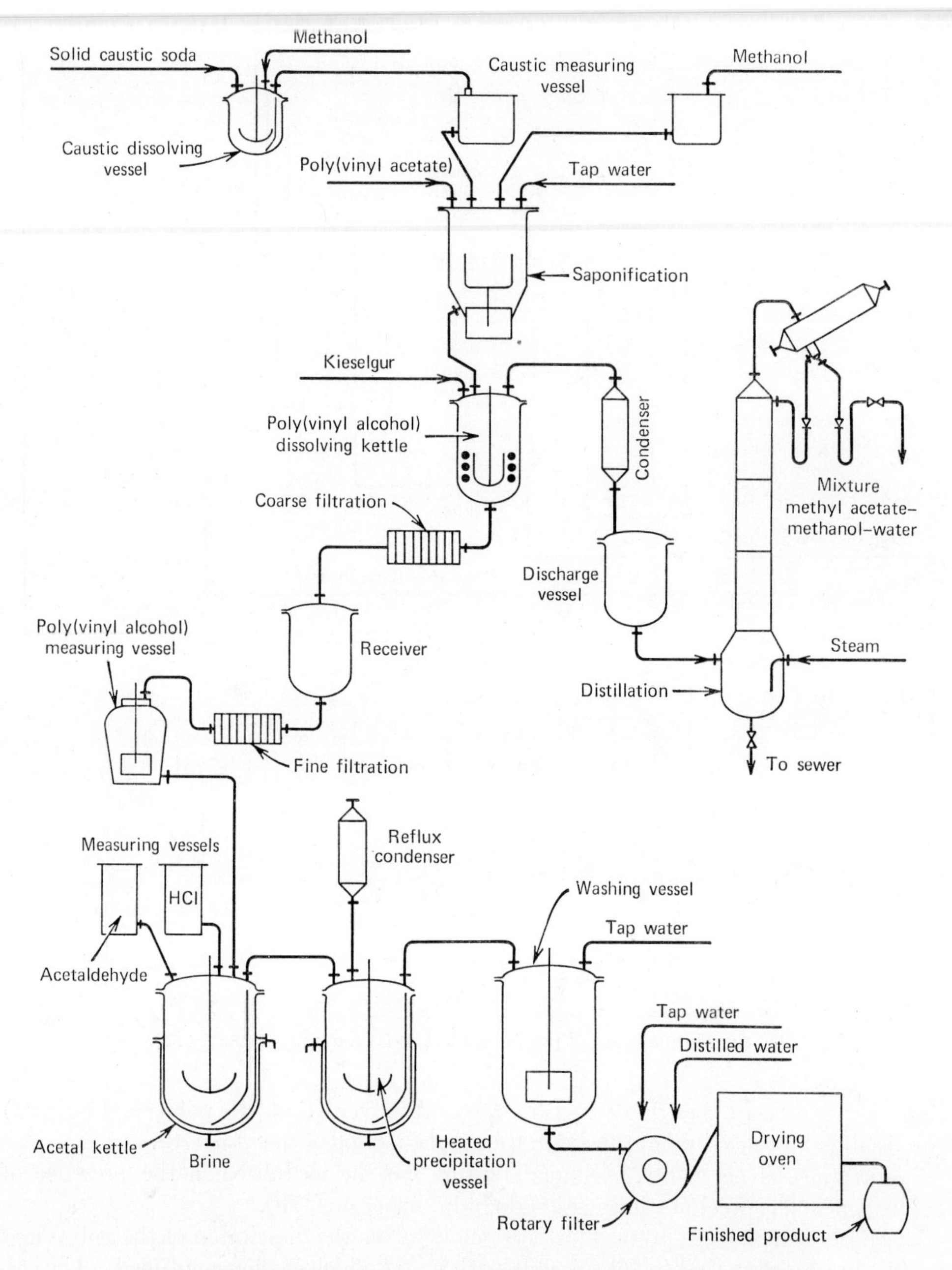

Fig. 5. Flowsheet for the two-step production of poly(vinyl acetal) (9)

acetalization of poly(vinyl alcohol). Poly(vinyl acetals) of the following structure have been investigated (83):

$$\begin{array}{l} \sim\!\!\sim\!\text{CH—CH}_2\text{—CH—CH}_2\text{—}\!\sim\!\!\sim \\ \quad\;\; | \qquad\qquad\quad | \\ \quad\;\; \text{O} \qquad\qquad\;\; \text{O} \\ \quad\;\; | \qquad\qquad\quad | \\ \quad\;\; \text{CH}_2\text{—O—R}\;\; \text{CH}_2\text{—O—R} \end{array}$$

It has been reported that a hemiacetal of formaldehyde could be produced by carefully reacting poly(vinyl alcohol) under mildly alkaline conditions with formaldehyde. The resulting polymeric hemiacetal is said to be useful as a formaldehyde donor since it releases the aldehyde slowly (27):

$$\sim\!\!\sim\text{CH}(\text{OCH}_2\text{OH})\text{—CH}_2\text{—CH}(\text{OCH}_2\text{OH})\text{—CH}_2\text{—}\sim\!\!\sim$$

Properties

The physical and chemical properties of poly(vinyl acetals) depend on the polymerization conditions of the parent poly(vinyl acetate), the hydrolysis conditions of the poly(vinyl alcohol), and the specific aldehyde which is used for acetalization. It can be seen that an almost infinite number of poly(vinyl acetals) can be made.

Physical Properties. A great many grades of poly(vinyl acetals) are available commercially in the United States and abroad. Table 4 lists some of the more common

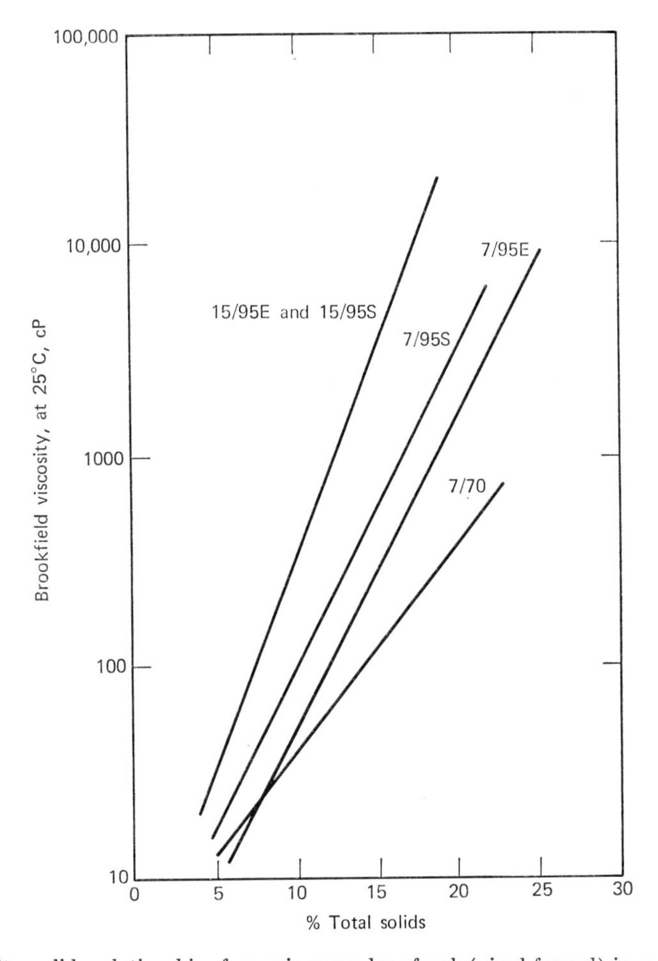

Fig. 6. Viscosity–solids relationships for various grades of poly(vinyl formal) in a toluene–ethanol (60:40) mixture (6).

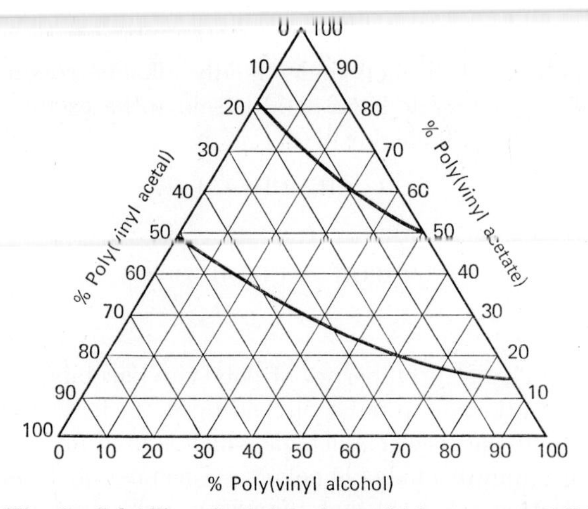

Fig. 7. Solubility of poly(vinyl acetals) in water at 0°C (89)

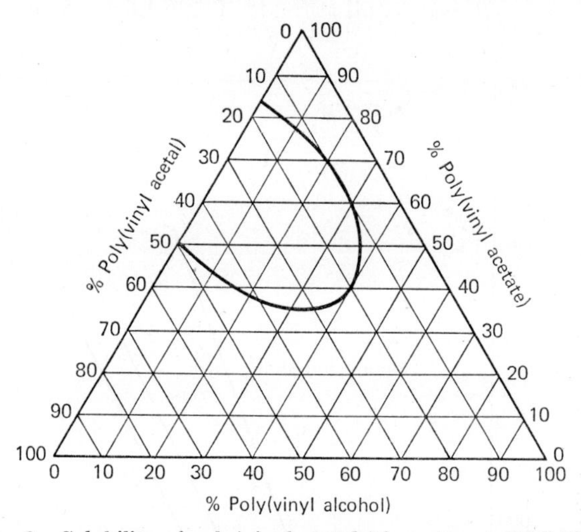

Fig. 8. Solubility of poly(vinyl acetals) in water at 40°C (89)

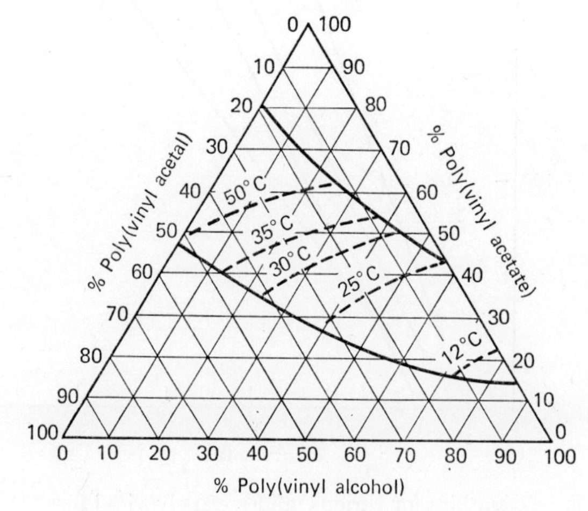

Fig. 9. Gelling of poly(vinyl acetals) (89).

Table 4. Some Trademarks for Poly(vinyl Acetals)

Manufacturers	Poly(vinyl formal)	Poly(vinyl acetal)	Poly(vinyl butyral)
Du Pont			Butacite
Monsanto	Formvar	Alvar	Butvar
Farbwerke Hoechst	Movital F	Movital O	Movital B
Wacker-Chemie	Pioloform F		Pioloform B
Nobel-Hoechst	Rhevyl F	Rhevyl A	Rhevyl B
Rhône-Poulenc	Rhovinal F	Rhovinal A	Rhovinal B
Union Carbide	Vinylite	Vinylite	Vinylite XYHL
PPG Industries			Vinal
Shin Nippon Chisso	Vinilex F		
Sekisui Chem. Co.			S-lec

trademarks. It can be seen that only the formal, acetal, and butyral are available. Each one of these is made in a number of molecular-weight grades.

The physical properties of poly(vinyl formals) now available in the United States are listed in Table 5. Poly(vinyl formals) are thermoplastic resins with a heat-distor-

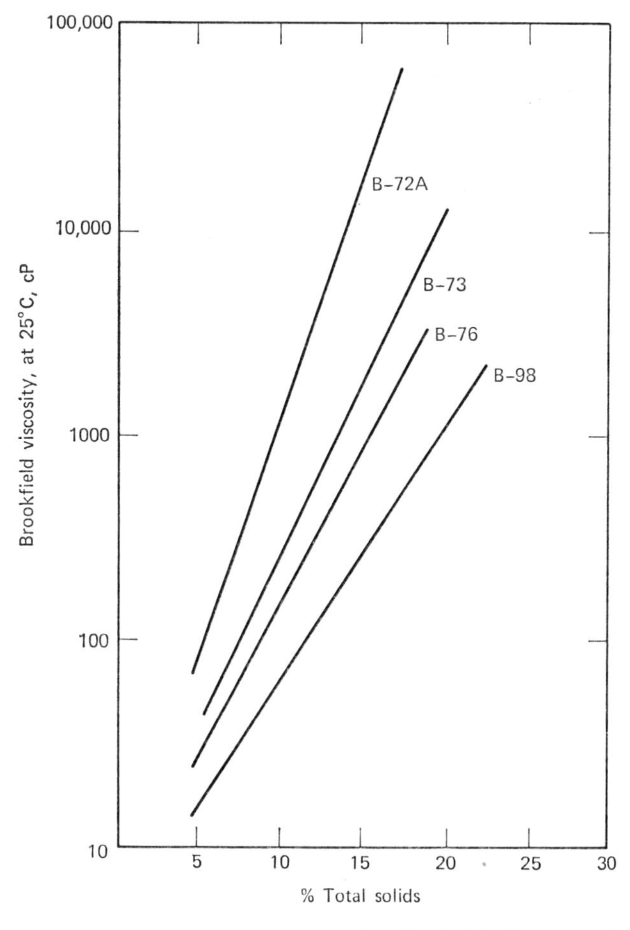

Fig. 10. Viscosity–solids relationships for various grades of poly(vinyl butyral) in a toluene–ethanol (60:40) mixture (6).

tion temperature of about 90°C. Increasing the acetate groups decreases the heat-distortion temperature, and also the tensile strength and the impact strength. Poly(vinyl formal) is soluble in acetic acid, dioxane, or a toluene–ethanol mixture, although it is insoluble in either toluene or ethanol. Figure 6 shows the viscosity–solids relationship for commercial resins. Table 6 lists some of the more common solvents for

Table 5. Properties of Poly(vinyl Formals) (84)

Property	ASTM test method	Formvar, type[e]					
		15/95E	15/95S	7/95E	7/95S	12/85	7/70
average degree of polymerization		500	500	350	350	430	350
poly(vinyl alcohol), %[a]		5–6	7–9	5–6.5	7–9	5–7	5–7
poly(vinyl acetate), %[b]		9.5–13	9.5–13	9.5–13	9.5–13	12–30	40–50
viscosity, cP[c]		40–60	50–80	15–20	15–22	18–22	7–11
viscosity, cP		3000–4000	3000–4500	300–500	300–500	500–600	100–140
stability (minimum), 1 hr, °C		120	150	120	150	120	120
specific gravity		1.227	1.229	1.227	1.229	1.219	1.214
refractive index, n_D^{25}	D 542-42	1.5	1.5	1.5	1.5	1.5	1.5
flow temperature, °C	D 569-48	160–170	160–170	140–145	140–145	145–150	
heat-distortion point, °C	D 648-49T	88–93	88–93	88–93	88–93	75–80	50–60
inflection temperature, °C	D 1043-49T	103–108	100–105	103–108	100–105	90–95	75–80
apparent modulus of elasticity,	D 1043-49T	3.5–6	3.5–6	3.5–6	3.5–6	4–6	4–6
10^5 psi	D 638-41T		2.5–3				
tensile strength, 10^4 psi	D 638-41T	1.0–1.1	1.0–1.1	1.0–1.1	1.0–1.1	1.0–1.1	1.0–1.1
elongation, %	D 638-41T	7–20	10–50	7–20	10–50	4–5	3–4
deformation, 4000 psi load, %							
at 50°C	D 621-48T	0.2–0.4	0.2–0.4	0.2–0.4	0.2–0.4	4–6	very high
at 70°C	D 621-48T	1–3	1–3	1–3	1–3	very high	very high
impact strength							
Izod notched,							
½×½ in., ft-lb	D 256-43T	1.2–2	1.2–2	1.0–1.4	1.0–1.4	0.5–0.7	0.4–0.6
½ × ⅛ in., ft-lb	D 256-43T				1.3–2		0.4–0.6
unnotched, ft-lb/in.	D 256-43T	>60	>60	>60	>60		
water absorption, %	D 570-40T	0.75	1.1	0.8	1.1	1.5	1.5

[a] By acetylation.

[b] By saponification.

[c] 5 g resin in 100 ml ethylene dichloride solution at 20°C.

[d] Viscosity of a 15% solution in a 60:40 toluene–ethanol mixture at 25°C using a Brookfield viscometer.

[e] Wire coating resin designated E, thermally stabilized resin designated S.

Table 6. Solvents for Poly(vinyl Formal)[a] (6)

Solvent	Formvar 15/95 (E&S) 7/95 (E&S)	Formvar 12/85	Formvar 7/70
acetic acid (glacial)	S	S	S
acetone	I	I	S
n-butanol	I	I	I
butyl acetate	I	I	I
carbon tetrachloride	I	I	I
chloroform	S	S	S
cresylic acid	S	S	S
cyclohexanone	I	S	S
diacetone alcohol	I	S	S
diisobutyl ketone	I	I	I
dioxane	S	S	S
ethanol, 95%	I	I	I
ethyl acetate, 99%	I	I	S
ethyl acetate, 85%	I	I	S
ethyl Cellosolve	I	I	S
ethylene chloride	S	S	S
furfural	S	S	S
hexane	I	I	I
2-propanol, 95%	I	I	I
methyl acetate	I	I	S
methanol	I	I	I
methyl Cellosolve	I	S	S
methyl Cellosolve acetate	I	S	S
methyl butynol	S	S	S
methyl pentynol	S	S	S
methyl ethyl ketone	I	I	S
methyl isobutyl ketone	I	I	I
nitropropane	I	S	S
toluene	I	I	I
toluene–ethanol (95%) (60:40 by wt)	S	S	S
xylene	I	I	I
xylene-*n*-butanol (60:40 by wt)	I	I	S

[a] S, completely soluble; I, insoluble or not completely soluble.
Solubility tests were run with solutions of 1 g of resin in 9 g of solvent for 24 hr at room temperature.

the poly(vinyl formals) (6). Figures 7, 8, and 9 show the change in solubilities of poly(vinyl acetals) as a function of acetal, acetyl, and hydroxyl contents. Figure 9 shows the compositional curves at which a solution of the given resin will gel when cooled to the indicated temperature (88,89).

Physical properties of poly(vinyl butyrals) are shown in Table 7 (6). They are being produced with various butyral contents and in various molecular weights grades depending on the application. Poly(vinyl butyrals) are much softer than are poly-(vinyl formals) since the C_4 chain internally plasticizes the resin.

Unplasticized poly(vinyl butyral) sheeting has a specific gravity of 1.07–1.10, a heat-distortion point of 50–60°C, a tensile strength (at break) of 4600–800 psi, an elongation (at break) of 70–110%, an impact strength (Charpy, notched) of 0.7–1.1 ft lb-in., and a dielectric strength (step-by-step) of 370–400 V/mil (6). Butyrals are soluble in the lower alcohols, dioxane, methyl Cellosolve, and in a toluene–ethanol

Table 7. Some Properties of Poly(vinyl Butyrals) (6)

Type	Butvar B-72A	Butvar B-73	Butvar B-76	Butvar B-90	Butvar B-98
form			white, free-flowing powder		
volatiles, % max	3.0	3.0	5.0	5.0	5.0
molecular weight	180,000	50,000	45,000	38,000	30,000
(weight average)	270,000	80,000	55,000	45,000	34,000
hydroxyl content, expressed as % poly(vinyl alcohol)	17.5–21.0	17.5–21.0	9.0–13.0	18.0–20.0	18.0–20.0
acetate content, expressed as % poly(vinyl acetate)	0–2.5	0–2.5	0–2.5	0–1.0	0–2.5
butyral content, expressed as % poly(vinyl butyral)	80	80	88	80	80
specific gravity	1.100	1.100	1.083	1.100	1.100
viscosity,[a] cP	ca 1,570	ca 400	ca 175	ca 195	ca 75
viscosity,[b] cP	8,000–18,000	1,000–4,000	500–1,000	600–1,200	200–450

[a] Viscosity of 10% solution in 95% ethanol at 25°C, using an Oswald viscometer.
[b] Viscosity of a 15% solution in 60:40 toluene–ethanol at 25°C using a Brookfield viscometer.

mixture. Poly(vinyl butyrals) are also soluble in concentrated soap solutions. They stay in solution even when diluted with water. The addition of methanol causes the dispersion to break (87). Table 8 lists the common solvents for the various commercial

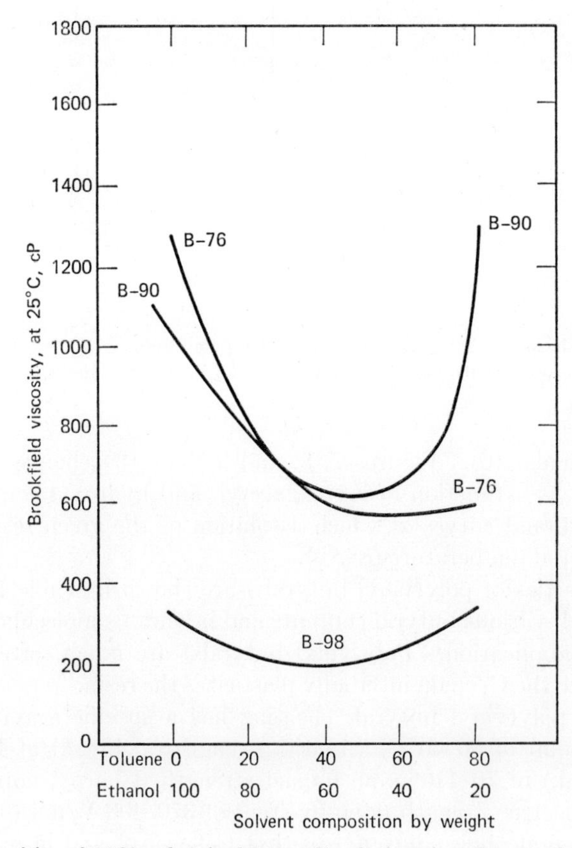

Fig. 11. Viscosities of 15% poly(vinyl butyral) solutions in toluene–ethanol mixture (6).

grades of poly(vinyl butyral) (6). Figures 10 and 11 show the viscosity of poly(vinyl butyral) solutions. In practice the choice of solvents is determined by the application. For example, in surface coatings when other resins are used in conjunction with butyrals the choice of suitable solvents is determined by compatibility studies. Solvent mixtures such as butyl acetate–methanol, ethyl acetate–methanol, and benzene–ethanol are often employed. Benzene, toluene, and methyl acetate often are used as diluents. Poly(vinyl butyrals) are insoluble in aliphatic hydrocarbons as well as in drying oils, waxes, and water. If the hydroxyl number of the resin is high, water will swell a film.

Table 8. Solvents for Poly(vinyl Butyral)[a] (6)

Solvent	Butvar[b] B-72A,B-73	Butvar[c] B-76	Butvar[c] B-90,B-98
acetic acid (glacial)	S	S	S
acetone	I	S	SW
n-butanol	S	S	S
butyl acetate	I	S	PS
carbon tetrachloride	I	PS	I
cyclohexanone	S	S	S
diacetone alcohol	PS	S	S
diisobutyl ketone	I	SW	I
dioxane	S	S	S
ethanol, 95%	S	S	S
ethyl acetate, 99%	I	S	PS
ethyl acetate, 85%	S	S	S
ethyl Cellosolve	S	S	S
ethylene chloride	SW	S	PS
ethylene glycol	I	I	I
isophorone	PS	S	S
2-propanol, 95%	S	S	S
isopropyl acetate	I	S	I
methanol	S	SW	S
methyl acetate	I	S	PS
methyl Cellosolve	S	S	S
methyl ethyl ketone	SW	S	PS
methyl isobutyl ketone	I	S	I
naphtha (light solvent)	I	SW	I
propylene dichloride	SW	S	SW
toluene	I	PS	SW
toluene–ethanol, 95% (60:40 by wt)	S	S	S
xylene	I	PS	SW

[a] S, soluble; PS, partially soluble; I, insoluble; SW, swells.
[b] 5% solids solution agitated for 24 hr at room temperature.
[c] 10% solids solution agitated for 24 hr at room temperature.

Coatings and films of poly(vinyl butyral) have excellent mechanical properties. It is found that the higher the molecular weight, the higher the tensile strength and elongation. Chemical resistance of both poly(vinyl formal) and poly(vinyl butyral) is excellent, especially in baked coatings. Poly(vinyl formal) and poly(vinyl butyral) possess a combination of properties such as high tensile strength, impact resistance, and elasticity which is not normally found in other polymers.

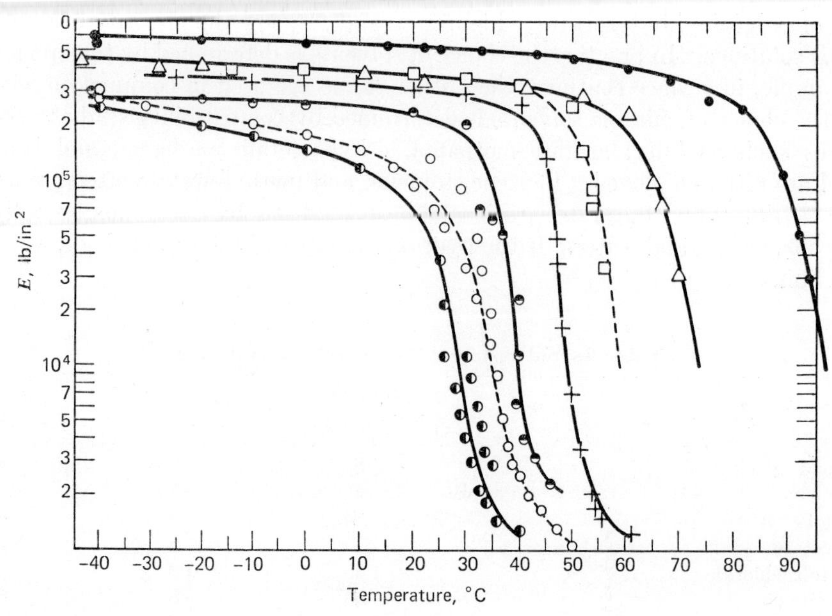

Fig. 12. Torsional modulus–temperature curves of poly(vinyl acetals) (90).

Symbol	*Aldehyde*	*Hydroxyl content expressed as poly(vinyl alcohol)*
●	acetaldehyde	11.8
△	propionaldehyde	13.7
□	isobutyraldehyde	12.9
+	butyraldehyde	12.0
◓	hexaldehyde	12.0
◐	heptaldehyde	13.0
○	2-ethylhexaldehyde	12.8

The electrical properties of poly(vinyl butyral) have been found to be intermediate between poly(vinyl acetate) and poly(vinyl chloride) (41). Dry films which have been conditioned at 20°C and 60% relative humidity show the following values:

dielectric constant	5–7 (50 Hz)
	3–5 (10^3 Hz)
	2–3 (10^6 Hz)
dielectric loss factor	0.7% (50 Hz)
volume resistivity	8×10^{14} Ω cm
electric field strength	400 kV/cm

The electrical properties of poly(vinyl formals) have been reviewed (85). The thermal stability of poly(vinyl acetals) is adequate in many applications (86). Stabilizers are often added, however, when the resins are formulated into wire coatings and baking finishes (153).

The specific heat for poly(vinyl butyral) with and without plasticizer was recently determined (95). The critical surface tension of poly(vinyl butyral) was found to be 24–25 dyn/cm (99).

Poly(vinyl acetals) have been made using higher aliphatic aldehydes (90). Figure 12 shows the torsional-modulus temperature curves of a series of acetals. It can be

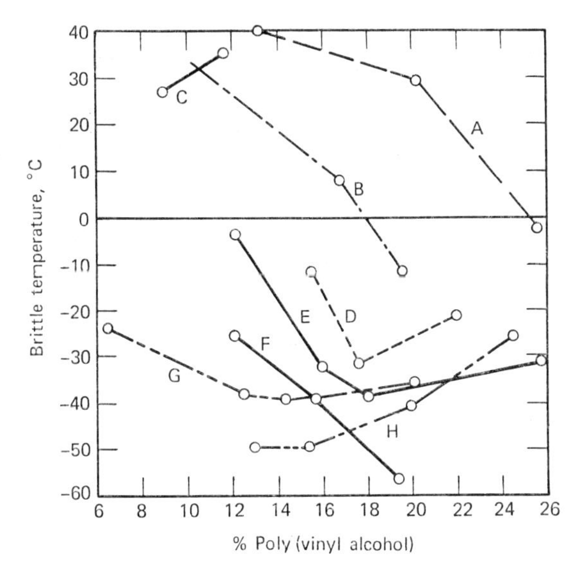

Fig. 13. Relation of brittle temperature to composition (90).

(A) Poly(vinyl isobutyral)
(B) Poly(vinyl propional)
(C) Poly(vinyl acetal)
(D) Poly(vinyl 2-ethylbutyral)
(E) Poly(vinyl butyral)
(F) Poly(vinyl hexanal)
(G) Poly(vinyl 2-ethylhexanal)
(H) Poly(vinyl heptanal)

seen that the long-side-chain acetals are effectively internally plasticized. This is also shown in Table 9 where the glass-transition temperatures (T_g) are shown and in Figure 13 where the brittle points are shown. The effect of residual hydroxyl groups on the torsional modulus of poly(vinyl butyral) is shown in Figure 14. Beyond about

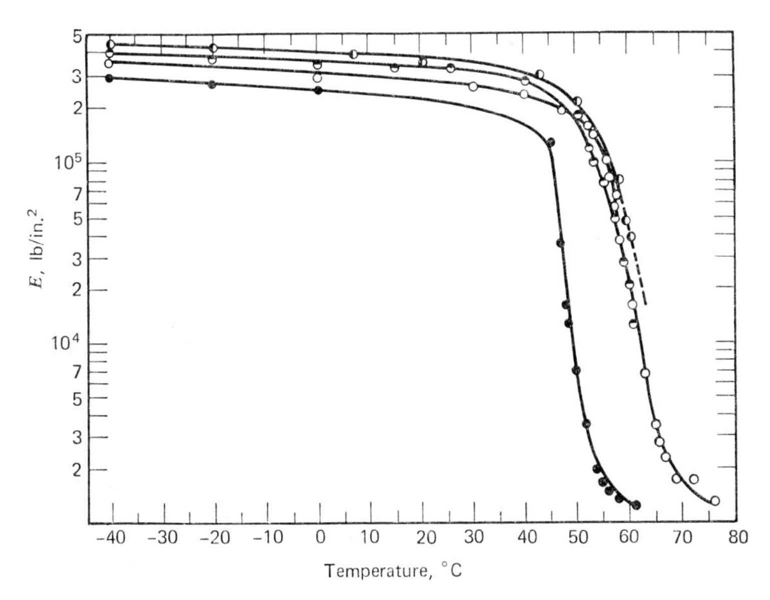

Fig. 14. Torsional modulus–temperature curves of poly(vinyl butyral) containing various amounts of alcohol groups expressed as % poly(vinyl alcohol) (90): ● 12.1%; ○ 16.0%; ◒ 18.0%; ◑ 25.6%.

12% OH groups the glass-transition temperature changes only slightly or, in other words, butyraldehyde is an effective internal plasticizer only when the hydroxyl group content of the final resin is low.

Typical mechanical properties are shown in Table 10. Although some of these acetals have interesting properties they have not been commercialized because of the higher price of higher aliphatic aldehydes.

Table 9. Glass-Transition Temperatures of Poly(vinyl Acetals) (90)

Aldehyde	Hydroxyl content[a]	T_g,[b] °C
acetaldehyde	11.9	86
	9.1	76
propionaldehyde	19.7	64
	16.7	54
	13.7	59
	10.3	59
isobutyraldehyde	25.6	63
	20.1	63
	13.3	53
butyraldehyde	25.6	54
	18.0	53
	16.0	51
	12.1	44
2-ethylbutyraldehyde	21.8	51
	17.7	51
	15.5	51
n-hexaldehyde	24.6	42
	19.3	
	19.6	42
	16.0	37
	12.2	29
n-heptaldehyde	20.2	31
	19.5	32
	15.4	20
	13.0	3
2-ethylhexaldehyde	19.7	29
	14.3	24
	12.8	14
	6.5	−10

[a] Expressed as % poly(vinyl alcohol).
[b] Determined by torsional modulus measurements.

Plasticizers. Many plasticizers for poly(vinyl acetal) resins have been disclosed in the patent literature (see Ref. 90a for an excellent survey). No general statement can be made because the wide variety of poly(vinyl acetals) and their different applications demand specific plasticizers and plasticizer combinations. Good plasticizers for poly(vinyl butyral) are (6) the following: dibutyl phthalate, dioctyl phthalate, tricresyl phosphate, tri(2-ethyl hexyl)phosphate, polyethylene glycol ethers, glyceryl monooleate, low-molecular-weight polyesters, dibutyl sebacate, and tributyl citrate. Many of these plasticizers are also suitable for poly(vinyl formal) in addition to the following (6): diphenyl phthalate, butyl benzyl phthalate, triphenyl phosphate, epoxidized soybean oil, polyesters, and rosin derivatives.

Table 10. Tensile Strength and Elongation of Plasticized Poly(vinyl Acetals) (90)

Aldehyde	Hydroxyl content[a]	DBP,[b] phr	Tensile strength, psi	Elongation, %
acetaldehyde	9.1	0	11,420	6
		54.9	5,840	215
isobutyraldehyde	19.7	0	8,280	5
		40.6	5,480	396
		43	5,060	395
	13.3	0	9,950	5.5
		30.0	4,820	267
butyraldehyde	25.6	0	10,020	8
	18.0	0	8,085	9
		15.0	5,190	10
		37.2	5,460	380
	12.1	0	8,630	10
		25.2	5,390	190
2-ethylhexaldehyde	14.3	0	4,360	10
		14.0	3,470	387
	6.5	0	2,600	434

[a] Expressed as % poly(vinyl alcohol) plasticizer.
[b] Dibutyl phthalate.

The mechanical properties of many acetals have been measured when plasticizer had been added (90). Dispersions of poly(vinyl butyral) generally contain 5–40 parts of plasticizer per 100 parts of resin.

Dilute Solution Properties. Only very few studies have been made concerning the dilute solution properties of poly(vinyl acetals). Intrinsic viscosity–molecular weight relationships seem to have been determined only for poly(vinyl formal) (94). Recently the dilute solution properties of poly(vinyl butyral) have been studied (91). It was shown that poly(vinyl butyral) assumes a more extended configuration than poly(vinyl alcohol). Chain flexibility for isotactic poly(vinyl butyral) is higher than for atactic poly(vinyl butyral). It was determined that the intrinsic viscosity of poly(vinyl acetals) depends not only on molecular weight and degree of acetalization, but also on the temperature of the polymerization of the parent poly(vinyl acetate) (92). The non-Newtonian behavior of poly(vinyl acetal) solutions in methyl ethyl ketone is pronounced (93).

The intrinsic viscosity–molecular weight relationship in dioxane at 30°C was determined for poly(vinyl formal) (94) with the following results:

$[\eta]$, *dl/g*	$\bar{M}_w$
1.190	337,000
0.927	234,000
0.625	131,000

Spectra. Infrared spectra of poly(vinyl formal) (96,103), poly(vinyl acetal) (96), and poly(vinyl butyral) (96) have been published (96) (Figs. 15–18).

One can see in the IR spectrum of poly(vinyl formal) five strong bands between 8 and 10 μm. They are characteristic of the acetal ring structure. Absorptions at 5.78 and 7.3 μm show residual vinyl acetate units, and at 2.9 and 6.95 μm show vinyl alcohol structure. For poly(vinyl acetal), bands at 7.45 μm, 8.8 μm, and 10.6 μm are characteristic for the cyclic acetal structure. Poly(vinyl butyral) spectra show the

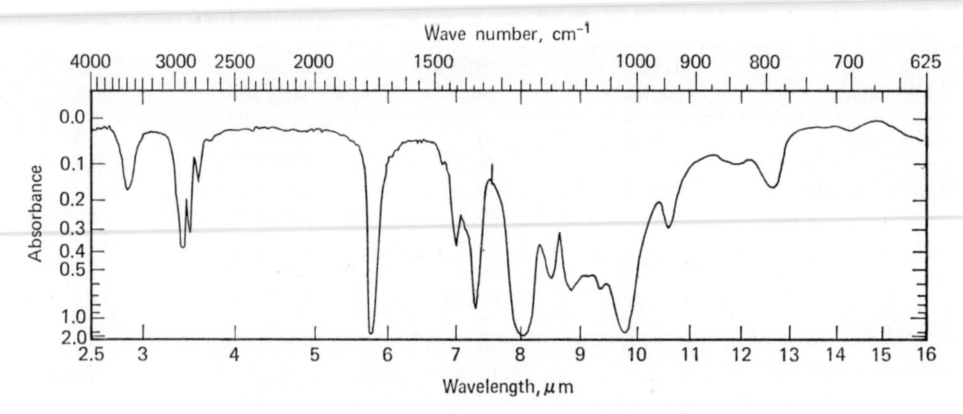

Fig. 15. Infrared spectrum of poly(vinyl formal) with 43–55% vinyl formal, 5–7% vinyl alcohol, and 40–50% vinyl acetate groups. Courtesy Dow Chemical Co.

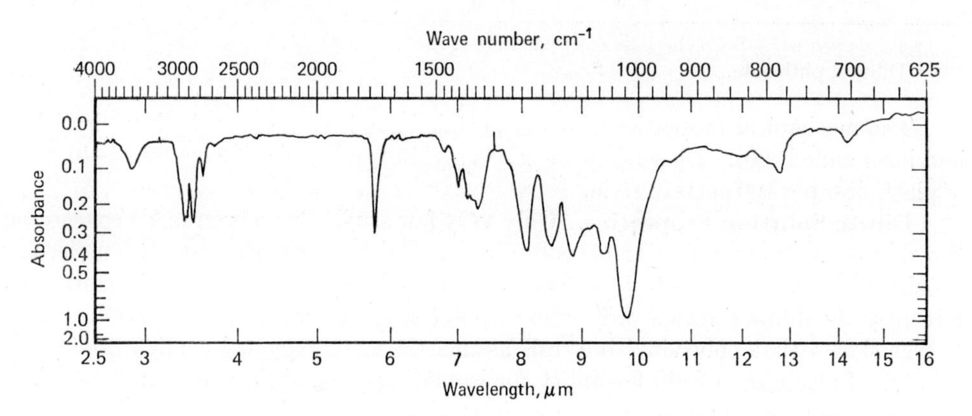

Fig. 16. Infrared spectrum of poly(vinyl formal) with 72–83.5% vinyl formal, 7–9% vinyl alcohol, and 9.5–13% vinyl acetate groups. Courtesy Dow Chemical Co.

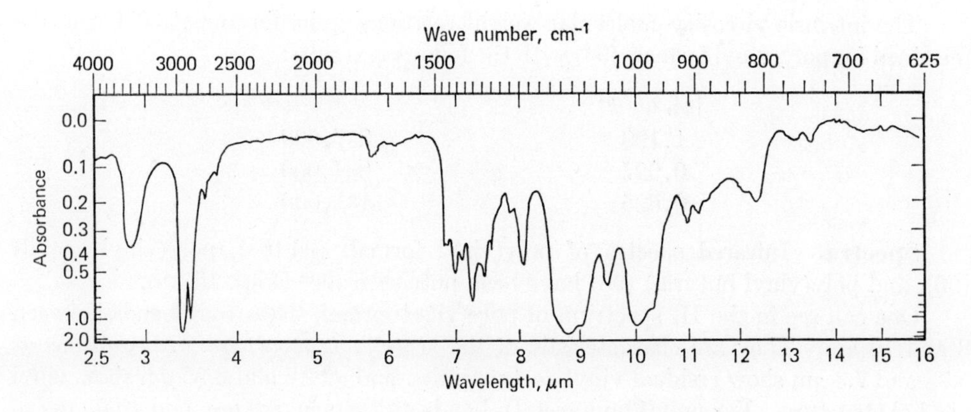

Fig. 17. Infrared spectrum of poly(vinyl butyral) with ca 88% vinyl butyral, 9–13% vinyl alcohol, and 2.5% max vinyl acetate groups. Courtesy Dow Chemical Co.

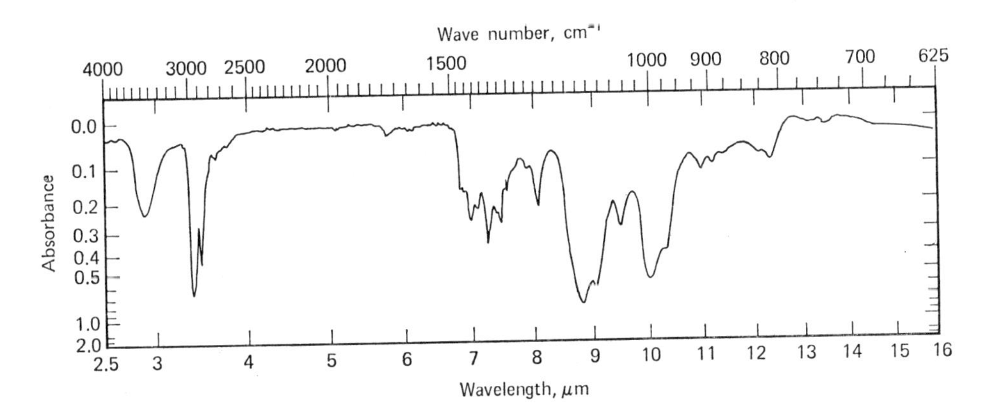

Fig. 18. Infrared spectrum of poly(vinyl butyral) with ca 80% vinyl butyral, 17.5–21% vinyl alcohol, and 2.5% max vinyl acetate groups. Courtesy Dow Chemical Co.

8.9 μm and 10.1 μm bands for the cyclic acetal. Also bands at 2.9 μm for the vinyl alcohol structure and sometimes a very weak band for the poly(vinyl acetate) structure can be seen.

Characteristic bands (96) for poly(vinyl formal) are 3.61, 8.07, 8.5, 8.8, and 9.8 μm; for poly(vinyl acetal), 7.46, 8.05, 8.8, 9.1, and 10.57 μm; for poly(vinyl butyral), 7.23, 8.8, 9.0, 9.5, 10.0, and 10.3 μm.

NMR spectra for poly(vinyl formal) have been studied extensively to elucidate the structure (97,103). It was found that the spectra were useful tools in the quantitative analysis of the steric configuration (98).

Table 11. Color of Poly(vinyl Acetal) Films After Treatment with Iodine–Iodide Solution (101)

Product	Type	Color after 30 sec[a]
Alvar 1570	acetal	green
Alvar 1580	acetal	green
Butvar B 76	butyral	yellow
Formvar 77	formal	blue
Formvar 1595	formal	black-violet
Mowital B 30 H	butyral	green
Mowital B 30 T	butyral	green
Mowital B 60 H	butyral	green
Mowital B 60 T	butyral	green
Mowital F 40	formal	blue
Pioloform B	butyral	green
Pioloform BL	butyral	green
Pioloform F	formal	black-violet
Revyl	butyral	yellow
Rhovinal B	butyral	yellow green
Vinilex FH	formal	black brown
Vinilex F	formal	blue violet
Vinylite XYHL	butyral	green
Vinylite VYNC	butyral	green

[a] Reagent: 10 parts of 50% acetic acid and 7 parts of iodine/iodide solution (prepared by mixing 1 part of KI and 0.9 part of iodine in 40 parts of water and 2 parts of glycerol).

Analysis. Poly(vinyl acetals) can be hydrolyzed by boiling a sample with 25–50% sulfuric acid. The aldehyde part can then be distilled off and qualitatively and quantitatively determined or can be determined in situ as the 2,4-dinitrophenylhydrazone of the aldehyde (100).

A very convenient qualitative color test can be performed using iodine reagent. Table 11 shows the color development when iodine–iodide solution is used with various poly(vinyl acetals) (101). It is still quite difficult to obtain accurate data on the acetal, acetyl, and hydroxyl contents. Other methods used are described in Ref. 101a.

Recently poly(vinyl formal) has been analyzed by pyrolysis gas chromatography (102). It is claimed that the analysis was rapid and accurate.

Chemical Properties. The chemical properties of poly(vinyl acetals) are similar to those of other hydroxyl-containing polymers. Reactions in solutions with isocyanates, epoxy compounds, phenolics, dialdehydes, anhydrides, *N*-methylol compounds such as those from melamine or urea, metal complexes, etc are possible for modification and crosslinking purposes.

The reaction of poly(vinyl acetals) with phosphoric acid has been studied (100). It was found that hydrolysis is the principal effect, although esterification of the hydroxyl groups of poly(vinyl butyral) also took place. The order of hydrolysis resistance of the acetals is as follows: poly(vinyl formal) > poly(vinyl propional) > poly(vinyl acetal) > poly(vinyl butyral). Poly(vinyl butyral) forms a complex with chromium salts in the presence of phosphoric acid (see Wash primers). Poly(vinyl formal) degrades at 150°C in oxygen, splitting off water and formaldehyde (104). When molded, a dish of poly(vinyl formal) resin is stable for 1 hr at 200°C.

Toxicity. The commercial poly(vinyl acetals) are nontoxic but the same precautions as with poly(vinyl alcohol) should be taken. Toxicological studies have been published on poly(vinyl formal) (see the section on Poly(vinyl Alcohol) in this article).

Poly(vinyl butyral) and poly(vinyl formal) are approved by the U.S. Food and Drug Administration as components in can coatings and paperboard coatings for food packaging use. Butvar B-76 poly(vinyl butyral) has also been approved as metal-can coating for meat packaging when used with thermosetting resins (6).

Applications

Poly(vinyl acetals) are mainly used in two application areas: coatings and adhesives. Owing to their relatively high cost only small amounts of resin are sold. Annual production of poly(vinyl formal) in the United States in 1969 was estimated at 5–6 million lb at an average selling price of about $0.90/lb; about 60 million lb of poly(vinyl butyral) was made in the same year and sold at an average price of $1.05/lb. The major suppliers to the coatings industry of these resins in the United States are Monsanto and Union Carbide whereas in the adhesive market (safety glass) Monsanto and Du Pont are the most important suppliers.

Coatings. Poly(vinyl formal) is extensively used in the manufacture of enamels for heat-resistant wire insulations for electrical applications (67). Other resins such as phenolics are combined with poly(vinyl formal) to give coatings of great toughness, adhesion, flexibility, and abrasion resistance which maintain their dielectric strength under the exacting conditions of wire manufacture and use (105–113). The oil and grease resistance of these coatings is excellent. Table 12 shows that the abrasion resistance, heat resistance, and chemical resistance of wire coatings made with poly(vinyl formal) are significantly better than those made of oleoresinous coatings (67).

Table 12. Comparison of Poly(vinyl Formal) and Oleoresinous Wire Enamels (67)

Test and description	Poly(vinyl formal)	Oleoresinous enamel
elongation, %	75	35
abrasion resistance		
revolving drum, 25°C, turns	200	30
revolving drum, 130°C, turns	75	5
Inca paddle, turns	2400	150
scrape, 0.009-in. blade, 25°C, oz	50	20
scrape, 0.016-in. blade, 105°C, oz	55	20
compression		
lb to cause failure at 25°C	5500	1800
20-lb load, 24 hr, 125°C	passes	fails
20-lb load, 24 hr, 150°C	passes	fails
heat aging, 105°C, hr	3000–3500	50–100
heat aging, 125°C, hr	400–700	12–24
heat shock, 2 hr at 105°C	passes	fails
solvent resistance		
kerosene	no effect	sl softened
naphtha	no effect	sl softened
toluene	no effect	fails
acetone	no effect	fails
ethanol	no effect	fails
power factor		
30°C	0.0076	0.0085
75°C	0.0046	0.036
dielectric constant		
30°C	3.6	3.0
75°C	3.4	3.6
loss factor		
30°C	0.028	0.025
75°C	0.025	0.12

The chemical interactions and the mechanism of cure in the poly(vinyl formal)–phenolic resin systems have been studied. It was found that the residual alcohol groups in the poly(vinyl formal) react with the phenolic methylol groups (114). Over the years the performance of the poly(vinyl formal) coatings has been significantly improved (115).

A small portion of the phenolic resin has been replaced by a blocked isocyanate resin ("Formetic" enamel). This resulted in a higher "cut-through" temperature for magnet wire and increased the resistance to refrigerants such as chlorodifluoromethane which is used in air conditioning applications. Operation at 130°C for 20,000 hr has been realized for magnet wire when epoxy resins were added to the standard poly(vinyl formal) coating (116–120).

Multicoat wires were developed to improve heat and overload resistance even further using a combination of poly(vinyl formal) with polyesters (121), polyimide (122–124), nylon (125), or poly(vinyl butyral). See also SURFACE COATINGS; WIRE AND CABLE COVERINGS.

Wash Primers. Poly(vinyl butyral) and to a lesser extent poly(vinyl formal) are important ingredients of so-called "wash-primer" formulations (126–131). They were developed by the U.S. Navy to provide a uniform surface preparation for applying antifouling vinyl coatings to ships (see also MARINE APPLICATIONS). The wash primers consist of an alcoholic solution of poly(vinyl butyral) and zinc tetraoxychromate and a second solution of phosphoric acid in alcohol. Although one-pack resin–phosphoric acid systems have been offered commercially they are not as effective as the two-pack systems. The essential components of the wash primer are poly(vinyl butyral), chromate ions, phosphoric acid, alcohol, and water.

A number of studies have been made to elucidate the mechanism of adhesion to metals and the corrosion resistance of the butyral coating (127,129,132). The phosphoric acid changes the chemical structure of poly(vinyl butyral) by rearranging the aldehyde, amounting probably to hydrolysis of hemiacetals and acetal bridges and formation of cyclic acetals. There is also a slight molecular-weight decrease due to chain scission at 1,2-glycol structures in the resin. It is most likely the alcohol reduces the chromium(VI) to chromium(III) and the resulting chromium(III) ions are then bound by complex formation to the poly(vinyl butyral). This complex, in turn, probably adsorbs phosphate ions. In this latter form it is adsorbed onto the metal surfaces. In effect, the product is an inorganic coating on the metal to which the organic coating is chemically attached (133). It was found that essentially no esterification of the free alcohol groups in poly(vinyl butyral) occurs.

Wash primers are widely used on a variety of metal structures such as storage tanks, ships, airplanes, etc. The following is a typical wash primer recipe (Butvar wash primer B-1030):

(a) To a solution of

	% by wt
Butvar B-76	1.24
ethanol (95%)	9.35
methyl ethyl ketone	9.97

add

	% by wt
basic zinc chromate pigment	11.52
Celite 266 (Johns Manville Corp.)	4.82

(b) Grind to Hegman fineness of 6, N. S. Scale.
(c) Add solution of

	% by wt
Butvar B-76	7.39
ethanol (95%)	23.08
methyl ethyl ketone	24.63
resinox P-97 (Monsanto Co.)	8.00

(d) Grind mixture A and package.

The phosphoric acid constituent is packaged separately (mixture B):

	% by wt
phosphoric acid (85%)	7.50
n-butanol	92.50

Equal weights of mixtures A and B are used. The raw materials cost is about $1.42/gal, the cost per square foot of painted surface about 26¢ and the pot life is about 8–12 hr. Other wash-primer recipes have been published (6,127). Weldable corrosion-preventive primers for iron or steel are based on poly(vinyl butyral) compositions which contain a phenolic resin and a rust inhibitor (137,138). Poly(vinyl butyral) is used fairly widely as a sealer in wood finishes where it confers outstanding "holdout," intercoat adhesion, moisture resistance, flexibility, toughness, and impact resistance (6,134).

Fabrics have been coated with poly(vinyl butyrals) both in solution (135) and dispersions (136). Butvar dispersions are used in the textile industry to impart increased abrasion resistance, durability strength, and slippage control. Table 13 lists the properties of typical plasticized poly(vinyl butyral) dispersions (6).

Table 13. Properties of Poly(vinyl Butyral) Dispersions (6)

Type	Butvar Dispersion BR	Butvar Dispersion FP
form	an aqueous dispersion of plasticized poly(vinyl butyral), milk-white in color	
total solids, %	50.0–51.0	50.0–51.0
viscosity, cP, max		
Brookfield, No. 3 spindle, 30 rpm, 25°C	1500	
Brookfield, No. 1 spindle, 30 rpm, 25°C		150
pH	8.0–10.5	8.0–10.5
particle size	most particles close to 0.5 μm; none larger than 1.0 μm	
particle charge	anionic	anionic
plasticizer content	40 parts per 100 parts of resin (28.6% of solids)	5 parts per 100 parts of resin (4.8% of solids)
lb/gal at 25°C	8.4	8.7

Poly(vinyl butyral) has also been proposed for heat-sealable coatings (41), temporary coatings for packaging applications (139), can coatings (163), and powder coatings (140).

Adhesive Applications. The largest use of poly(vinyl butyral) in 1970 (about 25–30 million lb in the U.S.) is for use as an interlayer in *automobile safety glass.* W. O. Herrmann was the first to propose this use in 1931 (141). The laminate consists of two layers of glass between which is sandwiched a film of plasticized poly(vinyl butyral). The adhesion of the poly(vinyl butyral) to glass is so strong that no glass splinters fly away when the glass laminate is broken on impact (142–144).

The plasticized poly(vinyl butyral) is extruded into a sheet and the glass panels laminated to it under heat and pressure. Typical plasticizers are triethylene glycol di-2-ethylbutyrate or dilaurate, dibutyl sebacate, triallyl phosphates, dialkyl phthalates, and ricinoleates. The glass is crystal clear and maintains impact resistance and clarity over wide temperature conditions (−40°C to +50°C). Even after exchanging the glass for 300 hr at 150°C there is only a 3% loss of light transmission (150). Figure 19 shows the impact resistance of safety glass laminated with various plastics. It can be seen that glass laminated with poly(vinyl butyral) is by far the best (151). The addition of metal salts to poly(vinyl butyral) has been studied to improve impact resistance (145–149).

Cellulose butyrate is laminated to poly(vinyl butyral) to produce shatterfree television tubes (152).

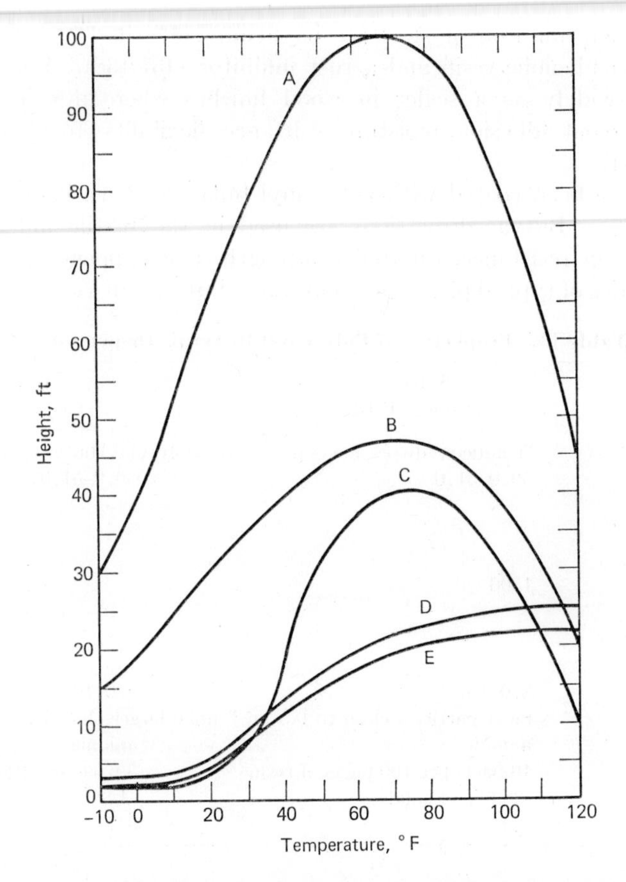

Fig. 19. Temperature vs resistance to break for safety glass constructed with different commercial types of plastic interlayer. (Breakthrough of 0.5 lb steel ball dropped upon a 12 × 12 in. glass plate supported around its periphery.) (151). (A) Poly(vinyl butyral), 0.025 in. thick; (B) poly(vinyl butyral), 0.015 in. thick; (C) acrylate resin, 0.026 in. thick; (D) cellulose acetate, 0.025 in. thick; (E) cellulose nitrate, 0.020 in. thick.

Structural Adhesives (154). Granular poly(vinyl formal) coated with a liquid phenolic resin will often form stronger bonds to metals than riveting or welding, when applied with heat and pressure. The process is known as the "Redux" process (156). The combination of poly(vinyl formal)–phenolic resin has been used in aircraft manufacture because of the high-shear-strength values at temperatures up to 250°C, coupled with high peel strength at low temperatures. Adhesive bonds have been durable in many varied climates (158,159) and the fatigue strength is very high (157). The poly(vinyl formal)–phenolic systems are now also used in honeycomb constructions (qv), brake linings, printed circuits, curtain walls (154) and other structural adhesives (160).

Poly(vinyl formal-acetal) has been used as adhesive to aluminum foil (155). Poly(vinyl formal) can also be used in combination with other resins, such as epoxy and melamine–formaldehyde resins. Detailed recipes are given in Ref. 154. The ratio of poly(vinyl butyral) to thermosetting components is optimal at 10:5 (161). Hot-melt adhesives are also formulated using poly(vinyl butyral) (162). See also ADHESION AND BONDING.

Miscellaneous Applications. Porous solid poly(vinyl formal) can be manufactured by foaming poly(vinyl alcohol) solution with air and then formalizing the foam (164–167). The foam can then be cut and used as sponges. These sponges have been sold under the trade name Ivalon. Poly(vinyl formal) foam has also been used as medical implants for the human body (164); however, toxicity complications arose (168) and these sponges are no longer used.

Poly(vinyl formals) can be molded with plasticizers and modifying agents. The impact strength can be increased markedly by blending in rubber compounds (169, 170). Polyblends with other plastics such aspo lyisobutylene, poly(ethyl acrylate), and chlorosulfonated polyethylene have properties that make them useful as solid plastics in electrical and other applications. Other potential applications are listed in Refs. 5 and 84. See also ELECTRICAL APPLICATIONS.

Poly(vinyl formal) fibers (partially formalized poly(vinyl alcohol)) have already been described in detail in this article on pp. 183–195.

Bibliography

General References

1. S. A. Miller, *Acetylene*, Vol. II, Academic Press, Inc., New York, 1966, pp. 296–304.
   General review of poly(vinyl acetals).
2. N. Platzer, *Modern Plastics* **28** (10), 142 (1951).
   Review of properties and uses of poly(vinyl acetals).
3. J. Scheiber, *Chemie und Technologie der Künstlichen Harze*, Vol. I, Wissenschaftliche Verlags Ges., Stuttgart, 1961, pp. 468–493.
   Very detailed review of poly(vinyl acetals) with the literature covered up to 1955 (in German).
4. P. Schneider, in E. Müller, ed., *Houben-Weyl Methoden der Organischen Chemie*, Vol. 14, Part 2, Georg Thieme Verlag, Stuttgart, 1963, pp. 717–723.
   Methods of laboratory preparation of poly(vinyl acetals).
5. H. Warson, in S. A. Miller, ed., *Poly(vinyl Acetals) in Ethylene*, E. Benn Ltd., London, 1969, pp. 1041–1051.
   General review of poly(vinyl acetals).
6. *Butvar, Poly(vinyl Butyral), and Formvar, Poly(vinyl Formal)*, Technical Bulletins Nos. 6070 and 6130, Monsanto Co., St. Louis, Mo., 1969.
7. F. Kainer, *Polyvinylalkohole*, Ferdinand Enke, Stuttgart, 1949 (in German); preparation of poly(vinyl acetals), pp. 63–86; safety glass applications, pp. 217–224.
8. I. Sakurada, ed., *Poly(vinyl Alcohol), First Osaka Symposium, Kobunshi Gakkai, Tokyo, 1955*, Chaps. 9, 10, 11 and 31.
   Discussion of poly(vinyl butyral) and poly(vinyl formal) applications. Also review of chemistry of acetalization and dilute solution properties of poly(vinyl acetals).
9. R. D. Dunlop, *Fiat Final Rept. No. 1109* (1947).
   Review of production and applications of poly(vinyl acetals) in Germany before 1945.
10. K. Toyoshima, *Kobunshi Kako* **18,** 182–712 (1969) (in Japanese).
    Detailed review on the mechanism of poly(vinyl acetal) formation.

References

11. W. O. Herrmann and W. Haehnel (to Consortium für Elektrochemische Industrie), Ger. Pat. 480,866 (1924).
12. G. O. Morrison, F. W. Skirrow, and K. G. Blaikie (to Canadian Electroproducts), U.S. Pat. 2,036,092 (1935).
13. W. Haehnel and W. O. Herrmann (to Consortium für Elektrochemische Industrie), Ger. Pat. 507,962 (1927).
14. H. Hopff (to I. G. Farben Industrie A.G.), U.S. Pat. 1,955,068 (1934).
15. H. Bauer, J. Heckmaier, H. Reinecke, and E. Bergmeister (to Wacker-Chemie), Ger. Pat. 929,643 (1952).

16. G. F. D'Alelio (to Allgem. Elektrizität Gesellschaft), Ger. Pat. 912,399 (1941).
17. W. O. Kenyon and W. F. Fowler, Jr. (to Eastman Kodak Co.), U.S. Pat. 2,397,548 (1942).
18. F. Winkler and J. W. Zimmermann (to Farbwerke Hoechst A.G.), Ger. Pat. 1,116,905 (1959).
19. W. H. Sharkey (to E. I. du Pont de Nemours & Co., Inc.), U.S. Pat. 2,396,209 (1942).
20. C. A. Sperati (to E. I. du Pont de Nemours & Co., Inc.), U.S. Pat. 2,421,971 (1944).
21. K. H. Kahrs, F. Winkler, and J. W. Zimmermann (to Farbwerke Hoechst A.G.), Ger. Pat. 1,086,435 (1959).
22. W. O. Herrmann (to Chemische Forschungs Gesellschaft m.b.H.), Ger. Pat. 690,332 (1941).
23. Compagnia Generale di Electricita, Ital. Pat. 395,170 (1941).
24. A. Voss, E. Dickhauser, and W. Starck (to I. G. Farben Industrie A.G.), Ger. Pat. 592,233 (1934).
25. Compagnia Generale di Electricita, Ital. Pat. 394,607 (1941).
26. O. M. Klimova, *Zh. Prikl. Khim.* **40**, 930 (1967).
27. H. Grafje (to BASF), Ger. Pat. 1,142,237 (1963).
28. K. Noma and T. Sone, *Kobunshi Kagaku* **4**, 50 (1947).
29. A. F. Fitzhugh (to Shawinigan Resins Corp.), U.S. Pat. 2,527,495 (1947).
30. E. Imoto and R. Motoyama, *Kobunshi Kagaku* **11**, 251 (1954).
31. S. Okamura and T. Motoyama, *Kogyo Kagaku Zasshi* **55**, 774 (1952).
32. S. Okamura, T. Motoyama, and K. Uno, *Kogyo Kagaku Zasshi* **55**, 776 (1952).
33. Eastman Kodak Co., Brit. Pat. 513,119 (1941).
34. Ref. 8, p. 81.
35. K. F. Beal and C. J. B. Thor, *J. Polymer Sci.* **1**, 540 (1946).
36. W. Starck (to Farbwerke Hoechst A.G.), Ger. Pat. 849,006 (1951).
37. E. T. Cline and H. B. Stevenson (to E. I. du Pont de Nemours & Co., Inc.), U.S. Pat. 2,606,803 (1953).
38. G. A. Schroeter and P. Riegger, *Kunststoffe* **44**, 228 (1954).
39. T. Motoyama and S. Okamura, *Kobunshi Kagaku* **7**, 265 (1950).
40. K. Noma, C. Y. Huang, and K. Tokita, *Kobunshi Kagaku* **6**, 439 (1949).
41. G. Schulz and G. Mueller, *Kunststoffe* **42**, 298 (1952).
42. T. Minami and T. Obata, *Kogyo Kagaku Zasshi* **57**, 826 (1954).
43. E. L. Martin (to E. I. du Pont de Nemours & Co., Inc.), U.S. Pat. 2,929,710 (1954).
44. J. O. Corner and E. L. Martin, *J. Am. Chem. Soc.* **76**, 3593 (1954).
45. A. Hachihama, M. Imoto, and C. Asao, *Kogyo Kagaku Zasshi* **47**, 919 (1944).
46. G. Kränzlein and U. Campert (to I. G. Farben Industrie A.G.), Ger. Pat. 729,774 (1937).
47. G. Kränzlein, A. Voss, and W. Starck (to I. G. Farben Industrie A.G.), Ger. Pat. 661,968 (1930).
48. K. Noma and C. Y. Huang, *Kobunshi Kagaku* **4**, 123 (1947).
49. H. Sonke (to I. G. Farben Industrie A.G.), Ger. Pat. 681,346 (1936).
50. J. D. Ryan (to Libbey-Owens-Ford Glass Co.), U.S. Pat. 2,425,568 (1947).
51. J. D. Ryan and F. B. Shaw (to Libbey-Owens-Ford Glass Co.), U.S. Pat. 2,447,773 (1948).
52. P. J. Flory, *J. Am. Chem. Soc.* **61**, 1518 (1938).
53. P. J. Flory, *J. Am. Chem. Soc.* **72**, 5052 (1950).
54. G. O. Morrison and A. F. Price (to Shawinigan Chemicals, Ltd.), U.S. Pat. 2,179,051 (1939).
55. J. G. McNally and R. H. van Dyke (to Eastman Kokak Co.), U.S. Pat. 2,269,216 (1942).
56. I. Sakurada and O. Yoshizaki, *Kobunshi Kagaku* **10**, 306 (1953).
57. *Ibid.*, **10**, 310 (1953).
58. *Ibid.*, **10**, 315 (1953).
59. I. Sakurada and K. Fuchino, *Riken Iho* **20**, 898 (1941).
60. Y. Oyanagi (to Kurashiki Rayon Co.), Japan. Pat. 446,472.
61. Ref. 10, p. 437.
62. Y. Ogata, M. Okano, and T. Ganke, *J. Am. Chem. Soc.* **78**, 2962 (1956).
63. G. Smets and B. Petit, *Makromol. Chem.* **33**, 41 (1959).
64. K. Shibatani, K. Fujii, Y. Oyanagi, J. Ukida, and M. Matsumoto, *J. Polymer Sci.* [C], *No. 23*, Part 1, 647 (1968).
65. K. Fujii, J. Ukida and M. Matsumoto, *Makromol. Chem.* **65**, 86 (1963).
66. K. Shibatani, K. Fujii, J. Ukida, and M. Matsumoto, *IUPAC Symposium on Macromolecular Chemistry, Tokyo, 1966, No. 5-5-03.*

67. A. F. Fitzhugh, E. Lavin, and G. O. Morrison, *J. Electrochem. Soc.* **100,** 351 (1953).
68. G. O. Morrison and A. F. Price (to Shawinigan Resins Corp.), U.S. Pat. 2,168,827 (1939).
69. E. Bergmeister, J. Heckmaier, and H. Zoebelein (to Wacker-Chemie GmbH), Ger. Pat. 1,071,343 (1957).
70. G. S. Stamatoff (to E. I. du Pont de Nemours & Co., Inc.), U.S. Pats. 2,400,957 (1943), and U.S. Pat. 2,422,754 (1947).
71. L. H. Rombach (to E. I. du Pont de Nemours & Co., Inc.), U.S. Pat. 3,153,009 (1964).
72. *Chem. Eng.* **61,** 122 (Feb. 1954).
73. R. D. Dunlop, *Fiat Report No. 1109,* 15.
74. T. P. G. Shaw and C. Monfet, *Alvar Bulletin,* Shawinigan Chemicals Ltd., 1952.
75. A. Voss and W. Starck (to I. G. Farben Industrie A.G.), Ger. Pat. 737,630 (1930).
76. H. Hopff and E. Kuehn (to I. G. Farben Industrie A.G.), Ger. Pat. 683,165 (1929).
77. W. H. Bromley (to Shawinigan Resins Corp.), U.S. Pat. 2,611,755 (1952).
78. F. Winkler (to Farbwerke Hoechst A.G.), Ger. Pat. 1,056,369 (1959).
79. K. Rosenbush, W. Pense, and F. Winkler (to Farbwerke Hoechst A.G.), Ger. Pat. 1,069,385 (1957).
80. H. M. Collins (to Shawinigan Chemicals Ltd.), U.S. Pat. 2,443,893 (1948).
81. S. G. Matsoyan, *J. Polymer Sci.* **52,** 189 (1961).
82. T. L. Tolbert (to Monsanto Company), U.S. Pat. 3,285,969 (1966).
83. U. Beyer, F. H. Müller, and H. Ringsdorf, *Makromol. Chem.* **101,** 74 (1967).
84. G. O. Morrison, in A. Standen, ed. *Kirk-Othmer Encyclopedia of Chemical Technology,* Vol. 21, 2nd ed., Interscience Publishers, a division of John Wiley & Sons, Inc., New York, 1970, pp. 304–317.
85. *Table of Dielectric Materials,* Vol. 2, National Defense Research Commission, Washington, D.C.
86. T. D. Callinan, *Thermal Evaluation of Formvar Resins, N.R.L. Report C-3397,* Naval Research Laboratory, Washington, D.C., 1948.
87. T. Isemura and Y. Kimura, *J. Polymer Sci.* **16,** 92 (1955).
88. Ref. 7, pp. 42–45.
89. T. F. Murray and W. O. Kenyon (to Eastman Kodak Co.), Ger. Pat. 728,445 (1942).
90. A. F. Fitzhugh and R. N. Crozier, *J. Polymer Sci.* **8,** 225 (1952); **9,** 96 (1952).
90a. Ref. 3, p. 477.
91. H. Matsuda, K. Yamano, and H. Inagaki, *Kogyo Kagaku Zasshi* **73,** 390 (1970).
92. M. Matsumoto and Y. Oyanagi, *J. Polymer Sci.* **37,** 132 (1959).
93. M. Matsumoto and Y. Oyanagi, *J. Appl. Polymer Sci.* **4,** 243 (1960).
94. W. R. Moore and H. R. Pathan, *Indian J. Technol.* **3,** 392 (1965).
95. H. Wilski, *Angew. Makromol. Chem.* **6,** 101 (1969).
96. O. Hummel, *Atlas der Kunststoff-Analyse,* Vol. 1, Part II, Carl Hanser Verlag, München, 1968, Spectra No. 1011–1017.
97. J. C. Woodbrey, in A. D. Ketley, ed., *The Stereochemistry of Macromolecules,* Vol. 3, Marcel Dekker, Inc., New York, 1968, Chap. 2, pp. 61–136.
98. K. Fujii, *J. Polymer Sci.* [B] **4,** 787 (1966).
99. S. Newman, *J. Colloid Interface Sci.* **25,** 341 (1967).
100. R. A. Barnes, *J. Polymer Sci.* **27,** 285 (1958).
101. K. Brockmann and G. Muller, *Farbe Lack* **61,** 217 (1955).
101a. Ref. 3, p. 493.
102. I. Takahashi, T. Yamane, T. Susuki, and T. Mukoyama, *Kogyo Kagaku Zasshi* **73,** 303 (1970).
103. K. Shibatani, T. Fujiwara, and K. Fujii, *J. Polymer Sci.* [A-1] **8,** 1693 (1970).
104. H. C. Beachell, P. Fortis, and J. Hucks, *J. Polymer Sci.* **7,** 353 (1951).
105. R. W. Hall (to General Electric Co.), U.S. Pat. 2,114,877 (1938).
106. R. H. Thielking (to Schenectady Varnish Co.), U.S. Pat. 2,154,057 (1939).
107. E. H. Jackson and R. W. Hall (to General Electric Co.), U.S. Pat. 2,307,588 (1943).
108. W. Patnode, E. J. Flynn, and J. A. Weh, *Ind. Eng. Chem.* **31,** 1063 (1939).
109. R. J. Anderson (to Monsanto Co.), U.S. Pat. 2,574,313 (1951).
110. K. Benton (to General Electric Co.), U.S. Pat. 2,215,996 (1941).
111. E. I. du Pont de Nemours & Co., Inc., Brit. Pat. 580,275 (1946).
112. J. G. Ford (to Westinghouse Electric Corp.), U.S. Pat. 2,372,074 (1945).
113. M. M. Sprung (to General Electric Co.), U.S. Pat. 2,531,169 (1950).

114. S. M. Cohen, R. E. Kass, and E. Lavin, *Ind. Eng. Chem.* **50**, 229 (1958).
115. E. Lavin, A. H. Markhart, and R. W. Ross, *Insulation* **8** (4), 25 (1962).
116. R. B. Young and J. R. Learn, "High Temperature Varnish-Wire Enamel Systems," *Paper Natl. Conf. on Application of Electrical Insulation, Cleveland, Ohio, 1958.*
117. A. F. Fitzhugh, E. Lavin, and A. H. Markhart, "New 130°C Varnished Formvar-Based Magnet Wire Systems," *Paper Natl. Conf. on Application of Electrical Insulation, Cleveland, Ohio, 1958.*
118. C. T. Straka, *Insulation* 7 (9), 17 (1961).
119. E. W. Daszewski, "Thermal Endurance of Epoxy Encapsulated Magnet Wire," *Paper Natl. Conf. on Application of Electrical Insulation, Washington, D.C., 1959.*
120. H. Lee, "Compatibility of Magnet Wire Insulation and Epoxy Encapsulating Resins," *Paper Natl. Conf. on Application of Electrical Insulation, Washington, D.C., 1959.*
121. R. G. Flowers and C. A. Winter (to General Electric Co.), U.S. Pat. 3,442,834 (1969).
122. "Test Procedures for Evaluation of the Thermal Stability of Enameled Wire in Air," *AIEE No. 57* (1959).
123. E. Lavin, A. H. Markhart, and C. F. Hunt, U.S. Pat. 3,105,775 (1963).
124. C. F. Hunt, A. F. Fitzhugh, and A. H. Markhart, *Electro-Technology* **60** (3), 131 (1962).
125. R. V. Carmer and E. W. Daszewski, "Overload Resistance of Film Insulated Magnet Wires," *Paper Natl. Conf. on Application of Electrical Insulation, Chicago, 1960.*
126. R. L. Whiting and P. F. Wangner (to United States, Navy Dept.), U.S. Pat. 2,525,107 (1950).
127. G. Müller, R. Bock, K. Hoffmann, and R. Kreinhofner, *Angew. Chem.* **68**, 746 (1956).
Excellent summary of the chemistry and technology of wash primers.
128. R. H. Chandler, "Wash Primers," in *Bibliographies in Paint Technology,* No. 8, Translation and Technical Information Services, London, 1964.
129. K. Weigel, *Farbe Lack* **68**, 309 (1962).
130. V. L. Larson, *Offic. Dig.* **26**, 837 (1954).
131. L. J. Coleman, *J. Oil Colour Chem. Assoc.* **42**, 10 (1959).
132. R. Ullman and F. R. Eirich, *Am. Chem. Soc., Div. Paint, Plastics, Printing Ink Chem., Preprints 18, No. 2,* 279 (1958).
133. S. M. Hirschfeld and E. R. Allen, *Offic. Dig.* **31**, 544 (1959); **31**, 629 (1959).
134. B. M. Brill, *Am. Paint J. Reprint,* Monsanto Co., St. Louis, Mo.
135. P. S. Plumb, *Ind. Eng. Chem.* **36**, 1035 (1944).
136. W. H. Bromley, Jr. (to Shawinigan Resins Corp.), U.S. Pat. 2,611,755 (1952).
137. M. D. Kellert and R. V. DeShay (to Monsanto Co.), U.S. Pat. 3,325,432 (1967).
138. W. Borkenhagen, R. Pohlmann, and W. Bauer (to VEB Lack-und Kunstharz fabrik), Brit. Pat. 1,093,200 (1967).
139. G. Schulz, G. Muller, R. Huth, and F. Reichert (to Farbwerke Hoechst A.G.), Ger. Pat. 852,298 (1952).
140. A. D. Yakoulev, I. A. Tolmachev, and I. S. Okrimenko, *Plast. Massy,* **1967** (6), 69.
141. W. O. Herrmann (to Chemische Forschungs Gesellschaft m.b.H.), Ger. Pat. 690,332 (1940).
142. E. W. Reid (to Union Carbide), U.S. Pat. 2,120,628 (1937).
143. J. D. Ryan (to Libbey-Owens-Ford Glass Co.), U.S. Pat. 2,232,806 (1937).
144. A. Weihe, *Kunststoffe* **31**, 52 (1941).
145. R. Houwink and G. Salomon, *Adhesion and Adhesives,* Vol. 1, American Elsevier Publishing Co., New York, 1965, pp. 318–326.
146. P. T. Mattimoe (to Libbey-Owens-Ford Glass Co.), U.S. Pat. 3,231,461 (1966).
147. F. T. Buckley and J. S. Nelson (to Monsanto Co.), U.S. Pat. 3,249,487 (1966).
148. E. Lavin, G. E. Mont, and A. F. Price (to Monsanto Co.), U.S. Pat. 3,262,837 (1966).
149. E. Lavin, G. E. Mont, and A. F. Price (to Monsanto Co.), U.S. Pat. 3,271,235 (1966).
150. Z. I. Bronstein and E. A. Zhukouskaya, *Stekl, Inform. Byul. Vses. Yuz. Nauch. Issled. Inst. Stekla* **10** (2), 8 (1957); through *Chem. Abstr.* **55**, 7784b (1961).
151. R. L. Wakeman, *The Chemistry of Plastics,* Reinhold Publishing Corp., New York, 1947, p. 384.
152. *Modern Plastics* **40**, 90 (Oct. 1962).
153. Ref. 3, pp. 476–477.
154. E. Lavin and J. A. Snelgrove, "Poly(vinyl Acetal) Adhesives," in I. Skeist, ed., *Handbook of Adhesives,* Reinhold Publishing Corp., New York, 1962, Chap. 29, p. 329.

155. A. I. Kislov, N. G. Drozdov, S. V. Yakubovich, and A. V. Uvarov, *Lakokrasochnye Materialy i ikh Primenenie* **6,** 32 (1967).
156. Aero Research Ltd., now Ciba (A.R.L.) Ltd., Brit. Pat. 577,823 (1942).
157. *Technical Note No. 144,* Ciba (A.R.L.) Ltd., 1954.
158. "The Tropical Durability of Metal Adhesives," *R.A.F. Tech. Note No. Chem. 1349* (1959).
159. H. W. Eichner, "Environmental Exposure of Adhesive-Bonded Metal Lapjoints," *W.A.D.C. Tech. Rept. 59,* Part I, 564, (1960)
160. J. L. Been and M. M. Grover (to Rubber and Asbestos Corp.), U.S. Pat. 2,920,990 (1960).
161. W. Whitney and S. C. Herman, *Adhesive Age* **34** (1), 22 (1960).
162. Ref. 154, p. 395.
163. American Can Co., Brit. Pat. 1,148,402 (1967).
164. R. Braun, *Kunststoffe* **50,** 729 (1960).
165. C. L. Wilson, U.S. Pat. 2,609,347 (1952).
166. H. G. Hammon (to C. L. Wilson), U.S. Pat. 2,653,917 (1953).
167. C. L. Wilson, U.S. Pat. 2,846,407 (1958).
168. V. Pardo and A. P. Shapiro, *Lab. Invest.* **15,** 617 (1966).
169. C. F. Fisk (to United States Rubber Co.), U.S. Pat. 2,684,352 (1954).
170. C. F. Fisk (to United States Rubber Co.), U.S. Pat. 2,775,572 (1956).

Martin K. Lindemann
Chas. S. Tanner Co.

VINYL ALKYL ETHER POLYMERS. See VINYL ETHER POLYMERS

N-VINYL AMIDE POLYMERS

Both cyclic and open-chain *N*-vinyl amides can be prepared, although at the present time only *N*-vinyl-2-pyrrolidone is commercially available. Poly-*N*-vinylpyrrolidone is manufactured and sold in the United States by GAF Corporation under the trademarks Plasdone and Polyclar. It is manufactured in Germany by Badische Anilin- und Soda-Fabrik as Luviskol K or Kollidon, Albigen A, and Igecoll. Most of the early work on vinyl amides was part of the Reppe study of the use of acetylene. Pyrrolidone and *N*-vinylpyrrolidone are still commercially made by a Reppe process using the reactions shown in equations 1–5.

$$HC{\equiv}CH + 2\,HCHO \longrightarrow HOCH_2C{\equiv}CCH_2OH \qquad (1)$$

$$HOCH_2C{\equiv}CCH_2OH \xrightarrow{\text{hydrogenation}} HOCH_2CH_2CH_2CH_2OH \qquad (2)$$

$$HOCH_2CH_2CH_2CH_2OH \xrightarrow[\Delta]{\text{Cu catalyst}} \text{(γ-butyrolactone)} \qquad (3)$$

$$\text{(γ-butyrolactone)} + NH_3 \longrightarrow \text{(2-pyrrolidone, N–H)} \qquad (4)$$

$$\text{(2-pyrrolidone, N–H)} + HC{\equiv}CH \xrightarrow{\text{base}} \text{(N-vinyl-2-pyrrolidone, N–}HC{=}CH_2) \qquad (5)$$

These reactions were first discovered during the 1930s at I. G. Farben in Germany. Polyvinylpyrrolidone was used extensively by that country as a blood-plasma extender during World War II. Starting in 1950 GAF Corporation began active commercial development of polyvinylpyrrolidone and built plants for manufacture at Calvert City, Kentucky (1956) and Texas City, Texas (1969). See also ACETYLENE AND ACETYLENIC POLYMERS.

The homopolymers of the *N*-vinyl amides are generally water soluble and on this property hinge many of the uses of this class of polymers. Some of the areas of use for polyvinylpyrrolidone are in adhesives, cosmetics, toiletries, pharmaceuticals, and detergents. It is widely used in the textile industry as a dye-stripping assistant and to improve the dyeability of certain fibers.

Monomers

Properties. In general the *N*-vinyl amide monomers are liquids or low-melting solids at room temperature and soluble in water and organic solvents. Table 1 lists properties of both open-chain and cyclic *N*-vinyl amides.

Table 1. Physical Properties of *N*-Vinyl Amides

	bp(mmHg)/mp, °C	n_D^{20}	References
	Open chain		
N-vinyl-*N*-			
methylacetamide	70(25)	1.4806	1
ethylacetamide	69(18)	1.4750	2
phenylacetamide	101(3)/52		1
methylpropionamide	69(13)	1.4797	2
ethylpropionamide	74(12)	1.4737	2
methylisobutyramide	75(19)	1.4739	2
methylbenzamide	137–139(14)	1.5670	2
	Cyclic		
N-vinyl-			
2-pyrrolidone	96(14)/13.5		3,4
2-piperidone	109–11(12)/42.5		5
ε-caprolactam	110.8–111.2(10)/34.5	1.5057	4,6
5-methyl-2-pyrrolidone	90–93(5)	1,4996	7
3,3,5-trimethyl-2-pyrrolidone	50–52(0.5)	1.4837	8
3-methyl-2-pyrrolidone	86(10)	1.4989	8

In the presence of dilute aqueous acid *N*-vinyl amides are readily hydrolyzed to acetaldehyde and the amide. A study of the hydrolysis of *N*-vinylpyrrolidone in aqueous acid at room temperature showed that at pH 5 only 1% of the monomer hydrolyzed in 6 hr (9).

Reaction of *N*-vinyl-2-pyrrolidone with dry hydrogen chloride (10) leads to formation of a dimer (eq. 6).

$$\underset{\text{N-2-pyrrolidone}}{HC{=}CH_2} \xrightarrow{HCl} \underset{\text{N-2-pyrrolidone}}{HC{=}CH}{-}\underset{\text{N-2-pyrrolidone}}{CHCH_3} \qquad (6)$$

Amides can be added to the double bond of *N*-vinylpyrrolidone to form the α-amidoethylpyrrolidones (11). Phenols add readily to the double bond (12,13). Alcohols and other active hydrogen compounds may be added to the double bond of *N*-vinyl amides (14).

N-vinylpyrrolidone is usually stabilized with 0.1% flake caustic soda to ensure that the monomer remains basic, to prevent hydrolysis as well as polymerization.

Manufacture. The most important industrial process for monomer manufacture is the direct vinylation of the amide with acetylene under basic conditions (2,4). Handling acetylene under pressure requires special precautions in limiting free space in reactors as well as in special compressors and pumps. Dilution of acetylene with nitrogen also greatly reduces danger of acetylene detonation (15). The usual process entails feeding a mixture of pyrrolidone and potassium hydroxide to a makeup tank, stripping water, and then charging this catalyst mixture with additional pyrrolidone to a tower maintained at 150–160°C under pressure (16). A mixture of nitrogen and acetylene is introduced at the bottom of the tower. Additional pyrrolidone is fed continuously to the bottom of the tower, and at the top of the tower crude *N*-vinylpyrrolidone is condensed and the gas mixture enriched and recycled. The crude monomer is first flash distilled to separate it from by-products; the yield in vinylation is on the order of 70%. The monomer is further purified by careful fractional distillation.

Vinylation using ethylene with a copper–palladium catalyst has also been reported (17). Transvinylation using vinyl ethers or vinyl acetate with mercuric salts can be used to prepare the *N*-vinyl amides (18,19). Pyrolysis of *N*-β-acetoxyethyl compounds can be used easily on a laboratory scale (1).

Polymerization

The *N*-vinyl amides are polymerized readily by free-radical initiators. The preferred free-radical catalysts are azobisisobutyronitrile and hydrogen peroxide (eqs. 7–9) (16,55). Peroxides such as benzoyl peroxide or lauroyl peroxide are not efficient catalysts for polymerization; they apparently take part with the monomer in a redox reaction that destroys catalyst at a faster rate than that of polymerization (6). Since the vinyl amides are hydrolyzed under acid conditions, polymerizations are best done at neutral or basic pH in water.

$$H_2O_2 \xrightarrow{NH_4OH} 2HO\bullet \qquad (7)$$

$$HO\bullet + CH_2{=}CH(NC_4H_6O) \longrightarrow HO{-}CH_2{-}\overset{H}{C}\bullet(NC_4H_6O) \qquad (8)$$

$$HO{-}CH_2{-}\overset{H}{C}\bullet(NC_4H_6O) + n\,CH_2{=}CH(NC_4H_6O) \longrightarrow HO{-}CH_2{-}\overset{H}{C}(NC_4H_6O){-}\!\left(CH_2{-}CH(NC_4H_6O)\right)_{n-1}\!{-}CH_2{-}CH\bullet(NC_4H_6O) \qquad (9)$$

(In eqs. 8 and 9 each NC_4H_6O group is a 2-oxopyrrolidin-1-yl ring attached through N.)

Open-chain *N*-alkyl-*N*-vinyl amides can also be polymerized cationically using iodine or boron trifluoride etherate (1). *N*-Vinyl-2-pyrrolidone has also been reported to polymerize with boron trifluoride etherate as a catalyst (20,21). No extensive work on cationic polymerization has been reported, though the *e* value of -1.14 would suggest that cationic polymerization of *N*-vinyl-2-pyrrolidone would be likely (22).

Poly-*N*-vinyl-2-pyrrolidone is the only *N*-vinyl amide polymer produced on a commercial scale, and is available in various viscosity ranges. Polymerization in water with hydrogen peroxide and ammonia in the temperature range 50–80°C has been used extensively to make polyvinylpyrrolidone (3,23–25). The role of ammonia has been shown to be more than simply that of a buffer, since the rate of polymerization using alkali metal hydroxides at the same pH is less than that with ammonia or amines (16). Very likely ammonia is taking part in a complicated redox system.

The polymerization is done in water at 20–60% concentration depending upon the desired viscosity range. The finished polymer may be spray dried. Polymerization in water using azobisisobutyronitrile at 50–60°C also readily yields polymer.

N-Vinyl-2-pyrrolidone copolymerizes with a variety of other vinyl monomers. Table 2 lists the reactivity ratios for copolymerization with various other monomers.

Table 2. Reactivity Ratios in Copolymerization of *N*-Vinylpyrrolidone (M_1)

Comonomer (M_2)	r_1	r_2	Reference
acrylonitrile	0.06 ± 0.07	0.18 ± 0.07	26
allyl alcohol	1.0	0.0	26
allyl acetate	1.6	0.17	26
ethylene	2.1	0.16	27
maleic anhydride	0.16	0.08	26
methyl methacrylate	0.02 ± 0.02	5.0	26,28
styrene	0.045 ± 0.05	15.7 ± 0.5	28
vinyl acetate	3.3 ± 0.15	0.20 ± 0.015	28
vinyl chloride	0.38	0.53	29
vinylene carbonate	0.4	0.7	30
vinyl cyclohexyl ether	3.84	0.0	31
vinyl phenyl ether	4.43	0.22	31

From the reactivity data, it can be seen that *N*-vinylpyrrolidone should copolymerize well with many of the vinyl monomers. The properties of the copolymer vary with the monomer ratios. Increasing the *N*-vinylpyrrolidone content increases the hydrophilic character and also the adhesiveness in the copolymers. Surface-active properties usually increase with increasing vinylpyrrolidone content.

Copolymers of *N*-vinylpyrrolidone and vinyl acetate are commercially available in varying grades with different vinylpyrrolidone contents (32). The copolymers can be prepared using solution polymerization in alcohols with azo catalysts (3). Emulsion polymerization using various catalysts is also a possible method of preparation. The copolymer is used in hair-spray resins, as a suspending agent in water, in tablet coating, and in shoe polishes.

Vinyl chloride is copolymerized with *N*-vinylpyrrolidone in butanone with an azo catalyst at 60°C (33). The soluble copolymers can be used as a sizing agent for glass fibers to promote adhesion of glass to laminates.

A number of copolymers made by grafting vinyl monomers into polyvinylpyrrolidone are commercially available. The monomers used in the grafting include ethyl acrylate, 2-ethylhexyl acrylate, styrene, and vinyl acetate (34–36). The grafting onto polyvinylpyrrolidone is done in emulsion using an anionic surfactant with ammonium peroxydisulfate as catalyst, with the temperature controlled by reflux. Such grafts are used in adhesives, textiles, and paper finishes, and as opacifiers. Some of the earliest copolymers reported for *N*-vinylpyrrolidone were with acrylic acid or sodium acrylate. These copolymers were prepared in water solution with hydrogen peroxide at 90°C (16,37).

Copolymerization of vinylpyrrolidone either to a random copolymer or as a graft onto polyacrylonitrile has been reported to improve the dyeability, water absorption characteristics, and solubility of polyacrylonitrile (38–42). Copolymerization may be done in emulsion or solution using azo, hydrogen peroxide, or peroxydisulfate catalysts.

Properties of Poly(vinyl Amides)

Polyvinylpyrrolidone. The homopolymer of *N*-vinyl-2-pyrrolidone is readily soluble in water and many organic solvents (43). It has been found that 0.5 mole of water is associated per monomer unit in the polymer (44) much like the hydration reported for various proteins. Polyvinylpyrrolidone is hygroscopic, the equilibrium water content being equal to approximately one third the relative humidity.

Polyvinylpyrrolidone powder is relatively stable when stored under ordinary conditions. Aqueous polyvinylpyrrolidone when protected from molds is stable for extended periods. Protection against molds is afforded by benzoic acid, dichlorophene, hexachlorophene, sorbic acid, and esters of *p*-hydroxybenzoic acid (see also BIOCIDES) as well as steam sterilization (15 lb pressure for 15 min).

Heating in air to 150°C or in the presence of strong alkali at 100°C will permanently insolubilize the polymer. Ammonium peroxydisulfate will gel polyvinylpyrrolidone in 30 min at about 90°C (45). Under the action of light the polymer can be crosslinked by diazo compounds or oxidizing agents such as dichromate (46–48).

One of the most unusual properties of polyvinylpyrrolidone is its great tendency to form complexes with many different substances. It forms such a tight complex with iodine that the iodine cannot be extracted with chloroform and there is no appreciable vapor pressure of iodine above the complex. The complex, however, retains the excellent germicidal properties of the iodine but with a greatly reduced toxicity and staining tendency (49). Insoluble complexes are formed on addition of polybasic acids such as poly(acrylic acid), tannic acid, or the copolymer of methyl vinyl ether and maleic acid to polyvinylpyrrolidone in aqueous solution. Although these products are insoluble in water, alcohol, or acetone, the reaction can be reversed by neutralizing the polyacid with base. This phenomenon is thought to be a hydrogen bonding effect of the polyvinylpyrrolidone, much as is observed in proteins. See also POLYELECTROLYTES.

In addition to iodine and polyacids, polyvinylpyrrolidone complexes with certain toxins, drugs, and toxic chemicals to reduce their toxicity (50,51). Many phenols such as resorcinol or pyrogallol precipitate polyvinylpyrrolidone from aqueous solution, although they will redissolve upon the addition of more water. Some dyes also strongly complex with polyvinylpyrrolidone and are the basis of the use of polyvinylpyrrolidone as a dye-stripping agent.

Polyvinylpyrrolidone is very soluble in water, being limited in a practical sense only by the viscosity of the resulting solutions at high concentration. There is no change in viscosity over the pH range of 1–10, although it increases in concentrated hydrochloric acid. The constants for the Mark-Houwink equation relating molecular weight to viscosity in water and other solvents have been reported as shown in Table 3.

Table 3. Molecular Weight–Intrinsic Viscosity Relationship, $[\eta] = KM^a$

Solvent	Temp, °C	K	a	Determined by light scattering[a]	Reference
water	25	6.76×10^{-2}	0.55	fractionated	52
water	25	5.65×10^{-2}	0.55	unfractionated	52
methanol	30	2.32×10^{-2}	0.65	fractionated	52,53
chloroform	25	1.94×10^{-2}	0.64	fractionated	53

[a] K and a are determined by measuring the molecular weight of fractionated or unfractionated polyvinylpyrrolidone whose intrinsic viscosity was measured. Concentrations are in g/dl.

Polyvinylpyrrolidone can be cast from a number of solvents such as methanol, water, or butyrolactone to give on drying clear, hard, and glossy films. Water acts as a plasticizer when not completely removed or if taken up by absorption from the air. Compatible resins such as carboxymethylcellulose, cellulose acetate, or shellac can be used to decrease the hygroscopicity. Polyvinylpyrrolidone has been found to be compatible with many natural and synthetic resins as well as with many inorganic salts (43).

Other Poly(vinyl Amides). Both poly-*N*-vinylcaprolactam and poly-*N*-vinylpiperidone precipitate from water solution on raising the temperature (54,55). Apparently this phenomenon is related to ring size, as polyvinylcaprolactam precipitates at 35°C whereas polyvinylpiperidone precipitates at 64–65°C and polyvinylpyrrolidone forms no precipitate up to 100°C. It has been found that the water solubility of polyvinylcaprolactam increases with molecular weight (55). The cloud-point phenomenon observed for the polymer is independent of concentration over the range of 0.05–5.0% in water.

Polyvinylcaprolactam is soluble in aromatic hydrocarbons, alcohols, ketones, dioxane, and chlorinated aromatic and aliphatic hydrocarbons (55).

The open-chain *N*-vinyl amide polymers are soluble in water and in some organic solvents, eg, alcohols and chlorohydrocarbons. Colorless hard films are obtained from such solutions (56).

Uses

Cosmetics and Toiletries (57–64). The properties of polyvinylpyrrolidone, such as its water solubility, its surface activity, and its complexing ability, make it useful as a thickening agent, emulsion stabilizer, desensitizer (eg, in detergents or germicidal products), and dye solubilizer. These properties are utilized in shampoos, hair tints, shaving creams, and emollients. Polyvinylpyrrolidone has a strong affinity for hair and is a film former so that it can be used in aerosol hair sprays. It has been shown to have antisoil redeposition properties and to be compatible with liquid heavy-duty detergents. It is also a loose color scavenger and reduces skin irritation of detergents.

Textiles and Dyes (61–73). Polyvinylpyrrolidone forms complexes with many dyes and also has surfactant properties that make it useful for fiber dyeing and pigment dispersion (see also DYEING). It has been used to improve the dye receptivity of hydrophobic fibers such as polyacrylonitrile or polypropylene. Many dyes become more water soluble in the presence of polyvinylpyrrolidone; the same property makes it useful for dye stripping cloth or paper. A product of GAF Corporation utilizing polyvinylpyrrolidone, Peregal ST, is sold to the textile trade for dye stripping and related uses. Polyvinylpyrrolidone is a suspending agent for titanium dioxide and organic pigments. It promotes better gloss because of its film-forming ability and is used in inks and polishes.

Pharmaceuticals (47,51,74–80). Polyvinylpyrrolidone has been studied extensively as a plasma volume expander for emergency use in the control of shock due to excessive blood loss, extensive burns, dehydration, or mechanical injury (see also MEDICAL APPLICATIONS). Plasdone C (GAF Corporation) pharmaceutical-grade polyvinylpyrrolidone is marketed for this purpose. In addition, Plasdone K is marketed for the pharmaceutical industry for use in tablet manufacture, as a dispersant and stabilizer for liquid dosage forms, and as a film former in topical preparations. The complexing ability of polyvinylpyrrolidone has been utilized to improve the release properties of drugs and anesthetics in topical preparations. Its properties as a suspending agent and detoxificant make for its use in injectable preparations of antibiotics, hormones, and analgesics.

Polyvinylpyrrolidone–iodine complex is a useful germicide with greatly reduced toxicity to mammals. It is marketed in gargles, tinctures, and ointments.

Adhesives (81–86). Polyvinylpyrrolidone has good adhesion to glass, metal, and plastics and thus is used in adhesive systems. It imparts high initial tack and strength, and its solubility allows systems to be based on either water or organic solvents. It can be used to give better bonding between glass fiber and other plastics, as in glass-fiber-reinforced composites. It has been used in both pressure-sensitive and water-remoistenable adhesives.

Beverage Clarification (87–92). As described above, polyvinylpyrrolidone forms complexes with many phenols and polyacids, including some tannins, to give insoluble products. In many cases such phenols and tannins are the cause of lack of clarity and poor stability in fruit juices, beer, wine, and other food products. Thus polyvinylpyrrolidone is used to improve clarity and to chillproof both alcoholic and nonalcoholic beverages. It is commercially available for such uses in soluble and insoluble forms (Polyclar L and Polyclar AT clarifiers, GAF Corporation) and is used to improve the clarity of such beverages as beer, wine, whiskey, vinegar, tea, and fruit juices. In removing the phenolic compounds by complexation, polyvinylpyrrolidone acts as a flavor and aroma stabilizer, which is especially important in the instant or concentrated beverages.

Miscellaneous Uses (93–100). Polyvinylpyrrolidone is an effective suspending agent, either as a primary protective colloid or as a secondary dispersant in polymerization. It is used in paper as a pigment dispersant. In paper coating, it is used as a leveling agent and also to improve the dye receptivity. Crosslinked polyvinylpyrrolidone films have been suggested for desalination membranes and for membranes for artificial kidney dialysis.

Economic Aspects

Polyvinylpyrrolidone is produced in the United States only by GAF Corporation. No production or sales figures are made public. 1969 prices for the various grades of polyvinylpyrrolidone are shown in Table 4.

Specifications and Standards

Poly-*N*-vinylpyrrolidone is available in four basic viscosity grades, K-15, K-30, K-60, and K-90. Table 5 gives some of the pertinent data for the various grades.

Table 4. Price of Poly-*N*-vinylpyrrolidone

Grade	Bulk price, $/lb
K-15	1.55
K-30	1.30
K-90	1.75
K-90 soln	1.50[a]
K-60 soln	1.12
Plasdone K	2.92
Plasdone C	3.50
Polyclar AT	2.50

[a] 100% basis.

Table 5. Commercial Types of Polyvinylpyrrolidone[a]

	K-15	K-30	K-60	K-90	K-90
physical form	off-white powder	off-white powder	clear, viscous aqueous solution	off-white solid	clear, viscous aqueous solution
K value[b]	15–21	26–35	50–62	80–100	80–100
average molecular weight	10,000	40,000	160,000	360,000	360,000
active content, % minimum	95	95	45	95	20
ash, % maximum	0.02	0.02		0.02	
unsaturation, % maximum as monomer	1.0	1.0		1.0	
packaging	150-, 25-lb drums; 5-lb pkg	100-, 25-lb drums; 5-lb pkg	450-, 100-, 50-lb drums; 10-lb pkg	100-, 50-, 25-lb drums; 5-lb pkg	450-, 100-, 50-lb drums; 10-lb pkg

[a] GAF Corporation.
[b] See the section on Analytical and Test Methods.

In addition to the four basic grades, Plasdone and Plasdone C pharmaceutical-grade polyvinylpyrrolidone and Polyclar AT and L beverage-clarification-grade polyvinylpyrrolidone are available. Table 6 lists the specifications for the pharmaceutical grade (for nonplasma use), which is available in three viscosity grades K-26-28, K-29-32, and K-33-36.

Table 7 lists the more stringent specifications for polyvinylpyrrolidone used as a blood-plasma expander.

Polyvinylpyrrolidone K-15, K-30, and K-90 powders are packed in polyethylene bags inside fiber-pack drums (100 lb net weight). Solutions of K-60 and K-90 are shipped in aluminum tank cars and 55-gal chrome-lined steel drums.

Table 6. Specifications for Pharmaceutical-Grade Polyvinylpyrrolidone

nitrogen content	12.6 ± 0.4%
moisture content	<15%
ash	<0.02%
unsaturation (calcd as monomer)	<1%
heavy metals content	<20 ppm
arsenic content	<2 ppm
packaging	100-lb drums; 25-, 5-, and 1-lb pkgs

Table 7. Specifications for Polyvinylpyrrolidone Plasma Expander[a]

K value[b]	30 ± 2
K-value distribution	not more than 15% greater than K-41
	not more than 25% lower than K-16
ash	<0.02%
moisture content	<5%
unsaturation (calcd as monomer)	<1%
heavy metals content	<20 ppm
arsenic content	<2 ppm
nitrogen content	12.6 ± 0.4%
acetaldehyde content	<0.5%
packaging	100-lb drums; 25-, 5-, and 1-lb pkgs

[a] Plasdone C, GAF Corporation.
[b] See the section on Analytical and Test Methods.

Analytical and Test Methods

The K value is a measure of molecular weight based on viscosity. The K value of polyvinylpyrrolidone is derived from the relative viscosity of a 1% aqueous solution by equation 10 (101), where c is concentration in g/100 ml solution, η_{rel} is ratio of solution viscosity to solvent viscosity, and the K value is 1000 k.

$$\log \frac{\eta_{rel}}{c} = \frac{75\,k^2}{1 + 1.5\,kc} + k \qquad (10)$$

Nitrogen content is determined by either Kjeldahl or Dumas methods; arsenic, by the standard USP method; and acetaldehyde, by the hydroxylamine method. Moisture is determined by Karl Fischer reagent for the solid products, and with the Cenco moisture balance for aqueous solutions. Residual monomer (N-vinyl-2-pyrrolidone) is determined by standard iodometric titration. Ash is determined by ignition and heavy metals by spectrographic emission.

Health and Safety Factors

Because of the use of polyvinylpyrrolidone in pharmaceutical applications as well as its history as a blood-plasma expander, the polymer has been studied extensively for any signs of toxicity. The polymer has been shown to be inert. The acute toxicity is very low and long-term administration does not demonstrate any ill effects. Polyvinylpyrrolidone has been shown to be essentially nontoxic by oral ingestion, inhalation, or injection. It is not a primary irritant, skin-fatiguing material, or sensitizer.

Acute toxicity studies show that the intravenous LD_{50} value is 12–15 g/kg (50,51, 102). Long-term studies including radioactive-tracer studies in rats and dogs conducted by independent research laboratories (102) have demonstrated that over 99% of the polyvinylpyrrolidone ingested is excreted by way of the intestinal tract. Animals fed up to 10% polyvinylpyrrolidone K-30 in their diets for up to two years showed no organic disturbances or toxic effects. Studies on the injection of polyvinylpyrrolidone into the body showed no harmful effects. A follow-up study of people who had received polyvinylpyrrolidone solutions as a blood-plasma expander showed no evidence of permanent kidney or liver damage (102).

In studies using rabbits, even after 36 days of exposing the eye membranes to polyvinylpyrrolidone solutions, no irritation or bonding to ophthalmic tissue was observed (103). Exposure of animals and human volunteers (104,105) to polyvinylpyrrolidone containing aerosols and of factory workers (106) to polyvinylpyrrolidone dusts of small particle size produced no evidence of ill effects.

Polyvinylpyrrolidone forms molecular complexes with many other substances and by such action may be a detoxifying agent. Iodine readily forms a complex, in which it retains all its germicidal properties but its oral toxicity to mammals is greatly reduced. Studies of the detoxifying properties of polyvinylpyrrolidone to the toxins produced in infections such as tetanus, diphtheria, and botulism have been reported (50,51). Potassium cyanide and nicotine in polyvinylpyrrolidone solution show reduced toxicity.

Pharmacological action may be enhanced and duration prolonged for certain antibiotics, drugs, and anesthetics by complexing with polyvinylpyrrolidone (50,51).

Bibliography

1. W. E. Hanford and H. B. Stevenson (to E. I. du Pont de Nemours & Co., Inc.), U.S. Pat. 2,231,905 (Feb. 18, 1941); *Chem. Abstr.* **35,** 3267 (1941).
2. H. Bestian and H. Jensen (to Farbwerke Hoechst A.G.), U.S. Pat. 3,324,177 (June 6, 1967); corresponds to Belg. Pat. 634,033 (Dec. 24, 1963); *Chem. Abstr.* **61,** 4227 (1964).
3. *Vinyl Pyrrolidone,* GAF Corporation Bulletin 7543–037.
4. W. Reppe, H. Krizkalla, O. Dornheim, and R. Sauerbier, U.S. Pat. 2,317,804 (April 27, 1943); *Chem. Abstr.* **37,** 6057 (1943).
5. Farbwerke Hoechst A.G., Fr. Pat. 1,340,350 (Oct. 18, 1963); *Chem. Abstr.* **61,** 1837 (1964).
6. O. F. Salomon, D. S. Vasilescu, and V. Tararescu, *J. Appl. Polymer Sci.* **12,** 1843 (1968).
7. Y. Hachihama and I. Hayashi, *Technol. Rept. Osaka Univ.* **4,** 173 (1954); through *Chem. Abstr.* **49,** 6626 (1955).
8. J. J. Nedwick (to Rohm & Haas Co.), U.S. Pat. 2,806,847 (Sept. 17, 1957); *Chem. Abstr.* **52,** 2931 (1958).
9. Ref. 3, p. 19.
10. J. W. Breitenbach, F. Galinousky, H. Nesvadba, and E. Wolf, *Naturwissenschaften* **42** (155), 440 (1955).
11. W. Reppe, *U.S. Dept. Commerce, Office Tech. Serv., PB Rept.* **18,** 852-S-18; *Bibliog. Tech. Repts.* **10,** 348 (1948); through Ref. 3, p. 17.
12. W. Reppe and H. Krizkalla (to Badische Anilin- und Soda-Fabrik), Ger. Pat. 851,197 (Oct 2, (1952); *Chem. Abstr.* **52,** 10179 (1958).
13. W. Reppe et al., *Ann. Chem.* **601,** 81 (1956).
14. R. A. Hickner, C. I. Judd, and W. W. Backe, *J. Org. Chem.* **32,** 729 (1967).
15. W. E. Hanford and D. L. Fuller, *Ind. Eng. Chem.* **40,** 1171 (1948).
16. C. E. Schildknecht, *Vinyl and Related Polymers,* John Wiley & Sons, Inc., New York, 1952, p. 662.
17. J. E. McKean and P. S. Starcher (to Union Carbide Corp.), U.S. Pat. 3,318,906 (May 9, 1967); *Chem. Abstr.* **68,** 39466 (1968).
18. W. E. Walles, W. F. Tousignant, and T. Houtman, Jr. (to Dow Chemical Co.), U.S. Pat. 2,891,058 (June 16, 1959); *Chem. Abstr.* **54,** 2359 (1960).
19. W. J. Peppel and J. D. Watkins (to Jefferson Chemical Co.), U.S. Pat. 3,019,231 (Jan. 30, 1962); *Chem. Abstr.* **54,** 2359 (1960).
20. C. E. Schildknecht, A. O. Zoss, and F. Grosser, *Ind. Eng. Chem.* **41,** 2891 (1949).
21. R. Bacskai, *Magy. Kem. Folyoirat* **61,** 97 (1955); through *Chem. Abstr.* **50,** 1360 (1956).
22. J. Brandrup and E. H. Immergut, *Polymer Handbook,* Interscience Publishers, a division of John Wiley & Sons, Inc., New York, 1966.
23. H. Fikentscher and K. Herrle, *U.S. Dept. Commerce, Office Tech. Serv., PB Rept.* 25,652, 73491; *Modern Plastics* **23,** 157 (1945).
24. J. W. Breitenbach and A. Schmidt, *Monatsh. Chem.* **83,** 833, 1288 (1952).
25. J. W. Copenhaver and M. H. Bigelow, *Acetylene and Carbon Monoxide Chemistry,* Reinhold Publishing Corp., New York, 1949, pp. 67–74.
26. F. Cech, "Polymerization and Copolymerization of Vinylpyrrolidone," Doctoral Thesis, University of Vienna, 1957.

27. R. A. Terteryan, E. E. Braudo, and A. I. Dintses, *Usp. Khim.* **34,** 666 (1965); *Chem. Abstr.* **63,** 1875 (1965).
28. J. F. Bork and L. E. Coleman, *J. Polymer Sci.* **43,** 413 (1960).
29. J. W. Breitenbach and H. Edelhauser, *Ric. Sci.* **25A,** 242 (1955).
30. K. Hayashi and G. Smets, *J. Polymer Sci.* **27,** 275 (1958).
31. F. P. Sidelkovskava, M. S. Askarov, and F. Ibragimov, *Vysokomolekul. Soedin.* **6,** 1810 (1964); *Chem. Abstr.* **62,** 6563 (1965).
32. *PVP/VA Copolymers,* GAF Corporation Bulletin 7543–031.
33. W. M. Perry (to GAF Corp.), U.S. Pat. 2,958,614 (Nov. 1, 1960); *Chem. Abstr.* **55,** 5899 (1961).
34. F. Grosser and M. R. Leibowitz (to GAF Corp.), U.S. Pat. 3,244,057 (April 5, 1966); corresponds to Ger. Pat. 1,156,985 (Nov. 7, 1963); *Chem. Abstr.* **60,** 1905 (1964).
35. F. Grosser and M. R. Leibowitz (to GAF Corp.), U.S. Pat. 3,244,658 (April 5, 1966); corresponds to Ger. Pat. 1,156,238 (Oct. 24, 1963); *Chem. Abstr.* **60,** 14,706 (1964).
36. *Polectron Emulsion Copolymers,* GAF Corporation Bulletin 7543–029.
37. C. Schuster, R. Sauerbier, and H. Fikentscher, U.S. Pat. 2,335,454 (Nov. 30, 1944); *Chem. Abstr.* **38,** 2770 (1944).
38. G. E. Ham (to Chemstrand Corp.), U.S. Pat. 2,643,990 (June 30, 1953); *Chem. Abstr.* **47,** 9025 (1953).
39. G. E. Ham and A. C. Craig (to Chemstrand Corp.), U.S. Pat. 2,971,937 (Feb. 14, 1961); *Chem. Abstr.* **55,** 12,870 (1961).
40. C. A. Levine (to Dow Chemical Co.), U.S. Pat. 2,979,447 (April 11, 1961); *Chem. Abstr.* **55,** 24,123 (1961).
41. Dow Chemical Co., Brit. Pat. 837,982 (June 22, 1960); *Chem. Abstr.* **54,** 25,868 (1960).
42. Dow Chemical Co., Brit. Pat. 843,063 (Aug. 4, 1960); *Chem. Abstr.* **55,** 6050 (1961).
43. *PVP,* GAF Corporation Bulletin 7583–033.
44. L. E. Miller and F. A. Hamm, *J. Phys. Chem.* **57,** 110 (1953).
45. C. E. Schildknecht (to GAF Corp.), U.S. Pat. 2,658,045 (Nov. 3, 1953); *Chem. Abstr.* **48,** 2413 (1954).
46. P. W. Dorst, *U.S. Dept. Commerce, Office Tech. Serv., PB Rept. 4116* (1945); *Bibliog. Tech. Repts., U.S. Dept. Com.* **1,** 327 (1945).
47. C. E. Rose and R. D. White, *U.S. Dept. Commerce, Office Tech. Serv., PB Rept. 1308* (1945); *Bibliog. Tech. Repts., U.S. Dept. Com.* **1,** 223 (1945).
48. S. C. Slifkin, *U.S. Dept. Commerce, Office Tech. Serv., PB Rept. 78256*; *Bibliog. Tech. Repts., U.S. Dept. Com.* **7,** 616 (1947).
49. *PVP-Iodine,* GAF Corporation Bulletin 7543–004.
50. *PVP—Preparation, Properties, and Applications in the Blood Field and Other Branches of Medicine,* GAF Corporation, New York, 1951.
51. *PVP Bibliography 1951–1966,* GAF Corporation, New York.
52. G. B. Levy and H. B. Frank, *J. Polymer Sci.* **17,** 247 (1955).
53. G. B. Levy and H. B. Frank, *J. Polymer Sci.* **10,** 371 (1953).
54. M. F. Shostakovski and F. P. Sydelkovskaya, *Vestn. Akad. Nauk. SSSR* **7,** 1957; through Ref. 55.
55. O. F. Solomon, M. Corciovei, I. Ciuta, and C. Boghina, *J. Appl. Polymer Sci.* **12,** 1835 (1968).
56. H. Bestian and D. Ulmschneider (to Farbwerke Hoechst A.G.), U.S. Pat. 3,316,224 (April 25, 1967).
57. F. J. Prescott, E. Hahnel, and D. Day, *Drug Cosmetic Ind.* **93,** 443, 540, 629, 702, 739 (1963).
58. M. Freifeld, J. R. Lyons, and A. J. Martinelli, *Am. Perfumer* **77,** 25 (1962).
59. *PVP Formulary,* GAF Corporation Bulletin 7543–066.
60. H. A. Shelanski, M. V. Shelanski, and A. Cantor, *J. Soc. Cosmetic Chemists* **5,** 129 (1954).
61. R. J. Holmes and D. B. Witwer, *Am. Dyestuff Reptr.* **44,** 702 (1955).
62. W. Fong, W. H. Ward, and H. P. Jundgren (to U.S. Sec. Agriculture), U.S. Pat. 3,000,830 (Sept. 19, 1961); *Chem. Abstr.* **56,** 4889 (1962).
63. *Chem. Eng. News* **34,** 500 (1956).
64. *Drug Trade News* **29,** 53 (1954).
65. E. C. Hansen, C. A. Bergman, and D. B. Witwer, *Am. Dyestuff Reptr.* **43,** 72 (1954).
66. H. W. Coover, Jr. (to Eastman Kodak Co.), U.S. Pat. 2,790,783 (April 30, 1957); *Chem. Abstr.* **51,** 11729 (1957).

67. S. A. Murdock, T. G. Traylor, and T. B. Lefferdink (to Dow Chemical Co.), U.S. Pat. 3,036,-033 (May 22, 1962); *Chem. Abstr.* **57,** 8750 (1962).
68. H. R. Mautner (to GAF Corp.), U.S. Pat. 2,955,008 (Oct. 4, 1960); *Chem. Abstr.* **55,** 4977 (1961).
69. E. Kaplan and I. Von (to American Cyanamid Corp.), U.S. Pat. 2,970,880 (Feb. 7, 1961); *Chem. Abstr.* **55,** 11867 (1961).
70. R. C. Riegel (to GAF Corp.), U.S. Pat. 2,955,011 (Oct. 4, 1960); *Chem. Abstr.* **55,** 4974 (1961).
71. F. W. Posselt and L. N. Stanley (to GAF Corp.), U.S. Pat. 2,953,422 (Sept. 20, 1960); corresponds to Brit. Pat. 815,458 (June 24, 1959); *Chem. Abstr.* **54,** 3974 (1960).
72. A. E. Moran and E. Kaplan (to American Cyanamid Corp.), U.S. Pat. 2,971,812 (Feb. 14, 1961); *Chem. Abstr.* **55,** 19,258 (1961).
73. S. A. Murdock, C. W. Davis, and F. A. Ehlers (to Dow Chemical Co.), U.S. Pat. 3,029,220 (April 10, 1962); *Chem. Abstr.* **57,** 8750 (1962).
74. *Plasdone C,* GAF Corporation Bulletin 7543–018.
75. *Plasdone in The Manufacture of Pharmaceutical Dosage Forms,* GAF Corporation, New York.
76. M. A. Lesser, *Drug Cosmetic Ind.* **75,** 1 (1954).
77. C. J. Endicott, T. A. Prickett, and A. A. Dellavis (to Abbott Laboratories), U.S. Pat. 2,820,741 (Jan. 21, 1958); *Chem. Abstr.* **52,** 6728 (1958).
78. H. A. Shelanski (to GAF Corp.), U.S. Pat. 2,739,922 (March 27, 1956); *Chem. Abstr.* **50,** 11,623 (1956).
79. R. C. Bogash, *Bull. Am. Soc. Hosp. Pharmacists* **13,** 226 (1956).
80. S. Siggia, *J. Am. Pharm. Assoc.* **46,** 201 (1957).
81. K. P. Plitt (to U.S. Dept. of Commerce), U.S. Pat. 3,028,351 (April 3, 1963); *Chem. Abstr.* **57,** 4890 (1962).
82. J. D. Russo and R. A. Weidener (to National Starch and Chemical Co.), U.S. Pat. 2,978,343 (April 4, 1961); *Chem. Abstr.* **55,** 17,106 (1961).
83. G. Robinson (to GAF Corp.), U.S. Pat. 2,941,980 (June 21, 1960); *Chem. Abstr.* **54,** 26,007 (1960).
84. A. Marjocchi and N. S. Janetos (to Owens-Corning Fiberglass Corp.), U.S. Pat. 2,931,739 (April 5, 1960); *Chem. Abstr.* **54,** 16,921 (1960).
85. J. Werner (to GAF Corp.), U.S. Pat. 2,853,465 (Sept. 23, 1958); *Chem. Abstr.* **53,** 5753 (1959).
86. J. Werner, R. Steckler, and F. A. Hessel (to GAF Corp.), U.S. Pat. 2,813,844 (Nov. 19, 1957); *Chem. Abstr.* **52,** 6850 (1958).
87. *Polyclar AT,* GAF Corporation Bulletin 7543–085.
88. *Polyclar AT Flavor Stabilizer and Clarifier,* GAF Corporation Bulletin 7543–158.
89. I. Stone (to Baxter Laboratories Inc.), U.S. Pat. 3,061,439 (Oct. 30, 1962); *Chem. Abstr.* **58,** 3860 (1963).
90. P. P. Gray, I. Stone, and J. E. Wylie (to Baxter Laboratories Inc.), U.S. Pat. 2,943,941 (July 10, 1960); *Chem. Abstr.* **54,** 25,564 (1960).
91. H. Tiedemann (to GAF Corp.), U.S. Pat. 3,022,173 (Feb. 20, 1962).
92. W. D. McFarlane, P. D. Bayne, J. L. Azorlosa, and A. J. Martinelli (to GAF Corp.), U.S. Pat. 3,117,004 (Jan. 7, 1964); *Chem. Abstr.* **60,** 7418 (1964).
93. H. Ohlinger and R. Fricker (to Badische Anilin- und Soda-Fabrik), U.S. Pat. 2,895,938 (July 21, 1959); *Chem. Abstr.* **53,** 18,554 (1959).
94. D. G. McNulty and R. I. Leininger (to Diamond Alkali), U.S. Pat. 2,890,199 (June 9, 1959); *Chem. Abstr.* **53,** 17,579 (1959).
95. F. Stastny (to Badische Anilin- und Soda-Fabrik), U.S. Pat. 2,787,809 (April 9, 1957).
96. F. Stastny and K. Buchholz (to Badische Anilin- und Soda-Fabrik), U.S. Pat. 2,744,291 (May 8, 1956); *Chem. Abstr.* **50,** 12,530 (1956).
97. H. Fikentscher, K. Herrle, and J. Jousset (to Badische Anilin- und Soda-Fabrik), Ger. Pat. 801,233 (Dec. 28, 1950); *Chem. Abstr.* **45,** 1811 (1951).
98. H. Fikentscher, K. Herrle, and J. Jousset (to Badische Anilin- und Soda-Fabrik), Ger. Pat. 801,746 (June 22, 1951); *Chem. Abstr.* **45,** 3651 (1951).
99. A. S. Douglass, M. Tagami, and H. K. Lansdale, *Polymer Preprints* **10** (2), 1161 (Sept. 1969).
100. R. A. Markle, R. D. Falb, and R. I. Leininger, *Rubber Plastics Age* **45,** 800 (1964).
101. H. Fikentscher, *Cellulosechemie* **13,** 60 (1932).
102. L. W. Burnette, *Proc. Sci. Sect. T.G.A.* **38,** 1 (1962).

103. E. Lederman, *Eye, Ear, Nose, Throat Monthly* **35**, 785 (1956).
104. J. Calandra and J. A. Kay, *Drug. Cosmetic Ind.* **84**, 174, 252 (1959).
105. J. H. Draye, A. A. Nelson, S. H. Newburger, and E. A. Kelley, *Soap Chem. Specialties* **35**, 91 (1959).
106. M. V. Shelanski, *Soap Chem. Specialties* **34**, 64, 87 (1958).

Donald H. Lorenz
GAF Corporation

VINYLAMINE POLYMERS

Poly(vinylamine) is generally prepared by the conversion of poly(*N*-substituted vinylamines). These polymers can be prepared from their respective monomers by conventional polymerization techniques. Vinylamine, however, has never been isolated. Attempts to prepare it have resulted only in rearrangement (eq. 1) (1,2) or decomposition (eq. 2) (3).

$$BrCH_2CH_2NH_2 \xrightarrow{-HBr} \underset{\text{(aziridine ring: } CH_2\text{—}CH_2 \text{ bridged by NH)}}{CH_2\text{—}CH_2} \quad (1)$$

$$CH_2{=}CH{-}NCO \xrightarrow{H_2O} CH_3CHO + NH_3 + CO_2 \quad (2)$$

Substituted poly(vinylamines) can be prepared in great variety either directly from their monomers (4,5) or by reactions on poly(vinylamine) (6–8,11).

Since only poly(vinylamine) itself or its derivatives seem to have any degree of technological interest, this article will concentrate on its chemistry and technology. See also POLYAMINES.

Preparation

The most suitable methods of preparation of poly(vinylamine) may be classified into three groups. Each of these methods starts with the preparation of a polymeric precursor followed by (a) hydrolysis or some similar cleavage reaction, (b) reduction, or (c) Hofmann degradation of poly(acrylamide).

Polymerization Followed by Hydrolysis. Historically, this was the first method employed to obtain poly(vinylamine). Hanford and Stevenson (9) prepared poly(*N*-vinyl phthalimide) and poly(*N*-vinyl succinimide) but could not obtain complete hydrolysis.

Jones and co-workers (10) have investigated a number of synthetic routes but could never isolate the pure polymer. One that proved somewhat successful used vinyl isocyanate as the monomer This was prepared from acrylyl chloride by reaction with sodium azide followed by a spontaneous Curtius rearrangement. Although the reaction did not go to completion, it was possible to obtain some poly(vinyl isocyanate) with the correct nitrogen analysis. Hydrolysis of this polymer in dilute hydrochloric acid gave what was assumed to be the hydrochloride of poly(vinylamine), but the pure polymer could not be obtained from it either by treatment with sodium hydroxide or silver oxide. Hydrolysis in acetic acid gave poly(vinyl acetamide) and in alcohol–hydrochloric acid the urethan, but attempts to hydrolyze these proved fruitless. See also ISOCYANATE POLYMERS.

Reynolds and Kenyon (11) were the first to isolate the desired polymer (**2**) via the hydrazine hydrolysis of poly(vinyl phthalimide). The reaction is considered to proceed as shown in equations 3 and 4.

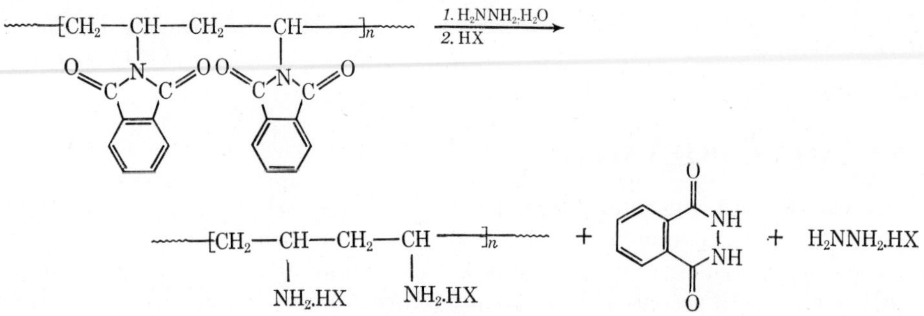

$$\sim\!\!\sim[CH_2-\underset{\text{N-phthalimido}}{CH}-CH_2-\underset{\text{N-phthalimido}}{CH}]_n\!\sim\!\sim \xrightarrow[2.\ HX]{1.\ H_2NNH_2.H_2O}$$

$$\sim\!\!\sim[CH_2-\underset{NH_2.HX}{CH}-CH_2-\underset{NH_2.HX}{CH}]_n\!\sim\!\sim \ + \ \text{phthalhydrazide} \ + \ H_2NNH_2.HX \quad (3)$$

(**1**)

$$(\mathbf{1}) + NaOH \longrightarrow \sim\!\!\sim[CH_2-\underset{NH_2}{CH}-CH_2-\underset{NH_2}{CH}]_n\!\sim\!\sim \ + \ NaX \ + \ H_2O \quad (4)$$

(**2**)
poly(vinylamine)

It is also possible to prepare derivatives of poly(vinylamine) directly from the intermediate resulting from the reaction of the phthalimide with hydrazine hydrate (step *1* in equation 3). This method was also employed by Nikolaev and Bondarenko (12).

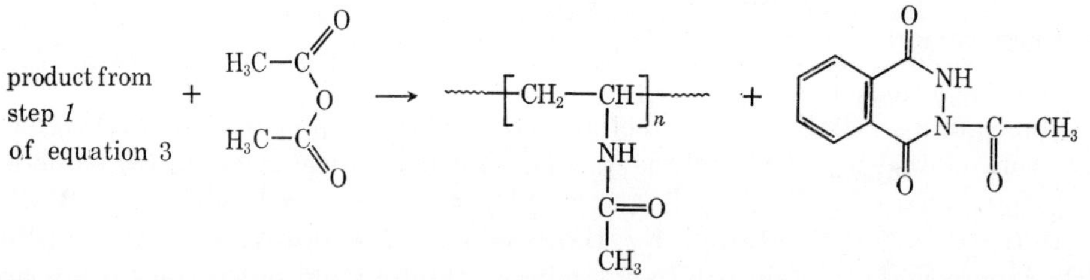

$$\text{product from step } \mathit{1} \text{ of equation 3} + (H_3C-CO)_2O \longrightarrow \sim\!\!\sim[CH_2-\underset{NH-C(=O)-CH_3}{CH}]_n\!\sim\!\sim \ + \ \text{N-acetylphthalhydrazide } (N-C(=O)-CH_3) \quad (5)$$

An alternative method employed by Reynolds and Kenyon (13) starts with poly(vinyl alcohol). Reaction with various sulfonyl chlorides produces poly(vinyl sulfonates). These can react with various amines to produce *N*-substituted poly(vinylamine), but the reactions do not go to completion.

Scheibler and Scheibler (14) prepared esters of vinylamine *N*-dicarboxylic acid by the following interesting synthetic route (eqs. 6–10). These can be polymerized by conventional free-radical catalysts and hydrolyzed.

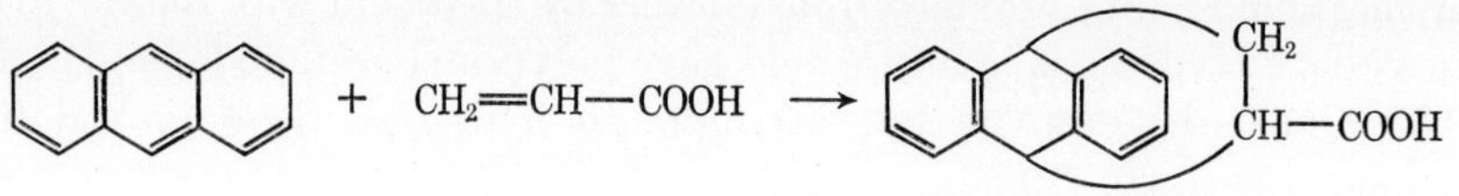

$$\text{anthracene} + CH_2{=}CH-COOH \longrightarrow \text{9,10-dihydro-9,10-ethanoanthracene with } CH_2-CH-COOH \text{ bridge} \quad (6)$$

(**3**)

$$(3) \xrightarrow[\substack{2.\ NaN_3 \\ 3.\ C_2H_5OH}]{1.\ SOCl_2} \text{(dibenzobicyclic)}\ \begin{matrix} CH_2 \\ | \\ CH{-}NH{-}COOC_2H_5 \end{matrix} \quad (4) \tag{7}$$

$$(4) \xrightarrow[2.\ ClCOOC_2H_5]{1.\ K} \text{(dibenzobicyclic)}\ \begin{matrix} CH_2 \\ | \\ CH{-}N(COOC_2H_5)_2 \end{matrix} \quad (5) \tag{8}$$

$$(5) \xrightarrow{\Delta} \text{(anthracene)} + CH_2{=}CH{-}N(COOC_2H_5)_2 \quad (6) \tag{9}$$

$$(6) \xrightarrow{Ba_2O_2} {-}[CH_2{-}CH]{-}\ \text{with } N(COOC_2H_5)(COOC_2H_5) \xrightarrow{HX} {-}[CH_2{-}CH]{-}\ \text{with } NH_2.HX \tag{10}$$

The isocyanate route afforded some more variations. Hart (15) prepared vinylureas and carbamates and found that while the ureas polymerized only with difficulty, the vinyl carbamates could be polymerized and the polymer hydrolyzed to give poly(vinylamine). Hart also found (16) that, while it was difficult to hydrolyze the methyl and ethyl esters, the hydrolysis proceeded very smoothly with the benzyl or the *t*-butyl esters.

It is indeed possible to prepare poly(vinylamine) directly from poly(vinyl isocyanate). Rath and Hilscher (17) react poly(acrylic acid) with hydrazoic acid in acidic media. This is added to a sodium hydroxide solution and, after the nitrogen evolution stops, the sulfate salt of poly(vinylamine) is precipitated with methanol.

Poly(*N*-vinyl diglycolimide) can be rapidly hydrolyzed to poly(vinylamine) under mild conditions in aqueous hydrochloric acid (18).

Polymerization Followed by Reduction. Jones and co-workers (10) were also among the first to investigate this method. They prepared nitroethylene by the dehydrochlorination of 1-chloro-2-nitroethane. The monomer could be polymerized to form a brown powder, but could not be reduced. Blomquist and co-workers (19) reasoned that Jones' polymer could not be reduced because it was crosslinked by reactions involving the active hydrogen on the carbon carrying the nitro group. Poly(2-nitropropene), lacking an active hydrogen, afforded them a convenient way to test this theory and permit the preparation of a carbon-substituted poly(vinylamine). The monomer was prepared by pyrolysis of 2-nitropropyl acetate or benzoate and polymerized anionically. The reduction was carried out catalytically, typically using Raney nickel as the catalyst in an autoclave at 1750 psig and 90°C in dioxane. Under such drastic conditions deamination accompanied the reaction, although a low-molecular-weight (3000–5000), completely reduced polymer with the correct nitrogen analysis could be isolated. See also NITROETHYLENE POLYMERS.

Hofmann Degradation of Poly(acrylamide). Jones and co-workers (10) tried the hypobromite degradation of poly(acrylamide) (eq. 11) but they did not ob-

$$\sim\!\left(CH_2{-}\underset{CONH_2}{\underset{|}{CH}}\right)_n\!\sim \xrightarrow{NaOBr} \sim\!\left(CH_2{-}\underset{NH_2}{\underset{|}{CH}}\right)_n\!\sim \tag{11}$$

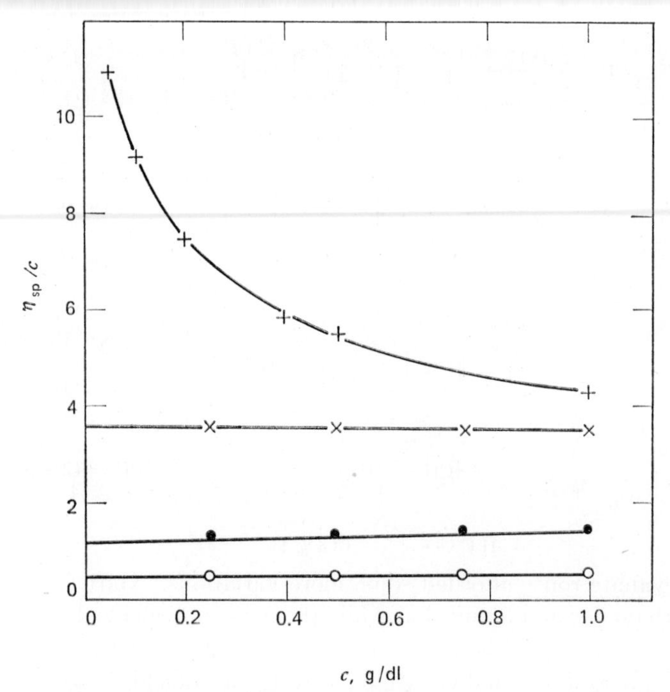

Fig. 1. Viscosities of poly(vinylamine hydrochloride) solutions in various aqueous media. Key: (+) H_2O; (×) NaCl 0.01*N*; (●) NaCl*N* 0.1*N*; (O) NaClN. Courtesy *Journal of Polymer Science* (15).

tain a pure product. A polymer containing 53% poly(vinylamine) units was reported to result from the reaction of poly(acrylamide) with hypohalite in methanol followed by hydrolysis (20).

Polymerization Kinetics. There are a number of routes leading to the preparation of poly(vinylamine), but there is very little information available about the polymerization kinetics of these various precursor monomers. Nikolaev and Bondarenko investigated the polymerization of the potassium salt of *N*-vinyl succinimide in aqueous solution by free-radical polymerization (21). The rate equation for the reaction was found to be equation 12, where [I] and [M] denote initiator and monomer concentrations, respectively. The activation energy was found to be 20 kcal/mole.

$$R_{pol} = k\,[I]^{0.7}[M]^2 \qquad (12)$$

Copolymers of Vinylamine. The copolymers are prepared by the same methods as the homopolymers. They are restricted, however, to systems where the post-polymerization reaction either would not affect the other component or its reaction is also desirable. Examples of the former case might be the reaction of an acrylonitrile–acrylamide copolymer with a hypochlorite (22) or the preparation of divinylbenzene–vinylamine copolymers via the hydrazine hydrolysis of a copolymer of divinylbenzene and *N*-vinyl phthalimide (23,24). Cases where reaction of both components is desirable would include copolymers of *N*-substituted vinylamines with vinylamine prepared from the appropriate *N*-vinyl carbonates followed by hydrolysis (25). Copolymers of vinyl alcohol–vinylamine can be prepared similarly (16,26). Of course, incomplete post-polymerization reactions in the preparation of poly(vinylamine) also yield a variety of copolymers.

Properties

Since pure poly(vinylamine) is difficult to prepare and examine under ambient conditions because of rapid absorption of carbon dioxide from the air (11), no information is available on its bulk properties. It is soluble in water, dilute acids, alcohol, and acetic acid, but insoluble in ether.

The relationship between its viscosity and molecular weight is not reported, although Katchalsky and co-workers (27) have determined the molecular weight of a sample of poly(vinylamine hydrobromide) and found it to be 16,500 by light scattering. The viscosity of poly(vinylamine hydrobromide) solution is greatly dependent on the ionic strength. Figure 1 shows the results obtained by Hart in solutions of varying ionic strength (15).

The polyelectrolyte properties of poly(vinylamine) have been investigated by Hart (15) and by Katchalsky and co-workers (27). Hart found that titration is possible in a water–pyridine mixture, but not in water alone. Katchalsky observed that the dependence of pH on the degree of ionization diverged from the known behavior of polyelectrolytes (qv).

In the infrared spectrum of the poly(vinylamine) free base, the 3400, 3325, and 1600 cm^{-1} absorption bands are assigned to NH_2, the 2900 and 1420 cm^{-1} bands to CH_2, and the 2815 and 1200 cm^{-1} bands to the CH (12). A peak found at 845 cm^{-1} is presumably due to a small amount of phenyl groups present because of incomplete hydrolysis of the phthalimide.

The NMR spectrum of poly(vinylamine hydrochloride) in D_2O at 100°C shows broad signals at approximately 2.2 ppm for the CH_2 group and at approximately 5.3 ppm for the CH (41).

Chemical Reactions of Poly(vinylamine)

The polymer undergoes the same kind of reactions that a monomeric amine might undergo. Thus it is possible to prepare poly(*N*-vinyl *p*-toluenesulfonamide), poly(*N*-vinylurethan), and poly(*N*-vinyl-*N′*-phenylurea) from it (11). The *N*-phenyl thiourea and *N*-*p*-bromophenylurea derivatives have been prepared (19). Reactions with aldehydes such as benzaldehyde, salicylaldehyde, or furfural yield the respective Schiff bases (6). Of particular interest are the polymeric chelates prepared from the salicylaldehyde (28) and 1,3-pentanedione adducts (29); all the chelated polymers showed a considerable improvement in thermal stability over the parent polymer. Poly(vinylamine) itself also forms chelates quite readily. The copper chelates are water soluble and, by adjusting the ionic strength, it is possible to keep them dissolved over a wide range of pH values (30). The chelation is reversible and results in a decrease in length of the polymer chain.

Uses

Although the utility of poly(vinylamine) or its derivatives has been demonstrated in a number of applications, it is doubltful that it will be commercially exploited except in a few special cases because of the difficulty of its preparation. It can be used as a constituent of anion-exchange resins (23,31) and it might be particularly useful in the selective binding of metals as chelates. Teyssie demonstrated that it will selectively bind copper in the presence of nickel and cobalt and will selectively bind those three metals in the presence of magnesium (32). Similarly, its use as a deflocculant has been suggested (33). In photography it is used as a constituent of the emulsion

(34–37). Vinylamine polymers act as dye receptors in synthetic fibers (22,26) and as carrier for photosensitive dye in photopolymerization (38). In the pharmaceutical field, poly(vinylamine) has been investigated as a possible salt-forming compound with penicillin to prolong the latter's therapeutic effect (39,40).

Bibliography

1. S. Gabriel, *Ber.* **21**, 1049, 2664 (1888).
2. C. C. Howard and W. Marckwald, *Ber.* **32**, 2036 (1899).
3. R. Hart, *Bull. Soc. Chim. Belges* **65**, 291 (1956).
4. I. G. Farbenindustrie A.G., Brit. Pat. 451,444 (1936).
5. K. H. Meyer and H. Hopff, *Ber.* **54**, 2274 (1921).
6. A. F. Nikolaev and V. M. Bondarenko, *Vysokomolekul. Soedin.* **6**, 1825 (1964).
7. V. M. Bondarenko, A. F. Nikolaev, and K. A. Makarov, *Vysokomolekul. Soedin.* **6**, 1829 (1964).
8. A. F. Nikolaev and V. M. Bondarenko, *Vysokomolekul. Soedin.* **7**, 1743 (1965).
9. W. E. Hanford and H. B. Stevenson (to E. I. du Pont de Nemours & Co., Inc.), U.S. Pats. 2,231,905 (1941), 2,276,840 (1942), and 2,365, 340 (1944).
10. G. D. Jones, J. Zomlefer, and K. Hawkins, *J. Org. Chem.* **9**, 500 (1944).
11. D. D. Reynolds and W. O. Kenyon, *J. Am. Chem. Soc.* **69**, 911 (1947).
12. A. F. Nikolaev and V. M. Bondarenko, *Vysokomolekul. Soedin.* **1964**, 146.
13. D. D. Reynolds and W. O. Kenyon, *J. Am. Chem. Soc.* **72**, 1591 (1950).
14. H. Scheibler and U. Scheibler, *Chem. Ber.* **87**, 379 (1954).
15. R. Hart, *Compt. Rend. 27° Congr. Intern. Chim. Ind., Brussels, 1954*, **20** Spec. No., 403 (1955); *J. Polymer Sci.* **29**, 629 (1958). Belg. Pats. 550,515 (1956) and 560,219 (1957).
16. R. Hart, *Makromol. Chem.* **32**, 51 (1959).
17. H. Rath and E. Hilscher (by E. Hilsher), Ger. Pat. 1,153,528 (1963).
18. Hercules Powder Co., Brit. Pat. 772,345 (1957).
19. A. T. Blomquist, W. J. Tapp, and J. R. Johnson, *J. Am. Chem. Soc.* **67**, 1519 (1945).
20. Gevaert Photo-Producten, Belg. Pat. 550,514 (1956).
21. A. F. Nikolaev and S. G. Bondarenko, *Vysokomolekul. Soedin.* **7**, 1822 (1965).
22. D. W. Chaney and H. G. Sommar (to American Viscose Corp.), U.S. Pat. 2,641,524 (1953).
23. I. Skondac and A. F. Nikolaev, *Vysokomolekul. Soedin.* **7**, 101 (1965).
24. *Ibid.*, p. 1835.
25. Gevaert Photo-Producten, Belg. Pat. 540,976 (1956).
26. A. F. Nikolaev, S. N. Ushakov, L. P. Vishnevetskaya, and N. A. Voronova, *Vysokomolekul. Soedin.* **5**, 547 (1963).
27. A. Katchalsky, J. Mazur, and P. Spitnik, *J. Polymer Sci.* **23**, 513 (1957).
28. V. M. Bondarenko, A. F. Nikolaev, and K. A. Makarov, *Vysokomolekul. Soedin.* **6**, 1829 (1964).
29. V. M. Bondarenko and A. F. Nikolaev, *Vysokomolekul. Soedin.* **7**, 2104 (1965).
30. P. Teyssie, C. Delorne, and M. T. Teyssie, *Makromol. Chem.* **84**, 51 (1965).
31. J. Skondak and A. F. Nikolaev, *Vysokomolekul. Soedin.* **7**, 1835 (1965).
32. P. Teyssie, *Makromol. Chem.* **66**, 133 (1963).
33. C. Kajisaki and K. Hori, *Bull. Govt. Res. Inst. Ceram. (Kyoto)* **6**, 25 (1952); *Chem. Abstr.* **49**, 12801h (1955).
34. D. D. Reynolds and W. D. Kenyon (to Kodak, Ltd.), Brit. Pat. 618,175 (1949).
35. Gevaert Photo-Producten, Belg. Pat. 550,514 (1956).
36. Gevaert Photo-Producten, Belg. Pat. 560,219 (1957).
37. Gevaert Photo-Producten, Brit. Pat. 860,631 (1961).
38. G. Smets, W. deWinter, and G. Delzenne, *J. Polymer Sci.* **55**, 767 (1961).
39. K. I. Shumikina and E. F. Panarin, *Antibiotiki* **11**, 628 (1966); *Chem. Abstr.* **65**, 12725f (1966).
40. K. I. Shumikina, E. F. Panarin, and S. N. Ushakov, *Antibiotiki* **11**, 767 (1966); *Chem. Abstr.* **65**, 20698f (1966).
41. M. Murano and H. J. Harwood, *Macromol.* **3**, 605 (1970).

Thomas A. Augurt
Lederle Laboratories
American Cyanamid Company

VINYLARENE POLYMERS

This article covers polymers derived from monomers that have a vinyl group attached to an aromatic ring (**1**). It does not cover aromatic monomers having a heteroatom in the ring, styrenes, or, except for 4-vinylbiphenyl, substituted styrenes. See also STYRENE POLYMERS; VINYLCARBAZOLE POLYMERS; VINYLPYRIDINE POLYMERS.

Monomers

Vinylarene monomers are generally prepared by dehydration of the corresponding carbinol, which can usually be obtained by the acetylation of the corresponding hydrocarbon and reduction of the ketone. The carbinol can also be obtained by the reaction of the aryl Grignard reagent with acetaldehyde (eq. 1).

Table 1. Properties of Some Vinylarene Monomers

Compound	bp, °C (mmHg)	mp, °C	Refractive index, n_D^{25}	References
1-vinylanthracene	160–170 (1)	62–63		1,2
2-vinylanthracene	160–170 (1)	186–187		1,2
9-vinylanthracene[a]	160–170 (1)	64–65		1,2
1-vinylnaphthalene[a]	86–87 (2)		1.6436	3
2-vinylnaphthalene[a]	76–81 (2.5)	65–66		3
4-vinylbiphenyl[a]	136–138 (6)	119.0–119.5		4
1-vinylpyrene		87–89		5
acenaphthylene[a]	103–105 (3)	92–93		6
2-vinylfluorene		133–134		3
6-vinyl-1,2,3,4-tetrahydronaphthalene	89–90 (2)		1.5699	3
vinyldecahydronaphthalene	103 (3)		1.5443	7
6-chloro-2-vinylnaphthalene		111.5–112.6		8
4-chloro-1-vinylnaphthalene	118–120 (2)		1.6408	3
9-vinylphenanthrene		36.4–37.8		9
2-vinylphenanthrene		116.5–118.0		9
3-vinylphenanthrene		58.2–59.4		9

[a] Commercially available.

$$\mathrm{ArH} + \mathrm{CH_3{-}\overset{O}{\overset{\|}{C}}{-}Cl} \longrightarrow \mathrm{Ar{-}\overset{O}{\overset{\|}{C}}{-}CH_3} \longrightarrow \mathrm{Ar{-}\overset{OH}{\overset{|}{C}}H{-}CH_3} \searrow$$

$$\mathrm{Ar{-}CH{=}CH_2} \quad \textbf{(1)} \qquad (1)$$

$$\mathrm{ArBr} + \mathrm{Mg} \longrightarrow \mathrm{ArMgBr} \xrightarrow{\mathrm{CH_3{-}\overset{O}{\overset{\|}{C}}{-}H}} \mathrm{Ar{-}\overset{OH}{\overset{|}{C}}H{-}CH_3} \nearrow$$

In Table 1 are listed some vinylarene monomers and their physical properties.

Anionic Polymerization

Kinetics of the anionic homopolymerization of 1-vinylnaphthalene (10), 2-vinylnaphthalene (10), and 9-vinylanthracene (11) in tetrahydrofuran at 25°C have been determined and propagation rate constants of 500, 300, and 0.2 $\mathrm{l\text{-}mole^{-1}\text{-}sec^{-1}}$ found. The greater reactivity of 1- and 2-vinylnaphthalene as compared with that of styrene has been attributed to their lower localization energies (10).

The anionic polymerization of 9-vinylanthracene produces only low-molecular-weight polymers, and initiation by naphthalene or biphenyl radical anions or by butyllithium yields oligomers having a $\overline{\mathrm{DP}}$ of 4–12 (11,12). A study of the reaction has shown that, although the concentration of the living ends remains unchanged during the reaction, the degree of polymerization does not correspond to the concentration of initiator, indicating an efficient chain-transfer reaction. When additional monomer is supplied to the polymerized system, more polymer forms without affecting its molecular weight, thus indicating that no chain transfer to polymer takes place.

It has been shown that 9-vinylanthracene can polymerize both along the vinyl group and across the central ring of the anthracene system (13), and structural analysis has shown that material polymerized in the presence of lithium, potassium, and sodium contains a lower percentage of anthracene rings than material polymerized with cesium.

$$\text{9-anthryl}{-}\mathrm{CH{=}CH_2} + \mathrm{Na\cdot} \longrightarrow \text{9-anthryl}{-}\mathrm{\overset{\ominus}{C}H{-}CH_2\cdot}\ \mathrm{Na^{\oplus}}\ \textbf{(2)} \longleftrightarrow \mathrm{\cdot CH_2{-}CH{=}}\text{(9,10-dihydroanthracen-9-ylidene)}\text{-}10\text{-}\mathrm{\overset{\ominus}{C}H}\ \mathrm{Na^{\oplus}}\ \textbf{(3)} \qquad (2)$$

$$\textbf{(2)} + \text{9-anthryl}{-}\mathrm{CH{=}CH_2} \longrightarrow \mathrm{Na^{\oplus}}\ \text{9-anthryl}{-}\mathrm{\overset{\ominus}{C}H{-}CH_2{-}CH(}\text{9-anthryl}\mathrm{){-}CH_2\cdot} \qquad (3)$$

$$\textbf{(3)} + \text{9-anthryl}{-}\mathrm{CH{=}CH_2} \longrightarrow \mathrm{\cdot CH_2{-}CH{=}}\text{(anthracenylidene)}\text{-}10\text{-}\mathrm{CH_2{-}CH{=}}\text{(anthracenylidene)}\text{-}10\text{-}\mathrm{\overset{\ominus}{C}H}\ \mathrm{Na^{\oplus}} \qquad (4)$$

The polymerization mechanism shown in equations 2–4 has been proposed (12,13). Accumulated physical and chemical evidence indicates that the predominant structure for the polymer is that resulting from a 1,6 across-the-ring addition. To account for the low molecular weight of the polymer, the chain-transfer reaction shown in equation 5 has been proposed (13).

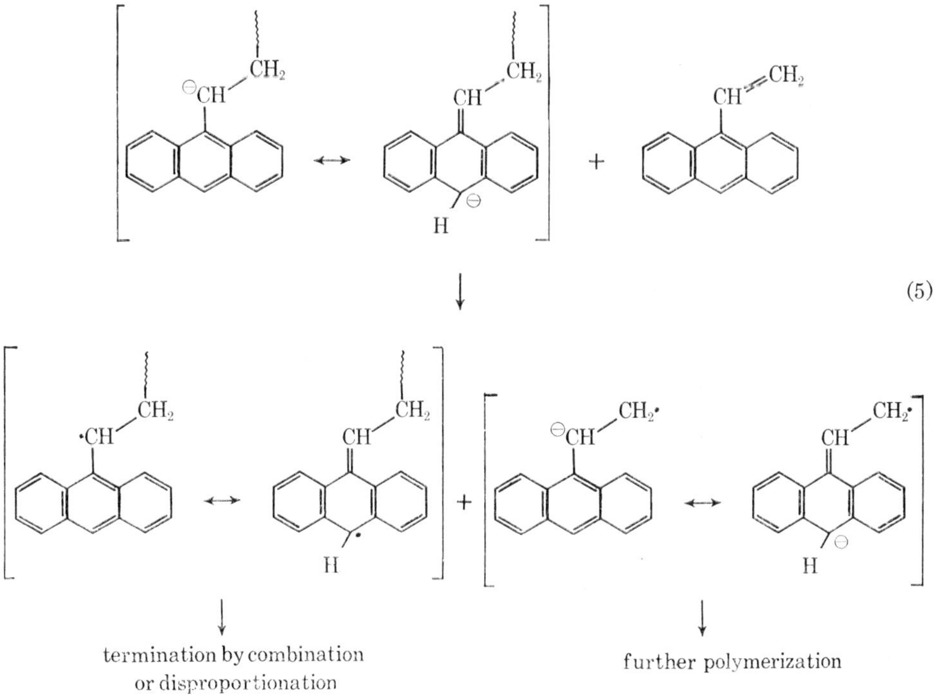

(5)

termination by combination or disproportionation

further polymerization

A kinetic study of the anionic polymerization of acenaphthylene has shown that the reaction follows pseudo-first-order kinetics and that a chain-transfer reaction to monomer similar to that observed for 9-vinylanthracene takes place (14). The highest molecular weight that could be obtained by anionic polymerization was 8000, although thermal polymerization in bulk produced polymers having very high molecular weight (ca 2,000,000). Although the chain-transfer mechanism has not been established, it probably involves electron transfer to monomer coupled with hydrogen abstraction from solvent.

The copolymerization of 1-vinylnaphthalene with 2-vinylpyridine and with styrene has been investigated in both sequential and simultaneous polymerizations (15), and good yields of copolymers were obtained when 1-vinylnaphthalene was initiated with a polystyrene anion. Interesting results are reported when styrene is initiated with a poly(1-vinylnaphthalene) anion (10): addition of two or three equivalents of styrene to "living" poly(1-vinylnaphthalene) leads to the disappearance of the characteristic 558-mμ absorption maximum of the poly(1-vinylnaphthalene) anion, but the expected 340-mμ maximum of the polystyrene anion does not appear. Instead, a new absorption peak at 440 mμ appears, but on standing for 24 hr the original 558-mμ peak of poly(1-vinylnaphthalene) reappears. When a large excess of styrene, twentyfold or more, is added the characteristic spectrum of polystyrene appears permanently.

The observations were explained by assuming that the reaction involves three steps (eqs. 6–8) (16).

$$\sim\!\!\sim CH_2-CH^{\ominus}\,Na^{\oplus}(\text{1-naphthyl}) + CH_2{=}CH-C_6H_5 \xrightarrow{k_1} \sim\!\!\sim CH_2-CH(CH_2)(H)(CH-C_6H_5)\ \ominus\ H\ Na^{\oplus} \tag{6}$$

$$\sim\!\!\sim CH_2-CH(CH_2)\,H\,CH-C_6H_5\ \ominus Na^{\oplus}\ H + CH_2{=}CH-C_6H_5 \xrightarrow{k_2}$$

$$\sim\!\!\sim CH_2-CH(\text{1-naphthyl})\ CH_2\ CH-C_6H_5\ CH_2\ CH-C_6H_5\ \ominus Na^{\oplus} \tag{7}$$

$$\sim\!\!\sim CH_2-CH\ CH_2\ CH\text{---}CH_2CH^{\ominus}-C_6H_5\ \ C_6H_5\ \ Na^{\oplus} \xrightarrow[k_3]{CH_2=CH-C_6H_5} \text{"living" polystyrene} \tag{8}$$

The addition of the first styrene molecule produces a benzyl-type anion that forms a bond with the preceding naphthalene group. The product resembles the adduct of "living" polystyrene and anthracene (17) and the product very slowly adds a second molecule of styrene. The addition of the second molecule destroys the complexing with naphthalene and the resulting polymer propagates as ordinary polystyrene does.

The complex formed on addition of a small excess of styrene to "living" poly(1-vinylnaphthalene) must be unstable because the spectrum of the poly(1-vinylnaphthalene) reappears within 24 hr. It has been concluded that the formation of the complex is reversible and that the equilibrium concentration of styrene is given by the reaction shown in equation 9. The reaction mixture must, however, contain some "liv-

$$\sim\!\!\sim CH(CH_2)\,H\,CH-C_6H_5\ \ominus\ H \rightleftarrows \sim\!\!\sim CH^{\ominus}(\text{1-naphthyl}) + CH_2{=}CH-C_6H_5 \tag{9}$$

ing" polystyrene anions since some segments have added two or more styrene units. Hence another equilibrium is established (eq. 10). These three equilibria are coupled

$$-\!\!\left[CH_2-\underset{C_6H_5}{CH}\right]_n\!-CH_2-\underset{C_6H_5}{CH^{\ominus}} \rightleftarrows -\!\!\left[CH_2-\underset{C_6H_5}{CH}\right]_{n-1}\!-CH_2\underset{C_6H_5}{CH^{\ominus}} + CH_2{=}\underset{C_6H_5}{CH} \tag{10}$$

in the overall process and the equilibrium of the overall process favors the right side (eq. 11). This scheme has been tested with α-methylstyrene, the propagation of which

$$\sim\text{CHCH}_2\overset{\ominus}{\text{C}}\text{H}-\text{C}_6\text{H}_5 \text{ (1-naphthyl on first CH)} + -[\text{CH}_2\underset{\text{C}_6\text{H}_5}{\text{CH}}]_n-\text{CH}_2\underset{\text{C}_6\text{H}_5}{\text{CH}}^{\ominus} \rightleftarrows \sim\text{CH}^{\ominus}\text{ (1-naphthyl)} + -[\text{CH}_2\underset{\text{C}_6\text{H}_5}{\text{CH}}]_{n+1}-\text{CH}_2\underset{\text{C}_6\text{H}_5}{\text{CH}}^{\ominus} \quad (11)$$

is thermodynamically unfavorable, and a stable complex was formed when this monomer was added to living poly(1-vinylnaphthalene) (18).

ABA block copolymers of 4-vinylbiphenyl and isoprene have been prepared using "living"-polymer techniques (19). Because of difficulties in achieving a rigorous purification of 4-vinylbiphenyl, a coupling technique was used whereby the A monomer was polymerized first, the B monomer was then added, and the AB anion was next coupled with a reactive dihalide. Using this technique the residual impurities in the A monomer only destroy some initiator; by estimating the degree of purity it is easy to use a slight excess of initiator to compensate for the amount destroyed by the impurities. Coupling of the AB anions was achieved by using phosgene, which was allowed to diffuse very slowly into a vigorously agitated polymer solution.

The block copolymers were characterized by gel permeation chromatography and, from a knowledge of the ratio of the refractive-index increments of the two homopolymers and the overall composition, a quantitative analysis was carried out (19). See also "LIVING POLYMERS."

Polymer Reactions

The transfer of an electron from alkali metals to an aromatic hydrocarbon such as naphthalene or biphenyl is well known (20). The same reaction occurs when the vinylarene group is attached to a polymer chain (21,22). The products have been referred to as polyradical anions (21), and are formed experimentally in all-glass, high-vacuum systems (23) by the reaction of the polymer in tetrahydrofuran with a sodium mirror at temperatures ranging from -80 to 30°C.

The reaction products have been characterized by viscometric, spectrophotometric, and electron-spin-resonance measurements (24). It was found that the viscosity of the solution decreases with time and that the final viscosity depends essentially on the alkali metal concentration. Spectrophotometric data have shown that with time the spectrum becomes almost identical to "living" polymer dianions, and electron-spin-resonance studies have indicated the presence of unpaired electrons in concentrations proportional to the sodium content. The disappearance of the signal to practically zero, the formation of anions, and the decrease in viscosity with time are consistent with a cleavage mechanism in which an electron migrates from the aromatic

$$\sim\text{CH}_2-\text{CH}-\text{CH}_2-\text{CH}\sim \text{ (each CH bearing a naphthyl radical anion, } \cdot\ominus\ \text{Na}^{\oplus}) \longrightarrow \sim\text{CH}_2-\text{CH}^{\ominus}\text{Na}^{\oplus}\text{ (1-naphthyl)} + \text{Na}^{\oplus}\overset{\ominus}{\text{C}}\text{H}_2-\text{CH}\sim\text{ (1-naphthyl)} \quad (12)$$

ring to the α carbon of the aliphatic chain with formation of a negatively charged end (eq. 12). The same mechanism has been proposed for poly(N-vinylcarbazole), poly(1-vinylnaphthalene), poly(2-vinylnaphthalene), and poly(4-vinylbiphenyl). Poly(acenaphthylene) degrades so fast that it is not possible to follow changes in viscosity as a function of time (25). It has also been found that monomeric fragments are produced (eq. 13).

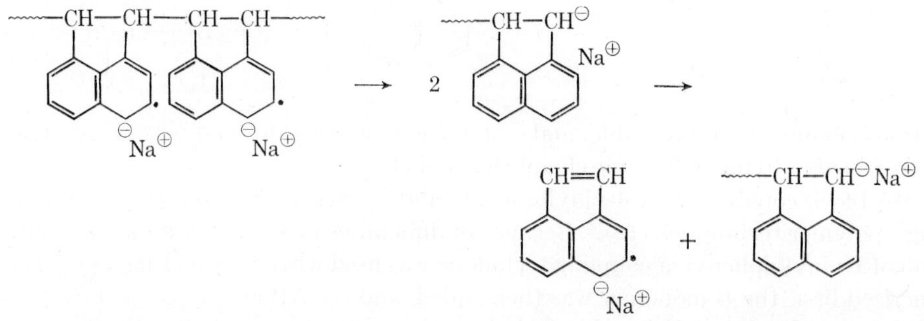

(13)

Polyradical anions have been used to initiate graft polymerization reactions (26–28). The reaction is not applicable to monomers that polymerize by an electron-transfer mechanism, where only homopolymerization is achieved (28). However, monomers such as cyclic ethers that cannot polymerize by an electron-transfer process but do polymerize anionically, do form graft copolymers. The mechanism of the poly-

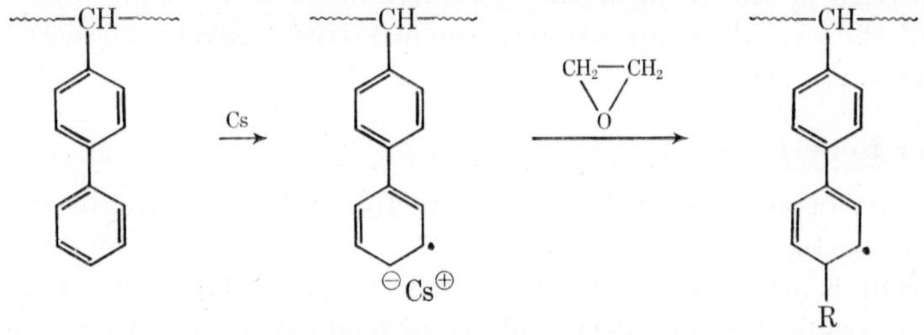

(14)

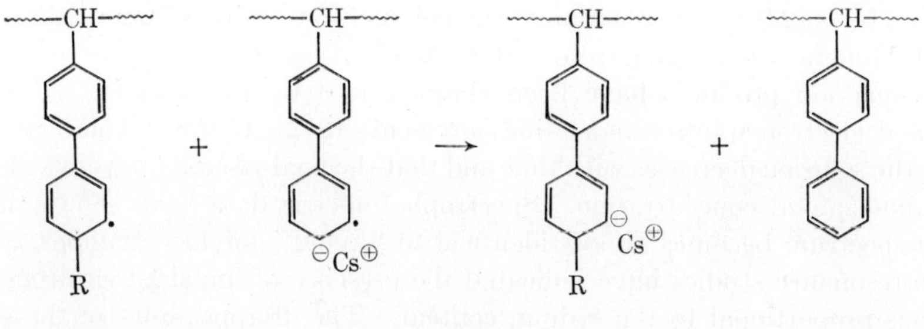

(15)

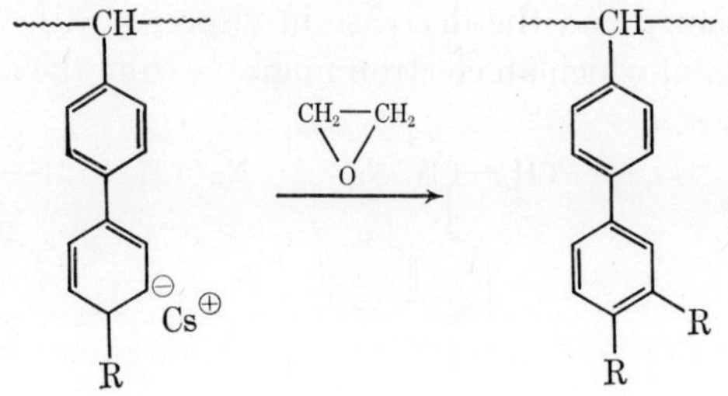

where R is $-\!\!(CH_2CH_2O)_n^{\ominus}\,Cs^{\oplus}$ (16)

merization is similar to that proposed for the carbonation of the naphthalene radical anion (eqs. 14–16) (29).

Poly(2-vinylfluorene) has been metalated with metallic sodium or lithium or with the corresponding naphthalene radical anions (eq. 17) (30), and graft copolymers with

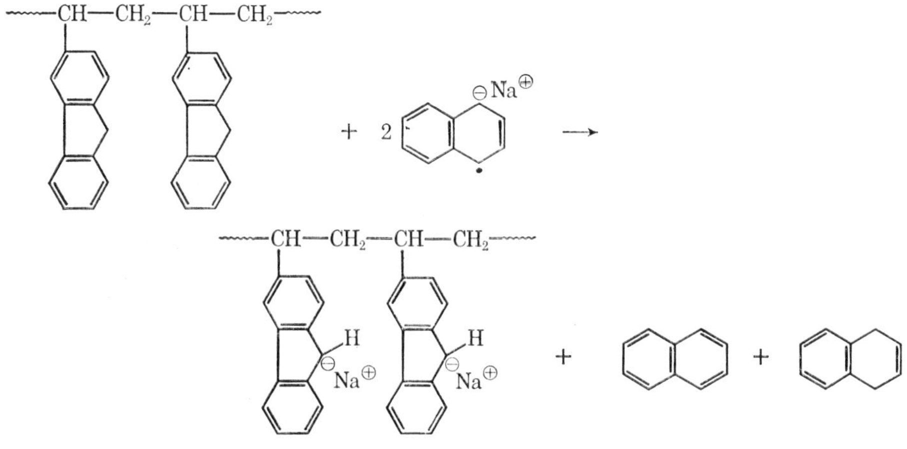

(17)

a variety of vinyl monomers such as styrene, methyl methacrylate, or vinylpyridine in addition to ethylene oxide have been prepared.

Metalation of 2-vinylnaphthalene units incorporated into a copolymer has also been used to provide sites for anionic grafting reactions (31). Thus, a copolymer of butadiene containing small proportions of 2-vinylnaphthalene has been prepared by free-radical copolymerization techniques, the resulting copolymer metalated with butyllithium, and styrene or 2-vinylnaphthalene graft copolymerized on the anionic sites. The resulting materials exhibited elastomeric properties similar to those of styrene–butadiene ABA block copolymers, provided the number of grafts per backbone were small.

Stereoregular Polymerization

Although the stereoregular polymerization of styrene and substituted styrenes has received considerable attention, other vinylarene monomers have been studied much less extensively. Natta and co-workers (32,33) have surveyed the stereoregular polymerization of over 20 vinyl aromatic monomers; among these were 1-vinylnaphthalene, 2-vinylnaphthalene, 1-vinyl-4-chloronaphthalene, 1,2,3,4-tetrahydro-6-vinylnaphthalene, 4-vinylbiphenyl, 9-vinylphenanthrene, and 9-vinylanthracene. This study established that Ziegler–Natta polymerizations are very sensitive to steric hindrance about the double bond and when the steric hindrance is excessive, such as in 9-vinylanthracene, no polymerization takes place. Although one study does report a polymerization of 9-vinylanthracene in yields from 20 to 90% depending on the Al/Ti ratio with an $Al(C_2H_5)_3$–$TiCl_4$ catalyst system (34), the results indicate a cationic polymerization.

Stereoregular polymers of 1-vinylnaphthalene, 2-vinylnaphthalene, and 4-vinylbiphenyl have been prepared using a $(C_2H_5)_3Al$–$TiCl_4$, $(C_2H_5)_2AlCl$–$TiCl_3$, or $(C_2H_5)_3$-Al–$TiCl_3$ catalyst system (35). The latter catalyst gave polymers in 75–95% conversion that were at least 90% isotactic. The atactic fraction could be separated from the isotactic ones by extraction with methyl ethyl ketone. The isotactic polymers were also characterized by infrared and nuclear-magnetic-resonance spectroscopy (35).

Not all stereoregular polymers could be crystallized. In polymers in which steric factors lead to a crystalline phase that would have a lower density than the amorphous phase, no crystallization took place. Thus, only 1-vinylnaphthalene produced a crystallizable polymer (32,33,35). An x-ray diffraction study on this polymer has been carried out (36). The Bragg distances in the unit cell are $a = b = 21.20$ Å and $c = 8.12$ Å, and the specific gravity is 1.12.

The stereoregular ionic polymerization of acenaphthylene has been investigated in some detail (37). Although four stereoisomers can be written, eg, cis and trans isotactic and cis and trans syndiotactic, a study of molecular models has shown that only the trans-isotactic and trans-syndiotactic conformations can exist in polymers. The trans-isotactic poly(acenaphthylene) forms a helix and the trans-syndiotactic poly(acenaphthylene) forms a "stair-stepped" rigid rod. These stereoisomers were obtained by *n*-butyllithium or boron trifluoride polymerizations and characterized by infrared and nuclear-magnetic-resonance spectroscopy.

Acenaphthylene has also been polymerized with an $Al(C_2H_5)_3$–$Ti(OC_3H_7)_4$ catalyst system, but no mention of stereoregularity was made (38).

Cationic Polymerization

The cationic polymerization of vinylarene monomers other than styrene is not well understood and little reliable quantitative information is available.

Acenaphthylene readily forms polymers of high molecular weight (6,40,41), although a dimer can be obtained when a solution of acenaphthylene in glacial acetic acid is treated with a small quantity of concentrated hydrochloric acid (39). The kinetics of the cationic polymerization catalyzed by boron trifluoride (42) and iodine (43) has also been studied. In the first case, a second-order reaction with respect to boron trifluoride was observed, and in the second case a high-order reaction with respect to iodine concentration and cocatalysis by hydrogen iodide was noted.

Unlike free-radical polymerization, the cationic polymerization of 9-vinylanthracene with initiators such as stannic chloride leads to very rapid polymerization rates (2). Early studies assumed a normal vinyl polymerization but it was later shown (44) that the normal addition takes place to only a very minor extent and that polymerization across the ring similar to that already discussed in the anionic polymerization (13) takes place. A wide variety of catalyst systems and solvents was also investigated (44).

Very little information is available on the cationic polymerization of other vinylarene monomers. 1-Vinylnaphthalene apparently can be polymerized to a high-molecular-weight product but monomers substituted in the α or β position of the vinyl group yield mainly dimers (45). The polymerization of 4-vinylbiphenyl with Friedel-Crafts catalysts has been reported (46) and 1-vinylpyrene (47) and 2-vinylfluorene (48) have also been polymerized with BF_3.

Stable carbonium ions such as tropylium hexachloroantimonate ($C_7H_7^{\oplus}$ $SbCl_6^{\ominus}$) and tetrafluoroborate ($C_7H_7^{\oplus}$ $BF_4^{\ominus}$) have been used to initiate the polymerization of acenaphthylene and 1- and 2-vinylnaphthalene (49).

Free-Radical Polymerization

The kinetics of the 2,2′-azobisisobutyronitrile-initiated bulk polymerization of 1-vinylnaphthalene have been reported (50). The polymerization rate is proportional to the ½ power of the initiator concentration and the first power of the monomer concentration. The molecular weight of the polymer was shown to be controlled by a chain-transfer reaction with the monomer, and a chain-transfer constant of 0.03, about 300 times that for styrene, was found. As a consequence, only low-molecular-weight polymers (2000–6000) were obtained. The bulk polymerization of 2-vinylnaphthalene leads to a product having a molecular weight of about 66,000. Emulsion polymerization techniques yielded a poly(1-vinylnaphthalene) having a molecular weight of 25,000 and a poly(2-vinylnaphthalene) having a molecular weight of 115,000 (51).

The relative ease of bulk polymerization of 1-vinylnaphthalene, 2-vinylnaphthalene, 6-vinyl-1,2,3,4-tetrahydronaphthalene, and vinyldecahydronaphthalene has been compared (7); 1- and 2-vinylnaphthalenes were the easiest to polymerize, 6-vinyl-1,2,3,4-tetrahydronaphthalene had polymerization rates comparable with those of unsubstituted styrene, and vinyldecahydronaphthalene did not polymerize during 30 days at 100°C.

The solid-state postpolymerization of ^{60}Co γ-irradiated 2-vinylnaphthalene has been studied (52). The monomer was irradiated at −78°C and then postpolymerized at temperatures ranging from −20 to 41°C. A limiting conversion of about 40% was obtained. The solid-state polymerization under pressure has also been investigated (53).

The polymerization rates of 1- and 9-vinylanthracene and 9-vinylphenanthrene have also been compared (54). The highest reactivity was shown by 9-vinylphenanthrene and the lowest by 9-vinylanthracene. The reactivities were explained on the basis of steric hindrance to conjugation between the ring system and the vinyl group and the nonaromatic character of the 9,10 double bond in phenanthrene. The free-radical polymerization of 9-vinylanthracene proceeds so slowly that it holds little promise as an acceptable polymerization technique. No studies have been reported in which the structure of this polymer has been examined.

Acenaphthylene can be polymerized to a high-molecular-weight polymer using free-radical initiators, and a molecular weight of over 150,000 has been reported (6). The kinetics of the thermal polymerization of a highly purified sample have been studied dilatometrically and a high activation energy for both initiation and propagation was found (55).

The effect of high pressure on the free-radical polymerization of acenaphthylene has also been investigated (56). It was found that the rate of polymerization is not increased as much by pressure as is that of other olefinic monomers such as styrene. The effect of pressure on molecular weight was also less than for polystyrene, and the molecular weight of the polymer increased by a factor of 2.6 between 1 and 2880 atm.

The solid-state polymerization of acenaphthylene initiated by x rays has been studied in air, in nitrogen, and under vacuum (57). The results indicate that the molecular weight is essentially independent of the total dose rate and the rate of polymerization is proportional to the first power of the dose rate. The polymer was amorphous as indicated by x-ray diffraction.

The polymerization rates of a series of substituted vinylbiphenyls have been found to be first order in monomer, and they were claimed to increase with increased conjugation and polarity of the substituents (58,59). Reactivity ratios for various vinylarene monomers are shown in Table 2.

As indicated by the $1/r_1$ values, all vinylarene monomers shown, with the exception of 9-vinylanthracene, are more reactive in copolymerization than is monomer M_1.

A series of copolymers of 4-vinylbiphenyl with styrene and vinylchlorobiphenyl and vinylfluorobiphenyl each with α-methylstyrene or α,p-dimethylstyrene have been prepared by mass and emulsion copolymerization (63). The 4-vinylbiphenyl–styrene copolymer was claimed to have improved resistance to heat distortion.

The effect of styrene on bulk polymerization rates and molecular weights of copolymers with various vinylnaphthalenes has received considerable attention (51,60, 61,64,65). Thus, the bulk polymerization rate of 1-vinylnaphthalene is decreased by the addition of styrene, and the rate reaches a minimum with 60 mole % styrene in the feed (60). In general, the addition of styrene to vinylnaphthalene increases the molecular weight of the copolymer (51). With 1-vinylnaphthalene, addition of styrene had little effect until about 60% had been added, and then the molecular weight increased almost linearly from 20,000 to 110,000. The increase in molecular weight of poly(2-vinylnaphthalene) by addition of styrene was in general more gradual but was more

Table 2. Reactivity Ratios for Various Vinylarene Monomers

M_1	M_2	r_1	r_2	$1/r_1$	Reference
styrene	1-vinylnaphthalene	0.67	1.35	1.49	60
	2-vinylnaphthalene	0.50	1.40	2.00	61
	4-chloro-1-vinylnaphthalene	0.85	0.80	1.18	61
	6-chloro-2-vinylnaphthalene	0.40	1.50	2.50	61
	1-vinylanthracene	0.81	0.57	1.24	54
	9-vinylanthracene	2.12	0.25	0.47	54
	9-vinylphenanthrene	0.58	2.36	1.72	54
	acenaphthylene	0.33	3.81	3.04	62
methyl methacrylate	2-vinylnaphthalene	0.4	1.0	2.50	61
	4-chloro-1-vinylnaphthalene	0.7	0.7	1.43	61
	6-chloro-2-vinylnaphthalene	0.45	1.6	2.22	61
methyl acrylate	2-vinylphenanthrene	0.1	2.0	10.0	61
	3-vinylphenanthrene	0.8	1.75	1.25	61

rapid at low styrene concentration. The same effect was also noted in methyl methacrylate–2-vinylnaphthalene copolymerization (61). As in homopolymerization, emulsion copolymerizations produce copolymers having higher molecular weights relative to those prepared by bulk polymerization (51).

Addition of styrene to 6-chloro-2-vinylnaphthalene leads to increasing rates with increasing styrene content in the feed, whereas the opposite is true with 4-chloro-1-vinylnaphthalene (61). The addition of methyl methacrylate has little or no effect on 4-chloro-1-vinylnaphthalene but decreases the rate of copolymerization of 6-chloro-2-vinylnaphthalene.

The copolymerization behavior of anthracene and phenanthrene derivatives with styrene has been investigated (2,54). The same order of decreasing activity (9-vinylphenanthrene > 1-vinylanthracene > 9-vinylanthracene) as in homopolymerization is also noted in copolymerization. Although the rate of copolymerization of 9-vinylanthracene with styrene is faster than that of 9-vinylanthracene alone, 9-vinylanthracene feeds greater than 25% by weight inhibit the polymerization of styrene (2). 2- And 3-vinylphenanthrenes have been copolymerized with methyl acrylate (61). Even though both monomers are more reactive than styrene toward methyl acrylate radicals, the addition of methyl acrylate to either of the phenanthrenes reduced both the molecular weight of the polymer and the rate of copolymerization.

Various copolymers of 1-vinylpyrene have been prepared and their softening points determined (47,66,67).

The copolymerization of acenaphthylene with other vinyl monomers has been described (41). Of these, the most extensively investigated was the copolymerization of styrene with acenaphthylene. Mass polymerizations using peroxide initiators or thermal polymerizations at 120–125°C for as long as 10 days yielded only low-molecular-weight copolymers. However, emulsion polymerization at 30°C with redox catalyst systems gave excellent yields and high-molecular-weight products (68). Terpolymers of acenaphthylene, styrene, and butadiene have also been prepared. Acenaphthylene has been copolymerized with divinylbenzene (69) and the crosslinked network sulfonated (70). Strongly acidic ion-exchange series were thus produced.

Solid-state, γ-radiation-induced copolymerization studies of acenaphthylene with acrylamide (71) and maleic anhydride (72) have been carried out. Only polyacrylamide homopolymers could be obtained in attempted copolymerizations of eutectic mixtures with acenaphthylene. Solid-state copolymerizations of maleic anhydride with acenaphthylene produced 1:1 copolymers. The same alternating copolymer was also obtained in free-radical solution copolymerization (73).

Graft copolymers of acenaphthylene onto polyethylene have been prepared by roll-mixing polyethylene, acenaphthylene, and benzoyl peroxide in air at 100°C (74). Maximum grafting was obtained at 30 min, and thereafter the amount grafted decreased because the grafted branches were selectively masticated. No grafting was obtained in the absence of benzoyl peroxide.

Polymer Properties

Characterization. In Table 3 are collected the parameters for the Mark-Houwink equation for some vinylarene polymers, correlating intrinsic viscosity with molecular weight.

Table 3. Parameters for the Mark-Houwink Equation[a]

Polymer	Solvent	Temp, °C	$K \times 10^4$	a	Reference
poly(4-vinylbiphenyl)	dimethoxyethane	25	3.70	0.53	75
	benzene	25	0.92	0.69	75
	2-methoxyethanol–dimethoxyethane	30	4.40	0.50	75
poly(acenaphthylene)	benzene	25	0.72	0.67	76
	1,2-dichloroethane	30	4.56	0.50	76
	tetrahydrofuran	20	9.66	0.87	77
poly(2-vinylnaphthalene)	benzene	20	0.66	0.71	78
	Decalin–toluene	30.2	5.20	0.50	78

[a] $[\eta] = KM^a$, in dl/g.

Light-scattering studies have shown that the coil size of poly(2-vinylnaphthalene) exceeds that of polystyrene by a factor of 1.4, indicating that substitution of benzene by a naphthalene ring increases the thermodynamic stiffness of the polymer (79). However, another study has shown that even though considerable hindrance to rotational motion of chain segments should be expected in poly(acenaphthylene), its dilute-solution behavior indicates that it has a hydrodynamic volume comparable with that of polystyrene (76). It has also been shown that poly(4-vinylbiphenyl), poly(1-vinylnaphthalene), and poly(2-vinylnaphthalene) can be represented by a common plot of intrinsic viscosity times the molecular weight of the repeat unit versus weight-average degree of polymerization, and that they also exhibit a common gel permeation chromatography calibration plot (80). These results lead to the somewhat surprising conclusion that all these vinylarene polymers have similar hydrodynamic volumes. Poly(acenaphthylene) could not be included in these studies because it has been found to be unstable in solution and to degrade by a free-radical mechanism that is at least partially an "unzipping" process (81) (see also DEPOLYMERIZATION).

A number of studies have dealt with a polarographic determination of monomer in polymers and copolymers of various vinylarene compounds (82–84). The polymers have also been characterized by spectrometric methods. Thus, poly(9-vinylanthracene) has been studied by UV, IR, and NMR (44); poly(2-vinylnaphthalene) by UV (85), IR, and NMR (35); poly(1-vinylnaphthalene) by UV(86), IR, and NMR (35); poly(4-vinylbiphenyl) by UV (19), IR, and NMR (35,87); and poly(acenaphthylene) by UV and IR (76).

Table 4. Conductivity Data for Some Vinyl Aromatic Polymers

Polymer–Donor[a]	Color	$\sigma_{298}/\Omega^{-1}\text{-cm}^{-1}$
[atactic PVN].TCNE	dark green	3.2×10^{-15}
[atactic PVN]$_2$.TCNE	dark green	7.4×10^{-15}
[atactic PVN]$_4$.TCNE	dark green	9.6×10^{-15}
[isotactic PVN].TCNE	dark green	2.8×10^{-15}
[atactic PVN].chloranil	dark brown	7.2×10^{-15}
[atactic PVN]$_4$.chloranil	dark brown	1.6×10^{-14}
[isotactic PVN].chloranil	dark brown	1.1×10^{-14}
[isotactic PVN]$_4$.chloranil	dark brown	1.3×10^{-14}
[atactic PVN]$_4$.DDQ	black	1.1×10^{-13}
[atactic PACN]$_4$.TCNE	green	3.7×10^{-15}

[a] PVN = poly(1-vinylnaphthalene); PACN = poly(acenaphthylene); TCNE = tetracyanoethylene; DDQ = 2,3-dichloro-5,6-dicyano-*p*-benzoquinone; chloranil = tetrachloro-*p*-benzoquinone.

Electrical Properties. A number of charge-transfer complexes have been prepared in which the electron donor is a vinylarene polymer. They are of interest because the complexes are known to show semiconductive properties in the solid state (see also POLYMERS, CONDUCTIVE).

The conductivity of charge-transfer complexes of both atactic and isotactic poly(1-vinylnaphthalene) and poly(acenaphthylene) with various electron acceptors are shown in Table 4 (88). The data indicate that the conductance does not change markedly with the proportion of electron acceptor in the complex. The electrical properties of poly(1-vinylanthracene) and poly(9-vinylanthracene) complexes with iodine, trinitrobenzene, and bromine were investigated and, depending on the method of preparation, the electrical resistivities of the iodine complexes were found to be between 10^6 and 10^{10} Ω-cm (34). The semiconductive properties of anionically prepared poly(9-vinylanthracene)–iodine charge-transfer complex have also been investigated (89).

Poly(vinylpyrene) and its charge-transfer complexes with iodine, tetracyanoethylene, and 7,7′,8,8′-tetracyanoquinodimethane have been studied (90). The polymer had a resistivity of 3×10^{15} Ω-cm; the iodine complex, 1.3×10^{8} Ω-cm; the tetracyanoethylene complex, 4.5×10^{12} Ω-cm; and the 7,7′,8,8′-tetracyanoquinodimethane complex, 1.1×10^{14} Ω-cm. The photoconduction of these compounds was also measured.

Using catalysts of the Ziegler-Natta type, a polymer from α-ethynylnaphthalene was prepared and a complex prepared with iodine in warm *o*-dichlorobenzene (91). The charge-transfer complex had a resistivity of 10^{10} Ω-cm and the polymer 10^{12} Ω-cm.

A film of a styrene–poly(2-vinylnaphthalene) copolymer has been treated with bromine at room temperature or iodine at 67–90°C (92). The resistivity of the polymer containing 50–60% bromine was claimed to have decreased by 7 to 9 orders of magnitude from that of the original polymer.

Utility and Application

Vinylarene monomers or their polymers have not found appreciable utility in a commercial sense. All the vinylarene monomers and a large number of their homo-, co-, and terpolymers have been investigated for applicability in numerous commercial fields, but only minimal acceptance has been realized.

In the area of fibers, modifications of poly(vinyl chloride)–acrylonitrile fibers with poly(vinylbiphenyl), poly(acenaphthylene), poly(vinylnaphthalene), and poly(vinylfluorene) have been reported to produce fibers of generally low cost and good thermal stability (93). Another report notes that the use of poly(vinylnaphthalene) in a fiber composition can replace poly(vinylcarbazole) as an effective promoter of dye receptivity (94).

A number of observations on the scintillation activity of 9-vinylanthracene and 9-vinylbiphenyl have been reported. These reports covered not only the basic studies on the nature of the excited states and molecular configuration as related to their optical and fluorescent properties (95–98), but also the commercial aspects that were illustrated, for example, by the discovery that 9-vinylanthracene was indeed a high-grade 1° and 2° scintillator, capable of replacing the widely used 1,4-bis(5-phenyl-2-oxazolyl)benzene (99).

Poly(9-vinylanthracene) (100), poly(acenaphthylene) (101–103), and poly(1-vinylnaphthalene) (103) have also been shown to be suitable as organic photoconductors. Electrophotographic printing plates incorporating some of these materials have been manufactured and tested (101,102).

One of the greatest applications for these materials is as a crosslinked polymeric matrix useful for the preparation of ion-exchange resins. Almost exclusive interest has been shown in the acenaphthylene–divinylbenzene system as a starting matrix (104, 105), although some work on a vinylnaphthylene–divinylbenzene system has been reported (106). After preparation of these crosslinked systems, they are further reacted through sulfonation (107) or chloromethylation and amination (108) to form either anionic (109) or cationic (110) exchange resins. Other miscellaneous applications reported for the polymers of the vinylarene monomers include their use as binders for electrical insulation products (111,112), molding compounds (113,114), organic semiconductors (115), resins for lacquers and coatings (116–119), and components of synthetic asphalt (120).

Bibliography

1. E. G. E. Hawkins, *J. Chem. Soc.* **1957,** 3858.
2. E. D. Bergmann and D. Katz, *J. Chem. Soc.* **1958,** 3216.
3. D. T. Mowry, M. Renoll, and W. F. Huber, *J. Am. Chem. Soc.* **68,** 1105 (1946).
4. W. F. Huber, M. Renoll, A. G. Rosson, and D. T. Mowry, *J. Am. Chem. Soc.* **68,** 1109 (1946).
5. R. G. Flowers and F. S. Nichols, *J. Am. Chem. Soc.* **71,** 3104 (1949).
6. R. G. Flowers and H. F. Miller, *J. Am. Chem. Soc.* **69,** 1388 (1947).
7. M. M. Koton, *J. Polymer Sci.* **30,** 331 (1958).
8. C. C. Price and G. H. Schilling, *J. Am. Chem. Soc.* **70,** 4265 (1948).
9. C. C. Price and B. D. Halpern, *J. Am. Chem. Soc.* **73,** 818 (1951).
10. F. Bahsteter, J. Smid, and M. Szwarc, *J. Am. Chem. Soc.* **85,** 3909 (1963).
11. A. Eisenberg and A. Rembaum, *J. Polymer Sci.* [B] **2,** 157 (1964).
12. R. H. Michel and W. P. Baker, *J. Polymer Sci.* [B] **2,** 163 (1964).
13. A. Rembaum and A. Eisenberg, in A. Peterlin, M. Goodman, S. Okamura, B. H. Zimm, and H. F. Mark, eds., *Macromolecular Reviews*, Vol. 1, Interscience, a division of John Wiley & Sons, Inc., New York, 1967, p. 57.
14. J. Moacanin and A. Rembaum, *J. Polymer Sci.* [B] **2,** 979 (1964).
15. E. Franta and P. Rempp, *Compt. Rend.* **254,** 674 (1962).
16. M. Szwarc, *Carbanions, Living Polymers and Electron Transfer Processes*, Interscience, a division of John Wiley & Sons, Inc., New York, 1968, p. 547.
17. S. N. Khanna, M. Levy, and M. Szwarc, *Trans. Faraday Soc.* **58,** 747 (1962).
18. J. Stearne, J. Smid, and M. Szwarc, *Trans. Faraday Soc.* **60,** 2054 (1964).
19. J. Heller, J. F. Schimscheimer, R. A. Pasternak, C. B. Kingsley, and J. Moacanin, *J. Polymer Sci.* [A] **1** (7), 73 (1969).
20. M. Szwarc, M. Levy, and R. Milkowitch, *J. Am. Chem. Soc.* **78,** 2656 (1956).
21. A. Rembaum and J. Moacanin, *J. Polymer Sci.* [B] **1,** 41 (1963).
22. J. Gole and G. Goutiere, *Compt. Rend.* **254,** 1067 (1962).
23. L. J. Fetters, *J. Res. Natl. Bur. Std. A* **70,** 421 (1966).
24. A. Rembaum, J. Moacanin, and R. Haack, *J. Macromol. Chem.* **1,** 657 (1966).
25. A. Rembaum, R. F. Haack, and A. M. Hermann, *J. Macromol. Chem.* **1,** 673 (1966).
26. A. Rembaum, J. Moacanin, and E. Cuddihy, *J. Polymer Sci.* [C] **4,** 529 (1964).
27. E. Cuddihy, J. Moacanin, and A. Rembaum, *J. Appl. Polymer Sci.* **9,** 1385 (1965).
28. G. Goutiere and J. Gole, *Bull. Soc. Chim. France* **1965,** 153.
29. D. Paul, D. Lipkin, and S. Weissman, *J. Am. Chem. Soc.* **78,** 116 (1956).
30. G. Goutiere and J. Gole, *Bull. Soc. Chim. France* **1965,** 162.
31. J. Heller and D. B. Miller, *J. Polymer Sci.* [B] **7,** 141 (1969).
32. G. Natta, F. Danusso, and D. Sianesi, *Makromol. Chem.* **28,** 253 (1958).
33. G. Natta, F. Danusso, D. Sianesi, and A. Macchi, *Chim. Ind.* (*Milan*) **41,** 964 (1959); through *Chem. Abstr.* **54,** 6182c (1960).
34. H. Inoue, K. Noda, T. Takiuchi, and E. Imoto, *Kogyo Kagaku Zasshi* **65,** 1286 (1962).
35. J. Heller and D. B. Miller, *J. Polymer Sci.* [A-1] **5,** 2323 (1967).
36. P. Corradim and P. Ganis, *Nuovo Cimento* **15,** 104 (1960).
37. V. M. Story and G. Canty, *J. Res. Natl. Bur. Std. A* **68,** 165 (1964).
38. M. Imoto and I. Soematsu, *Bull. Chem. Soc. Japan* **34,** 26 (1961).
39. J. Dolinsky and K. Dziwonski, *Chem. Ber.* **48,** 1917 (1915).
40. M. Kaufman and A. F. Williams, *J. Appl. Chem.* **1,** 589 (1951).
41. J. I. Jones, *J. Appl. Chem.* **1,** 568 (1951).
42. P. Giusti and F. Andruzzi, *Gazz. Chim. Ital.* **96,** 1563 (1966).
43. P. Giusti, G. Puce, and F. Andruzzi, *Makromol. Chem.* **98,** 170 (1966).
44. R. H. Michel, *J. Polymer Sci.* [A] **2,** 2533 (1964).
45. S. A. Zomis, *Zh. Obshch. Khim.* **9,** 119 (1939).
46. Brit. Pat. 606,364.
47. R. G. Flowers and F. S. Nichols, *J. Am. Chem. Soc.* **71,** 3104 (1949).
48. U.S. Pat. 2,476,737.
49. C. E. H. Bawn, C. Fitzsimmons, and A. Ledwith, *Proc. Chem. Soc.* **1964,** 391.
50. S. Lashaek, E. Broderick, and P. Bernstein, *J. Polymer Sci.* **39,** 223 (1959).
51. A. V. Golubeva, N. F. Usmanova, and A. A. Vansheidt, *J. Polymer Sci.* **52,** 63 (1961).

52. C. F. Parrish, W. A. Trinler, and C. K. Burton, *J. Polymer Sci.* [A-1] **5,** 2557 (1967).
53. S. Yura and S. Ono, *J. Chem. Soc. Japan* **53,** 268 (1950).
54. D. Katz, *J. Polymer Sci.* [A] **1,** 1635 (1963).
55. H. G. Gelhaar and K. Überreiter, *Kolloid Z.Z. Polymere* **209,** 136 (1966).
56. M. N. Romani and K. E. Weale, *Trans. Faraday Soc.* **62,** 2264 (1966).
57. C. S. H. Chen, *J. Polymer Sci.* **62,** S38 (1962).
58. P. A. El'tsova, M. M. Koton, O. K. Mineeva, and O. K. Surnina, *Vysokomolekul. Soedin.* **1,** 1369 (1959); through *Chem. Abstr.* **54,** 17954c.
59. A. V. Chernobai, L. I. Dimitrievskaya, Zh. S. Tirak'yants, and R. Ya. Delyatskaya, *Vysokomolekul. Soedin.* **7,** 1221 (1965); through *Chem. Abstr.* **63,** 11711h.
60. S. Loshaek and E. Broderick, *J. Polymer Sci.* **39,** 241 (1959).
61. C. C. Price, B. D. Halpern, and S. T. Voong, *J. Polymer Sci.* **11,** 575 (1953).
62. K. Überreiter and W. Krull, *Z. Physik Chem. (Frankfurt)* **12,** 303 (1957).
63. R. B. Seymour and J. M. Butler, U.S. Pat. 2,471,785 (1949).
64. M. M. Koton, *J. Gen. Chem. USSR (Engl. Transl.)* **9,** 1626 (1939).
65. N. F. Usmanova, A. V. Golubeva, A. A. Vansheidt, K. A. Sivograkova, and S. N. Dionikova, *Soviet Plastics (English Transl.)* **1961,** 3.
66. R. G. Flowers, U.S. Pat. 2,496,867 (1950).
67. R. G. Flowers, Can. Pat. 508,754 (1955).
68. K. R. Dunham, J. Vandenberghe, J. W. F. Faber, and W. F. Fowler, Jr., *J. Appl. Polymer Sci.* **7,** 143 (1963).
69. V. P. Li, A. B. Pashkov, and N. S. Shamis, *Plasticheskie Massy* **1963,** 6; through *Chem. Abstr.* **61,** 7104 (1964).
70. G. Manecke and J. Danhäuser, *Makromol. Chem.* **48,** **117** (1961).
71. M. Nishii, H. Tsukamoto, K. Hayashi, and S. Okamura, *Nippon Hoshasen Kobunshi Kenkyu Kyokai Nenpo* **5,** 115 (1963–64); through *Chem. Abstr.* **63,** 7116 (1965).
72. A. Shimizu, K. Hayashi, and S. Okamura, *Nippon Hoshasen Kobunshi Kenkyu Kyokai Nenpo* **5,** 129 (1963–64); through *Chem. Abstr.* **63,** 7117 (1965).
73. L. Strzelecki, *Bull. Soc. Chim. France* **1967,** 2659.
74. M. Akiyama, S. Kawakubo, and M. Kondo, *Rev. Elec. Commun. Lab. (Tokyo)* **12,** 693 (1964); through *Chem. Abstr.* **62,** 14888 (1964).
75. J. Moacanin, A. Rembaum, and R. K. Laudenslager, *Polymer Preprints* **4,** 179 (1963).
76. J. Moacanin, A. Rembaum, R. K. Laudenslager, and R. Adler, *J. Macromol. Sci.* **A1,** 1497 (1967).
77. N. Vene and G. Mohorcic, *Rept. J. Setfan Inst. (Ljubljana)* **5,** 71 (1958).
78. V. N. Cvetkov, S. I. Klenin, S. A. Frenkel, O. V. Fomitcheva, and A. G. Zhuse, *Vysokomolekul. Soedin.* **4,** 540 (1962).
79. V. E. Eskin and O. Z. Korotkina, *Vysokomolekul. Soedin.* **2,** 272 (1960); through *Chem. Abstr.* **54,** 20299.
80. J. Heller and J. Moacanin, *J. Polymer Sci.* [B] **6,** 595 (1968).
81. L. Utracki, N. Eliezer, and R. Simha, *J. Polymer Sci.* [B] **5,** 137 (1967).
82. V. D. Berzuglyi, V. N. Dmitrieva, and T. A. Batovskaya, *Zh. Analit. Khim.* **17,** 109 (1962).
83. T. A. Alekseeva and B. D. Bezuglyi, *Zh. Obshch. Khim.* **36,** 2054 (1966); through *Chem. Abstr.* **66,** 65912f.
84. V. D. Bezuglyi, T. A. Alekseeva, L. I. Dmitrievskaya, A. V. Chernobai, and L. P. Kruglyak, *Vysokomolekul. Soedin.* **6,** 125 (1964); through *Chem. Abstr.* **60,** 10789d.
85. H. A. Laitinen, F. A. Miller, and T. D. Parks, *J. Am. Chem. Soc.* **69,** 2707 (1947).
86. L. H. Klemm, H. Ziffer, J. W. Sprague, and W. Hodes, *J. Org. Chem.* **20,** 190 (1955).
87. E. E. Genser, *J. Mol. Spectry.* **16,** 56 (1965).
88. W. Slough, *Trans. Faraday Soc.* **58,** 2360 (1962).
89. A. Rembaum, A. Henry, and H. Waits, *Polymer Preprints* **4,** 109 (1963).
90. M. Ishizuka, K. Suzuki, and H. Mikawa, *14th Polymer Chemistry Meeting Tokyo*, 1965, [3F12] Preprint, p. 154.
91. S. Kambara, M. Hatano, N. Sera, and K. Shimamura, *J. Polymer Sci.* [B] **5,** 233 (1967).
92. Yu I. Vasilenok, B. E. Davydov, B. A. Kreutsel, and B. I. Sazhin, *Vysokomolekul. Soedin.* **7,** 626 (1965).
93. G. E. Ham, U.S. Pat. 2,749,321 (1956).

94. G. E. Ham, U.S. Pat. 2,769,793 (1956).
95. M. I. Dovgosheya, *Zh. Prikl. Spektro. Skojii* **1,** 260 (1964).
96. A. S. Cherkasov and K. G. Voldaikina, *Izv. Akad. Nauk. SSSR Ser. Fiz.* **27** (5), 628 (1963); through *Chem. Abstr.* **59,** 7057d (1963).
97. A. Heller and G. Rio, *Bull. Chim. Soc. France* **1963,** 1707.
98. V. D. Bezvglyi, A. V. Chernobai, L. I. Dmitrievskaya, R. S. Mil'ner, and M. I. Dovgosheya, *Stsintillyatoryi i Stsintillyats Materialy Sb.* **1963** (3), 72; through *Chem. Abstr.* **62,** 14139c (1965).
99. A. Heller and D. Katz, *J. Chem. Phys.* **35,** 1987 (1961).
100. K. Morimoto, E. Ishida, and A. Inami, *J. Polymer Sci.* [A-1] **5,** 1699 (1967).
101. H. Hoegel, Ger. Pat. 1,133,976 (1962).
102. A. G. Kalle, Brit. Pat. 952,906 (1964).
103. K. Morimoto and A. Inami, *Kogyo Kagaku Zasshi* **67,** 1938 (1964); through *Chem. Abstr.* **62,** 13996d (1965).
104. G. Manecke and J. Danhäuser, *Ionenaustaucher Einzeldarstell* **1,** 331 (1961).
105. G. Manecke and J. Danhäuser, *Makromol. Chem.* **56,** 208 (1962).
106. R. M. Wheaton and D. F. Harrington, U.S. Pat. 2,642,417 (1963).
107. D. Dolar, G. Mottorcic, and M. Pris, *Makromol. Chem.* **84,** 108 (1965).
108. A. B. Pashkov, V. S. Titov, and M. I. Itkina, USSR Pat. 126,265 (1960); through *Chem. Abstr.* **54,** 15766f (1960).
109. R. J. Phillips, Brit. Pat. 859,282 (1961).
110. G. S. Petrov, A. B. Pashkov, and A. V. Titov, USSR Pat. 105,382 (1957); through *Chem. Abstr.* **51,** 11616h (1957).
111. Belg. Pat. 651,120 (1965); through *Chem. Abstr.* **64,** 11437b (1966).
112. R. N. Cooper, Jr., U.S. Pat. 3,030,346 (1962).
113. D. A Tester, Brit. Pat. 933,696 (1963).
114. L. M. Debing, U.S. Pat. 2,454,250 (1948).
115. W. Slough, Brit. Pat. 1,009,361 (1965).
116. S. Schneider, Ger. Pat. 829,220 (1952).
117. A. G. Rütgerswerke, Brit. Pat. 725,459 (1955).
118. B. B. Kine, P. W. McWherter, and H. A. Alps, U.S. Pat. 2,828,222 (1958).
119. P. E. Marling, U.S. Pat. 2,603,611 (1952).
120. T. Edstrom, I. C. Lewis, and C. V. Mitchell, Fr. Pat. 1,358,003 (1964).

J. Heller
Pharmetrics, Inc.
T. Anyos
Stanford Research Institute

VINYLATION. See ACETYLENE AND ACETYLENIC POLYMERS

VINYLBENZENE POLYMERS. See STYRENE POLYMERS

VINYL BROMIDE POLYMERS

Vinyl bromide (bromoethene), $CH_2{=}CHBr$, was one of the earliest known vinyl compounds. It was initially prepared in 1835 by treating ethylene dibromide (1) with alkali and later in 1872 by the reaction of acetylene with hydrogen bromide (2). Accordingly, poly(vinyl bromide) was one of the first vinyl polymers discovered (3,4). Unfortunately, the investigators were not able to explain their observations that light-colored solids were produced from the exposure of vinyl bromide to sunlight and that these products were little affected by solvents and acids (3,4).

The first detailed study of poly(vinyl bromide) was made by Ostromysslenski in 1912 (5). The primary objective of this investigation appears to have been the elucidation of the molecular structure of the polymer, but commercial aspects were considered even at this early date. He concluded that the structure was best represented by a head-to-head as opposed to a head-to-tail configuration. The head-to-tail configuration $\text{-(}CH_2{-}CHBr\text{)-}_n$ was later shown to be correct as based primarily upon chemical reactions of the polymer. Ostromysslenski also reported that large quantities of hydrogen bromide were liberated upon heating to 220°C. Later studies have shown that the polymer is unstable even at room temperatures, and consequently, it is of little commercial significance (6,7).

Monomer

Properties. Some early data (8) were interpreted as evidence for resonance or hybridization of the ordinary structure (**1**) with excited structures such as (**2**), which are similar to those reported for phenyl halides (9,10). A study of electron diffraction and dipole moment yielded a C—Br distance of 1.86 or 1.87 Å, a Br—C—C angle of

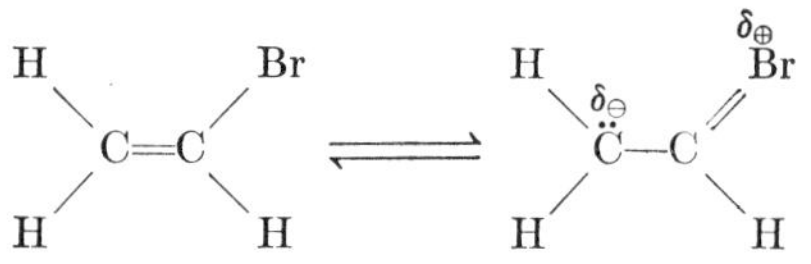

$$\underset{(1)}{H_2C{=}CHBr} \rightleftharpoons \underset{(2)}{H_2\overset{\delta\ominus}{\ddot{C}}{-}CH{=}\overset{\delta\oplus}{Br}}$$

122°, and a dipole moment of 1.41 D (11). It was found that neither bond contractions nor electric dipole moment anomalies could give accurate measures of the part that excited structures play in the actual molecule. The investigators concluded that their data supported structures (**1**) and (**2**), and an approximate 9:1 ratio of (**1**) to (**2**) was postulated. The infrared spectrum of vinyl bromide has been reported (12) and, when compared with spectra of vinyl chloride and iodide, shows, with three exceptions, a definite decrease in the frequency of a given vibration as the mass of the halogen atom increases. Additional infrared data as well as Raman spectra are described in Ref. 19.

Important physical properties of vinyl bromide are listed in Table 1.

Because the carbon–bromine bond in vinyl bromide is quite different from other alkyl-carbon–bromine bonds, there exist significant differences in the reactivity of this center in nucleophilic substitutions; in general, vinyl bromide resembles aryl bromides in its reactions with nucleophiles. It is unreactive under normal conditions, ie, with $I^{\ominus}$–acetone, $Ag^{\oplus}$, $OH^{\ominus}$, $OR^{\ominus}$, NH_3, $CN^{\ominus}$, and ArH–$AlCl_3$. Reactions of vinyl bromide with organometallic reagents are discussed in Refs. 20 and 21.

Table 1. Physical Properties of Vinyl Bromide

Property	Value	References
formula weight	106.260	
melting point, °C	−139.54	13
boiling point, °C	15.80	14–17
vapor pressure, mmHg		
−80°C	4.2	16
−60°C	18.5	16
−40°C	65	16
−20°C	178	16
0°C	422	16
20°C	895	16
40°C	1672	16
density, g/ml		
−40°C	1.6507	16
−20°C	1.6055	16
0°C	1.5610	16, 17
20°C	1.5152	16
30°C	1.4920	16
index of refraction, n_D		
20°C	1.441	18
25°C	1.435	18
dielectric constant	5.623	13
dipole moment, D	1.41 D	12
viscosity, cSt		
−20°C	0.2759	13
−10°C	0.2528	13
0°C	0.2393	13
molar refraction, R_D, cm^3 (obs)	18.917	13
molar refraction, R_D, cm^3 (calcd)	18.734	13

The bromine atom exerts an anomalous effect on reactivity and orientation in the electrophilic reactions of vinyl bromide, that is, on addition to the carbon–carbon double bond. For example, ethylene was found to brominate at least an order of magnitude faster than vinyl bromide, and the orientation of the attacking group was found to be influenced by the bromine atom on vinyl bromide (22).

Although vinyl bromide cannot be employed as an alkylating agent in Friedel-Crafts reactions, it will undergo acylation or alkylation in the presence of Friedel-Crafts catalysts and the appropriate halide (23). With highly reactive dienes, vinyl bromide can serve as the dienophile in the Diels-Alder reaction if the reaction is carried out at elevated temperatures and pressure.

Preparation. One of the older methods of preparing vinyl bromide which is still the choice of most laboratories consists of the dehydrobromination of 1,2-dibromoethane with alkali (eq. 1) (24). A study of the effects of alcohol and alkali upon the

$$CH_2Br{-}CH_2Br \xrightarrow[95\%\ C_2H_5OH]{20\%\ KOH} CH_2{=}CHBr + KBr + H_2O \qquad (1)$$

reaction revealed the following: if aqueous potassium carbonate was used, the product consisted of bromoethane and a small amount of vinyl bromide. When sodium hydroxide in 50% aqueous ethanol was used, vinyl bromide was the predominant product, and with sodium hydroxide in absolute ethanol, vinyl bromide was the exclusive product (25). In an interesting study of the decomposition of 1,2-dibromoethane into vinyl bromide and hydrogen bromide, the author concluded that 1,1-dibromoethane was formed as an intermediate when 1,2-dibromoethane was passed over Lewis-acid-type catalysts at elevated temperatures (26).

The commercial preparation of vinyl bromide stems from the early work of Reboul (27), who obtained vinyl bromide from the reaction of acetylene with hydrogen bromide (eq. 2). The primary disadvantage of this method is that significant quantities of

$$HC{\equiv}CH + HBr \xrightarrow{Hg^{2+}} CH_2{=}CHBr \qquad (2)$$

1,2-dibromoethane are obtained (28,29). Subsequent patents (30–34) point out the need for a catalyst and it appears that the halides of mercury, cerium, and copper are effective in decreasing by-product formation.

In the reaction of ethylene with bromine, vinyl bromide may be produced directly (35,36). Alternatively, 1,2-dibromoethane may be produced, depending upon the conditions of the reaction. The latter may be decomposed into vinyl bromide and hydrogen bromide by passing the gas through a solution of alcoholic potassium hydroxide (37). Additional methods of preparing vinyl bromide have been reported (38–42), but they are of little commercial significance.

Polymerization

Vinyl bromide is unexceptional with respect to the polymerization of most monosubstituted vinyl compounds. Typical initiators include peroxides, azobisisobutyronitrile, and photochemical means. Although the bromine is relatively inert in comparison with most alkyl bromides, the carbon–bromine bond is not impervious to organoalkali catalysts such as *n*-butyllithium or sodium naphthenate. In these cases side reactions leading to alkali vinyl compounds, vinyl carbene, and lithium acetylides are possible. The following section discusses those systems yielding normal poly(vinyl bromide).

Homopolymerization. Vinyl bromide may be polymerized under a variety of conditions and in various forms. The latter includes bulk, solution, emulsion, and suspension systems. The initial studies by Baumann (4) and Hoffman (3) were carried out in bulk, and sunlight was used as the initiator. These early investigators (3,4) (1860–1872) were primarily concerned with the phenomenon itself, which they found quite puzzling. Under their conditions, the rates of polymerization were slow, requiring 12–35 days before light-colored solids separated from the reaction mixture. Their concern and conclusions are quite reasonable with respect to the state of the art at that time. The first systematic investigation of the mechanism of vinyl bromide polymerization was conducted in 1957 and covered those factors that influence the photopolymerization rate including (*1*) solvents, (*2*) wavelength of the ultraviolet radiation used as the initiator, and (*3*) exposure time (44,45). The results indicated that the reaction was best characterized by a free-radical chain-polymerization mechanism. The inhibiting effect of dissolved oxygen upon the polymerization rate was investigated by Staudinger (43); his results probably account for the different rates reported by other investigators (6,46). The photochemical initiation may proceed by way of a diradical (eq. 3), which may initiate the reaction directly, or indirectly by de-

$$CH_2{=}CHBr + h\nu \longrightarrow \dot{C}H_2{-}\dot{C}HBr \qquad (3)$$

composing to vinyl and bromine radicals (eq. 4). Alternatively, the initiation may proceed by way of a dimeric diradical which has been postulated for thermal polymerizations (eq. 5).

$$\dot{C}H_2—\dot{C}HBr \longrightarrow CH_2=\dot{C}H + Br\cdot \quad (4)$$

$$2\ CH_2=CHBr \longrightarrow H\dot{C}Br—CH_2—CH_2—\dot{C}HBr \quad (5)$$

Some of the more common radical initiators used in both bulk and solution polymerizations include benzyl and di-*tert*-butyl peroxides and azobisisobutyronitrile (47).

Emulsion and suspension polymerizations of vinyl bromide have been reported (6,50). Potassium peroxydisulfate ($K_2S_2O_8$) is generally the initiator of choice, but azobisisobutyronitrile has been employed (6). For the former, the radical is generated by way of a redox mechanism and Fe^{2+} is commonly used as the reducing agent (eq. 6).

$$S_2O_8^{2-} + Fe^{2+} \longrightarrow SO_4\cdot + SO_4^{2-} + Fe^{3+} \quad (6)$$

Typical conditions for the suspension polymerization of vinyl bromide are as follows: monomer–water ratio of 1:2; initiator concentration of approximately 0.1% monomer and suspending agent concentration of 0.25% water; and a temperature of 50°C. Typical conditions for an emulsion polymerization are the following: monomer–water ratio of 1:1; initiator concentration of 0.25% monomer; emulsifier concentration of 0.02% water; and a temperature of 40°C.

The cationic polymerization of vinyl bromide has been reported (48,49). A strong Lewis acid is required to initiate the reaction; examples of these acids include $(n\text{-}C_4H_9)_3B$ (49) and $BF_3\cdot O(C_2H_5)_2$ (48). In either case, it appears that a trace of water is necessary (eqs. 7 and 8). Formation of the secondary carbonium ion, $CH_3—\overset{\oplus}{C}HBr$, is favored.

$$BF_3.O(C_2H_5)_2 + H_2O \longrightarrow (BF_3OH^{\ominus}.O(C_2H_5)_2)H^{\oplus} \quad (7)$$

$$(BF_3OH^{\ominus}.O(C_2H_5)_2)H^{\oplus} + CH_2=CHBr \longrightarrow (BF_3OH^{\ominus}.O(C_2H_5)_2) + CH_3—\overset{\oplus}{C}HBr \text{ or } \overset{\oplus}{C}H_2—CH_2Br \quad (8)$$

Copolymerization. The copolymerization of vinyl bromide with vinyl chloride (51), vinyl acetate (52), styrene (53,54), and acrylates (55,56) has been reported. The primary objective has been to enhance the stability of poly(vinyl bromide). In theory, the stability should improve if the monomer units of the second monomer are placed between the vinyl bromide monomer units. This conclusion was based on the observation that the loss of one molecule of hydrogen bromide was followed by the loss of another molecule at an adjacent group until a polyene was formed. The stabilities of the copolymers of methyl methacrylate and styrene with vinyl bromide, as illustrated in Table 2, are not good, however, or at least not exceptional. Moreover, a

Table 2. Thermal Stability of Various Copolymers (54)

Comonomer	% Vinyl bromide	% Loss of wt after 28 hr at 80°C	Color after 8 hr at 80°C
methyl methacrylate	59.0	1.9	red-brown
methyl methacrylate	3.6		white
styrene	10.7	0.7	white
styrene	3.2	0.0	white
none	100.0	0.2	pink

copolymer consisting of 59% vinyl bromide and 41% methyl methacrylate is less heat and color stable than pure poly(vinyl bromide) itself (54).

The reactivity ratios for the copolymerization of vinyl bromide with methyl methacrylate and styrene were determined at 0 and 28°C (54). The following values were reported:

	Irradiation at 0°C	*Irradiation at 28°C*
M_1: vinyl bromide	$r_1 = 0.05$	$r_1 = 0.05$
M_2: methyl methacrylate	$r_2 = 25$	$r_2 = 20$
M_1: vinyl bromide	$r_1 = 0.05$	$r_1 = 0.06$
M_2: styrene	$r_2 = 20–25$	$r_2 = 18$

In all cases the rates of copolymerization decrease when the mole fraction of vinyl bromide is increased; the most pronounced effect was observed for the photopolymerization with methyl methacrylate.

Polymer Properties

Data on the chemical and physical properties of poly(vinyl bromide) are somewhat meager, particularly when compared to the data existing for most vinyl polymers. This lack of information is, however, reasonable in terms of the instability of the polymer. Furthermore, this instability restricts both the physical and chemical measurements to the extent that most of the results are considered to be only qualitative in nature.

Physical Properties. White products are obtained from polymerizations initiated by either hydrogen peroxide (46) or benzoyl peroxide (6), both with and without solvent. However, the products were reported to darken upon drying under vacuum at 30°C (6). Similar results were also reported for the photochemical polymerization in ether solutions in sealed quartz tubes placed in front of a mercury arc.

Solubility studies (6) of poly(vinyl bromide) obtained from solution polymerization using potassium peroxydisulfate as catalyst (57) show that the polymer is insoluble in tri-*n*-butylamine, petroleum ether, carbon tetrachloride, nitromethane, and ethyl and methyl alcohol, is swelled in toluene and benzene, and is dissolved in methyl ethyl ketone, dioxane, and nitrobenzene. These data suggest that the cohesive-energy density of poly(vinyl bromide) is about 90, approximately the same as that of poly(vinyl chloride). The following statement concerning the structure of vinyl polymers is typical of most studies and illustrates the lack of physical data on poly(vinyl bromide). "The polymer . . . was left as a brittle powder which turned dark on standing. It was soluble in benzene and dioxane, only after refluxing, and the solutions were colored. Because of these difficulties, no attempts were made to determine viscosity and molecular weight" (48).

The effect of polymerization temperature on the molecular weight of poly(vinyl bromide) (49) is shown in Figure 1. The polymerizations were carried out using tri-*n*-butylborane as the initiator; the data show a maximum at about −35°C. The infrared spectra of poly(vinyl bromide) have been reported (47,50), but owing to the fragile nature of the cast film and the instability of the polymer, the resulting data are for the unoriented film only. The infrared spectrum of poly(vinyl bromide) is shown in Table 3.

Table 3. Infrared Absorption Bands of Poly(vinyl Bromide) (50)

Frequency, cm^{-1}	Intensity[a]	Frequency difference between PVC and PVBr	Frequency, cm^{-1}	Intensity[a]	Frequency difference between PVC and PVBr
537	s	78	1180	sh	16
563	s	72	1206	s	49
628	m	62	1274	sh	
646	m	62	1320	m	13
754	vw sh	10	1345	w	6
830	w	7	1376	w	5
905	sh	21	1430	s	−2
936	m	33	2845?	w	−23
1089	m	5	2878	w	−32
1119	vw sh	6	2937	m	−31
			3000	w	−6

[a] s, strong; m, medium; w, weak; vw, very weak; sh, shoulder.

In general, the absorption bands of poly(vinyl bromide) have lower frequencies than do the corresponding bands of poly(vinyl chloride). There are, however, a number of exceptions, namely, those assigned to the CH_2 bending mode and the CH and CH_2 stretching modes. See also VINYL CHLORIDE POLYMERS.

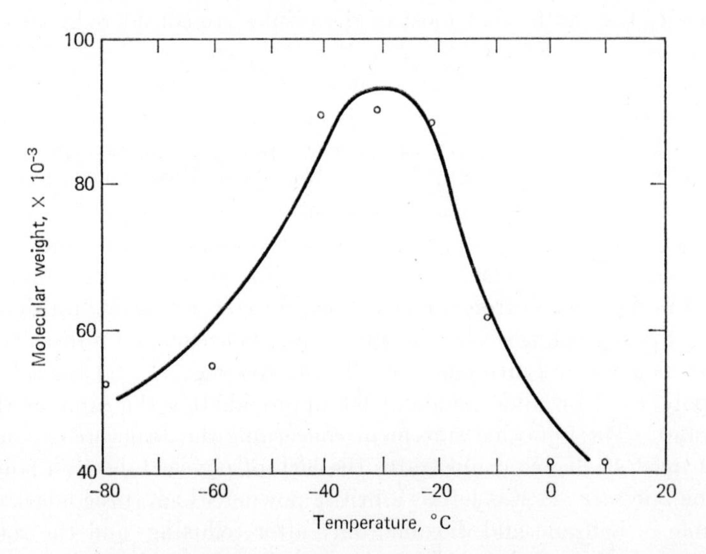

Fig. 1. Correlation of polymerization temperature with molecular weight of poly(vinyl bromide) (46).

Structure. After various early studies (5,58,59) Staudinger, Brunner, and Feisst (43) concluded that poly(vinyl bromide) is best represented by structure (**3**),

$$\left[CH_2{-}\underset{\displaystyle Br}{\underset{|}{CH}}{-}CH_2{-}\underset{\displaystyle Br}{\underset{|}{CH}} \right]_n$$

(**3**)

their results being based on the chemical reactions of the polymer. Further proof of the head-to-tail structure of poly(vinyl bromide) was based on both chemical and physical methods to characterize the product (7).

X-ray analysis showed that the presence of an aldehyde solvent increased the crystallinity of the polymer (47). A corresponding decrease in molecular weight was also noted as indicated by a 20:1 decrease in intrinsic viscosity (47). Further evidence on the molecular structure of poly(vinyl bromide) was obtained from NMR studies (60, 61). The spectra obtained at 60 and 100 MHz suggest that the free-radical polymerization of vinyl bromide at or near room temperature results in an atactic polymer, ie, one containing approximately equal concentrations of isotactic and syndiotactic diads (see also MICROTACTICITY). A small but distinct change in tacticity as a function of polymerization temperature has been reported (49). Results are depicted in Table 4

Table 4. Tacticity (Diads) of Poly(vinyl Bromide) as a Function of Polymerization Temperature (60)

Polymerization temp, °C	Isotactic, %	Syndiotactic, %
80	0.50	0.50
50	0.47	0.53
25	0.47	0.53
0	0.45	0.55
−40	0.40	0.60
−50	0.40	0.60
−60	0.38	0.62
−78	0.36	0.64

and, as expected, the data are quite similar to those reported for poly(vinyl chloride) (59).

Chemical Properties. The instability of poly(vinyl bromide) was first reported in 1912 (5); it was noted that the polymer turned black on heating to 130°C, and that much hydrogen bromide was evolved during decomposition at 220°C. The polymer decomposes at a much lower temperature, 25°C, when dissolved in pyridine, as shown by a rapid rise in conductance (6). The polymer reacts with triethylamine in methanol to produce a red-to-dark brown precipitate; a tertiary amine hydrobromide was found in the supernatant liquid. It was therefore concluded that tertiary amines dehydrohalogenate poly(vinyl bromide) (62) in preference to adding to it to form a polymeric quaternary salt. The elimination of hydrogen bromide by interaction of tertiary amines with the polymer chain is sterically favored, and is further enhanced by the production of allylic bromine atoms, —CHBrCH═CH—, which are more reactive than the original bromine atoms. Consequently, the tendency is to produce conjugated segments in the polymer chain.

The few reports of chemical reactions of poly(vinyl bromide) have been directed primarily toward structure elucidation. For example, poly(vinyl bromide) does not liberate iodine from peroxide-free dioxane solution of potassium iodide (7), a reaction that is characteristic of a 1,2-dihalide (63). In general, however, it is reasonable to conclude that most of the reactions of poly(vinyl chloride) are also possible for poly(vinyl bromide). See also VINYL CHLORIDE POLYMERS.

Bibliography

1. V. Regnault, *Ann. Chem.* **15,** 63 (1835); *Ann. Chim.* **59,** 358 (1835).
2. N. E. Reboul, *Jahresber.,* 304 (1872).
3. A. W. Hoffmann, *Ann. Chem.* **115,** 271 (1860).
4. E. Baumann, *Ann. Chem.* **163,** 312 (1872).
5. I. I. Ostromysslenski, *J. Russ. Phys. Chem. Soc.* **44,** 204, 240 (1912).
6. D. Edelson and R. M. Fuoss, *J. Am. Chem. Soc.* **71,** 3548 (1949).
7. G. S. Marvel, J. H. Sample, and M. F. Roy, *J. Am. Chem. Soc.* **61,** 3241 (1939).
8. L. O. Brockway, J. Y. Beach, and L. Pauling, *J. Am. Chem. Soc.* **57,** 2693 (1935).
9. L. E. Sutton, *Trans. Faraday Soc.* **30,** 789 (1934).
10. L. O. Brockway and K. J. Palmer, *J. Am. Chem. Soc.* **59,** 2181 (1937).
11. J. A. Hugill, I. E. Coop, and L. E. Sutton, *Trans. Faraday Soc.* **34,** 1518 (1938).
12. P. Torkington and H. W. Thompson, *J. Chem. Soc.* **1944,** 303, 597; *Proc. Roy. Soc. (London)* **184,** 31 (1945).
13. R. Dreisbach, *Advances in Chemistry Series No. 22,* American Chemical Society, Washington, D.C., 1959, p. 406.
14. M. Lecat, *Rec. Trav. Chim.* **46,** 240 (1927).
15. J. Timmermans, *Bull. Soc. Chim. Belges* **27,** 334 (1913).
16. A. Guyer, H. Schutze, and M. Weidenmann, *Helv. Chim. Acta* **20,** 936 (1937).
17. A. Juvala, *Chem. Ber.* **63,** 1989 (1930).
18. A. Amschutz, *Ann. Chem.* **221,** 141 (1883).
19. M. Hemptinne, *Trans. Faraday Soc.* **42,** 5 (1946).
20. M. Kharasch and C. Fuchs, *J. Am. Chem. Soc.* **65,** 604 (1943).
21. H. Gilman and A. Haubein, *J. Am. Chem. Soc.* **67,** 1420 (1945).
22. G. Ingold and S. Anantakrishman, *J. Chem. Soc.* **1935,** 1396.
23. L. Schmerling, *J. Am. Chem. Soc.* **71,** 698 (1949).
24. M. S. Kharasch, M. C. McNale, and F. R. Mayo, *J. Am. Chem. Soc.* **55,** 2521 (1933).
25. A. Berruoulli and W. Kambli, *Helv. Chim. Acta* **16,** 1187 (1938).
26. M. Kryszewski, *Roczniki Chem.* **26,** 350 (1952); **28,** 8634-b (1954); *Chem. Abstr.* **48,** 86346 (1954).
27. N. E. Robul, *Jahresber.,* 609 (1861).
28. R. Flid, V. Mironov, B. Ostrovskaya, and N. Avonova, *Zh. Fiz. Khim.* **33,** 119 (1959); *Chem. Abstr.* **53,** 12805 (1959).
29. R. Flid, *Izv. Vysshikh Uchebn. Zavedenii Khim. i Khim. Tekhnol.* **2,** 946 (1959); *Chem. Abstr.* **54,** 10478 (1960).
30. W. Jones and R. Barker, Brit. Pat. 628,731 (1949).
31. F. Schouteden and R. Tritsmans (to Gevaert Photo-Producten), U.S. Pat. 2,566,146 (1951).
32. Monsanto, Brit. Pat. 600,785 (1948).
33. T. Boyd (to Monsanto), U.S. Pat. 2,446,124 (1948).
34. W. Bauer (to Rohm and Haas), U.S. Pat. 1,414,852 (1922).
35. A. Cherniovsky (to N. V. de Bataafsche Petroleum Maatschappij), Brit. Pat. 635,013 (1950).
36. G. Schafer (to United Kingdom Chem. Ltd.), Brit. Pat. 670,523 (1952).
37. Escher. Wyss. Maschinen Fabriken A.G., Swiss Pat. 159,148 (1933).
38. Shell International, Fr. Pat. 1,441,233 (1966).
39. S. Rosenberg and A. Gibbson, *J. Am. Chem. Soc.* **70,** 2138 (1957).
40. S. Gelblum (to Dominion Tar and Chem. Co.), Brit. Pat. 774,125 (1957).
41. F. Hoffmann, *J. Org. Chem.* **14,** 105 (1949); **15,** 425 (1950).
42. A. Titov and F. Maklyaev, *Zh. Obshch. Khim.* **24,** 1631 (1954); *Chem. Abstr.* **50,** 12341 (1955).
43. H. Staudinger, M. Brunner, and W. Feisst, *Helv. Chim. Acta* **13,** 805 (1930).
44. M. Kryszewski, *Roczniki Chem.* **31,** 147, 893 (1957).
45. M. Kryszewski and J. Mandral, *J. Polymer Sci.* **29,** 103 (1958); *Roczniki Chem.* **29,** 567 (1955); *Chem. Abstr.* **50,** 1360 (1955).
46. A. Guyer and A. Schutze, *Helv. Chim. Acta* **17,** 1544 (1934).
47. G. Burnett, F. Ross, and J. Hay, *J. Polymer Sci.* [*B*] **5,** 271 (1967).
48. C. S. Marvel and E. Riddle, *J. Am. Chem. Soc.* **62,** 2666 (1940).
49. G. Talamini and G. Vidotto, *Makromol. Chem.* **100,** 45 (1967).
50. S. Narita, S. Ichinohe, and S. Enomoto, *J. Polymer Sci.* **37,** 273 (1959).

51. W. Scott and R. B. Seymour (to Wingfoot Corp.), U.S. Pat. 2,361,504 (1944).
52. F. Mayo, C. Walling, F. Lewis, and W. Hulse, *J. Am. Chem. Soc.* **70,** 1523 (1948).
53. P. Volans, *Plastics Inst. (London) Trans. J., Conf. Suppl.* **2,** 47 (1967); *Chem. Abstr.* **67,** 23710 (1967).
54. G. Blouer and L. Goldstein, *J. Polymer Sci.* **25,** 19 (1957).
55. F. Welch (to Union Carbide Co.), U.S. Pat. 3,321,417 (1967).
56. Sumitomo Chem. Co., Fr. Pat. 1,458,331 (1966).
57. I. M. Kolthoff and W. J. Dale, *J. Am. Chem. Soc.* **67,** 1672 (1945).
58. A. Harris, *Ann. Chem.* **395,** 216 (1913).
59. I. Ostromysslenski, Ger. Pat. 264,123 (1913).
60. M. Frata, C. Vidotto, and G. Talamini, *Chim. Ind. (Milan)* **48,** 42 (1966).
61. K. C. Ramey, D. C. Lini, and W. B. Wise, *J. Polymer Sci.* [B] **6,** 523 (1968).
62. C. R. Noller and R. Dinsmore, *J. Am. Chem. Soc.* **54,** 1025 (1932).
63. T. L. Davis and R. Heggie, *J. Org. Chem.* **2,** 470 (1937).

Kermit C. Ramey and David C. Lini
ARCO Chemical Company

VINYL BUTYL ETHER POLYMERS. See VINYL ETHER POLYMERS

VINYL BUTYRAL POLYMERS. See VINYL ALCOHOL POLYMERS

VINYLCARBAZOLE POLYMERS

This article deals with the polymers of vinylcarbazole and especially of 9-vinylcarbazole, or *N*-vinylcarbazole (**1**).

Preparation of Monomers

The preparative procedure differs depending upon the position of the vinyl group attached to the carbazole ring. The preparation of the 9-vinyl compounds has been more extensively investigated than that of the 3-vinyl compounds. Synthesis and properties of various vinylcarbazole derivatives are given in Table 1.

Reppe Synthesis. Direct vinylation of carbazole, which is readily available as a by-product from coal-tar distillation, is a practically important method (2–8). The reaction (eq. 1) was originally conducted at about 190°C with an acetylene pressure of 10–20 atm using a catalyst comprising an alkali metal hydroxide together with an

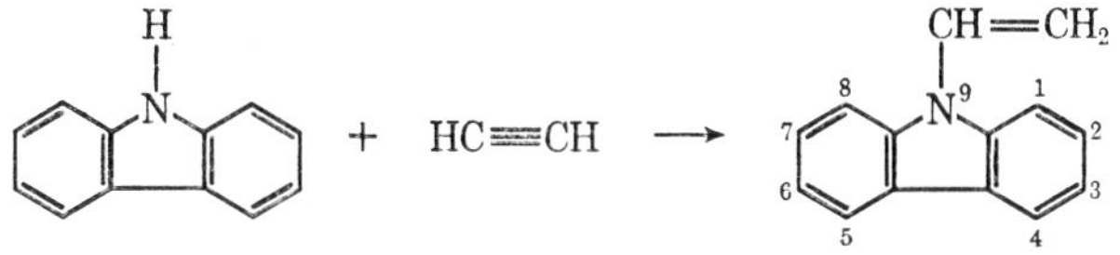

$$\text{carbazole (N–H)} + HC\equiv CH \longrightarrow \text{9-}(CH{=}CH_2)\text{carbazole} \qquad (1)$$

(**1**)

alkaline earth or zinc oxide activator (2,3). Lower reaction temperatures were later found to be favorable. Thus, the yield of 9-vinylcarbazole is improved to 95–99% for the vinylation at 145°C and 170 psi with cyclohexane as diluent (4).

Table 1. Synthesis and Properties of Vinylcarbazoles

Carbazole derivative	Bp, °C/mm Hg	Mp, °C	Method of synthesis[a]	Yield, %	Ref.
9-vinyl	170–180/15 140/1	64–67[c]	I, II, IV	max 99	1,4,8
9-vinyltetrahydro	125–130/0.5		I		1
3-amino-9-vinyl		105–106[d]	IV	66	26
3-chloro-9-vinyl		68–69[c]	IV	75	16,25
3,6-dibromo-9-vinyl		74–75[e]	IV	74	26
3,6-dichloro-9-vinyl		136[f]	IV	82	25
3-dimethylamino-9-vinyl	220–224/3		IV	55	26
3-methylamino-9-vinyl		95–96[d]	IV	73	26
3-nitro-9-vinyl		112–113[g]	IV	46	16,26
9-allyl		54–55[h]	V	58.0	27
9-Δ^3-butenyl		43.5–45[c]	[b]	76.5	27
9-Δ^4-pentenyl		46.6–47[c]	V	68	27
9-Δ^5-hexenyl		76–76.5[c]	V	56.5	27
3-vinyl-9-methyl	192–194/4–5	70–71[c]	III	50	18,19
3-vinyl-9-ethyl	180–183/2–3	66–67[c]	III	65	18,19
3-vinyl-9-propyl	198–200/5–6		III	70	18
3-vinyl-9-isopropyl	205–207/8–9	69–70	III	76.6	18
3-vinyl-9-butyl	206–209/5–6		III	46.0	18
3-vinyl-9-amyl	203–204/5–6		III	52.5	18
3-vinyl-9-isoamyl	213–215/5–6		III	53.5	18
3,6-divinyl-9-ethyl	195–198/2–3		III	10.0	17
3-vinyl-6-chloro-9-ethyl					22
3-vinyl-6-nitro-9-ethyl					22

[a] Key: I, Reppe synthesis; II, dehydration of a 2-hydroxyethylcarbazole; III, dehydration of ring-substituted 1-hydroxyalkylcarbazole; IV, dehydrochlorination of a 2-chloroalkylcarbazole.

[b] Carbazole → 9-(4-bromo(or chloro)butyl)carbazole → 9-(4-*N*,*N*-dimethylaminobutyl)-carbazole → amine oxide and pyrolysis.

[c] In methanol.

[d] In ligroin.

[e] In *n*-hexane.

[f] In acetone.

[g] In ligroin and ethanol.

[h] Sublimation.

The rate of vinylation is expressed by equation 2 (9). Direct vinylation with ethylene using Group VIII metal halide as catalyst has also been reported (10).

$$\text{rate} = k[C_2H_2][\text{carbazole}][\text{alkali}]^2 \tag{2}$$

Dehydration of 9-(2-Hydroxyethyl)carbazole. This process has been proposed for use on an industrial scale. Hydroxyethylation of carbazole is carried out by the reaction of potassium carbazole (carbazole plus potassium hydroxide) (eq. 3) with either ethylene oxide (95% yield) (11,12) shown in equation 4 or ethylene chloro-

$$\text{carbazole (N–H)} + \text{KOH} \longrightarrow \text{carbazole (N–K)} + H_2O \tag{3}$$

hydrin (about 40% yield) (13) shown in equation 5 in a ketone (12,13) or xylene (14,15) medium. 9-(2-Hydroxyethyl)carbazole (2) (mp, 83–83.5°C; bp, 234°C/

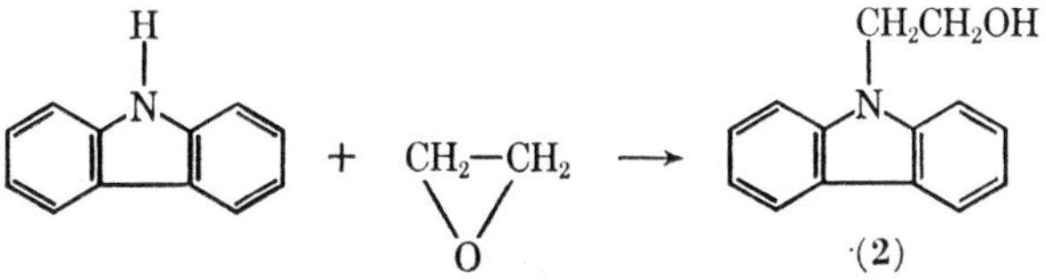

$$\text{carbazole (N–H)} + \underset{\text{O}}{CH_2{-}CH_2} \longrightarrow \text{9-}(CH_2CH_2OH)\text{carbazole } (2) \quad (4)$$

$$(2) \xrightarrow{\text{KOH}} (1) \quad (5)$$

10 mm Hg) is dehydrated by gradual heating with potassium hydroxide and the product distilled at 190–200°C/10 mm Hg (75–76% yield). The use of alumina or $KHSO_4$ as dehydrating agent is disadvantageous since they both catalyze the polymerization of 9-vinylcarbazole, and the yield of this compound decreases (12).

Dehydration of 1-Hydroxyethylcarbazoles. Compound (4), obtainable by Meerwein-Ponndorf reduction of the corresponding ketone (3) (19,55) (eq. 6) can be dehydrated by sodium hydroxide to give the corresponding vinyl compound (50–70%

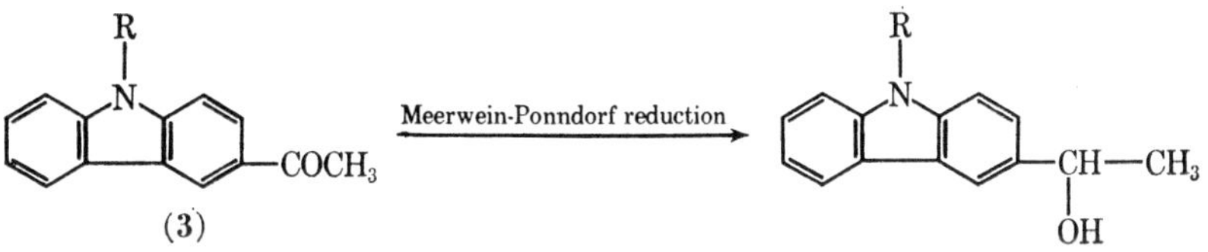

$$\text{9-R-carbazole-}COCH_3\ (3) \xrightarrow{\text{Meerwein-Ponndorf reduction}} \text{9-R-carbazole-}CH(OH){-}CH_3\ (4) \quad (6)$$

yield). Dehydration over alumina or $KHSO_4$ leads to polymer formation (20). 3,6-Divinyl-9-alkylcarbazole (17,21), 3-vinyl-6-chloro-9-ethylcarbazole, and 3-vinyl-6-nitro-9-ethylcarbazole (22) have been prepared in a similar way.

Dehydrochlorination of 9-(2-Chloroethyl)carbazole. This is a convenient laboratory process. 9-(2-Chloroethyl)carbazole is obtained by chloroethylation of carbazole with β-chloroethyl *p*-toluenesulfonate or by chlorination of the hydroxyethyl compound (23–25). The chloroethyl compound is readily converted to 9-vinylcarbazole by heating with alkali metal hydroxide. A variety of 3-substituted or 3,6-disubstituted 9-vinylcarbazoles have been prepared by this method. Substituents are introduced at the stage of chloroethylcarbazole (26).

Reaction of an Alkenyl Halide with a Carbazole Salt. This process is suitable for the preparation of higher homologs of 9-vinylcarbazole such as 9-allylcarbazole. Reaction of the sodium salt of carbazole with allyl bromide (27) was reported to be more convenient than the reaction of potassium salt of carbazole with allyl iodide (28). Production of 9-vinylcarbazole by condensation of the potassium salt of carbazole with vinyl chloride has been claimed (29).

Purification of 9-Vinylcarbazole. Carbazole, as the starting material of vinylcarbazole synthesis, often contains a high level of sulfur compounds. Complete elimination of these impurities is difficult although it is not impossible by repeated recrystallizations of 9-vinylcarbazole at a sacrifice in yield. Various purification methods have been surveyed with moderate success (8) and the use of sulfurfree carbazole is recommended. The inhibitory effect of impurities has been discussed for free-radical polymerization. Impurities not only retard but also accelerate the polymerization. The insoluble materials after extracting crude 9-vinylcarbazole prepared by the Reppe process with ether were shown to be good catalysts for polymerization (30). The action of oxygen during purification and storage of 9-vinylcarbazole seem to be responsible for the enhanced rate of thermal polymerization. For example, 9-vinylcarbazole recrystallized in an inert atmosphere polymerizes slowly at 120°C whereas the monomer recrystallized in air polymerizes within 5 minutes at 120°C (30). Little is known about the impurity effect on cationic polymerization. As judged from the extreme sensitivity of 9-vinylcarbazole to electrophilic attack, cationic polymerization would not be subject to inhibition by impurities such as sulfur-containing heterocycles or higher aromatic homologs.

Examples of purification are (a) treatment of crude monomer with Raney nickel and distillation or recrystallization (8); (b) heating with a free-radical initiator (azo compound) and recrystallization from methanol (31); (c) washing with methanol at 10°C after treating the crude monomer with activated charcoal and a small amount of benzene (32).

The purified monomer is thermally stable at room temperature for over a year.

Polymerization

Polymerization studies reported in the literature are almost exclusively for 9-vinylcarbazole. Very few, and only qualitative, data are available for substituted 9-vinylcarbazole (80) and 3-vinylcarbazole (109). 9-Vinylcarbazole is readily polymerized either by a free-radical mechanism or by a cationic mechanism, but not by an anionic mechanism. A characteristic feature of the polymerization is the initiation by electron-accepting compounds that have not been recognized as polymerization initiators for other vinyl compounds.

Free-Radical Polymerization. The rate of free-radical polymerization of 9-vinylcarbazole is very much higher than that of styrene. The overall rate constant of polymerization at 70°C in cyclohexanone, initiated by 0.01 mole of azobisisobutyronitrile, is determined by equation 7, where M_0 and M are monomer concentration at 0

$$k = [(2.303/t) \log M_0/M]$$

and t reaction time, respectively (33). The values of k are $94.5 \pm 2.4\ sec^{-1}$ for 9-vinylcarbazole and $19.3 \pm 3.5\ sec^{-1}$ for 3-vinyl-9-methylcarbazole, respectively. These rates of polymerization are quite large compared with styrene ($k = 9.4 \pm 0.4$). The heat of polymerization is 15.2 ± 0.3 kcal/mole (34).

Absolute rate constants were measured by means of a rotating sector (61). The value of k_p is $1.2 \times 10^6 \exp(-6900/RT)$, but the Arrhenius plots of k_t do not fall on a straight line. The values of k_t between 10 and -30°C are 30.6×10^4 (10°C), 11.0×10^4 (-10°C), and 0.92×10^4 (-30°C) liter-mole^{-1}-sec^{-1}. The rapid fall of k_t with decreasing temperature would seem to indicate the importance of a diffusion. The molecular shape of poly(9-vinylcarbazole) is a stiff rod, since the viscosity is independent of temperature below -30°C. The stiffness of the polymer would disturb diffusion of growing chains, resulting in a decrease in k_t.

Suspension (35–37) and emulsion polymerizations (38) have also been practiced. The di-*tert*-butyl peroxide, azobisisobutyronitrile, or a combination of both may be used as initiator. The polymerization is retarded and the molecular weight of the polymer lowered by amounts of anthracene comparable to or somewhat larger than the initiator concentration, but these effects show a saturation at higher anthracene concentrations. On the other hand, phenanthrene is a weak inhibitor which brings about the complete inhibition of polymerization at sufficiently high levels (39). When the amount of sulfur in monomer is over about 100 ppm, polymerization initiated by azobisisobutyronitrile (0.2 wt % of monomer at 100°C) is seriously interrupted (8).

Cationic Polymerization. Vinylcarbazole can be polymerized cationically even by the weakest initiators. The electron-releasing effect of nitrogen in the carbazole ring induces high nucleophilicity on the vinyl group. Furthermore, the cationic propagating species is stabilized by conjugation with the carbazole group. Consequently, the generally accepted concept that basic solvents inhibit cationic polymerization cannot be applied to the 9-vinylcarbazole. Even water does not inhibit the polymerization and acetone, ethyl acetate, tetrahydrofuran, and other basic solvents can be used (40,41). Protonic acids are generally not good cationic initiators for vinyl polymerization. However, 9-vinylcarbazole polymerizes readily in the presence of trace amounts of protonic acids, such as trichloroacetic acid (42); the molecular weight of polymer is low in comparison with that of polymer prepared by aprotic catalysts (43).

Besides the conventional cationic initiators, many other acidic compounds initiate polymerization of 9-vinylcarbazole, including oxidizing metal salts (41,44–47), perchlorates (47), rhenium pentachloride, molybdenum pentachloride (48), boron trifluoride chelated with 2-methyl-4,6-nonanedione (49), and stable carbonium or oxonium ion salts such as triphenylmethyl hexafluorophosphate and tropylium hexachloroantimonate (50–52). Many of them are suspected of inducing electron-transfer initiation and this problem, which is so closely connected to the chemistry of carbazoles, will be discussed later.

Reaction with Anionic Initiators. Attempts to polymerize 9-vinylcarbazole with anionic catalysts, such as sodium metal *n*-butyllithium or "living" polystyrene, were unsuccessful (53). Reduction of 9-vinylcarbazole on a sodium surface by one-electron transfer is observed by esr spectroscopy, but the anion radical is not capable of initiating propagation reaction of 9-vinylcarbazole. However, anionic polymerization of styrene could be initiated by the anion radical. The anion radical is unstable and eventually decomposes to carbazole and ethylene.

Polymerization with Organometallic Catalysts. There has been considerable discussion on the nature of Ziegler-Natta catalysts when used with 9-vinylcarbazole. An early report (54) indicated that coordinate polymerization of 9-vinylcarbazole was induced by *n*-butyllithium–titanium tetrachloride ratio (Li/Ti = 1.5–1.7) at 30°C in ligroin and that the stereospecificity ranged between 60 and 80%. Later, the reproducibility of the previous results was suspected (27). Since even weak Lewis acids such as titanium trichloride are effective catalysts for the polymerization of 9-vinylcarbazole, the claimed coordinate polymerization might, in fact, be a cationic polymerization. Studies of the polymerization of higher homologs of 9-vinylcarbazole with the Ziegler-Natta type catalysts provide indirect evidence against coordinate polymerization. 9-Allylcarbazole and 9-Δ^3-butenylcarbazole scarcely polymerize. Since a Ziegler-Natta catalyst does not require conjugation of a vinyl group with an electron-donor system, 9-allyl and 9-Δ^3-butenylcarbazole should be polymerized by the catalyst, regardless of the conjugation between the carbazole group and the olefinic double bond. Consequently, the combinations of titanium tetrachloride–triisobutylaluminum, titanium tetrachloride–triethylaluminum, titanium trichloride–triisobutylaluminum, and titanium tetrachloride–*n*-butyllithium behave essentially the same as the metal halides by themselves (56).

Stereospecific polymerization of 9-vinylcarbazole, however, seems to be possible using MR_mX_n (M = metal, R = organic radical, X = halogen), such as $(C_2H_5)_2AlCl$ (57). The crystallinity of polymer, as determined by x-ray diffraction analysis, is between 35 and 50%. The polymer is stable at 250°C and the reproducibility of the polymerization was confirmed (27).

Other examples of organometallic catalyst are $(C_2H_5)_2AlCl$ (0.01 mole) plus $VOCl_3$ (0.003 mole) in toluene at −20°C (58); $C_2H_5AlCl_2$, $(C_2H_5)_2AlCl$, $Ti(OC_4H_9)_2Cl_2$, $Ti(OAc)_2Cl_2$, $Fe(OAc)_2Cl$, or $Sn(OAc)_2Cl_2$ in toluene at −78 to +80°C (59); titanium tetrachloride absorbed on silica plus triisobutylaluminum in toluene (60).

9-Δ^4-Pentenylcarbazole and 9-Δ^5-hexenylcarbazole are polymerized to crystalline polymers by the titanium trichloride–$(C_2H_5)_2AlCl$ catalyst (27). The polymerization is believed to be one of coordinate polymerization.

Solid-State Polymerization. Solid-state polymerization of 9-vinylcarbazole may be carried out by the action of high-energy radiation (62–67,78), cationic catalysts (in some cases, electron acceptors) (68–75), and free-radical initiators (76,77). An induction period was observed in radiation-induced solid-state polymerization (62). The mechanism of propagation (eq. 8) has been suggested to be free-radical in nature.

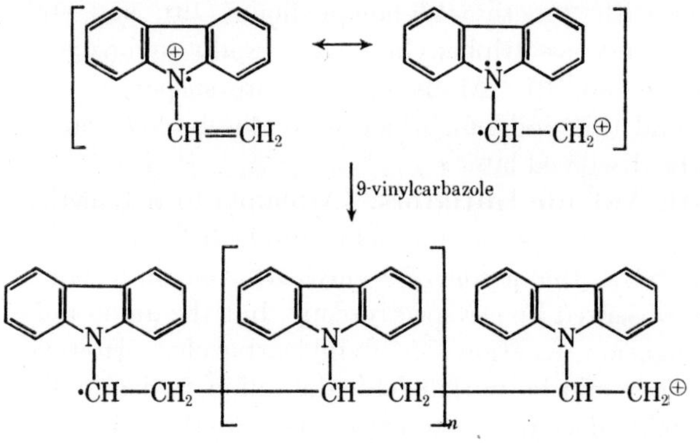

(8)

Electron-spin resonance study of irradiated solid 9-vinylcarbazole indicates that the initial formation of a cation radical which is trapped at 77°K (78). On warming up to 90°K, the ion radicals presumably initiate polymerization to an oligomeric state. The carbonium end group is neutralized at this stage and the remaining neutral radicals are the likely cause of post-polymerization (62). The size of the monomer crystals is a determining factor in the rate of solid-state polymerization; large crystals polymerize faster than small ones (65,67). The direction of propagation is considered to coincide with the *z* axis of the monomer crystals (66).

Solid-state cationic polymerization is mostly initiated by gaseous acidic compounds. Examined initiators are $AlBr_3$, $AlCl_3$, $SnCl_4$, BF_3, Cl_2, Br_2, I_2 (68), HCl, Cl_2, N_2O_4 (69); I_2, Cl_2, SO_2, (70,72,73,79); and BX_3 (X = F, Cl, Br) (71). There are some contradictory results with BF_3 gas; one report claims a polymerization (74), whereas another describes its inertness as an initiator (69). Peculiar features of cationic solid-state polymerization are that the molecular weight of the polymer is always low and independent of the nature of the catalyst (68). The effectiveness of a catalyst does not necessarily run parallel to its catalytic activity in homogeneous liquid phase. An important factor seems to be the size of the catalyst molecule.

Thus, HBr and NOCl are inert since they are not absorbed onto 9-vinylcarbazole (69).

Contrary to the general trend of solid-state vinyl polymerization, which produces only amorphous polymer, poly(9-vinylcarbazole) prepared by solid-state polymerization is partially crystalline (74). The crystallinity disappears by reprecipitation, however. Spontaneous thermal polymerization of 3,6-dibromo-9-vinylcarbazole also leads to the formation of crystalline polymer (80) which, likewise, is converted to amorphous polymer by reprecipitation.

Solid-state free-radical polymerization is induced by azobisisobutyronitrile impregnated in monomer crystals (77) or by the action of a water-soluble redox initiator system (ammonium peroxydisulfate plus sodium hydrogen sulfite) on monomer crystals suspended in water (76). In the latter case, the molecular weight of the polymer is much higher than in a cationic system and increases with temperature and catalyst concentration. The polymer thus obtained is also crystalline, as observed by means of a polarizing microscope.

Polymerization via Charge-Transfer Interaction. Since active monomers in cationic polymerization are good electron donors having low ionization potential, such monomers are capable of interacting with various electron acceptors. In the extreme case, the donor–acceptor interaction leads to the dissociation of the pair to a cation radical and an anion radical as a result of complete electron transfer. Then it is obviously possible that the cation radical could initiate polymerization either via a cationic or a free-radical path. When strong acceptors, such as iodine, are added to 9-vinylcarbazole, definitive evidence of free-spin generation is obtained by esr measurements (70). Consequently, the polymerization of 9-vinylcarbazole by strong electron transfer may be elucidated in terms of cation-radical formation. As will be discussed later, even in those presumably clear-cut systems, there still remains a serious problem of verifying the identity of the paramagnetic species and the actual initiating species. Furthermore, still more complicated phenomena are the initiation of polymerization of 9-vinylcarbazole by very weak acceptors such as acrylonitrile and methyl methacrylate (81).

Total ionization of these donor–acceptor pairs is energetically impossible and as an alternative explanation, the sophisticated concept of polymerization via mesomeric polarization (eq. 9) has been proposed (81). According to this mechanism, the propagating step should also be subject to the control of acceptor. However,

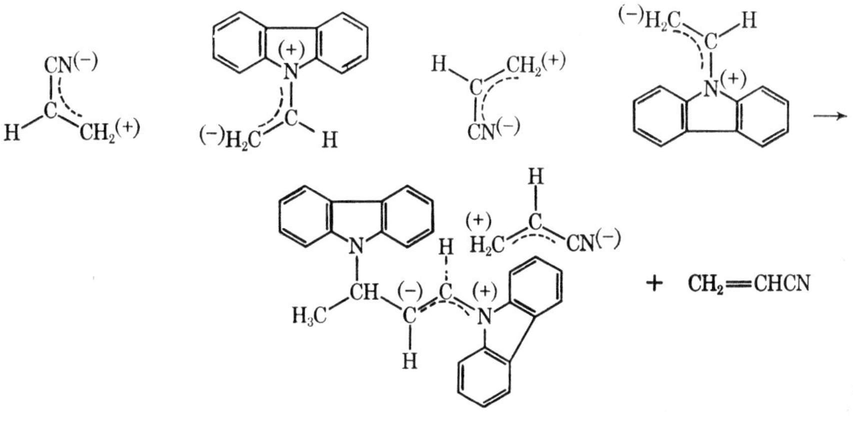

$$+ \; CH_2{=}CHCN \qquad (9)$$

Table 2. Polymerization of 9-Vinylcarbazole by Unconventional Initiators

Initiator	Remarks	Ref.
p-chloranil, 1,3,5-trinitrobenzene, tetranitromethane, chloro-2,4-dinitrobenzene, maleic anhydride, trichloroethylene, tetrachloroethylene, acrylonitrile, methyl methacrylate, carbon tetrachloride, ethyl cyanoacetate, cyclopentadiene	polymerization by mesomeric polarization is suggested	81, 98
p-chloranil, *o*-chloranil, *p*-bromanil, *p*-iodoanil, 2,3-dichloro-5,6-dicyanobenzoquinone, 1,4,-5,8-tetrachloroanthraquinone, 7,7,8,8-tetracyanoquinodimethane, tetracyanoethylene, chlorine, bromine, iodine, nitrogen dioxide, trichloroacetonitrile, pyromellitic anhydride, *p*-benzoquinone, the bromide salts of the Wurster ions of 1,4-diaminodurene and 1,6-diaminopyrene	cation-radical initiation is suggested	99
carbon tetrachloride	free-radical polymerization	62
	cationic polymerization	88
	initiation by chlorine?	90, 100
tetranitromethane	charge-transfer initiation is suggested; kinetic study	101
	initiation via nitroform is suggested	102
chloranil[a]	$R_p = K\,k_p[\mathrm{A}]_0[\mathrm{D}]_0^2$ A: acceptor, D: monomer	103
acrylonitrile	photopolymerization, radical + cation	82, 83
carbon tetrabromide	photopolymerization (very rapid)	114, 115
nitrobenzene	photopolymerization; cationic	45
maleic anhydride	a esr signal is observed; cationic	104
mixture of maleic anhydride and 3,3-bis(chloromethyl)oxetane	mixture is irradiated with γ rays and is mixed with monomer; cationic	105
phenolic phosphate or phosphite	acceleration by aromatic thiols	106
pyrylium salts, triphenylmethyl cation (anion: ClO_4^-, $SbCl_6^-$, BF_4^-)	charge-transfer initiation[b] or initiation by simple addition of carbonium ion	107, 108
sulfur dioxide, chlorine, bromine, iodine	esr signal is observed	70, 72, 73
	solid state	79, 110
chlorine, N_2O_4	solid state	69
sulfur dioxide	liquid phase	111, 120
sulfur dioxide + oxygen	instantaneous polymerization[c] at −75°C	112
$Fe(NO_3)_3$, $Cu(NO_3)_2$, $Ce(NH_4)_2(NO_3)_6$		
in methanol	polymerization (doubtful results)	113
in tetrahydrofuran, dimethylformamide, acetone, ethylene dichloride	polymerization	41, 44
$Fe(NO_3)_3$ in methanol	dimerization to *trans*-1,2-dicarbazylcyclobutane[d]; the residual monomer is recovered as 9-(1-methoxyethyl)-carbazole	91, 92
$Hg(CN)_2$ in acetonitrile	electrolytically induced polymerization, polymer + *trans*-1,2-dicarbazylcyclobutane	93
vanadate (Na^+, NH_4^+, Ag^+, Pb^{2+}, Co^{2+}, Cu^{2+})	cationic	116
diethylphosphoric acid	R_p = [catalyst] [monomer]^2/[H_2O] in ethylene dichloride	117
alkaline Na_2CrO_4	aqueous dispersion polymerization via complex?	118, 119

Table 2 (*continued*)

Initiator	Remarks	Ref.
$LiClO_4.H_2O$, $Mg(ClO_4)_2.H_2O$, $Ba(ClO_4)_2.H_2O$	in polar solvent, R_p = [catalyst] [monomer]^0	94,95 96
$AgNO_3$, $AgClO_4$	for $AgClO_4$, explosive polymerization in bulk	96
$AgClO_4$	in dilute benzene solution, the polymerization is photosensitive	121
LiCl, LiBr, LiI	catalytic activity: LiCl < LiBr < LiI; polymerization via mesomeric polarization suggested[e]	122
$NaAuCl_4.2H_2O$	R_p = [Au(III)] [monomer] in nitrobenzene; electron-transfer initiation suggested[f]	46
	extremely photosensitive	45,97,130

[a] A recent study indicates that a small amount of 2-hydroxy-3,5,6-trichloro-1,4-benzoquinone contained in chloranil is an effective initiator because of the high acidity of the phenolic hydroxyl group. This impurity cannot be removed by recrystallization or sublimation. Consequently, earlier works on polymerization by chloranil should be reexamined. When the impurity is removed by treating chloranil with alkaline substance, polymerization of 9-vinylcarbazole by the purified chloranil is very slow (123).

[b] Tropylium salts are known to form charge-transfer complexes with aromatic molecules.

[c] Proposed mechanism:

$$\text{9-vinylcarbazole} + SO_2 \rightleftharpoons \text{9-VCZ}\cdot^{\oplus} + SO_2\cdot^{\ominus}$$

$$\text{9-VCZ}\cdot^{\oplus} \longrightarrow \text{polymer} \qquad SO_2\cdot^{\ominus} \underset{O_2}{\rightleftharpoons} O_2\cdot^{\ominus}(SO_2)\ \text{stabilize}$$

[d] Structure is:

N—□—N (1,2-di-9-carbazolylcyclobutane)

[e] Suggested mechanism:

$$\rangle N{-}CH{=}CH_2 + Li^{\oplus}I^{\ominus} \longrightarrow \rangle \overset{(+)}{N}{=\!=}CH{=\!=}\overset{(-)(\cdot)}{CH_2}\ \cdots Li^{\oplus}\ \cdots I^{\ominus}$$

[f] Suggested electron-transfer initiation:

$$Au(III) \xrightarrow[\text{reducing agent}]{h\nu \text{ or}} Au(II)$$

$$Au(II) + \text{9-vinylcarbazole} \longrightarrow Au(I) + (\text{9-vinylcarbazole})\cdot^{\oplus}$$

In all investigations to date, the propagating species have shown no difference from cationic polymerization, except in polymerization under special conditions of photoirradiation or in the presence of various additives where simultaneous occurrence of radical and cationic propagations has been observed (82,83). The copolymerization behavior of 9-vinylcarbazole with *p*-methoxystyrene initiated by an acceptor (tetracyanoethylene) is almost identical with that initiated by boron trifluoride etherate, a typical cationic initiator (84). The mechanism of initiation is therefore still puzzling. Certainly, a more elaborate technique of detecting small differences in the nature of the propagating species is required, since the propagating species derived from organic donor–acceptor interaction should carry quite different counterions from those found in conventional cationic polymerization. However, there seems to be a considerable overlap between conventional cationic initiation, involving cation attack on the vinyl group, and charge-transfer initiation. For example, iodine is known as a weak cationic initiator, but during the polymerization of 9-vinylcarbazole by iodine, paramagnetic species are observed, indicating that a charge-transfer process is involved (79). The polymerization of 9-vinylcarbazole with nitroparaffins is cationic (85). It is tempting to consider charge-transfer interaction as the cause of initiation. Nevertheless, charge-transfer spectra are not detected and the dissociation of a proton from the nitro compound is the most likely initiation process. When aromatic polynitro compounds are employed (81), protonic dissociation (eq. 10) is not expected and charge-transfer initiation has been

$$CH_3NO_2 \rightleftarrows H^{\oplus} + {}^{\ominus}CH_2NO_2 \qquad (10)$$

suggested. When nitrobenzene is the acceptor as well as the solvent, spontaneous thermal polymerization is scarcely observed. This is quite understandable since nitrobenzene is a weaker acceptor than (polynitro)benzene.

The irradiation of the nitrobenzene–9-vinylcarbazole mixture brings about immediate cationic polymerization (45). Interpretation of the initiation mechanism is presented on the basis of a photosensitized charge-transfer process. The initiation may be caused either by the excitation of the ground-state charge-transfer complex (eq. 11) or by charge-transfer interaction of the excited donor or acceptor with the ground-state acceptor or donor (eq. 12) (86). In these equations D is the donor and A the acceptor.

$$D + A \rightleftarrows [D\text{---}A] \xrightarrow{h\nu} [D^{\oplus}\text{---}A^{\ominus}] \qquad (11)$$

$$D \text{ (or A)} \xrightarrow{h\nu} D^{*}(\text{or } A^{*}) \xrightarrow{A \text{ (or D)}} (D\text{---}A)^{*} \qquad (12)$$

It can readily be seen that charge-transfer phenomena and photochemistry are closely connected. A number of examples of photosensitized charge-transfer polymerization involving 9-vinylcarbazole have been reported (82,83,87). Acrylonitrile may be used as a polymerizible acceptor. Then, the photopolymerization system is all solid and even this requires no catalyst (82,83). The nature of the propagating species should be somewhat different from that of the thermal system. Thermal polymerization of the 9-vinylcarbazole–acrylonitrile system produces only homopolymer of 9-vinylcarbazole (81), possibly by a cationic path, whereas simultaneous occurrence of radical and cationic polymerization was confirmed in photopolymerization.

Halogenated hydrocarbons are also good photosensitizers for the polymerization of 9-vinylcarbazole. Carbon tetrachloride has been a subject of much discussion as to whether it is capable of initiating polymerization of 9-vinylcarbazole (62,88–90). Since the system was found to be very sensitive to light (81), the discrepancy of experimental results by different researchers may be due to photochemical effects. Photopolymerization of 9-vinylcarbazole by halogenated hydrocarbons is generally accompanied with coloration of the polymerization system; this is of interest in image-recording systems.

Metal salts are also good initiators for the polymerization of 9-vinylcarbazole. Oxidizing metal salts initiate polymerization presumably by an electron-transfer mechanism, as evidenced by esr spectroscopy (44,46). Initiation by protonic acids formed from hydrolysis of the metal salts is, however, an alternative possibility. Oxidation of monomer does not always initiate polymerization. When methanol is the solvent for the reaction, cyclic dimer (*trans*-1,2-dicarbazylcyclobutane) is formed (91,92). Polymerization and cyclodimerization may proceed simultaneously (93). Nontransition metal salts such as $LiClO_4.H_2O$, $Mg(ClO_4)_2.H_2O$, and $Ba(ClO_4)_2.H_2O$ are capable of initiating polymerization (94–96). All polymerization systems initiated by metal salts are of cationic nature and photoirradiation greatly accelerates the polymerization in some cases (30,45,97).

Table 3. Copolymerization of 9-Vinylcarbazole (M_1)

M_2	r_1	r_2	Temp, °C	Remarks[a]	Ref.
acrylonitrile	0.04 ± 0.02	0.28 ± 0.02	30		124
	0.05 ± 0.01	0.40 ± 0.10	30		83
allyl chloride	very large	0	70		125
2-chloroallyl chloride	very large	0	70		125
p-chlorostyrene	0.023 ± 0.003	7.0 ± 0.2	30		124
cyanoacetylene	0.075 ± 0.005	0.030 ± 0.005	30		124
2,5-dichlorostyrene	0.016 ± 0.002	8 ± 0.5	70		125
isobutyl vinyl ether	13.2 ± 0.5	2.1 ± 0.1	30	$NaAuCl_4.2H_2O$, dark	130
p-methoxystyrene	21.4 ± 0.81	0.13 ± 0.005	25	BF_3OEt_2, daylight	126
	23.2 ± 1.9	0.09 ± 0.01	0	BF_3OEt_2, daylight	126
	20.1 ± 1.48	0.095 ± 0.01	25	TCNE,[b] dark	126
	15.1 ± 1.2	0.13 ± 0.01	25	TCNE, daylight	126
	14.9 ± 3.2	0.16 ± 0.08	25	no catalyst in ethylene dichloride, daylight	126
methyl acrylate	0.050	0.50	75		127
	0.11 ± 0.02	0.43 ± 0.02	30		124
methyl methacrylate	0.20 ± 0.03	2.0 ± 0.3	70		128
	0.040	2.0	75		127
	0.07 ± 0.01	2.7 ± 0.1	30		124
styrene	0.035	5.7	75		127
	$0.012 \pm 15\%$	$5.5 \pm 15\%$	70		128
	>7	0.01 ± 0.01		$CuCl_2$ in THF	41
vinyl acetate	2.68 ± 0.10	0.126 ± 0.32	65		129
	3.02 ± 0.24	0.152 ± 0.018	100		129
vinyl acetate	3.9 ± 0.2	0.13 ± 0.03	30		124
vinyl butyrate	1.28 ± 0.06	0.059 ± 0.020	100		129
vinyl formate	4.22 ± 0.16	0.196 ± 0.004	100		129
vinylidene chloride	3.7	0.020			129
vinyl propionate	1.68 ± 0.140	0.076 ± 0.018	100		129

[a] Unless specified, conventional free-radical initiators were used.
[b] Tetracyanoethylene.

Table 4. Copolymers of 9-Vinylcarbazole

Comonomer	Conditions	Remark	Ref.
formaldehyde	Friedel-Crafts catalysts, 0–75°C		131,132
trioxane	Lewis acids; 30°C $< T <$ mp of trioxane	thermally stable polyoxymethylene	133
ethylene	$(C_2H_5)_3Al$–$TiCl_4$, room temp; ethylene/9-VCZ > 100/5 in feed; azobisisobutyronitrile, 110°C	good high-temperature resistance, good low-temperature flexibility, Carbathene	134,135, 138
ethylene–propylene, 1:2	organometallic catalyst, $(C_2H_5)_2AlCl$–$COCl_3$; −20°C		136
ethylene (85–99%)–acrylamide		the product is blended with polypropylene fiber to improve dyeability	137
vinyl thiolacetate	azobisisobutyronitrile		139
vinyl chloride			140
triphenylethylene	suspension polymerization	electrostatic capacitor	141
trivinyltrichlorobenzene			142
p-vinyldiphenyl sulfide	comonomer 75 parts, benzoyl peroxide		143
2-vinyldibenzofuran			144
p-fluorostyrene or 3,4-difluorostyrene		electrical insulator	145
vinyl chloride–1,4,5,6,7,7-hexachloro-5-norbornene-2,3-dicarboxylic acid	vinyl chloride > 80% 9-vinylcarbazole: 0.5–10%	heat-stable poly(vinyl chloride)	146

Polymerization reactions of 9-vinylcarbazole initiated by unconventional catalysts are recorded in Table 2.

Copolymerization. 9-Vinylcarbazole has a very large and negative *e* value (−1.40) as the result of conjugation (eq. 13). The cationic polymerizability of 9-

$$\left[\rangle N{-}CH{=}CH_2 \longleftrightarrow \rangle \overset{\oplus}{N}{=}CH{-}\overset{\ominus}{C}H_2 \right] \qquad (13)$$

vinylcarbazole is so high that its cationic copolymerization with all other monomers examined up to now yields an extremely large value of the reactivity ratio (Table 3).

The monomer reactivity ratios in cationic polymerization or in charge-transfer polymerization seem to depend upon the reaction conditions and the kind of initiator. In particular, the charge-transfer polymerization which is essentially cationic is influenced by photoirradiation. The differences in the values found for *r* for the copolymerization of 9-vinylcarbazole with *p*-methoxystyrene (126) are apparently beyond experimental error. It may be suggested that the photoeffect is to excite growing active species, or monomer, prior to the propagation reaction, or to produce active species of a different kind.

Although the monomer reactivity ratios are not given, various copolymerization systems are also found in patent literatures as shown in Table 4.

Little is known about vinylcarbazoles other than 9-vinylcarbazole. Copolymers of 3-vinyl-9-alkylcarbazole with divinylbenzene and other heterocyclic nitrogen-containing monomers have been reported (147).

Properties of Poly(9-vinylcarbazole)

Viscosity–Molecular Weight Relation. Several viscosity equations have been reported. For poly(9-vinylcarbazole), Ueberreiter and Springer (148) used equation 14 for molecular weights between 10^4 and 3×10^6 at 25°C in benzene solution.

$$[\eta] = (3.35 \times 10^{-2})\, \bar{M}_n^{0.58} \tag{14}$$

A recent study of poly(9-vinylcarbazole) under theta conditions (37 ± 1°C in toluene) provided equation 15 for the molecular-weight range between 4×10^4 and 2.30×10^6 (149).

$$[\eta]_\Theta = (7.62 \times 10^{-2})\, \bar{M}_n^{0.50} \quad \text{(in dl/g)} \tag{15}$$

The dimensions of the molecular chain in the ideal state $(\langle L^2\rangle_0/M)^{1/2}$ and the conformational parameter σ were calculated to be 6.33×10^{-11} cm and 2.85, respectively. The value of σ is large due to the repulsion between carbazyl groups. A semiempirical calculation of K_Θ based on specific chain stiffness and molecular weight per main chain atom gave $K_\Theta = 7.00 \times 10^{-2}$, which agreed fairly well with the observed value of K_Θ in equation 15 (150).

Table 5. Mechanical Properties of Poly(9-vinylcarbazole) (153)

Property	Polectron[a]	Luvican[b]
modulus, lb/in.^{2}	5.8×10^5	4.7×10^5
tensile strength, lb/in.^{2}	1,800	2,050
	18,000–20,000 (oriented fibers)	
elongation, %	0.32	
impact strength, ft-lb/in.	0.19	2.2
	0.51–1	
impact bending stress, cm/kg-cm		10–15
flexural strength, lb/in.^{2}	1,500–8,000 (increasing with fiber content)	5,500–6,900
	4,500–5,500 (−20°C to +80°C)	
	8,200–11,000	
hardness, kg/mm^2		14
Rockwell R	125	
Rockwell	113	
compressive strength, lb/in.^{2}		4,800
shear strength, lb/in.^{2}	3,500	
specific heat, cal/°C/g	0.3	
thermal conductivity, cal/cm/sec-°C		6.0×10^{-3}
coefficient of linear expansion, $(°C)^{-1}$	$45–55 \times 10^{-6}$	57×10^{-6} (20–100°C)
heat-distortion temperature (ASTM method), °C	100–150	
heat resistance, °C		
Martens		150
Vicat		190
flow temperature, °C		270
permissible continuous temperature, °C		120
specific gravity	1.2	1.2
refractive index at 20°C		1.69
water absorption	0.1%	12 mg/100 cm^2/week

[a] Trademark, General Aniline and Film Corp.
[b] Trademark, Badische Anilin- und Soda-Fabrik.

Mechanical and Thermal Properties. The characteristic property of poly(9-vinylcarbazole) is its extreme brittleness. The rigid and bulky side group which brings about brittleness is also reflected in the glass-transition temperature, 205°C, which is the highest among known vinyl polymers (151). The T_g increases with molecular weight (152).

A review (153) of the mechanical properties of poly(9-vinylcarbazole) indicates that the mechanical properties measured by different workers agree fairly well. However, the method and conditions of sample preparation for the measurements considerably affect the determined values. Table 5 shows data for Luvican, made in Germany by Badische Anilin- und Soda-Fabrik, and Polectron, made in the United States by General Aniline and Film Corp., both of which have been in quantity production.

Electrical Properties. The electrical properties, including photoconductivity of poly(9-vinylcarbazole), are extremely interesting and unique. The most important application of poly(9-vinylcarbazole) has been as a dielectric material because the changes of loss factors and permittivity are small over wide temperature and frequency ranges. The bulk and surface resistivity values are high in the dark. The electrical properties are shown in Table 6. All data are in relatively good agreement irrespective of the source of polymer.

Table 6. Electric Properties of Poly(9-vinylcarbazole) (153)

Property	Value
dielectric strength, mV/cm	1.1–0.86 over temperature range 25–150°C
	0.4 for samples 2.5 mm thick
	1.2 for samples 0.225 mm thick
resistivity, ohm-cm	$>10^{14}$ at neutron doses up to 5×10^{10} *nvt*[a]
	$8.0–0.5 \times 10^{18}$ over temperature range 25–150°C
	$10^{14}–10^{15}$
loss factor, %	0.04–0.1 at 25°C (1 kcps–100 Mcps)
	0.02–0.06 at 20°C (50 cps)
	0.08 at 20°C (800 cps)
	0.10 at 100°C (800 cps)
	0.16 at 150°C (800 cps)
	0.07 at 800 cps
	0.15 at 1 Mcps
	1 at 210°C
permittivity	3.1 at 20°C (50 cps)
	3.0 between 800 cps and 1 Mcps
arc resistance (ASTM D 495-42)	20 sec for 0.150-in. thick sample

[a] Note: *nvt* implies an integrated neutron flux of *n* neutrons/cm³ with velocity *v* cm/sec over a period of *t* sec.

There are scattered data on the dependence of electrical properties on temperature and frequency of the current. A 10% increase in capacitance over the temperature range 20–140°C was reported, whereas the insulating resistance of 2.000–3.200 megaohms at 60°C dropped to 21–25 megaohms at 120°C. Loss factor values increased from 0.7% at 30°C to 2% at 150°C (154). On the other hand, films by solvent casting were reported to have a constant capacitance up to 170°C and the power factor was almost unaffected up to 200°C, provided traces of solvent were removed (155). A capacitor prepared from poly(9-vinylcarbazole) in solution showed very small changes in capacitance and loss factor over the temperature range −50°C to +60°C at frequencies from 60 cps to 1 Mcps (156).

The photoconductivity of poly(9-vinylcarbazole) and its sensitization have been the interest of many researchers (157–161). Poly(9-vinylcarbazole) is at present the sole photoconductive polymer that has been used for electrophotography. The dark resistivity of poly(9-vinylcarbazole) was measured to be 1.9×10^{15} ohm-cm with an activation energy of 0.00 eV. Under photoirradiation the activation energy was 0.14 eV (161). The maximum photoconductivity was observed by the irradiation at ~360 mμ, which is the long-wavelength limit of absorption of poly(9-vinylcarbazole). The photocurrent was nearly proportional to the light intensity. The measurement of photoconductivity using a sandwich-type cell indicated that the charge carrier was the positive hole since the photocurrent was always greater if the photoirradiation was made on the positively charged electrode. The enhanced photocurrent in the presence of air was interpreted as a result of donor (poly(9-vinylcarbazole)) with acceptor (oxygen) promoting the formation of positive holes. The hole mobility was determined by transient photoconductivity as about 10^{-3} $cm^2/V/sec$ at room temperature (162). The thermal activation energy was 0.12 eV, which agreed fairly well with the temperature dependence of photocurrent determined from steady-state photoconductivity.

The addition of an electron acceptor is an efficient way of sensitization. Quantitative data are scanty, but the hole mobility of the poly(9-vinylcarbazole)–iodine charge-transfer complex has been reported to be about 0.5 $cm^2/V/sec$ (163).

Tetracyanoquinodimethane, tetracyanoethylene, chloranil, tetrabromonaphthodiquinone, and 2,5-dichloro-3,6-dihydroxy-1,4-benzoquinone were tested for their effects on the relative decay constant of charged poly(9-vinylcarbazole). These electron acceptors are strong sensitizers for goth negatively and positively charged poly(9-vinylcarbazole) (159). See also POLYMERS, CONDUCTIVE.

Practical Applications of Poly(9-vinylcarbazole)

Although various derivatives of vinylcarbazole were discussed in the section on Preparation of Monomers, only 9-vinylcarbazole is considered here from the viewpoint of practical applications, since the monomer is too expensive, in comparison with starting materials for general-purpose plastics, and the polymer is too brittle. There is no possibility of employing poly(9-vinylcarbazole) on a large scale. On the other hand, poly(9-vinylcarbazole) has many properties such as high thermal stability, unique dielectric properties, and photoconductivity which are hardly rivaled. Its high refractive index has interested manufacturers of contact lenses. Poly(9-vinylcarbazole) and copolymers containing 9-vinylcarbazole will therefore find their applications in specialized fields.

Capacitors and Insulators. Poly(9-vinylcarbazole) and copolymers with styrene and its derivatives have been most extensively investigated. Incorporation of 0.1–2% of styrene with 9-vinylcarbazole is sufficient to improve the workability of poly(9-vinylcarbazole) while retaining the electrical properties of the base polymer (164,165). As comonomers, divinylbenzene (166), divinyltetrachlorobenzene (167), and trivinyltrichlorobenzene (142) were reported. Radiation-induced polymerization of 9-vinylcarbazole, in the presence of plasticizer such as dibutyl phosphate has been reported. By this procedure, 9-vinylcarbazole alone or mixtures of 9-vinylcarbazole with various unsaturated hydrocarbons were polymerized. The process is useful for encapsulation, by irradiation of the articles when immersed in the molten monomer (168). Mechanical and electrical properties of polymers prepared by this procedure are comparable to ordinary poly(9-vinylcarbazole). The combination of poly(9-vinylcarbazole) with triphenyl phosphate or tritolyl phosphate has good thermal stability and is suitable for electric insulators (169). Poly(9-vinylcarbazole) prepared by heating 9-vinylcarbazole containing 0.01–1% of thio compound (thio-β-naphthol or thiourea) was reported to have a dielectric constant of 2.8, a loss angle of 0.001 at 1 Hz/sec, and a softening temperature of 160–180°C. This material is therefore useful for capacitor dielectrics (170).

Besides direct impregnation or film making, poly(9-vinylcarbazole) can be compounded with other polymeric materials. Although the uses of these compounded materials are not necessarily limited to the electric industry, the main applications are as capacitors and insulators. Poly(tetrafluoroethylene) was impregnated with molten 9-vinylcarbazole under high pressure and at elevated temperature; the monomer was polymerized in situ. The film has good receptivity in metallizing, and its use as dielectric spacer is suggested (171). An insulating sheet composed of poly(9-vinylcarbazole) layers and poly(isobutylene) has been described (172). A paper-based insulator may be prepared by impregnating the base with 9-vinylcarbazole and peroxide. After the impregnated base is dried, the paper is coated with bisphenol epoxide and dimethylaniline and finished by heating (173). The combination of an epoxy resin, phthalic anhydride, poly(9-vinylcarbazole), and filler is a good insulator which has low water absorption and an electric resistivity of 5 kV/mm (174). Epoxy resin may be hardened by an ordinary catalyst in the presence of poly(9-vinylcarbazole) (175). Combination of epoxy resin, polyester, and poly(9-vinylcarbazole) is suggested for the impregnation of coiled microcondensers (176). Patents also cover the use of poly(9-vinylcarbazole) doped with electrolytes as capacitor dielectrics for high-voltage use (177) and of lacquered plate containing poly(9-vinylcarbazole) as insulator (178).

Foams. The high softening of poly(9-vinylcarbazole) makes the foam useful as a thermal insulator at temperatures up to 150°C. The preparation of foamed poly(9-vinylcarbazole) by molding a mixture of the polymer powder and a blowing agent (179) has been claimed. In other processes (180,181) the incorporation of the blowing agent into the polymer is during polymerization or into polymer under pressure. Carbon tetrachloride and low-boiling hydrocarbons have been recommended. In another process, the polymer is dissolved in a low-boiling solvent (methylene chloride), the solution is then dispersed in water, and most of the solvent is removed. The partly dried granules are molded hot, yielding poly(9-vinylcarbazole) foam of densities down to 0.03 g/cc (182). Styrene–9-vinylcarbazole copolymers can be expanded in a similar manner (180,181,183).

A detailed study (37) indicates that blowing by a nonsolvent (acetone) alone is unsatisfactory, in particular when a foam of low density (<0.2 g/cc) is required. It was found, however, that impregnation of the polymer granules with dioxane or benzene and with azobisisobutyronitrile together with acetone results in greatly improved molding properties.

Properties of poly(9-vinylcarbazole) foam are summarized in Table 7. These properties are functions of density as well as of the shape and size of the bubbles.

9-Vinylcarbazole as a Minor Component in Copolymers. The copolymer of ethylene with 9-vinylcarbazole (ethylene/9-vinylcarbazole = 100/5) by a Ziegler-

Natta catalyst is resistant to ozonolysis and has high elongation (134). A ternary copolymer of ethylene (85–99%), 9-vinylcarbazole, and acrylamide was claimed to be useful as a blending material for polypropylene fiber to improve its dyeability (137). Radiation-induced grafting of 9-vinylcarbazole onto polyethylene has been reported (185,186). Grafting of 9-vinylcarbazole onto peroxidized polypropylene makes the fiber receptive to acid dyes (187). Two-stage polymerization of olefin in the presence of 9-vinylcarbazole was described as a method of producing dyeable polyolefins (188).

Table 7. Properties of Poly(9-Vinylcarbazole) Foam (184)

Item	Sample[a]	Density, g/cc	Result
weight loss on baking up to 170°C for 3–5 days	A, B	0.065–0.182	weight loss 3–6% distortion
dimensional stability on heating up to 200°C	A	0.05–0.07	<1%
	A	0.10–0.12	>3%
	A	0.16–0.18	>5%
maximum compressive modulus			
at 20°C	A	0.201	8×10^3 psi
	A	0.055	8×10^2 psi
at 180°C	A	0.04–0.06	$2–4 \times 10^2$ psi
tensile strength at 150°C	A	0.06	25.9 psi
	A	0.11	35.9 psi
	A	0.18	37.3 psi
elongation at failure			
at 20°C	A	0.06–0.18	<2%
at 180°C	A	0.06–0.18	<6%
thermal conductivity	A	0.053–0.18	8×10^{-5} kcal/cm²
dielectric constant, ϵ	A	0.061–0.135	1.078–1.186
loss factor, tan δ	A	0.060–0.179	$4–25 \times 10^{-4}$
effect of humid atmosphere on dielectric properties			no significant effect
flammability			flammable in air

[a] Key: A, granule (d = 0.5–5 mm) based foam; B, bead (d = 0.5–1.5 mm) based foam.

The use of 9-vinylcarbazole or its polymer as comonomer or blending component for acrylic fibers has also been described in patents (189,190). Copolymerization with acrylates (191), formaldehyde, or trioxane has appeared in the patent literature (131–133).

Image-Recording Systems. There are two important applications of 9-vinylcarbazole for imaging systems. The extreme sensitivity of 9-vinylcarbazole to charge-transfer photopolymerization permits its application to photopolymer systems. Although the mechanism of the polymerization has not yet been clarified, there seems to be no oxygen effect on this polymerization. By contrast, conventional photopolymerization via free-radical intermediates is generally retarded or inhibited by air during the initial period. The oxygen effect is particularly troublesome for imaging systems which are used as a thin layer having a large area exposed to air. An additional advantage of 9-vinylcarbazole as an imaging material is in the use of the color reaction of tertiary amines with halogenated hydrocarbons (192).

The use of 9-vinylcarbazole imaging systems was first described by Wainer (193). A series of patents covers several modifications of the 9-vinylcarbazole–halogenated hydrocarbon system such as the use as a photoresist (194); sensitization by alkaline earth iodide, iodates, or periodates (195); the combination of the 9-vinylcarbazole system with other arylamines in a polymer matrix (196); the sensitization by leuco triarylmethane dyes (197); and the use of hydrophilic polymer as matrix (198). These systems are all homogeneous and the sensitivity is very much less than in a silver emulsion.

A camera-speed photopolymerization system was recently announced (114,115, 199,200). It is constructed by the mixture of 9-vinylcarbazole and carbon tetrabromide finely dispersed in a gelatin matrix. Photopolymerization to yield an image is carried out by irradiation with blue or visible light followed by a more intense overall white-light exposure and heat. The 9-vinylcarbazole that was unreacted during the initial image-forming exposure is brought to a color reaction by the overall exposure, resulting in the production of a direct positive photograph. The system is sensitive enough to be exposed in a camera with shutter speed of 1 to 5 sec at *f*/4.7 in bright sunlight. This sensitivity can be enhanced by a factor of 10^2 by the process of dye sensitization and is comparable with that of a silver halide emulsion; exposure in a camera with shutter speeds of 1/50 and 1/200 sec at *f*/5.6 to *f*/11 in bright sunlight is sufficient for yielding an image. Developing and fixing of this system does not require any chemical reagents but only heat and white light under dry conditions.

Another application of poly(9-vinylcarbazole) is as a photoconductive material for electrophotography. At present, this field of application is probably much more important than the photopolymerization system. The advantages of organic (particularly polymer) photoconductors such as polyvinylcarbazole over inorganic photoconductors are transparency, high resolution, flexibility (film-forming property), light weight, panchromaticity, and the possibility of producing both positive and negative charges. In the case of poly(9-vinylcarbazole), an additional advantage is the larger photoconductivity in comparison with any other carbazole derivatives (201). As has been described in the section on properties, the photoconductivity of the polymer can be enhanced by the addition of electron acceptors. Dye sensitization also appears frequently in the literature. Thus, the patents given in Refs. 202–206 cover a variety of dyes and a great number of additives, which are mostly electron acceptors such as Lewis or Brønsted acids (207,208).

Modifications of photoconductive polymer by copolymerization of 9-vinylcarbazole with other monomers, and by the use of 9-vinylcarbazole analogs or derivatives (210) have been described; copolymers of 9-vinylcarbazole with 9-vinylanthracene (209) and poly(vinyl butyral) grafted with 9-vinylcarbazole are examples of this type. Substituted 9-vinylcarbazole derivatives have been extensively studied and patents have been issued covering homopolymers and copolymers containing 3,6-dibromo-9-vinylcarbazole (211–213); 3-iodo-9-vinylcarbazole (214); 3-amino-, 3-monoalkylamino-, 3-dialkylamino-, 3-benzalamino-9-vinylcarbazole (215); and 3-nitro-9-vinylcarbazole (216). Substituents at the 3- or 6- position can be introduced starting from poly(9-vinylcarbazole). The polymer reaction may be more practical than polymerization of substituted monomers since the preparation of substituted 9-vinylcarbazole is not an efficient process. The 3-nitro or 3,6-dibromo derivatives of poly(9-vinylcarbazole) are known to exhibit better photoconductivity than the parent polymer (217,218). The methods of bromination and nitration have also been patented (219,220).

In addition to electrophotography, the photoconductive property of poly(9-vinylcarbazole) has been applied to the formation of electrostatic wrinkle images (221). For this purpose, the imaging layer was composed of copolymer of 9-vinylcarbazole with a higher-alkyl acrylate (C_4–C_{18}) mixed with any other photoconductor, a dye, and a styrene–butadiene copolymer.

Bibliography

1. W. Reppe and E. Keyssner (to I. G. Farbenindustrie A.G.), Ger. Pat. 618,120 (Sept. 2, 1935); through *Chem. Abstr.* **30,** 110 (1936).
2. W. Reppe, "Acetylene Chemistry," *U.S. Dept. Comm., B. D. Rept.* **18852-S** (1949).
3. G. M. Kline, *Mod. Plastics* **24,** 157 (1946).
4. H. Beller, R. E. Christ, and F. Wuerth (to General Aniline & Film Corp.), Brit. Pat. 641,437 (Aug. 9, 1950); through *Chem. Abstr.* **45,** 8044 (1951).
5. O. Solomon, C. Ionescu, and I. Ciutǎ, *Chem. Tech.* (*Berlin*) **9,** 202 (1957); through *Chem. Abstr.* **51,** 15493 (1957).
6. K. Yamamoto (to Mitsui Chemical Industries Co.), Japan. Pat. 1714 (March 30, 1951); through *Chem. Abstr.* **A7,** 4917 (1953).
7. M. Amagasa, I. Yamaguchi, and R. Shioya, Japan. Pat. 954 (April 26, 1962); through *Chem. Abstr.* **58,** 3399 (1963).
8. H. Davidge, *J. Appl. Chem.* **9,** 241 (1959).
9. S. Otsuka and S. Murahashi, *Kogyo Kagaku Zasshi* **59,** 511 (1956).
10. Pullman Inc., Brit. Pat. 1,017,004 (Jan. 19, 1966); through *Chem. Abstr.* **64,** 14098 (1966)
11. H. Otsuki, I. Okano, and T. Takeda, *Kogyo Kagaku Zasshi* **49,** 169 (1946).
12. V. P. Lopatinskii, E. E. Sirotkina, I. P. Zherebtsov, and M. A. Lehman, *Tr. Tomsk. Gos. Univ., Ser. Khim.* **170,** 29 (1964); through *Chem. Abstr.* **63,** 565 (1965).
13. R. G. Flowers, H. F. Muller, and L. W. Flowers, *J. Am. Chem. Soc.* **70,** 3019 (1948).
14. H. Otsuki, I. Okano, and T. Takeda, *Kogyo Kagaku Zasshi* **49,** 169 (1946).
15. H. Otsuki, H. Funabashi, and K. Sakuma, *Kogyo Kagaku Zasshi* **50,** 51 (1947).
16. V. P. Lopatinskii and Yu. P. Shekirev (to Polytechnical Institute, Tomsk), U.S.S.R. Pat. 173,770 (Sept. 7, 1965); through *Chem. Abstr.* **64,** 2060 (1966).
17. V. P. Lopatinskii, E. E. Sirotkina, and M. M. Anasova, *Tr. Tomsk. Gos. Univ., Ser. Khim.* **170,** 49 (1964); through *Chem. Abstr.* **63,** 563 (1965).
18. V. P. Lopatinskii and E. E. Sirotkina, *Metody Polucheniya Khim. Reaktivov Preparatov* **11,** 40 (1964); through *Chem. Abstr.* **65,** 2203 (1966).
19. V. P. Lopatinskii and E. E. Sirotkina, *Izv. Tomsk. Politekh. Inst.* **136,** 26 (1965); through *Chem. Abstr.* **65,** 16930 (1966).
20. V. P. Lopatinskii and E. E. Sirotkina, *Izv. Tomsk. Politekh. Inst.* **148,** 73 (1967); through *Chem. Abstr.* **70,** 114938 (1969).
21. V. P. Lopatinskii and E. E. Sirotkina, *Izv. Tomsk. Politekh. Inst.* **126,** 67 (1964); through *Chem. Abstr.* **64,** 3457 (1966).
22. V. P. Lopatinskii, E. E. Sirotkina, and S. D. Pukalskaya, *Izv. Tomsk. Politekh. Inst.* **148,** 70 (1967); through *Chem. Abstr.* **70,** 87441 (1969).
23. G. R. Clemo and W. H. Perkin, Jr, *J. Chem. Soc.* **125,** 1804 (1924).
24. A. Inami, K. Morimoto, and Y. Murakami, *Nippon Kagaku Zasshi* **85,** 880 (1964).
25. V. P. Lopatinskii and I. P. Zherebtsov, *Izv. Tomsk. Politekh. Inst.* **136,** 23 (1965); through *Chem. Abstr.* **65,** 18550 (1966).
26. A. Inami, K. Morimoto, and Y. Murakami, *Nippon Kagaku Zasshi* **85,** 880 (1964).
27. J. Heller, D. J. Lyman, and W. A. Hewett, *Makromol. Chem.* **73,** 48 (1964).
28. B. Levy, *Mh. Chem.* **33,** 182 (1912).
29. W. Reppe, E. Keyssner, and F. Nicolai (to I. G. Farbenindustrie A.G.), Ger. Pat. 646,995 (April 13, 1937); through *Chem. Abstr.* **31,** 6258 (1937).
30. D. L. Nicol, M. Kaufman, and S. A. Miller (to British Oxygen Co. Ltd.), Brit. Pat. 718,912 (Nov. 24, 1954).
31. S. A. Miller and H. Davidge (to British Oxygen Co. Ltd.), U.S. Pat. 2,830,059 (April 8, 1958); through *Chem. Abstr.* **52,** 14696 (1958).

32. W. Schmidt and K. Jost (to Badische Anilin- & Soda-Fabrik A.G.), Ger. Pat. 1,025,875 (March 13, 1958); through *Chem. Abstr.* **54,** 9716 (1960).
33. A. V. Chernobai, Zh. S. Tirakyants, and R. Ya. Delyatitskaya, *Vysokomolekul. Soedin.* **9,** 664 (1967).
34. R. M. Joshi, *Makromol. Chem.* **55,** 33 (1962).
35. H. Davidge (to British Oxygen Co. Ltd.), Brit. Pat. 831,913 (April 6, 1960); through *Chem. Abstr.* **54,** 16925 (1960).
36. H. Davidge, *J. Appl. Chem.* **9,** 553 (1959).
37. L. P. Ellinger, *J. Appl. Polymer Sci.* **10,** 551 (1966).
38. Badische Anilin- & Soda-Fabrik A.G., Brit. Pat. 739,438 (Oct. 26, 1955); through *Chem. Abstr.* **50,** 17532 (1956).
39. L. P. Ellinger, *J. Appl. Polymer Sci.* **9,** 3939 (1965).
40. O. F. Solomon, I. Z. Ciuta, and N. Cobianu, *Polymer Letters* **2,** 311 (1964).
41. S. Tazuke, K. Nakagawa, and S. Okamura, *Polymer Letters* **3,** 923 (1965).
42. S. Tazuke, *Chem. Comm.* **1970,** 1277.
43. M. Biswas and I. Kar, *Indian J. Chem.* **5,** 119 (1967).
44. S. Tazuke, T. B. Tjoa, and S. Okamura, *J. Polymer Sci.* [A-1] **5,** 1911 (1967).
45. S. Tazuke, M. Asai, S. Ikeda, and S. Okamura, *Polymer Letters* **5,** 453 (1967).
46. S. Tazuke, M. Asai, and S. Okamura, *J. Polymer Sci.* [A-1] **6,** 1809 (1968).
47. O. F. Solomon, N. Cobianu, D. S. Vasilescu, and C. Boghinǎ, *Polymer Letters* **6,** 551 (1968).
48. H. Miyama, J. Tsuji, M. Morikawa, M. Kamachi, and T. Nogi (to Toyo Rayon Co. Ltd.), Japan. Pat. 16,049 (Sept. 2, 1967); through *Chem. Abstr.* **68,** 3461 (1968).
49. Ciba (A.R.L.) Ltd., Fr. Pat. 1,397,538 (April 30, 1965); through *Chem. Abstr.* **63,** 7131 (1965).
50. Distillers Co. Ltd., Neth. Appl. 6,612,244 (March 8, 1967); through *Chem. Abstr.* **67,** 64851 (1967).
51. C. E. H. Bawn, C. Fitzsimmons, and A. Ledwith, *Proc. Chem. Soc.* (*London*) **1964,** 391.
52. A. Ledwith, *J. Appl. Chem.* **17,** 344 (1967).
53. A. Rembaum, A. M. Hermann, and R. Haack, *Polymer Letters* **5,** 407 (1967).
54. O. F. Solomon, M. Dimonie, K. Ambrozh, and M. Tomescu, *J. Polymer Sci.* **52,** 205 (1961).
55. V. P. Lopatinskii and E. E. Sirotkina, U.S.S.R. Pat. 158,883 (Nov. 22, 1963); through *Chem. Abstr.* **60,** 11990 (1964).
56. J. Heller, D. O. Tieszen, and D. B. Parkinson, *J. Polymer Sci.* [A] **1,** 125 (1963).
57. Montecatini, Brit. Pat. 914,418 (Jan. 2, 1963); through *Chem. Abstr.* **58,** 9252 (1963).
58. J. Obloj, M. Nowakowska, and J. Pielichowski (to Instytut Ciezkiej Syntezy Organicznej), Polish Pat. 53,813 (Aug. 30, 1967); through *Chem. Abstr.* **68,** 60059 (1968).
59. G. Dall'Asta and A. Casale, *Atti Accad. Nazl. Lincei, Rend. Classe Sci. Fis., Mat. Nat.* **39,** 291 (1965); through *Chem. Abstr.* **65,** 3968 (1966).
60. J. C. MacKenzie and A. Orchechowski (to Cabot Corp.), U.S. Pat. 3,285,892 (Nov. 15, 1966); through *Chem. Abstr.* **66,** 19020 (1967).
61. J. Hughes and A. M. North, *Trans. Faraday Soc.* **62,** 1866 (1966).
62. A. Chapiro and G. Hardy, *J. Chim. Phys.* **59,** 993 (1962).
63. A. Chapiro, *U.S. At. Energy Comm.* **TID-7643,** 136–149 (1962); through *Chem. Abstr.* **58,** 6933 (1963).
64. A. Chapiro, *Magy. Kem. Lapja* **18,** 152 (1963); through *Chem. Abstr.* **59,** 2312 (1963).
65. J. Kroh and W. Pekala, *Bull. Acad. Polon. Sci., Ser. Sci. Chim.* **12,** 419 (1964); through *Chem. Abstr.* **62,** 1240 (1965).
66. Z. Geldecki, J. Karolak, W. Pekala, and J. Kroh, *Bull. Acad. Polon. Sci., Ser. Sci. Chim.* **15,** 209 (1967); through *Chem. Abstr.* **67,** 73959 (1967).
67. J. Kroh and W. Pekala, *Bull. Acad. Polon. Sci., Ser. Sci. Chim.* **14,** 55 (1966); through *Chem. Abstr.* **64,** 19795 (1966).
68. S. Okamura, T. Higashimura, and T. Matsuda, *Kobunshi Kagaku* **23,** 269 (1966).
69. R. A. Meyer and E. M. Christman, *J. Polymer Sci.* [A-1] **6,** 945 (1968).
70. M. Nishii, K. Tsuji, K. Hayashi, S. Okamura, and K. Takakura, *Kobunshi Kagaku* **23,** 445 (1966).
71. T. Matsuda, T. Higashimura, and S. Okamura, *Kobunshi Kagaku* **24,** 165 (1967).
72. K. Tsuji, K. Takakura, M. Nishii, K. Hayashi, and S. Okamura, *Ann. Rept. Japan. Assoc. Rad. Res. Polymers* **6,** 179 (1964–5).

73. M. Nishii, K. Tsuji, K. Takakura, K. Hayashi, and S. Okamura, *Ann. Rept. Japan. Assoc. Rad. Res. Polymers* **6,** 181 (1964–5).
74. S. Okamura, T. Higashimura, and T. Matsuda, *Kobunshi Kagaku* **22,** 180 (1965).
75. O. F. Solomon, N. Cobianu, and I. Z. Ciuta, *Inst. Bull. Politech. Bucuresti* **27,** 65 (1965). through *Chem. Abstr.* **64,** 3693 (1966).
76. T. Matsuda, T. Higashimura, and S. Okamura, *J. Macromol. Sci. Chem.* **2,** 43 (1968).
77. B. Boros-Gyevi, *Magy. Kem. Foly.* **75,** 87 (1969); through *Chem. Abstr.* **70,** 88284 (1969).
78. P. B. Ayscough, A. K. Roy, R. G. Groce, and S. Munari, *J. Polymer Sci.* [A-1] **6,** 1307 (1968).
79. K. Tsuji, K. Takakura, M. Nishii, K. Hayashi, and S. Okamura, *J. Polymer Sci.* [A-1] **4,** 2028 (1966).
80. K. Morimoto and H. Mitsuda, *Polymer Letters* **5,** 27 (1967).
81. L. P. Ellinger, *Polymer* **5,** 559 (1964).
82. S. Tazuke and S. Okamura, *Polymer Letters* **6,** 173 (1968).
83. S. Tazuke and S. Okamura, *J. Polymer Sci.* [A-1] **6,** 2907 (1968).
84. J. M. Barrales-Rienda, G. R. Brown, and D. C. Pepper, *Polymer* **10,** 327 (1969).
85. O. F. Solomon, M. Dimonie, and M. Tomescu, *Makromol. Chem.* **56,** 1 (1962).
86. S. Tazuke, *Yuki Gosei Kagaku Kyokaishi* **27,** 507 (1969).
87. S. Tazuke, *Adv. Polymer Sci.* **6,** 321 (1969).
88. J. W. Breitenbach and Ch. Srna, *Polymer Letters* **1,** 263 (1963).
89. J. W. Breitenbach and O. F. Olaj, *Polymer Letters* **2,** 685 (1964).
90. H. Scott, T. P. Konen, and M. M. Labes, *Polymer Letters* **2,** 689 (1964).
91. C. E. H. Bawn, A. Ledwith, and Yang Shin-Liu, *Chem. Ind.* (*London*) **1965,** 769.
92. S. McKinley, J. V. Crawford, and C. H. Wang, *J. Org. Chem.* **31,** 1963 (1966).
93. J. W. Breitenbach, O. F. Olaj, and F. Wehrmann, *Monatsh.* **95,** 1007 (1964).
94. O. F. Solomon and M. Dimonie, *J. Polymer Sci.* [C] **4,** 969 (1964).
95. O. F. Solomon, I. Z. Ciuta, N. Cobianu, and M. Georgescu, *Bull. Inst. Politech. Bucuresti* **27,** 59 (1965); through *Chem. Abstr.* **64,** 2169 (1966).
96. O. F. Solomon, N. Cobianu, D. S. Vasilescu, and C. Boghină, *Polymer Letters* **6,** 551 (1968).
97. S. Tazuke, M. Asai, and S. Okamura, *Kogyo Kagaku Zasshi* **72,** 1841 (1969).
98. L. P. Ellinger, *Polymer* **6,** 549 (1965).
99. H. Scott, G. A. Miller, and M. M. Labes, *Tetrahedron Letters* **17,** 1073 (1963).
100. H. Scott and M. M. Labes, *Polymer Letters* **1,** 413 (1963).
101. J. Pác and P. H. Plesch, *Polymer* **8,** 237, 252 (1967).
102. R. Gumbs, S. Penzek, J. Jagur-Grodzinski, and M. Szwarc, *Macromolecules* **2,** 77 (1969).
103. H. Nomori, M. Hatano, and S. Kambara, *Polymer Letters* **4,** 261 (1966).
104. K. Takakura, E. Kawa, K. Hayashi, and S. Okamura, *Ann. Rept. Japan. Assoc. Rad. Res. Polymers* **6,** 205 (1964–5).
105. K. Takakura, K. Hayashi, and S. Okamura, *Polymer Letters* **3,** 565 (1965).
106. E. H. Cornish (to Standard Telephone and Cables Ltd.), Brit. Pat. 1,003,910 (Sept. 8, 1965); through *Chem. Abstr.* **63,** 16491 (1965).
107. A. Ledwith, *J. Appl. Chem.* **17,** 344 (1967).
108. C. E. H. Bawn, R. Carruthers, and A. Ledwith, *Chem. Comm.* **1965,** 522.
109. E. S. Budnikova, E. E. Sirotkina, V. P. Lopatinskii, and M. A. Igumnova, *Vysokomolekul. Soedin.* **B10,** 447 (1968).
110. S. Okamura, T. Higashimura, and T. Matsuda, *Kobunshi Kagaku* **23,** 273 (1966).
111. O. F. Solomon, N. Cobianu, and V. Kucinschi, *Makromol. Chem.* **89,** 171 (1965).
112. T. Nagai, T. Miyazaki, and N. Tokura, *Polymer Letters* **6,** 345 (1968).
113. Chi-Hua Wang, *Chem. Ind.* (*London*) **1964,** 751.
114. Y. Yamada, T. H. Garland, Jr., and P. Bruck, *SPSE 1969 Ann. Conf.*, preprint.
115. Y. Yamada, T. H. Garland, Jr., and B. R. Tarr, *SPSE 1969 Ann. Conf.*, preprint.
116. M. Biswas, M. M. Maiti, and N. D. Ganguly, *Makromol. Chem.* **124,** 263 (1969).
117. H. Scott, T. P. Konen, and M. M. Labes, *Polymer Letters* **2,** 689 (1964).
118. W. Reppe, "Acetylene Chemistry," *U.S. Dept. Comm., O.I.S., B. D. Rept.* **18,852** (1949).
119. L. P. Ellinger, *Chem. Ind.* (*London*) **1963,** 1982.
120. W. Freudenberg (to General Aniline & Film Corp.), Brit. Pat. 624,819 (June 16, 1949); through *Chem. Abstr.* **44,** 2283 (1950).
121. Y. Takeda, M. Asai, and S. Tazuke, *Polymer Symp. 1970, Kyoto*, preprint.

122. L. P. Ellinger, *Polymer* **10**, 531 (1969).
123. T. Natsuume, Y. Akana, K. Tanabe, M. Fujimatsu, M. Shimizu, Y. Shirota, H. Hirata, S. Kusabayashi, and H. Mikawa, *Chem. Comm.* **1969**, 189.
124. A. M. North and K. E. Whitelock, *Polymer* **9**, 590 (1968).
125. T. Alfrey, Jr., J. Bohrer, and H. Mark, *Copolymerization*, Vol. VIII in *High Polymer* Series, Interscience Publishers, Inc., New York, 1952.
126. J. M. Barrales-Rienda, G. R. Brown, and D. C. Pepper, *Polymer* **10**, 327 (1969).
127. R. Hart, *Makromol. Chem.* **47**, 143 (1961).
128. T. Alfrey, Jr., and S. L. Kapur, *J. Polymer Sci.* **4**, 215 (1949).
129. S. N. Ushakov and A. F. Nikolaev, *Izv. Akad. Nauk SSSR, Otd. Khim. Nauk* **1956**, 83; through *Chem. Abstr.* **50**, 13867 (1956).
130. M. Asai, S. Tazuke, and S. Okamura, *Polymer Symp., 1968*, Matsuyama Preprint.
131. E. I. du Pont de Nemours & Co., Inc., Brit. Pat. 911,960 (Dec. 5, 1962); through *Chem. Abstr.* **58**, 5809 (1963).
132. N. Brown (to E. I. du Pont de Nemours & Co., Inc.), U.S. Pat. 3,194,790 (July 13, 1965); through *Chem. Abstr.* **63**, 10089 (1965).
133. Skanska Attikfabriken Aktiebolag, Brit. Pat. 991,538 (May 12, 1965); through *Chem. Abstr.* **63**, 4472 (1965).
134. E. L. Bush and M. M. Kumar (to Standard Telephone and Cables Ltd.), Brit. Pat. 1,083,894 (Sept. 20, 1967); through *Chem. Abstr.* **67**, 117578 (1967).
135. E. H. Cornish, E. L. Bush, and K. Kumar, *Plastics (London)* **31**, 1578 (1966).
136. J. Obloj, J. Polaczek, J. Pielichowski, M. Nowakowska, and K. Fraczek (to Institut Ciezkiej Syntezy Organicznej), Polish Pat. 52,434 (Nov. 30, 1966); through *Chem. Abstr.* **68**, 30887 (1968).
137. Sumimoto Chemical Co. Ltd. and Toyo Spinning Co. Ltd., Neth. Appl. 6,502,104 (Aug. 23, 1965); through *Chem. Abstr.* **64**, 8380 (1966).
138. G. D. Buckley and L. Seed (to Imperial Chemical Industries Ltd.), Brit. Pat. 706,412 (March 31, 1954); through *Chem. Abstr.* **48**, 12462 (1954).
139. Gy. Hardy, J. Varga, K. Nyitrai, I. Czajbik, and L. Zubonyai, *Vysokomolekul. Soedin.* **6**, 758 (1964).
140. G. Steinbach von Gaver and K. Magni (to Compagnie de Saint-Gobain), Fr. Pat. 80,489 (May 3, 1963); through *Chem. Abstr.* **59**, 6537 (1963).
141. P. Robinson (to Sprague Electric Co.), U.S. Pat. 2,725,369 (Nov. 29, 1955); through *Chem. Abstr.* **50**, 9787 (1956).
142. M. Markarian, S. D. Ross, and M. Nazzewski (to Sprague Electric Co.), U.S. Pat. 2,695,900 (Nov. 30, 1954); through *Chem. Abstr.* **49**, 14029 (1955).
143. L. A. Brooks, M. Markarian, and M. Nazzewski (to Sprague Electric Co.), U.S. Pat. 2,636,022 (April 21, 1953); through *Chem. Abstr.* **47**, 7826 (1953).
144. E. A. Kern (to General Electric Co.), U.S. Pat. 2,527,223 (Oct. 24, 1950); through *Chem. Abstr.* **45**, 392 (1951).
145. L. A. Brooks and M. Nazzewski (to Sprague Electric Co.), U.S. Pat. 2,406,319 (Aug. 27, 1946); through *Chem. Abstr.* **41**, 316 (1935).
146. K. Kamio, S. Nakada, and H. Takamatsu (to Japan Carbide Industries Co.), Japan. Pat. 12,184 (June 16, 1965); through *Chem. Abstr.* **63**, 16556 (1965).
147. V. P. Lopatinskii, E. E. Sirotkina, and M. P. Grosheva (to S. M. Kirov Polytechnic Institute, Tomsk), U.S.S.R. Pat. 204,590 (Oct. 20, 1967); through *Chem. Abstr.* **68**, 115426 (1968).
148. K. Ueberreiter and J. Springer, *Z. Physik. Chem. (Frankfurt)* **36**, 299 (1963).
149. N. Kuwahara, S. Higashide, N. Nakata, and M. Kaneko, *J. Polymer Sci.* [A] 7 (2), 285 (1969).
150. D. W. van Krevelen and P. J. Hoftyer, *J. Appl. Polymer Sci.* **11**, 1409 (1967).
151. L. E. Nielsen, *Mechanical Properties of Polymers*, Reinhold Publishing Corp., New York, 1962; Japanese translation, p. 15.
152. K. Ueberreiter and W. Bruns, *Ber. Bunsenges. Physik. Chem.* **68**, 541 (1964).
153. E. H. Cornish, *Plastics* **28**, 61 (1963).
154. W. M. Shine, *Mod. Plastics* **25**, 130 (1947), cited in Ref. 153.
155. W. E. Busse, I. M. Lambert, C. Mackinley, and H. R. Davidson, *Ind. Eng. Chem.* **40**, 2271 (1948), cited in Ref. 153.
156. L. Borsody, *IRE Trans. Component Pts.* **7** (1), 15 (1960), cited in Ref. 153.
157. A. I. Lakatos and J. Mort, *Phys. Rev. Letters* **21**, 1444 (1968).

158. H. Hoegl, *J. Phys. Chem.* **69,** 755 (1965).
159. S. Oka, T. Mori, S. Kusabayashi, Y. Yamamoto, M. Ishiguro, and H. Mikawa, *Electrophotography* (*Japan*) **5,** 152 (1964).
160. A. Inami, K. Morimoto, and Y. Hayashi, *Natl. Tech. Rept.* (*Japan*) **12,** 79 (1966).
161. M. Hayashi, M. Kuroda, K. Imura, and A. Inami, *Kobunshi Kagaku* **21,** 577 (1964).
162. A. Szymanski and M. M. Labes, *J. Chem. Phys.* **50,** 3568 (1969).
163. A. M. Hermann and A. Rembaum, *J. Polymer Sci.* [C] **17,** 107 (1967).
164. K. Geist (to Badische Anilin- & Soda-Fabrik A.G.), Ger. Pat. 1,097,680 (Jan. 19, 1961); through *Chem. Abstr.* **55,** 22920 (1961).
165. A. Von Hippel and L. G. Wesson, *Ind. Eng. Chem.* **38,** 1121 (1946).
166. A. S. Cummin and J. R. Hutzler (to General Electric Co.), U.S. Pat. 2,872,630 (Feb. 3, 1959); through *Chem. Abstr.* **53,** 8705 (1959).
167. M. Markarian (to Sprague Electric Co.), U.S. Pat. 2,528,445 (Oct. 31, 1950); through *Chem. Abstr.* **45,** 1813 (1951).
168. E. H. Cornish (to Standard Telephone and Cables Ltd.), Brit. Pat. 1,004,074 (Sept. 8, 1965); through *Chem. Abstr.* **63,** 13443 (1965).
169. Standard Electrica, S.A., Span. Pat. 290,150 (Nov. 14, 1963); through *Chem. Abstr.* **61,** 13497 (1964).
170. E. H. Cornish (to Standard Telephone and Cables Ltd.), Brit. Pat. 1,007,040 (Oct. 13, 1965); through *Chem. Abstr.* **64,** 836 (1966).
171. P. H. Netherwood (to Sprague Electric Co.), U.S. Pat. 2,930,714 (March 29, 1960); through *Chem. Abstr.* **54,** 16021 (1960).
172. I. G. Farbenindustrie A.G., Fr. Pat. 851,455 (Jan. 9, 1940); through *Chem. Abstr.* **36,** 2049 (1942).
173. P. Nowak (to Licentia Patent Verwaltungs G.m.b.H.), Ger. Pat. 1,124,570 (March 1, 1962); through *Chem. Abstr.* **57,** 1132 (1962).
174. G. Gahn and F. Schlezel (to Siemens A.G.), Ger. Pat. 1,290,275 (March 6, 1969); through *Chem. Abstr.* **70,** 98044 (1969).
175. P. Nowak (to Licentia Patent Verwaltungs G.m.b.H.), Ger. Pat. 1,070,823 (Dec. 10, 1959); through *Chem. Abstr.* **55,** 13715 (1961).
176. C. M. Hofbauer (to Ernst Roederstein Specialfabrik für Kondensatoren G.m.b.H.), Ger. Pat. 1,073,107 (Jan. 14, 1960); through *Chem. Abstr.* **55,** 10158 (1961).
177. J. Kleffner and W. Kliebisch (to Siemens & Halske A.G.), Ger. Pat. 974,880 (Appl. Sept. 14, (1949); through *Chem. Abstr.* **56,** 4521 (1962).
178. H. Schill (to Siemens & Halske A.G.), Ger. Pat. 1,089,024 (Appl. June 7, 1955); through *Chem. Abstr.* **56,** 3666 (1962).
179. E. Dumont and H. Reinhardt, Ger. Pat. 1,001,488 (Jan. 24, 1959); through *Chem. Abstr.* **53,** 23081 (1959).
180. F. Stastny and H. Gerlich (to Badische Anilin- & Soda-Fabrik A.G.), Ger. Pat. 934,692 (Nov. 3, 1955); through *Chem. Abstr.* **52,** 15960 (1958).
181. F. Stastny and K. Buchfolz (to Badische Anilin- & Soda-Fabrik A.G.), Ger. Pat. 951,299 (Oct. 25, 1956); through *Chem. Abstr.* **53,** 3774 (1959).
182. H. E. Knobloch, F. Meyer, and F. Stastny (to Badische Anilin- & Soda-Fabrik A.G.), Brit. Pat. 933,621 (Aug. 8, 1963); through *Chem. Abstr.* **59,** 11734 (1963).
183. F. Stastny (to Badische Anilin- & Soda-Fabrik A.G.), Ger. Pat. 956,808 (Jan. 24, 1957); through *Chem. Abstr.* **53,** 8718 (1959).
184. L. P. Ellinger, *J. Appl. Polymer Sci.* **10,** 575 (1967).
185. W. K. W. Chen. R. B. Mesrobian, D. S. Ballantine, D. J. Metz, and A. Gline, *J. Polymer Sci.* **23,** 903 (1957).
186. A. Chapiro, *J. Polymer Sci.* **29,** 321 (1958).
187. Montecatini, Brit. Pat. 850,471 (Oct. 5, 1960); through *Chem. Abstr.* **55,** 9894 (1961).
188. Y. Shin and O. Fukumoto (to Toyo Rayon Co. Ltd.), Japan. Pat. 26,585 (Dec. 29, 1963); through *Chem. Abstr.* **60,** 6979 (1964).
189. R. Sakurai, T. Tanabe, and H. Nagao, U.S. Pat. 2,902,335 (Sept. 1, 1959); through *Chem. Abstr.* **54,** 917 (1960).
190. G. E. Ham (to Chemstrand Corp.), U.S. Pat. 2,769,793 (Nov. 6, 1956); through *Chem. Abstr.* **55,** 9894 (1961).
191. Rohn & Haas Co., Brit. Pat. 959,775 (June 3, 1964); through *Chem. Abstr.* **61,** 5908 (1964).

192. R. H. Sprague, *Phot. Sci. Eng.* **5,** 98 (1961).
193. E. Wainer (to Horisons Inc.), U.S. Pat. 3,042,517 (July 3, 1962).
194. E. Wainer (to Horisons Inc.), U.S. Pat. 3,046,125 (July 24, 1962).
195. E. Wainer (to Horisons Inc.), U.S. Pat. 3,042,518 (Appl. Jan. 8, 1960).
196. Horisons Inc., Fr. Pat. 1,303,075 (Sept. 7, 1962).
197. Kalle A.G., Neth. Appl. 6,515,555 (June 13, 1966).
198. Kalle A.G., Neth. Appl. 6,516,580 (July 1, 1966).
199. N. T. Notley, *1969 Ann. Conf., SPSE,* preprint.
200. J. H. Jacobs, *1969 Ann. Conf., SPSE,* preprint.
201. K. Kinjo, S. Nagashima, and K. Yoshitake, *Electrophotography (Japan)* **3,** 29 (1961).
202. H. Hoegl, O. Sus, and W. Neugebauer (to Kalle A.G.), U.S. Pat. 3,037,861 (June 5, 1962).
203. Kalle and Co. A.G., Brit. Pat. 856,770 (Dec. 21, 1960); through *Chem. Abstr.* **55,** 20741 (1961).
204. Kalle and Co. A.G., Brit. Pat., 946,108 (Jan. 8, 1964); through *Chem. Abstr.* **60,** 11526 (1964).
205. Y. Hayashi, M. Kuroda, and A. Inami, *Bull. Chem. Soc. Japan* **39,** 1660 (1966).
206. H. Hoegl, E. Lind, and H. Schlesinger (to Azoplate Corp.), U.S. Pat. 3,232,755 (Feb. 1, 1966); through *Chem. Abstr.* **64,** 13614 (1966).
207. F. A. Levina, V. S. Myl'nikova, G. I. Rybalko, I. Sidaravicius, A. M. Sladkov, and A. N. Terenin, U.S.S.R. Pat. 169,395 (Dec. 18 1965); through *Chem. Abstr.* **64,** 16889 (1969).
208. Gevaert Photo Production N. V., Brit. Pat. 988,363 (April 7, 1965); through *Chem. Abstr.* **65,** 1662 (1966).
209. H. Ishida, K. Morimoto, and A. Inami (to Matsushita Electric Industrial Co.), Japan. Pat. 43-24750 (Oct. 25, 1968).
210. P. M. Cassiers and R. M. Hart (to Gevaert Photo Production N. V.), U.S. Pat. 3,155,503 (Nov. 3, 1964).
211. A. Inami, Y. Murakami, and K. Morimoto (to Matsushita Electric Industrial Co.), Japan. Pat. 42-19751 (Oct. 4, 1967).
212. A. Inami, Y. Murakami, and K. Morimoto (to Matsushita Electric Industrial Co.), Japan. Pat. 42-21875 (Oct. 27, 1967).
213. Matsushita Electric Industrial Co., Brit. Pat. 1,046,058 (Oct. 19, 1966).
214. A. Inami, A. Minobe, and K. Morimoto (to Matsushita Electric Industrial Co.), Japan. Pat. 43-7591 (March 22, 1968).
215. K. Morimoto, Y. Murakami, and A. Inami (to Matsushita Electric Industrial Co.), Japan. Pat. 42-9639 (May 18, 1967).
216. K. Morimoto, A. Inami, and Y. Murakami (to Matsushita Electric Industrial Co.), Japan. Pat. 41-14508 (Aug. 15, 1966).
217. K. Morimoto and A. Inami, *Kogyo Kagaku Zasshi* **67,** 1938 (1964).
218. K. Morimoto, Y. Yamamoto, and M. Kuroda (to Matsushita Electric Industrial Co.), Japan. Pat. 43-24753 (Oct. 25, 1968).
219. Y. Murakami, K. Morimoto, and A. Inami (to Matsushita Electric Industrial Co.), Japan. Pat. 42-25230 (Dec. 2, 1967).
220. K. Morimoto, A. Inami, and A. Monobe (to Matsushita Electric Industrial Co.), Japan. Pat. 42-22049 (Oct. 30, 1967).
221. Gevaert-Agfa N. V., Neth. Appl. 6,701,697 (April 25, 1967); through *Chem. Abstr.* **67,** 65027 (1967).

Shigeo Tazuke and Seizo Okamura
Kyoto University

VINYL CHLORIDE–ACRYLONITRILE COPOLYMER FIBERS. See MODACRYLIC FIBERS

VINYL CHLORIDE FIBERS. See Fibers under VINYL CHLORIDE POLYMERS

VINYL CHLORIDE POLYMERS

Vinyl chloride polymers, containing the repeating unit —CH_2CHCl—, include in addition to the homopolymer copolymers with such monomers as vinyl acetate, propylene, or vinylidene chloride. The polymers are used in both flexible and rigid form as moldings, foams, fibers, and films and sheeting. They also form the basis of plastisols, or vinyl dispersions, used extensively as coatings and molded articles. The homopolymers and the polymers with minor amounts of a comonomer are collectively referred to as PVC or PVC-type polymers in the literature. Chlorinated poly(vinyl chloride) is also a material of considerable interest.

INTRODUCTION

Historical Background

The first mention of poly(vinyl chloride) was in 1872 by Baumann (1) when he described the formation of a white powder by the action of sunlight on vinyl chloride contained in a sealed tube. He had no idea of the composition of the new product but his examination showed that it was unaffected by a wide range of solvents. The formation of a compound with the formula C_2H_3Cl had been reported earlier in 1835 by Regnault (2), but, apart from these two references, there was no further interest in vinyl chloride for forty years.

Renewed interest was brought about by the overcapacity which had arisen in Europe in the production of calcium carbide. The potential of acetylene as an illuminant had been greatly overestimated and this had led to the construction of many new carbide plants. This resulted in a substantial price drop in the early part of the twentieth century; in 1905 the price of calcium carbide was £28 per ton, but in 1909 it had dropped to £9. The position was particularly acute in Germany and it was here that an extensive research program was set under way to investigate possible chemical uses of acetylene. In 1912, Klatte (3) of Griesheim-Elektron filed a patent claiming the manufacture of vinyl chloride monomer by the reaction between acetylene and hydrogen chloride in the presence of a mercuric chloride catalyst. The formation of the polymer by the action of ultraviolet radiation was confirmed and in 1914 the use of organic peroxides as accelerators for the polymerization was known (4).

At the same time work on vinyl halides was being carried out in Russia by Ostromislensky (5). The polymer was being used as an intermediate in an attempt to produce synthetic rubber by dehydrochlorination, using alcoholic and aqueous potash.

Table 1. Consumption, in 10^3 Metric Tons,

	1958	1959	1960	1961	1962
Austria	6	8	12	14	15
Belgium-Luxembourg	9	10	8	11	18
Denmark	6	7	8	9	10
Finland	3	3	[a]	[a]	5
France	61	64	87	95	120
Italy	35	44	70	92	108
The Netherlands	18	23	28	28	30
Norway	2	4	6	7	8
Portugal	1	1	2	3	3
Spain	6	6	9	13	16
Sweden	11	13	17	19	25
Switzerland	6	7	11	13	18
United Kingdom	67	80	98	101	120
West Germany	88	119	157	177	206
Western Europe—*total*	*319*	*389*	*513*	*582*	*702*

[a] Not available.

Although there were some interesting observations on the solubility of various forms of poly(vinyl chloride) due undoubtedly to differences in molecular weight, Ostromislensky did not continue the work until after he emigrated to the United States some years later. The significance of Klatte's work was not realized and there was no follow-up after the initial patents, which were allowed to lapse in 1926.

In 1928 the publication of a number of patents revealed that several industrial groups were working in the field, and three independent companies had discovered that the processing of vinyl chloride polymers could be realized by copolymerizing vinyl chloride with vinyl acetate to improve handling characteristics (6–8). The real breakthrough was made by Semon of B. F. Goodrich (9), who showed that PVC could be processed and converted into a rubbery product by mixing and heating it with a high-boiling solvent such as tritolyl phosphate.

Industrial development ran parallel in Germany and the United States during the 1930s. Production in Europe got under way in Germany in 1931 and at about the same time Imperial Chemical Industries in England started their investigations into a process for the manufacture of vinyl chloride and methods of polymerization. In the United States production was started in the late 1930s by both Union Carbide and B. F. Goodrich. In conjunction with General Electric, B. F. Goodrich developed plasticized PVC as an insulator for wire and cable. The outbreak of World War II put a high pressure on this type of new insulating material which aged well, was less bulky, and had good chemical and flame resistance. During those years most of the production was put on allocation for military purposes. Both the copolymer and PVC were also used for military rainwear and waterproofing materials. It was only after the war that the development of new products for the consumer market grew and the industry really mushroomed.

In the United Kingdom, Imperial Chemical Industries had followed up their development with the construction of an 85 ton/yr pilot plant which started operations at the end of 1940. In 1942 this was followed by the first production plant with a capacity of 450 ton/yr. During the war a number of compounding plants were con-

of Vinyl Chloride Polymers in Western Europe

1963	1964	1965	1966	1967	1968	1969	1970 (estd)
[a]	[a]	20	22	23	27	30	
22	23	23	35	43	45	52	60
[a]	[a]	[a]	16	16	16	17	18
[a]	11	11	13	13	15	16	17
145	160	182	215	223	257	290	340
138	157	163	190	220	240	260	300
38	53	60	70	75	80	91	103
9	[a]	11	12	13	18	20	23
4	4	7	9	10	11	12	14
22	35	36	48	58	76	84	92
25	31	40	42	47	50	60	69
18	20	22	23	25	26	26	28
145	184	201	197	222	283	307	345
249	371	416	433	458	550	630	700
815	*1049*	*1192*	*1325*	*1446*	*1694*	*1895*	*2101*

structed to use home-produced resin and material which was imported from the United States under the Lendlease agreement.

In Europe all the first PVC plants had used the emulsion technique for the polymerization of monomer following the general practice of synthetic rubber manufacture (10). In Germany this practice was followed throughout World War II. The suspension polymerization process was followed from the start by the early American plants, but it was not adopted in the United Kingdom until about 1943, when the need for PVC with good electrical properties became critical.

The manufacture of copolymers was followed in the United States by Union Carbide using the solution process. These materials had been developed to reduce the difficulties in processing the homopolymer. A range of copolymers with acrylic esters was produced in Germany during World War II for solution applications.

Developments since the end of World War II have been too rapid to survey even briefly; suffice to say that whereas in 1945 worldwide production of PVC was 50,000 tons, by 1967 it had risen to about 4,000,000 tons.

Economic Aspects

The increase in total world production capacity of poly(vinyl chloride) to about 4 million tons in 1967 took place over a period of about 30 years, although the real progress has occurred in more recent years. In 1945 production was about 50,000 tons; in 1955 this had risen to 520,000 tons, and to 1.5 million tons in 1960. Most forecasters predict that the growth in consumption in Europe and the United States will continue at 9–12% per year.

Detailed figures for the consumption of poly(vinyl chloride) since 1958 in the countries of Western Europe are shown in Table 1. Western European producers are listed in Table 2. West Germany is outstanding in Europe for its high usage and dominates all its partners in the European Economic Community (EEC). France and Italy have about the same consumption, and this is almost identical with that of the United Kingdom, which far outstrips its partners in the European Free Trade Association

Table 2. Western European Producers of Vinyl Chloride Polymers in 1970[a] (11)

Producer	Location	Capacity, 10^3 metric ton/year
	West Germany	
Chemische Werke Huels	Marl	270
Wacker Chemie	Burghausen; Cologne	175
Badische Anilin- & Soda-Fabrik	Ludwigshafen	100
Farbwerke Hoechst	Gendorf; Knapsack	100
Deutsche Solvay	Burghausen	70
Dynamit Nobel	Troisdorf; Rheinfelden	60
	France	
Péchiney-Saint-Gobain	Saint-Auban; Saint-Fons	250
Solvic	Lavaux	140
Plastimer	Brignoud	70
Rhône-Poulenc	Péage de Roussillon	40
Aquitaine-Organico	Balan	30
	Italy	
Montecatini Edison	Porto Marghera; Brindisi	300
Società Italiana Resine	Porto Torres	70
ANIC	Ravenna	50
Solvic	Ferrara	50
Rumianca	Cagliari	50
	The Netherlands	
DSM	Beek	[b]
Shell Nederland Chemie	Pernis	50
	Belgium	
Solvic	Jemeppe	90
Badische Anilin- & Soda-Fabrik	Antwerp	50
	United Kingdom	
Imperial Chemical Industries	Hillhouse; Runcorn	180
BP Chemicals	Baglan Bay; Barry	140
	Spain	
Hispavic	Barcelona	50
Reposa	Miranda de Ebro	45
Monsanto	Monzon	30
	Scandinavia	
Fosfatbolaget, Sweden	Stockholm	85
Norsk Hydro, Norway	Heroya	50
Pekema Oy., Finland	Naantaly	[c]

[a] Courtesy *Chemical and Engineering News.*
[b] 75,000 ton/yr capacity planned for 1972.
[c] 50,000 ton/yr capacity planned for 1972.

Table 3. Per-Capita Consumption of Vinyl Chloride Polymers for 1967

Country	Consumption, lb
France	10.9
West Germany	18.3
Italy	9.8
Netherlands	16.6
Belgium and Luxembourg	10.1
EEC	*13.2*
United Kingdom	9.3
Austria	7.3
Denmark	7.8
Finland	6.5
Norway	7.9
Portugal	2.4
Switzerland	10.6
Sweden	14.0
EFTA	*8.7*
United States	11.1
Japan	12.9

(EFTA). West Germany has also a higher per-capita consumption than its Common Market partners (Table 3), although The Netherlands are not far behind. Within EFTA, Sweden has the highest per-capita usage with a figure of 14.0 lb. The figure for the United States is rather below that for the EEC but exceeds that for the EFTA block.

Production has also risen very rapidly in the United States; at the same time, the price has fallen sharply, as indicated in Figure 1. The capacity of the major producers is indicated in Table 4 for the monomer and in Table 5 for the polymers. The distribution of vinyl chloride polymers in various end uses is shown in Table 6.

Table 4. Capacity of United States Producers of Vinyl Chloride Monomer[a,b] (13)

Producer	Location	Capacity, 10^6 lb/yr
Allied Chemical Corporation	Geismar, La.	300
American Chemical, Inc.	Watson, Calif.	175
Continental Oil Company	Lake Charles, La.	600
Dow Chemical Company	Freeport, Texas	180
	Oyster Creek, Texas	800
	Plaquemine, La.	300
Diamond Shamrock Corporation	Deer Park, Texas	100
Ethyl Corporation	Baton Rouge, La.	300
	Houston, Texas	200
B. F. Goodrich Chemical Company	Calvert City, Ky.	1000
Monochem	Geismar, La.	300
PPG Industries	Lake Charles, La.	300
Tenneco	Houston, Texas	255
	Total	*4810*

[a] Courtesy *Chemical and Engineering News*.
[b] Early in 1970.

Table 5. Capacity of United States Producers of Vinyl Chloride Polymers[a,b] (14)

Company	Capacity, 10^6 lb/yr
Air Reduction Chemical Company	120
Allied Chemical Corporation	125
American Chemical, Inc.	70
Atlantic Tubing and Rubber	100
Borden Chemical Company	150
Diamond Shamrock Corporation	250
Escambia Chemical	50
Ethyl Corporation	125
Firestone Tire & Rubber Company	210
General Tire & Rubber Company	75
B. F. Goodrich Chemical Company	600
Goodyear Tire & Rubber Company	130
Great American Plastics Company	70
Hooker Chemical Corporation	90
Keysor Chemical	50
Monsanto Company	150
Olin Corporation	170
Pantasote Company	140
Stauffer Chemical Company	80
Tenneco Inc.	190
Thompson Plastics	150
Union Carbide Corporation	350
Uniroyal, Inc.	135
Total	*3580*

[a] Mid-1970.
[b] Courtesy *SPE Journal*.

Table 6. United States Domestic Consumption and Exports of Poly(vinyl Chloride) and Copolymers, million lb

Product	1967[a]	1968[a]	1969[b]	1970[b,c]
film and sheeting				
calendered	380	432	472	441
extruded	103	135	161	207
floor coverings				
calendered	250	267	267	240
coated	52	72	66	74
cable and wire	196	259	371	426
other extruded products	297	353	461	561
paper and textile coatings	101	125	99	93
phonograph records	104	107	132	150
plastisols for molding compounds	78	103	111	107
protective coatings and adhesives	72	94	84	85
injection and blow molding	72	94	102	132
other uses	302	278	340	603
exports	71	90	153	211
total	*2078*	*2409*	*2819*	*3330*

[a] Ref. 12.
[b] Ref. 15.
[c] Estimate.

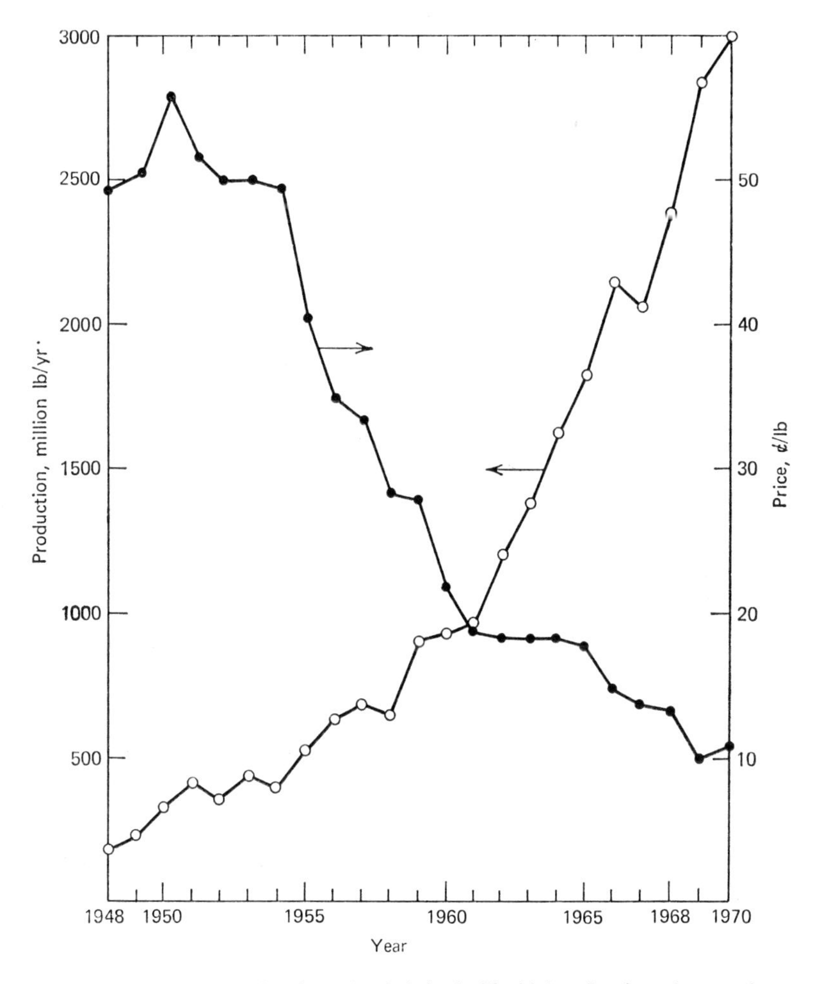

Fig. 1. Total United States production of poly(vinyl chloride) and price of general-purpose resin (calculated to equivalent 1968 dollars) (12).

The majority of countries with any industrial development have erected plants for the manufacture of PVC during recent years. These plants have been designed in most cases with the use of imported technology. Japan has been very active in this field besides having a sizeable export of manufactured PVC. The production and price history of vinyl chloride polymers for the last twelve years in Japan is summarized in Table 7. The major producers of PVC are listed in Table 8. A dramatic increase in the domestic market between 1966 and 1967 resulted in an increase in the per-capita consumption from 7.6 to 12.9 lb.

In the Soviet Union figures for the production and consumption are not readily available, but a recent article (17) gives some useful information. Production started in 1940 using suspension techniques for polymerization. Between 1950 and 1960 production increased by 200%, and then by 300% during the period from 1960 to 1965. It is claimed that the production of PVC in the Soviet Union is greater than that of any other plastic; further increases in production capacity are envisaged in the current five-year plan. It is estimated that in 1970 there are at least ten large manu-

Table 7. Production and Prices of Vinyl Chloride Polymers in Japan[a] (16)

	PVC (total)		Copolymer		Paste resin	
Year	Volume, ton	Price, yen/kg	Volume, ton	Price, yen/kg	Volume, ton	Price, yen/kg
1958	91,609	127	12,034	146	1,501	252
1959	180,091	122	15,193	150	3,014	244
1960	258,081	121	15,105	154	3,696	215
1961	308,933	115	19,577	157	4,175	193
1962	303,503	104	20,798	142	5,958	172
1963	348,848	104	26,193	134	8,269	171
1964	473,835	104	30,868	131	10,684	171
1965	482,973	102	26,883	132	10,606	171
1966	485,386	99	28,336	127	10,462	171
1967	697,967	96	32,281	125	11,544	166
1968	941,838	94	32,587	122	16,192	156
1969	1,047,075	91	32,289	111	25,526	142

[a] Courtesy *Japan Plastics Age.*

Table 8. Capacity of Japanese Producers of Vinyl Chloride Resins in 1970[a] (16)

Company	Capacity, metric ton/yr
Denki Kagaku Kogyo K.K.	54,780
Gunma Kagaku K.K.	54,780
Kanegafuchi Chemical Ind. Co., Ltd.	114,648
Kureha Chemical Ind. Co., Ltd.	87,384
Mitsui Toatsu Chemicals, Inc.	74,880
Mitsubishi Monsanto Chemical Co.	79,116
The Japanese Geon Co., Ltd.	122,820
Nippon Carbide Ind. Co., Ltd.	84,792
Nissin Chemical Ind. Co., Ltd.	57,612
Shin-Etsu Chemical Ind. Co., Ltd.	91,380
Chisso Corporation	87,300
Sumitomo Chemical Co., Ltd.	75,204
Tekkosha Co., Ltd.	66,216
Toa Gosei Chemical Ind. Co., Ltd.	54,780
Tokuyama Sekisui Co., Ltd.	30,804
Total	*1,136,496*

[a] Courtesy *Japan Plastics Age.*

facturing units in the U.S.S.R. The record of production is indicated below (the year 1965 being given an arbitrary rating of 100):

1965	1966	1967	1968	1969	1970
100	127	142	250	290	450

Another source (18) places Soviet, Polish, and total Eastern European capacity in 1970 at 300,000, 200,000, and 735,000 metric ton/yr, respectively.

As with all plastics materials, the price of PVC has fallen dramatically and this has contributed considerably to its high rate of growth. During the same period there has been a steady increase in price in nearly all basic raw materials so that PVC is now able to compete in a number of large-scale applications. Figure 1 shows the price structure in the United States over the last two decades. The downward price trend appears to have been halted in the latter half of 1967 when there were some small increases in both Europe and the United States. During 1968 the demand in Europe began to catch up with and, in the latter part of the year, exceed production capacity so that there were further small price increases. There have been many press announcements of new plants coming into production so that one can expect to see an excess of capacity over demand in the early 1970s.

Monomer

Production

Until the late 1950s, the production of monomeric vinyl chloride was generally carried out by following the classical route from acetylene and hydrogen chloride using a mercuric chloride catalyst (eq. 1). The mercuric chloride is supported on active

$$CH{\equiv}CH + HCl \xrightarrow{HgCl_2} CH_2{=}CHCl + 22.8 \text{ kcal/mole} \quad (1)$$

carbon and is contained in a series of reactors which are designed for the effective removal of the heat produced by the reaction. The catalyst is frequently modified by addition of potassium chloride and a number of heavy metal salts. It is essential that both gases be absolutely dry and free from impurities, such as phosphine in the case of acetylene. The catalyst is maintained at 80–150°C, the lower temperature being used when the catalyst is first brought into service; the temperature is gradually raised as its efficiency drops off. It is claimed that this loss in efficiency is due in part to volatilization of mercuric chloride and that the life of the catalyst can be prolonged if the reacting gases are passed first in one direction through the reactor tubes and then are reversed from time to time. A number of other catalysts have been studied but the majority of manufacturers use mercuric chloride as the main component. The gases are fed into the reactor tubes in equimolar proportions or with perhaps a slight excess of hydrogen chloride and the reaction goes substantially to completion. The hydrogen chloride is removed by distillation together with the light-end impurities, or by scrubbing. The monomer is dried and then distilled either under pressure or at atmospheric pressure using a refrigerated system. The monomer made by this process can contain as impurities very small amounts of acetylene, methylacetylene, 1,1-dichloroethane, allene, butadiene, or acetaldehyde.

Vinyl chloride is also produced by dehydrochlorination of dichloroethane (eq. 2).

$$ClCH_2{-}CH_2Cl \xrightarrow{\Delta} CH_2{=}CHCl + HCl \quad (2)$$

In the laboratory the reaction can be carried out using alcoholic caustic soda at about 60°C. On an industrial scale, vinyl chloride is produced from dichloroethane by pyrolysis at 300–600°C using a contact catalysis and at elevated pressures. The cracking reaction is taken to 55–60% conversion with about 95% efficiency. After cooling the gases, the vinyl chloride monomer of high purity is separated using a three-column purification system. The unchanged dichloroethane is returned to the cracking furnace while the hydrogen chloride is obtained as a pure anhydrous product which can then be reacted with acetylene using a mercuric chloride catalyst to produce a further quantity of vinyl chloride. This is known as the "*balanced route*" (see Fig. 2). Alternatively, the hydrogen chloride can be oxidized to chlorine using the Deacon or electrochemical processes, and this is then reacted with ethylene to produce dichloro-

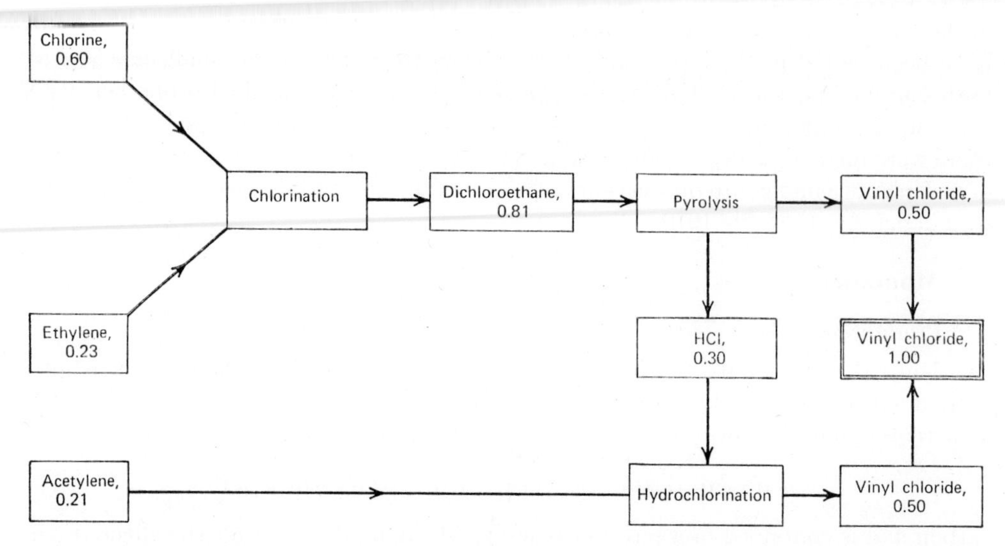

Fig. 2. Flow diagram of the "balanced route" process for production of vinyl chloride (weight balance).

ethane. This basic process has undergone considerable development with the need to use cheaper starting materials and reduce the cost of the monomer.

The most important development in vinyl chloride synthesis is the so-called *oxychlorination process* (19). In this process, ethylene is reacted with hydrogen chloride and oxygen to give dichloroethane, as shown in equation 3. A typical cata-

$$CH_2{=}CH_2 + 2\,HCl + \tfrac{1}{2}\,O_2 \longrightarrow ClCH_2CH_2Cl + H_2O \qquad (3)$$

lyst for this reaction consists of copper chlorides, potassium chloride, and lanthanum chloride on a support. The dichloroethane produced in this way as well as by chlorination of ethylene is pyrolyzed to vinyl chloride (eq. 2). The hydrogen chloride liberated is returned to the oxychlorination reactor where it undergoes the reaction indicated in equation 3. A flowsheet for such a process is shown in Figure 3. A typical analysis of vinyl chloride manufactured by oxychlorination is given in Table 9.

A further method using ethylene is the so-called *one-step process* (20,21). Ethylene, hydrogen chloride, and oxygen in the ratio of 2:1:1 are passed over a catalyst comprised of a mixture of cupric chloride and potassium chloride supported on diatomaceous earth at 470°C to give some 60% yield of vinyl chloride directly.

Table 9. Typical Impurity Levels in Vinyl Chloride

Impurity	Level, ppm
water	<100
acidity, as HCl	5
acetylene	1
other acetylenic compounds	7
ethylene	0.5
propylene	1
conjugated dienes	5
aldehydes, as acetaldehyde	3
nonvolatiles	50

In all the above processes from production of vinyl chloride it is essential that the hydrocarbon, whether it be acetylene or ethylene, is first produced in a pure form. When cracking processes are used, a greater part of the installation is concerned with the extraction and purification of the hydrocarbon—considerable saving can therefore be made if a mixture of acetylene and ethylene can be used without separation.

A process has been described in which a mixed dilute stream of acetylene and ethylene is used as a feedstock (22). It is produced directly from a hydrocarbon pyrolysis using naphtha. The mixture of acetylene and ethylene is purified only by removal of higher hydrocarbons; the other impurities, carbon monoxide, hydrogen, and methane, are left in as inert diluents. The acetylene in the mixture is reacted with hydrogen chloride in the presence of a mercuric chloride catalyst, and the vinyl chloride is then removed by washing out with dichloroethane. The ethylene in the remaining

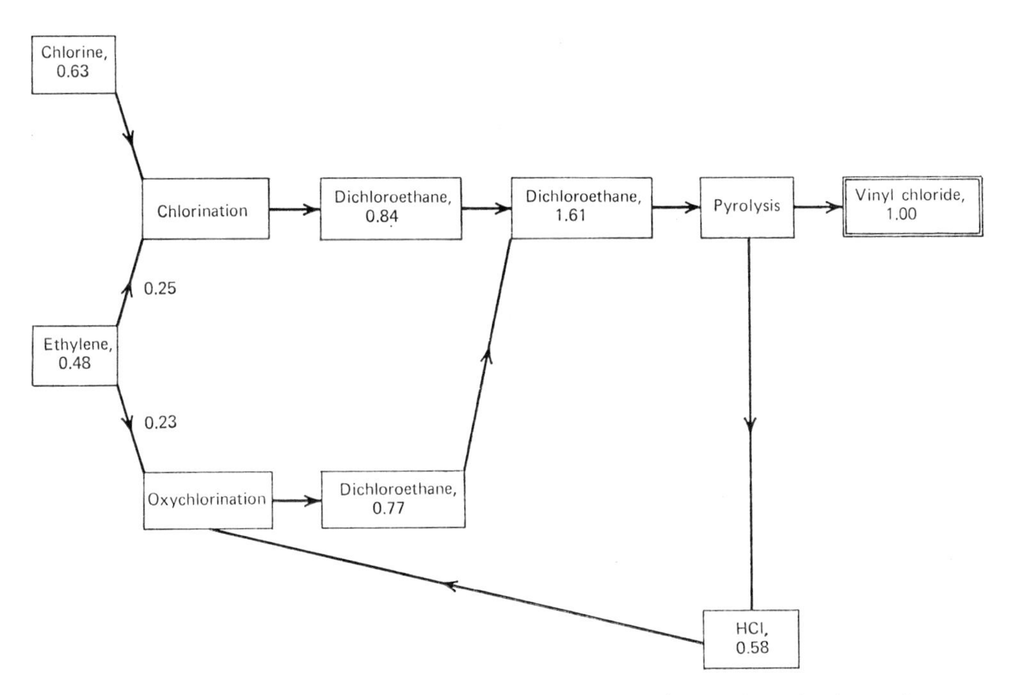

Fig. 3. Flow diagram of the oxychlorination process for production of vinyl chloride (weight balance).

gaseous mixture is then reacted with chlorine in the vapor phase over a fixed-bed catalyst. After removal of the dichloroethylene by condensation, the remaining gases are used as fuel gas in the cracking furnace. The dichloroethylene is cracked under pressure to give a mixture of vinyl chloride, hydrogen chloride, and unconverted dichloroethylene, which are separated in a three-column system. The hydrogen chloride is taken off the top of the first column and reacted in the first stage with the acetylene. The ratio of the acetylene and ethylene from the naphtha plant is controlled so as to keep the two reactions in balance and prevent any excess or shortage of hydrogen chloride. The pressure in the process falls progressively over each step so that there is need for only one compression stage. It is claimed that this is the cheapest process for the manufacture of the vinyl chloride, and particularly for smaller-scale operators (23).

Another recent development is the direct synthesis of vinyl chloride from ethane, air, and chlorine (24). This process, if successful on a commercial scale, could well result in a substantially decreased price for the vinyl chloride monomer.

It is impossible to decide precisely which process will produce vinyl chloride monomer at the lowest price. Local factors can have a significant effect; if hydrogen chloride is available as a waste gas, should its cost be accounted at zero value, or should the monomeric vinyl chloride process be credited with the cost of disposing of the hydrogen chloride if there was no other use for it? In the United States, oxychlorination in very large plants (in excess of 600 million lb/yr) seems to be the most economical route. The relative merits of processes based on acetylene or ethylene are examined critically in Ref. 25.

Properties

Vinyl chloride has a boiling point of approximately −14°C and is normally stored as a liquid under pressure. It can be transported safely without the addition of a polymerization inhibitor although as an added safeguard very small amounts of phenolic derivatives are sometimes added. This is frequently the case where the monomer may reach high temperatures due to climatic conditions; and these additives are removed before polymerization by washing with dilute caustic soda. Where the monomer is stored in cylinders over long periods, safety blow-off valves are usually provided as a precautionary measure. It has been generally assumed that vinyl chloride does not form peroxides by autoxidation as readily as do many other monomers and particularly such as vinylidene chloride. It has been reported that vinyl chloride can, under certain conditions, form a peroxide which can be isolated as a yellow oil (26); this is decomposed readily to form hydrogen chloride, carbon monoxide, carbon dioxide, formaldehyde, and glyoxal.

Vinyl chloride has only limited solubility in water (Fig. 4) so that in aqueous polymerization systems there are initially only two immiscible condensed phases

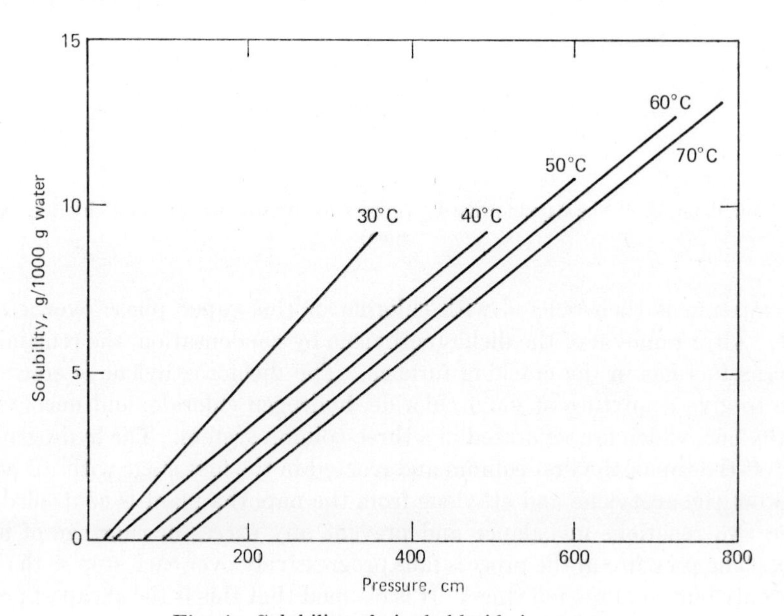

Fig. 4. Solubility of vinyl chloride in water.

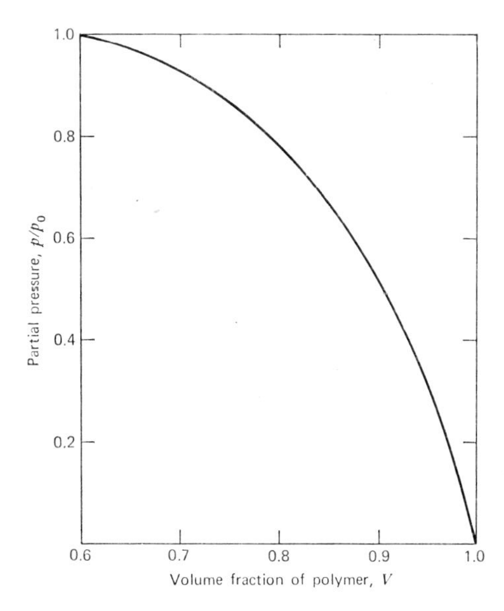

Fig. 5. Solubility of vinyl chloride in poly(vinyl chloride).

present in the system. Within a short time polymer separates out to form a third phase because it is totally insoluble in the monomer and precipitates completely. The monomer is soluble in the polymer to an extent as shown in Figure 5 where V is the volume fraction of polymer in the swollen phase and p/p_0 is the ratio of the actual pressure of the vinyl chloride to the saturation vapor pressure. This curve is for a temperature of 50°C which is a typical value at which polymerization is carried out. The solubility when expressed as a function of p/p_0 is reasonably independent of temperature.

Figure 6 shows several published sets of data (27–30) for the vapor pressure of vinyl chloride as a function of temperature. The full line represents what are believed to be the most accurate data, the maximum deviations of other data included in the diagram are of the order of −10% at 60°C. An empirical expression has been derived from these data, the vapor pressure, p, being given by equation 4,

$$\log p = a + \frac{b}{T} + 1.75 \log T + cT \tag{4}$$

where $a = 0.8420$
$b = -1150.9$
$c = -0.002415$
p = vapor pressure in atmospheres, absolute
T = temperature in °K

Other physical properties of the monomer are given in Table 10.

Much of the above information is taken from the manufacturers' technical literature (30) and has been obtained for the technical grade of vinyl chloride, which is only about 99.5% pure. The impurities comprise acetylene, (only a few parts per million if the polymerization is not to be affected), acetaldehyde, and higher chlorinated hydrocarbons if the monomer has been manufactured by the classical route using acetylene

Table 10. Physical Properties of Vinyl Chloride Monomer

molecular weight	62.501
boiling point at 760 mm, °C	−13.37
freezing point, °C	−153.79
flash point, °C	−78
liquid density, g/ml	
at −20°C	0.98343
at −25°C	0.99176
at −30°C	0.99986
viscosity, cP	
at −10°C	0.248
at −20°C	0.274
at −30°C	0.303
at −40°C	0.340
surface tension, dyne/cm	
at −10°C	20.88
at −20°C	22.27
at −30°C	23.87
refractive index, n_D^{15}	1.398
vapor pressure, mm	
at 25°C	3000
at −13.37°C	760
at −55.8°C	100
at −73.9°C	30
at −87.5°C	10
at −109.4°C	1
specific heat of liquid, cal/g/°C	0.38
specific heat of vapor, cal/g/mole/°C	10.8–12.83
latent heat of vaporization, cal/g	
at 25°C	71.26
at −13.37°C	79.53
latent heat of fusion, cal/g	18.14
critical temperature, °C	158.4
critical pressure, atm	52.2
absolute entropy at 25°C, cal/°K/g-mole	61.68
heat of formation, cal/g-mole	9000
dielectric constant at 10^5 Hz and 17.2°C	6.26

and hydrogen chloride, whereas hydrocarbons such as butadiene may be present when cracking techniques are used.

Handling and Safety Measures

Handling of vinyl chloride presents certain hazards because it is normally gaseous at room temperature and forms explosive mixtures with air; the explosive limits are 4.0% by volume in air (lower explosive limit) and 22.0% by volume in air (upper explosive limit). When large leakages of liquid vinyl chloride occur, it is difficult to wash the area clear because the monomer floats on the water and the vapor, which is heavier than air, does not disperse readily. Thus fires involving vinyl chloride are difficult to extinguish using water alone and carbon dioxide or chemical-type fire extinguishers should be available.

In high concentrations, the gaseous monomer has a sweet ethereal odor and produces anesthesia. Continual working in an atmosphere containing 500 ppm is considered sufficient to produce symptoms of drowsiness with an inability to concentrate.

A detailed study using rats (31) has shown that repeated daily exposure to concentrations of 2–5% of vinyl chloride causes no permanent damage. It has been confirmed from 5 minute exposures of human subjects to vinyl chloride in concentrations of up to 2.0% that a concentration of about 600 ppm is required to produce minimum effects of anesthesia under continuous exposure. Those affected by mild doses recover quickly when removed from the affected atmosphere but in the cases of more severe exposure, the administration of oxygen together with artificial respiration may be necessary if respiration is affected. The patient should be kept quiet and comfortably warm and special attention be given to those with uncertain heart action. In case of spillage, contaminated garments should be removed at once and the skin areas well washed with soapy water. If the skin is seriously affected, treatment as for frostbite should be given. A full account of all procedures to be used for handling vinyl chloride is given in Ref. 32.

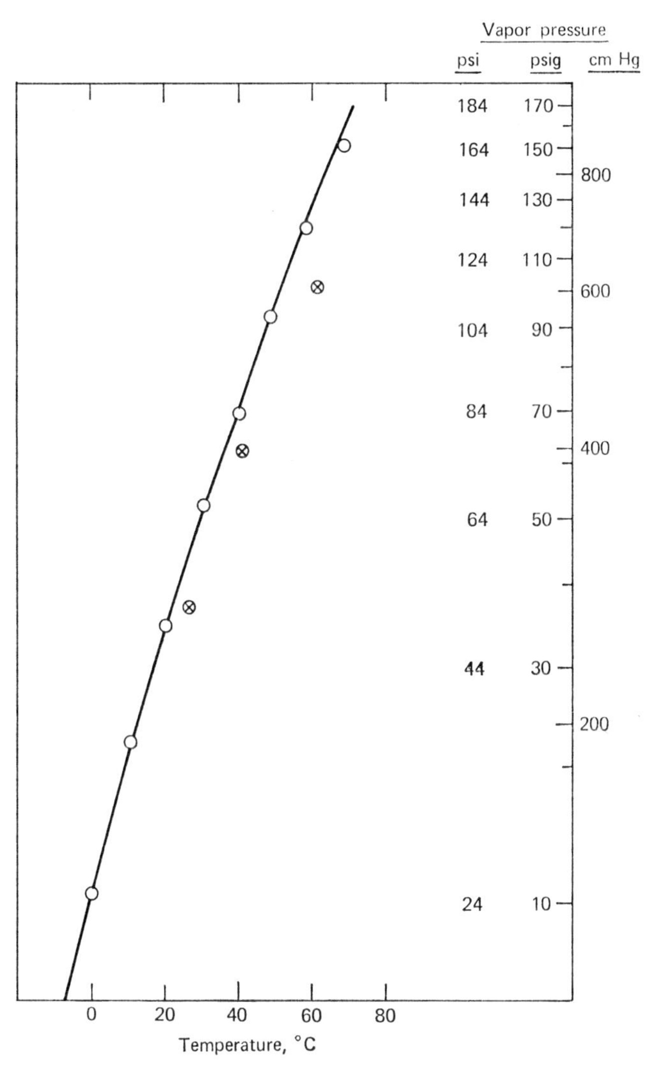

Fig. 6. Vapor pressure of vinyl chloride (27–29). Points marked ⊙ are from Ref. 28; those marked ⊗ are from Ref. 30.

Bibliography

1. E. Baumann, *Liebigs Ann.* **163,** 308 (1872).
2. V. Regnault, *Ann. Chim. (Phys.)* **58,** 307 (1835).
3. Chemische Fabrik Griesheim-Elektron, Ger. Pat. 278,249 (Oct. 11, 1912).
4. F. Klatte and A. Rollett (to Chemische Fabrik Griesheim-Elektron), U.S. Pat. 1,241,738 (Oct. 2, 1917).
5. I. Ostromislensky, *Chem. Zentralblatt* **1,** 1980 (1912).
6. E. W. Reid (to Carbide and Carbon Chemicals Corp.), U.S. Pat. 1,935,577 (Nov. 14, 1933).
7. A. Voss and E. Dickhauser (to I. G. Farbenindustrie A.G.), U.S. Pat. 2,012,177 (Aug. 20, 1935).
8. W. E. Lawson (to E. I. du Pont de Nemours & Co., Inc.), U.S. Pat. 1,867,014 (July 12, 1932).
9. W. L. Semon (to B. F. Goodrich Co.), U.S. Pat. 2,188,396 (Jan. 30, 1940); see also U.S. Pat. 1,929,453 (Oct. 10, 1933).
10. H. Fikentscher and C. Keuck (to I. G. Farbenindustrie A.G.), Ger. Pat. 654,989 (Jan. 6, 1938).
11. *Chem. Eng. News* **48,** 31 (Aug. 3, 1970).
12. M. J. R. Cantow in A. Standen, ed., *Kirk-Othmer Encyclopedia of Chemical Technology,* 2nd ed., Vol. 21, Interscience Publishers, a division of John Wiley & Sons, Inc., New York, 1970, pp. 369–412.
13. *Chem. Eng. News* **47,** 19 (Aug. 4, 1969).
14. *SPE J.* **26** (12), 38 (1970).
15. *Plastics World* **29,** 11 (Jan. 1971).
16. *Japan Plastics Age* **8,** 28 (Nov. 1970).
17. I. B. Kotlyav, *Plast. Mas.* p. 17 (Nov. 1967).
18. *Chem. Eng. News* **48,** 47 (Oct. 19, 1970).
19. *Chem. Week,* p. 93 (Aug. 22, 1964).
20. *European Chem. News,* p. 27 (Aug. 21, 1964).
21. Imperial Chemical Industries of Australia and New Zealand, Fr. Pat. 1,330,367 (June 6, 1962).
22. *European Chem. News,* p. 35 (Sept. 27, 1963).
23. S. Gomi, *Hydrocarbon Process. Petrol. Refiner* **43** (11), 165 (1964).
24. *Chem. Week,* p. 59 (March 3, 1971).
25. P. G. Caudle, *Chem. Ind.,* p. 1551 (Nov. 9, 1968).
26. G. A. Razuvaev and K. S. Minsker, *Zh. Obshch. Khim.* **28,** 983–991 (1958).
27. A. R. Berens, private communication.
28. J. Timmermans, *Physico-Chemical Constants of Pure Organic Compounds,* Elsevier Publishing Co., London, Vol. 1, 1950, p. 276; Vol. 2, 1965, p. 234.
29. E. R. Blout, W. P. Hohenstein, and H. Mark, *Monomers,* Interscience Publishing Co., Inc., New York, 1949.
30. *Vinyl Chloride Monomer,* bulletin, The Dow Chemical Co., 1954.
31. *Industrial Hygiene J.* **24,** 265 (1963).
32. Chemical Safety Data Sheet SD-56, Manufacturing Chemists Association Inc.

C. A. Brighton
BP Chemicals International Ltd.

POLYMERIZATION

Since 1946 kinetic studies of the free-radical polymerization of vinyl chloride have been carried out by many workers (1–21). The field has recently been reviewed (23). The majority of the early detailed studies were carried out in bulk using dilatometry to follow rates of polymerization. Dilatometry (qv) affords a sensitive technique for monitoring the conversion of this monomer because of the large volume contraction (36 percent contraction corresponds to 100 percent polymerization). However this approach has some serious limitations:

1. The range of investigation is restricted to conversions of about 20 percent or below, because separation of an insoluble polymer phase gradually obscures the meniscus at higher conversions.

2. Maintenance of isothermal conditions can be compromised by heat-transfer problems in a static or unstirred system, because of the high heat of polymerization (the most reliable values range from 23 to 26 kilocalories for the conversion of one gram-mole of liquid monomer to one monomer gram-mole of polymer (24,25)).

Although many important features of the polymerization were established in the decade 1946–1956, a universally acceptable mechanism did not emerge. More recent studies have concentrated on establishing the form of the precipitated polymer phase (11,15,17) defining the solubility relationships in the phases (26) and confirming that the kinetic features of the suspension process are essentially similar to those encountered in the bulk process (20,27). This last observation is very important since efficient heat transfer is more readily achieved using the suspension system, thus permitting collection of conversion–time data up to nearly 100 percent conversion (28). The new facts which are emerging from these combined approaches are already helping to restrict the choice of possible mechanisms.

Claims that the monomer can be polymerized by an anionic mechanism using organometallic or Ziegler-Natta type catalysts (29) have not been substantiated, initiation being attributed to extraneous free radicals (30). This conclusion is supported by the fact that reactivity ratios in copolymerizations involving vinyl chloride, and based on these catalysts, are the same as those derived with genuine free-radical initiators (30–32).

It has recently been suggested that polymerization can be effected anionically using alkyllithium or certain Grignard compounds (33,34). Polymers of high crystallinity can be isolated from the reaction products. However, the molecular weights of these polymers are extremely low (34) and it has been inferred that the stereoregularity is not substantially greater than that of free-radical polymers of the same molecular weight prepared at the same temperature (30,35,36).

Bulk Polymerization

There is general agreement among investigators that the bulk polymerization is characterized by the following features which must be taken into account in formulating a mechanism:

1. At a very early stage of the reaction the polymer precipitates from the reaction medium due to the insolubility of the polymer in the monomer. The system is therefore heterogeneous for a significant portion of the whole conversion.

2. With initiators which have half-lives considerably longer than the duration of the experiment, the rate is that of a polymerization auto-accelerating with conversion (1–7) (Fig. 1).

3. At a fixed conversion the dependence of the rate of polymerization on initiator concentration exhibits a fractional order in the vicinity of 0.5–0.6 (2,4,7,8,13).

4. At constant temperature the average molecular weight of the polymer is apparently independent of conversion or initiator concentration, at least for conversions below about 70 percent and over a wide range of initiator concentrations (2,3,6,7,

Table 1. Elementary Reaction Steps in Kinetic Schemes for Bulk Polymerization of Vinyl Chloride[a]

Equation no.	Reaction	Description	Rate
Reactions in the liquid phase			
(1)	$I \rightarrow R\cdot$	initiation	$R_i = fk_d[I]$
(2)	$R\cdot + M \rightarrow R\cdot$	propagation	$R_p = k_p[M][R\cdot]$
(3)	$R\cdot + M \rightarrow P + M\cdot$	chain transfer to monomer	$k_{tr,m}[M][R\cdot]$
(4)	$R\cdot + R\cdot \rightarrow P$	termination	$k_t[R\cdot]^2$
(5)	$M\cdot + M \rightarrow R\cdot$	propagation by radical from monomer transfer	$k_p[M\cdot][M]$
Reactions in polymer particles			
(6)	$R\cdot + P^x \rightarrow P^x + P^x\cdot$	trapping of radicals in polymer particles	$k_{trap}[R\cdot]F([P^x])$
(7)	$P^x\cdot + M^x \rightarrow P^x\cdot$	propagation	$R_p = k_p[P^x\cdot][M^x]$
(8)	$P^x\cdot + M^x \rightarrow P^x + M^x\cdot$	chain transfer to monomer	$k_{tr,m}[P^x\cdot][M^x]$
(9a)	$P^x\cdot + P^x\cdot \rightarrow P^x$	termination	$k_t'[P^x\cdot]^2$
(9b)	$P^x\cdot + M^x\cdot \rightarrow P^x$	termination	$k_t''[P^x\cdot][M^x\cdot]$
(9c)	$M^x\cdot + M^x\cdot \rightarrow M_2$	termination	$k_t'''[M^x\cdot]^2$
(10)	$M^x\cdot + M^x \rightarrow P^x\cdot$	propagation by radical from monomer transfer	$k_p[M^x\cdot][M^x]$
(11)	$M^x\cdot + P^x \rightarrow M^x + P^x\cdot$	chain transfer to polymer	$k_{tr,p}[M^x\cdot][P^x]$
(12)	$M^x\cdot \rightarrow M\cdot$	escape of monomer radicals to liquid phase	$k_{esc}[M^x\cdot]$

[a] Glossary of symbols:

I	initiator molecule
[I]	initiator concentration
f	initiator efficiency, fraction of primary radicals starting chains
F	function of
R_i	initiation rate, mole/liter-sec
R_p	propagation or polymerization rate, mole/liter-sec
R_t	termination rate, mole/liter-sec
k_d	initiator decomposition rate constant, sec^{-1}
$k_{tr,m}$	monomer transfer rate constant
$k_{tr,p}$	polymer transfer rate constant
k_p	propagation rate constant
k_t	termination rate constant
k_{esc}	radical escape rate constant
k_{trap}	radical trapping rate constant
M	monomer molecule
[M]	monomer concentration in liquid phase, mole/liter
$[M^x]$	average monomer concentration in active volume of a polymer particle
$M\cdot$	monomer radical in liquid phase
$[M\cdot]$	monomer radical concentration in liquid phase, mole/liter
$M^x\cdot$	monomer radical in polymer phase
$[M^x\cdot]$	average monomer radical concentration in the active volume of a particle
N_p	number of polymer particles per unit volume of suspension
P, P^x	polymer molecule
[P]	polymer concentration, monomer g-mole/liter suspension
$P^x\cdot$	chain radicals in polymer particles
$[P^x\cdot]$	average chain-radical concentration in active volume of a polymer particle
$R\cdot$	chain radical in liquid phase
$[R\cdot]$	average radical concentration in liquid phase
v_p	active volume of a polymer particle containing radicals
V_p	volume of polymer per unit volume of slurry
r	average radius of polymer particles
C	fractional conversion

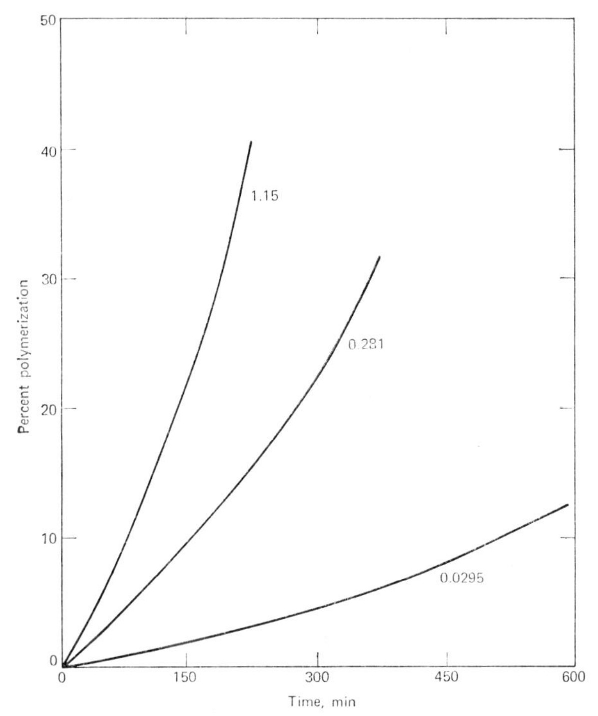

Fig. 1. Typical conversion–time curves at 47°C for polymerization of vinyl chloride at three different levels, in mole percent, of benzoyl peroxide and showing the auto-accelerating characteristic.

9,37). Less extensive evidence indicates that the molecular-weight distribution is equally insensitive to conversion and initiation rate in isothermal polymerization (38–40), although the kinetic chain length increases by a factor of threefold or more over the first 40 percent of conversion before levelling out (17,37).

5. Studies of the molecular weight of the polymer as a function of temperature of polymerization show a maximum molecular weight at about −30°C (13,31).

6. Addition of preformed polymer to the system produces an increase in the rate of polymerization. The addition to the bulk system of small amounts of liquids which are solvents for the polymer eliminates the auto-acceleration, at least in the range of conversion where the accumulated polymer is below its solubility limit in the system. Beyond this conversion the auto-acceleration reasserts itself (2,11,21,28). In other words, the auto-accelerating feature manifests itself only when polymer is present in a precipitated form but not when polymer is present in soluble form. (As pointed out above, the polymer is completely insoluble in monomer.)

Many of the postulated mechanisms represent the overall polymerization as a composite of two additive reactions—a liquid-phase and a solid-phase process. The various kinetic schemes which have been proposed for the bulk polymerization are summarized in Table 1. See also FREE-RADICAL POLYMERIZATION; KINETICS OF POLYMERIZATION.

Reaction 6 in Table 1 is a key reaction in determining the kinetic behavior. Bengough and Norrish proposed a surface immobilization theory (2). They assumed that the trapping of radicals in the polymer particles (reaction 6) is due to chain transfer of R· with polymer P^x on the surface of the particles, and that the effective surface area of the suspended polymer was proportional to the weight of polymer raised to the two-thirds power (eq. 13). They also assumed that [M], the monomer concentration

$$k_{trap}F[(P^x)] = Kk_{tr,p}[P^x]^{2/3} \tag{13}$$

in the liquid phase, was equal to $[M^x]$, the monomer concentration in the particles, that $k_t' = k_t'' = k_t''' = 0$, and that reaction 8 is always followed by reaction 12, so that the rate of trapping of radicals was equated to the rate of escape of radicals. The final rate expression takes the form given by equation 14, in agreement with observations that at

$$-\frac{d[M]}{dt} = K[I]^{1/2}([M] + K''[P^x]^{2/3}) \tag{14}$$

equal conversions ($[P^x]$ constant) the rate is proportional to the square root of the catalyst concentration. They confirmed that the rate was proportional to the conversion raised to the two-thirds power over the conversion range 8–20%.

The scheme proposed by Mickley and co-workers (11) is similar in many respects to that proposed by Bengough and Norrish. Thus they assume $k_t' = k_t'' = k_t''' = 0$. However, the radical trapping reaction (eq. 6) is either by shallow penetration by diffusion of R· radicals from the liquid phase into polymer particles, or by precipitation of the growing chain radicals, R·, onto the surface of the polymer particles. Kinetically indistinguishable, these reactions are written as equation 15, where D_L' is the diffusion

$$k_{trap}F([P^x]) = 4\pi r N_p D_L'[R\cdot] \tag{15}$$

coefficient of the chain radicals, R·, in the liquid phase. It is assumed that a unit volume of suspension contains N_p spherical polymer particles of average radius r. The radicals within the polymer particles, $P^x\cdot$ and $M^x\cdot$, are considered as being confined to an active volume, V_p, of each particle. When the conversion is low the polymer particles are assumed to be small and the active volume is equated with the total volume of the particles (eq. 16). As the conversion increases, the polymer particles are

$$v_p = 4(\pi r^3)/3 \tag{16}$$

assumed to increase in size so that radical activity is confined to the outer peripheral shell, of thickness δ, of the particles (eq. 17).

$$v_p = 4(\pi r^2 \delta) \tag{17}$$

The escape of monomer transfer radicals from polymer particles to liquid phase (reaction 12) is written as equation 18, the subscript s referring to surface concentration

$$k_{esc}[M^x\cdot] = 4\pi r N_p D_L'[M^x\cdot]_s \tag{18}$$

in the polymer particle. It is also assumed that the rate of reaction 11 is much greater than the sum of the rates of reaction 6 and reaction 4. The overall rate of polymerization then becomes equation 19.

$$-\frac{d[M]}{dt} = K'''[I]^{1/2}([M] + K''''N_p v_p[P^x]) \tag{19}$$

If N_p, the number of particles per unit volume of suspension, and $[P^x]$, the polymer concentration in the particles are constant, then equation 20 holds. The rate increment

$$-\frac{d[M]}{dt} = K'''[I]^{1/2}[M] + K'''''[I]^{1/2}v_p \tag{20}$$

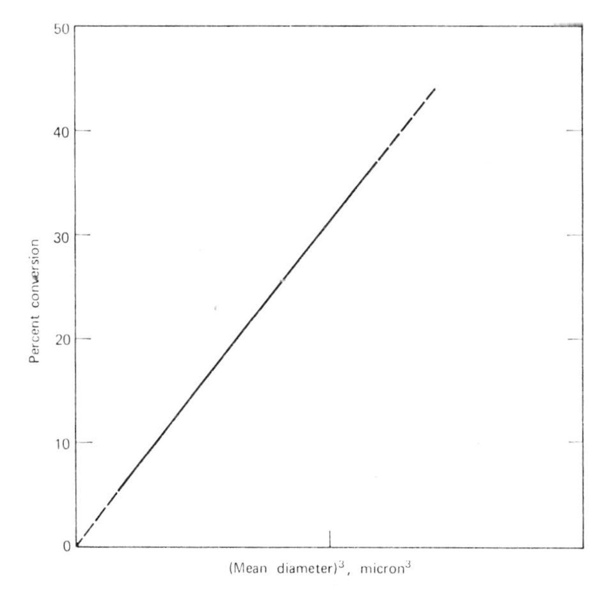

Fig. 2. Variation of microgranule size with conversion in bulk polymerization of vinyl chloride.

due to the presence of the polymer phase $K'''''[\mathrm{I}]^{1/2}v_p$ will therefore be proportional to the total polymer volume V_p when the conversion is low ($v_p = V_p/N_p$) and the total polymer surface, S, or $(V_p)^{2/3}$ at higher conversion (eq. 21). These authors (11)

$$v_p = \frac{4\pi r^2 \delta N_p}{N_p} = \frac{\delta S}{N_p} = \frac{(V_p)^{2/3}}{N_p} \tag{21}$$

observed a first-power law on polymer up to 5 percent conversion and a two-thirds power law on polymer from 8 to 30 percent conversion.

The validity of both mechanisms proposed so far (2,11) depends on a strong disposition to monomer and polymer transfer in the system. The invariance of molecular weight with catalyst concentration at a given temperature supports high transfer activity, but to achieve conformity with the observed independence of molecular-weight average and distribution with conversion implies constancy of the relative concentration of monomer and polymer at the locus of the polymerization.

The surface of the precipitated polymer plays a vital role in both mechanistic theories. Mickley appears to have been the first worker to consider the kinetics of nucleation, and the manner in which new surface is generated, as a function of conversion based on the aggregation of the insoluble polymer chains, generated homogeneously in solution. Using the treatment of von Smoluchowski (12) for the flocculation of particles in a suspension, he concluded that relatively early in the polymerization a stable situation will arise, which is characterized by a large and nearly constant number of aggregates of narrow size distribution suspended in the monomer. These aggregates will increase in volume but not in number as conversion increases, ie, N_p remains constant. He speculated that in the early stages of conversion the aggregates would be significantly less than 1 μ in size and their whole volume would have free access to monomer and radical activity in the suspension, leading to the first-order dependence on polymer concentration. Later only the outer shells would be involved owing to the increase in particle size and the two-thirds power dependence would take over.

Several of Mickley's ideas on the formation of polymer particles and the subsequent growth of surface with conversion have been completely vindicated in recent publications (15,17,22). In particular, a detailed study of the microstructure of the precipitated polymer phase as a function of conversion has been made (40). Some observations in the very early stages (less than 1 percent conversion) were consistent with nucleation phenomena and were associated with a discontinuity in the conversion–time plot. Beyond this stage and up to 30 percent conversion the precipitated polymer was composed of spheres of uniform size and constant in number. The results are shown in Figure 2 which illustrates the linear relationship between the conversion and the cube of the diameter of the particles. At conversions higher than 30 percent analysis was complicated by the incidence of nondispersible, fused particle clusters and an occasional small particle, suggesting limited secondary growth. The author made the interesting observation that the rate of polymerization was some 34 percent faster than for the pure monomer when the latter was diluted 40 volume percent with methanol, whereas the aggregates were flocculent and irregularly shaped.

Considerable depression of the rate constant for the mutual termination of macro radicals in the viscous polymer phase (reaction 9a) in comparison with the corresponding termination in the free monomer phase (reaction 4) is quite plausible. The complete absence of termination in this phase (eg, by reactions 9b and 9c) is less plausible. Breitenbach and Schindler (4) have considered this situation by assuming $k_t' < k_t$, and allowing the overall termination constant to decrease with conversion in accordance with equation 22, where C is the degree of conversion and a is a constant. According to their mechanism the rate of polymerization is given by equations 23–24.

$$k_{t(\text{overall})} = k_t/(1 + aC) \tag{22}$$

$$\frac{dC}{dt} = K''''''[\text{I}]^{1/2}[\text{M}](1 + K'''''''C)^{1/2} \tag{23}$$

$$C = K'''''''V_0^2t^2/4 + V_0t \tag{24a}$$

where

$$V_0 = K'''''''[\text{I}]^{1/2}[\text{M}] \tag{24b}$$

The authors claim that their experimental conversion curves are consistent with this equation up to 20 percent conversion. However, Cotman (40) found that the constant K''''''' was not single-valued but increases with the initiator-to-monomer ratio.

Attention has been drawn by Talamini (23) to an interesting empirical rate equation which can be derived from an observation of Danusso (7). The latter noted that conversion–time curves of experiments carried out at the same temperature with different initiator concentrations superimpose if graphed as conversion versus $[\text{I}]^{1/2}t$ instead of t. This implies a kinetic equation of the form in equation 25.

$$C = F([\text{I}]^{1/2}t) \tag{25}$$

This is developed into equation 26. With α constant and $n = 2$ this becomes equation 27. Some resemblance exists between this function and that of Breitenbach (equation 6) and also that of Magat (10) derived on the basis of non-steady-state kinetics.

$$C = \sum_{i=1}^{n} \alpha_i [\text{I}]^{1/2} t^i \tag{26}$$

$$C = \alpha_1[\text{I}]^{1/2}t + \alpha_2([\text{I}]^{1/2}t)^2 \tag{27}$$

Many authors (11,17,44) have drawn attention to the similarities and contrasts in the polymerization characteristics of vinyl chloride and of acrylonitrile (see ACRYLONITRILE POLYMERS). Auto-acceleration can be eliminated in both systems in the presence of suitable solvents for the polymers, although abnormal kinetic orders with respect to monomer and catalyst can still occur, particularly at high monomer concentrations. Under heterogeneous conditions, in bulk or in the presence of nonsolvents for the polymers, both show auto-acceleration, but while vinyl chloride retains relatively normal kinetic features ($R_p \propto [I]^{\sim 0.5}[M]^{\sim 1.0}$), acrylonitrile gives abnormal kinetics ($R_p \propto [I]^{\sim 0.9}[M]^{\sim 2}$). Bamford and Jenkins (44,45) have accounted for the behavior of acrylonitrile by the radical occlusion theory. Because the monomer does not swell the polymer, and because the transfer constant to the monomer is very low, polymer radicals which diffuse into or precipitate onto polymer particles have a very restricted opportunity of terminating by mutual interaction or of transferring their activity to monomer.

Apart from the proposals put forward by Breitenbach (4), the other mechanisms dealt with so far attempt to account for the normal kinetic orders observed in vinyl chloride by postulating that the potentially occluded radicals in the polymer phase can readily transfer their activity back to the liquid phase by transfer to monomer.

There is almost unanimous agreement (but see Ref. 38) that the monomer transfer constant is high in the vinyl chloride system although there is a fair spread in the published values (see Table 2).

Table 2. Chain-Transfer Constant to Monomer for Vinyl Chloride

Temp, °C	$C_{tr,m} \times 10^4$	Reference
25	4.3	21
30	6.25	47
40	7.7	46
50	13.5	47
50	14.8	37
50	50.0	11
60	10.8–12.8	47
70	23.8	47

However, the distinction of deepest significance between the two systems from the kinetic and mechanistic viewpoint has only recently been appreciated. Whereas acrylonitrile is virtually insoluble in its polymer (44), recent measurements have demonstrated (26) that the equilibrium swelling of the polymer by the monomer is quite marked in the vinyl chloride system. In the temperature range 20–70°C, which embraces the range normally employed for the manufacture of commercial polymers, the composition of the swollen polymer phase is approximately 60:40 polymer: monomer (v/v) in the presence of a liquid monomer-rich phase. There is good justification for the belief that this equilibrium condition is met during polymerization, at least under most conditions of commercial manufacture (28). Under these circumstances the polymerization in the swollen polymer phase can be regarded as quasi-normal, the unusual features being that:

1. Initiation is probably provided by diffusion and deposition of radicals from the free monomer phase.

2. The relative concentrations of monomer and polymer in the swollen phase is constant while a free monomer phase persists.

3. The normal termination reaction 9a may be severely depressed because of the high viscosity. It seems far less likely that the terminations involving a small radical (reactions 9b and 9c) can be ignored.

Thus the overall process can be regarded as the sum of two reactions (reactions 1–5 in the liquid phase and 6–12 in the polymer particles) each of which obeys the normal kinetic laws valid for homogeneous free-radical polymerization (qv). However, serious difficulties arise in attempting to express the stationary radical concentrations in either phase in terms of known quantities where termination occurs in both phases. By way of illustration, suppose that in the general scheme all radicals entering the polymer phase from the monomer phase terminate ultimately, and by and large indirectly, by reaction 9c so that reactions 9a, 9b, and 12 are absent. Further, for simplicity, suppose there is initially unit volume of monomer containing M_0 moles of monomer and I moles of catalyst and that no volume change accompanies polymerization, so that there is unit volume of slurry at any fractional conversion C. Concentrations are expressed in moles per unit volume of the phase in which the reactant is present.

Assuming steady-state conditions, then the overall rate of initiation per unit volume of slurry is equal to the overall rate of termination per unit volume of slurry, so that equations 28 or 29 hold, where V_p' is the total volume of a particle. Without

$$(1 - N_pV_p')fk_d\left(\frac{\mathrm{I}}{1 - N_pV_p}\right) = (1 - N_pV_p')k_t[\mathrm{R}\cdot]^2 + N_pV_p'k_t'''[\mathrm{M}^x\cdot]^2 \quad (28)$$

$$fk_d\mathrm{I} = (1 - N_pV_p')k_t[\mathrm{R}\cdot]^2 + N_pV_p'k_t'''[\mathrm{M}^x\cdot]^2 \quad (29)$$

making assumptions about the proportion of chain radicals R that enter the polymer phase in terms of some function of that phase, such as its volume or surface area, there appears to be no simple means of expressing the stationary concentration of radicals, $\mathrm{R}\cdot$ or $\mathrm{M}^x\cdot$, in either phase in explicit terms. Although one could arbitrarily define a stationary radical concentration $[\mathrm{X}\cdot]$ (eq. 30) akin to a root-mean-square average

$$[\mathrm{X}\cdot] = \sqrt[2]{(1 - N_pV_p')[\mathrm{R}\cdot]^2 + N_pV_p'(k_t'''/k_t)[\mathrm{M}^x\cdot]^2} \quad (30)$$

radical concentration throughout the whole suspension, and reproducing the behavior of the actual system, such an approach does not appear to advance a solution.

Recently, Talamini (20) has attempted to evade this problem in a rather ingenious way. He assumes equilibrium between the polymer-rich phase and monomer-rich phase, so that each phase has constant composition throughout the range of coexistence of the phases. Writing A for the monomer–polymer ratio in the concentrated phase, and R_d and R_c for the specific rates in the dilute and concentrated phase, respectively, (expressed as fractional conversion per unit time per unit weight of phase), and

$$Q = R_c/R_d \quad (31)$$

Talamini obtains equation 32,

$$\frac{dC}{dt} = R_d + CqR_d \quad (32)$$

where

$$q = AQ - A - 1 \quad (33)$$

Integration and replacement of R_d by $K''''''''[I]^{1/2}$ leads to equation 34

$$C = (1/q) \exp (qK''''''''[I]^{1/2}t - 1) \tag{34}$$

or the corresponding series expanded form (eq. 35).

$$C = (1/q) \sum_{n=1}^{\infty} [(qK'''''''')^n/n!][I]^{n/2}t^n \tag{35}$$

There are reservations regarding the development of the equation and the validity of this integration, in view of the arbitrary assumption that R_d and q are constants. However, the justification for this approach must be gaged by the fact that the author succeeds in fitting to equation 35 a variety of experimental conversion curves covering a range of initiators over a range of concentrations (see Fig. 3) and up to 60 percent conversion. Since the monomer-rich phase disappears at 70 percent conversion the original postulates would not hold beyond this point.

The author finds that q, "the autocatalysis factor," is essentially independent of the initiator for the three initiators studied, indicating that the quantitative aspects of auto-acceleration arising from phase separation are inherent properties of the monomer–polymer system. His treatment leads to values of $k_p/k_t^{1/2}$ for the monomer phase in the range 0.7–0.5 at 50°C; these are in reasonable agreement with other published values derived from homogeneous solution polymerization (Table 3). Identity of the mechanisms for bulk and suspension polymerization is also confirmed (20,27).

Table 3 is a collection of the published kinetic constants for the polymerization. No reliable values for the chain-transfer constant to terminated polymer appear to have been published.

Branching indexes can be used to calculate indirectly chain-transfer constants to terminated polymer on the assumption that branching arises exclusively by this

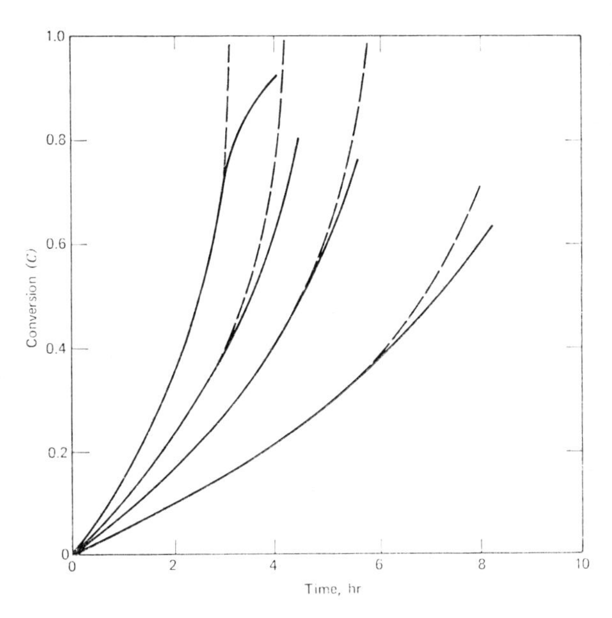

Fig. 3. Conversion curves for polymerization of vinyl chloride. Solid curves: polymerization of vinyl chloride in bulk initiated by four different levels of 2,2′-azobisisobutyronitrile. Dashed curves: theoretical curves based on equation 34 and with q and K given by 6.07 and 1.22, respectively.

Table 3. Propagation and Termination Constant for Vinyl Chloride Polymerization

k_p, (liter/mole-sec)	$k_t \times 10^{-6}$, (liter/mole-sec)	$(k_p/k_t) \times 10^6$	$k_p/(k_t)^{1/2}$	Temp, °C	Remarks	Ref.
6,200	1,100		0.2	25	soln	47
11,000	2,100		0.24	50	soln	47
		71	0.24	25	bulk	47
			0.7	50	soln	48
			0.6	50	soln	19

route. Values from this source range from 8.0×10^{-4} for polymer prepared at -50°C to 20×10^{-4} for polymer prepared at $+60$°C (12).

Effect of Oxygen

The inhibitory effect of oxygen in the polymerization of vinyl chloride has been investigated by several workers (1,40a,41–43). The authors of Ref. 40a showed that inhibition was due to copolymerization of the oxygen with the monomer to give polyperoxides which accumulated during the induction period. They studied the properties of the isolated peroxides and claimed that ideal 1:1 copolymers with oxygen could be formed. In a kinetic study of the effect of oxygen on the bulk polymerization (42) it was found that the rate of reaction with oxygen was nearly independent of the initial concentration of oxygen and proportional to the square root of the initiator concentration.

In a recent study the effect of initial oxygen pressures in the range 150–1200 mm Hg on the emulsion polymerization system was explored (43). The length of the induction period was inversely proportional to the square root of the initiator concentration, directly proportional to the square root of the initial amount of oxygen, and nearly independent of the emulsifier concentration. It was less firmly established that the duration of the induction period was inversely dependent on the volume of water phase, proportional to the 0.4 power of the volume of the gas phase, and independent of the volume of the monomer phase. The authors account for their observations by a cross termination reaction between normal and peroxy polymer chain radicals (eq. 36).

$$R\cdot + RO_2\cdot \rightarrow P \tag{36}$$

Solution Polymerization

One must distinguish between pseudo-solution polymerizations, in which the monomer–solvent combination gives rise to a swollen, precipitated, polymer-rich phase, and true solution polymerization in which the monomer and added solvent create conditions for true solution of the polymer. Solvents in the former class are methanol and cyclohexane, whereas true solution can be achieved with tetrahydrofuran, chlorobenzene, 1,2-dichloroethane, diethyl oxalate, and 2,4,6-trichloroheptane, provided the solvent concentrations in the systems are above a certain limit. In the case of cyclohexane (11) the kinetic behavior is identical with the bulk polymerization, although the polymerization rate is significantly lower. In the presence of methanol an enhanced rate of polymerization has been observed (17). On the basis of the mechanism proposed in Ref. 20 for the bulk polymerization, these rate differences could be accounted for by promotion (cyclohexane) or depression (methanol) of the swelling, and hence of the termination rate constant, in the polymer-rich phase.

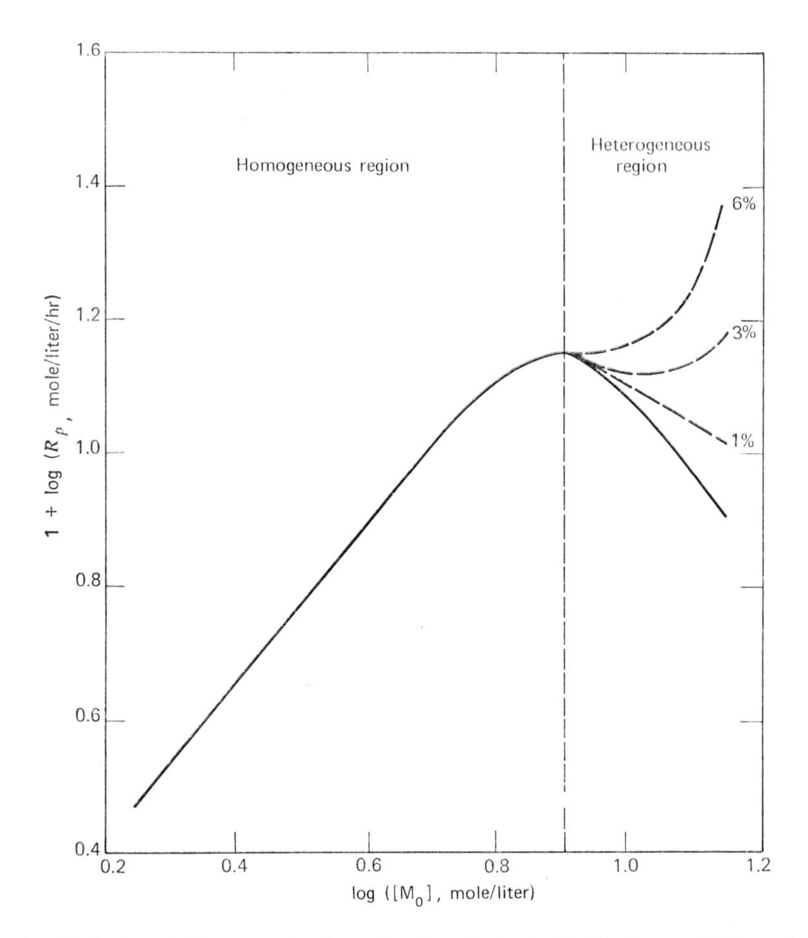

Fig. 4. Variation of the rate of polymerization of vinyl chloride in 1,2-dichloroethane as a function of monomer concentration at a fixed concentration of lauroyl peroxide. Solid curves: initial polymerization rates. Dashed curves: rates at indicated degrees of conversion.

In true solution systems polymerization rates are linear with time, ie, auto-acceleration is absent. However, auto-acceleration reappears when the monomer–solvent ratio is increased to the point where a granular precipitate manifests itself. Below this monomer concentration the rate of polymerization increases with the monomer concentration. The precise kinetic orders with respect to monomer and catalyst depend not only on the diluent but also on the investigator. Values ranging from 1 to 2 with respect to monomer have been quoted in the case of tetrahydrofuran (11,19,46). In conjunction with the effect discussed in the next paragraph it seems likely that these conflicting results could arise partly through the use of rate data acquired at different conversions by different workers, although solvent transfer may also complicate the kinetics. Molecular weights are low in both tetrahydrofuran (11) and chlorobenzene (19) and in the latter solvent the transfer appears to be degradative in character (19,46).

Many workers have observed an unexpected feature over the range of monomer–solvent ratios embracing the critical region of phase separation or precipitation, ie, that initial rates of polymerization decrease with increasing monomer concentration (11,19,

21). Because of the auto-accelerating effect in this domain the inverse dependence of rate on increasing monomer concentration becomes less pronounced with increasing conversion. At sufficiently high conversions a more normal increasing rate with increasing monomer concentration has been observed. Some of these features are summarized in Figure 4.

Most of these apparently anomalous features are capable of explanation on the basis of the mechanism proposed for the bulk polymerization (20). Controlling factors include the composition and the relative volumes of the polymer-rich and monomer-rich phases, the manner in which the propagating chain radicals distribute themselves between the two phases, and the ratio of the rate constants for termination in the two phases (19a,19c). Using either 1,2-dichloroethane or chlorobenzene as solvent, a significant observation has been made (19a); the molecular weight of the polymers, isolated at equivalent conversions, varies in the same way with monomer concentration, as does the average rate of polymerization, averaged over the same conversions (see Fig. 4).

Chain Transfer and Molecular-Weight Control

Sometimes it may be necessary to prepare a low-molecular-weight version of a given type of PVC polymer, and preservation of the required physical properties of the polymer powder can be obtained only by using the same polymerization temperature as used in the preparation of the parent. Since, in general, catalyst type and concentration do not exert any control over molecular weight in the vinyl chloride system, this situation would call for the use of a chain-transfer agent. See also CHAIN TRANSFER.

This mode of molecular-weight control could also be used if there were a temperature ceiling, due to pressure limitations, associated with the use of the polymerization vessel. The high reactivity of the propagating chain radical R· or $P^x\cdot$ in the vinyl chloride system implies that many additives will undergo transfer reactions in the system and reduce the molecular weight. However, choice can be severely restricted by the tendency for many additives to cause retardation effects and increase reaction times (see also INHIBITION AND RETARDATION). Halogenated methanes and ethylene

Table 4. Chain-Transfer Constants for Additives in Vinyl Chloride Polymerization at 50°C

Additive	$C_{tr} \times 10^4$	Reference
acetaldehyde	110	47
dodecyl vinyl ether	156	47
tetrahydrofuran	24	47
1,1,2,2-tetrachloroethane	60	46
trichloroethylene	850	46
toluene	48	46
carbon tetrabromide	50×10^4	47
	54×10^4	[a]
n-butyl mercaptan	250×10^4	[a]
chloroform	160	[a]
benzene	7.2	[a]
isopropylbenzene	270	[a]
cyclohexane	20	[a]
triethylamine	1128×10^4	[a]

[a] Estimates based on the *Q-e* scheme.

derivatives are the most favored chain-transfer agents. A variety of published chain-transfer constants for the vinyl chloride system are given in Table 4.

Early observations (5) made with very low concentrations of very active chain-transfer agents such as tetrabromomethane and *n*-dodecyl mercaptan revealed some startling effects in addition to the very low-molecular-weight polymers anticipated: (a) autocatalysis was eliminated; and (b) initial rates of polymerization were markedly faster than observed for the bulk polymerization without the chain-transfer agent, and increased with concentration of transfer agent to give finally a shallow maximum or plateau value.

Linear rates or absence of auto-acceleration are not inconsistent with the two-phase theory proposed earlier (19,20) when due allowance is made for the much lower viscosity in the polymer swollen phase when the polymer formed is of very low molecular weight, as in the presence of carbon tetrabromide. The kinetic distinctions between the two phases would tend to disappear; in particular the termination rate constant, k_t', in the swollen polymer phase would approach the value in the monomer-rich phase, k_t, and from the kinetic viewpoint the system would tend to manifest the characteristics of a homogeneous system.

The original explanation by the authors for the second phenomenon, which has recently been re-affirmed and confirmed for the heterogeneous bulk case and extended to embrace similar behavior in the solution polymerization of the monomer in 1,2-dichloroethane, has been the cause of considerable controversy. The authors propose (49) that in the normal bulk or solution polymerization of vinyl chloride, degradative chain transfer to monomer plays a significant role (eqs. 37–39).

Transfer to monomer

$$R\cdot + M \longrightarrow M\cdot \tag{37}$$

Termination

$$M\cdot + R\cdot \longrightarrow P \tag{38}$$

$$M\cdot + M\cdot \longrightarrow \text{inactive products} \tag{39}$$

They quote the frequently observed catalyst exponent of slightly greater than 0.5, which describes the influence of catalyst concentration on rate, as evidence for their proposal. In the presence of increasing amounts of active chain-transfer agent, T, termination by the relatively inactive monomer transfer radicals, M·, is gradually eliminated by transformation of the M· radicals into the more active radicals, T·, derived from the transfer agent (eqs. 40 and 41).

Transfer to transfer agent

$$M\cdot + T \longrightarrow T\cdot \text{ and inactive product} \tag{40}$$

Propagation

$$T\cdot + M \longrightarrow R\cdot \tag{41}$$

Rate promotion by the transfer agent requires that the T· radical derived from the transfer agent have a greater propensity to propagate rather than terminate than does the monomer transfer radical from which it sprang. An alternative, and probably more plausible, explanation was recently offered (50) which is consistent with the two-phase theory (19,20).

Copolymerization

Kinetic studies of copolymerization systems involving vinyl chloride have generally been restricted to one objective, namely, derivation of reactivity ratios. In most copolymerization reactions the composition is primarily determined by propagation, since initiation and termination reactions occur far less frequently than do propagation reactions. Thus, under steady-state conditions the consumption of monomers is given by equation 42, where $[M_1]$ and $[M_2]$ refer to monomer concentrations and r_1 and r_2 are

$$\frac{d[M_1]}{d[M_2]} = \frac{[M_1]}{[M_2]}\left(\frac{r_1[M_1] + [M_2]}{r_2[M_2] + [M_1]}\right) \qquad (42)$$

the monomer reactivity ratios as conventionally defined (see COPOLYMERIZATION). At low conversions $d[M_1]/d[M_2]$ may be replaced by m_1/m_2, where m_1 and m_2 refer to polymer composition in mole fractions; substituting $n = m_1/m_2$ and $x = M_1/M_2$, the equation rearranges to equation 43 and a plot of the lefthand side against x^2/n

$$x(n - 1)/n = r_1\left(\frac{x^2}{n}\right) - r_2 \qquad (43)$$

enables r_1 and r_2 to be abstracted as the slope and the intercept, respectively. A selection of reactivity ratios for copolymer systems involving vinyl chloride is given in Table 5. It will be appreciated that in copolymer systems that follow the phase-

Table 5. Reactivity Ratios for Copolymerization of Vinyl Chloride (M_1)

M_2	r_1	r_2	Temp, °C
acrylonitrile	0.02	3.28	60
	0.52	3.6	50
	0.04	2.7	40
butadiene	0.035	8.8	50
butyl acrylate	0.07	4.4	45
1-chloro-1-propene	1.13	0.24	50
2-chloro-1-propene	0.75	0.58	50
ethylene	3.6	0.24	90[a]
	1.85	0.20	70[b]
isobutylene	2.05	0.08	60
maleic anhydride	0.30	0.01	75
methyl acrylate	0.12	4.4	50
methyl methacrylate	0.02	15	45
n-octadecyl vinyl ether	2.1	0.1	50
n-octyl vinyl ether	1.9	0.1	50
propylene	2.27	0.3	60
styrene	0.035	5.7	75
	0.067	35	50
vinyl acetate	1.8, 1.35	0.6, 0.65	40
	1.68	0.23	60
	2.1	0.3	68
vinylidene chloride	0.2	1.8	45
	0.2	4.5	50
	0.23	3.15	50
	0.30	3.2	60
vinyl propionate	1.35	0.65	40

[a] At 1000 atm.
[b] At 300 atm.

separation pattern of vinyl chloride homopolymerization, published r_1 and r_2 values may be strictly erroneous. The monomer composition at the locus of reaction (the swollen polymer phase) will generally differ from the overall monomer composition, and M_1/M_2 in that swollen phase may not vary linearly with the initial M_1/M_2 ratio.

In principle it is possible to calculate the sequence distribution of the two monomer units in a copolymer on the basis of the reactivity ratios and the feed composition. The tediousness of the computation can now be avoided, since the publication of numerical tables of sequence distribution functions for a range of feed compositions and a range of reactivity ratios (50a). For the commercially most important copolymers of vinyl chloride these tables enable one to estimate that for monomer composition maintained constant at 80 moles of vinyl chloride and 20 moles of vinyl acetate by proper feedstock adjustment, over 70 percent of the vinyl chloride units present in the copolymer will be present in continuous sequences of 5 or more units and about 35 percent of the total vinyl chloride units will be present in continuous sequences of 10 or more units. However, rather than 10 sequences of length 10 units in every 100 monomer units in the homopolymer, this particular copolymer would have only 1 sequence of vinyl chloride units of length 10 for every 200 vinyl chloride units in the copolymer.

Industrial Manufacture

Most PVC resins are produced by polymerization of vinyl chloride in aqueous systems containing an emulsifying agent or a suspension stabilizer. In 1966 it was estimated that in the United States output was divided among the liquid-phase processes according to the following proportions: suspension, 82%; emulsion (for paste resins), 15%; and solution, 3%. However, the processes in which vinyl chloride is polymerized in solution or in bulk are assuming major importance, and at least one manufacturer in the United States in 1970 was using a bulk polymerization technique.

Emulsion Polymerization

Early Processes. The emulsion polymerization was developed mainly in Germany and the bulk of PVC produced there during World War II employed this technique. A typical example is that used at Bittefeld, where vinyl chloride was added to an equal weight of a 4% solution of Mersolat (sodium sulfonate of C_{15} hydrocarbon) and 0.4 parts of 40% hydrogen peroxide. Polymerization was carried out in closed autoclaves rotating at 4 rpm; these were made either of nickel or of glass-lined steel, were 12 meters long and 1 meter in diameter, and contained a charge of 3500 lb of monomeric vinyl chloride. The reaction temperature was maintained at 40–50°C for 24 hours, after which time the pressure had started to fall. The contents of the autoclave were discharged without any attempt to recover the unpolymerized vinyl chloride (10%) and the polymer emulsion was then spray dried to give the final product. There was little instrumentation in these early plants, with the result that polymerization conditions varied considerably and polymers of widely differing characteristics were produced. In fact, the product was graded according to its processing behavior in the laboratory. The PVC resins were classified according to their molecular weight into one of three ranges and also according to their processing characteristics. It is interesting to note that one grade was classified as a paste-making resin because of its ability to form a fluid mix with an equal weight of plasticizer.

An improved version of this process (51) was operated at Schkopau where vinyl chloride was polymerized continuously in a plant consisting of two tall glass-lined polymerizers connected in series. Both vessels were provided with jackets for heating and cooling and also with simple blade-type agitators which extended only to the upper third of their length. Both vessels were 7 meters high but the first of the two reactors had a diameter of 1.75 meters and the second a diameter of only 0.84 meter. The ingredients of the polymerization, consisting of emulsified solution, catalyst, and deionized water, were fed into the top of the first reactor and controlled by use of rotameters. The vinyl chloride monomer was pumped in through a separate feed system operated by a control valve The flow to the polymerization when operating at a steady state was:

	Flow, gal (U.S.)/hr
deionized water	41
emulsifier (10% soln of Mersolat)	14
potassium peroxydisulfate (1% soln)	5.1
vinyl chloride	43.4

The reactor was operated at 40–50°C depending on the molecular weight of the polymer required. In the first vessel, the heat was removed by circulating refrigerated brine at −20°C. The temperature was controlled in the second vessel by the use of cold water. The emulsion progressed down the first reactor and then passed to the top of the second, from which the final polymer emulsion was drawn off at the base and discharged into one of two tanks maintained at the same pressure as that of the polymerizers. After recovery of the unpolymerized monomer, the polymer was isolated either by spray drying, by coagulation with aluminum sulfate followed by washing filtration and drying, or by use of a drum dryer. Starting up the process required a special technique in which the vinyl chloride was polymerized in the first vessel with increasing amounts of monomer until the specific gravity of the emulsion reached 1.084. At this stage the continuous flow of ingredients was started and the flow of emulsion to the second reactor commenced. The process required close supervision, the amounts in the two reactors being adjusted so that the liquid levels came just to the top of the agitators. The speed of the agitators was variable, but was run so as to produce emulsification of the monomer without creating turbulence in the flow of the emulsion down the reactor.

At the bottom of the first vessel the conversion had reached about 88%, and when the reaction mixture passed from the second vessel into the discharge tanks had increased to 92%. The rate of polymerization was followed by taking specific gravity measurements of the polymer emulsion at various points in the system.

Polymer with improved electrical properties was produced by using hydrogen peroxide as catalyst instead of potassium peroxydipersulfate solution. This, however, led to a reduced rate of polymerization so that the through-put had to be lowered in order to maintain the required conversion values in the polymer emulsion at the bottom of the first reactor and in discharge from the second.

The process was used to produce polymers with a range of molecular weights and of electrical properties. The physical form of the product, obtained by spray drying, was small uniform spherical particles that gave plastisols on mixing with plasticizers,

Production of Paste-Forming Polymers. The emulsion process is now generally used for the manufacture of paste-forming polymers. The main essential of a good paste polymer is its ability to form a low-viscosity plastisol with the minimum of plasticizer, and this is generally achieved with polymers having a particle size distribution of 0.5–2 μ. The production techniques by which such properties are achieved have not been fully described in the technical literature, but some indication of the process is given in a number of patents (52–57) and articles (58–61). Vinyl chloride is polymerized in aqueous emulsion using two emulsifying agents, one of which is soluble in the aqueous phase and the other in the monomer. The polymer is then isolated either by coagulation and filtration or by spray drying the latex. The recipe given is as follows:

	Parts by weight
vinyl chloride	100
water	158
azobisisobutyronitrile	1
sodium dodecyl sulfate	1.58
cetyl alcohol	2

The polymerization is carried out at 50°C and taken substantially to completion. There are many instances in which the particle size is controlled by the addition of electrolytes to the emulsion system so that the reaction is carried out in a state of incipient coagulation.

Control of Particle Size. Strict control of particle size can be obtained by carrying out the polymerization with limitation of the emulsifier so that there are no micelles present in which initiation can take place. The number of polymer particles present in the system is determined by the initial addition of a seed latex containing particles of a small, uniform diameter (62,63).

The quantity of emulsifier must be sufficient to prevent coalescence of the polymer particles; it is normally 25–40% of the amount necessary for total soap coverage of the surface of the polymer particles. Additional vinyl chloride and emulsifier are added during the course of the reaction, which is initiated by use of a water-soluble catalyst such as potassium peroxydisulfate. The monomer is absorbed by the polymer particles present and polymerizes, thereby increasing the particle size, while the rate of addition of the emulsifier is maintained to prevent coalescence and at the same time to avoid the formation of micelles that could lead to the initiation of new particles (64). The process is complicated because the rates of addition of the ingredients are not constant over the whole period of the polymerization. This process is capable of producing PVC emulsions in which the particle size can have an average diameter of about 0.3–0.5 μm; the diameter of the seed particles may be as low as 0.02 μm but at this size the seed latex is quite unstable and it is an advantage to start with a seed of about 0.05 μm diameter. Besides the advantage of being able to produce PVC emulsions with a predetermined particle size, the solids content can be as high as 65%. The emulsifier content needed to maintain a stable latex is only of the order of 1%, which is considerably less than for most emulsion polymers.

This technique can be used to produce resins of known particle sizes, which can be blended to give any particle-size distribution and consequently differing paste-forming properties. At least one commercial resin is made by this process and appears, from the particle-size analysis, to consist of a blend of two polymers of particular particle size.

A modification of this process has been described in which the seed latex used in the polymerization consists of two or more latexes having average particle diameters different from one another (65). In the example the following recipe was used:

vinyl chloride	34 liters
water	51 liters
sodium dioctyl sulfosuccinate	45 g
PVC latex (average diameter 0.6 μ)	1.5% calcd on monomer
PVC latex (average diameter 0.25 μ)	2.5% calcd on monomer
sodium bicarbonate	68 g
potassium peroxydisulfate	6.8 g
sodium bisulfite	0.68 g

The reaction was carried out at 50°C with the addition after a period of 3, 6, and 9 hours of 45 g of dioctyl sodium sulfosuccinate on each occasion. When the pressure began to drop after 11 hours, a further equal quantity of emulsifier was added and the latex discharged from the reactor and the polymer isolated by spray drying. The resin formed a paste on the addition plasticizers in the ratio of 60:40 PVC:dioctyl phthalate, having a viscosity of 4200 cP after one day.

All the above techniques involve the use of water-soluble emulsifying agents, which normally remain in the resin after drying and make the material unsuitable for use in critical electrical applications. The use of water-insoluble soaps (66) overcomes this problem and at the same time has a beneficial effect on the heat stability of the resin. The soap has the general formula of ACOOM where A may vary between C_7H_{15} and $C_{21}H_{43}$ and M is a metal that may be lithium, barium, calcium, or magnesium. The soap may be produced in the reactor from the metal hydroxide and the fatty acid, and this has the advantage that homogenization is not required, whereas it is essential when the insoluble soap is added directly. The polymerization is carried out using a monomer–water ratio of about 1:2 with potassium peroxydisulfate as initiator. Low-speed agitation is used and the temperature is maintained at 40–50°C, the reaction taking about 15 hours to reach a conversion of 92%.

The final product is normally in the form of a stable latex from which the polymer can be obtained by coagulation or by spray drying. The process does not give the range of particle size that is required for the best paste-forming properties but normally produces resins with a fairly uniform particle diameter of about 3.5 μ.

The polymers are generally made from pure vinyl chloride but a number of paste resins based on copolymers are commercially available. Small amounts of vinyl acetate (up to 5%) are copolymerized to give resins with lower softening points.

Mechanism. When vinyl chloride is polymerized in an aqueous emulsion using a water-soluble catalyst, the reaction mechanism does not appear to follow the classical Smith-Ewart theory for this type of polymerization (67). Some initiation must occur within the soap micelles since the polymerization rate does increase to some extent with an increase in emulsifier concentration. This increase varies with different emulsifiers. The monomer is readily soluble in the polymer so that transfer from the solubilized monomer to the growing particles can occur easily. Some polymer will be formed from the monomer dissolved in the aqueous phase because of the relatively high solubility of monomeric vinyl chloride in water. When oil-soluble catalysts are used the mechanism becomes more complex. Some initiation will occur in the larger droplets of monomer so that the polymer formed in these particles will be similar to that produced in a suspension polymerization, but initiation in soap micelles with monomer transfer from the aqueous phase can also occur. These various mechanisms account for the broad particle-size distribution of the polymer obtained from emulsion polymerization using oil-soluble and mixed catalyst systems.

Suspension Polymerization

An early description of the suspension polymerization process is also contained in *German Plastics Practice* (68), which records the activities of the plastics industry during World War II. The monomer was polymerized in stainless-steel reactors using the following recipe:

	Quantity, kg
water (deionized)	6000
vinyl chloride	3000
benzoyl peroxide	4
poly(vinyl alcohol), 5% (80–85% hydrolyzed high-viscosity poly(vinyl acetate) with saponification number 100–130)	100

The agitation was provided by a paddle stirrer rotating at 40 rpm and the polymerization was carried out at 40°C, the reaction being stopped at 65% conversion (after 50 hours) to prevent the formation of low-molecular-weight polymers. After recovery of the excess monomer, the slurry was filtered, the polymer was washed and then finally centrifuged to give a cake containing 35% moisture which was dried in a Buttner drier. The PVC produced in this process had a high molecular weight (K value 80–85).

The suspension process (69) has become the most important in the commercial manufacture of PVC; it is essentially a batch process although attempts have been made to develop a continuous-flow system. It is now generally carried out in glass-lined reactors having a capacity of 2000–6000 U.S. gal. Stainless-steel polymerizers have been used, but these tend to produce more polymer build-up on the surface of the reactor during the reaction. Glass-lined polymerizers can be cleaned effectively by high-pressure water hoses, although entry into the vessel is required from time to time to remove the build-up that occurs in the gaseous phase on the vessel inlets. Because of the possibility of damaging the glass lining, solvent cleaning is adopted frequently as an effective means of removal of polymer build-up. Various types of agitators are fitted to the polymerizers and are used in conjunction with baffles to increase the degree of dispersion. The heat of polymerization is removed from the system by circulating cooling water through a jacket fitted to the reactor. The capacity of the unit can be improved by using chilled water for the cooling so that the initial charge of vinyl chloride monomer can be increased. An average value that is considered typical for the heat-transfer coefficient is 46 Btu/hr-sq ft-°F. The heat of polymerization for vinyl chloride is taken as 650 Btu/lb.

The recipes used at present do not appear to differ substantially from that developed in Germany but in addition to poly(vinyl alcohol) (partially saponified poly-(vinyl acetate)), gelatin, methylcellulose, and vinyl acetate–maleic anhydride copolymers are frequently used as the dispersant (70–73). The shape of the polymer particles can be varied by adding small amounts of secondary emulsifiers such as ethylene oxide condensation products and sulfonated oils to the recipe. In the use of gelatin, buffers are added to maintain the pH above the isoelectric point. The final properties of resin are very dependent on the system in which it is polymerized (74,75). Gelatin normally produces glassy spherical particles which have poor plasticizer absorption characteristics, whereas poly(vinyl alcohol) gives porous particles which readily absorb plasticizers to give dry powder blends.

The polymerization is carried out by first charging to the reactor the required amount of deionized water, then the dispersing agent, buffer, and initiator, such as lauryl peroxide or azobisisobutyronitrile, which are very often used. After sealing the vessel it is evacuated to remove oxygen, and then the monomer is fed in from a weighing tank. The agitator is then started and the contents of the reactor brought up to the polymerization temperature (45–60°C) by injecting steam into the jacket of the vessel. There is a short induction period, usually about an hour, before the polymerization gets under way and the controls governing the supply of cooling water to jacket take over to maintain the temperature at the predetermined value.

The maximum rate at which the polymerization can be allowed to proceed depends on the concentration of the initiator and on the amount of vinyl chloride present, ie, the ratio of monomer to water. The reaction does not proceed at a constant rate but increases gradually during the polymerization until it reaches a maximum at 60–80% conversion. This means that the amount of vinyl chloride that can be converted to polymer is limited by the rate of heat evolution at the reaction peak, if the heat transfer by the cooling water is to maintain the reaction temperature at its correct value. During the earlier stages of the polymerization the reaction rate is lower than the maximum so that the capacity of the reaction vessel is not fully utilized.

A more constant rate of polymerization can be obtained by using diisopropyl percarbonate (DIPP) as initiator and a number of manufacturers in the United States have adopted this practice. This initiator has the added advantage of reducing the induction period; it enables the output from a given polymerization plant to be increased by 20–50%. It is found that the percarbonate affects the structure of the polymer particle, particularly the porosity, so that the use of this type of initiator becomes limited to certain grades of polymer. In order to obtain high-porosity resins the conversion must be limited to about 50% so that the economic advantages of an increased polymerization rate are offset by the reduced amount of polymer obtained from each batch and the large amounts of unpolymerized vinyl chloride which must be recovered. These disadvantages have been overcome to some extent by a technique in which 10–20% of the suspension is withdrawn from the reactor when the conversion reaches 45–50%. The sudden drop in pressure increases the porosity of the polymer particles of this portion as the monomeric vinyl chloride is flashed off from the suspension. The polymerization of the remaining portion is then continued to 90–95% conversion. It is claimed that a larger initial charge of vinyl chloride can be used because the reaction peak is not reached until after the withdrawal of part of the suspension.

The molecular weight of the polymer is mainly determined by the temperature at which the polymerization is carried out. The rate increases as the temperature rises, the maximum in general use being about 60°C. Higher polymerization temperatures improve the heat-transfer process so that higher-plant outputs are obtained with the low-molecular-weight resins that are produced at the highest polymerization temperatures.

If there are any pressure limitations on the polymerization equipment, chain-transfer agents are sometimes used in the manufacture of low-molecular-weight polymers. These include chlorinated hydrocarbons and isobutylene.

The pressure in the system remains constant during the polymerization (at 55°C it is 115 psig) until the conversion reaches about 70%, after which it decreases rapidly as the reaction continues. Up to 75% conversion, free vinyl chloride is present in the system in the form of droplets in the aqueous phase, but at higher conversions the monomer is completely absorbed in the polymer particles. The polymer has a higher density than the monomer so that the volume of the organic phase shrinks by about 35% during the polymerization. This shrinkage has been utilized to reduce heat-transfer problems at the reaction peak, when about a 25% reduction in volume will have taken place, by injecting cold water into the polymerization system.

An average time for the entire polymerization process is about 16 hours, of which about 13 hours is taken up by the actual reaction. When the polymerization is substantially complete, the very much reduced reaction rate makes it uneconomical to take the conversion above 95%. The contents of the reactor are dumped into a large vessel located directly underneath (it is usual practice to have one dump tank servicing several polymerizers) and the excess vinyl chloride is then removed by a recovery system which is made up of a number of water-sealed compressors and a condenser to recover the monomeric vinyl chloride. The slurry may need heating during the recovery to facilitate the recovery of the monomer, and this is done by direct injection of live steam, taking care to avoid local heating and degradation of the polymer. When copolymers are being manufactured, the recovered monomer will contain some quantities of monomers other than vinyl chloride. This requires isolation of these monomers and reuse in the same type of polymerization.

The slurry is then blended with that from other reactors to reduce any small variations, after which it is dewatered using a superdecanter or continuous centrifuge to give a polymer cake containing about 20% moisture. Although the quantities of dispersants used in the reaction are very low (about 0.1% based on the weight of the monomer), improvements of the polymer can be made by washing. Sometimes this is done in the centrifuge or, alternatively, by redispersing the polymer cake in water and again centrifuging.

The process for suspension polymerization cannot be easily operated on a continuous basis because of the difficulty in preventing the polymer from separating out from the aqueous phase unless adequate agitation is maintained. One technique involves use of three continuous-flow stirred tank reactors, two of which are operated in parallel and in which the polymerization is taken to about 70% conversion. The suspension at this stage passes into the third reactor, where the reaction is taken substantially to completion. Settling of polymer in the connecting pipework still appears to create problems, and the build-up of a polymer film on the reactor walls reduces the heat transfer and gives rise to insoluble particles (fish-eyes) in the product. If these problems can be overcome, a continuous process could yield substantial savings by a reduction of labor costs and of capital investment.

Drying of the polymer is normally effected in a flash dryer in which the air stream is heated to about 80°C. The polymer particles are isolated by passage through a cyclone separator. A single stage will generally reduce the moisture content to around 5%. The drying is completed either by passage through the second stage of a flash dryer or, alternatively, by use of a rotary dryer. Fluidized-bed techniques have also been developed and are being used extensively. The most important feature of the dryer is the avoidance of areas in which the polymer can stick to the dryer walls and produce degraded material. Although the air is heated to about 80°C, the temperature of the polymer reaches only 60°C and normally no degradation will occur if the particles pass unimpeded through the drying system. The losses of polymer from the cyclone separators are less than 0.2%, and these can be eliminated by installation of bag filters on the outlet to the dryers. The dry polymer is screened to remove coarse particles, generally using a 40-mesh screen, and packed in multiwall paper bags.

A digital control system has been used for control of the polymerization plant (76). Computer control of the blending and of the polymerization temperature improves the quality of the product and reduces the cost by minimizing reactor downtime and increasing the output from the polymerization plant. A dual computer system performs the duties of calculating batch sizes and regulates the flow of additives accordingly. The mixer and the reactor charge at discharge points are controlled. The polymerization is regulated by adjusting steam and cooling water in the jackets of the reactors. The system provides alarms in unusual situations and a continuous indication of the polymerization conditions. It is claimed that the performance is much better with direct digital control than with analog control systems.

Solution and Bulk Polymerization

Although the use of aqueous dispersants accounts for the greater part of the manufacture of PVC, single-phase polymerization carried out either in the presence of or without an organic diluent represents an important manufacturing procedure. The process in which monomeric vinyl chloride is diluted with a solvent and polymerized by the addition of organic peroxide is usually referred to as solution polymerization, and is used exclusively for copolymers containing vinyl acetate. The basic process covers copolymers containing 10–25% of vinyl acetate. The mixed monomers are dissolved in a solvent such as *n*-butane or cyclohexane to form a 20% solution with 0.5% of an organic peroxide such as lauryl peroxide or acetyl benzoyl peroxide as initiator. The polymerization is carried out in a stirred autoclave at 40°C (77,78).

The stage at which the polymer begins to precipitate depends on the diluent used. When cyclohexane is the solvent a granular precipitate occurs right at the start of the polymerization. The contents of the autoclave are circulated through a filter press to remove the polymer and the filtrate is returned. Further amounts of the comonomers are added to maintain the concentrations of the vinyl chloride and vinyl acetate at the original values. The concentration of the peroxide is also maintained by further additions dissolved in the monomers. The polymer in the filter press is washed to remove residual diluent and any traces of organic peroxide which would have a detrimental effect on the heat stability. A certain amount of unpolymerized vinyl chloride is also removed from the polymer by the washing fluid and this is then recovered by fractional distillation.

Little information is available on the up-to-date techniques used in this process, but it is likely that there have been substantial improvements since the description in the original patents.

Bulk Polymerization (qv). The process in which vinyl chloride is polymerized without the addition of solvent is usually referred to as the "bulk polymerization" method. It has been developed almost exclusively by Péchiney-St. Gobain, who have licensed it all over the world. According to the latest information the polymerization is carried out by a two-step method. Vinyl chloride containing the initiator is charged to a small reactor fitted with a high-speed agitator and the reaction is carried out at 50–60°C. Two recipes have been disclosed for the process:

	Parts by weight	
vinyl chloride	100	100
azobisisobutyronitrile	0.016	0.02
diisopropyl percarbonate		0.00005

The polymer begins to precipitate when less than 1% has been formed and the reaction is continued until about 10% conversion after about 3 hours, when the entire contents of the reactor are charged into a large horizontal autoclave and further vinyl chloride is added. When the conversion reaches 20% the polymerization mass is in the form of a wet powder. At 40% conversion, the residual monomer is completely absorbed by the polymer so that the mass takes on the form of a dry powder. The autoclave is therefore specially designed to prevent agglomeration and to provide good temperature control so that the conversion can be taken up to 90% while retaining the polymer in a powder form. The residual vinyl chloride is removed under vacuum and the polymer then discharged and screened. The process has the advantage that no further drying is needed. A certain amount of oversize material is separated by the screens and is ground and disposed of as a substandard product. The capital cost of this type of plant is less than that needed for aqueous polymerization because of the absence of drying equipment.

Some problems are encountered in making copolymers by this route because the lower softening point of the product leads to the production of larger amounts of oversize material. Undoubtedly these problems will be overcome and copolymers containing at least small amounts of comonomer should become available from this process.

Bibliography

1. M. Prat, *Jean Mem. Serv. Chim. Etat (Paris)* **32,** 319 (1946).
2. W. I. Bengough and R. G. W. Norrish, *Proc. Royal Soc. (London), Ser. A* **200,** 301 (1950).
3. J. W. Breitenbach and A. Schindler, *Monatsh. Chem.* **80,** 429 (1949).
4. A. Schindler and J. W. Breitenbach, *Ric. Sci. Suppl.* **25,** 34 (1955).
5. J. W. Breitenbach and A. Schindler, *Monatsh. Chem.* **86,** 437 (1955); *J. Polymer Sci.* **18,** 435 (1955).
6. E. J. Arlman and W. M. Wagner, *J. Polymer Sci.* **9,** 581 (1951).
7. F. Danusso and G. Perugini, *Chim. Ind. (Milan)* **35,** 881 (1953).
8. F. Danusso, *Ric. Sci. (Suppl.)* **25,** 46 (1955).
9. F. Danusso and D. Sianesi, *Chim. Ind. (Milan)* **37,** 695 (1955).
10. M. Magat, *J. Polymer Sci.* **14,** 491 (1955).
11. H. S. Mickley, A. S. Michaels, and A. L. Moore, *J. Polymer Sci.* **60,** 121 (1962).
12. M. von Smoluchowski in H. R. Kruyt, ed., *Colloid Science,* Vol. 1, Elsevier, Amsterdam, 1952, p. 279.
13. G. Talamini and G. Vidotto, *Makromol. Chem.* **50,** 129 (1961); **53,** 21 (1962).
14. G. Talamini and G. Vidotto, *Chim. Ind. (Milan)* **46,** 371 (1964).
15. D. N. Bort, Ye. Ye. Rylov, N. A. Okladnov, B. P. Shtarkman, and V. A. Kargin, *Vysokomolekul. Soedin.* **7,** 50 (1965); **9,** 303 (1967).

16. V. V. Mazurek, *Vysokomolekul. Soedin.* **8,** 1174 (1966).
17. J. D. Cotman, M. F. Gonzales, and G. G. Claver, *J. Polymer Sci.* **A-1,** 1137 (1967).
18. G. Talamini, *J. Polymer Sci.* [A-2] **4,** 535 (1966).
19. G. Vidotto, A. Crosato-Arnaldi, and G. Talamini, *Makromol. Chem.* (a) **111,** 123 (1968); (b) **114,** 217 (1968); (c) **122,** 91 (1969).
20. A. Crosato-Arnaldi, P. Gasparini, and G. Talamini, *Makromol. Chem.* **117,** 140 (1968).
21. M. Ryska, M. Kolinsky, and D. Lim, *J. Polymer Sci.* **16C,** 621 (1967).
22. J. C. Thomas, *SPE J.* **67,** 61 (Oct. 1967).
23. G. Talamini and E. Peggion, Chap. 5 in G. Ham, ed., *Vinyl Polymerization,* Vol. 1, Part 1, Marcel Dekker Inc., New York, 1967.
24. R. M. Joshi and B. J. Zwolinski, *J. Polymer Sci.* [B] **3,** 779 (1965).
25. J. Benton, unpublished estimate.
26. H. Gerrens, W. Fink, and E. Köhnlein, *J. Polymer Sci.* **16C,** 2781 (1967).
27. E. Farber and M. Koral, *SPE Tech. Papers* **13,** 398 (1967).
28. M. R. Meeks, *Paper, 26th SPE Antec, May 6–10, 1968.*
29. Solvic S.A., Belg. Pat. 566,530–566,536 (July 29, 1960); Solvay et Cie., Belg. Pat. 545,968 (Oct. 9, 1959); Australian Pat. Appl. 26889.
30. U. Giannini and S. Cesca, *Chim. Ind. (Milan)* **44,** 371 (1962).
31. I. W. Fordham and C. L. Sturm, *J. Polymer Sci.* **33,** 803 (1958).
32. N. A. Shikari and A. Nishimura, *J. Polymer Sci.* **31,** 249 (1958).
33. A. Guyot and Pham Quang Tho, *J. Polymer Sci.* [C] **4,** 299 (1964); *Paper, Symp. Chem. Polymerisation Processes, London, 1965; Compt. Rend.* **256,** 165 (1963).
34. A. Guyot, M. Bert, and Pham Quang Tho, *J. Appl. Polymer Sci.* **12,** 639 (1965).
35. I. Rosen, P. N. Burleigh, and J. F. Gillespie, *J. Polymer Sci.* **54,** 31 (1961).
36. O. C. Böckman, *J. Polymer Sci.* **3A,** 3399 (1965).
37. F. Danusso, G. Pataro, and D. Sianesi, *Chim. Ind. (Milan)* **41,** 1170 (1959).
38. G. Talamini and G. Vidotto, *Chim. Ind. (Milan)* **46,** 16 (1964).
39. M. Kolinsky, M. Ryska, M. Bohdanecky, P. Kratochvil, K. Solc, and D. Lim, *J. Polymer Sci.* **C16,** 485 (1967).
40. G. Pezzin, G. Talamini and G. Vidotto, *Makromol. Chem.* **43,** 12 (1961).
40a. G. A. Razuvaev and K. S. Minsker, *Zh. Obshch. Khim.* **28,** 983 (1958).
41. E. Jenckel, J. Eckmans, and B. Rumbach, *Makromol. Chem.* **41,** 15 (1949).
42. Z. Machacek and F. Cermak, *Chem. Prûc* **15** (40), 484 (1965).
43. P. C. Mørk, *European Polymer J.* **5,** 261 (1969).
44. A. D. Jenkins, Chap. 6 of Ref. 23.
45. C. H. Bamford and A. D. Jenkins, *Proc. Royal Soc. (London), Ser. A* **228,** 220 (1955).
46. G. S. Park, University of Wales, private communication.
47. J. Brandrup and E. H. Immergut, eds., *Polymer Handbook,* Interscience Publishers, a division of John Wiley & Sons, Inc., New York, 1966.
48. G. V. Tkachenko, P. M. Khomikovski, and S. S. Medvedev, *Zh. Fiz. Khim.* **25,** 823 (1951).
49. J. W. Breitenbach, O. F. Olaf, H. Reif, and A. Schindler, *Makromol. Chem.* **122,** 51 (1969).
50. G. Talamini, *Makromol. Chem.* **140,** 249 (1971).
50a. C. Tosi, *Adv. Polymer Sci.* **5,** Part 4, 451 (1968).
51. R. D. Dunlop and F. E. Reese, *Ind. Eng. Chem.* **40** (4), 654 (1948).
52. G. E. A. Pears and A. Pajacykowski (to Imperial Chemical Industries Ltd.), Brit. Pat. 978,875 (Dec. 23, 1964).
53. Wacker-Chemie G.m.b.H., Brit. Pat. 1,007,867 (Oct. 22, 1965).
54. G. Kühne (to Farbwerke Hoechst A.G.), U.S. Pat. 3,208,965 (Sept. 28, 1965).
55. Kanegafuchi Chemical Industry Company Ltd., Fr. Pat. 1,509,473 (Jan. 26, 1967).
56. I. Harris (to Imperial Chemical Industries Ltd.), Brit. Pat. 698,359 (April 30, 1951).
57. D. E. Lintala (to Wingfoot Corp.), U.S. Pat. 2,890,211 (June 9, 1959).
58. C. Corso and E. Ferrari, *Materie Plastiche* p. 13 (Jan. 1962).
59. L. F. Albright, *Chem. Eng.* **74** (14), 85 (1967).
60. G. Marea and V. Cosma, *Materie Plastiche* **3** (2), 81 (1966).
61. O. Berger and W. Braun, *Plaste Kautschuk* **14** (9), 663 (Sept. 1967).
62. G. Benetta, V. Bresquar, G. Gatta, and F. Testa (to Sisedison s.P.a.), Brit. Pat. 984,487 (Feb. 24, 1965).
63. Società Edison, Brit. Pat. 1,050,625 (Dec. 18, 1963).

64. J. R. Powers (to B. F. Goodrich Co.), U.S. Pat. 2,520,959 (Sept. 5, 1950).
65. Montecatini, Brit. Pat. 928,556 (March 19, 1962).
66. B. F. Goodrich Co., Brit. Pat. 985,948 (June 15, 1962).
67. K. Giskehaug, *The Chemistry of Polymerisation Processes,* Society of the Chemical Industry Monograph No. 20, London, 1966, p. 235.
68. *German Plastics Practice,* Debell & Richardson, Springfield, Mass.
69. C. E. Schildknecht, *Vinyl and Related Polymers,* John Wiley & Sons, Inc., New York, 1952.
70. F. K. Schoenfeld and W. L. Semon (to B. F. Goodrich Co.), Brit. Pat. 641,442 (Aug. 1, 1947).
71. E. M. Jankowiak and A. A. Nelson (to The Dow Chemical Corp.), U.S. Pat. 3,042,665 (July 3, 1962).
72. Wacker-Chemie G.m.b.H., Brit. Pat. 1,017,483 (Jan. 19, 1966).
73. M. Baer (to Monsanto Chemical Co.), U.S. Pat. 2,476,474 (July 19, 1949).
74. F. H. Winslow and W. Matreyek, *Ind. Eng. Chem.* **43**, 1108 (1951).
75. P. Gluck and I. Gluck, *Materie Plastiche* **4** (6), 281 (1967).
76. *European Chem. News,* p. 52, June 14, 1968.
77. S. D. Douglas (to Union Carbide and Carbon Corp.), U.S. Pat. 2,075,429 (March 30, 1937).
78. E. W. Reid (to Carbide & Carbon Chemicals Corp.), U.S. Pat. 1,935,577 (Nov. 14, 1933).

J. L. Benton and C. A. Brighton
BP Chemicals International Ltd.

COPOLYMERS

The copolymerization of vinyl chloride with other monomers, and particularly vinyl acetate, was studied at a very early stage in the commercial development of PVC. The improved processing characteristics were recognized as essential for the fabrication of this class of materials. The discovery of the effect of certain high-boiling esters as plasticizers diverted attention away from this early work, but it was not long before the improved solution characteristics and other properties, such as moldability of rigid compounds, fulfilled certain areas of applications which could not be satisfied by the homopolymer.

Copolymers can be manufactured by either the emulsion, the solvent, or the suspension process; the latter is now the most widely used. The bulk polymerization is not suitable for copolymers except when only minimal quantities of comonomer are used. The comonomer does not necessarily polymerize at the same rate as the vinyl chloride so that the composition of the polymer which is formed at any particular instant during the reaction is not the same as the composition of monomeric phase. Some monomers polymerize at a faster rate than vinyl chloride, in which cases the copolymer formed in the early stages will be rich in comonomer and that formed in the later stages will tend to be almost pure PVC containing only small amounts of comonomer (see also COPOLYMERIZATION). Monomers that react in this manner include vinylidene chloride, acrylic esters, and maleic esters. A copolymer of uniform composition can be obtained only by feeding the comonomer into the polymerization at the same rate as it is used up, so as to maintain the constant monomer ratio that

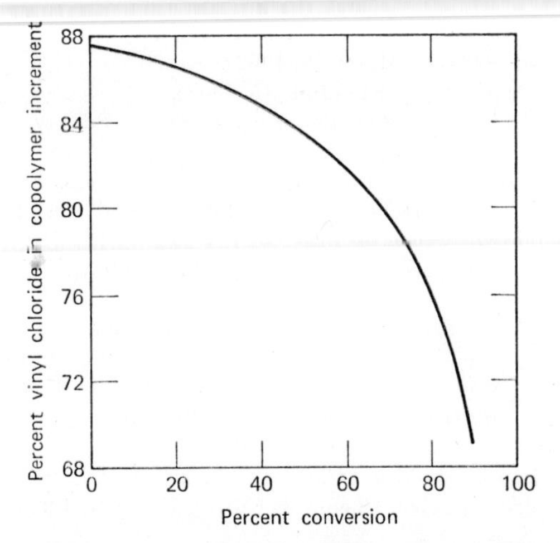

Fig. 1. Variation in composition of copolymer increment with conversion Initial charge 80/20 wt/wt vinyl chloride–vinyl acetate. Courtesy *British Plastics.*

produces the required copolymer composition. Vinyl acetate polymerizes at a slower rate than vinyl chloride so that the reaction requires the addition of further quantities of vinyl chloride in order to obtain a copolymer of uniform composition.

The changes in composition of the polymer that occur during the polymerization reaction in which all the monomers have been charged initially have been calculated for an 80:20 wt:wt vinyl chloride–vinyl acetate mixture polymerized at 60°C (1). This relationship was derived using the relative reactivity ratios determined for copolymerization in mass in 60°C (2) and is shown in Figure 1. Figure 2 shows the variations in the average composition of the copolymer as the reaction proceeds starting with 20% by weight vinyl acetate in the initial monomer mixture. The copolymer formed in the early stages contains more vinyl chloride than does the monomer mixture, so that the concentration of vinyl acetate in the residual monomer phase increases as the reaction proceeds and the copolymer becomes progressively richer in vinyl acetate. The relative reactivity ratios for these two monomers have been determined for polymerization at 60°C in an aqueous suspension system using an oil-soluble catalyst, and were found to be substantially the same as those reported above. Since the two monomers have widely differing vapor pressures, the changes in monomer composition can be followed experimentally by the progressive decrease in the pressure in the reactor. These data can be used to formulate a procedure for the progressive addition

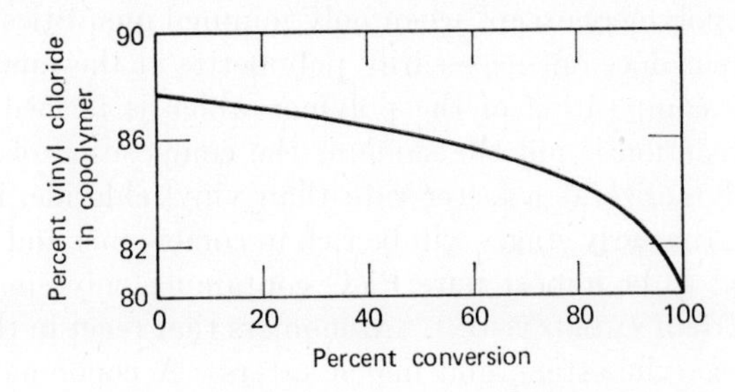

Fig. 2. Variations in average composition of copolymer with conversion. Initial charge 80/20 wt/wt vinyl chloride–vinyl acetate. Courtesy *British Plastics.*

Table 1. Theoretical Monomer Ratios and Operating Conditions for Forming Chemically Homogeneous Copolymer at 60°C[a]

Desired copolymer composition, %		Initial monomer charge[b]		Vinyl chloride to be added during polymerization[b]	Equilibrium pressure, lb/in.^{2}
Vinyl chloride	Vinyl acetate	Vinyl chloride	Vinyl acetate		
97	3	57	3	40	142
94	6	54	6	40	136
91	9	51	9	40	131
88	12	48	12	40	125
85	15	46	15	39	120
82	18	44	18	38	115
80	20	43	20	37	112

[a] Reproduced by permission of *British Plastics*.
[b] All monomer figures expressed as part by weight.

of vinyl chloride to the system in order to produce a copolymer of constant composition throughout the whole course of the polymerization. This is given for a range of vinyl acetate content in Table 1.

Many commercial products are manufactured by a batch process in which the whole of the comonomers is charged initially and then polymerized without any attempt being made to produce a uniform product. This can produce some advantages in physical properties, such as impact strength, but is deleterious if good solution properties are required. The solution process is used almost exclusively for the manufacture of uniform copolymers of vinyl acetate and gives products with particularly good solution characteristics. The solubility improves as the amount of vinyl acetate in the copolymer increases (3).

With Vinyl Esters

Vinyl Acetate. Copolymers of vinyl chloride with vinyl acetate undoubtedly account for the greater part of vinyl chloride copolymers available commercially. The vinyl acetate content varies from 2% up to about 20%. The lower-acetate resins are used for calendering and other processes, where some improvement in speed of processing is obtained over the homopolymers. The higher-acetate resins are used for solution applications, for the manufacture of phonograph records (4), and for vinyl asbestos tiles.

Such resins are used in solution in methyl ethyl ketone, tetrahydrofuran, or methyl isobutyl ketone. They have only limited solubility in aromatic hydrocarbon

Table 2. Solution Viscosity of Suspension Copolymer[a]

Solvent	Solution viscosity at 25°C,[b] cP
acetone	78
methyl ethyl ketone	95
tetrahydrofuran	130
methyl isobutyl ketone	167

[a] Escambia PVC 6145.
[b] Brookfield, 20% solids.

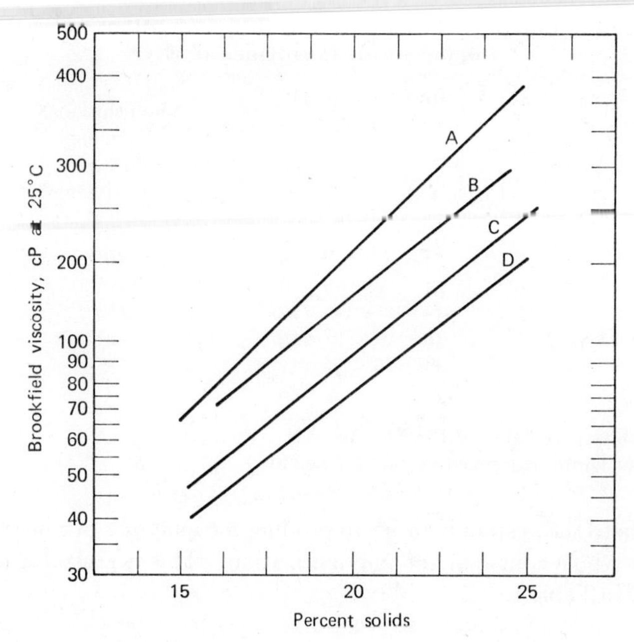

Fig. 3. Solution viscosity of PVC 6145 in methyl isobutyl ketone (curve A), tetrahydrofuran (curve B), methyl ethyl ketone (curve C), and acetone (curve D). Courtesy Escambia Chemical Corp.

solvents; these solvents are often used to lower the cost of the solvent system and control the viscosity of the solution. The resins produced by solution polymerization have better solubility characteristics than those produced in suspension systems; that is, the solutions have better clarity and lower viscosity at the same solids content. The suspension polymers vary appreciably in these properties, although commercial grades are now available that approach the quality of the solution-polymerized materials. The solution viscosity of such a copolymer (Escambia PVC 6145) containing 15% vinyl acetate is given in Table 2 for a range of solvents.

The variation in viscosity with solids content for the same resin is shown in Figure 3. Other properties of similar resins with differing vinyl acetate content are given in Table 3.

It was found that when this type of copolymer is used for lacquer applications, adhesion to metal is improved by the introduction of carboxyl groups into the polymer molecule. This can be achieved by copolymerizing maleic anhydride with the vinyl chloride and vinyl acetate. A description of the process given in a patent example discloses that a copolymer of 86% vinyl chloride, 13% vinyl acetate, and 1% maleic anhydride is formed by heating in dioxane solution with finely divided sodium hydroxide for 48 hours at 85–90°C. The solution was neutralized with acetic acid, then filtered and the resin precipitated in water. Carboxyl groups can also be introduced into the copolymer molecule by use of acrylic or methacrylic acid instead of maleic anhydride.

Copolymers are also available in which the acetate groups have been partially saponified to produce hydroxyl groups in the molecule. This improves the adhesion to metal substrates and also makes the copolymer highly compatible with alkyd, epoxy, acrylate, and urea–formaldehyde resins. The presence of hydroxyl groups can be used

Table 3. Properties of Vinyl Chloride Vinyl Acetate Copolymers[a]

Property	Test method	Value for grade number		
		R.65/81	R.51/83	R.46/88
vinyl acetate content (by wt), %		2	10	15
K value	relative viscosity at 25°C in ethylene dichloride, 0.5 g polymer in 100 ml soln	65	51	46
reduced viscosity, dl/g	ISO R 174	125	73	59
apparent density, g/ml	ISO R 60	0.45	0.48	0.78
packing density, g/ml		0.52	0.56	0.83
particle size, wt %	passing BS 60 mesh	99.9	99.9	99.0
	passing BS 200 mesh	5.8	21.8	9.0
volatile content, %	wt loss after 1 hr at 135°C	0.3	0.9	2.1

[a] Corvic copolymers; by permission of Imperial Chemical Industries Ltd.

to produce crosslinked products. This type of resin has a higher softening point and also a higher solution viscosity than the unhydrolyzed copolymer; they find use in the following applications: paint lacquers and printing inks; beverage-can coatings; paper and fabric coatings; strippable coatings; heat-sealable or nonblocking food packaging; and adhesion modifiers for organosol metal coatings.

The physical properties of the copolymers depend mainly on their vinyl acetate content; as this increases, the softening point of copolymers is reduced and processing is made easier than for homopolymers of similar molecular weight. The effect is due in part to the effects of the acetate side groups in the polymer chains; these reduce the tendency for ordered arrangement within the molecules and act in a manner comparable to that of plasticizers. This effect is responsible for their extensive use for molding phonograph records (where good flow is essential to fill the microgrooves) and for the manufacture of flooring compositions. In this latter application the copolymers of higher acetate content have good plasticity, which makes high filler acceptance possible; the resin content in the final flooring composition is usually only of the order of 20% for vinyl-asbestos tiles (see also FLOORING MATERIALS). An acetate copolymer is particularly suitable for making rigid sheet for subsequent vacuum-forming operations. The properties of such sheets are given in Table 4.

A copolymer containing about 5% vinyl acetate is available in the form of a paste-forming resin. It is made by polymerization in emulsion and then by spray drying the polymer latex. The plastisol has a lower fusion temperature than is normally obtained with homopolymers and is particularly suitable for applications where heat-sensitive materials such as textiles are involved. The copolymer has a

Table 4. Properties of Rigid Acetate Copolymers

Property	Value for different vinyl acetate contents, % by wt	
	3%	15%
tensile strength, psi	8200	8500
heat-distortion temperature at 66 psi, °C	67	57
Rockwell hardness	60	50
notched impact strength, ft-lb/in. notch	0.65	0.20

greater tendency to solvate in the presence of plasticizers and consequently the plastisol has rather poor aging characteristics. The effect of varying the vinyl acetate content of the polymer has been investigated (5). Polymers were formed from monomer mixtures containing 2, 4, and 6% vinyl acetate and these were found by infrared analyses to contain 1.4, 2.8, and 4.2% of incorporated acetate, respectively. A number of characteristics were assessed in this study.

Other Vinyl Esters. Copolymers have been manufactured using vinyl esters other than vinyl acetate, but generally these have been of only limited commercial interest. Vinyl propionate produces copolymers very similar to those obtained with vinyl acetate except it is claimed that smaller amounts of the vinyl propionate are needed to give comparable properties. This limited technical advantage has been outweighed by the lower price of the vinyl acetate monomer.

Copolymers with vinyl esters of stearic and mixed higher fatty acids have been produced in the United States and it is claimed that they have all the properties of externally plasticized PVC without its disadvantages. The vinyl stearate polymerizes at a much slower rate than the vinyl chloride so that it is essential to maintain the correct proportion of vinyl chloride in the system by continuously introducing this monomer during the polymerization. This is done by maintaining a constant pressure in the polymerizer, which ensures that the vinyl chloride and vinyl stearate are kept in constant ratio so long as the temperature does not vary (6). The vinyl ester used for the manufacture of this type of copolymer is generally made from mixed C_{12}–C_{18} fatty acids. One patent claims that a copolymer of this type containing only about 40% chlorine (about 30% vinyl ester) is useful as a release agent for pressure-sensitive adhesives (7).

Terpolymers based on vinyl chloride, vinyl stearate, and vinyl epoxystearate can be produced in a normal suspension system (8). An initial monomer composition with a mole ratio of 90:2:8 gives an 80% yield (after 48 hours at 50°C) of a copolymer containing the monomers in the ratio of 92.4:1.52:6.11. The polymer may be cross-linked by polyfunctional reagents that react with epoxy groups.

There are a number of references in the literature to the copolymerization of vinyl chloride with vinyl benzoate (9), vinyl chloracetate, and other vinyl esters (10). However, these copolymers have not exhibited any properties of note and were not produced commercially. See also VINYL ESTER POLYMERS.

With Olefins

Ethylene. There are many patents that relate to the copolymerization of vinyl chloride with olefins such as ethylene, propylene, and isobutylene. The majority of these pertain to ethylene; typical of these is one claiming the copolymerization of vinyl chloride and ethylene in the presence of trialkylboron (11). High conversion of monomer to polymer is obtained by adding the trialkylboron to the mixture of vinyl chloride and ethylene before introducing the dispersing agent. In one example, an 82% yield of dry polymer containing 2% by weight of ethylene units was obtained. When the water, catalyst, and dispersing agent were mixed prior to the addition of the monomers, only 0.8% of copolymer was formed. The copolymers containing small amounts of ethylene (up to 5%) have better processability than the homopolymers, but this is gained at the expense of a lower heat-distortion temperature. The heat stability is improved, owing, in part, to the lower temperatures at which processing can be carried out.

Polymers with greater amounts of ethylene can be produced as stable aqueous emulsions, which on drying give soft and flexible films devoid of any adhesiveness and with high transparency and strength (12). In the first step of this process, vinyl chloride is copolymerized with ethylene in aqueous emulsion at elevated pressure and in the presence of a water-soluble initiator to form a copolymer containing 50–75% vinyl chloride; this copolymer is then further polymerized, in a second stage, with 5–25% of vinyl chloride in the absence of ethylene but with the further addition of initiator, if necessary, so that the final product contains 65–85% of vinyl chloride. The water-soluble initiator used in the polymerization is sodium, potassium, or ammonium, peroxydisulfate or perborate. The emulsifier used is 2–10% (based on the weight of monomer) of a salt of a half ester of maleic acid and an aliphatic or cycloaliphatic monohydric alcohol (up to C_{24}) or a salt of a polymerizable unsaturated sulfonic acid which is copolymerized as a third monomer during the first stage.

There is an extensive range of patents describing methods for the copolymerization of ethylene and vinyl chloride. These vary from techniques involving the use of high pressures (13) of at least 20,000 psi to those in which relatively low pressures (100–1,000 psi) are used (14). The copolymers produced normally contain 3–8% of ethylene; they exhibit improved processability and better impact strength while retaining the dimensional stability and other properties of ordinary rigid PVC. See also Polar Copolymers under ETHYLENE POLYMERS.

Other Olefins. The majority of patents covering the copolymerization of vinyl chloride and ethylene also include propylene and other olefins. The commercial exploitation of copolymer containing these higher olefins has almost been entirely restricted to those containing propylene. They are made by charging propylene into a suspension polymerization of vinyl chloride containing a free-radical initiator, and then by adding further quantities of vinyl chloride to maintain a constant ratio of the two monomers. The propylene, by its high chain-transfer effect in the polymerization, reduces the molecular weight of the copolymer and gives rise to a high preponderance of propylene units at the end of the polymer chains. Copolymers with a range of molecular weights and containing various amounts of propylene are commonly available. The effects of polymerization temperature and propylene concentration on the intrinsic solution viscosity of the copolymer are given in Table 5.

The increased termination by propylene reduces the effect of the thermal instability of the end groups so that the copolymer is more stable than the homopolymer of equivalent molecular weight (15).

The copolymers have a low melt viscosity, which means that they can be processed at lower temperatures than PVC and with reduced amounts of stabilizer. The

Table 5. Viscosity of Vinyl Chloride–Propylene Copolymers

Polymerization temperature, °C	Intrinsic solution viscosity at various concentrations of propylene in the polymer[a]				
	0%	5%	7.5%	10%	15%
120	1.15	0.85	0.75	0.70	0.57
130	0.98	0.72	0.62	0.57	0.75
140	0.80	0.60	0.50	0.45	
150	0.62	0.48			

[a] ASTM D 1243-60 (method A).

Table 6. Rheological Properties of Rigid Vinyl Chloride–Propylene Resin Formulations[a]

Resin[b]	Inherent viscosity,[c] dl/g	Melt viscosity at 375°F, in poises, at various shear rates		
		10^2 sec^{-1}	10^3 sec^{-1}	10^4 sec^{-1}
Airco 401	0.87	83,000	9,200	1,000
Airco 405	0.75	33,000	5,800	800
Airco 420	0.66	21,000	4,800	660
Airco 470	0.60	13,000	3,600	530
vinyl chloride	0.93	92,000	9,800	1,100
homopolymer	0.63	31,000	5,700	760
vinyl chloride–vinyl acetate copolymer	0.50	6,000	2,800	520

[a] By permission of Airco Chemicals and Plastics Co.
[b] Compounds based on 3 phr of organotin stabilizer.
[c] ASTM D 1243-60A.

rheological properties of the Airco 400 series of vinyl chloride polymers in rigid formulations are given in Table 6.

The copolymers have proved to be of particular interest for making rigid bottles for use in food packaging since the amounts of additives needed are small and the nontoxic stabilizers based on calcium–zinc soaps can be used in spite of their reduced stabilizing efficiency (see also STABILIZATION).

The Airco 400 series resins have a degree of thermal stability that gives them a processing life two to three times that of vinyl chloride homopolymers of similar

Table 7. Physical Properties of Vinyl Chloride–Propylene Copolymers[a]

Property	ASTM method	Unmodified[b]				Impact-modified	
		Airco 401	Airco 405	Airco 420	Airco 470	Airco 401	Airco 470
specific gravity	D 792-64T	1.40	1.40	1.39	1.38	1.35	1.34
tensile strength, psi	D 638-64T (0.2 in./min)	8,700	8,400	8,200	8,000	6,200	5,800
tensile modulus, psi	D 638-64T (0.2 in./min)	490,000	450,000	440,000	415,000	350,000	350,000
ultimate tensile strength, psi	D 638-64T (0.2 in./min)	7,400	6,500	5,500	5,400	5,700	5,600
ultimate elongation, %	D 638-64T (0.2 in./min)	80	100	90	125	110	120
flexural yield strength, psi	D 790-66	15,400	14,700	14,300	13,800	9,500	9,200
flexural modulus, psi	D 790-66	505,000	480,000	450,000	455,000	410,000	390,000
compressive yield strength, psi	D 695-63T	11,500	11,400	11,100	10,700	8,000	7,800
deflection temperature, °F	D 648-56	159	158	158	155	160	158
Izod impact strength, ft-lb/in. notch	D 256-56 (Method A)	0.70	0.65	0.65	0.55	25	6.0

[a] By permission of Airco Chemicals and Plastics Co.
[b] The approximate propylene contents are 3% for Airco 401, 5% for 405, 6% for 420, and 7% for 470.

molecular weight. The physical properties vary according to the propylene content of the polymer (see Table 7).

The copolymerization of vinyl chloride and isobutylene can be carried out in aqueous emulsion using peroxydisulfate as initiator (16). The polymerization rate is slow and the polymer formed is of low molecular weight.

With Vinylidene Chloride

A range of copolymers of vinyl chloride and vinylidene chloride have been developed commercially. These are available with the vinylidene chloride content varying from 4 to 40%. The processing characteristics of the copolymers improve as the vinylidene chloride content increases, and there is also an increase in solubility, although this is not as good as that of the corresponding vinyl acetate copolymers. There is an extensive range of patents pertaining to the aspects of the copolymerization of vinyl chloride and vinylidene chloride. An early patent (17) describes the manufacture of a copolymer containing 10% vinylidene chloride in emulsion at 40°C using benzoyl peroxide as initiator. Vinylidene chloride enters the polymer chain preferentially and consequently, in order to obtain a homogeneous copolymer, vinylidene chloride is added during the course of the polymerization. The copolymers now available are manufactured using the suspension process, as developed for homopolymers, with the addition of monomeric vinylidene chloride during the reaction. This technique has the usual advantage of ease of isolation of the polymer from the aqueous slurry.

The development of the copolymers of vinyl chloride and vinylidene chloride arose because of the patent position covering the vinyl acetate copolymers. As these patents have lapsed, the majority of the vinylidene chloride copolymers have gone out of production. The main exception to this is the copolymer containing 40% vinylidene chloride. This resin is soluble in cheap hydrocarbon solvents and is used extensively as a beer-can lacquer. See also VINYLIDENE CHLORIDE POLYMERS.

With Acrylic Esters

These copolymers were developed and manufactured extensively in Germany during World War II. They were used for rigid applications such as the manufacture of press polished sheets for dials for instruments and windows for aircraft. Methyl acrylate was used as the comonomer, typical compositions containing 14 and 20% of the ester. Terpolymers were also manufactured with the addition of a maleic ester as the third monomer; in all cases the emulsion polymerization system was used with equipment similar to that employed for the homopolymer. The monomers were emulsified in a 1.25% solution of the sodium salt of oxo-octadecanesulfonic acid using a water-to-monomer ratio of 4:1. The aqueous phase contained 0.1% of potassium peroxydisulfate as initiator and the reaction was carried out at 40°C. The whole of the vinyl chloride, together with part of the comonomer, was charged at the start of the process and the remainder of the acrylate was added during the polymerization in order to maintain a constant copolymer composition.

The reaction took 24–36 hours to produce a polymer emulsion of 25–27% solids content which was coagulated by the addition of aluminum sulfate. The resin, after washing and neutralization of the coagulant, was heated to 80–85°C in an aqueous suspension to agglomerate and increase the size of the polymer particles. After

Table 8. Softening Points of Acrylic Ester–Vinyl Chloride Copolymers

	Softening point, °C, at various comonomer contents		
Comonomer	5% by wt	10% by wt	20% by wt
methyl acrylate	75	73	63
ethyl acrylate	72	67	61
n-butyl acrylate	70	65	60
octyl acrylate	68	64	56

filtration the polymer was washed with sodium carbonate to stabilize the resin and then dried.

The properties of the copolymer depend not only on the amount of comonomer present but also on the number of carbon atoms in alkyl radical of the acrylic ester. For example, the softening points of a number of copolymers based on different acrylic esters are given in Table 8.

Impact-Resistant Polymers. In the majority of methods used to produce copolymers, special techniques are used to ensure that the final products are homogeneous in composition because such materials have better properties than heterogeneous copolymers. However, in certain cases in copolymerization with acrylic esters, certain desirable properties can be obtained by producing heterogeneous polymers (18). When the whole of the acrylic monomer is introduced at the start of the polymerization the polymer formed in the first stages has a much higher acrylic ester content than the average composition. When 2-ethylhexyl acrylate is used as the comonomer, the poly-

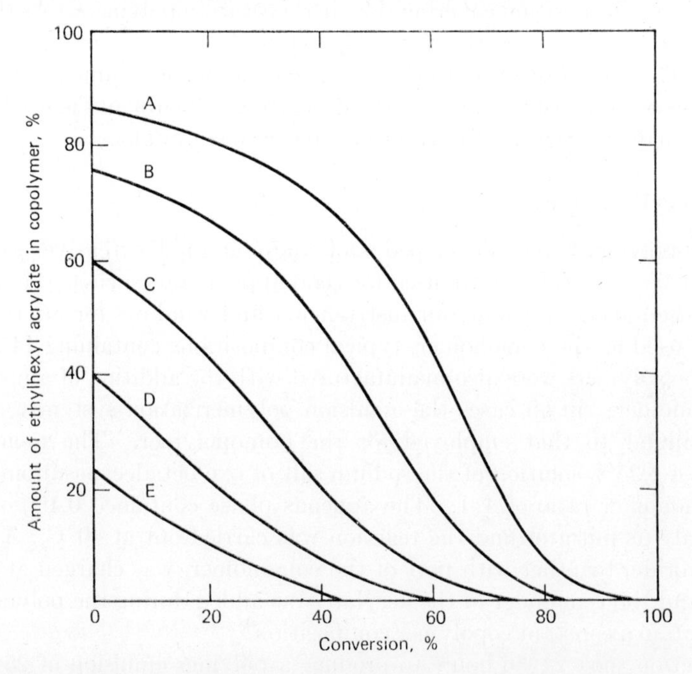

Fig. 4. Ester content of polymer as formed during polymerization. Key: curve A, 50% initial ester content in monomer; curve B, 35% initial ester content in monomer; curve C, 20% initial ester content in monomer; curve D, 10% initial ester content in monomer; curve E, 5% initial ester content in monomer.

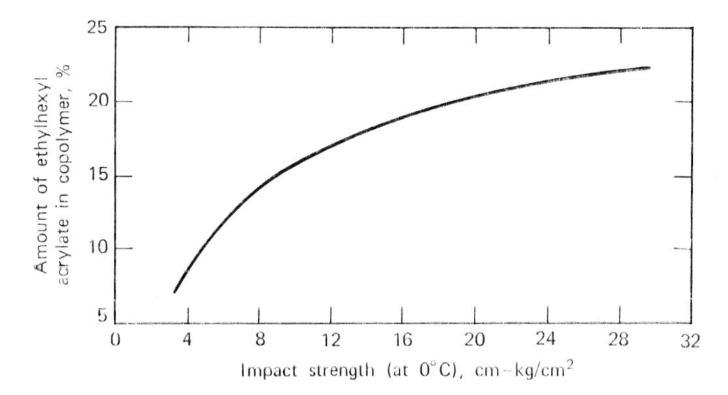

Fig. 5. Impact strength at 0°C of vinyl chloride–2-ethylhexyl acrylate copolymer.

mer containing high amounts of the ester will have a low second-order transition temperature, and the presence of these copolymer components in the composition lead to improved impact properties in rigid formulations and particularly at low temperatures. The relationship of the composition of the polymer to the initial composition of the monomer mixture is shown in Figure 4. It is evident that the polymer formed in the initial stages has a high acrylate content, whereas almost pure PVC is obtained toward the end of the process. When 20% of acrylate is present, about 15% of the polymer has an average 2-ethylhexyl acrylate content of about 55%. The second-order transition temperature of poly(2-ethylhexyl acrylate) is −65°C, and that of PVC is 80°C, so that the presence of 15% of a copolymer containing high amounts of acrylic ester is sufficient to impart improved low-temperature impact properties. Figure 5 shows the impact strength at 0°C for a range of copolymers. As the acrylate content increases the softening point falls, and is only about 40°C for a polymer containing 20% acrylate.

The properties of the copolymer, particularly the impact strength, depend on the polymerization technique. Copolymers containing 80% vinyl chloride and 20% ethylhexyl acrylate were prepared using both the emulsion and suspension processes, carrying out the reaction at 45°C and to 90% conversion. The suspension resin had a much higher impact strength (20 cm kg/cm²) than the emulsion product (4.5 cm-kg/cm²). This appears to be due to the difference in particle size; the best impact properties are obtained from polymers having a particle size between 0.5 and 5 μ.

With Vinyl Ethers

Copolymers of vinyl chloride with the lower vinyl ethers were introduced commercially at an early stage in the development of PVC in Germany. Both *n*-butyl vinyl and isobutyl vinyl ether were copolymerized with vinyl chloride in an emulsion system (19,20). The vinyl ethers are readily hydrolyzed under acid conditions so that buffers such as calcium carbonate, lead carbonate, and phosphates were added. Since the vinyl ethers polymerize more slowly than vinyl chloride, the polymerization was started with the entire quantity of the ether and only part of the vinyl chloride present. The rest of the vinyl chloride was added stepwise during the course of the reaction. In the emulsion copolymerization of 70 parts of vinyl chloride and 30 parts of *n*-butyl vinyl ether, it was found that the amount of ether in the copolymer increased as the emulsifier concentration was raised (see Table 9).

Table 9. Copolymerization of Vinyl Chloride and Vinyl *n*-Butyl Ether

Sodium 2-hydroxy-octadecanesulfonate emulsifier, %	Vinyl chloride content of copolymer, %
0	97
0.2	89
1.0	81.5
2.0	76

The copolymer containing isobutyl vinyl ether was used in Germany during World War II for blending with cellulose nitrate lacquers and as the adhesive for binding iron oxide particles to PVC films for a magnetic recording process. Copolymers of long-chain vinyl ethers and vinyl chloride are reported to have exceptional clarity and excellent mechanical properties (21–26). The best results in rigid composition appear to be obtained by blending PVC resin (above 90%) with a copolymer of vinyl

Table 10. Properties of a Modified Vinyl Chloride–Cetyl Vinyl Ether Copolymer

	Blend of copolymer with PVC	Homopolymer containing 1% dioctyl phthalate
impact strength, kg-m	4.1	3.0
brittle point, °C	−12	−8
flex temperature, °C	69	67

Table 11. Properties of Vinyl Chloride–Vinyl Ether Copolymer Blend

Property	Test method	Blend of copolymer with PVC	Poly(vinyl chloride)
specific gravity	ASTM D 792	1.39	1.40
tensile strength, kg/mm^2	Instron tester (full speed)	5.5	5.5
elongation, %	(50%/mm, 20°C)	150	150
impact strength, $kg\text{-}cm/cm^2$	Charpy method, V-notch	3.9	4.2
hardness	Rockwell	63	65
brittle temperature, °C	ASTM D 746-55T	−10	−12
softening temperature, °C	ASTM D 1043	64	70

chloride and a higher vinyl ether (up to 40%). The vinyl ethers specifically mentioned are cetyl vinyl ether and lauryl vinyl ether (see also VINYL ETHER POLYMERS). A copolymer containing 5 parts by weight of cetyl vinyl ether and 95 parts of vinyl chloride, and blended with a commercial-grade PVC (*K* value 65) in the proportion 1:29, has the same rheological behavior as a material formed by plasticizing the homopolymer with 1% dioctyl phthalate (25). The impact strength, brittle temperature, and flex temperature for the two compositions are given in Table 10.

The superior properties of the copolymer blend make the product suitable for applications such as the extrusion of clear rigid sheet. Properties of commercial grades of this type of blend are listed in Table 11.

With Maleic and Fumaric Esters

Copolymers with the lower alkyl maleates have been produced commercially both in Germany and in the United States. Diethyl maleate copolymerizes readily with vinyl chloride, entering the polymer chain at the faster rate so that the maleic ester content of the polymer is higher than that of the monomer mixture from which it is formed. A copolymer of uniform composition is produced by added portions of maleic ester during the polymerization (27). The copolymer containing 20% dimethyl maleate is reported to have a higher softening point than the copolymer based on 20% methyl acrylate, and also to have better processing properties.

The fumaric and maleic esters have fairly high boiling points so that it is difficult to remove any unpolymerized monomer from the resulting copolymer. This can present some problems because the esters have objectionable properties, particularly if they volatilize during compounding operations. For this reason and also because of the relatively high cost of maleic and fumaric esters compared with vinyl acetate, this type of copolymer is no longer available commercially, with one exception. A copolymer of vinyl chloride with di-2-ethylhexyl fumarate is used for blending with PVC in rigid compositions to produce good impact and good weathering properties. See also ACIDS, MALEIC AND FUMARIC.

With Acrylonitrile

Acrylonitrile copolymerizes with vinyl chloride to produce polymers in which the acrylonitrile content is lower than that of the monomer mixture from which they are formed (see also ACRYLONITRILE POLYMERS). Results obtained from polymerizations carried out over a range of conditions in both emulsion and single-phase systems show that the amount of acrylonitrile occurring in the polymer is only one half that in the monomer mixture from which it was produced, and it is necessary to feed acrylonitrile continuously during the polymerization in order to obtain resins of the degree of compositional uniformity desired for fiber applications (28). The copolymers become more soluble in acetone as the acrylonitrile content increases; those which find particular use in fibers have an acrylonitrile content of about 40% (see also MODACRYLIC FIBERS). The copolymer is made up into a 25% solution in acetone that also contains both light and heat stabilizers to prevent degradation during processing; is dry spun, the solvent being evaporated by a countercurrent stream of hot air as the fibers leave the spinnerets. The fibers are characterized by moderate stiffness and good resiliency, and are particularly useful for applications requiring these properties but only modest tensile strength. Drawing at elevated temperatures produces improved solvent resistance and a higher softening temperature. Stretching as high as 2500%, based on spun yarn, and filament deniers as low as 0.35 have been obtained experimentally, but in practice the filament denier of stretched yarn is usually of the order of 0.75–2.0. The stretched fibers have some degree of crystallinity but this is low compared with the more crystalline polymers, such as nylon.

One main disadvantage of this type of copolymer is that the presence of acrylonitrile produces a yellowish color in the product, but, in contrast to PVC, the color is bleached by strong sunlight. At elevated temperatures degradation takes place readily, with the result that crosslinking occurs and the copolymers become less soluble.

The fibers are particularly useful for making filter cloths and in other applications where their inherent fire retardancy (qv) is an advantage (29). See also the section on Fibers in this article (pp. 468 ff.).

Bibliography

1. C. M. Thomas and J. R. Hinds, *Brit. Plastics* **31** (12), 522 (1958).
2. F. R. Mayo et al., *J. Am. Chem. Soc.* **70,** 1523 (1948).
3. W. H. McKnight, *Mater. Protection* **4** (8), 32 (1965).
4. C. J. Martin, *SPE J.* **18,** 392 (April 1962).
5. C. W. Johnston and C. H. Brower, *SPE J.* **23** (6), 106 (June 1967).
6. A. J. Foglia (to Air Reduction Co., Inc.), U.S. Pat. 3,056,768 (Oct. 2, 1962).
7. W. S. Port and E. F. Jordan, Jr. (to U.S.A. as represented by the Secretary of Agriculture), U.S. Pat. 2,876,895 (March 10, 1959).
8. D. Swern (to U.S.A. as represented by the Secretary of Agriculture), U.S. Pat. 2,993,034 (July 18, 1961).
9. G. V. Trachenko, L. V. Stupen, L. P. Kofman, and L. A. Karacheva, *Zh. Fiz. Khim.* **32,** 2492 (1958).
10. W. Reppe, W. Starck, and A. Voss (to I. G. Farbenindustrie A.G.), U.S. Pat. 2,118,945 (May 31, 1938).
11. N. L. Zutty (to Union Carbide Corp.), U.S. Pat. 3,051,689 (Aug. 28, 1962).
12. H. Bartl, D. Kardt, and D. Clabisch (to Farbenfabriken Bayer A.G.), Ger. Pat. 1,211,395 (Sept. 18, 1963).
13. F. P. Reading, E. W. Wise, and J. H. Hodge (to Union Carbide Corp.), Can. Pat. 674,142 (Dec. 28, 1961).
14. Air Reduction Co., Inc., Brit. Pat. 1,096,887 (Aug. 6, 1965).
15. R. D. Deanin, *SPE J.* **23,** 50 (May 1967).
16. A. H. Gleason (to Standard Oil Development Co.), U.S. Pat. 2,379,292 (June 26, 1945).
17. C. H. Alexander and H. Tucker (to B. F. Goodrich Co.), U.S. Pat. 2,245,742 (June 17, 1941).
18. W. Albert, *Kunststoffe* **53** (2), 86 (1963).
19. H. Fikentscher and J. Hengstenberg (to I. G. Farbenindustrie A.G.), U.S. Pat. 2,100,900 (Nov. 30, 1937).
20. H. Fikentscher and R. Gaeth (to I. G. Farbenindustrie A.G.), U.S. Pat. 2,016,490 (Oct. 8, 1935).
21. Oxygen, Hydrogen & Oil Industries Co., Japan. Pat. 17,594.
22. Shin-Estu Chemical Industry Co. Ltd., Brit. Pat. 877,100 (Sept. 3, 1961).
23. *Kogyo Kagaku Zasshi* **63,** 592 (1962).
24. *Kogyo Kagaku Zasshi* **65,** 1001 (1962).
25. Y. Hoshi and M. Onozuka (to Kureha Kasei Co. Ltd.), U.S. Pat. 3,168,594 (Feb. 2, 1965).
26. Kureha Kasei Co. Ltd., Brit. Pat. 928,799 (June 12, 1963).
27. J. J. P. Staudinger and C. A. Brighton (to Distillers Co. Ltd.), Brit. Pat. 581,995 (Oct. 31, 1946).
28. E. W. Rugeley, T. A. Feild, and G. H. Fremon, *Ind. Eng. Chem.* **40,** 1724 (1948).
29. E. Stowell, *Papers, Am. Assoc. Textile Technol.* **3,** 30 (1947).

C. A. Brighton
BP Chemicals International Ltd.

PROPERTIES

Structure

The determination of the average chemical structure of individual PVC chains has been beset with great difficulties. The extent and type of stereoregularity and branching, and the factors which control these properties, are of paramount importance since they reflect the polymerization mechanism and can often dominate certain polymer properties, eg, thermal and photochemical stability, crystallinity, and rheology of polymer melts and solutions. It is convenient and appropriate to consider both branching and stereoregularity together since attempts to define and estimate the variation of stereoregularity have frequently been compromised by the concomitant variation in chain branching. See also MICROTACTICITY; STEREOREGULARITY.

Reduction with zinc dust leads to the conclusion that in PVC prepared at normal temperatures the monomer units are incorporated predominantly in a head-to-tail sequence (1).

Branching. Branching in PVC has generally been determined by reduction of the polymer in tetrahydrofuran solution by lithium aluminum hydride to a polyethylene structure, followed by estimation of the $CH_3:CH_2$ ratio in the molten product by infrared measurements (2), usually through the absorbances at 1378 cm^{-1} and 1368 cm^{-1}. Although a calibration curve for $CH_3:CH_2$ against ratio of optical densities can be relatively accurate at high $CH_3:CH_2$ ratios, using a range of *n*-paraffins, the biggest source of error arises in extrapolating this curve into the range of ratios encountered in the reduction products of PVC. It has not yet been resolved whether the branching in PVC is predominantly short or long chain in nature and this factor will also influence the estimates.

Although most investigators (3–5) agree that the branching decreases significantly as the temperature of polymerization decreases and becomes trivial below about −40 to −60°C, there are wide discrepancies among absolute values. The results published in two detailed studies on branching may be quoted to illustrate the extent of the variation. The numerical values for branching incidence are expressed as the number of methyl groups per 100 methylene groups in the polyethylene structures obtained by reduction of the PVC. For example, polymers prepared at 90, 50, −15, and −75°C to undisclosed conversions yielded branching indexes of 0.27, 0.20, 0.05, and 0, respectively. A commercial polymer, presumably a high-conversion sample, made at 55–60°C gave a value of 0.20 (5). In another study, polymers prepared at low conversion (~5%) with alkylboron and oxygen as initiator at 25, −20, and −60°C had indexes of 1.78, 1.30, and 0.65, respectively. A commercial polymer made at 50°C to 95% conversion gave a value of 1.80 (4). Some of the discrepancies may be associated with varying interference by initiator end groups.

If the mechanism of bulk polymerization of vinyl chloride put forward in Ref. 6 is correct (see p. 326), no significant variation of branching with conversion would be anticipated at least up to 70–80% conversion, irrespective of whether the branching arose intramolecularly (short branches) or intermolecularly (long branches). One author claimed that branching was constant up to 95% conversion (4). The same author failed to find evidence for sufficient terminal $CH_2{=}CH{-}$ groups in the reduced polymers to account for the general claim that monomer transfer controls the molecular weight in PVC.

Based on the measured transfer constant to 2,4,6-trichloroheptane (7), an analog of PVC, one can calculate a branching incidence of one branch per molecule at 50°C. In establishing solution viscosity–molecular weight relationships for PVC, it has frequently been observed (8,9) that the double logarithmic plot of intrinsic viscosity versus molecular weight for polymer fractions was linear for molecular weights up to about 150,000–170,000, but that significant deviations from linearity occurred in the direction of low viscosity for the fractions with molecular weights in excess of this value. The departure is consistent with a higher incidence of long-chain branches per molecule in the higher-molecular-weight fractions. The Zimm and Kilb function (10) has been applied to one set of data to derive a branching index of approximately 0.15 for certain polymers made at 50°C (8). Another estimate of branching is based on the assumption that the branch points will be initiated predominantly at the carbon–hydrogen bonds of the CHCl groups leading to tertiary carbon–chlorine bonds in the branched polymer (11).

A recent publication (12) discloses branching indexes varying from 0.33 to 0.82 for PVC samples prepared over the temperature range −50°C to +60°C.

Stereoregularity and Crystallinity. Commercial poly(vinyl chloride) is a largely amorphous polymer but with a small proportion (about 5%) of small and imperfect crystallites. The rather diffuse x-ray diffraction pattern has a repeat distance of about 5 Å and agrees with a syndiotactic planar zigzag structure (13,14). A reduction of the polymerization temperature gives sharper, but still diffuse, x-ray patterns and the polymer becomes less soluble in solvents. Estimated melting points range from 155°C for polymers made at 125°C to 310°C for those made at −75°C. There is a corresponding increase in glass-transition temperature from 68 to 105°C (5,15). Commercial poly(vinyl chlorides) have T_g values of about 80–85°C (15).

Significant changes also occur in the infrared spectra. There are a number of absorbances in the region 700–600 cm^{-1} and the relative intensities of several of these vary with polymerization temperature. Using frequencies at 635 and 692 cm^{-1}, the ratio of intensities was found to change from 1.3 at 70°C to 2.5 at −70°C (16). The assignments for these bands were considered from the spectra of model compounds to correspond to the stable conformations of syndiotactic and isotactic configurations of adjacent monomer units, and it was concluded that there was an increase in syndiotacticity with decreasing polymerization temperature. Other investigators using different bands calculated that the syndiotacticity changed from 0.52 at 50°C to 0.86 at −60°C (17). These figures were based, however, on an assumption made to fit the observed temperature dependence of the infrared absorption ratio, that $\Delta H_s^i = 1$ kcal/mole and $\Delta S_s^i = 0$, and they cannot be regarded as absolute values. (ΔH_s^i and ΔS_s^i are the differences in enthalpy and entropy of isotactic and syndiotactic placement, respectively.)

Conducting the polymerization in aliphatic aldehydes as solvents (18,19) led to an even more marked increase in crystallinity, accompanied by an increase in the 635 cm^{-1}:692 cm^{-1} intensity ratio to a value in the region of 3.5–4, independent of the polymerization temperature. The explanation was thought to be the solvation of the growing chain end by the aldehyde, favoring approach of the monomer for syndiotactic addition. Alternative, and more reasonable, explanations for the effect of the aldehyde are given on p. 362. The effect of aldehydes in increasing polymer crystallinity has been consistently confirmed.

The conclusion that PVC produced at low temperatures or in aldehyde solution is highly syndiotactic has been challenged (20,21). NMR studies indicated that the polymer structure was of almost constant composition, with about two syndiotactic units to each isotactic unit, irrespective of the temperature of polymerization. The value of ΔH_s^i was estimated to be 0.2 kcal/mole at most and the increased crystallinity was attributed to reduced branching (21). Two other infrared frequencies were used for assessing the chain stereochemistry (22); the results quoted have been in fair agreement with NMR data, but unlike the results of later workers a moderate dependence of syndiotacticity on temperature was found.

Significant contributions have been made to the assignment of the infrared bands, previously subject to some doubt, by examining stereoisomers of 2,4,-dichloropentane (DL and meso forms) and 2,4,6-trichloroheptane (DLD, DLL, and DDD isomers) (25–27). Dipole moment measurements have been used to supplement the data and resolve dubious assignments. Frequencies at 608 and 641 cm^{-1} were found to correspond to the planar zigzag (*tttt**) syndiotactic isomer; the heterotactic form absorbs at 610, 627, and 687 cm^{-1} (probably from the *tttg* with some *ttgt* conformation) and the isotactic structure, probably tg_n or gt_n, at 619 and 688 cm^{-1}. From this it would appear that the steric assignments ascribed to the spectra of poly(vinyl chloride) are correct, although the quantitative estimates may be subject to some error. Recent NMR estimates (23,24) agree with the conclusions of Ref. 21 that the stereochemistry is almost independent of polymerization temperature. On the basis of the two kinds of accessible rotational isomers assumed by heterotactic 2,4,6-trichloroheptane (*tttg*, *ttgt*), it has been suggested that the flexible joints in the PVC chain are located in the atactic regions (25–27).

Earlier published data on poly(vinyl chloride) are given in Table 1. From these values a definite conclusion about the effect of polymerization temperature on configuration cannot be reached. The infrared data, which are thought to be less reliable (25), give considerably higher values than NMR; and in one study (22) a distinct temperature dependence is observed.

The last value in Table 1, showing the absence of any influence of *n*-butyraldehyde on the stereochemistry of poly(vinyl chloride) is in agreement with the previously expressed views (21) and in contrast with the infrared data (19). The generally high estimates obtained from infrared spectra force the conclusion that the band intensities may be affected by factors other than chain configuration, for example, by changes in crystallinity. It has also been shown that the infrared spectra in the solid state of a given sample of PVC can be very dependent on the pretreatment given to the sample.

Table 1. Structure of Poly(vinyl Chloride)

Polymerization temp, °C	Syndiotacticity		Ref.
	By NMR[a]	By infrared	
50	0.60		24
	0.50 (0.55)		23
		0.72	25
30	0.51	0.54	22
0	0.59	0.59	22
−20	0.54 (0.65)		23
		0.69	23
−25	0.54		22
−30	0.68		24
		0.73	25
−35	0.73	0.72	22
−50	0.56 (0.68)		23
−70	0.85	0.80	22
−75	0.60 (0.61)		23
−78		0.77	25
50[b]	0.53 (0.56)		23

[a] The figures in parentheses are results obtained from proton resonances.
[b] Butyraldehyde solvent.

It is suggested that significant but variable concentrations of high-energy chain conformations can be generated by the constraints existing in the solid state (28).

Ambiguities in the interpretation of the NMR spectra now seem to have been resolved. The meso methylene groups in PVC or the model compounds are heterosteric and the protons of such groups are generally magnetically nonequivalent. Substantial nonequivalence was confirmed for meso methylene protons in the 2,4-dichloropentane model compounds (27,29,30). However, the differences between the methylene protons of the isotactic or meso–meso form of the higher homolog 2,4,6-trichloroheptane are very small (25–27). As the isotactic chain is lengthened this difference would presumably become insignificant, leading to the conclusion that the protons of meso methylene groups in PVC are fortuitously equivalent. Differences in the methylene protons in PVC should be detected more easily in the α,β-dideutero polymers which possess neither vicinal nor geminal proton–proton spin couplings. Poly(α-*cis*,β-d_2-vinyl chloride) has been examined (31), and although eight of the ten expected methylene proton resonances were displayed, it was concluded that not only are the average shieldings of all the meso methylene protons nearly the same, but also the shieldings of all the racemic or syndio protons are nearly the same. More highly resolved spectra of the poly(α-d_2-vinyl chloride) obtained recently have supported these findings and also indicated Bernouillian distributions of tactic placements in PVC (22). The final conclusion is that PVC prepared at 100°C is virtually atactic and that decreasing the polymerization temperature even to −78°C does not promote the syndiotactic placement content very significantly. In numerical terms, if σ is the probability that a polymer chain will add a monomer unit to give the same configuration as that of the last unit at the growing end, in other words generation of an isotactic diad, then the deduced values of σ are 0.46 at 100°C, falling to 0.37 at −78°C. The revised values of ΔH_s^i and ΔS_s^i are 310 ± 20 cal/mole and 0.6 ± 0.01 esu/mole, respectively (32); the former is higher than the original value of 200 cal/mole (21) but substantially less than the early estimate of 600 cal/mole (16).

The increase in size and perfection of crystallites, along with the increase in melting point and glass-transition temperature, with reduced polymerization temperature seems to be much more pronounced than can be accounted for by the increase in syndiotacticity. A reduction in other forms of structural irregularity, such as head-to-head addition or branching, with lower polymerization temperatures, could help to relieve the dilemma. Complexing with an aldehyde solvent may be equivalent to a lowering of the temperature in this context. An alternative explanation for the effect of aldehyde solvents would be reduction of molecular weight which would facilitate crystallization. A resonable correlation was observed between crystallinity and degree of polymerization (18). Poly(vinyl chlorides) with low molecular weights (DP 32) have been prepared (33) using various solvents (including butyraldehyde), and it was observed that the degree of crystallinity increased with decrease in molecular weight. Even for the highest branching indexes reported (4) a significant proportion of molecules will be branchfree when M_n falls to around 1800.

There could also be differences in the distribution of steric sequences in polymers of the same average tacticity. It has been suggested (34) that the crystallites in PVC are formed from short syndiotactic blocks of about four or five units, which would not occur frequently in a nearly random polymer of 50–60% syndiotacticity.

Mention should be made of a novel free-radical polymerization of vinyl chloride carried out in urea canal complexes or clathrates (35). By analogy with the geometrical

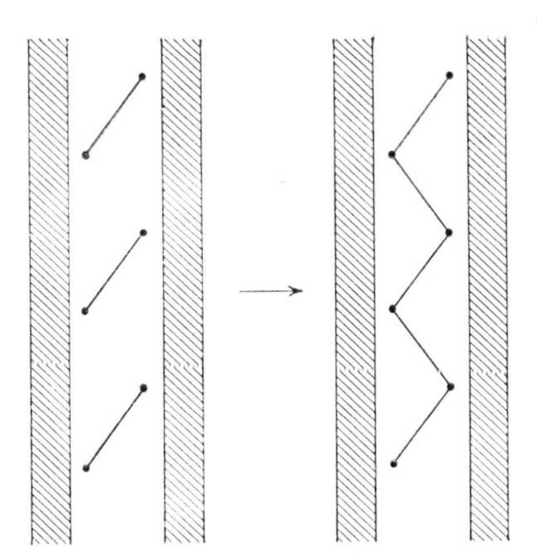

Fig. 1. Schematic representation of polymerization of vinyl chloride in urea clathrate.

proposals for the related polymerization of substituted butadienes in thiourea canal complexes it is assumed that the vinyl chloride molecules are stacked in the channel in the appropriate spatial disposition for propagation but protected from termination or branching with adjacent sequences by the walls of the canal (Fig. 1). Initiation was by radiation and the urea was removed with aqueous acetone. Infrared evidence was advanced indicating that the product was highly syndiotactic (28); however, a more detailed examination by the recent sophisticated NMR techniques should be extremely revealing.

Property Values

Poly(vinyl chloride) is available commercially in a wide range of particle sizes, the particular range depending on the method of polymerization. Emulsion polymers, which include those used for making plastisols, have a particle size of 0.2–2 μ; the polymers that are not paste-forming fall in the lower end of the range and have a more regular size than the paste-forming resins, which have a particle-size distribution across this whole range. Some paste-forming polymers have a bimodal size distribution with about 75% of the particles around 0.8 μ and 25% around 0.2 μ. Suspension resins have a particle size range of 5–300 μ.

The shape of the particles can vary from almost perfect spheres to irregular structures that have a surface area at least eight times that of a spherical particle of the same weight. The increased surface area improves the rate of interaction with plasticizers so that dry powder blends can be produced by simple mixing in a stirred vessel. The powder mixing properties depend on the particle-size distribution and are best for those resins with a uniform particle size. This is to be expected since particles of approximately the same diameter and shape absorb plasticizers at much the same rate. The inner structure of the polymer particles is also a determining factor in the rate at which interaction takes place (36).

Typical property ranges for vinyl chloride homopolymers and copolymers are given in Table 2.

Molecular Weight. The number-average molecular weight ($\bar{M}_n$) for most commercial PVC resins lies in the range 50,000–120,000. Processing characteristics can be varied by producing polymers of different molecular weights. The higher-molecular-weight polymers give finished products, both plasticized and rigid, with better physical properties than would be obtained with the lower-molecular-weight polymers. These latter, however, produce compounds with a lower melt viscosity, so that they are essential for some fabrication techniques, eg, the injection molding of thin-walled articles of rigid PVC. The softening point of the resins is reduced in the low-molecular-weight products, which also have improved solubility in organic solvents.

The use of dilute-solution viscosity measurements is an effective means of determining the molecular weight, although the value obtained, $\bar{M}_v$, will correspond to $\bar{M}_w$ or $\bar{M}_n$ only if the polymer is homogeneous and monodisperse. The Mark-Houwink relationship, $[\eta] = KM_v^a$, where $[\eta]$ is the intrinsic viscosity in dl/g, is most commonly used. Values of K and a have been determined by a number of workers and considerable variations have been found even when the same solvent is used. The results shown in Table 3 were obtained with cyclohexanone as solvent.

Table 2. Typical Properties of Vinyl Chloride Homopolymers and Copolymers[a]

Property	ASTM test method	Rigid	Flexible
	Processing		
molding qualities		fair to good	good
compression-molding temp, °F		285–400	285–350
compression-molding pressure, psi		750–2,000	500–2,000
injection-molding temp, °F		300–415	320–385
injection-molding pressure, psi		10,000–40,000	8,000–25,000
compression ratio		2.0–2.3	2.0–2.3
mold shrinkage (linear), in./in.		0.001–0.005	0.010–0.050 (varies with plasticizer)
specific gravity	D 792	1.35–1.45	1.16–1.35
specific volume, cu in./lb	D 792	20.5–19.1	23.8–20.5
	Mechanical		
tensile strength, psi	D 638, D 651	5,000–9,000	1,500–3,500
elongation, %	D 638	2.0–40	200–450
tensile modulus, 10^5 psi	D 638	3.5–6.0	
compressive strength, psi	D 695	8,000–13,000	900–1,700
flexural yield strength, psi	D 790	10,000–16,000	
impact strength, ft-lb/in. of notch (½ × ½ in. notched bar, Izod test)	D 256	0.4–20	varies depending on type and amount of plasticizer
hardness, Shore	D 785	65–85D	50–100A
	Thermal		
thermal conductivity, 10^{-4} cal/(sec)(cm²)(°C/cm)	C 177	3.0–7.0	3.0–4.0
specific heat, cal/°C/g (25°C)		0.2–0.28	0.3–0.5
thermal expansion, 10^{-5} in./in./°C	D 696	5.0–18.5	7.0–25
resistance to heat, °F (continuous)		150–175	150–175
deflection temp, °F	D 648		
at 264 psi fiber stress		130–175	
at 66 psi fiber stress		135–180	

Table 2 (*continued*)

Property	ASTM test method	Rigid	Flexible
		Electrical	
volume resistivity, ohm/cm (at 50% rh and 23°C)	D 257	$>10^{16}$	10^{11}–10^{15}
dielectric strength, ⅛-in. thickness, volts/mil	D 149		
short-time		425–1,300	300–1,000
step-by-step		375–750	275–290
dielectric constant	D 150		
at 60 Hz		3.2–3.6	5.0–9.0
at 10^3 Hz		3.0–3.3	4.0–8.0
at 10^6 Hz		2.8–3.1	3.3–4.5
dissipation (power) factor	D 150		
at 60 Hz		0.007–0.020	0.08–0.15
at 10^3 Hz		0.009–0.017	0.07–0.16
at 10^6 Hz		0.006–0.019	0.04–0.14
arc resistance, sec	D 495	60–80	
		Optical	
refractive index, n_D	D 542	1.52–1.55	
clarity		..........transparent to opaque..........	
		Resistance characteristics	
water absorption, 24 hr, ⅛-in. thick, %	D 570	0.07–0.4	0.15–0.75
burning rate (flammability), in./min	D 635	self-extinguishing	slow to self-extinguishing
effect of weak acids	D 543	none	none
effect of strong acids	D 543	none to slight	none to slight
effect of weak alkalies	D 543	none	none
effect of strong alkalies	D 543	none	none
effect of organic solvents	D 543	resistant to alcohols, aliphatic hydrocarbons, and oils; soluble or swells in ketones and esters; swells in aromatic hydrocarbons	

[a] Courtesy *Modern Plastics*, Encyclopedia Issue 1970–1971, McGraw-Hill Book Co., New York.

Table 3. Parameters for Mark-Houwink Equation, Determed in Cyclohexanone

K	a	Temperature, °C	Type of mol wt average	Remarks	Ref.
1.95×10^{-4}	0.79	25	$\bar{M}_n$	unfractionated polymer	a
1.16×10^{-4}	0.85	20	$\bar{M}_n$	unfractionated polymer	b
2.04×10^{-3}	0.56	25	$\bar{M}_n$	fractionated polymer	c
2.4×10^{-4}	0.77	25	$\bar{M}_n$	unfractionated polymer	d
1.1×10^{-5}	1.0	25	$\bar{M}_w$	unfractionated polymer	e
1.3×10^{-3}	0.63	25	$\bar{M}_n$	fractionated polymer	f

[a] J. Hengstenberg, *Angew. Chem.* **62,** 26 (1950).
[b] J. Breitenbach et al., *Kolloid Z.* **127,** 1 (1952).
[c] Z. Mencik, *Chem. Listy* **49,** 1958 (1955).
[d] G. Danusso et al., *Chem. Ind.* (*Milan*) **36,** 883 (1954).
[e] G. Ciampa and M. Schwindt, *Makromol. Chem.* **21,** 169 (1956).
[f] G. Bier and H. Kramer, *Makromol. Chem.* **28–29,** 151 (1956).

More recently both number- and weight-average molecular weights were determined for PVC fractions of a commercial emulsion resin precipitated from tetrahydrofuran (37). Special precautions were taken to avoid diffusion of low-molecular-weight fractions, a complication that may account for some of the variations found in the data of Table 3. Determinations were made in cyclohexanone at 25°C and the following values were obtained.

$K = 2.08 \times 10^{-3}$ $a = 0.56 \pm 0.02$ number-average mw
$K = 1.74 \times 10^{-3}$ $a = 0.55 \pm 0.02$ weight-average mw

Another possible cause for the variations obtained by different workers may be due to the use of different types of polymer; the conditions prevailing in an emulsion system would favor the formation of more branched polymer than in suspension-polymerized PVC, and this would tend to give lower values for a.

Solubility. The compatibility of polymers with solvents and plasticizers can be predicted by using solubility parameters (see also PLASTICIZERS; COHESIVE-ENERGY DENSITY). Values of the solubility parameter of PVC have been calculated and determined experimentally, and fall in the range 9.48–9.7.

Solubility of polymers containing functional groups, such as the acetate groups in vinyl chloride–vinyl acetate copolymers, is also determined by hydrogen bonding and to some extent by dipole interactions. In the case of PVC, it is considered that the C–H is sufficiently negative to form weak hydrogen bonds. To predict the solubility in solvents it is necessary to take account of both the solubility parameter and the proton-attracting power of the media. See also SOLUBILITY OF POLYMERS.

The best solvents for PVC are tetrahydrofuran and cyclohexanone, but a wide range of other organic solvents can also be used. These include ketones, halogenated hydrocarbons, and aromatic compounds such as nitrobenzene. Alcohols and aliphatic hydrocarbons have very little solvent action and can be used as precipitants for retrieving the polymer from solutions. This treatment can be extended to predict the effectiveness of plasticizers since the process of plasticization is simply a matter of the polymer becoming completely compatible with a solvent. The dielectric constant of the plasticizer can also be used to predict compatibility and is particularly useful when used with the solubility parameter (38).

The molecular-weight distribution can be established by subjecting PVC to fractionation by progressive precipitation of the polymer from a dilute solution. The polymer is dissolved in cyclohexanone or tetrahydrofuran to a concentration of up to about 1%; successive precipitation is obtained by addition of a nonsolvent such as butyl or amyl alcohol, or, in the case of tetrahydrofuran, water can be used. It is essential to maintain effective agitation of the solution during the addition of precipitant to obtain the polymer in a finely divided form which can be easily filtered. The polymer fractions produced with each successive addition of prescribed volumes of precipitant are isolated, and after weighing are used to measure the intrinsic viscosity. The higher-molecular-weight polymer precipitates in the first fraction and progressively lower-molecular-weight material is obtained with successive additions of precipitant. The precipitation can be followed continuously by turbidimetric titration so long as the particle size of the polymer precipitated does not vary. The increase in absorbance is measured with the progressive addition of the nonsolvent. Some anomalies have been reported (39) in the use of this method for fractionating PVC, because the fractions are not pure and tend to carry down polymer of other molecular weights. This has occurred in systems using cyclohexanone or tetrahydrofuran as solvent and aliphatic monohydric alcohols as precipitants; the effect was independent of the initial polymer concentration. This overlap results in a molecular-weight-distribution curve that is too narrow, as has been reported using the system cyclohexanone–butyl alcohol (40). It is apparently due to association of the polymer molecules into independent clusters so that only part of the tangled molecules can interact with the solvent–precipitant mixtures.

Electrical Properties (qv). The electrical properties of PVC are dependent on the amount and type of plasticizer, on the additives, eg, stabilizers, lubricants, and pigments, and also on the processing which the material undergoes during fabrication. The electrical properties of the compositions also depend on the polymerization additives remaining in the resin after the finishing process. The quantities of dispersants that remain in suspension resins are very low, and for this reason this type is most widely used for electrical applications. Many of these polymerization additives, for example, gelatin used as a protective colloid, improve the electrical properties, and consequently suspension polymers are usually better than polymers produced by bulk polymerization. The polymerization residues that remain in the resin vary among manufacturers and this can result in different effects when adding the same ingredient. An example of this can be found in the calcined clay used in some compounds to produce higher values of volume resistivity; resins from some manufacturers show a considerable improvement, whereas resins of corresponding molecular weight from other suppliers may show hardly any change.

Polymers produced by emulsion polymerization using relatively high amounts of anionic emulsifiers are generally isolated by spray drying the polymer latex so that the dried polymer contains 2–3% of material which drastically reduces the electrical properties; therefore plastisols are used only in applications in which the electrical performance is not critical.

The electrical properties of compounded PVC also depend on its thermal history. During compounding and fabrication, hydrogen chloride is split off from the PVC and reacts with the stabilizer present; since it is usual to use lead compounds as stabilizers, lead chloride will be formed. The conductance of the PVC compound increases as the heating is prolonged, owing to the formation of chloride ions. When the effectiveness of the stabilizer comes to an end, there is a more sudden increase in the conductance (41). The point at which this occurs depends on the initial quantity of stabilizer present. In the case of plasticizers such as dibenzyl sebacate, which require more extensive compounding to produce a homogeneous product, the conductance decreases in the initial stages until complete homogeneity is obtained. See also STABILIZATION.

Unlike nonpolar polymers, such as polystyrene and polyethylene, the electrical properties of PVC vary with temperature and frequency (see also ELECTRICAL PROPERTIES).

The change in *dielectric constant* over a wide frequency range for flexible PVC contaning tricresyl phosphate is shown in Figure 2 (42). The dielectric constant decreases as the frequency increases but the rate of change depends on the quantity of plasticizer present. The dielectric constant of compositions containing only small amounts of tricresyl phosphate is fairly low and varies little with increasing frequency. The dependence of dielectric constant on plasticizer content is illustrated in Figure 3. The most rapid changes occur in the region where the compound becomes flexible; this

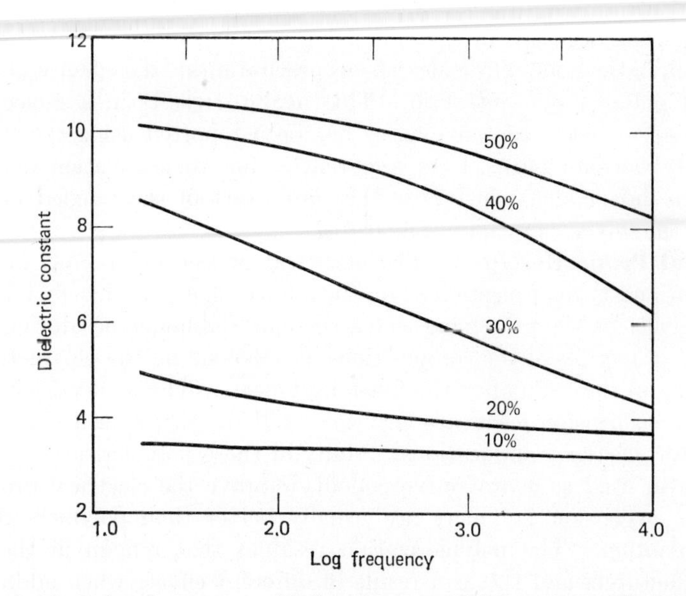

Fig. 2. Variation in dielectric constant with frequency at 40°C for compounds containing tricresy phosphate.

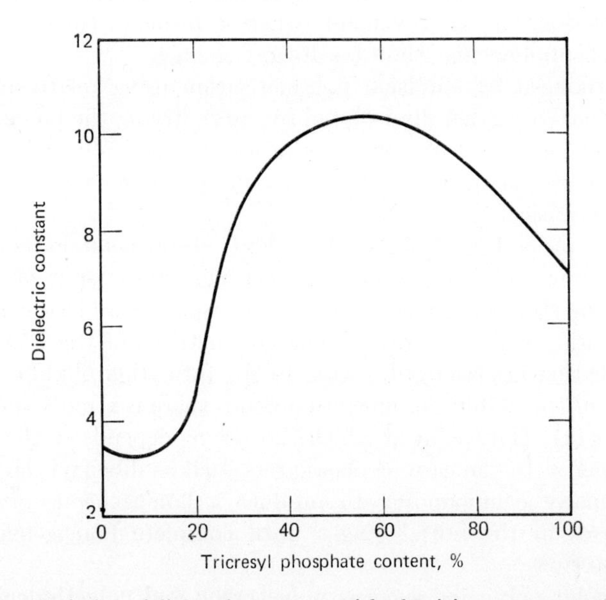

Fig. 3. Variation of dielectric constant with plasticizer content at 60 Hz.

corresponds to the region in which the *loss factor* passes through a sharp maximum (Fig. 4).

The changes in the electrical properties accompanying changes in plasticizer content arise through the effect of plasticization on the mobility of the polar groups in the polymer molecule. Increasing softness, ie, increasing plasticizer concentration, enables the polar groups to follow the forces arising from the imposition of an electric field. The high value of the loss factor corresponds to the maximum phase difference between the dipole motion and the field. When the amount of plasticizer is increased,

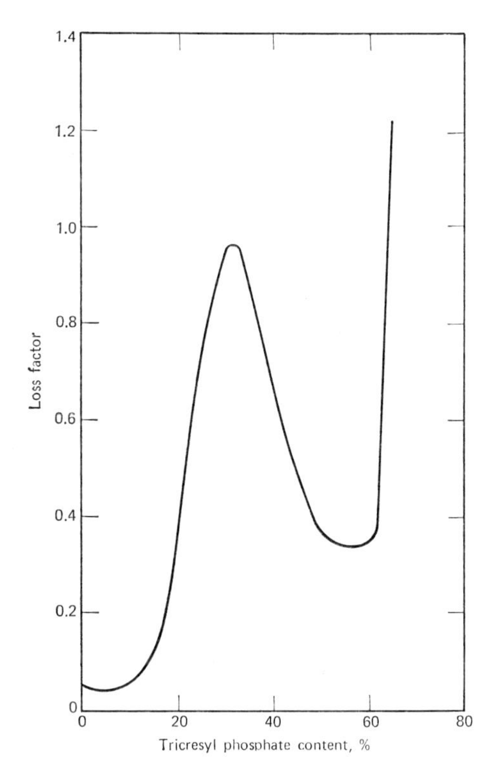

Fig. 4. Variation of loss factor with plasticizer content at 60 Hz.

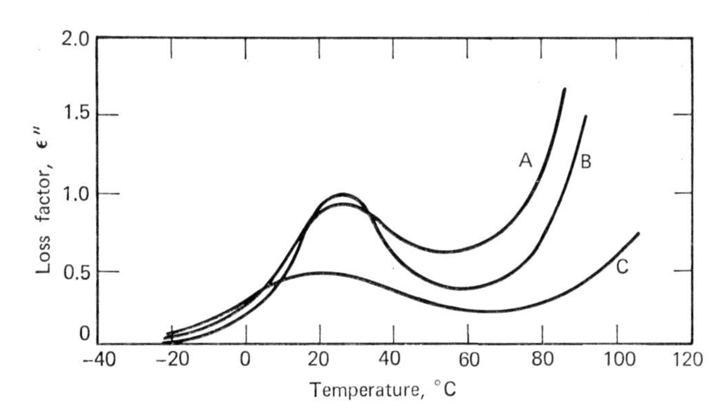

Fig. 5. Variation of loss factor with temperature at 60 Hz for PVC containing various plasticizers: curve A, butyl phthalyl butyl glycolate (39%); curve B, tricresyl phosphate (40%); curve C, dioctyl phthalate (40%).

so that the viscosity of the system becomes lower, the phase difference between the dipole motion and the field is reduced, resulting in a lower loss factor and an increased dielectric constant. When the temperature is increased, the effect on the viscosity is the same as increasing the plasticizer concentration, so that the maximum in the loss factor shifts to lower concentrations (43,44). Data obtained for several plasticizers indicate that the maximum in the loss factor occurs at the temperatures at which the compounds have comparable hardness. The effect of different plasticizers on the loss

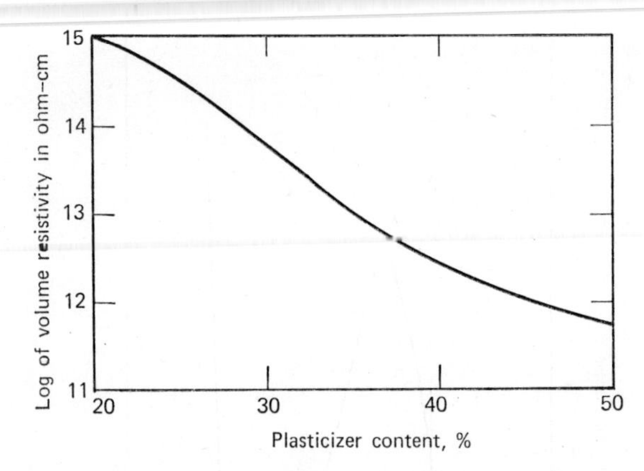

Fig. 6. Effect of dioctyl phthalate on volume resistivity at 23°C. Formulation: suspension resin (K = 65), 100 parts by weight; white lead paste, 4.5 pts; calcium stearate, 1.0 pts.

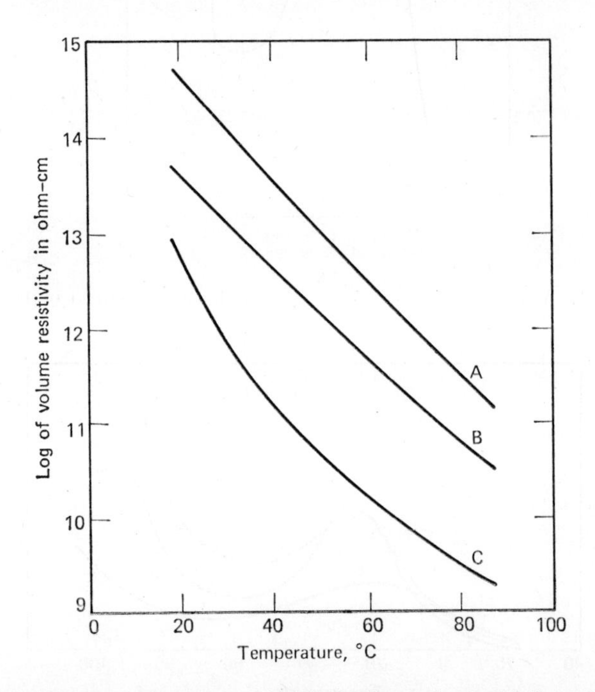

Fig. 7. Effect of temperature on volume resistivity. Curve A, 28.5% dioctyl phthalate; curve B, 33% dioctyl phthalate; curve C, 40% tricresyl phosphate.

factor is shown in Figure 5 in which the compounds tested were made up with different amounts of plasticizer so as to obtain materials of comparable hardness.

Volume resistivity is dependent on both the formulation and the temperature. The effect of the variation of the plasticizer content on a compound containing dioctyl phthalate is shown in Figure 6. This property decreases steadily (Fig. 7) as the temperature is increased.

Plastisols. The properties of plastisols formed by mixing PVC resins specifically designed for the purpose with plasticizer depend on the proportions of the two components, on the type of plasticizer used, and on the particle-size distribution and sur-

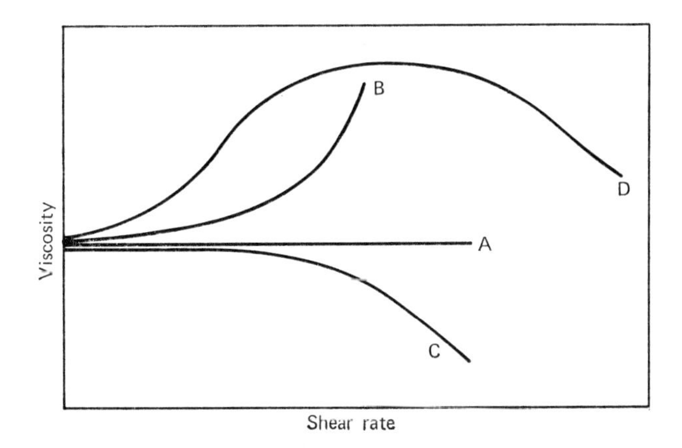

Fig. 8. Types of viscosity behavior as a function of shear rate. Key: curve A, Newtonian; curve B, dilatant; curve C, thixotropic; curve D, mixed flow.

face characteristics of the resin. Paste-forming polymers are usually manufactured by spray drying a PVC latex, and it has been found that the properties of the plastisol are very dependent on the temperature conditions in the dryer. When the moisture from the latex has evaporated, the polymer particles, which are formed as agglomerates, tend to sinter and fuse together; the extent to which this occurs depends on the severity of the drying conditions. A resin dried at high temperatures undergoes no change on storage in dibutyl phthalate for several days (45), whereas a resin dried at low temperatures disperses easily and breaks down, with very little agitation, into the discrete particles.

The flow behavior of plastisols varies according to the shear rate, and different resins can produce quite different effects. It is unusual for a plastisol to exhibit Newtonian flow and have a constant viscosity over a wide shear rate range. Many plastisols exhibit thixotropy at high shear rates and a reduced viscosity. Dilatancy is characterized by a rapid increase in viscosity with increasing shear rate. Some plastisols can have combinations of different flow properties within the range of shear rates encountered; for example, a material may exhibit dilatancy up to a critical shear rate beyond which the viscosity remains constant over a narrow range of shear rate. With further increases in shear rate, the plastisol exhibits a thixotropic effect and the viscosity decreases. The causes of these different types of flow behavior, which are illustrated in Figure 8, have been discussed by a number of authors (46,47). See also MELT VISCOSITY.

Thermal Properties. The glass-transition temperature for PVC will vary with the temperature at which polymerization is carried out but practically all commercial resins have a T_g of between 80 and 85°C. PVC produced at −75°C has a T_g of 105°C and this is reduced to 68°C when the temperature of polymerization is raised to 125°C. The glass-transition temperature is reduced by plasticization and normal flexible compounds have a T_g below room temperature. Although PVC is generally regarded as an amorphous material, it has crystalline regions within the clusters of molecules and for normal commercial products the value for the melting point for these regions is about 220°C.

Other thermal properties are as follows: coefficient of thermal expansion, $5\text{–}18.5 \times 10^{-5}$ cm/cm/°C; thermal conductivity, 3.9×10^{-4} cal/sec/cm^2/°C/cm; specific heat,

0.32–0.51 cal/°C/cm. These values will vary according to the formulation; for instance, the coefficient of thermal expansion for plasticized PVC is slightly higher than that for the plasticized material.

Testing

The techniques and methods by which the characteristics of PVC resins are determined vary widely among the numerous manufacturers, although there are moves within the International Organisation for Standardisation to agree on standard methods of assessment of basic properties.

Particle Size. The size of the large-particle resins (made by the suspension and bulk techniques) can be determined by the use of a series of standard sieves; a number of different standards are available but these are being standardized by the ISO (48). The recommended weight of powder varies from 25 to 10 g according to its density. The sieving operation may be carried out either wet or dry with the use of a machine or by hand, for a fixed period of time. Since PVC resins tend to pick up an electrostatic charge, wet sieving is more frequently used. This method is rather tedious, particularly if the fractions of PVC have to be dried after wet sieving. The micromerograph, which is a two-layer sedimentation balance, has proved to be useful for measurement of the particle size of suspension resins since its recommended range is 1–250 μ. The instrument is largely automatic and reproducible results are obtained easily and rapidly without the need for highly trained operators.

The measurement of the particle size of emulsion resins is much more difficult since the average size is less than 2 μ. Since this type of polymer is used almost exclusively for making plastisols, the important criterion is that the resin is free of large particles which will produce graininess in paste-spreading applications. This can be determined readily by converting the resin into plastisol using a standard recipe; the North fineness number is then determined using a Hegman grind gage which has a taped channel with a depth from 0.0000 to 0.0040 in. After filling the channel with paste and then removing the excess with a scraper, the point on the channel at which streaks begin to appear is given as the North fineness number. A good paste resin gives a reading of at least 4 (the scale reading varies from 0 at the end of maximum depth to 8 where the channel depth is zero). This measurement is normally adequate for all paste applications.

The size of the polymer particles in PVC latexes can be determined by using the Coulter counter. In this method the number and size of particles suspended in an electrolyte are measured by causing them to pass through a small orifice, and the changes in resistance as they flow through generate voltage pulses whose amplitudes are proportional to the volume of the particles. The pulses are amplified, sized, and counted, and from the derived data the particle size of the disperse phase can be determined.

Thermal Stability. Since polymers and copolymers of vinyl chloride tend to decompose at elevated temperatures, it is useful to have a test which can differentiate one resin from another. The decomposition of the polymers causes a *change in color*, and the rate at which this occurs is very dependent on the formulation of the compound. The test which is recommended by the International Organisation for Standardisation (49) depends on assessment of the color changes that occur when the test specimens in the form of disks (14 mm diam by 1 mm thick) are heated to 120–200°C. The disks are removed at 5 minute intervals and sufficient specimens are used so that the test will be of long enough duration to result in complete blackening of the test specimen. In general, the test time is 60 to 120 minutes. The test specimens are fixed on a card in the order of their time of exposure, and the time is recorded of (a) the first observed change in color, and (b) complete blackening, together with the test temperature.

The *evolution of hydrogen chloride* when PVC resins are heated above 100°C has been studied extensively; however, this method is of limited use for the assessment of thermal stability because in fabrication techniques where the PVC is compounded with stabilizers and other additives, the stability of the compound can be varied over a wide range and these variations are much greater than that of the original polymer and bear little relation to differences in the evolution of hydrogen chloride. A method has been used, however, for the assessment of the thermal stability of PVC based on the initial evolution of hydrogen chloride. A fixed quantity of the resin is heated in a test tube into the top of which is suspended a piece of congo red paper. The heating is carried out in an oil bath maintained at 150°C and the time is recorded when the congo red paper first shows signs of a change in color.

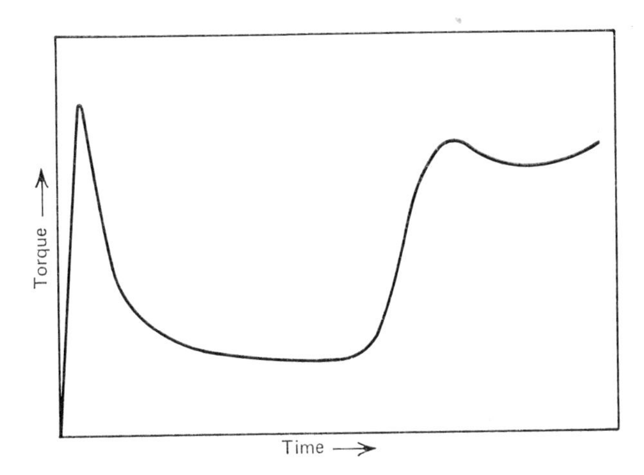

Fig. 9. Typical torque–time curve for PVC recorded on a Brabender plastograph.

In addition to the evolution of hydrogen chloride, which in the case of compounded materials is neutralized by the stabilizer and only becomes detectable when the stabilizer is exhausted, and the formation of color, the *melt viscosity* of the compound undergoes a rapid increase due to crosslinking of the polymer. The effect can be determined very effectively by use of the Brabender plastograph. The compound in the form a powder mix is introduced into the heated mixing chamber of the plastograph with the machine running. As the PVC compound begins to fuse, the torque rises rapidly and reaches a maximum when the compound is fully gelled. It then falls and the temperature of the material continues to rise until the torque reaches a steady value when the temperature of the PVC compound is constant; this will depend on the temperature to which the mixing chamber is heated and also the rate of mixing and is measured by a thermocouple which protrudes through the mixer into the contents. After a period of mixing, crosslinking of the PVC takes place with a sudden increase in the melt viscosity which causes the torque to rise. The inflection on this section of the curve indicates the stage at which the crosslinking has reduced the hot strength of the material to the point at which it starts to disintegrate into crumb form. Figure 9 shows a typical chart as recorded and the time from the stage at which the torque begins to rise (50) at the commencement of fusion to the point at which it rises a second time, ie, when degradation sets in, is recorded as the dynamic thermal stability. See also the article LUBRICANTS for a further discussion of this technique.

When using constant-stability data, it should be noted that the PVC compound with the lower torque (lower melt viscosity) can probably be processed at a lower temperature than a polymer with a higher torque. This will have the effect of greatly increasing the thermal stability and invalidating the comparison at constant temperature. Nevertheless, data on the efficiency of different stabilizers are best obtained from experiments carried out under constant temperature. The stability of PVC can be determined with the plastograph over a wide range of conditions representative of those encountered in most processing techniques. This can be achieved by means of varying the torque, the speed of rotors, the temperature of the PVC compound, and the temperature of the mixer jacket.

Molecular Weight. The molecular weight of PVC can be determined precisely using any of the standard techniques which are available for high polymers (see MOLECULAR-WEIGHT DETERMINATION). All commercial resins have a wide molecular-weight distribution so that the results obtained are only average values. Osmotic-pressure measurements can be used to give number-average molecular weights, whereas weight-average values are obtained by using light-scattering techniques. The intrinsic viscosity is also a useful indication of molecular weight (discussed above in the section on Property Values; see also VISCOMETRY).

For technological applications, simpler methods can be used to obtain a figure related to the molecular weight and giving an adequate indication of those properties that are molecular-weight dependent. All commercial producers use dilute-solution viscosity measurements as a means of differentiating their various grades of polymer. The polymer is dissolved in a solvent such as cyclohexanone, ethylene dichloride, nitrobenzene, or tetrahydrofuran, to a concentration of 0.2–1 g in 100 ml of solution. The viscosity of the solution is then determined, usually at 25°C, using an Ubbelohde viscometer, after which the time of flow of the pure solvent is also measured. The result is then expressed as the relative viscosity or the specific viscosity.

In Europe, Fikentscher K values are derived from the relative viscosity (51) (eq. 1) where $K = 1000k$ for a concentration of 1.0 g/dl of solution.

$$\log \eta_{\text{rel}} = \left[\frac{75\,k^2}{1 + 1.5\,kc} + k\right]c \tag{1}$$

The value of K depends on the solvent used for the viscosity measurement. In an attempt to overcome this difficulty, the ISO has decided that the molecular weight should be represented by its *viscosity index* and not by its K value. The viscosity index is determined at 25°C using a solution of 0.5 g PVC in 100 ml of solution with cyclohexanone as solvent.

The viscosity index is then given by equation 2, where t is efflux time of the solution, t_0 is efflux time of the solvent, and c is concentration in g/dl.

$$\text{viscosity index} = \frac{(t/t_0) - 1}{c} \tag{2}$$

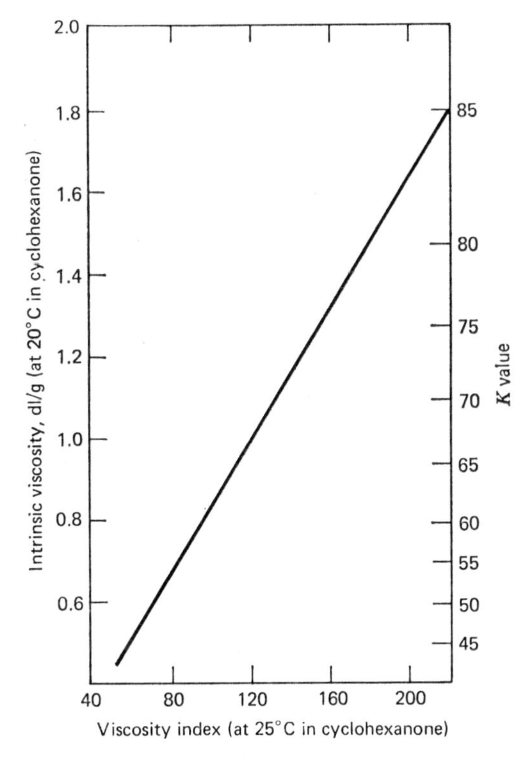

Fig. 10. Correlation of viscosity index equivalence with K value and intrinsic viscosity.

Figure 10 shows the relationships between viscosity index, intrinsic viscosity, and K value in cyclohexanone at 20°C.

The determination of the solution viscosity can be exacting and time consuming, and does not necessarily give an accurate indication of processing characteristics. However, useful information can be obtained by an empirical method using the Brabender plastograph (52).

The melt viscosity of a plasticized compound, as expressed by the torque, is shown to be directly proportional to the K value as determined by solution viscosity measurements. In one experiment, the torque of a plasticized PVC compound containing 40% of dioctyl phthalate was measured in the plastograph at 160°C and 56 rpm using the Banbury type jacketed-roller mixing chamber. When the mixing is carried out, the torque falls rapidly from the start and reaches a plateau after about 20

Table 4. Correlation of Plateau Value of Torque with K Value and Relative Viscosity[a]

K value	Relative viscosity	Torque, m-kg/sec
65.2	2.27	0.71
69.5	2.48	0.90
70.0	2.50	0.90
72.0	2.63	0.96
73.7	2.74	1.00
80.5	3.30	1.30

[a] Measurements on 1% solution in cyclohexanone at 25°C.

minutes when the compound is homogeneous and fully gelled. When a number of polymer samples of differing K value were tested, it was found that the plateau value of the torque was in proportion to the K value, as shown in Table 4.

It is essential to ensure that the conditions in the mixing chamber of the plastograph are constant for reproducible results to be obtained. In view of the extensive use of this type of equipment, this method has advantages over the more usual procedures and is particularly suitable for raw material testing of PVC polymers in control laboratories associated with compounding plants.

Fish Eyes. When PVC resins are compounded with plasticizers, the ease and speed with which homogeneous compounds are obtained are of considerable importance to the fabricator. Suspension resins invariably contain particles that are plasticized much less quickly than the bulk of the material. These particles show up in transparent films as ungelled spots which, in pigmented compositions, remain colorless. They are generally referred to as "fish eyes." The rate at which these ungelled particles disperse on compounding depends on the viscosity of the mix after gelation; in harder compounds containing small amounts of plasticizer the fish eyes will be subjected to higher shearing in the process, so that they will disappear more readily. The majority of commercial resins contain a number of particles that prove exceptionally difficult to disperse; these give rise to persistent fish eyes, which may arise from any one of the following causes (53): (a) dust or fibers; (b) blending of a PVC of higher molecular weight; (c) blending with a less porous PVC; (d) blending a fine PVC resin with a coarser one; (e) blending plasticized PVC with PVC resin; and (f) nonuniform plasticizer concentration. It is apparent, therefore, that fish eyes may appear in the final product as a result of unsatisfactory compounding conditions, in addition to being present in the original PVC resin.

The rate of interaction of the resin particles with plasticizer varies according to particle size, and more uniform rates are obtained with polymers of narrow particle-size distribution and uniform porosity. It has been observed that fish eyes may arise because the particles of polymers produced by the suspension process have a skin that is insoluble in plasticizers at 150°C, whereas the core is readily compatible with plasticizer at this temperature. This skin is probably a graft copolymer formed by interaction with the suspending agent, and those particles containing the larger amounts of the latter will be less soluble in plasticizers and may be responsible for the presence of persistent fish eyes.

The assessment of the PVC resins for fish eye content can be carried out by milling a black plasticized compound on an ordinary laboratory mill (54). A sample is taken off each minute, drawn down to a standard thickness, and examined for transparent colorless spots against the light. A large number of fish eyes is observed in the early samples, but the number decreases rapidly as milling is continued. A good resin produces a compound which after a relatively short time contains no fish eyes. The usual product has a few persistent fish eyes that remain even on prolonged milling. The test is reproducible provided constant milling conditions are maintained. The method can also be used to assess the rate of gelation with different plasticizers.

Bulk Density. The bulk density of suspension resins is of considerable importance because of its effect on the rate of output of many fabrication techniques. It is determined by the measurement of the weight of a fixed volume of PVC resin according to ASTM D 1895 (55). The ISO method (56) differs in some details but gives values which differ by only -0.01 to $+0.03$ g/cm^3 based on the average of five tests for each of three materials. With increasing use of PVC powder compounds, the bulk density has appreciable significance in extrusion performance. The majority of commercial polymers recommended for extrusion have a bulk density of at least 0.5 g/cm^3. The general range of all resins lies between 0.4 and 0.55 g/cm^3.

There has been encouragement to raise the bulk density because of the increasing use of bulk transport for the delivery of polymers. Severe competition has cut profit margins drastically so that manufacturers have seized upon the opportunity to cut transport costs by getting a bigger load into the bulk containers. The advantages of higher bulk density lie mainly in uses of unplasticized products because these resins generally have reduced porosity and poor plasticizer acceptance. The resins with lower bulk density are normally used in techniques using plasticized compositions.

Impurities and Foreign Particles. Although the degree of contamination of the polymer by foreign particles is, strictly speaking, not a basic property of PVC, it is of considerable practical importance. There is an increasing use of PVC in glass-clear applications where dark particles can be seen readily; these particles consist either of extraneous dirt or of burnt particles caused by overheating in the drying system. The degree of contamination can be determined by counting, using an illuminating magnifier, the dark particles in a specified area of polymer contained as a thin layer between two glass plates, the top one of which is etched with a grid of ten one-inch squares. The number of foreign particles counted in the total area is recorded as the foreign particle count. (This technique can be varied in detail (57) but is generally accepted by all major manufacturers of PVC.) It is suitable, however, for only suspension resins, and no comparable method exists for plastisol resins.

Porosity. The porosity of the resin has a considerable influence on the rate of interaction with plasticizers under any given set of conditions. There are a number of test procedures by which this property is assessed, but perhaps the most realistic is that in which the absorption rate is measured directly. This rate depends on several variables besides the porosity or surface area of the polymer particles, eg, on the temperature, type and amount of plasticizer, and rate of agitation. One method of measurement uses a planetary-motion paddle mixer which can be heated (58). A fixed amount of PVC resin is charged into the mixer at the prescribed temperature, after which a quantity of plasticizer is added and the agitator is set in motion. The absorption of plasticizer is considered to be complete when a sample from the mixer shows no plasticizer stain if pressed between two pieces of absorbent paper for a period of one minute. The plasticizer absorption time is the period of time from the point of addition of plasticizer to the end point as determined above. Other methods for studying the interaction between polymers and plasticizers include the use of a microscope equipped with a heated stage whereby the dissolution of PVC particles in plasticizers is followed visually (59,60).

The Brabender plastograph has proved to be a useful tool for determining the rate of plasticizer uptake and a method has been developed (61) which bears a close relation to the techniques used in large-scale compounding. A quantity of PVC is introduced into a small planetary mixer connected to the dynamometric part of a Brabender rheometer; the mixer is heated (88°C) and the resin is stirred for a time to reach temperature equilibrium; a quantity of plasticizer is added (at normal temperature) and the mixing is continued until a dry mixture is obtained. The torque, which is a measure of the consistency of the blend, shows first a sudden increase due to the addition of the plasticizer, then levels off for a period, and finally decreases to a minimum followed by a leveling off. The time required to produce a dry blend is measured by the point at which the torque reaches the minimum; where this cannot be clearly identified it can be determined by the intersection of extrapolated portions of the curves. A plot of this mixing time against the amount of plasticizer gives a reasonably straight line, the slope of which depends on the speed of the stirrer, the temperature at which the test is run, and also on the nature of the plasticizer used.

The above methods do not give a true measure of porosity because in the process of mixing the plasticizer not only penetrates into the pores but also diffuses into the body of the polymer particle. The first takes place fairly quickly and is largely independent of temperature, but this does not produce a dry blend unless the quantity of plasticizer is sufficiently small. The migration of the plasticizer from the surface to the mass of the polymer is rather slower and depends on the temperature and speed of mixing. From the practical viewpoint the speed at which dry blends can be produced controls the output of a compounding unit, and the significance of the factors which influence it can be overlooked.

An effective method for the measurement of the actual pore sizes of polymer particles is the use of a high-pressure mercury porosimeter. The instrument measures the size of pores ranging from 75 to 75,000 Å by subjecting the material to mercury over a wide pressure range and measuring the resulting changes in volume. The volume of mercury which penetrates the pores of the sample is measured dilatometrically and is recorded automatically as a function of the pressure. The sample is placed in a glass tube with a graduated capillary system, and then deaerated by application of a high vacuum. It is filled with mercury and then subjected to increases in pressure up to 1000 kg/cm^2. The necessary pressure is produced by a hydraulic system using two pistons which operate over the range 0–100 kg/cm^2 and from 0 to 1000 kg/cm^2. The recorder indicates the pressure existing in the system over the two ranges.

The relationship between pore size and exerted pressure under equilibrium conditions is expressed by equation 3 where r is pore radius, σ is mercury surface tension, θ

$$pr = 2\sigma \cos \theta \tag{3}$$

is wetting angle, and p is absolute pressure exerted. When using mercury and assuming for σ a value of 480 dyne/cm and a wetting angle θ of 141.3 and also that all the pores are cylindrical in shape, we obtain the relationship in equation 4 where r

$$r = \frac{75{,}000}{p} \tag{4}$$

is pore radius in Å and p is absolute applied pressure in kg/cm^2. If the pores are not exactly circular, but are of irregular form, the relationship does not strictly apply, but remains correct from a relative point of view.

Electrical Properties. The electrical properties are normally determined by measurement of the volume resistivity of the PVC in a standard plasticized formulation. The conditions under which the samples are prepared must be carefully controlled for reproducible results to be obtained. Methods differ according to the procedure used for conditioning the samples and the conditions under which the electrical measurements are made. According to ASTM method D 257 (62) the volume resistivity is measured on the dry sheets at 50°C, whereas the British Standard Method (63) specifies 23°C after the samples have been aged for 24 hours in distilled water. The values obtained at the lower temperature (23°C) are higher than those obtained

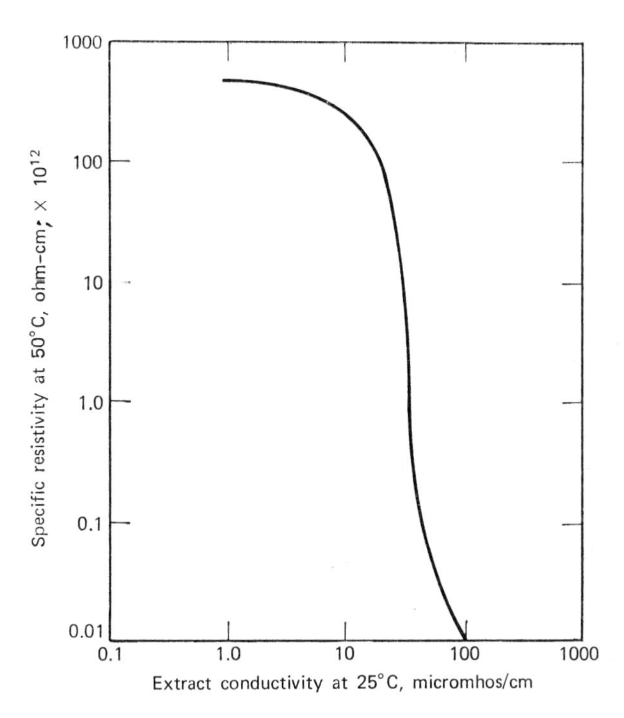

Fig. 11. Conductivity of water–isopropyl alcohol extract of PVC resins versus specific resistivity of the resin. The point marked Δ represents dry slab resistivity.

at 50°C, and consequently differences in the electrical properties are more easily measured at the lower temperature. The need for careful control of the processing conditions has already been emphasized (64); in addition stocks of raw materials must be maintained so that the same ingredients are used for each determination. Even so, it is necessary to carry out measurements in duplicate and also to have a standard sample included in each set of experiments. This technique does not always measure the difference in the electrical properties due to the resin.

Attempts have been made to measure electrical properties by a technique that does not involve converting the resin into a plasticized compound (65). The method is based on the principle that the resulting electrical properties of a plasticized PVC compound are determined by the ion content of the resin. It is assumed that the ionic contaminants are not part of the polymer structure, which is covalent and not readily ionized. If, therefore, the resin is extracted with a suitable solvent, such as water or isopropyl alcohol, the increase in the conductivity of the extractant should be a measure of the ionic impurities of the PVC and, consequently, give an indication of the insulation properties of plasticized compounds made from that resin. A number of extraction methods using both isopropyl alcohol and high-purity water as well as mixtures of the two have been used. The conductivity of the resin extracts were then measured using a cell and a Wheatstone bridge, and the results were compared with resistivity measurements obtained by the conventional technique on plasticized compositions. Figure 11 shows the comparison of the resistivity of compounds made from a range of different resins with the extract conductivity obtained using a mixture of water and isopropyl alcohol. The resistivity of a typical plasticized compound was found to decrease sharply when resins containing large amounts of ionizable materials were used, but there does not appear to be a direct relationship between the two although the extraction method is capable of distinguishing between electrical and nonelectrical grades.

Plastisols. Paste viscosity and gelation behavior are considered.

Paste Viscosity. The measurement of the viscosity over a range of shear rates is important for predicting the performance of plastisols in the applications in which they are used. Two basic types of viscometer are used: In the extrusion rheometer the paste is extruded through an orifice under standard conditions. This method (66) has the advantage that the viscosity can be measured under high shear rates such as are encountered in spreading opeations in which the plastisol is run at high speed. The plastisol is made up using, eg, 100 parts of resin with 60 parts of plasticizer in a simple stirred mixer under strict temperature control; it is essential to remove any air from the plastisol either by letting it stand at room temperature or by keeping it for a while under vacuum. The plastisol is then brought to the test temperature (23 ± 0.2°C) by allowing it to remain in a thermostat for 2½ hours, after which it is filled into the barrel of a Castor-Severs viscometer and extruded through the orifice under a range of pressures, the quantity of paste being measured over a convenient time period.

The shear stress at which the test is carried out is calculated according to equation 5, where P is the pressure in the rheometer (in psi), R is the radius of the orifice (in cm), and L is the length of the orifice (in cm).

$$\text{shear stress (in psi)} = \frac{PR}{2L} \tag{5}$$

The shear rate is given by equation 6, where W is the weight of the exudate (in

$$\text{shear rate (in sec}^{-1}\text{)} = \frac{4W}{\pi R^3 dT} \tag{6}$$

grams), d is the density of the sample (ASTM designation D 1475) (in g/ml), and T is the efflux time (in sec).

The viscosity is then given by equation 7.

$$\text{viscosity (in poises)} = \frac{\text{shear stress}}{\text{shear rate}} \times 6.895 \times 10^4 \tag{7}$$

The results are given for each of the following pressures: 100, 70, 40, and 10 pounds per square inch.

Wide differences in flow behavior have been found in the extensive range of commercial paste resins. For example, the viscosities of four polymers that produce dilatant plastisols have been measured at different pressures (67), as shown in Figure 12. The constant-pressure lines are also included on the graph, and show that the measurement of viscosity at equivalent pressures does not correspond to the same rates of shear.

The second type of instrument used for the measurement of viscosity is the Brookfield viscometer, which employs a rotating spindle and is limited to low rates of shear. It is useful for evaluating performance in applications such as dipping and rotational molding. After making up the plastisol and conditioning it, the viscosity is measured by using a No. 6 spindle if the Brookfield viscometer model RVF is used or a No. 4 in the case of the model LVF. Readings are taken at a number of revolutions at each speed (68).

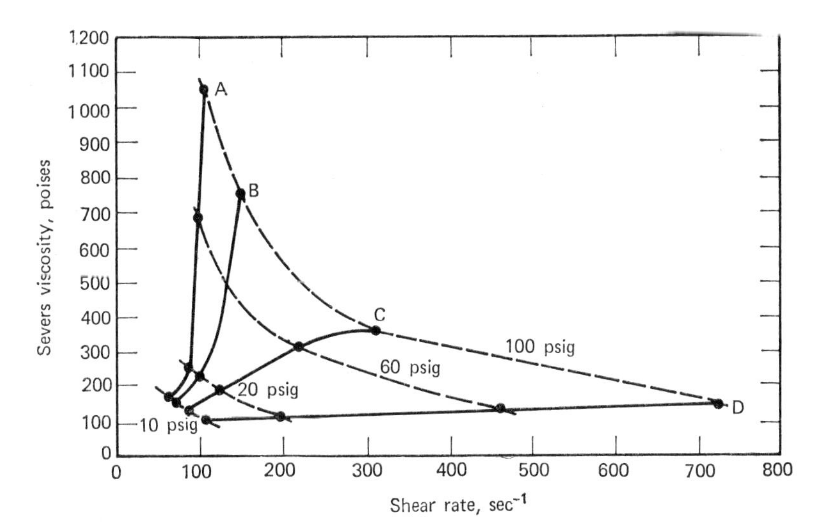

Fig. 12. Variation of viscosity with shear rate shows constant-pressure lines (dashed curves) (67). Test formula: resin, 100 parts; dioctyl phthalate, 60 parts. Points marked A, B, C, and D are plastisols based on four different paste resins. Courtesy *Modern Plastics.*

The viscosity, in poises, is then calculated according to equation 8, where D_{100}

$$\text{viscosity (in poises)} = \frac{100 \times D_{100}}{\psi} \quad \text{or} \quad \frac{100 \times D_{500}}{\psi} \tag{8}$$

is the dial reading on the 100 scale; D_{500} is the dial reading on the 500 scale; and ψ is the spindle speed (in rpm). The viscosity of the plastisol tends to change on standing so that it is usual also to give results taken after 24 hours and 7 days.

Gelation Behavior. The viscosity of the plastisol is very dependent on the temperature of the system; if this is raised, the viscosity is reduced, as is normal with most fluids, but as the temperature is raised still further, the polymer and plasticizer begin to interact and the viscosity of the system rises sharply. Many empirical methods are available for establishing the temperature at which plastisols are gelled; these are compared in Ref. 69. The method dependent on viscosity measurements necessitates taking readings on a plastisol while it is being heated and using a Brookfield viscometer. A thermocouple positioned in the plastisol near the viscometer spindle is used to measure the temperature, which is raised by immersion of the sample in an oil bath maintained at a constant temperature of around 100°C.

The gelation characteristics are also measured by tipping a trough filled with plastisol onto a flat temperature gradient plate which is then tilted at a small angle in a direction perpendicular to the temperature gradient. The plastisol flows down the incline over varying distances dependent on the temperature of the plate and the time necessary to raise the viscosity of the plastisol to a point of no flow. The various methods give different values according to the stage to which gelation has taken place; the viscosity increase as determined by the viscometer indicates the first stages of interaction between the resin and the plasticizer when the plastisol is said to have set and the tensile strength of the compound has reached only a low value.

Other methods correspond to more advanced stages of polymer–plasticizer interaction with corresponding increases in the tensile strength, the maximum value being reached when the resin has become fully gelled. This can be assessed readily by using the Brabender plastograph in which the plastisol is heated in a Z-blade mixer and the torque is recorded as the temperature increases. When the fusion stage is reached, the torque shows a rapid increase. It is essential that the mixer be fitted with a thermocouple which protrudes through the side directly into the plastisol so that its actual temperature can be determined. This will generally be higher than the temperature of the oil circulating through the jacket because of the frictional heat generated by the action of the mixer blade.

Specifications

Each PVC manufacturer produces a wide range of polymers which vary according to molecular weight, particle size, bulk density, resin porosity, etc. Standard methods for the estimation of many of these properties have not been universally adopted and the potential purchaser of poly(vinyl chloride) can be confused by the technical literature setting out the characteristics of competitive products. In order to standardize the specifications for the basic properties both the American Society for Testing and Materials (ASTM) and the International Standards Organization (ISO) have developed classification systems.

The ASTM system (D 1755-66) recognizes two types of resin, *general-purpose,* which is used for dry blending or processing with plasticizers, and *paste or dispersion resins,* used for making organosols and plastisols. The former are identified by the designation GP and the latter by a designation D. The letter is followed by a series of six numbers for the classification of the general-purpose resins, and a series of three numbers for the classification of paste-making resins. The two classification systems are shown in Tables 5 and 6. The classification numbers are given in the precise order shown in the tables. Each number is used to denote the specific range of the designated property; if no value is given for any one of the properties, a zero is included in place of the range number. This system does not attempt to cover all the relevant properties of PVC resins, but it is of considerable value and can be easily extended.

A similar method of classification has been adopted by the International Standards Organization (ISO Recommendation R 1060). It subdivides the properties into principal and secondary classes and also includes a suffix to describe the method of polymerization. The recommendation, shown in Table 7, is intended to cover both general-purpose and paste-making resins and the suffixes G and P are used to differentiate between them. However, the first edition, adopted in April 1969, does not set any limits for the properties of plastisols and leaves these for agreement between the purchaser and the supplier. The two classifications are similar in their general approach but there are some important differences. The ISO recommendation is aimed at giving a wider coverage of the resin properties but some of them are left for agreement with the purchaser whereas others, for example, the ash content, have only little effect on the compounding properties or the final performance of the material. In addition to the suffix indicating the type of resin, the coding includes a second suffix to describe the method of manufacture of the polymer; emulsion resins, e; suspension, s; mass (bulk) m; solution, sc; and sp for any intermediate method and new techniques.

An ASTM classification scheme has recently been adopted to describe copolymer resins. The specifications cover copolymers with the following monomers: vinyl acetate, vinylidene chloride, maleic esters, and acrylonitrile, and can be extended

Table 5. Classification of General-Purpose Resins[a] of Poly(vinyl Chloride) According to ASTM D 1755–66

Order of classification numbers[b]	Property	Test method	Property range for each classification number									
			0	1	2	3	4	5	6	7	8	9
first	dilute-solution viscosity	D 1243	unspec.	<0.70	0.70–0.79	0.80–0.89	0.90–0.99	1.00–1.09	1.10–1.19	1.20–1.29	1.30–1.39	$\geqslant$1.40
second	sieve analysis, % through No. 200 (74 μm) sieve	D 1705	unspec.	0–9	10–19	20–29	30–39	40–49	50–59	60–79	80–99	100
third	apparent (bulk) density, g/liter	D 1895	unspec.	<144	144–232	233–328	329–425	426–520	521–616	617–712	713–808	$\geqslant$809
fourth	plasticizer sorption, parts dioctyl phthalate per hundred of resin	D 1755	unspec.	<50	50–74	75–99	100–124	125–149	150–174	175–199	200–224	$\geqslant$225
fifth	dry flow, S/400 cm³	D 1755	unspec.			<2.0	2.0–3.9	4.0–5.9	6.0–7.9	8.0–9.9	$\geqslant$10	
sixth	conductivity (max), micromhos/cm-g	D 1125	unspec.			6	12	18	>18			

[a] Courtesy American Society for Testing and Materials.
[b] All numbers are preceded by the designation GP.

Table 6. Classification of Dispersion Resins[a] of Poly(vinyl Chloride) According to ASTM D 1755–66

Order of classification numbers[b]	Property	Test method	Property range for each classification number									
			0	1	2	3	4	5	6	7	8	9
first	dilute-solution viscosity	D 1243	unspec.	<0.90	0.90–0.99	1.00–1.09	1.10–1.19	1.20–1.29	1.30–1.39	1.40–1.49	1.50–1.59	$\geqslant$1.60
second	Brookfield viscosity (RVF), poises	D 1824	unspec.	0–24	25–49	50–74	75–99	100–124	125–149	150–174	175–199	>199
third	Severs viscosity, poises	D 1823	unspec.	0–49	50–99	100–149	150–199	200–299	300–499	500–999	1000–1499	>1499

[a] Courtesy American Society for Testing and Materials.
[b] All numbers are preceded by the designation D.

Table 7. Classification of Poly(vinyl Chloride) Resins According to ISO Recommendation R 1060

Order of classification numbers	Property	Test method	Property range for each classification number: 0	1	2	3	4	5	6	7	8	9
			Principal									
first	viscosity number, ml/g	ISO R 174	unspec.	<70	70 to 80	>80 to 90	>90 to 105	>105 to 120	>120 to 135	>135 to 155	>155 to 175	>175
second	apparent bulk density,[a] g/ml	ISO R 60	unspec.	<0.25	0.25 to 0.35	>0.35 to 0.45	>0.45 to 0.55	>0.55 to 0.65	>0.65			
	sieve analysis	ISO DR 1624	unspec.									
third	% retained on 0.063-mm sieve			<0.5	0.5 to 5	>5 to 20	>20 to 50	>50 to 90	>90			
fourth	% retained on 0.250-mm sieve			<0.5	0.5 to 5	>5 to 20	>20 to 50	>50 to 90	>90			
fifth	ash as sulfates, %	ISO DR 1270 method B	unspec.	<0.20	0.20 to 0.40	>0.40 to 0.80	>0.80 to 1.60	>1.60				
			Secondary									
sixth	volatile matters (including water), %	ISO DR 1269	unspec.	<0.30	0.30 to 1	>1 to 3	>3					
seventh	pH of aqueous extract	ISO DR 1264	unspec.	<6.5	6.5 to 8.5	>8.5 to 10.5	>10.5					
			Particular									
	impurities and foreign particles	for agreement between purchaser and supplier										
	behavior in the presence of plasticizer											
	ability to form dry blends		values to be agreed between purchaser and supplier									
	properties of plastisol											
	electrical properties											
	thermal stability											

[a] This property is of no interest for paste resins. A zero must be placed for this property in the designation.

Table 8. Classification of Vinyl Chloride Copolymer Resins[a] According to ASTM D 2474–69

Order of classification numbers[b]	Property	Property range for each classification number									
		0	1	2	3	4	5	6	7	8	9
first	type of copolymer	unspec.	vinyl acetate	vinylidene chloride	maleic ester	acrylonitrile					polycomponent systems
second	concn of comonomer, % by wt of resin	unspec.	0–5	6–10	11–15	16–20	21–25	26–30	31–35	36–40	41–45
third	dilute-solution viscosity	unspec.	<0.40	0.40–0.49	0.50–0.59	0.60–0.69	0.70–0.79	0.80–0.89	0.90–0.99	1.00–1.10	>1.10
fourth	bulk density, g/liter	unspec.	<144	141–232	233–328	329–425	426–520	521–616	617–712	713–808	≥809
fifth	specific gravity, 23/23°C	unspec.	<1.350	1.350–1.359	1.360–1.369	1.370–1.379	1.380–1.389	1.390–1.399	1.400–1.409	1.410–1.419	1.420–1.429
sixth	specific gravity, 23/23°C	unspec.	1.430–1.439	1.440–1.449	1.450–1.459	1.460–1.469	1.470–1.479	1.480–1.489	1.490–1.499	1.500–1.509	>1.509

[a] Courtesy American Society for Testing and Materials.
[b] All numbers are preceded by the designation C.

easily to include other comonomers. The designation C is used to indicate that the resin is a copolymer. This letter precedes the classification numbers shown in Table 8.

Degradation and Stabilization

It will be appropriate to define the terms "degradation" and "stabilization" as applied to PVC before beginning this section. On exposure to various forms of energy, several properties of PVC change. Hydrogen chloride is evolved, color develops, and ultimately changes in physical properties occur. The term degradation has been applied indiscriminately to all these reactions and, as a result, confusion has arisen in certain cases where authors have applied different criteria for their degradation studies. To avoid ambiguity, it is essential to define more precisely the nature of the degradation. Likewise, the term stabilization may acquire differing significance in different applications. For example, whereas evolution of hydrogen chloride and accompanying corrosion problems may be of greatest significance in processing deeply colored products, slight color change would be more significant during outdoor aging of lightly colored PVC. In the first instance, the requirement of the stabilizer would be to prevent evolution of hydrogen chloride and in the second case, to retain the original color, in addition to trapping the hydrogen chloride. See also DEGRADATION; STABILIZATION.

The relative instability of PVC compared with simple alkyl halides (70) has led to speculation concerning the structure of PVC. More particularly, such irregularities as branch points (with tertiary chlorine atoms), unsaturation, and head-to-head monomer units arising from probable polymerization side-reactions have been suggested. Arising from such speculation, a range of model compounds for PVC has been synthesized, incorporating various features of proposed irregularities.

Thermal Degradation of Model Compounds. Pyrolysis of alkyl halides was studied by Asahina and Onozuka, who like most earlier workers, relied upon gas-phase pyrolysis, using infrared techniques to follow the reaction (71). In all cases the products were hydrogen chloride and olefin. First-order kinetics were observed for compounds containing secondary and tertiary chlorine and both terminal and nonterminal unsaturation. The least stable compounds were those containing allylic or tertiary chlorine. The presence of double bonds in nonallylic (to chlorine) structures was less deleterious. Conversely, the presence of a vicinal double bond (in 3-chloro-2-pentene) imparted significant improvement in stability. This result is in direct contrast to the results of Erbe and co-workers (72) for 3-chloro-2-hexene, although these latter workers claimed the analogous 4-chloro-3-heptene to be substantially more stable.

Appreciating some of the limitations imposed by the techniques of earlier workers, Mayer and co-workers studied the degradation of allylic structures in the liquid phase (73). In this case, first-order kinetics were observed only during the early stages of degradation. Subsequent stages of the reaction were described as "complex" and the authors speculated on the probability that the reaction may also involve both rearrangements and back additions of hydrogen chloride. The complexity of the reaction in liquid phase compared with that occurring in the gas phase illustrates the need to study model polymers.

The effect of branching was studied in copolymers of vinyl chloride with a variety of suitable comonomers, eg, 2-chloropropene (74). It was concluded that branch points represent the greatest weakness in PVC. Similar conclusions were reached based on model copolymers containing both tertiary hydrogen (copolymer with 1-chloropropene) and tertiary chlorine (copolymer with 2-chloropropene) of which the latter was least stable (75). However, in all copolymer studies, the doubt must persist that the copolymer may not consist entirely of the predicted structure.

Head-to-head PVC has been produced by chlorination of both *cis*- and *trans*-1,4-polybutadiene (76,77). There is some indication in these publications that head-to-head PVC is less stable than conventional PVC, which is believed to be largely head-to-tail (78). However, Murayama (77) also showed that whereas head-to-head PVC starts to degrade at lower temperatures than the head-to-tail PVC, the rate of dehydrochlorination is higher for the head-to-tail form. This result led to the suggestion that the vicinal arrangement of the chlorine atoms leads to instability but that the direction of subsequent dehydrochlorination could lead to a more stable product (**1**) analogous to polychloroprene (eq. 9).

$$\sim\!\!\sim CH_2{-}\underset{\displaystyle Cl}{\underset{|}{CH}}{-}\underset{\displaystyle Cl}{\underset{|}{CH}}{-}CH_2 \sim\!\!\sim$$

$$\swarrow \qquad\qquad \searrow$$

$$\underset{(\mathbf{1})}{{-}CH_2{-}CH{=}\underset{\displaystyle Cl}{\underset{|}{C}}{-}CH_2 \sim\!\!\sim} \qquad\qquad \underset{(\mathbf{2})}{\sim\!\!\sim CH{=}CH{-}\underset{\displaystyle Cl}{\underset{|}{CH}}{-}CH_2 \sim\!\!\sim} \qquad (9)$$

The production of (**1**) might account for the observed rates of dehydrochlorination. However, it is difficult to reconcile substantial formation of (**1**) with the formation of long sequences of conjugated polyenes, which would be anticipated to derive from (**2**). Bailey observed that the color of degraded head-to-head PVC was more marked than that of the head-to-tail PVC, which would suggest a greater concentration of long conjugated polyene units (76). However, it is not clear whether Bailey's observations refer to polymers degraded for the same period of time or to the same extent of dehydrochlorination.

Thermal Degradation of PVC. Although hydrogen chloride is the main volatile product of the degradation process below about 300–400°C, traces of other volatiles, notably benzene, have been recorded (79). Earlier reports suggested that little benzene was evolved below about 300°C, but with increasing sophistication and resolution of experimental techniques, this product has now been detected at temperatures as low as 180°C (80,81).

Color can be detected in the residue at extremely low degrees of degradation, corresponding closely to the earliest detection of hydrogen chloride being liberated from the polymer. This color has been shown to consist of broad absorption ranging from the ultraviolet portion of the spectrum (about 200 nm) well into the visible, up to or beyond about 600 nm (82,83). The maximum absorption is generally observed in the ultraviolet region.

Color. There can be little doubt that the color in degraded PVC arises from the presence of sequences of conjugation, ie, of structures of the type

$$-CH_2(CH{=}CH)_nCHCl-$$

Since the simple polyenes have absorption spectra consisting of a few well-defined peaks with absorption wavelengths that increase with increasing values for n (84) it might be imagined that the spectrum of degraded PVC would be readily predictable. Unfortunately, insufficient data are available for long polyenes and this fact, coupled with various other uncertainties, such as possible bathochromic shifts, solvent effects, etc, make accurate analysis of the absorption spectra impossible at the present time. The situation is further complicated by the presence of oxygen during degradation, when some bleaching may occur as a result of partial oxidation of the polyenes. Nevertheless, several attempts have been made on a semiempirical basis to resolve these spectra in terms of polyene distribution (85,86).

Elimination occurring at random along the polymer chain would be expected to produce initially short polyenes, which would lengthen with increasing extent of reaction. In practice, a constant distribution is observed, and only the number, not the sequence length, of conjugated polyenes increases with increasing extent of reaction (87).

Absorption spectra very similar to those found in thermally degraded PVC are obtained when PVC is degraded chemically in solution at room temperature in the presence of suitable bases (88,89). On the other hand, suitably modified treatment with basic catalysts can produce absorption spectra with a single broad maximum at about 550 nm (88). It has been suggested that under these latter conditions the polyene sequences run the length of the polymer chain, whereas the former treatment gives rise to relatively short polyene sequences (88). Thus it may be deduced that in thermally degraded PVC the sequences are also short. This conclusion could not be drawn unequivocally from direct analysis of the spectra, in view of uncertainties surrounding possible absorption behavior of very long polyenes. Most estimates of maximum polyene sequence length are in the range 10–20 (90).

During the early stages of degradation, the concentration of polyenes increases in direct proportion to the volume of hydrogen chloride evolved (85,87) but after evolution of a few percent hydrogen chloride the rate of increase of polyene concentration fails to match the rate of hydrogen chloride evolution (85,87,89).

Dehydrochlorination. The rate at which dehydrochlorination occurs during the early stages of degradation may increase, decrease, or remain approximately constant as a function of time, depending on the choice of degradation temperature (85,90). The autoretarding temperature range is below about 170°C and the auto-accelerating region above about 200°C. At greater extents of dehydrochlorination the rate approaches zero, giving apparent limiting levels of dehydrochlorination well below 100% of the theoretical value, eg, about 5% at 170°C, and 65% at 230°C. These limiting values increase with temperature until at about 250°C approximately 100% dehydrochlorination can be obtained (90). Further dehydrochlorination beyond the limiting value results when the temperature is subsequently raised (90). This phenomenon has been attributed to the existence of a series of metastable states. It is argued that a number of irregular structures exist, either present initially or else produced during degradation at any given temperature. On raising the temperature, these irregular structures are activated and give rise to additional hydrogen chloride.

Estimates of the overall energy of activation, E_a, for dehydrochlorination, in vacuum (91) or in nitrogen (79,85,92) lie in the approximate range 25–35 kcal/mole (cf E_a for simple secondary alkyl halides of approximately 45–55 kcal/mole (70)). However, in oxygen (92) or at temperatures below about 180–200°C (90,93) significantly lower values have been reported (12–20 kcal/mole).

On the basis that dehydrochlorination may be initiated at chain ends, it would be anticipated that the rate of dehydrochlorination should be inversely proportional to molecular weight. Such inverse relationships have been observed (92), but in most cases the extrapolated rate at infinite chain length is significantly different from zero. It has been shown that whereas an inverse relationship is observed for lower-molecular-weight fractions, the rate of dehydrochlorination is independent of molecular weight above some limiting value (94). This result could be consistent with the estimate of a higher incidence of branch points in lower-molecular-weight fractions of PVC (95).

Opinion concerning the role of liberated hydrogen chloride in the degradation of PVC has undergone several distinct changes. For many years it was suspected that hydrogen chloride could catalyze the reaction, until 1954 when Arlman (96) discounted such catalysis. Regrettably, this conclusion was not challenged for several years, until Rieche et al. (97) and, most convincingly, Talamini (98), demonstrated clearly that hydrogen chloride is indeed a catalyst. Subsequently, it has been shown that hydrogen chloride is a catalyst not only for dehydrochlorination, but also discoloration (85,99) and crosslinking (85,99).

Mechanism. In formulating possible mechanisms for the degradation of PVC, it is logical to consider first the pyrolysis of simple alkyl halides (70). Alkyl halides are capable of decomposing by unimolecular elimination reactions, but in many cases a radical chain process may also occur, and when it does this latter mechanism generally predominates. The most widely accepted radical chain process for alkyl halides consists of three basic steps: (a) generation of halide atom, (b) abstraction by halide atom of hydrogen atom from alkyl halide, and (c) rearrangement of radical generated in (b) to give an olefin plus another halide atom which may continue the process.

Extension of this scheme to PVC gives the following propagation reactions (eqs. 10–12) (100). In this scheme, the chain can be continued only if the abstracted hy-

$$R\cdot + \sim\!\!\sim CH_2-\underset{Cl}{\underset{|}{CH}}-\sim\!\!\sim \longrightarrow RH + \sim\!\!\sim \dot{C}H-\underset{Cl}{\underset{|}{CH}}-\sim\!\!\sim \quad (10)$$

$$\sim\!\!\sim \dot{C}H-\underset{Cl}{\underset{|}{CH}}-\sim\!\!\sim \longrightarrow \sim\!\!\sim CH{=}CH-CH_2-\underset{Cl}{\underset{|}{CH}}-\sim\!\!\sim + Cl\cdot \quad (11)$$

$$-CH{=}CH-CH_2-\underset{Cl}{\underset{|}{CH}}-\sim\!\!\sim + Cl\cdot \longrightarrow \sim\!\!\sim CH{=}CH-\dot{C}H-\underset{Cl}{\underset{|}{CH}}-\sim\!\!\sim + HCl \quad (12)$$

drogen atom is from the secondary (methylenic) position. The radical produced by abstraction of a tertiary hydrogen has no obvious provision for rearrangement and is likely to be capable only of crosslinking by combination with another similar radical. Highly favored secondary relative to tertiary hydrogen atom abstraction is not a feature of many abstraction reactions, but it is known that the normal order of susceptibility to abstraction is often inverted in chlorination reactions (101), notably in the chlorination of PVC to give largely 1,2-dichloroethylene units (102).

A more serious obstacle to this (or any other) radical mechanism for PVC is the difficulty in accounting for the initial radical to give an overall energy of activation on the order of 30 kcal/mole. This obstacle is generally overcome by postulating the existence of structural irregularities in the PVC. Some of these defects have been discussed above. Another objection to a radical process is that it is difficult to account in such terms for the catalytic effect of hydrogen chloride.

No absolute proof in favor of a radical process has yet been offered. The strongest evidence in support of such a mechanism is the detection of radicals in degrading PVC (75,80) although such detections are recorded at relatively high degrees of dehydrochlorination (greater than 10%). Other evidence quoted in support of a radical mechanism includes the ability of degrading PVC to initiate the polymerization of methyl methacrylate (103) and the degradation of poly(methyl methacrylate) (104).

Alternatively, two other classes of mechanism have been proposed, molecular and ionic. It is doubtful if any of the authors supporting an ionic mechanism (85,97) have been in favor of true (separate) ion formation rather than extensive polarization of the carbon–chlorine bond. Thus the elimination step might be represented as a four-center reaction (eq. 13).

$$\sim\!\!\sim\!\text{CH}(\text{Cl})\text{—CH}(\text{H})\!\sim\!\!\sim \longrightarrow \underset{\textbf{(3)}}{\sim\!\!\sim\!\text{CH}\text{==}\text{CH}\!\sim\!\!\sim \atop {}^{\delta-}\text{Cl}\text{----}\text{H}^{\delta+}} \longrightarrow \sim\!\!\sim\!\text{CH}\text{=}\text{CH}\!\sim\!\!\sim + \text{HCl} \qquad (13)$$

In this initial elimination step the difference between ionic and unimolecular reactions depends solely on the formation of the transition state (**3**). However, if distinction is made between polarized and neutral transition-state formation, the difference between these mechanisms is more significant in relation to subsequent elimination steps. Spectroscopic evidence has been cited previously suggesting that conjugated polyene sequences containing up to about twenty double bonds are formed at very low extents of dehydrochlorination (significantly less than 0.1%). Taking an average degree of polymerization for PVC of about 800, it follows that at the stage of degradation at which color can first be detected the mean loss of hydrogen chloride is less than one molecule per polymer chain. Since visible discoloration requires conjugated sequences of double bonds in excess of six, it is evident that after the elimination of the first molecule of hydrogen chloride from a PVC chain subsequent eliminations are very much more likely to occur from adjoining monomer units rather than at random throughout the chain. In the case of the ionic mechanism above, the substantially increased reactivity of allylic chlorine toward heterolytic scission relative to that of the corresponding saturated chloride is well known (105). This increased reactivity would enable several propagation steps of the above type to occur to give conjugated unsaturation. Conversely there is no clear literature evidence in support of substantially decreased stability of allylic chlorine toward four-center homolytic scission in a unimolecular process, which would be necessary to make possible the formation of conjugated double bonds rather than isolated double bonds.

As in the case of radical reactions, no absolute proof has yet been offered in support of ionic reactions. Among the points in support of an ionic mechanism the catalytic effect of hydrogen chloride and the influence of dielectric strength (85) or polarity (106) of the medium have been cited. Further support has been based on theoretical considerations (107,108).

The possibility must not be ignored that several mechanisms may operate concurrently. Rieche, for example, has suggested that the increased rate of dehydrochlorination in oxygen may be due to an oxygen-catalyzed initial radical reaction, occurring in addition to the ionic reaction prevalent in the absence of oxygen (97).

Stabilization. Several general reviews on this topic have appeared (109–111) (see also the article STABILIZATION). A wide range of materials has been employed as stabilizers for PVC, their only common property apparently being an ability to react with hydrogen chloride. In many cases this reaction with hydrogen chloride continues until all the stabilizer is converted to the corresponding chloride (85,112,113). Thus induction periods are observed in the presence of stabilizers, during which no hydrogen chloride is liberated from the system, although it continues to be evolved from the polymer. Depending on whether the resulting chloride is a catalyst for the degradation (85,113) the ultimate rate of dehydrochlorination may be increased. Both Nagatomi (110) and Onozuka (113) have explained synergism in terms of a reduction in the rate of generation of the more harmful metallic chloride in a system containing two or more metal-containing stabilizers.

The question whether the stabilizer merely reacts with hydrogen chloride or in addition, reacts with the PVC, is unresolved. If the stabilizer fails to react directly with the PVC, then the stabilizing action must derive (a) from the reduction in concentration of hydrogen chloride that could catalyze the degradation, and (b) by reduction of the probability of forming harmful metal chlorides, eg, ferric chloride, by contact with processing machinery. It would seem reasonable to suppose that some of the simpler stabilizers, eg, calcium stearate, act in this way. Using labeled stabilizers it was shown that a portion of some metal soap and dialkyltin dicarboxylate stabilizers is retained in the polymer after partial thermal degradation (114–117). By labeling different groups in the stabilizers, it was observed that the soap or carboxylate moieties were retained to a greater extent than the metal atom or alkyl groups. It was concluded that the stabilizer is involved in some coordinate complex with the polymer chain. Similar conclusions have been reached by other workers (85,118). The ability of mono- and dialkyltin stabilizers to react with five or with four moles, respectively, of hydrogen chloride (82,85) taken in conjunction with reports of the stability of the alkyltin bonds (117) has led to the suggestion that these stabilizers generate Friedel-Crafts type catalysts during the degradation of PVC (85). It has been suggested that several organotin stabilizers may act as dienophiles in Diels-Alder type reactions (97). Both Friedel-Crafts and Diels-Alder reactions have been postulated to account for the considerable reduction in visible absorption that occurs when PVC is degraded in the presence of organotin stabilizers.

Photochemical Degradation. Most studies of the effects of photochemical decomposition have been carried out to determine the suitability of compounded PVC in outdoor applications, using various combinations of lamps and filters designed to simulate natural sunlight. Very few publications exist in which it is possible to define precisely the photochemical parameters pertaining to given results on aging.

General conclusions are that the reactions occurring in the photochemical degradation are similar to those occurring during the thermal degradation of PVC (119–127). However, the variation in results reported is considerable (128,129). The rates of dehydrochlorination are too low to measure conveniently or accurately, and the emphasis has shifted to study of oxidative and crosslinking processes, and their effects on physical properties. Several reports of optimum wavelengths around 270–280 nm (130,131) or around 300–310 nm (132,133) have been contradicted by reports that the reaction is independent of wavelengths below about 340 nm (127).

Mechanism. There is widespread support for a radical chain process occurring during the photochemical degradation of PVC (100,127,130,132) although the exact nature of the initiation process is still a matter for conjecture. It is unlikely that carbon–chlorine bond rupture (estimated energy about 75 kcal/mole) will occur with radiation of wavelength greater than about 300 nm. Simple alkyl chlorides are stable at wavelengths down to 220 nm (130). The most probable initiation site would seem to be unsaturated structures produced during thermal processing, but this topic has received little attention (132).

Bibliography

1. C. S. Marvel, J. H. Sample, and M. F. Roy, *J. Am. Chem. Soc.* **51**, 3241 (1939).
2. J. D. Cotman, Jr., *Ann. N.Y. Acad. Sci.* **57**, 417 (1953); *J. Am. Chem. Soc.* **77**, 2790 (1955).
3. M. H. George, R. J. Grisenthwaite, and R. F. Hunter, *Chem. Ind.* **1958**, 1114.
4. G. Boccato, A. Rigo, G. Talamini, and Zilio-Grandi, *Makromol. Chem.* **108**, 218 (1967).
5. A. Nakajima, M. Hamada, and S. Hayashi, *Makromol. Chem.* **95**, 40 (1966).
6. A. Crosato-Arnaldi, P. Gasparini, and G. Talamini, *Makromol. Chem.* **117**, 140 (1968).
7. D. Lim and M. Kolinsky, *J. Polymer Sci.* **53**, 173 (1961).
8. M. Kolinsky, M. Ryska, M. Bohdanecky, P. Kratochvil, K. Solc, and D. Lim, *J. Polymer Sci.* [C] **16**, 485 (1967).
9. M. Freeman and P. P. Manning, *J. Polymer Sci.* [A] **2**, 2017 (1964).
10. B. H. Zimm and R. W. Kilb, *J. Polymer Sci.* **37**, 19 (1959).
11. A. Caraculacu, O. Wichterle, and B. Schneider, *J. Polymer Sci.* [C] **16**, 495 (1967).
12. M. Carrega, C. Bonnebat, and G. Zednik, *Anal. Chem.* **42**, 1807 (1970).
13. C. S. Fuller, *Chem. Rev.* **26**, 143 (1940).
14. G. Natta and P. Corradini, *J. Polymer Sci.* **20**, 251 (1956).
15. F. R. Reding, E. R. Walter, and F. J. Welch, *J. Polymer Sci.* **56**, 225 (1962).
16. J. W. L. Fordham, P. H. Burleigh, and C. L. Sturm, *J. Polymer Sci.* **41**, 73 (1959).
17. M. Takeda and K. Iimura, *J. Polymer Sci.* **57**, 383 (1962).
18. I. Rosen, P. N. Burleigh, and J. F. Gillespie, *J. Polymer Sci.* **54**, 31 (1961).
19. P. N. Burleigh, *J. Am. Chem. Soc.* **82**, 749 (1960).
20. U. Giannini and S. Cesca, *Chim. Ind. (Milan)* **44**, 371 (1962).
21. F. A. Bovey and G. V. D. Tiers, *Chem. Ind. (London)* **1962**, 1826.
22. H. Germar, K. H. Hellwege, and U. Johnsen, *Makromol. Chem.* **60**, 106 (1963). U. Johnsen, *J. Polymer Sci.* **54**, 56 (1961).
23. W. C. Tincher, *Makromol. Chem.* **85**, 20 (1965); *J. Polymer Sci.* **62**, S148 (1962).
24. R. Chujo, S. Satoh, T. Ozeki, and E. Nagai, *J. Polymer Sci.* **61**, 512 (1962). S. Satoh, *J. Polymer Sci.* [A] **2**, 5221 (1964).
25. T. Shimanouchi, M. Tasumi, and Y. Abe, *Makromol. Chem.* **86**, 43 (1965).
26. T. Shimanouchi, *Special Lecture, IUPAC Symp., Prague, 1965;* published as *Macromolecular Chemistry,* Vol. 2, Butterworths, London, 1966.
27. Y. Abe, M. Tashumi, T. Shimanouchi, S. Satoh, and R. Chujo, *J. Polymer Sci.* [A] **4**, 1413 (1966).
28. S. Krimm, J. J. Shipman, V. L. Folt, and A. R. Berens, *J. Polymer Sci.* **3**, 2755 (1965).
29. D. Doskocilova, *J. Polymer Sci.* [B] **2**, 421 (1964).
30. H. Pivcova and B. Schneider, *Coll. Czech. Chem. Commun.* **31**, 3154 (1966).
31. Y. Yoshino and J. Komiyama, *J. Polymer Sci.* [B] **3**, 311 (1965).
32. F. A. Bovey, F. P. Hood, E. W. Anderson, and R. L. Kornegay, *J. Phys. Chem.* **71**, 312 (1967).
33. O. C. Böckman, *J. Polymer Sci.* [A] **3**, 3399 (1965).
34. G. Talamini and G. Vidotto, *Polymer Previews* **2**, 227 (1966).
35. D. M. White, *J. Am. Chem. Soc.* **82**, 5678 (1960).
36. S. Poorvik, *Plastvärlden* **4**, 47 (1967).
37. W. R. Moore and R. J. Hutchinson, *Nature* **200**, 1095 (1963).
38. J. R. Darby, N. W. Touchette, and K. Sears, *SPE Ann. Tech. Conf.* **13**, 405 (1967).
39. Z. Mencik, *J. Polymer Sci.* **17**, 147 (1955).
40. P. Doty, H. Wagner, and S. Singer, *J. Phys. Colloid Chem.* **51**, 32 (1947).
41. R. M. Fuoss, *J. Am. Chem. Soc.* **61**, 2329 (1939).
42. *Ibid.*, p. 2334.
43. J. M. Davies, R. F. Miller, and W. F. Busse, *J. Am. Chem. Soc.* **63**, 361 (1941).
44. J. M. Davies, R. F. Miller, and W. F. Busse, *J. Appl. Phys.* **21**, 605 (1950).
45. I. A. Mukhina, B. P. Shtarkman, and I. N. Vishnevskaya, *Mekhanism Protsessov Plenkoobrazov. Polim. Rastvorov Dispersii, Akad. Nauk SSSR, Sb. Statei* **1966**, 124.
46. H. Green, *Industrial Rheology and Rheological Structures,* John Wiley & Sons, Inc., New York, 1949.
47. J. J. Hermans, *Flow Properties of Disperse Systems,* North Holland, New York, 1953.
48. ISO DR 1624, International Standards Organization.
49. ISO Recommendation R 305, International Standards Organization.
50. J. W. Watts, "Measurement of the Stability of Poly(vinyl Chloride) Using the C. W. Brabender Plastograph," *Paper, SPE Conf., 1963.*

51. H. Fikentscher, *Cellulose Chem.* **13,** 58 (1932).
52. W. T. Blake, *Plastics Technol.* **4,** 909 (1958).
53. M. R. Rector, *Plastics Technol.* **2,** 390 (1956).
54. O. Leuchs, *Kunststoffe* **50** (4), 227 (1960).
55. ASTM Method D 1895-65T, American Society for Testing and Materials.
56. ISO R 60, International Standards Organization.
57. ISO 1265, International Standards Organization.
58. A. W. M. Coaker and M. W. Williams, *Mod. Plastics* **33** (2), 160 (1955).
59. C. E. Anagnostopoulos, A. Y. Coram, and H. R. Gamrath, *Appl. Polymer Sci.* **4** (11), 181 (1960).
60. H. Luther, F. O. Glander, and E. Schleese, *Kunststoffe* **52** (1), 7 (1962).
61. E. Berger, C. Drap, and R. Malavoi, *SPE J.* **24** (7), 37 (1968).
62. ASTM D 257-58, "Test for Electrical Materials," American Society for Testing and Materials.
63. *Brit. Standard Spec.* 2782.
64. C. E. Balmer and R. F. Conyne, *Rubber Age* **79** (1), 105 (1956).
65. J. B. DeCoste and B. A. Stiratelli, *Rubber Age* **83** (2), 279 (1960).
66. ASTM D 1823-61T, American Society for Testing and Materials.
67. A. C. Werner, *Mod. Plastics* **34** (6), 139 (1957).
68. ASTM D 1824-61T, American Society for Testing and Materials.
69. J. A. Greenhoe, *SPE J.* **17** (12), 1314 (1961).
70. A. Maccoll, *Chem. Rev.* **69,** 33 (1969).
71. M. Asahina and M. Onozuka, *J. Polymer Sci.* [A] **2,** 3505, 3515 (1964).
72. F. von Erbe, T. Grewer, and K. Wehage, *Angew. Chem.* **74,** 985 (1962).
73. Z. Mayer, L. Michailov, B. Obereigner, and D. Lim, *Chem. Ind. (London)* **12,** 508 (1965).
74. D. Braun and M. Thallmaier, *J. Polymer Sci.* [C] **16,** 2351 (1967).
75. V. P. Gupta and L. E. St. Pierre, *IUPAC Symp. Macromol. Chem., Budapest, 1969,* Preprint 11/73.
76. F. E. Bailey, Jr., J. P. Henry, R. D. Lundberg, and J. M. Whelan, *J. Polymer Sci.* [B] **2,** 447 (1964).
77. N. Murayama and Y. Amagi, *J. Polymer Sci.* [B] **2,** 115 (1966).
78. C. S. Marvel, J. H. Sample, and M. F. Roy, *J. Am. Chem. Soc.* **61,** 3241 (1939).
79. R. R. Stromberg, S. Straus, and B. G. Achhammer, *J. Polymer Sci.* **35,** 355 (1959).
80. I. Ouchi, *J. Polymer Sci.* [A] **3,** 2685 (1965).
81. M. B. Neiman, R. A. Papko, and V. S. Pudov, *Vysokomolekul. Soedin.* **10A,** 841 (1968).
82. W. C. Geddes, *European Polymer J.* **3,** 747 (1967).
83. V. W. Fox, J. G. Hendricks, and H. J. Ratti, *Ind. Eng. Chem.* **41,** 1774 (1949).
84. F. Sondheimer, D. A. Ben-Efraim, and R. Wolovsky, *J. Am. Chem. Soc.* **83,** 1675 (1961).
85. G. C. Marks, J. L. Benton, and C. M. Thomas, *S.C.I. Monograph* **26,** 204 (1967).
86. T. Kelen, G. Galambos, F. Tudos, and G. Balint, *IUPAC Symp. Macromol. Chem., Budapest, 1969,* Preprint 11/07.
87. M. Thallmaier and D. Braun, *Makromol. Chem.* **108,** 241 (1967).
88. J. P. Roth, P. Rempp, and J. Parrod, *Compt. Rend.* **251,** 2970 (1960).
89. W. I. Bengough and I. K. Varma, *European Polymer J.* **2,** 61 (1966).
90. A. Guyot, J. P. Benevise, and Y. Trambouze, *J. Appl. Polymer Sci.* **6,** 103 (1962).
91. N. Grassie, *Chem. Ind. (London)* **1954,** 161.
92. G. Talamini and G. Pezzin, *Makromol. Chem.* **39,** 26 (1960).
93. T. Kelen, G. Balint, F. Tudos, and G. Galambos, *IUPAC Symp. Maromol. Chem., Budapest, 1969,* Preprint 11/08.
94. A. Crosato-Arnaldi, G. Palma, E. Peggion, and G. Talamini, *J. Appl. Polymer Sci.* **8,** 747 (1964).
95. W. I. Bengough and M. Onozuka, *Polymer* **6,** 625 (1965).
96. E. J. Arlman, *J. Polymer Sci.* **12,** 543 (1954).
97. A. Rieche, A. Grimm, and H. Mucke, *Kunststoffe* **52,** 265 (1962).
98. G. Talamini, G. Cinque, and G. Palma, *Mater. Plast. Elast.* **30,** 317 (1964).
99. T. Morikawa, *Chem. High Polymers (Japan)* **25,** 505 (1968).
100. D. E. Winkler, *J. Polymer Sci.* **35,** 3 (1959).
101. N. Colebourne and E. S. Stern, *J. Chem. Soc.* **1965,** 3599.
102. J. Petersen and B. Ranby, *Makromol. Chem.* **102,** 83 (1967).
103. C. H. Bamford and D. F. Fenton, *Polymer* **10,** 63 (1969).

104. I. C. McNeil and D. Neil, *Makromol. Chem.* **117,** 265 (1968).
105. D. Bethell and V. Gold, *Quart. Rev. Chem. Soc. London* **12,** 173 (1958).
106. T. Morikawa, *Kagaku To Kogyo (Osaka)* **38,** 672 (1964).
107. M. Imoto and T. Nakaya, *Polymer Rept.* **93,** 42 (1966).
108. L. Valko and I. Tvaroska, *IUPAC, Symp. Macromol. Chem., Budapest, 1969,* Preprint 11/01.
109. E. L. Scalzo, *Materie Plastiche* **28,** 682 (1962).
110. R. Nagatomi and Y. Saeki, *Japan Plastics Age* **5,** 51 (1967).
111. F. Chevassus and R. de Broutelles, *Stabilization of Poly(vinyl Chloride),* Edward Arnold, London, 1963.
112. E. W. J. Michell and D. G. Pearson, *J. Appl. Chem.* **17,** 171 (1967).
113. M. Onozuka, *J. Polymer Sci.* [A] **1,** 2229 (1967).
114. A. H. Frye and R. W. Horst, *J. Polymer Sci.* **45,** 1 (1960).
115. A. H. Frye, R. W. Horst, and M. A. Paliobagis, *J. Polymer Sci.* [A] **2,** 1765 (1964).
116. A. H. Frye, R. W. Horst, and M. A. Paliobagis, *J. Polymer Sci.* [A] **2,** 1785 (1964).
117. A. H. Frye, R. W. Horst, and M. A. Paliobagis, *J. Polymer Sci.* [A] **2,** 1801 (1964).
118. T. Morikawa and K. Yoshida, *Kagaku To Kogyo* **38,** 667 (1964).
119. G. P. Mack, *Mod. Plastics* **31,** 150 (1953).
120. B. Dolezel, *Chem. Prumysl* **13,** 160 (1963).
121. V. E. Gray and I. R. Wright, *J. Appl. Polymer Sci.* **8,** 1505 (1964).
122. G. V. Hintzenstern, *Plaste Kautschuk* **10,** 16 (1963).
123. T. Kimura, *Kobunshi Kagaku* **20,** 257 (1963).
124. D. H. Weichert and K. Buehler, *Plaste Kautschuk* **12,** 664 (1965).
125. M. Helmstedt, *Plaste Kautschuk* **13,** 394 (1966).
126. H. H. Frey, *Kunststoffe* **53,** 103 (1963).
127. M. A. Golub and J. A. Parker, *Makromol. Chem.* **85,** 6 (1965).
128. L. B. Weisfeld, C. A. Thacker, and L. I. Nass, *SPE J.* **21,** 649 (1965).
129. F. Baxmann and K. Kopetz, *Kunststoffe* **52,** 465 (1962).
130. A. S. Kenyon, *Natl. Bur. Std. Circ.* **525,** 81 (1953).
131. B. Dolezel and J. Stepek, *Chem. Prumysl* **10,** 381 (1960).
132. W. Jasching, *Kunststoffe* **52,** 458 (1962).
133. R. C. Hirt, N. Z. Searle, and R. G. Schmitt, *SPE Trans.* **1,** 21 (1961).

C. A. Brighton, G. C. Marks, and J. L. Benton
BP Chemicals International Ltd.

COMPOUNDING

Compound Ingredients

Only in a few applications can poly(vinyl chloride) be used without a number of additives, which must meet certain specific requirements to enable the PVC compound to be processed and converted into the finished product. The conversion of resin into rubbery compounds by heating with tricresyl phosphate formed the basis of the original patent (1) and established the potential of PVC for the manufacture of substitutes for natural rubber in electrical applications. A wide range of high-boiling materials is now available as *plasticizers,* which can produce PVC compounds with various properties. High-boiling liquids compatible with plasticizers are often used in the formulation in order to reduce the cost; attempts are made to use any high-boiling waste product that is available at low cost; these are generally referred to as *extender plasticizers.* See also PLASTICIZERS.

Since the resin is unstable at the elevated temperatures required to process PVC compounds, *heat stabilizers* are added in order to avoid the evolution of hydrogen chloride and prevent discoloration (see the section on Degradation and Stabilization). Materials that are effective in this capacity are generally capable of chemically reacting with the hydrogen chloride as it is formed and thus preventing the autocatalytic effect of the decomposition products from becoming apparent. Stabilizer is consumed during the whole time the ingredients in a formulation are being converted into a fully fused (or gelled) compound, and an adequate amount must still be left to allow the final processing to be carried out in safety. Metallic salts based on lead, barium, and cadmium are most effective, and synergistic effects can be produced by using mixed compounds. See also STABILIZATION.

Lubricants are incorporated in PVC compounds to serve two purposes: first, to improve the flow of polymer chains past one another, and second, to help the flow of the molten PVC compound through processing equipment and to prevent it from sticking to the metal surfaces. Generally two types of lubricant are used for the two purposes: *internal lubricants* are required for the first, and *external lubricants* for the second. See also LUBRICANTS.

Fillers are used generally to extend the PVC resin and produce lower-cost compounds. This result is obtained only at the expense of properties such as strength, clarity, resilience, and light weight. In general the reduction in quality of these properties is directly proportional to the quantity of filler present in the compound. See also FILLERS.

Pigments are used to produce colored materials, but only special types are satisfactory because most pigments have a detrimental effect, eg, on the thermal stability of the compound. *Dyes* are used less frequently than pigments because they are more likely to bleed to the surface. It is found that pigments that are satisfactory for plasticized compounds may not necessarily be suitable for use in unplasticized or rigid compounds.

The formulation of compounds having all these various additives is based on PVC resins made by the suspension or bulk process (see the section on Polymerization). The choice of the resin is determined first by the process by which the compound is being fabricated (eg, a low-molecular-weight resin would be needed for injection molding unplasticized PVC), and second, by the ultimate physical properties required. It is often necessary to compromise between two competing objectives. Generally, PVC compounds based on high-molecular-weight resins are more difficult to process but produce better physical properties in the finished article.

The formulation of *plastisols* is largely determined by the properties required for the conditions under which the material is to be applied, eg, the viscosity under high rates of shear such as would be encountered in high-speed spreading operations. The extent to which viscosity characteristics of the paste change during storage would also be relevant. The physical properties of the fused product do not play a large part in controlling the additives and the quantities in which they are used, except perhaps with regard to the softness of the compound, which is controlled by the type of plasticizer and ratio of plasticizer to resin. There are limits to which these can be varied if a satisfactory plastisol is to result. The addition of lubricants is not generally required, because the effects for which these compounds are incorporated do not apply in processes involving plastisols.

Plasticized Poly(vinyl Chloride)

Plasticizers. The main classes of compounds used as plasticizers are phthalates, phosphates, esters of aliphatic dibasic acids (low-temperature plasticizers), and polyesters. Chlorinated paraffins as well as aliphatic and aromatic oils are classed as extenders. A number of other compounds are used for special properties, but these form only a small portion of the total plasticizer use.

Phthalates. The phthalates have the best all-round properties and are also the cheapest primary plasticizers available; a number of isomers of dioctyl phthalate are produced. Di-2-ethylhexyl phthalate and diisooctyl phthalate are the most widely used because they give good overall performance for a wide range of uses at low cost. Dialphanol phthalate, an ester from mixed C_7–C_9 alcohols, is used extensively in the United Kingdom because of the availability of the alcohols at low cost.

The physical properties of PVC compounds based on phthalic acid esters of the various isomers of octyl alcohol do not vary appreciably, and Figure 1 can be taken as a reasonable guide to indicate variation of tensile strength and 100% modulus with plasticizer content. These plasticizers are used in a wide range of applications in which the temperature does not exceed 60°C; for temperatures in excess of this, eg, in electric-blanket wiring, the use of ditridecyl phthalate is satisfactory. (This plasticizer, together with suitable antioxidant, meets the Underwriters' Laboratories' requirements for 105°C wiring in thick-section (0.050 cm) products but fails with thinner sections.)

Other mixed alcohols are now finding use for the manufacture of phthalate plasticizers; straight-chain alcohols containing 6–10 carbon atoms can be produced from coconut oil or by synthesis, although the distribution of the C_8 and C_{10} alcohols varies in the two sources. These phthalates have lower volatility than the octyl esters and somewhat better low-temperature flexibility. Since these mixed alcohols are essentially unbranched, they are more stable to oxidation than are the isomers of octyl phthalate.

Phosphates. Although unplasticized PVC is self-extinguishing, the addition of plasticizers generally reduces its flame-retardant properties; the phosphate plasticizers and chlorinated paraffins, however, behave differently in this respect. Tritolyl phosphate was used extensively for reducing the flammability of PVC compounds, but has now been largely superseded on economic grounds by trixylyl phosphate; this plasticizer has the disadvantage that the low-temperature properties of the compounds

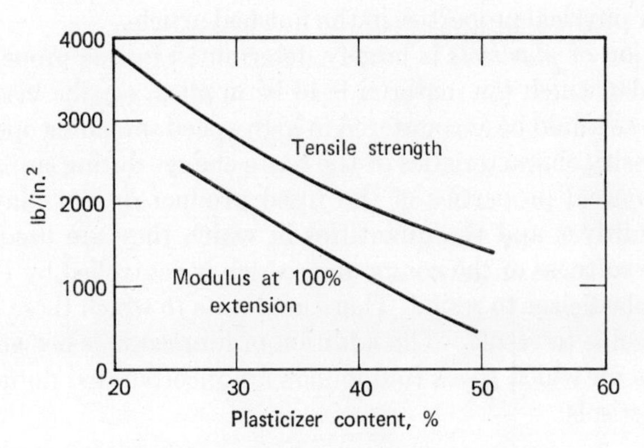

Fig. 1. Effect of plasticizer content (dioctyl phthalate) on tensile strength and 100% modulus of PVC.

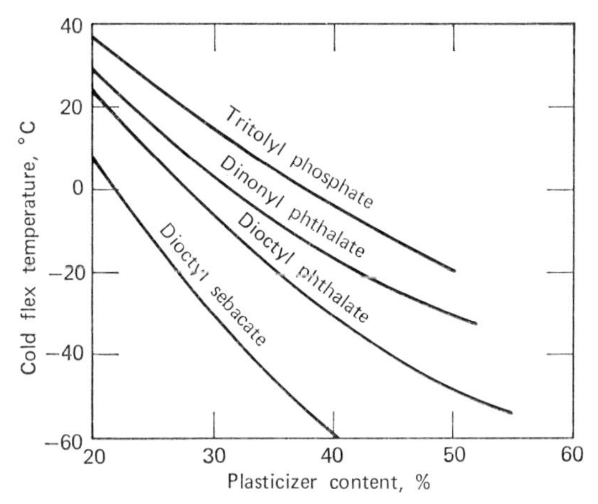

Fig. 2. Effect of plasticizer on cold flex temperature.

are much inferior to those based on phthalic acid plasticizers (see Fig. 2) (see also FIRE RETARDANCY). The continuing effort to reduce the cost of plasticizers has led to the introduction of new products based on a mixed feedstock of phenol and *o*- and *p*-chlorophenol, which is reacted with phosphorus oxychloride to form phosphate esters. These plasticizers give compounds which are of poorer heat stability than are normally obtained with tritolyl and trixylyl phosphates, and which produce undesirable odors that limit their use. A further development has been the use of a feedstock of phenol and isopropylated phenol to give phosphate plasticizers with greatly improved light stability and an almost complete lack of odor.

Esters of Aliphatic Dibasic Acids. In Figure 2 it can be seen that at a plasticizer content of 30% the cold flex temperature of compounds based on dioctyl sebacate is about 25°C lower than that of those based on dioctyl phthalate. Esters based on aliphatic dibasic acids generally give improved low-temperature flexibility; the sebacates give the best properties followed by the azelates, the adipates, and finally the nylonates, which are made from mixed adipic, glutaric, and succinic acids. The higher-molecular-weight compounds have the added advantage that they are less volatile and are more resistant to extraction by aqueous media.

Polyesters. The need for compounds with special properties, such as are used in contact with hydrocarbon fuels, has resulted in the development of a wide range of polyester plasticizers. The first polyesters to be introduced were unmodified polyesters, a typical example being poly(propylene sebacate). The disadvantage of unmodified polyesters is their high viscosity with consequent difficulty in handling, high processing temperatures for optimum physical properties, poor plasticizing efficiency, and poor low-temperature properties. This has led to modification of these polyesters by addition to the reaction mixture of a portion of a monocarboxylic acid and/or a monohydric alcohol, which results in the termination of the polymer molecules to give lower-molecular-weight products. The different types of polyester that can be produced can be listed as follows:

unmodified polyester $(A\text{-}G)_n$
alcohol-modified polyester $R\text{-}(A\text{-}G)_n\text{-}A\text{-}R$
acid-modified polyester $M\text{-}G\text{-}(A\text{-}G)_n\text{-}M$

where A = dibasic acid, G = glycol, R = monohydric alcohol, and M = monobasic acid.

There are continuing developments in producing polyester plasticizers with improved resistance to extraction, particularly for compositions that can be safely dry-cleaned with chlorine-containing solvents. Some progress has been made in this direction but no composition can yet be claimed as entirely satisfactory for this application.

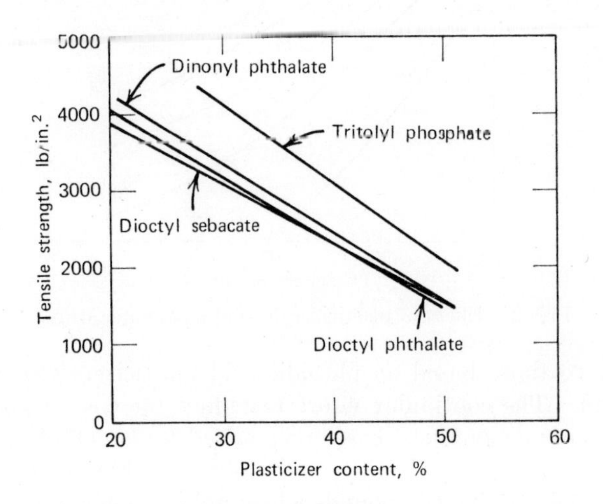

Fig. 3. Effect of different plasticizers on tensile strength.

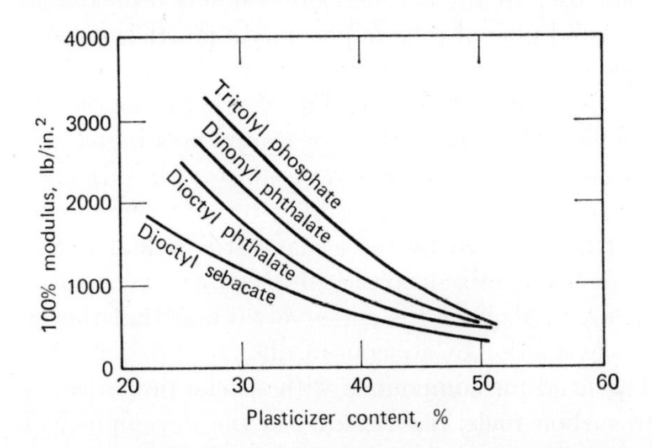

Fig. 4. Effect of different plasticizers on 100% modulus.

Chlorinated Paraffins. Chlorinated paraffins are now used extensively for partial replacement of primary plasticizers to give compositions that are cheaper, and have better flame retardancy and improved low-temperature properties. The chlorinated paraffins usually have a chlorine content of 42–56% and an average carbon chain length of C_{15}–C_{25}. The materials with the higher chlorine content are more compatible with PVC and can be used in larger proportions in the plasticizer blend. Whereas the phosphate plasticizers are normally used in flame-retardant compositions, chlorinated paraffin can be incorporated to give better low-temperature properties with a reduction in cost. The effect of this type of plasticizer has been studied in blends with a number of the more widely used primary plasticizers (2).

Effects on Properties. Apart from the resin, the plasticizer is the most important ingredient in a flexible vinyl compound. The type and amount determines the flexibility and influences the other properties such as tensile strength, modulus, specific gravity, hardness, cold flex, and elongation at break. The effect of different plasticizers and of variation of concentration on tensile strength is shown in Figure 3, and on the 100% modulus in Figure 4. The electrical properties of PVC compounds are also dependent on the type of plasticizer and the amount present. The effect with dioctyl phthalate on the volume resistivity is shown in Figure 5. The most suitable plasticizers for the achievement of good electrical properties are characterized by stability toward hydrolysis, a strong hydrophobic nature, and nonpolarity as far as is consistent with compatibility.

The formulation of a PVC compound to achieve a given set of properties is usually established in the laboratory by making a series of compounds based on previous results. This can be costly and time consuming, and the formulator can never be sure that he has obtained the optimum plasticizer blend at the minimum cost. Hence

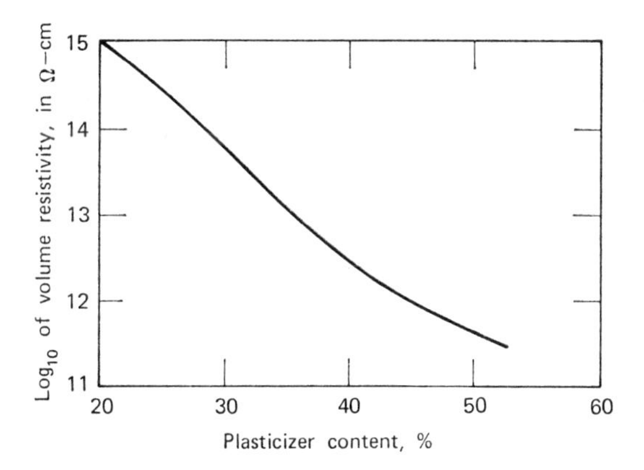

Fig. 5. Effect of dioctyl phthalate content on volume resistivity.

there is need for a simple and reliable method of accurately formulating a PVC compound to have a given set of properties. A number of systematic approaches to this problem have been reported; in the first the performance data are correlated with concentration for binary plasticizer systems and then reference charts are prepared (3). The method is limited to binary plasticizer systems, and the number of charts becomes large if all available plasticizers are included (eg, 435 charts for 30 plasticizers).

The properties of a vinyl compound that contains a blend of plasticizers can be predicted by assuming that the value of any measurable property, eg, tensile strength, is proportional to the "efficiency concentration" of each of the plasticizers present (4). The efficiency concentration is defined as the concentration of that plasticizer which, when used as the sole plasticizer, gives the same hardness as that which will be obtained with the blend of plasticizers. Hardness is a convenient property to use and tensile strength can be just as effective. The value P_{Bi} of any property i is then given by

$$P_{Bi} = \sum_{j=1}^{n} \frac{c_j P_{ij}}{c_j^0}$$

where P_{Bi} is the *i*th property of the blend, n is the number of plasticizers in the blend, c_j is the concentration of the *j*th plasticizer, in parts per hundred of resin (phr), c_j^0 is the "efficiency concentration" of the *j*th plasticizer, in phr, and P_{ij} is the value of the *i*th property of the *j*th plasticizer when used alone at C_j^0.

The equation has been used also to formulate a computer program whereby it is possible to retrieve data on the effect of plasticizers on a range of properties, and process them to solve each of the following three types of problem: estimating the performance of a given plasticizer blend; selecting the optimum blend for a given set of specifications; and plotting performance and cost diagrams for a given ternary plasticizer system (5). The optimization problem is generally limited to the selection of the optimum combination of three or fewer plasticizers, because experience has shown that practically all requirements can be met with such a number.

Numerous individual compounds used for the plasticization of poly(vinyl chloride) and the process of plasticization are discussed in the article on PLASTICIZERS.

Stabilizers. The development of satisfactory stabilizers for vinyl compounds has led to a wide range of chemicals that are available for the many different applications of PVC. They may be divided into five main groups: (*1*) amines; (*2*) organotin compounds; (*3*) basic lead compounds; (*4*) metallic salts of phenols, and of aromatic and aliphatic acids; and (*5*) additives such as antioxidants, chelating agents, ultraviolet-radiation absorbers, and epoxidized oils. This wide range of stabilizers with their different chemical structures suggests that there may be differing mechanisms in the stabilization process. In many aspects the different stabilizers are complementary, and no one stabilizer fulfills all requirements.

The largest class of stabilizers is based on *compounds of lead*, such as white lead, basic lead carbonate, and basic lead sulfates; in addition, dibasic lead phosphite, dibasic lead phthalate, dibasic and monobasic lead stearates, and basic lead silicate are used for certain special properties. Many other lead compounds have become commercially available at one time or another, but the compounds listed above are the most widely used. It is found that the lead compounds differ in at least one specific property. White lead is used extensively for stabilization of plasticized extrusion compounds for cable covering, but if the cable is required to work at elevated temperatures then basic lead phthalate is frequently used because of its antioxidant properties; this particular lead stabilizer also improves the electrical properties of the compound. The major limitations to the use of lead stabilizers in flexible compounds are the opacity, which makes it impossible to produce a transparent material, the toxic properties, which prevent their inclusion in materials intended for use in contact with food, and the sulfur-staining characteristics.

The *long-chain aliphatic acid salts of cadmium and barium* find extensive use as stabilizers, particularly in blends of the two. Cadmium soaps give excellent heat stability for a fairly short time, after which the compound darkens rapidly; barium soaps are less effective but longer lasting. Mixtures of the two produce a synergistic effect that makes them particularly suitable for flexible compounds where clarity or translucency and freedom from staining are required. Besides the aliphatic acid salts, barium phenoxide is sometimes included, as well as zinc soaps in amounts ranging from one eighth to one third of the amount of the barium–cadmium stabilizer. A single salt of this class is hardly ever used as a stabilizer, and most of the materials are single-package stabilizing systems, the compositions of which are jealously guarded by the manufacturers who claim special properties for their product. This class of stabilizers cannot be used for compounds intended for food packaging because of the toxic properties of both the barium and the cadmium salts. This type of stabilizer, which is rich in cadmium and phosphites, is also very effective as a light stabilizer.

Organotin stabilizers are generally used where the most effective stabilizing action, together with high clarity, is required. This class of compound is the most expensive of the stabilizers available, but the cost is offset to some extent because the quantities used are much less than for other stabilizers (1–2% instead of about 5%). The compounds used are generally to be found within the following classes of compounds: di-*n*-alkyltin mercaptides, di-*n*-alkyltin dilaurates, and dibutyltin dimaleate. The high clarity obtained with this class of stabilizer is due to the fact that the organotin compounds react with hydrogen chloride produced during processing to form dialkyltin dichlorides, which are highly soluble in both plasticized and rigid vinyl compounds.

Other types of material have found use as stabilizers, either alone or mixed with other stabilizing compounds in package systems. *Epoxy resins* are frequently used with other types because of their synergistic effect and because they are not sufficiently effective when used alone. *Epoxidized oils* and esters are also frequently used as auxiliary stabilizers. Mixtures of calcium and zinc carboxylates now form the basis of a number of stabilizers used in applications where low toxic hazard is required. Esters of aminocrotonic acids are also used in similar applications.

The choice of the best stabilizing system is difficult because the test methods available generally do not bear much relation to the conditions encountered in processing equipment. The system that is ultimately developed is frequently derived as a result of long experience by the formulator with the process under examination. As a rule, when flexible PVC compounds are processed by any of the normal processing techniques, such as Banbury mixing, calendering, extruding, or injection molding, the best results are obtained with barium–cadmium or lead stabilizers rather than with organotin compounds, unless particularly high temperatures (above 190°C) are used.

Attempts have been made by several producers of vinyl resins to incorporate a stabilizing system during the polymerization process. It was common practice in the manufacture of emulsion resins to add sodium carbonate to the aqueous emulsion after completion of polymerization, and this produced better stability in the final product. In suspension polymerization the addition of lead acetate to the aqueous system leads to the incorporation of the lead salt in the polymer particle, which results in a resin of improved stability. The most interesting work in this field has been the attempts to introduce stabilizing comonomers into the polymer chain. Glycidyl methacrylate and styrene oxide are two epoxy derivatives which, it is claimed, produce copolymers of greatly improved thermal stability. The use of a vinyl organotin compound as comonomer is an elegant method of producing a stable vinyl resin. Although these various methods do improve the stability of PVC, it is still necessary to add conventional stabilizers, and the advantage gained by the use of such resins does not justify the increased price of the materials.

The article on STABILIZATION lists numerous individual stabilizers and discusses the mechanism of their action. See also DRIERS AND METALLIC SOAPS.

Lubricants. The addition of lubricants to plasticized compositions is essential to ensure that the materials can be processed satisfactorily. The quantities required, however, are much less than those required for rigid compositions, and the use of metal soaps as stabilizers often imparts sufficient lubricity without the addition of further lubricant. In addition to metal soaps, the following are normally used as lubricants:

stearic acid, glyceryl monostearate, ethyl diaminostearate, and paraffin and low-molecular-weight waxes.

The waxes and stearic acid are only partially compatible with plasticized PVC, and this determines their efficiency as lubricants; if excessive quantities are used, therefore, they will migrate to the surface of the composition and cause blooming. Metallic soaps such as calcium, barium, and lead stearate are less critical and can be used at concentrations of 0.5–1.5% of the polymer. An increasing number of new lubricants have appeared but these are generally chemically similar to those listed above. The measurement of the efficiency of a particular lubricant system has been carried out using a high-pressure capillary viscometer (6), and from the results of a range of lubricants can be classified according to their chemical structure. The Brabender plastograph has also been used to determine the fusion time of a PVC composition (7,8). See also LUBRICANTS.

Fillers. The prime objective in adding fillers to plasticized PVC compositions is to reduce the cost. There is no chemical interaction between the filler and the resin so that reinforcement, as in the case of the addition of carbon black to rubber compounds, does not occur. In general, the addition of fillers produces a reduction in both the physical and chemical properties, although there may be some advantages, such as improved electrical properties, resistance to ultraviolet radiation, drier polymer–plasticizer blends, and control of the surface finish. The choice of filler is made so as to achieve the desired improvements with the minimum reduction of those properties that are affected; these include tensile strength, elongation, low-temperature performance, and abrasion resistance. Fillers commonly used are various forms of calcium carbonate, and because of their oil-absorption characteristics, the addition of such materials increases the hardness of compounds at the same level of plasticizer; this gives rise to an increase in the melt viscosity during processing so that output rates are reduced. If satisfactory dispersion in the compound is to be obtained, the particle size must be in the range of 0.2–15 μ for the majority of applications, although for asbestos–vinyl floor tiles, larger-particle-size materials can be used. The falloff in mechanical properties is generally lower with the finer fillers, although these have higher oil absorption and consequently produce harder compounds.

The calcium carbonate fillers are available in both natural and precipitated types; the former are made by grinding materials, whereas the latter are produced chemically. The material forms are usually described as whitings and are predominantly amorphous, whereas the precipitated product has a much finer particle-size distribution (0.2–5 μ) and is crystalline in nature. The crystalline grades are more easily dispersed during compounding than are the amorphous whitings, which have a tendency to agglomerate. Other fillers that find applications in plasticized compositions are calcined clays, talc, asbestos, slate dust, calcium silicate, and barytes, but these form only about 10% of the total quantity used. The calcium carbonate fillers are available also as grades in which the particles are coated with calcium stearate; this coating can be applied to both the natural and precipitated forms so that the plasticizer absorption is reduced. The calcium stearate coating contributes to the lubrication of the compound, thus reducing the requirement of additional lubricants.

In assessing the cost reduction that can be obtained by using fillers, the following factors must be considered. (*1*) The cost of the filler must obviously be taken into account; uncoated calcium carbonate fillers vary in price from 0.5 to 1.2 ¢/lb, the coated whitings cost around 2.5 ¢/lb, and coated precipitated calcium carbonate costs around

6 ¢/lb. (*2*) The specific gravity can vary from 2.63 for calcined clay to 4.46 for barytes. Calcium carbonate fillers have a specific gravity of 2.65–2.71, so that the specific gravity of the compounded PVC will rise and affect the cost on a volume basis accordingly. (*3*) The last factor to be considered is the extent to which the mechanical properties of the compound can be reduced and still give an acceptable performance, not only as regards the end-use application, but also the output on the processing equipment. In this latter context a raw-material cost advantage using one filler may be more than offset by the use of a more expensive filler that permits a faster output and consequently a lower processing cost.

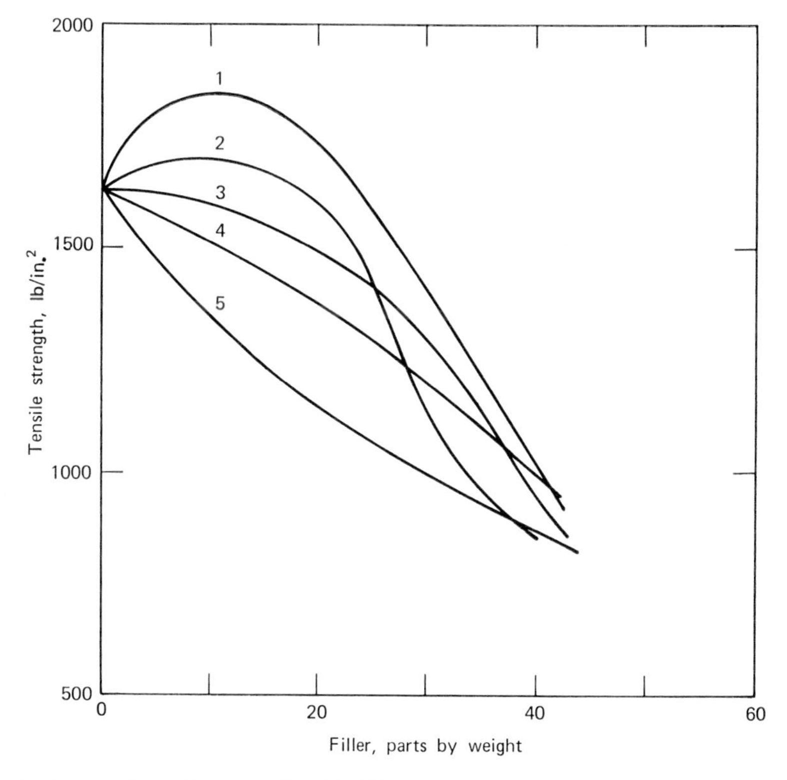

Fig. 6. Variation of tensile strength with filler content. 1, finely divided china clay; 2, lightly colored china clay; 3, coated natural whiting; 4, precipitated whiting; 5, asbestos.

Figure 6 shows the change in tensile strength that occurs with the addition of a number of fillers. At low concentrations, some clay fillers give an apparent increase, but this is exceptional; at the normal levels used (20–30% by weight of the total), the tensile strength is only about two thirds of the value for the unfilled compound. These results were obtained by adding the fillers to the following composition:

	parts by weight
PVC (high mol wt)	50
tritolyl phosphate	50
white lead (dispersed in dioctyl phosphate)	4.5
calcium stearate	1

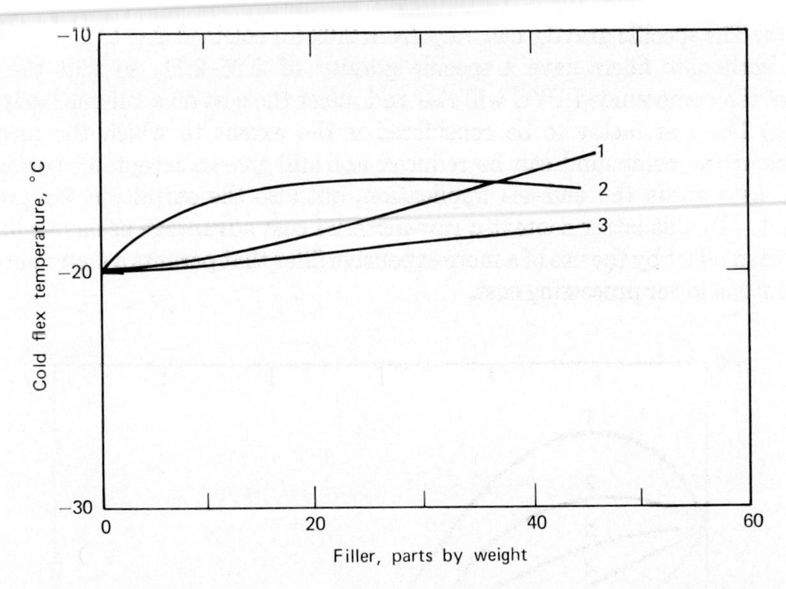

Fig. 7. Variation of cold flex with filler content. 1, fine china clays; 2, calcium silicate; 3, whitings.

The effects on other properties of the addition of filler is shown in Figure 7 (cold flex temperature) and Figure 8 (compound hardness). Whitings can be added in fairly large quantities without having much effect on the cold flex temperature. The coated whitings also have little effect on the hardness of the compound if the quantity does not exceed 20 parts.

The calcium carbonate fillers reduce the acid resistance of PVC compounds and have an adverse effect on resistance to many chemical products; they also produce an increase in water absorption, although this is less than for other fillers.

Many waste products have been assessed as fillers; these are often materials available at a low cost, but they find only limited application. Many trace metal compounds have a deleterious effect on the thermal stability of PVC and the presence of such impurities can make a material unsuitable as a filler. See also FILLERS.

Pigments. When selecting pigments for coloring PVC compounds it is essential to make sure that the colorant is stable at the processing temperature (120–180°C) and that it will not be affected by the trace amounts of hydrogen chloride evolved. Good light stability, resistance to the action of soaps and detergents, and fastness to wet and dry rubbing are also necessary. Suitable inorganic pigments include titanium dioxide, cadmium yellows and reds, and pigment grades of carbon black. Many organic colorants can also be used, including anthraquinone and indathrene blues, yellows, and oranges, and phthalocyanine blues and greens. Organic pigments that may be soluble in the plasticized compound should not be used since migration of the pigment is likely to occur.

Some pigments reduce the thermal stability of the compound, but those colorants based on metals whose derivatives are used as stabilizers may actually improve the stability. See also PIGMENTS.

Other Additives. *Flame Retardants.* The flammability of flexible PVC compositions is determined largely by the plasticizers that are used. Aryl phosphates and chlorinated paraffins can be included with other plasticizers to give materials with

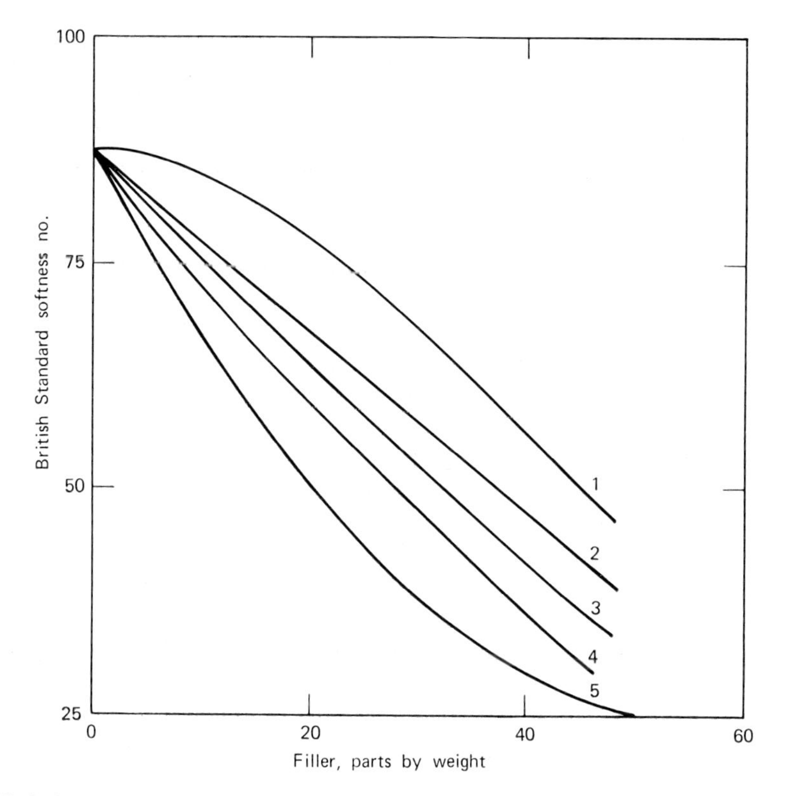

Fig. 8. Variation of British Standard softness with filler content. 1, coated natural whiting; 2, precipitated whiting; 3, coated precipitated whiting; 4, lightly colored china clay; 5, asbestos.

acceptable flame resistance but this is achieved only at the expense of low-temperature properties. If flame resistance is required along with the physical properties that can be obtained only with phthalate plasticizers, then a special retardant, such as antimony oxide, is used in concentrations of up to about 5%. Other materials that find some use, although they are not as effective as antimony oxide, are barium metaborate, zinc borate, zinc antimony polyalkylphosphate, and zinc boron polyalkylphosphate. See also FIRE RETARDANCY.

Ultraviolet-Radiation Stabilizers. Because of the degradative effect of UV radiation on the polymer molecule, stabilizers are usually incorporated in plasticized compositions when the materials are used outdoors, to prevent discoloration and embrittlement or loss of tensile strength. Such applications include rainwear, vinyl-coated canopies, garden hose, and automobile seat covers, as well as floor tiles that are subjected to fluorescent illumination. There are several types of UV-radiation stabilizers that can be used, but their effectiveness depends to some extent on the heat stabilizer used; for example, the hydroxybenzophenones give the best results when used with barium cadmium stabilizers and epoxy plasticizers. A concentration of about 0.5% by weight is usually sufficient to give the required protection. Benzotriazoles can produce a significant improvement in the UV stability of plasticized vinyl compositions when used in amounts as low as 0.1% by weight. Other products that are effective include benzylidene malonates, salicylates, and substituted acrylonitriles. Lead stabilizers are not normally used for flexible compounds for outdoor applications, but dibasic lead phosphite gives some measure of protection for electrical insulation and outdoor upholstery. Pigmented compositions that contain titanium dioxide or carbon black have improved light stability owing to their opacity to ultraviolet rays. See also ULTRAVIOLET-RADIATION ABSORBERS.

Antioxidants. The degradation of the resin in PVC compounds is controlled largely by the addition of heat and UV stabilizers. Some falloff in properties can also result, however, due to oxidation of the plasticizer (9). A study of the oxidation of alkyl sebacates has shown that peroxides are formed in the initial stages, and these are followed by the formation of acids and carbon dioxide as the principal products. The formation of peroxides during the early stages of the oxidation can also have a detrimental effect on the stability of resin, so that the addition of an antioxidant can have a profound effect on the stability of plasticized PVC compounds, both during processing and aging. The use of antioxidants to reduce plasticizer oxidation has been pursued more actively in the United States than elsewhere, and many commercial plasticizers also contain an antioxidant; a product that is often used is 2,2-bis(4-hydroxyphenyl)propane, more commonly known as diphenylolpropane. This more general use of antioxidants in the United States is probably due to the inclusion of stringent thermal aging tests in American specifications; an example of such a specification is the Underwriters' Laboratories Standard UL 83 entitled "Thermoplastic Insulated Wires," which requires a sample to retain 65% of its original elongation at break after aging at 100°C for 7 days. By contrast, British Standard 2746, "P.V.C. Insulation and Sheath of Electric Cables," specifies a 3% loss in weight after heating in an oven at 82°C for 5 days; such a specification need not require the inclusion of an antioxidant in the formulation.

The calendering of transparent flexible PVC sheet involves the generation of frictional heat because of the high linear speed used; this results in the need for the inclusion of an antioxidant in the formulation if the original color is to be maintained; the amount required is usually on the order of 0.1% by weight.

There is an extensive range of products that are effective in maintaining the physical properties of plasticized PVC on aging at elevated temperatures; these include the alkylated phenols, bisphenols, hydroquinone derivatives, and other phenolic compounds, aromatic amines, esters, and organic phosphites and phosphates. See also ANTIOXIDANTS.

Antistatic Agents. The superior electrical properties of flexible PVC compounds result in the ready development of static buildup during both manufacture and use. This can present such problems as dirt pickup, discomfort, and hazard from sparks, etc. Most products added to compounds to reduce static are hydrophilic substances with poor compatibility with the PVC composition, so that they tend to migrate slowly to the surface; a conducting film results which prevents the formation of static charges. A similar effect can be obtained by applying a thin film to the surface of the final product; this produces only temporary protection but is sufficient to ensure dispersion of the static charge, eg, during film processing.

The types of antistatic agents incorporated in PVC compounds include fatty amines and amides, phosphate esters, quaternary ammonium salts, and esters of poly(oxyethylene) glycol. The quantities used vary between 0.1 and 1.0%; if this is exceeded the migration can produce a surface bloom and cause difficulties with heat-sealing operations.

Many antistatic agents give colored products and can also reduce the thermal stability, so that special care is needed in formulating the recipe. See also ANTISTATIC AGENTS.

Fungicides. Although the polymer is not susceptible to attack, plasticizers can be degraded by various microorganisms so that plasticized compositions for use in the tropics, and particularly in hot and humid climates, need to be suitably stabilized. The degree to which attack can occur depends to a very large extent on the type of plasticizer used. The phthalates, phosphates, and chlorinated hydrocarbons are resistant, but the adipates, sebacates, epoxidized oils, and polymeric plasticizers are known to be subject to attack. The initial attack is always on the surface and, if it is not very severe, is limited to discoloration and change of appearance; however, when the attack goes further, there will be a deterioration in the physical properties.

There are a variety of products that can be used to give protection including organotin compounds, brominated salicylanilides, mercaptans, quaternary ammonium compounds, arsenic compounds, and copper compounds. There are some materials that have been developed for specific applications, eg, (tri-*n*-butyltin) sulfosalicylate for flexible PVC film at levels of 0.5%; it is extremely toxic and must be handled with special care. *N*-(Trichloromethylthio)phthalimide is used for PVC-coated films in concentrations of 0.25–0.5% (or higher if very susceptible plasticizers are included). Copper 8-quinolinolate is useful for compositions such as coated fabrics, electrical insulation, and other products which must be kept free from mildew growth. Many of these products are very toxic and their use is limited to those applications where possible migration cannot give rise to any hazard. See also BIOCIDES.

Plastisols

A plastisol, or vinyl dispersion, is normally formed using 30–50% plasticizer and is formulated to meet the special requirements of the process in which it is used. The final physical properties of the fused material are of secondary consideration and are, in fact, subject to limits imposed by the rheological properties required in the initial plastisol, which depend on the chemical and physical nature of the plasticizers and the ratio in which they are used. Besides the amounts and types of plasticizers used, the rheological properties also depend on the age of the plastisol, its temperature, and the rate of shear at which the viscosity is measured. The viscosity starts to increase when a plastisol is stored after manufacture. When the temperature is raised the viscosity initially falls, but as the rate of solvation between the polymer and plasticizers increases, the viscosity undergoes a rapid increase, which is followed by setting and complete fusion. The effect of the rate of shear varies according to the composition of the plastisol.

The effects of various plasticizers on the rheological properties of the system have been studied extensively (10). The phthalates give a good all-round performance at relatively low cost; they give pastes of low viscosity, with good shelf life and good physical properties in the final product. The octyl phthalates, both straight-chain and branched, are widely used in paste applications. The phosphates have a high solvating power, which enables them to be used where lower fusion temperatures are required. The use of phosphate plasticizers makes possible the incorporation of larger amounts of fillers and extenders before incompatibility limits are reached; they impart a degree of flame retardancy and good resistance to extraction by oils, but are subject

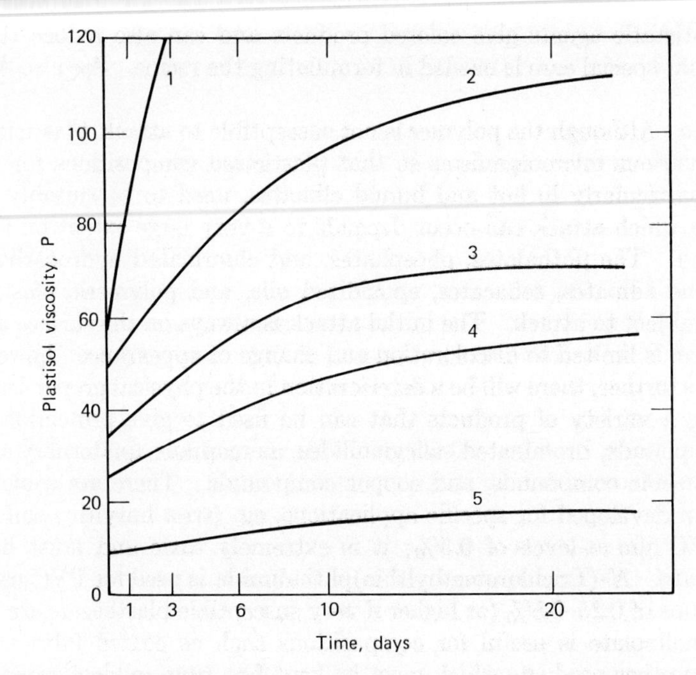

Fig. 9. Viscosity of PVC plastisols. 1, dibutyl phthalate; 2, di-2-ethylhexyl phthalate; 3, straight-chain C_6–C_{10} phthalate; 4, isooctyl isodecyl esters of mixed adipic, glutaric, and succinic acids; 5, triethylene glycol dicaprylate. The mixtures were made using 70 parts of plasticizer with 100 parts of resin, then stored at 23°C, at which temperature the viscosities were determined using a Ferranti Portable Viscometer, model VH.

to extraction by water and detergent solutions. The aliphatic esters of adipic and sebacic acids impart very good low-temperature performance, but they have low solvating power and require higher fusion temperatures than the phthalates. The chlorinated hydrocarbons, which are used generally to reduce costs and improve flame retardancy, are high-viscosity liquids and tend to produce relatively thick pastes with dilatant characteristics.

The viscosities of plastisols obtained with a number of plasticizers and using a typical commercial paste-forming resin are shown in Figure 9. The butyl phthalates, which have a strong solvating action on the polymer particles, give plastisols that age rapidly, whereas a nonsolvating plasticizer such as triethylene glycol caprylate produces a plastisol that undergoes little change on storage at room temperature. The rate at which the viscosity of a plastisol increases on storage is of considerable importance to the user, and attempts have been made to predict these changes from accelerated results obtained at elevated temperatures (11,12).

Reasonably accurate results for natural aging can be predicted by comparing the absolute increase in the viscosity values for natural aging with the absolute increase in the viscosity for accelerated aging at 45°C. The method does not appear to be very satisfactory for pastes that undergo rapid changes in viscosity at room temperature.

When the temperature is raised the viscosity falls at first and reaches a minimum at about 45°C; as the temperature is raised further the viscosity rises sharply at about 60°C; the exact temperature varies according to both the resin and the plasticizer. The increase in viscosity on heating is due to solvation of the polymer by the plasticizer and can be followed by examining the system under the electron microscope. This can be seen clearly in Figure 10. This change of viscosity at increased temperatures is of considerable importance and determines the performance of the plastisol in applications such as rotational casting and slush molding. It has been studied extensively for a number of plasticizers (13). The temperatures at which the rise in viscosity takes place generally lie within the range 55–70°C for all the plasticizers used with paste-making resins.

Stabilizers. The types of stabilizers used for plasticized PVC compounds are also effective for the formulation of plastisols. Basic lead carbonate is used extensively and is a very effective heat stabilizer but does not impart much light stability; tribasic lead sulfate and dibasic lead phosphite are effective and give better light stability;

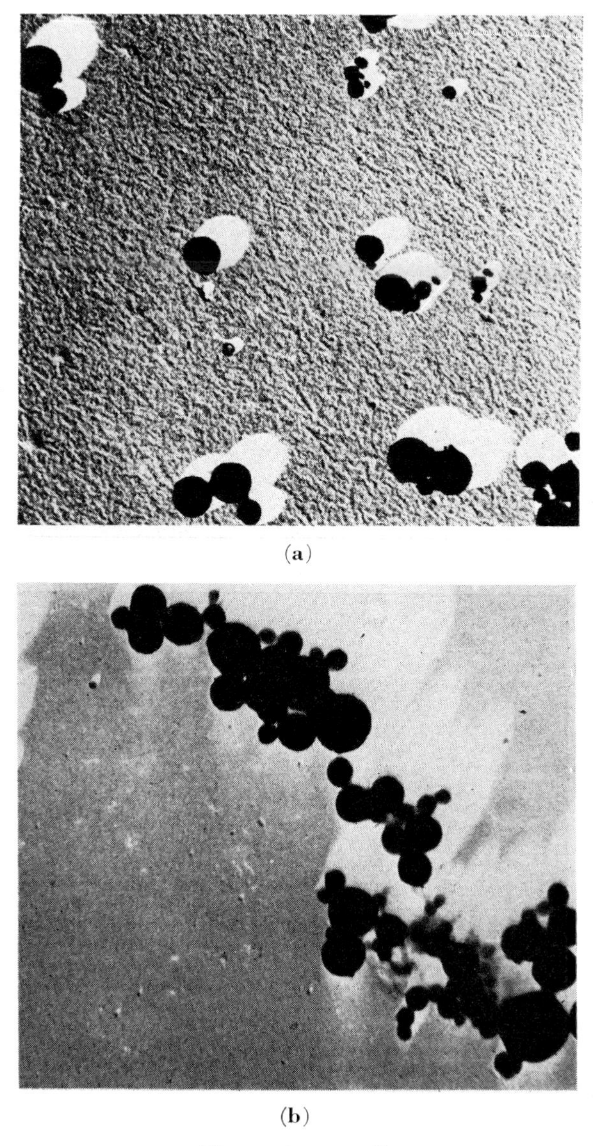

(a)

(b)

Fig. 10 *(continued)*

the quantities used are normally 3–5 parts per hundred of polymer. The use of lead stabilizers is limited because they give opaque products and are not suitable for food applications or other uses where a toxic hazard may arise. For such applications certain calcium–zinc or octyltin stabilizers should be used, especially in conjunction with 2–3 phr of an epoxy plasticizer to impart greater heat and light stability. An effective improvement in light stability can be obtained by the addition of an ultraviolet-radiation absorber such as 2-(2′-hydroxy-5′-methylphenyl)benzotriazole; 0.2 phr is normally sufficient.

Fillers (qv). Fillers are frequently incorporated in plastisols to improve the weathering characteristics, to reduce surface tack, and to increase the hardness of fused films. The most commonly used filler is calcium carbonate, either as precipitated

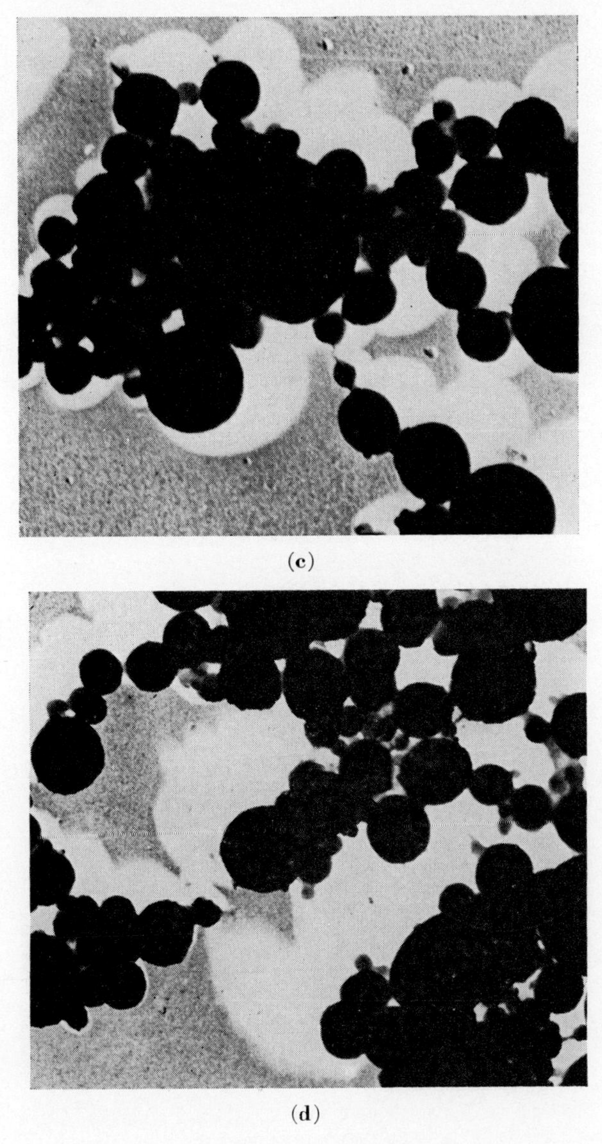

(c)

(d)

Fig. 10 (*continued*)

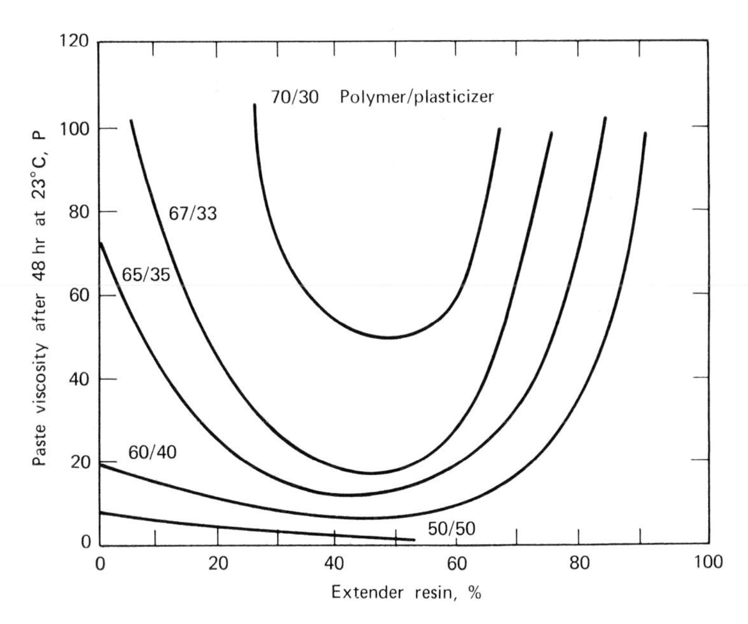

Fig. 11. Viscosity of blends of paste and extended resins.

characteristics, such as a slightly lower fusion point. This effect is obtained only with polymers composed of spherical particles with low oil-absorption characteristics; with the addition of resins that readily form powder mixes, the paste viscosity increases and continues to rise on aging. The variation of viscosity obtained by the blending of an extender polymer with a paste resin at various plasticizer concentrations is shown in Figure 11; this practice has proved useful in spreading applications, in rotational casting, and in spraying.

Viscosity Modifiers. Viscosity modifiers are used when a given set of physical properties is required in the final product but the viscosity of the initial plastisol does not meet the requirements of the conversion process. In practice, a lower viscosity is more frequently required, and a number of viscosity depressants are available. A typical product of this class is hexylene glycol. Although a lower initial viscosity is obtained, some depressants lead to a more rapid increase on aging so that the effect is nullified after a certain time.

Other Types of Plastisols. The greater part of paste-making resins is used in the type of formulations described above. Other types of dispersions are also encountered, and are generally referred to as organosols, rigisols, or plastigels.

Organosols are pastes in which the liquid phase consists of plasticizers and volatile diluents. Normally the volatile component is a nonsolvent for the polymer and forms only a small portion of the total composition. The formulation of organosols depends on finding the right balance between the plasticizer and the organic solvent used; this has been studied extensively (14). *Rigisols* are plastisols that are so compounded that the fused end products are hard and inflexible at normal temperatures. This effect is achieved by using plasticizers that give low-viscosity plastisols, so that the quantity can be kept to a minimum, and also by adding volatile components that are nonsolvents for the resin. *Plastigels* are puttylike plastisols which can be molded at ordinary pressures and retain their shape during fusion.

Unplasticized, or Rigid, PVC

The formulation of rigid PVC compounds is determined largely by the equipment in which they are processed; they may be used in the form of granules or chips, produced from fused compounds made in compounding equipment such as the Banbury mixer, or used as powder blends produced in a high-speed mixer. The latter requires more critical formulation than is necessary when material is completely fused before being fed to the processing equipment. The basic ingredients for both forms of the compound are the same: polymer, stabilizer, lubricants (internal and external), and pigments.

If special properties are required during processing or special physical properties in the finished product, processing aids and impact improvers may also be included.

Polymer. Both the final processing technique and the ultimate application will determine not only the type of resin to be used, ie, homopolymer or copolymer, but also its molecular weight. The copolymers have better flow properties and lower softening temperatures than the homopolymers, so that they are used in applications such as phonograph records, where the melt viscosity of the composition must be low during molding. The vinyl ether copolymers and low vinylidene chloride copolymers (4%) are used extensively for making glass-clear rigid sheet. The homopolymers are used wherever possible because they are lower in price and give better physical properties to the finished product. Developments in processing equipment over the years have permitted an increasing use of homopolymers at the expense of copolymers. This tendency has been more pronounced in Europe than in the United States, where copolymers, eg, vinyl chloride–propylene copolymers, are more widely used.

PVC is available in a wide range of molecular weights in order to satisfy the various fabrication techniques; for calendering or pipe extrusion, a polymer with a number-average molecular weight of about 50,000 (K value of 65) will give optimum properties; for sheet extrusion, a resin of number-average molecular weight of about 45,000 ($K = 60$) is preferred; for injection molding or blow molding, a resin of about 40,000 ($K = 55$) is needed.

If the compound is supplied as granule or chip, then the particle size and shape do not have a significant effect, except in that the amount of resin that can be fed into the compounding equipment depends on the bulk density. When using the compound as a powder blend, the choice of the resin is much more important (15). Emulsion resins are not used for powder blends because they have a small particle size and are too light (see also the section on Manufacture). Suspension resins are used widely; they can be obtained in a range of particle sizes and shapes with varying fusion characteristics. Bulk-polymerized PVC is composed of porous particles that can be readily impregnated by stabilizers and lubricants and are particularly suitable for producing powder blends. The interaction between the polymer and the additives takes place in the final processing operation so that each component plays a critical part, and it is difficult to generalize about the effect of the various ingredients. This means that the formulation of a powder blend is specific to the process and, indeed, sometimes to the particular item of equipment being used. When a powder blend is used for extrusion, the packing density of the polymer should be as high as possible, consistent with good powder flow and gelation behavior. The gelation properties can be determined using the Brabender plastograph, and from the results obtained the polymer can be selected to suit the particular item of equipment for which the compound is being designed.

Stabilizers. The same types of stabilizers are used for rigid compounds as for plasticized, and the choice is largely determined by the final application. Lead stabilizers are used extensively for pipe extrusion and profiles. There are various legal requirements throughout the world regarding the use of lead stabilizers for pipe for conveying potable water; in some countries they are not permitted, whereas in others (British Standard 3505) they can be used so long as the extractability does not exceed a certain value.

Where clarity is required, then, the organotin stabilizers and the tin mercaptides are generally used. If the final product is to be used for food-packaging applications, stabilizers such as di(*n*-octyl)tin maleate are used (16) because the dibutyltin derivatives are more toxic and would be unacceptable at the level of migration usually observed with this class of stabilizer. The tin mercaptides are also unsuitable for food contact because of their disagreeable odor. The dibutyltin stabilizers are more effective stabilizers and are used for applications where good outdoor performance is required.

Stabilizing systems based on mixed salts of calcium and zinc are finding increasing use for rigid compounds for bottles, in spite of their inferior stabilizing properties, because their residual odor and toxic properties are lower. Amino crotonic esters have also been used for PVC bottles because of their low toxicity, but they tend to leave a residual odor that is not acceptable for many foodstuffs. The increasing use of rigid PVC for building applications has required the development of compounds with improved weathering characteristics. Exposure studies on clear sheet show that the thio-organotin stabilizers used in conjunction with an ultraviolet stabilizer give the best results (17). It has been claimed, however, that a combination of a barium–cadmium soap, a phosphite chelating agent, an epoxy plasticizer, and an ultraviolet-radiation absorber gives a rigid compound with excellent heat and light stability (18). Conflicting results obtained with organotin stabilizers suggest that dibutyltin maleate gives better results than a dibutyltin bismercapto ester derivative, especially when used in conjunction with an ultraviolet-radiation absorber. When pigmented products are used, the addition of titanium dioxide greatly improves the weathering properties and reduces the falling off of physical properties, particularly impact strength.

Lubricants (qv). Lubricants in rigid PVC have a decisive influence both on the behavior during processing and on the properties of the finished product. The primary requirements of a lubricant are to improve the flow properties of the fused compound, and to help movement through equipment by lubrication of the interface between the PVC and the metal surfaces.

Lubricants are generally divided into two classes, internal and external (19). An internal lubricant should be easily compatible with PVC so that it can reduce the melt viscosity of the compound, improve its flow characteristics, and thereby reduce the frictional heat developed by shear forces. An external lubricant must act as a lubricating layer on the surface of the PVC where it comes into contact with the metal surfaces of the processing equipment. It should therefore have only limited compatibility with the PVC, particularly at processing temperatures, so that it exudes readily to the surface where it should form a continuous film. Studies have been carried out by several workers in an attempt to determine the efficiency of lubricating systems by measurements of their physical properties (20,21).

Among commonly used external lubricants are synthetic waxes; fatty esters, ethers, and alcohols; low-molecular-weight polyethylenes; and lead stearate. Internal lubricants include long-chain fatty acids, calcium stearate, alkylated fatty acids, and long-chain alkyl amines.

All lubricants should be in readily dispersable form and should be thoroughly premixed for optimum efficiency. Although the above classification is generally recognized, it has been shown that the majority of lubricants exhibit both internal and external behavior (22). In selecting the appropriate lubricant system, account must be taken of the other ingredients. The organotin and nonmetallic stabilizers have little lubricating effect, so that the addition of both types of lubricant is required. Metal stearates are both stabilizers and lubricants, thus requiring little further lubrication. The barium–cadmium soaps introduce a certain amount of internal lubrication, so that in these cases only a small amount of external lubricant is needed. The exact lubrication needed for a specific application can be determined only by experience and no publication on the subject can do more than act as a general guide.

Fillers (qv). Fillers are used in rigid compositions in order to reduce the cost of the product and, in some instances, improve physical properties. The materials generally used are those used for plasticized compounds and for plastisols: calcium carbonate; whitings, calcites, and dolomite; china clay; titanium dioxide; calcium metasilicate; and barium sulfate.

The wide range of products based on calcium carbonate have proved useful; the use of precipitated and coated natural whitings has not only speeded the processing of unplasticized compounds but has enabled the impact strength to be improved substantially through use of loadings of up to 15%. Titanium dioxide is used where opacity and optimum brightness are required; this filler also improves the UV stability of rigid compounds and is used where good outdoor performance is specified (23). The technique of reinforcing resins by inclusion of glass fibers has now been extended to rigid PVC (24); 25% by weight of glass fiber in a rigid PVC compound increases the ultimate tensile strength from about 8,000 to 15,800 psi when determined at a strain rate of 0.05 in./min, while at a high strain rate (400 in./min), the tensile strength increases to 25,500 psi. The use of long fibers generally results in lower ultimate tensile strengths than does that of shorter fibers.

Pigments (qv). The pigments and colorants used in rigid PVC compounds are generally the same as used for flexible PVC compounds. Since the compounding temperatures are higher (190°C compared to 140–160°C), however, the pigments are required to be more heat resistant. The extent to which the pigments migrate depends on their compatibility with the other components of the system, so that materials that are satisfactory in this respect for plasticized compounds may not be suitable for rigid products, and vice versa.

Since there is an increasing usage of unplasticized vinyl compounds for outdoor applications, the use of pigments that are stable to UV light is important. In practice, where products are subjected to long-term weather exposure, it is very difficult to obtain pigments that do not fade in due course; the color range is therefore limited to white and gray, which can be obtained by use of blends of titanium dioxide and carbon black. The use of carbon black alone gives a considerable improvement in UV stability of rigid compounds, so that the addition of separate UV stabilizers may not be required. When such black compounds are exposed to the sun, however, the temperature buildup can be considerable and can have an adverse effect on the performance of the material.

The final choice of pigments and colorants will depend on the formulation of the compound and the long-term performance to which it is to be subjected.

Other Additives. Antioxidants are not used in rigid compounds because they are included normally to prevent degradation of the plasticizer. The same generally applies to fungicides because attack by microorganisms is directed at the plasticizer component. Certain lubricants in rigid formulations may be susceptible to attack under very hot and humid conditions and some protection may be required. The classes of compounds used for plasticized compounds may also be effective with rigid compositions.

Ultraviolet stabilizers are used extensively in rigid compounds because of the increase in building applications (qv) for this type of material, and again the compounds mentioned previously have proved effective. Titanium dioxide, in amounts of up to about 10%, acts as a very effective UV stabilizer so that white rigid vinyls exhibit good light stability outdoors. Flame retardants are not used, normally because flammability is not a problem. The buildup of antistatic charges can be a serious problem and applications using rigid vinyl film require the inclusion of antistatic agents. In phonograph record compounds, quaternary ammonium compounds are used in amounts of up to 1.5%. The inclusion of carbon black helps also to reduce static buildup.

Impact Improvers. Ordinary rigid PVC compounds tend to fail under impact with brittle fracture (see IMPACT RESISTANCE); this has limited their use in certain critical applications and resulted in the development of a wide range of impact improvers. Many of these impact improvers have a glass-transition temperature below room temperature; ie, they are rubbery, and when added to the PVC compound in concentrations of about 5–10% they increase the (Izod) impact strength up to about 20 ft-lb/in. notch. This improvement in impact strength is achieved by increasing the ductility of the material, ie, by reducing the modulus. The tensile strength is reduced from about 8,000 to 6,000 psi and the elongation at break usually increases by a factor of two. The chemical resistance of the modified compounds is not as good as that of ordinary rigid compounds. Experimental evidence available on rubber-toughened systems indicates that, in order to produce an improvement in impact strength, the rubber must be present as a separate phase of discrete particles. A rubber that is completely compatible acts as a plasticizer and gives some increase in ductility but not to the same extent as one with only limited compatibility. It has been suggested that in two-phase systems the rubber particles tend to absorb some of the propagation energy of a crack and thus limit its growth in the material (25). Impact improvers are available with widely varying compositions, but acrylonitrile–butadiene–styrene (ABS) terpolymers are the most widely used (see also ACRYLONITRILE POLYMERS). The relative merits of a number of rubbery impact improvers have been discussed (26). The types of modifiers which are at present commercially available include acrylonitrile–butadiene–styrene (ABS) terpolymers, methyl methacrylate–butadiene–styrene terpolymers, chlorinated polyethylene, ethylene–vinyl acetate copolymers (EVA), vinyl chloride–EVA graft polymer, polyethylene, and copolymers of vinyl chloride with octyl acrylate or octyl fumarate. Besides these products there are a large number of other compositions covered by patents.

The acrylonitrile–butadiene–styrene terpolymers are well established as impact improvers (27) and a concentration of only 5% can produce good impact properties. The final product is slightly yellow because of the acrylonitrile content of the terpolymer, and the majority of this type of modifier give a hazy effect because of the difference in refractive indexes of the two components. For this reason the methacrylate–styrene–butadiene terpolymers have found increasing use, particularly in applications where good color and clarity are required; such properties are particularly useful for compounds specially designed for blow molding rigid bottles. The impact improvement obtained with the methyl methacrylate terpolymer is less than with the acrylonitrile terpolymer, but the advantages of better clarity and color outweigh this disadvantage.

In Europe, chlorinated polyethylene is used extensively as an impact improver and, because of its excellent light stability, has gained wide acceptance in compositions used for external building applications (28). The impact improver is not sold as a separate additive but as a component of a granulated compound, whereas in the United States a chlorinated high-molecular-weight polyethylene containing 40% chlorine by weight is available separately. The quantity of chlorinated polyethylene required to give an impact strength (Izod) of about 15 ft-lb/in. is about 10–15%, whereas this value can be obtained with smaller amounts of ABS modifiers. See also ETHYLENE POLYMERS.

Copolymers of ethylene and vinyl acetate are also claimed to improve impact strength when used in blends with PVC. The compound has good processing characteristics combined with very good mechanical properties and excellent weathering resistance. A further development in this field has been the introduction of a graft polymer based on vinyl chloride with 10% of an ethylene vinyl acetate copolymer, which has proved to be particularly useful for rigid compounds for building components and also for blown bottles. A graft polymer containing 30–50% of the ethylene–vinyl acetate copolymer is claimed to have the same flexible characteristics as a plasticized compound.

Additions of 1–5% of very fine powdered polyethylene are claimed to give good flow and impact properties in rigid and semirigid PVC formulations. The polyethylene is made up of spherical particles less than 30 μ in diameter and is said to form a completely homogeneous mixture with the PVC, acting as a lubricant for processing and as an impact improver. This type of impact modifier is claimed to produce the same increase in impact strength when used in 1% concentration as would be obtained with 5% of a more usual type.

The use of an impact modifier frequently necessitates the addition of a processing aid to improve the processing characteristics of the component and help in the dispersion of the ingredients. These products enhance flow behavior during compounding and give better surface smoothness and gloss to the finished product without affecting the hardness; they also make it possible to operate the equipment at a faster rate. They also improve the hot strength of rigid vinyl, a particularly beneficial effect when extruding sheet for vacuum forming. The type of compound used for this purpose is sometimes a styrene–acrylonitrile copolymer or an acrylic copolymer. The quantities used in the compound are usually about the same as for those of the impact modifier.

The impact strength can be improved by copolymerization of vinyl chloride with octyl esters, such as octyl fumarate or acrylate; this has been referred to above. The effectiveness of the comonomer depends on the production of a heterogeneous copolymer which contains components high in ester content and with rubbery characteristics. The addition of about 20% of 2-ethylhexyl acrylate raises the impact strength from 4.5 to 20 kg/cm^2. This technique has the disadvantage that the softening point is reduced rather more than when 10–15% of an external impact modifier is added. The Vicat softening point of the above copolymer is about 42°C, as compared with 77°C for a high-impact grade containing 10% of ABS modifier.

Table 3. Effect of Impact Modifier on Properties of Rigid Poly(vinyl Chloride)

Property	ASTM test	Type I, normal impact	Type II, high impact
flammability	D 635	self-extinguishing	
impact strength (Izod), ft-lb/in. notch	D 256	0.6	18.0
tensile strength at yield, psi	D 638	7000	6000
elongation at yield, %	D 638	2.8	2.8
tensile modulus, psi	D 638	430,000	390,000
hardness, Rockwell R	D 785	115	110
heat distortion temperature under load (264 psi), °F	D 648	166	160
specific gravity	D 792	1.40	1.36

The properties of rigid compounds are given in Table 3 below; type I is the unmodified material and type II contains an impact modifier. The values can be varied according to the other ingredients in the formulation and are also dependent on the amount of processing (29).

Typical formulations which give these properties are as follows:

Normal impact

	parts by weight
PVC (*K* 61–64)	100
tribasic lead sulfate	3.0
dibasic lead stearate	2.0
acrylic copolymer	3.0
calcium stearate	0.5

High impact

	parts by weight
PVC (*K* 68–70)	100
acrylic copolymer	2.0
ABS impact modifier	5.0
tribasic lead sulfate	3.0
dibasic lead stearate	0.8
calcium stearate	1.5

Techniques

In order that PVC be processed satisfactorily, it is necessary that all the additives, ie, plasticizers, stabilizers, lubricants, etc, be fully dispersed to yield a homogeneous material. This can be done in a variety of ways and by using any of the several types of equipment available. The compound may be powdered in the form of a free-flowing powder or as a granular product; the latter can be obtained only by heating and mixing the ingredients above the fluxing temperature, after which the compound can be converted into granules or pellets. The compounding process is complicated and has not yet been fully analyzed; however, the following separate processes can all take place:

(*i*) Simple mixing of all the ingredients (resin, plasticizers, stabilizers, lubricants, etc).
(*ii*) Absorption of liquid ingredients onto the surface layer of resin particles.
(*iii*) Complete plasticization of the solvated resin particles.
(*iv*) Cohesion between plasticized resin particles.
(*v*) Loss of identity of the individual particles.
(*vi*) Chemical interaction of the polymer with some of the ingredients (eg, stabilizers).
(*vii*) Degradation of polymer by action of heat and shear forces.

The manufacture of powder blends involves only the first three stages, and stage *vi* to some extent. The manufacture of a completely fluxed compound involves all the stages enumerated and, because the temperatures that the material reaches are much higher, the extent of chemical interaction between the polymer and the stabilizer is greater than in the manufacture of a powder blend. None of these various processes has been considered theoretically, although the simple mixing operation has been examined by a number of authors (30–35). The interaction between the polymer particles and the plasticizer has been the subject of a number of interesting papers (36–47). During the compounding operation all the above processes take place simultaneously, although simple mixing must be substantially complete before any appreciable temperature rise occurs. This is necessary to ensure that the stabilizer is completely dispersed throughout the resin so that excessive local degradation does not occur when the mixture comes into contact with the hot surface of the equipment. Complete and uniform distribution of the plasticizer before fusion occurs is no less important; otherwise an excessive amount of unplasticized particles may remain dispersed in the fused mass, and these become even more difficult to disperse uniformly. This condition is particularly responsible for the presence of "fish eyes" in the flexible products (see the discussion of "fish eyes," p. 376).

It has been general practice with both flexible and rigid compounds to use techniques involving dispersion of the ingredients in a mixer, and then to complete the fusion on a two-roll mill from which the material is stripped and granulated; this granulated compound is then used as the feedstock for subsequent extrusion and molding operations. The technique has the disadvantage that the PVC compound goes through two separate heating cycles and must be stabilized accordingly, although it is still used extensively by manufacturers who produce compounds for sale to fabricators not equipped to do their own compounding. During recent years, the introduction of suspension resins with high porosity has made it possible to make powder blends; these are free flowing and can be fed directly to fabrication equipment. This technique can be used with compounds containing large amounts of plasticizer so long as the resin has a sufficiently high porosity to give a free-flowing powder that does not stick and cause bridging in the feed mechanism of the equipment. Powder blends have economic advantages over granular compounds because the compounding process is less expensive to operate, and because the reduced heat history allows the use of smaller amounts of stabilizer.

A wide range of equipment is now available for compounding PVC; this is indicative of the rapid developments that have taken place in this branch of PVC technology. The majority of such compounding equipment requires the feed material to be in the form of a dry blend containing all the ingredients of the composition. This preblending can be carried out in simple jacketed internal mixers, such as ribbon or tumble blenders and dough-type mixers. The resin is charged to the blender and heated to 80–100°C with agitation; the mixed plasticizers are heated to about the same temperature and then added slowly through a spray head. This may take up to 20 min, depending on the amount of plasticizer being used. The elevated temperatures improve the rate of solvation of the resin by the plasticizer, and help to remove any traces of moisture that had been present. The stabilizer and colorants are added, preferably as dispersions in the plasticizer. This is followed by other dry ingredients, such as fillers. When the plasticizer has become fully absorbed and the mixture has dried up, the lubricant is dispersed in plasticizer and mixed thoroughly. The plasticized blend is then cooled below 60°C and screened before it passes to the next stage of the compounding process.

This procedure must be varied according to the ingredient; for example, polymeric plasticizers require higher mixing temperatures because of the slower rate of solvation of the polymer by the plasticizer. The quality of the premix depends on careful control of the operating procedure and ensuring that adequate mixing is obtained, by not making the additions too quickly. If the blends are not used immediately after mixing, then the temperature must be reduced to not more than 30°C because it is difficult to dissipate the heat of the material stored in bulk, and degradation of the resin can occur.

Although this blending process is slow, consistent products can be obtained with no restriction on the types of resin used. See also COMPOUNDING.

Two-Roll Mill. This piece of equipment is still the standard for rubber compounding and it was logical that it should be the first to be used for compounding PVC. Two rolls, each measuring 16–24 in. in diameter and up to 84 in. in length, are arranged in parallel in a horizontal plane with a gap of 1–2 mm between them. The rolls rotate at different speeds, the ratio being about 1.4:1, with the slower roll being located at the front. An important feature is the control equipment that regulates the temperature of the mill rolls by adjusting the flow of steam and of cooling water. The material in the form of a powder blend is fed onto the rolls maintained at temperatures of 150–160°C, and the compound is completely mixed in about 10 min, the exact time being dependent on the batch size. The mixing operation is aided by stripping the banded material from the mill roll and returning it so that it moves along the roll surface and is more effectively kneaded by the shearing action of the rolls rotating at different speeds. The optimum conditions for compounding vary according to the composition of the compound, but the following is an example of general mixing procedure:

batch mixing	depends on mill size
mill temperature	150–160°C
front roll steam pressure	varies between 25 lb/in.^{2} for 40-in. mills to 150 lb/in.^{2} for 84-in. mills
cycle time	approximately 10 min
roll speed	12 rpm front roll, and a higher speed on back roll to give the maximum friction ratio

The compounded material is stripped off as a continuous band after the specified time. If it is to be fed to an extruder, it is then passed to a granulator where the strip of material is cut into small cubes, usually about $\frac{1}{8}$–$\frac{1}{4}$ in. The compound is still warm at this stage, so it is then cooled by an air blast before storage or packaging.

Banbury Mixer. The Banbury mixer was designed originally for rubber compounding but it has proved to be very effective for providing fully fused and completely homogeneous PVC compounds, both rigid and flexible. The mixer consists of a jacketed mixing chamber fitted with two grooved rotors that rotate in opposite directions (see also COMPOUNDING; RUBBER COMPOUNDING). The material is kept in close contact with the rotors by a pressure ram, which maintains pressure on the charge during the compounding operation. The Banbury mixer is available in a wide range of sizes, as listed in Table 4.

The Banbury mixer is very effective for compounding PVC because the combined action of the rotors and the ram produces more vigorous kneading than is obtainable

Table 4. Banbury Mixers

Type	Capacity, liters	Rotor speeds	Kneading capacity, lb/hr	hp
00	5	40:45	experimental	15
1A	30	30:35	300–450	50–75
3A	96	30:35	850–1100	100–150
9A	160	30:35	1500–2000	200–300
11	240	21:28, 36:42	2500–3000	600–800

with ordinary open equipment. Considerable frictional heat is generated, and this makes it difficult to control the stock temperature and avoid degradation of the polymer. The operation of internal mixers has been described extensively and the important aspects of their use for the manufacture of PVC compounds have been discussed in the literature (48–50).

The main disadvantage of the Banbury mixer is that it can be operated only on a batch basis, and this can lead to batch-to-batch variations because of possible differences in temperatures produced in the mix. The discharge doors and the glands are very prone to leakage, which occurs particularly with very hard or very soft compounds. A further disadvantage lies in the difficulty of cleaning the equipment, so that if a change of composition is planned, a prolonged downtime for cleaning is required, and even so there is always a possibility of contamination on resuming compounding with the new composition. The mixed material is discharged from the Banbury as a large doughlike mass which must be converted to a sheet for granulation. This is usually done by dropping the charge directly onto a two-roll mill from which the material is stripped after a number of passes through the nip.

Precise processing conditions depend on the composition of the compound. However, those listed in Table 5 will serve as a guide to cycle times and temperatures.

Table 5. Processing Conditions for Banbury Mixer

internal mixer, batch weight	rated capacity of machine × specific gravity of the compound (approximately 125 lb for a 3A Banbury)
temperature of Banbury	
flexible compounds	jacket and rotor 70–90 lb/in.^{2} steam pressure
rigid compounds	jacket 20–30 lb/in.^{2} steam pressure
cycle time	normally 4–7 min
dropping temperature	130–170°C; should be the lowest possible to obtain a fully fused compound
ram	
flexible compounds	normal mix with ram pressure on
rigid compounds	mix with ram floating
mill temperature	
front roll	60–80 lb/in.^{2} steam pressure
back roll	15–30 lb/in.^{2} steam pressure
cycle time	must correlate with the internal mixer to give a continuous flow in production; size of the mill should be sufficient to handle the batch weight within the mixer cycle time; for a 3A Banbury, a 60-in. mill is required
roll speed	approximately 15–20 rpm with a speed differential between the rolls from 1.05:1 to 1.1:1

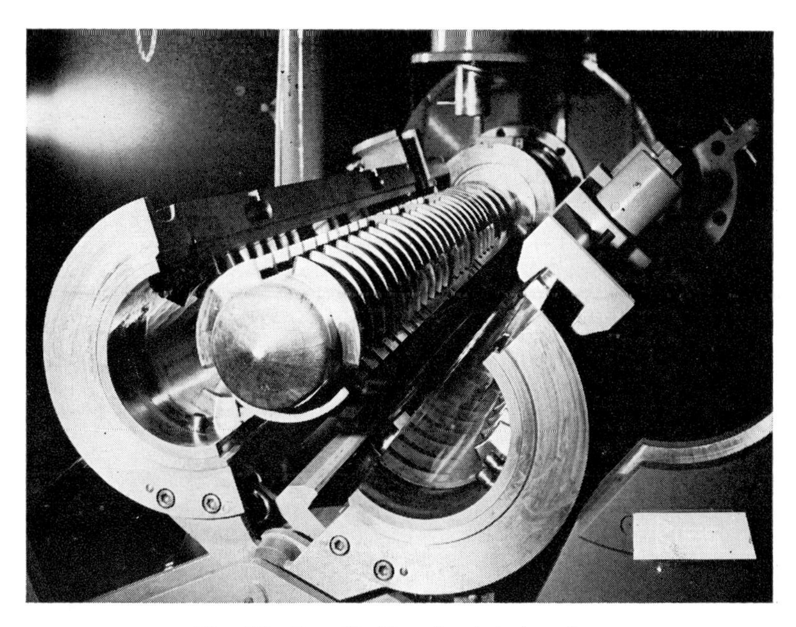

Fig. 12. Buss Ko-Kneader, interior view.

As mentioned previously, the overall mixing time and temperature are dependent on the composition of the compound, but the process is also governed by a number of factors such as the plasticizer absorption characteristics of the resin, its molecular weight, and fusion temperature.

The performance of the Banbury mixer can be improved by the use of *automatic weighing and feeding devices.* The PVC can be fed directly into the mixer by means of a weighing conveyor, which is supplied from a storage bunker. The weighing conveyor is mounted on load cells that transmit signals to the main control panel. The plasticizer can be injected directly into the chamber by pumps, the output of which can be varied. The other additives must be blended either with the resin, in the case of dry components, or with the plasticizer, in the case of liquid components. The program is controlled by an electronic device that determines the feed of resin blend and plasticizer, the operation of the ram and hopper door, the total mixing time, and the discharge door. A counter on the control panel can be set so that a predetermined number of batches can be made.

An *automatic feeding system,* which employs batch bins and a tipping mechanism to feed the Banbury mixer, has also been developed. The plasticizer is injected directly into the mixer by the same pump system as that used above. The other dry ingredients are weighed separately into individual bins, but other liquids would preferably need to be blended with the plasticizers. The system is so programmed that the operator need only ensure that the batch bins are delivered in the correct sequence to the Banbury.

Continuous Mixers. During recent years there have been active developments in the design of equipment for the continuous compounding of plastics, with particular reference to PVC. This type of equipment has considerable advantages over the batch-type mixer in that a more consistent compound can be produced, and generally, because of a shorter heat history the compound can be formulated with lesser amounts of stabilizer. The material does not receive as much kneading as in the batch-type mixer; consequently, special care must be taken to ensure that the feed is completely homogeneous since the working of the compound in a continuous mixer is not sufficient to eliminate any variations. Most continuous compounding machines are basically single- or twin-screw extruders of modified design; the ordinary extruder is rather ineffective and the modifications are aimed to increase the working of the compound.

The Buss Ko-Kneader (51) is a single-screw machine (Fig. 12) in which the worm is interrupted at intervals and the wall of the barrel is provided with three rows of kneading teeth. The characteristic feature of the Ko-Kneader is that the screw, while rotating, also reciprocates, as a result of which the compound undergoes vigorous shearing by the interaction of the specially profiled flights on the screw with the teeth in the casing (Fig. 13). In the usual arrangement, the machine may be heated or cooled in three zones, ie, the middle zone, the outlet zone, and the screw itself, all of

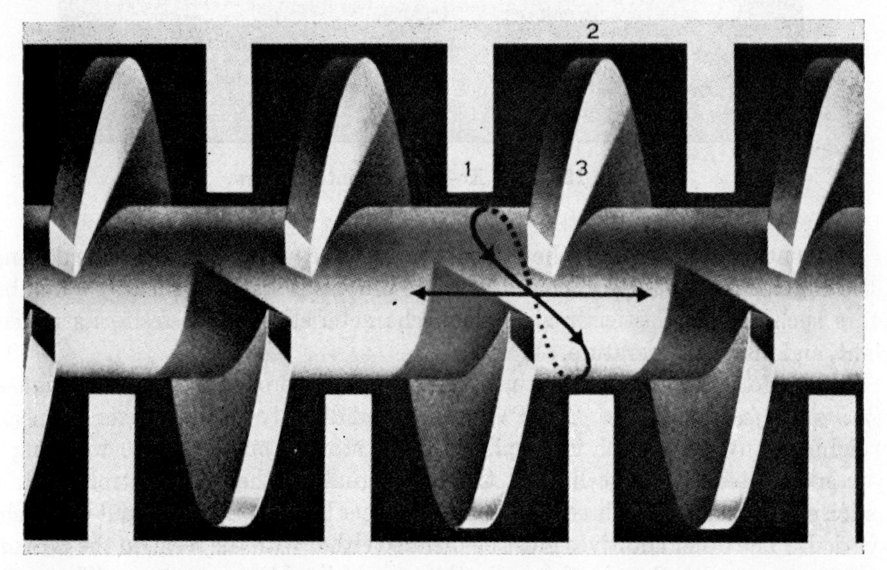

Fig. 13. Buss Ko-Kneader screw. 1, kneading tooth; 2, kneader casing; 3, screw flight.

which may be operated at different temperatures if necessary, according to the requirements of the material. The barrel of the machine is split about the vertical center line and arranged for rapid opening so that the whole interior can be rendered completely accessible for cleaning. The Ko-Kneader fitted with a die plate to extrude a circular or flat section is frequently used for the direct feeding of calenders. If face-cut granulate is required, a short crosshead extruder is fitted to the machine and the compound is extruded through a suitable die plate where it is cut to the required thickness by means of a rotating knife. Both the extruder and the cutting mechanism are supplied with variable-speed drives. The granulate produced when using the die-face cutter is cooled by either air or water and then conveyed pneumatically via a cyclone separator to storage bins. Four main sizes of the PT type of Ko-Kneader, the type suitable for processing PVC compounds, are available with barrel diameters of 100, 140, 200, and 300 mm, and are capable of outputs ranging from 200 to 3000 kg/hr, in addition to which a laboratory model with a 46-mm diameter barrel is produced.

The Werner and Pfleiderer Plastificator (Fig. 14) is another version of the single-screw compounder, in which mixing is effected by friction of the material between a conical rotor fitted with a number of fillets and a corresponding conical zone in the extruder barrel. The degree of compounding can be varied by axial adjustment of the screw. The premix is fed to the main screw by the action of variable-speed screws which convey the material from the hopper. Mixing takes place in the short conical section, from which the compound was originally granulated directly. The time available for complete compounding was inadequate in the early model unless it was

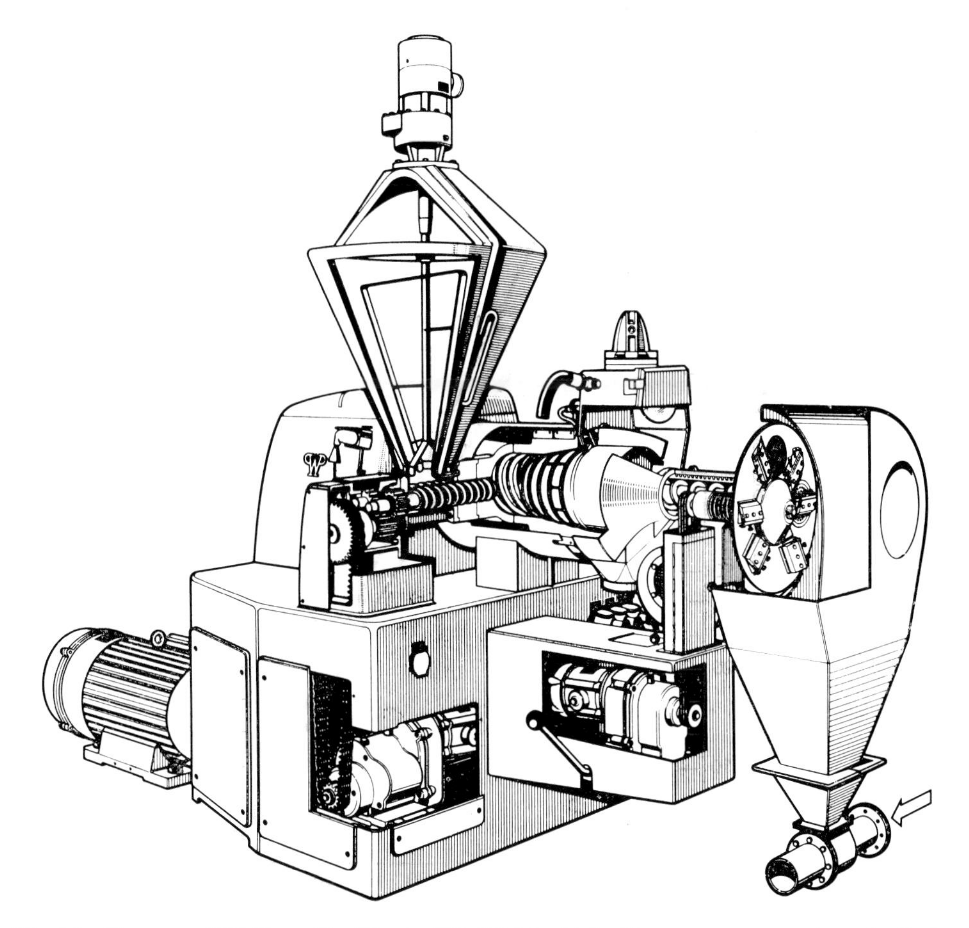

Fig. 14. Werner and Pfleiderer Plastificator PK 400/11 with feed hopper, vacuum unit, and pelletizer.

operated at low outputs, but this has been largely overcome by the provision of a second screw section, to which the compounded material passes as it leaves the conical section, and which extrudes it through a die with a die-face cutter to produce a uniform granule. The plastificator is capable of producing a PVC compound at rates up to about 200 kg/hr. It gives best results on the type of compounds used for cable covering; these are plasticized materials with a softness of 20–40 in the British Standard range. The softness number is part of the British Standard specification for testing plastics. Figure 15 shows the relationship of British Standard softness to the Shore A scale (see HARDNESS).

A recent development of interest is the Schalker Eisenhütte planetary screw extruder (Fig. 16), which incorporates a multiple-screw system in which six or eight screws are arranged in planetary fashion around a single larger-diameter central driven screw (52). The main screw and planetary screw flights are arranged at an angle of 45°. The inside of the barrel is also flighted to the same depth as the screw, which increases the surface area and ensures that as the main screw rotates the planetary screws counter-rotate around it. Because of the very effective mixing and compounding of

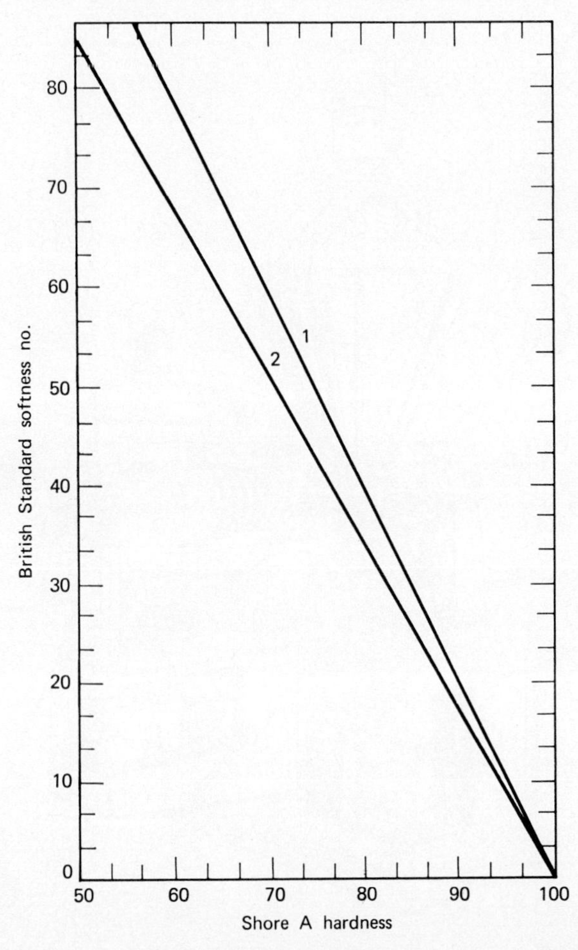

Fig. 15. Comparison of British Standard softness and Shore A hardness. 1, 3-sec indentation (DIN 53505, ASTM D 676). 2, 15-sec indentation (ASTM D 1432). Temperature, 23 ± 0.1°C.

the planetary screws, the plasticizing section of the machine is extremely short. Four models are available; the output of Model P 150 is claimed to be up to 400 kg/hr for rigid PVC and up to 500 kg/hr for plasticized compounds. This machine can also be used for direct extrusion of powder blends if the main central screw is extended to eliminate the turbulence created in the polymer melt by the action of the planetary screws before it enters the die zone. It is not entirely satisfactory for unplasticized or very soft compounds because of the high frictional heat developed in the conical section of the machine.

There is a wide range of single-screw compounders of varying efficiency, although these have been largely superseded by the multiscrew machines.

The Farrel continuous mixer (Fig. 17) has been developed by the manufacturers of the Banbury mixer, which it resembles in its compounding characteristics. The machine is fitted with twin rotors, which are readily accessible for cleaning by moving the mixing chamber forward and rotating it clear of the rotors. The mixer operates on

Fig. 16. Planetary screw extruder (Schalker Eisenhütte).

Fig. 17. Farrel continuous mixer.

a powder blend feed, which is fed on a continuous basis. The temperature of the compounded material can be effectively and accurately controlled over a wide range of feed rates; variations in temperature can be made readily by adjusting the discharge opening by a hydraulically operated mechanism. The machine is available in four sizes, the 4CM model having an output up to 540 kg/hr for plasticized PVC and up to 270 kg/hr for rigid compounds. The largest machine, the 14CM, has corresponding outputs of 10,800 kg/hr and 5,400 kg/hr. This type of equipment does not depend upon tight clearances within close tolerances to provide efficient mixing, and therefore its operation is not affected by relatively small amounts of abrasive or corrosive wear.

The main advantage of the continuous mixer over an automated Banbury mixer lies in its versatility; the output of a continuous mixer can be adjusted over a range of about 10:1, whereas the output of the Banbury can be altered only by varying the number of batches processed. The equipment needed for making the premix for feeding the continuous mixer can be simpler than that required for the batch mixer so that the overall installation is simpler and cheaper to install.

Multiscrew Compounders. The extent to which a twin-screw extruder can act as a compounding unit depends on the interaction between the two screws, and this is determined by the shape of the flights, the degree to which they intermesh, and the relative direction of rotation of the screws. The usefulness of a conventional extruder for the compounding of PVC has led to the development of a number of multiscrew machines designed specifically for this operation. A typical example of such a machine is the Werner and Pfleiderer ZKS 53 twin-screw disc kneader, in which screw sections and kneading discs delivering forward or backward can be mounted on the shafts, which rotate in the same direction. The advantage of this machine is that the screw and kneader configurations can be changed to meet the specific requirments of the material being processed. Different lengths of screw can also be supplied, and the kneading discs are shaped approximately in the form of equilateral triangles. The output varies according to the type of compound being processed; it is 100–200 kg/-hr for unplasticized PVC and 200–450 kg/hr for plasticized material.

The Kombiplast Type 83/200 (Werner and Pfleiderer) is a special machine developed for processing and face-cutting rigid PVC at outputs in the range 340–400 kg/hr. It consists of two separate units which can be operated quite independently of each other; the first unit is a twin-screw extruder mixer and the second is a single-screw extruder which operates at comparatively slow speeds. The transfer of the material from the twin screw to the single screw provides a venting post at neutral pressure.

High-Speed Mixers. The compounding equipment described above is used to convert the blend of PVC and additives into a compound, which is then shaped into cubes or some other form. This was always regarded as essential for the final fabrication process, but now both injection molding and extrusion equipment have been redesigned so that powder blends can be processed satisfactorily. The earlier problems of erratic and low production rates and of poor physical properties of the final product have been overcome. It is essential, however, that the powder blend be made in a high-speed mixer in which the temperature of the mix is raised by the development of frictional heat. Such mixers employ rotors that can be operated at speeds from 500 to several thousand revolutions per minute. The Henschel Fluid Mixer (Henschel-Werke GmbH) is a typical example, and is so called because the rotors impart a fluid-like motion to the contents during the mixing operation (53). It is produced in a

Fig. 18. Papenmeier Universal Mixer.

range of sizes from that suitable for use in the laboratory to one that will mix a batch weighing 100 kg. The action of the high-speed rotors causes the temperature of the contents to rise fairly quickly, and it is the usual practice to raise the temperature of the powder blend to 80–120°C. The time required for this varies according to the composition, and mixing cycles can take from a few minutes up to 20 min. The time required depends on the speed at which the rotors are operated. Complete reliance on frictional heat for raising the temperature of the powder mix can lead to some overheating when larger batches are compounded; this can be overcome by supplying some heat from the outside. The mixers can be supplied with a jacket suitable for the heating medium. The injection of heat through the outside of the mixer allows the

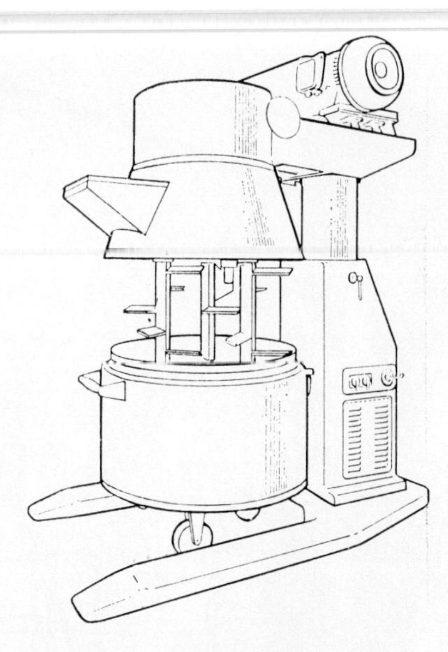

Fig. 19. Vertical paddle mixer.

rotor speed to be reduced so that the frictional heat buildup is less, and the same mixing cycles can be maintained with less danger of overheating. It is now common practice to include a cooling mixer with the high-speed equipment so that the batch can be discharged as soon as it reaches the mixing temperature and cooling is allowed to take place while the next batch is being made up (54).

Other high-speed mixers include the Dierks Diosna R 200 with a working capacity of about 50 kg per batch, the Papenmeier Universal Dry Mixer (G. Papenmeier KG) with a capacity range of 200–350 kg (Fig. 18), and the Vaterland M 160 (Vaterland-Werke) with a working capacity of about 35 kg. The Novax RS 4 and RS 120 (Pentax Maschinen) are mixers with a design different from those described above. The high-speed impeller is mounted on a horizontal shaft and the casing is provided with lugs that can be adjusted to vary the rate at which frictional heat is generated. When this type of equipment is used to produce a powder blend to a rigid formulation, all the ingredients can be added to the resin at the start of the mixing cycle. With plasticized compositions, and particularly with those producing soft compounds, the whole amount of the PVC together with all the additives and half the amount of plasticizer are added to the mixer initially. The temperature rises rapidly as mixing takes place and the plasticizer is absorbed; the point at which this is complete coincides with a drop in the current consumption of the rotor motor. The rest of the plasticizer is then added and the temperature of the powder blend is raised to 120°C, at which stage the plasticizer is completely absorbed and partial fusion has occurred. The extent to which this takes place depends on the speed of the rotors and the rate at which the temperature rises. With slower speeds and longer cycle times, the amount of fused material will be less than at higher speeds of the rotor. It is necessary to establish the optimum conditions for the blending operation according to the quality of the product produced in the final process (55).

Wet Granulation. This is a recent development (56) in the compounding of PVC, and although it is claimed to have substantial advantages it does not appear to

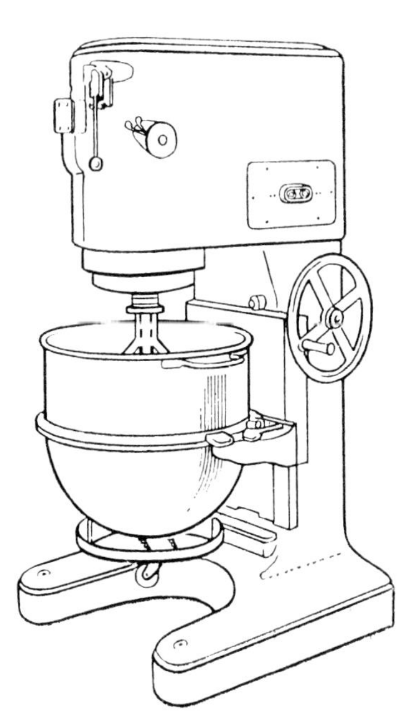

Fig. 20. Vertical planetary mixer.

be used to any great extent. The PVC is used directly from the polymerizer while it is still in an aqueous suspension. A certain quantity of the material is introduced into the granulating vessel, to which part of the other additives, with the exception of the plasticizer, are also added. The plasticizer is heated separately in a vessel to which the remainder of the stabilizer and the lubricants are also added, and this mixture is then fed into the granulator, which already contains the PVC suspension. Under the conditions of addition, the resin particles are rapidly plasticized; the compound is produced in the form of granules in a very short time and these separate out readily from the aqueous suspension. The agglomerates can be filtered easily by means of a screen and the compound is dried either in a drum or fluidized-bed drier. The advantage of the process lies with the drying operation; the granulated compound already contains the stabilizer, so that it must be less liable to degradation at the elevated temperatures required to evaporate the water than is the unstabilized polymer as it dries directly after polymerization. Excessive temperatures must be avoided, however, to minimize the loss by volatilization of the plasticizer. Besides giving a product of improved properties, the process is said to be cheaper to operate because of the reduced energy requirements; however, it is limited to polymer manufacture since it would be uneconomic to redisperse the polymer after it had been dried once.

Mixing of Plastisols. PVC pastes can be produced readily using simple paddle-type mixers, which provide sufficient shear to disperse the polymer completely in the plasticizer without causing any rise in temperature (57). The more simple types of mixer may tend to produce some lumpiness, so further treatment on a single- or triple-roll mill may be needed to disperse the agglomerates. Equipment in most general use is the vertical paddle mixer (Fig. 19), the vertical planetary mixer (Fig. 20), and the horizontal Z-blade mixer (Fig. 21). These mixers are ideal for operation under vacuum, which is essential to avoid inclusion of air in the plastisol; they can handle a

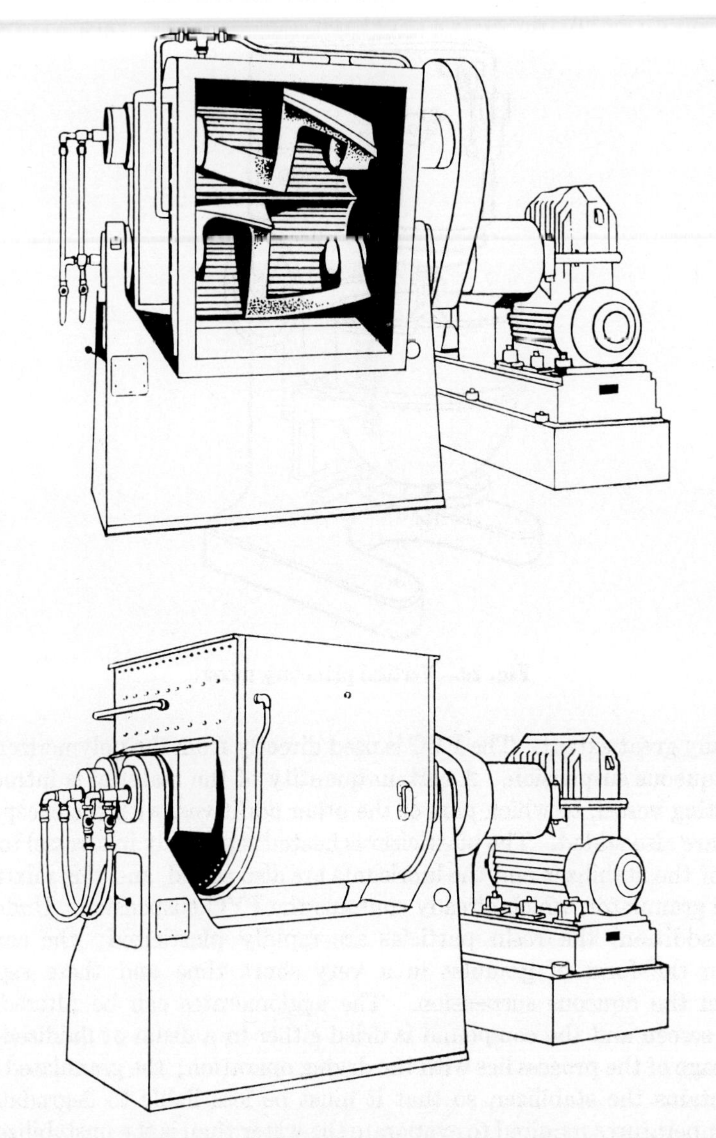

Fig. 21. Horizontal dough or Z-blade mixer.

batch size of up to about 1000 l, and are easily cleaned. The horizontal Z-blade mixer is the most efficient for economical large-scale production, capital costs are fairly low, and the machine can be adapted for automatic operation. High-speed mixers are also used for making plastisols; they are equipped with variable-speed agitators mounted with toothed discs. The mixer can be operated under vacuum. It must be operated with caution, however, because, although the average temperature of the plastisol may be kept below the critical limit during the mixing, some overheating may occur in close vicinity to the agitator disc.

The PVC resin is charged to a paddle mixer and plasticizer just sufficient to form a paste is added. Fillers, stabilizers, and other ingredients in suitably prepared form are then incorporated into the mixture. The viscosity is adjusted by further addition of plasticizer to give the consistency that will produce maximum shear from the blades of the mixer. This is known as the thick-mixing stage and is necessary in order to disperse fully all the ingredients. Care is needed to ensure that the temperature of the plastisol does not rise above 32°C, since higher temperatures cause an increase of viscosity. After mixing at this stage for some time, the remainder of the plasticizer and diluents are added slowly and mixing is continued until a smooth, homogeneous paste is obtained. When a high-speed mixer is used it is usual to charge all the polymer and then add the plasticizer rapidly so that the whole mixing cycle takes place within a short time, usually about two minutes. If the mixing procedure has not been properly carried out, it may be necessary to disperse any agglomerates by passing the paste through a single- or triple-roll mill.

For many applications it is essenial that the plastisol be completely free of absorbed air bubbles; this can be achieved by carrying out the mixing under vacuum. If deaeration is required, various techniques are available, but the ease with which this can be done depends on the viscosity and flow characteristics of the plastisol. If the paste viscosity is low, entrapped air can be removed by allowing the paste to stand for a few hours. Medium-viscosity and thixotropic pastes may require gentle agitation and subjection to a vacuum. For deaerating highly viscous pastes, it may be necessary to pass a thin layer of the paste through an evacuated chamber.

Bibliography

1. W. L. Semon (B.F. Goodrich Chemical Co.), U.S. Pat. 2,188,396.
2. K. M. Bell, B. W. McAdam, and H. J. Caesar, *Kunststoffe* **59,** 272 (May 1969).
3. J. McBroom, *Mod. Plastics* **43** (5), 145 (1966).
4. L. H. Wartman, *Mod. Plastics* **32** (6), 139 (1955).
5. Y. P. Tang and E. B. Harris, *SPE (Soc. Plastics Engrs.) J.* **23** (11), 91 (1967).
6. G. Illman, *SPE (Soc. Plastics Engrs.) J.* **23** (6), 71 (1967).
7. G. Menges and J. Muller, *Plastverarbeiter* **17,** 398 (1966).
8. G. J. van Veersen, *Kunststoffe* **59** (3), 180 (1969).
9. H. C. Murfitt, *Brit. Plastics* **33** (12), 578 (1960).
10. E. T. Severs and J. M. Austin, *Ind. Eng. Chem.* **46** (11), 2369 (1964).
11. F. Bohme and K. Muller, *Plaste Kautschuk* **7,** 124 (1960).
12. I. Tomaszewska, J. Raczynska, and Z. Roszkowski, *Plaste Kautschuk* **15,** 353 (1968).
13. W. D. Todd, D. Esarove, and W. M. Smith, *Mod. Plastics* **34** (1), 159 (1956).
14. G. M. Powell, R. W. Quarles, C. I. Spessard, W. H. McKnight, and T. E. Mullen, *Mod. Plastics* **28** (10), 129 (1951).
15. D. R. Jones and J. C. Hawkes, *Plastics Inst. (London) Trans. J.* **35** (120), 773 (1967).
16. *Plastics World* **26,** 45 (March 1968).
17. L. B. Weisfeld, G. A. Thacker, and L. I. Nass, *SPE (Soc. Plastics Engrs.) J.* **21** (7), 649 (1965).
18. N. L. Perry, *Mod. Plastics* **40** (9), 156 (1963).
19. V. Oakes and B. Hughes, *Plastics* **31** (348), 1132 (1966).
20. U. Jacobson, *Brit. Plastics* **34** (6), 328 (1961).
21. G. J. van Veersen, *Kunststoffe* **59** (3), 19 (1969).
22. B. I. Marshall, *Brit. Plastics* **42** (8), 70 (1969).
23. C. H. Kuist and L. D. Maxim, *SPE (Soc. Plastics Engrs.) J.* **24** (7), 46 (1968).
24. J. E. Theberge and N. T. Hall, *Mod. Plastics* **46** (7), 114 (1969).
25. E. H. Mertz, C. H. Claver, and M. Baer, *J. Polymer Sci.* **22,** 325 (1956).
26. J. E. Bramfitt and J. M. Heaps, *Advances in PVC Compounding and Processing,* Maclaren, London, 1961, p. 41.
27. A. W. Carlson, T. A. Jones, and J. R. Martin, *Mod. Plastics* **44** (9), 155 (1967).
28. J. L. O'Toole, A. A. Reventas, and T. R. von Toerne, *Mod. Plastics* **41** (7), 149 (March 1964).
29. R. Phillips, D. S. Dixler, and C. A. Heiberger, *SPE (Soc. Plastics Engrs.) J.* **21** (11), 1305 (1965).

30. E. C. Bernhardt, ed., *Processing of Thermoplastic Materials*, Reinhold Publishing Corp., New York, 1959, Chap. 3.
31. P. V. Danckwerts, *Chem. Eng. Sci.* **2,** 1 (1953).
32. P. M. C. Lacey, *J. Appl. Chem.* **4,** 257 (1954).
33. W. D. Mohr, R. L. Saxton, and C. H. Jepson, *Ind. Eng. Chem.* **49,** 1855 (1957).
34. R. S. Spencer and R. M. Wiley, *J. Colloid Sci.* **6,** 133 (1951).
35. J. F. E. Adams and A. G. Baker, *Trans. Inst. Chem. Engrs. London* **34** (1), 91 (1956).
36. A. Hartmann and F. E. Glander, *Kolloid-Z.* **137,** 79 (1954).
37. A. Hartmann, *Kolloid-Z.* **142,** 123 (1955).
38. J. F. Ehlers and K. R. Goldstein, *Kolloid-Z.* **118,** 137 (1950).
39. G. Beck, *Kunststoffe* **42,** 37 (1952).
40. P. Schmidt, *Kunststoffe* **41,** 23 (1951).
41. P. Schmidt, *Kunststoffe* **42,** 142 (1952).
42. J. F. Ehlers, *Chem. Tech. (Berlin)* **1,** 138 (1949).
43. W. Birnthaler, *Kunststoffe* **38,** 11 (1948).
44. F. Wurstlin, *Kolloid-Z.* **113,** 18 (1949).
45. F. Wurstlin and H. Klein, *Kunststoffe* **42,** 445 (1952).
46. T. Shiramatsu and N. Ueda, *Nippon Gomu Kyokaishi* **31,** 97 (1958).
47. A. Cittadini and R. Paolillo, *Materie Plastische* **26** (3), 219 (1960).
48. H. S. Bergen and J. R. Darby, *Ind. Eng. Chem.* **43,** 2404 (1951).
49. R. Hammond, *Trans. P.I.* **26,** 49 (1958).
50. B. S. Dyer, *Trans. P.I.* **27,** 84 (1959).
51. *Plastics* **34,** 378,329 (April 1969).
52. *Brit. Plastics* **40** (4), 104 (April 1967).
53. Wick and Koenig, *Plastics* **46,** 583 (1956).
54. *Intern. Plast. Eng.*, **6** (1), 14 (Jan. 1966).
55. W. Brehm, *Kunststoffe* **58,** 337 (1968).
56. *Kunststoffe* **58,** 750 (1968).
57. D. O. Senft, *Chem. Process Eng.* **50** (12), 78 (1969).

C. A. Brighton
BP Chemicals International Ltd.

FABRICATION

The conversion of both plasticized and rigid PVC into finished products can be carried out by the usual processing techniques, such as extrusion, injection molding, compression molding, blow molding, and calendering. Plasticized compounds can be processed more easily than rigid products because the reduced softening point makes it possible to use lower temperatures, 140–160°C, compared with 185–190°C. The high processing temperature for unplasticized compounds requires that equipment be designed so that the material flows through without holdup; otherwise stagnation results in thermal decomposition, which will be evident by partial discoloration of the product and in more severe cases, by an explosive generation of hydrogen chloride.

Extrusion

Both flexible and rigid vinyl formulations can be extruded into a variety of shapes, including profiles, tubes, and sheet, as well as wire sheathing by use of a crosshead die. The compound can be processed from either a powder blend or granulated material. Generally the granulated material can be handled on extruders of shorter screw length; in the case of plasticized PVC an extruder with a screw of 16:1 length-to-diameter (L/D) ratio is satisfactory, whereas if powder blends are being extruded, equipment with a screw of 20:1 L/D ratio is required for pellets and 25:1 for powder. Single-screw machines are normally used for plasticized compounds, the compression ratio being 1.5:1 to 2.5:1 for granules and 2.5:1 to 4:1 for powder blends. Rigid compounds can also be processed on single-screw extruders, but with increasing use of powder blends higher outputs of better-quality extrudates are obtained with multiscrew machines. The twin-screw extruder has been developed extensively for the extrusion of rigid compounds, and is widely used in Europe where it is generally considered to be the best type of machine for this type of material (1); the single-screw machine, however, is still favored in the United States. The temperatures at which extrusion is carried out depend on the type of compound being processed, and can be determined only by careful experimentation; to obtain the best physical properties in the finished product, the temperature of the compound as it leaves the die must be maintained at its correct value. The temperature gradient on the extruder must be adjusted so that the head temperature approximates the temperature at which the compound emerges from the breaker plate. The quality of surface finish is determined by the correct adjustment of head temperature, which should correlate as closely as possible with the temperature of the compound as it passes through to the die. Typical extruder temperatures for processing vinyl compounds are given in Table 1.

Table 1. Typical Extruder Temperatures for Poly(vinyl Chloride)

	Temp, °C
Plasticized compounds	
feed end of screw	120–140
front end of screw	140–160
head	150–170
die	160–180
Unplasticized (rigid) compounds	
feed end of screw	140–150
front end of screw	155–165
head	165–175
die	170–190

The temperatures used for powder blends are at the upper end of the scale, and those for granulated compounds at the lower end. Since higher-compression-ratio screws are used for powder blends, there is an increase in back pressure, which will increase the shear forces; this in turn will lead to faster fusion and greater development of heat. The control of temperature along the barrel may become a problem and it is not unusual for the frictional heat developed along the screw to be sufficient for maintaining the temperatures listed. The screw speed determines the temperatures that develop along the screw and if it is maintained at too high a value, it may prove impossible to hold the temperatures within the limits given. Cooling the screw by water circulation

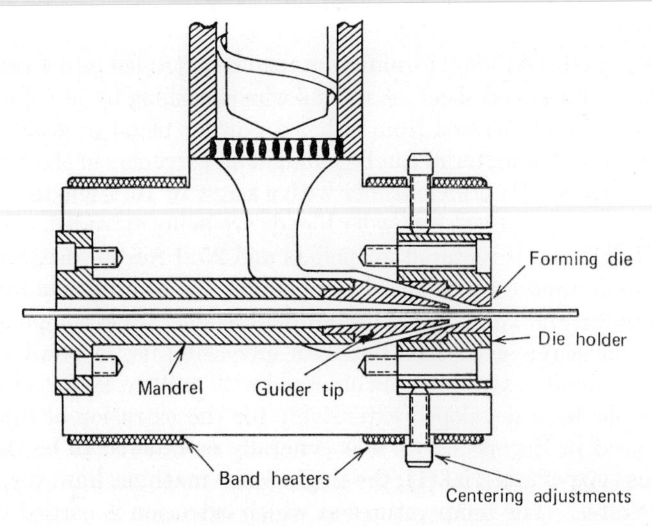

Fig. 1. Crosshead die for wire coating.

increases the shearing effect; this may be necessary to ensure adequate fusion and mixing of the compound in the extrusion of rigid products from powder blends.

Dies for extruding any vinyl compound must be highly streamlined to minimize burning during long production runs; a hard polished chrome plating on the interior gives additional protection against decomposition, improves the surface appearance of the extrudate, and extends the useful life of the die. Unplasticized PVC has a very high melt viscosity at processing temperatures, and therefore it is important that the machinery be sufficiently robust to contain the considerable pressures induced by the high carrying efficiency of the screw. The screw speeds employed for rigid PVC are generally between 5 and 40 rpm but the larger machines operate only between 5 and 15 rpm. Such low speeds require particularly strong bearings and low gearing of the motor drive. Completely variable screw speed is considered a necessity for the extrusion of rigid PVC. See also MELT EXTRUSION. An extensive study of the effect of the various components of a plasticized PVC compound on its extrusion performance has been carried out (2), from which it is evident that for a compound of given softness, ie, fixed polymer–plasticizer ratio, the type and concentration of lubricant play a very important role.

Wire Coating. A significant part of extruded plasticized PVC is used for wire and cable insulation, in which high rates of output are of paramount importance. The wire to be coated passes straight through a crosshead die at right angles to the length of the extruder. The polymer melt enters the crosshead from the extruder and is directed through 90° before it surrounds the wire and emerges through the die (Fig. 1), which controls the final thickness of material around the wire. The temperature of the die must be carefully controlled and the centering of the wire is critical. Since the insulation covering of the electric wire must be kept free from foreign particles that may produce a breakdown of the insulation, a fine meshed screen is used in front of the breaker plate to remove any contaminants.

The wire, which is pulled through the die by action of a capstan, may be preheated electrically or flamed to remove lubricants and to improve adhesion. It is essential to maintain the wire speed within very close limits to ensure that a uniform thickness of

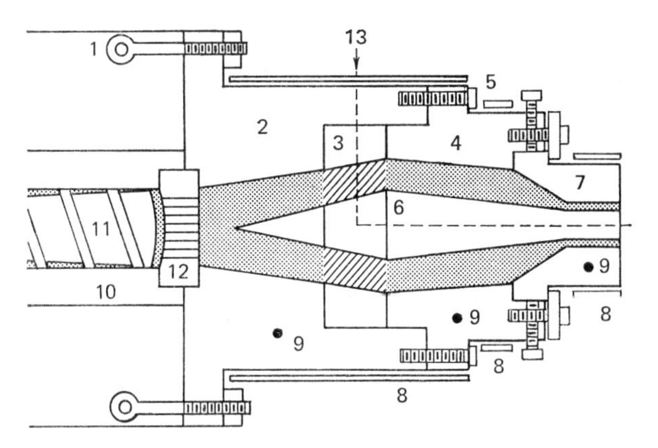

Fig. 2. Suitable head design for the extrusion of rigid PVC in hollow profiles. 1, Six swing bolts; 2, head; 3, spider; 4, adaptor; 5, six bolts; 6, mandrel; 7, die; 8, heater bands; 9, thermocouple; 10, barrel; 11, screw; 12, breaker plate; 13, air-pressure entry.

the PVC coating is maintained. Both granules and powder blends can be used as feed to the extruder, but the latter are prone to produce porosity in the wire covering unless a vented extruder is used (3). The rate at which the wire can be coated satisfactorily depends on the composition of the PVC compound, the surface finish being the controlling factor. See also WIRE AND CABLE COVERINGS.

Rigid Pipe and Tubing. The extrusion of rigid PVC pipe and tubing has contributed in a very large degree to the expansion of the use of PVC. Heavy-walled pipes can be extruded up to a diameter of 1 m and are used for water and gas distribution, for drainage, and for industrial applications. Thin-walled pipes and tubes are used for gutters and downspouts, ventilation ducting, waste disposal, and electrical conduit. The extrusion of hollow profiles can be carried out using either granular or powder compounds. Single-screw extruders are used mainly with granular material, but the economic advantages obtained with powder blends have resulted in a considerable extension of the use of multiscrew extruders (4). In-plant compounding of powder blends with the use of a multiscrew extruder can produce savings in the cost of raw material of 1–3 ¢/lb over buying the granulated compound.

The extrusion of hollow profiles requires a special die with an internal mandrel design as shown in Figure 2. The quality of the extruded product depends on correct die design; one of the main requirements is for a smooth finish with an adequate land length, ordinarily about twenty times the wall thickness of the pipe. Land length cannot be used alone, and another factor to be taken into consideration is that the volumetric capacity of the head should be at least thirty times that of the die land. Crosshead dies may also be used, but careful design of the radius of entry into the die is required to prevent stagnation on the radius which should be as great as possible. For small pipe and profiles, the use of polytetrafluoroethylene-coated dies and mandrels helps to produce high-quality extrusions.

Various cooling and sizing devices are used to adjust the internal and external diameters after the tube emerges from the die. Either pressure or vacuum sizing may be employed; in the former the most satisfactory means of maintaining pressure in a pipe is by use of a floating bung. This allows lengths of pipe to be removed without losing pressure and prevents unnecessary scrap. Vacuum sizing is achieved by applying a vacuum, through separate sections in the cooled sizing die to the hollow section during extrusion. This draws the section against the wall of the sizing die, ensuring that the external dimension is maintained. Two separate vacuum zones are required in the sizing die to maintain the exact dimensions of heavier sections.

Another method of maintaining shape is by using an extended mandrel that is water cooled. After leaving the sizing equipment the pipe passes through a cooling bath; refrigerated water is often necessary for rapid cooling to prevent distortion. The best results in sizing and cooling devices are obtained with minimum friction or drag. The haul-off equipment, which must provide a steady pull, may be comprised of an electrically driven double belt with fully adjustable speed and height.

There is a growing trend to manufacture pipes of larger diameters with greater thicknesses. Machines that can produce pipe of up to 40-in. diameter are capable of a production rate of about 1000 lb/hr; the thickness of the pipe may be up to 2 in.

Blown Film. Blown PVC film, both rigid and plasticized, is obtained by extrusion through a special tubular die in a manner similar to that which is used for polyethylene. Plasticized film must be extruded vertically upward using a right-angle die on a normal horizontal extruder. This die must avoid the occurrence of points of stagnation or regions of decompression. Best results are obtained from dies of "gooseneck" design, in which the compound is extruded through a cylindrical section with a gradual change in direction from the horizontal to the vertical. Air pressure is introduced through the die, increasing the diameter of the bubble up to about three times the diameter at the die. Temperature control is required to maintain a constant film thickness and bubble diameter. The blown tube passes through pinch rolls, ensuring the maintenance of the air pressure within the bubble. A perforated ring for cooling is located above the die, and its position affects the film thickness, which is also determined by the rate of haul-off of the film. Orientation can be introduced into the film by stretching to produce a product suitable for shrink-film applications; this orientation is controlled by adjustment of the blow ratio and the speed of windup (see also BIAXIAL ORIENTATION).

Rigid film can be produced using a similar technique, except that the greater stiffness of the extrudate allows the process to be operated in a completely horizontal setup. This makes die construction simpler in that the polymer melt does not have to travel through 90°, a very important feature in design when processing unplasticized compounds. Rigid PVC films are normally unplasticized, although in Europe there is a large production of film usually described as rigid PVC and containing about 17% plasticizer. Film thicknesses can be as great as 0.02 in. but the usual value is about 0.001 in. For greater thicknesses a flat-sheet die is used; this technique has now been developed for thin film as well (5,6). See also FILMS AND SHEETING.

Sheet Extrusion. There has been a rapid increase in the market for unplasticized PVC sheet owing to its use for building applications. A large portion of the product is produced as transparent corrugated sheet with a thickness of $\frac{1}{16}$–$\frac{1}{8}$ in. by extrusion from a flat-sheet (7) or circular die, the tube in the latter case being cut at the die exit and opened out into a flat sheet. After the sheet has been extruded, it is corrugated by passage through forming rollers. Both single- and twin-screw machines can be used satisfactorily for the extrusion with flat-sheet dies, whereas circular dies are normally employed only with a twin-screw extruder. Best results are obtained from compounds using PVC resins with a K value of 55–60; it is now usual to extrude directly from dry blends using high-bulk-density polymers with relatively nonporous particles. The design of the flat-sheet die is complicated because the extrudate enters the manifold as a cylindrical slug, which must be formed into a flat sheet in such a way that the rate of flow must be the same at all points from the outside edges of the die to the center. The length of the flow path to the extreme edges of the die is much longer than to the center, and uniformity of flow of the polymer melt is obtained by the combined use of a profiled restrictor bar and a temperature gradient across the die lips.

The shape of the manifold is largely determined by the melt flow characteristics of the PVC compound so that the formulation is critical (and the die will probably be unsuitable for other plastics). This complication is much less evident with circular dies, but, on the other hand, slitting and opening of the extrudate may give rise to distortion in the final product.

When the sheet leaves the equipment it is usually passed through polishing rolls operating at 90–100°C to improve the surface finish. If it is to be corrugated, the sheet is reheated on leaving the polishing rolls, using infrared heaters before going through the corrugating equipment. Corrugation is normally done longitudinally with a marked reduction in the overall width of the sheet.

Thinner rigid sheet than the above for packaging applications can be produced using the flat-sheet die and reducing the thickness by stretching the sheet as it leaves the die lips. This produces orientation in the direction of the machine operation. This technique is particularly suitable for producing heavier-gage sheet (>0.02 in.). The sheet is water cooled and passed through polished steel rolls; this method produces flat sheet of maximum clarity and free from distortion. See also FILMS AND SHEETING.

Molding

Injection Molding. The molding of PVC has developed rapidly with the availability of the screw-type injection molding machines; prior to this, when the ram-type injection machine was standard, only plasticized PVC could be molded and it required special attention to prevent holdup of material with consequent degradation. Unplasticized PVC can be handled only on screw-type machines, which have the advantage of being able to deliver a low-viscosity homogeneous melt into the mold cavity with the minimum possibility of stagnation and with the maximum possible pressure on the melt. The working life of rigid PVC at the temperatures required for injection molding necessitates the introduction of the polymer melt into the mold in the shortest possible time once the optimum temperature has been reached; at most this should take only two cycles. This was almost impossible to achieve with ram machines, but the reciprocating-screw injection molding machine is similar to a single-screw extruder in that a homogeneous melt is obtained by shear forces set up by the rotation of the screw. As the material accumulates at the front of the screw, the latter retracts until a preset position is reached at which the volume of accumulated molten material is sufficient to fill the mold; the screw then stops rotating and moves forward to inject the melt into the mold. The screw then starts to rotate again and prepares the melt for the next shot. The screw is ordinarily of constant pitch with a compression ratio of about 2:1, has a long tapering cone with a very sharp tip, and rotates at a speed of 20–50 rpm. The molding cycle can be varied over a wide range by adjustment of the speed of rotation of the screw, and the times for the various stages of the process. The molten compound has a high melt viscosity compared with the majority of moldable thermoplastics, and there is no need to use a nonreturn flow valve at the outlet of the injection molding machine. In fact, the use of such a valve must be avoided; otherwise, holdup with subsequent degradation occurs.

The pressures during injection of rigid compounds are approximately 12,000–15,000 lb/in.^{2}; for plasticized compounds it is considerably less. (These pressures pertain to rigid compounds based on low-molecular-weight PVC with a K value of about 55.) High-molecular-weight resins such as used for extrusion are not suitable for the injection molding of rigid PVC compounds because their melt viscosity creates pressures too high for most machines. Some variations in molding performance can be obtained by varying the composition of the compound, particularly the concentration of the lubricant. The feedstock may be in the form of either granules or powder blend; the latter, as in other processing techniques, offers economic advantages but generally poses some problems. See also Injection Molding in MOLDING.

Blow Molding. The blow molding of rigid PVC is generally limited to the manufacture of bottles. These are produced by extrusion of a tubular parison, which is fed directly to a blow-molding device and blown while it is still hot. Some of the first production of PVC bottles used a two-stage process in which the parison was extruded as a continuous tube, which was then cut into the required lengths; these were then reheated prior to feeding to the blow-molding operation. This process had some advantages, eg, more efficient extruder operation, but the uniform reheating of the parisons required close control, which could be attained only with expensive equipment specially designed for the operation. The main advantage that better quality control could be achieved in the extrusion of the parisons has now largely disappeared with the general improvements in blow-molding machinery, and the process has been largely superseded.

There is now a wide range of equipment suitable for blowing PVC bottles; the design takes account of the need to avoid holdup or exposure of the compound to a high temperature for too long a time. Thus most of the equipment involves either continuous extrusion of a parison or the use of a reciprocating-screw system to form the parison. Reciprocating-screw systems are not yet widely used but appear to have some advantages, such as production of less scrap, the molding of special shapes giving better control of wall thickness, and more uniform parison temperature. The equipment has the disadvantage that it gives lower production rates than the extrusion process. Normally continuous-extrusion equipment is operated in conjunction with a multiple-mold indexing system in which the previously blown bottles are cooled as additional parisons are formed, clamped, and blown from a continuously extruded tube. Further details of blow-molding techniques can be found under Blow Molding in MOLDING.

The production of PVC bottles is usually limited to about 2500/hr. A recent development (8) claims that it is now possible to design a machine with an output of at least 10,000/hr in sizes from 8 to 32 oz.

The size of blown PVC containers is significantly less than the dimensions that can be obtained with polyolefins. This is due mainly to the fact that the thermal instability of PVC prevents the use of the accumulator type of equipment that is essential for blow-molded items of large capacity. Two-liter PVC containers are in general production, and it has proved technically possible to manufacture containers of up to two-gallon capacity.

Machines can operate both with powder blends and with granular compounds, although the latter have now been superseded because of their high production costs. The compounds are normally based on PVC with a K value of 55; if glass-clear bottles are required, organotin stabilizers must be used for the best results. Copolymers of vinyl chloride and propylene have been developed primarily for the manufacture of PVC bottle compounds and are used for this application to some extent in the United States. They can be processed at lower temperatures than would be needed for homopolymers so that reduced amounts of stabilizer are required (9). With the increasing use of PVC for food packaging there is need for a nontoxic stabilizer; calcium–zinc stabilizers are now widely used although the clarity of the final product is not as good as with the tin stabilizers. Changes in the Food and Drug Administration regulations now permit the use of certain organotin stabilizers for packing some foods as long as extractability into the food does not exceed a certain limit (10,11).

Although strength of a bottle depends largely upon its design, impact resistance can be improved by the incorporation of impact improvers, which can be used without substantially reducing the clarity of the bottle. The effects of the various components in PVC compounds on the properties of the bottles are discussed in detail in Ref. 12.

Compression Molding. Compression molding has only limited use with PVC, but it is used for processing rigid material into phonograph records with good surface properties. In the manufacture of phonograph records, a rigid compound based on a copolymer containing about 15% vinyl acetate is used in order to have a low-viscosity melt that gives an accurate reproduction of the details of the grooves of the record. The compound is mixed and fused, and passed to a two-roll mill from which it is stripped off and cut into biscuits of the requisite weight. These are then compression molded in a record press under a pressure of about 2000 psi. Some cooling is required at the end of the cycle to ensure that the record can be removed from the mold without distortion. Cycle times can be as low as 15 sec. The initial compound can also be prepared in pellet form using a continuous compounder. This is then fed to the press as a preheated agglomerate by heating a preweighed amount of pellets to a temperature slightly above their softening point, usually in a container through which a heated air stream can pass. This technique has been superseded largely by a method in which a powder blend is fed directly to a small extruder mounted in a vertical position over each record press; this is known as the "boomer" method. The equipment can be operated to deliver the requisite amount of material for each pressing cycle in the form of a molten extrudate directly into the mold. Other techniques such as injection molding are now replacing compression molding; this has occurred because of the improvements in molding machinery that allow higher pressures to be used.

Compression molding is the only way of making thick high-quality rigid sheet. A number of sheets of calendered material are pressed together to form a heavy sheet up to 1 in. thick. The compound used for this process is usually based on a vinyl chloride copolymer containing a small amount of vinylidene chloride. The product has excellent clarity and good impact properties. See also Compression and Transfer Molding in MOLDING.

Calendering

Calendering is used extensively for the production of both plasticized and rigid PVC sheet; it is capable of producing high-quality material at very high rates of output. Calenders first used for poly(vinyl chloride) were designed for processing rubber; the latest equipment incorporates improvements resulting in increased efficiency, high production rates, wider sheets with improved thickness control, and better equipment performance. See also CALENDERING.

Although the cost of a calender together with the auxiliary equipment is very high, this is the most economical method for the production of PVC film and sheeting, particularly for materials of relatively great thicknesses (13). A calender can operate on thin plasticized PVC sheeting in widths up to 84 in. and thicknesses of 0.003–0.004 in. with a tolerance of ±0.0001 in. at speeds in excess of 100 yd/min. At this speed the quantity of PVC compound being processed is about 3500 lb/hr; with heavier sheets the output will probably be on the order of 5000 lb/hr. It is important, if good-quality sheeting is to be produced, that the compound fed to the calender be homogeneous and at the correct temperature; this temperature must be closely controlled and is different for different types of compound. It is essential that the compound be subjected to the minimum heating cycle during the process because of the high temperatures encountered at the final calender nip in the manufacture of thin sheet; these temperatures may exceed 200°C. The normal method for feeding the compound to the calender is by mixing the polymer with all the other ingredients in a ribbon-type blender from which the powder blend is fed to a Banbury mixer, which discharges the mixed compound onto a two-roll mill. This compounding step ensures that batch-to-batch variation is minimized. From the mill the compound is fed to the calender either as rolls of sheet or continuously by strip feeding. Continuous compounders such as the Buss Ko-Kneader or the Werner and Pfleiderer Plastificator are used to feed the calender directly.

Because of the high temperatures encountered in the nips of the calender rolls where the compound is subjected to high shear, it is possible to use high-molecular-weight polymers (*K* value about 70) as the basis of the formulation. The various types of sheeting produced at present require the development of a special compound for each particular end use; special stabilizers are required to avoid sulfur staining and plate-out, and to give good weatherability, whereas special plasticizers are used for low-temperature performance, fire retardancy, and resistance to extraction. Acrylic polymers and other polymeric processing aids are used to improve calendering performance, particularly in rigid formulations (see the section on Compound Ingredients, above). The composition affects the surface properties of the sheet, such as the ease with which it can be printed. All these variations in formulation require specific operating conditions on the calender, which must be established by almost empirical means on production equipment, since laboratory equipment is capable of giving only a broad guide to start-up conditions.

Two general formulations for transparent unsupported film with good processing characteristics and good stability against sunlight and natural aging are given as follows:

	(*a*)	(*b*)
	parts by weight	
suspension resin (*K* value = 70)	00.0	100.0
dioctyl phthalate	50.0	38.0
barium–cadmium stabilizer	2.0	2.0
lubricant	≤0.5	≤0.5
phenyl salicylate	1.0	1.0

The calender can also be used for coating fabric and paper with plasticized PVC sheet; the substrate is fed into the nip of the last roll to carry out the lamination. It is necessary to preheat the fabric immediately before it enters the nip and to ensure that its temperature approximates that of the stock on the calender. For paper coating it is important to dry the roll of paper stock for 24 hr before processing.

An increasing demand for thin rigid sheet has necessitated developments in calendering techniques because the material is difficult to process and requires extreme pressures and high roll temperatures. The stock temperature generally approaches 200°C for compounds based on homopolymers, whereas vinyl chloride–vinyl acetate or vinyl chloride–vinylidene chloride copolymers can be handled about 25°C lower. The higher temperatures can be obtained only by using oil heating or high-pressure steam on the rolls, and in addition, higher horsepower may be needed because of the higher melt viscosity of the compounds. Normally, rigid materials are fluxed in the same way as plasticized compounds before being fed to the calender, but equipment has been developed that eliminates the prefluxing step. Pellets or powder can be fed directly to the hot rolls of the calender and the resulting molten material is carried through the finishing rolls to produce rigid sheets or material laminated to a substrate. However, this technique gives low production rates, and for the highest possible outputs it is essential to feed the calender with prefluxed material.

Thermoforming

Rigid PVC sheet can be thermoformed to produce thin-walled articles that would be very difficult to make by injection molding. The equipment is basically the same as that used for other materials; the process is particularly suitable for the economical production of a limited number of components since mold costs are low and the equipment is inexpensive. Rigid PVC sheet can be formed in thicknesses from 0.002 to 0.25 in.; the formulation of the compound affects the thermoforming characteristics, and sheet based on vinyl chloride–vinyl acetate copolymers is frequently used to improve forming behavior. In general, however, any standard type of thermoforming equipment is sufficiently versatile to be capable of thermoforming rigid PVC sheet by adjustment of the operating conditions. Some additional care is needed, however, in this process; during heating of the sheet both time and temperature must be controlled carefully, and the shortest possible cycles must be adopted to reduce the possibility of thermal degradation. The most efficient method is to use infrared radiation and to heat both surfaces of the sheet at the same time. Thick sheets can pose more problems than thin and special attention is required during both the heating and forming operations. The optimum temperature for thermoforming is that at which the elongation at break is at a maximum; for rigid PVC this is limited to a relatively narrow temperature range, which varies with the composition of the sheet. Oriented sheets, particularly those that are biaxially oriented, may present problems owing to the high tension set up by the tendency to shrink on heating. This may require the provision of special clamping devices to counteract the forces created within the sheet. It can be avoided by heating the sheet to higher temperatures for longer times to allow relaxation to occur, but this can present problems when using thicker materials. Thermoforming produces a relatively large amount of scrap, up to 50%; since it is not practicable to reconvert this into sheet, some provision must be made for its use in other processes. This further processing will extend the thermal history of the PVC compound, and so it must be stabilized accordingly. The quantity of scrap can be reduced by using multiple-cavity molds. See also THERMOFORMING.

Bonding

Hot Gas Welding. This technique can be used for joining rigid PVC components; it closely resembles the technique used for welding metals except, of course, that comparatively low temperatures are used. At approximately 180°C pressure fusion welding becomes possible, although careful temperature control is necessary since fusion occurs within the range of thermal decomposition. The technique employs a hot gas torch, which may be heated by either gas or electricity; both air and nitrogen are used as welding gases. When compressed air is used, care must be taken to remove any contaminants, such as oil or water vapor, which would otherwise impair weld strength. Accurate control of gas pressure is essential since variations will influence the outlet gas temperature. Welding rigid PVC normally requires a flow rate on the order of 45 l/min and a gas outlet temperature approximately 270–280°C about ¼ in. away from the nozzle. Welding is accomplished using the torch in conjunction with a filler rod having a small preformed bend. The rod is laid in the groove between the abutting sections and softened by the torch, taking care that softening does not occur too far above the groove. Excessive softening reduces the vertically applied pressure, which is essential if strong welds are to be achieved. Stretching the filler rod should also be avoided since contraction on cooling will tend to promote cracking. Overlying runs must be on surfaces free from decomposed material, and if slightly plasticized grades of filler rod are used, a reduction in gas temperature may be necessary.

Solvent Cementing. Solvent cementing is often more convenient than welding, although the bond strength of the joint may not be as good. Cemented joints have a cleaner finish and are used when the additional material applied in welding is objectionable or unsightly. All surfaces should be thoroughly cleaned with methyl ethyl ketone or carbon tetrachloride before the cement, which is a concentrated solution of PVC in a suitable solvent such as tetrahydrofuran, is applied to the dry surfaces. Full load should not be applied to a cemented structure for at least 48 hr. The rate of evaporation of solvents from a joint is necessarily slow and tests show that bond strength increases progressively during this period. The time taken for the development of full bond strength depends on the solvent system. A better joint is obtained when using cements in which the solvents used are those most effective for dissolving PVC. Bond strength is improved by the use of pressure. When joints fit loosely adhesives based on epoxy resins can be used with advantage; since no solvents are used no shrinkage problems are presented. See also ADHESION AND BONDING.

Foams

Flexible Foams. Several processes are in use for the manufacture of flexible PVC foams. These may be divided roughly into mechanical blowing and chemical blowing; in both types the PVC is converted into a plastisol as a preliminary operation. See also the section Plastisol Processing below.

In the process in most general use in Western Europe, the plastisol is fed continuously to a refrigerated cooler where it is mixed with an inert gas, such as carbon dioxide, which dissolves under pressure in the mixture; this is then fed through spray nozzles under carefully controlled conditions onto a moving conveyor that carries the slowly expanding foam into ovens where the temperature is gradually raised, by infrared or dielectric heating, to the fusion value (14). After the foam has been fully formed it is cooled and then slit into the required thickness.

In an alternative process air is whisked into the paste by means of a mechanical whisk and the resulting froth is stabilized by a special stabilizing agent. This is added to the paste during or immediately after mixing (15). The mixture is then either fed into molds and fused by heating to 175°C, or fed continuously onto a conveyor belt that takes the material through a series of ovens. This technique is particularly suitable for coating thin layers of foam onto fabric.

Carbon dioxide is used as the expanding gas in the process shown in Figure 3. The plastisol passes slowly down through the absorption tower which is a packed column, while carbon dioxide is fed in at the bottom and flows countercurrent to the paste (16). The pressure inside the system is maintained at 30–90 psi and the gas flow is adjusted so that the plastisol is saturated with carbon dioxide by the time it reaches the reservoir. When high production rates or low foam densities are required a two-stage absorber may be needed. The gas-saturated plastisol is transferred by means of an air pump and is fed through a needle-type gun into the molds, in which the material is fused in an oven. Other techniques based on this principle have also been described (17).

In the above processes the properties of the foam depend not only on the formulation of the plastisol but also to a large extent on the operating conditions. The size of the bubbles formed in the foam as it is released from the pressure maintained inside the equipment is naturally determined by the physical conditions prevailing. When the wet foam is heated there is a reduction in surface tension and an increase in the volume of the gas inside the bubble. The reduction in surface tension tends to decrease the size of the bubble whereas the increase in the volume of gas has the opposite effect. The individual bubbles of the foam are unstable at this stage, and the rate of fusion of the PVC determines the final structure. If the plastisol fuses too rapidly, expansion of the foam is arrested and a high-density foam is obtained. If the fusion is slow, then expansion is excessive and breakdown of the cellular structure takes place. The correct balance between fusion time and time for maximum expansion is essential for the production of good-quality foam. The density of the products generally ranges from about 4 to 40 lb/ft^3 and higher, and they have an open cell structure.

Molded cellular products can be manufactured readily using an organic blowing agent as an ingredient of the plastisol. These blowing agents are the same as those used in the manufacture of expanded leatherlike cloth (see the section on Coating, under Plastisol Processing, below). Three different methods are used: (*i*) using a normal-size mold and expanding after fusion; (*ii*) using an oversized mold and allowing expansion during fusion; and (*iii*) fusing at atmospheric pressure.

In the first method the mold is filled with the plastisol containing the blowing agent and is fused under a pressure of approximately 1–3 tons/in.^{2} A typical formulation used for this process is as follows:

	parts by weight
PVC (paste forming)	100
dioctyl phthalate	140
4,4′-oxybis(benzenesulfonyl hydrazide)	27
calcium stearate	6

After the heating cycle the mold is allowed to cool under pressure to about 50°C, after which the finished product is removed and heated to about 100°C to complete the expansion. Closed-cell foams of very uniform expansion and pore size are obtained by this method. The process requires strong molds and heavy-duty presses to accommodate the high pressures used.

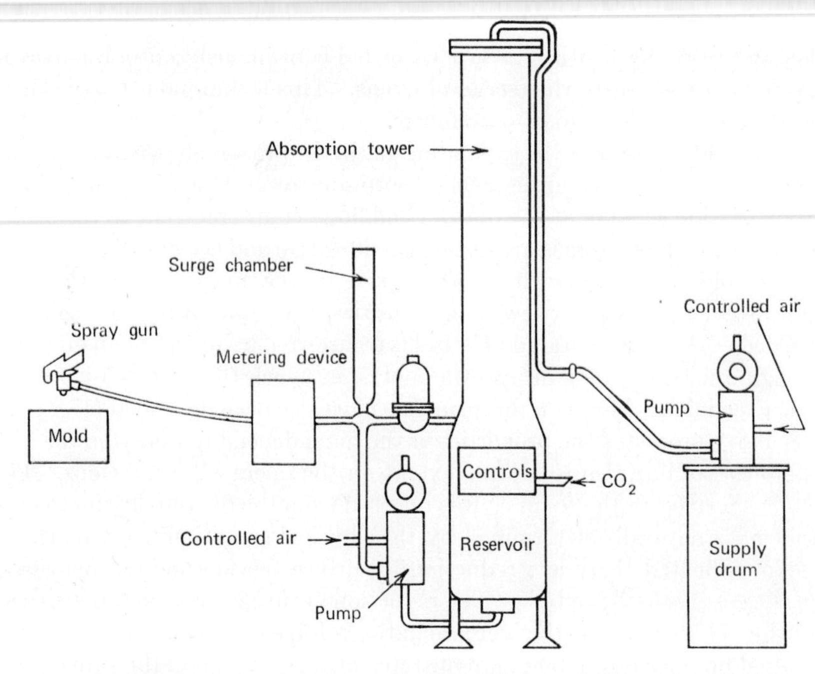

Fig. 3. Diagram for process for producing PVC foamed with carbon dioxide.

The second method, in which an oversized mold is used, requires the establishment of the exact operating conditions in order to obtain good-quality products. If too much plastisol is used, a relatively thick hard outer skin is formed. The addition of too much blowing agent can produce an unstable foam structure and cause the cells to break down. The mold must be adequately vented so that the air can be displaced when full expansion occurs during the pressing period. It is not necessary to reheat the molded product after it has been removed from the mold. The cell size is generally larger than that obtained by fusing under pressure and expanding afterward.

Expanding PVC pastes under atmospheric pressure gives a foam with interconnecting cells. A consistent product requires strict control of both formulation and operating conditions. The foam is made by first heating the paste to the decomposition temperature of the blowing agent; this should be below the fusion temperature of the compound. After expansion has taken place, the temperature is raised to fuse the plastisol. A formulation suitable for this type is the following:

	parts by weight
paste resin	100
didecyl phthalate	100
calcium oxide	1
epoxidized soya bean oil	3
Mellite 101 (a barium–cadmium stabilizer produced by Albright and Wilson)	2
neutral calcium Petronate (a petroleum sulfonate produced by Witco Chemical Corp.)	3.6
blowing agent	10

The blowing agent sodium borohydride operates by a completely different mechanism; this produces expansion by interaction with water under acidic conditions at room temperature to liberate hydrogen, which is responsible for the expansion (eq. 1). (The hydrogen, however, does not remain long in the cells, but diffuses out rapidly and is replaced by air. Safety measures must be observed to keep the concentration of

$$NaBH_4 + 2\,H_2O \rightarrow NaBO_2 + 4\,H_2 \quad (1)$$

hydrogen around processing equipment low, generally below 4%.) Although sodium borohydride is expensive, it produces such a very large volume of gas that it is economically competitive with the least expensive organic chemical blowing agents. This blowing agent is used in a two-component system, which reacts shortly after mixing; therefore, the two components are metered into the mold where expansion takes place prior to fusion. When large articles are being produced or low density is required, the blowing with borohydride must be retarded by the addition of other components (18).

Calendering. Flexible PVC foam can be produced as a thin sheet by incorporation of a blowing agent into the calendered sheet. Azodicarbonamide is added during the preliminary compounding stage and the sheet is then calendered at a temperature lower than the decomposition temperature of the blowing agent. This sheet is usually laminated as a sandwich with fabric below and an unfoamed calendered sheet on top as a protective coat. The decomposition of the blowing agent is then brought about by passing the laminate through an oven at a temperature higher than that used for calendering and lamination. See also the section on Plastisol Processing below.

Extrusion. Flexible PVC foam can also be produced by extrusion, although this technique is generally limited to the wire and cable industry where the vinyl foam is used as a cable filler, cable jacket, and primary insulation. There are two different methods of extrusion; in the first, the blowing agent is compounded with all the other ingredients, as in calendering at a temperature below the decomposition temperature of the blowing agent. The compounded material is diced and then extruded onto the cable; expansion is brought about by raising the temperature of the extrudate in a heated tunnel after the cable emerges from the die. In the second method the compound is made as granules without the inclusion of the blowing agent, which is added by tumbling just before extrusion. The temperatures in the equipment are maintained so that the blowing agent decomposes in the machine and expanded material emerges from the die. The product is cooled immediately to prevent collapse. This last technique requires careful control of the operating conditions, and the extruder must have close tolerance between the barrel and the screw so that the pressure generated by the blowing agent is not allowed to dissipate. High temperatures are generally necessary to obtain satisfactory decomposition of the blowing agent, and the compound must be formulated accordingly. The hot-melt viscosity must be high enough to prevent surging of the extrudate from the die owing to pressure developed within the equipment. In cable applications the foamed section is preferred with a solid skin of PVC; this is obtained by running the temperature of the die considerably below the temperature of the compound so that the outer layer forms a cool skin that limits the expansion to the core of the section. See also the section on Wire Coating above.

Rotational Molding. This process is used for the production of toys and automobile accessories by incorporation of a blowing agent such as azodicarbonamide into the plastisol. A compound is charged to the mold, which is heated in the usual way while being rotated in two planes; the temperature is then raised above the decomposition point of the blowing agent to produce a cellular structure. If the amount of blowing agent is too great; however, mottled effects are produced and a satisfactory surface is not obtained.

Injection Molding. All types of injection-molding machines can be used for the production of cellular moldings of PVC, although a screw-type machine is preferred and close control of the operating conditions is needed to obtain the correct degree of expansion (19). The process differs, however, from the injection molding of conventional PVC compounds in that the injection speed must be high to eliminate foaming in both the gates and the nozzle. The injection pressure and the locking pressure are much reduced and the pressure required to fill the mold is derived almost entirely from the gas. The amount of blowing agent and the mold closure time largely determine the degree of expansion and the density of final molding; the longer the closure time, the higher is the product density. Density reductions of 15% in moldings can be obtained by the inclusion of 0.15–0.4% blowing agent with little change in physical properties. The maximum reduction obtainable with conventional molding equipment is about 30%; lower densities can be obtained on machines specially designed for the production of cellular moldings (20).

Rigid Foam. Developments in the use of rigid PVC foams have been restricted, largely because of the much higher price compared with expanded polystyrene and rigid polyurethan. Its use has been limited to those fields in which its superior structural properties and other special characteristics such as low flammability and good resistance to chemicals and abrasion are essential. Restrictions on use imposed by the relatively low softening point of PVC have been overcome to some extent by the development of a crosslinked foam. This is produced by reacting PVC with maleic anhydride, and hydrolyzing the anhydride groups. This produces a crosslinked product on reaction with a diisocyanate. The crosslinked compound is then immersed in hot water or steam to produce carbon dioxide, which functions as a blowing agent and forms a strong network structure (21). The foams produced by this process range in density from 2 to 6 lb/ft^3 and are available in sheets from $\frac{1}{4}$ to $1\frac{1}{2}$ in. thick. It has about 90% closed-cell structure.

Plastisol Processing

Coating. A very large percentage of paste-making resins are used for spreading plastisols on various substrates such as textiles, sheet steel, and paper. It is reported that PVC consumption for coating in the United States in 1968 was 108 million pounds, which was almost completely used for coating textiles (22). Commercial growth continues in this field in spite of competition from coated fabrics produced by lamination of calendered PVC film, because of the superior feel and appearance of the plastisol-coated material.

Coating equipment generally consists of an adjustable doctor knife or spreading blade supported over a steel plate or roller; however, there are many elaborations of this setup. Variations have been made on the means of support for the fabric and there are also numerous types of doctor blades for ensuring that the desired amount of plastisol is metered onto the fabric. Rollers are sometimes used instead of doctor blades. The paste is fed by hand, gravity, or pump to the front of the doctor blade, which may be arranged so that it forms one side of a trough over the fabric to ensure that there is a uniform area of paste in contact with the fabric. Details of the various arrangements of coating equipment are discussed and illustrated in the article COATING METHODS.

Paste viscosities vary appreciably according to the nature of the substrate and the type of product being made; it is possible to spread pastes ranging from a free-flowing liquid to a fairly heavy dough and with viscosities ranging from 1,000 to 40,000 cP. The rheology of the paste is important. Pseudoplastic pastes tend to give heavier coatings as the viscosity falls with increasing rates of shear and the plastisol flows more readily under the blade. Dilatant pastes resist the spreading action of the blade and may even fail to coat the substrate under extreme conditions. Pastes that are pseudoplastic at slow rates of shear may become dilatant at higher rates of shear, so that adequate information on the paste viscosity over a wide range of shear rates is necessary for production control. Coating speeds vary considerably; the normal range lies between 2 and 30 m/min. Thick heavy doughs with high filler and pigment content can be used only at low spreading speeds, whereas low-viscosity pastes are required for spreading at high speeds. Coating thickness also influences rate of coating, thin coats tending to the higher rates and thicker coats to lower.

The coated material passes through an oven to set and then to fuse the plastisol; the temperature and the length of the heating oven depends on the rate of coating and on the formulation. When the paste contains a volatile solvent a low-temperature preheating oven is also needed. It is usual practice to use two separate coats for fabrics, the undercoat being specially designed for adhesion and at the same time to avoid excessive penetration of the substrate, and the top coat producing the decorative finish.

Foamed Coatings. There is increasing use of an expandable layer to make coated fabrics that have improved softness and hand and are able to stretch around double curvatures. The material resembles natural leather (see LEATHER-LIKE MATERIALS). It ordinarily consists of a continuous outer surface, an expanded inner layer, and a knitted cotton backing. It may be manufactured by either the direct or the "upside-down" method. In the former the knitted fabric, which may require a supporting material, is fed to the coating machine where an expandable paste is coated directly onto it; it is then passed through an oven to set the PVC but at a temperature that is not high enough to cause complete fusion of the PVC or to decompose the blowing agent. The fabric is cooled and re-rolled, after which further layers of expandable paste may be applied until the required thickness is obtained. After application of a nonexpandable top coat, the material is passed through a heating oven at a temperature high enough to produce complete fusion of the PVC and to decompose the blowing agent in the intermediate layers.

The alternative method has some advantages, because of the difficulty of handling the knitted fabric in the initial stages and the need to avoid excessive permeation without losing the improved handling characteristics. In the "upside-down" method the layers are built up in the reverse order on a carrier belt, which must have a uniform surface to which PVC paste will adhere but from which it can be stripped off cleanly without damage. A heavy-weight paper of about 150 g/m^2 with a surface coating of a release agent is generally used.

The plastisol used for the top coat is formulated for the special properties required in the final application. The plastisol contains a chemical blowing agent that may be activated by certain types of stabilizer. Stabilizers that catalyze the decomposition of the blowing agent include cadmium–zinc complexes (liquid), zinc complexes (liquid), and dibasic lead phthalate (solid). Blowing agents used include azodicarbonamide, azobisisobutyronitrile, 4,4′-oxybis(benzenesulfonyl hydrazide), dinitrosopentamethylenetetramine, aminoguanidine bicarbonate, and benzil monohydrazone. The use of these blowing agents is discussed in detail in the article BLOWING AGENTS.

Belting (qv). Conveyor belting made by laminating together layers of cotton duck coated with PVC paste is used in coal mines, particularly in the United Kingdom. The interply coats are based on a flexible formulation, and the outer coats are antistatic and resistant to abrasion. The lamination is carried out either by pressing the plies together in an ordinary press or by rotocure. The press method can be used only batchwise, whereas the rotocure operates continuously. Layers of coated fabric come together on the surface of a rotating drum, which is heated to assist lamination. A simpler method of making a PVC belt consists of encapsulating a fabric core in PVC paste. The core is drawn through a tank containing the plastisol, then is heated to partially fuse and dry the PVC. It is then passed through a bath containing a plastisol with a lower plasticizer content, to give a hard surface on heating. Methods of coating fabrics are also discussed in detail in the article FABRICS, COATED.

Dip Coating. This process consists basically of coating forms with a PVC paste, which is then fused in an oven. There are two techniques: one is to obtain a product that is stripped from the form after fusion; in the second, the coated layer is intended to form a protective layer for the base. The paste may be applied to the substrate by either hot or cold dipping. In hot dipping, the form is heated and immersed at a controlled speed into the plastisol; this may be repeated several times to build up the required thickness, after which the coated form is inverted to prevent drips and fused by heating in an oven. In order to maintain a uniform thickness it may be necessary to rotate the form on a drum to overcome flow problems.

Although hot dipping is the usual method of dip molding, cold dipping is also used, but this technique is much more difficult to control and problems of flow are encountered. It is particularly useful, however, for objects with varying metal thickness because hot dipping could in such instances result in reduced thickness of the coating in the thinner metal areas, which cool more rapidly when dipped into the plastisol.

Molding Methods. *Rotational Casting.* This is one of the major methods of processing plastisols. A measured amount of paste is charged into a sealed split mold which is attached to a plate; this plate is secured to a spindle that can be rotated through two planes, at right angles to each other, by a series of gears. The speed of rotation is comparatively slow so that the plastisol flows by the effect of gravity to form a layer of uniform thickness over the whole of the mold cavity. The complete mold assembly is in an oven so that it can be heated during rotation while the PVC is fused. The mold is then removed from the oven and cooled by water after which it is opened to remove the molded article. The mold assembly is usually designed to carry two or more molds; these are charged alternately by stopping the operation in the middle of the heating cycle of the first mold to clamp the second mold in position. It is essential that the plastisol should have fused sufficiently at this stage that it does not flow to the bottom of the mold when it stops rotating. Applications for rotational moldings include flexible toys and automobile arm rests and crash pads. These latter are produced by rotating the mold containing a normal plastisol until a skin is formed on the inside. The mold is then opened during the heating cycle; a PVC paste formulated with a blowing agent and with or without a metal insert is introduced, after which the mold is closed and the cycle is continued until the whole system has completely fused.

The process has the advantage that the equipment is relatively cheap and molds are inexpensive and easy to make. The molding is carried out at atmospheric pressure. Aluminum is suitable for the molds, which can be cast or machined if the shape of the cavity is not complicated. The mold should be mechanically strong but the metal should be as thin as possible to ensure that there is rapid heat transfer for the fusion and cooling of the paste. See also Rotational Molding in the article MOLDING.

Slush Molding. This process consists essentially of pouring a plastisol into a metal mold, which is then heated to fuse or partially fuse the material in close contact with the surface of the mold; the excess plastisol is then poured off and the heating of the mold is continued to complete the fusion of the PVC paste. Fusion is carried out in an oven maintained at 170–220°C, although higher temperatures, even up to 270°C, are frequently employed.

In the "two-pour" method, the mold is filled with PVC paste and then drained by standing upside down. Since no heat is applied at this stage, the thin layer of paste remaining on the walls of the mold is unfused. The mold is then passed quickly through an oven at a temperature of 177–215°C so that partial fusion of the layer takes place. When the mold emerges from the oven it is fairly hot and it is then refilled with plastisol; the mold temperature causes fusion of a further layer of paste in contact with the initial layer. After the plastisol is emptied from the mold the combined layers of material adhering to the surface are fused by reheating in an oven and then cooled before removing the final article.

Low-Pressure Injection Molding. This technique can be used for molding PVC shoe soles. As with all the other techniques for processing plastisols, the costs of the machine and the molds are lower than for injection molding of other types of PVC compound. For fast molding cycles, the viscosity of a plastisol must be fairly low to give good flow behavior and rapid mold fusion; the compound must attain high tensile strength at a relatively low fusion temperature. A suitable formulation for such an application would be as follows:

	parts by weight
PVC paste resin	80
suspension (diluent) resin	20
dioctyl phthalate	50
dibutyl phthalate	20
cadmium–barium stabilizer	1
lubricant	0.5

The use of low-pressure injection transfer of plastisol caulking compounds into joint cavities has been common practice in the motor industry for some time. The technique has also been used for producing ring seals for tobacco and food container lids.

Bibliography

1. *Mod. Plastics* **46** (11), 73 (Nov. 1969).
2. C. M. Thomas, *Plastics Inst. (London) Trans. J.* **35** (120), 793 (Dec. 1967).
3. L. G. P. Smith, *SPE (Soc. Plastics Engrs.) J.* **20,** 999 (1964).
4. A. Domininghaus, *Brit. Plastics* **39** (8), 475 (Aug. 1966).
5. B. Georgii, *Kunststoff-Rundschau* **14** (7), 317 (July 1967).
6. J. J. Barney, *Mod. Plastics* **46** (12), 116 (Dec. 1969).
7. D. J. Dowrick, *Plastics* **30,** 67 (Oct. 1965).
8. *Mod. Plastics* **46** (10), 114 (Oct. 1969).
9. *Mod. Plastics* **45** (12), 78 (Aug. 1968); *SPE (Soc. Plastics Engrs.) J.* **24** (6), 23 (June 1968).

10. *Plastics World* **26,** 45 (March 1968).
11. H. C. Carr, *SPE (Soc. Plastics Engrs.) J.* **25,** 72 (Oct. 1969).
12. W. B. Sisson, *Plastics & Polymers* **36** (126), 453 (Oct. 1968).
13. N. Stackhouse, *Plastics Inst. (London) Trans. J.* **32** (99), 227 (June 1964).
14. Elastomer Corp., U.S. Pat. 2,666,036 (1953).
15. R. T. Vanderbilt Co., Inc., U.S. Pat. 3,288,729 (1962).
16. I. Dennis, U.S. Pat. 2,763,475 (1954).
17. O. Fuchs, F. Heckhelt, and A. Herz, *Kunststoffe* **55** (9), 717 (1965).
18. B.F. Goodrich Chemical Co., U.S. Pat. 3,084,127 (1963).
19. K. T. Collington, *Plastics* **34** (380), 724 (1969).
20. *Plastics Technol.* **14** (13), 11 (1968).
21. *Mod. Plastics* **45** (2), 91 (1967).
22. *Mod. Plastics* **46** (5), 26 (May 1969).

C. A. Brighton
BP Chemicals International Ltd.

APPLICATIONS

At present, in 1970, the commercial markets for poly(vinyl chloride) are expanding rapidly, with the building industry and packaging attracting most attention. The use of rigid compounds is increasing more rapidly than that of plasticized materials, which now account for about two thirds of PVC used. Sales of PVC can be divided into the major markets outlined in Table 1.

Table 1. Major Markets for PVC in the United States

Market	Usage in 1969, 10^6 lb	% of total
building and construction gutters and downspouts, windows, pipes and fittings, cladding and sidings, waterstops	470	17.6
flooring tiles, continuous flexible flooring materials	347	13.0
electrical wire and cable coverings, conduit	405	15.2
clothing rainwear, shoes, baby pants	161	6.0
home furnishings wall coverings, garden hose, curtains, upholstery	501	18.8
packaging film, bottles, thermoformed containers	208	7.5
toys and leisure phonograph records, balls and dolls, golf bags	213	8.0
transportation upholstery and seat covers, wire covering, crash pads	224	8.4
miscellaneous tubing, credit cards, paints and lacquers	141	5.3
total	2,670	

Building and Construction

This industry is one of the major users of PVC (see also BUILDING AND CONSTRUCTION APPLICATIONS). Applications of unplasticized materials are at present being increasingly developed. Flooring accounts for the greatest use, although its rate of commercial growth has slowed considerably during recent years. It is expected that the major increase in demand for PVC products for general building applications in the near future will take place in the unplasticized grades. These materials have good physical properties and fire retardancy, and can be readily fabricated. The weathering of rigid compounds can be improved by the use of special formulations so that at least a twenty-year outdoor service can be expected with practically no maintenance. In order to achieve this performance only a limited color range can be used, and most manufacturers supply products only in white and light gray. Dark colors are avoided in exterior applications to minimize temperature increase when the product is exposed to the sun. This good weather resistance has led to the use of rigid PVC for sidings and cladding, window frames, shutters, building panels, roofing, and gutters and downspouts. Waterstops are another application; they are flexible extrusions used to prevent water ingress at junctions of cast concrete slabs.

A wide range of extruded and thermoformed PVC sections are available for outside cladding of buildings; these are produced in weather-tested compounds and are usually white. The rapid increase of this application is due to the increasing cost of maintenance of wooden sidings and the decreasing cost in the installation in comparison with traditional materials. In the United Kingdom an injection-molded PVC panel has been developed (1); each side measures 9 ft × 3 ft and weighs approximately 28 lb. The panels are joined back to back by a number of brackets made of polypropylene and are then mounted vertically and filled with sand and gravel or some other more permanent filling. The system is designed so that the service pipes can be incorporated into the interior of the panel.

Use of PVC window frames has increased considerably in both Europe and the United States during recent years. In Germany, in 1969, about 8% of the total number of window frames installed were made from rigid PVC. The majority of these were made from extrusions by heat welding mitered sections; in some instances the corners are reinforced with metal, and the possibility of deflection in the straight sections is reduced by sliding a metal section into the PVC profile. In the United States a casement window based on a PVC sheathed wooden section has been introduced (2). The sash is produced by extruding PVC over a wooden section using a crosshead die. The frame is vacuum formed and fabricated over a wooden core. Windows produced from profiles have the advantage that there is no limit to the number of sizes that can be produced, but the joints at the corners are areas of potential weakness unless they are reinforced. This has created an interest in injection molding (3,4) although mold costs are high and only a very limited number of sizes could be produced. With high-volume production the costs of the injection-molded windows should be less than the same sizes made from extruded profiles.

Rainwater gutters and downspouts of PVC have almost completely replaced cast iron and asbestos in the United Kingdom. Their main advantage is in reduced maintenance and in ease of installation because of their light weight. In countries with heavy snowfalls and low winter temperatures there has been much less use of PVC gutters, although products are now specially designed for such locations.

Another growing market, particularly in the United States, is the use of vinyl coatings and paints. Copolymers of vinyl chloride have proved to give superior performance as binders for high-quality paints for exterior woodwork. This type of product gives a performance equivalent to acrylic-based paints but at a lower cost. See also the section on Coatings, below.

Floor Coverings

Vinyl–asbestos floor tiles are based on compositions containing about 20% of a vinyl chloride–vinyl acetate copolymer and about 80% of asbestos and mineral fillers. The copolymer most suitable for this application contains about 15% vinyl acetate. Plasticizers and resins (eg, coumarone) are often added to improve filler wetting and ease of calendering. These tiles are relatively inexpensive and give floors that stand up well under hard wear. See also FLOORING MATERIALS.

Compositions containing smaller amounts of filler are used to make flexible floor coverings available as tiles or sheet. Flexibility is improved by the addition of plasticizer; this product has greater resilience than a vinyl–asbestos flooring and therefore is quieter and more comfortable. It is essential to formulate compounds in such a way that traffic staining is minimized; this is a problem with light-colored flooring. A hard surface provides the best protection; this can be obtained by laminating a hard top coat, based on a homopolymer, to the flooring material. The use of lead stabilizers should be avoided as these increase susceptibility to sulfur staining.

For laminated flooring a base material, such as felt, thick paper, or asbestos, is combined with a PVC sheet printed with the final design, and a protective top layer of clear PVC is them laminated to the surface. Plastisols can be used at any stage for the manufacture of this type of flooring. It has limited durability because the protective coat must be thin so that the pattern is not obscured by light reflection.

Foam flooring is made of a base sheet to which a plastisol containing a chemical blowing agent is applied. After fusing the plastisol without decomposing the blowing agent, the material is printed with the final design and then a further layer of a plastisol is applied as a top coat; this is fused and the intermediate layer is expanded in one operation. Decorative effects can be produced by the use of special agents in the printing ink that deactivate the blowing agent in specific areas. This type of flooring is very much more resilient than others; it is very comfortable to walk on and has good recovery from indentation.

Pipe

PVC pipes account for a large part of the applications of this polymer in the construction industry (see also PIPE). They are available in two types: the first is capable of operating at service temperatures up to 65°C; this is normal impact-resistant pipe used for pressure applications. The second type is suitable only for service temperatures up to 60°C; it is a high-impact grade for special applications. In Europe there is less interest in high-impact formulations. Improved formulations have made it possible for PVC pipe to meet practically all national specifications. In the United Kingdom there are two British Standard specifications (5,6) covering use for cold-water services and for industrial applications. Thickness and outside diameter vary according to the country of manufacture, although there have been moves toward some degree of standardization; fittings are designed accordingly. The pressure at which pipes can operate depends on the circumferential, or hoop, stress, on the thickness of the pipe wall, and on its diameter. The internal pressure is related to these factors by the equation

$$P = 2ts/D$$

where P is pressure, s is hoop stress, t is thickness, and D is outside diameter.

The value of the hoop stress that can be maintained indefinitely, ie, 50 yr, is obtained by extrapolating the curve giving time at which pipes fail by rupture versus hoop stress. To establish the long-term pressure that can be applied, the value taken for the hoop stress must be less than the limiting value as determined above. The safety factor applied to the hoop stress varies with different national specifications; the value of the hoop stress is normally given at a temperature of 23°C, and decreases as the temperature rises. The Plastics Pipe Division of the Society of the Plastics Industry recommends a safety factor of one half the hoop stress at 100,000 hr, or 11.4 yr. Pipe considered suitable for operating at 150 psi in municipal water service has an instantaneous burst pressure of 800–1000 psi. Pipe rated for 100 psi at 23°C can be used at 128 psi at 0°C and only 75 psi at 40°C.

In general, PVC pipe requires two to four times the number of supports required by steel pipe, owing to the tendency of the pipe to sag. A continuous channel support is better than individual clamps and is essential if the pipe is carrying fluids at elevated temperatures. The high coefficient of expansion of PVC must also be taken into account when designing pipe systems, and adequate provision of expansion joints is essential. For a pipe of 1000 ft in length, the change in length is about 3.3 in. for every 5°C change in temperature. Pressure loss in PVC piping is much less than for traditional materials.

Joints between pipes and fittings are usually made by the use of solvent cement. A properly made joint should be as strong as the pipe itself. This type of joint uses a tapered socket; the gap between the pipe and the fitting is filled with material that has been softened by the solvent and displaced when the pipe is forced into the socket. The solvent cement joint can be difficult to make effectively if unsatisfactory conditions prevail, or if the temperature is so low that the solvent evaporates only very slowly. A number of mechanical joining systems are available; these all use a gasket to prevent leaks and a push-to seat method of joining. This type of joining can be carried out satisfactorily and quickly under practically all weather conditions; other advantages include wider acceptable tolerances and ample allowance for expansion and contraction owing to the length of the socket. It is essential that the gasketing material be suitable for use with the fluid being conveyed in the pipe.

Rigid pipe has gained wide acceptance for water mains and service lines, for drainage, and for soil pipes. The chief advantage over metal pipes rests in corrosion resistance and lower installation costs. Another growing application is in the relining of existing substandard sewers. Although a pipe of smaller diameter must be used for this purpose, the smoother bore creates a higher flow velocity, and in most instances there is little difference in net capacities. In Europe there has been considerable interest in the potential of rigid PVC for gas distribution. This use at present is limited by the presence of aromatic hydrocarbons in the gas (7–9), but with increasing replacement of gas produced from coal by natural gas, the economic advantages of rigid PVC pipe might lead to a large increase in usage. Rigid PVC pipes are at present extruded in diameters up to 1 m.

An interesting development in drainage is a system that depends on the use of a rigid PVC bowl of 19-in. diameter to replace the traditional manhole built in brickwork (10). The injection-molded bowl is hemispherical; the top is located at ground level and is covered by a cast-iron cover. Up to six inlet connectors may be joined to the bowl at any required angle. Access to the drain line from the bowl is achieved by leading down a 4-in. pipe instead of constructing a masonry shaft. The system can result in considerable savings in cost by reducing the amount of brickwork and the time required for installation.

Electrical Insulation

Electrical insulation was the first major application for PVC and still forms a substantial portion of the total usage (see also ELECTRICAL APPLICATIONS); the wide range of properties obtainable make PVC compounds particularly suitable for electrical

Table 2. Some Typical Compounds Used for Electrical Applications

Category	British Standard softness no.	Components	Parts
general dielectric	18	suspension resin ($K = 65$)	71.5
		dioctyl phthalate	28.5
		calcined clay	5.0
		tribasic lead stearate	5.5
		ethylene stearamide	0.6
general dielectric	33	suspension resin ($K = 65$)	66
		dioctyl phthalate	34
		calcined clay	5
		tribasic lead stearate	4
		ethylene stearamide	0.5
thin-walled dielectric	12	suspension resin ($K = 60$)	76
		dioctyl phthalate	24
		tribasic lead stearate	4
		ethylene stearamide	0.5
sheathing	42	suspension resin ($K = 65$)	57
		dioctyl phthalate	29
		chloroparaffin (chlorine content 42%)	14
		whiting	30
		tribasic lead stearate	4
		barium stearate	0.5
	38	suspension resin ($K = 65$)	62.5
		dioctyl phthalate	25.5
		chloroparaffin (chlorine content 42%)	12
		whiting	30
		tribasic lead stearate	4
		calcium stearate	1

insulation for temperatures from −30 to +105°C. Present applications include household wiring, electrical and radio appliances, and automobile electrical systems. Applications requiring the use of compounds specially formulated for high temperatures (11) range from electric-blanket wire and under-floor heating to high-voltage cables in which the conductor is normally operating at temperatures up to 100°C. This last use is still very much in the development stage; in Europe the use of PVC insulation for 15 kV cables is permitted, and it can be expected that higher voltages will be possible as compounds with better heat stability and higher heat-distortion temperatures are developed. PVC compounds differ in their electrical behavior from nonpolar polymers such as polystyrene and polyethylene. The power factor passes through a maximum as the frequency changes, and the best insulating properties are obtained at frequencies of about 60 cycles and in excess of 10^6 cycles. The frequency range over which the loss factor is at a maximum is the best for high-frequency welding. General effects of compound composition on electrical properties are covered in the section on Properties (see also ELECTRICAL PROPERTIES). Some typical compounds used for electrical applications are given in Table 2.

Packaging

PVC is used in packaging as flexible and rigid film, thermoformed containers, blow-molded bottles, and coatings for cans and paper. Its moisture resistance and low gas permeability (12) make PVC particularly suitable for food packaging, as is evident in the rapid increase in its consumption for blown bottles. There has been extensive work on the use of PVC bottles for carbonated drinks, and market trials have been run in Germany on bottled beer. The bottle must withstand pressures of up to 60 psi and be of such a thickness that the carbon dioxide is not dissipated for at least three months (13). High clarity is obtained with PVC bottles by use of organotin stabilizers, which are permitted in most countries for food applications. When such stabilizers are not permitted, calcium–zinc and aminocrotonic ester stabilizers must be used, although they are not as effective as the organotin compounds. A demand for bottles with good impact properties has necessitated the inclusion of acrylic and ABS impact modifiers in the compound. This type of additive reduces the clarity of PVC compounds, and special grades with a refractive index closer to that of PVC have been developed for clear bottles. These new modifiers do not give such effective impact improvement as the more usual types, and larger amounts are needed to obtain the same impact strength. Bottles containing impact modifiers also suffer from the disadvantage that stress whitening can occur if the material is creased or flexed. A bottle based on unmodified PVC can be dropped a distance of about one foot without breaking, and this increases to about five feet for those based on compounds containing 10% ABS modifier.

In the majority of countries the composition of plastics packages used for foodstuffs is controlled by legislation. This legislation varies considerably from one country to another, with the consequence that certain additives may be permitted in some countries and not in others; the use of organotin stabilizers is a case in point. The majority of plasticizers used for flexible PVC are extracted by alcohol and fats, and plasticized compositions are accordingly limited to use with foods into which plasticizer migration is very low. Certain grades of plasticized PVC are used for meat wrapping, because flexible films have a much higher permeability to oxygen, which is essential if the meat is to keep its color.

Plasticized PVC film can be produced with good clarity, desirable for extensive use in packaging. It can also be produced in a biaxially oriented form for shrink wrapping, with varying degrees of shrinkage in the two directions. Rigid film and sheet is used in food packaging because of its good physical properties and vapor permeability characteristics. Thin sheet can be thermoformed easily for containers for margarine, cakes, and dairy products. See also PACKAGING MATERIALS; FOOD AND DRUG APPLICATIONS.

Coatings

Plastisols are used extensively for coating fabric, paper, and steel sheet; they are also widely used for dip coating metal objects. There is considerable use of copolymers dissolved in organic solvents for solution coating where thin films are required; the solvents ordinarily used include esters, ketones, and chlorinated hydrocarbons. Diluents such as toluene or xylene may be added to reduce costs and control the evaporation rate. Homopolymers are not satisfactory for this application because of their limited solubility in the above solvents, and because of the problem of adhesion to metal substrates. Copolymers containing up to about 20% vinyl acetate are in most general use; they are modified by partial hydrolysis to introduce hydroxyl groups which improve both compatibility with other resins and adhesion to substrates. Carboxyl groups are introduced into the polymer chain by the incorporation of a third monomer such as maleic acid, and these materials are used for coatings for metal surfaces such as automobiles. Copolymers of vinyl chloride and vinylidene chloride that are soluble in aromatic hydrocarbons can also be used for metal coating.

Multicoat systems are at present required for coatings without pinholes; a primer coat that gives good adhesion is used as a base coat and a top coat of a hydrolyzed vinyl chloride–vinyl acetate copolymer is added. This type of material is particularly suitable for marine applications because of its antifouling properties. In order to withstand physical damage a coating of 6–10 mils thickness is needed, and this can only be obtained by using a number of coats. Poly(vinyl butyral) and zinc-rich primers are often used with a top coat of a vinyl acetate copolymer. See also MARINE APPLICATIONS.

A recent development in coatings is the application of plasticized PVC compounds by fluidized bed, electrostatic spraying, and powder spraying techniques. The item to be coated is first preheated to about 250°C and then brought into contact with the powder compound, which forms a covering over the surface. The whole is then reheated to complete fusion and produce a continuous coating. Special materials have for some time been produced for this application method by freeze grinding a fused compound, but recently resins have been produced that can be compounded by ordinary powder-blending techniques (14). In order to obtain good adhesion to metal surfaces it is sometimes necessary to use a coating of a solution of a copolymer as a primer. Fluidized bed and electrostatic or powder spraying produce thicker coatings than would be obtained by coating with a solution, and have the advantage that no solvents are needed. Costs are much lower than for solution coating. The use of a dry blended powder requires higher temperatures than are needed for the material produced by freeze grinding a fused PVC compound.

Latexes are used extensively for coating substrates such as paper and textiles. Ordinarily the polymer or copolymer is plasticized. Latexes based on copolymers of vinyl chloride and ethyl acrylate are particularly suitable for coating heat-sensitive materials. Coatings of this type produce a decorative effect and provide a water-resistant, strong, and easily cleaned surface. Coated papers include wall coverings, shelf lining, book covers, and packaging materials. Coated fabrics include tarpaulins, tablecloths, and upholstery. Plastisols are often used for such purposes, although a latex is frequently used as a precoat. See also SURFACE COATINGS.

Fibers

Fibers based on vinyl chloride copolymers have been available for some time in the United States; they have the advantages of good weather resistance, low flammability, good abrasion resistance, and low cost. This makes them particularly useful for a number of industrial applications, such as protective clothing and filter fabrics. Other uses include blankets, curtains, imitation hair, and moldable fabrics. Copolymers containing 40% acrylonitrile and 60% vinyl chloride (see MODACRYLIC FIBERS) or 15% vinyl acetate and 85% vinyl chloride (vinyon fibers) are used extensively for this type of fiber.

In Europe and Japan fibers are produced from vinyl chloride homopolymers, which have the advantage over copolymer fibers of better resistance to heat and to dry-cleaning solvents. The polymers used have a higher crystallinity for fibers than the usual vinyl chloride polymers; this is obtained by polymerization at lower temperatures than usual. Fibers produced from commercial PVC shrink at 60–70°C; when they are heated in boiling water, the shrinkage can be as much as 50%. It has been reported that a heat-resistant product can be produced by heterogeneous polymerization of vinyl chloride in methanol with monomer recirculation at about 0°C (15). The transition temperature of this polymer is about 20°C higher than that of ordinary PVC; this markedly improves the heat resistance of thermally stabilized fibers, shrinkage being reduced to about one fifth of that equivalent to fibers produced from ordinary PVC.

Fibers based on PVC are discussed at greater length in the section on Fibers, p. 468.

Foams

PVC can be produced in expanded form by any of the usual blowing techniques. The materials have a number of advantages, both technical and economic, that have led to a notable increase in production over recent years. Flexible foams are normally made by the expansion of plastisols (see the section on Fabrication); they are widely used in upholstery for mattresses, for wire and cable coatings, in automobiles, and for carpet underlays.

PVC rigid foams are used in structural sandwich cores, aircraft bulkheads, roof insulation, and as insulation for marine applications. Superior structural properties give PVC foam a considerable potential for sandwich panels in the building industry if the price structure can be reduced to the level of other foams, such as polyurethan and polystyrene. Cost reductions may possibly be achieved by the development of continuous extrusion techniques, which have been studied extensively, in which organic blowing agents are introduced into the compound. Expanded products can be produced (16) by using an extruder that has a central torpedo arranged so as to produce a hollow profile, which is allowed to expand toward the center. The products formed have a continuous skin with a density gradient decreasing toward the center. The average density of such sections can be as low as 25 lb/ft^3. Hollow sections can be produced, and such a technique would be suitable for making pipes of expanded material.

Bibliography

1. *Plastics* **33** (370), 836 (1968).
2. *Plastics World* **27** (10), 22 (1969).
3. S. C. Hart-Still, Brit. Pat. 1,104,687 (1964).
4. Aro Plastic Building Supplies, Fr. Pat. 1,498,290 (1965).
5. *Brit. Standard Spec. 3505*, 1968.
6. *Brit. Standard Spec. 3506*, 1962.
7. W. F. Gardner and I. Kalushner, *J. Inst. Gas Eng.* **2** (3), 184 (1962).
8. J. L. Benton and C. A. Brighton, *J. Inst. Gas Eng.* **5** (3), 185 (1965).
9. W. Altmann, *Plaste Kautschuk* **13** (7), 418 (1966).
10. J. S. Donovan, *Brit. Plastics* **43** (3), 77 (1970).
11. I. Phillips, *Brit. Plastics* **37** (5,6), 261, 325 (1964).
12. M. V. Maneral, *SPE (Soc. Plastics Engrs.) J.* **25** (11), 31 (1969).
13. *Chem. Eng. News* **46** (30), 35 (1968).
14. W. Adams, *Advances in PVC Compounding and Processing*, Maclaren, 1962, p. 82.
15. A. N. Zay'yalov, Yu. V. Glazhovskii, L. N. Zubov, B. K. Kruptsov, and V. D. Fikhman, *Khim. Volokna* **1969** (2), 29.

C. A. Brighton
BP Chemicals International Ltd

CHLORINATED POLY(VINYL CHLORIDE)

Early work on the chlorination of poly(vinyl chloride) was carried out in Germany at about the same time as the industrial development of the base polymer was taking place. The product with a chlorine content of about 65% by weight is more soluble than PVC in organic solvents and was used as a lacquer and for the manufacture of fibers by the wet-spinning process. During World War II the production of this material in Germany rose to about five million pounds a year. This type of polymer was superseded commercially by copolymers of vinylidene chloride (see VINYLIDENE CHLORIDE POLYMERS), which proved more satisfactory for the applications mentioned although chlorinated PVC has continued to be available in Europe for specialized uses in lacquer. Interest in chlorinated PVC was renewed in 1962 with the publication of a patent (1) in which it was disclosed that if PVC was chlorinated in an aqueous suspension, polymers with increased softening point could be obtained, and that these could be compounded to give products particularly suitable for the manufacture of pipe for high-temperature applications.

Manufacture

In one early method of manufacture a 10% solution of PVC in tetrachloroethane was chlorinated at 90°C for about 24 hr, after which the polymer was isolated by precipitation with an equal volume of methanol (2). The product had a chlorine content of 63–64% and was completely soluble in acetone. An alternative method consisted of treating a 10% solution of PVC in chloroform with liquid chlorine under pressure; the reaction was initiated by heating the reactants to about 50°C, at which temperature the heat of reaction caused the temperature to rise to about 120°C. This procedure reduced the reaction time to about 5 hr and produced a polymer with about the same chlorine content as did the earlier method using gaseous chlorine. This process is difficult and expensive to operate because of the large quantities of solvent required to form the initial polymer solution, and an alternative technique has been disclosed in which PVC is chlorinated in suspension in carbon tetrachloride at 60°C (3). As the reaction proceeds the polymer passes into solution so that the final product must be isolated by precipitation with alcohol.

Recent developments have shown that PVC can be chlorinated under certain conditions to give products that have higher softening points than the original polymer, as well as improved thermal stability. To an aqueous dispersion of PVC polymerized by the suspension process is added about 10% by weight of chloroform in order to swell the polymer. The reaction mixture is subjected to ultraviolet radiation and maintained at 55°C while gaseous chlorine is introduced. A vigorous reaction takes place with the evolution of hydrogen chloride, and after 5 hr the chlorine content of the polymer has risen to about 66%. The solvent in the system is stripped off by steam distillation, after which the aqueous medium is neutralized by the addition of alkali, and the polymer is filtered off, washed, and dried. This method has been modified by using a free-radical initiator such as benzoyl peroxide instead of ultraviolet radiation. This process is comparatively expensive because the majority of the swelling agent is lost (in the case of chloroform it is converted to carbon tetrachloride), and also because of the purification and isolation techniques for the final product. Modifications of this process involve carrying out the chlorination in an aqueous dispersion that contains at least 20% hydrogen chloride at the outset and maintaining the reaction temperature in the range 0–30°C (4). It is claimed that improved thermal stability is obtained if chloroform or some other swelling agent is introduced in the vapor phase with the chlorine (5). It is also claimed that dry PVC powder can be chlorinated directly at 180°C (6), but at this temperature the chlorinated product could not be expected to have as good thermal stability as that produced in dispersion or solution at lower temperatures.

Properties

Most chlorinated PVC resins available commercially have a chlorine content in the range 62–67% by weight and a specific gravity of 1.50–1.575. The softening point of the resin increases as the chlorine content is raised, and Figure 1 shows a plot of the chlorine content against heat-distortion temperature.

It has been found experimentally that it is not possible to introduce more than one chlorine atom for every monomeric unit, on the average, into the polymer molecule. The maximum chlorine content that can be obtained is therefore about 73%; this corresponds to a polymer that consists essentially of poly(1,2-dichloroethylene) with a specific gravity of 1.69–1.71 (theoretical value 1.70), and a heat-distortion temperature of about 115°C. Such a material is most difficult to process because, at the temperatures required to obtain adequate melt flow, decomposition—with the evolution of hydrogen chloride—cannot be controlled; however, the sintering methods used for fluorinated polymers can be employed.

Processing of the partially chlorinated resins requires temperatures about 10°C higher than for PVC. The most suitable stabilizers include barium–cadmium laurate,

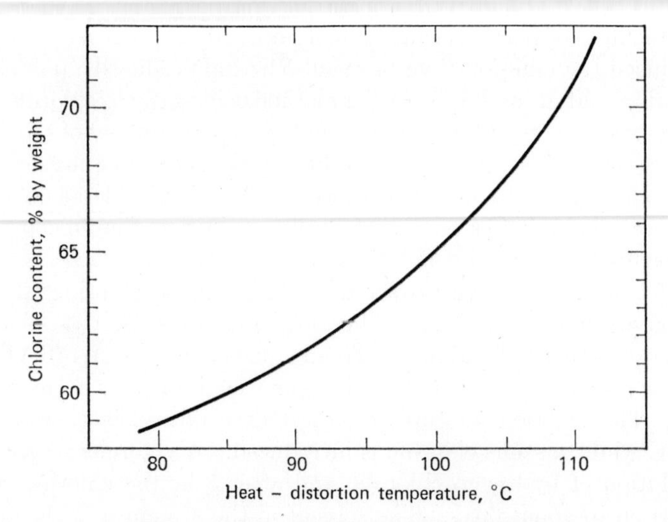

Fig. 1. Softening point of chlorinated PVC.

dibutyltin maleate, and lead stearate; the amount of stabilizer needed is no greater than usual for PVC (7). See the sections on Compounding and Fabrication.

The thermal stability is improved by chlorination, but this improvement is offset by the need to heat the compound to higher temperatures in order to process it. A study of the effects of the variables in the chlorination reaction on the thermal stability has been made (8). The use of chlorobenzene as the swelling agent in the suspension chlorination produces a polymer with better thermal stability than is obtained using chloroform (although the patent literature suggests that the latter is more generally used commercially). This may be due to a greater tendency by chloroform to form free radicals, leading to the presence of trichloromethyl groups in the chlorinated polymer. The thermal stability shows a considerable improvement even at low amounts of chlorination when the reaction is carried out using chlorobenzene as the swelling agent. With further chlorination (up to a chlorine content of 64.5%) the thermal stability

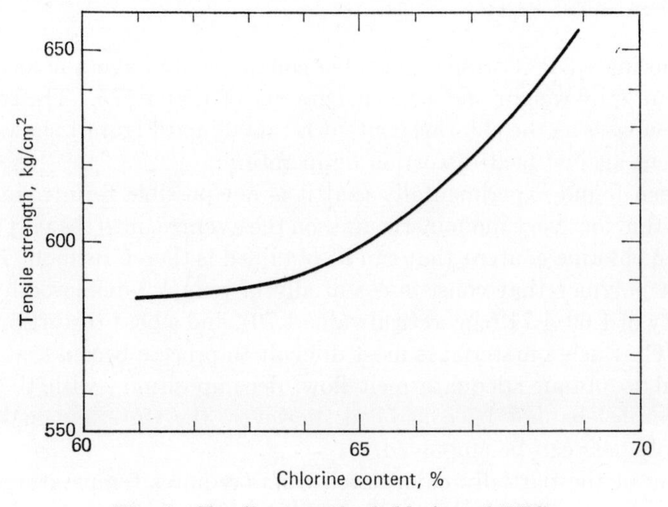

Fig. 2. Tensile strength of chlorinated PVC.

increases to a maximum, after which it falls off with further chlorination. When chloroform is used the heat stability falls sharply and then rises again.

The improvement in thermal stability, particularly at low levels of chlorination, is due undoubtedly to the addition of chlorine to the terminal double bonds of the PVC molecule, which are some of the main centers of instability. As chlorination is continued, some addition will occur to the other double bonds in the molecule; the molecular entanglement will make this second stage more difficult than the first, but should be lessened by the presence of swelling agents. The difference in the effect on the thermal stability in the presence of the two swelling agents, chlorobenzene and chloroform, must be due in part to the readiness of the latter to form free radicals and be converted into carbon tetrachloride.

The tensile strength increases with increasing chlorine content as shown in Figure 2. As the temperature rises the tensile strength falls steadily but still retains a useful value at the temperature of boiling water. The performance of chlorinated PVC compounds in applications at elevated temperatures and under stress is controlled by its creep characteristics; these are compared with those of PVC in Figure 3.

Physical properties of a typical compound suitable for the extrusion of pipe are given in Table 1.

Table 1. Physical Properties of a Typical Chlorinated PVC Compound for Pipe Extrusion[a]

		ASTM method
specific gravity	1.54	D 792
heat deformation temp, °C		
at 264 psi	102	D 648
at 66 psi	119	
impact strength (Izod), ft-lb/in. notch (½ in. × ⅛ in. bar)		D 256
at 82°C	20.5	
at 60°C	16.0	
at 23°C	6.3	
at 0°C	2.6	
tensile strength, psi		
at 100°C	2,100	D 638
at 82°C	3,200	
at 60°C	5,000	
at 23°C	7,300	
at 0°C	9,000	
tensile modulus at 23°C, psi	365,000	D 638
elongation at break at 23°C, %	65	
flexural strength at 23°C, psi	14,500	D 790
flexural modulus at 23°C, psi	380,000	D 790
repeated flexural stress (fatigue), psi	1,900	D 671
coefficient of thermal expansion, cm/cm-°C	7.9×10^{-5}	D 696
thermal conductivity, cal/(sec)(cm²)(°C/cm)	3.3×10^{-4}	C 177
water absorption, 24 hr at 23°C, % weight increase	0.11	D 570
volume resistivity at 23°C, Ω-cm	8×10^{15}	D 257
dielectric strength, V/mil	1,250	D 149

[a] Hi-Temp Geon, B.F. Goodrich Chemical Co.

Structure

A number of investigators have studied the structure of chlorinated polymers by various methods. Early studies by x-ray diffraction suggested that the substitution

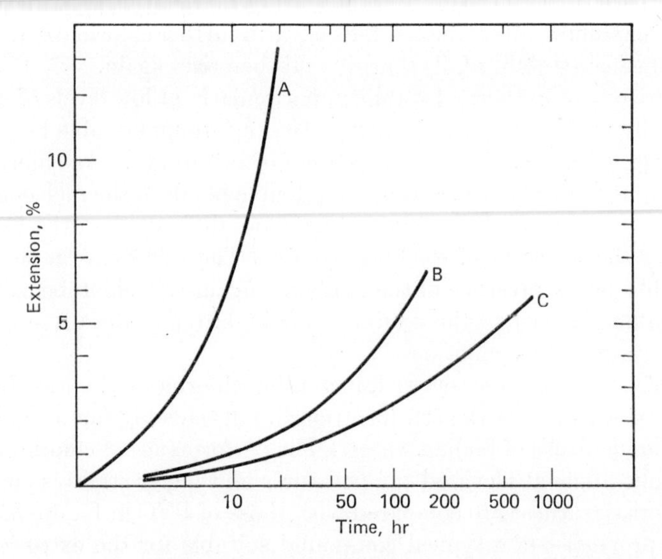

Fig. 3. Creep characteristics of chlorinated PVC at 80°C. A, PVC, 1000 psi; B, chlorinated PVC, 1500 psi; C, chlorinated PVC, 1000 psi.

by chlorine takes place on the carbon atoms that are already chlorinated, so that —CCl_2— groups are formed (9); this would produce a structure equivalent to a copolymer of vinyl chloride and vinylidene chloride. Other authors using infrared spectroscopy found chlorination to take place mainly at the methylene groups (10,11). The reaction is not restricted to these groups, and some substitution also occurs at the —CHCl— groups. The extent to which the two types of chlorination occur has been investigated for a number of different types of reaction, and it is confirmed that chlorination takes place preferentially on the methylene groups (12). The distribution between the two different types of chlorinated structure is independent of the chlorination method.

Accordingly, structures (**1**), (**2**), and (**3**) will occur in chlorinated PVC. In further

—CH_2—CHCl— (**1**) —CHCl—CHCl— (**2**) —CH_2—CCl_2— (**3**)

studies (13) samples of chlorinated PVC were prepared by chlorination at 30°C according to the method described above (1) and the NMR spectra were examined; the composition of chlorinated polymers thus determined is given in Table 2.

Table 2. Structural Composition of Chlorinated PVC

	Chlorine content	Ratio CH_2CHCl^a	Structure (1)	Structure (2)	Structure (3)
PVC	56.6	2.000	100	0	0
chlorinated PVC	57.5	1.864	97.2	2.8	0
	61.2	1.421	81.1	17.5	1.4
	63.2	1.175	71.1	26.8	2.1
	69.5	0.766	31.1	49.9	19.0
	72.6	0.626	5.5	60.5	34.0

[a] Ratio of the number of hydrogen atoms in —CH_2— groups to those in —CHCl— groups.

The results suggest that the chlorinated PVC available commercially with a chlorine content of about 65% contains only very small amounts of consecutive —CH_2—CCl_2— units along the chain. In the early stages of the chlorination reaction, the chlorine reacts preferentially with the —CH_2— groups, and substantial substitution with —CHCl— groups takes place only when the total chlorine content approaches 70%. In the interpretation of the NMR spectra no account has been taken of the presence of steric regularities in the polymer chains, but it is evident that tacticity does influence the spectra. The presence of head-to-head monomer units in the polymer chain may also affect the interpretation of the data. It has been reported that when the chlorine content exceeds 67%, small amounts of the structure —$CHClCCl_2$— can be detected (14).

Applications

The main use for chlorinated PVC in its early development was as textile fibers resistant to hot water. The polymer containing 62–64% chlorine was dissolved in acetone at 40°C to form a 28% solution, which was stabilized by the addition of 0.5% dihydroxy diethyl sulfide. After cooling, the fiber was produced by spinning into water to give fibers ranging from 90 to 1000 den. The fibers were used for fabrics requiring good fungal and chemical resistance, as well as for filter cloths to be used under acidic conditions and other applications, such as ropes, for which natural fibers were unsuitable. Although chlorinated PVC is no longer used for making fibers, it still forms the base for adhesives and lacquers for special applications.

The products from chlorination of high-molecular-weight suspension polymers are used for pipes and fittings, and as sheet for high-temperature applications. Compounds are produced using stabilizers and lubricants similar to those for unplasticized PVC (see the section on Compounding), but because of the higher softening point, higher processing temperatures are required. For this reason, conditions are rather more critical than for PVC and the majority of fabricators use materials compounded by the resin manufacturer. Compounds of various molecular weights are available, those based on higher-molecular-weight PVC being used for extrusion of pipe, whereas the lower-molecular-weight materials are used for injection molding of fittings. The compounds are processed on conventional equipment and by all the techniques common to thermoplastic materials (15). The melt viscosity of the chlorinated polymer is much higher than that of unplasticized PVC, and in a normal extrusion process, there is a tendency for the generation of an excessive amount of heat from shear forces, as well as the production of higher pressures in the die head (16). Screw recommendations are the same as those for rigid PVC and should have a length-to-diameter ratio of 24:1. Hard chrome plating of the screw and the interior die surfaces are recommended for the production of high-quality extrudate. In order to obtain optimum physical properties, the extruder should be operated under such conditions as to give a stock temperature of 195°C. The material in the form of granules is slightly hydroscopic and must be dried by heating for 2 hr at about 100°C before extrusion or molding. Pipe installation can be assembled readily from the various components by normal solvent converting techniques using adhesives based on tetrahydrofuran. Ducts, tanks, hoods, and other structures are easily fabricated from calendered sheet by conventional hot gas plastic welding equipment using a hand gun and prewarmed rod at 260–315°C. Products can also be made by thermoforming techniques.

Properties of Pipes. There has been considerable activity both in Europe and the United States on the replacement of copper by pipe made from chlorinated PVC for domestic hot water and central heating systems (see also PIPE). The performance of pipe under stress at elevated temperatures has been determined experimentally by a number of investigators (17). The impact resistance of chlorinated PVC decreases with the chlorine content and the softening point is raised. This reduction in impact strength has necessitated the use of impact improvers for certain applications and the addition of these additives leads to a reduction of the softening point. Where the optimum heat resistance is required, such as in chemical engineering equipment, no impact improvers are added. The majority of manufacturers, therefore, supply two grades of material for the extrusion of pipe, depending on the impact resistance and the softening point required.

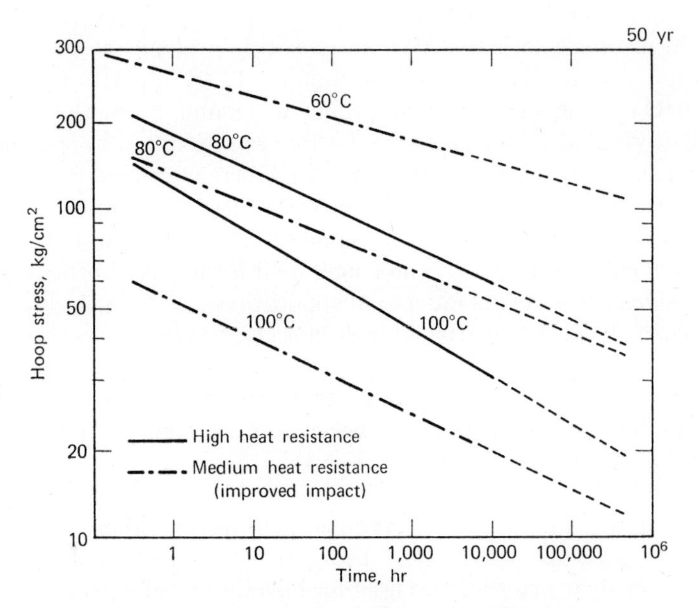

Fig. 4. Burst strength of chlorinated PVC pipes.

Figure 4 shows stress versus time to failure for two types of chlorinated PVC at elevated temperatures. By extrapolating the results obtained for the high-heat-resistant product at 80°C, the hoop stress for failure after a period of 50 yr is determined as 40 kg/cm^2 (570 lb/in.^{2}); the value for pipe made from the material with lower heat resistance (but with improved impact strength) is marginally lower than this. The results at 100°C show a much larger difference, the corresponding values being 22 kg/cm^2 and about 14 kg/cm^2. The extrapolated value for the hoop stress at 60°C for a life of 50 yr for the less heat-resistant material is about 100 kg/cm^2 (1400 lb/in.^{2}), which corresponds to that of ordinary rigid PVC pipe at 20°C and may therefore be used under the same conditions.

In Europe, PVC has been used extensively for the extrusion of pipes used for the conveyance of liquid wastes from the sanitary units to the drainage system. A number of failures have occurred due to excessively high temperatures of the waste water, and some manufacturers have raised the softening point of the pipes by using material based on a blend of PVC and chlorinated PVC. The compounds are prepared by the usual compounding techniques with stabilizers and lubricants. The following table shows the softening point for a range of compounds based on blends of PVC and a chlorinated polymer with a chlorine content of 65%:

chlorinated PVC, parts by weight	100	80	60	40	20	
PVC (K value 65; $\bar{M}_n$ ca 50,000), parts by weight		20	40	60	80	100
softening point, Vicat, 5 kg, °C	112	101	92.5	87	83	80

A rise of about 10°C in the Vicat softening point is considered adequate to give the margin of safety that will eliminate the risk of failure; this is obtained with blends containing approximately equal amounts of the two polymers. There is by no means universal agreement that the small number of failures have occurred because of the low

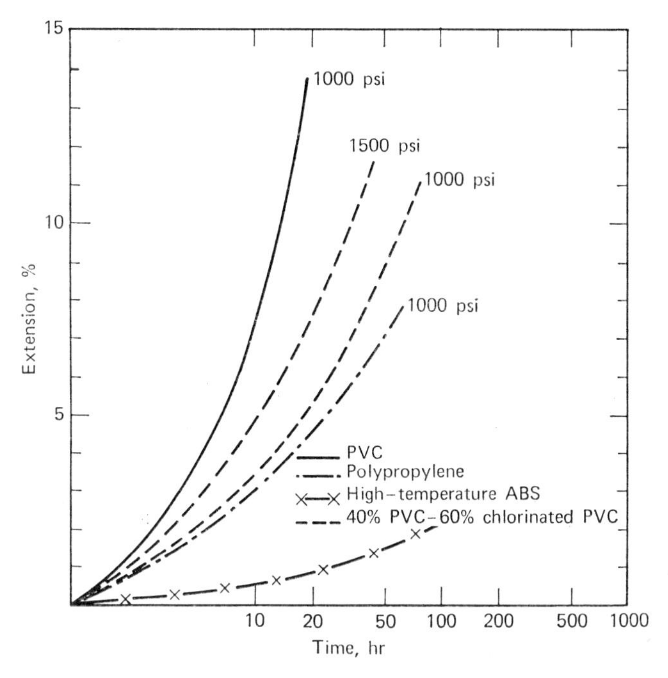

Fig. 5. Creep characteristics of PVC–chlorinated PVC blends at 80°C.

softening point of PVC rather than because of unsatisfactory installation. For this reason it is considered that PVC is quite adequate for this application so long as the thickness is at least 0.125 in. Although the blends of 40% of chlorinated PVC with 60% PVC have a Vicat softening point 7°C higher than that of PVC, this difference is not reflected in the creep characteristics of the material; Figure 5 shows that the blend has comparatively little advantage.

Bibliography

1. B.F. Goodrich Chemical Co., Brit. Pat. 893,288.
2. J. M. Bell, W. C. Coggin, and W. E. Gloor, *German Plastics Practice*, DeBell and Richardson, Springfield, Mass., 1946, p. 79.
3. C. Schonburg, U.S. Pat. 1,982,765.
4. Imperial Chemical Industries, Ltd., Brit. Pat. 1,120,769.
5. Dynamit Nobel, U.S. Pat. 3,362,896.
6. H. Jacqué, Ger. Pat. 801,304.
7. R. Huth, *Kunststoffe Plastics*, **55,** 319 (May 1965).
8. E. N. Zil'berman, P. S. Pyryalova, and F. A. Ekstrin, *Plasticheskie Massy* **8,** 10 (Aug. 1968).
9. O. Seipold, *Chem. Tech. (Berlin)* **5,** 467 (1953).
10. H. Tucks and D. Louis, *Makromol. Chem.* **22** (1), 1 (1957).
11. H. Germar, *Makromol. Chem.* **86,** 89 (1965).
12. O. Fredriksen and J. A. Crowo, *Makromol. Chem.* **100,** 231 (1967).
13. J. Petersen and B. Rånby, *Makromol. Chem.* **102,** 83 (1967).
14. S. Tsuge, T. Okumoto, and T. Takeuchi, *Macromolecules* **2,** 277–280 (May-June, 1969).
15. E. M. Faber, *SPE (Soc. Plastics Engrs.) J.* **22** (8), 49 (1966).
16. G. H. Arnold, *Plastics and Polymers* **38** (133), 21 (1970).
17. P. Decroly, *Plastics Paint and Rubber* 66 (March/April 1968).

C. A. Brighton
BP Chemicals International Ltd.

FIBERS

Historically, poly(vinyl chloride) (PVC) enjoys the distinction of being the first synthetic polymer to be fabricated into man-made fibers. A patent issued in 1913 (1) speaks of making artificial threads by coagulating a solution of poly(vinyl chloride) in chlorobenzene. This work was somewhat premature, however, and serious development of PVC fibers did not begin until the early 1930s in both Germany and the United States.

The initial obstacle to commercial development was the insolubility of the material in the solvents commercially available at that time. This problem was overcome in Europe in two ways. The Germans developed a technique of "postchlorination" of PVC which rendered it soluble in acetone (2), while the French discovered that PVC could be dissolved in mixed solvents, especially in acetone–carbon disulfide mixtures (3). The German fibers were designated PeCe fibers and the French Rhovyl. A subsequent German discovery was the solubility of PVC in tetrahydrofuran, which led to the development of "PCU" fibers which were not postchlorinated.

In the United States, the solubility problem was solved by the Carbide and Carbon Chemicals Company (now Union Carbide Corporation), by copolymerizing vinyl chloride with vinyl acetate and other monomers. In particular, the copolymer with vinyl acetate was found to be soluble in acetone and was developed into a commercial fiber by this company in cooperation with American Viscose Corporation. Another copolymer developed by Carbide and Carbon was that using acrylonitrile as comonomer. This material has also been a commercial success and is described in detail in the article on modacrylic fibers (qv) and hence will not be covered here.

The Textile Fiber Products Identification Act of 1960 defines as "vinyon" any fiber containing at least 85% by weight of vinyl chloride units. Under this definition all homopolymer PVC fibers as well as the vinyl chloride–vinyl acetate copolymer mentioned above (which is 86% vinyl chloride) are legally designated as vinyon. In the United States, however, the vinyl acetate copolymer is the only vinyl chloride fiber manufactured in reasonably large quantities and is known as Vinyon HH. It should be emphasized, however, that the generic term "vinyon fiber" may be applied to homopolymeric PVC or other copolymers containing 85% or more vinyl chloride.

Chemical Compositions

Homopolymer. The structure of vinyl chloride is $CH_2{=}CHCl$. Polymerization is via addition, and the structure of the polymer is symbolized by $\left(CH_2CHCl\right)_n$. In the past, there was considerable controversy over the question of whether the vinyl chloride units added in a head-to-tail fashion, eg,

$$\sim\!\!\sim CH_2—CHCl—CH_2—CHCl—\sim\!\!\sim$$

or in a head-to-head, tail-to-tail manner,

$$\sim\!\!\sim CH_2—CHCl—CHCl—CH_2—\sim\!\!\sim$$

The head-to-tail structure was conclusively demonstrated by the work of Marvel and co-workers (4) and Flory (5).

Examination of the structural formula given above shows that the carbon atoms containing the chlorine atoms are asymmetric and therefore may exist in two stereoisomeric forms. This gives rise to three possible conformations for the molecule as follows:

(*1*) All asymmetric atoms have the same configuration. This is called an isotactic conformation.

(*2*) The asymmetric atoms have alternate *d* and *l* configurations. This is called the syndiotactic conformation.

(*3*) The asymmetric atom configurations are randomly ordered. This is called the atactic conformation.

The exact conformation achieved is important because only structures (*1*) and (*2*) above would have sufficient regularity to permit the molecules to crystallize, and crystallinity is one of the two most important molecular parameters of a fiber that affect its physical properties. (The other parameter is the molecular orientation; see FIBERS).

X-ray investigations of poly(vinyl chloride) polymerized at room temperature or higher, in solution, suspension, or bulk, reveal that very little crystallinity is present in the polymer (6). However, in 1959 Fordham showed that, on the basis of energy considerations, the addition of one vinyl group to another requires slightly less energy for the syndiotactic configuration, ie, that in which the chlorine atoms are trans to each other, than for the isotactic configuration with both chlorine atoms on the same side of the chain (7). This means that at lower temperatures or in solvents of low dielectric constants, the syndiotactic addition is favored and crystallizable PVC may be produced. This has proved to be the case, and it is found that polymerizing PVC at temperatures substantially below $-20°C$ results in increased tacticity, giving increased crystallinity, higher softening temperatures, and higher glass-transition temperatures (8,9). The Italian firm Chatillon S.p.A. has commercialized a fiber, Leavil (or Leavin), based on this discovery and showing superior heat and solvent resistance (11) in the crystallized form. Burleigh (10) has demonstrated similar effects in PVC polymerized in aliphatic aldehydes (low dielectric constant) at 50°C, but the molecular weights obtained were extremely low (about 5000).

Postchlorinated PVC. As mentioned previously, one of the first commercial vinyl chloride fibers was made from PVC that had been further chlorinated. These fibers, called PeCe, were first produced commercially (12) in 1934 and are still being produced in limited quantity in East Germany. In the process believed to be used, a solution of 7–8% PVC in tetrachloroethane is saturated with chlorine gas at 70°C. The heat of reaction is continuously removed and the temperature kept below 120°C. After 24–40 hr the solution is cooled, excess chlorine and hydrochloric acid are removed under vacuum, and the chlorinated polymer is precipitated with methanol, filtered, washed with methanol, and dried. The chlorine content of the resulting polymer is raised from 56 to 62–65% by this process, and the resulting polymer is soluble in acetone and thus ready for dry or wet spinning.

The structure of chlorinated PVC is a subject of controversy, particularly in regard to the relative abundance of dichloroethylene groups (ie, —CHCl—CHCl—) and vinylidene chloride groups (ie, —CH_2—CCl_2—) (13–15). According to Petersen and Rånby (15) the dichloroethylene groups increase continuously with increasing chlorination, whereas the vinylidene chloride groups begin to form at about 60% chlorine and then continue to increase as the chlorine content increases. These workers (15) chlorinated PVC in a suspension of water and chloroform at room temperature. If their results can be translated to the commercial product (solvent chlorinated), there are about 27% dichloroethylene groups, 2% vinylidene chloride groups, and 71% vinyl chloride groups at a total chlorine content of 63% in the polymer.

Although these fibers never achieved great commercial success there appears to be increasing interest in this type of polymer (ca 68–70% Cl), owing to the fact that its heat-distortion temperature is about 20°C higher than the homopolymer and it shows an improved impact strength over homopolymer, which makes the material attractive for plastic molding and extrusion applications (16). This in turn may revive interest in the polymer as a fiber former. See also VINYLIDENE CHLORIDE POLYMERS.

Vinyl Chloride–Vinyl Acetate Copolymer. This material, Vinyon HH, is a copolymer containing 86% vinyl chloride and 14% vinyl acetate and is soluble in acetone and hence easily dry spun into fibers. The polymer is made by free-radical copolymerization of vinyl chloride and vinyl acetate in solution, using a catalyst such as benzoyl peroxide (17), at temperatures around 30°C. The molecular weight should be carefully controlled, for uniformity of fiber properties and for optimum solubility in acetone.

The resulting material consists of a random copolymer of vinyl chloride and vinyl acetate, linked in a head-to-tail fashion. Work by Douglas and Stoops (18) showed that the chlorine content of the copolymer, and hence the mole ratio of vinyl chloride to vinyl acetate, did not vary greatly in the molecular weight range from 5,800 to 15,800.

Graft Copolymers and Polymer Blends. Recent developments of PVC fibers in Japan have resulted in a vinyl chloride–vinyl alcohol fiber being produced under the names Cordela and Kohjin. The fiber is said to be 50% vinyl chloride and 50% vinyl alcohol, and is a mixture of a graft copolymer of the two monomers blended with the two homopolymers (19). The spinning technique is said to be emulsion spinning, ie, spinning of an emulsion of the polymer species which is coagulated in the spinning medium. The French PVC fiber Clevyl is reportedly produced from a blend of two homopolymers.

Manufacture

Polymerization. Vinyl chloride, manufactured from ethylene or acetylene (20), may be polymerized in a variety of ways, eg, in bulk, solution, emulsion, or suspension. The most important of these, accounting for the largest volume of polymer produced, is suspension polymerization, with emulsion a close second. Suspension polymerization is favored by the producers of PVC plastics because of its processing advantages, ie, better heat transfer and hence temperature control, and because of product quality, especially the particle size of the resulting white, powdery polymer. Emulsion polymerization is used for the preparation of PVC latexes, ie, colloidal suspensions of PVC which have many end uses. Solution polymerization is still used to prepare the polymer for Vinyon HH fibers because it lends itself to continuous production and to a careful control of molecular-weight distribution. Bulk polymerization was not used commercially until recently, because of the difficulty of controlling the heat of reaction, but has now become practical (see the other sections of this article dealing with commercial methods of polymerization).

All the above methods of polymerization depend on a free-radical reaction. An initiator (sometimes called a catalyst) is added to the monomer to provide free radicals, which then react with monomer to begin the polymerization reaction. The growing polymer chain is stopped by a chain-transfer reaction to monomer or by a chain-stopping reaction with impurities. The rate of the reaction is proportional to the square root of the initiator concentration and to the monomer concentration. The average molecular weight is inversely proportional to the temperature.

In a typical suspension polymerization the cooled monomer is charged to an autoclave, under pressure, containing water and a dispersing agent such as poly(vinyl alcohol) or gelatin. The vinyl chloride forms relatively large droplets, each of which becomes a miniature reactor when the initiator, a monomer-soluble material such as benzoyl peroxide, is added. The mixture is agitated and the temperature is carefully controlled during polymerization. Because of the low viscosity of the suspension, heat transfer is very efficient. Conversion efficiency is very close to 100%. After polymerization the white granules of PVC are separated from the suspension, the unreacted monomer is removed under vacuum and heat, and the polymer particles are washed and dried. The size of the granules can be very carefully controlled under this system and this enables the producer to offer material of better solubility for further processing into plastic, fibers, or film. See also SUSPENSION POLYMERIZATION.

Emulsion polymerization is superficially similar to this, but is believed to proceed by a different physical mechanism. The sizes of the monomer droplets are much smaller, being only several hundred angstroms in diameter, and the initiator is water soluble. It is believed that polymerization takes place in micelles of emulsifier, with the monomer serving as a reservoir supplying the emulsion micelle. The end result is a very stable emulsion (latex) which must be dried if it is desired to recover the polymer in solid form. See also EMULSION POLYMERIZATION.

Bulk polymerization (qv) may be used to prepare low-temperature polymerized, syndiotactic PVC. For practical purposes, however, a concentrated solution is normally used. Although there is a large body of technical information on the chemistry of this process, little is known of the commercial process used. Various organometallic compounds may be used as catalysts, usually in a redox system, which give rise to free radicals. Typical catalysts are tri-*n*-butylborane with hydrogen peroxide (21), tetraethyllead with ceric ammonium nitrate (22), and triethylborane with pyridine (23). Photoinitiated polymerization with uranyl nitrate catalyst has also been reported (24). Polymerization takes place at temperatures from 0 to −80°C, either in bulk or concentrated solution in solvents such as methanol or heptane. A temperature of −20°C seems to be a good compromise, giving good syndiotacticity and high molecular weight. The molecular weight of the polymer goes through a maximum at about −20°C as the temperature is reduced (25). Table 1, taken from the work of Nakajima and co-workers (24), shows the profound effect of the temperature of polymerization on the conformation and physical properties of the polymer. It has been reported that improved syndiotactic fibers can be prepared at −30°C with an initiator consisting of cumene hydroperoxide, sulfur dioxide, and sodium alkoxide (41).

Table 1. Effect of Polymerization Temperature on Properties of Poly(vinyl Chloride) (24)

Polymerization temp, °C	$\overline{DP}$	Degree of branching, $CH_3/100\ CH_2$	Syndiotacticity[a]	Density[b]	Crystallinity[c]	T_g, °C
90	380	27	0.51	1.391	11.3	
55–60	1200	0.20	0.53	1.391	11.3	85
50	1460	0.20	0.54	1.392	13.2	
20	2760	0.15	0.52	1.393	15.0	
−15	4700	0.05	0.64	1.416	57.3	105
−75	1700	0.00	0.77	1.431	84.2	

[a] Fraction of vinyl chloride groups attached in the syndiotactic configuration, as measured by infrared.

[b] Density after annealing at 85°C for 2 hr, measured at 30°C.

[c] Percent crystallinity calculated from density.

The effect of increasing syndiotacticity and decreasing branching is to make the polymer more crystallizable, and this has a marked effect on its ability to form good fibers.

The copolymer with vinyl acetate used for making Vinyon HH fiber is made by solution polymerization. Although the exact details of the process have not been revealed, an old patent (26) discloses a continuous process in butane solution at temperatures down to 30°C using acetyl benzoyl peroxide as initiator. Careful control of time and temperature is required to give a reproducible molecular weight and a narrow molecular-weight distribution. Control of color is also important for fiber applications, and excellent colorless polymers are said to be produced by this process.

The polymer used in the production of Cordela fibers, according to the patent literature (27), is a mixture of poly(vinyl alcohol) and a graft copolymer of vinyl chloride and vinyl alcohol. An emulsion of about ten parts of vinyl chloride and one part of poly(vinyl alcohol) is made containing a catalyst, eg, potassium peroxydisulfate. Polymerization takes place in an autoclave at 45°C, during which PVC is grafted onto the poly(vinyl alcohol). After polymerization additional poly(vinyl alcohol) is added to bring the mole ratio of poly(vinyl chloride) to poly(vinyl alcohol) to approximately unity. This resulting mixture then forms the spinning dope from which the fibers are produced.

Spinning. Of the three methods of forming fibers from polymers, wet, dry, and melt spinning, only the last has not been used to produce a commercial textile fiber from PVC. The reason for this is that at the melting point (or flow point) of PVC (about 175°C), thermal degradation of the polymer takes place. The molten polymer is also degraded by high shear rates commonly encountered in melt extruders. These two factors lead to off-color fibers, and black particles that plug jet holes. In extruding the polymer for plastic articles, film, and sheet, stabilizers are added in large quantities. These also act as plasticizers and would seriously degrade the physical properties if used to form fibers. Melt spinning has been used, however, in Japan to produce monofilaments and continuous filaments for fishing nets.

In wet and dry spinning of PVC the polymer is first dissolved in a suitable solvent before extrusion through the spinneret at moderate temperatures. In wet spinning, the emerging filaments come into contact with a coagulating bath, whereas in dry spinning the solvent is evaporated by means of hot air.

As mentioned previously, one of the initial problems of spinning PVC was the selection of a suitable solvent. In the past two or three decades, however, a large number of solvents have become commercially available for PVC. A patent survey of the years 1947–1964 has revealed the following solvents claimed: cyclohexanone; tetrahydrofuran; methyl ethyl ketone– or acetone–carbon disulfide mixtures; dimethylformamide; thionyl chloride; trichloroethylene–nitromethane mixture; diacetone alcohol mixed with various ketones; *N*-methyltetrahydrofurfurylamine; tetrahydrofurfuryl alcohol; tetrahydrofuran with lactones or lactams; and tetrahydrofuran with sulfones or sulfoxides. However, only two or three solvents or solvent systems are currently in commercial use. In dry spinning, a 50:50 or 60:40 solution of benzene–acetone is used in Japan, and a 50:50 mixture of carbon disulfide and acetone in Europe. In wet spinning regular PVC, a solution in tetrahydrofuran or tetrahydrofuran–acetone may be used, whereas for the syndiotactic, crystallizable PVC a solution in cyclohexanone is used. For chlorinated PVC and Vinyon HH the solvent is acetone, but Cordela is wet spun from an aqueous emulsion.

In all cases, the spinning dope is made as concentrated as possible, consistent with practical handling of highly viscous solutions, to minimize the amount of solvent recovery necessary per pound of fiber. Concentrations on the order of 20–30% by weight are normally used in dry spinning with lower concentrations (10–20%) in wet spinning. Various techniques may be used to dissolve the polymer in certain solvents, for example, contacting the polymer with solvent at low temperature to achieve good dispersion under minimum swelling conditions, and then raising the temperature to obtain dissolution (28). Another technique is to wet the polymer with one component of a two-component solvent before adding the second component. This also gives good dispersion before the polymer mass is allowed to dissolve. Corbiere, Terra, and Paris (29) have published a detailed study of the solvent power of mixed solvents for PVC.

The concentrated solution is very carefully filtered prior to spinning to remove gel particles and dirt which would plug holes in the spinnerets and thus shorten jet life. Since the polymer is stable at room temperature, the solution may be stored indefinitely prior to spinning.

Dry spinning is accomplished in a manner and in apparatus similar to those used in spinning cellulose acetate (see Cellulose esters, organic). The polymer solution is heated under pressure, filtered a second time, and extruded through the holes of the spinneret into a long tube or chamber containing heated air, which evaporates the solvent. The solvent-laden air may move either cocurrent or countercurrent to the filaments, but generally the latter is the case. The air emerging from the tube goes to the solvent-recovery system, which may be a carbon adsorption bed, water, or a condenser. The emerging filaments are oiled to provide lubrication and antistatic protection and wound on a spool or bobbin in a conventional manner, or else formed into a tow for further processing into staple.

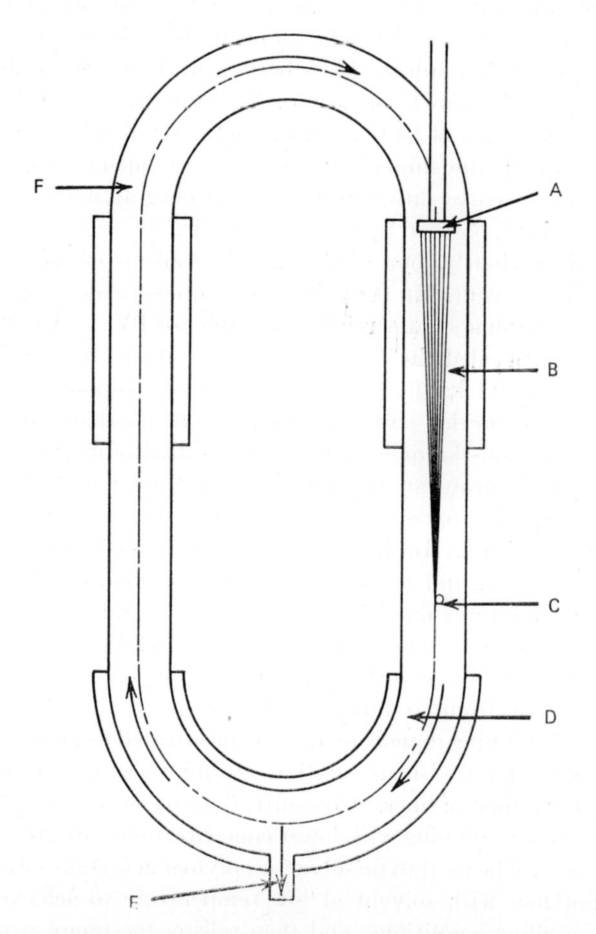

Fig. 1. Spinning cell. A, spinneret; B, heated section; C, to yarn takeup; D, cooled seetion; E, to solvent recycle; F, heated section (32).

Société Rhovyl, the largest European producer of PVC fibers, uses a variation on the conventional dry spinning system in which the spinning tube forms part of a system that is closed except for the entrance of the filaments through the spinneret and their exit at a small orifice near the bottom of the tube (30). The solvent-laden air passes cocurrent with the filaments to a section below their exit where it is condensed by a cooling section of the tube. The condensed solvent is continuously drawn off while the air, with considerably less solvent vapor, is reheated and again transferred to the area of the spinning head. See Figure 1 (32).

Spinning speeds in dry spinning range from about 100 to 500 m/min. However, the total denier of the spun yarn is limited by the rate of diffusion of solvent out of the yarn, which in turn is a function of the size of the tube, the velocity of the air, and the nature of the solvent. In general, only fine-denier yarns for continuous filament may be dry spun. In the case of staple, however, the tow may be made from a number of dry-spun filaments gathered together, or it may be wet spun. Spinning speeds for wet spinning may be only one tenth of those used in dry spinning, but the ability to spin many thousands of filaments simultaneously may be more economical in the long run; ie, the total spun denier per minute may be larger even though the individual filaments travel at a slower speed.

In wet spinning, the polymer solution emerging from the jet is contacted with a coagulating bath. This bath must be a nonsolvent for the polymer but miscible with the liquid used to dissolve the polymer in the spinning dope. The spin bath solidifies the stream of polymer into a filament and provides a medium for the diffusion of the solvent out of the thread. These two functions must be nicely balanced. Thus, too rapid a diffusion of solvent leads to weak, porous fibers containing large voids, and too rapid a coagulation may make removal of the solvent extremely difficult.

In the wet spinning of the syndiotactic fiber Leavil (31), the polymer is dissolved in hot cyclohexanone at 137°C. A hot, 18% solution is extruded through jet holes of 0.1 mm diameter into a cold (60°C) bath consisting of 50% water, 24% cyclohexanone, and 26% ethyl alcohol. The drop in temperature, as well as the presence of water, coagulates the polymer, while the ethyl alcohol solubilizes the cyclohexanone solvent in the spin bath. On emerging from the spin bath the filaments are washed with a solution of ethyl alcohol to remove all traces of cyclohexanone.

In the case of the vinyl chloride–vinyl alcohol graft copolymer mixture Cordela, the emulsion prepared as previously described, which contains about 60% solid material, is spun into a 40% (wt/vol) aqueous solution of sodium sulfate. The high salt content presumably breaks the emulsion, permitting the solid particles to coalesce into a continuous fiber which is then dried in air without tension. See also MAN-MADE FIBERS, MANUFACTURE.

Aftertreatment. PVC fibers immediately after spinning are weak and highly extensible. As is the case with most synthetic fibers, it is necessary to stretch the fiber severalfold in order to orient the molecules in the direction of the fiber axis and thus impart strength to the material, usually at a sacrifice of extensibility. Stretching is done in air or water at temperatures substantially above the glass-transition temperature of 75°C, and to 200–700% of the unstretched length. Although this improves the strength at room temperature, the stresses that are built into the molecular structure are relieved when the fiber is subsequently heated above the glass-transition temperature. The result is a shrinkage of the fiber, which is one of the major defects of PVC as a textile material, ie, lack of dimensional stability toward hot-water laundering.

Types of Products. Regular, atactic PVC is available in a wide range of deniers, in both continuous filament and staple. Typical sizes for staple are 1.8, 3.0, 3.5, 5.0, 6.0, 9.0, and 15 den/filament. Typical filament yarn constructions are 75/24, 120/38, 200/64, 800/256, and 3200/1024 den/filament count, which is about 3.1 den/filament. The fibers may also be characterized by their shrinkage at boiling-water temperatures, since this may be controlled by a relaxed shrinkage at high temperatures after the stretching operation. Thus, Rhovyl 55 will shrink 55%, Clevyl F will shrink 25%, and Rhovyl T will shrink 0% at 100°C (32).

The syndiotactic fiber Leavil is available only in staple form, probably because the wet-spinning process would not be economical for fine-denier continuous filament. A full range of deniers is available from 2.5 to 35.0 den, and shrinkage of these fibers is essentially zero at boiling-water temperatures.

Vinyon HH is available only as staple because the major end use, ie, as a fibrous binder in nonwoven fabrics (qv), demands it in that form even though the fiber is dry spun. Fibers are offered in 1.5, 3.0, and 5.5 den/filament and a variety of cut lengths from as short as ¼ to 4 in. Cordela, produced by a form of wet spinning, is available as staple or tow in deniers from 1.5 to 60.

Physical Properties

Mechanical Properties. The mechanical properties of a textile fiber are determined by the degree of orientation of the molecules along the fiber axis and by the crystallinity of the fiber, as well as by the inherent properties of the macromolecules, eg, the flexibility of the molecular backbone. In general, the tenacity and modulus are determined by the orientation, whereas the elongation to break, stiffness, and stability to heat are functions of the crystallinity. In ordinary atactic PVC the crystallinity is of a very low order; hence the material acts more or less as a highly viscous liquid or a glass, depending on the temperature. The demarcation temperature between the glassy and viscous liquid states is the glass-transition temperature (qv). For ordinary PVC this temperature is 75°C (32). Below this temperature the individual atoms and segments of the molecule can engage only in vibratory or limited rotational movement.

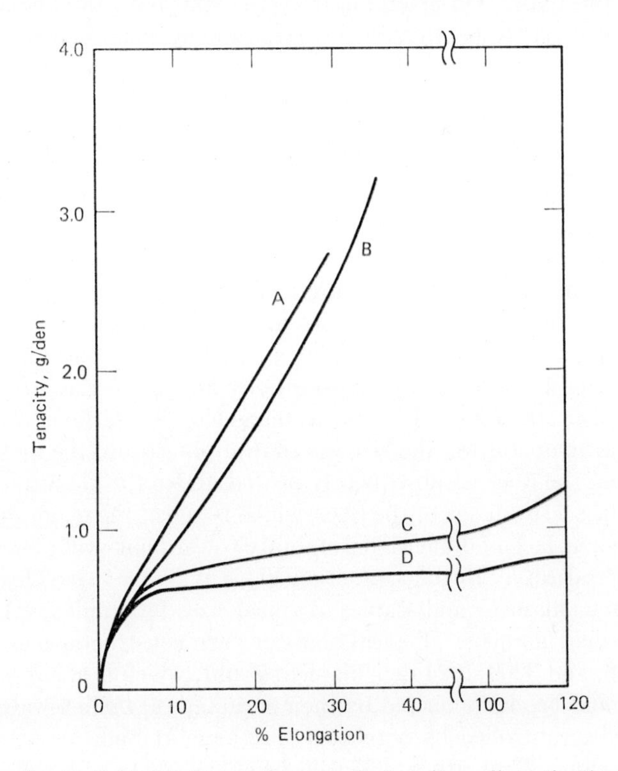

Fig. 2. A, C, atactic vinyl chloride homopolymers; B, Leavil syndiotactic fiber; D, Vinyon HH copolymer fiber.

Above the transition temperature large segments of the molecules can move and rotate in concert. Hence at temperatures approaching or above 75°C various strains that are built into the fiber may relax. The most important effect is that the orientation of the molecules becomes more random, the fiber shrinks, and the surface becomes soft and tacky. This effect is simultaneously the main drawback and one of the major advantages of ordinary PVC fibers.

It is a disadvantage in that it means that articles made of these fibers may not be laundered in hot water or ironed without serious shrinkage and softening. It is an advantage in the construction of various novelty yarns and fabrics, eg, bulked yarns, cloque fabrics, and "sculptured" pile fabrics, where the shrinkage is deliberately used to create special effects (32).

Ordinary PVC yarns and fibers may have tenacities ranging from 1.2 to 3.0 g/den and extensibilities from 12 to 180% with high tenacity corresponding to low extensibility and vice versa. Obviously, this range of properties spans the spectrum from wool-like to cottonlike properties. Figure 2 shows the stress–strain curves of atactic vinyl chloride homopolymer fibers at the two ends of the scale as well as of Leavil and Vinyon HH. It is to be noted that in all cases there is a yield point in the curve at a few percent elongation. This is the point beyond which the deformation is not recoverable.

The syndiotactic fiber Leavil is distinguished from ordinary PVC by the fact that it is crystallizable. This crystallinity, which develops during the hot stretching or later heat setting of the fiber or fabric, tends to stabilize the molecular configuration of the fiber to the point where much higher temperatures are required to relax the strains produced by stretching. Thus, whereas ordinary PVC, such as Rhovyl 55, begins to shrink at about 65°C when heated in silicone oil, stabilized Leavil does not start to shrink until 120°C is reached (33). Other fibers, such as Clevyl, have intermediate heat stability (75–110°C, depending on grade). The higher crystallinity of Leavil also makes it possible to develop a higher tenacity and elongation to break, such as 3.6 g/den tenacity at 40% extensibility for 2.5 den staple. A further advantage is decreased swelling in organic solvents, which makes syndiotactic PVC more resistant to drycleaning agents than regular PVC.

The copolymer fiber Vinyon HH shows the same lack of crystallinity as homopolymer, atactic PVC. In addition the presence of vinyl acetate makes the fiber liable to thermal shrinkage at even lower temperatures (50°C) than the homopolymer. This characteristic has been commercially exploited in that the major end use of Vinyon HH staple is as a fibrous binder for nonwoven fabrics. The Vinyon HH is dispersed and intimately mixed with the other fibers making up the nonwoven, and the application of heat enables the fabric to be sealed to itself. A simple example is in tea bags. Vinyon fibers as currently marketed have inadequate physical properties for textile uses: tenacity, about 0.6 g/den, and extensibility, 125%. These can be improved by stretching, however, if the end use warrants it.

Cordela fiber is said to have tenacities in the range of 2.8–3.3 g/den with extensibilities of 20–24% in the dry state. Wet strengths drop to 2.0–2.3 g/den at equivalent extensibilities (19). This sensitivity to water is undoubtedly due to the poly(vinyl alcohol) present in the fiber.

Nonflammability. PVC fibers, in all the forms discussed in this article, are completely nonflammable in the pure state, ie, unmixed with other fibers; in fact, PVC is the only organic synthetic fiber that has this property. When exposed to an open flame, PVC chars but does not propagate the flame, nor does it melt and drip, as is the case with many other synthetic fibers. PVC homopolymer fibers, moreover, impart nonflammability to blends with other fibers, provided the mixture contains 60% or more PVC.

Moisture Regain. Moisture regain is virtually nil for homopolymer PVC fibers, about 0.1% for Vinyon HH, and 3.2% for Cordela.

Chemical Resistance. Homopolymer PVC fibers are inert to all chemicals with the exception of chlorinated organic solvents and certain other organic chemicals, eg, dimethylformamide, cyclohexanone, and tetrahydrofuran, which tend to swell or dissolve the fiber. A greater range of organic chemicals will dissolve Vinyon HH and postchlorinated PVC, including most low-molecular-weight ketones and aldehydes, and the same general conclusion is probably true regarding Cordela, although no data are yet available on this fiber.

Biological Resistance. All PVC fibers show a high resistance to biological deterioration; ie, they do not mildew or rot.

Weathering. In a study of the effect of atmospheric conditions on the properties of synthetic fibers, PVC fibers demonstrated outstanding ability to withstand weathering (34).

Electrical Properties. The dielectric constant of PVC is 3.2 at 60 Hz and 2.9 at 10^6 Hz with corresponding loss factor tan δ of 0.013 and 0.016, respectively. The dielectric strength is 400–800 V/mil and the volume resistivity is 10^{14-16} Ω-cm (35). The dielectric constant of Vinyon HH is 3.45 at 1 Hz, with a loss factor of 0.016.

The low dielectric constant and high resistivity are responsible for the high static electrical charges built up on PVC fibers that are not protected by an antistatic agent (qv). This high charge in turn has been claimed to be responsible for the analgesic and antirheumatic properties of PVC fibers (36).

Dyeing

Because of their hydrophobicity and lack of specific polar groups on the molecule, PVC fibers are generally difficult to dye in deep or bright shades. For this reason the practice of "dope" or "solution" dyeing is often encountered in PVC fiber production. In this system, a colored pigment is added in finely dispersed form to the spinning solution and the resulting fiber retains the pigment. Such colored fibers are very fast and resistant to sunlight, laundering, etc. On the other hand, this method of achieving colored fibers is not very economical from a manufacturing standpoint unless large quantities of a given shade are made. Otherwise, the cost of cleaning equipment for each new color, matching colors, and maintaining large inventories of various shades becomes prohibitive.

PVC fibers can be dyed by using the same type of dyestuff originally developed for acetate fibers, ie, the dispersed dyes. In this system the dyestuff, which is insoluble in water, is dispersed in water and brought into contact with the fabric or fibers. Because of its greater solubility in the organic fiber phase, the dyestuff dissolves in the fiber to a greater or lesser extent, usually dependent on the temperature. Most regular and postchlorinated PVC fibers are temperature sensitive, as mentioned above, and cannot be heated above 60°C in the dyebath without shrinking, loss of strength, and deformation of the fabric. Some PVC fibers, eg, Rhovyl T, Clevyl T, and Movil, have been allowed to relax in boiling water during manufacture. These fibers may be dyed at the boil to achieve deeper shades. Fibers that have not had this heat treatment can be dyed only to pale shades at low temperatures unless a dyeing assistant or "carrier" is used. This is an organic solvent, eg, an aromatic halide, which is dispersed with the dyestuff and whose function is to swell the fiber and thus increase the penetration of the dyestuff. Of course, such swelling also impairs the mechanical properties of the fibers, but not to the same degree as heat.

Syndiotactic PVC is much more resistant to heat than the atactic form and can be satisfactorily dyed at the boil without the aid of a carrier. In fact, it is claimed that this fiber can be dyed under pressure at still higher temperatures than normal boiling to achieve even greater penetration of the dyestuffs (33). Dimensionally stable PVC, such as the homopolymer blend in Clevyl fibers, also has good stability at dyeing temperatures (Clevyl T does not shrink up to 110°C).

Vinyon HH is subject to the same problem in dyeing as atactic PVC with the added disadvantage of a slightly lower softening temperature, or about 50°C. Since this fiber is generally more soluble than homopolymer PVC, smaller quantities of carrier at lower temperatures are effective in dyeing. The 50% poly(vinyl alcohol) present in Cordela improves the dyeability markedly by providing active dyesites on the molecule (27). This fiber is said to be dyeable with sulfide, azoic, vat, disperse, and complex salt dyes (19). See also DYEING.

Applications

The history of PVC fibers is an interesting study of the effect of economic, political, and technological factors on the production and utilization of textile material. These fibers were initially produced in Germany in the years prior to World War II and undoubtedly exploited as part of the effort of the German government at this time to become self-sufficient and independent of imports of natural fibers, eg, cotton and wool. However, PVC could not compete with these fibers and the newer synthetics in the open market that came into existence at the end of the war, and PVC fibers consequently fell into disrepute. However, PVC is still one of the cheapest organic polymers, and this economic fact has kept interest in these fibers from lagging and has led to research and development into the best ways to utilize this material in the modern world.

The place of PVC fibers in today's textile market is strictly dictated by the economic and technical advantages and disadvantages of the fiber. PVC is an inexpensive, strong, nonflammable, chemically resistant, and weather-resistant fiber with an unfortunate sensitivity to temperatures below the boiling point of water. This latter fact has kept ordinary PVC out of the really large textile markets for woven materials that can be laundered in hot water, ironed, heat set, or resin treated. Thus, the fiber has sought markets where high-temperature resistance is not required and where its virtues give it a definite advantage.

The one unique quality that PVC provides is its nonflammability. This has led to a broad spectrum of end uses including curtains, furnishings, upholstering fabrics, women's and children's nightwear, blankets, mattresses, rugs, and industrial fire-protective clothing. Blends with wool or nylon are also used in various clothing applications; it is said that at least 75% PVC is required to impart sufficient nonflammability to blends (37). Automotive and aircraft upholstery fabrics are a growing end use. It is certain that the nonflammable property of PVC will be of increasing importance in the future as governments and consumers become more conscious of the flammability of most textile materials.

The weather and rot resistance of PVC has led to applications in the field of tents, tarpaulins, nets, and other exposed, outdoor textile articles such as flags. Because of its water repellency and its excellent thermal insulation PVC finds application in ski-wear, mountain climbing and hunting clothing, sleeping bags, and winter underwear. The chemical resistance of PVC is also an advantage in protective clothing for workers in the chemical industry.

The French firm Société Rhovyl has been most aggressive in developing markets for PVC fibers. These have included applications where the heat sensitivity of the fibers has been turned into an advantage. For example, spun yarns made with blends containing PVC staple fibers, when heated above the glass-transition temperature of PVC, become quite bulky owing to the retraction of the PVC fibers. By heating a fabric woven from PVC, the shrinkage of the warp and filling yarns results in a very tight weave. The resulting fabric can be used as a map substrate or for protective clothing for chemical workers. When PVC fibers form only part of the fabric, shrinkage produces novel "cloque" effects. PVC fibers in a pile fabric can be used to produce a sculptured or relief effect when heat is applied. Woven fabrics of PVC can also be molded to intricate forms under the application of heat and pressure (32).

PVC fibers are also finding increasing markets in the rapidly expanding field of nonwoven fabrics. Nonwovens made of pure PVC may be calendered, using heat and pressure to give a smooth surface, or by using embossing rollers, various designs may be impressed. Needle-punched batts made with PVC fibers in blends with other fibers may be given greater strength by the thermal retraction of the PVC after forming. PVC fibers may also be used as fibrous binders in nonwoven blends with other fibers. Mention has been made of the use of Vinyon HH in the preparation of teabag nonwovens, where the Vinyon also permits the bags to be heat-sealed, thus dispensing with a costly sewing operation.

The syndiotactic fiber Leavil has been commercially available in large quantities only since late 1967, and hence large markets for it have not yet developed. However, its superior thermal and drycleaning resistance promises more extensive applications than the regular PVC. It is noteworthy that Société Rhovyl has plans to commercialize a similar fiber, called Stavinyl. Teijin Ltd. of Japan has also investigated this fiber and has several patents in this field. Similarly the graft copolymer fiber Cordela, or Polychlal as it is sometimes called, has not been on the market long enough for assessment of its potential. In this case, the superior moisture absorbance and dyeability are expected to take it into applications beyond those available to atactic PVC homopolymer. The market potential of both these fibers will depend to a certain extent on their price, since the manufacturing process for both fibers is more complex and hence probably more costly than dry-spun regular PVC. In this respect they will be competing against the more extensively used fibers, eg, nylon, polyester, and above all acrylics, which they resemble.

Economic Aspects

Atactic PVC homopolymer is one of the most abundant and least costly synthetic materials in the world. These attractive raw material considerations have been one of the decisive factors in the continuing interest in PVC fibers. Unfortunately, the wet and dry spinning processes required for fiber production are more expensive than the melt spinning used for fibers such as nylon and polyester, enabling the latter fibers to rival PVC in price. Nonetheless, it is probably true that PVC fibers will always be somewhat less expensive than these fibers and thus will always have a certain advantage. This may not be true of the syndiotactic fibers where the more costly low-temperature polymerization is required and a complex wet spinning technology is involved. Of course, the nonflammability of PVC fibers also gives them a competitive advantage, although a great deal of research and development is currently being devoted to methods of conferring nonflammability to other fibers (see FIRE RETARDANCY).

Owing to their relatively small percentage of the total fibers market, production figures on PVC fibers are difficult to obtain, usually being reported in the category "other noncellulosics" along with glass, saran, etc. In 1964, however, the two major producers in the world were Société Rhovyl of France and Teijin, Ltd. of Japan (38). Table 2 lists the major producers of PVC fibers, as well as estimated capacities and trademarks associated with each producer.

Table 2. Major Producers of Poly(vinyl Chloride) Fibers (38, 39)

Company	Country	Trademarks	Type of fiber	Estimated capacity, millions of lb
Société Rhovyl	France	Rhovyl, Thermovyl, Fibrovyl, Retractyl	atactic homopolymer	>50
		Clevyl	polyblend of two PVC homopolymers	
Teijin, Ltd.	Japan	Teviron	atactic homopolymer	15
Polymer Ind.	Italy	Movil	atactic homopolymer	8
Fiberfabriken Wolfen	East Germany	PeCe	postchlorinated	6
Acetilen Fiber	Yugoslavia		atactic homopolymer	2
	U.S.S.R.	Khlorin	atactic homopolymer	
Chatillon S.p.A.	Italy	Leavil[a]	syndiotactic homopolymer	7
FMC Corp.	U.S.A.	Vinyon HH	vinyl acetate copolymer	4
Kohjin Co. Ltd.	Japan	Cordela[b]	vinyl alcohol graft copolymer	

[a] Also called Leavin.
[b] Also called Polychlal, Kohjin, and Super Enbilon.

Analysis and Testing

A quick screening test for PVC fibers can be made by attempting to burn the sample. If the material burns readily, it is obviously not PVC. However, a nonflammable fiber may be PVC, another nonflammable material, or a material that has been rendered flame-retardant by means of an additive or special treatment.

An organic chlorine determination by Schöniger combustion or by a similar method will distinguish between homopolymer, postchlorinated PVC, and copolymers. PVC homopolymer contains about 56% chlorine, postchlorinated PVC, about 62–65% chlorine, and copolymers, less than 56%. Vinyon HH contains 47% chlorine, and Cordela about 25% chlorine. Postchlorinated PVC and Vinyon HH are soluble in acetone, whereas homopolymer and Cordela are not. It must be borne in mind that there are also other chlorine-containing fibers, such as saran and modacrylics.

Infrared absorbance measurements may be used to distinguish between syndiotactic and atactic PVC. In particular, the ratio of the absorbance of the band at 646 cm^{-1} to that at 692 cm^{-1} increases linearly in proportion to the syndiotactic content of the polymer. For example, PVC polymerized at 50°C has an absorbance ratio of about 1.4, whereas that made at −50°C has a ratio of about 2.2 (8).

Health and Safety Factors

PVC is said to be completely nondermatitic and nonallergenic. Its nonflammable characteristic has encouraged its use in hospitals, sanatoriums, and the like. In this connection, the antirheumatic properties of PVC are of interest. It is claimed in both Japan and Europe that the wearing of undergarments or use of bed sheets made of PVC gives significant relief to patients suffering from rheumatism, arthritis, and related painful conditions. This was initially thought to be due to the high negative electric charge generated by PVC in contact with the skin, which is not found with any other fiber (36). However, the emphasis now seems to be on the superior warmth of the garments due to the low thermal conductivity of PVC (40).

Bibliography

1. F. Klatte (to Chemische Fabrik Griesheim-Elektron), Ger. Pat. 281,877 (1913).
2. E. Hubert, *Vierjahresplan* **4,** 222 (1940); abstr. *Kunststoffe* **30,** 244 (1940).
3. J. Corbiere, *Atomes,* 201 (1947).
4. C. S. Marvel et al., *J. Am. Chem. Soc.* **61,** 3241 (1939); *J. Am. Chem. Soc* **64,** 2356 (1942).
5. P. J. Flory, *J. Am. Chem. Soc.* **61,** 1518 (1939).
6. C. S. Fuller, *Chem. Rev.* **26,** 162 (1940).
7. J. W. L. Fordham, *J. Polymer Sci.* **39,** 321 (1959).
8. J. W. L. Fordham, P. A. Burleigh, and C. L. Sturm, *J. Polymer Sci.* **41,** 73 (1959).
9. F. P. Reding, E. R. Walter, and F. J. Welch, *J. Polymer Sci.* **56,** 225 (1962).
10. P. H. Burleigh, *J. Am. Chem. Soc.* **82,** 749 (1960).
11. *Man-Made Textiles* **43,** 26 (1966).
12. C. Schoenburg (to I. G. Farbenindustrie), Ger. Pat. 596,911 (1934).
13. V. F. Krska, J. Stamburg, and Z. Pelzbauer, *Angew. Makromol. Chem.* **3,** 149 (1968).
14. H. Germar, *Makromol. Chem.* **86,** 89 (1965).
15. J. Petersen and B. Rånby, *Makromol. Chem.* **102,** 83 (1967).
16. W. Pungs, *Kunststoffe* **57,** 317 (1967).
17. S. D. Douglas (to Carbide and Carbon Chemical Co.), U.S. Pat. 2,075,575 (1937).
18. S. D. Douglas and W. H. Stoops, *Ind. Eng. Chem.* **28,** 1152 (1936).
19. N. C. Heimbold, *Textile World,* 117 (June 1967).
20. C. F. Ruebensaal, *Chem. Eng.* **57,** 102 (Dec. 1950).
21. Monsanto Chemical Co., Brit. Pat. 958,969 (May 27, 1964).
22. Societa Edison, Belg. Pat. 661,932 (April 2, 1964).
23. Societa Edison, Belg. Pat. 609,856 (Nov. 3, 1961).
24. A. Nakajima, H. Hamada, and S. Hayashi, *Makromol. Chem.* **95,** 40 (1966).
25. G. Borsini and M. Cipolla, *J. Polymer Sci.* **B2,** 291 (1964).
26. S. D. Douglas (to Carbide and Carbon Chemical Co.), U.S. Pat. 2,075,429 (1937).
27. S. Okamura et al. (to Toyo Kogaku Co.), U.S. Pat. 3,111,370 (Nov. 19, 1967).
28. Teijin Ltd, Japan. Pat. 260,309 (Nov. 12, 1958).
29. J. Corbiere, P. Terra, and R. Paris, *J. Polymer Sci.* **8,** 101 (1952).
30. A. Mouchiroud and J. Trillat (to Societe Rhodiaceta), U.S. Pat. 2,472,842 (June 14, 1949).
31. Applicazioni Chimica S.p.A., Belg. Pat. 630,561 (April 3, 1963).
32. L. Gord, in H. Mark, S. Atlas, and E. Cernia, eds., *Man-Made Fibers: Science and Technology* Vol. 3, Interscience Publishers, a division of John Wiley & Sons, Inc., New York, 1968, pp. 327–355.
33. C. Mazzolini, *Fibre Colori* **17,** 214 (1967).
34. J. Lunenschloss and H. Stegherr, *Textil Praxis* **15,** 939 (Sept. 1960).

35. C. E. Schildknecht, *Vinyl and Related Polymers*, John Wiley & Sons, Inc., New York, 1952, p. 421.
36. C. Frager, *Textile Res. J.* **32**, 168 (1962).
37. *Skinner's Silk Record* **38**, 907 (1964).
38. Noyes Development Corp., *Man-Made Fiber Producers*, New York, 1964.
39. *Chem. Eng. News.* 36 (March 13, 1967).
40. *Textile Industries*, 78 (Aug. 1968).
41. *Chem. Eng. News*, 37 (Oct. 12, 1970).

J. P. Dux
American Viscose Division
FMC Corporation

VINYL CYANIDE POLYMERS. See ACRYLONITRILE POLYMERS

VINYLCYCLOALKANE POLYMERS

Vinylcycloalkanes are typical allylic olefins and cannot be polymerized in the presence of free-radical initiators. Thus there was very little interest in vinylcycloalkanes until the discovery of the Ziegler-Natta coordination catalysts, which could effect the polymerization of α-olefins. Polymers have now been prepared from all the vinylcycloalkanes with these catalysts. Vinylcyclohexane yields a high-melting, crystalline polymer which might be useful as an insulator; however, none of the vinylcycloalkane polymers is sold commercially. See also ALLYL POLYMERS.

Monomers

Physical Properties. Vinylcycloalkanes exhibit the properties of normal hydrocarbons; ie, they are colorless, flammable liquids with low densities. The compounds can be purified by fractional distillation and do not require involved separation procedures. The physical constants of several polymerizable vinylcycloalkanes are summarized in Table 1. The infrared spectra of vinylcyclopropane, vinylcyclopentane, and vinylcyclohexane along with several related cycloalkane derivatives have been compiled (5).

Chemical Properties. Vinylcycloalkanes undergo the usual electrophilic reactions of olefins. The reactivity of vinylcyclopropane is unique, because the cyclo-

Table 1. Properties of Vinylcycloalkanes

Monomer	bp, °C/mmHg	d_4^{25}	n_D^{25}	Yield, %	Reference
vinylcyclopropane	40.5–40.8/755	0.7157	1.4104	44.5[a]	1
vinylcyclobutane	68/753	0.7467	1.4210	85[b]	2
vinylcyclopentane	98.8–99/755	0.7850	1.4336	83[c]	1
vinylcyclohexane	127/760	0.8012 (d_4^{20})	1.4462 (n_D^{20})	80[c]	3
4-methylvinylcyclohexane[d]	142.3	0.7912	1.4403	15[c]	4
3-methylvinylcyclohexane[d]	144–145	0.7936	1.4428	17[c]	4
2-methylvinylcyclohexane[d]	149	0.8168	1.4510	60[b]	4
vinylcycloheptane	160/762		1.4589	28.5[a]	2

[a] Via cycloalkylethyl xanthate pyrolysis.
[b] Pyrolysis of cycloalkylethyl *N*-oxide.
[c] Pyrolysis of cycloalkylethyl acetate.
[d] Cis or trans isomer.

propane ring interacts with the vinyl group and 1,5 addition products are obtained under both ionic and radical conditions (eq. 1) (6).

$$CH_2{=}CH{-}\triangleleft \; + \; Br{-}CCl_3 \xrightarrow{h\nu} Cl_3CCH_2CH{=}CHCH_2CH_2Br \quad \text{(cis and trans)} \tag{1}$$

The most significant reaction of vinylcycloalkanes derived from stable rings is the isomerization of the double bond to form ethylidene- and 1-, 3-, or 4-ethylcycloalkenes (eq. 2). The ethylidene and the cycloalkene isomers are several orders of magnitude more stable than the vinylcycloalkanes. Although the ratio of 1-ethylcycloalkene to

$$(CH_2)_n\langle CH{-}CH{=}CH_2,\; CH_2\rangle \longrightarrow \quad n = 3, 4$$

$$(CH_2)_n\langle C{=}CH{-}CH_3,\; CH_2\rangle + (CH_2)_n\langle C{-}CH_2CH_3 \,\|\, CH\rangle + (CH_2)_{n-1}\langle CH{-}CH_2{-}CH_3,\; CH{=}CH\rangle \tag{2}$$

the other isomers varies with cycloalkane ring size and with the conditions employed for equilibration, the 1-ethylcycloalkene is definitely the most stable isomer in all cases. The equilibrium ratios were observed after treating the vinylcycloalkane at room temperature over a sodium-on-alumina catalyst for 24 hr (7). At 250°C on an activated alumina bed equilibrium was reached after 15–20 passes through a 120 × 10 mm column (7). Chromic oxide on alumina was also effective at 250°C (8). Vinylcyclohexane is completely isomerized in the presence of triiron dodecacarbonyl in 19 hr at 125°C (9). Thermal isomerization of vinylcyclopropane yields cyclopentene (10).

Synthesis. Most of the common techniques for synthesizing vinyl compounds have been applied to the preparation of vinylcycloalkanes. Pyrolysis techniques are favored because isomerization of the product mixture can be minimized. Dehydrohalogenations in the presence of strong bases yield ethylidene derivatives as the major products (11). The synthesis of vinylcyclopropane requires special techniques because of the reactivity of the cyclopropyl group.

There are three general classes of pyrolysis reactions that produce olefins: the pyrolysis of acetates, of xanthates (Chugaev), and of amine oxides (Cope). The acetate pyrolysis is the most common technique because the intermediates are easier to prepare. Addition of either acetaldehyde or ethylene oxide to a cycloalkylmagnesium bromide affords the corresponding 1-cycloalkylethanol (**1**) or 2-cycloalkylethanol (**2**) (eqs. 3a and 3b). Dehydration of the cycloalkylethanol derivatives over alumina at

$$(CH_2)_n\langle CHMgBr \xrightarrow[2.\ H_2O]{1.\ CH_3CHO} (CH_2)_n\langle CH{-}CH(OH){-}CH_3 \quad (\mathbf{1}) \qquad n = 4, 5, 6 \tag{3a}$$

$$\xrightarrow[2.\ H_2O]{1.\ \text{ethylene oxide}} (CH_2)_n\langle CH{-}CH_2CH_2{-}OH \quad (\mathbf{2}) \tag{3b}$$

350°C yielded ethylidenecycloalkanes as the major product (12). However, thoria-catalyzed dehydration was stereoselective, yielding α-olefins if extremely short contact times (0.1–0.5 sec) were employed (12a). Apparently the short contact time minimizes concomitant isomerization of the vinyl derivatives to the more thermodynamically stable isomers.

The acetates dervied from (**1**) or (**2**) can be pyrolyzed at 450–500°C under nitrogen in a column packed with glass helixes. Under these conditions a mixture of ethylidene- and vinylcycloalkanes is formed, but the vinyl compound can be isolated by careful fractional distillation. The ratio of vinylcycloalkane to each of the other components is much higher when 2-cycloalkylethyl acetate (15) is pyrolyzed because the double bond can form in the terminal position only (eqs. 4a and 4b). Although the vinyl-

$$(CH_2)_n\,CH{-}CH_2{-}CH_2OCOCH_3 \xrightarrow{450^\circ C} \underset{80\%}{(CH_2)_n\,CH{-}CH{=}CH_2} + CH_3COOH \quad (4a)$$

$$(CH_2)_n\,CH{-}\underset{OCOCH_3}{\underset{|}{CH}}{-}CH_3 \xrightarrow{450^\circ C} \underset{60{-}65\%}{(CH_2)_n\,CH{-}CH{=}CH_2} + \underset{35{-}40\%}{(CH_2)_n\,C{=}CHCH_3} \quad (4b)$$

cycloalkane is still the major product from the pyrolysis of 1-cycloalkylethyl acetates, a significant percentage of the product mixture is the ethylidenecycloalkane formed by abstraction of the tertiary hydrogen atom on the cycloalkane (eq. 4b). Some thermal isomerization of the products occurs, as evidenced by the presence of ethylcycloalkene isomers in product mixtures derived from either 1-cycloalkylethyl acetate or 2-cycloalkylethyl acetate.

In spite of the lower yield of vinylcycloalkane obtained from the pyrolysis of 1-cycloalkylethyl acetates, this is the most common technique employed because the 1-cycloalkylethanols are more readily available. Catalytic reduction of acetophenone derivatives affords the 1-cyclohexylethanol in about 40% yield (12). The major side reactions, dehydration or hydrogenolysis of the phenylmethylcarbinol intermediate, could not be prevented. However, cyclohexyl methyl ketones can be reduced cleanly either by hydrogenation or with complex metal hydrides. The cycloalkyl methyl ketones are available through direct acylation of cycloalkenes followed by reduction of the methyl cycloalkenyl ketones formed (13). Reduction of cyclohexyl methyl ketone with $LiAlD_4$ followed by treatment of the deuterated alcohol with acetyl chloride and pyrolysis afforded vinylcyclohexane-α-*d* in 65% overall yield (14). Recently a technique involving addition of ethylene to chlorocyclohexane in the presence of aluminum chloride, converting the β-chloroethylcyclohexane to the acetate, and then cracking led to the synthesis of vinylcyclohexane. The scope of this reaction has been expanded to include a large number of alkyl-substituted vinylcycloalkane derivatives (15).

Application of other pyrolysis techniques requires slight modifications in the syntheses of the monomer precursors, which generally increase the number of steps in the synthesis. However, xanthate pyrolysis affords a higher yield of pure vinylcycloheptane than the pyrolysis of the corresponding 2-cycloheptylethyl acetate (2). Several methylvinylcyclohexanes have been prepared by pyrolysis of the corresponding *N*,*N*-dimethyl-2-(cycloalkyl)ethylamine oxides (4), which were synthesized from xylyl cyanides through the reaction sequence shown in equation 5. The mild conditions required for the pyrolysis minimize isomerization, but problems in reduction of the xylyl cyanide reduce the overall yield from xylyl bromide to about 20%. The synthesis of vinylcyclobutane through the Cope reaction has been effected in 36% yield in six steps from cyclobutanecarboxylic acid (2).

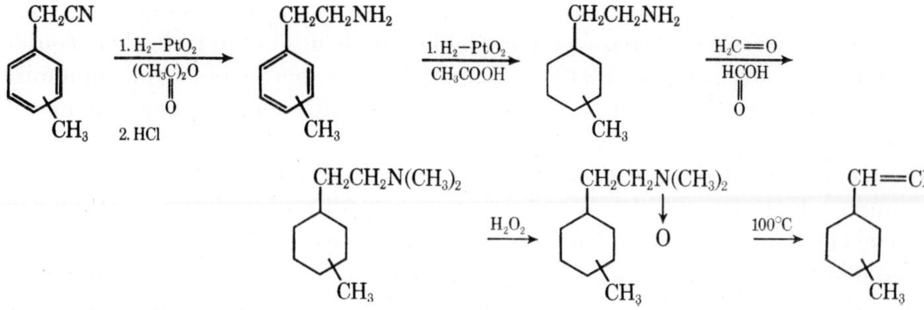

Pure vinylcyclohexane can be prepared in 98% yield by cracking the Diels-Alder adduct (4) at 333–390°C (16). Since the anthracene can be recovered quantitatively and 4-vinyl-1-cyclohexene is readily available from the thermal dimerization of butadiene, the scheme outlined represents the best route to vinylcyclohexane (eq. 6).

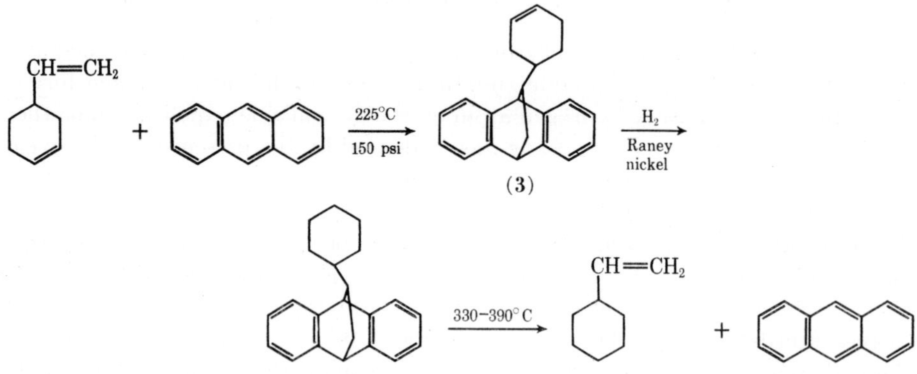

A similar technique for selective reduction of the internal double bond of 4-vinyl-1-cyclohexene requires that the vinylcyclohexene displace isobutene from triisobutylaluminum; the tris(cyclohexenylethyl)aluminum can be hydrogenated and then the vinylcyclohexane can be displaced with 1-dodecene (17). A monomer synthesis that may have industrial significance is the free-radical addition of acetylene to cycloalkanes (18). At 140°C under a partial pressure of 200 psi of acetylene, radicals generated from di-*tert*-butyl peroxide abstract a secondary or tertiary proton from the cycloalkane, and acetylene adds to this to form a vinyl radical, which in turn abstracts a proton from the reaction medium to form the vinylcycloalkane. The kinetic chain length for formation of the monoadduct is very short; thus the yield of vinylcycloalkane is comparable to the initial initiator concentration.

Since the preparation of vinylcyclopropane by pyrolysis techniques is accompanied by extensive rearrangement (10), several alternative approaches have been used. Generation of CH_2: from methylene iodide with a zinc–copper couple (Simmons-Smith reaction) in the presence of an excess of butadiene affords a 26% yield of vinylcyclopropane along with 23% dicyclopropyl (19). Since this mixture is difficult to separate, dibromocarbene addition to butadienes was used to prepare 2,2-dibromovinylcyclopropane (**5**), which undergoes reductive debromination to yield vinylcyclopropane (eq. 7) (20). Butadiene also reacts with dimethylsulfonium methylide to

$$CH_2{=}CH{-}CH{=}CH_2 + {:}CBr_2 \longrightarrow \underset{(5)\ (67\%)}{CH_2{=}CH{-}CH{-}CH_2\ (\text{ring } CBr_2)} \xrightarrow[\text{dioxane}]{Na,\ C_2H_5OH} \underset{(58\%)}{CH_2{=}C{-}CH{-}CH_2\ (\text{ring } CH_2)} \qquad (7)$$

produce vinylcyclopropane in 30% yield (21). The 2,2-dihalovinylcyclopropanes (**5**) formed by the addition of dihalocarbenes to butadiene can also be polymerized. Selective reduction of the *gem*-dihalocyclopropanes with tri-*n*-butyltin hydride yields the corresponding 2-halovinylcyclopropane (22). 2-Phenylvinylcyclopropane can be prepared from butadiene through either photolytic decomposition of phenyldiazomethane or treatment with benzal bromide in the presence of butyllithium (23).

Polymerization

Ziegler-Natta Techniques. Vinylcycloalkanes polymerize without rearrangement on the surface of heterogeneous Ziegler-Natta-type catalysts. The polymers are composed of both low-molecular-weight amorphous fractions and high-molecular-weight crystalline fractions. The molecular weight of the crystalline fraction varies upon changing the size of the cycloalkyl substituent, but this may be a function of monomer purity. The physical properties of the poly(vinylcycloalkanes) are summarized in Table 2. The polymerization of vinylcyclohexane has been effected by a variety of transition metal halide–organometallic combinations, which have been compiled in Table 3. According to the patent literature, the addition of carbonyl-containing additives increases the rate of polymerization and the molecular weight of the polymer formed.

cis-1,3-Divinylcyclohexane and *cis*-1,3-divinylcyclopentane cyclopolymerize in the presence of Ziegler-Natta catalysts to yield soluble, virtually saturated polymers (31). The linear polymers are composed of bicyclo[3.3.1]nonane and bicyclo[3.2.1]-octane repeat units, respectively (eq. 8). See also CYCLOPOLYMERIZATION.

$$\text{cis-1,3-divinylcyclohexane} \xrightarrow[Al(i\text{-}C_4H_9)_3{-}TiCl_4\ (1{:}1)]{n\text{-heptane}} \left[\text{bicyclo[3.3.1]nonane}{-}CH_2\right]$$

$$\text{cis-1,3-divinylcyclopentane} \longrightarrow \left[\text{bicyclo[3.2.1]octane}{-}CH_2\right] \qquad (8)$$

Topchiev et al. have also obtained crystalline poly(vinylcyclohexane) with a chromium oxide on aluminum silicate catalyst (Phillips catalyst) (26). By mixing small increments of triisobutylaluminum with the chromium oxide, the yield of polymer could be increased. The extent of crystallinity was not reported but the powder diffraction patterns of the samples prepared on chromium oxide were slightly more diffuse than those observed from samples prepared under similar conditions with Ziegler-Natta catalyst. The chromium oxide also catalyzes the isomerization of

Table 2. Physical Properties of Poly(vinylcycloalkanes) Prepared on Heterogeneous Catalyst

Monomer	Catalyst type[a]	Temp, °C	Time, days	Yield, %	% Crystallinity[b]	$[\eta]$, dl/g[c]	mp, °C[d]	Reference
vinylcyclopropane	A	68	2	35.4	63		>260 decomp	1
vinylcyclobutane	B	75	7	24.5	66		228	2
vinylcyclopentane	B	68	5	13.7	87.5	0.56	>260 decomp	1
vinylcyclopentane	C	40	0.5		100		292[e]	24
vinylcyclohexane	A	68	1.5	46.5	93.5		>300	1
vinylcyclohexane	C	40	0.5		93.2	5.02 (toluene)	383[e]	24
vinylcyclohexane	D	50	2	15.1	80.1	1.9	385	25
4-methylvinylcyclohexane	B	80	1.5	30.0		0.45	225–250	4
3-methylvinylcyclohexane	B	70	1.5	42.5	30.0	0.74	276–355	4
vinylcycloheptane	B	72	7	9.8	58.8	0.39	>300	2

[a] Solvent, heptane; A, freshly ground $TiCl_3$, $Al(i\text{-}C_4H_9)_3$ (mole ratio Ti/Al = 1:4); B, $TiCl_4$, $Al(i\text{-}C_4H_9)_3$ (Ti/Al = 1:2); C, "AA" $TiCl_3$, $Al(C_2H_5)_2Cl$ (Ti/Al = 1:1); D, $TiCl_4$, $Al(C_2H_5)_3$ (Ti/Al = 1:3).

[b] Insoluble fraction remaining after continuous extraction with ether for at least 24 hr.

[c] Measured in benzene unless otherwise noted.

[d] Estimated on Fisher-Johns melting-point block.

[e] Determined by differential thermal analysis.

Table 3. Heterogeneous Catalyst Systems for the Preparation of Crystalline Poly(vinylcyclohexane)

Catalyst	Cocatalyst	Additive	Mole ratio	Diluent	Reaction temp, °C	Reaction time, hr	Reference
$TiCl_3$	$Al(i\text{-}C_4H_9)_3$		1:4	heptane	68	36	1
"AA" $TiCl_3$[a]	$Al(C_2H_5)_2Cl$		1:1	heptane	40	12	24
$TiCl_4$	$Al(C_2H_5)_3$		1:3	cyclohexane	50	48	25
$TiCl_4$	$Al(i\text{-}C_4H_9)_3$		1:2	heptane	80	5	26
$TiCl_4$	Al	$Ti(OC_4H_9)_3$		heptane	90	8	27
$TiCl_4$	NaH	HMPA[b]	1:1:0.5	heptane	85	6	28
$TiCl_4$	$NaAl(C_2H_4)_4$	HMPA	1:1:0.5	heptane	55	6	28
$TiCl_4$	CH_2Cl_2 + Al	(tetrahydrofuran ring)–CH_2OCOCH_3	1:1:1	heptane	55	6	29
$TiCl_4$	Na	DMF[c]	1:1:0.25	heptane	75	6	28
$TiCl_3$	C_2H_5Li	$AlCl_3$	1:1:1	cyclohexane	70	4	30
$TiCl_3$	$C_2H_5AlCl_2$	$CH_3\overset{O}{\overset{\|}{C}}OCH_2CH_2O\overset{O}{\overset{\|}{C}}CH_3$					

[a] α-Titanium trichloride prepared by high-temperature reduction of titanium tetrachloride with aluminum. It contains one-third more aluminum chloride per mole $TiCl_3$.

[b] Hexamethylphosphorus triamide.

[c] Dimethylformamide.

vinylcyclohexane to ethylidenecyclohexane, which does not polymerize and tends to inhibit the polymerization.

Kinetics. The rate of vinylcyclohexane polymerization in the presence of $TiCl_3$–$Al(C_2H_5)_3$ catalyst has been studied in the temperature range 30–70°C in cyclohexane or *n*-heptane (25). The initial rates, obtained from the first 120 min at 50°C, were measured dilatometrically at monomer concentrations of 0.45–3.15 mole/l. The polymerization was first order in monomer for monomer concentrations greater than 0.8 mole/l, but in dilute solutions the reaction appeared to be second order. Catalyst variables were also explored: the initial rate of polymerization exhibited a linear dependence upon the titanium trichloride concentration, and a slight maximum was observed in the plot of rate vs aluminum-to-titanium ratio at Al/Ti = 0.28. The intrinsic viscosity of the polymer increased with reaction time and monomer concentration, and was essentially independent of the triethylaluminum concentration. The Arrhenius activation energy was calculated to be 5.2 kcal/mole, which is too high for a diffusion-controlled reaction.

Either the adsorption of monomer on the catalyst surface or the surface migration of hydrogen atoms to generate fresh catalyst sites was considered to be the rate-controlling step in the polymerization. At low monomer concentrations the rate of monomer adsorption is a factor in the overall polymerization rate, and a higher kinetic order in monomer is expected. However, at high monomer concentrations the reactive sites are saturated with adsorbed monomer and the polymerization rate is proportional to the rate of alkylation of monomer by migration of hydrogen atoms from the catalyst surface. Thus the hydrogen atoms serve as activators in the polymerization and under steady-state conditions the concentration of the surface alkyl groups generated by the monomer is governed by the rates of monomer reaction with hydrogen atoms, of desorption, and of removal by chain growth (propagation). If only that part of the surface containing adsorbed hydrogen atoms is considered and a steady state is assumed, then

$$k_a(1 - \theta)[\mathrm{M}] = k_d\theta + k_p\theta[\mathrm{M}]$$

where θ is the fraction of H atoms which have reacted with monomer and k_a, k_d, and k_p are the specific rate constants for alkylation of monomer at the catalyst surface, desorption, and propagation, respectively. The rate of polymerization is thus given by equation 9:

$$R_p = k\theta[\mathrm{M}] = \frac{kk_a[\mathrm{M}]^2}{k_d + [\mathrm{M}](k_p + k_a)} \tag{9}$$

Note that at low monomer concentrations the rate is proportional to $[\mathrm{M}]^2$ but it becomes first order at higher monomer concentrations. This is consistent with the experimental observations and indicates that the concentration of active catalyst sites is dependent upon the monomer concentration. Similar results were also observed in the polymerization of propylene and 1-pentene under the same conditions. A comparison of the relative rates of polymerization illustrates the reduced reactivity of vinylcyclohexane; ie, propylene is 300 times and 1-pentene is 100 times as reactive as vinylcyclohexane. See also STEREOREGULAR LINEAR POLYMERS.

Cationic Polymerization (qv). The polymerization of vinylcycloalkanes can be initiated by Friedel-Crafts catalysts such as $AlBr_3$ or $AlCl_3$. The molecular weight of the polymers increases when the polymerization temperature is lowered to

−100°C but no polymers with molecular weights greater than 10,500 have been isolated. The vinylcycloalkanes can be considered analogs of 3-methyl-1-butene, which isomerizes before propagation to yield a 1,3-enchainment (see ISOMERIZATION POLYMERIZATION). Indeed, the polymerization of vinylcyclohexane (24,33) and vinylcyclopropane (32) is accompanied by hydride transfer reactions which lead to rearranged polymers with cycloalkyl groups as integral parts of the chain (eq. 10).

$$CH_2{=}CH\text{–}C_6H_{11} \xrightarrow{R^{\oplus}} R\text{—}CH_2\text{—}\overset{\oplus}{C}H\text{–}C_6H_{11} \xrightarrow{\text{rearrangement}} RCH_2\text{—}CH_2\text{–}C_6H_{10}^{\oplus} \xrightarrow{CH=CH_2\text{–}C_6H_{11}} \left[CH_2CH_2\text{–}C_6H_{10} \right]_x \quad (10)$$

(6)

The structure of the rearranged polymer was assigned on the basis of infrared and NMR evidence. When the infrared spectrum of cationically initiated poly(vinylcyclohexane) was compared with spectra of Ziegler-Natta-initiated polymer, characteristic differences in the methylene rocking region (700–800 cm^{-1}) were observed. A similar comparison of NMR spectra was less conclusive due to the poorly resolved spectra, but the absorption bands were consistent with the proposed structures. Attempts to prepare model compounds by polymerization of 2,5-spirooctane (**6**) failed.

The polymerization of vinylcyclopropane may proceed by way of a nonclassical bicyclobutonium ion which leads to the incorporation of either olefin groups (**7**) or cyclobutane units (**8**) into the polymer backbone (eq. 11) (32). Vinylcyclopropane

$$CH_2{=}C(R)\text{–}C_3H_5 \xrightarrow{AlBr_3} \sim CH_2\text{—}\overset{\oplus}{C}(C_3H_5)\text{—}R \longrightarrow \sim CH_2\text{—}C(R)(C_3H_5)\sim$$

R = H, aryl

rearrangement →

$$\sim CH_2\text{—}C(R)\text{(bicyclobutonium ion)} \longrightarrow \sim CH_2\text{—}C(R){=}CH\text{—}CH_2CH_2\sim \quad (7) \quad (11)$$

$$\longrightarrow \sim CH\text{—}C(R)\text{—}CH_2\sim \ (\text{ring } CH_2\text{—}CH_2) \quad (8)$$

yields a polymer of 60% unrearranged repeat units, 10% structure (**7**), and 30% structure (**8**). 1-Arylvinylcyclopropanes yielded higher percentages of repeat units derived from a bicyclobutonium ion intermediate; the 1-*p*-cumyl derivative was quantitatively converted to a homopolymer of structure (**8**). The influence of the 1-aryl substituents on the repeat unit distribution was attributed to the large steric interaction involved in adding a monomer unit to the α carbon of a nonrearranged chain end.

Recently, the concept of a bicyclobutonium ion intermediate in vinyl cyclopropane polymerization has been challenged (36). Attempted cationic polymerization of several 1,1-disubstituted 2-vinylcyclopropane derivatives failed to produce polymers. Furthermore, *cis*-1-halo-2-vinylcyclopropanes yielded polymers containing greater than 70% 1,2-enchainments, whereas the corresponding trans isomers failed to polymerize. These results were interpreted on the basis of steric inhibition of the 1,2 propagation step, coupled with the failure of the cyclopropyl carbonium ion to relieve this hindrance by isomerizing to the bicyclobutonium ion.

Steric interactions may also account for the failure of vinylcyclopentane to yield a rearranged polymer. Since the formation of an sp^2 hybridized carbon is favored in five-membered ring systems, a hydride shift should occur readily. However, addition of monomer to the alkylcyclopentyl carbonium ion requires that it be converted to an sp^3 hybridized carbon with two large alkyl substituents. The steric compression involved in this conversion apparently prohibits this mode of addition. Thus, the formation of rearranged poly(vinylcyclohexane) is a demonstration of the enhanced flexibility of the cyclohexane ring, which enables it to accommodate this steric compression.

Free-Radical Polymerization (qv). There are a few exceptions to the general rule that vinylcycloalkanes do not polymerize in the presence of free-radical initiators. The polymerization of vinylcyclopropane derivatives with polar substituents in the ring can be initiated by benzoyl peroxide, azobisisobutyronitrile, ultraviolet radiation, or cumene hydroperoxide in an aqueous emulsion (34,35). The polymerization is accompanied by ring opening to yield a 1,5 polymer containing a carbon–carbon double bond in the backbone, as shown in equation 12 where $I\cdot$ is an initiator radical. The

$$CH_2{=}CH{-}\underset{\substack{\diagdown\ \diagup \\ C \\ R'\ \ R''}}{CH{-}CH_2} \xrightarrow{I\cdot} I{-}CH_2{-}\dot{C}H{-}\underset{\substack{\diagdown\ \diagup \\ C \\ R'\ \ R''}}{CH{-}CH_2} \longrightarrow$$

$R' = H$, $R'' = \underset{\substack{\| \\ O}}{C}OC_2H_5$

$R' = R'' = Cl$

$$I{-}CH_2CH{=}CH{-}CH_2{-}\underset{R''}{\overset{R'}{\overset{|}{\underset{|}{C}}}}\cdot \longrightarrow \sim\!\!\left[CH_2{-}CH{=}CH{-}CH_2{-}\underset{R''}{\overset{R'}{\overset{|}{\underset{|}{C}}}}\right]\!\!\sim \quad (12)$$

structure of the polymer was confirmed by titration of the main-chain unsaturation with benzoyl hydroperoxide as well as by infrared and NMR analysis. The availability of substituted vinylcyclopropanes through carbene intermediates, coupled with the isomerization occurring during propagation, enhances the theoretical interest in vinylcyclopropane polymerization. The following cyclopropane derivatives have been prepared and polymerized in the presence of azobisisobutyronitrile: vinyl, isopropenyl, 1-methyl-1-vinyl, 1,1-dichloro-2-methyl-2-vinyl, *cis*- and *trans*-1-chloro-2-vinyl, and 1,1-dibromo-2-vinyl (36). In every case 1,5 enchainment was observed. These results contrast sharply with the results obtained using cationic initiators, in which case copolymers of 1,2, 1,5, and cyclobutyl enchainments were obtained.

High-molecular-weight, crystalline 1,2 homopolymers were prepared by free-radical polymerization of 1,1-dichloro-2,2-difluoro-3-vinylcyclobutane, the cycloaddition product of 1,1-dichloro-2,2-difluoroethylene and butadiene (37). Polymerization could be effected in bulk with *t*-butyl peroxide or in emulsion with potassium peroxydisulfate to yield a polymer soluble in pyridine with a softening point at 180–190°C and a molecular weight above 20,000. Quenching the melt in cold water gave a clear, colorless, flexible film with a tensile strength of 7000 lb/in.^{2}

Copolymerization. Detailed investigation of the copolymerizability of vinylcycloalkanes has been restricted to Ziegler-Natta systems. The vinylcycloalkanes copolymerized with 1-butene (38), 3-methyl-1-butene, and 4-methyl-1-pentene (39) to yield crystalline copolymers. The most effective catalyst composition was a 2:1 mole ratio of diethylaluminum chloride to titanium trichloride. Since the vinylcycloalkanes are much less reactive in copolymerization than the comonomers, incremental addition of comonomer is required to obtain a copolymer with a uniformly random sequence distribution. A block copolymer of vinylcyclohexane and 4-methyl-1-pentene was prepared by a different mode of addition of monomers; each monomer was allowed to polymerize for 24 hr before the other monomer was added. The relative reactivity of vinylcyclohexane with 4-methyl-1-pentene is illustrated in Table 4. The high conversions reported preclude the calculation of meaningful reactivity ratios. The high-molecular-weight, heptane-insoluble copolymers could be spun from trichloroethylene into isopropanol to produce tough pliable fibers. Copolymers containing at least 66% vinylcyclohexane exhibited melting points comparable to those of nylon or polyester fibers. Thus the excellent mechanical properties of the copolymers are achieved without sacrificing the high-melting characteristics of the corresponding homopolymers.

Table 4. Copolymerization of Vinylcyclohexane (M_1) and 4-Methyl-1-pentene (M_2)[a] (39)

Comonomer composition, mole %		Copolymer composition, mole %		% Conversion	η_{inh}[b]	mp, °C[c]
M_1	M_2	M_1	M_2			
84.6	15.4	80	20	60	3.05	325–330
73.3	26.7	66	34	56	2.90	>250
48	52	44	66	70	3.10	210–300
24	76	17	83	69	3.30	230–235
13	87	9	91	70	3.75	220–225

[a] Polymerization conditions: 2 $Al(C_2H_5)_2Cl$:1 $TiCl_3$, 0.03–0.04 mole $TiCl_3$ to 1 mole monomer; slurried in heptane at 26–28°C for 16 hr.

[b] Inherent viscosity of 0.1% trichloroethylene solution, dl/g.

[c] Disappearance of birefringence.

Ziegler-Natta systems have been employed in the preparation of copolymers of vinylcyclohexane with ethylene (40), propylene (41), and styrene (42). The ethylene–vinylcyclohexane copolymers are useful in reinforcing paraffin coatings on paper, cardboard, fabrics, and foods. Free-radical initiators effect the copolymerization of vinylcyclohexane with acrylonitrile (43) and with vinyl chloride (44). Thiocarbonyl fluoride copolymerizes with vinylcyclohexane in the presence of a trialkylboron–oxygen redox couple to yield a tacky elastomer containing 24% sulfur (eq. 13) (45). 1,2-Divinylcyclohexane reacts with sulfur dioxide in the presence of free radicals to produce a linear copolymer (eq. 14) (46). Attempted cationic copolymerization of vinylcyclohexane with isobutylene failed (47).

$$F_2C{=}S + CH_2{=}CH{-}C_6H_{11} \xrightarrow{O_2-(C_4H_7)_3B\,(1{:}2)} \left[CF_2{-}S{-}CH_2{-}CH(C_6H_{11}) \right]_x \qquad (13)$$

$$C_6H_{10}(CH{=}CH_2)_2 + SO_2 \xrightarrow{R\cdot} \left[\text{bicyclic sulfone repeat unit } (SO_2) \right]_x \qquad (14)$$

Physical Properties

The most outstanding property of poly(vinylcycloalkanes) is their high crystallinity. Polymers derived from nonbranched cyclic α-olefins tend to be rubbery noncrystalline materials if the monomer contains more than five carbon atoms (see also OLEFIN POLYMERS). However, vinylcycloalkanes fall into a general class of branched 1-olefins which exhibit a high propensity to crystallize even when incorporated into copolymers. This suggests that vinylcycloalkanes polymerize in a stereospecific manner to yield highly ordered chains which pack very well and are stiffened by the favorable arrangement of alkyl groups.

X-ray analysis has demonstrated that Ziegler-Natta catalysts yield isotactic poly(vinylcycloalkanes) that become completely crystalline upon orientation and annealing at high temperatures. The drawing temperature determines the precise crystal structure. For example, poly(vinylcyclohexane) crystallizes in a 4_1 helix (form I) when oriented between 240 and 260°C (48) but a second crystal structure (form II) is obtained at drawing temperatures below 240°C (49). A compilation of the crystal structures of oriented poly(vinylcycloalkanes) is shown in Table 5. Most of the polymers form tetragonal unit cells containing four helical chains. Forms I and II vary only in the pitch and helix diameter; an increase in pitch produces a corresponding decrease in helix diameter to minimize the distortion of the axes. The space requirements of the cycloalkyl substituents determine the pitch of the helix. The fact that poly(vinylcyclopropane) crystallizes in both the 3_1 and 10_3 helixes suggests that the cyclopropyl group can barely be accommodated in the 3_1 helix and all larger substituents will be forced into the more extended 4_1 helix by side-chain interaction.

Single crystals of poly(vinylcyclohexane) have been grown from dilute trichloroethylene solutions (51). Crystals grown from 0.1% solutions were dendritic but at lower concentrations (<0.05%) the polymer precipitated almost entirely as single crystals. The electron diffraction pattern obtained from a platelet showed the orientation of the molecular chain to be perpendicular to the a–b plane. Since the height of the crystal, approximately 220 Å, is much less than the length of an individual chain, the crystal must be the folded chain type.

The dynamic mechanical properties of poly(vinylcyclohexane) at acoustic and low ultrasonic frequencies have been determined in conjunction with a study of polymers containing cycloalkyl groups (52). All polymers except poly(vinylcyclohexane)

Table 5. Summary of Unit-Cell Data for Poly(vinylcycloalkanes) (49)

Polymer	Helix type[a]	Space group[b]	a, Å	c, Å	Helixes per cell	Repeat length per monomer, Å	ρ, calc	ρ, obs
			Hexagonal, a = b					
vinylcyclopropane (I)[c]	3_1	$P3_1,P3_2$	13.6	6.48	3	2.16	0.9805	0.975
			Tetragonal, a = b					
vinylcyclopropane (II)[c]	10_3	$I\bar{4}$	15.21	20.85	4	2.09	0.926	
vinylcyclobutane	4_1		34.12	6.6	16	1.65		
vinylcyclopentane (I)	4_1	$I4_1/a$	20.14	6.50	4	1.625	0.970	
vinylcyclopentane (II)	12_3	$I4$ or $I\bar{4}$	20.14	19.5	4	1.625	0.970	0.954
vinylcyclopentane (III)	10_3	$P4/mmm$ or $P4/nmm$	37.3	19.8	16	1.98	0.931	0.955
vinylcyclohexane (I)	4_1	$I4_1/a$	21.99	6.43	4	1.605	0.942	0.9475
vinylcyclohexane (II)	24_7	$I4$ or $I\bar{4}$	20.48	44.58	4	1.86	0.940	
vinylcycloheptane	4_1	$I4_1/a$	23.4–23.5	6.5	4	1.625	0.926	

[a] The helix notation indicates the number of monomer units per unit of periodicity in the helix screw sense. For example, 10_3 describes a helix that requires ten monomer units and three turns to complete the repeating sequence.

[b] The space group notation is an abbreviation that characterizes the symmetry elements contained by the unit cell. A complete definition of each space group can be found in Ref. 56.

[c] Ref. 50.

exhibit a γ relaxation connected with the internal motion of the cycloalkyl substituent. The activation parameters of this relaxation depend exclusively on the nature of the cycloalkyl ring. However, poly(vinylcyclohexane) did exhibit a secondary relaxation effect (δ process) at lower temperatures. The δ process was observed for the entire series of polymers; the position on the temperature scale, peak heights, and activation parameters depend upon the nature of the group to which the cycloalkyl substituent is attached. The δ process was attributed to some complex motion of the entire cyclohexyl ring about the bonds linking it to the polymer chain. See also MECHANICAL PROPERTIES.

The dissipation factor, tan δ, and specific-volume electrical resistance were determined for poly(vinylcyclohexane) in the temperature range from $-180°C$ to $+160°C$ and from 50 to 10^6 Hz (53). The dielectric properties of poly(vinylcyclohexane) were found to be superior to those of polystyrene or polytetrafluoroethylene; the dielectric permeability coefficient is -0.2 in the glass phase and increases to -0.6 in the more elastomeric phase (at about 100°C). The dielectric coefficient combined with the high melting point of the polymer make it ideal for insulator applications.

Poly(vinylcycloalkanes) are generally soluble in aromatic solvents, Decalin, Tetralin, and chlorinated ethylenes. The intrinsic viscosity–molecular weight relationship for atactic poly(vinylcyclohexane), which was prepared by hydrogenation of polystyrene, has been determined in a theta solvent, tetrahydrofuran (54). Assuming that $a = 0.5$ in theta solvents, the value for K_θ was calculated to be 0.0788 ml/g by means of equation 15, where $[\eta]_\theta$ is in ml/g.

$$[\eta]_\theta = K_\theta \bar{M}_v^{0.5} \quad (15)$$

Chemical Properties

The alkanes are practically inert to chemical modification unless the cycloalkyl substituent is a strained ring. Treatment of a poly(vinylcyclohexane) film with atomic oxygen, which was generated by passing oxygen at low pressure through a radio-frequency coil, produces a moderate weight loss (55). The polymer is more resistant to the attack by oxygen atoms than is polypropylene but more sensitive than polystyrene or polytetrafluoroethylene. Although molecular oxygen is much less reactive than atomic oxygen, autoxidation would be expected to parallel these results, and poly(vinylcycloalkanes) probably exhibit good oxidative stability.

Poly(vinylcyclopropane) can be modified by treatment with either hydrogen bromide, bromine, *p*-toluenesulfonic acid, or a mixture of acetic acid and sulfonic acid. In each case, addition across the cyclopropyl ring occurs to only 85% of the theoretical ring openings and the crystallinity of the polymer is destroyed.

Bibliography

1. C. G. Overberger, A. E. Borchert, and A. Katchman, *J. Polymer Sci.* **44,** 491 (1960).
2. C. G. Overberger, H. Kaye, and G. Walsh, *J. Polymer Sci.* [A] **2,** 755 (1964).
3. J. R. Van Der Bij and E. C. Kooyman, *Rec. Trav. Chim.* **71,** 837 (1952).
4. C. G. Overberger and J. E. Mulvaney, *J. Am. Chem. Soc.* **81,** 4697 (1959).
5. A. E. Borchert and C. G. Overberger, *J. Polymer Sci.* **44,** 483 (1960).
6. E. S. Huyser and L. R. Munson, *J. Org. Chem.* **30,** 1436 (1965); S. Sarel and E. Breuer, *J. Am. Chem. Soc.* **81,** 6522 (1959); I. S. Lishanskii, A. M. Guliev, A. G. Zak, O. S. Formina, and A. S. Khachaturov, *Dokl. Akad. Nauk. SSSR* **170,** 1084 (1966).
7. J. Herling, J. Shabtai, and E. Gil-Av, *J. Am. Chem. Soc.* **87,** 4107, 4111 (1965).
8. R. Ya Levina, W. W. Mezentsova, and P. A. Akishin, *Zh. Obshch. Khim.* **23,** 562 (1953).

9. T. A. Manuel, *J. Org. Chem.* **27**, 394 (1962).
10. C. G. Overberger and A. E. Borchert, *J. Am. Chem. Soc.* **82**, 1007 (1960); M. R. Willcott and V. H. Cargle, *J. Am. Chem. Soc.* **89**, 723 (1967).
11. A. Maccioni and M. Secci, *Ann. Chim. (Rome)* **54**, 266 (1964); *Chem. Abstr.* **61**, 4237h (1969).
12. D. V. Mushenko, E. G. Lebedeva, V. P. Khimich, V. S. Chagina, and N. S. Barinov, *Zh. Prikl. Khim.* **39**, 2596, 2769 (1966).
12a. A. J. Lundeen and R. VanHoozer, *J. Am. Chem. Soc.* **85**, 2180 (1963).
13. C. D. Nenitzescu and A. T. Balaban, "Aliphatic Acylation" in G. A. Olah, ed., *Friedel-Crafts and Related Reactions*, Vol. III, Part 2, Interscience Publishers, a division of John Wiley & Sons, Inc., New York, 1964.
14. C. G. Overberger, A. Katchman, and J. E. Mulvaney, *J. Org. Chem.* **25**, 271 (1960).
15. Sh. G. Sadykhov, S. D. Meichtiev, V. A. Soldatova, M. A. Salimov, F. A. Mamedov, and N. S. Guseinov, *Neftekhimiya* **8**, 655 (1968); *Chem. Abstr.* **70**, 57233c (1969); *Chem. Abstr.* **58**, 11228a (1963).
16. L. H. Slaugh and E. F. Magoon, *J. Org. Chem.* **27**, 1037 (1962).
17. I. E. Levine (to California Research Corp.), Ger. Pat. 1,156,784 (Nov. 7, 1963).
18. N. F. Cywinski and H. J. Hepp, *J. Org. Chem.* **30**, 3814 (1965); N. I. Shuikin, B. L. Lebedev, and V. G. Wikol'skii, *Neftekhimiya* **6**, 544 (1966); *Dokl. Akad. Nauk SSSR* **158**, 692 (1964).
19. C. G. Overberger and G. W. Halek, *J. Org. Chem.* **28**, 867 (1963); G. Wittig and F. Wingler, *Chem. Ber.* **97**, 2146 (1964).
20. S. Nishida, I. Moritani, K. Ito, and K. Sakai, *J. Org. Chem.* **32**, 939 (1967).
21. J. Kiji and M. Iwamoto, *Tetrahedron Letters* **1966** (24), 2749.
22. D. Seyferth, H. Yamazaki, and D. L. Alleston, *J. Org. Chem.* **28**, 703 (1963).
23. I. S. Lishawskii and A. B. Zvyagina, *Zh. Org. Khim.* **4**, 184 (1968).
24. A. D. Ketley and R. J. Ehrig, *J. Polymer Sci.* [A] **2**, 4461 (1964).
25. W. H. McCarty and G. Parravano, *J. Polymer Sci.* [A] **3**, 4029 (1965).
26. E. A. Mushina, A. I. Perel'man, A. V. Topchiev, and B. A. Krentsel, *J. Polymer Sci.* **52**, 199 (1961); *Doklady Akad. Nauk SSSR* **130**, 1344 (1960); A. V. Topchiev, *J. Polymer Sci.* **53**, 195 (1961).
27. H. W. Coover, Jr. (to Eastman Kodak Co.), U.S. Pat. 2,948,712 (Aug. 9, 1960); *Chem. Abstr.* **55**, 1079i (1961).
28. H. W. Coover, Jr., and F. B. Joynèr (to Eastman Kodak Co.), U.S. Pat. 2,967,856 (Jan. 10, 1961); *Chem. Abstr.* **55**, 8941e (1961); U.S. Pat. 2,973,348 (Feb. 28, 1961); *Chem. Abstr.* **55**, 12930b (1961); U.S. Pat. 3,287,340 (Nov. 22, 1966); *Chem. Abstr.* **66**, 29362t (1967); U.S. Pat. 3,230,208 (Jan. 18, 1966); *Chem. Abstr.* **64**, 14309b (1966).
29. N. H. Shearer, Jr., and H. W. Coover, Jr. (to Eastman Kodak Co.), U.S. Pat. 3,397,196 (Aug. 13, 1968); *Chem. Abstr.* **69**, 67899t (1968).
30. B. Gmuender, Fr. Pat. 1,336,744 (Sept. 6, 1963); *Chem. Abstr.* **60**, 5713e (1964).
31. G. C. Corfield, A. Crawshaw, G. E. Butler, and M. L. Miles, *Chem. Comm.* **238** (1966).
32. A. D. Ketley, A. J. Berlin, and L. P. Fisher, *J. Polymer Sci.* [A-1] **5**, 227 (1967).
33. J. P. Kennedy, J. J. Elliott, and W. Naegele, *J. Polymer Sci.* [A] **2**, 5029 (1964).
34. I. S. Lishanskii, A. G. Zak, Ye. F. Fedorova, and A. S. Khachaturov, *Vysokomol. Soedin.* **7**, 966 (1965).
35. T. Takahashi and I. Yamashita, *J. Polymer Sci.* [B] **3**, 251 (1965).
36. T. Takahashi, *J. Polymer Sci.* [A-1] **6**, 403, 3327 (1968).
37. V. L. Folt (to B. F. Goodrich Co.), U.S. Pat. 3,111,509 (Nov. 19, 1963); *Chem. Abstr.* **60**, P4276g (1964); P. D. Bartlett, L. K. Montgomery, and B. Seidel, *J. Am. Chem. Soc.* **86**, 616 (1964).
38. Imperial Chemical Industries, Ltd., Neth. Pat. Appl. 6,508,723 (Jan. 10, 1966); *Chem. Abstr.* **64**, 17739h (1966).
39. E. V. Kirkland and N. S. Millington (to Celanese Corp.), U.S. Pat. 3,376,248 (April 2, 1968); *Chem. Abstr.* **68**, 96739a (1968).
40. "Shell" International Research Maatachappij N.V., Neth. Pat. Appl. 6,516,554 (June 23, 1966); *Chem. Abstr.* **65**, Pc 18863g (1966).
41. Dow Chemical Co., Neth. Pat. Appl. 6,400,976 (Aug. 7, 1964); *Chem. Abstr.* **62**, P1761c (1965).
42. G. F. D'Alelio (to Dal Mon Research Co.), U.S. Pat. 3,075,957 (Jan. 29, 1963); *Chem. Abstr.* **58**, P11482e (1963).

43. C. S. Y. Kim, E. O. Hook, F. Veatch, and E. C. Hughes, *Kunststoff-Rundschau* **12**, 65 (1965); *Chem. Abstr.* **63,** 8496c.
44. F. P. Reding, E. W. Wise, and J. H. Hoge (to Union Carbide Corp.), U.S. Pat. 3,256,256 (June 14, 1966); *Chem. Abstr.* **65,** Pc9051h (1966).
45. A. L. Barney, J. M. Bruce, Jr., J. N. Coker, H. W. Jacobson, and W. H. Sharkey, *J. Polymer Sci.* [A-1] **4,** 2617 (1966).
46. Peninsular Chem. Research, Inc., Brit. Pat. 1,013,230 (Dec. 15, 1965); *Chem. Abstr.* **64,** P6787c (1966).
47. J. P. Kennedy, S. Bank, and R. G. Squires, *J. Macromol. Sci. Chem.* **1,** 961 (1967).
48. G. Natta, P. Corradini, and I. W. Bassi, *Makromol. Chem.* **33,** 247 (1959).
49. H. D. Noether, *J. Polymer Sci.* [C] **16,** 725 (1967).
50. H. D. Noether, C. G. Overberger, and G. Halek, *J. Polymer Sci.* [A] **7,** 201 (1969).
51. J. D. Hutchinson, *J. Polymer Sci.* [A] **3,** 2710 (1965).
52. V. Frosini, P. Magagnini, E. Butta, and M. Baccaredda, *Kolloid Z. Z. Polym.* **213,** 115 (1966).
53. S. T. Barsamyan, A. S. Appesyan, V. L. Kleiner, and L. L. Stotskaya, *Plasticheskie Massy* **3,** 13 (1968); *Chem. Abstr.* **69,** 3281c.
54. H-G. Elias and O. Etter, *J. Macromol. Chem.* **1,** 431 (1966).
55. R. H. Hansen, J. V. Pascale, T. DeBenedictis, and P. M. Rentzepis, *J. Polymer Sci.* [A] **3,** 2205 (1965).
56. *International Tables for X-ray Crystallography,* Vol. 1, The Kynoch Press, Birmingham, England, 1965.

William H. Daly
Louisiana State University

VINYL DISPERSIONS. See VINYL CHLORIDE POLYMERS

VINYLENE CARBONATE POLYMERS

Vinylene carbonate is an unsaturated, cyclic 1,2-substituted ethylene derivative (**1**). Hydrolysis of the vinylene carbonate rings produces a polyglycol containing vicinal hydroxyl groups. The structure of the hydrolyzed homopolymer is indicated in structure (**2**). The presence of the numerous hydroxyl groups is reflected in the physical properties of the polymers and in its high reactivity.

```
HC═══CH
 |     |
 O     O              ~~~( CH—CH )~~~
  \   /                    |   |
    C                     OH  OH   n
    ‖
    O
   (1)                    (2)
```

Monomer

Newman and Addor (1) described the first synthesis of the monomer. A stream of chlorine was passed through freshly distilled ethylene carbonate at 63–70°C, under the influence of ultraviolet radiation. After 24 hr, the gain in weight corresponded to monochloro substitution. Vacuum distillation yielded 5% of 1,2-dichlorethylene carbonate and 69% of monochloroethylene carbonate: bp, 106–107°C at 10–11 torr; n_D^{25} 1.4530; d_4^{25}, 1.5082. Strong strained-ring carbonyl absorption was found at 5.45 μm. The purified monochloroethylene was then treated at reflux in dry ether with triethylamine. Distillation yielded 59% of a colorless liquid, bp 76–79°C at 37 torr.

Pure vinylene carbonate has the following properties: bp, 73–74°C at 32 torr, 102°C at 135 torr; mp, 22°C; n_D^{25}, 1.4190; d_4^{25}, 1.3541; C–H absorption, 3.12 μm; C=O absorption, 5.48 μm.

Several detailed studies have been carried out on the microwave (2,3), infrared, and Raman (4,5) spectra of vinylene carbonate. It has been established that the molecule can be characterized by a planar C_{2v} symmetry, and most of the fundamental modes of vibration have been assigned.

The Diels-Alder reaction can be carried out successfully with vinylene carbonate and the following dienes: 1,3 butadiene, 2,3 dimethyl 1,3 butadiene, cyclopentadiene, hexachlorocyclopentadiene, furan, and anthracene (6).

Polymerization

Smets and Hayashi (8,9) investigated the polymerization kinetics of vinylene carbonate initiated by azobisisobutyronitrile (AIBN) in acetone, ethyl benzoate, ethylene carbonate, benzene, and methyl benzoate. The overall activation energy of the azobisisobutyronitrile-initiated polymerization is 22.2 kcal/mole. Chain transfer to the monomer has been ruled out because of the lack of a labile hydrogen atom in the molecule. Haas and Schuler (10) confirmed that the vinylene carbonate does not polymerize under the influence of cationic and anionic catalysts, but does so under the influence of benzoyl peroxide and other free-radical initiators. The production of high-molecular-weight poly(vinylene carbonate) has been described by Field and Schaefgen (11) who used azobisisobutyronitrile initiation in bulk and in solution, the latter with dimethyl sulfoxide and ethylene carbonate. They established that the polymer is formed through the carbon–carbon double bond and that it is amorphous. This was confirmed by Kazanskaya and Klimova (12). The polymerization process has been established to be free radical in nature and the radical species responsible for it are probably formed in a redox reaction (13,14). Free radicals are also capable of initiating copolymerizations in aqueous solution with water-soluble monomers (15,16). A possible mechanism for the process in acidic media—by analogy with the reactions of ascorbic acid and dihydroxymaleic acid with oxygen—involves the following redox reaction between vinylene glycol (a hypothetical compound formed by hydrolysis of the carbonate) and atmospheric oxygen (14):

$$\begin{array}{c} HC{-}OH \\ \| \\ HC{-}OH \end{array} + O_2 \rightarrow \begin{array}{c} HC{-}O\cdot \\ \| \\ HC{-}OH \end{array} + HO_2\cdot$$

$$\begin{array}{c} HC{-}O\cdot \\ \| \\ HC{-}OH \end{array} + O_2 + H_2O \rightarrow 2\,HCOOH + HO\cdot \qquad (1)$$

In the azobisisobutyronitrile-initiated liquid-phase bulk polymerization of vinylene carbonate, Hardy and Nyitrai (17) found the average degree of polymerization for the product to be independent of the initiator concentration. This indicates that chain termination is predominantly by chain transfer to the monomer molecule. The process may be described by equation 2:

$$k_p/k_t^{1/2} = 740 \exp\left(\frac{-7400}{RT}\right) \qquad (2)$$

The values obtained in this way are different from the analogous values of vinyl compounds that form reactive radicals but do agree, in order of magnitude, with those of maleic anhydride.

Kinetics of the γ-ray-initiated liquid-state polymerization of vinylene carbonate gave a value of 0.16 for the G_{init} radiation chemical yield. This value is very low, or about one order of magnitude smaller than the corresponding value of vinyl acetate. Hardy and co-workers studied the γ-ray-initiated solid-state post- and in-source polymerization of vinylene carbonate (18,19) using electron-spin resonance (qv). The spectrum of the radical detectable in the system undergoing post-polymerization differs significantly from the spectra of the primary irradiation products. The rate of solid-state polymerization is initially linear, then decelerates and, finally, accelerates. The molecular weight of the polymer formed in the linear stage is independent of the conversion, whereas in the accelerating stage it increases with the conversion. Molecular-weight distribution of the polymers, as determined by ultracentrifugation (qv), is rather narrow in the linear stage but broad in the accelerating one. Schulz and Vollkommer (20) fractionated poly(vinylene carbonate) using dimethylformamide as a solvent and methanol or n-butanol as precipitating agents.

Derivatives

Poly(vinylene carbonate) can be easily hydrolyzed in basic (10,11,21,22) and in acidic (9,23) media to produce poly(vinylene glycol) (**2**). This polymethylol is not soluble in water, because of the great frequency of hydrogen bonds, but is soluble in dimethyl sulfoxide (21), in hot 20% sodium hydroxide, and in the aqueous 35% solution of benzyltrimethylammonium hydroxide (10). The heat of hydrolysis in dilute sulfuric acid is 9 kcal/mole (9). In concentrated aqueous ammonia the following reaction takes place (eq. 3) (11):

$$\sim\!\!\text{CH}{-}\text{CH}\!\!\sim \text{(cyclic carbonate: O–C(=O)–O bridging both CH)} \xrightarrow[H_2O]{NH_3} \sim\!\!\text{CH(OH)}{-}\text{CH(O–C(=O)–NH}_2)\!\!\sim \qquad (3)$$

A systematic study of the aminolysis with n-butylamine in dimethylformamide was made by Nemirovsky and Skorokhodov (24–26) utilizing both poly(vinylene carbonate) and ethylene carbonate, as a model compound. The rate of the poly(vinylene carbonate) aminolysis was found to be considerably higher than that of ethylene carbonate. These data indicate that the catalysis occurs by the opened monomeric unit that is adjacent to the reacting one. The catalytic action is probably connected with the carbonate groups. Syntheses and properties of the following products were prepared (24):

$$\sim\!\!\text{CH}{-}\text{CH}\!\!\sim \text{(cyclic carbonate: O–C(=O)–O bridging both CH)} + RNH_2 \longrightarrow \sim\!\!\text{CH(O–C(=O)–NH–R)}{-}\text{CH(OH)}\!\!\sim \qquad (4)$$

where R = CH_3, C_2H_5, n-C_4H_9, n-C_6H_{13}, n-$C_{10}H_{21}$, β-hydroxyethyl, and cyclohexyl.

Further products derived from reactions of the β-hydroxyl group have been studied (27), including those resulting from nitration (eq. 5), acylation (eq. 6), and tosylation (eq. 7) (28):

$$\begin{array}{ccc} \sim\text{CH—CH}\sim & & \sim\text{CH—CH}\sim \\ |\qquad | & & |\qquad | \\ \text{O}\quad\ \text{OH} & \xrightarrow{\text{HNO}_3} & \text{O}\quad\ \text{ONO}_2 \\ | & & | \\ \text{C=O} & & \text{C=O} \\ | & & | \\ \text{N—R}' & & \text{N—R}' \\ | & & | \\ \text{R}'' & & \text{R}'' \end{array} \tag{5}$$

$$\begin{array}{ccc} \sim\text{CH—CH}\sim & & \sim\text{CH—CH}\sim \\ |\qquad | & & |\qquad | \\ \text{O}\quad\ \text{OH} & \xrightarrow{\text{RCOCl}} & \text{O}\quad\ \text{O—C—R} \\ | & & |\qquad\quad \| \\ \text{C=O} & & \text{C=O}\quad\ \text{O} \\ | & & | \\ \text{N—R}' & & \text{N—R}' \\ | & & | \\ \text{R}'' & & \text{R}'' \end{array} \tag{6}$$

$$\begin{array}{ccc} \sim\text{CH—CH}\sim & & \sim\text{CH—CH}\sim \\ |\qquad | & & |\qquad | \\ \text{O}\quad\ \text{OH} & \xrightarrow[\text{pyridine}]{\text{TsCl}} & \text{O}\quad\ \text{OTs} \\ | & & | \\ \text{C=O} & & \text{C=O} \\ | & & | \\ \text{NH} & & \text{NH} \\ | & & | \\ \text{R} & & \text{R} \end{array} \tag{7}$$

where TsCl = p-$CH_3C_6H_4SO_2Cl$ and R = CH_3, n-C_4H_9, or $—CH_2CH_2OH$. These chemically modified products are good fiber- and film-forming polymers.

Other derivatives of hydrolyzed poly(vinylene carbonate) include the acetates (11,21), cinnamates, and other esters (22), phenylurethans (21), alkyl ethers, and acetals resulting from reaction with formaldehyde, butyraldehyde, and benzaldehyde (23).

Copolymerization

The sequence length of the built-in vinylene carbonate units can be determined by hydrolysis followed by cleavage of the glycol with periodate (29); an isolated vinylene carbonate unit yields no formic acid. For copolymerizations with vinyl acetate, good agreement was found between the theoretically calculated and the experimentally obtained sequence-length distributions (30). The numerical results of copolymerization investigations are summarized in Table 1. The copolymerization studies indicate unambiguously the relatively low reactivity of vinylene carbonate, which can be explained by the 1,2-substitution. Resonance stabilization has been invoked as the cause for the negative e value (32).

Vinylene carbonate copolymers have been reported with many comonomers, for example, acrylonitrile (39,40), methyl methacrylate (40), ethylene (41,43,44,46), vinyl chloride (42,45), and tetrafluoroethylene (42).

The redox system formed in aqueous medium by atmospheric oxygen can be utilized to prepare copolymers with methyl methacrylate, vinylpyrrolidone, vinyl acetate, acrylic acid, and methacrylic acid (36,37). The copolymer of vinylene carbonate and vinyl acetate may also be produced in aqueous medium with ammonium peroxydisulfate–ascorbic acid redox system (41). Copolymerization of vinylene car-

Table 1. Copolymerization Parameters of Vinylene Carbonate (M_1)

Comonomer (M_2)	Initiator	Temp, °C	r_1	r_2	Q_1	e_1	Reference
vinyl acetate	AIBN[a]	70	0.27	3.0	0.012	−0.6	(9,31)
vinyl acetate	BPO[b]	70	0.15	4.0	0.008	−1.18	(10)
vinyl acetate	AIBN	60	0.13	7.3	0.002	−0.73	(32)
vinyl acetate	BPO	55	0.058	3.71			(30)
vinyl thiolacetate	AIBN	60	0.04	12.9			(33)
vinyl chloride	AIBN	80	0.09	5.2	0.012	−0.6	(31)
vinylpyrrolidone	AIBN	60	0.4	0.7	0.012	−0.6	(31)
acrylonitrile	AIBN	60	0.085	14.9			(38)
acrylamide	atm O_2	20	0.065	14.2	0.012	−0.89	(34)
acrylic acid	atm O_2	20	pH 1.5–2.0 0.020	32.5			
			pH 3.0 0.014	16.3			(35)
			pH 7.0 0.185	8.7			
methacrylic acid	atm O_2	20	pH 1.5–2.0 0.011	26.5			
			pH 3.0 0.027	10.0			(35)
			pH 7.0 0.33	3.5			
methyl methacrylate	AIBN	70	0.005	70.0	0.012	−0.6	(31)
methyl methacrylate	atm O_2	20	0.01	67.0	0.026	−0.23	(36)
styrene	AIBN	60	0	8–20			(32)
isobutyl vinyl ether	AIBN	50	0.160	0.185			(37)
methyl vinyl sulfide	AIBN	60	0.05	10.6	0.01	−0.71	(32)

[a] Azobisisobutyronitrile.
[b] Benzoyl peroxide.

bonate and methyl methacrylate takes place under the influence of β-radiation (49). In this case the copolymer contains more vinylene carbonate units than when using free-radical initiators for the copolymerization. Vinylene carbonate can also be grafted onto poly(methyl methacrylate) with γ-irradiation. The polymer obtained is more resistant to radiation than the original poly(methyl methacrylate). The rate of grafting is significantly increased in the presence of methyl alcohol; this effect may be the result of improved diffusion of the vinylene carbonate or of radiosensitization (50).

Uses

Vinylene carbonate polymers, including homopolymers, copolymers, and derivatives, have been suggested for fibers (28,39), films (28), plastics (40,41,52,53), impregnants (51), adhesives (52), and as membranes for water desalination (54). In the latter case, high salt rejections together with high flux have been reported.

Bibliography

1. M. S. Newman and R. W. Addor, *J. Am. Chem. Soc.* **75,** 1263 (1953).
2. G. R. Slayton, J. W. Simmons, and J. H. Goldstein, *J. Chem. Phys.* **22,** 1678 (1954).
3. K. L. Dorris, C. O. Britt, and J. E. Boggs, *J. Chem. Phys.* **44,** 1352 (1966).
4. J. L. Hales, J. J. Idris, and W. Kynaston, *J. Chem. Soc.* **1957,** 618.
5. K. L. Dorris, J. E. Boggs, A. Danti, and L. L. Altpeter, *J. Chem. Phys.* **46,** 1191 (1967).
6. M. S. Newman and R. W. Addor, *J. Am. Chem. Soc.* **77,** 3789 (1955).
7. M. S. Newman, *J. Org. Chem.* **26,** 2630 (1961).
8. G. Smets and K. Hayashi, *J. Polymer Sci.* **29,** 257 (1958).
9. K. Hayashi, *Kyoto Daigaku Nippon Kagakuseni Kenkyusho Koenshu* **15,** 69 (1958); *Chem. Abstr.* **54,** 958e (1960).
10. H. C. Haas and N. W. Schuler, *J. Polymer Sci.* **31,** 237 (1958).

11. N. D. Field and J. R. Schaefgen, *J. Polymer Sci.* **58,** 533 (1962).
12. V. F. Kazanskaya and O. M. Klimova, *Zh. Prikl. Khim.* **38,** 432 (1965).
13. V. F. Kazanskaya and O. M. Klimova, *Izv. Vys. Utsebn, Zavedenij SSSR, Khim. i Khimits. Technol.* **9,** 641 (1966).
14. V. F. Kazanskaya and O. M. Klimova, *Vysokomol. Soedin.* **A9,** 1889 (1967).
15. O. M. Klimova and V. F. Kazanskaya, *Zh. Prikl. Khim.* **38,** 434 (1965) (*J. Appl. Chem.*, in Russian).
16. T. F. Baskova and O. M. Klimova, *Vysokomol. Soedin.* **B10,** 358 (1968) (in Russian).
17. G. Hardy and K. Nyitrai, *Acta Chimica A. Sc. Hung.* **56,** 39 (1968).
18. G. Hardy, K. Nyitrai, G. Nagy, J. Erö, and M. Kisbényi, *Acta Chimica A. Sc. Hung.* **56,** 61 (1968).
19. G. Hardy, K. Nyitrai, F. Cser, and A. Nagy, *European Polymer J.* **4,** 289 (1968).
20. R. C. Schulz and N. Vollkommer, *Makromol. Chem.* **116,** 288 (1968).
21. C. C. Unruh and D. A. Smith, *J. Org. Chem.* **23,** 625 (1958).
22. M. G. Krakovyak, S. I. Klenin, and S. S. Skorokhodov, *Vysokomol. Soedin.* **7,** 1576 (1965) (in Russian).
23. O. M. Klimova, A. M. Kuras, V. V. Stepanov, and N. I. Harlamova, *Zh. Prikl. Khim.* **37,** 1152 (1964) (*J. Appl. Chem.*, in Russian).
24. V. D. Nemirovsky, M. A. Pavlovskaja, V. V. Stepanov, and S. S. Skorokhodov, *Vysokomol. Soedin.* **7,** 1580 (1965) (in Russian).
25. V. D. Nemirovsky and S. S. Skorokhodov, *J. Polymer Sci.* [C] *No. 16,* 1471 (1967).
26. V. D. Nemirovsky and S. S. Skorokhodov, *Vysokomol. Soedin.* **A9,** 2142 (1967) (in Russian).
27. V. D. Nemirovsky, S. S. Skorokhodov, and K. K. Kalninsh, *Vysokomol. Soedin.* **A9,** 15 (1967) (in Russian).
28. V. B. Lustsik, V. D. Nemirovsky, and S. S. Skorokhodov, *Vysokomol. Soedin.* **B9,** 840 (1967).
29. C. Schuerch, *J. Polymer Sci.* **13,** 405 (1954).
30. H. L. Marder and C. Schuerch, *J. Polymer Sci.* **44,** 129 (1960).
31. K. Hayashi and G. Smets, *J. Polymer Sci.* **27,** 275 (1958).
32. J. M. Judge and C. C. Price, *J. Polymer Sci.* **41,** 435 (1959).
33. C. G. Overberger, H. Biletch, and R. G. Nickerson, *J. Polymer Sci.* **27,** 381 (1958).
34. T. F. Baskova and O. M. Klimova, *Vysokomol. Soedin.* **B10,** 63 (1968) (in Russian).
35. T. F. Baskova, O. M. Klimova, and L. G. Sutulova, *Vysokomol. Soedin.* **B10,** 220 (1968) (in Russian).
36. N. K. Maratova and O. M. Klimova, *Vysokomol. Soedin.* **B10,** 87 (1968) (in Russian).
37. R. C. Schulz and R. Wolf, *Kolloid-Z.* **220,** 148 (1967).
38. E. A. Rassolova, M. A. Zharkhova, G. I. Kudrjavcev, and V. S. Klimenkov, *Khim. Volokna* **11,** 50 (1969) (*Chemical Fibers,* in Russian).
39. J. A. Price and J. J. Padbury (to American Cyanamid Co.), U.S. Pat. 2,722,525 (1955); *Chem. Abstr.* **50,** 3011h (1956).
40. E. W. Gluesenkamp and J. D. Calfee (to Monsanto Chemical Co.), U.S. Pat. 2,847,402 (1958); *Chem. Abstr.* **52,** 21251g (1958).
41. E. W. Gluesenkamp and J. D. Calfee (to Monsanto Chemical Co.), U.S. Pat. 2,847,398 (1958); *Chem. Abstr.* **52,** 19258b (1958).
42. E. W. Gluesenkamp and J. D. Calfee (to Monsanto Chemical Co.), U.S. Pat. 2,847,401 (1958); *Chem. Abstr.* **52,** 21251d (1958).
43. I. O. Salyer and J. A. Herbig (to Monsanto Chemical Co.), U.S. Pat. 2,957,847 (1960); *Chem. Abstr.* **55,** 6939i (1961).
44. I. O. Salyer and J. A. Herbig (to Monsanto Chemical Co.), U.S. Pat. 2,945,836 (1960); *Chem. Abstr.* **54,** 26005h (1960).
45. I. O. Salyer and J. D. Calfee (to Monsanto Chemical Co.), U.S. Pat. 2,934,514 (1960); *Chem. Abstr.* **54,** 16927i (1960).
46. T. V. Krejcer and R. A. Terteryan, *Vysokomol. Soedin.* **B11,** 345 (1969) (in Russian).
47. O. M. Klimova and V. F. Kazanskaya, *Zh. Prikl. Khim.* **38,** 434 (1965), (*J. Appl. Chem.*, in Russian).
48. L. E. Klubikova, O. M. Klimova, and A. V. Jaros, *Zh. Prikl. Khim.* **38,** 1188 (1965), (*J. Appl. Chem.*, in Russian).
49. N. K. Maratova, O. M. Klimova, and I. K. Karpov, *Vysokomol. Soedin.* **B10,** 566 (1968) (in Russian).

50. N. K. Zaiceva, O. M. Klimova, I. K. Karpov, and A. D. Zhsurbenko, *Vysokomol. Soedin.* **B11**, 196 (1969).
51. E. K. Drechsel (to American Cyanamid Co.), U.S. Pat. 2,930,779 (1960); *Chem. Abstr.* **54**, 17967g (1960).
52. E. K. Drechsel (to American Cyanamid Co.), U.S. Pat. 2,794,013 (1957); *Chem. Abstr.* **51**, 13465c (1957).
53. F. M. Rugg, J. J. Smith, J. E. Potts, and E. F. Bonner (to Union Carbide Corp.), Brit. Pat. 814,393 (1959); *Chem. Abstr.* **53**, 18554c (1959).
54. *Saline Water Conversion Report for 1967*, Office of Saline Water, U.S. Department of the Interior, Washington, D.C.

G. Hardy
Research Institute for the Plastics Industry (Hungary)

VINYL ESTER POLYMERS. See VINYL ESTER POLYMERS, Vol. 15

VINYL ETHER POLYMERS

Vinyl ethers were prepared as early as 1927 by reaction of acetylene with alcohols as part of the study of acetylene reactions by Reppe and his associates. Polymers were available in Europe from I. G. Farben (Badische Anilin) in the 1930s and introduced in the United States in the early 1940s.

The vinyl ether monomers are readily polymerized and copolymerized to many materials of varied use. Some of the more important uses are in adhesives, processing aids, lubricants and greases, paints, hair grooming aids, molding compounds, fibers, and films.

Vinyl ether monomers, as well as homopolymers and copolymers, are produced commercially in the United States by GAF Corporation (1). Union Carbide Corporation also supplies some vinyl ethers in the United States. Badische Anilin- und Soda-Fabrik A.G. supplies vinyl ethers in Europe (2). Imperial Chemical Industries has reported a method of making vinyl ethers from ethylene and alcohols but in 1970 had not entered commercial production.

MONOMERS

Table 1 lists the physical properties of the C_1–C_4 vinyl ethers. The C_1–C_4 vinyl ethers are soluble in solvents such as hexane, carbon tetrachloride, methanol, ethanol, ether, dioxane, ethyl acetate, benzene, and ethoxyethanol.

Tables 2–6 list some of the pertinent data on alkyl, aryl, divinyl, α- and β-substituted, and functionally substituted vinyl ethers.

The chemical reactions of the vinyl ethers are typical of most vinyl compounds. Extensive discussion of their reactions can be found in Refs. 5, 9, and 27. Vinyl

Table 1. Physical Properties of the C_1–C_4 Vinyl Ethers (3–5)

	Methyl	Ethyl	Isopropyl	*n*-Butyl	Isobutyl
odor	sweet, pleasant	pleasant	pleasant	pleasant	pleasant
boiling point, °C	5.5	35.6	55–56	94.3	83
freezing point, °C	−222	−115.3	−140	−112.7	−132.3
specific gravity at 20/4°C	0.7511	0.753	0.753	0.778	0.767
refractive index, n_D^{25}	1.3947	1.3734	1.3829	1.3997	1.3946
solubility in water, 20°C, wt %	0.97	0.9	0.6	0.3	0.2
flash point, °C	20[a]	−18[b]		0[b]	−10[b]
heat of vaporization, at 1 atm, cal/g		87.8		75.6	77.2

[a] Cleveland open cup.
[b] Tag open cup (ASTM D 1310).

Table 2. Properties of Alkyl Vinyl Ethers

Alkyl group	Bp, °C (mm Hg)	n_D^{25}	d_4^{20}	Reference
s-butyl	81	1.3970 (20°C)	0.7715	5
t-butyl	75	1.3922 (20°C)	0.7691	5
n-amyl	111			7
isoamyl	110	1.4070 (20°C)	0.7833	8
1,2-dimethylpropyl	102			9
n-hexyl	143.5	1.4171 (20°C)	0.7966	5
1,2,2-trimethylpropyl	115			9
2-ethylbutyl	132.2	1.4185 (20°C)	0.8011	5
1,3-dimethylbutyl	47 (50)			9
2,2-dimethylbutyl	124.5	1.4125	0.7929	10
diisopropylmethyl	139			9
n-octyl	58 (4)	1.4268 (20°C)	0.8024	5
2-ethylhexyl	174; 61–63 (11)	1.4247	0.8088	5,9
1-methylheptyl	78 (19)	1.4211		9
2,2-dimethylhexyl	167.5	1.4232	0.8031	10
n-decyl	101 (10)	1.4346		9
2,2-dimethyloctyl	51 (1.1)	1.4303	0.8102	10
2,2-dimethyldecyl	93 (2)	1.4373	0.8176	10
n-tetradecyl	140–145 (4)			9
2,2-dimethyldodecyl	127 (1.8)	1.4445	0.8246	10
n-hexadecyl	160–165 (4)			9
2,2-dimethyltetradecyl	154 (1)	1.4515	0.8319	10
n-octadecyl	182 (3) (mp 30°C)	1.4515		9
oleyl	170–175 (2)			9

Table 3. Properties of Aryl Vinyl Ethers (11)

Aryl group	Bp, °C (mm Hg)	n_D^{20}
phenyl	155–157	1.5226
o-cresyl	170	
p-cresyl	174	1.5186
p-chlorophenyl	194	
2,4-dichlorophenyl	105 (15)	
2,4,6-trichlorophenyl	118 (14) (mp, 35°C)	
α-naphthyl	259	1.6152
β-naphthyl	264 (mp, 64°C)	

Table 4. Properties of Divinyl Ethers

	Bp, °C (mm Hg)	n_D^{20}	d_4^{20}	Reference
$CH_2{=}CHOCH{=}CH_2$	28.5 (fp, −101°C)	1.3989	0.773	12
$CH_2{=}CHOCH_2CH_2OCH{=}CH_2$	126	1.4338		12
$CH_2{=}CHO(CH_2CH_2O)_2CH{=}CH_2$	110 (18)			12
$CH_2{=}CHO(CH_2CH_2O)_3CH{=}CH_2$	110–120 (10)			13
$CH_2{=}CHO(CH_2)_4OCH{=}CH_2$	60–65 (20)			13
$CH_2{=}CHOCH_2OCH{=}CH_2$	90			12
$CH_2{=}CHOCH(CH_2CH_2CH_3)OCH{=}CH_2$	58 (32)			12

Table 5. Properties of α- and β-Substituted Vinyl Ethers

Vinyl ether	Bp, °C (mm Hg)	n_D^{20}	d_4^{20}	Reference
methyl α-methylvinyl ether	32.5–33.5	1.3788	0.73	14
methyl α-chlorovinyl ether	122; 123		1.02	15
methyl β-methylvinyl ether				
(*cis*)	45.0	1.3917		16
(*trans*)	48.5	1.3899		
methyl β-chlorovinyl ether	100–118			17
ethyl α-ethylvinyl ether	87	1.4018		15
ethyl β-methylvinyl ether				
(*cis*)	69.0 (755)	1.3986		16
(*trans*)	75.0 (758)	1.3978		
ethyl α-phenylvinyl ether	96 (12.5)			15
isopropyl β-methylvinyl ether				
(*cis*)	83 (750)	1.400		16
(*trans*)	90.5 (750)	1.4018		
n-butyl-β-methylvinyl ether				
(*cis*)	119; 121 (765)	1.4131		16,18
(*trans*)	126; 127.5 (765)	1.4143		
isobutyl-β-methylvinyl ether				
(*cis*)	110.5 (755)	1.4079		16
(*trans*)	119.0	1.4100		
t-butyl-β-methylvinyl ether				
(*cis, trans*)	185–187 (746)	1.440	0.8800	19

ethers in the presence of acid are hydrolyzed to the corresponding alcohol and acetaldehyde (eq. 1) (28–31). A kinetic study of the reaction showed the following order of

$$ROCH{=}CH_2 + H_3^{\oplus}O \longrightarrow ROH + CH_3{-}\overset{\overset{\large O}{\|}}{C}{-}H + H^{\oplus} \tag{1}$$

reactivities: *t*-butyl > isopropyl > ethyl > methyl (28). This is the same order observed in simple cationic polymerization and is related to the basicity of the oxygen rather than electron release to the double bond.

Reaction with halogen (chlorine and bromine) at low temperatures gives the normal addition product. In the presence of alcohols β-haloacetates are produced (32). One of the more important reactions involving vinyl ethers is the transesterification (transvinylation) reaction, which makes possible preparation of base-sensitive vinyl ethers. This reaction is catalyzed by mercuric salts (6) or can be carried out over a diatomaceous earth catalyst at 250–320°C.

Table 6. Functionally Substituted Vinyl Ethers ($CH_2{=}CHOR$)

R	Bp, °C (mm Hg)	n_D^{25}	d_{20}^{20}	Reference
CH_2CH_2OH	140	1.4532		20
$CH_2CH_2CH_2CH_2OH$	10_2 (50)	1.4427	0.939_4^{20}	21
CH_2CH_2Cl	108 (fp, −70°C)	1.4361	1.0493	9
$CH_2CH_2OCH_3$	109 (fp, −83°C)	1.4072	0.8967	9
$CH_2CH_2N(CH_3)_2$	124	1.4215		9
CF_3	−18–−15			22
$CF(CF_3)_2$	28–30			23
$CF(CF_3)(CF_2Cl)$	58–60	1.3298		23
$CF(CF_2Cl)_2$	101.5	1.3559		23
$CH_2CH_2OCH_2CH_2OH$	108 (12)			13
$CH_2CH_2CHOHCH_3$	117–120			13
$CH_2CH_2OCH_2CH_2OCH_3$	177	1.4252		20
$CH_2CH_2OCH_2CH_2OC_2H_5$	125 (10)			20
$CH_2CH{=}CH_2$	67	1.4095	0.805_4^{20}	12
CH_2COOCH_3	76.8 (49.5)	1.4232		24
$CH_2\underset{\diagdown O \diagup}{CH{-}CH_2}$	138–139	1.4458 (11°C)	1.105_{11}^{11}	25
$CH_2CH_2{-}N$ (pyrrolidone ring, C=O)	135 (3)			26

In addition to the normal reactions of a vinyl compound, vinyl ethers can act as a source of acetaldehyde (eq. 2) (33).

$$CH_2{=}CHOR + HCN \xrightarrow[H_2O]{\text{ion-exchange resin}} CH_3CHOHCN + ROH \qquad (2)$$

Reaction of ammonia with vinyl ethers in the presence of copper or ammonium salts gives 2-methyl-5-ethylpyridines (34–37).

Since the vinyl ethers are prone to hydrolysis and to polymerization by acidic materials, acids and water are to be avoided. Stabilizers ordinarily used are solid alkalies (potassium hydroxide pellets), or high-boiling amines such as triethanolamine. No radical-trap stabilizers are ordinarily required with adequate removal of air during handling and distillation.

Manufacture

The major commercial process for the manufacture of vinyl ethers is the reaction of acetylene with alcohols under basic conditions at temperatures of 120–180°C (7,13). A likely mechanism for the reaction involves addition of the metal alcoholate to the triple bond (eq. 3a) in the rate-controlling step followed by metal–alcohol exchange (eq. 3b) (7).

$$HC{\equiv}CH + ROM \longrightarrow ROCH{=}CHM \qquad (3a)$$

$$ROCH{=}CHM + ROH \longrightarrow ROM + ROCH{=}CH_2 \qquad (3b)$$

A typical procedure for a batch vinylation would be conducted in a high-pressure autoclave with a suitable safety barricade. The reactor would be half charged with alcohol containing either alkali metal hydroxide or alkoxide, purged of air with nitrogen, and the temperature raised to 120–180°C. A gas mixture of acetylene and nitrogen (usually 1 part acetylene to 2 parts nitrogen) is introduced at an operating pressure of 100 psi or above and held at that pressure by continuous feed until there is no further uptake of acetylene. For more sluggish vinylations higher acetylene ratios are used, but special autoclaves capable of withstanding detonation pressures are required. Normally the vinylation is carried to a minimum of 80% completion, after which the reaction mixutre is cooled, acetylene vented, and the product isolated by distillation. See also ACETYLENE AND ACETYLENIC POLYMERS.

Another important commercial process for the preparation of vinyl ethers is the thermal cracking of acetals. The cracking reactions are usually at 250–400°C over catalysts such as palladium on asbestos (38), thoria (39), or metal sulfates on alumina (40). The acetal precursors are prepared by reaction of acetaldehyde with appropriate alcohols using acid catalysts. Calcium chloride is an especially advantageous catalyst since it also functions to remove water and drive the reaction to completion. Acetals are also formed in good yields by reaction of acetylene with alcohols in the presence of acidic mercuric catalysts (eq. 4) (41).

$$HC{\equiv}CH + 2\,ROH \xrightarrow[HgO]{BF_3} CH_3CH(OR)_2 \qquad (4)$$

Reaction of primary alcohols with vinyl acetate in the presence of a similar type of catalyst produces acetals in 80–90% yield (42). Reaction of ethylene with alcohols using platinum group metals can also give acetals (43).

If the vinyl ethers that cannot withstand the hot basic conditions of the vinylation reaction or if the acetal is difficult to prepare, the exchange or transvinylation reaction can be used (eq. 5). This reaction is catalyzed by mercuric salts and is carried out in the vapor phase (44–46).

$$\underset{\displaystyle OR}{CH_2{=}\overset{}{C}H} + R'OH \xrightarrow{Hg^{2+}} \underset{\displaystyle OR'}{CH_2{=}\overset{}{C}H} + ROH \qquad (5)$$

Vinyl ethers have also been reported to be produced by reaction of ethylene with alcohols in the presence of palladium(II) chloride complexes or palladium metal (47, 48). Certain vinyl ethers have been produced by dehydrochlorination reactions both commercially and in laboratory preparations. Reaction of 2,2′-dichlorodiethyl ether with sodium hydroxide is commercially used to manufacture 2-chloroethyl vinyl ether (eq. 6) (49,50).

$$ClCH_2CH_2OCH_2CH_2Cl \xrightarrow{NaOH} CH_2{=}CHOCH_2CH_2Cl + NaCl + H_2O \qquad (6)$$

Monomer Analysis

The original Reppe method of analysis was hydrolysis of the vinyl ether to acetaldehyde followed by typical aldehyde analysis (13,51,52). This suffered from the disadvantage that aldehydes and acetals, typical impurities, act as interfering impurities and could not be determined. A major improvement in the determination of vinyl ethers was the development of the iodometric method (52). Excess iodine in the presence of methanol gives the iodoacetal with the vinyl ether and is back titrated with thiosulfate (eq. 7).

$$CH_2{=}CHOR + CH_3OH + I_2 \longrightarrow ICH_2CHOROCH_3 + HI \qquad (7)$$

In this method aldehydes and acetals are not interfering. The pH must not be allowed to go below 2 if acid impurities are present, as this leads to iodine formation from the HI. High pH must also be avoided since under basic conditions vinyl ethers and acetaldehyde impurities can give iodoform by reaction with iodine.

Another method is based on the reaction of vinyl ethers with mercuric acetate in the presence of methanol (eq. 8).

$$CH_2{=}CHOR + Hg(OAc)_2 + CH_3OH \longrightarrow AcOHgCH_2CH(OR)(OCH_3) + HOAc \qquad (8)$$

The acetic acid produced can then be titrated with standard methanolic base after the excess mercuric acetate is destroyed by reaction with sodium chloride (53) or sodium bromide (eq. 9) (54).

$$Hg(OAc)_2 + 2\,NaX \longrightarrow HgX_2 + 2\,NaOAc \qquad (9)$$

One variation on this method uses a nonaqueous hydrochloric acid titration (55) and is based on the difference between a blank and the one with double bonds. The following reactions (eqs. 10–12) explain the method.

$$Hg(OAc)_2 + 2\,HCl \longrightarrow HgCl_2 + 2\,HOAc \qquad (10)$$

$$CH_2{=}CHOR + Hg(OAc)_2 + CH_3OH \longrightarrow AcOHgCH_2CH(OR)(OCH_3) + HOAc \qquad (11)$$

$$AcOHgCH_2CH(OR)(OCH_3) + HCl \longrightarrow ClHgCH_2CH(OR)(OCH_3) + HOAc \qquad (12)$$

Since the acetoxy mercuriacetals are thermally unstable, the reaction must be kept below −10°C, and the titration should not be done above 15°C (54). One of the advantages is that the higher vinyl ethers which are very water insoluble can be more readily determined.

Gas–liquid chromatography is also an easily adaptable method for analysis of the monomers. Using a column packed with Carbowax 400 (10%) on 60/80 Chromosorb P a linear correlation of peak area with concentration was reported (56).

Bibliography

1. *Chem. Eng. News* **45** (24), 32 (1967).
2. *Chem. Eng. News* **43** (12), 42 (1965).
3. C. E. Schildknecht, O. A. Zoss, and C. McKinley, *Ind. Eng. Chem.* **39,** 180 (1947).
4. *Vinyl Monomers,* Bulletins F-41210 (1965) and F-40800A (1965), Union Carbide Corp.
5. *Alkyl Vinyl Ethers,* Bulletin 7543-055, GAF Corporation.
6. R. L. Adelman, *J. Am. Chem. Soc.* **77,** 1669 (1955).
7. W. Reppe (to I. G. Farbenindustrie A.G.), U.S. Pat. 1,959,927 (May 22, 1934); *Chem. Abstr.* **28,** 4431 (1934).
8. A. E. Favorskii and M. F. Shostakovskii, *J. Gen. Chem. (USSR)* **13,** 1 (1943).
9. C. Schildknecht, *Vinyl and Related Polymers,* John Wiley & Sons, Inc., New York, 1952, p. 597.
10. H. J. Hagemeyer, Jr., A. E. Blood, and J. D. Heller (to Eastman Kodak), U.S. Pat. 3,265,675 (Aug. 9, 1966); Fr. Pat. 1,352,522 (Feb. 14, 1964); *Chem. Abstr.* **61,** 9604 (1965).
11. Ref. 9, p. 623.
12. Ref. 9, p. 618.
13. W. Reppe, et al., *Ann. Chem.* **601,** 81 (1956).
14. M. Goodman and Y. Fan, *J. Am. Chem. Soc.* **86,** 4922 (1964).
15. Ref. 9, p. 616.
16. M. Farina, M. Peraldo, and G. Bressan, *Chem. Ind. (Milan)* **42,** 967 (1960); through *Chem. Abstr.* **55,** 11284 (1961).

17. G. Diab and M. Crauland (to Saint-Gobain), Fr. Pat. 1,090,421 (March 30, 1955); *Chem. Abstr.* **53,** 7987 (1959).
18. A. Mizote, S. Kusudo, T. Higashimura, and S. Okamura, *J. Polymer Sci.* [A-1], **5,** 1727 (1967).
19. C. M. Hill, L. Haynes, D. E. Simmons, and M. E. Hill, *J. Am. Chem. Soc.* **80,** 3624 (1958).
20. Ref. 9, p. 615.
21. *Mono and Divinyl Ether of Butanediol,* GAF Corporation Preliminary Data Sheet.
22. P. E. Aldrich (to E. I. du Pont de Nemours & Co., Inc.), U.S. Pat. 3,162,622 (Dec. 22, 1964).
23. A. G. Pittman, B. A. Ludwig, and D. L. Sharp, *J. Polymer Sci.* **6,** 1741 (1968).
24. D. D. Coffman, G. H. Kalb, and A. B. Ness, *J. Org. Chem.* **13,** 223 (1948).
25. W. Kawai and S. Tsutsumi, *Nippon Kagaku Zasshi* **80,** 88 (1959); through *Chem. Abstr.* **55,** 4466 (1961).
26. Ref. 9, p. 621.
27. N. D. Field and D. H. Lorenz, in E. C. Leonard, ed., *Vinyl and Diene Monomers,* John Wiley & Sons, Inc., New York, 1970, Chap. 7.
28. A. Ledwith and H. J. Woods, *J. Chem. Soc., Ser. B* **1966,** 753.
29. J. Wislicenus, *Ann. Chem.* **192,** 106 (1878).
30. T. L. Jacobs and S. Searles, *J. Am. Chem. Soc.* **66,** 684 (1944).
31. M. F. Shostakovskii, *J. Gen. Chem. (USSR)* **16,** 1143 (1946); *Chem. Abstr.* **41,** 2691 (1947).
32. J. W. Copenhaver (to GAF Corp.), U.S. Pat. 2,550,637 (April 24, 1951); *Chem. Abstr.* **45,** 8029 (1951).
33. C. J. Schmidle and R. C. Mansfield (to Rohm and Haas Co.), U.S. Pat. 2,736,743 (Feb. 28, 1956); *Chem. Abstr.* **50,** 10761 (1956).
34. H. Krzikalla (to Badische Anilin- und Soda-Fabrik A.G.), Ger. Pat. 896,648 (Nov. 12, 1953); *Chem. Abstr.* **52,** 9221 (1958).
35. H. Krzikalla and E. Woldar (to Badische Anilin- und Soda-Fabrik A.G.), Ger. Pat. 890,957 (Sept. 24, 1953); *Chem. Abstr.* **52,** 14707 (1958).
36. J. E. Mahan (to Phillips Petroleum Co.), U.S. Pat. 2,706,730 (April 19, 1955); *Chem. Abstr.* **50,** 2683 (1956).
37. E. Kobayashi (to Mitsubishi Chemical Industries Co.), Jap. Pat. 1,134 (Feb. 17, 1956); through *Chem. Abstr.* **51,** 5845 (1957).
38. K. Baur (to I. G. Farbenindustrie A.G.), U.S. Pat. 1,931,858 (Oct. 24, 1934); *Chem. Abstr.* **28,** 485 (1934).
39. M. Calanac, *Compt. Rend.* **190,** 881 (1930).
40. M. Kitabatake, S. Owari, and K. Kuno, *Soc. Synth. Org. Chem. Japan* **25,** 70 (1967).
41. J. A. Neuwland, R. R. Vogt, and W. L. Foohey, *J. Am. Chem. Soc.* **52,** 1018, 2892 (1930).
42. W. J. Croxall, F. J. Glavis, and H. T. Neher, *J. Am. Chem. Soc.* **70,** 2805 (1948).
43. W. D. Schaeffer (to Union Oil), U.S. Pat. 3,285,970 (Nov. 15, 1966); *Chem. Abstr.* **66,** 37437 (1967).
44. R. I. Hoaglin and D. H. Hersch (to Union Carbide Corp.), U.S. Pat. 2,566,415 (Sept. 15, 1951); *Chem. Abstr.* **46,** 2562 (1952).
45. W. H. Watanabe and L. E. Conlon, *J. Am. Chem. Soc.* **79,** 2828 (1957).
46. H. Yuki, K. Hatada, K. Nagata, and K. Kaziyama, *Bull. Chem. Soc., Japan* **42,** 3546 (1969).
47. E. W. Stern and M. L. Spector, *Proc. Chem. Soc.* **1961,** 370.
48. Imperial Chemical Industries, Neth. Pat. 6,411,879 (April 15, 1965); through *Chem. Abstr.* **63,** 9813 (1965).
49. L. H. Cretcher, J. A. Koch, and W. H. Pittenger, *J. Am. Chem. Soc.* **47,** 1175 (1925).
50. C. E. Rehberg, *J. Am. Chem. Soc.* **71,** 3247 (1949).
51. S. Siggia, *Anal. Chem.* **19,** 1025 (1947).
52. S. Siggia and R. L. Edsberg, *Anal. Chem.* **20,** 762 (1948).
53. R. W. Morten, *Anal. Chem.* **21,** 921 (1949).
54. J. B. Johnson and J. P. Fletcher, *Anal. Chem.* **31,** 1563 (1959).
55. M. N. Das, *Anal. Chem.* **26,** 1086 (1954).
56. A. Ledwith and H. J. Woods, *J. Chem. Soc. Ser. B* **1966,** 753.

Donald H. Lorenz
GAF Corporation

POLYMERS

Preparation

The first polymerization of a vinyl ether was reported in 1878 (1); the addition of iodine to ethyl vinyl ether produced a violent reaction resulting in the formation of a resinous material. However, the major impetus for the study of vinyl ethers began with the discovery by Reppe and co-workers at I.G. Farbenindustrie of a simple method of producing the monomers (see the section on Monomers, p. 504, as well as ACETYLENE AND ACETYLENIC POLYMERS). Lewis acids, such as boron trifluoride, were found to be excellent catalysts for the polymerization and a rubbery poly(isobutyl vinyl ether) produced in this way was sold in Germany under the trademark Oppanol C even before World War II. Shostakovskii and co-workers, in the Soviet Union, also studied cationic vinyl ether polymerization at about this time (for reviews, see Refs. 2 and 3).

It is now known that vinyl ethers produce only low-mw homopolymers (viscous liquids) when initiation is carried out with free-radical catalysts, ultraviolet radiation, or heat alone. Anionic initiation apparently fails completely, as would be expected from the electron-donating nature of the ether substituent (see also ANIONIC POLYMERIZATION). By far, the preferred catalysts are those known to produce cationic polymerization (qv). Proper choice of reaction conditions even permits the preparation of stereoregular poly(vinyl ethers). High-energy radiation (β and γ rays) also causes polymerization and gives at room temperature essentially quantitative yields of high-mw polymers (4,5).

Formation of Atactic Homopolymers. Atactic homopolymers are best made by the use of acidic catalysts, especially of the Friedel-Crafts type (eg, boron trifluoride, aluminum trichloride, stannic chloride, etc, and their complexes). Polymerization is effected in bulk or in an inert, dry solvent. The reaction may be excessively vigorous, so it is important to moderate it for safety considerations and in order to control the molecular weight. The extraordinarily rapid polymerization brought about, eg, by boron trifluoride on the lower alkyl vinyl ethers has been termed "flash polymerization" (6). Hydrocarbons, such as propane, are frequently employed as solvents but some chlorinated solvents, which have also been used, are reported to be chain-transfer agents (7). Impurities in the monomer or the solvent (eg, alcohols, aldehydes, water) tend to reduce the rate of reaction and the molecular weight. Much greater control of the reaction can be obtained if polymerization is carried out at -40 to -70°C and if the Lewis acid is complexed (eg, boron trifluoride etherate). The propagation step probably proceeds as shown in eq. 1, thereby producing a head-to-tail

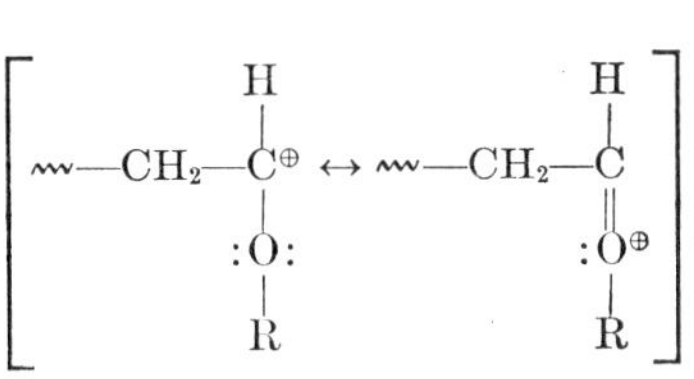

$$\left[\begin{array}{c} \text{H} \\ | \\ \sim\!\!\sim\text{—CH}_2\text{—C}^{\oplus} \\ | \\ :\ddot{\text{O}}: \\ | \\ \text{R} \end{array} \leftrightarrow \begin{array}{c} \text{H} \\ | \\ \sim\!\!\sim\text{—CH}_2\text{—C} \\ \| \\ :\text{O}^{\oplus} \\ | \\ \text{R} \end{array}\right] \qquad \text{(1a)}$$

(1)

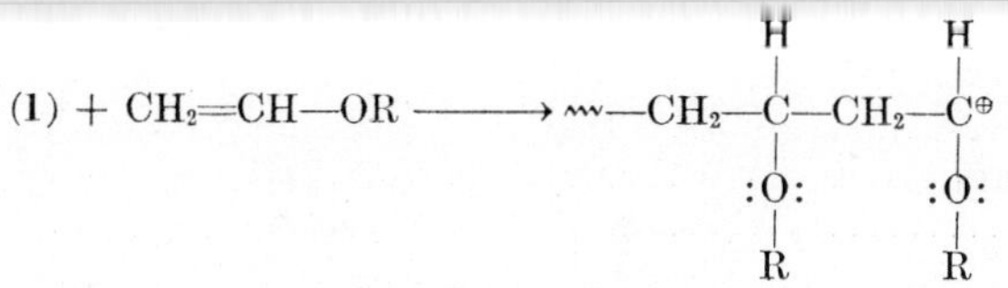

$$(1) + CH_2{=}CH{-}OR \longrightarrow \sim\!\!\sim CH_2{-}\underset{\underset{R}{|}}{\underset{:\ddot{O}:}{\underset{|}{\overset{\overset{H}{|}}{C}}}}{-}CH_2{-}\underset{\underset{R}{|}}{\underset{:\ddot{O}:}{\underset{|}{\overset{\overset{H}{|}}{C^{\oplus}}}}} \qquad (1b)$$

structure. The more highly branched the alkyl group R, the greater the reactivity of the monomer (8). Long-chain alkyl ethers are generally less reactive than the short-chain homologs. Aromatic vinyl ethers do not polymerize readily and are prone to side reactions, such as rearrangements and condensations (2). If the boron trifluoride is complexed, the nature of the etherate affects the rate of polymerization as follows (3,25): anisole > isopropyl ether > ethyl ether > *n*-butyl ether > tetrahydrofuran.

Table 1. Typical Conditions for the Homopolymerization of Vinyl Ethers (12)

Catalyst	Solvent	Reaction temp, °C	Remarks
SO_2	bulk	−10	
$SnCl_2$	bulk	12	
RMgBr	bulk	20–30	crystalline
$BF_3.2H_2O$	bulk	3–5	
$Fe_2(SO_4)_3$	bulk	25–30	isotactic
$BF_3.Et_2O$	bulk	−78	isotactic
I_2	ethyl ether	25	
CrO_2Cl_2	hexane	0–8	
$BF_3.Et_2O$	propane	−70	
Et_2AlCl	toluene	−78	isotactic
$EtAlCl_2$	toluene	−80	crystalline
TiF_3	hexane–methylene chloride	60	crystalline
$Al(OR)_3 + HF$	heptane	0–25	crystalline
$Al_2(SO_4)_3.3H_2SO_4$	heptane	5–100	crystalline
$Cr_2(SO_4)_3.H_2SO_4$	hexane	40	crystalline
$VOSO_4$	bulk	25–30	crystalline
AlOCl, AlOBr, AlOEt	toluene		crystalline
$AlR_3 + VCl_4$	heptane	30	crystalline
$AlR_3 + TiCl_4$	pentane		crystalline

Various chain-transfer or termination mechanisms have been proposed (3), resulting in the formation of end groups of the types indicated in structures (**2**), (**3**), and (**4**).

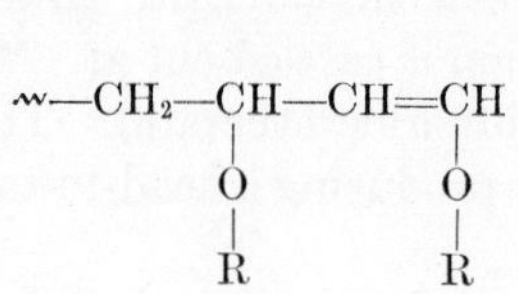

$$\sim\!\!\sim CH_2{-}\underset{\underset{R}{|}}{\underset{O}{\underset{|}{CH}}}{-}CH{=}\underset{\underset{R}{|}}{\underset{O}{\underset{|}{CH}}} \quad (\mathbf{2})$$

$$\sim\!\!\sim CH_2{-}\underset{\underset{R}{|}}{\underset{O}{\underset{|}{CH}}}{-}O{-}CH{=}CH_2 \quad (\mathbf{3})$$

$$\sim\!\!\sim CH_2CH \text{ (ring: } CH_2CH{-}CH_2{-}CHOR{-}CH_2{-}CH(OR){-}O{-}\text{)} \quad (\mathbf{4})$$

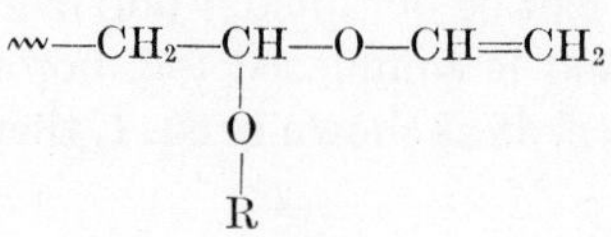

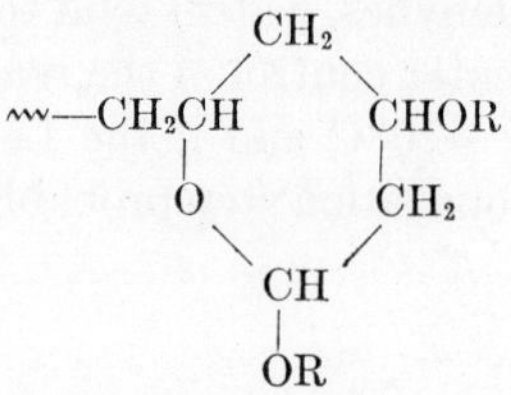

Formation of Stereoregular Homopolymers. The formation of stereoregular vinyl ether homopolymers produced by cationic catalysis has been the subject of numerous investigations. Historically, the poly(vinyl ethers) occupy an important

Table 2. Some Catalysts for the Stereospecific Polymerization of Alkyl Vinyl Ethers at 0°C or Above (11)

Catalyst	Alkyl group	Crystalline polymer, % insolubles
Metal Oxide Type		
CrO_3	isobutyl	22
CrO_2Cl_2	isobutyl	30
Metal Halide Type		
TiF_3	isobutyl	42
$TiF_4 + Ti(O\text{-}i\text{-}C_3H_7)_4$	methyl	66
$TiF_4 + Al(O\text{-}i\text{-}C_3H_7)_3$	methyl	70
Metal Sulfate Type		
$Al_2(SO_4)_3.H_2SO_4.7H_2O$	ethyl	some
	isopropyl	54
	n-butyl	some
	isobutyl	25
$Al(i\text{-}C_4H_9)_3 + H_2SO_4$	methyl	some
$Al(O\text{-}i\text{-}C_3H_7)_3 + H_2SO_4 + Al(i\text{-}C_4H_9)_3$	methyl	40
$Al_2(SO_4)_3 + Al(O\text{-}i\text{-}C_3H_7)_3$	methyl	some
	isopropyl	some
	butyl	some
$Al_2(SO_4)_3 + Al(i\text{-}C_4H_9)_3.THF$	*t*-butyl	some
$Al_2(SeO_4)_3 + Al(O\text{-}i\text{-}C_3H_7)_3$	isobutyl	50
$Ti(O\text{-}i\text{-}C_3H_7)_4 + H_2SO_4$	methyl	68
$MgSO_4 + H_2SO_4$	isobutyl	20
$Cr_2(SO_4)_3 + H_2SO_4$	isobutyl	20
$Fe_2(SO_4)_3.6H_2O$	ethyl	0
$Fe_2(SO_4)_3.H_2SO_4.3\text{–}4H_2O$	ethyl	some
	isopropyl	some
$Fe(OC_2H_5)_3 + H_2SO_4 + Al(i\text{-}C_4H_9)_3.THF$	isopropyl	some
$Fe_2(SO_4)_3 + Al(O\text{-}i\text{-}C_3H_7)_3$	methyl	some
$Fe_2(SO_4)_3 + Al(i\text{-}C_3H_7)_3$	methyl	some
Modified Ziegler-Natta Catalysts		
PSV[a] + $Al(i\text{-}C_4H_9)_3.THF$	methyl	40
	neopentyl	1
PSV[a] + $Al(i\text{-}C_4H_9)_3$	ethyl	1
	n-butyl	trace
	isobutyl	4
	t-butyl	1[b]
$VCl_2.VCl_2.AlCl_3 + Al(i\text{-}C_4H_9)_3.THF$	methyl	3
ground $VCl_2 + Al(i\text{-}C_4H_9)_3$	methyl	3
ground $VCl_3 + Al(i\text{-}C_4H_9)_3$	methyl	3
dibenzene Cr + $TiCl_4$	isobutyl	some
$TiCl_3 + Al(i\text{-}C_4H_9)_3$	methyl	trace
$TiCl_4 + Al(i\text{-}C_4H_9)_3$	ethyl	0
	isobutyl	0[b]
	allyl	0[b,c]

[a] See eq. 2 for description.
[b] Some crystalline polymer at −78°C.
[c] Polymerized the vinyl double bond at −78°C.

place in the development of polymer science, since crystalline polymers, later recognized to be isotactic, were already prepared in 1948. These investigations have been reviewed in Refs. 2, 3, 9–11. Typical conditions for the homopolymerizations of vinyl ethers are shown in Table 1 (12). The table indicates the broad range of polymerization systems that may be employed, although not all of these produce stereoregular polymers. Some catalysts that are clearly stereospecific are shown in Table 2 (11). The preparation of the so-called PSV (pretreated stoichiometric vanadium) catalyst in Table 2 is shown in eq. 2 (13). Its analysis is given in Table 3. This modified Ziegler-

$$\tfrac{1}{3}\,(C_2H_5)_3Al + VCl_4 \xrightarrow[\text{heptane, 90°C}]{\text{16 hr}} \underset{(5)}{VCl_3.\tfrac{1}{3}AlCl_3} \qquad (2a)$$

$$(5) + 2\,(i\text{-}C_4H_9)_3Al.THF \xrightarrow[\text{heptane}]{\text{20 hr}} \text{PSV catalyst} \qquad (2b)$$

Natta catalyst has the advantage of being effective at room temperature and of yielding highly crystalline polymer. The stereospecific propagation step has been represented as shown in eq. 3 (13), where Cl′ is the chloride counterion, Cl is a chlorine

$$\begin{array}{c} M' \leftarrow Cl'^{\delta-} \\ \uparrow \\ OR \\ | \\ X\;\; CH{=}CH_2\;\; \overset{\delta+}{CH}{-}CH_2 \\ Cl \qquad | \qquad | \\ | \qquad OR \quad \wr \\ M'' \overset{A}{\leftarrow} \end{array} \longrightarrow \begin{array}{c} M' - Cl' \\ \uparrow \\ OR \\ | \qquad\quad B \\ X\;\; \delta^{+}CH{-}CH_2 \quad CH{-}CH_2 \\ \delta^{-}\,Cl \qquad | \qquad | \\ \downarrow \qquad OR \quad \wr \\ M'' \end{array} \qquad (3)$$

atom, M′ is one type of metal ion (presumably vanadium), M″ is another metal center, X is a bridging group (Cl or OR), *A* is a coordinate bond that is broken and replaced by bond *B*. M″ is thereby freed to coordinate with another monomer molecule. Both metal centers are located at the surface of the insoluble component of the catalyst.

Table 3. Composition of the Pretreated Stoichiometric Vanadium (PSV) Catalyst (13)

Heptane-insoluble part	Heptane-soluble part
highly crystalline	
(not VCl_2, VCl_3, or $AlCl_3$)	
99% vanadium	1% vanadium
88% V(II)	
12% V(III)	
0.2 Al per vanadium	$(i\text{-}Bu)_3Al$
$(i\text{-}Bu)_2AlCl$	$(i\text{-}Bu)_2AlCl$
$(i\text{-}Bu)AlCl_2$	$(i\text{-}Bu)AlCl_2$?
$AlCl_3$?	
probably no V—R bonds	
(R = Et or *i*-Bu)	

Another catalyst system that yields stereoregular poly(vinyl ethers) at room temperature consists of an insoluble metal sulfate–sulfuric acid complex. A proposed mechanism for coordinate cationic polymerization is shown in Figure 1. The boron trifluoride etherate catalyst is stereospecific at low temperatures, provided the rate of polymerization is slow (10); but it is less stereospecific than some other catalysts (see Table 4).

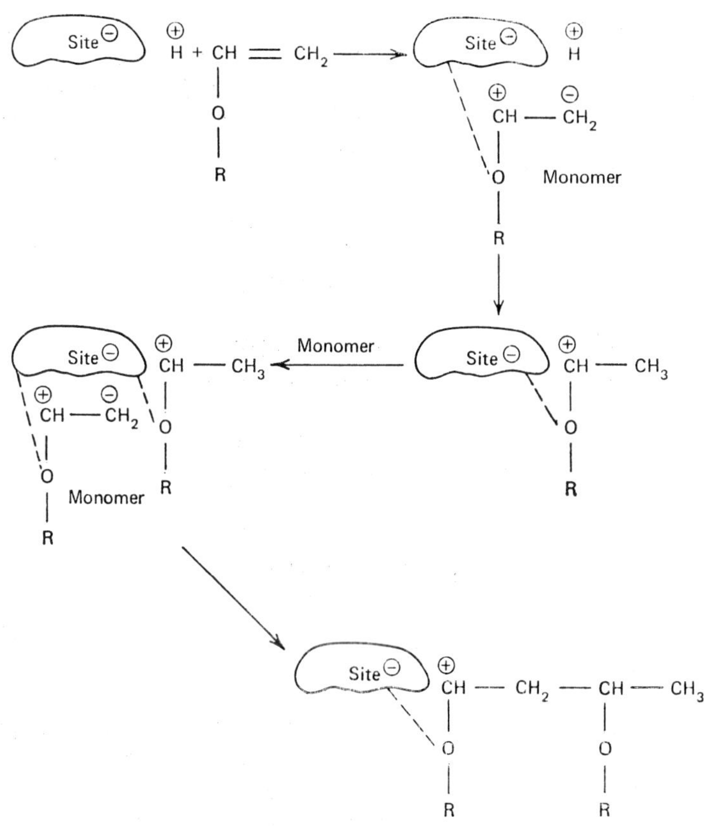

Fig. 1. Proposed mechanism for the coordinate cationic polymerization of alkyl vinyl ethers with a metal sulfate–sulfuric acid complex (3).

Homogenous catalysts have also been proposed (11,15). They are effective because there are two centers of coordination in the monomer: the double bond and the oxygen atom. A suitable catalyst is $(C_5H_5)_2TiCl_2AlX_2$ (see COORDINATE POLYMERIZATION). Homogenous catalysts give highest stereoregularity, if the polymerizations

Table 4. Comparison of Various Catalyst Systems for the Polymerization of Isobutyl Vinyl Ether (9)

Catalyst	BF_3.etherate	Et_2AlCl	$Al_2(SO_4)_3$-H_2SO_4	$(i\text{-}Bu)_3Al$-VCl_3-$\frac{1}{3}AlCl_3$
polymerization				
solvent	propane	toluene	hexane	*n*-heptane
temperature, °C	−80 to −60	−78	30	30
time, hr	0.5–2.0	3–6	1–2	18–24
conversion, %	high	85	60–80	95
product, % insolubles				
boiling acetone		95	15–25	
cold benzene	0	0		5
properties, insolubles				
inherent viscosity, dl/g	0.4–10	2.1	0.5–1.5	2.8
melting point, °C	95	110		170

proceed at low temperatures. The counterion also plays an important part (11,15). Some systems appear to be homogeneous at first, but then proceed in a gel phase as the conversion increases (14). Proliferous polymerization (qv) also seems to enhance stereoregularity of the product (10). Various polymerization systems are compared in Table 4 for the case of isobutyl vinyl ether, a monomer that has been widely studied because of its ready conversion to an isotactic polymer (16).

Grignard compounds are another type of catalyst that polymerizes vinyl ethers; eg, isobutyl vinyl ether can be polymerized to a crystalline product, mp 107°C, provided a trace of oxygen or carbon dioxide is present (9). It is of interest to note that many of the stereospecific catalysts (eg, the Ziegler-Natta types) do not polymerize the allyl group in allyl vinyl ether, thus providing support for the concept that a cationic coordinate mechanism is involved (the allyl group is not subject to cationic polymerization) (26).

Copolymerization. Vinyl ethers copolymerize readily with each other. As in the case of homopolymerization, cationic catalysis is preferred over free-radical initiation.

Of greater practical interest is the copolymerization of vinyl ethers with other types of monomers. In this case, cationic polymerization is less desirable because of the tendency of the vinyl ethers to homopolymerize. Vinyl ethers with branched alkyl groups seem to be most susceptible to cationic copolymerization. The use of Ziegler-Natta catalysts has been claimed to permit the preparation of copolymers with olefins (17,18).

Free-radical polymerization (qv) is by far the preferred method of preparing the industrially important copolymers. It is especially suitable for copolymerization with monomers having electron-withdrawing groups, such as the acrylates and methacrylates, maleic anhydride, vinyl chloride, vinyl acetate, etc. Solution, emulsion, suspension, or bulk polymerization techniques can be used, depending on the nature of the reactants. If aqueous systems are employed, it is desirable to maintain a pH of about 8 or above to prevent hydrolysis of the vinyl ether. Reactivity ratios are known for some comonomer systems; those of commercial interest are summarized in Table 5 (12). The r_2 values for the alkyl vinyl ethers are very low and approach zero. There is considerable evidence that charge-transfer complexes form between vinyl ethers and maleic anhydride; these complexes participate in the copolymerization (27).

Table 5. Reactivity Ratios[a] for Free-Radical Copolymerization of Alkyl Vinyl Ethers (M_2) with Other Vinyl Monomers (12)

Comonomer (M_1)	r_1	r_2
acrylonitrile	0.8–1	0
butyl maleate	0–0.1	0
maleic anhydride	0.0	0
methyl acrylate	2.7–3	0
methyl methacrylate	10	0
styrene	greater than 50	0
vinyl acetate	3.4–3.7	0
vinyl chloride	1.7–2.2	0
vinylidene chloride	1.3–1.5	0

[a] Bulk copolymerization with methyl, octyl, dodecyl, and octadecyl vinyl ethers using benzoyl peroxide as initiator at 40–100°C.

Terpolymers and quaterpolymers are also important industrially; for example, BASF markets a series of latexes trademarked Acronal that contain, in addition to isobutyl vinyl ether, up to three comonomers such as methyl acrylate, ethyl acrylate, butyl acrylate, acrylonitrile, styrene, or acrylic acid (2). See also COPOLYMERIZATION.

Properties and Uses

The physical properties of poly(alkyl vinyl ethers) depend on the molecular weight, the nature of the alkyl group, and the stereoregularity. Traditionally, the molecular weight of the polymers has been estimated by the so-called K value, a viscosity measurement originated by Fikentscher (19). It is obtained by making use of the relationship shown in eq. 4, where η_{rel} is the ratio of the viscosity of the solution to that of the solvent and c is the concentration of polymer, in g/dl.

$$\log_{10} \eta_{rel}/c = \frac{75k^2}{1 + 1.5kc} + k \tag{4}$$

The Fikentscher K value equals 10^3k. It has been found that the intrinsic viscosity in 2-butanone at 30°C of methyl and ethyl vinyl ether polymers can be related to molecular weight by eq. 5 (20), where $[\eta]$ is in dl/g.

$$[\eta] = 1.37 \times 10^{-3} M^{0.54} \tag{5}$$

The appearance of the homopolymers can range from viscous liquids to elastomeric solids. The long-chain polymers are waxy in nature. Table 6 gives the properties of some commercial amorphous homopolymers (10). Solubility characteristics of atactic homopolymers are given in Table 7. For comparison, the solubility properties of stereoregular polymers are shown in Table 8, together with melting points and glass-transition temperatures. It can be seen that, whereas the atactic poly(methyl vinyl

Table 6. Some Commercial Vinyl Ether Homopolymers (10)

Vinyl ether polymer[a]	Polymer viscosity[b]	Trademark	Fields of use
methyl			
viscous liquid (balsam-like)	$K = 50$	Lutonal M (BASF)	plasticizer for coatings; aqueous adhesive tackifier
viscous liquid (balsam-like)	$\eta_{inh} = 0.3–0.5$	Gantrez M (GAF)	
ethyl			
viscous liquid	$K = 60$ and lower grades	Lutonal A (BASF)	plasticizer for cellulose nitrate and natural resin lacquers
elastomeric solid high polymer	(solid and solutions supplied)	PVEE (Union Carbide)	pressure-sensitive adhesive base
isobutyl			
viscous liquid	$K = 60$	Lutonal I (BASF)	tackifier for adhesives
viscous liquid	$\eta_{sp} = 0.1–0.5$ $K = 25–50$	Gantrez B (GAF)	tackifier for adhesives
elastomeric solid	$K = 70–130$ $\eta_{sp} = 2–6$	Oppanol C (BASF)	pressure-sensitive adhesive base
octadecyl			
waxy solid	low DP, mp = 50°C	V-Wax (BASF)	polishing and waxing agents

[a] K values are those of Fikentscher (eq. 4); the η_{inh} values are for 1% solutions.

[b] Some viscous liquid polymers are also supplied as high-solids solutions, eg, 70% in toluene.

Table 7. Solubility Characteristics of Some Atactic Poly(Alkyl Vinyl Ethers)[a] (2)

	Nature of alkyl group		
Solvent	Methyl	Ethyl	Isobutyl
water	sol	insol	insol
methanol; ethanol	sol	sol	insol
2-propanol; *n*-butanol	sol	sol	sol
benzene; toluene	sol	sol	sol
petroleum ether	insol	sol	sol
methylene chloride; chloroform	sol	sol	sol
ethyl ether	insol	insol	sol
acetone; cyclohexanone	sol	sol	sol
ethyl acetate; butyl acetate	sol	sol	sol

[a] At room temperature for polymers with Fikentscher K values of 50–60.

Table 8. Properties of Some Crystalline Poly(Alkyl Vinyl Ethers) (2,21)

Alkyl group	Mp, °C	T_g, °C[a]	Insoluble in
methyl	144–150	−31	water, methanol, heptane
ethyl	86	−42	methanol, heptane
n-propyl	76	−49	heptane, acetone
isopropyl	190–191	−3	heptane, acetone, methanol
n-butyl	64	−55	heptane
isobutyl	165–170	−19	heptane, benzene
tert-butyl	238–260		heptane, benzene

[a] Of predominantly amorphous material.

Table 9. Some Commercial Vinyl Ether Copolymers

poly(methyl vinyl ether-*co*-maleic anhydride)	
Gantrez AN	GAF Corporation
Viscofras	ICI Ltd.
half-amide of poly(methyl vinyl ether-*co*-maleic anhydride)	
Gantrez AN-4651	GAF Corporation
ethyl half-ester of poly(methyl vinyl ether-*co*-maleic anhydride)	
Gantrez ES-225	GAF Corporation
poly(octadecyl vinyl ether-*co*-maleic anhydride)	
Gantrez AN-8194	GAF Corporation
poly(isobutyl vinyl ether-*co*-vinyl chloride)	
Gantrez VC	GAF Corporation
Vinoflex MP 400	BASF A.G.
poly(isooctyl vinyl ether-*co*-vinyl acetate)	
Gantrez AC-1810	GAF Corporation
poly[isobutyl vinyl ether(50%)-*co*-methyl acrylate (30%)-*co*-acrylonitrile (20%)]	
Acronal 430D	BASF A.G.

ether), for example, is soluble in water or methanol, the corresponding crystalline polymer is insoluble in these media.

Poly(vinyl ethers) must be stabilized by the addition of antioxidants (qv) or ultraviolet-radiation absorbers (qv). They exhibit outstanding tack characteristics and are, therefore, widely used in pressure-sensitive adhesives. They are often modified by the addition of hydrocarbon resins (qv), rosin esters (see ROSIN), phenolic resins (qv), or acrylic polymers (see ACRYLIC ESTER POLYMERS).

The homopolymers can be cured by radiation or peroxides to produce interesting vulcanizates. Copolymers with small amounts of vulcanizable groups (such as dienes, allyl vinyl ethers, 2-chloroethyl vinyl ether, etc) have also been studied (3). The vinyl ether can also be the minor component of a copolymeric elastomer (see ACRYLIC ELASTOMERS).

Some of the commercially important copolymers of vinyl ethers are listed in Table 9. These materials as well as the major homopolymers will be discussed in some detail.

Homopolymers. Among the commercially important homopolymers are those with methyl, ethyl, or isobutyl groups.

Poly(methyl Vinyl Ether) (22). The homopolymer is soluble in water at room temperature but solutions become gradually hazy as the temperature is increased. Eventually the polymer precipitates; the temperature at which this takes place is known as the "cloud point." The cloud point can be raised (up to 100°C) by the addition of a water-miscible solvent or a surfactant such as the sodium salt of the sulfated adduct of ethylene oxide to nonylphenol.

Poly(methyl vinyl ether) has extensive uses in adhesives and coatings, as a non-migrating plasticizer, and as tackifier. It improves adhesion to metals, glass, plastics, and other surfaces within a broad range of free energies (see ADHESION AND BONDING). Because of its hydrophilic character, monomolecular water layers on substrates do not interfere with its adhesion characteristics. Poly(methyl vinyl ether) is a modifier for acrylic pressure-sensitive adhesives for use in tapes, labels, and decals. It can be used as a hot-melt adhesive, as a pigment-wetting agent, and as a plasticizer for printing inks. The polymer also has uses as a textile-sizing agent and as a stabilizer in emulsion polymerization.

Poly(ethyl Vinyl Ether). The homopolymer of ethyl vinyl ether is available in a variety of grades, some of which are shown in Table 10 (10). It is useful as a base or modifier for pressure-sensitive adhesives, calking compounds, and laminating resins as well as in adhesives for paper, cloth, vinyls, metal, and wood (22).

Poly(isobutyl Vinyl Ether) (22). This polymer is used either in the dry state or as a solution in an organic solvent. It has excellent adhesion to plastics, metals, and coated surfaces and finds applications as an adhesion promoter and plasticizer in pressure-sensitive tapes and labels, as well as in various adhesive compositions and surface coatings. The polymer is also useful as a viscosity-index improver for lubricants and as a tackifier for elastomers.

Copolymers. The most important copolymers are those listed in Table 9.

Poly(methyl Vinyl Ether-co-Maleic Anhydride) (22). Vinyl ethers copolymerize readily with maleic anhydride to form alternating copolymers (see ACIDS, MALEIC AND FUMARIC). Several grades of the methyl vinyl ether copolymer are available ranging in specific viscosity (1 g/dl at 25°C in ethyl methyl ketone) from 0.1 to 3.5. The

Table 10. Properties of Some Commercial Poly(ethyl Vinyl Ethers)[a] (10)

Polymer type	Low viscosity		High viscosity		Extra-high viscosity	
grade designation	EHBC	EDBC	EHBM	EDBM	EHBN	EDBN
solids in hexane, %	80	98	28	98	25	98
approximate reduced viscosity, 20°C	0.3	0.3	4.0	4.0	5.0	5.0
specific gravity, n_{20}^{20}	0.908	0.973	0.747	0.968	0.725	0.968

[a] PVEE polymers, Union Carbide Corp.

polymer is soluble in water with hydrolysis of the anhydride groups and many of its applications involve its aqueous solutions, often mixed with other water-soluble polymers. It is used as a thickening and suspending agent (eg, for pharmaceutical preparations), protective colloid, sizing agent for textiles, flocculant in ore beneficiation, dispersing agent for pigments, beater additive in papermaking to improve sizing, etc. Like the vinyl ether polymers already discussed, it is a useful component of adhesives; it has Food and Drug Administration approval for inclusion in food-packaging adhesives. Numerous derivatives may be prepared from the copolymer, including various salts and half-esters. The ethyl, isopropyl, and butyl half-esters are useful in cosmetic applications, particularly as film formers in hair sprays. One of the most interesting derivatives is the *ammonium salt of the half-amide*, which is formed when anhydrous ammonia is bubbled into a slurry of the copolymer in benzene (eq. 6).

$$\sim\!\!\sim CH_2-\underset{H_3CO}{CH}-\underset{CO\diagdown O \diagup}{CH}-\underset{CO}{CH}-\sim\!\!\sim \xrightarrow{NH_3} \sim\!\!\sim CH_2-\underset{H_3CO}{CH}-\underset{\underset{O^{\ominus}\,NH_4^{\oplus}}{CO}}{CH}-\underset{CONH_2}{CH}-\sim\!\!\sim \quad (6)$$

(6)

Polymer (**6**), a polyelectrolyte (qv), is soluble in water and polar organic solvents. Its main use is as a thickener for latexes, cosmetics, adhesives, blasting compositions, etc. It is also useful as an emulsion stabilizer and in various photoreproduction applications (22).

Poly(octadecyl Vinyl Ether-co-Maleic Anhydride). This copolymer also has an alternating structure. It forms concentrated solutions in toluene from which waxy films may be deposited which serve as antiblocking agents and release coatings. The dry polymer is compatible with paraffins and polyethylene (10,22).

Copolymers with Vinyl Chloride. Copolymerization of vinyl ethers (eg, 20–30% isobutyl vinyl ether) with vinyl chloride results in products that are particularly useful as coatings. The copolymers are readily soluble in various organic solvents, have good adhesion and flexibility properties, adequate flame retardancy, and good resistance to alkali. Marine paints and other corrosion-resistant finishes are a major application (10,12,22).

Copolymerization of vinyl chloride with minor amounts of long-chain (C_{12}–C_{16}) vinyl ethers results in poly(vinyl chloride)-type polymers with improved processing characteristics (23). The polymers are said to be "internally plasticized" (see also VINYL CHLORIDE POLYMERS).

Copolymers with Vinyl Acetate. The copolymer of isooctyl vinyl ether with vinyl acetate is available as a viscous emulsion from which clear, heat-sealable films can be obtained. Typical uses are as a pigment binder for paper coatings, an adhesive, a release-coating modifier, and an adhesion promoter (22). The isobutyl vinyl ether–vinyl acetate copolymer has textile applications (12). (See also VINYL ESTER POLYMERS, Vol. 15.)

Miscellaneous Copolymers. There are numerous other interesting copolymers, including those with olefins and fluorinated olefins (2). The copolymer of methyl vinyl ether with ethylene can be produced at elevated temperatures and pressures in the presence of free-radical initiators (12). Fluorinated vinyl ethers have also been studied; for example, the copolymer of tetrafluoroethylene with trifluoromethyl trifluorovinyl ether has shown promise as a heat-resistant elastomer (24).

Bibliography

1. J. Wislicenus, *Ann. Chem.* **192,** 106 (1878).
2. W. Kern and V. Jaacks in E. Müller, ed., *Makromolekulare Stoffe,* Part I (Vol. 14 of *Houben-Weyl Methoden der Organischen Chemie*), Georg Thieme Verlag, Stuttgart, 1961, pp. 921–972.
   A very detailed review with emphasis on preparation and properties.
3. J. Lal in J. P. Kennedy and E. Törnqvist, eds., *Polymer Chemistry of Synthetic Elastomers,* Part I, Interscience Publishers, a division of John Wiley & Sons, Inc., New York, 1968, pp. 331–376.
   An excellent review of polymerization mechanisms and kinetics and uses of the polymers as elastomers.
4. J. G. Fee, W. S. Port, and L. P. Witnauer, *J. Polymer Sci.* **33,** 95 (1958).
5. S. H. Pinner and R. Worrall, *J. Appl. Polymer Sci.* **2,** 122 (1959).
6. M. Otto, H. Güterbock, and A. Hellemans (to Jasco), U.S. Pat. 2,311,567 (1943); *U.S. Dept. Commerce OTS Rept. PB 67694* and *FIAT Rept. 944* (post-World War II Allied investigative reports of German developments).
7. E. J. Duffek, *J. Am. Oil Chemists Soc.* **37,** 37 (1960).
8. C. E. Schildknecht, A. O. Zoss, and F. Grosser, *Ind. Eng. Chem.* **41,** 2894 (1949).
9. L. Reich and A. Schindler, *Polymerization by Organometallic Compounds,* Interscience Publishers, a division of John Wiley & Sons, Inc., New York, 1966, pp. 672–679.
10. C. E. Schildknecht in A. Standen, ed., *Kirk-Othmer Encyclopedia of Chemical Technology,* Vol. 21, 2nd ed., Interscience Publishers, a division of John Wiley & Sons, Inc., New York, 1970, pp. 412–426.
11. R. W. Lenz, *Organic Chemistry of Synthetic High Polymers,* Interscience Publishers, a division of John Wiley & Sons, Inc., New York, 1967, pp. 523–527; 640–644.
12. N. D. Field and D. H. Lorenz in E. C. Leonard, ed., *Vinyl and Diene Monomers,* Part I, Wiley-Interscience, a division of John Wiley & Sons, Inc., New York, 1970, pp. 365–411.
13. E. J. Vandenberg, *J. Polymer Sci.* [C] *No. 1,* 207 (1963).
14. G. J. Blake and A. M. Carlson, *J. Polymer Sci.* [A-1] **4,** 1813 (1966).
15. T. Higashimura, T. Watanabe, K. Suzuoki, and S. Okamura, *J. Polymer Sci.* [C], *No. 4,* 361 (1963).
16. J. R. Elliott, ed., *Macromolecular Syntheses,* Vol. 2, John Wiley & Sons, Inc., New York, 1966, pp. 27–32.
17. E. W. Gluesenkamp (to Monsanto), U.S. Pat. 3,026,290 (1962).
18. Toyo Rayon Co., Brit. Pat. 1,063,040 (1967); Fr. Pat. 1,401,605 (1965); through *Chem. Abstr.* **64,** 19828 (1966).
19. H. Fikentscher, *Cellulosechemie* **13,** 60 (1932).
20. J. A. Manson and G. J. Arquette, *Makromol. Chem.* **37,** 187 (1960).
21. J. Lal and G. S. Trick, *J. Polymer Sci.* [A] **2,** 4559 (1964).
22. Based on information supplied by F. B. Lane, GAF Corp., New York.
23. Y. Hoshi and M. Onozuku (to Kureha Kasei), U.S. Pat. 3,168,594 (1965).
24. A. L. Barney, *Am. Chem. Soc. Polymer Preprints* **10,** 1483 (1969).
25. D. D. Eley and A. Seabrooke, *J. Chem. Soc.* **1964,** 2226.
26. J. Lal, *J. Polymer Sci.* **31,** 179 (1958).
27. S. Iwatsuki and Y. Yamashita, *J. Polymer Sci.* [A-1] **5,** 1753 (1967).

Norbert M. Bikales
Consultant

VINYL ETHYL ETHER POLYMERS. See VINYL ETHER POLYMERS

VINYL FLUORIDE POLYMERS

The vinyl fluoride polymers, which include both homopolymers and copolymers, have the base unit shown by structure (1). Prior to the early 1940s, the physical

$$\sim\!\!\sim \overset{\displaystyle H}{\underset{\displaystyle H}{\overset{|}{\underset{|}{C}}}} - \overset{\displaystyle H}{\underset{\displaystyle F}{\overset{|}{\underset{|}{C}}}} \sim\!\!\sim$$

(1)

properties of poly(vinyl fluoride) were virtually unknown. This may be due for the most part to the difficulty of polymerizing the monomer. Vinyl fluoride has a high critical temperature and a low boiling point; thus, its polymerization generally requires high-pressure techniques similar in many respects to those employed to prepare low-density polyethylene (see ETHYLENE POLYMERS). More recent advances in the use of Ziegler-Natta type catalysts have resulted in the preparation of vinyl fluoride homopolymers and copolymers at low temperatures and greatly reduced pressures. The first commercial form of poly(vinyl fluoride) was placed on the market by Du Pont in the early 1960s under the trade name Tedlar.

Poly(vinyl fluoride) in some respects resembles poly(vinyl chloride) (see VINYL CHLORIDE POLYMERS). Both polymers have a tendency to split off hydrogen halides when heated to elevated temperatures. Poly(vinyl fluoride) burns slowly, whereas poly(vinyl chloride) does not burn. The tendency of poly(vinyl fluoride) to crystallize is far greater than that of poly(vinyl chloride). Homopolymers and copolymers made from vinyl fluoride have exceptional resistance to chemical attack and to water adsorption, and have unusual solvent resistance. Poly(vinyl fluoride) is quite stable at relatively high temperatures and has exceptional resistance to degradation by sunlight. This unusual combination of properties makes these homopolymers and copolymers quite suitable for coatings or laminates for outdoor exposure, the field in which they have their greatest application.

Monomer

Preparation. Both laboratory and commercial syntheses of vinyl fluoride are considered.

Laboratory Synthesis. Vinyl fluoride was first prepared in 1901 by the reaction between 1,1-difluoro-2-bromoethane and zinc (1,2). Phenylmagnesium bromide in ether and alcoholic potassium iodide were also used in place of the metal for removing halogen. It was subsequently reported that vinyl fluoride could be prepared by the pyrolysis of 1,1-difluoroethane at 275°C over a chromium fluoride catalyst contained in a platinum tube (3). Pyrolysis of 1,1-difluoroethane resulting in the formation of vinyl fluoride may be catalyzed at 400°C in the presence of oxygen

Table 1. Physical Properties of Vinyl Fluoride[a]

chemical formula	$CH_2{=}CHF$
molecular weight	46.046
boiling point, °C	−72.0
freezing point, °C	−160
critical temperature, °C	54.7
critical pressure, psia	760
critical density, g/cc	0.320
vapor pressure at 21°C, atm	25.2
liquid density at 21°C, g/cc	0.636

[a] By permission of E. I. du Pont de Nemours & Co., Inc.

Table 2. Thermal Properties of Vinyl Fluoride[a]

latent heat, cal/g-°C			
at −28.9°C		41.05	
−17.8°C		38.58	
−6.7°C		36.12	
4.4°C		33.48	
15.6°C		30.60	
26.7°C		27.28	
	Liquid		Vapor (1 atm)
heat capacity, cal/g-°C			
at −28.9°C	0.484		0.224
−17.8°C	0.492		0.229
−6.7°C	0.500		0.235
4.4°C	0.510		0.241
15.6°C	0.542		0.246
26.7°C	0.650		0.252
thermal conductivity, cal/sec-cm²-(°C/cm), × 10³			
at −28.9°C			0.0248
−17.8°C	0.293		0.0268
−6.7°C	0.297		0.0289
4.4°C	0.305		0.0305
15.6°C	0.347		0.0322
26.7°C	0.417		0.0338

[a] By permission of E. I. du Pont de Nemours & Co., Inc.

(4). The literature also describes the preparation of vinyl fluoride by the addition of hydrogen fluoride to acetylene without a catalyst (5–7) and with mercuric oxide as the catalyst (8). This synthesis is not different in its essential aspects from that of the synthesis of vinyl chloride. However, in the case of vinyl fluoride, lower temperatures are employed. Acetylene and gaseous hydrogen fluoride are passed over contact catalysts such as carbon pellets or charcoal containing mercuric chloride or fluoride (eq. 1) (9). The by-products of this reaction are difluoroethane and vinylidene fluoride.

$$CH{\equiv}CH + HF \xrightarrow[\text{on charcoal}]{HgCl_2} CH_2{=}CHF \qquad (1)$$

An integrated process for the preparation of vinyl fluoride from 1,1-difluoroethane has been reported (51). In this case, 1,1-difluoroethane is converted to vinyl fluoride via a disproportionation reaction whereby acetylene is reacted with 1,1-difluoroethane at elevated temperatures in the presence of aluminum fluoride catalyst.

Commercial Synthesis. The preparative details for the commercial synthesis of vinyl fluoride monomer have not been described in the literature and are believed to be of proprietary nature. It is presumed that the addition of hydrogen fluoride to acetylene represents a commercial approach to the preparation of vinyl fluoride monomer (10).

Commercially supplied vinyl fluoride monomer is usually stabilized with terpenes, such as *d*-limonene, to inhibit spontaneous polymerization.

Properties. Vinyl fluoride is a very low-boiling monomer. It is a useful intermediate for the introduction of fluorine atoms into organic compounds. The presence of such fluorine atoms often increases the chemical stability of organic compounds, decreases their solubility, and retards their degradation by light or other physical agents.

Table 3. Solubility Relationships for Vinyl Fluoride[a]

Solvent	Solubility of vinyl fluoride, cc gas/cc solvent	Henry's law constant at 30°C, atm	Vinyl fluoride pressure, atm	Solubility in water, g/100 g water, at 27°C	79°C	100°C
in organic solvents						
ethyl alcohol	4					
diethyl ether	5.5					
adiponitrile	4	50.8				
propionitrile	10	32.9				
acetonitrile	10.5	41.9				
butyrolactone	5.2	57.0				
dimethylformamide	8.9	33.4				
in water[b]			8.5		0.3	
			17.3		0.5	0.4
			27.2	1.1		
			34.0	1.5	0.9	0.8
			68.0		1.5	1.5

[a] By permission of E. I. du Pont de Nemours & Co., Inc.
[b] Vinyl fluoride forms a hydrate at 15.6°C and 22 atm (324 psia).

Tables 1–3 summarize some of the physical properties of vinyl fluoride monomer (11).

Flammability. Vinyl fluoride is flammable in air between the limits of 2.6 ± 0.5% and 21.7 ± 1.0% vinyl fluoride by volume. It is classified as flammable by the Interstate Commerce Commission. Preliminary studies indicate a minimum ignition temperature of about 400°C for mixtures of air and vinyl fluoride.

Toxicity. A maximum allowable concentration of vinyl fluoride for a single short exposure for human beings has been proposed at 20% by volume (12).

Polymerization

Owing to its low boiling point (−72°C) and high critical temperature (54.71°C) vinyl fluoride is quite difficult to polymerize and generally requires high-pressure processes (50–1000 atm). In this respect it tends to resemble ethylene to a greater extent than it does the other vinyl halides.

Poly(vinyl fluoride) was first prepared in 1934 by polymerizing vinyl fluoride in toluene at 67°C and 6000 atm for 16 hr (13). However, not until 1946 was vinyl fluoride effectively polymerized and the poly(vinyl fluoride) characterized (14); the polymerization was conducted in quartz capillary tubes under autogenous pressures in the presence of peroxides or under catalysts such as ultraviolet radiation.

In general, organic peroxides are believed to be used in the commercial preparation of poly(vinyl fluoride). Catalyst systems derived from azo compounds, alkali metals, and ultraviolet radiation have also been described in the literature but are generally less effective. The molecular weight of the resultant polymer is affected by the amount of catalyst employed. High catalyst concentrations may cause a reduction in efficiency and result in polymer with a lower viscosity.

The mechanism of addition of vinyl fluoride monomer to a propagating chain appears to be essentially head-to-tail (67). However, it is also indicated (34,67) that, contrary to other vinyl monomers, vinyl fluoride does give rise to a significant degree of head-to-head addition at higher polymerization temperatures. As the temperature is decreased, the probability of monomer adding to form $—CH_2CHF-CHFCH_2—$ enchainment is diminished. Thus, the lower the polymerization temperature, the greater will be the degree of chemical regularity. This will result in a higher degree of crystallinity in the lower-temperature product. Reactions for polymerization at temperatures between 0 and 50°C can be given as follows:

Radical formation

$$R—R \xrightarrow{\Delta} R\cdot + R\cdot$$

Initiation

$$R\cdot + CH_2{=}CHF \rightarrow RCH_2CHF\cdot$$

Propagation

$$RCH_2CHF\cdot + n\ CH_2{=}CHF \rightarrow R(CH_2CHF)_{n+1}\cdot$$

Termination

$$R(CH_2CHF)_{n+1}\cdot + R(CH_2CHF)_m\cdot \rightarrow R(CH_2CHF)_{n+1}(CHFCH_2)_mR$$

$$\downarrow$$

$$R(CH_2CHF)_nCH{=}CHF + R(CH_2CHF)_mH$$

Transfer

$$R(CH_2CHF)_{n+1} + AB \rightarrow R(CH_2CHF)_{n+1}A + B\cdot$$

Since the polymerization of vinyl fluoride with peroxides and azo compounds is affected by temperature, increased temperatures lower the molecular weight of the polymer.

Ziegler-Natta type catalysts allow vinyl fluoride to be polymerized at temperatures below its critical temperature (0–50°C). These catalyst systems, in addition to their great commercial implication, make it possible to polymerize vinyl fluoride in the laboratory without extensive high-pressure equipment.

Polymerization Techniques. The first commercial polymerization of vinyl fluoride in the late 1940s was carried out under high pressure. In a typical process, polymerization was carried out in an aqueous suspension using benzoyl peroxide catalyst at 80°C and under 900 atm pressure for 6.5 hr (15–18).

Modified Suspension Polymerization. In the early 1950s a new, lower-pressure polymerization technique was described which markedly reduced the pressure requirements for the suspension polymerization of vinyl fluoride. Azo catalyst systems were employed in place of the benzoyl peroxide used in previous high-pressure polymerizations (19). The reactions with azo catalysts were conducted at 25–100°C and between 25 and 100 atm pressure over 18–19 hr. In a typical procedure, a stainless-steel reactor is flushed with oxygen-free nitrogen and then charged with 150 parts of deoxygenated distilled water, 0.150 part of 2,2′-azobisisobutyronitrile, and 150 parts of vinyl fluoride monomer (acetylene-free). The reactor is then agitated and the temperature brought to 70°C within one hour. The pressure is maintained at 82 atm, and the reaction continued under these conditions for 18 hr. The unreacted monomer is then bled off and the reactor opened. The product, 75.8 parts of poly(vinyl fluoride), is collected in the form of a white cake (14).

Emulsion Polymerization (66). Vinyl fluoride has generally been found to be more difficult to polymerize by emulsion processes than are other monomers. This is believed to be due to extensive chain transfer with the surfactants. However, emulsion polymerization is quite desirable since it can be conducted at lower pressures.

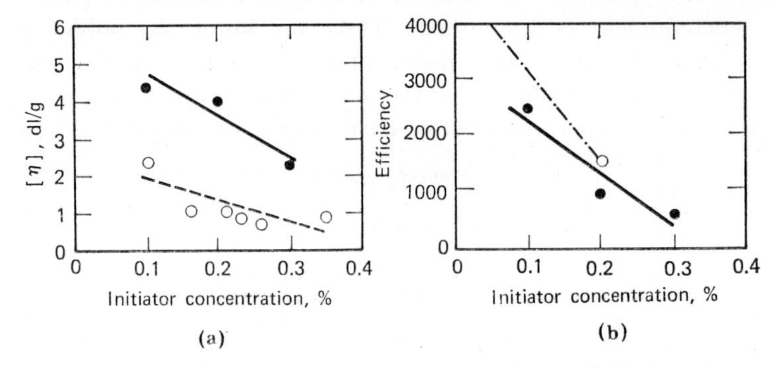

Fig. 1. Effects of initiator concentration on (**a**) viscosity of polymer (in dimethylformamide at 100°C), and (**b**) initiator efficiency (in moles monomer polymerized/moles initiator). Key: ●, benzoyl peroxide, 250 atm, 80°C, absorption stops; ○ in (**a**), azoamidine hydrochloride, 100 atm, 70°C, constant pressure drop; ○ in (**b**), azobisisobutyronitrile, 75 atm, 70°C, to total pressure drop.

Conventional surfactants are ineffective, and only perfluorinated carboxylic acids containing between seven and eight carbon atoms are suitable emulsifiers. The use of such perfluorinated emulsifiers results in high conversion of vinyl fluoride to polymer.

A typical recipe for emulsion polymerization consists of 200 parts by weight of water, 0.2 part of ammonium persulfate, 1.2 parts of perfluorinated carboxylic acid emulsifier (L 1159, 3M Corp.), and 100 parts of vinyl fluoride. Polymerization takes place at 46°C, with stirring, under the autogenous pressure of 42.5 atm, for 12 hr, to 85% conversion of monomer. The polymer is isolated by precipitation with an electrolyte.

Effects of Polymerization Variables. An extensive study has been made of the effects of various polymerization variables on the polymerization of vinyl fluoride and on the physical and chemical properties of the resulting polymer (20). It covered polymerizations over the temperature range between 35 and 165°C, and pressures from 1 to 1000 atm, as well as a number of catalyst systems including redox

Table 4. Initiators for Vinyl Fluoride Polymerization[a]

Initiator	Temp, °C	Pressure, atm	Reaction time, hr	Conversion	Mol wt
	Between 35 and 100°C				
2,2′-azobis-2,4,4-trimethylvaleronitrile	37	1000	4	high	high
2,2′-azobis-2,4-dimethylvaleronitrile	50	500	8	moderate	medium
2,2′-azobisisobutyronitrile	70	70	19	90%	high
2,2′-azobis(isobutyramidine hydrochloride)	70	100	4	40%	high
benzoyl peroxide	80	250	13	37%	very high
ammonium persulfate	85	250	2	33%	medium
	Above 100°C				
2,2′-azobis(cyclohexane carbonitrile)	115	175	1.5	32%	moderately high
diethyl peroxide	122	600	14.5	good	medium
tert-butyl peroxide	140	500	0.5	good	medium
2(2′-hydroxyethylazo)-2-ethylbutyronitrile	145	250		good	medium

[a] Ref. 20. Initiator concentration is specified only as constant; exact value is not given.

systems, alkali metals, and thermally generated free radicals. The following discussion is drawn largely from this study.

Initiator. High initiator concentrations lower the initiator efficiency (in terms of mass of monomers polymerized per mole of initiator) and produce polymer with a lower molecular weight (Fig. 1). Table 4 lists a number of initiators with their effects on conversion and molecular weight.

Temperature. In general, polymerization temperature has a pronounced effect on the molecular weight of the polymer. Polymers prepared at elevated temperatures are reported to be more branched; this effect is probably due to more rapid chain transfer. As the temperature of polymerization is increased, for a given initiator concentration, the polymer formed has a lower molecular weight. With peroxides as well as azo compounds, the initiator efficiency increases with temperature to a maximum and then rapidly decreases (Fig. 2).

Pressure. Increasing the pressure increases the rate of polymerization. Increased pressure also favors the formation of higher-molecular-weight polymer and more initiator efficiency (Figs. 3 and 4). However, in the case of azobisisobutyronitrile, the maximum efficiency appears to be independent of pressure.

Polymerization Media. Organic solvents used in place of water as polymerization media are reported to result in extensive telomerization with vinyl fluoride, causing drastic reduction in molecular weight.

Ziegler-Natta Alkylboron Catalysts. In the early 1960s the polymerization of vinyl fluoride by Ziegler-Natta type catalysts was described in the literature. For example, a patent granted in 1963 describes the polymerization of vinyl fluoride employing a catalyst system composed of diethylaluminum bromide–titanium tetrachloride–carbon tetrachloride (21). In a study of the polymerization and copolymerization of vinyl fluoride reviewing the use of Ziegler-Natta and alkylboron catalysts over the temperature range of 0–50°C, a Ziegler-Natta system based on vanadyl acetylacetonate and AlR(OR)Cl compounds was reported to have good activity (22).

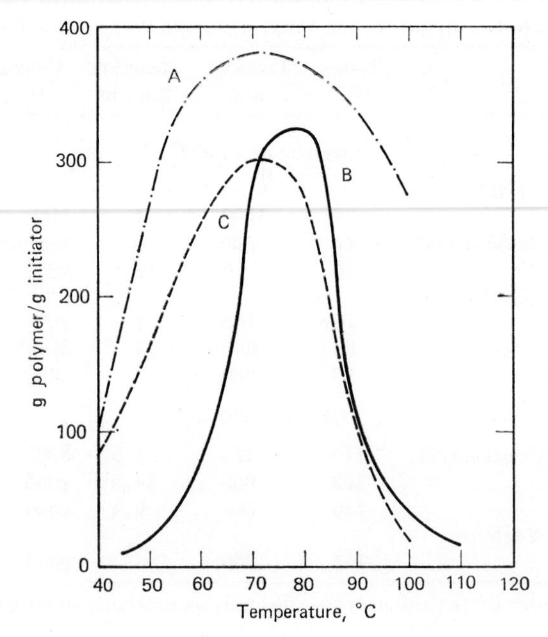

Fig. 2. Effect of temperature on initiator efficiency. Key: curve A, 0.1% 2,2′-azobisisobutyronitrile at 250 atm pressure; curve B, 0.1% benzoyl peroxide at 250 atm pressure; curve C, 0.2% 2,2′-azobisisobutyronitrile at 100 atm pressure.

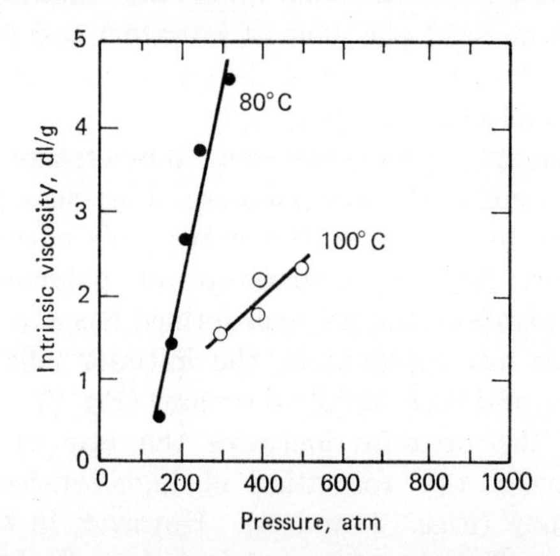

Fig. 3. Effect of polymerization pressure on viscosity using 0.2% benzoyl peroxide in aqueous medium. (Viscosity determined in dimethylformamide at 100°C.)

Increased reaction rates and higher degrees of polymerization were reported for oxygen-activated alkylboron catalysts. In all cases studied, the polymerization proceeded by a free-radical mechanism. However, crystallinity and melt temperature of the polymer were higher than those of poly(vinyl fluoride) prepared by high-pressure processes.

Since the heterogeneous catalyst systems of the type described in Refs. 21 and 22 are known to result in stereospecific polymerization of certain monomers, Sianesi (22) attempted to determine whether this was the case with vinyl fluoride. His studies, employing alkylboron and Ziegler-type catalysts, gave no evidence of increased

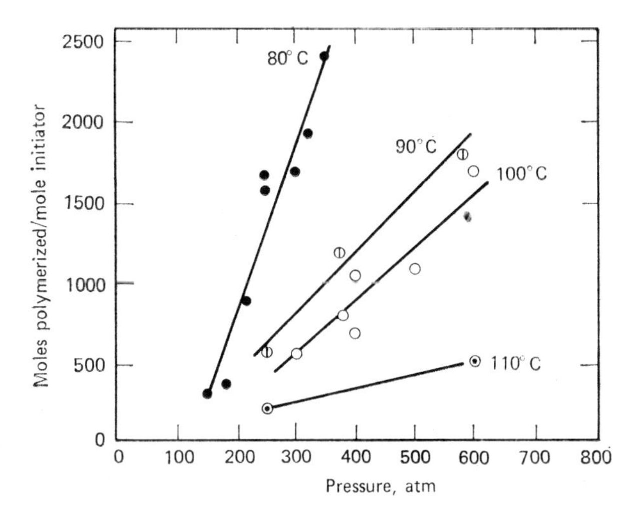

Fig. 4. Effect of polymerization pressure on initiator efficiency using 0.2% benzoyl peroxide in aqueous medium.

stereoregularity in the polymer formed. The polymers had the same structures as the atactic polymers prepared with conventional catalysts at higher temperatures. The absence of stereospecificity confirmed that the mechanism was indeed free radical. Thus, the reaction path can be presumed to be the same as that of conventional free-radical-initiated polymerization conducted with peroxy or azo catalysts.

X-ray diffraction patterns of polymers prepared with Ziegler-Natta catalyst systems at low temperatures are similar to those of polymers prepared by conventional free-radical processes (22); only minor differences exist, which are attributable to differences in the degree of polymerization. Thus the crystal structure of

Table 5. Variation of the Melting Point and Degree of Crystallinity of Poly(vinyl Fluoride) with the Polymerization Temperature Using Alkylboron Catalyst[a]

Polymerization temperature, °C	Melting point, °C	Degree of crystallinity,[b] % ± 2
85	197–205	37
40	205–215	44
30	218–225	45
20	222–230	45
10	220–235	48
0	225–235	50

[a] Polymerization pressure, 300 atm.

[b] From the heat of fusion (differential scanning calorimeter) and the averaged $\Delta H_u = 1813$ cal/monomer unit.

polymers prepared at low temperatures using Ziegler-Natta catalysts must be the same as that of the randomly oriented poly(vinyl fluoride) prepared at elevated temperatures, although a greater degree of crystallinity exists. Polymerization temperature affects the degree of crystallinity and melting point (Table 5). Presumably the higher degree of crystallinity of polymer prepared at lower temperature is due to an increase in chain regularity. In other references to the use of alkylboron catalysts for vinyl fluoride polymerization, the polymerizations were conducted in the range of −30 to 55°C and yielded polymers with melting points around 225°C (23,24).

Continuous Polymerization. A continuous process for the commercial polymerization of vinyl fluoride has been described in which a mixture of monomer, water, and free-radical initiator is stirred at 100°C and 265 atm. A small amount of low-molecular-weight olefin (C_3–C_7) chain regulator is added to help avoid the formation of very high-molecular-weight poly(vinyl fluoride) which is claimed to be objectionable due to its poor processing characteristics (25). A continuous polymerization of vinyl fluoride, for use as a coating for laminating structures designed for outdoor exposure, is carried out by continuously passing a water-soluble initiator, water, and vinyl fluoride into a first reaction zone maintained at 60–90°C and 1,500–9,000 psi; further polymerization subsequently takes place in a second reaction zone maintained at 90–140°C and 4,000–14,000 psi (26).

Copolymerization

In general, the conditions employed for homopolymerization (ie, media, temperature, pressure, and catalysts) apply to copolymerization as well. A wide variety of monomers can be copolymerized with vinyl fluoride (15–18). Copolymer systems described in the early patent literature include vinyl fluoride–vinylidene carbonate (27); vinyl fluoride–ethyl acrylate–acrylonitrile–acrylic acid (28); vinyl fluoride–vinyl formate (29); vinyl fluoride–vinyl acetate–tetrafluoroethylene (30); vinyl fluoride–vinyl chloride (31); and vinyl fluoride–chlorotrifluoroethylene (32).

Conventional free-radical initiators are used in the above copolymerizations; however, vinyl fluoride can be copolymerized by Ziegler-Natta and alkylboron cata-

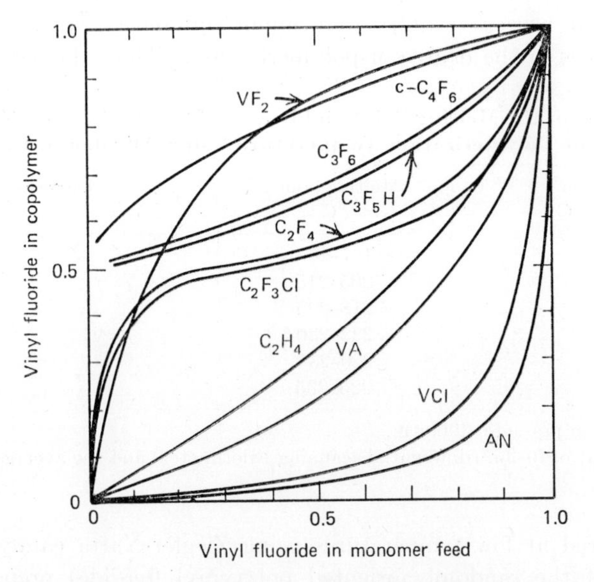

Fig. 5. Monomer–copolymer composition in the copolymerization of vinyl fluoride at 30°C with $O_2/B(i\text{-}C_4H_9)_2$ = 0.5 initiator (based on total monomer charged) and ethyl acetate as solvent. Comonomers: vinyl chloride (VCl), acrylonitrile (AN), vinyl acetate (VA), chlorotrifluoroethylene, ethylene, chlorotrifluoroethylene, tetrafluoroethylene, *cis*-1-hydropentafluoropropene (C_3F_5H), hexafluoropropene (C_3F_6), hexafluorocyclobutene(c-C_4F_6), and vinylidene fluoride (VF_2).

lysts. As in the case of homopolymerization of vinyl fluoride, copolymerization proceeds by a free-radical mechanism. Copolymerizations of vinyl fluoride with a number of different comonomers are described and Q and e values given in Ref. 22.

Reactivity Ratios. Besides brief references to the reactivity ratios of vinyl fluoride and a limited number of monomers (33), a fairly extensive study has been published covering a number of different monomers over a range of compositions, with either a Ziegler-Natta or alkylboron catalyst (22). (See also the article COPOLYMERIZATION for a discussion of reactivity ratios with Ziegler-Natta type catalysts.) Figure 5 represents monomer–copolymer composition curves for the alkylboron-catalyzed copolymerizations. The copolymerization reactivity ratios determined are listed in Tables 6 and 7. Q and e values of 0.010 ± 0.005 and −0.8 ± 0.2 were also found.

Table 6. Reactivity Ratios in the Copolymerization of Vinyl Fluoride (M_1) at 30°C with Ziegler-Natta Catalysts

Initiator	M_2	r_1	r_2	r_1r_2
$\frac{(i\text{-}C_4H_9O)ClAl\text{-}(i\text{-}C_4H_9)}{VO(acetylacetonate)_2} = 2$	vinyl chloride	0.07 ± 0.002	9 ± 1	0.63
	vinylidene fluoride	4.2 ± 0.4	0.18 ± 0.02	0.75
	hexafluoropropene	1.1 ± 0.05	0	0
$\frac{Al(i\text{-}C_4H_9)_3}{Ti(O\text{-}i\text{-}C_3H_7)_4} = 3$	hexafluoropropene	25 ± 5	0.04 ± 0.02	1.0

Table 7. Reactivity Ratios in the Copolymerization of Vinyl Fluoride (M_1) at 30°C with $B(i\text{-}C_4H_9)_3\text{-}O_2$ Catalyst

M_2	r_1	r_2	r_1r_2
$CH_2{=}CH_2$	0.3 ± 0.03	1.7 ± 0.1	0.51 ± 0.08
$CH_2{=}CHCl$	0.05 ± 0.005	11.0 ± 1	0.55 ± 0.10
$CH_2{=}CF_2$	5.5 ± 0.5	0.17 ± 0.03	0.93 ± 0.20
$CF_2{=}CF_2$	0.27 ± 0.03	0.05 ± 0.02	0.013 ± 0.007
$CFCl{=}CF_2$	0.18 ± 0.02	0.006 ± 0.02	0.011 ± 0.005
$CF_3CF{=}CF_2$	1.01 ± 0.01	0	0
$CF_3CF{=}CFH$ (*cis*)	0.9 ± 0.05	0	0
cyclo-C_4F_6	3 ± 0.6	0	0
$CH_2{=}CHOCOCH_3$	0.16 ± 0.01	2.9 ± 0.2	0.46 ± 0.05
$CH_2{=}CHCN$	~1×10^{-3}	24 ± 2	~0.024

The literature does not indicate that any copolymers of vinyl fluoride are at present in commercial use. However, since vinyl fluoride copolymerizes with a variety of monomers, the products are potentially useful for specific applications in coatings. Copolymer can be prepared that will coalesce into films at room temperature. These copolymers have good weather, solvent, and abrasion resistance, provided the fluorine content is not too low.

Properties

Typical physical and chemical properties of a poly(vinyl fluoride) film are listed in Table 8.

Table 8. Typical Properties of Poly(vinyl Fluoride) Film[a]

Property	Test method	Value
	Physical properties	
bursting strength at 23°C, psi	Mullen, ASTM D 774	19–70
coefficient of friction (film/metal) at 23°C	Instron	0.15–0.30
density at 23°C, g/cc	weighed samples	1.38–1.57
impact strength at 23°C, kg-cm/mil	Du Pont pneumatic tester	2.7–5.6
moisture absorption at 23°C	water immersion	<0.5% for all types
moisture-vapor transmission at 39.5°C, g/(100 m^2)(hr)(mil) (53 mm Hg)		157–205
moisture-vapor transmission at 39.5°C and 80% rh, g^{-1}/24 hr-sq m-mm Hg	ASTM E 96-58T	1.5
refractive index at 30°C, n_D	ASTM D 542, Abbé refractometer	1.46
tear strength at 23°C, g/mil		
propagated	Elmendorf	12–100
initial (Graves)	ASTM 1004	450–620
tensile modulus at 23°C, psi	ASTM D 882, method A (100% elongation/min—Instron)	250–375 × 10^3
ultimate tensile strength at 23°C, psi	ASTM D 882, method A (100% elongation/min—Instron)	7.0–18.0 × 10^3
ultimate elongation at 23°C, %	ASTM D 882, method A (100% elongation/min—Instron)	115–250
ultimate yield at 23°C, psi	ASTM D 882, method A (100% elongation/min—Instron)	6,000–4,900
	Chemical properties	
chemical resistance at 25°C, after 1 yr immersion in acids, bases, or solvents		no visible effect
chemical resistance at boiling temp, after 2 hr immersion in acids, bases, or solvents		no visible effect
chemical resistance after soil burial for 5 yr		strength and appearance not affected
gas permeability at 23.5°C, cc/(100 sq in.)(24 hr)(atm)(mil)	ASTM D 1434	
carbon dioxide		11.1
helium		150
hydrogen		58.1
nitrogen		0.25
oxygen		3.2
vapor permeability at 23.5°C, g/(100 m^2)(hr)(mil)	ASTM E 96, modified	
acetic acid		45
acetone		10,000
benzene		90
carbon tetrachloride		50
ethyl acetate		1,000
ethyl alcohol		35
hexane		55

Table 8 (*continued*)

Property	Test method	Value	
water-vapor permeability at 39.5°C, g/(100 m²)(hr)(mil)(53 mm Hg)	Du Pont permeability method	180	
weatherability, facing south at 45° to horizontal, Florida exposure		excellent	
	Thermal properties		
aging at 150°C, hr	circulating air oven	3,000	
flammability		slow burning to self-extinguishing	
linear coefficient of expansion, in./in./°F		2.8×10^{-5}	
shrinkage, %	air oven, 30 min		
Type 20, MD and TD[b]		4% at 130°C	
Type 30, TD only		4% at 170°C	
Type 40, TD only		2.5% at 170°C	
temperature range for continuous use, °C		−72 to 107	
zero strength, °C	hot bar	260–300	
	Electrical properties		
		Film type	
		200 SG40TR	200 AM30WH
corona endurance at 60 cps, 1000 V/mil, hr	ASTM suggested T method	2.5	6.2
dielectric constant, 1 kc at 23°C	ASTM D 150	8.5	9.9
dielectric strength, 60 cps, kV/mil	ASTM D 150	3.4	3.5
dissipation factor, %	ASTM D 150		
1000 cps, 23°C		1.6	1.4
1000 cps, 70°C		2.7	1.7
10 kc, 23°C		4.2	3.4
10 kc, 70°C		2.1	1.6
volume resistivity, ohm-cm	ASTM D 257		
at 23°C		4×10^{13}	7×10^{14}
at 100°C		2×10^{10}	1.5×10^{11}

[a] Tedlar, Du Pont.
[b] MD, machine direction; TD, transverse direction.

Molecular Weight and Solubility. Poly(vinyl fluoride) has very low solubility in ordinary solvents, a property that has created some difficulty in characterization, especially in light scattering. High-molecular-weight poly(vinyl fluoride) is insoluble in ordinary solvents at temperatures below 100°C, but above 100°C dissolves in *N*-substituted amides, dinitriles, ketones, tetramethylene sulfone, and tetramethylurea. Solvents that have been used for characterization include dimethylformamide, dimethylacetamide, γ-butyrolactone, and hexamethyl phosphoramide. Molecular weights have been determined by sedimentation velocity and osmotic pressure. In the case of poly(vinyl fluoride) in dimethylformamide, the ratio of change in refractive index to concentration change, dn/dc, is about 0.02 cc/g, so light-scattering and sedimentation-equilibrium measurements are not possible in this solvent (36). Light-scattering measurements have been made using a solvent pair. Dimethylformamide to which 0.1N LiBr was added to suppress polyelectrolyte interference has been used to determine intrinsic viscosity and osmotic number-average molecular weight at 90°C; and sedimentation-velocity molecular weight at 100°C. Samples encompassed the range M_n = 76,000–234,000; M_s = 143,000–654,000; M_s/M_n = 2.5–5.6. The equation $[\eta] = 6.42 \times 10^5 M_s^{0.80}$ was obtained where $a = 0.80$ corresponds to an extended polar polymer in a good solvent. Sedimentation-velocity measurements showed some low-molecular-weight material and a high-molecular-weight tail. The sedimentation constant, S_0, was 0.0929 $M_s^{0.40}$. The unperturbed dimension A ($= \sqrt{\bar{r}_0^2/M}$ where $\bar{r}^2$ is the root-mean-square end-to-end distance) was 0.786. The steric factor σ ($= A/A_F$) was similar to that of other poly(vinyl halides), indicating similar steric hindrance to free rotation of the halogen atoms. Table 9 shows that the similarities in σ can be explained by the compensating effects of an increase in molar volume (tending to increase A) and a corresponding decrease in electronegativity difference (tending to decrease A).

Table 9. Steric Factors of Vinyl Halide Polymers (36)

	Poly(vinyl fluoride)	Poly(vinyl chloride)	Poly(vinyl bromide)
steric factor	1.72	1.83	1.82
molar volume of substituent, cc	12	28	35
electronegativity difference of C–X bond atoms	1.5	0.5	0.3

High-viscosity poly(vinyl fluoride) ($[\eta]$ = 3–4 dl/g) was soluble to 10–15% at 110°C in dimethylformamide. Polymers prepared with water-soluble catalysts were less likely to form gels on cooling the solution. Molecular weights were determined by tagged initiator and by osmotic pressure at 100°C in dimethylformamide, dimethylacetamide, and γ-butyrolactone, as were intrinsic viscosities; number-average molecular weights from 20,000 to 180,000 were found (20). In determinations of inherent viscosities using 0.05 g poly(vinyl fluoride) in 100 ml hexamethyl phosphoramide dissolved at high temperature and run at 30°C, a typical resin had an inherent viscosity of 2.6 dl/g (37).

Structure. The infrared and Raman spectra of poly(vinyl fluoride) indicate that the polymer is atactic (38) although earlier studies based on less information indicated an isotactic structure (39). NMR studies showed that one monomer unit in six was added in the reverse direction, ie, head-to-head (40). A decrease in the amount of head-to-head addition corresponded to a decrease in the temperature of polymerization. For example, the head-to-head addition at 180°C was one unit in six, whereas at 20°C it was one unit in eight.

X-ray investigation has shown a high degree of crystallinity. This can be reconciled by considering the fluorine atom similar in size requirements to the hydrogen atom. The crystal lattice could form with no differentiation between the two atoms. Strong hydrogen bonding would decrease solubility and produce solution viscosities larger than expected; both are observed with poly(vinyl fluoride).

Dimensions of the unit cell of poly(vinyl fluoride) by x-ray methods were found to be: $a = b = 4.93$ Å, $c = 2.53$ Å; if the crystal structure is similar to that of poly(vinyl alcohol), which has a planar zigzag structure with an O–O repeat distance of 2.52 Å, then the F–F repeat distance of 2.53 Å would seem reasonable (41). It

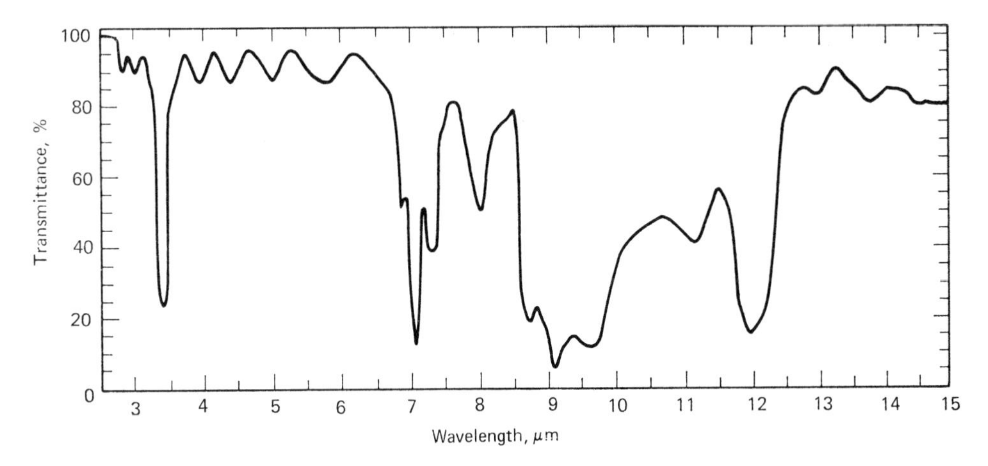

Fig. 6. Infrared absorption spectrum of poly(vinyl fluoride) film (Tedlar, Type 50 SG20TR, Du Pont). Reprinted from *Tedlar Bulletin TD-5* by permission of E. I. du Pont de Nemours & Co., Inc.

was concluded that the unit cell was hexagonal (theoretical density 1.44; observed poly(vinyl fluoride) density 1.38). Another study showed great crystallinity, but an irregular pattern (39); a unit cell was found to have orthogonal axes and to contain two monomer units, whose cell dimensions are: $a = 8.57$ Å, $b = 4.95$ Å, and $c = 2.52$ Å. From these data and assuming similarities to poly(vinylidine fluoride), it was concluded that the true unit cell was an orthorhombic space group Cm2m (42).

Thermal Transitions. In measurements of dielectric properties, two loss processes were observed: a high-temperature loss for which $\Delta H^{\ddagger} = 21$ kcal/mole, and $\Delta S^{\ddagger} = 13$ eu (entropy units), and a low-temperature loss in which $\Delta H^{\ddagger} = 23$ kcal/mole (43). This latter was assigned to a glass transition (T_g). Free torsion-pendulum studies showed transition at 41°C (1.7 Hz) and −20°C (5.4 Hz), with T_g assigned to 41°C (44); forced torsion-pendulum studies gave a T_g value at 20 Hz of 70°C (45). Thermal stabilities of fluorinated hydrocarbon polymers have been determined by DTA (46); the order of decreasing stability was polytetrafluoroethylene > polyethylene > poly(vinylidene fluoride) > poly(vinyl fluoride). Poly(vinyl fluoride) gave an endotherm at 192°C and decomposed at 355°C with release of sufficient energy to rupture the 2800 psi rupture disc employed in the apparatus. The melting point of poly(vinyl fluoride) has been reported as 210°C (47).

Optical Properties. Poly(vinyl fluoride) film is exceptionally transparent to light. A film of Tedlar 100 BG30TR (Du Pont) 1 mil thick had over 90% transmission from 400 mμ to above 1.4 μ and over 80% transmission from 230 to 400 mμ. The light transmitted decreases rapidly below 200 mμ (48). Poly(vinyl fluoride) absorbs strongly in the 7–13 μ region of the infrared (49) (see Fig. 6). Three optical axes can be distinguished in oriented films; the lowest refractive index is 1.444 through the film thickness, the highest 1.464 in the plane of the film. The birefringence measured in highly oriented film was 0.0186; for film of low orientation, 0.006.

Effects of Radiation. When poly(vinyl fluoride) film 0.003 in. thick was subjected to electron-beam radiation, tensile strength and elongation decreased with increasing radiation dose, but less rapidly than does that of polytetrafluoroethylene

(50). At a 100 Mrad dose, the film has lost virtually all of its structural properties; it does retain them, however, when irradiated at 32 Mrad. When heated to 255°C, unirradiated film contracted longitudinally to 50% of its initial value; when irradiated at 8 Mrad, it contracted to 80% of its initial value; and, when irradiated to 32 Mrad, it retained its initial dimension.

Effects of Plasticizers. Although a number of materials are compatible with the polymer at an elevated temperature (above 100°C), they exude on cooling to room temperature. Dibutyl phthalate and tricresyl phosphate are retained in amounts of 5–10% (20). This amount of plasticizer lowers the softening temperature of the polymer without imparting rubbery characteristics. The surface tension of poly(vinyl fluoride) film has been reported as 28 dyne/cm (47).

Fabrication

Commercial poly(vinyl fluoride) resin is fabricated into films and coatings. At present a self-supporting film is sold by Du Pont, and a dispersion-grade resin by Diamond Shamrock. Fabrication of the film, Tedlar, involves laminating it to a substrate. Literature references to molded articles do exist (2,52), but none appears to be commercially significant. There are no commercially practical room-temperature solvents for poly(vinyl fluoride). There are, however, "latent" solvents which solvate the resin at elevated temperatures; these latent solvents are used to prepare a dispersion of the small resin particles, similar to an organosol of poly(vinyl chloride). Film extrusion has been shown possible with modified poly(vinyl fluoride) (35). A dispersion of the resin in a latent solvent may be heated to fuse the poly(vinyl fluoride): fusion may take place either on a substrate, on a heated continuous belt, or as the polymer is extruded into a hot liquid such as mineral oil (53).

Du Pont supplies four types of Tedlar (formerly Teslar), types 15, 20, 30 and 40, which presumably differ in degree of orientation, the higher number having lesser orientation. The more highly oriented films have better tensile strength and impact resistance and greater resistance to flexural fatigue. The less oriented films have high tear strength, greater extensibility at high temperature, and less shrinkage (49). A more dimensionally stable product may result from vulcanization (54). Two types of film surface are available, a strippable surface and an adherable surface for use with adhesives, inks, and coatings. The latter may be produced by an electric discharge process (55). The untreated (strippable) surface may be heat sealed at 205–220°C and 20 psi with a dwell time of 1–2 sec and a cooling time of 3–4 sec (56). Gloss variations, colored, transparent, and ultraviolet-absorbing films are then laminated to substrates; specific adhesives are supplied for various substrates. Acrylic adhesive has been used to bond Tedlar to steel (57). Rubber and epoxy adhesives are also used. When used as a protective film for glass-reinforced thermoset resins, the thermosetting resin itself is used as the adhesive.

The problems of fabricating poly(vinyl fluoride) stem from the high degree of order and the large number of hydrogen bonds in the polymer, which result in a high melting point near the decomposition temperature. Successful processing requires the use of latent solvents to disrupt the polymer order and, therefore, lower the minimum temperature for film forming. The latent solvent must not only produce a continuous film of polymer, but must also have the correct volatility. If the solvent is too volatile it will evaporate before the resin particles completely coalesce; if it has insufficient volatility it will not be completely removed in an economical baking cycle. Other components that are usually added to the formulation include pigments, usually inorganic, in amounts of 5 to 15%, stabilizers in small amounts, often polyhydroxy compounds in combination with sulfur-containing (eg, thiodialkanoic acid diester (58)) or phosphorus-containing compounds, and polymeric viscosity modifiers to improve film leveling. A typical dispersion of Diamond Shamrock resin would have the following properties (47):

solids content, 30–35 wt %, 23–26 vol %
efflux time, number 4 Ford cup, 60–80 sec
viscosity, plate and cone, 0 sec^{-1}, 200 cP
10^4 sec^{-1}, 120 cP
density, 9.7–9.9 lb/gal
coverage (dry) per gallon, 370–420 mil ft^2
fineness of grind, Hegman scale, 7+
shelf stability, soft settling after 30 days

Dispersions are applied commercially by air, airless, and electrostatic spray equipment, and by reverse roll coating (see COATING METHODS). Fluidized-bed coatings have also been employed (35).

Applications

Protective Films and Coatings. Poly(vinyl fluoride) is used principally in films and coatings. It is a moderately expensive coating.

The major amount of poly(vinyl fluoride) resin used is purchased as a film and laminated to a substrate. The largest commercial use of poly(vinyl fluoride) is to provide a durable, weather- and stain-resisting surface to building products. Over twenty-five years of outdoor weathering tests have shown remarkable retention of appearance and physical properties (60,61). It is used as a protective surfacing material or decorative coating for prefabricated building panels and siding; as a weather-resistant skin for reinforced plastic panels, increasing the outdoor life of the component by preventing erosion of the resin and subsequent moisture penetration; as a weather-resistant surface for industrial roofing; as a protective covering for pipe insulation; and as an insulation jacketing for storage tanks (59). Uniformity of the coating is assured since it is applied as factory-inspected film to the substrate. Other advantages are freedom from pinholes and voids, and high resistance to peeling, blistering, cracking, crazing, or chalking even in hostile environments. Resistance to abrasion, chemicals, solvents, and staining lead to the use of poly(vinyl fluoride) film as a top coat on vinyl and other wall coverings, as an interior finish to commercial aircraft, and externally as a surface for rigid plastic sheet for highway markers, housing for lawn mowers and air-conditioning units, and in home accessories such as shutters and gutters. Poly(vinyl fluoride) can be bonded to galvanized steel, plywood, timber hardboard, insulation board, aluminum, reinforced plastics, decorative laminates, roofing, pipe insulation, and jacketing foils and paper. See also BUILDING AND CONSTRUCTION APPLICATIONS.

Poly(vinyl fluoride) film has superior abrasion resistance when used as a surfacing material because of its exceptional toughness. Pigmented film does not chalk, craze, or erode appreciably even after prolonged exposure to weathering. Film does not fracture when the building material to which it is bonded expands or contracts. Usually, poly(vinyl fluoride) film has much greater flexibility and formability than the substrates to which it is bonded; thus, operations such as drawing, embossing, and vacuum forming of the substrate can be employed without damaging the film. It also has excellent resistance to solvent and chemical attack and other outstanding properties, as indicated in Table 8.

Du Pont is, in 1970, marketing four types of Tedlar poly(vinyl fluoride) film. Types for building applications include a high-shrinkage grade intended for use as a surfacing film for polyester–glass panels, another grade designed for surfacing vinyls for wall coverings and other interior building products, and a film with medium tensile strength and medium elongation for lamination to a wide variety of substrates including woods, metals, building felts, and plastics. A fourth type of film with good formability, high elongation, and good heat-sealing properties is used primarily as a release agent for molding reinforced plastics. As well as differing grades of film, different surfaces are available designed to give the maximum bonding with various adhesives. Other surfaces are designed to have antistick properties and are used when the film serves as a mold-release agent (59).

Diamond Shamrock markets a poly(vinyl fluoride) dispersion resin, Dalvor 720. This is formulated with a latent solvent and used as a premium performance coating, applied by air, airless, and electrostatic spray, or reverse roll coating techniques and then oven baked to fuse the polymer (see COATING METHODS). These dispersions are used to line steel shipping containers, taking advantage of their high chemical and solvent resistance, for coil coatings, and as an additive to other resin systems for improved chemical resistance and lower frictional losses.

Conventional equipment used for paint and lacquer prefinishing of sheet materials can, in many cases, be utilized for finishing or surfacing with poly(vinyl fluoride). Hydrolytic stability, low oxygen permeability, grease resistance, toughness, and inherent chemical inertness make poly(vinyl fluoride) film suitable for packaging greasy substances, such as oil, and corrosive chemicals.

Release Film. Because of its excellent release properties, poly(vinyl fluoride) film is suitable for use as release sheeting in low- and high-pressure molding of phenolic, epoxide, polyester, and other reinforced plastics. Poly(vinyl fluoride) contains neither plasticizers nor moisture which could leach into the resins at high temperatures during curing. Poly(vinyl fluoride) is particularly suitable as a shrink tape, parting sheet, or as a release film for bag molding. It resists embrittlement and strips cleanly from the mold. It conforms easily to complex shapes and holds a vacuum under severe curing conditions in bag molding because of its flexibility and good elongation. At elevated temperatures at all degrees of humidity and in contact with common solvents, poly(vinyl fluoride) film maintains its toughness. However, it is not recommended for resin systems with cure temperatures above 205°C. The extremely low gas permeability of the film does, however, provide protection against air inhibition of resins while the film itself does not inhibit curing (59,62).

Other Applications. Poly(vinyl fluoride film has also been used as the covering for solar stills (63), and has been considered for use in spacecraft (64) and as a backing for photographic emulsion (65).

Economic Aspects

The poly(vinyl fluoride) market is dominated by Du Pont, which produced the first commercial resin as Teslar (now Tedlar) film in 1963. Sales figures are not available, but it has been reported that sales were 50% larger in 1968 than they were in 1967 (66). Du Pont is currently marketing film in four thicknesses: 0.0005, 0.001, 0.0015, and 0.002 in. Prices depending upon thickness, surface treatment, and pigmentation range from $2.75 to $5.00 per pound which is equivalent to from $0.02 to $0.07 per ft². The grade used for exterior siding lamination, 0.0015 in. pigmented, sells for $0.03–$0.035 per ft² but sold at $0.06 per ft² when first introduced (61).

Bibliography

1. F. Swartz, *Bull. Acad. Royale Belg.* **1901,** 383.
2. F. Swartz, *Bull. Acad. Royale Belg.* **1909,** 728.
3. D. D. Coffman and R. D. Cramer (to E. I. du Pont de Nemours & Co., Inc.), U.S. Pat. 2,461,523 (1948).
4. O. W. Cass (to Du Pont), U.S. Pat. 2,442,993 (1947).
5. H. Plauson (to Plauson's Ltd.), U.S. Pat. 1,425,130 (1922).
6. A. V. Grosse and C. B. Linn, *J. Am. Chem. Soc.* **64,** 2289 (1942).
7. A. L. Henue, *Organic Reactions,* Vol. 2, John Wiley & Sons, Inc., New York, 1944, p. 66.
8. J. Soll (to I. G. Farbenindustrie A.G.), U.S. Pat. 2,118,901 (1938).
9. L. F. Salisbury (to Du Pont), U.S. Pat. 2,437,307 (1948).
10. K. H. Mieglitz (to Kali-Chemie), U.S. Pat. 3,230,692 (1966).
11. *Fluoromonomers,* Brochure No. FM-1, E. I. du Pont de Nemours & Co., Inc., Wilmington, Del., 1965.
12. D. Lester and L. A. Greenberg, *Arch. Ind. Hygiene Occup. Med.* **2,** 335 (1950).
13. H. W. Starkweather, *J. Am. Chem. Soc.* **56,** 1870 (1934).
14. A. E. Newkirk, *J. Am. Chem. Soc.* **68,** 2467 (1946).
15. D. D. Coffman and T. A. Ford (to Du Pont), U.S. Pat. 2,419,008 (1947).
16. D. D. Coffman and T. A. Ford (to Du Pont), U.S. Pat. 2,461,523 (1948).
17. D. D. Coffman and T. A. Ford (to Du Pont), U.S. Pat. 2,419,009 (1947).
18. D. D. Coffman and T. A. Ford (to Du Pont), U.S. Pat. 2,419,010 (1947).
19. F. L. Johnston (to Du Pont), U.S. Pat. 2,510,783 (1950).
20. G. H. Kalb et al., *J. Appl. Polymer Sci.* **4** (10), 55–61 (1960).
21. G. H. Crawford (to Minnesota Mining & Manufacturing Co.), U.S. Pat. 3,084,144 (1963).
22. D. Sianesi and G. Caporiccio, *J. Polymer Sci.* [A-1] **6** (2), 335–352 (1968).
23. Montecatini, Brit. Pat. 1,029,635 (1966).
24. Montecatini, Brit. Pat. 1,039,914 (1966).
25. J. L. Hecht (to Du Pont), U.S. Pat. 3,265,678 (1966).
26. Du Pont, Brit. Pat. 1,077,728 (1967).
27. E. W. Gluesenkamp (to Monsanto Chemical Co.), U.S. Pat. 2,847,401 (1958).
28. J. R. Straughan (to Union Carbide Corp.), U.S. Pat. 3,057,812 (1962).
29. C. A. Thomas (to Du Pont), U.S. Pat. 2,406,717 (1947).
30. C. A. Thomas (to Monsanto Chemical Co.), U.S. Pat. 2,362,960 (1944).
31. R. M. Joyce (to Du Pont), U.S. Pat. 2,479,367 (1949).
32. A. L. Dittman (to M. W. Kellogg Co.), U.S. Pat. 2,689,241 (1954).
33. G. E. Ham, ed., *Copolymerization,* Interscience Publishers, a division of John Wiley & Sons, Inc., New York, 1964, p. 644.
34. T. J. Dougherty, *J. Am. Chem. Soc.* **86,** 460 (1964).
35. J. P. Stallings and R. A. Paradis, *J. Appl. Polymer Sci.* **14,** 461 (1970).
36. M. L. Wallach and M. A. Kabayama, *J. Polymer Sci.* [A-1] **4,** 2667–2674 (1966).
37. J. A. Proctor (to Du Pont), U.S. Pat. 3,096,299.
38. J. L. Koenig and F. J. Boerio, *Makromol. Chem.* **125,** 302–305 (1969).
39. G. Natta, I. Bassi, and G. Allegra, *Atti Accad. Nazl. Lincei-Rend. Classe Sci. Fis. Mat. e Nat.* **31,** 350–356 (1961).
40. C. W. Wilson, III, and E. R. Santee, Jr., *J. Polymer Sci.* [C] **1,** 97 (1965).
41. R. C. Golike, *J. Polymer Sci.* **42,** 582 (1960).
42. J. B. Lando, H. G. Olf, and A. Peterlin, *J. Polymer Sci.* [A] **4,** 941 (1966).
43. E. Sacher, *J. Polymer Sci.* [A-2] **6,** 1813 (1968).
44. K. Schmeider and K. Wolf, *Kolloid Z.* **134,** 149 (1953).

45. T. Honjo and S. Ogawa, *Kogyo Gijutsuin Sen'i Kogyo Shinkensho Kenkyo Hokoku* **65,** 1 (1963).
46. K. L. Paciorek, W. G. Lajiness, R. G. Spain, and C. T. Lenk, *J. Polymer Sci.* **61,** S-42 (1962).
47. J. J. Dietrich, T. E. Hedge, and M. E. Kucsma, *Paint Varnish Prod.* **56,** 75 (1966).
48. *Technical Bulletin TD-5,* E. I. du Pont de Nemours & Co., Inc., Wilmington, Del., March 1967.
49. V. L. Simril and B. A. Curry, *J. Appl. Polymer Sci.* **4,** 62–68 (1960).
50. R. Timmerman, and W. Greyson, *J. Appl. Polymer Sci.* **6,** 456–460 (1962).
51. T. E. Hedge (to Diamond Shamrock Corp.), U.S. Pat. 3,317,619 (1967).
52. A. L. Barney and W. Honsbere (to Du Pont), Fr. Pat. 1,578,405 (Aug. 1969).
53. I. E. Wolinski (to Du Pont), U.S. Pat. 3,255,099 (June 1966).
54. *Technical Information Bulletin TD-14,* E. I. du Pont de Nemours & Co., Inc., Wilmington, Del., April 1966.
55. *Adhesives Age,* p. 38, October 1962.
56. Deutsche Solvay Werke G.m.b.H., Brit. Pat. 1,004,172 (1965).
57. J. S. Proctor (to Du Pont), U.S. Pat. 3,110,692 (1963).
58. C. Neros (to Diamond Alkali (now Diamond Shamrock)), Brit. Pat. 1,059,597 (Feb. 1967).
59. *Brit. Plastics* **41,** 75 (1968).
60. G. R. McKay, *Rubber Plastics Age* **48,** 719–721 (1967).
61. J. Kestler, *Mod. Plastics* **46** (9), 82 (1969).
62a. A. Yazgi, *Poliplasti* **16,** 5–9 (1968).
62b. A. Yazgi, *Plast. Mod. Elast.* **20,** 68–70 (1968).
63. R. D. Johnson and C. H. M. van Bauel, *Sci.* **149,** 1377–1379 (1965).
64. *NASA Tech. Rept. 32-1411,* Jet Propulsion Lab, Pasadena, California, June 1969.
65. D. Wang (to Du Pont), U.S. Pat. 3,497,357.
66. J. G. Frielink, Brit. Pat. 1,161,958.
67. J. L. Koenig and J. J. Mannion, *J. Polymer Sci.* [A-2] **4,** 401–414 (1966).

Fredric S. Cohen
Polaroid Corporation
Paul Kraft
Stauffer Chemical Company

VINYL FORMAL POLYMERS. See Vinyl alcohol polymers

VINYL FORMATE POLYMERS. See Vinyl ester polymers, Vol. 15

VINYLIDENE CHLORIDE POLYMERS

Two important characteristics of polymers of vinylidene chloride, $CH_2{=}CCl_2$, are thermal stability and impermeability to gases and vapors. The present commercial success of these materials is due to the fact that the instability problem was overcome so that the barrier properties could be exploited. The techniques, copolymerization and plasticization, were developed by Ralph Wiley and co-workers in the laboratories of The Dow Chemical Company during the period of 1933–1939. The commercialization of these polymers under the trademark Saran began in 1939. The polymerization of vinylidene chloride had been observed nearly a century before this and the structure of the polymer had been correctly deduced before 1930, but, until the above-described development began, poly(vinylidene chloride) was primarily a laboratory curiosity (1).

The homopolymer (Saran A), although it has valuable properties, was not commercialized because of the difficulty in fabricating it. A great many copolymers were prepared, but only three types have stood the test of time: vinylidene chloride–vinyl chloride copolymers (Saran B), vinylidene chloride–alkyl acrylate copolymers (Saran C), and vinylidene chloride–acrylonitrile copolymers (Saran F).

Saran is now a generic term in the United States for polymers with high vinylidene chloride content. (However, elsewhere, Dow continues to maintain its proprietary rights in the Saran trademark and its registrations.) Unfortunately, the practice of using poly(vinylidene chloride) and Saran synonymously in the United States developed even in the very early days. As a consequence, many of the materials identified in the literature as poly(vinylidene chloride) (or "PVDC") were actually copolymers of unknown composition.

The chemistry and properties of vinylidene chloride polymers have been treated in several reviews. Reinhardt (1) describes the early work done at The Dow Chemical Company. Other general treatments include those of Staudinger (2) and Schildknecht (3).

Monomer

Physical Properties. Pure vinylidene chloride (1,1-dichloroethylene) is a colorless, mobile liquid with a characteristic "sweet" odor. Its properties are summarized in Table 1. Vinylidene chloride is soluble in most polar and nonpolar organic solvents. An azeotrope forms with 6% methanol (5). Distillation of the azeotrope, with subsequent water extraction of the methanol, is a method of purification. For safe handling procedures of this monomer, see Ref. 4.

Manufacture. Vinylidene chloride monomer may be conveniently prepared in the laboratory by the reaction of 1,1,2-trichloroethane with aqueous alkali (eq. 1).

$$2\,ClCH_2{-}CHCl_2 + Ca(OH)_2 \xrightarrow{90°C} 2\,CH_2{=}CCl_2 + CaCl_2 + H_2O \qquad (1)$$

This general method was used by early investigators and was the subject of later patents (6,7). Other methods are based on bromochloroethane (8), trichloroethylacetate (9), tetrachloroethane (10), and catalytic cracking of trichloroethane (11). Mixed dichloroethylenes are prepared by introducing acetylene into a liquid acid copper catalyst solution containing both cupric chloride and cuprous chloride (12). The addition of an alkali chloride and/or alkaline earth chloride modifies the process to make larger quantities of 1,1-dichloroethylene in addition to 1,2-dichloroethylene. Vinylidene chloride is prepared commercially by the dehydrochlorination of 1,1,2-trichloroethane by lime or caustic used in slight excess (2–10%) (13). A liquid-phase reaction, operated continuously at 98–99°C, yields approximately 90%. Performance with caustic is better than with lime.

Vinylidene chloride is purified by washing with water, drying, and fractional distillation. It is normally inhibited at this point. Two commercial grades are available: one with 200 ppm of methyl ether of hydroquinone and the other with 0.6–0.8 weight percent phenol. A typical analysis is shown in Table 2. Many inhibitors for the polymerization of vinylidene chloride have been described in patents (14–17), but the two listed above are most often used. For many polymerizations, the low level of methyl ether of hydroquinone need not be removed; it can be overcome by polymerization initiators. The higher level of phenol must be removed, either by alkali extraction or distillation. Vinylidene chloride with inhibitor removed should be refrigerated at $-10°C$, under nitrogen atmosphere, in the dark, in a nickel-lined or baked phenolic-lined storage tank. If not used within one day, it should be reinhibited.

Table 1. Physical Properties of Vinylidene Chloride (4)

Property	Value
molecular weight	96.95
boiling point at 760 mm Hg, °C	31.56
freezing point, °C	−122.5
density	
at −20°C	1.2902
at 0°C	1.2517
at 20°C	1.2132
refractive index, n_D	
at 10°C	1.43062
at 15°C	1.42777
at 20°C	1.42468
latent heat of vaporization, kcal/mole	
at 25°C	6.328
at boiling point	6.257
dielectric constant at 16°C	4.67
vapor pressure, mm Hg, $\log P$ mm $= 6.98200 - 1104.29/(T + 237.697)$	
at −20°C	80
at −10°C	135
at 0°C	215
at 20°C	495
at 30°C	720
specific heat, cal/g	0.275
heat of combustion, kcal/mole	261.93
heat of polymerization, kcal/mole	−18.0 ± 0.9
absolute viscosity, cP	
at −20°C	0.4478
at 0°C	0.3939
at 20°C	0.3302
flash point, °F	
Tag closed cup	55
Cleveland open cup	5
explosive mixture with air, % by vol	
lower limit	7.3
upper limit	16.0
autoignition temperature, °F	1058
solubility of monomer in water, wt % (25°C)	0.021
solubility of water in monomer, wt % (25°C)	0.035

Vinylidene chloride is produced commercially in the United States by The Dow Chemical Company, PPG Industries, and Vulcan Materials; in England, by Imperial Chemical Industries; in Germany, by Badische Anilin- und Soda-Fabrik and Chemische Werke-Huels; in Japan, by Asahi-Dow Chemical Company and Kureha Chemical Company.

Table 2. Typical Analysis of Vinylidene Chloride (4)

vinylidene chloride (excluding inhibitors), wt %	99.8
vinyl chloride, ppm	850
cis-1,2-dichloroethylene, ppm	10
trans-1,2-dichloroethylene, ppm	900
1,1-dichloroethane, ppm	<10
ethylene dichloride, ppm	<10
1,1,1-trichloroethane, ppm	150
trichloroethylene, ppm	<10
inhibitor, ppm	
methyl ether of hydroquinone	200
phenol	6000–8000

Homopolymerization

Vinylidene chloride polymerizes by both ionic and free-radical reactions (1,2,3,18). Processes based on the latter are, by far, the more common. Vinylidene chloride is of average reactivity when compared to other unsaturated monomers. The pair of chlorine substituents act to stabilize radicals in the intermediate state of an addition reaction. Since they are also strongly electron-withdrawing, they polarize the double bond making it susceptible to anionic attack. For the same reason, a carbonium-ion intermediate would not be favored.

The 1,1-disubstitution causes significant steric interactions in the polymer (19). This is evident from the heat of polymerization (see Table 1). When corrected for the heat of the fusion of the semicrystalline polymer produced, it is significantly less than the theoretical value of 20 kcal/mole for the process of converting a double bond to two single bonds. However, the steric strain is insufficient to prevent the easy polymerization of vinylidene chloride since depolymerization is favored only at high temperatures; the estimated ceiling temperature for poly(vinylidene chloride) is about 400°C (20).

The free-radical polymerization of vinylidene chloride has been carried out in slurry, suspension, and emulsion. Solution polymerization in a medium that dissolves both monomer and polymer has not been studied. Slurry polymerizations are normally used only at the laboratory level. They can be carried out in bulk or in common solvents like benzene. Poly(vinylidene chloride) is insoluble in these media and separates from the liquid phase as a crystalline powder. The heterogeneity of the reaction makes stirring and heat transfer difficult; as a consequence, these reactions cannot be easily controlled on a large scale. (Aqueous emulsion or suspension reactions are preferred for any large-scale operation.) Slurry reactions are usually initiated by the thermal decomposition of organic peroxides or azo compounds. Purely thermal initiation can occur, but rates are very slow (21). The spontaneous polymerization of vinylidene chloride so often observed when the monomer is stored at room temperature is due to peroxides formed by the reaction of vinylidene chloride with oxygen. Very pure monomer does not polymerize under these conditions. Ultraviolet (22) or gamma (23) radiation will induce polymerization.

Bulk Polymerization. The heterogeneous nature of the bulk polymerization of vinylidene chloride is apparent from the rapid development of turbidity in the reaction medium following initiation. The turbidity is due to the presence of minute poly(vinylidene chloride) crystals. As the reaction progresses, the crystalline phase grows at the expense of the liquid. Eventually, a point is reached where the liquid slurry

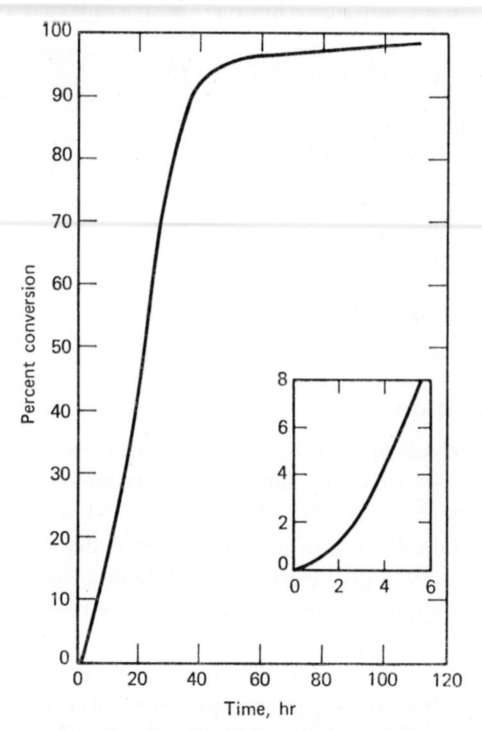

Fig. 1. Bulk polymerization of vinylidene chloride (1); 0.5% benzoyl peroxide at 45°C. The insert shows the behavior in the early stages.

solidifies into a solid mass. A typical conversion–time curve is shown in Figure 1 for a benzoyl peroxide-catalyzed bulk polymerization (1). The reaction can be divided into three stages. The first stage is characterized by a rapidly increasing rate. The rate levels off in the second stage to a fairly constant value, often called the steady-state region. Throughout the first two stages, monomer concentration remains constant because the polymer separates into another phase. The third stage shows a gradual decrease in rate to zero, owing to depletion of monomer supply. Since the mass solidifies while monomer is still present (usually at conversions below 20%), further polymerization generates void space. The final solid, therefore, is opaque and quite porous. A similar pattern of behavior is observed when vinylidene chloride is polymerized in solvents like benzene that do not dissolve or swell the polymer. In this case, of course, the reaction mixture may not solidify if the monomer concentration is low.

Heterogeneous polymerization is characteristic of a number of monomers, including vinyl chloride and acrylonitrile. A completely satisfactory mechanism for these reactions has not yet evolved. In the case of vinylidene chloride this is at least partly due to a lack of experimental data. Only two kinetic studies have been reported: in the work of Burnett and Melville (22), ultraviolet-radiation initiation in bulk and in cyclohexane was studied; Bengough and Norrish (24) investigated the benzoyl peroxide-initiated reaction. In neither case was the investigation broad enough to establish the mechanism.

Considerably more effort has been devoted to the study of the other monomers. A rather complete qualitative description of these cases is available in a recent review by Jenkins (25). Experimental studies of heterogeneous polymerization have convinced most investigators that the mechanism involves a two-phase reaction. That in the liquid follows the usual homogeneous kinetics, but with the possibility also of transferring radicals to the solid phase. This is a consequence of the low solubility of the polymer in the liquid phase. The interesting feature of this mechanism is the suggestion that further polymerization takes place in the solid phase. Radicals can continue to propagate the chains, but the rate of termination is reduced. The autoacceleration observed in these systems is attributed to such a solid-state reaction.

There is little doubt that the solid polymer phase is the key factor. No abnormalities are observed in the polymerization kinetics of these monomers when measured in a homogeneous system. Even when the amount of solvent present is only sufficient to convert the solid phase into a highly swollen gel, the kinetics are normal. This suggests that, even though the radicals in the solid may be immobilized, monomer is still free to diffuse to the reaction site.

Trapped Radicals. Radicals can be trapped in the solid phase under certain conditions. This phenomenon has been qualitatively described by the "occlusion" theory. When radicals become "occluded," or buried, in the solid phase, they enjoy unusually long life times. These radicals are responsible for the long "aftereffects" observed in heterogeneous acrylonitrile polymerizations. They can be made available for further reactions by swelling or heating the polymer. This leads to the so-called "fast" reaction. The presence of radicals in the solid phase was first suspected because of these anomalies. However, they have since also been detected by electron-spin resonance signals (25).

As pointed out above, there have been many attempts to derive rate equations for heterogeneous polymerization. None of these can be classified as completely satisfactory. Until recently, the structure of the solid phase was ignored, but experiments by Cotman and co-workers (26) now show this to be a significant factor. These authors have made a thorough comparison of the various theories proposed before 1967. They conclude that none is adequate; they suggest that the problem is probably unresolvable because of the complexity of the solid phase. The extension of theories derived for vinyl chloride or acrylonitrile to vinylidene chloride must be considered with caution. The similarity between these cases may be only superficial. The plots of conversion versus time follow the same characteristic pattern; but one cannot draw any conclusions from this fact alone.

Although evidence for trapped radicals in poly(vinyl chloride) and polyacrylonitrile is strong, the situation with poly(vinylidene chloride) is not so clear. Burnett and Melville, in their study using ultraviolet initiation, observed no abnormal photochemical aftereffects, and Bamford and co-workers could detect no radicals in poly(vinylidene chloride). Bawn and co-workers were able to generate trapped radicals by polymerizing with gamma rays (21). They found a very low post-irradiation rate and could only induce a slight "fast" reaction by heating. They concluded that the radicals were present in low concentration and could react only if the crystalline phase was partially destroyed. However, the experiments reported by Bengough and Norrish (24) demonstrate the presence of radicals in the solid phase.

Crystallinity and Solubility. The major differences between the polymerization of vinylidene chloride and the above-mentioned vinyl monomers involve both the respective polymer structures and the nature of the solid phase. Poly(vinylidene chloride) is well above its glass-transition temperature over the range of temperatures in which kinetics have been studied. The vinyl polymers, on the other hand, are converted to glassy solids. Poly(vinylidene chloride) is a hard solid only by virtue of its high crystallinity; poly(vinyl chloride) and polyacrylonitrile are only slightly crystalline.

Poly(vinylidene chloride) is highly crystalline because it has a symmetrical structure. This structure and the solubility of the polymer are independent of the temperature of polymerization. Talamini and Vidotto (27), in a study of the effect of polymerization temperature on molecular weight, found no abnormalities. The solubility of poly(vinylidene chloride) in its monomer or in common solvents in the temperature range normally used for polymerization studies (0–60°C) is virtually zero. The particles remain essentially unswollen and should contain no monomer except that adsorbed on their surfaces. The interior of the crystals ought to remain inaccessible unless the polymer is heated to a high enough temperature for diffusion into the crystalline phase to occur.

Kinetics. The kinetic studies, complicated by polymer precipitation, do not shed much light on the mechanism of reaction. In the photoinitiated reaction (22), however, the kinetics are normal in other respects. The rate is proportional to the half power of the light intensity and is first order in the monomer (reaction carried out in cyclohexane). The overall activation energy is 5 kcal/mole and molecular weight increases with reaction temperature. Abnormalities were observed only in the temperature dependence of the derived rate constants. This was attributed to heterogeneity.

Emulsion and Suspension Polymerizations. In emulsion and suspension reactions, the polymer is insoluble in the monomer, and the polymer and monomer are both insoluble in water. Suspension reactions are similar in behavior to slurry reactions. Oil-soluble initiators are employed so that conditions in the monomer–polymer droplet are analogous to a small mass reaction. Emulsion polymerizations are more complex. Since the monomer is insoluble in the polymer particle, the simple Smith-Ewart theory should not apply (28). This theory assumes that polymerization takes place in monomer-swollen polymer particles. It is more likely, in the case of vinylidene chloride, that polymerization occurs on the particle surface. See EMULSION POLYMERIZATION.

The first emulsion polymerization of vinylidene chloride was carried out before 1940. During the following decade, extensive kinetic studies were undertaken by Moll and co-workers at The Dow Chemical Company. These data have never been published, although some of the conclusions drawn from them were discussed by Rubens and Boyer (29). The first kinetic studies to appear in the literature were included in reviews by Staudinger in 1947 (2). He showed conversion–time curves for emulsion polymerization using both oil-soluble and water-soluble initiators. Much faster rates were obtained with the latter. The rate was found to increase with the initiator concentration in either case.

In 1951, Wiener (30) published a detailed account of the peroxydisulfate-initiated polymerization of vinylidene chloride, both in soap solution and in emulsion. He measured the solubility of vinylidene chloride in soap solutions and in latexes at 25°C

Table 3. Solubility of Vinylidene Chloride in the Presence of Soap at 25°C (30)

potassium laurate concentration, mole/liter	0	0.05	0.10	0.20
solubility of vinylidene chloride, mole/liter	0.066	0.079	0.098	0.144
mole vinylidene chloride per mole soap at saturation		0.26	0.32	0.39

by a vapor-pressure method. The values he obtained (Table 3) are surprisingly high. Vinylidene chloride appears to be somewhat more soluble than styrene in pure water.

Wiener's emulsion polymerization kinetic data fitted the empirical equation

$$R_p = k[\mathrm{I}]^{1/2} \tag{2}$$

where $k \propto [\mathrm{S}]^{3/5}$ and R_p is the polymerization rate and [I] and [S] are initiator and soap concentrations, respectively. These results are in agreement with the Smith-Ewart theory. A kinetic study of the reaction in soap solution showed that the rate was first order in monomer concentration. Surprisingly, this study uncovered no abnormalities in behavior.

A more extensive study of the batch emulsion polymerization of vinylidene chloride was reported by a group at Olin Mathieson Corporation (31–33). They used a redox initiator system, $(NH_4)_2S_2O_8$–$Na_2S_2O_5$, in a stirred kettle at reflux (~32°C)

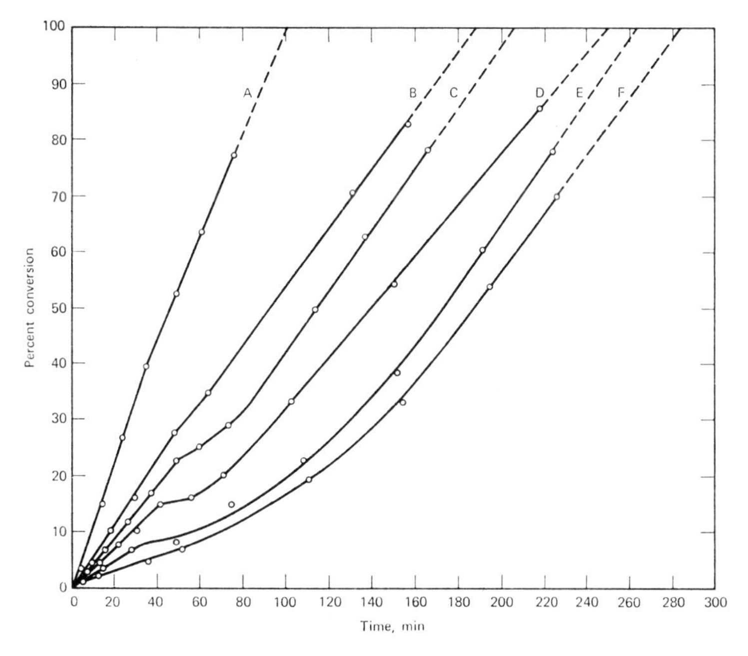

Fig. 2. Emulsion polymerization of vinylidene chloride. Effect of surfactant concentration on emulsion polymerization of vinylidene chloride at 36.0°C (31). Initial catalyst charge: 0.15 g $(NH_4)S_2O_3$, 0.15 g $Na_2S_2O_2$/100 g monomer. Sodium lauryl sulfate concentrations: curve A, 10.0 g; curve B, 5.0 g; curve C, 3.0 g; curve D, 2.0 g; curve E, 1.0 g; curve F, 0.50 g/100 g.

with sodium lauryl sulfate as emulsifier. A typical recipe used in this study was 100 g vinylidene chloride, 0.15 g $(NH_4)_2S_2O_8$, 0.15 g $Na_2S_2O_5$, 109.7 g water, 10 g soap, at a stirring speed of 285 rpm. The initiator level, soap concentration, and rate of stirring were varied. A typical set of conversion–time curves, showing the effect of emulsifier level, is plotted in Figure 2. The dependence of initial rates on the variables was in agreement with the Smith-Ewart theory (33).

Redox initiator systems are normally used in the emulsion polymerization of vinylidene chloride in order to get high rates at low temperatures. Reactions must be carried out below about 80°C to prevent degradation of the polymer. Poly(vinylidene chloride) in emulsion is also attacked by aqueous base. Therefore, reactions should be carried out at low pH.

The instability of poly(vinylidene chloride) is one of the reasons that ionic initiation of vinylidene chloride polymerization has not been used extensively. Many of the common catalysts either react with the polymer or catalyze its degradation. For example, butyllithium will polymerize vinylidene chloride by an anionic mechanism; but the product is a low-molecular-weight, discolored polymer with a low chlorine content (34). Cationic polymerization of vinylidene chloride seems unlikely in view of its structure (35). However, in the low-temperature radiation-induced copolymerization of vinylidene chloride with isobutylene, the reactivity ratios vary markedly with temperature, indicating a change from a simple free-radical mechanism (36).

Copolymerization

The homopolymer of vinylidene chloride has little commercial value; the importance of vinylidene chloride as a monomer is due to the utility of its copolymers with other vinyl monomers. It is most easily copolymerized with acrylates, but also reacts, though more slowly, with monomers like styrene that form highly resonance-stabilized radicals. This is in accord with its Q–e values (38,39) (see COPOLYMERIZATION). The double substitution with chlorine should place vinylidene chloride in the center of the Q–e map, where, in fact, it does appear (35). Its values are $Q = 0.22$, $e = 0.36$ (37). However, there is some question as to whether the Q–e scheme should apply to vinylidene chloride. Until recently experimental evidence favored the view that it should (39).

Reactivity ratios (r_1, r_2) for vinylidene chloride with the important comonomers are listed in Table 4. Values for other comonomers may be found in Ref. 37. These values are for free-radical copolymerizations. They are all based on the assumption that the simple copolymerization theory is applicable to vinylidene chloride. Deviations from the simple theory have been studied by spectroscopic analysis of the copolymers (36,40–50). A listing of polymers for which no copolymerization parameters have been determined may be found in Ref. 51.

Table 4. Reactivity Ratios of Vinylidene Chloride (M_1) in Copolymerization (37)

Monomer, M_2	r_1	r_2	Temp, °C
styrene	0.14 ± 0.05	2.0 ± 0.1	60
vinyl chloride	3.2	0.3	60
acrylonitrile	0.37 ± 0.10	0.91 ± 0.10	60
methyl acrylate	1.0	1.0	60–70
methyl methacrylate	0.24 ± 0.03	2.53 ± 0.01	60
vinyl acetate	6	0.1	68

The commercially important copolymers include those with vinyl chloride, acrylonitrile, or various alkyl acrylates, but many commercial saran polymers contain three or more components, vinylidene chloride being the major one. Usually, one component is introduced to improve the processability or solubility of the polymer. The others are added to modify specific end-use properties. Most of these compositions have been described in the patent literature. A partial listing is also given in Ref. 51.

Bulk copolymerizations yielding copolymers high in vinylidene chloride content are normally heterogeneous. Two of the most important pairs, vinylidene chloride–

vinyl chloride and vinylidene chloride–acrylonitrile, are heterogeneous over the entire composition range. In both cases, at either composition extreme, the product separates initially in a powdery form, but for intermediate compositions, the reaction mixture may only gel. Copolymers in this composition range are swollen but not dissolved completely by the monomer mixture at normal polymerization temperatures. Copolymers containing greater than 10–15 mole % acrylate are normally soluble. These reactions are therefore homogeneous, and if carried to completion, yield clear solid castings of the copolymer. Most copolymerizations can be carried out in solution because of the greater solubility of the copolymers in common solvents.

During copolymerization, one monomer may add to the copolymer more rapidly than the other. Thus, except for the unusual case when reactivity ratios are both unity, batch reactions carried to completion yield polymers of broad composition distribution. This is generally an undesirable result.

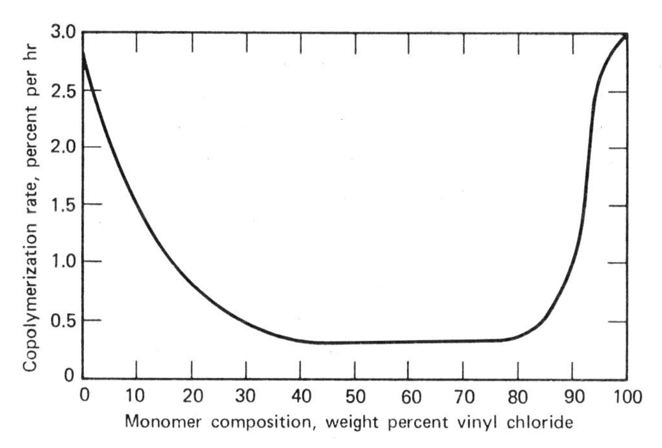

Fig. 3. Copolymerization rate of vinylidene chloride and vinyl chloride as a function of monomer composition (1). Catalyst, 0.50% benzoyl peroxide; temperature, 45°C; dark.

Vinylidene chloride copolymerizes randomly with methyl acrylate and nearly so with other acrylates. Very severe composition drift occurs, however, in copolymerizations with vinyl chloride or methacrylates. Several methods have been developed to produce homogeneous copolymers regardless of reactivity ratio. These methods are applicable mainly to emulsion and suspension methods where adequate stirring can be maintained (see the section on Commercial Methods of Polymerization, pp. 550 ff.).

The rates of copolymerization of vinylidene chloride with small amounts of a second monomer are normally lower than the rate of homopolymerization. The dependence of rate on monomer composition for the vinylidene chloride–vinyl chloride system is shown in Figure 3. For details of the kinetics of this system see Refs. 52 and 53.

Copolymers of vinylidene chloride can also be prepared by methods other than free-radical polymerization. The possible cationic copolymerization by irradiation of vinylidene chloride and isobutylene at low temperatures has already been mentioned. On the other hand, irradiation of vinylidene chloride and styrene at low temperatures yields copolymers with compositions characteristic of free-radical-initiated polymerizations (23). Copolymers have been prepared by various kinds of organometallic catalysts (54–57). The butyllithium-initiated reactions are clearly anionic on the basis of the observed reactivity ratios (55). The mechanism by which more complex catalyst systems operate is not well established (56,57).

Ziegler-type catalysts can be used to make block copolymers of vinylidene chloride (58). This method is of recent origin and has not yet been extensively developed. Graft copolymers containing vinylidene chloride can be made, but have also been little studied. They are normally prepared by free-radical methods. Irradiation of vinylidene chloride vapor in the presence of a polymeric substrate produces a graft (59). Polymerization of vinylidene chloride in the presence of vinyl acetate copolymers using chemical initiators also yields grafted products (60). Vinyl chloride can be grafted to a vinylidene chloride–olefin copolymer in the same way (61). None of these materials has been well characterized.

Commercial Methods of Polymerization

Industrial processes are essentially modifications of the laboratory methods already described, to allow operation on a larger scale. The nature of the end use dictates, to some extent, the method of manufacture. Emulsion polymerization (qv) and suspension polymerization (qv) are the preferred industrial processes. Either process is carried out in a closed, stirred reactor that is glass-lined and jacketed for heating and cooling. The reactor must be purged of oxygen and the water and monomer must be free of metallic impurities to avoid an adverse effect on the thermal stability of the polymer.

Emulsion polymerization is used commercially to make vinylidene chloride copolymers, which are utilized in two ways. In some applications, the latex which results is used directly (usually with additional stabilizing ingredients) as a coating vehicle; in others, the polymer is first isolated from the latex as a dry powder before use. In applications where the polymer is not used in latex form, the emulsion process is used because it gives advantages over alternative methods that outweigh the disadvantages. In those cases, the polymer is recovered, usually by coagulating the latex with electrolyte, followed by washing and drying.

The principal advantages of emulsion polymerization as a process for preparing vinylidene chloride polymers are: (a) High-molecular-weight polymers can be produced in reasonable reaction times (especially pertinent to copolymers with vinyl chloride). The initiation and propagation steps can be controlled more independently than is the case for the suspension process. (b) Monomer can be added during the polymerization to maintain copolymer composition control.

The disadvantages of emulsion polymerization essentially result from the relatively high concentration of additives in the recipe. The water-soluble initiators, activators, and surface-active agents generally cause the polymer to have greater water sensitivity, poorer electrical properties, and poorer heat and light stability. Recovery of the polymer by coagulation, washing, and drying will, to some extent, improve these properties over those of the polymer deposited in latex form.

A typical recipe for batch emulsion polymerization is given in Table 5 (62). Reaction times of 7–8 hr at 30°C are required for 95–98% conversion. A latex is produced with an average particle diameter of 1000–1500 Å. In addition to the ingredients given in Table 5, other modifying ingredients may be present, such as other colloidal protective agents (eg, gelatin or carboxymethylcellulose), initiator activators (redox types), chelates, plasticizers, stabilizers, and chain-transfer agents.

Table 5. Typical Recipe for Batch Emulsion Polymerization (62)

	Parts by weight
vinylidene chloride	78
vinyl chloride	22
water	180
potassium persulfate	0.22
sodium bisulfite	0.11
dihexyl sodium sulfosuccinate, 80%	3.58
nitric acid, 69%	0.07

Surfactants used commercially are generally anionic, alone or in combination with nonionic types. Representative anionic emulsifiers are the sodium alkylarylsulfonates, the alkyl esters of sodium sulfosuccinic acid, and the sodium salt of fatty alcohol sulfates. Nonionic emulsifiers are of the ethoxylated alkylphenol type.

Free-radical sources other than persulfates may be used, such as hydrogen peroxide, organic hydroperoxides, perborates, and percarbonates. Many of these are used in redox pairs where an activator promotes the decomposition of the peroxy compound (qv). Examples are persulfate or perchlorate activated with bisulfite, hydrogen peroxide with metallic ions, and organic hydroperoxides with sodium formaldehyde sulfoxylate. The use of activators causes the decomposition of initiator to occur at lower reaction temperatures, which allows the preparation of higher-molecular-weight polymer within reasonable reaction times. This is an advantage in the particular case of copolymers of vinylidene chloride with vinyl chloride. The use of oil-soluble initiators has been reported, but they are usually effective only when activated by a water-soluble activator or reducing agent.

Suspension polymerization is used commercially for vinylidene chloride polymers that are utilized as molding and extrusion resins. The principal advantage of the suspension process over the emulsion process is the use of fewer ingredients that can detract from the polymer properties. Stability is improved and water sensitivity is decreased. Extended reaction times and the practical difficulty of preparing higher-molecular-weight polymers are disadvantages of the suspension as compared to the emulsion process, particularly with copolymers with vinyl chloride.

A typical recipe for suspension polymerization is given in Table 6 (63). At a reaction temperature of 60°C, the polymerization proceeds to 85–90% conversion in 30–60 hr. Unreacted monomer is removed by vacuum pumps, condensed, and reused after processing. The polymer is obtained in the form of small (30–100 mesh) beads that are dewatered by a centrifuge and then dried in a flash dryer or fluidized-bed dryer. Suspension polymerization employs monomer-soluble initiators and polymerization occurs inside suspended monomer droplets which are formed by the shearing action of the agitator and are prevented from coalescence by the protective colloid.

Table 6. Typical Recipe for Suspension Polymerization (63)

	Parts by weight
vinylidene chloride	85
vinyl chloride	15
deionized water	200
(methyl)(hydroxypropyl) cellulose, 400 cP	0.05
lauroyl peroxide	0.3

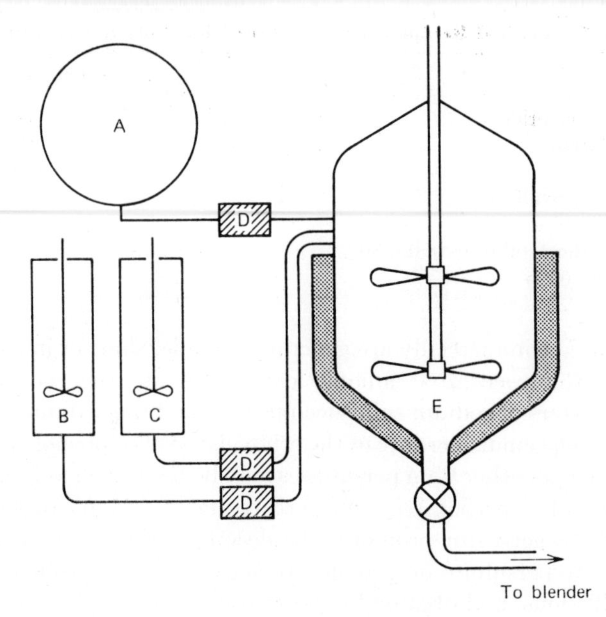

Fig. 4. Apparatus for continuous addition emulsion polymerization of vinylidene chloride–acrylate mixture (65). Key: A, monomer feed; B, catalyst feed; C, activator feed; D, metering pump; E, reactor.

It is important that the initiator be uniformly dissolved in the monomer before droplet formation. Unequal distribution of initiator will cause some droplets to polymerize faster than others, leading to monomer diffusion from slowly polymerizing droplets to rapidly polymerizing droplets. The latter will form polymer beads that are dense, hard, and glassy, and extremely difficult to fabricate, owing to their inability to accept stabilizers and plasticizers. Common protective colloids that prevent droplet coalescence and control particle size are poly(vinyl alcohol), gelatin, and methylcellulose. Organic peroxides, percarbonates, and azo compounds are used as initiators for vinylidene chloride suspension polymerization.

The above-described processes do not compensate for composition drift. Constant-composition processes have been designed using either emulsion or suspension reactions. It is more difficult to design controlled composition processes using suspension methods. In one approach (64), the less reactive component is removed continuously from the reaction in order to keep the unreacted monomer composition constant. This method has been used effectively in a case like the vinylidene chloride–vinyl chloride copolymerization where the more slowly reacting component is a gas and can be bled off during the reaction to maintain constant pressure. In many other cases, there is no practical way of removing the more slowly reacting component.

The method of continuous-addition emulsion polymerization is quite general. In this process, one or more components are metered continuously into the reaction. If the system is properly balanced, a steady state is reached in which a copolymer of uniform composition is produced. Woodford (65) has described a process of this type in some detail. The apparatus is shown in Figure 4. A typical recipe is given in Table 7.

The monomers are charged to the weigh tank A which is kept under a nitrogen blanket. The soaps (components *3* and *4*), initiator (*5*), as well as part of the water

Table 7. Typical Recipe for Continuous-Addition Emulsion Polymerization

	Parts by weight
1. vinylidene chloride	468
2. comonomer	52
3. Tergitol NP 35[a] nonionic wetting agent	12
4. sodium lauryl sulfate, 25%	12
5. ammonium persulfate, 5%	10
6. sodium metabisulfite, 5%	10
7. water	436

[a] Registered trademark of Union Carbide Corp.

are charged to tank B, and the reducing agent and some water to tank C. The remaining water is charged to the reactor and the system is sealed and purged. The temperature is raised to 40°C and one-tenth of the monomer and initiator charges are added. Then one-tenth of the activator is pumped in. When the reaction is underway, as indicated by an exotherm and pressure drop, feeds of A, B, and C are started at programmed rates, slowly at first and then at gradually increasing rates. The emulsion is maintained at constant temperature during the run by pumping cooling water through the jacket. When all components have been fed in and the exotherm begins to subside, an end shot of initiator and reducing agent are added to complete the reaction. This process produces a latex suitable for coating paper, plastic film, and other substrates.

Structure

The chemical composition of poly(vinylidene chloride) has been confirmed by various techniques including elemental analysis, x-ray diffraction analysis, infrared, raman, and NMR spectroscopy, and degradation studies. The polymer chain is made up of vinylidene chloride units added head to tail (**1**). Since the repeat unit is sym-

$$\sim\!\!\sim CH_2-CCl_2-CH_2-CCl_2-CH_2-CCl_2-\sim\!\!\sim$$

(**1**)

metrical, no possibility exists for stereoisomerism. Variations in structure can come about only by head-to-head addition, branching, or degradation reactions that do not cause chain scission. This includes such reactions as thermal dehydrochlorination, which creates double bonds in the structure (**2**) or a variety of ill-defined oxida-

$$\sim\!\!\sim CH_2-CCl_2-CH{=}CCl-CH_2-CCl_2-\sim\!\!\sim$$

(**2**)

tion and hydrolysis reactions which generate carbonyl groups. The infrared spectra of poly(vinylidene chloride) often show traces of both unsaturation and carbonyl absorption. The slightly yellow tinge of many of these polymers comes from the same source; the pure polymer is colorless. Elemental analyses for chlorine are normally slightly lower than the theoretical value, which is 73.14%.

The high crystallinity of poly(vinylidene chloride) indicates that significant amounts of head-to-head addition, or branching, cannot be present. This has been confirmed by an examination of the NMR spectra (41). There is only one peak in the spectrum due to the methylene hydrogens. Either branching or another mode of addition would produce nonequivalent hydrogens and a more complicated spectrum. However, the method would not be able to detect small amounts of such structures.

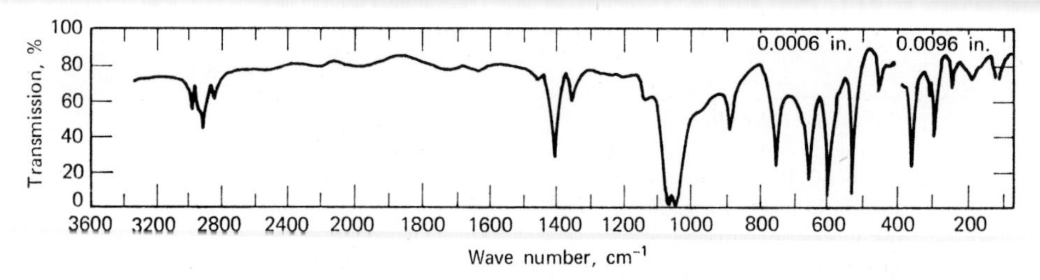

Fig. 5. Infrared spectrum of unoriented poly(vinylidene chloride) (67).

The infrared spectra can also be interpreted in terms of a simple head-to-tail structure. A typical spectrum, obtained with highly crystalline poly(vinylidene chloride) powder, is shown in Figure 5. Band assignments have been proposed (66–69). An interpretation of the spectra is complicated by the fact that many bands are sensitive to polymer crystallinity, particularly those at 1360, 1046, 887, and 754 cm^{-1}. The ratio of bands, such as 1046 cm^{-1} to a band proportional to film thickness such as 1070 cm^{-1}, can be used as a measure of relative crystallinity (70). The assignment of many bands has been made by comparison of crystalline and amorphous copolymer spectra. Poly(vinylidene chloride) itself is difficult to prepare in the form of an amorphous film and, as a consequence, cannot be studied directly.

Although the chemical composition of the poly(vinylidene chloride) chain is well-established, almost nothing is known about its size or size distribution. No absolute measurements of molecular weights have ever been reported. Viscosity studies indicate, however, that the chain is unbranched and can be made with degrees of polymerization which are estimated to lie in the range from 100 to well over 10,000 ([η] from 0.01 to >2.0 dl/g) (71).

The interpretation of viscosity is based mainly on comparison with copolymers which can be studied by conventional dilute-solution methods such as light scattering. (There is no suitable solvent for poly(vinylidene chloride) in which these measurements could be made.) Both molecular weights and distributions are typical of what would be expected from a free-radical polymerization.

Crystal Structure. The crystal structure of poly(vinylidene chloride) has not been completely solved. Several unit cells have been proposed. The data are collected in Table 8. The best representation is probably that of Okuda (72,74). The unit cell contains four monomer units with two monomer units per repeat distance. The calculated density is higher than the experimental values, which range from 1.80 to 1.94 g/cc (25°C) depending on sample history. This is usually the case with crystalline polymers because samples of 100% crystallinity cannot normally be obtained. A direct calculation of the polymer density from volume changes during polymerization yields a value of 1.97 g/cc (75). If this value is correct, the unit cell densities may be on the low side.

Table 8. Crystallographic Data for Poly(vinylidene Chloride) (37)

Crystal system	Space group	*a*, Å	*b*, Å	*c*, Å	*β*, degrees	Repeat units per cell	Density, g/cc
monoclinic		13.69	4.67	6.296	123.8	4	1.949
monoclinic		22.54	4.68	12.53	84.2	16	1.959
monoclinic	C2-2	6.73	4.68	12.54	123.6	4	1.96

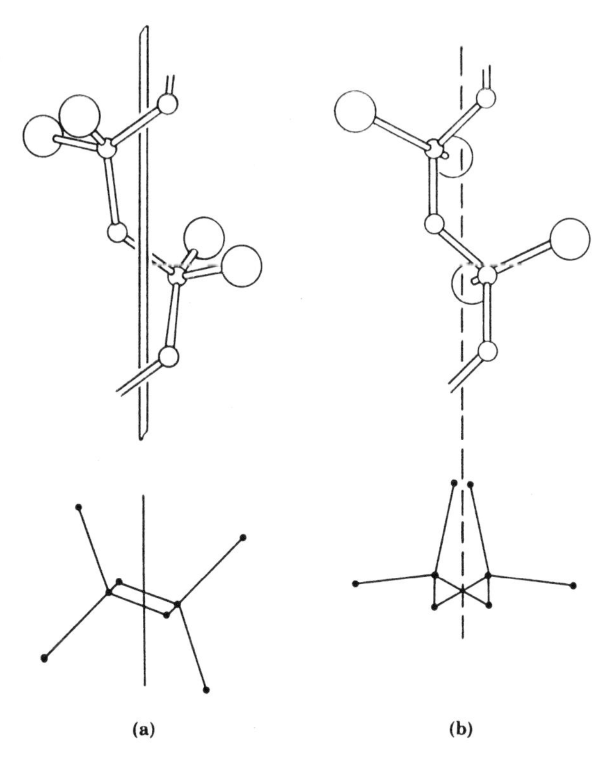

Fig. 6. Chain model of poly(vinylidene chloride): (**a**) twofold helix; (**b**) conformation with a glide plane as proposed by Miyazawa and Ideguchi (80).

The repeat distance along the chain axis, 4.68 Å, is significantly less than that calculated for a planar zigzag structure. Therefore, the polymer must be in some other conformation; a number of conformations has been proposed, but there is no agreement as yet on which is correct. Frevel (73) proposed a 2_1 helix with expanded bond angles. This structure gave the best agreement with his x-ray diffraction data. The conformation is sketched out in Figure 6. Fuller (76) suggested a staggered planar zigzag, but noted that a helical structure would also be consistent with his x-ray data. DeSantis and co-workers (77,78) conclude from x-ray data, conformational energy analysis, and optical Fourier transforms of the various models that the helical conformation is correct. Okuda (79) has also carried out a partial structure analysis using x-ray diffraction methods and concludes that the *tgtg'* (a modified version of the Fuller structure) conformation proposed by Miyozawa and Ideguchi (80) is more likely. The latter authors deduced this conformation from a study of the infrared spectra. Other infrared spectroscopists have reached a similar conclusion (66). Additional support for this conformation has come from a recent analysis of the laser-excited Raman spectra of poly(vinylidene chloride) (81).

Morphology and Transitions. The highly crystalline particles of poly(vinylidene chloride) that precipitate during polymerization are aggregates of thin lamellar crystals (82). The substructures are 50–100 Å thick and 100 or more times larger in the other dimensions. In some respects, they resemble the lamellar crystals grown from dilute solution (72). Not much is known about either type. Both appear to be built up of folded chains oriented normal to the lamellar surface.

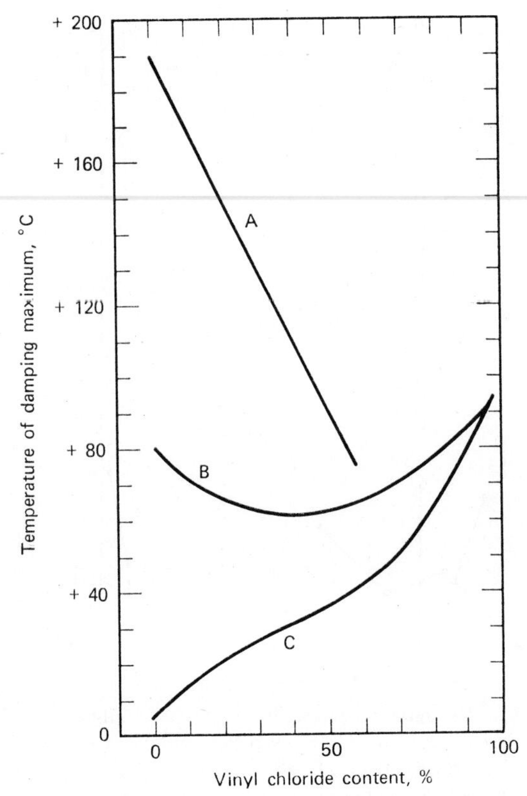

Fig. 7. Temperature positions of the damping maxima of a series of vinylidene chloride–vinyl chloride copolymers as function of composition (by weight) (87). Curve A, mean temperature of the melting range of the crystalline portions; curve B, mean temperature of the softening range of the strained amorphous portions; curve C, mean temperature of the softening range of the amorphous portions.

Melting points of "as-polymerized" powders are high, ranging from 198 to 205°C when measured by differential thermal analysis (qv) or hot-stage microscopy. A lower melting point is often observed at high heating rates, indicating that the powders may be annealing during the measurement. "As-polymerized" poly(vinylidene chloride) does not show a well-defined glass-transition temperature owing to its high crystallinity, but a sample can be melted at 210°C and quenched rapidly to an amorphous state at −20°C. The amorphous polymer shows a glass transition at −17°C by dilatometry (83). Values ranging from −19 to −11°C have been observed, depending both on methods of measurement and on sample preparation.

Once melted, poly(vinylidene chloride) does not regain its "as-polymerized" morphology when subsequently crystallized, but recrystallizes into a spherulitic habit (84). Quenching and low-temperature annealing generate many small nuclei which, on heating, grow rapidly into small spherulites. Slow crystallization at higher temperatures produces fewer, but much larger, spherulites (85).

The melting point and degree of crystallinity of recrystallized poly(vinylidene chloride) also depends on crystallization conditions. T_m increases with crystallization temperature, but the "as-polymerized" value cannot be achieved. There is no reason to believe that even these values represent the true melting point of poly(vinylidene

Table 9. Transition Temperatures and Associated Properties of Poly(vinylidene Chloride)

Property	Preferred value	Range
melting point, °C	202	198–205
glass-transition temperature, °C	−17	−19 to −11
alpha transition, °C	80	
density (amorphous), 25°C, g/cc	1.775	1.67–1.775
density (unit cell), g/cc	1.96	1.949–1.96
density (crystalline) at 25°C		1.80–1.97
refractive index (cryst), n_D^{25}	1.63	
heat of fusion, ΔH, cal/mole	1500	1100–1900

chloride). It may be as high as 220°C, but slow, high-temperature recrystallization and annealing experiments are not feasible owing to the thermal instability of the polymer (86).

Other transitions in poly(vinylidene chloride) have been observed by dynamic methods. Schmieder and Wolf (87), using a torsion pendulum, observed a higher temperature relaxation process in poly(vinylidene chloride) and copolymers with vinyl chloride which they associated with strained amorphous regions. It appears at about 80°C in poly(vinylidene chloride) and decreases with vinyl chloride content (Fig. 7). McCrum (88) identifies this as an α transition. The β transition in crystalline poly(vinylidene chloride) occurs around 12–20°C depending on the frequency of the measurement. The β peak corresponds to T_g, but these values are substantially higher than T_g values obtained by slower rate measurements. This is too large a difference to be accounted for solely by rate differences, indicating a possible dependence of T_g on crystallinity. Both α and β transitions can be detected by broad-line NMR (89) and by dielectric measurements (90), and are found at approximately the same temperature as the mechanical-loss peak, when compared at the same frequencies. Collected data on transitions are given in Table 9.

Crystallinity has been measured by various methods including x-ray, calorimetry, and from densities; the results are in substantial agreement (91). Infrared can also be used to measure changes in crystallinity. This method gives a relative degree of crystallinity and has not been calibrated with absolute methods. See also CRYSTALLINITY.

Copolymers. The properties of poly(vinylidene chloride) are usually modified by copolymerization. Copolymers of high vinylidene chloride content have lower melting points and higher glass-transition temperatures than the homopolymer. Copolymers containing more than about 15 mole % acrylate or methacrylate are amorphous. Substantially more acrylonitrile, ~25%, or vinyl chloride, ~45%, is required to destroy crystallinity completely. The effect of different types of comonomers on T_m is shown in Figure 8. The vinylidene chloride–methyl acrylate copolymers obey Flory's melting-point-depression theory and give a reasonable value for the heat of fusion. The theory does not seem to hold for the vinylidene chloride–vinyl chloride system (72). This is most likely caused by the fact that vinyl chloride units can enter into the poly(vinylidene chloride) crystal structure as defects. Consequently, vinyl chloride is less effective in lowering T_m.

The glass-transition temperatures of saran copolymers have been studied by several authors (65,83,92–94). Some of the data are shown in Figure 9. In every case, T_g

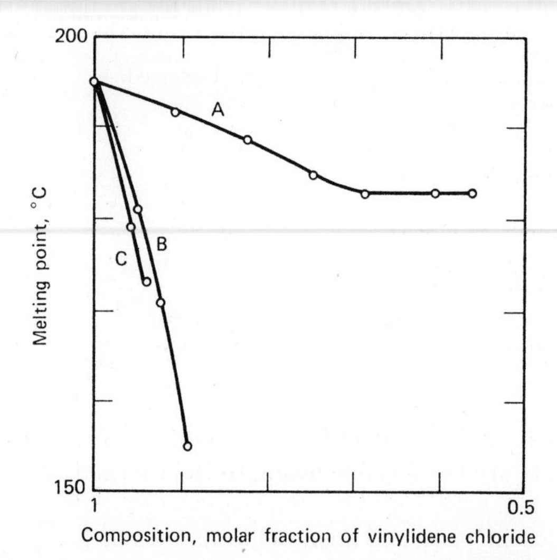

Fig. 8. Melting points plotted against composition for vinylidene chloride–vinyl chloride (curve A), vinylidene chloride–methyl acrylate (curve B), and vinylidene chloride–octyl acrylate (curve C) copolymers (72).

increases with comonomer content at low comonomer levels even in cases where the T_g of the homopolymer of the other component is lower. In some cases, a maximum T_g is observed at intermediate compositions. In others, where the T_g of the homopolymer is much higher than the T_g of poly(vinylidene chloride), the glass-transition tempera-

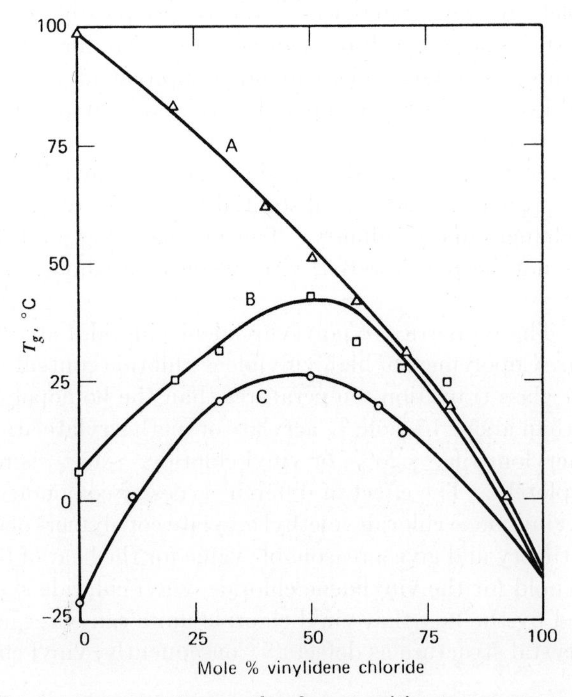

Fig. 9. Effect of comonomer structure on the glass-transition temperatures of saran copolymers (94). Key: curve A, vinylidene chloride–acrylonitrile; curve B, vinylidene chloride–methyl acrylate; curve C, vinylidene chloride–ethyl acrylate.

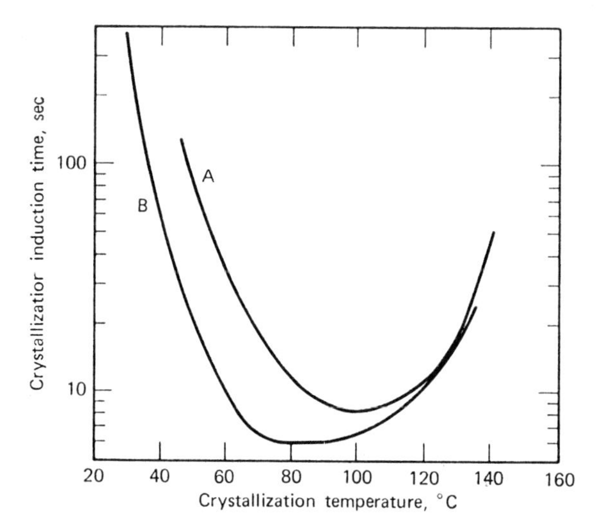

Fig. 10. Typical crystallization induction period curves for a normally crystalline vinylidene chloride–vinyl chloride copolymer (1). Curve A, no plasticizer; curve B, 7% plasticizer. Specimen thickness, 0.050 in.

tures of the copolymers increase over the entire composition range. Even here, there are differences in behavior; T_g increases most rapidly at low acrylonitrile levels, but changes most slowly at low vinyl chloride levels. This suggests that polar interactions affect the former; but the increase in T_g in the vinylidene chloride–vinyl chloride copolymers may be simply due to loss of chain symmetry. Because of these effects, the temperature range in which copolymers can crystallize is drastically narrowed. Crystallization induction times are prolonged and subsequent crystallization takes place at a low rate over long periods of time. As shown in Figure 10, plasticization, which lowers T_g, decreases crystallization induction times significantly. The 10 mole % butyl acrylate copolymer has a T_g of 8°C; the 10% octyl derivative, −3°C; and the 10% octadecyl derivative, −16°C. The rates of crystallization of these copolymers are inversely related to the glass-transition temperatures (95). Apparently, the long alkyl side chains act as internal plasticizers, lowering the melt viscosity of the copolymer even though the acrylate group acts to raise it. See also KINETICS OF CRYSTALLIZATION.

The more common crystalline copolymers show their maximum rates of crystallization in the range of 80–120°C. In many cases, these have broad composition distributions containing both fractions of high vinylidene chloride content which crystallize rapidly, and other fractions that do not crystallize at all. Poly(vinylidene chloride) itself probably crystallizes at a maximum rate at 140–150°C, but the process is difficult to follow because of severe polymer degradation. The copolymers may remain amorphous for considerable periods of time if quenched to room temperature. The induction time before the onset of crystallization depends on both the type and amount of comonomer; the homopolymer crystallizes within minutes at 25°C.

Orientation or mechanical working accelerates crystallization and has a pronounced effect on morphology. Okuda (72) found that in uniaxially oriented filaments the crystals are oriented along the fiber axis. The long period, by small-angle scattering, is 76 Å and decreases with comonomer content. The fiber is 43% crystalline and has a melting point of 195°C with an average crystal thickness of 45 Å. The crystal size is not greatly affected by comonomer content, but both crystallinity and melting point fall.

Copolymerization also affects morphology under other crystallization conditions. Copolymers in the form of cast or molded sheets are much more transparent because of the small spherulite size. In extreme cases, crystallinity cannot even be detected optically, but its effect on mechanical properties is pronounced. Before crystallization, films are soft and rubbery with low modulus and high elongation. After crystallization, they are leathery and tough with a higher modulus and lower elongation.

Significant amounts of comonomer also reduce the ability of the polymer to form lamellar crystals from solution. In some cases, the polymer merely gels the solution as it precipitates rather than forming distinct crystals. At somewhat higher vinylidene chloride content, it may precipitate out but in the form of aggregated, ill-defined particles and clusters.

Properties

Solubility. Poly(vinylidene chloride), like many high-melting polymers, does not dissolve in most common solvents at ambient temperatures. Copolymers, particularly those of low crystallinity, are much more soluble. Nevertheless, one of the outstanding characteristics of saran polymers is their resistance to a wide range of solvents and chemical reagents.

The insolubility of poly(vinylidene chloride) is due less to its polarity than to its high melting point. It dissolves readily in a wide variety of solvents at temperatures above 130°C (86). The polarity of the polymer is important only in mixtures with specific polar aprotic solvents; many solvents of this general class are able to solvate poly(vinylidene chloride) strongly enough to depress the melting point more than 100°C.

Solubility is normally correlated with cohesive-energy densities (qv) or solubility parameters (see SOLUBILITY). Burrell has reported a value of 12.2 for the solubility parameter of poly(vinylidene chloride) (96). This is much higher than a value estimated from solubility studies in nonpolar solvents, 10.1 ± 0.3, or calculated from Small's relationship, 10.25. The latter values are more likely to be correct. In any case, the use of the solubility parameter scheme for polar crystalline polymers like poly(vinylidene chloride) is of limited value.

Poly(vinylidene chloride) will dissolve in a variety of nonpolar solvents of matching solubility parameter at 130°C or above (82). The specific polar aprotic solvents, on the other hand, can dissolve poly(vinylidene chloride) at much lower temperatures. A list of both good nonpolar solvents and specific solvents is given in Table 10. The relative solvent activity is characterized by the temperature at which a 1% mixture of polymer in the solvent becomes homogeneous when heated rapidly.

Poly(vinylidene chloride) also dissolves readily in certain solvent mixtures (97). One component must be a sulfoxide or amide of the type described above. Effective cosolvents are less polar and have cyclic structures. They include both aliphatic and aromatic hydrocarbons, ethers, sulfides, and ketones. Acidic or hydrogen-bonding solvents have an opposite effect, rendering the polar aprotic component less effective. Both hydrocarbons and hydrogen-bonding solvents are nonsolvents for poly(vinylidene chloride).

Table 10. Solvents for Poly(vinylidene Chloride)

Solvents	Temperature,[a] °C
Nonpolar	
1,3-dibromopropane	126
bromobenzene	129
α-chloronaphthalene	134
2-methylnaphthalene	134
o-dichlorobenzene	135
tetrahydronaphthalene[b]	142
Polar aprotic	
hexamethylphosphoramide	−7.2
tetramethylene sulfoxide	28
N-acetylpiperidine	34
N-methylpyrrolidone	42
N-formylhexamethylenimine	44
trimethylene sulfide	74
N-*n*-butylpyrrolidone	75
isopropyl sulfoxide	79
N-formylpiperidine	80
N-acetylpyrrolidine	86
tetrahydrothiophene	87
N,*N*-dimethylacetamide	87
cyclooctanone	90
cycloheptanone	96
n-butyl sulfoxide	98

[a] Temperature at which a 1% mixture of polymer in solvent becomes homogeneous.
[b] Mixed isomers.

As polymerized, poly(vinylidene chloride) is not in its most stable state; annealing and/or recrystallization can raise the temperature at which it dissolves. Polymers of low crystallinity can be dissolved at a lower temperature to form metastable solutions. On standing at the solution temperature, they will gel or become turbid, indicating precipitation.

Copolymers with a high enough vinylidene chloride content to be quite crystalline, behave in a manner similar to that of poly(vinylidene chloride). They are more soluble, however, because of their lower melting points. The solubility of amorphous copolymers is much higher. The selection of solvents, in either case, varies somewhat with the type of comonomer. Some of the more common types are listed in Table 11.

Table 11. Common Solvents for Saran Copolymers

Solvent	Copolymer type	Range, °C
tetrahydrofuran	all	<60
methyl ethyl ketone	low crystallinity	<80
1,4-dioxane	all	50–100
cyclohexanone	all	50–100
cyclopentanone	all	50–100
ethyl acetate	low crystallinity	<80
chlorobenzene	all	100–130
dichlorobenzene	all	100–140
dimethylformamide	high acrylonitrile	<100

Solvents that dissolve poly(vinylidene chloride) will also dissolve the copolymers, but at lower temperatures.

Dilute-Solution Properties. Solution properties of poly(vinylidene chloride) have not been studied to any extent. The few data that are available have been obtained at high temperatures. Molecular weights of copolymers have been measured in tetrahydrofuran, methyl ethyl ketone, and *o*-dichlorobenzene, among others. Techniques include viscometry (qv), osmometry (qv), light scattering (qv), and gel-permeation chromatography (see FRACTIONATION). NMR spectra are obtained in either *o*-dichlorobenzene or bromobenzene at high temperatures. Acrylate copolymers behave like typical flexible-backbone, vinyl-type polymers. The length of the acrylate ester side chain has little effect on properties (98). Some are shown in Table 12.

Table 12. Chain Dimensions of Saran Copolymers

Copolymer, mole %	Theta temp, °C	Solvent	K_θ [a]	$\left(\frac{r_0^2}{M_w}\right)^{1/2}$ [b]	$\left(\frac{r^2}{r_0^2}\right)^{1/2}$ [c]
ethyl acrylate, 14.9%	49.6	ethyl acetoacetate	6.64×10^{-4}	6.81×10^{-9}	2.18
n-butyl acrylate, 16.7%	44.0	benzyl alcohol	7.30×10^{-4}	7.03×10^{-9}	2.31
n-hexyl acrylate, 14.5%	56.8	benzyl alcohol	5.50×10^{-4}	6.04×10^{-9}	2.14
n-octyl acrylate, 15.6%	77.9	benzyl alcohol	5.87×10^{-4}	6.53×10^{-9}	2.23

[a] $[\eta]_\theta = K_\theta M_w^{1/2}$, dl/g.
[b] $(r_0^2)^{1/2}$ = unperturbed root-mean-square end-to-end distance.
[c] $(r^2)^{1/2}$ = measured end-to-end distance of polymer chain in methyl ethyl ketone.

Molecular weights of poly(vinylidene chloride) have not been measured directly, but the constants in the Mark-Houwink relation, $[\eta] = KM^a$, for solutions in trichlorobenzene have been estimated from kinetic studies (eq. 3) (22).

$$[\eta] = 0.45 \times 10^{-4} M^{0.7} \quad \text{in dl/g} \qquad (3)$$

This is not too different for the relationships obtained for copolymers in methyl ethyl ketone. A 15% ethyl acrylate copolymer gives equation 4, where $\bar{M}_v$ is the viscosity-average molecular weight.

$$[\eta] = 2.88 \times 10^{-4} \bar{M}_v^{0.6} \quad \text{in dl/g} \qquad (4)$$

Poly(vinylidene chloride) in good solvents like tetramethylene sulfoxide at 25°C is estimated to have an intrinsic viscosity–molecular weight relation given approximately by equation 5.

$$[\eta] \cong 1 \times 10^{4} \bar{M}_v^{0.7} \quad \text{in dl/g} \qquad (5)$$

In many cases, the molecular weight of poly(vinylidene chloride) and saran copolymers are characterized by the absolute viscosity of a 2% solution in *o*-dichlorobenzene at 140°C. The exact correlation between this number and molecular weight is not known.

Mechanical Properties. Because of the difficulty of fabricating poly(vinylidene chloride) into suitable test specimens, very few direct measurements of its mechanical properties have been made. In many cases, however, the properties of copolymers have been studied as a function of composition and the properties of poly(vinylidene chloride) can be estimated by extrapolation. Table 13 lists some properties characteristic of unplasticized copolymers with high vinylidene chloride content. The actual

Table 13. Estimated Mechanical Properties of Poly(vinylidene Chloride)

Property	Range
tensile strength, psi	
unoriented	5,000–10,000
oriented	30,000–60,000
elongation, %	
unoriented	10–20
oriented	15–40
softening range (heat distortion), °C	100–150
flow temperature, °C	>185
brittle temperature, °C	−10 to +10
impact strength, ft-lb/in.	0.5–1.0

Table 14. Effect of Stretch Ratio upon Tensile Strength and Percent Elongation[a] (99)

Stretch ratio	Tensile strength, psi	Percent elongation
2.50:1	34,080	23.2
3.00:1	43,960	26.3
3.50:1	45,820	19.2
4.00:1	46,380	19.7

[a] Instron tests at 2 in./min with 90% vinylidene chloride–10% vinyl chloride.

performance of a given specimen is very sensitive to morphology, including amount and kind of crystallinity, as well as orientation. Tensile strength increases with crystallinity, while toughness and elongation drop. Orientation, however, improves all three properties. Table 14 shows the effect of orientation on properties of a vinylidene chloride–vinyl chloride copolymer.

The effect of temperature on the dynamic mechanical modulus and the loss factor is shown in Figure 11. The incorporation of vinyl chloride units in the polymer results in a drop in dynamic modulus owing to the reduction in crystallinity (see also Fig. 7). At the same time, however, the glass-transition temperature is raised; the softening effect observed at room temperature, therefore, is accompanied by increased brittleness at lower temperatures. Saran B copolymers are normally plasticized in order to avoid this difficulty. Small amounts of plasticizer (2–10%) can depress T_g significantly without loss of strength at ambient temperatures.

At higher vinyl chloride content, the T_g of the copolymer falls above room temperature and the modulus rises again. A minimum in modulus (or maximum in softness) is normally observed in copolymers where T_g falls above room temperature. A thermomechanical analysis of both vinylidene chloride–acrylonitrile and vinylidene chloride–methyl methacrylate copolymer systems shows a minimum in softening point at 79.4 and 68.1 mole % vinylidene chloride, respectively (100).

In cases where the copolymers have substantially lower glass-transition temperatures, the modulus falls with increasing comonomer content and is therefore accompanied by an improvement in low-temperature performance. At low acrylate levels (<10%), however, T_g increases with comonomer content. The brittle points in this range may, therefore, be higher than that of poly(vinylidene chloride).

The long side chains of the acrylate ester group can apparently act as internal plasticizers. Substitution of a carboxyl group on the polymer chain, in itself, increases

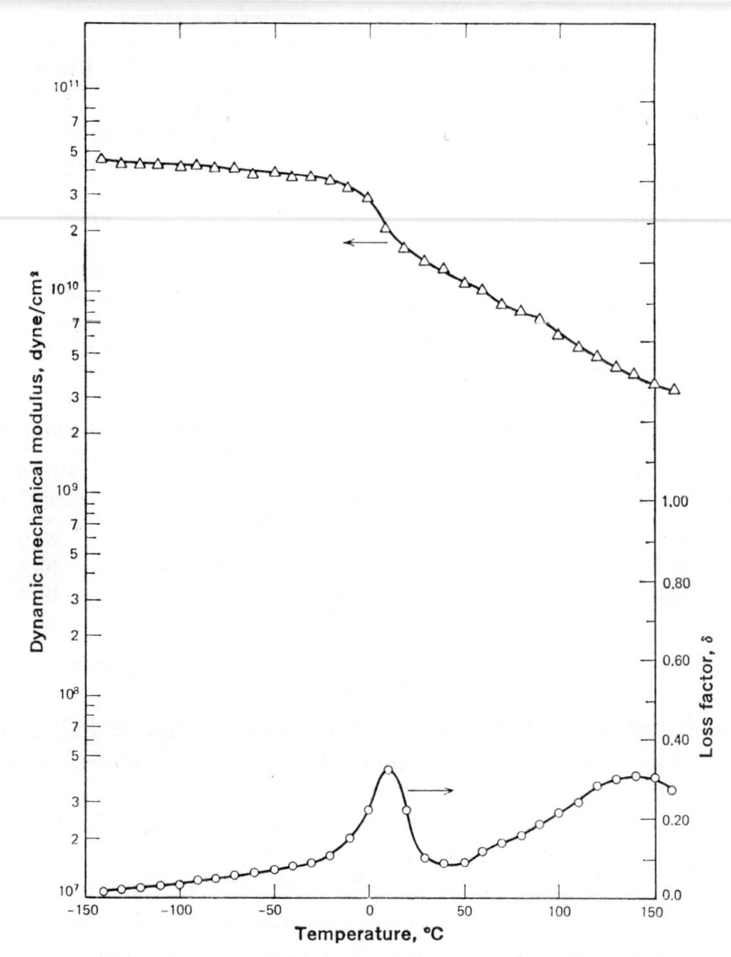

Fig. 11. Dynamic mechanical modulus and loss factor of poly(vinylidene chloride) as a function of temperature. Courtesy S. G. Turley, The Dow Chemical Company.

brittleness. A more polar substituent, such as an *N*-alkyl amide group, is even less desirable. Copolymers of vinylidene chloride with *N*-alkylacrylamide are more brittle than the corresponding acrylates even when the side chains are long (101). Side-chain crystallization may be a contributing factor.

Barrier Properties. Vinylidene chloride polymers are, in comparison with other polymers, very impermeable to a wide variety of gases and liquids. Without a doubt, this is a consequence of the combination of high density and high crystallinity in the polymer. An increase in either tends to reduce permeability. But a more subtle factor may be the symmetry of the polymer structure. Lasoski (102) has shown that both polyisobutylene and poly(vinylidene chloride) have unusually low permeabilities to water compared to their monosubstituted counterparts, polypropylene and poly(vinyl chloride).

Table 15 gives the permeability of a saran film to a variety of gases. Permeability is affected by both kind and amounts of comonomer as well as by crystallinity. In contrast to mechanical behavior, however, orientation seems to have only a minor effect. Permeability of semicrystalline polymers can be expressed as a function of crystallinity (eq. 6) (103) where D_a and S_a are the coefficient of diffusion and the

$$P = D_a S_a (1 - \phi)^2 \tag{6}$$

solubility of the penetrant in the amorphous polymer and ϕ is the volume fraction of the crystalline phase. S_a should be related to the interaction between polymer and diffusing species and, as a consequence, be dependent on the copolymer composition. In practice, it is observed that a more polar comonomer, like acrylonitrile, increases

Table 15. Permeability of Saran Film to Various Gases

Gas	Temp, °C	Activation energy, E_a, kcal/mole	$P \times 10^{10}$ [a]
N_2	30	16.8	0.00094
O_2	30	15.9	0.00053
CO_2	30	12.3	0.03
He	34		0.31
H_2O	25	11.0	0.5
H_2S	30	17.8	0.03

[a] Units: cm^3 (STP)-cm/cm^2-sec-cm Hg.

moisture-vapor transmission rate more than vinyl chloride, other factors being constant. For the same reason, acrylonitrile copolymers are more resistant to penetrants of low cohesive-energy density. D_a can also be influenced by copolymerization. Comonomers that lower T_g and increase the free volume in the amorphous phase will increase permeability. This seems to be the way in which the higher acrylates act, though they undoubtedly influence S_a as well. Plasticizers increase permeability for similar reasons. See also BARRIERS, VAPOR; PERMEABILITY.

Degradation

Poly(vinylidene chloride) has a marked tendency to eliminate hydrogen chloride. The reaction is normally described as a two-step process (eqs. 7 and 8).

Formation of a conjugated polyene

$$-\!(CH_2-CCl_2)\!-_n \xrightarrow{\text{fast}} -\!(CH=CCl)\!-_n + n\,HCl \tag{7}$$

Carbonization

$$-\!(CH=CCl)\!-_n \xrightarrow{\text{slow}} 2n\,C + n\,HCl \tag{8}$$

Although this simplified description serves to illustrate the main features of the reaction, it by no means should be accepted as a mechanism. In fact, poly(vinylidene chloride) can degrade by several distinct mechanisms. Degradation can be effected by heat, ultraviolet radiation, ionizing radiation (x rays, γ rays), basic reagents, and catalytic metals or salts. The common feature of these reactions is that chlorine is removed from the polymer either as Cl or HCl, depending on the medium. But the carbonaceous residue remaining after various extents of decomposition is very much dependent on the method of degradation. In addition, most poly(vinylidene chloride) degradation reactions are heterogeneous, with the polymer present as a solid phase. In such cases, the course of the reaction is also influenced by the polymer morphology and the interaction between phases.

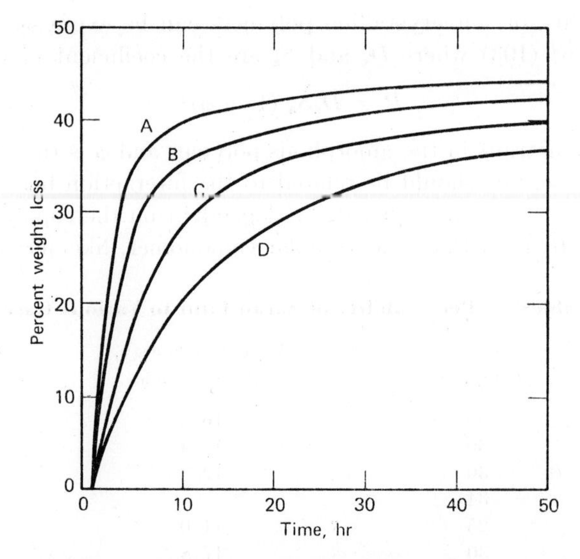

Fig. 12. Weight loss observed for various samples (107). The upper four curves are for pure poly(vinylidene chloride) decomposed at the following temperatures: curve A, 190°C; curve B, 180°C; curve C, 170°C; and curve D, 160°C.

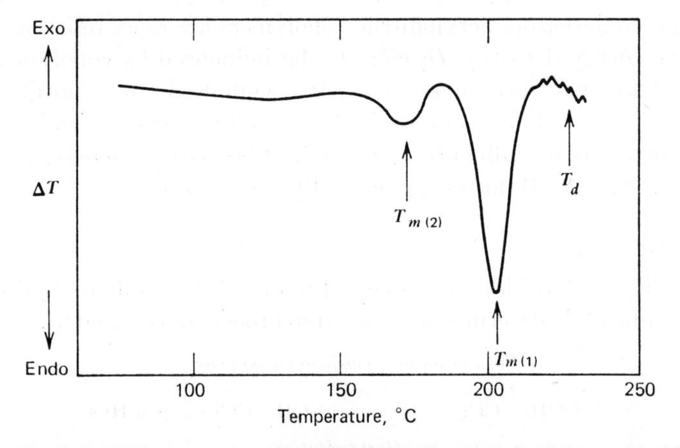

Fig. 13. Typical DTA trace for "as-polymerized" poly(vinylidene chloride) (111).

Thermal degradation has been most widely studied (104–110). Poly(vinylidene chloride) begins to decompose at about 125°C. Some typical kinetic data for the isothermal reactions at temperatures below the melting point are shown in Figure 12. The extent of reaction is measured in terms of the hydrogen chloride loss (assuming this to be the only volatile product).

In many cases, the rate accelerates, initially, but reaches a constant value in the range from 10 to 45% reaction. The reaction slows down drastically beyond 45–50%. Much higher temperatures (>700°C) are needed to completely remove all chlorine from the polymer.

In the very early stages of the thermal decomposition (<1%), poly(vinylidene chloride) begins to discolor and becomes insoluble. The melting point drops and bands corresponding to unsaturation appear in the infrared spectrum. As the reaction pro-

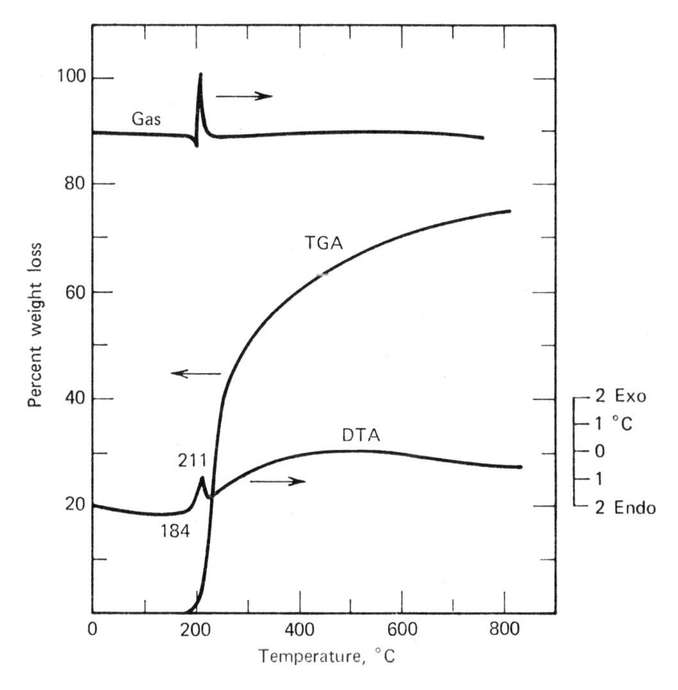

Fig. 14. DTA, TGA, and gas evolution for poly(vinylidene chloride) (112).

ceeds into the constant-rate region, conjugated double bonds are detected. The polymer becomes infusible and the poly(vinylidene chloride) crystal structure is destroyed. An increasing concentration of free radicals can be detected by esr. If further reaction is brought about by raising the temperature, aromatic structures are formed. Finally, at very high temperature, graphitization of the carbonaceous residue takes place (108). Since there is a significant induction period at temperatures below the melting point, the polymer can be heated rapidly to the melting point without degradation. Figure 13 shows a typical differential thermal analysis (111) for an "as-polymerized" poly(vinylidene chloride) sample. The double peak is associated with the fusion process; it is characteristic of polymers that undergo recrystallization during the heating process.

A combination DTA–TGA–gas evolution experiment (Fig. 14) illustrates further that no significant decomposition occurs below T_m. Once melting begins, however, the decomposition reaction takes off almost explosively. The polymer foams and turns black while releasing a burst of hydrogen chloride gas. Again, only about 50% of the available hydrogen chloride is released in the initial reaction. More hydrogen chloride comes off as the polymer is heated to higher temperature, but at a much lower rate. The reactions that take place when decomposition is initiated above the melting point are not well-defined. The products are similar, ie, conjugated polymers, but may be less highly crosslinked.

Poly(vinylidene chloride) does not appear to degrade at a measurable rate in the dark at temperatures below ~100°C. When exposed to ultraviolet or sunlight, however, it discolors. Hydrogen chloride is eliminated in the process and the polymer becomes crosslinked. Quantitative studies of this process have been carried out mainly with copolymer films. The color cannot be accounted for by the introduction of random double bonds. Boyer and co-workers (104,113) proposed instead the formation of conjugated polyene structures. This was based on the assumption that allylic chlorine in a unit adjacent to an already formed double bond would be more reactive. Hence, elimination would occur sequentially. In support of this theory, they observed a shift in light absorption to longer wavelengths with increasing amounts of degradation and a positive test for polyene structure. But the samples they used were degraded chemically rather than photolytically.

Oster and co-workers (114–116) observed the formation of a species absorbing at 285 mμ with no shift to longer wavelengths with increasing exposure to light (Fig. 15). This suggests the formation of a single species or, at least, a narrow distribution of conjugated sequence lengths. Again, the irradiated films gave a positive test for polyene structure and were also observed to be fluorescent. In these experiments, absorbance increased linearly with exposure time; the concentration of free electrons, the gel fraction of the polymer, and the conductivity also increased with exposure. Irradiated samples were also less thermally stable. In later studies, an increase in T_g with exposure time was observed. This was accompanied by a drop in film density and an increase in the insoluble fraction. Unlike ultraviolet, higher energy irradiation with γ rays causes chain scission in poly(vinylidene chloride). Copolymers of vinylidene chloride and vinyl chloride undergo both crosslinking and chain scission (poly(vinyl chloride) itself crosslinks) but the net result is dependent on polymer morphology as well as on copolymer composition (117).

The role of polymer morphology was not considered in the above-described studies. One might expect, however, that this should be a major variable in any solid-state reaction. Everett and co-workers have recently proposed a new mechanism that does include this aspect (110,134,135). They found that if degradation was carried out below 190°C, the polymer morphology was largely retained, indicating that the crystalline solid is decomposed without softening. The polymers studied had lamellar structures with thicknesses of 100 Å. The carbon particles had a similar appearance, but

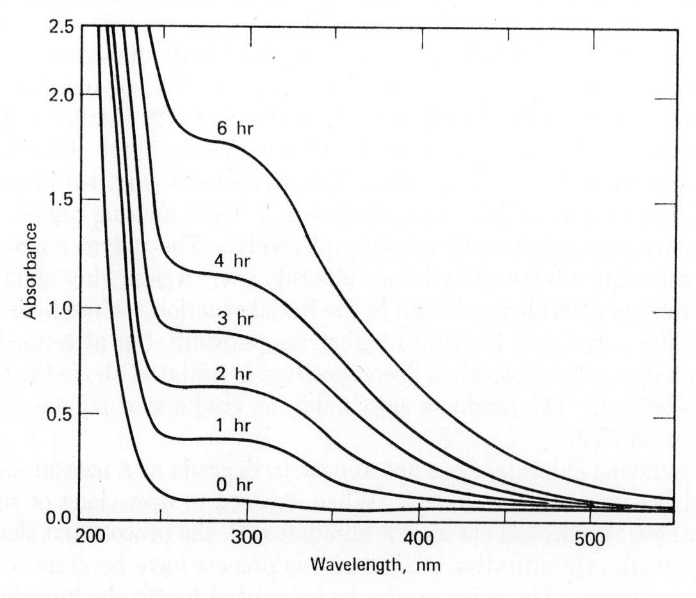

Fig. 15. Absorption spectra of saran film irradiated at 254 nm for different times (0–6 hr as indicated on graph) (114). Intensity, 5×10^{18} quanta/cm²/hr.

the thickness was increased to 200–500 Å. They suggested that this was due to crosslinking between lamellae during degradation. The reaction was found to be first order following an induction period of accelerating rate. No molecular-weight dependence on kinetics could be found, as was reported by Burnett and co-workers (109), but an effect of particle size was observed.

Additional work was concerned with irradiation and degradation in various atmospheres (136). The overall evidence pointed to a free-radical chain reaction initiated at the particle surface. Based on the assumption of a folded-chain lamellar crystal structure, Everett and co-workers proposed the following mechanism: a monomer unit in a chain fold is more reactive because of steric strain. It dissociates thermally into a radical pair as shown by equation 9.

$$\sim\!\sim CH_2-\overset{\displaystyle Cl}{\underset{\displaystyle Cl}{\overset{|}{\underset{|}{C}}}}-\sim\!\sim \xrightarrow{\Delta} \sim\!\sim CH_2-\overset{\cdot}{\underset{\displaystyle Cl}{\underset{|}{C}}}-\sim\!\sim + Cl\cdot \qquad (9)$$

The local chain conformations determine whether the chain radical crosslinks by coupling with a neighboring fold, or whether it propagates into the crystal by initiating a chain reaction. This mechanism yields a crosslinked polychloroacetylene product containing trapped radicals. The latter serve to initiate Diels-Alder condensation when the temperature is raised above 200°C.

Effect of Bases and Salts. The decomposition of poly(vinylidene chloride) brought about by the action of strong bases, is quite clearly an ionic reaction. But details of the mechanism are not known. The ultimate products are carbon and chloride ion. At intermediate stages, conjugated polyene structures and acetylenic bonds are formed. The reaction is very fast even at liquid ammonia temperatures (−33°C) (118). There are at least two possible mechanisms, given by equations 10 and 11. The former is most likely in view of the behavior of low-molecular-weight compounds in elimination reactions (119).

Concerted elimination

$$B\!: \; H-\overset{|}{\underset{|}{C}}-\overset{|}{\underset{|}{C}}-Cl \longrightarrow BH^{\oplus} + \;\rangle C=C\langle\; + Cl^{\ominus} \qquad (10)$$

Carbanion intermediate

$$B\!: + H-\overset{|}{\underset{|}{C}}-\overset{|}{\underset{|}{C}}-Cl \rightleftharpoons BH^{\oplus} + {}^{\ominus}\overset{|}{\underset{|}{C}}-\overset{|}{\underset{|}{C}}-Cl \longrightarrow BH^{\oplus} + \;\rangle C=C\langle\; + Cl^{\ominus} \qquad (11)$$

The reaction can be effected by various strong bases including alcoholic potassium hydroxide, alkylmetal compounds, metal alkoxides, metal amides in liquid ammonia, active metals, lithium chloride in dimethylformamide, etc. The reaction is normally heterogeneous because poly(vinylidene chloride) is not soluble in the usual solvents for such reactions. The rate is, therefore, probably controlled by the surface area. Aqueous bases have a limited effect on poly(vinylidene chloride) primarily because the polymer is scarcely wetted or swollen by water; hot concentrated sodium hydroxide solution will decompose the polymer over a long period of time, however. Weak bases such as ammonia, amines, or polar aprotic solvents also accelerate the decomposition of poly(vinylidene chloride). The products from these reactions do not appear to have a simple polyene structure. In many cases, substitution reactions leading to bound nitrogen can occur competitively (120). These reagents, in addition, can also swell the polymer. Pyridine, for example, would be a fairly good solvent for saran if it did not attack the polymer chemically. In a nonsolvent mixture, however, pyridine does not penetrate into the polymer phase (121). Studies on single crystals indicate that it removes hydrogen chloride only from the surface. Both kinetic studies and product characterization suggest that the reaction of two units in each chain fold can take place easily; further reaction is greatly retarded either by inability of the pyridine to diffuse into the crystal, or by steric factors.

Various metal salts such as ferric chloride, aluminum chloride, and zinc chloride, catalyze the thermal decomposition of poly(vinylidene chloride) (104). This problem is of great practical importance. If saran polymers come in contact with a metal surface, the metal chloride forms and catalyzes further decomposition. The reaction is thereby tremendously accelerated. As a consequence, attempts to extrude unstabilized saran in conventional steel equipment lead to almost explosive decomposition. Very little is known about the mechanism of the reactions; the major industrial emphasis has been on preventive measures. Metal parts intended to be used with saran are fabricated from acid-resistant alloys or nickel (nickel salts are much less active as catalysts). In addition, the polymers are usually stabilized with some type of metal-ion scavenger.

Stabilization. The art of stabilizing saran-type polymers is highly developed. Although not much is understood about mechanisms in detail, some general principles have been established. The ideal stabilizer system should: (a) absorb, or combine, with hydrochloric acid gas in an irreversible manner under the conditions of use, but not have such strong affinity as to strip hydrogen chloride from the polymer chain; (b) act as a selective ultraviolet-radiation absorber (qv) to reduce the total ultraviolet energy absorbed in the polymer itself; (c) contain a reactive dienophilic molecule capable of destroying the discoloration by reacting with and breaking up the color-producing, conjugated polyene sequences; (d) possess antioxidant activity in order to prolong the induction period of the oxidation process and prevent the formation of carbonyl groups and other chlorine-labilizing structures resulting from oxidation of polymer molecules; and (e) have the ability to chelate metals, such as iron, and prevent the formation of metallic chlorides which act as catalysts for polymer degradation.

Acid acceptors are of three general types: alkaline-earth and heavy-metal salts of weak acids, such as barium, cadmium, or lead fatty acid salts (see DRIERS AND METALLIC SOAPS); epoxy compounds, such as epoxidized soybean oil or glycidyl ethers and esters (see EPOXIDATION); and organotin compounds, such as salts of carboxylic acids and organotin mercaptides (see STABILIZATION). The principal compounds of commercial interest as ultraviolet-radiation absorbers (qv) are derivatives of salicylic acid, resorcylic acid, benzophenone, and benzotriazole. Examples of dienophiles that have been used are maleic anhydride and dibasic lead maleate.

Antioxidants (qv) generally fall into two classes: those that react with a free radical to stop a radical chain and those that reduce hydroperoxides to alcohols. Phenolic antioxidants, such as 2,6-di-*t*-butyl-4-methylphenol and substituted bisphenols, fall into the first class. The second class is exemplified by organic sulfur compounds and organic phosphites. The phosphites have the ability to chelate metals as do the Versenes (Dow Chemical Co.) and citric acid or citrates. The performance of organic phosphites in functioning as antioxidants and as chelating agents illustrates the dual ability of many of the stabilizing compounds. It is common practice to use a combination of stabilizing compounds to achieve optimum benefit. Synergism of stabilizing action is frequently observed. An excellent list of patents on stabilization of vinyl chloride and vinylidene chloride polymers and copolymers has been tabulated by Chevassus and de Broutelles (137).

Processing and Uses

Molding (122). Vinylidene chloride copolymer resins are fabricated by injection, compression, and transfer molding (see MOLDING). The fairly sharp crystalline melting temperature of the resins makes the injection-molding process unique. Whereas, with other thermoplastics, cold molds will hasten the cooling of molded parts and shorten the molding cycle, with vinylidene chloride copolymers a cold mold will supercool the resin and produce soft, flexible, amorphous pieces. Rapid hardening is achieved by using heated molds, to induce crystallinity; the molded pieces can be ejected from the mold at temperatures as high as 100°C in a strainfree, dimensionally stable form. Very rapid cycles can be realized with heavy sections. Sink marks are not usually a problem since the outer skin of the molded piece cools quickly below the crystalline melting temperature and crystallization takes place rapidly. The range of molding temperatures is rather narrow owing to the crystalline nature of the resin and its thermal sensitivity. All crystallites must be melted to obtain low polymer melt viscosity, but prolonged or excessive heating must be avoided to prevent dehydrochlorination.

Thermal degradation is a problem, even when the resin is formulated with the very best thermal stabilizers. Molding equipment is designed to alleviate this problem by having all passages through the heating cylinder streamlined to prevent plastic buildup. Any plastic that remains in the heating cylinder for longer than a few minutes will decompose, releasing hydrochloric acid gas and forming black degradation products. It is especially important that an injection-molding heating cylinder not be shut down when loaded with molten resin. The cylinder must be purged with a more stable resin, such as polystyrene.

The metal parts of the injection molder (liner, torpedo, nozzle) that contact the hot molten resin must be of the noncatalytic type to prevent an accelerated decomposition of the polymer. In addition, they must be resistant to corrosion by acid gas. Nickel alloys, such as DuraNickel and Hastelloy B brands, are recommended. The injection mold need not be made of noncatalytic metals; any high-grade tool steel may be used, since the plastic is cooling in the mold and undergoing little decomposition. However, the mold requires good venting to allow the passage of small amounts of acid gas as well as air. Vents tend to corrode and have to be cleaned periodically.

Vinylidene chloride copolymer resins can be compression molded by conventional heating and cooling of the mold as is done with other thermoplastics. Hot plastic from an extruder or dielectrically heated preforms may be used. The preferred method is dielectric heating of compressed powder preforms, where gentle heating can be applied to a preform practically free of air. A 5-kW generator at about 27 megacycles frequency will heat approximately 1 lb of resin per minute in a 6-in. diameter preform. The heated preform will have a powder skin which should be kneaded into the hot plastic to yield a smooth compression-molded surface. The mold can have a noncatalytic surface provided by chrome plating, although slight etching may occur. The mold should be well cored to allow an adequate supply of water for quick cooling. The plastic may be molded in conventional transfer molds at temperatures of 150–

200°F. Dielectrically heated preforms, unkneaded, may be used and material remaining in the transfer cylinder may be collected for reheating to make a full cylinder charge. Practically no sprue scrap is generated. See also DIELECTRIC HEATING.

Molded parts of vinylidene chloride copolymer plastics are used to satisfy the industrial requirements of chemical resistance (qv). They are used in gasoline filters, spinnerets, valves, pipe fittings, containers, and chemical-process equipment, etc. Complex articles are constructed from molded parts by welding; the welds are practically as strong as the other portions of the molded part and are as chemically resistant. The weld is strong because of the high fluidity of the hot plastic which, in turn, is due to the sharp melting crystalline character. Hot-air welding at 400–500°F is an excellent method; in addition, hot plate welding, frictional welding, or radiant heating may be used. Molded parts have good physical properties, but have lower tensile strength than films or fibers, since crystallization is random in molded parts but well-developed by orientation in films and fibers.

Table 16 shows physical properties of a typical molded vinylidene chloride plastic.

Table 16. Properties of Molded Saran 281[a]

Property	Value
tensile strength, psi	3,500–4,500
elongation at yield, point, %	15–25
modulus of elasticity, tension, psi	$0.7–2.0 \times 10^5$
impact strength, ½ × ½ in. notched bar, Izod, ft-lb/in.	2–8
flexural strength, psi	15,000–17,000
compressive strength at yield point, psi	7,500–8,300
resistance to heat, continuous, °F	approx 170
resistance to heat, intermittent, °F	220
specific heat, cal/°C/g	0.32
thermal expansion, per °F	8.78×10^{-5}
thermal conductivity, cal/sec/cm²/°C/cm	2.2×10^{-4}
hardness, Rockwell M	50–65
specific gravity	1.68–1.75
refractive index, n_D	1.61
power factor, 60, 10^3, and 10^6 cycles	0.03–0.15
breakdown voltage, 60 cycles, instantaneous, V/mil	
0.125-in. thickness	500
0.020-in. thickness	1,500
0.001-in. thickness	300
volume resistivity at 50% rh and 25°C, ohm-cm	$10^{14}–10^{16}$
water absorption, ASTM D 570-40T	less than 0.1%
burning rate	self-extinguishing
mold shrinkage (injection-molded), in./in.	0.005–0.030
injection-molding temperature, °F	300–350
injection-molding pressure, psi	10,000–30,000
compression-molding temperature, °F	250–350
compression-molding pressure, psi	250–5,000

[a] The Dow Chemical Co.

Extrusion (99,122). Extrusion (qv) of vinylidene chloride copolymers, which is the major fabrication technique for filaments, films, rods and tubing or pipe, requires the same concerns for thermal degradation, streamlined flow, and noncatalytic materials of construction that were described above for injection molding. The plastic leaves the extrusion die in a completely amorphous condition and is maintained in this state by quench-cooling in a water bath at about 50°F, inhibiting the recrystallization.

In this state, the plastic is soft, weak, and pliable. If it is allowed to remain at room temperature, it will gradually harden and partially recrystallize at a slow rate with a random crystal arrangement. Heat treatment can be used to recrystallize at controlled rates.

Crystal orientation is developed in the supercooled extrudate by stretching and heat treatment. In the manufacture of *filaments*, the stretching produces orientation in a single direction, developing unidirectional properties of high tensile strength, flexibility, long fatigue life, and good elasticity. The filaments are taken from the supercooling tank, wrapped several times around smooth take-off rolls, and then wrapped several times around orienting rolls which have a linear speed about four times that of the take-off rolls. The difference in roll speeds produces the mechanical stretching and causes orientation of crystallites along the longitudinal axis during partial recrystallization. Heat treatment may be used during, or after, stretching to affect the degree of crystallization and control the physical properties of the oriented filaments.

A variation of the orientation process is used to produce vinylidene chloride copolymer *films*. The plastic is extruded into tube form, which is supercooled and subsequently biaxially oriented (see BIAXIAL ORIENTATION) in a continuous process known as the *bubble process*. The supercooled tube is flattened and passed through two sets of pinch rolls arranged so that the second set of rolls is traveling faster than the first set. Between the two sets of rolls, air is injected into the tube to create a bubble which is entrapped by the pinch rolls. The entrapped air bubble remains stationary, while the extruded tube is oriented as it passes around the bubble. Orientation is produced in the transverse direction, as well as the longitudinal direction, creating in the film excellent tensile strength, elongation, and flexibility (123).

Extruded monofilaments in diameters of 5–15 mils have been widely used as furniture and automobile upholstery, drapery fabric, venetian-blind tape, filter cloths, etc (see p. 575). Chemically resistant tubing and pipe liners are also extruded. Pipe liners are used in steel pipes to protect against corrosion. In a completely lined system, all components, including valves and fittings, are protected with preformed inserts.

The biaxially oriented extruded films are used in packaging applications where their excellent resistance to water vapor and most gases makes them an ideal transparent barrier. Being highly oriented, these films exhibit some shrinkage when exposed to higher than normal temperatures. Preshrinking or heat setting may be performed to minimize residual shrink, or the shrinkage may be used to advantage in the heat shrinking of packaging overwraps. The films are used in tube form or flat, or converted to bags. Electronic or dielectric heating is the most satisfactory method for welding the film to itself, although hot-plate sealers or cements can also produce airtight seals.

Lacquer Resins (124). Vinylidene chloride polymers have several properties that are valuable in the coatings industry: their excellent resistance to gas and moisture vapor transmission, their good resistance to attack by solvents and by fats and oils, their high strength, and their ability to be heat sealed. These characteristics derive from the highly crystalline nature of the very high vinylidene chloride composition of the polymer, which ranges from about 80% to more than 90% by weight. Minor constituents in these copolymers are generally from the group: vinyl chloride, alkyl acrylates, alkyl methacrylates, acrylonitrile, methacrylonitrile, and vinyl acetate. Small concentrations of carboxylic acids, such as acrylic acid, methacrylic acid, or itaconic acid, are sometimes included to enhance adhesion of the polymer to the substrate. Crystallization is reduced with increasing concentration of the comonomers; some polymers used commercially fail to crystallize at all.

Acetone, methyl isobutyl ketone, dimethylformamide, ethyl acetate, methyl ethyl ketone, and tetrahydrofuran, especially the two latter, are solvents for vinylidene chloride polymers used in lacquer coatings. Toluene may be used as a diluent. Lacquers prepared at 10–20% polymer solids by weight in a solvent blend of 2 parts ketone and 1 part toluene will have a viscosity in the range of 200–1000 cP. With polymers of very high vinylidene chloride content, lacquers can be prepared and stored at room temperature in the tetrahydrofuran–toluene system. Methyl ethyl ketone lacquers must be prepared and maintained at 60–70°C or the lacquer will form a solid gel. It is critical in the manufacture of polymers for lacquer application to maintain a fairly narrow compositional distribution in the polymer in order to achieve good dissolution properties.

The lacquers are applied commercially by roller coating, dip and doctor, knife coater, and spraying (see COATING METHODS). Spraying is useful only with low-viscosity lacquers and solvent balance is important to avoid webbing from the spray gun. Solvent removal is difficult from heavy coatings, and multiple coatings are recommended where a heavy film is desired, allowing sufficient time between coats to avoid solvent lifting of the previous coat. In the machine coating of the flexible substrata (paper, plastic films), the solvent is removed by infrared heating or forced-air drying at temperatures of 90–140°C. Temperatures in the range of 60–95°C promote the recrystallization of the polymer, after the solvent has been removed. Failure to recrystallize the polymer will leave a soft, amorphous coating that blocks or adheres between concentric layers in a rewound roll. A recrystallized coating can be rewound without blocking. Handling properties of the coated film are improved by adding small amounts of wax as a slip agent and of talc or silica as an antiblocking agent (qv) to the lacquer system. The concentration of additives is kept low to prevent any serious detraction from the vapor-transmission properties of the vinylidene chloride polymer coating. For this reason, plasticizers are seldom, if ever, used.

A primary use of vinylidene chloride polymer lacquers is the coating of films made from regenerated cellulose (125), polyamides, polyester, polyethylene, polypropylene, poly(vinyl chloride), and polyethylene-coated board or paper to impart resistance to fats, oils, oxygen, and water vapor. These are used, mainly, in the packaging of foodstuffs, where the additional features of inertness, lack of odor or taste, and nontoxicity are required. Vinylidene chloride polymers have been used extensively as interior coatings for ship-tanks (126), railroad tank cars, and fuel storage tanks (127), and for coating of steel piles (128) and structures. The high chemical resistance and good adhesion have resulted in excellent long-term performance of the coating. Brushing and spraying are suitable methods of application.

The excellent adhesion to primed films of polyesters, combined with good dielectric properties and good surface properties, make the vinylidene chloride polymers very suitable as binders for iron-oxide-pigmented coatings of magnetic tapes (129). They have shown very good performance in audio tapes, video tapes, and computer tapes.

Saran Latex (130–132). Vinylidene chloride polymers are often made in emulsion, but normally they are isolated, dried, and used as conventional resins. In the last decade, however, stable saran latexes have been prepared which can be used directly for coatings and other applications. They have the desirable characteristic of forming an extremely water-resistant coating.

Not surprisingly, the major applications for these materials are as barrier coatings on paper products and, recently, on plastic films. The heat-sealing characteristics of saran coatings are also valuable in many applications. Saran latexes have been used in combination with cement to make high-strength mortars and concretes. They are also used as binders for paints and nonwoven fabrics; these latexes yield composites that are both water-resistant and nonflammable.

Poly(vinylidene chloride) latexes can be easily prepared by the same methods, but have few uses because of their inability to form films. Copolymers of high vinylidene chloride content form films when freshly prepared, but soon crystallize and lose this desirable characteristic. Since crystallinity in the final product is very often desirable, in barrier coatings for example, a major problem has been to prevent crystallization in the latex during storage but to induce rapid crystallization of the polymer after coating. This has been accomplished by using the proper combination of comonomers with vinylidene chloride.

Most latexes are made with varying amounts of acrylates, methacrylates, acrylonitrile, and vinyl chloride, the total amount of comonomer ranging from about 7 to 20%. The properties of a typical latex used for paper coating are listed in Table 17.

Table 17. Typical Properties of a Saran Latex

total solids, %	60–62
viscosity at 25°C, cP	25
pH	4.5–5.0
color	white–cream
particle size, μ	approx 0.25
weight per gal, lb	11.5
mechanical stability	excellent
storage stability	excellent
chemical stability	not stable to divalent or trivalent ions

Latexes are usually formulated with antiblocking, slip, and wetting agents. They can be deposited in conventional coating processes such as with a doctor blade. Coating speeds in excess of 1000 ft/min can be attained and the coating can be dried in forced air or radiant heat ovens. Two coats are normally applied to reduce pinholing. A precoat is often used on porous substrates to reduce the quantity of the more expensive saran latex needed (133).

Fibers (99)

Saran filaments have the desirable properties of high strength and flexibility and are chemically resistant and self-extinguishing. The monofilaments have been widely used in automotive seat cover, outdoor furniture, agricultural shade cloth, filter fabrics, insect screening, window awning fabrics, venetian blind tape, and brush bristles. Extruded multifilament yarns have been used for draperies, upholstery, doll hair, dust mops, and various industrial fabrics.

Vinylidene chloride copolymers are fabricated into fibers by melt-spinning (extrusion) or by solvent-spinning (wet or dry) methods. Saran monofilaments are extruded in sizes of 5–15 mils in diameter, the most commonly used sizes being 10–12 mils. Multifilaments are used mostly as staple fiber, but very little as continuous yarns. Staple yarn is commercially available at 16 and 22 denier and has been produced experimentally at 10 denier. Filaments extruded and cut for staple yarn have a curl

Table 18. Maximum Stretch Ratio at Various Temperatures

Temperature, °C	Maximum stretch ratio
20	3.4
30	4.1
40	4.8
50	5.5
60	6.2
70	7.0

Table 19. Typical Properties of 0.010-in. Diameter Round Saran Monofilament

tensile strength, psi	40,000
elongation, %	15–25
water absorption, 27-hr immersion, %	<0.1
specific gravity	1.65–1.72
softening point, °C	115–140
flammability	self-extinguishing
shrinkage, %	depends upon the formulation and after-treatment (2–60% is possible)
heat resistance, intermittent, °C	77
heat resistance, continuous	shrinks at 100°C
flexural strength, psi	15,000–17,000
modulus of elasticity, tension, psi	700–2,000

imparted during the extrusion process. The solvent resistance of vinylidene chloride copolymers limits the preparation of spinning dopes—use of such solvents as tetrahydrofuran is necessary to achieve 30–40% resin solids in the dope. Solvent removal from the fibers and solvent recovery are difficult.

Physical properties of the monofilaments are determined by the stretch ratio applied during orientation of the amorphous vinylidene chloride copolymer. (The orientation process is described on p. 573.) The data in Table 14 show that the elongation decreases and the tensile strength increases with increasing stretch ratio. In addition, knot strength, stiffness, and shrinkage increase with the stretch ratio. The maximum stretch ratio which can be applied to the supercooled, amorphous filament is dependent upon the stretch temperature as shown in Table 18. Typical physical properties of the saran monofilaments are shown in Table 19.

Bibliography

1. R. C. Reinhardt, *Ind. Eng. Chem.* **35,** 422 (1943).
2. J. J. P. Staudinger, *Brit. Plastics* **19** (9), 381 (1947); **19** (10), 453 (1947).
3. C. E. Schildknecht, *Vinyl and Related Polymers,* John Wiley & Sons, Inc., New York, 1952, Chap. VIII.
4. *Vinylidene Chloride Monomer,* Inorganic Chemicals Department Bulletin, The Dow Chemical Company, 1966.
5. F. L. Taylor and L. H. Horsley (to The Dow Chemical Company), U.S. Pat. 2,293,317 (1942).
6. I. G. Farbenindustrie A.G., Ger. Pat. 529,604 (1931).
7. W. N. Howell (to Imperial Chemical Industries Ltd.), Brit. Pat. 534,733 (1941).
8. M. L. Henry, *Bull. Soc. Chim.* **42** (2), 262 (1884).
9. A. Jocitsch, *J. Russ. Phys. Chem. Soc.* **30,** 998 (1898).
10. Compagnie des Produits Chimiques et Electrométallurgiques, Fr. Pat. 786,803 (1935).
11. A. W. Hanson and W. C. Goggin (to Dow), U.S. Pat. 2,238,020 (1947).

12. A. Jacobowsky and K. Sennewald (to Knapsack-Griesheim A.G.), U.S. Pat. 2,915,565 (1959).
13. P. W. Sherwood, *Ind. Eng. Chem.* **54** (12), 29–33 (1962).
14. E. C. Britton and W. J. LeFevre (to Dow), U.S. Pats. 2,121,009–2,121,012 (1938).
15. G. H. Coleman and J. W. Zemba (to Dow), U.S. Pats. 2,136,333 and 2,136,334 (1938).
16. G. H. Coleman and J. W. Zemba (to Dow), U.S. Pat. 2,160,944 (1939).
17. R. M. Wiley (to Dow), U.S. Pats. 2,136,347–2,136,349 (1938).
18. G. Talamini and E. Peggion in G. E. Ham, ed., *Vinyl Polymerization,* Vol. 1, Part 1, Marcel Dekker, Inc., New York, 1967, Chap. 5.
19. P. J. Flory, *Principles of Polymer Chemistry,* Cornell University Press, Ithaca, N.Y., 1953, Chap. VI.
20. B. V. Lebedev, I. B. Rabinovich, and V. A. Budarina, *Vysokomolekul. Soedin.* **A9,** 488 (1967).
21. C. E. Bawn, T. P. Hobin, and W. J. McGarry, *J. Chim. Phys.* **56,** 791 (1959).
22. J. D. Burnett and H. W. Melville, *Trans. Faraday Soc.* **46,** 976 (1950).
23. W. J. Burlant and D. H. Green, *J. Polymer Sci.* **31,** 227 (1958).
24. W. I. Bengough and R. G. W. Norrish, *Proc. Royal Soc.* (*London*), *Ser. A* **218,** 149 (1953).
25. A. D. Jenkins in G. E. Ham, ed., *Vinyl Polymerization,* Vol. 1, Part 1, Marcel Dekker, Inc., New York, 1967, Chap. 6.
26. J. D. Cotman, M. F. Gonzalez, and G. C. Claver, *J. Polymer Sci.* [A-1] **5,** 1137 (1967).
27. G. Talamini and G. Vidotto, *Chim. Ind.* **46,** 371 (1964).
28. B. M. E. Van Der Hoff in *Polymerization and Poly Condensation Processes,* American Chemical Society, Washington, D.C., 1962.
29. L. C. Rubens and R. F. Boyer in R. H. Boundy and R. F. Boyer, eds., *Styrene,* American Chemical Society, Monograph No. 115, Reinhold Publishing Corp., New York, 1952.
30. H. Wiener, *J. Polymer Sci.* **7,** 1 (1950).
31. P. M. Hay, J. C. Light, L. Marker, R. W. Murray, A. T. Santicola, O. J. Sweeting, and J. G. Wepsic, *J. Appl. Polymer Sci.* **5,** 23 (1961).
32. J. C. Light, L. Marker, A. T. Santicola, and O. J. Sweeting, *J. Appl. Polymer Sci.* **5,** 31 (1961).
33. C. P. Evans, P. M. Hay, L. Marker, R. W. Murray, and O. J. Sweeting, *J. Appl. Polymer Sci.* **5,** 39 (1961).
34. A. Konishi, *Bull. Chem. Soc., Japan* **35,** 197 (1962).
35. A. Konishi, *Bull. Chem. Soc. Japan* **35,** 193 (1962).
36. A. P. Sheinker et al., *Dokl. Akad. Nauk SSSR* **124,** 632 (1959).
37. J. Brandrup and E. H. Immergut, eds., *Polymer Handbook,* Interscience Publishers, a division of John Wiley & Sons, Inc., New York, 1966.
38. T. Alfrey, Jr., and J. J. Bohrer, and H. Mark, eds., *Copolymerization,* Interscience Publishing Co., New York, 1954.
39. T. Alfrey, Jr., and L. J. Young in G. E. Ham, ed., *Copolymerization,* Interscience Publishers, a division of John Wiley & Sons, Inc., New York, 1964, Chap. 2.
40. V. Johnsen and K. Kolbe, *Kolloid* Z. **232,** 712 (1969).
41. T. Fisher, J. B. Kinsinger, and C. W. Wilson, *Polymer Letters* **4,** 379 (1966).
42. J. B. Kinsinger, T. Fisher, and C. W. Wilson, *Polymer Letters* **5,** 285 (1967).
43. K. H. Hellwege, V. Johnsen, and K. Kolbe, *Kolloid* Z. **214,** 45 (1966).
44. V. Johnsen and W. Lesch, *Kolloid* Z. **233,** 863 (1969).
45. J. L. McClanahan and S. A. Previtera, *J. Polymer Sci.* [A] **3,** 3919 (1965).
46. R. Chujo, S. Satoh, and E. Nagai, *J. Polymer Sci.* [A] **2,** 895 (1964).
47. V. Johnsen, *Kolloid* Z. **210,** 1 (1966).
48. K. Ito, S. Iwase, and Y. Yamashita, *Makromol. Chem.* **110,** 233 (1967).
49. Y. Yamashita, K. Ito, S. Ikuma, and H. Kada, *Polymer Letters* **6,** 219 (1968).
50. H. Germar, *Makromol. Chem.* **84,** 36 (1965).
51. J. F. Gabbett and W. M. Smith in G. E. Ham, ed., *Copolymerization,* Interscience Publishers, a division of John Wiley & Sons, Inc., New York, 1964, Chap. 10.
52. H. W. Melville and L. Valentine, *Proc. Royal Soc.* (*London*), *Ser. A* **200,** 358 (1950).
53. W. I. Bengough and R. G. W. Norrish, *Proc. Royal Soc.* (*London*), *Ser. A* **218,** 155 (1953).
54. N. Yamazaki, K. Sasaki, T. Nisiimura, and S. Kambara, *Am. Chem. Soc. Polymer Preprints* **5,** 667 (1964).
55. A. Konishi, *Bull. Chem. Soc., Japan* **35,** 395 (1962).

56. B. L. Erusalimskii, I. G. Krasnosel'skaya, V. V. Mazurek, and V. G. Gasan-Zade, *Dokl. Akad. Nauk SSSR* **169,** 114 (1966).
57. Chisso Corp., Brit. Pat. 1,119,746 (1967).
58. H. J. Hagemeyer and M. B. Edwards (to Eastman Kodak Co.), U.S. Pat. 3,453,346 (1969).
59. A. V. Vlasov, L. G. Tokareva, D. Ya. Tsvankin, B. L. Tsetlin, and M. V. Shablygin, *Dokl. Akad. Nauk SSSR* **161,** 857 (1965).
60. R. Buning and W. Pungs (to Dynamit Nobel), Can. Pat. 98,905 (1968).
61. M. Baer (to Monsanto), U.S. Pat. 3,366,709 (1968).
62. P. K. Isacs and A. Trafimow (to W. R. Grace & Co.), U.S. Pat. 3,033,812 (1962).
63. L. C. Friedrich, Jr., J. W. Peters, and M. R. Rector (to Dow), U.S. Pat. 2,968,651 (1961).
64. J. Heerema (to Dow), U.S. Pat. 2,482,771 (1944).
65. D. W. Woodford, *Chem. Ind.* **1966,** 316; to Scott Bader & Co., U.S. Pat. 3,291,769 (1966).
66. S. Krimm, *Fortschr. Hochpolymer.-Forsch.* **2,** 51 (1960).
67. S. Krimm and C. Y. Liang, *J. Polymer Sci.* **22,** 95 (1956).
68. S. Narita, S. Ichinohe, and S. Enomoto, *J. Polymer Sci.* **37,** 251 (1959).
69. S. Narita, S. Ichinohe, and S. Enomoto, *J. Polymer Sci.* **37,** 263 (1959).
70. J. G. Cobler, M. W. Long, and E. G. Owens in O. J. Sweeting, ed., *The Science and Technology of Films,* Vol. I, Interscience Publishers, a division of John Wiley & Sons, Inc., New York, 1968, Chap. 15.
71. R. A. Wessling, Dow Chemical Co., unpublished results.
72. K. Okuda, *J. Polymer Sci.* [A] **2,** 1749 (1964).
73. L. K. Frevel, unpublished results, 1938, reported in Ref. 1.
74. S. Narita and K. Okuda, *J. Polymer Sci.* **38,** 270 (1959).
75. E. J. Arlman and W. M. Wagner, *Trans. Faraday Soc.* **49,** 832 (1953).
76. C. S. Fuller, *Chem. Rev.* **26,** 143 (1940).
77. P. DeSantis, E. Giglio, A. M. Liquori, and A. Ripamonti, *J. Polymer Sci.* [A] **1,** 1383 (1963).
78. V. M. Coiro, P. DeSantis, A. M. Liquori, and A. Ripamonti, *Polymer Letters* **4,** 821 (1966).
79. K. Okuda, private communication.
80. T. Miyazawa and Y. Ideguchi, *Polymer Letters* **3,** 541 (1965).
81. P. T. Hendra, J. R. Mackenzie, and P. Holliday, *Spectrochim. Acta* **25A,** 1349 (1969).
82. R. A. Wessling, unpublished results.
83. R. F. Boyer and R. S. Spencer, *J. Appl. Phys.* **15,** 398 (1944).
84. G. Schuur, *Some Aspects of the Crystallization of High Polymers,* Rubber-Stichting, Delft, 1955, p. 18.
85. R. M. Wiley, Dow Chemical Co., unpublished results.
86. R. A. Wessling, *J. Appl. Polymer Sci.* **14,** 1531 (1970).
87. K. Schmieder and K. Wolf, *Kolloid Z.* **134,** 149 (1953).
88. N. G. McCrum, B. E. Read, and G. Williams, *Anelastic and Dielectric Effects in Polymeric Solids,* John Wiley & Sons, Inc., New York, 1967, Chap. 11.
89. T. Hideshima and M. Kakizaki, *Rept. Progr. High Polymer Phys. Japan* **7,** 271 (1964).
90. Y. Ishida, M. Yamamoto, and M. Takayanagi, *Kolloid Z.* **168,** 124 (1960).
91. V. P. Lebedev, N. A. Okladnov, and M. N. Shlykova, *Vysokomolekul. Soedin.* **A9,** 495 (1967).
92. K. H. Illers, *Kolloid Z.* **190,** 16 (1963).
93. E. F. Jordan, W. E. Palm, L. P. Witnauer, and W. S. Port, *Ind. Eng. Chem.* **49,** 1695 (1957).
94. E. Powell, M. J. Clay, and B. J. Sauntson, *J. Appl. Polymer Sci.* **12,** 1765 (1968).
95. G. R. Riser and L. P. Witnauer, *Am. Chem. Soc. Polymer Preprints* **2,** 218 (1961).
96. H. Burrell, *Offic. Dig.* **1955,** p. 726.
97. R. A. Wessling, *J. Appl. Polymer Sci.* **14,** 2263 (1970).
98. M. Asahina, M. Sato, and T. Kobayashi, *Bull. Chem. Soc. Japan* **35,** 630 (1962).
99. E. D. Serdensky in H. Mark, S. Atlas, and E. Cernia, eds., *Man-Made Fibers: Science and Technology,* Vol. 3, Interscience Publishers, a division of John Wiley & Sons, Inc., New York, 1968, p. 319.
100. G. S. Kolesnikov, L. S. Fedorova, B. L. Tsetlin, and N. V. Klimentova, *Izv. Akad. Nauk SSSR, Otd. Khim. Nauk* **1959,** 731–735.
101. E. F. Jordan, G. R. Riser, B. Artymyshyn, W. E. Parker, J. W. Pensabene, and A. N. Wrigley, *J. Appl. Polymer Sci.* **13,** 1777 (1969).
102. S. W. Lasoski, *J. Appl. Polymer Sci.* **4,** 118 (1960).

103. S. W. Lasoski and W. H. Cobbs, *J. Polymer Sci.* **36,** 21 (1959).
104. L. A. Matheson and R. F. Boyer, *Ind. Eng. Chem.* **44,** 867 (1952).
105. C. B. Havens in "Polymer Degradation Mechanisms," *Natl. Bur. Std. Circ.* **525,** 107–122 (1953).
106. J. R. Dacey and D. A. Cadenhead, *Proc. Fourth Conf. on Carbon,* Buffalo University Press, 1960, pp. 315–319.
107. J. R. Dacey and R. G. Barradas, *Can. J. Chem.* **41,** 180 (1963).
108. F. H. Winslow, W. O. Baker, and W. A. Yager, *Proc. First and Second Conf. on Carbon,* Buffalo University Press, 1956, pp. 93–102.
109. G. M. Burnett, R. A. Haldon, and J. N. Hay, *European Polymer J.* **3,** 449 (1967).
110. A. Bailey and D. H. Everett, *J. Polymer Sci.* [A-2] **7,** 87 (1969).
111. I. R. Harrison and R. A. Wessling, to be published.
112. D. Dollimore and R. R. Heal, *Carbon* **5,** 65 (1967).
113. R. F. Boyer, *J. Phys. Colloid Chem.* **51,** 80 (1947).
114. G. Oster, G. K. Oster, and M. Kryszewski, *J. Polymer Sci.* **57,** 937 (1962).
115. G. Oster, *Nature* **191,** 164 (1961).
116. M. Kryszewski and M. Mucha, *Bull. Acad. Polon. Sci. (Ser. Sci. Chim.)* **13,** 53 (1965).
117. D. E. Harmer and J. A. Raab, *J. Polymer Sci.* **55,** 821 (1961).
118. E. Tsuchida, C-N. Shih, I. Shinohara, and S. Kambara, *J. Polymer Sci.* [A] **2,** 3347 (1964).
119. D. J. McLennan, *Quart. Rev.* **21** (4), 490 (1967).
120. T. Nakagawa, *Kogyo Kagaku Zasshi* **71,** 1272 (1968).
121. I. R. Harrison and E. Baer, *J. Colloid Interface Sci.* **31,** 176 (1969).
122. W. C. Goggin and R. D. Lowry, *Ind. Eng. Chem.* **34,** 327 (1942).
123. W. T. Stephenson (to Dow), U.S. Pat. 2,452,080 (1948).
124. *Saran F Resin,* Technical Bulletin, Dow Chemical-Europe, 1969.
125. P. M. Hauser (to E. I. du Pont de Nemours & Co., Inc.), U.S. Pat. 2,462,185 (1949).
126. W. W. Cranmer, *Corrosion* **8** (6), 195–204 (1952).
127. J. E. Cowling, I. J. Eggert, and A. L. Alexander, *Ind. Eng. Chem.* **46,** 1977–1985 (1954).
128. R. L. Alumbaugh, *Mater. Protection* **3** (7), 34–36, 39–45 (1964).
129. J. P. Talley (to Ampex Corporation), U.S. Pat. 3,144,352 (1964).
130. L. J. Wood, *Mod. Packaging* **33,** 125 (1960).
131. R. F. Avery, *Tappi* **45,** 356 (1962).
132. A. D. Jordan, *Tappi* **45,** 865 (1962).
133. E. A. Chirokas, *Tappi* **50,** 59A (1967).
134. D. H. Davies, D. H. Everett, and D. J. Taylor, *Trans. Faraday Soc.,* to be published, 1971.
135. A. Bailey and D. H. Everett, *Nature* **211,** 1082 (1966).
136. D. H. Everett and D. J. Taylor, to be published *Trans. Faraday Soc.,* 1971.
137. F. Chevasus and R. de Broutelles, *The Stabilization of Polyvinyl Chloride,* St. Martin's Press, Inc., New York, 1963.

R. A. Wessling and F. G. Edwards
The Dow Chemical Company

VINYLIDENE CYANIDE POLYMERS

Vinylidene cyanide, $CH_2{=}C(CN)_2$, is an extremely reactive monomer that undergoes rapid ionic polymerization in the presence of almost any weak base to form a hydrolytically unstable homopolymer. The monomer polymerizes readily with a wide variety of comonomers such as vinyl acetate, styrene, and dienes to form alternating rather than random copolymers. Many of these copolymers have crystalline structures and unusually high melting or softening points.

Although a considerable effort was made to develop moldable or extrudable plastics based on vinylidene cyanide, the effort was not successful commercially. A fiber from the 1:1 copolymer of vinylidene cyanide and vinyl acetate was developed by the B. F. Goodrich Company and for a time produced commercially under the trade name Darvan. (The rights to the fiber are now owned by The Celanese Corporation.)

Monomer

Synthesis. Vinylidene cyanide can be prepared by the following methods (1–9).

1. Pyrolysis of 4,4-dicyanocyclohexene (2,3).

$$H_2C{=}C(COOC_2H_5)_2 + CH_2{=}CH{-}CH{=}CH_2 \longrightarrow \text{4,4-bis(ethoxycarbonyl)cyclohexene} \xrightarrow[KOH]{\text{hydrolysis}} \text{cyclohexene-4,4-(COOH)}_2 \xrightarrow{SOCl_2} \text{cyclohexene-4,4-(COCl)}_2$$

$$\text{cyclohexene-4,4-(COCl)}_2 \xrightarrow{NH_3} \text{cyclohexene-4,4-(CONH}_2)_2 \xrightarrow{P_2O_5} \text{cyclohexene-4,4-(CN)}_2 \xrightarrow{\text{heat}} H_2C{=}C(CN)_2 + CH_2{=}CH{-}CH{=}CH_2$$

2. Pyrolysis of 1,1,3,3-tetracyanopropane (1,4).

$$2\,H_2C(CN)_2 + H_2CO \longrightarrow H{-}C(CN)_2{-}CH_2{-}C(CN)_2{-}H \xrightarrow{150\text{–}200^\circ C} H_2C{=}C(CN)_2 + H_2C(CN)_2$$

3. Pyrolysis of 1-acetoxy-1,1-dicyanoethane (1,1-dicyanoethyl acetate, "di-(acetyl cyanide)" (DAC)) (1,5–7).

$$H_2C{=}C{=}O \text{ or } (CH_3CO)_2O + HCN \longrightarrow H_3C{-}C(CN)_2{-}O{-}C(=O){-}CH_3 \xrightarrow[C]{600\text{–}650^\circ} H_2C{=}C(CN)_2 + CH_3COOH$$

4. Pyrolysis of poly(vinylidene cyanide) (9).

$$-\!(CH_2-C(CN)_2)_n\!- \xrightarrow{>160^\circ C} H_2C\!\!<\!\!\begin{smallmatrix}CN\\CN\end{smallmatrix}$$

Commercially, the monomer is prepared by pyrolysis of 1-acetoxy-1,1-dicyanoethane. This method utilizes relatively inexpensive, readily obtainable starting materials. The addition of HCN to ketene or to acetic anhydride is a base-catalyzed reaction (1,8). Suitable catalysts are tertiary amines such as triethylamine, trimethylamine, or amine ion-exchange resins. Yields of 90% or more of 1-acetoxy-1,1-dicyanoethane can be obtained. The addition is conducted in either the liquid or vapor phase in a temperature range of 0–150°C.

It is desirable to use high purity 1-acetoxy-1,1-dicyanoethane for the pyrolysis since vinylidene cyanide is rapidly polymerized by many impurities. Purification of the 1-acetoxy-1,1-dicyanoethane can be accomplished by vacuum distillation followed by recrystallization. The pyrolysis of 1-acetoxy-1,1-dicyanoethane to vinylidene cyanide can be carried out at reduced or at atmospheric pressure using either glass or metal apparatus (1,5–7). In one method of pyrolysis 1-acetoxy-1,1-dicyanoethane vapor at 10 mm pressure is passed through a 3-ft-long, ⅝-in. inside-diameter brass tube packed with a tight roll of brass screen. The tube is heated to 600–650°C. The pyrolysis product is trapped in a receiver cooled with dry ice–acetone and containing a small amount of phosphorus pentoxide as a stabilizer. This pyrolysis product is purified by vacuum distillation (54–57°C at 15 mm), followed by crystallization from dry chloroform at −30°C, washing with dry toluene (cooled to −30°C), and flash distillation to yield pure monomer.

Use of a large variety of surfaces and diluents has little apparent effect upon increasing the yields of vinylidene cyanide from the pyrolysis. Diluents for 1-acetoxy-1,1-dicyanoethane in the pyrolysis help reduce the formation of polymer and carbon in the pyrolysis tube. Among the effective diluents are chlorobenzene, toluene, hydrogen, nitrogen, and natural gas. Heat-transfer studies indicate that optimum conversion of 1-acetoxy-1,1-dicyanoethane to monomer occurs when the 1-acetoxy-1,1-dicyanoethane is heated rapidly to pyrolysis temperature and the products are removed immediately from the hot area. Optimum yields of monomer are obtained at high conversions of 1-acetoxy-1,1-dicyanoethane. A cracking temperature of about 560°C and residence time of about 3.5–5.5 seconds appear to be optimum for good yields (80–85%). Rapid quenching of the pyrolysis product is extremely important.

Purification. The pyrolysis of 1-acetoxy-1,1-dicyanoethane has been found to yield a large number of compounds. Among the compounds identified in the pyrolysis product are: vinylidene cyanide, acetic acid, acetonitrile, pyruvonitrile, malonitrile, dimethylmalonitrile, succinonitrile, acrylonitrile, glutaronitrile, butenonitrile, 1-acetoxy-1,1-dicyanoethane, 2-methylvinylidene cyanide, and acetone. Many methods for purification of this crude pyrolysis product have been investigated. Vinylidene cyanide can be separated by distillation, crystallization, or a combination of both. Distillation requires an efficient fractionating column operated under vacuum if reasonable purity is to be obtained.

The pyrolysis product can also be roughly fractionated, then crystallized from chloroform. When the monomer is about 95% pure it is possible to crystallize it fractionally, without added solvent to obtain pure monomer.

Properties (1). Some physical properties of vinylidene cyanide are listed in Table 1. Vinylidene cyanide is miscible with benzene, nitromethane, and trichloroethylene, and is substantially insoluble in aliphatic hydrocarbons. It polymerizes instantaneously in the cold upon contact with water, alcohols, amines, amides, ketones (except ketones without enol forms, such as benzophenone), and a large number of other organic and inorganic substances. Infrared and Raman spectra have also been determined (10,11).

Stabilization (12–17). Efforts to stabilize vinylidene cyanide have been quite extensive. Many compounds have been tested as possible stabilizers for vinylidene cyanide. Among the effective stabilizers are phosphorus pentoxide, sulfur trioxide, and concentrated sulfuric acid. Negatively substituted sulfonic acids are an extremely effective class of stabilizer; among the sulfonic acids found effective are 2,4-dinitro-1-naphthol-7-sulfonic acid, *m*-nitrobenzenesulfonic acid, *m*-benzenedisulfonic acid, 4-sulfophthalic anhydride, and 2,5-dichlorobenzenesulfonic acid.

Table 1. Physical Properties of Vinylidene Cyanide

boiling point, °C	
at 5 mm Hg	40.0
at 6 mm Hg	42.5
at 8 mm Hg	46.8
at 10 mm Hg	50.5
at 760 mm Hg	154.0
melting point, °C	−9.7
density, d_4^{23}	0.992
refractive index, n_D^{26}	1.4411
molar refraction	
MR_D (obs)	20.7
MR_D (calcd)	19.7
cryoscopic constant, K_f	3.6
latent heat of vaporization, L_v, cal/mole	13,900
heat of fusion, H_f, cal/mole	3,400

Analysis. Analysis for vinylidene cyanide is based on Diels-Alder adduct formation with a measured amount of anthracene. Details of the procedure are given in Refs. 18 and 18a.

Toxicity (1). Vinylidene cyanide is a very strong lacrimator. It has a delayed action on the mucous membranes, slowly developing a formaldehyde sting. Exposure of test animals to vinylidene cyanide in concentrations of 0.96–1.6 ppm for a few hours resulted in some fatalities. The dead animals exhibited diffuse degeneration of the brain, liver, and kidney, and acute chemical pneumonitis.

Homopolymerization

Vinylidene cyanide polymerizes in the cold upon contact with water to form a hard, white, infusible polymer (19–21). This type of initiation occurs with alcohols, amines, and ketones. Anionic polymerization appears to proceed owing to the polarization of π electrons of the double bond and the presence of unshared electrons on the nitrile groups giving the polarized structure (**1**) (20).

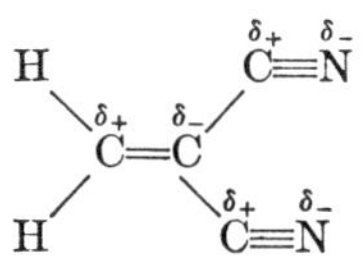

$$\begin{array}{c} \text{H} \quad\quad \overset{\delta_+}{\text{C}}\equiv\overset{\delta_-}{\text{N}} \\ \diagdown\,\overset{\delta_+}{\text{C}}=\overset{\delta_-}{\text{C}}\,\diagup \\ \diagup \quad\quad \diagdown \\ \text{H} \quad\quad \overset{\delta_+}{\text{C}}\equiv\overset{\delta_-}{\text{N}} \end{array}$$

(1)

Initiation would proceed as shown in equation 1.

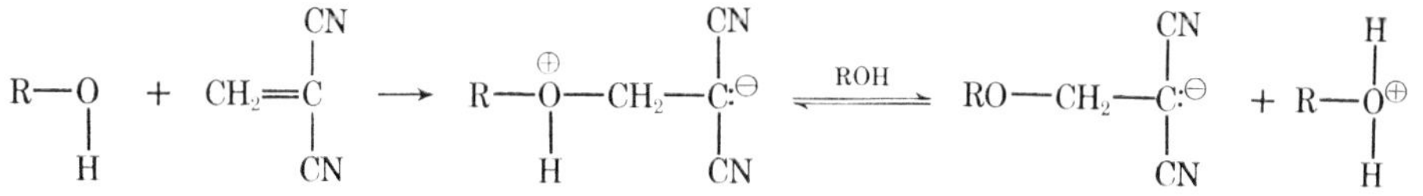

$$R{-}\underset{H}{\underset{|}{O}} + CH_2{=}\underset{CN}{\underset{|}{\overset{CN}{\overset{|}{C}}}} \longrightarrow R{-}\underset{H}{\underset{|}{\overset{\oplus}{O}}}{-}CH_2{-}\underset{CN}{\underset{|}{\overset{CN}{\overset{|}{C}}}}{:}^{\ominus} \xrightleftharpoons{ROH} RO{-}CH_2{-}\underset{CN}{\underset{|}{\overset{CN}{\overset{|}{C}}}}{:}^{\ominus} + R{-}\underset{H}{\underset{|}{\overset{H}{\overset{|}{O}}}}{}^{\oplus} \quad (1)$$

Polymerization then proceeds ionically (eq. 2).

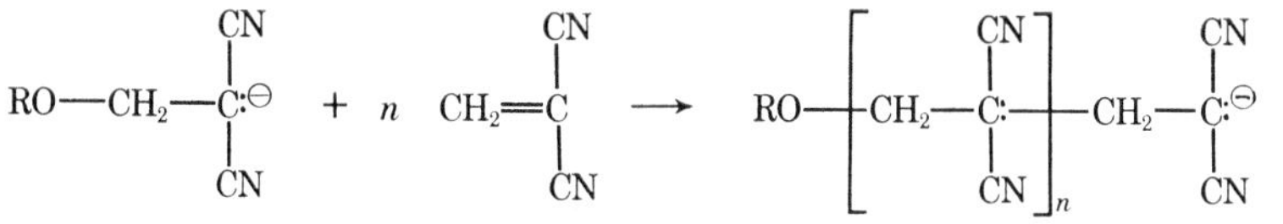

$$RO{-}CH_2{-}\underset{CN}{\underset{|}{\overset{CN}{\overset{|}{C}}}}{:}^{\ominus} + n\ CH_2{=}\underset{CN}{\underset{|}{\overset{CN}{\overset{|}{C}}}} \longrightarrow RO{-}\left[CH_2{-}\underset{CN}{\underset{|}{\overset{CN}{\overset{|}{C}}}}{:}\right]_n{-}CH_2{-}\underset{CN}{\underset{|}{\overset{CN}{\overset{|}{C}}}}{:}^{\ominus} \quad (2)$$

Homopolymerization of vinylidene cyanide with free-radical initiators proceeds very slowly owing to a slow propagation rate. This slow rate is attributed to a high electrostatic repulsion between growing chain and polarized monomer and to resonance stabilization of the chain end.

High-molecular-weight polymers can be prepared by carefully regulating the rate of the polymerization (19). If one such procedure a benzene solution of vinylidene cyanide is added to dimethylformamide benzene solution at 0°C. See also ACRYLONITRILE POLYMERS; ANIONIC POLYMERIZATION; CYANOETHYLATION.

Properties of the Homopolymer

Poly(vinylidene cyanide) is a hard, white, infusible polymer with a density of 1.31 (19). Solvents include dimethylformamide, tetramethylene sulfone, tetramethylurea, triethyl phosphate, diethylcyanamide (19,22–24). The molecular weight of the polymer is related to the specific viscosity by equation 3 (21) where $K_m = 1.5 \times 10^{-4}$ and

$$\bar{M} = \eta_{sp}/K_m c \quad (3)$$

concentration is in moles of monomer per liter of solution. No sharp melting point is observed for the polymer, and x-ray diffraction patterns show little crystallinity. Depolymerization begins at 160°C and the rate increases as the temperature is raised. The polymer is sensitive to moisture, turning dark on standing in moist air and degrading in contact with water or, especially, on contact with bases (19).

Fibers and films can be prepared from solutions of the polymer in appropriate solvents. Tenacities of 77,000 psi at 22% elongation are reported for fibers spun from dimethylformamide, dried, and hot stretched (19). Films cast from dimethylformamide are clear and flexible (21).

The suggested mechanism for the hydrolytic degradation of poly(vinylidene cyanide) involves attack along the chain followed by chain scission (eqs. 4 and 5).

$$\sim\!CH_2-\underset{CN}{\overset{CN}{C}}-CH_2-\underset{CN}{\overset{CN}{C}}\!\sim \xrightarrow{HOH} \sim\!CH_2-\underset{CN}{\overset{CN}{CH}} + HO-CH_2-\underset{CN}{\overset{CN}{C}}\!\sim \quad (4)$$

$$HO-CH_2-\underset{CN}{\overset{CN}{C}}\!\sim \rightleftharpoons CH_2O + H-\underset{CN}{\overset{CN}{C}}\!\sim \quad (5)$$

This mechanism accounts for the appearance of formaldehyde when the homopolymer remains in contact with water. Poly(vinylidene cyanide) reacts with active hydrogen compounds in the presence of bases to give products containing one or more $—CH_2CH(CN)_2$ groups (25–28). The mechanism of this reaction is believed to be according to equations 6–8. These reactions appear to be generally applicable to all

$$AH + B: \rightleftharpoons A^{\ominus} + BH^{\oplus} \quad (6)$$

$$\sim\!\left(CH_2-\underset{CN}{\overset{CN}{C}}\right)_n CH_2-\underset{CN}{\overset{CN}{CH}} + A^{\ominus}$$

$$\rightleftharpoons$$

$$\sim\!\left(CH_2-\underset{CN}{\overset{CN}{C}}\right)_{n-1} CH_2-\underset{CN}{\overset{CN}{C^{\ominus}}} + A-CH_2-\underset{CN}{\overset{CN}{CH}}$$

$$\rightleftharpoons$$

$$\sim\!\left(CH_2-\underset{CN}{\overset{CN}{C}}\right)_{n-1} CH_2-\underset{CN}{\overset{CN}{C}}-H + A-CH_2-\underset{CN}{\overset{CN}{C^{\ominus}}} \quad (7)$$

$$A-CH_2-\underset{CN}{\overset{CN}{C^{\ominus}}} + BH^{\oplus} \xrightarrow{HCl} A-CH_2-\underset{CN}{\overset{CN}{CH}} + BH^{\oplus}Cl^{\ominus} \quad (8)$$

active hydrogen compounds except those that would be expected to give products in which the $—CH_2CH(CN)_2$ group is attached to nitrogen or oxygen.

Copolymerization

Vinylidene cyanide copolymerizes readily with a wide variety of common monomers. Owing to the tendency of vinylidene cyanide to homopolymerize, almost explosively, upon contact with water, sulfones, amides, and alcohols, copolymerizations are usually carried out in bulk or in inert solvents such as benzene and other aromatic hydrocarbons (19).

Three distinct polymerization mechanisms have been observed: (a) autocatalytic, (b) anionic, and (c) free-radical. The autocatalytic copolymerization (ie, with styrene), however, may in reality be ionic with the comonomer acting as the initiator. A charge-transfer mechanism similar to those reported in Ref. 38 also appears possible.

Reactivity Ratios. Table 2 lists monomer reactivity ratios (r_1 and r_2), the products r_1r_2, and the relative reactivities ($1/r_1$) of vinylidene cyanide with several comonomers (29). All of the values were not determined at the same temperature, but no effect of temperature upon reactivity ratios was found in the range studied (0–50°C) (30). The most active comonomers (styrene, 2-chloro-1,3-butadiene, 2,5-dichlorostyrene, and methyl methacrylate) all have electron-rich double bonds (31). The least active comonomers (acrylic acid, vinyl chloride, dichloroethylene, and maleic anhydride) all have electron-poor double bonds.

Table 2. Free-Radical Monomer Reactivity Ratios of Various Comonomers with Vinylidene Cyanide

Comonomer (M_2)	r_1[a]	r_2	r_1r_2	$1/r_1$
styrene	0.001 (0.008–0.004[b])	0.005 (0.003–0.009[b])	5.0×10^{-6}	1000
2-chloro-1,3-butadiene	0.0048 (0.0032–0.010)	0.016 (0.0097–0.048)	7.7×10^{-5}	208
2,5-dichlorostyrene	0.0092 (0.0057–0.010)	0.031 (0.024–0.042)	2.8×10^{-4}	110
methyl methacrylate	0.031 (0.023–0.077)	0.046 (0.035–0.11)	1.4×10^{-3}	32
vinylidene chloride	0.049 (0.043–0.17)	0.012 (0.007–0.25)	5.9×10^{-4}	20
methyl α-chloroacrylate	0.091 (0.064–0.14)	0.41 (0.27–0.55)	3.7×10^{-2}	11
vinyl benzoate	0.10 (0.081–0.51)	0.008 (0.008–0.07)	8.0×10^{-4}	10
vinyl acetate	0.11 (0.073–0.20)	0.0054 (0.0026–0.013)	5.9×10^{-4}	9.1
vinyl chloroacetate	0.13 (0.08–0.18)	0.00 (0.00–0.004)		7.7
2-chloropropene	0.20 (0.14–0.26)			5.0
acrylic acid	0.29 (0.21–0.37)	0.26 (0.20–0.32)	7.5×10^{-2}	3.5
vinyl chloride	0.54 (0.23–0.72)	0.017 (0.007–0.040)	9.2×10^{-3}	1.9
cis-dichloroethylene	30 (20–50)	0.0		0.03
trans-dichloroethylene	30 (20–50)	0.0		0.03
maleic anhydride	45 (35–65)	0.0		0.02

[a] The subscript 1 refers to vinylidene cyanide.

[b] These values are not probable errors, but rather maximum deviations. An objective and quantitative method for calculating the probable errors of r values is simply not available. The differential form of the copolymer equation was solved by the curve-fitting method, or by the method of least squares.

The data for the dichloroethylenes and for maleic anhydride show them to fit very poorly into the copolymerization equation. The mother liquors from these systems were examined carefully for soluble polymers but none was found.

No r values are given for vinylidene cyanide–vinyl ether systems since rapid ionic polymerization apparently precludes the possibility of free-radical initiation. The vinyl ethers are as sensitive to cationic initiation (32,33) as the vinylidene cyanide is to anionic initiation (19). Evidently the ether acts as a base toward vinylidene cyanide, whereas vinylidene cyanide acts as an acid toward the ether. The net result is a novel situation wherein the mixture of the two monomers results in simultaneous cationic and anionic polymerizations. Two polymers were always isolated from these mixtures: a benzene-soluble polymer high in vinyl ether content, and a benzene-insoluble polymer high in vinylidene cyanide content. The formation of the soluble polymer can be inhibited completely by the addition of triethanolamine to the vinyl ether prior to the addition of vinylidene cyanide. The formation of the insoluble polymer can be stopped by the addition of phosphorus pentoxide to vinylidene cyanide prior to the addition of the vinyl ether. Triethanolamine is a known inhibitor for the anionic polymerization of vinylidene cyanide.

The strong tendency of vinylidene cyanide to alternate during copolymerization is immediately apparent from the data in Table 2. Six of the copolymer systems have r_1r_2 products of less than 9.0×10^{-4}. The only copolymers with little tendency to alternate are the dichloroethylenes and maleic anhydride. Monomers having polarities similar to that of vinylidene cyanide would, of course, not be expected to alternate with vinylidene cyanide.

The vinylidene cyanide–styrene and the vinylidene cyanide–dichlorostyrene systems exhibit a rapid "cross" initiation of a type predicted for uncatalyzed copolymerization of more strongly alternating pairs of monomers (34). The system vinylidene cyanide–styrene is very strongly alternating, and free-radical copolymerization does proceed readily at room temperature in the absence of catalysts.

Based on an analysis of the copolymerization data in Table 2, it has been proposed that monomers such as acrylonitrile and vinylidene cyanide exhibit repulsive effects between adding monomer and copolymer containing high proportions of the same monomer (35). Such an effect in free-radical polymerizations should be characteristic of monomers possessing a high order of reactivity in anionic copolymerizations. This type of reactivity depends on the electron deficiency of double bonds arising from electron-withdrawing characteristics of attached groups. Vinylidene cyanide possesses a pronounced dipole due to two nitrile groups conjugated with a methylenic double bond. This type of repulsion is not accounted for in the Mayo-Lewis copolymerization equation, and significant deviations from this equation occur in vinylidene cyanide copolymerizations (36). Equation 9 was derived, assuming negligible r_2 and r_2' values,

$$n - 1 = r_1'x\ (1 + r_1x)/(1 + r_1'x) \tag{9}$$

where x is the molar ratio of monomer 1 to monomer 2 in the reaction mixture, n is the molar ratio of monomer 1 to monomer 2 in the copolymer, r is the reactivity ratio for a radical with the same penultimate and terminal unit, and r' is the ratio for a radical with different units. In this equation the entire deviation is ascribed to the effects of the penultimate unit in the polymer radical on the adding monomer.

The validity of this treatment applied to vinylidene cyanide copolymerizations depends on negligible addition of vinylidene cyanide monomer to a radical ending in vinylidene cyanide at concentrations up to 50 mole percent vinylidene cyanide. Where this effect is slight but significant, r_1' can be estimated from the slope of the tangent drawn from the origin to the copolymer composition–reaction mixture composition curve. Estimation of r_1 remains the same (37). The copolymerization data involving vinylidene cyanide (30) were analyzed for the above effect by plotting $n - 1$ against x. The experimental curves exhibited slope r' at $x = 0$ and slope r as $x \rightarrow \infty$. Systems examined in this way are listed in Table 3. In all cases reactivity of vinylidene cyanide for a growing free radical ending in the other monomer is seen to diminish as the probability increases of finding vinylidene cyanide in the penultimate position. Further discussion of the effects of the penultimate and of more remote units can be found in the article COPOLYMERIZATION.

Q and e Values. Average values of e and log Q based on independent calculations from several comonomer systems have been given as 2.7 and 0.91, respectively (30).

Table 3. Reactivity Ratios from the Barb Equation[a]

M_2	r_2'	r_2
methyl α-chloroacrylate	0.245 ± 0.030	0.40 ± 0.002
acrylic acid	0.168 ± 0.003	0.193 ± 0.003
methyl methacrylate	0.04 ± 0.001	0.082 ± 0.002
2,5-dichlorostyrene	0.022 ± 0.001	0.32 ± 0.001
vinylidene chloride	0.010 ± 0.0005	0.012 ± 0.005
2-chlorobutadiene	0.000	0.057

[a] Equation 9.

The high e and Q values of vinylidene cyanide are consistent with the structure of the monomer, since the two nitrile groups on the one carbon atom cause considerable polarization of the electrons of the carbon–carbon double bond while at the same time providing sites for an odd electron on the vinylidene cyanide radical.

Properties of the Copolymer

Alternating 1:1 copolymers are, with some exceptions (eg, vinylidene cyanide–methyl α-chloroacrylate), distinguished by their resistance to attack by alkali. Vinylidene cyanide homopolymer, as previously stated, degrades when treated with dilute aqueous alkali. Copolymers containing adjacent vinylidene cyanide units have also been found to be alkali sensitive.

The following discussion is concerned with a description of the polymerization and properties of a cross section of vinylidene cyanide–comonomer systems. These descriptions should also, to a limited extent, characterize the copolymerization behavior of vinylidene cyanide with like monomers. That is, α-methylstyrene and vinyltoluene, for example, may be somewhat similar to styrene in their copolymerization behavior. The only vinylidene cyanide copolymer whose preparation has been studied in large-scale equipment is the vinylidene cyanide–vinyl acetate copolymer. This polymer formed the basis for the fiber Darvan, which was first sold commercially by the B. F. Goodrich Co. and subsequently taken from the market (39).

Alternating 1:1 Copolymers. Properties of some alternating copolymers of vinylidene cyanide are listed in Table 4. Many of the alternating copolymers listed are either too brittle or are extremely difficult to process by normal mixing, milling, or calendering operations to form sheets and films. Cyanoethyl esters, cyanoethyl ethers, esters of γ-keto acids, *N*-alkyl-, and *N,N*-dialkylsulfonamides have been recommended as plasticizers that reduce the melt viscosity and facilitate molding (67–69).

Styrene (20,40–42). Vinylidene cyanide copolymerizes rapidly with styrene in the absence of added catalysts to form a perfect alternating copolymer which is alkali resistant. The autocatalytic copolymerization appears to proceed by a free-radical mechanism since hydroxyaromatic compounds can be used as inhibitors or retarders. Radicals necessary for the polymerization may be generated by the interaction of one comonomer with the other (34). At least two competing reactions occur during the autocatalytic styrene–vinylidene cyanide reaction, one giving the copolymer in 50–60% yields, the other a crystalline material in 40–50% yields, composed of one molecule of styrene and two molecules of vinylidene cyanide, which melts sharply at 160–161°C. The copolymers exhibit the physical properties shown in Table 4. Cast films and dry spun fibers were found to be extremely brittle. Substituted styrenes can also be co-

Table 4. Properties of Alternating 1:1 Copolymers of Vinylidene Cyanide

Comonomer	Softening or melting point, °C	Inherent viscosity[a]	Solubility	Melt viscosity	Melt stability	Distinguishing characteristics	Refs.
vinyl acetate	180	4.6–6.5[b,c]	dimethylformamide; tetramethylene sulfide; nitromethane; acetone/H_2O = 85/15; acetonitrile; acetonitrile/H_2O = 85/15; dimethyl sulfone; diethylcyanamide	high	poor	hydrophobic fibers with excellent physical properties; not crystallizable	20,39,42, 49–55
isobutylene	140–148 (115)	1.03–1.76[c]	ketones; dimethylformamide; carbon tetrachloride; tetrahydrofuran; nitromethane	high	good	crystallizable; very brittle molded objects	20,39,42,63
vinylidene chloride	225–235 (205)	1.8–2.2[b,d]	cyclopentanone; cyclohexanone; dimethylformamide; dimethylacetamide; ethylene carbonate, γ-butyrolactone	high	poor	crystallizable; alkali-sensitive	20,39,65
vinyl chloride	200–220	1.2–1.6[b,c]	dimethylformamide; nitromethane	high	poor	not crystallizable; stable toward dilute alkali	20,39,42,66
styrene	180–190	0.85–0.95	acetone; methyl ethyl ketone; cyclohexanone; acetone; γ-butyrolactone; dimethylformamide; nitromethane	very high	good	not crystallizable; very brittle molded objects	20,39–42
butadiene	267–270	too insoluble to determine	hot dimethylformamide	low	very poor	crystallizable	20,39,43–45
2-chloro-1,3-butadiene	240 (155)		dimethylformamide	low	poor	crystallizable	20,46
methyl methacrylate	170–180	6.0–7.0[b]	ketones; dimethylformamide; nitromethane	high	excellent	very brittle molded objects; not crystallizable	20,39,42,47

[a] All inherent viscosity values determined with 0.4 g polymer per 100 ml solvent at 24.2°C.
[b] Intrinsic viscosity values (g/dl).
[c] In dimethylformamide.
[d] In butyrolactone.

polymerized with vinylidene chloride under conditions similar to those employed for styrene (40).

Butadiene (20,45). The conjugated dienes react readily with vinylidene cyanide to give either a Diels-Alder adduct or a copolymer. The polymerization reaction is enhanced if the reaction is run in dilute solution or at a low temperature. Both these conditions reduce the polymerization rate. However, catalysts exist that accelerate the rate of polymerization relative to the rate of adduct formation; such catalyst systems include (a) sulfur dioxide and an organic thiol (43), (b) an inorganic acid (preferably hydrochloric) and an organic thiol (44), and (c) a hydroperoxide initiator activated with hydrogen chloride or sulfur dioxide.

2-Chloro-1,3-butadiene (20,46). Autocatalytic bulk polymerization of the monomers ordinarily proceeds so rapidly and exothermically from room temperature that the resulting copolymer becomes slightly charred. When more than 90% by weight of solvent is used, however, a free-radical initiator is necessary since under these conditions a Diels-Alder adduct is formed in large quantities and only a very small portion of copolymer is obtained. The two monomers enter the polymer in essentially equimolar ratio regardless of the degree of conversion and the charging ratio. The properties of a typical alternating copolymer are listed in Table 4. Patent literature describes these polymers as hard, resinous, nonrubbery materials which are insoluble in benzene, toluene, ethers, and alcohols. They melt at approximately 240°C with slight decomposition and can be melt spun into fibers possessing high tensile strength and low elongation. Filaments are said to be cold drawable.

Methyl Methacrylate (20,42,47). Copolymerization of vinylidene cyanide with methyl methacrylate does not appear to be autocatalytic. Polymerization in bulk, using a feed containing 60 weight % vinylidene cyanide, produces an alternating copolymer. The reaction mixture becomes viscous very rapidly and an accelerated reaction takes place as the viscosity increases. The percent conversion versus time is shown in Figure 1.

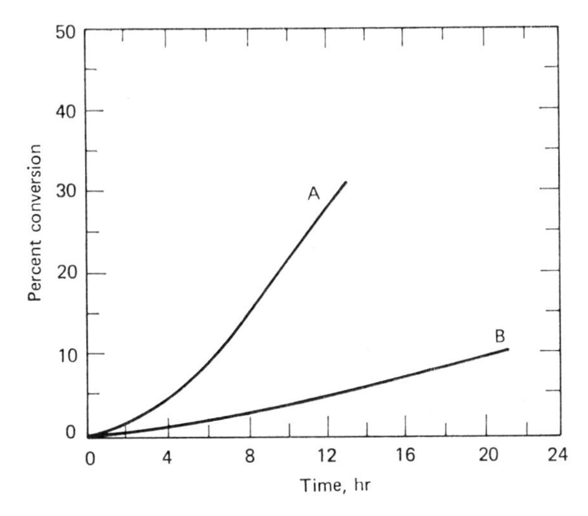

Fig. 1. Monomer conversion versus copolymerization time for (curve A) vinylidene cyanide–methyl methacrylate (60 wt % vinylidene cyanide in feed) at 50°C, under nitrogen, using 0.05% bis-(*o*-chlorobenzoyl) peroxide initiator, and (curve B) vinylidene cyanide–vinylidene chloride (60 wt % vinylidene cyanide in feed), under the same reaction conditions.

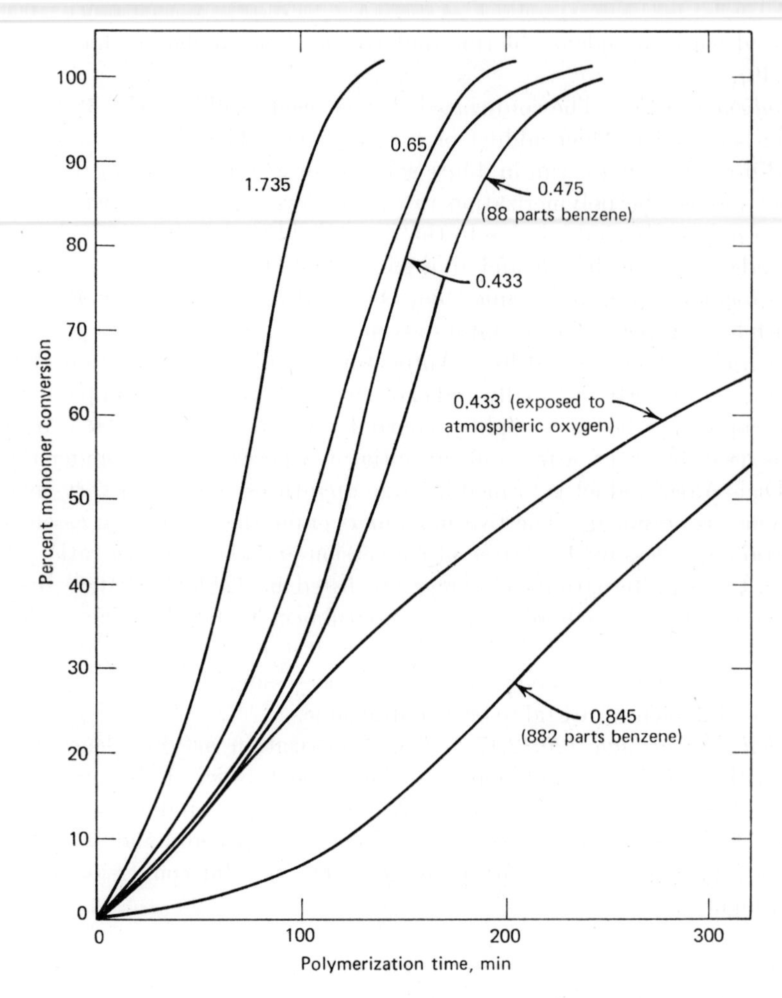

Fig. 2. Percent monomer conversion versus polymerization time at 40°C for vinylidene cyanide–vinyl acetate copolymerization. Parts by weight of bis(*o*-chlorobenzoyl) peroxide are as indicated on the curve.

The vinylidene cyanide–methyl methacrylate system (or vinylidene cyanide with any other comonomer that homopolymerizes) potentially lends itself to the formation of copolymers low in vinylidene cyanide. Methyl methacrylate can theoretically add to its own terminal radical to give copolymers high in methyl methacrylate. A high charging rate of methyl methacrylate, however, does not ensure equal distribution of vinylidene cyanide units throughout all polymer chains. The strong tendency of vinylidene cyanide to alternate will allow the formation of polymer high in vinylidene cyanide in the early stages of the polymerization and polymer low in vinylidene cyanide in the late stages of polymerization. This is especially true when the polymerization is run to high conversion.

Bulk copolymerization of vinylidene cyanide with other acrylates has also been studied (48).

Vinyl Acetate (20,42,49–55). This copolymer formed the basis for the only vinylidene cyanide polymer that has enjoyed commercial status. The fiber, Darvan, was

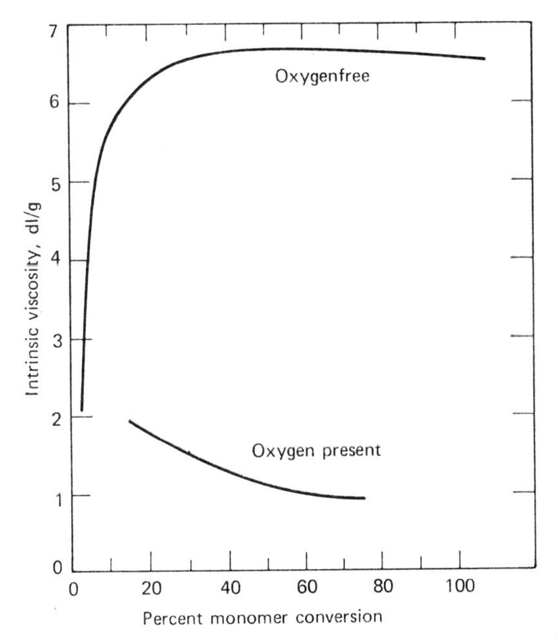

Fig. 3. Intrinsic viscosity (in dimethylformamide at 24.2°C) versus monomer conversion for vinylidene cyanide–vinyl acetate copolymerization.

first offered commercially by the B. F. Goodrich Company and subsequently by the Celanese Corporation and by Hoechst in Germany.

The commercial polymer was prepared by bulk copolymerization of vinylidene cyanide (100 part) and vinyl acetate (807 parts), with bis(*o*-chlorobenzoyl) peroxide (0.433 parts) as initiator, at 40°C. The copolymer is insoluble in the monomer mixture and forms a gelatinous, viscous slurry up to about 35% monomer conversion (about 8.5% total solids) which transform into a semisolid and finally a friable solid at 100% monomer conversion or 23.4% total solids. The polymerization reaction has no prolonged induction period and proceeds to high monomer conversion (Fig. 2). Oxygen inhibits the polymerization and greatly lowers both conversion and the intrinsic viscosity of the resulting polymer (Figs. 2 and 3). The intrinsic viscosity of the polymer is inversely dependent upon the temperature of polymerization. The rate of polymerization is proportional to the square root of the catalyst concentration in the range studied (Figs. 2 and 3). Addition of benzene as diluent retards the polymerization rate (Fig. 2) and lowers the intrinsic viscosity of the polymer.

Benzoyl peroxide may be satisfactorily substituted for bis(*o*,*o*′-dichlorobenzoyl) peroxide to produce comparable vinylidene cyanide–vinyl acetate copolymers; this polymerization, however, is slower and the intrinsic viscosities are higher than that produced with bis(*o*,*o*′-dichlorobenzoyl) peroxide at a given temperature.

Information concerning the molecular-weight distribution of vinylidene cyanide–vinyl acetate copolymer and the relationship of intrinsic viscosity (in dimethylformamide at 25°C) to molecular weight (Fig. 4) has been reported (56).

Values for $\bar{M}_n$ were calculated from osmotic-pressure measurements using the equation $M = (RT/\pi c)_{c\rightarrow 0}$ (see OSMOMETRY) (56,57). Using the method of least squares, the best straight line determined for this relationship is given by the equation $[\eta]_0 = 1.536 \times 10^{-4} M^{0.78}$.

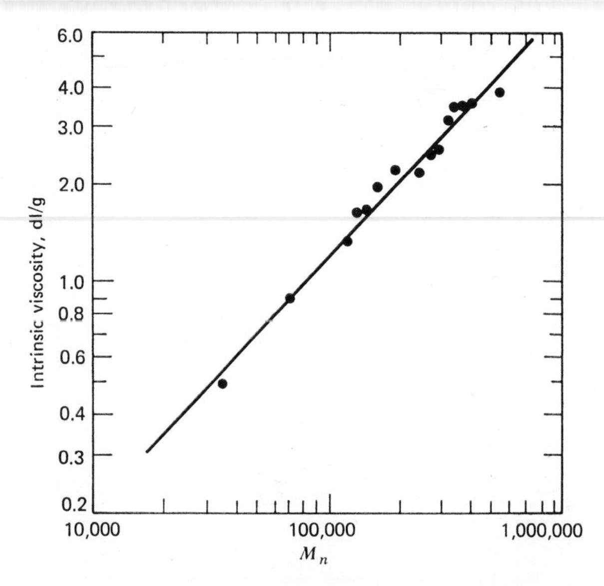

Fig. 4. Intrinsic viscosity versus molecular weight for vinylidene cyanide–vinyl acetate copolymer fractions dissolved in dimethylformamide at 25°C.

X-ray diffraction studies on fibers and films of vinylidene cyanide–vinyl acetate copolymer have also been published (58,59). Spinning solutions of vinylidene cyanide–vinyl acetate copolymers in dimethylformamide apparently require stabilization against degradation and/or discoloration. Various acidic materials such as phosphorus pentoxide and carboxylic acid anhydrides have been recommended (60,61). Sodium nitrite and other inorganic salts are claimed to be stabilizers for acetone–water and acetonitrile–water solutions (62).

Isobutylene (20,42,63). Gaseous isobutylene (or propylene) is sufficiently soluble in a benzene reaction medium to facilitate copolymerization with vinylidene cyanide. It is interesting that the lowest mono-olefin homolog, ethylene, does not form an alternating copolymer with vinylidene cyanide and polymerizes randomly regardless of the charging ratio of the two monomers. 2-Chloropropene forms alternating vinylidene cyanide copolymers softening at about the same temperature as the isobutylene copolymers.

Vinylidene Chloride (20,65). Rate of monomer conversion in vinylidene cyanide–vinylidene chloride bulk copolymerization is indicated in Figure 1. The rate of conversion is greatly increased in going from 45 to 60°C. The softening temperature of the polymer decreases with an increase in polymerization temperature. Conversion, polymer softening temperature, and intrinsic viscosity go through a maximum when the monomer charge is approximately 40 mole percent vinylidene cyanide. If benzene is added as a diluent, a progressive increase in benzene concentration causes a corresponding reduction in conversion for a given reaction time and a lowering of the intrinsic viscosity and softening temperature of the polymer.

Regardless of the method of preparation, all copolymers of vinylidene cyanide and vinylidene chloride undergo a color change when placed in 1% aqueous sodium hydroxide. The intensity of the color developed depends upon the mole percent of vinylidene cyanide in the copolymer and the immersion time.

One of the shortcomings of polymers containing chloroaliphatic repeating units is their heat sensitivity. Vinylidene cyanide–vinylidene chloride alternating copolymers appear to be slightly more stable than unstabilized poly(vinyl chloride).

Vinyl Chloride (20,42,66). Copolymerizations of vinylidene cyanide with vinyl chloride are usually carried out in bulk since the monomers are mutually soluble. Essentially 1:1 alternating copolymers are produced regardless of the monomer charging ratio. The products exhibit excellent resistance to attack by dilute alkaline solutions. The polymers are also quite stable in dimethylformamide solutions from which flexible, almost colorless, films have been cast.

Random Copolymers. Characteristics of some random copolymers of various monomers with vinylidene cyanide are listed in Table 5.

Table 5. Random Copolymers of Various Monomers with Vinylidene Cyanide

Comonomer	Mole percent charged	Mole percent combined	Comment	Refs.
methyl acrylate	5.5 90.8	38.5 82.3	hard resinous copolymers completely insoluble in acetone	70
n-butyl acrylate	15.4 89.1	66.7 92.0	hard resinous copolymers completely insoluble in acetone	70
acrylonitrile	10.0 70.0	10.2 29.2	hard, infusible solid, soluble in dimethylformamide and tetramethylene sulfone	71
acrylic acid	10.0 90.0	27.7 77.2	copolymers with methacrylic acid and sorbic acid are also obtained; physical properties are not reported	72
methyl vinyl ketone	8.2 50.0	40.2 44.6	resinous copolymers softening at 135–140°C	73
vinyl isobutyl ether			two fractions obtained: benzene-soluble (3 mole % $V(CH)_2$) and benzene-insoluble (69.5 mole % $V(CH)_2$)	74,75
ethylene	0.96 3.3	49.6 73.7	alkali-sensitive polymers with poor heat stability; product depolymerizes below 250°C	64
allyl acetate	1.3 92.0	52.5 78.5	cast films obtained from dimethylformamide; intrinsic viscosity varies from 0.4 to 0.6 in dimethylformamide	76

The alkyl acrylate–vinylidene cyanide copolymers are reported to be hard resinous materials that are more thermoplastic than homopolymers of vinylidene cyanide and harder than homopolymers of the alkyl acrylates.

Copolymerization with ethylene yields a mixture containing polyethylene, poly(vinylidene cyanide), and polymers containing indeterminable combinations of the monomers. The products are alkali sensitive and have poor heat stability; they darken at comparatively low temperature and begin to depolymerize below 250°C.

Random copolymers of vinylidene cyanide have also been obtained from allyl chloride (76), allyl cyanide (76), methallyl chloride (76), 1,1-difluoro-2,2-dichloroethylene (77), 1,2-difluoro-1,2-dichloroethylene (77), *cis*- and *trans*-1,2-dichloroethylenes (78), itaconic anhydride (79), maleic anhydride, diethyl maleate, and diethyl fumarate.

Fibers from Vinylidene Cyanide Copolymers

Although a considerable number of the polymers mentioned above are able to form fibers, the vinylidene cyanide–vinyl acetate copolymer is best suited for fiber formation. Many of the copolymers were eliminated because their melting or softening points were below 170–180°C. Likewise brittle copolymers such as vinylidene cyanide–styrene and those having poor melt or alkali sensitivity such as vinylidene cyanide copolymers with dienes, vinyl chloride, or vinylidene chloride are of less interest than the more stable vinyl acetate copolymer.

Fibers have been spun from the vinyl chloride, vinylidene chloride, and vinyl acetate copolymers with vinylidene cyanide by solution techniques, and from the butadiene–styrene–vinylidene cyanide terpolymer by a modified melt-spinning technique.

Fiber properties of three of these copolymers are listed in Table 6; properties of poly(vinylidene cyanide-*co*-vinyl acetate) are discussed in greater detail below.

Fibers of Vinylidene Cyanide–Vinyl Acetate Copolymer. The 1:1 alternating copolymer has been produced commercially as Darvan (formerly Darlan) (20); the fibers are also known by the generic name nytril. Properties of these fibers are listed in Table 7.

The softness, elasticity, resilience, and resistance to moisture of the nytril fiber are attributable to the fine structure. The low degree of orientation and lateral order together with the large number of dipole–dipole contact points provided by the cyano groups result in a strong flexible network structure. These same factors also tend to enhance the hot wet properties of the fiber.

The strength of the nytril fiber, while somewhat lower than that of acrylic and modacrylic fibers, is adequate for textile applications, and under hot wet conditions is superior to that of the acrylic fibers.

The effect of temperature on wet and dry moduli is shown in Figure 5. Treatment of the nytril fiber in saturated steam reveals that considerable shrinkage sets in at 120–130°C, indicating the presence of a wet glass-transition temperature in this region. This is reinforced by the fact that dye penetration is very drastically accelerated by pressure dyeing above 120°C. The setting of impressed configurational geometry such as crimp can be achieved best by treatment in steam at 127°C (20 lb/in.^{2} pressure).

Table 6. Mechanical Properties of Fibers Prepared from Copolymers of Vinylidene Cyanide

Property	Copolymer with butadiene (40%)–styrene (10%)	Copolymer with vinylidene chloride (50%)	Copolymer with vinyl chloride (50%)
tenacity, g/den	3.0	2.8	1.6
elongation, %	14	6.5	19
modulus, g/den	40	34	45
elastic recovery from 3% elongation, dry			
work recovery (resilience), %	37	33	45
tensile strain recovery, %	51	55	79
elastic recovery from 3% elongation, wet			
work recovery (resilience), %	28		38
tensile strain recovery, %	43		68

Table 7. Properties of Nytril Fibers

Property	Value for nytril	Value for steam-set nytril
	Physical properties	
specific gravity[a]	1.21	
sticking temperature,[a] °C	1.70	
crystallinity[b]	none	
orientation	slight	
lateral order	slight	
	Mechanical properties	
tenacity, g/den		
at 23°C, 65% rh	2.2	2.1
at 23°C, wet	1.8	1.8
at 70°C, wet	1.5	1.4
elongation, %		
at 23°C, 65% rh	26	35
at 23°C, wet	30	36
at 70°C, wet	31	40
modulus, g/den		
at 23°C, 65% rh	25	22
at 23°C, wet	20	17
at 70°C, wet	11	10
energy to uncrimp, g-cm/den/cm × 10^4	1.8	0.8
	Elastic properties[c]	
work recovery, %		
at 23°C, 65% rh	22	30
at 23°C, wet	16	33
at 70°C, wet	16	35
permanent set, %		
at 23°C, 65% rh	24	14
at 23°C, wet	34	15
at 70°C, wet	43	9

[a] Ref. 54.
[b] By x-ray studies.
[c] ASTM D 1774-61-T, 5% extension.

The elastic behavior has also been examined by sonic modulus techniques (80). Figure 6 indicates the proportion of strain which is immediately recoverable; the nytril fiber compares very favorably with silk and wool in contrast to the acrylic fiber. Steam setting improves the performance slightly.

Dyeability. The nytril fiber Darvan displays affinity for disperse or nonionic dyes but is penetrated by the dye only with difficulty. Indeed this is one of the serious drawbacks of nytril fiber, necessitating the use of carriers or elevated temperatures and pressures to achieve deeper shades. The improvement in dyeability experienced under pressure at 120°C is believed to result from an increase in diffusion rate that normally occurs above the wet glass-transition temperature. Aside from this, little can be said for the mechanism of dyeability with disperse dyes.

The use of promoters in the dye bath has been shown to effect a decrease in wet glass transition so that pressure dyeing is unnecessary (81–83).

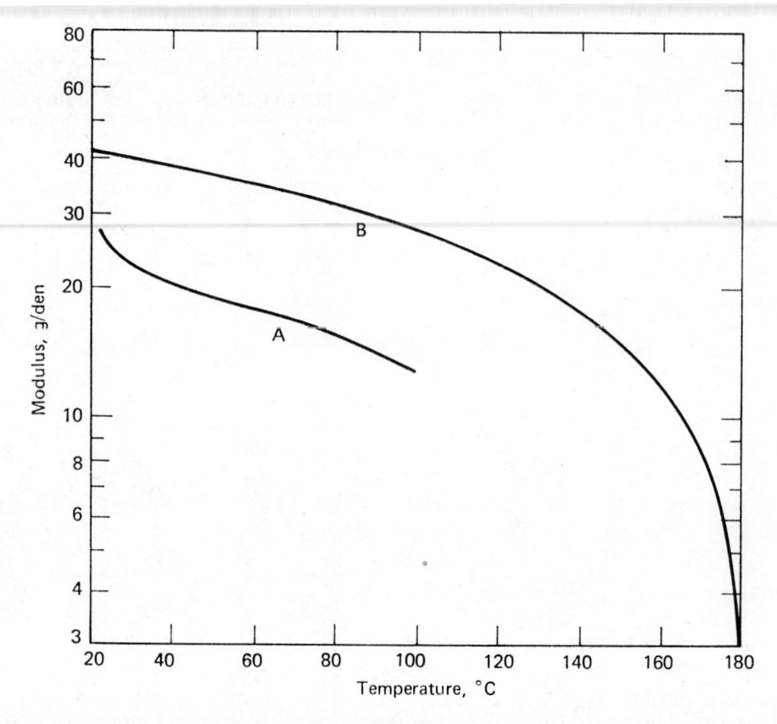

Fig. 5. Modulus–temperature relationship of nytril fiber, wet (curve A) and dry (curve B), measured at 1% strain.

A more chemical approach to improving dyeability involved attempts to convert cyano or acetate groups into more functionally dyeable moieties. Modification of the substrate has been attempted with various reagents (see, eg, Refs. 84–88), all of which resulted in only marginal improvements of no commercial importance.

Table 8. Composition and Properties of Vinylidene Cyanide–Butadiene–Styrene Terpolymers

Monomer	Mole % composition of polymer			
vinylidene cyanide	50	50	50	50
butadiene	50	34	27	
styrene		16	23	50
Property	**Value**			
melting point, °C	270			
softening point, °C		170	180	186
		180	195	186
T_1,[a] °C	140		120	152
T_2,[b] °C	270		195	186
solubility in dimethyl-formamide	insoluble	soluble	soluble	soluble
film properties	brittle	flexible	tough	brittle

[a] T_1 is closely related to the second-order transition temperature or to the softening point, whichever is higher.

[b] T_2 is the temperature at which a fixed constant rate of extrusion is obtained.

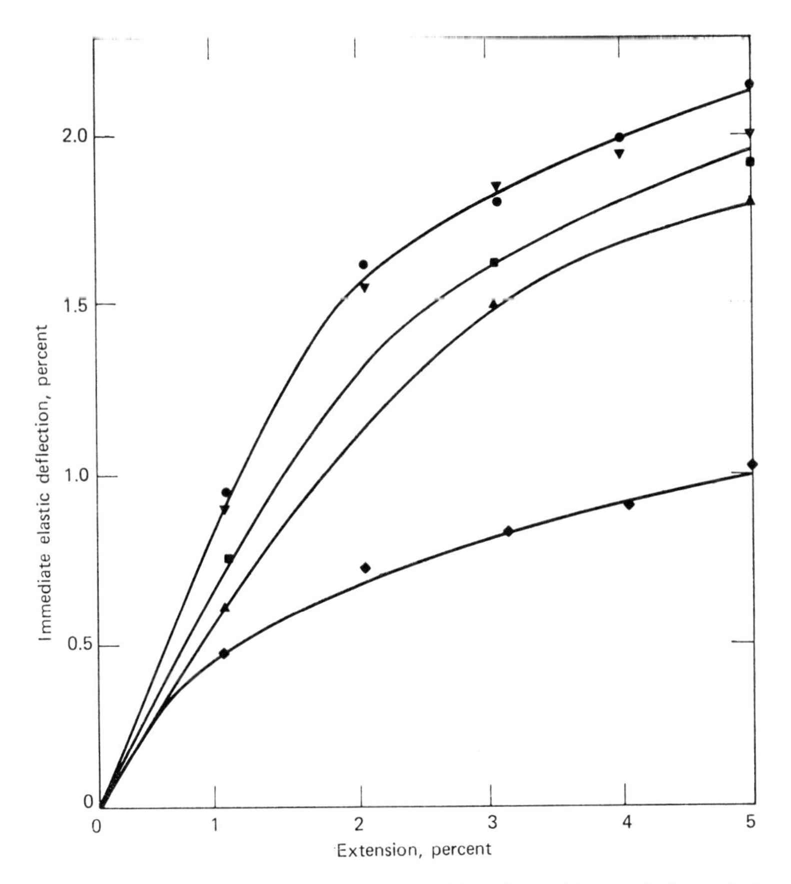

Fig. 6. Proportion of strain immediately recoverable. Key: (▲) nytril fiber; (■) nytril fiber, steam-set; (●) silk; (▼) wool; (◆) acrylic (acrylonitrile and less than 10% polar vinyl comonomers).

Co-spinning of polymeric and nonpolymeric additives with nytril fiber proved to be a more successful approach. Of a great number of additives tested for this purpose, however, only a very few successfully opened the structure enough to accelerate dyeability without significantly reducing the physical properties. The most effective opener was about 5–10% of a copolymer of *N,N*-dimethylacrylamide with vinyl acetate (89,90). Other additives that have been reported to be effective openers were poly(vinyl formal) (91) and poly(alkylene glycols) and other derivatives (92). The primary effect of co-spun additives is to accelerate diffusion of disperse dyes into the fiber rather than promote affinity for acid or cationic dyes. An acid dyeable composition was obtained by co-spinning poly(vinylidene cyanide–vinyl acetate) with a copolymer of styrene and a vinylpyridine (93). Basic dyeability of poly(vinylidene cyanide–vinyl acetate) can be achieved by terpolymerization with strongly acid monomers (94), most notably with styrenesulfonic acid (95,96).

Terpolymers of Vinylidene Cyanide for Plastics

A considerable number of terpolymers of vinylidene cyanide have been evaluated or use as plastics. The principal objectives were to modify solubility, overcome brittleness, and raise the melting of softening point of the copolymers to an optimum range for the extrusion of clear plastics. Property improvements, however, are not as readily achieved for plastic end uses as for the fiber modifications mentioned above. The influence of terpolymerization of vinylidene cyanide with various monomers has been studied (44). Properties of some vinylidene cyanide–butadiene–styrene terpolymers are summarized in Table 8.

Terpolymers have also been prepared from combinations of monomers including vinyl acetate, vinylidene chloride, and acrylonitrile, as well as with two dienes (butadiene, isoprene, 2,3-dimethylbutadiene, and/or piperylene) (96). In general, terpolymerization did not overcome the poor melt stability of the diene copolymers. Vinylidene cyanide–vinyl acetate terpolymerized with acrylonitrile, methyl methacrylate, styrene, and vinyl chloride has also been studied (96).

In summary, it can be said that in an extensive effort to develop moldable or extrudable plastics based in vinylidene cyanide and a great number of monomer combinations, the one that showed the greatest promise was the vinylidene cyanide–butadiene–styrene terpolymer, but because of inadequate properties it was never produced commercially.

Bibliography

1. A. E. Ardis, S. J. Averill, H. Gilbert, F. F. Miller, R. F. Schmidt, F. D. Stewart, and H. L. Trumbull, *J. Am. Chem. Soc.* **72,** 1305 (1950).
2. A. E. Ardis, S. J. Averill, H. Gilbert, F. F. Miller, R. F. Schmidt, F. D. Stewart, and H. L. Trumbull, *J. Am. Chem. Soc.* **72,** 3127 (1950).
3. A. E. Ardis (to B. F. Goodrich & Co.), U.S. Pat. 2,502,412 (April 4, 1950).
4. H. Gilbert (to B. F. Goodrich & Co.), U.S. Pat. 2,514,387 (July 11, 1950).
5. A. E. Ardis (to B. F. Goodrich & Co.), U.S. Pat. 2,467,270 (July 19, 1949).
6. L. F. Reuter and R. D. Smith (to B. F. Goodrich), U.S. Pat. 2,663,721 (Dec. 22, 1953).
7. A. E. Ardis (to B. F. Goodrich), U.S. Pat. 2,663,726 (Dec. 22, 1953).
8. A. E. Ardis (to B. F. Goodrich), U.S. Pat. 2,623,062 (Dec. 23, 1952).
9. A. E. Ardis and H. Gilbert (to B. F. Goodrich), U.S. Pat. 2,535,827 (Dec. 26, 1950).
10. M. L. Dannis and J. J. Shipman, *J. Chem. Phys.* **19,** 382 (1951).
11. A. Rosenberg and J. P. Devlin, *Spectrochim. Acta* **21** (9), 1613 (1965).
12. H. Gilbert (to B. F. Goodrich), U.S. Pat. 2,467,398 (April 19, 1949).
13. H. Gilbert and A. E. Ardis (to B. F. Goodrich), U.S. Pat. 2,614,117 (Oct. 14, 1952).
14. F. F. Miller (to B. F. Goodrich), U.S. Pat. 2,614,118 (Oct 14, 1952).
15. F. F. Miller (to B. F. Goodrich), U.S. Pat. 2,535,861 (Oct, 14, 1952).
16. A. E. Ardis (to B. F. Goodrich), U.S. Pat. 2,665,298 (Jan. 5, 1954).
17. A. E. Ardis (to B. F. Goodrich), U.S. Pat. 2,665,299 (Jan. 5, 1954).
18. W. P. Tyler, D. W. Beesing, and S. J. Averill, *Anal. Chem.* **26,** 674 (1954).
18a. S. J. Averill and H. L. Trumbull, *J. Am. Chem. Soc.* **76,** 1159 (1954).
19. H. Gilbert, F. F. Miller, S. J. Averill, R. F. Schmidt, F. D. Stewart, and H. L. Trumbull, *J. Am. Chem. Soc.* **76,** 1074 (1954).
20. B. S. Sprague, H. E. Green, L. F. Reuter, and R. D. Smith, *Angew. Chem.* (*Intern. Ed. Eng.*) **1,** 425 (1962).
21. R. F. Schmidt, A. E. Ardis, and H. Gilbert (to B. F. Goodrich), U.S. Pat. 2,589,294 (March 18, 1952).
22. A. E. Ardis (to B. F. Goodrich), U.S. Pat. 2,574,369 (Nov. 6, 1951).
23. R. F. Schmidt (to B. F. Goodrich), U.S. Pat. 2,594,353 (April 29, 1952).
24. A. E. Ardis (to B. F. Goodrich), U.S. Pat. 2,600,180 (June 10, 1952).
25. J. C. Westfahl (to B. F. Goodrich), U.S. Pat. 2,864,850 (Dec. 16, 1958).
26. J. C. Westfahl (to B. F. Goodrich), U.S. Pat. 2,804,469 (Aug. 27, 1957).
27. J. C. Westfahl and T. C. Greshem, *J. Am. Chem. Soc.* **78,** 2588 (1956).
28. J. C. Westfahl, *J. Am. Chem. Soc.* **80,** 871(1958).

29. F. R. Mayo and F. M. Lewis, *J. Am. Chem. Soc.* **66,** 1594 (1944).
30. H. Gilbert, F. F. Miller, S. J. Averill, E. J. Carlson, V. L. Folt, H. J. Heller, F. D. Stewart, R. F. Schmidt, and H. L. Trumbull, *J. Am. Chem. Soc.* **78,** 1669 (1956).
31. C. C. Price, *Mechanisms of Reactions at Carbon–Carbon Double Bonds,* Interscience Publishers, Inc., New York, 1945, p. 96.
32. E. R. Blout and H. Mark, *Monomers,* Interscience Publishers, Inc., New York, 1951, pp. 16, 37.
33. D. D. Eley and A. W. Richards, *Trans. Faraday Soc.* **45,** 425 (1949).
34. C. Walling, *J. Am. Chem. Soc.* **71,** 1930 (1949).
35. G. E. Ham, *J. Polymer Sci.* **24,** 349 (1957).
36. W. G. Barb, *J. Polymer Sci.* **11,** 117 (1953).
37. G. E. Ham, *J. Polymer Sci.* **14,** 87 (1954).
38. N. G. Gaylord and A. Takahashi, "Addition and Condensation Polymerization Processes," *Advan. Chem.* **99,** pp. 94–125.
39. *Chem. Week,* Jan. 16, 1960, p. 21.
40. F. F. Miller (to B. F. Goodrich), U.S. Pat. 2,615,868 (Oct. 28, 1952).
41. M. Hirooka, *Kobunshi* **16** (188), 1172 (1967); through *Chem. Abstr.* **68,** 50374z (1968).
42. H. Gilbert and F. F. Miller (to B. F. Goodrich), U. S. Pat. 2,745,814 (May 15, 1956).
43. H. Gilbert and F. F. Miller (to B. F. Goodrich), U.S. Pat. 2,740,769 (April 3, 1956).
44. H. Gilbert and F. F. Miller (to B. F. Goodrich), U.S. Pat. 2,740,770 (April 3, 1956).
45. S. J. Averill (to B. F. Goodrich), U.S. Pat. 2,615,873 (Oct. 28, 1952).
46. S. J. Averill (to B. F. Goodrich), U.S. Pat. 2,615,872 (Oct. 28, 1952).
47. V. L. Folt (to B. F. Goodrich), U.S. Pat. 2,615,871 (Oct. 28, 1952).
48. V. L. Folt and H. Gilbert (to B. F. Goodrich), U.S. Pat. 2,615,879 (Oct. 28, 1952).
49. J. B. Dickey (to Eastman Kodak), U.S. Pat. 2,466,395 (April 5, 1949).
50. H. Gilbert and F. F. Miller (to B. F. Goodrich), U.S. Pat. 2,615,866 (Oct. 28, 1952).
51. G. V. Wooton and H. M. Hoxie (to B. F. Goodrich), U.S. Pat. 2,871,214 (Jan. 27, 1959).
52. H. Gilbert and F. F. Miller (to B. F. Goodrich), U.S. Pat. 2,649,426 (Aug. 18, 1953).
53. G. V. Wooton (to B. F. Goodrich), U.S. Pat. 2,862,903 (Dec. 2, 1958).
54. A. E. Ardis (to B. F. Goodrich), U.S. Pat. 2,600,180 (June 10, 1952).
55. B. F. Goodrich, Brit. Pat. 809,346 (Feb. 25, 1959).
56. J. A. Yanko, *J. Polymer Sci.* **22,** 153 (1956).
57. J. A. Yanko, *J. Polymer Sci.* **19,** 437 (1956).
58. J. A. Yanko and J. W. Born, *J. Polymer Sci.* **27,** 145 (1958).
59 T Yurugi, A. Yamaguchi, and T. Ogihara, *Seni Gakkaishi* **17,** 1197 (1961).
60. Kurashiki Rayon Co., Japan, Brit. Pat. 951,219 (March 4, 1964).
61. S. J. Averill (to B. F. Goodrich), U.S. Pat. 2,614,090 (Oct. 14, 1952).
62. G. V. Wooton and H. M. Hoxie (to B. F. Goodrich), U.S. Pat. 2,871,214 (Jan. 27, 1959).
63. A. E. Ardis (to B. F. Goodrich), U.S. Pat. 2,615,865 (Oct. 28, 1952).
64. G. Gilbert and F. F. Miller (to B. F. Goodrich), U.S. Pat. 2,615,874 (Oct. 28, 1952).
65. V. L. Folt (to B. F. Goodrich), U.S. Pat. 2,615,870 (Oct. 28, 1952).
66. V. L. Folt (to B. F. Goodrich), U.S. Pat. 2,615,869 (Oct. 28, 1952).
67. H. J. Heller (to B. F. Goodrich), U.S. Pat. 2,952,653 (Sept. 13, 1960).
68. D. G. Dobay (to B. F. Goodrich), U.S. Pat. 2,838,467 (June 10, 1958).
69. D. G. Dobay (to B. F. Goodrich), U.S. Pat. 2,855,375 (Oct. 7, 1958).
70. V. L. Folt (to B. F. Goodrich), U.S. Pat. 2,615,880 (Oct. 28, 1952).
71. H. Gilbert (to B. F. Goodrich), U.S. Pat. 2,628,954 (Feb. 17, 1953).
72. E. J. Carlson (to B. F. Goodrich), U.S. Pat. 2,657,197 (Oct. 27, 1953).
73. H. Gilbert and F. F. Miller (to B. F. Goodrich), U.S. Pat. 2,654,724 (Oct. 6, 1953).
74. B. F. Goodrich, Brit. Pat. 756,839 (Sept. 12, 1956).
75. F. F. Miller and H. Gilbert (to B. F. Goodrich), Ger. Pat. 953,660 (Dec. 6, 1956).
76. H. Gilbert, F. F. Miller, and V. L. Folt (to B. F. Goodrich), U.S. Pat. 2,650,911 (Sept. 1, 1953).
77. H. Gilbert and F. F. Miller (to B. F. Goodrich), U.S. Pat. 2,654,728 (Oct. 6, 1953).
78. V. L. Folt (to B. F. Goodrich), U.S. Pat. 2,615,878 (Oct. 28, 1952).
79. W. Goeltner and H. Hoyer (to Farbwerke Hoechst), Ger. Pat. 1,137,553 (Oct. 4, 1962).
80. W. J. Hamburger et al., *Textile Res. J.* **22,** 695 (1952).
81. H. Heller (to B. F. Goodrich), U.S. Pat. 2,848,296 (Aug. 19, 1958).

82. H. G. Sommar (to B. F. Goodrich), U.S. Pat. 2,921,831 (Jan. 19, 1960).
83. M. Kallman, and V. Salvin (to Celanese Corp.), U.S. Pat. 3,159,519 (Nov. 10, 1964).
84. S. M. Davis and G. Gateff (to B. F. Goodrich), U.S. Pat. 2,819,253 (Jan. 7, 1958).
85. H. Gilbert and G. V. Wooton (to B. F. Goodrich), Ger. Pat. 1,069,567 (Nov. 26, 1959).
86. H. Gilbert and G. V. Wooton (to B. F. Goodrich), U.S. Pat. 3,031,253 (April 24, 1962).
87. O. Yamamoto (to Bureau of Industrial Technics), Japan. Pat. 5,292 (1962).
88. W. Happe, M. Lederer, and G. Messwarb (to Farbwerke Hoechst A. G.), Ger. Pat. 1,226,303 (Oct. 6, 1966).
89. J. C. McCarty and M. C. Pastorelle, Fr. Addn. 83,141 (June 19, 1964), U.S. Application, Feb. 28, 1962.
90. F. D. Stewart (to Celanese Corp.), U.S. Pat. 3,139,621 (1964).
91. O. Fukushima and K. Matsubayashi (to Kurashiki Rayon Co. Ltd.), U.S. Pat. 3,137,675 (June 16, 1964).
92. J. R. Adams, Jr., U. S. Salvin, and C. L. Smart (to Celanese Corp.), U.S. Pat. 3,206,420 (Sept. 14, 1965).
93. J. C. McCarty and F. D. Stewart (to Celanese Corp.), U.S. Pat. 3,103,497 (Sept. 10, 1963).
94. H. Heller (to B. F. Goodrich), U.S. Pat. 2,838,476 (June 10, 1958).
95. A. B. Conciatori and C. L. Smart (to Celanese Corp.), U.S. Pat. 3,180,857 (April 27, 1965).
96. H. Gilbert and F. F. Miller (to B. F. Goodrich), U.S. Pat. 2,716,104 (Aug. 23, 1955).

A. B. Conciatori, L. E. Trapasso, and R. W. Stackman
Celanese Research Company

VINYLIDENE FLUORIDE POLYMERS

PLASTICS

Poly(vinylidene fluoride) is the name given to the polymer of 1,1-difluoroethylene. The structure of the repeating unit is $—CH_2—CF_2—$. The vinylidene fluoride monomer $CH_2{=}CF_2$ is readily polymerized by free-radical initiators to give high-molecular-weight crystalline polymers containing over 59% fluorine by weight. Hydrogen and fluorine atoms are spatially symmetrical, which optimizes cross-bonding forces between polymer molecules. The polymer has application in many areas of the chemical processing industry, in the electrical and electronics industry, as long-lasting exterior coating, and in other fields where high-performance plastics are required.

Monomer

Vinylidene fluoride, or 1,1-difluoroethylene, is a colorless, nearly odorless gas boiling at −82°C. It can be prepared by dehydrohalogenation of 1-chloro-1,1-difluoroethane (1–6), 1-bromo-1,1-difluoroethane (7), or 1,1,1-trifluoroethane (8), or by dehalogenation of 1,2-dichloro-1,1-difluoroethane (9–12).

$$CH_3CClF_2 \xrightarrow{-HCl} CH_2{=}CF_2$$

$$CH_3CBrF_2 \xrightarrow{-HBr} CH_2{=}CF_2$$

$$CH_3CF_3 \xrightarrow{-HF} CH_2{=}CF_2$$

$$CH_2Cl{-}CClF_2 \xrightarrow{-Cl_2} CH_2{=}CF_2$$

The preferred monomer precursor, 1-chloro-1,1-difluoroethane, has been prepared from acetylene, vinylidene chloride, and 1,1,1-trichloroethane.

$$CH{\equiv}CH + 2\ HF \longrightarrow CH_3{-}CHF_2 \xrightarrow{Cl_2} CH_3{-}CClF_2 + HCl$$

$$CH_2{=}CCl_2 + 2\ HF \longrightarrow CH_3{-}CClF_2 + HCl$$

$$CH_3{-}CCl_3 + 2\ HF \longrightarrow CH_3{-}CClF_2 + 2\ HCl$$

The monomer can also be continuously prepared by the pyrolysis of trifluoromethane in the presence of a catalyst or with either methane or ethylene (13–16). Using either methanol or dichloromethane as a source of carbene moiety, vinylidene fluoride can be continuously prepared from difluorochloromethane (17, 18). Pyrolysis of dichlorodifluoromethane with either methane or methyl chloride yields the monomer (19,20). Methane with either trifluorobromo- or trifluorochloromethane can also yield vinylidene fluoride (21) on pyrolysis.

Vinylidene fluoride is stored and shipped without polymerization inhibitor. If desired, terpenes or quinones can be added to inhibit polymerization. E. I. du Pont de Nemours & Co., Inc., and Pennwalt Corporation are producers of vinylidene fluoride in the United States. The properties of the monomer are given in Table 1.

Table 1. Properties of Vinylidene Fluoride

		Reference
formula	$CH_2{=}CF_2$	
molecular weight	64.04	
melting point, °C	−144	
boiling point, °C	−84	
critical constants		
T_c, °C	30.1	103
P_c, psig	643	103
D_c, g/ml	0.417	103
density, liquid, at 23.6°C, g/cm^3	0.617	103
explosive limits, vol % in air	5.8–20.3	104
toxicity	reported to be nontoxic	105
heat of formation, $\Delta H_f°$(g), kcal/mole	-80 ± 0.8	106
activation energy, E_a, kcal/mole	38.54	41
radiation-chem. yield, moles/100 eV	4.3×10^4	40
heat of vaporization, at 25°C, cal/g-mole	980	103
specific heat, gas, at C_p 25°C, cal/(mole)(°C)	14.37	103

Polymerization

Vinylidene fluoride readily polymerizes by a radical-initiated reaction. Recent reports reveal that the monomer can also be polymerized with coordination catalysts of the Ziegler-Natta type (22, 23).

The first successful polymerization of vinylidene fluoride was disclosed in 1944 (24). This patent mentions aqueous recipes utilizing radical initiators such as benzoyl peroxide, ammonium persulfate and sodium bisulfite, and oxygen. The recommended reaction pressure was above 300 atmospheres at a temperature of 20–250°C.

Subsequent studies have shown that vinylidene fluoride can be polymerized at lower pressures and that the reaction can be carried out in suspension, emulsion, or solution. In the polymerization of vinylidene fluoride, the proper selection of the radical initiator, dispersing agent, and reaction medium is important. Polymerization pressures in most processes range from 10 to 300 atmospheres and temperatures from 10 to 150°C.

For the emulsion polymerization of vinylidene fluoride, as in the case of most fluorine-containing monomers, a chemically stable fluorinated surfactant must be employed. Buffers may be used. Both inorganic peroxy compounds such as persulfates and organic peroxides (25–27) may be used as polymerization initiators. Organic percarbonate compounds were also found to be useful initiators for vinylidene fluoride polymerization in emulsion (28). Vinylidene fluoride polymerizations in suspension have been carried out using an aqueous recipe with and without colloidal dispersants. As initiators, organic percarbonate and peroxy compounds are used (29,30). Solution polymerization of vinylidene fluoride in various solvents and initiated by radical initiators (31,32) or by ionizing irradiation has also been reported (33,34).

Alkylboron compounds activated by oxygen or oxygen compounds initiate vinylidene fluoride polymerization in the presence of water with or without surface-active substances or in organic solvents at temperatures between −50 and +50°C (35). Also di-*tert*-alkyl hyponitrites were found to initiate vinylidene fluoride polymerization (36). The gamma-ray (^{60}Co) induced polymerization has been investigated by several authors (33,34,37–40).

Many copolymers of vinylidene fluoride have been reported. Significant are copolymers with tetrafluoroethylene (42), chlorotrifluoroethylene (43,44), hexafluoropropene (45), or pentafluoropropene (46, 47). The last three produce high-performance elastomers with excellent high-temperature properties. In this class belong terpolymers of vinylidene fluoride, tetrafluoroethylene, and hexafluoropropene (48) or pentafluoropropene (49). Waxes have also been reported from vinylidene fluoride (50). In general, other monomers are capable of being polymerized with vinylidene fluoride. Prime interest has been concentrated on the mentioned fluorocarbon monomers which help retain or enhance the desirable thermal, chemical, and mechanical properties of the vinylidene structure.

Properties

Poly(vinylidene fluoride) is a crystalline polymer with about 68% crystalline content (34), high mechanical strength, and high impact strength, especially when compared to poly(tetrafluoroethylene) or poly(chlorotrifluoroethylene). Oriented poly(vinylidene fluoride) films and fiber exhibit exceptional mechanical strength. In addition, poly(vinylidene fluoride) maintains superior resistance to elastic deformation under load (as creep), as well as exceptional ability to withstand repeated flexure or fatigue. Its resistance to abrasion by a variety of tests is excellent, especially considering its hardness. Typical physical properties of poly(vinylidene fluoride) are shown in Table 2. The thermochemical reaction of poly(vinylidene fluoride) has been described by equation 1 (51):

$$1/n\ (CH_2{-}CF_2)_n \text{ solid} + CO_2\ (g) + 20\ H_2O\ (liq) = 2\ CO_2\ (g) + 2\ [HF.10\ H_2O\ (liq)] \quad (1)$$

Heat of combustion ($\Delta H_c°$), -225.99 kcal/mole
Heat of formation ($\Delta H°_{f298.15}$), -113.3 kcal/mole

The characteristics of the polymer depend on molecular weight, molecular-weight-distribution, extent of irregularities along the polymer chain, and crystalline form. Vinylidene fluoride adds primarily to the polymer chain through the CH_2 group, forming $RCH_2{-}CF_2{-}$ units (R is the polymer or copolymer residue) (52). High-resolution NMR has established (53–55) that approximately 5% of the monomer units are re-

Table 2. Properties of Poly(vinylidene Fluoride)

Property	Value
clarity	transparent to translucent
melting point, crystalline, °C	158–197
specific gravity	1.75–1.80
refractive index, n_D^{25}	1.42
molding temperatures, °C	200–275
mold shrinkage, average, cm/cm	0.020
color possibilities	unlimited
machining qualities	excellent
flammability	self-extinguishing, nondripping
tensile strength, psi	
at 25°C	6100–8500
at 100°C	5000
elongation, %	
at 25°C	50–500
at 100°C	400
yield point, psi	
at 25°C	5500–7500
at 100°C	2500
creep, at 2000 psi and 25°C for 10,000 hr, cm/cm	0.02
compressive strength, at 25°C, psi	10,000
modulus of elasticity, at 25°C, psi	
in tension	1.2×10^5–2.4×10^5
in flexure	2.0×10^5
in compression	1.2×10^5
Izod impact, at 25°C, ft-lb/in.	
notched	2.6–3.8
unnotched	30
durometer hardness, Shore, D scale	77–80
heat-distortion temperature, °C	
at 66 psi	141–168
at 264 psi	73–128
abrasion resistance, Tabor CS-17, 0.5 kg load, mg/1000 cycles	17.6
coefficient of sliding friction to steel	0.14–0.17
thermal coefficient of linear expansion, per °C	1.5×10^{-6}
thermal conductivity, at 25–160°C, cal/(hr)(cm²)(°C/cm)	1.0
specific heat, cal/(g)(°C)	0.33
thermal degradation temperature, °C	>315
low-temperature embrittlement, °C	<−62
water absorption, %	0.04
moisture vapor permeability, for 1 mil thickness, g/(24 hr)(m²)	1.0
radiation resistance (Co^{60}), megarad (retains tensile strength of about 85% of its original value)	1000

versed (head-to-head) along the polymer chain, but it is believed that tail-to-tail addition takes place following this head-to-head structure.

Polymorphism of poly(vinylidene fluoride) has been the subject of numerous papers (34,56–74). Poly(vinylidene fluoride) exists in at least two crystalline forms. There is some controversy about structural detail of the poly(vinylidene fluoride) forms and about the mechanism of their formation. The structure of the planar zigzag conformation defined as Phase I (beta form) has been well established (52,53,57,64). The chain molecules of Phase II (alpha form) have a *trans*-gauche-*trans*-gauche conformation as proposed by Myazawa for poly(vinylidene chloride) (75). Phase III (59–61, 66,72,73) is less known. Thermomechanical conditions for transitions among the three modifications are summarized in Ref. 72. It is inferred that polymorphism of poly(vinylidene fluoride) is affected by thermomechanical treatment (55,61,71,72), polymerization conditions (34,68,69,74), and solvents used to cast films (60,67,70). Crystallization from the melt, by either slow or rapid cooling, yields only Phase II (61,62,64).

Due to the polymorphism of poly(vinylidene fluoride), various melting points and densities have been noted for the polymer (34,50,72). The *crystalline melting point* varies from 158 to 197°C as defined by differential scanning calorimeter (DSC). The *glass-transition temperature* of poly(vinylidene fluoride) is −40°C (77–79).

The *electrical properties* of poly(vinylidene fluoride) are listed in Table 3. Dielectric strength and volume resistivity are quite high. Dielectric properties of poly(vinylidene fluoride) are sensitive to the structure of the molecular chain, because poly(vinylidene fluoride) has a distinct dipole in the main polymer chain and two types of crystal structures which differ chemically, thermally, and mechanically from one another. The dielectric absorption spectrum of poly(vinylidene fluoride) depends on the crystalline modification and is affected by molecular orientation (80–86). The contribution of dipole orientation to the dielectric constant of poly(vinylidene fluoride) is markedly affected by passage between the rolls of a calender (84).

Table 3. Electrical Properties of Poly(vinylidene Fluoride) Homopolymer

	60 Hz	10^3 Hz	10^6 Hz	10^9 Hz
dielectric constant	8.40	7.72	6.43	2.98
dissipation factor	0.0497	0.0191	0.159	0.11
volume resistivity, Ω-cm				2×10^{14}
dielectric strength, short time, V/mil				
125-mil thickness				260
8-mil thickness				1280

Unlike most vinylidene polymers, poly(vinylidene fluoride) undergoes *crosslinking* when exposed to ionizing radiation (87,88). The dose tolerance of poly(vinylidene fluoride) is very high. A tensile strength of about 85% of the original value is retained to doses as high as 1000 megarads. The solubility behavior, indicating crosslinking, is a function of dose. Even at doses as low as 2 megarads solubility is reduced to 74%, and the solubility decreases rapidly until it is only 4% after a dose of 590 megarads (88). This property of poly(vinylidene fluoride) is the basis of several technical applications (76,89–92).

The *thermal behavior* of poly(vinylidene fluoride) has been studied by oven pyrolysis (93). The predominant reaction appears to be the elimination of hydrogen fluoride

(94). Poly(vinylidene fluoride) is a good heat insulator: it can dissipate intense thermal radiation through decomposition without flaming (95).

Recent studies of polymer electrets have shown that the depolarization currents on heating poly(vinylidene fluoride) have peaks that correspond to transition temperatures associated with various types of molecular motions (96–102). A piezoelectric effect stable for many months was found in poly(vinylidene fluoride) films drawn heated to 90°C, and cooled while a static electric field was applied (102).

Poly(vinylidene fluoride) is resistant to attack or penetration by most corrosive chemicals and organic compounds, including acids, alkalies, strong oxidizing agents, and halogens at room temperature. Some highly polar solvents, ketones, and ethers are absorbed and soften the polymer to varying degrees depending upon time and temperature of exposure. Good *solvents* for poly(vinylidene fluoride) are dimethylacetamide, dimethylformamide, hexamethylphosphoramide, and dimethyl sulfoxide. Acetone is strongly absorbed at room temperature, and causes swelling and softening of the polymer. The polymer is attacked by fuming sulfuric acid and strongly basic primary amines, otherwise its *chemical resistance* is outstanding. The U.S. Food and Drug Administration has approved poly(vinylidene fluoride) for use in processing food; molded parts, such as gaskets, seals, valves, pumps, tubing, and lined and solid pipe, can now be used in food processing plants and equipment, especially where chemical resistance is required.

Other adverse environmental conditions produce little or no effect upon poly(vinylidene fluoride). Like other fluorine-containing polymers (qv), it is essentially unaffected by sunlight or other ultraviolet radiation. Exposure to high vacuum (10 torr) does not cause changes in its properties. Because the backbone of the molecule is shielded by the fluorine atoms with strong bonds to the carbon chain, oxidative attack of the polymer does not occur as is the case with many thermoplastic polymers. The operating temperature range is considered to be from –40 to 150°C. Severe bending at −40°C does not cause cracking, even after exposure to 150°C for long periods. Although it will char when exposed to direct flame, the polymer is quickly self-extinguishing and does not drip while the flame is applied. Poly(vinylidene fluoride) has been classified "Self Extinguishing, Group I" by Underwriters' Laboratories, Inc. When exposure to temperatures over 250°C is anticipated, provision should be made for dissipating or absorbing any hydrogen fluoride that might be liberated.

Fabrication and Processing

Poly(vinylidene fluoride) can be processed by practically any of the normal methods known for thermoplastics. Fabrication (107,108) by compression molding, transfer molding, injection molding, blow molding, extrusion, coining, vacuum or pressure forming, thermoforming, welding, and machining of poly(vinylidene fluoride) has been successfully performed using conventional equipment. (See MELT EXTENSION; MOLDING; THERMOFORMING.) Rotational molding, fluidized-bed coating, flame spraying, and plasma-jet spraying have also been evaluated, but because of the high melt viscosity and lack of attraction for particles to each other (except under pressure) these methods of fabrication are not feasible at this time. Until new technologies are developed, coatings for architectural finishes and corrosion control for the chemical process industry are being applied as dispersions in the form of modified organisols. Coatings of poly(vinylidene fluoride) can be applied by spraying (conventional and various electrostatic approaches), roller coating, dipping, slush coating, impregnating, and wet lay-up laminating. (See COATING METHODS; LAMINATES.)

The most efficient method for service evaluation for poly(vinylidene fluoride) is usually compression molding. Quite often existing molds originally designed for thermosetting plastics or modified rubber compounds can be used. Otherwise slugs or billets are molded from which prototypes can be machined. Poly(vinylidene fluoride) pellets are most suitable for molding. They should be heated using a high-frequency dielectric preheater followed by oven heating so that the polymer is uniformly heated to 175–215°C. Melted polymer is then charged into the mold, which is best heated to 120–160°C. Since poly(vinylidene fluoride) has a high shrinkage factor, the lowest melt temperature and mold temperature at which a finished part can be successfully formed are recommended. Molding pressures vary between 1000 and 10,000 psi, depending on melt temperature, and shape and size of the article being formed.

Transfer-molding conditions are closely related to those described for compression molding, except that higher temperatures and pressures are usually required, depending on the intricacy of the parts and especially if inserts are included. For best results, inserts should be coated with poly(vinylidene fluoride) dispersion, cured to dissipate solvent, and charged to the mold heated to approximately 140–160°C to ensure a tenacious bond.

Reciprocating-screw-injection is preferred over straight ram-injection molding, although many parts have been successfully formed using the latter-type equipment. Melt temperatures can vary from 200 to 270°C, and pressure up to 20,000 psi may be required. Because of the high melt viscosity and heat retention of the molten polymer, runnerless molds and central gating are most desirable. If runners are needed, then shortest routings and absolutely balanced layout must be observed. "Family" molds which incorporate various sizes and geometry should be avoided.

Poly(vinylidene fluoride) is readily extrudable into any form including sheet, film, pipe, and wire insulation. Extruders should have a 20/1 L/D ratio or more, and include a screw which has between four and six metering flights. Of most significant importance is the need for a long, gradual transition (compression stage) from feed to metering sections. Certain types of high-shear-mixing screws have proved advantageous. On the other hand, those types that incorporate raised sections, such as pins or island-type sections, cause serious degradation of the polymer. Long shallow entrance angles and well-polished, chrome-plated surfaces are required for all die parts. Extrusion temperature should be between 215 and 270°C. Pressures as high as 10,000 psi may be required. Pressure transducers of the flush type are preferred so as not to present any "hang-up" spot in the flow stream. Any area where stagnation of molten polymer can occur will result in a decomposition site.

Applications

The balance of properties of poly(vinylidene fluoride) has led to extensive applications in three major markets: a base for long-life finishes for exterior metal siding; electrical insulation; and a variety of chemical process equipment. In addition, there are several specialized uses for poly(vinylidene fluoride) in smaller, but significant quantities.

Coatings. One of the most rapidly growing markets for poly(vinylidene fluoride) is in its use as a base for durable, long-life coatings for exterior finishes. The resin, blended with suitable heat-stable pigments and other additives, is mixed with a solvent system, and is factory-applied by coil coating or spray techniques. Metal substrates for the most part are either aluminum or galvanized steel. Applied to coil stock, finishes based on high percentages of poly(vinyl fluoride) can be postformed for a variety of aesthetic and functional applications without chipping or peeling. Projected life of such coatings is in a range of twenty to thirty years. Typical metal applications include roofing, louvers, panel siding, window frames, and a variety of special components.

Electrical Applications. As an electrical insulating material, poly(vinylidene fluoride) serves in a variety of applications, including primary insulation on wire used in computers, and as jacketing on aircraft wire and geophysical cable. Poly(vinylidene fluoride) is used as a primary insulation on hook-up wire used in computer back panels. Computers require not only satisfactory electrical properties, but also exceptional physical toughness of a wire insulation. Also it is the computer manufacturer's constant objective to reduce wire diameter so that additional wires may be placed in the same amount of space. Naturally, creep (cold flow) must be avoided to prevent electrical shorts. Additionally, automatic wiring of back panels requires an insulation with a high degree of mechanical strength (109).

In other insulation applications, use of poly(vinylidene fluoride) heat-shrinkable tubing is at present growing rapidly. Poly(vinylidene fluoride) has excellent resistance to abrasion and cut-through. The standard product shrinks 50% of its supplied diameter, and greater reduction ratios are possible for specialized uses. One such special use involves the incorporation of a ring of solder within the heat-shrinkable tubing, forming a so-called "solder sleeve." This device is used by the millions in the electronic, aircraft, and aerospace industries. In other electrical and electronic applications, poly(vinylidene fluoride) is used in coil bobbins, connector blocks, potting boots, wire bundling devices, and anode caps.

Chemical Processing Equipment. The last major market for poly(vinylidene fluoride) involves the use of material in chemical processing equipment. This includes piping systems, both lined and solid, and a variety of solid and lined valves and pumps. In addition, poly(vinylidene fluoride) film is used for valve diaphragms and for packaging for corrosive chemicals. Owing to its unique physical properties, poly(vinylidene fluoride) is the only fluorocarbon polymer available as a rigid piping. Field applications have demonstrated that poly(vinylidene fluoride) is particularly suitable for halogens and strong oxidizing agents, although it is not recommended for such highly polar solvents as ketones and esters. In addition to pipes, valves, and pumps, special devices are molded or otherwise fabricated from the resin. These include spargers, dutchman, large-diameter ducts, tower packing, and filter cloth woven from poly(vinylidene fluoride) monofilament.

Although the above markets are certainly the largest users from the volume point of view, poly(vinylidene fluoride) in a specially prepared porous form has gained wide acceptance in one consumer product, fine-line marking pens. Poly(vinylidene fluoride) is suitable in this case because of its combination of mechanical toughness, durability, and chemical resistance to marking inks.

Economic Aspects

Although the first patent for poly(vinylidene fluoride) was issued in 1948 to Ford and Hanford, it was not until 1961 that quantities of the polymer were fabricated into commercial products. In 1962, Pennwalt Corporation constructed a small plant with a capacity of several hundred thousand pounds per year. In 1965, a large commercial monomer and polymer plant was completed by Pennwalt, and in 1970, was expanded. The base price of the Kynar homopolymer powder in 1971 was $3.25/lb in truckload quantities. Polytetrafluoroethylene and polychlorotrifluoroethylene at present sell for $3.25–6.60/lb; thus, they are costlier on a volume basis because of their higher specific gravity, 2.2, versus 1.76 for poly(vinylidene fluoride).

Bibliography

1. F. B. Downing, A. F. Benning, and R. C. McHarness (to E. I. du Pont de Nemours & Co., Inc.), U.S. Pat. 2,551,573 (1951).
2. C. B. Miller (to Allied Chemical Corp.), U.S. Pat. 2,628,989 (1953).
3. R. Mantell and W. S. Barnhart (to M. W. Kellogg Co.), U.S. Pat. 2,774,799 (1956).
4. B. P. Zverev, A. L. Goldinov, Yu. A. Panshin, L. M. Borovnev, and N. S. Shirokova, U.S.S.R. Pat. 216,699 (1958).
5. F. Kaess, K. Leinhard, and H. Michaud (to Sueddeutsche Kalkstickstoff-Werke A.G.), Ger. Pat. 1,288,085 (1969).
6. F. Kaess, K. Leinhard, and H. Michaud (to Sueddeutsche Kalkstickstoff-Werke A.G.), Ger. Pat. 1,288,593 (1969).
7. Produits Chimiques Pechiney Saint-Gobain, Fr. Pat. 1,337,360 (1963).
8. H. Ukihashi and M. Ichimura (to Asahi Glass Co., Ltd.), Japan. Pat. 68–29,126 (1968).
9. A. F. Benning, F. B. Downing, and R. J. Plunkett (to E. I. du Pont de Nemours & Co., Inc.), U.S. Pat. 2,401,897 (1946).
10. M. E. Miville and J. J. Earley (to Pennwalt Corporation), U.S. Pat. 3,246,041 (1966); Neth. Appl. 6,508,619 (1967); Ger. Pat. 1,253,702 (1967).
11. Kureha Chem. Ind. Co., Ltd., Japan. Pat. 68–11,202 (1968).
12. J. C. Calfee and C. B. Miller (to Allied Chemical Corp.), U.S. Pat. 2,734,090 (1956).
13. F. Olstowski (to Dow Chemical Co.), U.S. Pat. 3,047,637 (1962).
14. A. E. Pavlath and F. H. Walker (to Stauffer Chemical Co.), Fr. Pat. 1,330,146 (1963).
15. M. Hauptschein and A. H. Fainberg (to Pennwalt Corporation), U.S. Pat. 3,188,356 (1965).
16. S. Okazaki and N. Sakauchi (to Kureha Chemical Industry Co., Ltd.), Japan. Pat. 65–22,453 (1965).
17. D. M. Marquis (to E. I. du Pont de Nemours & Co., Inc.), U.S. Pat. 3,073,870 (1963).
18. Y. Kometani (to Daikin Kogyo Co., Ltd.), Japan. Pat. 68–10,602 (1968).
19. H. Madai, Ger. (East) Pat. 42,730 (1966).
20. J. R. Soulen and W. F. Schwartz (to Pennwalt Corp.), U.S. Pat. 3,428,695 (1969).
21. F. Olstowski and J. D. Watson (to Dow Chemical Co.), U.S. Pat. 3,089,910 (1963).
22. Dow Chemical Co., Fr. Pat. 1,566,523 (1969).
23. G. F. Helfrich and E. J. Rothermel (to Dow Chemical Co.), U.S. Pat. 3,380,977 (1968).
24. Th. A. Ford and W. E. Hanford (to E. I. du Pont de Nemours & Co., Inc.), U.S. Pat. 2,435,537 (1948).
25. M. Hauptschein (to Pennwalt Corp.), U.S. Pat. 3,193,539 (1965).
26. Daikin Kogyo Co., Ltd., Brit. Pat. 1,179,078 (1970).
27. H. Iserson (to Pennwalt Corp.), U.S. Pat. 3,245,971 (1966).
28. G. H. McCain, J. R. Semancik, and J. J. Dietrich (to Diamond Shamrock Corp.), U.S. Pat. 3,475,396 (1969).
29. S. Okazaki and N. Sakauchi (to Kureha Chemical Co., Ltd.), Japan. Pat. 43–4867 (1968); Fr. Pat. 1,419,741 (1965).
30. Daikin Kogyo Co., Ltd., Brit. Pat. 1,178,227 (1967).
31. P. D. Carlson (to E. I. du Pont de Nemours & Co., Inc.), Ger. Off. 1,806,426 (1969); Neth. Pat. 68–15,435 (1969).
32. Kali-Chemie A.G., Brit. Pat. 1,057,088 (1967).
33. Asahi Glass Co., Ltd., Brit. Pat. 1,188,889 (1970).
34. W. W. Doll and J. B. Lando, *J. Appl. Polymer Sci.* **14**, 1767–1773 (1970).
35. Deutsche Solvay-Werke G.m.b.H., Brit. Pat. 1,004,172 (1965); Belg. Pat. 620,986 (1963).
36. R. G. Foster (to Imperial Chemical Industries Ltd.), Brit. Pat. 1,149,451 (1969).

37. L. A. Bulygina and E. V. Volkova, *Radiats. Khim. Polim. Mater. Simp. Moscow, 1964*, pp. 122–126.
38. Daikin Kogyo Co., Ltd., Fr. Pat. 1,394,585 (1965).
39. E. V. Volkova, P. V. Zimakov, and A. V. Fokin, *Dokl. Akad. Nauk S.S.S.R.* **167** (5), 1057–1059 (1966).
40. E. V. Volkova, P. V. Zimakov, and A. V. Fokin, *At. Energ.* **26** (3), 240–245 (1969).
41. S. S. Dubov, M. A. Landau, E. V. Volkova, and L. A. Bulygina, *Zh. Fiz. Khim.* **43** (6), 1514–1515 (1969).
42. T. A. Ford (to E. I. du Pont de Nemours & Co., Inc.), U.S. Pat. 2,468,054 (1949).
43. A. Dittman, H. J. Passino, and W. O. Teeters (to M. W. Kellogg Co.), U.S. Pat. 2,738,343 (1956).
44. A. Dittman, H. J. Passino, and W. O. Teeters (to M. W. Kellogg Co.), U.S. Pat. 2,752,331 (1956).
45. D. R. Rexford (to E. I. du Pont de Nemours Co., Inc.), U.S. Pat. 3,051,677 (1962).
46. A. N. Bolstad (to 3M Co.), U.S. Pat. 3,163,628 (1964).
47. D. Sianesi, G. C. Bernardi, and A. Regio (to Montedison Co.), U.S. Pat. 3,331,823 (1967).
48. J. R. Pailthorp and H. E. Schroeder (to E. I. du Pont de Nemours & Co., Inc.), U.S. Pat. 2,968,649 (1961).
49. D. Sianesi, G. C. Bernardi, and G. Diotallevi (to Montedison Co.), U.S. Pat. 3,335,106 (1967).
50. C. B. Miller and J. D. Calfee (to Allied Chemical Corp.), U.S. Pat. 2,635,093 (1953).
51. W. D. Wood, J. L. Lacina, B. L. DePrater, and J. P. McCullough, *J. Phys. Chem.* **68** (3), 579–586 (1964).
52. L. L. Maksimov and E. G. Zotikov, *Vysokomol. Soedin., Ser. B* **11** (11), 818–821 (1969).
53. R. E. Naylor and S. W. Lasoski, *J. Polymer Sci.* **44**, 1 (1960).
54. C. W. Wilson, *J. Polymer Sci.* [A] **1**, 1305 (1963).
55. C. W. Wilson and E. R. Santee, Jr., *J. Polymer Sci.* [C] **8**, 97 (1965).
56. E. L. Galperin, Yu. V. Strogalin, and M. P. Mlenik, *Vysokomol. Soedin.* **7** (3), 933–939 (1965).
57. J. B. Lando, H. G. Olf, and A. Peterlin, *J. Polymer Sci.* [A-1] **4**, 941–951 (1966).
58. W. W. Doll and J. B. Lando, *J. Macromol. Sci. Phys.* **4** (2), 309–329 (1970).
59. G. Cortili and G. Zerbi, *Spectrochim. Acta* **23A**, 2216–2218 (1967).
60. N. I. Makarevich and N. I. Sushko, *Zh. Prikl. Spektrosk* **5** (11), 917–920 (1969).
61. J. B. Lando and W. W. Doll, *J. Macromol. Sci. Phys.* **2** (2), 205–218 (1968).
62. R. P. Teulings, J. H. Dumbleton, and R. L. Miller, *Polymer Letters* **6**, 441 (1968).
63. F. J. Boerio and J. L. Koenig, *J. Polymer Sci.* [A-2] **7**, 1489–1494 (1969).
64. E. L. Galperin and B. P. Kosmynin, *Vysokomol. Soedin. A* **11** (7), 1432–1436 (1969).
65. G. Natta, G. Allegra, I. W. Bassi, D. Sianesi, G. Caporiccio, and F. Torti, *J. Polymer Sci.* [A] **13**, 4263 (1965).
66. K. Sakoku and A. Peterlin, *J. Macromol. Sci. Phys. B* **1** (2), 401–406 (1967).
67. K. Okuda, T. Yoshida, M. Sugita, and M. Asahina, *Polymer Letters* **5**, 465–468 (1967).
68. S. G. Malkevich and L. I. Tarutina, *Vysokomol. Soedin. B* **10** (12), 881–882 (1968).
69. E. L. Galperin, S. S. Dubov, E. V. Volkova, M. P. Mlenik, and L. A. Bulygina, *Vysokomol. Soedin.* **8** (11), 2033 (1966).
70. S. Enomoto, Y. Kawai, and M. Sugita, *J. Polymer Sci.* [A-2] **6**, 861–869 (1968).
71. B. P. Kosmynin, E. L. Galperin, and D. Ya. Zvankin, *Vysokomol. Soedin. A* **12** (6), 1254 (1970).
72. R. Hasegawa, Y. Tanabe, M. Kobayashi, and H. Jadokoro, *J. Polymer Sci.* [A-2] **8**, 1073–1087 (1970).
73. E. L. Galperin, B. P. Kosmynin, and R. A. Bychbov, *Vysokomol. Soedin., Ser. B* **12** (7), 555–557 (1970); *Chem. Abstr.* **73**, 77754 (1970).
74. E. L. Galperin, B. P. Kosmynin, L. A. Aslanyan, M. P. Mlenik, and V. K. Smirnov, *Vysokomol. Soedin., Ser. A* **12** (7), 1654–1661 (1970); *Chem. Abstr.* **73**, 77746 (1970).
75. T. Miyazawa and Y. Ideguchi, *J. Polymer Sci.* [B] **3**, 541 (1965).
76. Raychem. Corp., Belg. Pat. 733,485 (1969).
77. L. Mandelkern, G. M. Martin, and F. A. Quinn, Jr., *J. Res. Natl. Bur. Std.* **58**, 137 (1957).
78. S. Yano, *J. Polymer Sci.* [A-2] **8**, 1057 (1970).
79. H. Kakutani, *J. Polymer Sci.* [A-2] **8**, 1177 (1970).
80. T. Wentink, Jr., *J. Appl. Phys.* **32**, 1063–1064 (1961).

81. S. P. Kabin, G. G. Harkevich, and G. P. Mikhailov, *Vysokomol. Soedin.* **3** (4), 618–623 (1961).
82. Y. Ishida, M. Watanabe, and K. Yamafuji, *Kolloid-Z.* **200,** 48 (1964).
83. A. Peterlin and J. D. Holbrook, *Kolloid-Z.* **203,** 68 (1965).
84. A. Peterlin and J. Elwell, *J. Mater. Sci.* **2** (1), 1–6 (1967).
85. H. Kakutani, *Kobunshi Kagaku* **26,** 83 (1969).
86. H. Sasabe, S. Saito, M. Asahina and H. Kakutani, *J. Polymer Sci.* [A-2] **7,** 1405 (1969).
87. R. Timmerman, *J. Appl. Polymer Sci.* **6,** 456–460 (1962).
88. G. D. Sands and G. F. Pezdirtz, *NASA Rpt. N66-13497* (1965).
89. R. Timmerman (to Radiation Dynamics, Inc.), U.S. Pat. 3,142,629 (1964).
90. V. L. Lanza and E. C. Stivers (to Raychem Corp.), U.S. Pat. 3,269,862 (1966).
91. V. L. Lanza (to Raychem Corp.), Canadian Pat. 838,674 (1970).
92. J. T. Robin Claburn (to Raychem Corp.), Ger. Offen. 1,909,489 (1969).
93. L. A. Osentyevich and A. N. Pravednikov, *Vysokomol. Soedin. B* **10** (1), 49–52 (1958).
94. J. T. Stapler and W. J. Barnes, *U.S. Clearing House for Sci. Tech. Inf. AD-672509* (1968).
95. H. Iserson and F. I. Koblitz (to Pennwalt Corp.), U.S. Pat. 3,510,429 (1970).
96. T. Takamatsu and E. Fukada, *Rika Gaku Kenkyusho Hokoku* **45** (3), 49–54 (1969).
97. T. Takamatsu and E. Fukada, *Rika Gaku Kenkyusho Kokoku* **45** (1), 1–9 (1969).
98. T. Takamatsu, *Sen-i To Kogyo* **2** (9), 649–655 (1969).
99. K. Euler, *J. Elektrotech. Z. Ausg. A.* **90** (23), 600–603 (1969).
100. T. Takamatsu and E. Fukada, *Polymer J.* **1,** 101–106 (1970).
101. E. Fukada and S. Takashita, *Jap. J. Appl. Phys.* **8** (7), 960 (1969).
102. H. Kawai, *Jap. J. Appl. Phys.* **8** (7), 975–976 (1969).
103. W. H. Mears, R. F. Stahl, S. R. Orfeo, R. C. Shair, L. F. Wells, W. Thompson, and H. Mc-Cann, *Ind. Eng. Chem.* **47** (7), 1449–1454 (1955).
104. A. N. Baratov and V. M. Kucher, *Zh. Prikl. Khim.* **38** (5), 1068–1072 (1965).
105. *Fed. Reg. Title 21. Cl. 1B, Designation 121. 2593.*
106. C. A. Neugebauer and J. L. Margrave, *J. Phys. Chem.* **60,** 1318 (1956).
107. A. A. Dukert, "Polyvinylidene Fluoride Resin–Kynar, Fabrication and Applications," *Soc. Plastics Eng. 18th Ann. Nat. Tech. Conf., Pittsburgh, Pa., Jan. 1962.*
108. W. S. Barnhart, N. Capron, A. A. Dukert, R. A. Ferren, H. Iserson, M. E. Miville, and L. E. Robb, "Polyvinylidene Fluoride–RC-2525 Resin Properties," *Soc. Plastics Eng. 17th Ann. Natl. Tech. Conf., Washington, D.C., Jan. 1961.*
109. A. A. Dukert and J. H. Houser, "Using Kynar Insulation in Computer Wiring," *Computer Design Magazine,* Jan. 1970.

J. E. Dohany, A. A. Dukert, and S. S. Preston III
Pennwalt Corporation

ELASTOMERS

Elastomeric copolymers containing vinylidene fluoride are manufactured in the United States by E. I. du Pont de Nemours & Company, Inc., under the trademark Viton and by Minnesota Mining and Manufacturing (3M) Company under the trademarks Fluorel and Kel-F. In Italy, Montecatini Edison S.p.A. manufactures a fluorocarbon rubber called Tecnoflon. Fluorocarbon rubbers, presumably copolymers containing vinylidene fluoride, also are manufactured in the Soviet Union and in Japan by SKF and Daikin, respectively. See also FLUORINE-CONTAINING POLYMERS; ELASTOMERS, SYNTHETIC.

Preparation

The elastomeric copolymers of vinylidene fluoride with hexafluoropropylene, chlorotrifluoroethylene, or pentafluoropropylene are prepared in free-radical polymerization systems, eg, with peroxide or azo-type catalysts or by irradiation. Patents to du Pont (1,2), to the 3M Company through the M. W. Kellogg Company (3), and to Montecatini (4) all indicate that an emulsion-polymerization process operated under several hundred pounds pressure at temperatures between 40 and 100°C, with a peroxy initiator such as an organic peroxide, hydrogen peroxide, a peroxydisulfate, or a peroxydisulfate/bisulfite initiation system, can be effectively used in the synthesis of the copolymers. Fairly careful control of composition is required to obtain optimum elastomeric properties. Polymers containing a very high proportion of vinylidene fluoride are classified as plastics rather than elastomers. Preparation of low-molecular-weight fluid elastomers, by carrying out the polymerization with organic-soluble sources of free radicals in a polymer solvent that also acts as a chain-transfer agent for the polymerization, has also been described (5).

Fluorocarbon elastomers currently of greatest commercial importance are copolymers of vinylidene fluoride and hexafluoropropylene in the approximate weight ratio of 60:40 and a terpolymer of vinylidene fluoride, hexafluoropropylene, and tetrafluoroethylene. Viton A, A-35, AHV, B, B-50, and E-60, as well as Fluorel 2140 and 2160, fit these descriptions. The text that follows pertains only to these types, which will be termed "fluoroelastomers" for convenience.

Processing

Vinylidene fluoride-based elastomers are available in a wide variety of viscosity grades. The low Mooney viscosity of Viton A-35 (ML-10 at 100°C of about 35) and the high Mooney viscosity of Viton AHV (ML-10 at 121°C of over 150) exemplify the range. Processing characteristics are closely related to viscosity. Viton fluoroelastomer differs from many other elastomers in that, for a given viscosity, it has relatively low "nerve," ie, a short-lived viscoelastic memory. Viscosity breakdown does not occur to any large extent on mastication.

Most of the fluoroelastomers described above may be mixed on open, two-roll mills or in internal mixers in much the same manner as other elastomers. Their compounds likewise are processed on equipment conventional to the rubber industry. With minor consideration to compounding, they are readily molded, extruded, or calendered. Extrudates of the higher-viscosity compounds tend to have a rough, sandpaper-like finish which is best overcome by using a cool, long land die. Water-jacketed aluminum dies have been used successfully. In fact, with the exception of molding, most processing operations are best carried out at lower temperatures than used with other elastomers.

Fluoroelastomer compounds are virtually devoid of tack. Building operations (eg, roll building) entail, therefore, the use of a solvent such as methyl ethyl ketone to wet the plies before laminating. See also RUBBER COMPOUNDING AND PROCESSING.

Curing. The best vulcanizate properties are achieved by a two-step curing process. The first step entails the application of both heat and pressure, such as molding in a press or fabric wrapping and curing in a pressurized open-steam autoclave. The second step is called a postcure and is carried out in an air oven. For thick sections, the temperature of the oven usually must be raised gradually or stepwise, to prevent fissuring of the part. A typical curing condition would be 10–30 min in a press at 325–350°F, depending on size of the molding, followed by 24 hr at 450°F in an air oven. Depending again on size of the part, the oven postcure may require a 4–8 hr rise to curing temperature. Postcuring is required for vulcanizates to have good heat and compression-set resistance. For solvent, oil, and chemical resistance alone, postcuring is not required.

The following theory of the curing of vinylidene fluoride polymers at least partially explains why these conditions are required.

Curing Chemistry. Both free-radical and ionic processes have been recommended for the curing of vinylidene fluoride copolymer elastomers. The free-radical crosslinking process, used comparatively little, involves the thermal decomposition of relatively large amounts, up to 5%, of a source of energy-rich radicals, such as benzoyl peroxide (6) or difluorodiazine (7). It has been envisioned that, as the initiator decomposes, the free radicals formed abstract hydrogen from the methylene groups along the polymer chain, and the resultant polymer radicals interact, directly or through the intermediacy of radical traps, to form the crosslink. Magnesium oxide, as an acid acceptor, and carbon black fillers are usually included in the recipe. A postcure of the molded articles is normally required to produce satisfactory vulcanizate properties. The good resistance of the vinylidene fluoride polymers to attack by free radicals makes this curing chemistry relatively inefficient; radiation curing, presumably by radical processes, has also not been practiced extensively.

The commonly used curing chemistry for the vinylidene fluoride copolymer elastomer is believed to involve initially the removal of the elements of hydrogen fluoride from the polymer chain by reaction of the polymer with base (8–11). The resulting highly polarized $\diagup CH{=}CF \diagup$ units will react with each other to form crosslinks if no other reagent is present, but will react with suitable bisnucleophiles (diamines, bisphenols, dimercaptans) much more readily.

A variety of bases have been used for the dehydrofluorination reaction, including primary, secondary, and tertiary amines, tetraalkylguanidines, and strongly alkaline metal alkoxides and hydroxides. In various studies the initial formation of the hydrogen fluoride reaction product with most of these bases has been established. Although these bases will convert the vinylidene fluoride polymers to a crosslinked condition relatively slowly, better vulcanizate properties are achieved if there is added to the recipe a bisnucleophile capable of reacting with two of the $\diagup CH{=}CF \diagup$ units postulated to have been created by the dehydrofluorination described above. Bis-(primary amines), which act as their own bases (11), bismercaptans (12), and bisphenols (13) have been utilized.

With all of these systems it is advantageous to provide a metal oxide acid acceptor to remove hydrogen fluoride from the system and to deactivate any source of easily available fluoride ion. Magnesium, calcium, and lead oxides are preferred, perhaps as a consequence of the insolubility and stability of the corresponding fluorides.

The function of the postcure step, usually necessary to produce optimum vulcanizate properties, is the removal of volatile by-products, in many cases water, from the molded articles, rather than to establish additional crosslinks (10). The enhanced physical properties seen after postcure may also reflect improved polymer-filler interaction as a result of the volatilization of moisture.

The course of the reaction with Viton fluoroelastomer and a diamine curing agent has been described by Smith (11), following the reaction sequence indicated below.

$$\underset{\text{hexamethylenediamine carbamate (HMDAC, DIAK No. 1)}}{H_3N^+\text{—}(CH_2)_6\text{—}NHCOO^-} \xrightarrow{\text{heat}} H_2N\text{—}(CH_2)_6\text{—}NH_2 + CO_2 \tag{1a}$$

$$\sim\!\!\sim\!\!(CH_2\text{—}CF_2\text{—}CF_2\text{—}\underset{|}{\overset{CF_3}{C}}F)_n\!\!\sim\!\!\sim + H_2N\text{—}(CH_2)_6\text{—}NH_2 \longrightarrow$$

$$\sim\!\!\sim\!\!(CH{=}CF\text{—}CF_2\text{—}\overset{CF_3}{C}F)_n\!\!\sim\!\!\sim + H_2N\text{—}(CH_2)_6\text{—}NH_3^+F^- \tag{1b}$$

$$2\,H_2N\text{—}(CH_2)_6\text{—}NH_3^+F^- + MgO \longrightarrow MgF_2 + H_2O + 2\,H_2N\text{—}(CH_2)_6\text{—}NH_2 \tag{1c}$$

$$\sim\!\!\sim\!\!(CH{=}CF\text{—}CF_2\text{—}\overset{CF_3}{C}F)_m\!\!\sim\!\!\sim + NH_2\text{—}(CH_2)_6\text{—}NH_2 + \sim\!\!\sim\!\!(CH{=}CF\text{—}CF_2\text{—}\overset{CF_3}{C}F)_n\!\!\sim\!\!\sim \longrightarrow$$

$$\begin{matrix} \sim\!\!\sim\!\!(CH_2\text{—}CF\text{—}CF_2\text{—}\overset{CF_3}{C}F)_m\!\!\sim\!\!\sim \\ | \\ NH \\ | \\ (CH_2)_6 \\ | \\ NH \\ | \\ \sim\!\!\sim\!\!(CH_2\text{—}CF\text{—}CF_2\text{—}\overset{CF_3}{C}F)_n\!\!\sim\!\!\sim \end{matrix} \longrightarrow$$

$$\begin{matrix} \sim\!\!\sim\!\!(CH_2\text{—}C\text{—}CF_2\text{—}\overset{CF_3}{C}F)_m\!\!\sim\!\!\sim \\ \| \\ N \\ | \\ (CH_2)_6 \\ | \\ N \\ \| \\ \sim\!\!\sim\!\!(CH_2\text{—}C\text{—}CF_2\text{—}\overset{CF_3}{C}F)_n\!\!\sim\!\!\sim \end{matrix} + 2\,HF \tag{1d}$$

$$2\,HF + MgO \longrightarrow MgF_2 + H_2O \tag{1e}$$

When bisphenols are used as the nucleophiles instead of the bis(primary amine), the crosslink is believed to contain ether linkages, in the place of the Schiff base structure shown.

Compounding. Practically all curing systems include a metal oxide. Magnesia (MgO) is by far the most commonly used but litharge (PbO), calcium oxide, and zinc oxide together with dibasic lead phosphite occasionally are used for special properties. Amounts used vary from 4 to 20 phr.

Either of two different types of crosslinking agent is currently used commercially. The older is based on diamines. These include DIAK No. 1 (hexamethylenediamine carbamate), DIAK No. 3 (*N*,*N*′-dicinnamylidene-1,6-hexanediamine), and DIAK No. 4 (an alicyclic amine salt). Amounts used vary from 1 to 3 phr. The second and more recently developed type is an aromatic dihydroxy compound of which DIAK No. 5, hydroquinone, is an example. Aromatic dihydroxy compounds require an alkaline activator such as calcium hydroxide and/or DIAK Super 6 to effect a crosslink. The aromatic dihydroxy–alkaline activator curing system provides an excellent balance between processing safety (Mooney scorch resistance) and rapid cure rate in the press. More importantly, it affords a means for obtaining vulcanizates having remarkably good resistance to compression set.

Filler Loading. One of the unique features in compounding Viton or Fluorel fluoroelastomers is the limitation in the amount and type of filler or reinforcing agent that can be used. As will be discussed subsequently, limitations on the permissible amount and type of plasticizer prohibit loading with large amounts, especially of highly reinforcing fillers. Excessive loading causes processing to become intolerably difficult, and the vulcanizate hardness and stiffness also become too high for all but few end-use applications. Generally, a loading of a soft carbon black is used; twenty to forty phr of MT Carbon black is common. A fibrous calcium or magnesium silicate (eg, Cab-O-Lite P-4) in the same amount also makes a good filler. With no filler, tensile strength is low.

Plasticizers. Choice of a plasticizer is limited by the fact that most of those that are useful in other elastomers are too volatile above 400°F and, therefore, are lost during the oven postcure or during subsequent service at high temperature. Furthermore, because fluoroelastomer vulcanizates are very solvent resistant, the polymer and its vulcanizates have limited compatibility with most fluids used as plasticizers. Fluorosilicone oils and certain high-molecular-weight aromatic hydrocarbons are the most useful. These can be used to aid processing or to reduce vulcanizate hardness. No plasticizers that improve low-temperature properties have so far been found that are practical to use.

Processing Aid. Viton LM, a low-molecular-weight vinylidene fluoride–hexafluoropropylene copolymer, may be used to lower the viscosity of uncured compounds for ease of processing. Generally, 10–20 phr suffice. It is retained during postcuring but the product may lose some weight as a result of volatiles loss during high-temperature service. A slight reduction in state of cure also results from the use of Viton LM.

Pigmentation. Viton fluoroelastomers may be compounded to give bright, color-stable vulcanizates if the dihydroxy curing system is used. Hitherto, color was limited to dark shades because the diamine curing agents discolored during postcure or subsequent heat service. If a filler other than black is desired, the commonly used inorganic pigments (qv) and titanium dioxide offer a wide range of colors.

Vulcanizate Properties

Fluoroelastomers in general are best known and used for their inherent resistance to the deteriorating influences of heat, oils, solvents, and chemicals, whether alone or in combination. Retention of mechanical properties after prolonged exposure to temperatures of 400°F and above is remarkable.

In reading the discussion that follows, it must be borne in mind that compounding can enhance certain properties, while compromising, to some degree, others. Unless the changes in compounding are drastic, however, the properties described are typical of most products made from fluoroelastomers.

Hardness. General-purpose fluoroelastomer vulcanizates have a durometer A hardness of about 70. Formulations can be provided over the range of 50 to 95. Diamine-cured vulcanizates drop in hardness 15–20 points as the temperature is raised from 75 to 400°F. Dihydroxy-cured vulcanizates do not change in hardness over this temperature range.

Tensile Properties. Tensile-strength values of 2000 psi at 75°F are typical. At higher temperatures, strength is less, eg, 1000 psi at 200°F or 300 psi at 500°F. Elongation at break varies greatly with the formulation and state of cure, but generally falls within the range of 125–300% at 75°F. It decreases as temperature is increased, as also does modulus.

Heat Resistance. Fluoroelastomers have excellent resistance to degradation by heat. Vulcanizates remain usefully elastic almost indefinitely when exposed to laboratory air oven aging up to 400°F or to intermittent test exposures up to 500°F. Continuous service limits are generally considered to be: >3000 hr at 450°F, 1000 hr at 500°F, 240 hr at 550°F, and 48 hr at 600°F.

Chemical Resistance. The resistance of fluoroelastomers is better than that of other commercially available elastomers with respect to aliphatic and aromatic hydrocarbons, chlorinated solvents, oils, fuels, lubricants, and most mineral acids. The elastomers are not suitable for use in ketones, certain esters and ethers, some amines, hot anhydrous hydrofluoric acid or chlorosulfonic acid, and a few proprietary fluids.

Low-Temperature Properties. Brittle temperature (ASTM D 746) varies with specimen thickness. Typical values are: 0.075 in. thick, −40°F; 0.025 in. thick, −55°F. The glass-transition temperature, as measured by the Clash-Berg stiffness test (ASTM D 1043), is about 0°F. Temperature retraction (ASTM D 1329) T_{10} is −5°F.

Compression-Set Resistance. Using the dihydroxy curing system, very low compression-set values are attainable over a wide temperature range, particularly with polymers recently developed specifically for this property. Values for a Viton E-60 compound measured by ASTM Method D 395 (Method B) are shown below:

Pellets (0.5 in. × 1.129 in.)		%
70 hr	75°F	10
22 hr	400°F	9
70 hr	400°F	19
168 hr	400°F	29
336 hr	400°F	36
O-Rings (1 in. × 0.139 in.)		%
70 hr	75°F	6
22 hr	400°F	12
70 hr	400°F	24
168 hr	400°F	38
336 hr	400°F	51
22 hr	450°F	26
70 hr	450°F	65
168 hr	450°F	82
22 hr	500°F	56
70 hr	500°F	94

Abrasion and Tear Resistance. Vulcanizates of the fluoroelastomers have good tear strength and excellent abrasion resistance compared to those of most general-purpose elastomers.

Resilience. Resilience is not particularly high, running about 50% at room temperature for a 70-durometer A-hardness vulcanizate, as measured by ASTM D 945.

Ozone and Oxygen Resistance. Fluoroelastomers are essentially ozoneproof. Aging in oxygen under pressure and at high temperature has, likewise, virtually no effect.

Steam Resistance. Special compounding is required but, nevertheless, the fluoroelastomers discussed herein are not particularly good in steam service.

Radiation Resistance. Resistance to both beta and gamma radiation is about in the middle range for elastomers in general.

Permeability. Gas permeability is low, being about equivalent to that of butyl rubber (see BUTYLENE POLYMERS).

Fungus Resistance. Fungus will not grow readily on typical vulcanizates.

Electrical Properties. Typical values for DC resistivity are on the order of 2×10^{13} Ω-cm, for specific inductive capacity around 15, for power factor about 5%, and for dielectric strength 500 V/mil.

Weather Resistance. Vulcanizates of Viton fluoroelastomer showed little or no change in physical properties or appearance after ten years' outdoor exposure in southern Florida.

Uses

The commercial use of copolymers of vinylidene fluoride and hexafluoropropylene, and those containing tetrafluoroethylene as well, are governed largely by the excellent resistance to heat, fluids, and compression set of their vulcanizates. Elastomeric seals for industrial, aerospace, and automotive equipment constitute the largest markets. These are mostly O-rings of all sizes, flat or lathe-cut gaskets, and to a lesser extent lip-type rotating- or reciprocating-shaft seals. Coated fabrics for diaphragms and sheet goods account for considerable consumption. Hose linings, tubing, and industrial gloves for chemical service, rolls for hot or corrosive service, and extruded goods for such purposes as autoclave and oven seals constitute examples of the varied end uses to which fluoroelastomers are put.

An amendment to the Federal Food, Drug and Cosmetic Act provides for the use of Viton A, Viton AHV, and Viton B vulcanizates containing magnesia and DIAK No. 1 or DIAK No. 4 in specified maximum amounts in the formulation of rubber articles intended for repeated food-contact use.

Fluoroelastomers are high-priced in relation to general-purpose elastomers. Viton, eg, costs $10.00/lb. In addition, its specific gravity is high (1.82–1.86) so that the materials cost of a product made of fluoroelastomer is a significant factor in its sales price.

Handling Precautions. The following is quoted from one supplier's literature: "Under ordinary handling conditions, Viton fluoroelastomer and its vulcanizates present no health hazards. However, it is known that some decomposition products are highly toxic. With the polymer and its conventional vulcanizates, decomposition has been noted at temperatures above 500°F (260°C). Therefore, vulcanization at temperatures approaching 500°F (260°C) as well as the use of vulcanizates at such temperatures should be carried out with adequate ventilation."

Vigorous exothermic decomposition of fluoroelastomer compounds during processing have occurred on rare occasions. These have been associated with exceptionally high processing temperatures, the addition of finely divided metals, (eg, aluminum powder), accidental use of grossly excessive amounts of diamine curing agents, or the use of dinitroso-type blowing agents.

Bibliography

1. D. R. Rexford (to E. I. du Pont de Nemours & Co., Inc.), U.S. Pat. 3,051,677 (1962).
2. J. R. Pailthorp and H. E. Schroeder (to E. I. du Pont de Nemours & Co., Inc.), U.S. Pat. 2,968,649 (1961).
3. A. L. Dittman, A. J. Passino, and J. M. Wrightson (to M. W. Kellogg Co.), U.S. Pat. 2,689,241 (1954).
4. D. Sianesi, G. C. Bernardi, and A. Reggio (to Montecantini), U.S. Pat. 3,331,823 (1967).
5. G. A. Gallagher (to E. I. du Pont de Nemours & Co., Inc.), U.S. Pat. 3,069,401 (1962).
6. A. S. Novikov, F. A. Galik-Ogly, and N. S. Gilinskaya, *Kauchuk i Rezina* **21** (2), 4–10 (1962).
7. J. F. Smith and J. R. Albin, *Ind. Eng. Chem., Prod. Res. Develop.* **2** (4), 284–286 (1963).
8. K. L. Paciorek, L. C. Mitchell, and C. T. Lenk, *J. Poly. Sci.* **45,** 405–413 (1960).
9. J. F. Smith, *Rubber World* **142** (3), 102–107 (1960).
10. J. F. Smith and G. T. Perkins, *Rubber Plast. Age* **41** (11), 1362 (1960).
11. J. F. Smith, *Proc. International Conf., Washington,* 575–581 (1959).
12. J. F. Smith, *Rubber World* **148** (2), 263–266 (1959).
13. R. P. Conger (to U.S. Rubber), U.S. Pat. 3,142,660 (1964).

General References

The manufacturers of fluoroelastomers supply valuable bulletins on processing and properties.

D. C. Thompson and A. L. Barney
E. I. du Pont de Nemours & Co., Inc.

VINYLIMIDAZOLE POLYMERS. See POLYIMIDAZOLES

VINYL ISOBUTYL ETHER POLYMERS. See VINYL ETHER POLYMERS

VINYL ISOCYANATE POLYMERS. See ISOCYANATE POLYMERS

VINYL KETAL POLYMERS. See VINYL ALCOHOL POLYMERS

VINYL KETONE POLYMERS

The polymerization of α,β-unsaturated ketones, $CH_2{=}CR'{-}\underset{\underset{O}{\|}}{C}{-}R''$, can be initiated by free-radical, cationic, and anionic catalysts. Both vinyl (R = H) and isopropenyl (R = CH_3) ketones are extremely reactive monomers which polymerize spontaneously upon exposure to heat or sunlight. Polymers derived from vinyl ketones have been known since 1903 (1), but no significant commercial application has been found. However, the investigation of vinyl ketone polymerization has been instrumental in the development of fundamental polymer chemistry. Staudinger showed that poly(methyl isopropenyl ketone) is a high-molecular-weight polymer by converting the polyketone to a series of polymeric derivatives (2). Marvel and Levesque demonstrated by cyclizing poly(methyl vinyl ketone) that vinyl monomers undergo head-to-tail addition (3). Koehler observed that the clear, tough polymers formed from methyl isopropenyl ketone exhibit an elasticity similar to that of ivory; thus the first attempt at commercialization involved the manufacture of billiard balls (4). This application, as well as most subsequent uses, was thwarted by the poor thermal and photochemical stability of vinyl ketone polymers. Commercial applications will depend upon the development of a stabilizer that will inhibit the rapid decolorization and degradation of the ketone polymers upon exposure to heat and light. Alternatively, applications that utilize these properties, ie, polymeric photosensitizers, may revitalize interest in these polymers. See also KETONE POLYMERS; KETENE POLYMERS; ACROLEIN POLYMERS.

Monomers

Physical Properties. Vinyl ketones are colorless, flammable liquids with strong irritating odors. With the exception of methyl vinyl ketone, the monomers exhibit limited (<5%) solubility in water and are miscible with most organic solvents (5). Methyl vinyl ketone is miscible with water in all proportions but the monomer can be isolated by saturation of the solution with salt. A methyl vinyl ketone–water mixture (85:15% by volume) forms an azeotrope with a boiling point of 75.8°C. Another azeotrope (bp 73–74°C) is formed by a mixture of methyl vinyl ketone, acetone, and water in a 3:3:1 composition. Pure methyl vinyl ketone can be isolated from the tertiary azeotrope by adding acetic anhydride, which removes the water, and fractionally distilling the mixture of vinyl ketone, acetone, and acetic acid (6). The physical constants of several polymerizable α,β-unsaturated ketones are summarized in Table 1.

Table 1. Physical Properties of Polymerizable α,β-Unsaturated Ketones

Ketone[a]	Mol wt	bp, °C/mmHg	d_4^{20}	n_D^{20}	Yield, %[b]	Reference
methyl vinyl (3-buten-2-one)	70.08	81.4/760 32/120	0.8636	1.4086	87	5
ethyl vinyl (1-penten-3-one)	84.12	102.2/740 38/60	0.8468	1.4192		5
isopropyl vinyl (4-methyl-1-penten-3-one)	98.16	49/77	0.847	1.4303	83.4	7
t-butyl vinyl (4,4-dimethyl-1-penten-3-one)	112.20	59–60/103		1.4222 (15°C)	61.7	8
divinyl (1,4-pentadien-3-one)	82.12	41/62	0.8839	1.4440	54.6	7
phenyl vinyl (acrylophenone)	132.2	108–112/13–14 58–60/0.2	1.060	1.5522 1.5580	82.2	7 10
methyl isopropenyl (3-methyl-3-buten-2-one)	84.12	98/760 97–98/760	0.8550_{20}^{20} 0.841	1.4220 1.4220	 94.5	5 7
isopropenyl vinyl (2-methyl-1,4-pentadien-3-one)	96.16	59/95	0.8906	1.4540	64.3	7
isopropenyl phenyl	146.2	62–64/4	1.025	1.5455	83.4	7
1-fluorovinyl methyl	88.08	71.3			58[c]	9

[a] Nomenclature accepted by *Chemical Abstracts* given in parentheses.
[b] Maximum yield obtained from the corresponding methyl or ethyl ketone via a Mannich reaction.
[c] Phosphoric acid dehydration of 1-fluoro-2-hydroxyethyl methyl ketone.

The ultraviolet spectra of the α,β-unsaturated ketones exhibit a $\pi-\pi^*$ transition in the region between 210 and 250 mμ. The λ_{max} and ϵ depend upon the nature of the substituents: alkyl (ca 220 mμ) (11) or aryl (ca 248 mμ) (12). A comparative study of ultraviolet and infrared spectra showed that alkyl vinyl ketones exist primarily in a planar *s*-cis conformation, which is consistent with their activity in Diels-Alder reactions (13). The infrared spectra of methyl and ethyl vinyl ketones contain a doublet in the carbonyl region, which is indicative of an equilibrium between the *s*-cis and *s*-trans conformations. The absorption of the vinylic protons of alkyl vinyl ketones is shown in the NMR spectra by a complex multiplet of the ABC type. A precise analysis of the spectra has been reported (14).

Chemical Properties. Vinyl ketones are very reactive compounds and serve as important chemical intermediates in the synthesis of steroids, vitamin A, bicyclic derivatives, and heterocyclic compounds (15). In addition to the normal reactions of the vinyl and the carbonyl groups, α,β-unsaturated ketones undergo 1,4 additions readily. Their activity in Diels-Alder reactions is unusual in that they behave as both

Table 2. Reactivity of Methyl Vinyl Ketone

Type of reaction	Reagent	Reaction conditions and catalysts	Product	Reference
Diels-Alder	cyclopentadiene	refluxing ether, exothermic	$-C(=O)-CH_3$ (bicyclic adduct); 62% *endo*, 38% *exo*	20
Diels-Alder	methyl vinyl ketone	145°C, 22 hr	CH_3, O, $C(=O)-CH_3$ (dihydropyran)	17
Michael 1,2 addition	phthalimide	refluxing dioxane, 30 min, $NaOCH_3$	$N-CH_2CH_2-C(=O)-CH_3$ (phthalimido, C=O, C=O)	21
1,4 addition	1-*N*-morpholinocyclohexene	(*1*) refluxing benzene, (*2*) acid hydrolysis	O (bicyclic enone); 72% α,β, 28% β,γ	22
addition to carbonyl	acetylene	Na, liquid NH_3	$CH_2{=}CH-C(CH_3)(OH)-C{\equiv}CH$	18

dienes and dienophiles (16). Thus dihydropyran derivatives, which are formed by the Diels-Alder dimerization of vinyl ketones, begin to contaminate pure monomer after long periods of storage at room temperature (17). This reaction cannot be inhibited, but the rate of dimerization is minimized at low temperatures. The activating influence of the carbonyl group facilitates the Michael-type addition of vinyl ketones to compounds containing activated hydrogen atoms (18) (see also CYANOETHYLATION). For example, methyl vinyl ketone reacts with water, alcohols, amines, malonic esters, or imides to yield either β-acetylethyl derivatives or 1,4 addition products. These reactions are generally base-catalyzed and can be inhibited by storing the vinyl ketones in the presence of weak acids. Finally, the ketone group behaves as a normal carbonyl function yielding oximes, phenylhydrazones, and other ketone derivatives. The selective addition of acetylene to the carbonyl group of methyl vinyl ketone is an important step in the synthesis of vitamin A (19). Table 2 illustrates a few typical examples of these reactions.

Monomer Synthesis. Alkyl vinyl ketones were synthesized in 1906 by heating β-chloroethyl ketones with diethylaniline (23). A large number of syntheses have been developed since then but only three or four have general applicability. Most of the syntheses are carried out at low pH to minimize the base-catalyzed condensation of the vinyl ketones. Alternatively vinyl ketones can be generated in situ from β-(diethylamino)ethyl ketones when the reaction must be run in basic media (24).

One useful technique for preparing both vinyl and isopropenyl derivatives involves a base-catalyzed condensation of formaldehyde with methyl or ethyl ketones, respectively. Thermal dehydration of the β-ketoalcohol intermediates in the presence of weak acid catalysts produces the α,β-unsaturated ketones in 50–60% yields. The best results are obtained when the initial pH is around 8.3–8.5 and the ketone-to-formaldehyde ratio is maintained at 4:1, to prevent the reaction of the methylol derivative with more formaldehyde, which leads to higher condensation products (6). Several variations of this procedure have been reported (25). The reaction of acetone with formaldehyde in the gas phase, by passing the reactants at 250–300°C over lead zeolite or silica gel impregnated with alkali metal hydroxides, is one of the most practical applications of this technique (26).

Methyl vinyl ketone is synthesized industrially by the hydration of vinylacetylene. The reaction is catalyzed by acetates, formates, or sulfates of mercury, silver, cadmium, copper, or zinc in the presence of acids (27,28). The oxidation of 1-butene to methyl vinyl ketone, in 72% yield, by the formation of olefin–mercuric salt complexes followed by the decomposition of these complexes with acid may become commercially feasible (29). Similar oxidation procedures using cupric salts have also been reported, but only 40% yields of vinyl ketone were obtained (30). The methyl vinyl ketone produced by these processes is generally yellow and is contaminated by acetone, water, acetaldehyde, 3-ketobutanol, and divinylacetylene. Extensive purification procedures have been developed (28).

Preparation of vinyl ketones via the Mannich reaction overcomes many of the drawbacks cited in the procedures described above. The α,β-unsaturated ketones are obtained in high yields (see Table 1) under mild conditions from readily available starting materials; thus, this is the best technique for laboratory preparation (7). The Mannich base, which is formed by heating equimolar quantities of ketone, formalin, and diethylamine hydrochloride for 1 hr at 95°C (eq. 1, where the ketone is methyl ethyl ketone), is isolated and pyrolyzed at 150–210°C under reduced pressure (eq. 2).

$$CH_3{-}\underset{\underset{O}{\|}}{C}{-}CH_2CH_3 + CH_2O + (CH_3CH_2)_2NH.HCl \longrightarrow CH_3{-}\underset{\underset{O}{\|}}{C}{-}\underset{\underset{\underset{HCl}{CH_2{-}N(CH_2CH_3)_2 \cdot}}{|}}{C}HCH_3 + H_2O \qquad (1)$$

$$\text{(1)} \quad \text{(1)} \xrightarrow[80\text{–}100\ \text{mmHg}]{150\text{–}200^\circ C} CH_3{-}\underset{\underset{O}{\|}}{C}{-}\overset{\overset{CH_3}{|}}{C}{=}CH_2 + (CH_3CH_2)_2NH.HCl \qquad (2)$$

The α,β-unsaturated ketone (methyl isopropenyl ketone in eq. 2) is distilled from the reaction mixture; the diethylamine hydrochloride can be recovered and recycled. A detailed kinetic study of this process showed that the pH must be maintained below 2 to obtain reasonable rates of Mannich-base formation (31). At 95°C both acetone and methyl ethyl ketone had reacted quantitatively in 6 hr. The substituted α-carbon reacts preferentially to an α-methyl group when a 1:1 mole ratio of formaldehyde and diethylamine is employed; ie, methyl isobutyl ketone yields α-isopropylvinyl methyl ketone. However, divinyl ketones can be prepared by using excess formaldehyde and diethylamine hydrochloride. This is shown in equation 3 for acetone.

$$CH_3\underset{\underset{O}{\|}}{C}CH_3 + 2\,CH_2O + 2\,(CH_3CH_2)_2NH.HCl \xrightarrow[10\ \text{hr}]{100^\circ C} \underset{\text{(2)}}{[(CH_3CH_2)_2N{-}CH_2CH_2]_2\underset{\underset{O}{\|}}{C}\cdot HCl}$$

$$\text{(2)} \xrightarrow[110\ \text{mmHg}]{\Delta} CH_2{=}CH{-}\underset{\underset{O}{\|}}{C}{-}CH{=}CH_2 + 2\,(CH_3CH_2)_2NH.HCl \qquad (3)$$

Most of the classical techniques of monomer synthesis have been applied to the preparation of α,β-unsaturated ketones. Among them one finds pyrolysis of 3-acetoxy-2-butanone (32), catalytic oxidation of 3-buten-2-ol (33) and 2,3-butanediol (34), and decarboxylation of γ-keto-α,β-unsaturated carboxylic acids (35). The addition of acid chlorides to ethylene yields β-chloroethyl alkyl ketones, which can be dehydrohalogenated with triethylamine. Isopropyl, isobutyl, *t*-butyl, neopentyl, and cyclohexyl vinyl ketones have been prepared by this technique (36). A synthesis of aromatic vinyl ketones, which should have general applications, involves a Friedel-Crafts acylation with α-chloropropionyl chloride followed by dehydrohalogenation of the chloroketone with triethylamine (37). Good yields were reported in spite of the basic conditions required in the dehydrohalogenation step.

α,β-Unsaturated ketones are usually inhibited by a mixture of acetic or formic acid and hydroquinone or catechol (5). These inhibitors can be removed by treating the monomer with anhydrous sodium carbonate, filtering, and distilling the filtrate several times under nitrogen. For kinetic investigations, a prepolymerization step is advisable. The monomer is refluxed under nitrogen in the presence of a free-radical catalyst until 10–15% has polymerized. The remaining monomer can then be distilled from the polymerizing mixture (38).

Polymerization Methods

Free-Radical Polymerization. Most of the common free-radical systems are effective in initiating vinyl ketone polymerization. Azobisisobutyronitrile is considered the best catalyst for producing polymers with good color stability; all other catalysts leave acid residues or degrade the polymer during the polymerization. γ-Irradiation is classified as a free-radical "initiator" because the polymerization can be inhibited by quinone. Although bulk polymerization is feasible, better results are obtained in solvents such as cyclohexane or petroleum ether, which dissolve the monomer but not the polymer (precipitation polymerization). Precipitation polymerization produces higher rates of reaction as well as higher molecular weights than homogeneous solutions under comparable conditions. This phenomenon has been attributed to the reduction of the rate of termination relative to the propagation rate, thus increasing the concentration of growing radicals (39) (see GEL EFFECT). A similar effect is observed in methyl methacrylate, acrylonitrile, vinyl acetate, and styrene polymerizations.

Since methyl vinyl ketone is completely miscible in water, an emulsion-type polymerization can be carried out without adding emulsifiers. A poly(methyl vinyl ketone) latex is formed which coagulates on standing (40). The latex can be stabilized by adding poly(vinyl alcohol), or other water-soluble polymers containing hydroxyl or carboxyl groups, to the aqueous monomer solution (41). The best catalyst system is an equimolar mixture of potassium peroxydisulfate–silver nitrate, but hydrogen peroxide, ammonium peroxydisulfate–silver nitrate, potassium peroxydisulfate–ferrous ammonium sulfate, and potassium permanganate–oxalic acid dihydrate have been used as initiators. Marvel and Casey found that higher-molecular-weight polymers could be obtained if the water solubility of the monomer was decreased by adding sodium chloride and an emulsifier, potassium caproate (42). However, optimum results were obtained with the potassium peroxydisulfate–silver nitrate system at low temperatures. Polymers with molecular weights on the order of $3–4 \times 10^5$ could be prepared at $-15°C$. Phenyl vinyl ketone was polymerized in an emulsion containing 7.5% soap and 0.2% potassium peroxydisulfate; high conversion to polymer with a maximum inherent viscosity of 0.65 was achieved (42). By running the redox polymerization of methyl vinyl ketone in alcohol–water mixtures containing 30–80% alcohol, the system remains homogeneous (38,43). The polymers obtained can be redissolved in alcohol–water mixtures but not in pure alcohol. Table 3 summarizes the free-radical polymerization conditions reported for α,β-unsaturated ketones.

Kinetics of Polymerization. Several detailed kinetic studies of the polymerization of methyl isopropenyl ketone have been reported. Haward investigated the precipitation polymerization in cyclohexane dilatometrically and observed a maximum rate at a cyclohexane concentration of 77.5% at 25°C, and a similar maximum at 57% when the temperature was raised to 40°C (39). The viscosity also increased fourfold as the diluent concentration was raised from 7 to 43%. Smets and Oosterbosch conducted a study of both the bulk and solution polymerizations and observed that the rate law was one-half order in initiator (benzoyl peroxide) in bulk and first order in monomer (benzene solution) (49). A similar rate law was reported earlier for the benzoyl peroxide-catalyzed polymerization of methyl vinyl ketone in toluene and cyclohexane (6). Smets calculated the energy of activation to be about 5 kcal/mole in the temperature range 64–100.5°C. This compares favorably with the value of 4.8 recently reported for the ^{60}Co-initiated polymerization of methyl vinyl ketone in the range from -78 to 20°C (44). The uncatalyzed thermal polymerization of methyl isopropenyl ketone was studied in bulk and in either aromatic or aliphatic hydrocarbons (46). No induction period was observed and the initial rates were not influenced by the nature of the solvent. Moreover, the polymerization was autocatalytic in cyclohexane; this is con-

Table 3. Free-Radical Polymerization Conditions for Polymers Derived from α,β-Unsaturated Ketones

Ketone	Catalyst[a]	Solvent	Ketone concn, mole/l	Temp, °C	Time, hr	Yield, %	$[\eta]$, dl/g[b]	Reference
methyl vinyl	BPO	none[c]	12.33	26	40	40	0.65 (MEK)	42
methyl vinyl	BPO	benzene[d]	3.5	25	100	57	1.08	6
methyl vinyl	BPO	cyclohexane[d]	3.5	25	100	98	1.8	6
methyl vinyl	BPO	acetone[c]	3.5	25	100	35	0.9	6
methyl vinyl	AIBN	ethanol–water (7:3)[c]	6.5	50	1.5	62		38
methyl vinyl	$K_2S_2O_8$[e]	water–$AgNO_3$[e]	1.43	30	3	91	1.28 (MEK)	40
methyl vinyl	$K_2S_2O_8$[e]	water–$AgNO_3$[e]	1.8	−15	42	67	2.99 (MEK)	42
methyl vinyl	γ rays	none	12.33	−78			0.3 (acetone)	44
methyl isopropenyl	AIBN	none[c]	10.18	75	24	91	0.33	45
methyl isopropenyl	BPO	none[c]	10.18	87	24	41	0.29	45
methyl isopropenyl	none	toluene[c]	3.8	80	5	3.1	0.41 (DMF)	46
methyl isopropenyl	BPO	cyclohexane[d]	5.6	25	200	60	1.5 (dioxane)	39
methyl isopropenyl	$B(n\text{-}C_4H_9)_3$	toluene[c]	5.8	0	48	38	1.2 ($CHCl_3$)	47
t-butyl vinyl	AIBN	benzene[c]		60	26	66		8
phenyl vinyl	BPO	toluene[c]	1.7	60	48	32		47
phenyl vinyl	AIBN	benzene[c]	1.2	55	10	73	0.26 (benzene)	42
phenyl vinyl	AIBN	toluene[c]		78			1.5 (benzene)	48
phenyl vinyl	$K_2S_2O_8$[e]	water–7.5% ORR soap[f]	1.5	26	24	69	0.65 (benzene)	42
fluorovinyl methyl	AIBN	benzene	1.8	65	17	64.3	3.1 ($CHCl_3$)	9
fluorovinyl methyl	AIBN	chloroform[c]	1.8	65	17	50	1.4 ($CHCl_3$)	9

[a] Approximately 1 mole % of monomer, BPO = benzoyl peroxide, AIBN = azobisisobutyronitrile.

[b] Solvents given in parentheses if they were reported.

[c] Homogeneous solution.

[d] Polymer precipitates at low conversions.

[e] 0.2 mole % of monomer, emulsion polymerization.

[f] Office of Rubber Reserve soap; a mixture of trimethylalkylammonium chlorides in which the alkyl group is a straight-chain C_{12}–C_{18} saturated hydrocarbon.

sistent with the gel effect observed in precipitation polymerization. The rate of thermal polymerization was found to be proportional to the square of the monomer concentration.

Ionic Polymerization. Ionic initiators can also effect the polymerization of vinyl ketones. Methyl vinyl ketone yielded a solid polymer when boron trifluoride etherate was added to a mixture of monomer and carbon dioxide in petroleum ether (50). Crystalline poly(ethyl, isopropyl, *t*-butyl, and cyclohexyl vinyl ketones) were prepared by precipitation polymerization using metallic lithium or alkyllithium catalysts (8,36). Amorphous polymers were obtained under similar conditions if the polymerization remained homogeneous. Phenyl Grignard reagents initiated the stereoregular polymerization of isopropenyl methyl ketone (51). Complexes formed by the coordination of alkylaluminum or alkylzinc compounds with alkali metal alkyls ("-ate" complexes) produced crystalline methyl vinyl and methyl isopropenyl ketone polymers, but phenyl vinyl ketone polymers prepared under similar conditions were amorphous (10,52). The catalysts always yield mixtures of amorphous (type I) and crystalline (type II) polymers, which must be fractionated by extraction with acetone or hot 3-heptanone. Crystalline poly(methyl vinyl ketone) is insoluble in most solvents but can be dissolved in formic acid. The x-ray diffraction patterns, infrared spectra, and melting points confirm that there are significant differences between type I and type II polymers. The polymerization conditions and properties of the polymers prepared in the presence of ionic catalysts are summarized in Table 4.

Recently vinyl ketone polymerizations accompanied by hydrogen transfer in the propagation step have been reported (54). Treatment of methyl vinyl ketone with sodium *t*-butoxide in toluene yielded a low-molecular-weight polymer with the structure $\left[CH_2—\overset{O}{\overset{\|}{C}}—CH_2—CH_2 \right]_x$. These experiments were conducted in the presence of *N*-phenyl-β-napthylamine to inhibit free-radical initiation. Although the choice of inhibitor was probably fortuitous, subsequent work showed that this inhibitor promotes the hydrogen-transfer mechanism (55). Polymerizations accompanied by cyclization have also been observed in the presence of alkyllithium and Grignard reagents (10,51). See also ISOMERIZATION POLYMERIZATION.

Copolymerization. The copolymerization of α,β-unsaturated ketones has been studied extensively in an effort to improve the poor chemical and thermal stability exhibited by the homopolymers. Methyl vinyl ketone is similar to styrene in its copolymerizability, as indicated by its copolymerization parameters, $e = 0.7$ and $Q = 1.0$. The resonance factor, Q, is identical but the electronic factor, e, has the opposite polarity (56) (see COPOLYMERIZATION). Thus, one would expect a high tendency toward alternation with styrene which would lead to isolated vinyl ketone units in the copolymers and to a corresponding increase in color stability. Schwann and Price have reevaluated the Q–e scheme to include temperature factors and report revised values of $Q = 2.8$ kcal/mole and $e = 0.19 \times 10^{-10}$ esu for the resonance and electronic factors, respectively (57). Further correlations of Q and e to Hammett substituent constants have been described (58). Price-Alfrey copolymerization parameters reported for other α,β-unsaturated ketones include phenyl vinyl ketone, $e = 0.66$ and $Q = 2.09$ (37); methyl isopropenyl ketone, $e = 0.53$ and $Q = 1.49$ (59) (recalculated by Wich and Brodoway to be $e = 1.23$ and $Q = 1.95$ (60)).

Although the vinyl ketones have been copolymerized with most of the common vinyl and diene monomers, this work is described in patents or articles that do not con-

Table 4. Ionic Polymerization Conditions and Physical Properties of Polymers Derived From α,β-Unsaturated Ketones

Ketone	Catalyst[a]	Solvent	Temp, °C	Time, hr	Yield[b]		$\eta_{sp/c}$, dl/g[c]	Melting point, °C	Reference
					Type I	Type II			
methyl vinyl	BF_3–$(C_2H_5)_2$	petroleum ether–carbon dioxide	−78	0.5					50
methyl vinyl	C_6H_5MgBr	toluene	0	168	63	6	1.39	165–170	10
methyl vinyl	$Al(C_2H_5)_3$	toluene	0	168	62	9	1.3	145–160	10
methyl vinyl	$CaZn(C_2H_5)_4$	toluene	0	96	64[d]	8	1.2	162–164	10
methyl isopropenyl	$(C_2H_5)_2Mg$–$CoCl_2$	ether	20					219–223	53
methyl isopropenyl	$Al(C_2H_5)_3$	toluene	0	48	62	13	1.5	160–190[e]	10
methyl isopropenyl	$LiAl(C_2H_5)_3Bu$	toluene	0	48	65	12		>140[e]	10
methyl isopropenyl	$CaZn(C_2H_5)_4$	toluene	0	48	61	20	1.3	210–215	10
methyl isopropenyl	$Zn(C_2H_5)_2$	toluene	0	48	43	22	1.6	200–205	10
methyl isopropenyl	C_6H_5MgI in toluene	ethylene chloride	−20	5		30		240	51
methyl isopropenyl	C_6H_5MgI in toluene	chloroform	−40	20		42		240	51
isopropyl vinyl	Li	chloroform	−25				4.0	220[f]	36
t-butyl vinyl	$LiAlH_4$ in THF	chloroform	0	3			0.7	154	36
t-butyl vinyl	Li	hexane	25	3	24	56		140–150	8
t-butyl vinyl	C_4H_9Li	hexane	25	0.66	25	58		140–150	8
t-butyl vinyl	C_4H_9Li	tetrahydrofuran	0	6	46.2		0.53	90[e]	8
phenyl vinyl	C_4H_9Li	toluene	0	48	31			89–90[e]	10
phenyl vinyl	$SrZn(C_2H_5)_4$	toluene	0	48	73	2		80–90[e]	10
phenyl vinyl	KCN	dimethylformamide	−78		28		0.10		42

[a] Initiator concentration is approximately 1 mole % of monomer.
[b] Total conversion is the sum of type I and type II.
[c] Values apply to type II fractions. Viscosities measured in chloroform at concentrations varying from 0.2 to 0.6 g/dl.
[d] Partially crystalline with melting point of 145–160°C.
[e] Amorphous x-ray diffraction pattern.
[f] Depolymerization occurred at this temperature.

Table 5. Reactivity Ratios for Vinyl Ketone Copolymerization

M_1	r_1	M_2	r_2	T, °C	Reference
methyl vinyl ketone	1.78 ± 0.22	acrylonitrile	0.61 ± 0.04	60	61
	1.6 ± 0.1	butyl acrylate	0.65 ± 0.07	50	62
	4.4	phenyl vinyl ether	0.1		63
	0.35 ± 0.02	styrene	0.29 ± 0.04	60	61
	0.26 ± 0.04	4,6-diamino-2-vinyl-*s*-triazine	1.2 ± 0.15	60	64
	7.00	vinyl acetate	0.05	70	65
	8.3	vinyl chloride	0.10	70	66
	1.8	vinylidene chloride	0.55	70	66
methyl isopropenyl ketone	0.70 ± 0.14	acrylonitrile	0.36 ± 0.08	80	66
	0.1 ± 0.05	2-chloro-1,3-butadiene	3.6 ± 0.2	40	60
	1.7	α-methylstyrene	0.3		66
	0.66	styrene	0.32	80	66
	0.29 ± 0.6	styrene	0.44 ± 0.10	80	66
	4.5 ± 0.1	vinylidene chloride	0.15 ± 0.02	60	59
phenyl vinyl ketone	1.10 ± 0.10	styrene	0.107		37

tain sufficient experimental data to derive the reactivity ratios. The experimental values reported in the literature are tabulated in Table 5.

Copolymerization of methyl vinyl ketone with styrene under anionic conditions yields copolymers containing a minimum of 98% methyl vinyl ketone for all initial monomer compositions (52). Apparently the growing ketone anion cannot add to styrene; the trace amounts of styrene found were formed from styryl anions generated by the initiator before a cross-propagation step occurred. Similarly, poly(*t*-butyl vinyl ketone) was isolated from the attempted copolymerization of *t*-butyl vinyl ketone with methyl methacrylate in the presence of metallic lithium or biphenyllithium (8).

β-Substituted α,β-unsaturated ketones can also be copolymerized. Marvel and co-workers showed that benzalacetophenone copolymerizes with butadiene, styrene, and methyl methacrylate using emulsion techniques (67). No homopolymer could be prepared, but 1,2-disubstituted ethylenes are known to be difficult to homopolymerize. Subsequent work in this area showed that benzalacetone, furfuralacetophenone, thiophene or pyridine analogs of benzalacetophenone, and *cis*- or *trans*-dibenzoylethylene form copolymers with butadiene, isoprene, or styrene (68). All these comonomers exhibit reactivity ratios very close to zero and, thus, yield alternating copolymers. A detailed description of the composition and properties of these copolymers has been published (69). Recently the copolymerizability of 4-cyclopentene-1,3-dione, a monomer similar in structure to maleic anhydride, was studied (70). Acrylonitrile and methyl methacrylate yielded normal monomer feed–copolymer composition relationships, but anomalous results were obtained for the styrene copolymers.

Copolymers with a wide range of physical properties have been prepared in an effort to commercialize vinyl ketones. Most of the work has been directed toward the preparation of oil- and solvent-resistant rubbers to replace styrene–butadiene rubber. Emulsion copolymerization of butadiene with either methyl vinyl ketone or methyl isopropenyl ketone yielded rubbers with good solvent resistance and low-temperature flexibility, but the products tended to harden on storage and were not compatible with natural rubber (71). Copolymerization of methyl isopropenyl ketone with styrene improved the injection-molding characteristics over those of the ketone homopolymer and retained most of the good properties (45). Low-molecular-weight copolymers of methyl vinyl ketone and ethylene or vinylidene chloride were evaluated as adhesives for cellulose (72). Although these copolymers fulfill most of the criteria for these applications, the presence of the reactive carbonyl function reduces their aging characteristics and imparts an undesirable sensitivity to alkaline reagents. This sensitivity can be exploited if crosslinked resins are required. Copolymers of butyl acrylate and methyl vinyl ketone can be crosslinked by treatment with hydrazine (62). The addition of primary and secondary amines to copolymers containing carbonyl groups produces crosslinked resins with anion-exchange capabilities (73). Crosslinked resins containing carbonyl groups can be prepared by using divinyl ketone as a comonomer (74).

Terpolymers with excellent clarity were prepared from monomer mixtures corresponding to the azeotropic copolymer compositions using the following monomer combinations: methyl vinyl ketone, styrene, acrylonitrile; methyl vinyl ketone, styrene, methyl methacrylate; methyl vinyl ketone, styrene, acrylic acid; and methyl vinyl ketone, methacrylonitrile, vinyltoluene (75). Terpolymers of methyl isopropenyl ketone with butadiene and acrylonitrile have limited application as specialty rubbers (76,77).

Physical Properties of Polymers

The physical properties of the polymers depend upon the polymerization conditions; poly(methyl vinyl ketone) ranges from a viscous oil to a hard plastic or rubbery mass. All polymers prepared with free-radical initiators are amorphous materials with low softening points and poor thermal or chemical stability. Extensive monomer purification, low polymerization temperatures, and neutral catalysts such as azobisisobutyronitrile are required to prepare colorless polymers. Exposure to light, heat, or traces of acid will accelerate the discoloration, which is accompanied by crosslinking. Alkali or amines also promote crosslinking via an aldol condensation; the polymers become intractable, brittle, highly colored resins.

Amorphous poly(methyl vinyl ketone) softens at 40–60°C and is soluble in the monomer, acetone, methyl ethyl ketone, tetrahydrofuran, dimethylformamide, acetic acid, ethyl acetate, dioxane, pyridine, and chloroform. It is insoluble in aromatic and aliphatic hydrocarbons, water, ethyl ether, and carbon tetrachloride. The density of the polymer is 1.12, which corresponds to a 25% contraction during polymerization (5). Colorless samples have a refractive index of 1.5 and exhibit a strong ultraviolet absorption around 290 mμ.

Poly(methyl isopropenyl ketone) exhibits physical properties very similar to poly(methyl methacrylate). A 24% contraction during polymerization leads to a polymer with a density of 1.116 (39). Brown and co-workers evaluated this polymer as a transparent molding compound; the softening point (ca 80°C) and refractive index (1.5212) were acceptable, but the thermal and photochemical stability was unsatisfactory. The Rockwell hardness, water absorption, tensile- and flexural-strength properties exceeded those reported for polystyrene and poly(methyl methacrylate) (45). The solubility properties are similar to those of poly(methyl vinyl ketone), with the exception of improved solubility in aromatic hydrocarbons.

Phenyl vinyl ketone has not been studied extensively; however, a softening point of 80–90°C has been reported for the polymer (44). 1-Fluorovinyl methyl ketone polymer exhibits a glass-transition temperature at 142°C (DTA) and a break in the TGA curve around 280°C (9). Thus, poly(1-fluorovinyl methyl ketone) is considerably more stable than the corresponding 1-chlorovinyl polymer, which degrades at room temperature (78).

The molecular weights of vinyl ketone polymers are relatively low owing to the lability of the protons alpha to the carbonyl group. The chain-transfer constant to methyl isopropenyl ketone (C_M) was found to be 4.0×10^{-4} at 80°C. Chain-transfer constants to initiator and several solvents were also reported (46). The molecular-weight–viscosity relationships used to estimate the degree of polymerization have recently been summarized (79). For poly(methyl isopropenyl ketone) the following equations apply:

$$\begin{aligned} \log\,[\eta] &= -4.558 + 0.81_7 \log \bar{M}_w \text{ in dimethylformamide at } 35^\circ\text{C} \\ &= -4.161 + 0.72 \log \bar{M}_w \text{ in dioxane at } 25^\circ\text{C} \\ &= -3.682 + 0.60 \log \bar{M}_w \text{ in methyl ethyl ketone at } 25^\circ\text{C} \end{aligned}$$

where $[\eta]$ is expressed in deciliters/gram. Comparable relationships have not been defined for poly(methyl vinyl ketone). However, Guillet and Norrish report that an intrinsic viscosity of 1.27 dl/g in methyl ethyl ketone corresponds to a number-average molecular weight of 254,000 (80).

Degradation Reactions

Thermal and Chemical Processes. Polymers of vinyl or isopropenyl alkyl ketones lose water at temperatures above 250°C to yield glassy, red, noncrosslinked products. Marvel proposed that an intramolecular aldol condensation (eq. 4) was responsible for this degradation and showed that 15–21% of the oxygen remained in the polymer (3,81). The residual oxygen content was in good agreement with the theoretical value of 18.4% expected for the proportion of unchanged monomer units that would become isolated if the water were eliminated from random pairs of adjacent units (82). The color is produced by the formation of a sequence of conjugated double bonds. The ultraviolet spectrum of the degraded polymer clearly indicates the pres-

$$\left[\begin{matrix} & H & H_2 & H & H_2 & \\ \sim & C & & C & & CH \sim \\ & C{=}O & & C{=}O & & C{=}O \\ & CH_3 & & CH_3 & & CH_3 \end{matrix} \right]_x \xrightarrow[-H_2O]{\Delta} \left[\begin{matrix} & H & H_2 & H & H_2 & \\ \sim & C & & C & & CH \sim \\ & C{=}C & & C{=}O & & C{=}O \\ & CH_3 & H & & & CH_3 \end{matrix} \right]_x \xrightarrow{-H_2O} \left[\begin{matrix} & H & H_2 & H & H_2 & \\ \sim & C & & C & & CH \sim \\ & C{=}C & & C{=}C & & C{=}O \\ & CH_3 & H & & H & \end{matrix} \right]_x \xrightarrow{etc} \quad (4)$$

ence of three different cyclic species originating from the condensation of two, three, or four methyl ketone units (83,84). A kinetic analysis of the reaction showed that the relative concentrations of these species can be derived by assuming that the condensation occurs between random pairs of ketone groups. No penultimate effect or chain reaction could be detected. The structure of the cyclic adducts was confirmed by infrared spectroscopy. The decrease in the absorption bands at 1710 ($>$C=O), 1356 (—CH_3), and 1160 cm^{-1} (—$COCH_3$) is accompanied by an increase in the intensity of the bands at 1650, 1615, and 1580 cm^{-1} (conjugated $>$C=O, —C=C—) (85). In contrast to the thermal degradation, amine- or alkali-catalyzed degradation occurs by a chain process. Treatment of poly(methyl vinyl ketone) with primary or secondary amines at 110°C produces a very rapid discoloration. Not only is the reaction accelerated, but the sample eventually turns black, indicating longer conjugated sequences than those obtained by purely thermal degradation. The chain reaction initiated by the amines is analogous to the degradation of polyacrylonitrile (see DEGRADATION). Indeed, the presence of methyl vinyl ketone units in acrylonitrile copolymers exerts a catalytic influence on the thermal degradation of these copolymers. Chlorinated solvents such as tetrachloroethane or chloroform also catalyze the aldol condensation of poly(methyl isopropenyl ketone) (84). Alkali is so effective that cyclized polymer is frequently the only product isolated from anionic polymerization (10,86). When solutions of the ketone polymers are treated with alkali, an exothermic reaction yielding black, brittle crosslinked products ensues. Traces of acids also catalyze the degradation at elevated temperatures (45).

The chemical properties of crystalline poly(methyl vinyl ketone) and poly(methyl isopropenyl ketone) are not outstanding. Although the stereoregular structure of these polymers produced an increase of 100°C in melting points, no improvement in the chemical or thermal stability was observed (84). This could be attributed to the presence of amorphous segments in the crystalline fractions; the thermal degradation of a sample of crystalline poly(isopropenyl methyl ketone) exhibited a more rapid initial loss in weight than the amorphous sample, but the total weight loss of the crystalline sample was less. The rate of intramolecular aldol condensation of the crystalline sample was also enhanced, which might be expected if the functional groups were juxtaposed in an isotactic structure.

Photochemical Processes. Photolytic degradation of poly(methyl vinyl ketone) is a more complicated process. Irradiation of dioxane solutions (80) or thin films (87) with monochromatic light at 313 mμ produces an initial rapid reduction in molecular weight followed by a slower process which is accompanied by the evolution of acetaldehyde, carbon monoxide, and methane. These results can be interpreted by assuming a concomitant occurrence of Norrish type I (eq. 5) and type II (eq. 6) ketone cleavages:

Type I

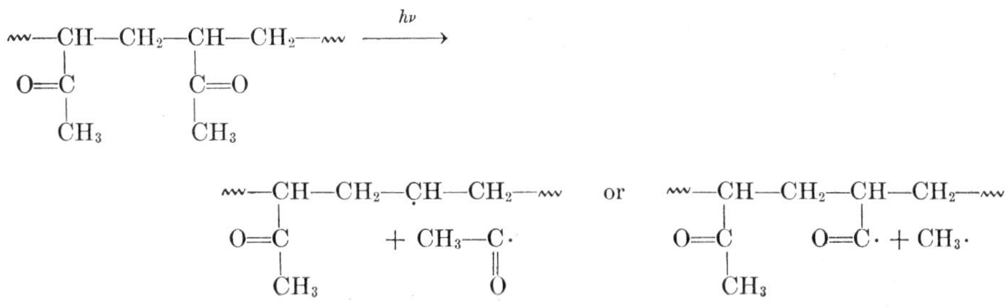

```
~~—CH—CH2—CH—CH2—~~   hν
    |        |        ——→
  O=C        C=O
    |        |
    CH3      CH3

~~—CH—CH2—ĊH—CH2—~~              or   ~~—CH—CH2—CH—CH2—~~        (5)
    |                                     |        |
  O=C            + CH3—C·               O=C      O=C· + CH3·
    |                  ||                 |
    CH3                O                  CH3
```

Type II

$$\sim\!\!\sim CH(C(=O)CH_3)-CH_2-CH(C(=O)CH_3)-CH_2\sim\!\!\sim \xrightarrow{h\nu}$$

$$\sim\!\!\sim C(C(=O)CH_3){=}CH_2 + CH_2(C(=O)CH_3)-CH_2\sim\!\!\sim \quad \text{or} \quad \sim\!\!\sim CH(C(=O)CH_3)-CH_3 + CH(C(=O)CH_3){=}CH\sim\!\!\sim \qquad (6)$$

Type II reactions are responsible for the decrease in molecular weight and the formation of double bonds conjugated with the carbonyl groups, which could be identified by both ultraviolet spectroscopy and relative rates of oxidation with peroxybenzoic acid. Type I reactions yield the gaseous by-products and introduce free radicals into the polymer backbone which ultimately form crosslinks. The evolution of carbon monoxide observed during the polymerization of methyl vinyl ketone with ultraviolet radiation is probably due to a simultaneous degradation of the polymer formed (88). Polyacrylonitrile, poly(methyl methacrylate), and poly(vinyl acetate) can be grafted to poly(methyl vinyl ketone) via the radicals formed by irradiation of the latter polymer dissolved in the suitable monomer (89).

Poly(methyl isopropenyl ketone) degrades almost quantitatively to monomer upon exposure to 313 $m\mu$ radiation at temperatures in the range 150–200°C (87). Very low yields of carbon monoxide and methane could also be detected. An unzipping mechanism similar to that proposed for poly(methyl methacrylate) would be consistent with these observations, but the complicated dependence of the reaction rate on molecular weight, temperature, and light intensity suggests that a more complex reaction sequence is occurring. In contrast to these results, Shultz observed that irradiation of dioxane solutions of poly(methyl isopropenyl ketone) at 23°C effected a random chain scission, as evidenced by the reduction in viscosity per unit of radiant energy absorbed (79). Similar results were obtained from γ-irradiation of solid polymer samples. A comparison of the absorbed energy per random chain scission for poly(methyl isopropenyl ketone) and poly(methyl methacrylate) enabled the author to conclude that the polyketone has a higher absorption coefficient (k = 526 cm^{-1}) and lower scission energy (E_d = 22 eV/scission) than poly(methyl methacrylate) (k = 18.4 cm^{-1}, E_d = 2100 eV/scission) under 2536 Å light irradiation.

Chemical Transformations. The future of poly(vinyl ketones) in polymer chemistry lies in their potential as reactive polymer substrates. By converting the carbonyl group to a less reactive function, one can increase the thermal and photochemical stability and introduce new functional groups that are not accessible by conventional polymerization techniques (see also ACROLEIN POLYMERS). Thus, Marvel and co-workers were able to introduce oxime groups into vinyl ketone polymers (42). Unfortunately, the reaction of poly(methyl vinyl ketone) with hydrogen cyanide, phenylhydrazine, semicarbazide, mercaptans, or amines is complicated by the simultaneous cyclization of adjacent methyl ketone groups (90). Although the expected derivative is formed to some extent, the major reaction is usually cyclization. A unique cyclization reaction is observed when poly(phenyl vinyl ketone) is treated with trityl perchlorate in refluxing nitromethane–acetic acid mixtures; a polymer composed of up to 50% 3,5-disubstituted 2,6-diphenylpyrylium perchlorate units is isolated (91) (eq. 7).

$$\left[-CH_2-\overset{H}{C}-\overset{H_2}{C}-\overset{H}{C}- \right]_x \text{ with } \underset{C_6H_5}{C=O} \text{ groups} + (C_6H_5)_3C^{\oplus}ClO_4^{\ominus} \xrightarrow[(1)\ CH_3NO_2,\ (2)\ CH_3COOH]{\Delta} \left[-CH_2-(\text{pyrylium ring}, C_6H_5, O^{\oplus}, C_6H_5, ClO_4^{\ominus})-\overset{H_2}{C}-\underset{\underset{C_6H_5}{C=O}}{CH}- \right]_y \qquad (7)$$

The cyclization reaction can be minimized by preparing copolymers containing a maximum of 30% vinyl ketone. The comonomer units isolate the ketone moiety and prevent intramolecular reactions. The copolymers exhibit normal ketone reactivity; a few of the more unusual reactions are described below.

Marvel and Wright were able to prepare phosphorus-containing copolymers of styrene, methyl methacrylate, acrylonitrile, and butadiene by treating the corresponding methyl vinyl or methyl isopropenyl ketone copolymer with phosphorus trichloride followed by hydrolysis (eq. 8). The copolymers containing the hydroxyphosphonic

$$\left[-\underset{\underset{CH_3}{C=O}}{CH}-CH_2- \right]_n \xrightarrow[(2)\ H_2O-CH_3COOH]{(1)\ PCl_3} \left[-\underset{HO-\underset{CH_3}{C}-PO_3H_2}{CH}-CH_2- \right]_n \qquad (8)$$

acid groups tended to crosslink readily, but were much less flammable than the starting materials (92). The Wittig reaction was employed to convert the carbonyl groups in methyl vinyl ketone–styrene copolymers to methylene and ethylidene substituents (93). The reaction is accompanied by crosslinking, as indicated by the formation of about 25% insoluble gel and an increase in the number-average molecular weight of the soluble fraction.

Polyketones can be reduced to secondary alcohols by catalytic hydrogenation over platinum oxide or reaction with lithium aluminum hydride or potassium borohydride (90,94). The products are poly(keto alcohols) since a maximum conversion of only 50% was attained. A mechanism involving a cyclic complex, analogous to that observed in Meerwein-Ponndorf-Verley reductions, was postulated to explain the 50% conversion limit. The reaction of ethylmagnesium bromide with the polyketones yields tertiary alcohols but cyclization of adjacent ketone groups becomes a serious side reaction. An interesting variation of the complex metal hydride reduction is the preparation of optically active poly(methyl vinyl carbinol) by the reduction of poly(methyl vinyl ketone) with lithium borneoxyl aluminum hydride (95). Kun and Cassidy report that treatment of either poly(methyl vinyl ketone) or a styrene–methyl vinyl ketone copolymer with aluminum isopropoxide in benzene failed to effect a Meerwein-Ponndorf-Verley reduction (96). However, they observed that lithium aluminum hydride in benzene–ether mixtures completely reduced the copolymers. The reduced reactivity of the polyketones toward aluminum isopropoxide was attributed to steric hindrance. Subsequent experiments with the same copolymer in the melt (ca 160°C) showed that aluminum isopropoxide can effect reduction at high temperatures (97), but the products are crosslinked.

Derivatives of poly(vinyl ketones) have also been used as precursors for the preparation of polypyridines (42). Poly(methyl vinyl ketone) was converted to a polyketoxime which yielded up to 75 mole % 2,6-dimethylpyridine structures upon treatment with ethanolic hydrogen chloride (eq. 9).

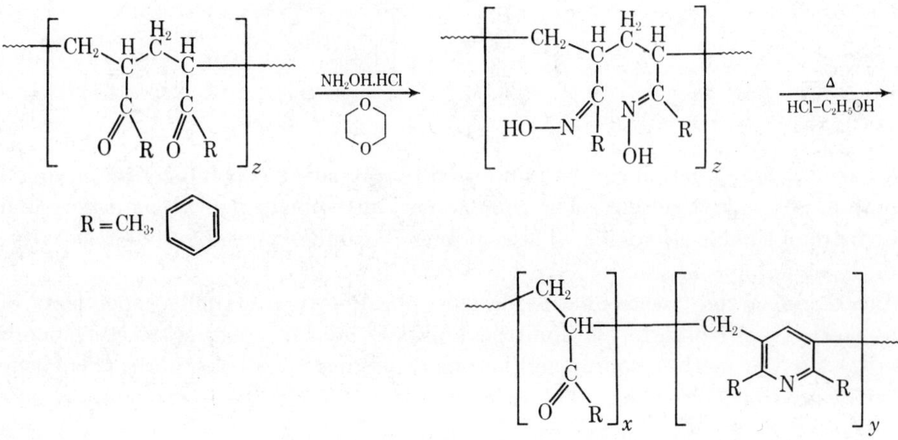

(9)

The most promising new application of polyketones is their use as polymeric photosensitizers. Cassidy and Moser report that poly(phenyl vinyl ketone), dissolved in deaerated benzene with an excess of *d,l*-α-phenylethanol, underwent nearly complete photoreduction to a polypinacol upon irradiation with a strong ultraviolet source (48). Poly(phenyl vinyl ketone) sensitized the cis–trans isomerization of *cis*-piperylene in benzene solution or in nitrogen–piperylene gas mixtures, which were sealed in a tube coated with polymer particles. The vapor-phase isomerization proceeded more rapidly, indicating that rapid diffusion of the diene to the activated polymer surface occurred and that the solvent plays a nonessential role in the energy-transfer step. Solid polymer in inert solvents or in the reactants sensitized the formation of quadricyclene from norbornadiene (eq. 10) as well as conversion of myrcene to 5,5-dimethyl-1-vinylbicyclo[2.1.1]hexane (eq. 11) (98). The emission spectrum of the polymer in an

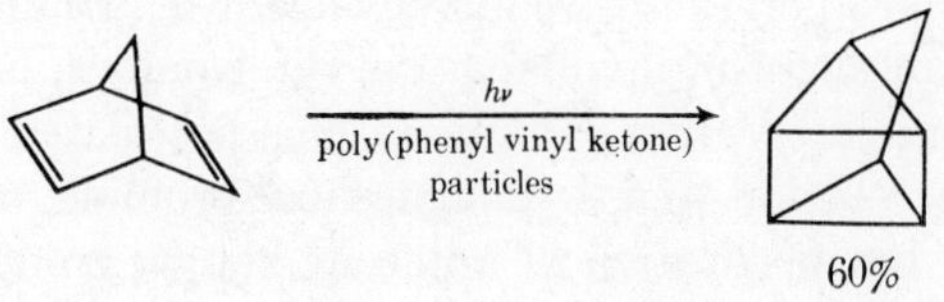

(10)

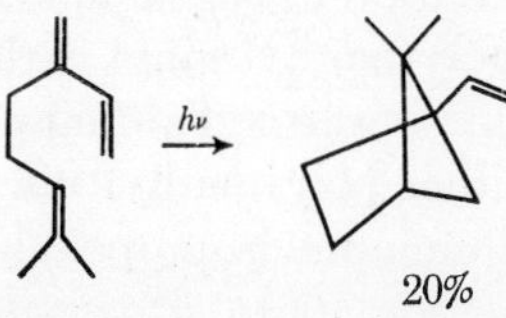

(11)

ether–tetrahydrofuran glass showed a well-resolved progression of bands with maxima at 395 (O—O band, triplet energy of 77.2 kcal), 420, 450, and 485 mμ, which is similar to the emission spectrum of acetophenone under the same conditions. Thus, the triplet state of poly(phenyl vinyl ketone) is closely analogous to acetophenone which is commonly used as a photosensitizer.

Poly(vinyl acetophenone) as well as copolymers of vinyl acetophenone and styrene have been prepared but their photosensitizing capabilities were not described (99,100). However, treatment of polymers containing acetophenone substituents with benzaldehyde in the presence of concentrated sulfuric acid introduced benzalacetophenone units, which dimerized on exposure to light (eq. 12) (101). Photosensitive crosslinking agents of this type are of interest in the preparation of photoresist resins for use in photoreproduction.

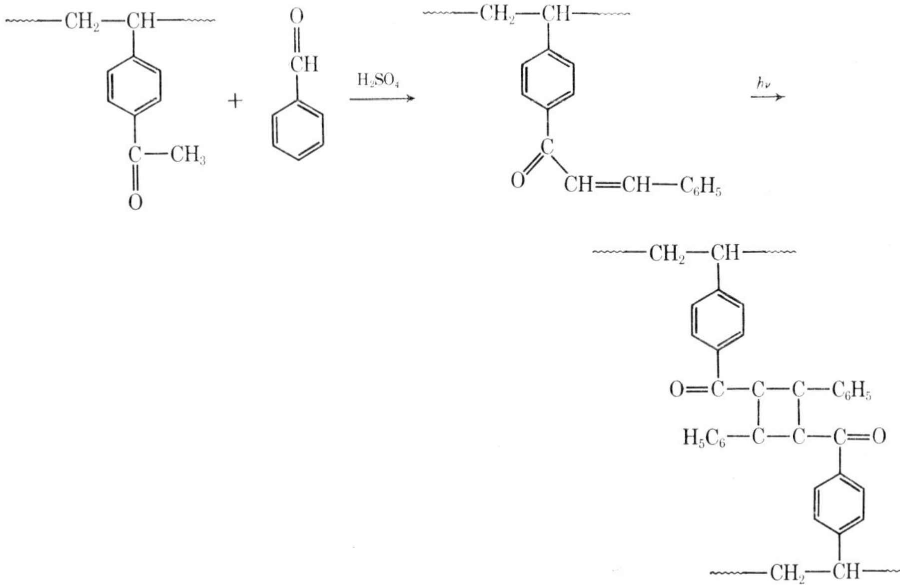

$\sim\sim CH_2—CH\sim\sim$ (C_6H_4—C(=O)—CH_3) + O=CH—C_6H_5 $\xrightarrow{H_2SO_4}$ $\sim\sim CH_2—CH\sim\sim$ (C_6H_4—C(=O)—CH=CH—C_6H_5) $\xrightarrow{h\nu}$

$\sim\sim CH_2—CH\sim\sim$ / C_6H_4 / O=C—C—C—C_6H_5 / H_5C_6—C—C—C=O / C_6H_4 / $\sim\sim CH_2—CH\sim\sim$ (12)

Analytical Techniques

α,β-Unsaturated ketones can be analyzed by quantitative hydrogenation (102), polarographic reduction (103), bromate–bromide titration (104), quantitative oxime formation (104), and gas chromatography (105). The polarographic determination is most selective and can be used in the presence of formaldehyde and acetone (106). However, the bromate–bromide or oxime titrations can be performed more rapidly and yield satisfactory results. Trace amounts of α,β-unsaturated ketones can be detected by a chromotropic acid spot test (107) or by gas chromatography. The retention times for methyl vinyl and methyl isopropenyl ketone on a variety of stationary phases have been reported (108).

The composition of methyl isopropenyl ketone copolymers has been determined by quantitative oxime formation, ultraviolet (109) and infrared techniques, and radiotracer methods (110).

Safety Precautions

Vinyl ketones are generally toxic, lacrimatory compounds. Dermatitis results from exposure to methyl vinyl ketone vapor or liquid (111). Methyl isopropenyl ketone is less irritating, but prolonged exposure to its vapor should be avoided. The physiological properties of the higher homologs have not been reported. The normal precautions for handling toxic, flammable liquids should be observed.

Bibliography

1. C. M. Van Marle and B. Tollens, *Chem. Ber.* **36,** 1351 (1903); H. Schaefier and B. Tollens, *Chem. Ber.* **39,** 2187 (1906).
2. H. Staudinger and B. Ritzenthaler, *Chem. Ber.* **67,** 1773 (1934).
3. C. S. Marvel and C. L. Levesque, *J. Am. Chem. Soc.* **60,** 280 (1938); **61,** 3234 (1939).
4. G. Merling and H. Koehler (to Bayer), U.S. Pat. 981,668.9 (1909).
5. C. E. Schildknecht, *Vinyl and Related Polymers,* John Wiley & Sons, Inc., New York, 1952, pp. 682ff.
6. T. White and R. N. Haward, *J. Chem. Soc.* **1943,** 25.
7. M. I. Farberov and G. S. Mironov, *Dokl. Akad. Nauk SSSR* **148,** 1095 (1963); *Chem. Abstr.* **59,** 5062a.
8. C. G. Overberger and A. M. Schiller, *J. Polymer Sci.* [C] **1,** 325 (1963); **54,** 530 (1961).
9. J. A. Sedlak and K. Matsuda, *J. Polymer Sci.* [A] **3,** 2329 (1965).
10. T. Tsuruta, R. Fujio, and J. Furukawa, *Makromol. Chem.* **80,** 172 (1964).
11. L. K. Evans and A. E. Gilliam, *J. Chem. Soc.* **1941,** 815.
12. K. Bowden, E. A. Braude, and E. R. H. Jones, *J. Chem. Soc.* **1946,** 948.
13. R. Mecke and K. Noack, *Chem. Ber.* **93,** 210 (1960).
14. S. Castellano and J. S. Waugh, *J. Chem. Phys.* **37,** 1951 (1962).
15. L. F. Fieser and M. Fieser, *Reagents for Organic Synthesis,* John Wiley & Sons, Inc., New York, 1967, p. 697.
16. J. Colonge and G. Descotes, "α,β-Unsaturated Carbonyl Compounds as Dienes," in J. Hamer, ed., *1,4-Cycloaddition Reactions, The Diels-Alder Reaction in Heterocyclic Syntheses,* Academic Press, Inc., New York, 1967.
17. J. Colonge and J. Dreux, *Compt. Rend.* **228,** 582 (1949).
18. E. D. Bergmann, D. Ginsburg, and R. Pappo, *Organic Reactions* **10,** 179–560 (1959).
19. O. Isler, W. Huber, A. Ronco, and M. Korler, *Helv. Chem. Acta* **30,** 1911 (1947).
20. J. G. Dinwiddie, Jr., and S. P. McManus, *J. Org. Chem.* **30,** 766 (1965).
21. H. Irai, S. Shima, and N. Murata, *Kogyo Kagaku Zasshi* **62,** 82 (1959); *Chem. Abstr.* **58,** 5659a.
22. G. Stork, A. Brizzolara, H. Landesman, J. Szmuszkovicz, and R. Terrell, *J. Am. Chem. Soc.* **85,** 207 (1963).
23. E. E. Blaise and M. Maire, *Compt. Rend.* **142,** 215 (1906).
24. A. L. Wilds, R. M. Nowak, and K. E. McCaleb, *Org. Syn. Coll. Vol.* **4,** 281 (1963).
25. G. Morgan, N. J. L. Megson, and K. W. Pepper, *Chem. Ind. (London)* **57,** 885 (1938); E. F. Landau and E. P. Irany, *J. Org. Chem.* **12,** 422 (1946); A. Wohl and A. Prill, *Ann. Chem.* **140,** 139 (1924).
26. E. M. McManon, J. H. Roper, W. P. Utermohler, R. H. Haser, R. C. Harris, and J. H. Brant, *J. Am. Chem. Soc.* **70,** 2971 (1948); J. H. Brant and R. L. Hasche (to Eastman Kodak Co.), U.S. Pat. 2,245,567 (1939); S. Malinowski, S. Benbener, J. Pasynkiewilzand, and E. Wosciechowsica, *Roczniki Chem.* **32,** 1089 (1958); *Chem. Abstr.* **53,** 7974.
27. A. S. Carter (to E. I. du Pont de Nemours and Co., Inc.), U.S. Pat. 1,896,161 (1933); R. F. Conaway (Du Pont), U.S. Pat. 1,967,225 (1934); Brit. Pat. 388,402 (1933); Fr. Pat. 719,309 (1931).
28. O. Nicodemus and W. Weibezahn (to I.G. Farben), Ger. Pat. 590,237 (1932); H. Lange and O. Horn (to I.G. Farben), U.S. Pat. 2,267,829 (1939).
29. Imperial Chemical Industries, Ltd., Belg. Pat. 660,006 (Aug. 19, 1965).
30. N. I. Popova, B. V. Kabakova, F. A. Milman, and E. E. Vermel, *Dokl. Akad. Nauk SSSR.* **155,** 149 (1964); *Chem. Abstr.* **60,** 13070c.
31. G. S. Mironov, M. I. Farberov, and I. M. Orlova, *Zh. Prikl. Khim.* **36,** 654 (1963); *Chem. Abstr.* **59,** 7413b.
32. J. R. Long (to Wingfoot Corp.), U.S. Pat. 2,256,149 (1942).
33. J. J. Kolfenbach, E. F. Tuller, L. A. Underkofler, and E. T. Fulmer, *Ind. Eng. Chem.* **37,** 1178 (1945).
34. W. J. Hale and L. A. Underkofler (to Natural Agrol Co.), U.S. Pat. 2,371,577 (1945).
35. S. O. Lawesson, E. H. Larsen, G. Sundstrom, and H. J. Jakobsen, *Acta Chem. Scand.* **17,** 2216 (1963).

36. P. R. Thomas, G. J. Tyler, T. E. Edwards, A. T. Radcliffe, and R. C. P. Cubbon, *Polymer* **5**, 525 (1964); P. R. Thomas and G. T. Tyler (to British Nylon Spinners, Ltd.), Brit. Pat. 878,898 (1961), 886,958 (1962).
37. T. Otsu, J. Ushirone, and M. Imoto, *Kogyo Kagaku Zasshi* **69**, 516 (1966); *Chem. Abstr.* **65**, 15516g.
38. R. C. Schulz, H. Cherdon, and W. Kern, in Houben-Weyl *Methoden der organischen Chemie*, Vol. XIV/1, Georg Thieme Verlag, Stuttgart, 1961, pp. 1090ff.
39. R. N. Haward, *J. Polymer Sci.* **3**, 10 (1948).
40. G. S. Whitby, M. D. Gross, J. R. Miller, and A. J. Costanza, *J. Polymer Sci.* **16**, 549 (1955).
41. W. Starck and H. Freudenberger (to I.G. Farben), Ger. Pat. 727,955 (1934).
42. C. S. Marvel and D. S. Casey, *J. Org. Chem.* **24**, 957 (1959).
43. W. W. Groves (to I.G. Farben), Brit. Pat. 498,383 (1937); Fr. Pat. 838,084 (1938).
44. Y. Goto, Y. Tabata, and H. Sobue, *Kogyo Kagaku Zasshi* **67**, 1276 (1964).
45. F. Brown, F. Berardinelli, R. J. Kray, and L. J. Rosen, *Ind. Eng. Chem.* **51**, 79 (1959).
46. A. K. Chaudhuri and S. Basu, *Makromol. Chem.* **29**, 48 (1959); A. K. Chaudhuri, *Makromol. Chem.* **31**, 214 (1959).
47. T. Tsuruta, R. Fujio, and J. Furukawa, *Makromol. Chem.* **80**, 172 (1964).
48. R. E. Moser and H. G. Cassidy, *J. Polymer Sci.* [B] **2**, 545 (1964).
49. G. Smets and L. Oosterbosch, *Bull. Soc. Chim. Belges* **61**, 139 (1952).
50. C. E. Schildknecht, A. O. Zoss, and F. Grosser, *Ind. Eng. Chem.* **41**, 2891 (1949).
51. H. Watanabe, R. Koyama, H. Nagal, and A. Nishioka, *J. Polymer Sci.* **62**, 575 (1962).
52. R. Fujio, T. Tsuruta, and J. Furukawa, *Makromol. Chem.* **52**, 233 (1962); *Kogyo Kagaku Zasshi* **66**, 1339 (1963); *Chem. Abstr.* **60**, 13332d.
53. P. A. Small and D. G. M. Wood (to Imperial Chemical Industries, Ltd.), Brit. Pat. 862,862 (1961).
54. S. Iwatsuki, Yuya Yamashita, and Yoshio Eshii, *J. Polymer Sci.* [B] **1**, 545 (1963).
55. N. G. Karapetyan, S. M. Voskanyan, O. A. Tonoyan, and G. N. Chukhadzhyan, *Izv. Akad. Nauk. Arm. SSR, Khim. Nauk.* **18**, 371 (1965); *Chem. Abstr.* **64**, 5216g.
56. C. C. Price, *J. Polymer Sci.* **3**, 772 (1948).
57. T. C. Schwann and C. C. Price, *J. Polymer Soc.* **40**, 457 (1959).
58. M. Charton and A. J. Capat, *J. Polymer Sci.* [A] **2**, 1321 (1964).
59. E. C. Chapin, G. E. Ham, and C. L. Mills, *J. Polymer Sci.* **4**, 597 (1949).
60. G. S. Wich and N. Brodoway, *J. Polymer Sci.* [A] **1**, 2163 (1963).
61. F. M. Lewis, C. Walling, W. Cummings, E. R. Briggs, and W. J. Wenison, *J. Am. Chem. Soc.* **70**, 1527 (1948).
62. W. Cooper and E. Catterall, *Can. J. Chem.* **34**, 387 (1956).
63. F. Ida, K. Uemura, and S. Abe, *Kagaku To Kogyo (Osaka)* **38**, 215 (1964); *Chem. Abstr.* **61**, 7105d.
64. C. G. Overberger and F. W. Michelotti, *J. Am. Chem. Soc.* **80**, 988 (1958).
65. H. C. Haas and M. S. Simon, *J. Polymer Sci.* **9**, 309 (1952).
66. L. J. Young, *J. Polymer Sci.* **54**, 411 (1961).
67. C. S. Marvel, J. E. McCorkle, T. R. Fukuto, and J. C. Wright, *J. Polymer Sci.* **6**, 776 (1951).
68. C. S. Marvel, W. R. Peterson, H. K. Inskip, J. E. McCorkle, W. K. Taft, and B. G. Labbe, *Ind. Eng. Chem.* **45**, 1532 (1953).
69. L. E. Coleman, Jr., and N. A. Meinhardt, *Fortschr. Hochpolym. Forsch.* **1**, 159 (1959).
70. A. Winston and F. L. Hamb, *J. Polymer Sci.* **2**, 4475 (1964); **3**, 583 (1965).
71. G. S. Whitby, *Synthetic Rubber*, John Wiley & Sons, Inc., New York, 1954, pp. 697, 726–727.
72. A. D. McLaren, *J. Polymer Sci.* **3**, 654 (1948); A. D. McLaren and C. J. Seiler, *J. Polymer Sci.* **4**, 63 (1952).
73. J. C. H. Hwa (to Rohm and Haas Co.), U.S. Pat. 2,597,491 (1952).
74. T. R. E. Kressman (to Perutit Co. Ltd.), Brit. Pat. 804,782 (1958).
75. R. J. Slocombe, *J. Polymer Sci.* **26**, 9 (1957).
76. G. B. Sterling, J. G. Cobler, D. S. Erley, and F. A. Blanchard, *Anal. Chem.* **31**, 1612 (1959).
77. M. Felden, D. R. Hammel, and R. N. Laundrie, *Ind. Eng. Chem.* **46**, 2248 (1954).
78. J. R. Catch, D. E. Elliott, D. H. Hey, and E. R. H. Jones, *J. Chem. Soc.* **1948**, 278; E. M. Kosower and G. S. Wu, *J. Org. Chem.* **28**, 633 (1963).
79. A. R. Shultz, *J. Polymer Sci.* **47**, 267 (1960).

80. J. E. Guillot and R. G. W. Norrish, *Proc. Royal Soc.* (*London*) *Ser. A* **233,** 153 (1955).
81. C. J. Marvel, E. H. Riddle, and J. O. Conner, *J. Am. Chem. Soc.* **64,** 92 (1942).
82. P. J. Flory, *J. Am. Chem. Soc.* **64,** 177 (1942).
83. J. N. Hay, *Makromol. Chem.* **67,** 31 (1963).
84. K. Matsuzaki and T. C. Lay, *Makromol. Chem.* **110,** 185 (1967).
85. N. Grassie and J. N. Hay, *Makromol. Chem.* **64,** 82 (1963).
86. S. Iwatsuki, Y. Yamashita, and Y. Ishii, *Kogyo Kagaku Zasshi* **66,** 1162 (1963); *Chem. Abstr* **60,** 9360f.
87. K. F. Wissbrun, *J. Am. Chem. Soc.* **81,** 58 (1959).
88. H. W. Melville, T. T. Jones, and R. F. Tuckett, *Chem. Ind.* (*London*) **59,** 267 (1940); *Proc. Royal Soc.* (*London*) **187,** 19 (1946).
89. J. E. Guillet and R. G. W. Norrish, *Proc. Royal. Soc.* (*London*) *Ser. A* **233,** 172 (1955).
90. R. C. Schulz, H. Vielhaber, and W. Kern, *Kunststoffe* **50,** 500 (1960).
91. L. Strzelecki, *Compt. Rend. Acad. Sci. Paris, Ser. C.* **265,** 1094 (1967).
92. C. S. Marvel and J. C. Wright, *J. Polymer Sci.* **8,** 495 (1952).
93. L. X. Mallavarapu and A. Ravve, *J. Polymer Sci.* [A] **3,** 593 (1965).
94. E. Selegny, L. Merle-Aubrey, and N. Thoai, *Bull. Soc. Chim. France* **1967,** 3166.
95. Y. Minoura and H. Yamaguchi, *J. Polymer Sci.* [A] **6,** 2013 (1968).
96. K. A. Kun and H. G. Cassidy, *J. Polymer Sci.* **44,** 383 (1960).
97. R. S. Gregorian and R. W. Bush, *J. Polymer Sci.* **32,** 481 (1964); Grace Chemical Co., U.S. Pat. 3,175,996 (1965).
98. P. A. Leermakers and F. C. James, *J. Org. Chem.* **32,** 2898 (1967).
99. W. O. Kenyon and G. P. Waugh, *J. Polymer Sci.* **32,** 83 (1958).
100. R. Beckerbauer and H. E. Baumgarten, *J. Polymer Sci.* [A] **2,** 823 (1964).
101. C. C. Unruh, *J. Polymer Sci.* **45,** 325 (1960); *J. Appl. Polymer Sci.* **2,** 358 (1959).
102. E. C. Dunlop, *Ann. N.Y. Acad. Sci.* **53,** 1087 (1951).
103. C. W. Johnson, C. G. Overberger, and W. J. Seagers, *J. Am. Chem. Soc.* **75,** 1495 (1953).
104. B. Buděšínský, K. Mňouček, F. Jančik, and E. Kraus, *Chem. Listy* **51,** 1819 (1957); *Chem. Abstr.* **52,** 1860f.
105. M. Kitahara and T. Konishi, *Rika Gaku Kenkyusho Hokoku* **38,** 90 (1962); *Chem. Abstr.* **58,** 5021e.
106. V. A. Devyatnin and I. A. Solunina, *Tr. Vses. Nauchn. Issled. Vitamin. Inst.* **7,** 104 (1961); *Chem. Abstr.* **59,** 29e.
107. T. Mitsui and Y. Miyatake, *Rika Gaku Kenkyusho Hokoku* **38,** 189, 446 (1962); *Chem. Abstr.* **58,** 1318d; **60,** 6216h.
108. C. E. R. Jones, *Gas Chromatog., Proc. Symp. 3rd, Edinburgh,* **1960,** 401.
109. J. J. Pepe, I. Kniel, and M. Czuha, Jr., *Anal. Chem.* **27,** 755 (1955).
110. G. B. Sterling, J. G. Cobler, D. S. Erley, and F. S. Blanchard, *Anal. Chem.* **31,** 1612 (1959).
111. N. I. Sax, *Dangerous Properties of Industrial Materials,* 3rd ed., Reinhold Publishing Corp., New York, 1968, p. 938.

William H. Daly
Louisiana State University

VINYL METHYL ETHER POLYMERS. See VINYL ETHER POLYMERS

VINYLNAPHTHALENE POLYMERS. See VINYLARENE POLYMERS

VINYLON FIBERS. See VINYL ALCOHOL POLYMERS

VINYL PLASTICS

In common usage, the term vinyl plastics (or "vinyls") refers especially to those plastics derived from vinyl chloride and vinyl acetate. See also VINYL CHLORIDE POLYMERS; VINYL ESTER POLYMERS, Vol. 15.

VINYL PLASTISOLS.

See VINYL CHLORIDE POLYMERS

VINYL POLYMERIZATION

The term vinyl polymerization is used to denote addition polymerization of vinyl compounds, $CH_2{=}CHX$, where X is an alkyl group, aryl group, OR, OCOR, Cl, COOR, CN, etc. See also ADDITION POLYMERIZATION.

VINYL PROPIONATE POLYMERS.

See VINYL ESTER POLYMERS, Vol. 15

VINYLPYRIDINE POLYMERS

A large number of vinylpyridine polymers, including both homopolymers and copolymers, have been prepared. The most common vinylpyridine monomers are 2-vinylpyridine (**1**), 4-vinylpyridine (**2**), and 2-methyl-5-vinylpyridine (**3**). Vinylpyridine polymers have a wide range of applications, some of the most interesting of which are based on their ability to act as polyelectrolytes and to form complexes at the basic nitrogen atom.

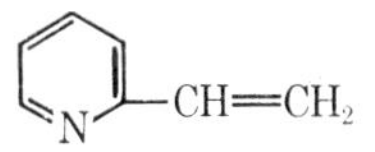

CH=CH₂ structures:

(1) 2-pyridyl–$CH{=}CH_2$ (2) 4-pyridyl–$CH{=}CH_2$ (3) $H_2C{=}CH$–pyridine–CH_3

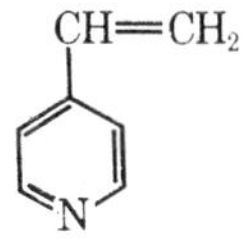

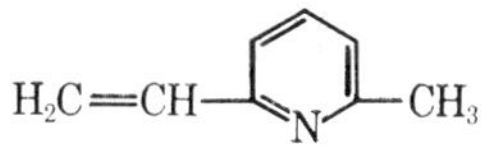

Monomers

Synthesis. Vinylpyridine monomers can be synthesized by a variety of methods. The suitable synthetic method depends largely on the position at which the vinyl group is to be introduced. The general synthetic process is given by equation 1, where X is H, OH, halogen, OOCR, or $^{\oplus}NRCl^{\ominus}$. Similar procedures can be applied for the

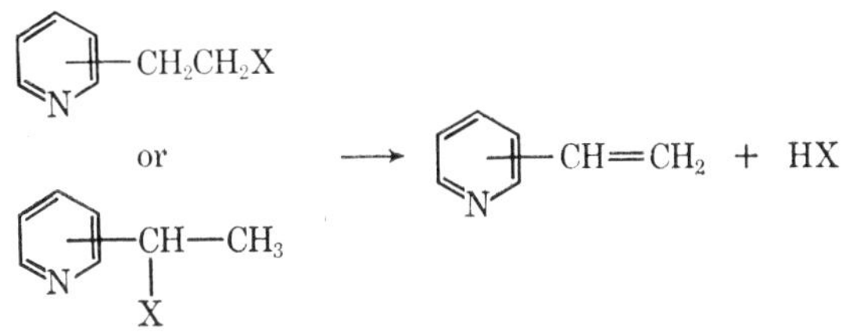

$$\text{Py}{-}CH_2CH_2X \;\text{ or }\; \text{Py}{-}\underset{X}{\underset{|}{CH}}{-}CH_3 \longrightarrow \text{Py}{-}CH{=}CH_2 + HX \qquad (1)$$

synthesis of α-substituted vinylpyridines. Among these various methods, the most economical and industrially important are vinyl group formation by dehydrogenation of the ethyl group, and by dehydration of the hydroxyethyl group.

Method I. Catalytic Dehydrogenation of Ethylpyridine. The catalytic dehydrogenation is carried out in the gas phase at 500–800°C and using catalysts similar to those for the production of styrene from ethylbenzene; the dehydrogenation is carried out in an atmosphere of nitrogen or superheated steam. Examples of catalysts and conditions include the following: (a) $Cr_2O_3 + Mo_2O_3$ alone or supported on a base such as alumina or bauxite; ThO_2 on alumina; Cr_2O_3 plus alkaline earth or alkali metal oxides or hydroxides (1). (b) 93% Fe_2O_3 + 5% Cr_2O_3 + 2% KOH at 540–700°C (2). (c) 50% ZnO + 40% Al_2O_3 + 10% CaO at 600–650°C (3,4). (d) 86–88% ZnO or

Fe_2O_3 + Cr_2O_3 (5). (e) 40–45% Cr_2O_3 + 50–55% K_2CO_3 + 3–7% Fe_2O_3 at 540–760°C (6). (f) 80% ZnO + 18% Fe_2O_3 + 1.2% K_2O + 0.8% Cr_2O_3 at 580–600°C (7). (g) ZnO with Cr_2O_3 at 670°C (8). (h) 55–80% MgO + 10–30% Fe_2O_3 + 5–10% CuO + about 5% K_2O at 600–650°C (9,10). (i) 25–40% ZnO + 35–50% MgO + 0–50% Al_2O_3 at 500–700°C (11). (j) 2-Methyl-5-ethylpyridine with 2 wt % of iodine pyrolyzed at 700–725°C in the presence of 0.6 mole oxygen (12).

The dehydrogenation method is directed chiefly to the production of 2-methyl-5-vinylpyridine since 2-methyl-5-ethylpyridine is the most readily available of the ethylpyridine derivatives. One difficulty in this process is in separating 2-methyl-5-vinylpyridine from a number of by-products. Although the amounts are small, pyridine, 2-picoline, 3-picoline, 2,5-lutidine, 3-ethylpyridine, 3-vinylpyridine, 2-ethyl-5-vinylpyridine, 2,6-dimethyl-3-ethylpyridine, 2-methyl-5-isopropenylpyridine, 2-vinyl-5-ethylpyridine, and 2-methyl-5-propen-1-ylpyridine have been reported as by-products (2,13). A number of patents cover various purification methods for 2-methyl-5-vinylpyridine on the basis of differences in solubility, melting point, or boiling point (14).

When 2- or 4-ethylpyridine is available, this method would be more advantageous for the production of 2- or 4-vinylpyridine than Method II starting from 2- or 4-picoline.

Method II. Dehydration of Hydroxyalkylpyridine. Dehydration of an α- or β-hydroxyalkyl group is a conventional preparative method for vinyl compounds. The dehydration process proceeds with relatively high yield, but the preparation of hydroxyalkylpyridine is a rather inefficient process. A common method is the reaction of 2- or 4-picoline with formaldehyde (eq. 2). The methyl group at the 2- and 4-positions is reactive and can be methylolated. Methylolation is conducted with various

$$C_5H_4N\text{–}CH_3 + HCHO \longrightarrow C_5H_4N\text{–}CH_2CH_2OH \tag{2}$$

catalysts, generally under high pressure. Some examples are (a) 2-picoline with formalin (37%)–H_3PO_4 at 150–160°C in a nitrogen atmosphere (15); (b) 2-picoline with paraformaldehyde at 140°C and 50 atm for 8 hr (yield = 29%) (16); (c) 2-picoline with formalin (37%)–formic acid at 240°C and 62–142 atm (yield = 53.6%) (17); and (d) 2-picoline, 2,6-lutidine, 2,4-lutidine with formalin (38%)–$K_2S_2O_8$–hydroquinone–trinitrobenzene at 220°C. Yields of corresponding vinyl compounds after dehydration with potassium hydroxide are 75–85% based on alkylpyridines consumed (25).

Other methods of preparing hydroxyalkylpyridines are given by equations 3–7 (18–20).

$$C_5H_4N\text{–}CH_3 + n\text{-}C_4H_9Li \longrightarrow C_5H_4N\text{–}CH_2Li + C_4H_{10} \xrightarrow{CH_2O} C_5H_4N\text{–}CH_2CH_2OH \tag{3}$$

$$C_5H_5N + (CH_3)_2CO \xrightarrow[Mg]{HgCl_2} C_5H_4N\text{–}C(CH_3)_2\text{–}OH \tag{4}$$

$$C_5H_4N\text{–}COOCH_3 \xrightarrow{CH_3MgI} C_5H_4N\text{–}C(CH_3)(OH)\text{–}CH_3 \tag{5}$$

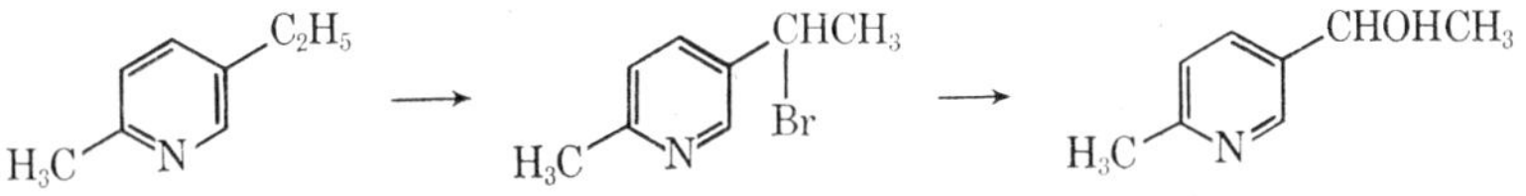

C_2H_5 ; H_3C N → CHCH$_3$ Br ; H_3C N → CHOHCH$_3$; H_3C N (6)

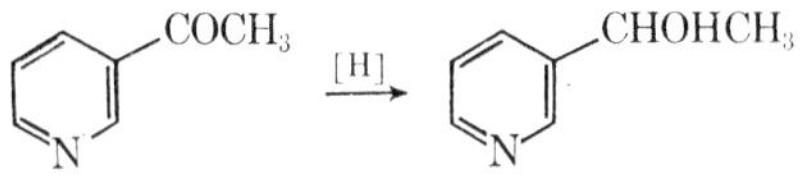

COCH$_3$; N $\xrightarrow{[H]}$ CHOHCH$_3$; N (7)

Dehydration of 2-hydroxyalkylpyridine gives 2-vinylpyridine with very good yield (15,16,23–26). 4-Vinylpyridine is obtained in moderate yield (56%) (22), whereas the yield of 3-vinylpyridine is much poorer (18–20%) (27). However, some 3-hydroxyalkyl groups are active, depending upon other substituents. Thus, 5-bromo-3-isopropenylpyridine is prepared from the corresponding β-hydroxypropyl compound with 89% yield (19). As dehydrating catalysts, alkali hydroxide, H_2SO_4, $KHSO_4$, and P_2O_5 are common for liquid-phase reactions, and at elevated temperatures Al_2O_3 is used for gas-phase dehydration.

Method III. Dehydrohalogenation. Vinylpyridines can be derived from dehydrohalogenation of β-haloethyl compounds (eq. 8) (28).

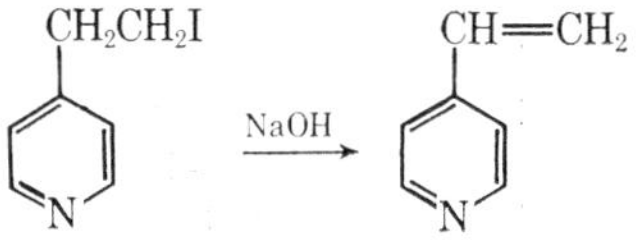

CH_2CH_2I ; N $\xrightarrow{NaOH}$ CH=CH$_2$; N (8)

Method IV. Pyrolysis of α-Acetoxyalkylpyridine. Pyrolysis of 2- or 4-(α-acetoxyethyl)pyridine at 500°C on silica or quartz gives the corresponding vinylpyridines in 40–83% yield (29,30).

Miscellaneous Methods. 3-(α-Chlorethyl)pyridine is converted to 3-vinylpyridine via the trimethylammonium salt (eq. 9) (31).

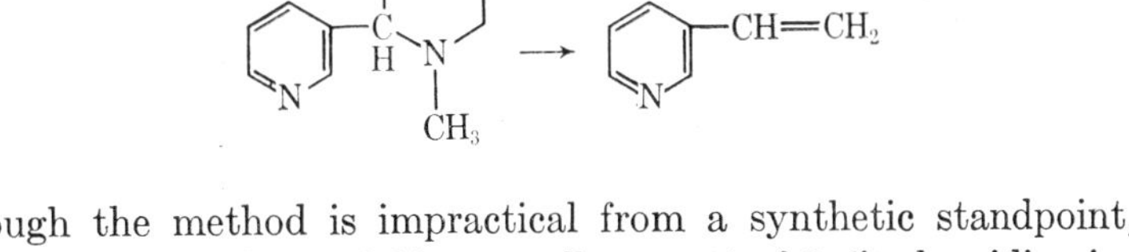

Cl ; CH—CH$_3$; N → $N^{\oplus}(CH_3)_3Cl^{\ominus}$; CH—CH$_3$; N → CH=CH$_2$; N (9)

Pyrolysis of nicotine gives 3-vinylpyridine (eq. 10) but the yield is poor (32).

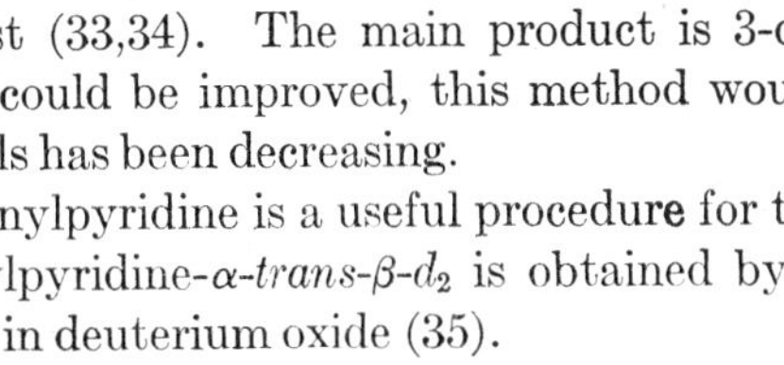

C ; H N ; N ; CH$_3$ → CH=CH$_2$; N (10)

Although the method is impractical from a synthetic standpoint, reaction of acrylonitrile with butadiene yields a small amount of 2-vinylpyridine in the presence of Al_2O_3–Cr_2O_3 catalyst (33,34). The main product is 3-cyanocyclohexene. If the selectivity of catalyst could be improved, this method would be attractive since the cost of the raw materials has been decreasing.

Reduction of ethynylpyridine is a useful procedure for the synthesis of deuterated vinylpyridine. 2-Vinylpyridine-α-*trans*-β-d_2 is obtained by reducing ethynylpyridine with chromium sulfate in deuterium oxide (35).

Table 1. Properties of Vinylpyridine Derivatives

Substituents	Bp/mm Hg, °C	n_D (at °C)	Mp of picrate, °C	Preparative methods	References
2-vinyl	158–159/760				16,17,30,33,34,
	69–71/30	1.5495(20)		I,II,IV,V	55–60,62
	42–44/10	1.5386(30)			
3-vinyl	67–68/18	1.5530(20)	143–144	II,V	27,32,35,55,58,61
4-vinyl	65/15	1.5499(20)	198–199	I,II,III,IV	28,29,30,55,58
	78–82/23	1.5450(20)			
3-methyl-2-vinyl	72–74/12		135–136	II	63
4-methyl-2-vinyl	80–82/20	1.5420(25)	167–168	II	26,64,65
	73/12				
6-methyl-2-vinyl	73/21,	1.5320	163	II	26,55,65–71
	65–70/12				
	72–74/18	1.5380	160.5		
3-ethyl-2-vinyl	85–90/18		115–111	II	72
5-ethyl-2-vinyl	71–72/5,		130.5–131.5	I	13
	96/20				
4-(2-hydroxyethyl)-2-vinyl	93–97/0.5	1.5503(20)	122	II	64
4-propoxy-2-vinyl	125–127/20				73
2-methyl-3-vinyl	96–100/15	1.5410(20)		I	4
5-bromo-3-vinyl	74–75/3	1.5810		II	19
2-methyl-5-vinyl	75/15	1.5454	−12[a]	I	1–3,5–12,20,53,
			157–158		55,61,73
2-ethyl-5-vinyl	93/20	1.5375	160–161	I	13
2-R-5-vinyl					
R = $(CH_2)_2N(CH_3)_2$	123–124/6	1.5345			
R = $CH(CH_2N(CH_3)_2)_2$	135–137/6	1.5345			74
R = $C(=CH_2)—CH_2N(CH_3)_2$	127–130/6	1.5570			
2-methyl-4-vinyl	76–77/20	1.5410(20)	166.5–167.5	II	64,71,75
	78–79/11	1.5394(25)	169		
2,4-dimethyl-6-vinyl	87/15	1.5382(20)	155–156	II	55,64
	81–82/16				
2-chloro-4-methyl-3-vinyl	89/6	1.5584(20)		III	76
2,6-dichloro-4-methyl-3-vinyl	142–143/16		(no picrate formation)	III	77
2,4-divinyl	95–97/15	1.5552(25)	151	II	64
2,5-divinyl	80/6	1.5920(20)		[b]	74
2,6-divinyl	88–89/16	1.5710(25)	140.5	II	67,70,78
		1.5620(20)			
2-methyl-4,6-divinyl	91–92/12	1.5643(25)	143	II	64
4-methyl-2,6-divinyl			165–166	II	64
2-isopropenyl	63–67/10	1.5241(25)		II	19
3-isopropenyl	75/10	1.5381(25)		II	19
5-chloro-isopropenyl	70–73/3	1.5554(25)		II	19
5-bromo-isopropenyl	74–75/3	1.5810(25)		II	19
4-(α-acetoxyvinyl)	76/0.5				37
2-ethynyl	85/32				43
3-ethynyl	87/30		38.5[a]		43
4-ethynyl			50[a]		43

Table 1 (*continued*)

Substituents	Bp/mm Hg, °C	n_D (at °C)	Mp of picrate, °C	Preparative methods	References
2-vinylpyridine *N*-oxide	100.5–102/0.42	1.6050(25)	112.5–113.5 45[a]		39,79,80
5-ethyl-2-vinylpyridine *N*-oxide	118–119/0.3	1.5801(26)	130.5–131.5		39,79
N-vinylpyridinium perchlorate			95.5–7.5[a]		41
N-vinylpyridinium fluoroborate			75.5–6.5[a]		41
N-vinylpyridinium hexachloroplatinate			211–3[a] 193[a]		41 42

[a] Mp of free monomer.
[b] Starting material is 2-methyl-5-vinylpyridine; the action of formaldehyde and diethylamine on the methyl group gives various derivatives.

α,α-Disubstituted Derivatives. A series of isopropenylpyridines can be obtained in a manner similar to Method II (eq. 11) (19). 4-(α-Acetoxyvinyl)pyridine is synthesized from 4-vinylpyridine as shown by equation 12 (37).

$$\text{2-Py}-C(CH_3)(OH)-CH_3 \longrightarrow \text{2-Py}-C(CH_3)=CH_2 \qquad (11)$$

$$\text{4-Py}-CH=CH_2 \xrightarrow[0\text{-}5°C]{KMnO_4} \text{4-Py}-CHOH-CH_2OH \xrightarrow[\text{heating overnight at }100°C]{(CH_3CO)_2O} \text{4-Py}-C(O-COCH_3)=CH_2 \qquad (12)$$

Vinylpyridine N-Oxide. Method II can be applied to β-hydroxyethylpyridine *N*-oxide (eq. 13) (38,39). The overall yields are 40–50% for R = H and 25–30% for R = C_2H_5.

$$\text{5-R-2-Py}-CH_2CH_2OH \xrightarrow[CH_3COOH]{H_2O_2} \text{5-R-2-Py(N}\rightarrow\text{O)}-CH_2CH_2OH \xrightarrow[\text{or NaOH}]{KHSO_4} \text{5-R-2-Py(N}\rightarrow\text{O)}-CH=CH_2 \qquad (13)$$

Ethynylpyridine. Ethynylpyridine is prepared by the action of phosphorus pentachloride on acetylpyridine followed by dehydrochlorination by potassium hydroxide (43). The starting material may be 2-vinylpyridine, which is brominated and then dehydrobrominated, yielding 2-ethynylpyridine (44).

Vinylpyridinium Compounds and Vinylpyridine Complexes. The reactivity of the nitrogen atom of the pyridine group of vinylpyridine is very much the same as that of pyridine or the alkylpyridines. Addition of an acidic component to a solution of vinylpyridine, however, often induces polymerization instead of formation of monomeric salts. 4-Vinylpyridine reacts with methyl *p*-toluenesulfonate to give

1-methyl-4-vinylpyridinium *p*-toluenesulfonate at −15°C (40). This salt can be readily polymerized.

Many transition metal salts form stoichiometric complexes with vinylpyridine. Salts and complexes of vinylpyridine are discussed below in connection with their polymerizability.

N-Vinylpyridinium salts are obtained by the reaction of pyridine with 1,2-ethylene dibromide followed by dehydrobromination with silver oxide (eq. 14). The pyridinium salt can be obtained as bromofluoride, perchlorate, or hexachloroplatinate (41,42).

$$\text{C}_5\text{H}_5\text{N} + \text{BrCH}_2\text{CH}_2\text{Br} \longrightarrow \text{C}_5\text{H}_5\overset{\oplus}{\text{N}}\text{–CH}_2\text{CH}_2\text{Br} \xrightarrow[\text{AgX}]{\text{Ag}_2\text{O}} \text{C}_5\text{H}_5\overset{\oplus}{\text{N}}\text{–CH=CH}_2\ \text{X}^{\ominus} \quad (\text{X} = \text{ClO}_4^{\ominus}, \text{BF}_4^{\ominus}) \tag{14}$$

Stabilization. Since the tendency of vinylpyridine to polymerize is high (about as great as that of styrene), stabilizers must be added to the monomer. Compounds reported to act as stabilizers include dinitrochloroaniline (45), phenothiazine derivatives (46), phthalein sulfone compounds such as phenol red (47), diarylamines (48), aromatic sulfur compounds such as thiols and polysulfides (49), β-phenylnaphthylamine, diphenylamine, trinitrobenzene (50), tetraalkylammonium borohydride (51), ammonium salt of *N*-nitroso-*N*-phenyl(or 1-naphthyl)hydroxylamine (52), sulfur, picric acid, and α-nitroso-β-naphthol (53).

Toxicity. 2-Vinylpyridine has a marked local and general resorptive action on contact with skin. Exposure to vapor (0.005 mg/liter) causes a retarded gain of weight, changes in the neuromuscular excitability, and reduction of the arterial pressure and of the leucocyte count in laboratory animals. Maximum permissible levels are 0.005 mg/liter in air (54). In these authors' experience, all vinylpyridine derivatives are irritating to the skin. The grade of dermatitis and its symptoms depend upon allergic sensitivity of individuals.

Physical Constants. Properties of vinylpyridines and related compounds are tabulated in Table 1.

Determination of Purity. Impurities in vinylpyridine can be determined by double-bond analysis (81–83), infrared spectroscopy (84), polarography (85), ultraviolet spectroscopy (86), and gas chromatography (86). The impurity level is particularly of concern for 2-methyl-5-vinylpyridine. The severe reaction conditions of Method I inevitably bring about migration or cracking of the alkyl group as side reactions. Also, the separation of 2-methyl-5-vinylpyridine from unreacted 2-methyl-5-ethylpyridine is difficult. Levels of 0.01% of 2-methyl-5-ethylpyridine in 2-methyl-5-vinylpyridine can be detected by gas chromatography.

Polymerization

Vinylpyridine undergoes both free-radical and anionic polymerization. Since the structure and electronic state of vinylpyridine are similar to those of styrene, the polymerizabilities (at least in radical polymerization) of these two monomers resemble each other. The presence of basic nitrogen in vinylpyridine, however, introduces unique and complicating factors into the polymerization. Complex or salt formation of monomer would alter the electronic state of the monomer entirely.

Cationic polymerization of vinylpyridine has not been reported. Although the vinyl group might have a high enough π electron density for cationic propagation,

Table 2. Rate Constants for Radical Polymerization of Vinylpyridine

Rate constant	2-Vinylpyridine	4-Vinylpyridine	2-Methyl-5-vinylpyridine	Styrene (Ref. 106)
k_p, liter-mole^{-1}-sec^{-1} at 25°C	96.5 (Ref. 105)[a] 186 (Ref. 105) E_p = 8.0 kcal/mole	12 (Ref. 106)	50 (Ref. 107)	25.1
k_t, liter-mole^{-1}-sec^{-1} at 25°C	8.9×10^6 (Ref. 105)[a] 3.3×10^7 (Ref. 105) E_t = 5.0 kcal/mole	3×10^6 (Ref. 106)	3.5×10^6(Ref. 107)	3.63×10^6
k_{tr}, liter-mole^{-1}-sec^{-1} at 25°C		8×10^{-3} (to (monomer) (Ref. 106)		8.81×10^{-4} (to monomer)
k_i (thermal), sec^{-1} at 25°C		5×10^{-13}	6.5×10^{-4} (Ref. 111)[b]	9.51×10^{-16}
$k_{overall}$,[c] liter-mole^{-1}-sec^{-1}	20.2 ± 1.5 (Ref. 108)	42.8 ± 4.5 (Ref. 108)	22.0 ± 3.1 (Ref. 108)	9.4 ± 0.4 (Ref. 108)
heat of polymerization, kcal/mole	17.6 ± 0.6 (Ref. 109)	18.7 ± 0.3 (Ref. 110)		17.0 ± 0.3 (Ref. 110)

[a] Probably more reliable.

[b] Thermal rate at 120°C.

[c] At 70°C in cyclohexanone, $[\text{monomer}]_0$ = 1 mole/liter, [azobisisobutyronitrile] = 0.01 mole/liter. $k = [(2.303/t) \log(A_0/A)] \times 10^4$ (min^{-1}), where A_0 and A are monomer concentrations at t = 0 and t, respectively.

cationic initiators would be deactivated by complex formation with the basic pyridyl group.

Free-Radical Polymerization. Kinetic constants for homogeneous liquid-phase free-radical polymerization are given in Table 2. These rate constants are of the same order of magnitude as those for styrene. The reactivity of vinylpyridine is, however, somewhat higher than that of styrene. The methyl affinities (rate constant of methyl radical addition to monomer/rate constant of hydrogen abstraction from isooctane by methyl radical) of 2-vinylpyridine, 4-vinylpyridine, and styrene are 1360, 1360, and 792, respectively (104). The reactivities of 2-vinylpyridine and 4-vinylpyridine toward styryl radical relative to styrene are 1.82 and 1.62, respectively.

The kinetics of the catalyzed thermal polymerization of 2-methyl-5-vinylpyridine do not satisfy the theoretical equation 15. The monomer exponent is higher than unity (1.3 and 1.5 in benzene for benzoyl peroxide and for azobisisobutyronitrile initia-

$$-d[\mathrm{M}]/dt = k[\mathrm{I}]^{1/2}[\mathrm{M}] \tag{15}$$

tors, respectively (111); and 2 in ethanol (112)). Surface-catalyzed polymerization is also observed (106); this phenomenon is related to the activation of monomer by absorption or complex formation. In this way clay acts as a catalyst for polymerization of vinylpyridine (113).

Emulsion polymerization is generally carried out in the presence of butadiene as comonomer. Emulsion homopolymerization of vinylpyridine (116–119) does not agree with theories based on the assumption that micelles are the sole propagating sites (120–124) (see also EMULSION POLYMERIZATION). For the monomer of relatively high solubility in the aqueous phase (1.04% at 60°C for 2-methyl-5-vinylpyridine), the aqueous phase is apparently important as the locus of polymerization (119). The situation is the same for butadiene–styrene–2-vinylpyridine emulsion copolymerization (125). The copolymerization kinetics of acrylonitrile with 2-methyl-5-vinylpyridine have been investigated from a technological standpoint (114,115).

Radiation-induced polymerization of 2-vinylpyridine hydrochloride (126) and of 2-methyl-5-vinylpyridine in the liquid and solid states (127) has been reported. In the case of 2-vinylpyridine hydrochloride the reaction is very rapid; only amorphous polymer is obtained even when crystalline monomer salt is irradiated. Polymerization of 2-methyl-5-vinylpyridine proceeds via radical intermediates in the liquid phase ($R_p \propto I_{abs}^{1/2}$, E = 5.7 kcal/mole), whereas ionic mechanism has been proposed on the basis of kinetic and scavenger studies ($R_p \propto I_{abs}$, E = 0.3 kcal/mole) for the solid-state polymerization at low temperatures.

Unusual catalyst systems such as $LiClO_4$ (128), cupric salts (129–133), and SO_2CCl_4 have been reported. A detailed study of cupric acetate-initiated polymerization of 4-vinylpyridine indicates that complex formation between monomer and cupric salt is the primary step followed by oxidation of monomer by an electron-transfer mechanism. The cation radical formed from monomer is claimed to be an active initiating species (132). Consequently, the presence of conjugation between vinyl group and Cu(II) (eq. 16) facilitates the electron-transfer process. The polymerization of 4-vinylpyridine by cupric acetate is indeed much faster than the polymerization of 2-methyl-5-vinylpyridine or of sterically hindered 2-vinylpyridine. The activation of monomer by complex or salt formation is a specific feature of polymerization of vinylpyridine.

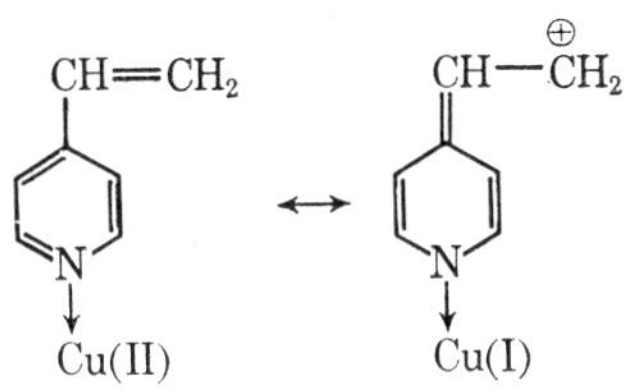

(16)

Anionic and Coordinate Polymerizations. The high electronegativity of the nitrogen atom activates the monomer toward attack by carbanions. Anionic polymerizability of vinylpyridine is considerably higher than that of styrene, as can be seen from the propagation rate constants in Table 3 (87). On the other hand, the carbanion formed from vinylpyridine is less reactive than the styryl anion. Measurement of the rate constant for reaction between ∿—(styrene)–(2-vinylpyridine)$^{\ominus}$ and 2-vinylpyridine indicates that the reactivity of this anion depends on n, and that the rate constant decreases with increasing n (see Table 3) (85). The relative rate constants are important in conducting block copolymerization. A complicating factor in the anionic polymerization of vinylpyridine is the fact that the pyridine ring may be attacked by the anion. Reactions of styryl, α-methylstyryl, and cumyl anions with poly(2-vinylpyridine) have been studied (eq. 17) (89). Consequently, there is a good possibility of preparing a graft copolymer by these anionic chain-transfer reactions.

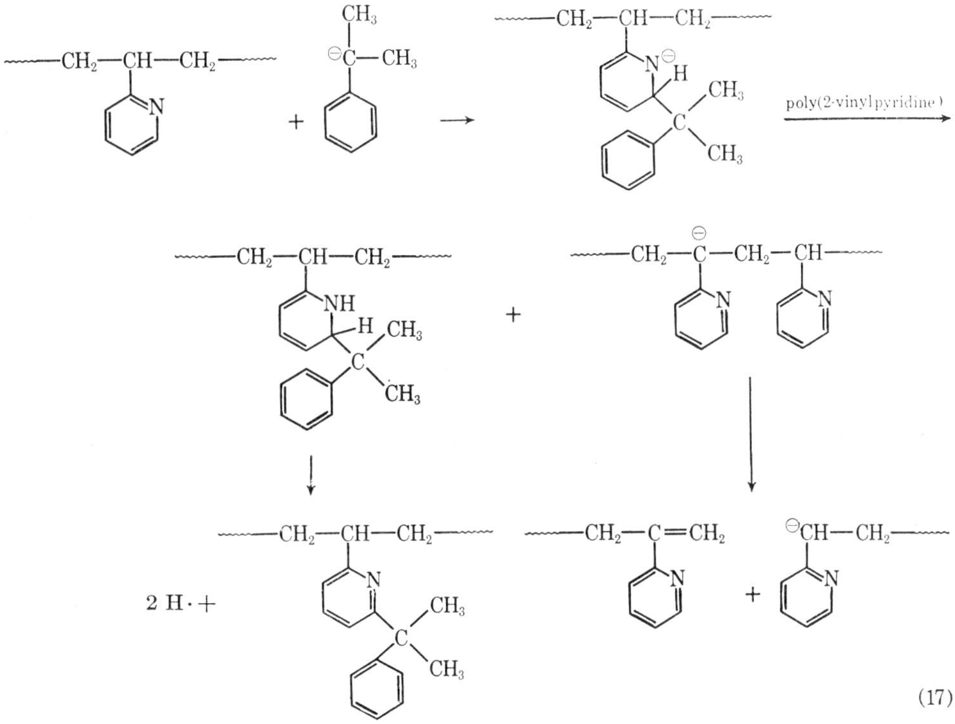

(17)

Both electron-transfer and carbanion initiators are capable of inducing polymerization. Examples of initiator systems include n-butyllithium (102), sodium compounds of biphenyl and of naphthalene, cumylpotassium (90), sodium metal in liquid ammonia (91), sodium adduct of vinylpyridine (92), phenylmagnesium bromide (93),

Table 3. Propagation Rate Constant of Vinylpyridine in Anionic Polymerization[a]

With anion of	Monomer	k_p, l-mole^{-1}-sec^{-1}	Reference
poly(2-vinylpyridine)	2-vinylpyridine	7,300	87
poly(4-vinylpyridine)	4-vinylpyridine	3,500	87
polystyrene	2-vinylpyridine	>30,000	87
polystyrene	4-vinylpyridine	>30,000	87
poly(2-vinylpyridine)	styrene	<1	87
~—(styrene)(2-vinylpyridine)$_n$			
$n = 1.1$	2-vinylpyridine	30,600	88
$n = 2.04$	2-vinylpyridine	11,000	88
$n = 3.2$	2-vinylpyridine	8,020	88
$n > 5$	2-vinylpyridine	ca 7,000	88

[a] Counterion $Na^{\oplus}$. In tetrahydrofuran solvent at 25°C.

butylmagnesium iodide (94), and many other organomagnesium or organoberyllium compounds, as well as lithium aluminum hydride (95–97). The magnesium, beryllium, and lithium compounds are of particular interest in that in the case of 2-vinylpyridine but not of 4-vinylpyridine, they act as coordination catalysts and result in stereospecific polymerization. Polymerization data are tabulated in Table 4. This difference between 2- and 4-vinylpyridine can be attributed to the fact that the nitrogen atom, in the case of 4-vinylpyridine, is too far apart from the vinyl group for orientation of monomer by coordination to catalyst to be possible. Even when the counterion is the noncoordinating sodium ion, it is more firmly bound by "living" poly(2-vinylpyridine) anion than by "living" poly(4-vinylpyridine), as indicated by the lower conductivity of "living" poly(2-vinylpyridine) anion than of "living" polystyrene or poly(4-vinylpyridine) anions (98).

These coordinating organometallic catalysts can also be used for alternating copolymerization of 2-vinylpyridine with stilbazole (99,100). Ziegler-Natta catalysts

Table 4. Stereospecific Polymerization of 2-Vinylpyridine[a]

Catalyst	% conversion	Acetone-insoluble crystalline polymer, % of total polymer	$[\eta]$,[b] dl/g	Monomer:catalyst ratio
$Be[N(CH_3)_2]_2$	40	40	0.27	ca 40
$Al[N(CH_3)_2]_3$	15	85	0.35	
$(C_2H_5)_2NMgCl$	96	88	0.30	ca 80
$(C_2H_5)_2NMgBr$	96	85	0.27	
$(C_2H_5)_2NMgI$	98	83	0.17	
$(CH_3)(C_6H_5)NMgBr$	93	88	0.49	
$(p\text{-}BrC_6H_4)(CH_3)NMgBr$	85	93	0.58	
$(p\text{-}BrC_6H_4)(CH_3)NMgC_2H_5$	98	85	0.15	
$Be(C_2H_5)_2$	97	65		ca 15
$LiAlH_4$	98	82		
$Mg(C_2H_5)_2$	97	90	0.28	ca 30
$Mg(C_6H_5)_2$	96	93	0.34	
C_6H_5MgBr	92	98	0.49	

[a] Polymerization at 45°C for 5 hr in toluene.
[b] In dimethylformamide at 30°C.

absorbed on silica have been described as catalysts of vinylpyridine copolymerizations (101).

Electrochemical polymerization of 4-vinylpyridine in liquid ammonia is apparently initiated by direct electron transfer to monomer. The red-orange color of the vinylpyridine radical-ion is observed at the cathode (103). See also ELECTROCHEMICAL INITIATION.

Polymerization of Activated Monomer. Interaction of vinylpyridine with acidic molecules induces polymerization. This phenomenon was described in a patent as early as 1958 (135), and recent detailed studies have opened a new and interesting field of polymerization (136–142). That the propagating species is anionic is shown by the fact that the polymerization is inhibited by carbon dioxide but not by radical scavengers. The general mechanism of polymerization can be shown by equations 18 and 19.

$$RX + CH_2{=}CH\text{–}C_5H_4N \longrightarrow CH_2{=}CH\text{–}C_5H_4\overset{\oplus}{N}\text{–}R\ X^{\ominus} \rightleftharpoons X\text{–}CH_2\text{–}\overset{\ominus}{C}H\text{–}C_5H_4\overset{\oplus}{N}\text{–}R \qquad (18)$$

$$X\text{–}CH_2\text{–}\overset{\ominus}{C}H\text{–}C_5H_4\overset{\oplus}{N}\text{–}R + CH_2{=}CH\text{–}C_5H_4\overset{\oplus}{N}\text{–}R\ X^{\ominus} \longrightarrow X\text{–}CH_2\text{–}CH(C_5H_4\overset{\oplus}{N}\text{–}R\ X^{\ominus})\text{–}CH_2\text{–}\overset{\ominus}{C}H\text{–}C_5H_4\overset{\oplus}{N}\text{–}R \qquad (19)$$

RX may be an alkyl halide (136,141), dimethyl sulfate (137,139,142), or a protonic acid (140). The anion produced by equation 18 must be very unreactive because of the positive charge on the nitrogen. Consequently, only monomers that are activated by quaternization can add to the growing chain end. When other monomers such as styrene or acrylonitrile are added or used as solvent, only quaternary poly(4-vinylpyridine) is produced (136,141), since these vinyl monomers are not reactive enough to add to the stable propagating species. In a manner similar to equation 17, the contribution of resonance structures is an important factor in monomer activation. Thus, the rate of polymerization decreases in the order 4-vinylpyridine > 2-methyl-5-vinylpyridine > 2-vinylpyridine. The slow rate of polymerization of 2-vinylpyridine is probably due to steric effects. The mechanism of polymerization of 2-vinylpyridine by methyl iodide is controversial. Since the polymer obtained is not completely quaternized and copolymerization of the quaternary salt of 2-vinylpyridine with styrene is possible, a charge-transfer copolymerization of a radical nature between 2-vinlypyridine and its quaternized derivative has been suggested (143).

The monomer molecules are considered to be spatially organized prior to polymerization. Consequently, there is an optimum temperature above which the organized monomer array is destroyed and the rate of polymerization decreases (142). The formation of stereoregular poly(4-vinylpyridine) by acid-catalyzed polymerization is attributed to the organization of monomer (140).

This polymerization technique has been extended to the use of polyacids as catalysts for the polymerization of vinylpyridine (144–148). The conformation of the polymer chain in solution (extended or entangled, helical or random coil) is an important factor in determining the activity of the polymer catalyst (149).

When vinylpyridine is complexed with nonoxidizing metal salts, the electronic state of the monomer is affected as a result of coordination bond formation, and this naturally also affects the reactivity of monomer (150–152). Free-radical polymerization of vinylpyridine complexed with metal salts of Group IIB indicates that the reactivity of the complexed monomer is very high for 4-vinylpyridine but not for 2-methyl-5-vinylpyridine (Table 5). This result is understandable if the π-type interaction between metal salt and pyridine group is considered. The monomer conjugation is enhanced more if resonance structures such as shown by equation 17 are possible. The contribution of resonance structures in determining the relative stabilities of complexes of 4-vinylpyridine, 2-vinylpyridine, and 2-methyl-5-vinylpyridine has been studied by infrared spectroscopy and by nuclear magnetic resonance (151,153).

Table 5. Copolymerization of Vinylpyridine with Styrene in Concentrated Solution of Zinc Acetate at 50°C (M_2 = Styrene)

M_1	r_1	r_2	Q_1	e_1
4-vinylpyridine[a]	0.52–0.7	0.54–0.62	0.82	−0.20
4-vinylpyridine–Zn(II)	2.7 ± 0.5	0.08 ± 0.03	4.7	+0.44
2-vinylpyridine[a]	0.9–1.81	0.55	1.30	−0.50
2-vinylpyridine–Zn(II)	3.35 ± 0.3	0.55 ± 0.15	r_1r_2 = 1.84	
2-methyl-5-vinylpyridine[a]	0.68–1.19	0.6–0.88	0.99	−0.58
2-methyl-5-vinylpyridine–Zn(II)	2.0 ± 0.2	0.35 ± 0.05	1.75	−0.26

[a] Data of copolymerization were taken from Ref. 161.

Theoretically derived reactivity indexes explain homo- and copolymerizability of metal-complexed vinylpyridine satisfactorily (154). The effect of complex formation on the rate of homopolymerization is shown in Figure 1. Catalyzed and spontaneous polymerizabilities of metal-complexed vinylpyridine correlate well. The complexed monomer active in propagation is also active in spontaneous initiation (155). Thus, 4-vinylpyridine absorbed on a solid surface polymerizes spontaneously (106,156), whereas 2-vinylpyridine is not susceptible to surface-catalyzed polymerization. The inertness of 2-vinylpyridine complexes must be at least in part due to steric effects, although theoretical calculations also predict the low reactivity of 2-vinylpyridine complexes. The NMR spectra of 2-vinylpyridine complexed with BF_3 or $AlBr_3$ indicate considerable rotation of the vinyl group due to interference between the vinyl

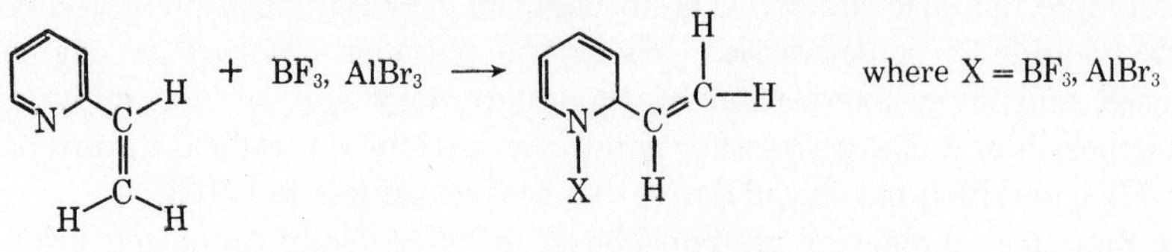

$$\text{(2-vinylpyridine)} + BF_3, AlBr_3 \longrightarrow \text{(N-X complex)} \qquad \text{where } X = BF_3, AlBr_3 \qquad (20)$$

$$\text{(2-vinylpyridine)} + H^{\oplus} \longrightarrow \text{(N-protonated 2-vinylpyridinium)} \qquad (21)$$

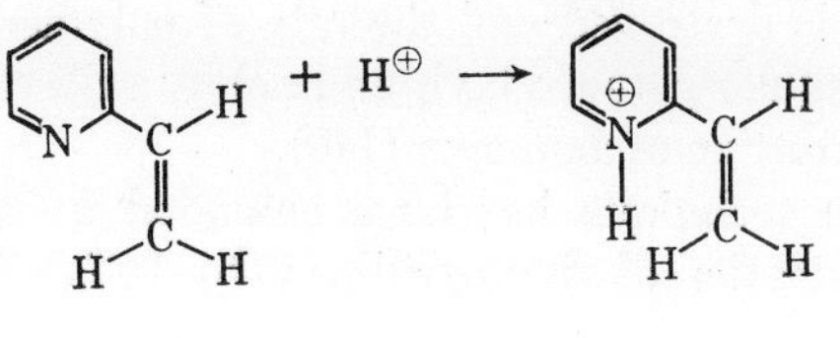

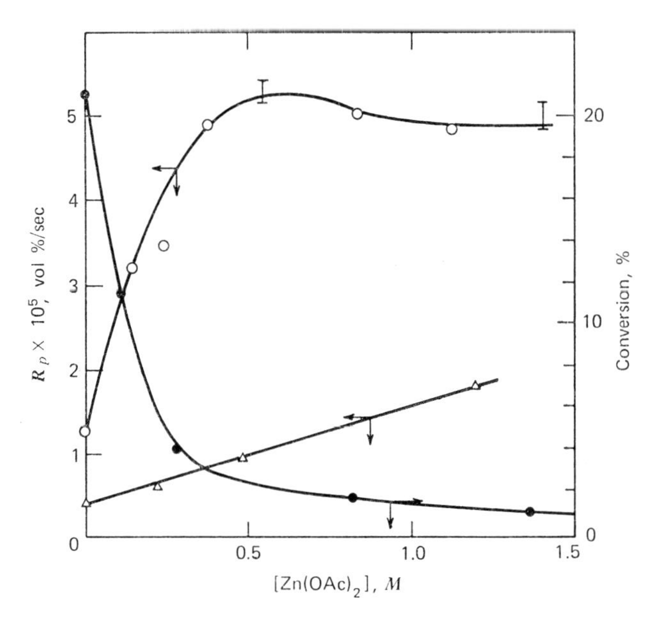

Fig. 1. Polymerization of vinylpyridine in the presence of $Zn(CH_3COO)_2$ in dimethylformamide at 50°C. O, [4-vinylpyridine] = 0.43 *M*, [azobisisobutyronitrile] = 1.22×10^{-2} *M*. △, 2-methyl-5-vinylpyridine, R_p measured by a dilatometer. ●, [2-vinylpyridine] = 0.85 *M*, [azobisisobutyronitrile] = 2.44×10^{-2} *M*. Polymerization time, 8 hr.

group and the Lewis acid (eq. 20) (158). When the acid attached to the nitrogen atom of pyridine is as small as a proton, the position of the vinyl group is normal (eq. 21). Such distortion of the vinyl group would certainly affect the reactivity of the monomer.

Copolymerization. Published data on copolymerization of vinylpyridines are largely limited to radical copolymerization. However, alternating copolymerization of vinylpyridine with trinitrostyrene has been suggested to proceed through a charge-transfer reaction (159).

Copolymerization reactivity ratios are tabulated in Table 6. Other data can be found in Ref. 160. Copolymerization of 2-vinylpyridine with *p*-benzoquinone and 2,5-dimethylbenzoquinone (160), and of 2-methyl-5-vinylpyridine with fumaric acid (207), vitamin A acrylate (208), propenyl isopropenyl ketone, and vinyl isobutenyl ketone (209) has been discussed.

The rates of copolymerization of various vinylpyridines with *p*-divinylbenzene decrease in the following order (80°C in bulk) (210): 4-vinylpyridine > 4-methyl-2-vinylpyridine > 2-vinylpyridine > 5-ethyl-2-vinylpyridine > 2-methyl-5-vinylpyridine > 6-methyl-2-vinylpyridine.

Block and Graft Copolymerization. This field of study is so closely connected to practical application that most of the information is published in the form of patents. The common methods of living anionic polymerization, polymerization with a polymer catalyst bearing radical or anion sources, radiation-induced reaction, and radical polymerization in the presence of polymer can all be applied to vinylpyridine systems. As a special method, the susceptibility of the pyridine ring toward carbanion attack is utilized for preparing block and graft polymers. The reaction mechanism can be visualized as in Figure 2. The choice of reaction path depends on reaction conditions such as solvent, counterion, and reactivity of attacking living anion (211,212).

Table 6. Reactivity Ratios for Copolymerization of Vinylpyridine

M_1	r_1	M_2	r_2	T, °C	Reference
		Binary system			
2-vinylpyridine	ca 0	acrolein	ca 4	50	162
	0.47 ± 0.03	acrylonitrile	0.113 ± 0.002	60	163
	2.13	acrylonitrile	0.086	40[a]	164
	21.9 ± 5.52	acrylonitrile	0.05 ± 0.01	60	165,166
	0.097 ± 0.04	butyl acrylate	2.51 ± 0.05	60	167
	0.06 ± 0.01	2-chloro-1,3-butadiene	5.19 ± 0.03	60	165,166
	0.63 ± 0.07	dichlorostyrene	0.11 ± 0.07	60	168
	1.1	2,5-dichlorostyrene	0.9	70	169
	0.836 ± 0.18	(2,2-dimethyl-1,3-dioxolan-4-yl)methyl acrylate	0.12 ± 0.1	50	170
	1.06 ± 0.10	4-(2,2-dimethyl-1,3-dioxolan-4-yl)methyl itaconate	0.16 ± 0.11	50	170
	0.76 ± 0.06	[(2,2-dimethyl-1,3-dioxolan-4-yl) methyl] methyl fumarate	0	50	170
	32.0	divinyl ether	ca 0 (assumed)	60	171
	0.23 ± 0.05	ethyl acrylate	0.19 ± 0.06	75	172
	0.47 ± 0.07	isoprene	0.59 ± 0.05	60	165,166
	1.65 ± 0.05	isopropenylacetylene	0.55 ± 0.01	60	173
	1.55 ± 0.10	methacrylic acid	0.58 ± 0.05	70	174
	1.38 ± 0.04	methacrylic acid	0.44 ± 0.02		175
	2.03 ± 0.49	methyl acrylate	0.20 ± 0.09	60	168
	1.58 ± 0.05	methyl acrylate	0.168 ± 0.003	60	176
	0.77 ± 0.02	methyl methacrylate	0.439 ± 0.002	60	176
	0.86 ± 0.06	methyl methacrylate	0.395 ± 0.025	60	177
	0.84 ± 15%	methyl methacrylate	0.20 ± 6.5%	25	184
	0.70 ± 15%	methyl methacrylate	0.33 ± 15%	70	178
	4.0 ± 0.07	phenylacetylene	0.2 ± 0.05	60	179
	0.9 ± 0.2	styrene	0.56 ± 0.02	60	176
	1.14	styrene	0.55	60	180
	1.135 ± 0.08	styrene	0.55 ± 0.025	60	181
	1.81 ± 0.05	styrene	0.55 ± 0.03	60	165,166
	1.33	styrene	0.57	50	182,183
	0.76–0.862	styrene	0.16–0.486	[b]	182,183
	0.36 ± 0.04	styrene	0.16 ± 0.02	22[b]	183
	1.27	styrene	0.50	25	184
	10	vinyl acetate	0.3		210
	30 ± 15	vinyl acetate	0	70	178
	1.5 ± 0.5	1-vinyl-2-ethylacetylene	0.6 ± 1	60	173
4-vinylpyridine	0.41 ± 0.09	acrylonitrile	0.113 ± 0.005	60	163
	5.15 ± 0.09	butyl acrylate	0.46 ± 0.09	60	185
	1.7 ± 0.2	methyl acrylate	0.22 ± 0.01	60	176
	0.79 ± 0.05	methyl methacrylate	0.574 ± 0.004	60	176
	0.52 ± 0.06	styrene	0.62 ± 0.02	80	186
	0.7 ± 0.1	styrene	0.54 ± 0.03	60	176
	0.30 ± 0.02	*N,N,N*-trimethyl-*N*-(2-methacryloyl)-ammonium iodide	0.61 ± 0.09	60	187
	23.4	vinyl chloride	0.02	60	188
	0.48	vinylhydroquinone dibenzoate	0.40	78	189

Table 6 (*continued*)

M_1	r_1	M_2	r_2	T, °C	Reference
2-methyl-5-vinyl-pyridine	0.01 ± 0.09	acrylamide	0.56 ± 0.09	60	190
	0.27 ± 0.04	acrylonitrile	0.116 ± 0.003	60	191
	1.10 ± 0.2	acrylonitrile	0.10 ± 0.05		192
	0.72 ± 0.03	butadiene	1.32 ± 0.01		193
	0.474 ± 0.031	butadiene	0.600 ± 0.006		194
	80	diallyl ether	0	70	195,278
	0.90 ± 0.10	*o*-divinylbenzene	0.45 ± 0.15	60[c]	196
	0.17 ± 0.09	ethyl cinnamate	1.49 ± 0.16	70	197
	0.88 ± 0.10	methyl acrylate	0.172 ± 0.007	60	198
	0.70 ± 0.10	methyl acrylate	0.35 ± 0.5		199
	0.85 ± 0.03	methacrylic acid	0.43 ± 0.2		200
	0.61 ± 0.08	methyl methacrylate	0.46 ± 0.02	60	198
	2.9 ± 0.3	styrene	0.5 ± 0.1	70[d]	201
	0.68 ± 0.1	styrene	0.6 ± 0.1	70	201
	2.9 ± 0.3	styrene	0.3 ± 0.1	70[e]	201
	0.91 ± 0.02	styrene	0.812 ± 0.005	60	198
	1.19 ± 0.12	styrene	0.88 ± 0.2	60	202
	0.32 ± 0.065	triethylene glycol dimethacrylate	0.59 ± 0.06		203
5-ethyl-2-vinyl-pyridine	0.43 ± 0.05	acrylonitrile	0.02 ± 0.02	60	163
	1.16 ± 0.08	methyl acrylate	0.179 ± 0.006	60	176
	0.69 ± 0.03	methyl methacrylate	0.395 ± 0.003	60	176
	1.2 ± 0.2	styrene	0.79 ± 0.03	60	176
2-vinylpyridine *N*-oxide	3.9 ± 0.8	methyl methacrylate	0.13 ± 0.09	60	39
	2.1 ± 0.6	styrene	0.11 ± 0.01	60	39
5-ethyl-2-vinyl-pyridine *N*-oxide	5.5 ± 0.5	methyl methacrylate	0.11 ± 0.01	60	39
	4.7 ± 0.6	methyl methacrylate	0.12 ± 0.02	60	39
	2.6 ± 0.3	styrene	0.10 ± 0.01	60	39
N-vinylpyridinium fluoroborate	0.20	acrylonitrile	1.06	60	41
	0.2	methyl acrylate	1.5	60	41
	0.008	methyl methacrylate	4.75	60	41

Ternary system

Comonomers	r_{12}	r_{21}	r_{13}	r_{31}	r_{23}	r_{32}	T, °C
M_1 = 2-vinylpyridine, M_2 = methyl acrylate, M_3 = dichlorostyrene (204)	2.03 ± 0.49	0.20 ± 0.09	0.63 ± 0.07	0.11 ± 0.07	0.25 ± 0.04	4.27 ± 0.28	60[f]
M_1 = 4-vinylpyridine, M_2 = methyl acrylate, M_3 = isoprene (205)	2.14	0.15	2.49	0.32	0.12	0.75	50
M_1 = 2-methyl-5-vinylpyridine, M_2 = styrene, M_3 = butadiene (206)	0.801 ± 0.030	0.73 ± 0.010	0.412 ± 0.026	1.30 ± 0.07	0.825 ± 0.086	1.39 ± 0.15	30[g]

[a] Aqueous phase.
[b] Hydroxy compounds are added (CH_3COOH, C_6H_5OH, CH_3OH, or C_2H_5OH).
[c] Based on simple treatment. If the cyclized radical R_3

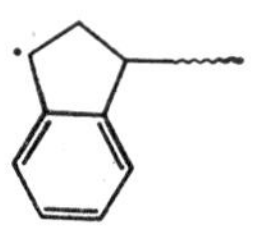

is distinguished, $r_3 = k_{32}/k_{31} = 0.2 \pm 0.3$, $r_1 = 0.78 \pm 0.15$, $r_2 = 0.62 \pm 0.05$, $r_c = k_c/k_{22} = 2.6$, and $k_c/k_{21} = 1.6$.
[d] Sodium metal as catalyst.
[e] $TiCl_4$–$Al(C_2H_5)_3$ as catalyst.
[f] Bulk.
[g] Emulsion.

Graft and block polymerizations involving vinylpyridine and its polymers are listed in Table 7.

Structure and Properties of Polyvinylpyridine

Tacticity. Stereoregular polymers can be prepared for 2-vinylpyridine (95–97). The crystallinity of isotactic poly(2-vinylpyridine) was originally confirmed by x-ray diffraction analysis (identity period about 6.7 Å). Crystalline poly(2-vinylpyridine) can be fractionated from amorphous polymer by extracting the latter polymer with acetone. Crystalline bands of poly(2-vinylpyridine) were observed by infrared spectroscopy at 928 cm^{-1} and 1180 cm^{-1} (255). Recent progress in NMR studies of poly(2-vinylpyridine) has made possible the quantitative analysis of tacticity of

Table 7. Block and Graft Copolymerization of Vinylpyridine

Backbone polymer	Grafting monomer	Remarks	Reference
	A. Radiation-induced graft copolymerization		
polyethylene	2-vinylpyridine		213
	4-vinylpyridine, 2-methyl-5-vinylpyridine		214
	2-methyl-5-vinylpyridine	preirradiation (peroxidation)	215
polypropylene	vinylpyridine	preirradiation	216,217
	2-methyl-5-vinylpyridine	slow neutron irradiation	218
polytetrafluoroethylene	4-vinylpyridine	aqueous phase	219
polytetrafluoroethylene	4-vinylpyridine		220
copolymer of vinylidene chloride and trifluorochloroethylene	4-vinylpyridine	emulsion	221
poly(vinyl chloride)	4-vinylpyridine	vapor phase	222
polyethylene	styrene + 4-vinylpyridine		223
poly(ethylene terephthalate)	4-vinylpyridine	preirradiation	224
nylon	4-vinylpyridine	preirradiation	225
cellulose	2-vinylpyridine	vapor phase	226
	B. Graft copolymerization by radical chain transfer		
copolymer of vinylidene fluoride and tetrafluoroethylene	2-methyl-5-vinylpyridine, 2-vinylpyridine, 4-vinylpyridine	benzoyl peroxide or azobisisobutyronitrile	227
chloroprene rubber	vinylpyridine	benzoyl peroxide	228
polybutadiene	vinylpyridine	azobisisobutyronitrile	229
	C. Radical graft polymerization initiated by functional groups in backbone polymer		
cellulose	2-vinylpyridine	ceric ammonium sulfate	230
polycyclopentadiene	2-methyl-5-vinylpyridine	the backbone polymer is preoxidized	231
cellulose acetate	2-methyl-5-vinylpyridine	preoxidation by hydrogen peroxide	232
poly(2-methyl-5-vinylpyridine)	acrylonitrile, styrene, methacrylic acid	diazo group in backbone polymer as end group	233
copolymer of acrylonitrile or 2-methyl-5-vinylpyridine with 4-(vinylsulfonyl)-2-aminoanisole	acrylonitrile, vinylpyridine	the amino groups in trunk polymer are converted to diazo compounds	234

Table 7 (*continued*)

Backbone polymer	Grafting monomer	Remarks	Reference
	D. Block copolymerization by the use of "living" polymers		
polybutadiene anion	2-methyl-5-vinylpyridine, 4-vinylpyridine, 2-vinylpyridine		
polystyrene anion	2-vinylpyridine, 4-vinylpyridine	copolymerization (styrene–4-vinylpyridine, 4-vinylpyridine–2-vinylpyridine, styrene–2-vinylpyridine) possible	235
poly(2-methyl-5-vinylpyridine) anion	acetaldehyde	backbone polymer is polymerized by dibarium benzophenone	236,237
poly(2-vinylpyridine) anion	styrene	polymers of five sequences are prepared	238,239
poly(2-methyl-5-vinylpyridine) anion	formaldehyde		240
poly(2-vinylpyridine) (α,ω-dicarbanionic)	divinylbenzene, glycol dimethacrylate, hexanediol dimethacrylate	crosslinking	241

E. Graft or block copolymerization initiated by functional groups attached to polymer

Polymer catalyst	Grafting monomer	Reference
poly(acrylonitrile–4-vinylpyridine)	β-propiolactone	242
poly(styrene–4-vinylpyridine) with sodium naphthalene	methyl methacrylate	243,244
poly(styrene–acrylonitrile) with sodium	4-vinylpyridine, 2-vinylpyridine	245
potassium alkoxide of starch or dextrin	4-vinylpyridine	246
poly(styrene–acrylonitrile) + dimethylsilylmethylmagnesium chloride	4-vinylpyridine	247
poly(3,3-diphenyl-1-propene) + sodium (or lithium) naphthalene	styrene-2-vinylpyridine, methyl methacrylate-2-vinylpyridine (sequenced graft)	248–250
poly(2-vinylpyridine) with butyllithium	2-vinylpyridine	251
poly(vinylbenzophenone sodiumketyl)	4-vinylpyridine	252

F. Graft or block copolymerization by reactions between polymers

Reaction	Reacting site	Reference
"living" polystyrene + poly(2-vinylpyridine) or "living" poly(α-methylstyrene)	pyridine ring or main chain of poly(2-vinylpyridine)	212
("living" polystyrene + difunctional nitriles) hydrolyzed to carbonyl compounds, then added with "living" poly(2-vinylpyridine)	carbonyl group	253
chloromethylated polyethylene + "living" poly(2-vinylpyridine)	chloromethyl group	254
"living" poly(2-vinylpyridine) + "living" polystyrene or living polyisoprene	pyridine ring or α-carbon of poly(2-vinylpyridine)	211

poly(2-vinylpyridine) (256–258). It is still uncertain whether or not poly(4-vinylpyridine) exists in crystalline form (140,259).

Viscosity–Molecular Weight Relationship. Constants for the Mark-Houwink equations determined under various conditions are tabulated in Table 8.

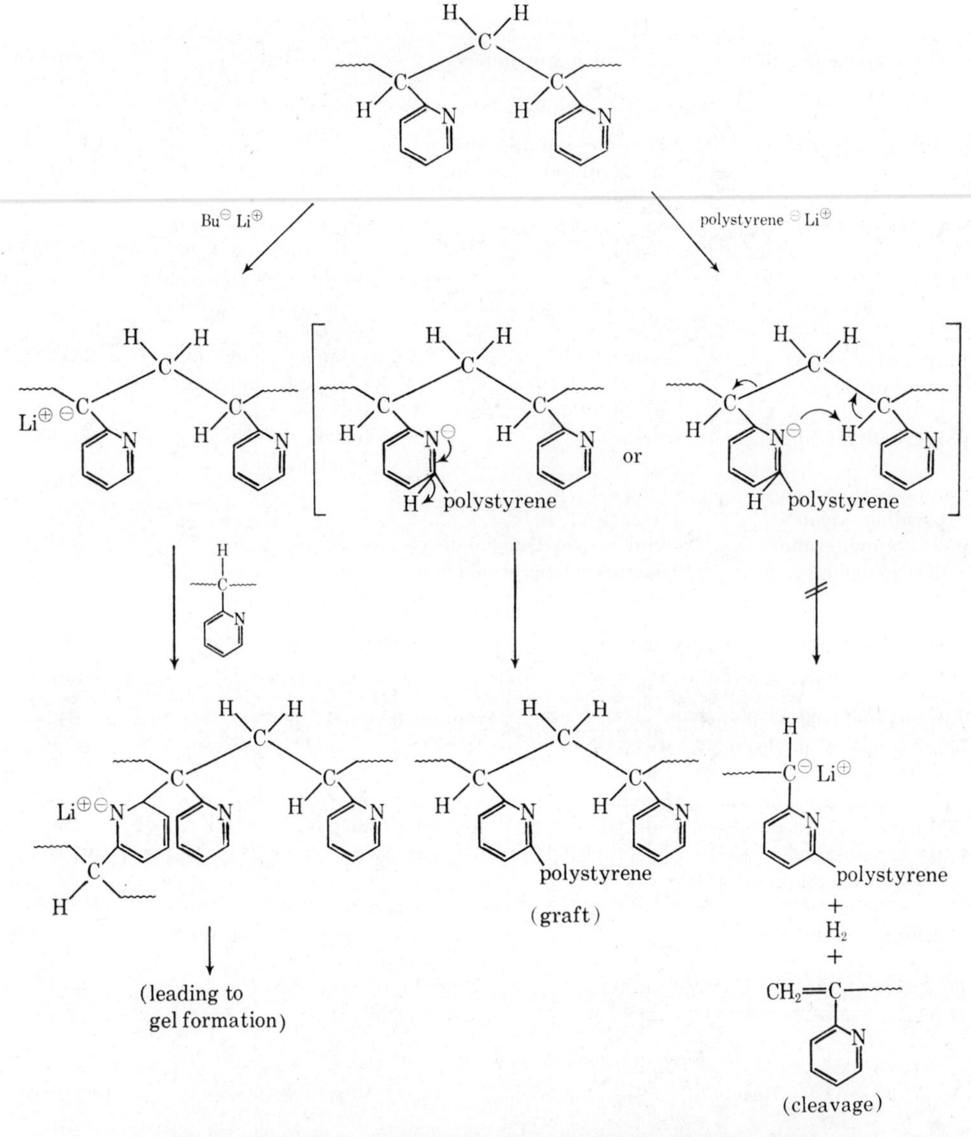

Fig. 2. Preparation of block and graft copolymers of vinylpyridine.

Complex Formation. Since pyridine is a good electron donor as well as a strong ligand capable of forming a coordination bond, polyvinylpyridine forms complexes with both inorganic and organic molecules (see also ELECTRON-TRANSFER POLYMERS; POLYELECTROLYTES). In particular, complexes with strong electron acceptors such as tetracyanoquinodimethane (TCNQ), tetracyanoethylene (TCNE), and iodine have been studied from the viewpoint of electric conductivity and behavior as semiconductors. The TCNQ complex has a conductivity as high as about 10^{-3} Ω^{-1}-cm^{-1}. The reaction of polyvinylpyridinium salt with TCNQ anion radical produces a 1:1 complex with a rather low conductivity (about 10^{-10} Ω^{-1} cm^{-1}). When this complex is treated with free TCNQ, the conductivity increases to 10^{-3} to 10^{-5} Ω^{-1}-cm^{-1} (271,

Table 8. Constants for Viscosity–Molecular Weight Relation[a] of Vinylpyridine Polymers

Polymer	K	a	Solvent	T, °C	Ref.
poly(2-vinylpyridine)	1.22×10^{-4}	0.73	92.01% ethanol	25	260
	2.8×10^{-4}	0.66	ethanol	25	261
	1.2×10^{-3}	0.50	heptane–propanol (Θ solvent)	25	261
	1.38×10^{-4}	0.69	pyridine	25	262
	9.72×10^{-4}	0.47	methyl ethyl ketone	25	262
	1.03×10^{-4}	0.746	ethanol	10	263
	1.09×10^{-4}	0.729	ethanol	25	263
	0.99×10^{-4}	0.733	ethanol	40	263
	1.06×10^{-4}	0.718	ethanol	65	263
	0.64×10^{-4}	0.74	ethanol + 2.7N ammonia	25	263
	1.603×10^{-2}	0.70	ethanol	25	264
	1.13×10^{-4}	0.73	methanol	25	260
	1.162×10^{-2}	0.69	dimethylformamide	25	264
	1.47×10^{-4}	0.67	dimethylformamide	25	260
	1.70×10^{2}	0.68	ethanol + sodium acetate 0.1M	25	264
	1.70×10^{-4}	0.64	benzene	25	260
	3.09×10^{-4}	0.58	dioxane	25	260
isotactic poly(2-vinylpyridine)	4.624×10^{-2}	0.59	ethanol + sodium acetate 0.1M	25	264
	4.467×10^{-2}	0.57	dimethylformamide + lithium chloride 0.1M	25	264
poly(4-vinylpyridine)	2.5×10^{-4}	0.68	ethanol	25	265
	3.80×10^{-4}	0.57	butanone–isopropanol (86:14)	25	266
	1.20×10^{-4}	0.73	92.01% ethanol	25	266
	3.055×10^{-5}	1.20	pyridine	25	106
poly(2-methyl-5-vinylpyridine)	19×10^{-3}	0.64	methyl ethyl ketone	25	267
	18.6×10^{-3}	0.70	methanol	25	267
	0.80×10^{-4}	0.76	methanol	25	268
	84×10^{-3}	0.50	n-butyl acetate (Θ solvent)	21.8	267
	80×10^{-3}	0.50	methyl isopropyl ketone (Θ solvent)	37.4	267
	83×10^{-3}	0.50	n-amyl acetate (Θ solvent)	48.2	267
	18×10^{-3}	0.83	dimethylformamide	25	269
	13×10^{-3}	0.76	methanol	25	269
	61.7×10^{-3}	0.615	methanol	25	270

[a] $[\eta] = KM^a$, in dl/g.

272). On heat treatment of the poly(4-vinylpyridine–TCNQ) complex, the resistivity was reported to decrease from 4.7×10^8 to 4.04×10^3 Ω-cm, whereas the temperature coefficient decreased from 1.01 to 0.15 eV (273). The high conductivity would be brought about by the $R^{\oplus}(TCNQ)_2^{\ominus}$ complex, as expected from the conductivity measurement of simple monomeric salt (274). See also Semiconductive Polymers under POLYMERS, CONDUCTIVE.

Complexes of polyvinylpyridine are listed in Table 9.

Although complex formation is not stoichiometric, interactions of poly(4-vinylpyridine) with chlorophyll or magnesium phthalocyanine (285), of poly(2-vinylpyridine–acrylic acid) with protein (286), and of poly(4-vinylpyridine), poly(4-vinylpyridine–acrylic acid), or poly(4-vinylpyridinium salt) with DNA (287) have been reported. Absorption of DNA on poly(4-vinylpyridine–styrene) or poly(4-vinylpyridine–methyl methacrylate) is also known (288). The natures of biologically interesting compounds are certainly affected by such interactions. Thus, the al-

Table 9. Complexes of Polyvinylpyridine

Polymer	Complexing agent	Remarks	Reference
poly(4-vinylpyridine), poly(2-vinylpyridine)	TCNQ	resistivity = 10^{3}–10^{10} Ω-cm depending upon preparative conditions	271,272,273, 275,276,277
	TCNE	conductivity = 10^{-8} to 10^{-10} Ω^{-1}-cm^{-1} ϵ = 1–1.3 eV	278
	iodine	conductivity = 10^{-4} to 10^{-7} Ω^{-1}-cm^{-1}	275,279
	bromine		273,280
	$SbCl_5$		273,275
poly(4-vinylpyridine), poly(2-vinylpyridine), poly(2-methyl-5-vinylpyridine)	benzoquinone		275,281
poly(2-methyl-5-vinylpyridine)	dinitrotoluene dinitroxylene	viscosity of polymer increases	282
poly(4-vinylpyridine)	$Ni(SCN)_2$	absorption of aromatic hydrocarbon by clathrate formation	283
polyvinylpyridine	various metal salts		284

lomerization of chlorophyll by oxygen is catalyzed by complex formation with poly-(4-vinylpyridine) (285). The presence of a complex-forming polymer affects the melting point of DNA (denaturation temperature) and prevents its enzymatic hydrolysis.

Metal complexes of polyvinylpyridine are generally insoluble and infusible when more than two pyridine groups are attached to a metal salt. The coordination bonds would crosslink the polymer, producing network structures. The formation of intermolecular coordination bonds is most prominent for poly(4-vinylpyridine) whereas poly(2-vinylpyridine) is much less crosslinked, as indicated by the solubility of the complexed polymer (150,151,289).

Quaternized Polyvinylpyridine

Polyvinylpyridine is readily quaternized by acids or alkyl halides and the resulting polyelectrolytes have been the subject of many studies. The rate of quaternization is subject to effects of neighboring groups (290) or steric hindrance (291), and consequently, the rate becomes slower than that expected from normal bimolecular kinetics during the course of reaction. The solution properties of polyvinylpyridinium salts have been investigated since the late 1940s using polymers of the lower alkylvinylpyridinium halides or polyvinylpyridine dissolved in acidic media (292–301). In the aqueous solution of polyion (eg, poly(4-vinylpyridinium hydrochloride)), the polyion expands at first with increasing dilution, owing to the decrease in the screening effect of the counterions on the intramolecular electrostatic repulsion. However, further increase brings about extensive hydrolysis of polyion, which decreases the intramolecular repulsion, and the polyion begins to contract (301). Consequently, the hydrodynamic properties of polyion show a discontinuity with respect to the polyion concentration. When the *N*-alkyl group is long-chain hydrocarbon, the behavior as a polyelectrolyte is converted to that of a polysoap (302–306). Dodecyl or nonyl bromide is the common alkyl halide for the preparation of polysoap. When the alkyl groups are very hydrophobic they apparently tend to aggregate, and the relation η_{sp}/c vs c becomes normal (306).

Amphoteric polyelectrolytes synthesized by copolymerization of vinylpyridine with an acidic monomer, in most cases with acrylic acid or methacrylic acid, form another interesting subject of study (307–313). There is evidence of complex formation between carboxylic acid and pyridine (311), and the viscosity is at a minimum at the isoelectric point (310). Salt formation by polyvinylpyridine depends on the configuration of the polymer. Optical rotation of *d*-tartaric acid in the presence of poly(2-vinylpyridine) is higher when the polymer is isotactic (314). See also POLYELECTROLYTES.

Applications

Vinylpyridine has practical use as a copolymerization component for synthetic rubbers and fibers. The homopolymer of 2-methyl-5-vinylpyridine has mechanical properties similar to those of polystyrene (315). There seems to be no advantage of the homopolymer of 2-methyl-5-vinylpyridine over polystyrene as a plastic material, however, since the vinylpyridine is more fragile than polystyrene.

Rubber and Adhesives. Vinylpyridine rubber (a copolymer of butadiene with vinylpyridine and, in some cases, other vinyl compounds such as styrene) has better flex cracking, tearing, resilience, and tensile properties than styrene–butadiene rubber. It was reported that the highest values for strength, elasticity, and abrasion resistance were achieved when the rubber contained butadiene (70%), styrene (25%), and 2-methyl-5-vinylpyridine (5%) (316). Since vinylpyridine has a high affinity for metal salts, inorganic salts such as zinc chloride, zinc oxide, stannous chloride, and cadmium chloride (317–321) can be used as vulcanizing agent. Alkylene dihalides may be used together with inorganic salts (321). Epoxy resin is also a good vulcanizing agent for vinylpyridine rubber (322).

Vinylpyridine monomers complexed with metal salts act as vulcanizing agents for ethylene–propylene or butadiene–styrene rubber (323,324). For butadiene–styrene rubber, dynamic properties are improved and scorch is retarded. For ethylene–propylene rubber, the tensile strength is increased from 32 to 78 kg/cm^2 when 15 parts of 2-methyl-5-vinylpyridine–zinc chloride complex is added to the rubber together with 10 parts of zinc oxide and 1 part of peroxide, and then the mixture is vulcanized.

Blending of vinylpyridine rubber with other polymers has been of considerable interest particularly from the viewpoint of a bonding material between rubber and tire cord. Vinylpyridine rubber itself has a high adhesivity for tire cord materials such as rayon, nylon, and polyester (325–327). Combinations of vinylpyridine rubber with resorcinol–formaldehyde resin (428–331), with epoxylated novolac and zein, casein, or caseinated salt (332), and with phenol-blocked diphenylurethan resin (333) are potential bonding materials.

Vinylpyridine may be used also as a bonding material between glass fiber and resin in fiber-reinforced plastics (334).

Textile Applications. Vinylpyridine is an important comonomer for improving dyeability of acrylic fiber. Besides this use, the following proposals can be found in the literature: grafting of poly(vinyl chloride) fiber with 4-vinylpyridine (355), addition of polymer latex containing vinylpyridine to the spinning solution of poly(vinyl alcohol) (356), grafting of polypropylene fiber with 2-methyl-5-vinylpyridine (357), blending of poly(4-vinylpyridine) (358) or the copolymer poly(propylene-*b*-4-vinylpyridine) (359,360) or the copolymer of 4-vinylpyridine and *N*-substituted acrylamide (357) with polypropylene fiber. When cellulose is treated with the iodo complex of poly(2-methyl-5-vinylpyridine), a bacteriostatic effect is produced (361).

Since the quaternized polymer is conductive, an antistatic effect is to be expected when the polymer is coated on fiber (362) or film (363,364).

Uses as a Polyelectrolyte. These uses include applications as an ion exchanger, flocculant, emulsifier, etc. Vinylpyridine copolymers are weak base resins that can be used as anion-exchange resins after quaternization or as selective ion exchangers for metal, using complex formation by the pyridine side group. Amphoteric ion exchangers are easily obtained by copolymerization with acidic monomers. Examples of ion-exchange polymers containing vinylpyridine are listed in Table 10. See also ION-EXCHANGE POLYMERS.

Polyvinylpyridinium salts can be used as flocculants (348–352). Combinations with nonionic water-soluble polymer may have synergistic effects on flocculation (353). See also FLOCCULATION.

Copolymers of vinylpyridine with acrylic or methacrylic acid are soluble in water over a wide range of pH; an amphoteric coating material for pharmaceutical use can be prepared from the copolymers (354).

Uses as a Catalyst. If monomeric pyridine acts as a catalyst in some given reaction, polyvinylpyridine could certainly be expected to exhibit a more or less

Table 10. Ion-Exchange Polymers Containing Vinylpyridine

Components	Remarks	Reference
2-methyl-5-vinylpyridine + dimethyl maleate or dimethyl fumarate converted to hydrazide after polymerization	effective for metal ion (Cu^{2+}, Ni^{2+}, Co^{2+}, Mg^{2+})	335
2,5-divinylpyridine + 2-(β-diethylamino-ethyl)-5-vinylpyridine	selective complexing resin	336
4-vinylpyridine–acrylic acid–divinylbenzene	selective for UO_2^{2+}	337
poly(vinyl alcohol) grafted with 2-methyl-5-vinylpyridine		338
polyethylene grafted with 4-vinylpyridine, polyethylene grafted with ethyl acrylate and blended polyethylene	sulfonated and *N*-methylated (amphoteric)	339
N-methyl-2-vinyl-5-ethylpyridine–styrene–divinylbenzene	may be prepared in rubber matrix	340
2-methyl-5-vinylpyridine–divinylbenzene or triethylene glycol dimethacrylate	quaternized with benzyl chloride or methyl *p*-toluenesulfonate	341
poly(4-vinylpyridine) or poly(4-vinylpyridine–styrene) + 1,4-diiodobutane	ion-exchange membrane formed on saran net	342
poly(2-vinylpyridine) + halomethylated toluene–formalin resin		343
2-vinylpyridine + 2,5-divinylpyridine		344
2-methyl-5-vinylpyridine + maleic anhydride (or fumaric anhydride) + divinylbenzene	amphoteric	345
vinylpyridine–styrene–divinylbenzene sulfonated	amphoteric	346
crosslinked poly(cyanovinylpyridine), hydrolyzed	amphoteric	347

Table 11. Reactions Catalyzed by Vinylpyridine Polymers

Reaction	Catalyst	Remarks	Reference
hydrolysis			
of 2,4-$(NO_2)_2C_6H_4OOCCH_3$,	poly(4-vinylpyridine)	aqueous	365
of 3-NO_2-4-$CH_3COOC_6H_4SO_3K$		comparison with γ-picoline	366
of 4-NO_2-$C_6H_4OOCCH_3$	poly(4-vinylpyridine)		367
	poly(4-vinylpyridine) partially quaternized by 2-(β-chloroethyl)pyridine		368
condensation of C_6H_5NCO with CH_3OH	poly(4-vinylpyridine)	rate is high when the reaction proceeds heterogeneously	369
	poly(4-vinylpyridine–styrene)		
condensation of 1,4-$(OCN)_2C_6H_4$ with HOC_2H_4OH	poly(4-vinylpyridine), poly(4-vinylpyridine–styrene)		370
sulfonation of starch	poly(2-vinylpyridine) + SO_3		371
chlorination of $(NO_2)_3C_6H_2OH$ by $POCl_3$	poly(2-methyl-5-vinylpyridine), poly(4-vinylpyridine)	reaction proceeds via polymer picrate	372
oxidation of $C_6H_5CH_2Br$ to C_6H_5CHO	poly(2-methyl-5-vinylpyridine *N*-oxide)	catalyst is reduced to poly(2-methyl-5-vinylpyridine)	372
reversible redox reaction of *N,N'*-ethylenebis(salicylideneiminato)cobalt(II) and *N,N'*-ethylenebis(salicylaldiminato)cobalt(II)	poly(4-vinylpyridine)		373
polymerization of *N*-carboxyanhydride of amino acids to polypeptides	poly(2-vinylpyridine)	local concentration effect if amino acid contains electron-acceptor groups	374
polymerization of acrylonitrile	poly(4-vinylpyridine), poly(5-ethyl-2-vinylpyridine) } + CCl_4	poly(4-vinylpyridine) is more active, whereas monomeric pyridine is inactive	375
polymerization of propylene	polypropylene grafted with vinylpyridine, or poly(2-vinylpyridine) + $(C_2H_5)_2AlCl$–$TiCl_4$	polymer catalyst is more active than pyridine	376

similar catalytic effect. However, since it is a polymeric catalyst, neighboring group effect, conformational as well as configurational effects, and problems of local concentration of reacting species might result in different reaction features not expected from the study of monomeric catalysts. The study of polymer catalysts as prototypes of enzymatic reactions is now attracting many researchers. From a practical viewpoint, fixation of catalytically active groups on insoluble polymer is advantageous for continuous operation in a factory. Examples of polyvinylpyridine as catalyst are listed in Table 11.

Uses as Poly(2-vinylpyridine *N*-Oxide). This polymer, obtained either by polymerization of 2-vinylpyridine *N*-oxide or by oxidation of poly(2-vinylpyridine) with hydrogen peroxide, is especially effective for treatment of silicosis (377–379). Among analogous polymers, only poly(2-vinylpyridine *N*-oxide) is effective, and there seems to be a correlation between the effectiveness of silicosis treatment and stability of the complex formed between polymer and silicic acid (380). The ineffective poly(4-vinylpyridine *N*-oxide) forms only a weak complex with silicic acid (381). Hydrogen bond formation between *N*-oxide and silicic acid was detected by NMR spectroscopy (392).

Miscellaneous Uses. Miscellaneous uses of polyvinylpyridine include the following: stabilizer or emulsifier in emulsion polymerization in acidic media (382, 383); copolymer of vinylhydroquinone as redox resin (384,385); coating material for electrophotography (386); film, mordant, or interlayer coating to prevent migration of dyestuff in photography (387,388); corrosion inhibitor of metal (389); coating on porous material used for hyperfiltration of salt water (390,391); and binder for rocket fuel.

Bibliography

1. C. R. Wagner (to Phillips Petroleum Co.), U.S. Pat. 2,732,376 (Jan. 24, 1956); *Chem. Abstr.* **50,** 9450 (1956).
2. J. E. Mahan (to Phillips Petroleum Co.), U.S. Pat. 2,769,811 (Nov. 6, 1956); *Chem. Abstr.* **51,** 7433 (1957).
3. F. Runge, G. Naumann, and M. Morgner (to VEB Farbenfabrik Wolfen), Ger. (East) Pat. 13,099 (April 25, 1957); *Chem. Abstr.* **53,** 5292 (1959).
4. F. Runge, G. Naumann, and M. Morgner (to VEB Farbenfabrik Wolfen), Brit. Pat. 828,205; *Chem. Abstr.* **54,** 13149 (1960).
5. M. I. Farberov, A. M. Kutin, B. F. Ustavshchikov, T. P. Vernova, and A. F. Frolov, *Zh. Prikl. Khim.* **34,** 632 (1961); *Chem. Abstr.* **55,** 17630 (1961).
6. Phillips Petroleum Co., Brit. Pat. 873,998 (Appl. May 6, 1959); *Chem. Abstr.* **56,** 7285 (1962).
7. G. Cevidalli, J. Herzenberg, and A. Nenz (to Sisedison Societa per Azioni), Ital. Pat. 596,924 (Aug. 11, 1959); *Chem. Abstr.* **57,** 12443 (1963).
8. T. Oga, *Koru Taru* **14,** 380 (1962).
9. S. Yoshioka, T. Ohmae, and N. Hasegawa, *Kogyo Kagaku Zasshi* **65,** 1995 (1962).
10. S. Yoshida, T. Ohmae, and M. Hasegawa (to Japan Synthetic Chemical Industries Ltd.), Japan. Pat. 6,536 (May 6, 1964); *Chem. Abstr.* **61,** 13289 (1964).
11. Y. Matsuda, Y. Nakahara, R. Kato, Y. Yasuda, Y. Moriyama, and Y. Ito (to Dainippon Celluloid Co. Ltd), Japan. Pat. 10,334 (Aug. 6, 1962); *Chem. Abstr.* **59,** 5140 (1963).
12. L. A. Burrows and G. H. Kalb (to E. I. du Pont de Nemours & Co., Inc.), U.S. Pat. 2,677,688 (May 4, 1954); *Chem. Abstr.* **51,** 15603 (1957).
13. G. B. Gechele, A. Nenz, C. Garbuglio, and S. Pietra, *Chim. Ind.* (*Milan*) **42,** 959 (1960).
14. M. Yoshida, *Yuki Gosei Kagaku Kyokai Shi* **16,** 571 (1958).
15. J. E. Mahan (to Phillips Petroleum Co.), U.S. Pat. 2,512,600 (June 27, 1950); *Chem. Abstr.* **44,** 9987 (1950).
16. K. Winterfeld and C. Heinen, *Chem. Ann.* **573,** 85 (1951).

17. E. Profft, *Chem. Tech. (Berlin)* **7,** 511 (1955).
18. J. Finkelstein and R. C. Elderfield, *J. Org. Chem.* **4,** 374 (1939).
19. G. B. Bachman and D. D. Micucci, *J. Am. Chem. Soc.* **70,** 2381 (1948).
20. M. Yoshida and H. Kumagae, *Kogyo Kagaku Zasshi* **59,** 196 (1956).
21. F. M. Strong, *J. Am. Chem. Soc.* **55,** 816 (1933).
22. R. L. Frank, *Ind. Eng. Chem.* **40,** 879 (1948).
23. J. E. Mahan (to Phillips Petroleum Co.), U.S. Pat. 2,534,285 (Dec. 19, 1950); *Chem. Abstr.* **45,** 3425 (1951).
24. L. F. Salisbury (to E. I. du Pont de Nemours & Co., Inc.), Brit. Pat. 632,661 (Nov. 28, 1949); *Chem. Abstr.* **44,** 4513 (1950).
25. H. L. Dimond, L. J. Fleckenstein, and M. O. Shrader (to Pittsburgh Coke & Chemical Co.), U.S. Pat. 2,848,456 (Aug. 19, 1958); *Chem. Abstr.* **53,** 1384 (1959).
26. S. Chrzczonowicz, J. Michalski, K. Studniarski, and H. Zajac, Polish Pat. 42,386 (Oct. 15, 1959); *Chem. Abstr.* **55,** 6501 (1961).
27. H. A. Iddles, E. H. Lang, and D. C. Gregg, *J. Am. Chem. Soc.* **59,** 1945 (1937).
28. *Beilsteins Handbuch der Organischen Chemie,* Vol. XX, Springer Verlag, Berlin, 1935, p. 170.
29. Yu. I. Chumakov and Yu. P. Shapovalova, *Metody Polucheniya Khim. Reaktivov i Preparatov, Gos. Kom. Sov. Min. SSSR po Khim.* **1964,** 43; *Chem. Abstr.* **64,** 15832 (1966).
30. Yu. I. Chumakov and Yu. P. Shapovalova, *Zh. Organ. Khim.* **1,** 940 (1965); *Chem. Abstr.* **63,** 6960 (1965).
31. W. E. Doering, *J. Am. Chem. Soc.* **69,** 2461 (1949).
32. C. F. Woodward, *J. Am. Chem. Soc.* **66,** 911 (1944).
33. G. J. Janz and N. E. Duncan, *J. Am. Chem. Soc.* **75,** 5389 (1953).
34. G. J. Janz and N. E. Duncan, *Nature* **171,** 933 (1953).
35. C. H. Jarboe and C. J. Rosene, *J. Chem. Soc.* **1961,** 2455.
36. K. Matsuzaki and T. Sugimoto, *J. Polymer Sci.* [A-2] **5,** 1320 (1967).
37. H. S. Haas (to Polaroid Corp.), U.S. Pat. 3,407,186 (Oct. 22, 1968); *Chem. Abstr.* **70,** 12141 (1969).
38. F. E. Cislak (to Reilly Tar & Chemical Corp.), U.S. Pat. 2,749,349 (June 5, 1956); *Chem. Abstr.* **51,** 4442 (1957).
39. T. Tamikado, T. Sakai, and K. Sagisawa, *Makromol. Chem.* **50,** 244 (1962).
40. D. D. Reynolds and T. T. M. Laakso (to Eastman Kodak Co.), U.S. Pat. 2,725,381 (Nov. 29, 1955).
41. I. N. Duling and C. C. Price, *J. Am. Chem. Soc.* **84,** 578 (1962).
42. E. Schmidt, *Arch. Pharm.* **251,** 183 (1913).
43. Y. Okamoto and D. Alia, *Chem. Ind. (London)* **1964,** 1311.
44. N. N. Dykhanov and T. S. Ryzhkova, *Metody Polucheniya Khim. Reaktivov i Preparatov, Gos. Kom. Sov. Min. SSSR Khim. Sb.* **1964,** 104; *Chem. Abstr.* **64,** 12638 (1966).
45. W. L. Smith, M. F. Potts, and P. S. Hudson (to Phillips Petroleum Co.), U.S. Pat. 2,874,159 (Feb. 17, 1959); *Chem. Abstr.* **53,** 12310 (1959).
46. K. Saruwatari and M. Matsushima (to Yoshitomi Drug Manufacturing Co.), Japan. Pat. 5,872 (July 4, 1959); *Chem. Abstr.* **54,** 14274 (1960).
47. A. Nenz and G. B. Gechele (to Sisedison Societa per Azioni), Ital. Pat. 601,911 (Feb. 16, 1960); *Chem. Abstr.* **55,** 12934 (1961).
48. G. B. Gechele, A Nenz, and G. Barberis (to Sisedison Societa per Azioni), Ital. Pat. 601,912 (Feb. 19, 1960); *Chem. Abstr.* **55,** 12934 (1961).
49. J. T. Dunn and D. T. Manning (to Union Carbide Corp.), U.S. Pat. 3,158,615 (Nov. 24, 1964); *Chem. Abstr.* **62,** 11972 (1965).
50. P-T. Li, H-F. Kao, and C-Y. Kung, *Communist Chinese Sci. Abstr.* **88,** 65 (1965); *Chem. Abstr.* **63,** 14984 (1965).
51. J. J. Arrigo (to Universal Oil Products Co.), U.S. Pat. 3,230,225 (Jan. 18, 1966); *Chem. Abstr.* **64,** 9838 (1966).
52. Copolymer Rubber and Chemical Corp., Neth. Pat. Appl. 6,511,747 (March 11, 1966); *Chem. Abstr.* **65,** 10697 (1966).
53. M. I. Farberov, B. F. Ustavshchikov, A. M. Kutin, T. B. Vernova, and E. V. Yorosh, *Izv. Vysshikh Uchebn. Zavedenii Khim. i Khim. Tekhnol. 1958* **5,** 92; *Chem. Abstr.* **53,** 11364 (1959).
54. A. I. Dukhovnaya, *Gigiena Truda i Prof. Zabolevaniya* **10,** 9 (1966); *Chem. Abstr.* **65,** 4523 (1966).

55. R. L. Frank, C. E. Adams, J. B. Blegen, P. V. Smith, A. E. Jure, C. H. Schroeder, and M. M. Goff, *Ind. Eng. Chem.* **40,** 879 (1948).
56. R. H. Linnell, *J. Org. Chem.* **25,** 290 (1960).
57. B. M. Kuindzki, L. D. Gluzman, M. A. Zepalova, R. M. Tsip, A. D. Val'kova, and A. A. Rok, U.S.S.R. Pat. 135,488 (Feb. 15, 1961); *Chem. Abstr.* **55,** 16571 (1961).
58. G. Favini, *Gazz. Chim. Ital.* **93,** 635 (1963); *Chem. Abstr.* **59,** 13457 (1963).
59. A. A. Rok and L. D. Gluzman, *Sb. Nauchn. Tr., Ukr. Nauchn. Issled. Uglekhim. Inst.* **1965,** 144; *Chem. Abstr.* **65,** 16818 (1966).
60. Z. Yu. Kokoshko, V. G. Kitaese, Z. V. Pushkareva, and V. E. Blokhin, *Zh. Obshch. Khim.* **37,** 58 (1967); *Chem. Abstr.* **66,** 89338 (1967).
61. D. H. White and J. M. Folz (to Phillips Petroleum Co.), U.S. Pat. 2,962,498 (Nov. 29, 1960); *Chem. Abstr.* **55,** 10478 (1961).
62. Ref. 28, p. 256.
63. F. Bohlmann, E. Winterfeldt, P. Studt, H. Laurent, G. Boroschewski, and K. M. Klein, *Chem. Ber.* **94,** 3151 (1961).
64. R. Bodalski, J. Michalski, and K. Studniarski, *Roczniki Chem.* **40,** 1505 (1966); *Chem. Abstr.* **66,** 94890 (1967).
65. H. F. Kauffman (to Allied Chemical & Dye Corp.), U.S. Pat. 2,556,845 (June 12, 1951); *Chem. Abstr.* **46,** 1456 (1952).
66. K. S. N. Prasad and R. Raper, *J. Chem. Soc.* **1956,** 217.
67. J. Michalski and K. Studniarski, *Roczniki Chem.* **29,** 1141 (1955); *Chem. Abstr.* **50,** 12044 (1956).
68. M. Maruoka, K. Isagawa, and Y. Fushizaki, *Kobunshi Kagaku* **18,** 751 (1961).
69. K. S. N. Prasad and R. Raper, *J. Chem. Soc.* **1956,** 217.
70. B. M. Kuindzhi, M. A. Zepalova, L. D. Gluzman, A. K. Va'kova, R. M. Tsin, I. V. Zaitseva, and A. A. Rokk, *Metody Polucheniya Khim. Reaktivov Prep.* **15,** 93 (1967); *Chem. Abstr.* **68,** 114378 (1968).
71. G. B. Gechele and S. Pietra, *J. Org. Chem.* **26,** 4412 (1961).
72. G. F. Smith and J. T. Wróbel, *J. Chem. Soc.* **1960,** 1463.
73. D. M. Haskell (to Phillips Petroleum Co.), U.S. Pat. 2,716,118 (Aug. 23, 1955); *Chem. Abstr.* **50,** 5770 (1956).
74. E. Sh. Kagan and B. I. Ardashev, *Khim. Geterotsikl. Soedin.* **1968,** 1066; *Chem. Abstr.* **70,** 77740 (1969).
75. R. W. Sudhoff (to Monsanto Co.), U.S. Pat. 3,344,143 (Sept. 26, 1967); *Chem. Abstr.* **68,** 87168 (1968).
76. L. N. Yakhontov and M. V. Rubtsov, *Zh. Obshch. Khim.* **31,** 3281 (1961); *Chem. Abstr.* **57,** 3426 (1962).
77. M. V. Rubtsov and L. N. Yakhontov, *Zh. Obshch. Khim.* **25,** 1820 (1955); *Chem. Abstr.* **50,** 7107 (1956).
78. R. Bodalski, J. Michalski, and K. Studniarski, *Roczniki Chem.* **38,** 1337 (1964); *Chem. Abstr.* **62,** 1627 (1965).
79. T. Tamikado, *Tokyo Kogyo Shikensho Hokoku* **57,** 241 (1962).
80. V. Boekelheide and R. Sharrer, *J. Org. Chem.* **26,** 3802 (1961).
81. R. W. Martin, *Ann.* **21,** 921 (1949).
82. J. T. Hays (to Hercules Powder Co.), U.S. Pat. 2,611,769 (Sept. 23, 1952).
83. B. Philipp and U. Bartels, *Acta Chim. Acad. Sci. Hung.* **32,** 19 (1962); *Chem. Abstr.* **58,** 933 (1963).
84. H. Kamio, M. Nishikawa, and T. Kanzawa, *Bunseki Kagaku* **10,** 851 (1961).
85. M. Yoshida, *Kogyo Kagaku Zasshi* **63,** 893 (1960).
86. Y. Takayama, *Kogyo Kagaku Zasshi* **62,** 658 (1959).
87. J. Smid and M. Szwarc, *J. Polymer Sci.* **61,** 31 (1962).
88. C. L. Lee, J. Smid, and M. Szwarc, *Trans. Faraday Soc.* **59,** 1192 (1963).
89. M. Fontanille and P. Sigwalt, *Bull. Soc. Chim. France* **1967,** 4095.
90. M. Fontanille and P. Sigwalt, *Bull. Soc. Chim. France* **1967,** 4083.
91. D. Laurin and G. Parravano, *J. Polymer Sci.* [A-1] **6** 1047 (1968).
92. K. Kuwata, H. Kawazura, and K. Hirota, *Nippon Kagaku Zasshi* **81,** 1770 (1960).
93. M. Fontanille and P. Sigwalt, *Compt. Rend. Ser. C* **263,** 624 (1966).
94. B. L. Erusalimskii and Yu. N. Ovsyannikov, *Vysokomolekul. Soedin.* **B10,** 725 (1968); *Chem. Abstr.* **70,** 20384 (1969).

95. G. Natta, G. Mazzanti, G. Dall'Asta, and P. Longi, *Makromol. Chem.* **37,** 160 (1960).
96. G. Natta, G. Mazzanti, P. Longi, G. Dall'Asta, and F. Bernardini, *J. Polymer Sci.* **51,** 487 (1961).
97. G. Natta, G. Mazzanti, P. Longi, G. Dall'Asta, and F. Bernardini (to Montecatini Societa Generale per l'Industria Mineraria e Chimica), Ger. Pat. 1,114,638 (Oct. 5, 1961); *Chem. Abstr.* **57,** 1076 (1962).
98. M. Tardi, D. Rouge, and P. Sigwalt, *European Polymer J.* **3,** 85 (1967).
99. G. Natta, P. Longi, and U. Nordio, *Makromol. Chem.* **83,** 161 (1965).
100. P. Longi, U. Nordio, and E. Pellino (to Montecatini), Ital. Pat. 726,445 (Nov. 15, 1966); *Chem. Abstr.* **69,** 36605 (1968).
101. J. C. MacKenzie and A. Orchechowski (to Cabot Corp.), U.S. Pat. 3,285,892 (Nov. 15, 1966); *Chem. Abstr.* **66,** 19020 (1967).
102. P. P. Spiegelman and G. Parravano, *J. Polymer Sci.* [A] **2,** 2245 (1964).
103. D. Lanrin and G. Parravano, *J. Polymer Sci.* [B] **4,** 797 (1966).
104. F. Leavitt, V. Stannett, and M. Szwarc, *J. Polymer Sci.* **31,** 193 (1958).
105. W. I. Bengough and W. Henderson, *Trans. Faraday Soc.* **61,** 141 (1965).
106. P. F. Onyon, *Trans. Faraday Soc.* **51,** 400 (1955).
107. A. F. Revzin and Kh. S. Bagdasar'yan, *Zh. Fiz. Khim.* **38,** 1020 (1964); *Chem. Abstr.* **61,** 3199 (1964).
108. A. V. Chernobai, Zh. S. Tirakyants, and R. Ya. Delyatitskaya, *Vysokomolekul. Soedin.* **A9,** 664 (1967); *Chem. Abstr.* **67,** 22260 (1967).
109. R. M. Joshi, *Makromol. Chem.* **55,** 35 (1962).
110. R. M. Joshi, *J. Polymer Sci.* **56,** 313 (1962)
111. V. G. Ostroverkov, I. S. Vakarchuk, and V. G. Sinyavskii, *Vysokomolekul. Soedin.* **3,** 1197 (1961); *Chem. Abstr.* **56,** 8921 (1962).
112. V. G. Karkozov, R. K. Gavurina, V. S. Polonskii, and A. I. Smirnova, *Vysokomolekul. Soedin.* **A10,** 1343 (1968); *Chem. Abstr.* **69,** 44217 (1968).
113. C. R. Frink, *Frontiers Plant Sci. (Conn. Agr. Exptl. Station, New Haven)* **15,** 5 (1963).
114. M. A. Zharkova, E. A. Rassolova, G. I. Kudryavtsev, and V. S. Klimenkov, *Khim. Volokna* **1961,** 13; *Chem. Abstr.* **56,** 3630 (1962).
115. S. Yuguchi, H. Kiuchi, and M. Watanabe, *Kobunshi Kagaku* **18,** 510 (1961).
116. J. E. Pritchard, M. H. Opheim, and P. H. Moyer, *Ind. Eng. Chem.* **47,** 863 (1955).
117. E. B. Fitzgerald and R. M. Fuoss, *Ind. Eng. Chem.* **42,** 1603 (1950).
118. A. Katchalsky, K. Rosenheck, and B. Altman, *J. Polymer Sci.* **23,** 955 (1957).
119. L. Crescentini, G. B. Gechele, and M. Pizzoli, *European Polymer J.* **1,** 293 (1965).
120. W. D. Harkins, *J. Chem. Phys.* **13,** 381 (1945).
121. W. D. Harkins, *J. Am. Chem. Soc.* **69,** 1428 (1947).
122. W. D. Harkins, *J. Polymer Sci.* **5,** 217 (1950).
123. W. V. Smith and R. H. Ewart, *J. Chem. Phys.* **16,** 592 (1948).
124. W. V. Smith, *J. Am. Chem. Soc.* **70,** 3695 (1948).
125. K. Ueno and S. Abe, *Kogyo Kagaku Zasshi* **68,** 401 (1965).
126. S. Fujioka, K. Hayashi, and S. Okamura, *Nippon Hoshasen Kobunshi Kenkyu Kyokai Nempo* **4,** 199 (1962).
127. Y. Tabata, H. Kitano, and H. Sobue, *J. Polymer Sci.* [A] **2,** 3639 (1964).
128. R. G. Hodgdon, Jr. (to Monsanto Research Corp.), U.S. Pat. 3,239,494 (March 8, 1966); *Chem. Abstr.* **64,** 16011 (1966).
129. M. F. Potts and P. F. Hudson (to Phillips Petroleum Co.), U.S. Pat. 2,767,159 (Oct. 16, 1956); *Chem. Abstr.* **51,** 4054 (1957).
130. Chi-Hua Wang, *Chem. Ind. (London)* **1964,** 751.
131. S. Tazuke and S. Okamura, *Polymer Letters* **3,** 135 (1965).
132. S. Tazuke and S. Okamura, *J. Polymer Sci.* [A-1] **4,** 141 (1966).
133. S. Tazuke, K. Nakagawa, and S. Okamura, *Polymer Letters* **3,** 923 (1965).
134. M. Matsuda, Y. Ishioroshi, and T. Hirayama, *J. Polymer Sci.* [B] **4,** 815 (1966).
135. J. E. Pritchard (to Phillips Petroleum Co.), U.S. Pat. 2,862,902 (Dec. 2, 1958); *Chem. Abstr.* **53,** 7670 (1959).
136. V. A. Kargin, V. A. Kabanov, K. V. Aliev, and E. F. Razvodovskii, *Dokl. Akad. Nauk SSSR* **160,** 604 (1965).
137. V. A. Kabanov, T. I. Patrikeeva, and V. A. Kargin, *Dokl. Akad. Nauk SSSR* **168,** 1350 (1966).

138. V. A. Kabanov, T. L. Patrikeeva, O. V. Kargina, and V. A. Kargin, *J. Polymer Sci.* [C] **23,** 357 (1966).
139. T. I. Patrikeeva, T. E. Nechaeva, M. I. Mustafaev, V. A. Kabanov, and V. A. Kargin, *Vysokomolekul. Soedin.* **A9,** 332 (1967).
140. V. A. Kabanov and V. A. Petrovskaya, *Vysokomolekul. Soedin.* **B10,** 797 (1968).
141. V. A. Kabanov, K. V. Aliev, and V. A. Kargin, *Vysokomolekul. Soedin.* **A10,** 1618 (1968).
142. V. A. Kabanov, T. I. Patrikeeva, O. V. Kargina, and V. A. Kargin, *Intern. Symp. Macromol. Chem. Tokyo* **1966,** preprint 2-2-09.
143. S. Iwatsuki, T. Kokubo, K. Motomatsu, M. Tsuji, and Y. Yamashita, *Makromol. Chem.* **120,** 154 (1968).
144. V. A. Kargin, V. A. Kabanov, and O. V. Kargina, *Dokl. Akad. Nauk SSSR* **161,** 1131 (1965).
145. O. V. Kargina, I. V. Adorova, V. A. Kabanov, and V. A. Kargin, *Dokl. Akad. Nauk SSSR* **170,** 1130 (1966).
146. O. V. Kargina, V. A. Kabanov, and V. A. Kargin, *J. Polymer Sci.* [C] **22,** 339 (1967).
147. O. V. Kargina, M. V. Ul'yanova, V. A. Kabanov, and V. A. Kargin, *Vysokomolekul. Soedin.* **A9,** 340 (1967).
148. V. A. Kabanov, V. A. Petrovskaya, and V. A. Kargin, *Vysokomolekul. Soedin.* **A10,** 925 (1968).
149. H. Morawetz, *Svensk Kem. Tidskr.* **79,** 309 (1967).
150. S. Tazuke, N. Sato, and S. Okamura, *J. Polymer Sci.* [A-1] **4,** 2461 (1966).
151. S. Tazuke and S. Okamura, *J. Polymer Sci.* [A-1] **5,** 1083 (1967).
152. S. Tazuke, K. Shimada, and S. Okamura, *J. Polymer Sci.* [A-1] **7,** 879 (1969).
153. S. Tazuke and S. Okamura, *Polymer Letters* **5,** 95 (1967).
154. S. Tazuke, K. Tsuji, T. Yonezawa, and S. Okamura, *J. Phys. Chem.* **71,** 2957 (1967).
155. S. Tazuke, T. Okada, and S. Okamura, *Makromol. Chem.* **115,** 213 (1968).
156. H. Z. Friedlander, *Polymer Preprints* **4,** 300 (1963).
157. H. Z. Friedlander and C. R. Frink, *J. Polymer Sci.* [B] **2,** 475 (1964).
158. H. H. Perkampus and U. Krueger, *Ber. Bunsenges. Phys. Chem.* **71,** 439 (1967).
159. N. C. Yang and Y. Gaoni, *J. Am. Chem. Soc.* **86,** 5022 (1964).
160. C. F. Hauser and N. L. Zutty, *Polymer Preprints* **8,** 369 (1967); *Chem. Abstr.* **67,** 11777 (1967).
161. G. E. Ham, ed., *Copolymerization*, Interscience Publishers, a division of John Wiley & Sons, Inc., New York, 1964.
162. R. C. Schulz, E. Kaiser, and E. Kern, *Makromol. Chem.* **58,** 160 (1962).
163. Y. Iwakura, T. Tamikado, M. Yamaguchi, and K. Takei, *J. Polymer Sci.* **39,** 203 (1959).
164. S. Yuguchi and M. Watanabe, *Makromol. Chem.* **49,** 243 (1961).
165. M. M. Koton, *J. Polymer Sci.* **30,** 331 (1958).
166. M. M. Koton, *Chem. Abstr.* **55,** 16546g (1961).
167. W. M. Thomas and M. T. O'Shaughnessy, *J. Polymer Sci.* **11,** 455 (1953).
168. S. L. Aggarwal and F. A. Long, *J. Polymer Sci.* **11,** 127 (1953).
169. T. Alfrey, Jr., and B. Magel, in T. Alfrey, Jr., J. Bohrer, and H. F. Mark, eds., quoted in *Copolymerization*, Vol. VIII in *High Polymer* Series, Interscience Publishers, Inc., New York, 1952.
170. G. F. D'Alelio and R. J. Caiola, *J. Polymer Sci.* [A-1] **5,** 287 (1967).
171. G. B. Butler, G. Vanhaeren, and M. F. Ramadier, *J. Polymer Sci.* [A-1] **5,** 1265 (1967).
172. G. van Paesschen and G. Smets, *Bull. Soc. Chim. Belges* **64,** 173 (1955).
173. C. C. Price and T. F. McKeon, **41,** 445 (1959).
174. T. Alfrey, Jr., and H. Morawetz, *J. Am. Chem. Soc.* **74,** 436 (1952).
175. A. V. Ryabov, Yu. D. Somchikov, and N. N. Slavnitskaya, *Tr. po Khim. i Khim. Tekhnol.* **1963,** 334; *Chem. Abstr.* **61,** 5771 (1964).
176. T. Tamikado, *J. Polymer Sci.* **43,** 489 (1960).
177. C. Walling, E. R. Briggs, and K. B. Wolfstirn, *J. Am. Chem. Soc.* **70,** 1543 (1948).
178. T. Alfrey, Jr., J. Bohrer, H. Haas, and C. Lewis, *J. Polymer Sci.* **5,** 719 (1950).
179. C. C. Price and C. E. Greene, *J. Polymer Sci.* **6,** 111 (1951).
180. C. G. Overberger and F. W. Michelotti, *J. Am. Chem. Soc.* **80,** 988 (1958).
181. C. Walling, E. R. Briggs, and K. B. Wolfstirn, *J. Am. Chem. Soc.* **70,** 1543 (1948).
182. A. V. Ryabov, Yu. D. Semchikov, N. N. Slavnitskaya, and V. N. Vakhrusheva, *Dokl. Akad. Nauk SSSR* **154,** 1135 (1964); *Chem. Abstr.* **60,** 13324 (1964).
183. A. V. Ryabov, Yu. D. Semchikov, and V. N. Vakhrusheva, *Tr. po Khim. i Khim. Tekhnol.* **1963,** 188; *Chem. Abstr.* **60,** 9448 (1964).

184. W. J. Burlant and D. Green, *J. Polymer Sci.* **31**, 227 (1960).
185. B. L. Funt and E. A. Ogryzlo, *J. Polymer Sci.* **25**, 279 (1957).
186. R. M. Fuoss and G. I. Cathers, *J. Polymer Sci.* **4**, 97 (1949).
187. C. G. Overberger, H. Biletch, and R. G. Nickerson, *J. Polymer Sci.* **27**, 381 (1958).
188. K. Matsuoka, M. Otsuka, K. Takemoto, and M. Imoto, *Kogyo Kagaku Zasshi* **69**, 137 (1966).
189. H. Kamogawa and H. G. Cassidy, *J. Polymer Sci.* [A] **1**, 1971 (1963).
190. Ref. 161, cited as private communication.
191. T. Tamikado, *Makromol. Chem.* **38**, 85 (1960).
192. T. Yamamoto, *Kogyo Kagaku Zasshi* **62**, 476 (1959).
193. V. L. Tsailingol'd, M. I. Farberov, and G. A. Burgova, *Chem. Abstr.* **54**, 5157c (1960).
194. Pin Ts'ai Li and K'ai-Kuo Wu, *K'o Hsueh Ch'u Pau She* **1963**, 151.
195. C. Aso and M. Sogabe, *Kogyo Kagaku Zasshi* **68**, 1970 (1965).
196. C. Aso and T. Nawata, *Kogyo Kagaku Zasshi* **68**, 549 (1965).
197. E. V. Kuznetsov, I. P. Prokhorova, and N. I. Avvakumova, *Tr. Kazansk. Khim.-Tekhnol. Inst.* **36**, 411 (1967); *Chem. Abstr.* **69**, 97219 (1968).
198. T. Tomikado, *Makromol. Chem.* **38**, 85 (1960).
199. V. D. Bezuglyi and T. A. Alekseeva, *Ukr. Khim. Zh.* **31**, 392 (1965); *Chem. Abstr.* **63**, 5750 (1965).
200. B. K. Basov, V. L. Tsailingol'd, and E. G. Lazaryant, *Prom. Sin. Kauch.* **1**, 17 (1967); *Chem. Abstr.* **69**, 77783 (1968).
201. I. Sakurada, *Kobunshi Kagaku* **18**, 496 (1961).
202. V. G. Ostroverkhov, I. S. Vakarchuk, and V. G. Sinyavskii, *Chem. Abstr.* **56**, 8921d (1962).
203. A. B. Dabankov, L. B. Znbakova, and A. A. Gurov, *Vysokomolekul. Soedin.* **6**, 235 (1964); *Chem. Abstr.* **60**, 14680 (1964).
204. S. L. Aggarwal and F. A. Long, *J. Polymer Sci.* **11**, 127 (1953).
205. F. Ida, K. Uemura, and S. Abe, *Kagaku To Kogyo (Osaka)* **39**, 565 (1965).
206. L. Crescentini, G. B. Gechele, and A. Zanella, *J. Appl. Polymer Sci.* **9**, 1323 (1965).
207. V. S. Lebedev and R. K. Gavurina, *Vysokomolekul. Soedin.* **6**, 1161 (1964); *Chem. Abstr.* **61**, 13441 (1964).
208. T. Ida, S. Takahashi, and I. Utsumi, *J. Vitaminol. (Kyoto)* **9**, 269 (1963); *Chem. Abstr.* **61**, 15931 (1964).
209. M. G. Avetyan, E. G. Darbinyan, and S. G. Matsoyan, *Izv. Akad. Nauk Arm. SSR, Khim. Nauk* **16**, 247 (1963); *Chem. Abstr.* **60**, 666 (1964).
210. S. Okamura and K. Uno, *Kobunshi Kagaku* **8**, 467 (1951).
211. A. B. Gosnell, J. A. Gervasi, D. K. Woods, and V. Stannett, *J. Polymer Sci.* [C] **22**, 611 (1969).
212a. A. Dondos and P. Rempp, *Compt. Rend. Acad. Sci., Paris, Ser. C* **264**, 869 (1967).
212b. M. Fontanille and P. Sigwalt, *Bull. Soc. Chim. France* **1967**, 4087.
213. V. Stannett, J. L. Williams, A. B. Gosnell, and J. A. Gervasi, *J. Polymer Sci.* [B] **6**, 185 (1968).
214. M. Yanagita, H. Kawabe, K. Shinohara, and T. Takamatsu, *Sci. Papers Inst. Phys. Chem. Res. (Tokyo)* **56**, 218 (1962).
215. T. V. Druzhinina, A. A. Konhin, and A. I. Birger, *Khim. Volokna* **1968**, 61; *Chem. Abstr.* **69**, 37905 (1968).
216. K. Sadakata, K. Ito, K. Kimura, T. Kishino, and Y. Nakamura (to Mitsubishi Rayon Co.), Japan. Pat. 6,818,144 (Aug. 1, 1968); *Chem. Abstr.* **70**, 29726 (1969).
217. G. Odor and F. Geleji, *Magy. Kem. Lapja* **17**, 221 (1962); *Chem. Abstr.* **57**, 11411 (1962).
218. M. Umezawa, K. Hirota, and Z. Takezaki, *Nippon Isotope Kaigi Hobunshu* **4**, 356 (1961).
219. A. Chapiro and P. Seidler, *European Polymer J.* **1**, 189 (1965).
220. G. Bex, A. Chapiro, M. B. Huglin, A. M. Jendrychowska-Bonamour, and T. O'Neil, *J. Polymer Sci.* [C] **22**, 493 (1967).
221. American Machine & Foundry Co., Brit. Pat. 952,452 (March 18, 1964); *Chem. Abstr.* **61**, 796 (1964).
222. T. Takamatsu, *Rika Gaku Kenkyusho Hokoku* **37**, 1 (1961).
223. G. G. Odian, A. Rossi, E. Ratchik, and T. Acker, *J. Polymer Sci.* **54**, S11 (1961).
224. M. R. Houlton and J. K. Thomas, *Intern. J. Appl. Radiation Isotopes* **11**, 45 (1961).
225. J. A. W. Sykes and J. K. Thomas, *J. Polymer Sci.* **55**, 721 (1961).
226. Kh. U. Usmanov, U. A. Azizov, and M. U. Sadykov, *Radiats. Khim. Polim. Mater. Simp., Moscow* **1964**, 153; *Chem. Abstr.* **66**, 96044 (1967).

227. G. S. Kolesnikov, A. S. Tevlina, S. E. Vasyukov, and V. S. Smirnov, *Plasticheskie Massy* **4,** 22 (1968); *Chem. Abstr.* **69,** 19914 (1968).
228. F. Ida and K. Uemura, *Kagaku To Kogyo (Osaka)* **35,** 326 (1961).
229. K. Uemura and F. Ida, *Kagaku To Kogyo (Osaka)* **35,** 374 (1961).
230. R. J. E. Cumberbirch and J. R. Holker, *J. Soc. Dyers Colourists* **82,** 59 (1966).
231. C. Aso, T. Kunitake, and S. Ushio, *Kobunshi Kagaku* **19,** 734 (1962).
232. R. Kavalyunas, G. D. Shershneva, R. M. Livshits, and Z. A. Rogovin, *Vysokomolekul. Soedin.* **8,** 240 (1966); *Chem. Abstr.* **64,** 19963 (1966).
233. Yu. G. Kryazhev and Z. A. Rogovin, *Vysokomolekul. Soedin.* **6,** 672 (1964); *Chem. Abstr.* **61,** 3206 (1964).
234. Yu. G. Kryazhev and Z. A. Rogovin, *Zh. Vses. Khim. Obshchestva im. D. I. Mendeleeva* **8,** 118 (1963); *Chem. Abstr.* **58,** 14130 (1963).
235. E. Franta and P. Rempp, *Compt. Rend.* **254,** 674 (1962).
236. H. Takida and K. Noro, *Kobunshi Kagaku* **21,** 459 (1964).
237. H. Takida and K. Noro, *Kobunshi Kagaku* **22,** 717 (1965).
238. M. Fontanille and P. Sigwalt, *Compt. Rend.* **251,** 2947 (1960).
239. G. Champetier, M. Foutanille, A. C. Korn, and P. Sigwalt, *J. Polymer Sci.* **58,** 911 (1962).
240. K. Noro, H. Kawazura, T. Moriyama, and S. Yoshioka, *Makromol. Chem.* **83,** 35 (1965).
241. P. Rempp, H. Benoit, and P. Weiss (to Centre National de la Recherche Scientifique), Fr. Pat. 1,552,263 (Jan. 3, 1969); *Chem. Abstr.* **71,** 39754 (1969).
242. T. Shiota, Y. Goto, and K. Hayashi, *J. Appl. Polymer Sci.* **11,** 773 (1967).
243. G. Gontiere and J. Gole, *Bull. Soc. Chim. France* **1965,** 153.
244. G. Gontiere, J. P. Leonetti, and J. Gole, *Compt. Rend.* **257,** 2485 (1963).
245. G. Greber and G. Egle, *Makromol. Chem.* **59,** 174 (1963).
246. S. Sasson and A. Zilkha, *European Polymer J.* **5,** 369 (1969).
247. G. Greber and G. Egle, *Makromol. Chem.* **62,** 196 (1963).
248. A. Dondos and P. Rempp, *Compt. Rend.* **256,** 4443 (1963).
249. A. Dondos, *Bull. Soc. Chim. France* **1963,** 2762.
250. A. Dondos and P. Rempp, *Bull. Soc. Chim. France* **1962,** 2313.
251. A. Dondos, *Bull. Soc. Chim. France* **1967,** 910.
252. G. Greber and G. Egle, *Makromol. Chem.* **54,** 136 (1962).
253. J. A. Gervasi, A. B. Gosnell, D. K. Woods, and V. Stannett, *J. Polymer Sci.* [A-1] **6,** 859 (1968).
254. J. A. Gervasi, A. B. Gosnell, and V. Stannett, *Polymer Preprints* **8,** 785 (1967).
255. S. Arichi, *Bull. Chem. Soc. Japan* **41,** 244 (1968).
256. G. Geuskens, J. C. Lubikulu, and C. David, *Polymer* **1,** 63 (1966).
257. K. Matsuzaki and S. Sugimoto, *J. Polymer Sci.* [A-2] **5,** 1320 (1967).
258. G. Weill and G. Hermann, *J. Polymer Sci.* [A-2] **5,** 1293 (1967).
259. P. Longi, I. W. Bassi, F. Greco, and M. Cambini, *Tetrahedron Letters* **1964,** 995.
260. S. Arichi, H. Matsuura, Y. Tanimoto, and H. Murota, *Bull. Chem. Soc. Japan* **39,** 434 (1966).
261. A. J. Hyde and R. B. Taylor, *Polymer* **4,** 1 (1963).
262. S. Arichi, *J. Sci. Hiroshima Univ., Ser. A-II* **29,** 97 (1965).
263. S. Arichi, *Bull. Chem. Soc. Japan* **41,** 548 (1968).
264. C. Loucheux and Z. Czlonkowska, *J. Polymer Sci.* [C] **16,** 4001 (1965).
265. J. B. Berkowitz, M. Yamin, and R. M. Fuoss, *J. Polymer Sci.* **28,** 69 (1958).
266. A. G. Boyes and U. P. Strauss, *J. Polymer Sci.* **22,** 463 (1956).
267. C. Garbuglio, L. Crescentini, A. Mula, and G. B. Gechele, *Makromol. Chem.* **97,** 97 (1966).
268. M. Miura, Y. Kubota, and T. Masuzukawa, *Bull. Chem. Soc. Japan* **38,** 316 (1965).
269. H. Sato and T. Yamamoto, *Nippon Kagaku Zasshi* **80,** 1393 (1949).
270. A. V. Topchiev, M. M. Kusakov, G. D. Kalyazhnaya, N. N. Kapstov, A. Yu. Koshevnik, and E. A. Razumovskaya, *Neftekhimiya* **3** (1), 90 (1963).
271. J. H. Lupinski, K. D. Kopple, and J. J. Hertz, *J. Polymer Sci.* [C] **16,** 1561 (1967).
272. H. Inoue, Y. Kida, M. Yamamoto, M. Okawara, and E. Imoto, *Kogyo Kagaku Zasshi* **69,** 774 (1966).
273. A. Mizoguchi, H. Moriga, T. Shimizu, and Y. Amano, *Natl. Tech. Rept. (Matsushita Elec. Ind, Co. Osaka)* **9,** 407 (1963).
274. L. R. Melby, R. J. Harder, W. R. Hertler, W. Mahler, R. S. Benson, and W. E. Mochel, *J. Am. Chem. Soc.* **84,** 3374 (1962).

275. K. Yagi and M. Hanai, *Kogyo Kagaku Zasshi* **69,** 881 (1966).
276. J. H. Lupinski and K. D. Kopple, *Science* **146,** 1038 (1964).
277. H. Nomori, M. Hatano, and S. Kambara, *Kogyo Kagaku Zasshi* **67,** 1600 (1964).
278. A. Berlin, A. I. Sherle, and N. A. Markova, *Vysokomolekul. Soedin.* **B11,** 21 (1969); *Chem. Abstr.* **70,** 88412 (1969).
279. S. B. Mainthia, P. L. Kronick, and M. M. Labes, *J. Chem. Phys.* **41,** 2206 (1964).
280. W. Slough, *Trans. Faraday Soc.* **55,** 1030 (1959).
281. H. Sugiyama and H. Kamogawa, *J. Polymer Sci.* [A-1] **4,** 2281 (1966).
282. A. V. Topchiev, N. N. Kaptsov, G. D. Kulyuzhnaya, A. I. Mityaeva, and I. E. Balitskaya, *Dokl. Akad. Nauk SSSR* **143,** 621 (1962); *Chem. Abstr.* **57,** 3621 (1962).
283. S. Smets, V. Balogh, and Y. Castille, *J. Polymer Sci.* [C] **4,** 1467 (1964).
284. S. Okamura, S. Tazuke, and Y. Kozima (to Research Institute for Production Development). Japan. Pat. 41-14672 (Aug. 18, 1966).
285. G. R. Seely, *J. Phys. Chem.* **71,** 2091 (1967).
286. H. Morawetz and N. L. Hughes, Jr., *J. Phys. Chem.* **56,** 64 (1952).
287. D. Bach and I. R. Miller, *Biochim. Biophys. Acta* **114,** 311 (1966).
288. I. R. Miller, *Biochim. Biophys. Acta* **103,** 219 (1965).
289. S. Tazuke, unpublished results.
290. L. Y. Chow and R. M. Fuoss, *J. Polymer Sci.* **27,** 569 (1958).
291. C. B. Arends, *J. Chem. Phys.* **39,** 1903 (1963).
292. R. M. Fuoss and U. P. Strauss, *J. Polymer Sci.* **3,** 246 (1948).
293. R. M. Fuoss, *J. Polymer Sci.* **3,** 603 (1948).
294. R. M. Fuoss and G. I. Cathers, *J. Polymer Sci.* **4,** 97 (1949).
295. R. M. Fuoss and W. N. Maclay, *J. Polymer Sci.* **6,** 305 (1951).
296. W. N. Maclay and R. M. Fuoss, *J. Polymer Sci.* **6,** 511 (1951).
297. F. T. Wall, J. J. Ondrejcin, and M. Pikramenou, *J. Am. Chem. Soc.* **73,** 2821 (1951).
298. H. Eisinberg and J. Pouyet, *J. Polymer Sci.* **13,** 85 (1954).
299. D. O. Jordan, A. R. Mathieson, and M. R. Porter, *J. Polymer Sci.* **21,** 463 (1956).
300. W. Slough, *Trans. Faraday Soc.* **55,** 1030 (1959).
301. D. O. Jordan, T. Kurucsev, and R. L. Darskus, *Polymer* **6,** 303 (1965).
302. U. P. Strauss and E. G. Jackson, *J. Polymer Sci.* **6,** 649 (1951).
303. U. P. Strauss and S. S. Slowata, *J. Phys. Chem.* **61,** 411 (1957).
304. U. P. Strauss, N. L. Gershfeld, and H. Spiera, *J. Am. Chem. Soc.* **76,** 5909 (1954).
305. U. P. Strauss, N. L. Gershfeld, and E. H. Crook, *J. Phys. Chem.* **60,** 577 (1956).
306. A. Yu. Chernikhov and S. S. Medvedev, *Dokl. Akad. Nauk SSSR* **180,** 913 (1968); *Chem. Abstr.* **69,** 44235 (1968).
307. T. Alfrey and H. Morawetz, *J. Am. Chem. Soc.* **74,** 436 (1952).
308. H. L. Wagner and F. A. Long, *J. Phys. Colloid Chem.* **55,** 1512 (1951).
309. A. Katchalsky and I. R. Miller, *J. Polymer Sci.* **13,** 57 (1954).
310. A. Katchalsky, *J. Polymer Sci.* **7,** 393 (1951).
311. A. V. Ryabov, Yu. D. Semchikov, and N. N. Slavnitskaya, *Tr. po Khim. i Khim. Tekhnol.* **1963,** 161; *Chem. Abstr.* **60,** 9372 (1964).
312. I. T. Slyusarov and S. S. Urazovskii, *Vysokomolekul. Soedin.* **4,** 481 (1962); *Chem. Abstr.* **59,** 14124 (1963).
313. R. P. Mitra, M. Atreyi, and R. C. Gupta, *J. Electroanal. Chem. Interfacial Electrochem.* **17,** 227 (1968).
314. R. C. Schulz and J. Schwaab, *Makromol. Chem.* **85,** 297 (1965).
315. G. B. Gechele and G. Convalle, *J. Appl. Polymer Sci.* **5,** 203 (1961).
316. V. N. Reikh, A. E. Kalaus, D. B. Bognslavskii, A. I. Opalev, L. I. Dubovik, Kh. N. Borodushkina, and Yu. I. Fedorova, *Kauchuk i Rezina* **20,** 2 (1961); *Chem. Abstr.* **55,** 24068 (1961).
317. W. F. Brucksch, Jr., *Rubber Chem. Technol.* **36,** 975 (1963).
318. W. F. Brucksch, Jr., *Rubber Chem. Technol.* **35,** 453 (1962).
319. E. P. Kopylov, V. G. Epshtein, E. G. Lazaryants, and V. L. Tsailingol'd, *Kauchuk i Rezina* **21** (10), 19 (1962); *Chem. Abstr.* **58,** 8116 (1963).
320. E. P. Kopylov, V. G. Epshtein, E. G. Lazaryants, V. L. Tsailingol'd, and L. N. Mantseva, *Kauchuk i Rezina* **22,** 9 (1963); *Chem. Abstr.* **59,** 12993 (1963).
321. W. B. Reynolds, J. E. Pritchard, M. H. Opheim, and G. Kraus, *Proc. Rubber Technol. Conf., 3rd, London, 1954,* p. 226; *Chem. Abstr.* **50,** 16166 (1956).

322. V. L. Makarova and N. D. Zakharov, *Kauchuk i Rezina* **25** (5), 14 (1966); *Chem. Abstr.* **65,** 7424 (1966).
323. A. A. Dontsov, G. K. Lobacheva, and B. A. Dogadkin, *Kauchuk i Rezina* **27,** 19 (1968); *Chem. Abstr.* **68,** 88003 (1968).
324. A. A. Dontsov et al., *Kauchuk i Rezina* **27,** 16 (1968); *Chem. Abstr.* **70,** 29871 (1969).
325. V. L. Tsailingol'd et al., *Yaroslavsk. Prom.* **1958** (5), 22; *Chem. Abstr.* **57,** 13940 (1962).
326. E. S. Popova, Z. P. Krillova, G. S. Savvateeva, and Z. N. Asovskaya, *Uch. Zap. Yaroslavsk. Technol. Inst.* **1960,** 171; *Chem. Abstr.* **57,** 1004 (1962).
327. I. L. Shmurak and R. V. Uzina, *Vysokomolekul. Soedin.* **8,** 2065 (1966); *Chem. Abstr.* **66,** 66437 (1967).
328. V. L. Tsailingol'd et al., *Kauchuk i Rezina* **18,** 6 (1959); *Chem. Abstr.* **53,** 19426 (1959).
329. D. B. Boguslavskii, V. G. Epshtein, T. E. Ognevskaya, L. A. Lyapina, and V. G. Lyubeznikov, *Kauchuk i Rezina* **19** (8), 13 (1960); *Chem. Abstr.* **55,** 24069 (1961).
330. N. V. Belyaeva, N. L. Garetovskaya, N. F. Simonova, B. M. Babkin, and Ya. A. Schmidt, *Kauchuk i Rezina* **26,** 33 (1967); *Chem. Abstr.* **66,** 66537 (1967).
331. Canadian Industries Ltd. (to Imperial Chemical Industries Ltd.), Brit. Pat. 1,082,531 (Sept. 6, 1967); *Chem. Abstr.* **68,** 50832 (1968).
332. H. R. Krysiak (to E. I. du Pont de Nemours & Co., Inc.), U.S. Pat. 3,419,452 (Dec. 31, 1968); *Chem. Abstr.* **70,** 48464 (1969).
333. W. L. Thompson, T. B. Marshall, and A. T. Sweet, *Adhesives Age* **2,** 30 (1959); *Chem. Abstr.* **59,** 816 (1963).
334. C. E. Wheelock, *Ind. Eng. Chem.* **49,** 1929 (1957).
335. E. M. Vasil'eva, E. K. Podval'naya, O. P. Kolomeitsev, and R. K. Gavurina, *Vysokomolekul. Soedin.* **A9,** 1499 (1967); *Chem. Abstr.* **68,** 30524 (1968).
336. E. Sh. Kagan, B. I. Ardashev, A. D. Garnovskii, V. F. Krammanovich, D. A. Osipov, and V. T. Panyushkin (to Rostov State University and Novocherkassk Polytechnic Institute), U.S.S.R. Pat. 197,946 (June 9, 1967); *Chem. Abstr.* **68,** 87879 (1968).
337. F. Wolf, R. Hauptmann, and D. Warnecke, *Z. Anal. Chem.* **238,** 432 (1968).
338. E. B. Trostyanskaya and A. S. Tevlina, *Vysokomolekul. Soedin.* **5,** 44 (1963); *Chem. Abstr.* **59,** 2988 (1963).
339. H. Kawabe and M. Yanagita, *Bull. Chem. Soc. Japan* **42,** 1029 (1969).
340. M. Nishimura and M. Sugihara, *Kogyo Kagaku Zasshi* **66,** 1461 (1963).
341. A. B. Davankov and L. B. Zubakova, *Vysokomolekul. Soedin.* **2,** 884 (1960); *Chem. Abstr.* **55,** 7900 (1961).
342. M. Nishimura, M. Sugihara, M. Morioka, and M. Okita, *Kogyo Kagaku Zasshi* **70,** 1040 (1967).
343. V. N. Laskorin, N. Ya. Lyubman, G. K. Imangazieva, F. T. Shostak, and S. M. Imanbekova (to Institute of Chemical Sciences, Academy of Sciences, Kazakh S.S.R.), U.S.S.R. Pat. 224,063 (Aug. 6, 1968); *Chem. Abstr.* **70,** 48216 (1969).
344. N. B. Galitskaya, A. B. Pashkov, and E. I. Lyustgarten, *Plasticheskie Massy* **1967,** 13; *Chem. Abstr.* **68,** 60158 (1968).
345. E. M. Vasil'eva and R. K. Gavurina, *Vysokomolekul. Soedin.* **8,** 713 (1966); *Chem. Abstr.* **65,** 2420 (1966).
346. A. V. Smirnov, Yu. A. Leikin, A. B. Davankov, V. V. Korshak, and V. Rataichak, *Vysokomolekul. Soedin.* **B9,** 657 (1967); *Chem. Abstr.* **68,** 13645 (1968).
347. W. E. Feely (to Rohm & Haas Co.), U.S. Pat. 3,320,186 (May 16, 1967); *Chem. Abstr.* **67,** 22434 (1967).
348. L. I. Gabrielova and A. K. Livshits, *Tsvetn. Metal.* **36,** 77 (1963); *Chem. Abstr.* **59,** 7181 (1963).
349. W. P. Shyluk, *J. Polymer Sci.* [A-2] **6,** 2009 (1968).
350. K. Sakaguchi and K. Nagase, *Kogyo Kagaku Zasshi* **69,** 1192 (1966).
351. W. P. Shyluk, *J. Polymer Sci.* [A-2] **7,** 27 (1969).
352. I. Hashida, T. Shinagawa, and M. Sugihara, *Kagaku To Kogyo* (*Osaka*) **41,** 373 (1967).
353. A. K. Livshits and L. I. Gabrielova, *Gorn. Zh.* **1960,** 67; *Chem. Abstr.* **54,** 11895 (1960).
354. T. Ida, S. Kishi, S. Takahashi, and I. Utsumi, *J. Pharm. Sci.* **51,** 1061 (1962).
355. J. R. Puig and G. Gaussens, *Bull. Inst. Textile Fr.* **21,** 865 (1967); *Chem. Abstr.* **68,** 88059 (1968).
356. K. Matsubayashi and Y. Hirano, *Sen-i Gakkaishi* **17,** 643, 650 (1961).
357. Y. Shinohara, *Sen-i Gakkaishi* **18,** 485 (1962).

358. H. W. Coover, Jr., and F. B. Joyner (to Uniroyal Inc.), U.S. Pat. 3,315,014 (April 18, 1967); *Chem. Abstr.* **67**, 12470 (1967).
359. K. Nakatsuka, F. Ide, and Y. Jo (to Mitsubishi Rayon Co.), Japan. Pat. 6,806,536 (March 11, 1968); *Chem. Abstr.* **69**, 52907 (1968).
360. K. Nakatsuka, F. Ide, and Y. Jo (to Mitsubishi Rayon Co.), Japan. Pat. 6,806,535 (March 11, 1968); *Chem. Abstr.* **69**, 78405 (1968).
361. T. A. Mal'tseva, A. D. Virnik, G. D. Pestereva, and Z. A. Rogovin, *Izv. Vysshikh Ucheb. Zavedenii Tekhnol. Tekstil'n. Prom.* **1966** (4), 92; *Chem. Abstr.* **66**, 56484 (1967).
362. B. Lionel and E. Collins, *J. Polymer Sci.* **28**, 359 (1958).
363. G. B. Gechele and G. Convalle, *J. Appl. Polymer Sci.* **8**, 801 (1964).
364. S. Porejko, L. M. A. Szepke-Wrobel, and J. Trepka, *Polimery* **12**, 77 (1967); *Chem. Abstr.* **67**, 86456 (1967).
365. R. L. Lestinger and T. J. Savereide, *J. Am. Chem. Soc.* **84**, 3122 (1962).
366. R. L. Lestinger and T. J. Savereide, *J. Am. Chem. Soc.* **84**, 114 (1962).
367. S. K. Pluzhnov, Yu. E. Kirsh, V. A. Kabanov, and V. A. Kargin, *Dokl. Akad. Nauk SSSR* **185**, 843 (1969); *Chem. Abstr.* **71**, 22442 (1969).
368. Yu. E. Kirsh, V. A. Kabanov, and V. A. Kargin, *Vysokomolekul. Soedin.* **A10**, 349 (1968); *Chem. Abstr.* **68**, 105580 (1968).
369. I. Massukh, V. S. Pshezhetskii, V. A. Kabanov, and V. A. Kargin, *Vysokomolekul. Soedin.* **A9**, 839 (1967); *Chem. Abstr.* **67**, 43493 (1967).
370. V. S. Pshezhetskii, I. Massukh, and V. A. Kabanov, *J. Polymer Sci.* [C] **22**, 309 (1967).
371. H. E. Smith, C. R. Russell, and C. E. Rist, *Cereal Chem.* **40**, 282 (1963); *Chem. Abstr.* **59**, 2998 (1963).
372. M. Okawara, Y. Kurusu, and E. Imoto, *Kogyo Kagaku Zasshi* **65**, 1658 (1962).
373. A. Misono, S. Koda, and Y. Uchida, *Bull. Chem. Soc. Japan* **42**, 580 (1969).
374. K. Suzuoki, Y. Imanishi, T. Higashimura, and S. Okamura, *Biopolymers* **7**, 917 (1969).
375. M. Imoto, K. Takamoto, T. Otsuki, N. Ueda, S. Tahara, and H. Azuma, *Makromol. Chem.* **110**, 37 (1947).
376. R. Backskai (to Chevron Research Co.), Brit. Pat. 1,148,650 (April 16, 1969); *Chem. Abstr.* **60**, 115688 (1969).
377. G. Mohn, *Beitr. Silikose-Forsch.* **88**, 1 (1965); *Chem. Abstr.* **64**, 10296 (1966).
378. F. J. Strecker, *Beitr. Silikose-Forsch. Sonderband* **6**, 437 (1963); *Chem. Abstr.* **64**, 10306 (1966).
379. H. Sakabe and K. Koshi, *Ind. Health (Kawasaki Jap.)* **5**, 181 (1967).
380. P. F. Holt, H. Lindsay, and E. T. Nasrallah, *Nature* **216**, 611 (1967).
381. P. F. Holt and E. T. Nasrallah, *J. Chem. Soc. B* **1968**, 233.
382. J. E. Pritchard, M. H. Opheim, and P. H. Moyer, *Ind. Eng. Chem.* **47**, 863 (1955).
383. S. Arichi, *J. Sci. Hiroshima Univ., Ser. A-II* **25**, 419 (1962).
384. M. Ezrin, I. H. Updegraff, and H. G. Cassidy, *J. Am. Chem. Soc.* **75**, 1610 (1953).
385. L. Luttinger and H. G. Cassidy, *J. Polymer Sci.* **20**, 417 (1956).
386. F. A. Levina, G. I. Rybalko, I. Sidaravicius, and A. M. Sladkov (to State Optical Institute, Institute of Elemental Organic Compounds, Sci.-Res. Institute of Electrography of the State Committee on Radio-electronics of the U.S.S.R.), U.S.S.R. Pat. 165,970 (Oct. 26, 1964); *Chem. Abstr.* **62**, 9980 (1965).
387. H. C. Haas (to Polaroid Corp.), U.S. Pat. 3,382,215 (March 7, 1968); *Chem. Abstr.* **69, 10944** (1968).
388. D. W. Heseltine, D. A. Brooks, R. C. Taber, and J. C. R. Williams (to Eastman Kodak Co.), Fr. Pat. 1,489,356 (July 21, 1967); *Chem. Abstr.* **69**, 27251 (1968).
389. V. I. Zavrazhina, Yu. N. Mikhailovskii, and P. I. Zubov, *Zashch. Metal.* **3**, 700 (1967); *Chem. Abstr.* **68**, 65089 (1968).
390. K. A. Krans, A. J. Shor, and J. S. Johnson, Jr., *Desalination* **2**, 243 (1967); *Chem. Abstr.* **68**, 53166, 1968).
391. J. S. Johnson et al., *Desalination* **5**, 359 (1969); *Chem. Abstr.* **70**, 80745 (1969).
392. P. F. Holt and H. Lindsey, *J. Chem. Soc. C* **1969**, 1012.

Shigeo Tazuke and Seizo Okamura
Kyoto University

N-VINYLPYRROLIDONE POLYMERS. See *N*-VINYL AMIDE POLYMERS

VINYL STEARATE POLYMERS. See VINYL ESTER POLYMERS, Vol. 15

VINYL SULFONE POLYMERS. See SULFUR-CONTAINING POLYMERS

VINYLSULFONIC ACID POLYMERS. See ETHYLENESULFONIC ACID POLYMERS

VINYL TILE. See FLOORING MATERIALS

VINYLTOLUENE POLYMERS. See STYRENE POLYMERS

VINYON FIBERS. See Fibers under VINYL CHLORIDE POLYMERS

VIRUSES. See POLYNUCLEOTIDES

VISCOELASTICITY

The Nature of Viscoelastic Behavior

The classical theory of elasticity deals with mechanical properties of elastic solids, for which, in accordance with Hooke's law, stress is always directly proportional to strain in small deformations but independent of the rate of strain. The classical theory of hydrodynamics deals with properties of viscous liquids, for which, in accordance with Newton's law, the stress is always directly proportional to rate of strain but independent of the strain itself. These categories are idealizations, however; although the behavior of many solids approaches Hooke's law for infinitesimal strains, and that of many liquids approaches Newton's law for infinitesimal rates of strain, under other conditions deviations are observed. Two types of deviations may be distinguished.

First, when *finite* strains are imposed on solids (especially those soft enough to be deformed substantially without breaking) the stress–strain relations are much more complicated (non-Hookean); similarly, in steady flow with *finite* strain rates, many fluids (especially polymeric solutions and melts) exhibit marked deviations from Newton's law (non-Newtonian flow). The dividing line between "infinitesimal" and "finite" depends, of course, on the level of precision under consideration, and it varies greatly from one material to another.

Second, even if both strain and rate of strain are infinitesimal, a system may exhibit behavior which combines liquidlike and solidlike characteristics. For example, a body which is not quite solid does not maintain a constant deformation under constant stress but goes on slowly deforming with time, or creeps. When such a body is constrained at constant deformation, the stress required to hold it diminishes gradually, or relaxes. On the other hand, a body which is not quite liquid may, while flowing under constant stress, store some of the energy input, instead of dissipating it all as heat; and it may recover part of its deformation when the stress is removed (elastic recoil). When such bodies are subjected to sinusoidally oscillating stress, the strain is neither exactly in phase with the stress (as it would be for a perfectly elastic solid) nor 90° out of phase (as it would be for a perfectly viscous liquid) but is somewhere in between. Some of the energy input is stored and recovered in each cycle, and some is dissipated as heat. Materials whose behavior exhibits such characteristics are called viscoelastic. If both strain and rate of strain are infinitesimal, we have *linear* viscoelastic behavior; then, in a given experiment the ratio of stress to strain is a function of time (or frequency) alone, and not of stress magnitude.

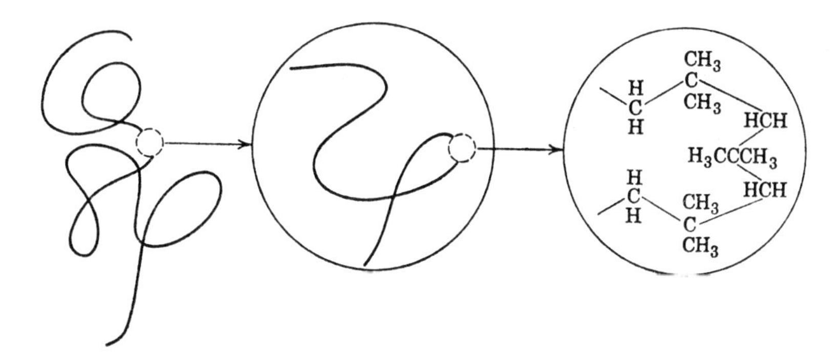

Fig. 1. Symbolic representation of long-range and short-range coutour relationships in a flexible polymer molecule (polyisobutylene).

The relations between stress, strain, and their time dependences are in general described by a "constitutive equation" or "rheological equation of state." If strains and/or rates of strain are finite, the constitutive equation may be quite complicated. If they are infinitesimal, however, corresponding to linear viscoelastic behavior, the constitutive equation is relatively simple, and most of the material in this article falls under its jurisdiction.

In many of the materials of interest in classical physics, as well as of practical importance in engineering, viscoelastic anomalies are negligible or of minor significance. Though the foundations of the phenomenological theory of linear viscoelasticity were inspired by creep and relaxation experiments on fibers of metal and glass (1–4) and the dissipation of energy in sinusoidally oscillating deformations has provided valuable information about the structure of metals (5), the deviations from perfect elasticity here are small. In polymeric systems, by contrast, mechanical behavior is dominated by viscoelastic phenomena which are often truly spectacular.

The prominence of viscoelasticity in polymers is not unexpected when one considers the complicated molecular adjustments which must underlie any macroscopic mechanical deformation. In deformation of a hard solid such as diamond, sodium chloride, or crystalline zinc, atoms are displaced from equilibrium positions in fields of force which are quite local in character; from knowledge of the interatomic potentials, elastic constants can be calculated (6). Other mechanical phenomena reflect structural imperfections involving distances discontinuously larger than atomic dimensions (5,6). In an ordinary liquid, viscous flow reflects the change with time, under stress, of the distribution of molecules surrounding a given molecule; here, too, the relevant forces and processes of readjustment are quite local in character, and from knowledge of them the viscosity can in principle be calculated (7). In a polymer, on the other hand, each flexible threadlike molecule pervades an average volume much greater than atomic dimensions and is continually changing the shape of its contour as it wriggles and writhes with its thermal energy. To characterize the various configurations or contour shapes which it assumes, it is necessary to consider (qualitatively speaking) gross long-range contour relationships, somewhat more local relationships seen with a more detailed scale, and so on, eventually including the orientation of bonds in the chain backbone with respect to each other on a scale of atomic dimensions, as symbolized in Figure 1 (8). Alfrey (8a) has referred to these spatial relationships, viewed over progressively longer ranges, as "kinks, curls, and convolutions." Rearrangements on a local scale (kinks) are relatively rapid, on a long-range scale (convolutions) very slow. Under stress, a new assortment of configurations is obtained; the response to the local aspects of the new distribution is rapid, the response to the long-range aspects is slow, and all told there is a very wide and continuous range of time scale covering the response of such a system to external stress.

Every polymeric system has a glass-transition temperature below which the writhing thermal motions essentially cease. Here, long-range convolutional readjustments are severely restricted; there is still a wide range of response rates to external stress, but different in nature.

From measurements of viscoelastic properties of polymers, information can be obtained about the nature and the rates of the configurational rearrangements, and the disposition and interaction of the macromolecules in both their short-range and their long-range interrelations. From the standpoint of the physical chemist, this provides a field of inquiry with unique features of interest. Investigation of viscoelastic properties of polymers has also been greatly stimulated, of course, by the practical importance of mechanical behavior in the processing and utilization of rubbers, plastics, and fibers. As a result, a very high proportion of all studies on viscoelasticity in the past three decades has been devoted to the viscoelasticity of polymers. See also DEFORMATION; FRACTURE; MECHANICAL PROPERTIES; MELT VISCOSITY.

Strain, Stress, and Linear Constitutive Equations for Simple Shear

The principal purpose of this article is to relate the viscoelasticity of polymers to molecular structure and modes of molecular motion, and to describe the dependence of viscoelastic properties on molecular weight, molecular-weight distribution, temperature, concentration, chemical structure, and other variables. However, it is necessary first to provide a phenomenological background with definitions of strain and stress and their interrelations in a medium regarded as a continuum.

Equations of Change. Experimental measurements of mechanical properties are usually made by observing external forces and changes in external dimensions of a body with a certain shape—a cube, disc, rod, or fiber. The connection between forces and deformations in a specific experiment depends not only on the stress–strain relations (the constitutive equation) but also on two other relations (9). These are the equation of continuity, expressing conservation of mass:

$$\frac{\partial}{\partial t}\rho = -\sum_{i=1}^{3}\frac{\partial}{\partial x_i}(\rho v_i) \tag{1}$$

and the equation of motion, expressing conservation of momentum, with three components of the form:

$$\rho\left(\frac{\partial}{\partial t}v_j + \sum_{i=1}^{3} v_i \frac{\partial}{\partial x_i} v_j\right) = \sum_{i=1}^{3}\frac{\partial}{\partial x_i}\sigma_{ij} + \rho g_j \tag{2}$$

Here ρ is the density, t the time, x_i the three Cartesian coordinates, and v_i the components of velocity in the respective directions of these coordinates. Equation 2 is repeated with $j = 1, 2, 3$ corresponding to these directions; g_j is the component of gravitational acceleration in the j direction, and σ_{ij} the appropriate component of the stress tensor (see below). (A third restriction, conservation of energy, can be omitted for a process at constant temperature; the discussion in this section is limited to isothermal conditions.) However, many experiments are designed so that both sides of equation 1 are zero, and that in equation 2 the inertial and gravitational forces represented by the first and last terms are negligible. In this case, the internal states of stress and strain can be calculated from observable quantities by the constitutive equation alone. For infinitesimal deformations, the appropriate relations for viscoelastic materials involve the same geometrical form factors as in the classical theory of equilibrium elasticity.

Infinitesimal Strain Tensor. In a viscoelastic as in a perfectly elastic body, the state of deformation at a given point is specified by a strain tensor which represents the relative changes in dimensions and angles of a small cubical element cut out at that position. The rate of strain tensor expresses the time derivatives of these relative dimensions and angles. Similarly, the state of stress is specified by a stress tensor which represents the forces acting on different faces of the cubical element from different directions. For details, the reader is referred to standard treatises (8–15).

For an infinitesimal deformation, the components of the infinitesimal strain tensor in rectangular coordinates with the three Cartesian directions denoted by the subscripts 1, 2, 3 are

$$\gamma_{ij} = \begin{pmatrix} 2\,\partial u_1/\partial x_1 & \partial u_2/\partial x_1 + \partial u_1/\partial x_2 & \partial u_3/\partial x_1 + \partial u_1/\partial x_3 \\ \partial u_2/\partial x_1 + \partial u_1/\partial x_2 & 2\,\partial\, u_2/\partial x_2 & \partial u_2/\partial x_3 + \partial u_3/\partial x_2 \\ \partial u_3/\partial x_1 + \partial u_1/\partial x_3 & \partial u_2/\partial x_3 + \partial u_3/\partial x_2 & 2\,\partial u_3/\partial x_3 \end{pmatrix} \tag{3}$$

where x_i and u_i are respectively the coordinates of the point where the strain is specified and its displacement in the strained state; ie, $u_i = x_i - x_i^0$, where the superscript 0 refers to the unstrained state. The rate of strain tensor, $\dot{\gamma}_{ij}$, is formulated similarly with u_i replaced by v_i, the velocity of displacement ($= \partial u_i/\partial t$).

In most treatises (12,15), the strain tensor is defined with all components smaller by a factor of ½ than in equation 3, so that $\gamma_{11} = \partial u_1/\partial x_1$ and $\gamma_{21} = \frac{1}{2}\,(\partial u_2/\partial x_1 + \partial u_1/\partial x_2)$. However, such a definition makes discussion of shear or shear flow somewhat clumsy: either a "practical" shear strain and "practical" shear rate must be introduced which are twice γ_{21} and $\dot{\gamma}_{21}$ respectively, or else a factor of two must be carried in the constitutive equations. Since most of the discussion in this article is concerned with shear deformations, we use the definition of equation 3 which follows Bird (16) and Carreau (17) and is comparable with the treatment of Lodge (13); the reader is referred to Table I of Ref. 17. This does cause a slight inconvenience in the discussion of compressive and tensile strain, where a "practical" measure of strain is subsequently introduced. In older treatises on elasticity (10,14), strains are defined without the factor of two appearing in the diagonal components of equation 3, but with the other components the same.

Stress Tensor. Associated with the strain and its time dependence is a state of stress

$$\sigma_{ij} = \begin{pmatrix} \sigma_{11} & \sigma_{12} & \sigma_{13} \\ \sigma_{21} & \sigma_{22} & \sigma_{23} \\ \sigma_{31} & \sigma_{32} & \sigma_{33} \end{pmatrix} \tag{4}$$

where σ_{ij} is the component, parallel to the j direction, of the force per unit area acting on the face of a cubical element which is perpendicular to the i direction. The normal stresses σ_{ij} are taken as positive for tension and negative for compression. (This convention is not universal; the opposite signs are used in some treatments (9).) The components of the stress tensor are illustrated graphically in Figure 2.

Constitutive Equation for Linear Viscoelasticity in Simple Shear. If the deformation is uniform (homogeneous), the stress and strain components do not vary with position and are independent of x_i. There are several specific types of uniform deformation for which the strain and stress tensors assume a relatively simple form. One that corresponds to a commonly used experimental geometry is simple shear, where two opposite faces of the cubical element are displaced by sliding, as illustrated

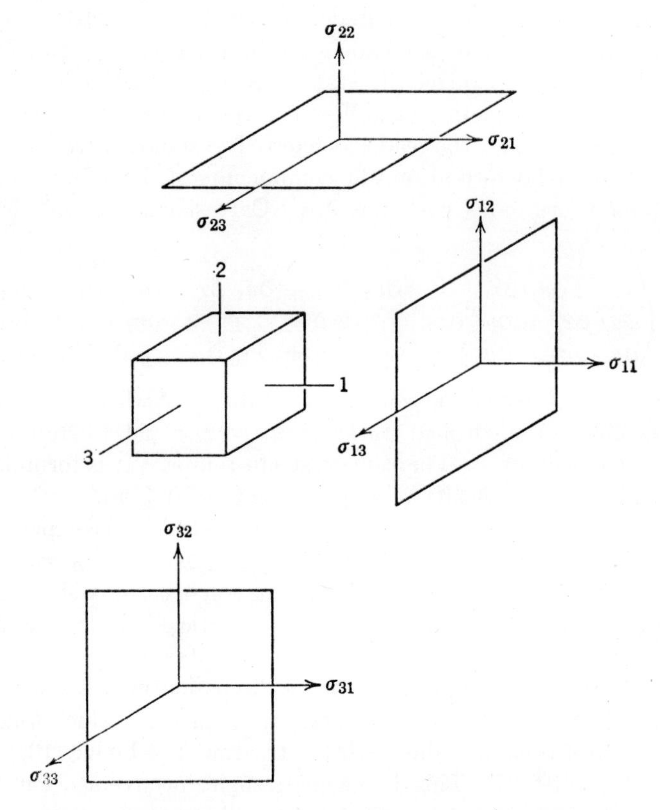

Fig. 2. Identification of the components of the stress tensor.

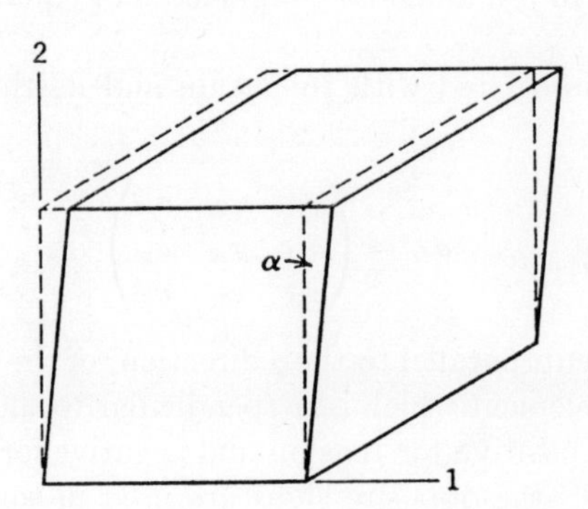

Fig. 3. Illustration of simple shear of a cubical element.

in Figure 3. Conventionally, the 13 plane slides in the 1 direction; the strains and stresses are then

$$\gamma_{ij} = \begin{pmatrix} 0 & \gamma_{12} & 0 \\ \gamma_{21} & 0 & 0 \\ 0 & 0 & 0 \end{pmatrix} \tag{5}$$

where $\gamma_{21} = \gamma_{12} = \partial u_1/\partial x_2 = \tan \alpha \cong \alpha$, and

$$\sigma_{ij} = \begin{pmatrix} 0 & \sigma_{12} & 0 \\ \sigma_{21} & 0 & 0 \\ 0 & 0 & 0 \end{pmatrix} \tag{6}$$

in which the two nonzero components can be identified in Figure 2. (If a constant hydrostatic pressure is imposed, the diagonal zeros are replaced by other components as discussed below.) The strain γ_{21} and stress σ_{21} are functions of time, and they are connected by a constitutive equation for linear viscoelasticity which for simple shear has a very simple form.

This linear constitutive equation is based on the principle that the effects of sequential changes in strain are additive (12,15):

$$\sigma_{21}(t) = \int_{-\infty}^{t} G(t - t')\dot{\gamma}_{21}(t')\, dt' \tag{7}$$

where $\dot{\gamma}_{21} = \partial\gamma_{21}/\partial t$ is the shear rate, $G(t - t')$ is called the relaxation modulus, with a physical significance which will be apparent later, and the integration is carried out over all past times t' up to the current time t. An alternative formulation (12) which can be obtained from equation 7 by integration by parts is

$$\sigma_{21}(t) = \int_{-\infty}^{t} m(t - t')\gamma_{21}(t')\, dt' + G(0)\gamma_{21}(t) \tag{7a}$$

where the integration covers the history of the strain γ_{21} (the displacements being referred to the positions of material particles at the current time t) instead of that of the rate of strain $\dot{\gamma}_{21}$; $m(t - t')$, the memory function, is $dG(t - t')/dt'$.

An alternative constitutive equation can be written to express the strain in terms of the history of the stress, in the form

$$\gamma_{21}(t) = \int_{-\infty}^{t} J(t - t')\dot{\sigma}_{21}(t')\, dt' \tag{8}$$

where $\dot{\sigma}_{21} = d\sigma_{21}/dt$ and $J(t - t')$ is called the creep compliance because of its physical significance to be described later.

From knowledge of the shear relaxation modulus, the memory function, or the creep compliance function of a particular material, its stress–strain relations for any kind of experiment in shear with a prescribed time dependence of stress or strain can be predicted as long as the motions are sufficiently small or sufficiently slow. Some time-dependent patterns of particular interest are described in the following section and analyzed in terms of equation 7. Equation 8 will be used to illustrate certain other experimental patterns below.

Linear Time-Dependent Experiments in Shear

Stress Relaxation Experiment. Suppose a shear strain γ_{21} is imposed within a brief period of time ξ by a constant rate of strain $\dot{\gamma}_{21} = \gamma_{21}/\xi$ (Fig. 4). Equation 7 can then be expressed as

$$\sigma_{21}(t) = \int_{t_0-\xi}^{t_0} G(t - t')(\gamma_{21}/\xi)\, dt' \tag{9}$$

where t_0 is the time at which the strain is complete, since the rate of strain is zero both before and after the interval represented by this integral. By the Theorem of the Mean, equation 9 can be written

$$\sigma_{21}(t) = (\gamma_{21}/\xi)\xi G(t - t_0 + \epsilon\xi), \qquad 0 \leq \epsilon \leq 1$$

Thus, if the strain is accomplished by the time $t = 0$,

$$\sigma_{21}(t) = \gamma_{21}G(t + \epsilon\xi) \tag{10}$$

and for times long compared with ξ, the loading interval, this is indistinguishable from

$$\sigma_{21}(t) = \gamma_{21}G(t) \tag{11}$$

The physical meaning of the relaxation modulus $G(t)$ is apparent in terms of this simple experiment. In general, the ratio of a stress to the corresponding strain is called a modulus, and for a perfectly elastic solid the equilibrium shear modulus G is defined

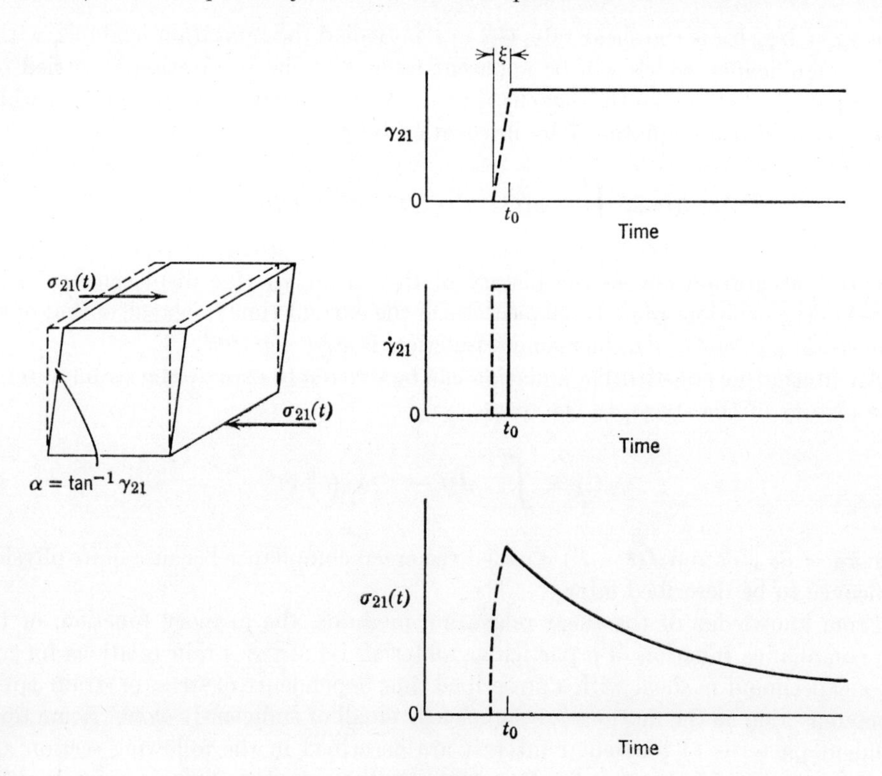

Fig. 4. Geometry and time profile of a simple shear stress relaxation experiment following sudden strain.

as σ_{21}/γ_{21}; $G(t)$ is its time-dependent analog as measured in an experiment *with this particular time pattern.*

There is another kind of stress relaxation experiment which must not be confused with that defined by the function $G(t)$. In a system which can undergo deformation

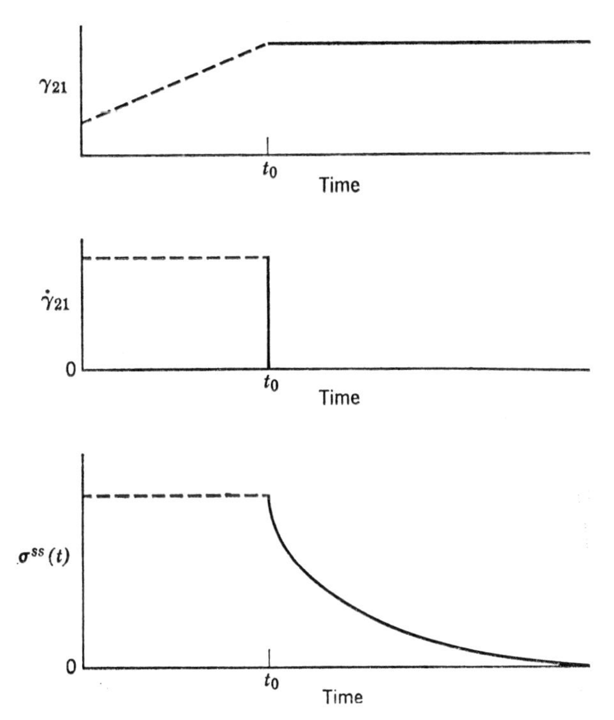

Fig. 5. Time profile of a simple shear stress relaxation experiment following cessation of steady-state flow.

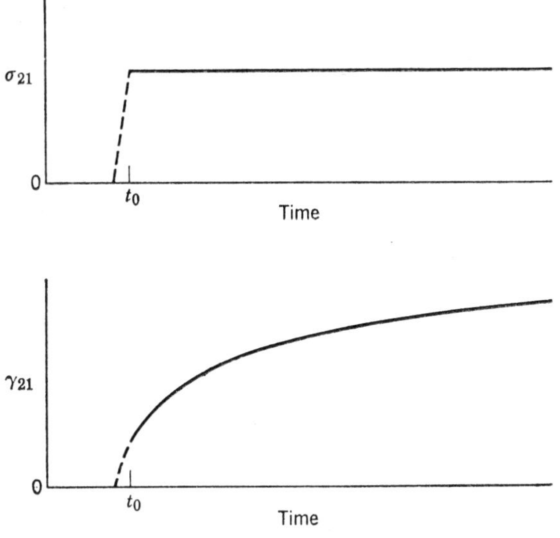

Fig. 6. Time profile of a shear creep experiment.

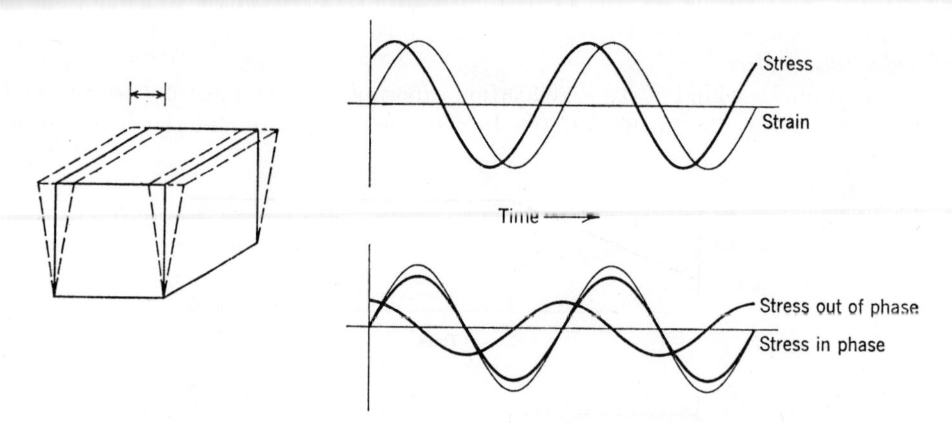

Fig. 7. Geometry and time profile of a simple shear experiment with sinusoidally varying shear.

in steady shear flow, the change in σ_{21} with time can be followed after abrupt cessation of such flow (Fig. 5). Such an experiment is much less commonly performed; the course of this type of stress relaxation, distinguished by the notation $\sigma_{21}^{ss}(t)$, and its relation to $G(t)$ are discussed in other treatises. The term "stress relaxation" without further qualification will always apply to the experiment depicted in Figure 4.

Creep Experiment. Alternatively, a shear stress σ_{21} may be applied within a very brief period before time 0 and then maintained constant (Fig. 6). The dependence of the strain γ_{21} on time can be derived from equation 8 exactly as equation 11 is obtained from equation 7, with the result

$$\gamma_{21}(t) = \sigma_{21}J(t) \tag{12}$$

This shows the physical meaning of the creep compliance $J(t)$, which has the dimensions of reciprocal modulus and is a monotonically nondecreasing function of time. For a perfectly elastic solid, $J = 1/G$. However, for a viscoelastic material, $J(t) \neq 1/G(t)$, because of the difference between the two experimental time patterns (17a).

Other Types of Transient Experiments. In deformation with a constant rate of strain, γ_{21} is increased linearly with time (at constant $\dot{\gamma}_{21}$); if the total deformation is small, $\sigma_{21}(t)$ can be rather simply related to $G(t)$. In deformation with a constant rate of stress loading, σ_{21} is increased linearly with time; then $\gamma_{21}(t)$ can be rather simply related to $J(t)$ (17a). Another pattern of interest involves removal of stress after it has been applied for a sufficiently long time to achieve either a constant strain or a constant rate of strain. This type of experiment will be discussed below, under Elastic Recovery.

Periodic or Dynamic Experiments. To supplement the above transient (ie, nonperiodic) experiments and provide information corresponding to very short times, the stress may be varied periodically, usually with a sinusoidal alternation at a frequency ν in cycles/sec (Hz) or ω $(= 2\pi\nu)$ in radians/sec. A periodic experiment at frequency ω is qualitatively equivalent to a transient experiment at time $t = 1/\omega$, as will be evident in the examples shown. If the viscoelastic behavior is linear, it is found that the strain will also alternate sinusoidally but will be out of phase with the stress (Fig. 7). This can be shown from the constitutive equation as follows. Let

$$\gamma_{21} = \gamma_{21}^0 \sin \omega t \tag{13}$$

where $\gamma_{21}{}^0$ is the maximum amplitude of the strain. Then

$$\dot{\gamma}_{21} = \omega\gamma_{21}{}^0 \cos \omega t \tag{14}$$

Substituting in equation 7, denoting $t - t'$ by s, we have

$$\sigma_{21}(t) = \int_0^\infty G(s)\omega\gamma_{21}{}^0 \cos [\omega(t - s)]\, ds$$

$$= \gamma_{21}{}^0 \left[\omega \int_0^\infty G(s) \sin \omega s\, ds\right] \sin \omega t + \gamma_{21}{}^0 \left[\omega \int_0^\infty G(s) \cos \omega s\, ds\right] \cos \omega t \tag{15}$$

The integrals converge only if $G(s) \to 0$ as $s \to \infty$; otherwise they must be formulated somewhat differently. It is clear that the term in $\sin \omega t$ is in phase with γ_{21} and the term in $\cos \omega t$ is 90° out of phase; σ_{21} is periodic in ω but out of phase with γ_{21} to a degree depending on the relative magnitudes of these terms. The quantities in brackets are functions of frequency but not of elapsed time, so equation 15 can be conveniently written

$$\sigma_{21} = \gamma_{21}{}^0 (G' \sin \omega t + G'' \cos \omega t) \tag{16}$$

thereby defining two frequency-dependent functions—the shear storage modulus G' and the shear loss modulus G''.

It is instructive to write the stress in an alternative form displaying the *amplitude* $\sigma_{21}{}^0$ of the stress and the *phase angle* δ between stress and strain. From trigonometric relations,

$$\sigma_{21} = \sigma_{21}{}^0 \sin (\omega t + \delta) = \sigma_{21}{}^0 \cos \delta \sin \omega t + \sigma_{21}{}^0 \sin \delta \cos \omega t \tag{17}$$

Comparison of equations 16 and 17 shows that

$$G' = (\sigma_{21}{}^0/\gamma_{21}{}^0) \cos \delta \tag{18}$$

$$G'' = (\sigma_{21}{}^0/\gamma_{21}{}^0) \sin \delta \tag{19}$$

$$G''/G' = \tan \delta \tag{20}$$

It is evident that each periodic, or dynamic, measurement at a given frequency provides simultaneously *two* independent quantities, either G' and G'' or else $\tan \delta$ and $\sigma_{21}{}^0/\gamma_{21}{}^0$, the ratio of peak stress to peak strain. (Over a range of frequencies, however, G' and G'' are not independent and can be interrelated (17a).)

It is usually convenient to express the sinusoidally varying stress as a complex quantity. Then the modulus is also complex, given by

$$\overset{*}{\boldsymbol{\sigma}}_{21}/\overset{*}{\gamma}_{21} = \mathbf{G}^* = G' + iG'' \tag{21}$$

$$|\mathbf{G}^*| = \sigma_{21}{}^0/\gamma_{21}{}^0 = \sqrt{G'^2 + G''^2} \tag{22}$$

This corresponds to a vectorial resolution of the components in the complex plane as shown in Figure 8. It is evident that G' is the ratio of the stress in phase with the strain to the strain, whereas G'' is the ratio of the stress 90° out of phase with the strain to the strain.

The data from sinusoidal experiments can also be expressed in terms of a complex compliance

$$\mathbf{J}^* = \overset{*}{\gamma}_{21}/\overset{*}{\boldsymbol{\sigma}}_{21} = 1/\mathbf{G}^* = J' - iJ'' \tag{23}$$

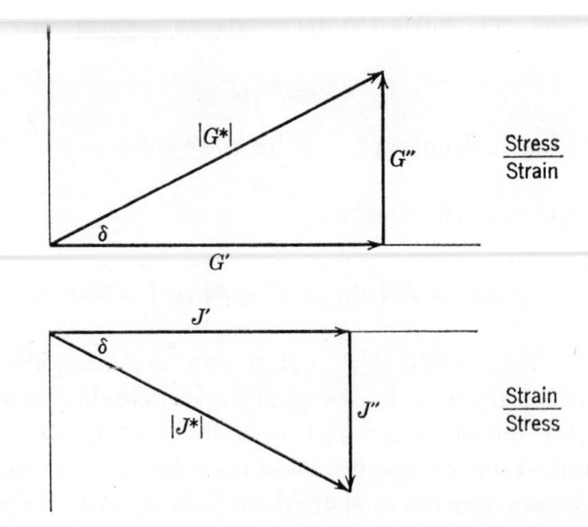

Fig. 8. Vectorial resolution of components of complex modulus and compliance in sinusoidal shear deformations.

The storage compliance J' is the ratio of the strain in phase with the stress to the stress, whereas the loss compliance J'' is the ratio of the strain 90° out phase with the stress to the stress (Fig. 8). The storage modulus G' and compliance J' are so named because they are associated with energy storage and release in the periodic deformation. The loss modulus G'' and compliance J'' are associated with the dissipation or loss of energy as heat. Note that $J''/J' = G''/G' = \tan\delta$.

Reference to the complex modulus $\mathbf{G}^*$ may be regarded as abbreviated language for talking about G' and G'' simultaneously, and similarly for $\mathbf{J}^*$. The quantities G'' and J'', though sometimes called the "imaginary" components of their complex counterparts, are of course real numbers.

Although $\mathbf{J}^* = 1/\mathbf{G}^*$, their individual components are not reciprocally related, but are connected by the following equations:

$$J' = \frac{G'}{(G'^2 + G''^2)} = \frac{1/G'}{1 + \tan^2\delta} \tag{24}$$

$$J'' = \frac{G''}{(G'^2 + G''^2)} = \frac{1/G''}{1 + (\tan^2\delta)^{-1}} \tag{25}$$

$$G' = \frac{J'}{(J'^2 + J''^2)} = \frac{1/J'}{1 + \tan^2\delta} \tag{26}$$

$$G'' = \frac{J''}{(J'^2 + J''^2)} = \frac{1/J''}{1 + (\tan^2\delta)^{-1}} \tag{27}$$

Periodic measurements can be made, depending on circumstances, at frequencies from 10^{-5} to 10^8 Hz; usually a given experimental method will cover only two to three powers of ten, but a great variety of methods is available (17b).

Correlation of Experimental Data to Provide Information over Wide Ranges of Time Scale. To cover a wide enough time scale to reflect the variety of molecular motions in polymeric systems, often ten to fifteen logarithmic decades, one must usually combine information from transient and sinusoidal experiments. It is then necessary to calculate from the results of one type of measurement what would have been observed in the other, in the same range of time (or inverse radian frequency). Fortunately, this is possible, provided the viscoelastic behavior is linear. In principle, knowledge of any one of the functions $J(t)$, $G(t)$, $G'(\omega)$, $G''(\omega)$, $J'(\omega)$, $J''(\omega)$ over the entire range of time or frequency (plus in certain cases one or two additional constants) permits calculation of all the others. Even if the values are not available over the entire range of the argument, approximation methods can be applied (17a).

Since each of the preceding functions can be calculated from any other, it is an arbitrary matter which is chosen to depict the behavior of a system and to correlate with theoretical formulations on a molecular basis. In fact, two other derived functions are sometimes used for the latter purpose—the relaxation and retardation spectra, H and L, which will be defined below. Actually, different aspects of the viscoelastic behavior, and the molecular phenomena which underlie them, have different degrees of prominence in the various functions enumerated above, so it is worthwhile to examine the form of several of these functions even when all are calculated from the same experimental data. A qualitative survey of their appearance will be presented below, under the illustrations of viscoelastic behavior.

Even when only one experimental method is available, covering perhaps two or three decades of logarithmic time or frequency scale, the viscoelastic functions can be traced out over a much larger effective range by making measurements at different temperatures, and using a sort of principle of viscoelastic corresponding states. In many cases, an increase in temperature is nearly equivalent to an increase in time or a decrease in frequency in its effect on a modulus or compliance, and a temperature range of 100° may provide an effective time range of ten logarithmic decades. This scheme must be used judiciously, with regard to the reservations in the theories which support it, and to previous experimental experience (18). When properly applied, it yields plots in terms of reduced variables which can be used with considerable confidence to deduce molecular parameters as well as to predict viscoelastic behavior in regions of time or frequency scale not experimentally accessible.

The Relaxation and Retardation Spectra

In addition to the basic viscoelastic functions described above, there are two useful derived functions, the relaxation spectrum and the retardation spectrum. These can be introduced most clearly in terms of mechanical models.

The form of the time dependence of $J(t)$ or $G(t)$, or of the frequency dependence of $G'(\omega)$, $G''(\omega)$, $J'(\omega)$, or $J''(\omega)$, can be imitated by the behavior of a mechanical model with a sufficient number of elastic elements (springs) and viscous elements (dashpots imagined as pistons moving in oil). The simplest mechanical model analogous to a

Fig. 9. The Maxwell element. **Fig. 10.** The Voigt element.

viscoelastic system is one spring combined with one dashpot, either in series (Fig. 9) or in parallel (Fig. 10). Here force applied to the terminals of the model is analogous to σ_{21}, the relative displacement of the terminals is analogous to γ_{21}, and the rate of displacement is analogous to $\dot{\gamma}_{21}$. Each spring element is assigned a stiffness (force/displacement) analogous to a shear modulus contribution G_i, and each dashpot is assigned a frictional resistance (force/velocity) analogous to a viscosity contribution η_i. The dimensions do not correspond (force/displacement is dynes/cm and modulus is dynes/cm^2) and the geometry looks like extension rather than shear, but the mathematical analogy is satisfactory.

Figure 9 represents a *Maxwell element.* If the spring corresponds to a shear rigidity $G_i = 1/J_i$ (we choose shear as the type of deformation to be worked out in detail, though any other deformation would do as well) and the dashpot to a viscosity η_i, then the *relaxation time* of the element is defined as $\tau_i = \eta_i/G_i$ and is a measure of the time required for stress relaxation. If η_i is in poises and G_i in dynes/ cm^2, τ_i is in seconds.

The viscoelastic functions exhibited by the Maxwell element can easily be derived and are summarized as follows (eqs. 28–35).

$$J(t) = J_i + t/\eta \tag{28}$$

$$G(t) = G_i e^{-t/\tau_i} \tag{29}$$

$$G'(\omega) = G_i\omega^2\tau_i^2/(1 + \omega^2\tau_i^2) \tag{30}$$

$$G''(\omega) = G_i\omega\tau_i/(1 + \omega^2\tau_i^2) \tag{31}$$

$$\eta'(\omega) = \eta_i/(1 + \omega^2\tau_i^2) \tag{32}$$

$$J'(\omega) = J_i \tag{33}$$

$$J''(\omega) = J_i/\omega\tau_i = 1/\omega\eta_i \tag{34}$$

$$\tan\delta = 1/\omega\tau_i \tag{35}$$

Figure 10 represents a *Voigt element.* If the spring and dashpot have the same significance as before, their ratio τ_i is defined as the retardation time and is a measure of the time required for the extension of the spring to its equilibrium length while retarded by the dashpot. (The model of Figure 10 was introduced by Kelvin before Voigt, and is sometimes called the Kelvin element, but the name of Voigt has come to be more generally used.)

The viscoelastic functions exhibited by the Voigt element are as follows.

$$J(t) = J_i(1 - e^{-t/\tau_i}) \tag{36}$$

$$G(t) = G_i \tag{37}$$

$$G'(\omega) = G_i \tag{38}$$

$$G''(\omega) = G_i\omega\tau_i = \omega\eta_i \tag{39}$$

$$\eta'(\omega) = \eta_i \tag{40}$$

$$J'(\omega) = J_i/(1 + \omega^2\tau_i^2) \tag{41}$$

$$J''(\omega) = J_i\omega\tau_i/(1 + \omega^2\tau_i^2) \tag{42}$$

$$\tan\delta = \omega\tau_i \tag{43}$$

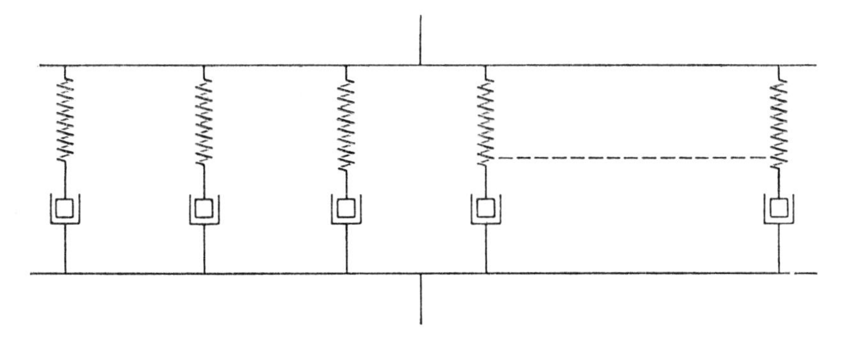

Fig. 11. Generalized Maxwell model.

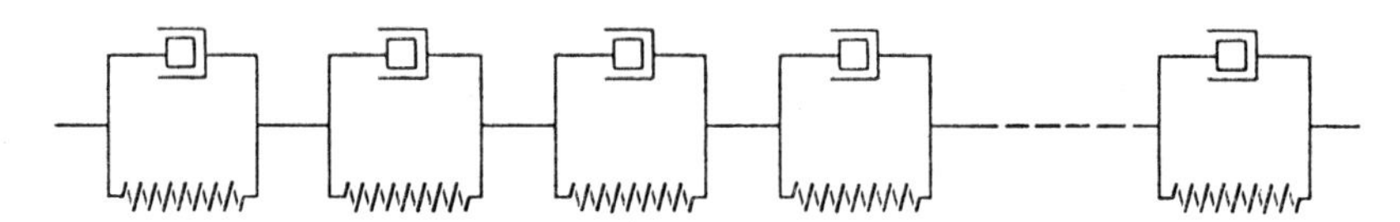

Fig. 12. Generalized Voigt model.

Graphs of these functions have been given in many places (11,19), but will not be reproduced here because they are too drastically simplified to correspond to any real viscoelastic behavior (with rare exceptions). It may be noted that the creep of the Voigt element and the stress relaxation of the Maxwell element are exponential functions of time. Most observed creep and relaxation processes in polymers progress more gradually than specified by these simple equations. The components of the complex modulus for the Maxwell element, and those of the complex compliance for the Voigt element, depend on frequency in a manner reminiscent of the Debye dispersion of the complex dielectric constant. Again, observed moduli and compliances change with frequency more gradually than this.

Generalized Maxwell and Voigt Models. The correct frequency and time dependences can be represented by combinations of Maxwell and Voigt elements, as illustrated in Figures 11 and 12.

For example, when the Maxwell model (Fig. 11) is subjected to a stress relaxation experiment as in Figure 4, the force on each spring–dashpot pair relaxes exponentially; expressed in terms of the modulus analogy, we have the contribution of the ith pair to the modulus, $G_i(t)$, given by

$$G_i(t) = G_i e^{-tG_i/\eta_i} \equiv G_i e^{-t/\tau_i} \tag{44}$$

in which the time constants τ_i, the relaxation times, are defined as η_i/G_i. Then

$$G(t) = \sigma_{21}(t)/\gamma_{21} = \sum_{i=1}^{n} G_i e^{-t/\tau_i} \tag{45}$$

The corresponding constitutive equation analogous to equation 7a is

$$\sigma_{21} = \int_{-\infty}^{t} \left\{ \sum_i (G_i/\tau_i) e^{-(t-t')/\tau_i} \right\} \gamma_{21}(t')\, dt' \tag{46}$$

in which the quantity in brackets corresponds to the memory function. With a sufficient number of elements the continuous functions $G(t - t')$ and $m(t - t')$ can be represented by the sums indicated. If appropriate values for all G_i and τ_i (or η_i) have been assigned, in principle all the viscoelastic functions can be calculated. In practice, such a procedure is rarely attempted, except for rough calculations. The chief value of a model is as a guide for qualitative thinking.

The models of Figures 11 and 12 are equivalent with an appropriate assignment of parameters subject to certain requirements which depend on whether the system is a viscoelastic liquid or a viscoelastic solid. To make such a dichotomy may seem paradoxical since all viscoelastic materials are intermediate between solids and liquids. However, it is usually possible to distinguish between those which in a creep experiment eventually achieve steady flow or linearly increasing deformation with time (liquids) and those which eventually closely approach an equilibrium deformation or zero strain rate (solids). For liquids, $G(t)$ approaches zero as $t \rightarrow \infty$; for solids, $G(t)$ approaches a constant finite value (G_e). Only in systems with exceedingly long relaxation times is the distinction uncertain. For a *liquid,* all viscosities in the Maxwell model of Figure 11 must be finite and one spring in the Voigt model of Figure 12 must be zero. For a *solid,* one viscosity in the Maxwell model must be infinite and all springs in the Voigt model must be nonzero. It should be noted that a viscoelastic liquid, in this sense, may have a very high viscosity with the superficial appearance of a solid.

It was pointed out by Poincaré (20) that if a physical phenomenon can be represented by a mechanical model it can also be represented by an infinite number of other models. Thus, springs and dashpots could be arranged in countless different patterns, as discussed by Kuhn (21), all of them equivalent. In particular, there are rules for interrelating the parameters of the equivalent Maxwell and Voigt models (8a).

As an example of the qualitative usefulness of mechanical models, the difference between the two stress relaxation experiments portrayed in Figures 4 and 5 can be easily understood in terms of the model of Figure 11. If the strain is suddenly imposed (Fig. 4), the individual forces on the spring–dashpot pairs of the model are initially distributed in proportion to the relative stiffnesses of the *springs.* Subsequently, each contribution to the total force relaxes according to its own relaxation time. By contrast, if the experiment begins with a state of steady flow (Fig. 5), the individual forces on the spring–dashpot pairs are initially distributed in proportion to the relative frictional resistances of the *dashpots.* In the subsequent relaxation, the relaxation times are the same, but the magnitudes of the relaxing force contributions are different, so the course of the total force as a function of time is entirely different.

The time-dependent behavior of the Maxwell and Voigt elements is exactly analogous to the time-dependent electrical behavior of combinations of resistances and capacities or resistances and inductances. There are several possible ways of setting up the analogy. In particular, if capacities are equated to springs and dashpots to resistances, the storage and dissipative units correspond correctly physically, but the topology is backwards—ie, parallel mechanical connections correspond to series electrical connections. If resistances are equated to springs and capacities to dashpots, the two topologies are identical but the physical analogy is less satisfactory. A large literature has been devoted to this subject (22–24). The analog of the Boltzmann superposition principle in the electrical case is known as the Hopkinson superposition principle.

Discrete Viscoelastic Spectra. Any number of Maxwell elements in series have the properties of a single Maxwell element with $J = \Sigma J_i$ and $1/\eta = \Sigma(1/\eta_i)$; any number of Voigt elements in parallel have the properties of a single Voigt element with

$G = \Sigma G_i$ and $\eta = \Sigma \eta_i$. However, Maxwell elements in parallel or Voigt elements in series, as in Figures 11 and 12, obviously have much more complicated properties.

A group of Maxwell elements in parallel represents a discrete spectrum of relaxation times, each time τ_i being associated with a spectral strength G_i. Since in a parallel arrangement the forces (or stresses) are additive, it can readily be shown that for the Maxwell model, Figure 11, the viscoelastic functions $G(t)$, $G'(\omega)$, $G''(\omega)$, and $\eta'(\omega)$ are obtained simply by summing the expressions in equations 29–32 over all the parallel elements; thus if there are n elements, $G(t)$ is given by equation 45.

For a viscoelastic solid, one of the relaxation times must be infinite and the corresponding modulus contribution is G_e, the equilibrium modulus. The functions $J(t)$, $J'(\omega)$, and $J''(\omega)$ can also be calculated, but not in a simple manner.

A group of Voigt elements in series represents a discrete spectrum of retardation times, each time τ_i being associated with a spectral compliance magnitude J_i. Since in a series arrangement the strains are additive, it turns out that for the Voigt model (Fig. 12) the viscoelastic functions $J(t)$, $J'(\omega)$, and $J''(\omega)$ are obtained by summing the expressions in equations 36, 41, and 42 over all the series elements; thus

$$J(t) = \sum_{i=1}^{n} J_i \left(1 - e^{-t/\tau_i}\right) \tag{47}$$

(to which a term t/η must be added if one of the springs has zero rigidity, as must be the case for an uncrosslinked polymer), etc. The functions $G(t)$, $G'(\omega)$, $G''(\omega)$, and $\eta'(\omega)$ cannot be simply expressed for this model.

Any experimentally observed stress relaxation curve which decreases monotonically can in principle be fitted with any desired degree of accuracy to a series of terms as in equation 45 by taking n sufficiently large, and this would amount to determining the discrete spectrum of "lines," each with a location τ_i and intensity G_i. Similarly, fitting creep data to equation 47 would amount to experimental determination of the discrete retardation spectrum. Certain molecular theories do indeed predict discrete spectra corresponding to equations 45 and 47. In the analysis of experimental data, however, it is difficult or impossible to resolve more than a few lines. The contributions with the longest relaxation times can in principle be separated from a stress relaxation experiment by a procedure analogous to the analysis of radioactive decay in a mixture of radioactive species (25–27). Beyond this, the empirical choice of parameters τ_i and G_i (or J_i) would be largely arbitrary. Although an arbitary set of parameters would suffice to predict macroscopic behavior, it would not be unique and would be of little value for theoretical interpretation. This difficulty can be avoided, however, by substituting continuous spectra.

The Relaxation Spectrum. If the number of elements in the Maxwell model of Figure 11 is increased without limit, the result is a continuous spectrum in which each infinitesimal contribution to rigidity $F\,d\tau$ is associated with relaxation times lying in the range between τ and $\tau + d\tau$. Actually, experience has shown that a logarithmic time scale is far more convenient; accordingly the continuous relaxation spectrum is defined as $H\,d\ln\tau$, the contribution to rigidity associated with relaxation times whose logarithms lie in the range between $\ln\tau$ and $\ln\tau + d\ln\tau$. (Evidently, $H = F\tau$.) For the continuous spectrum, equation 45 becomes

$$G(t) = G_e + \int_{-\infty}^{\infty} He^{-t/\tau}\,d\ln\tau \tag{48}$$

which may alternatively be taken as a mathematical definition of H in terms of the relaxation function $G(t)$ which is introduced in the basic constitutive equation, equation 7. The constant G_e is added to allow for a discrete contribution to the spectrum with $\tau = \infty$, for viscoelastic solids; for viscoelastic liquids (uncrosslinked polymers), of course, $G_e = 0$. (Here we continue to follow the notation recommended by the Committee on Nomenclature of the Society of Rheology. Various other symbols have been used for H, and in some cases a spectrum has been defined by an equation analogous to equation 48 with $d \log_{10} \tau$ instead of $d \ln \tau$, thereby differing by a factor of 2.303.)

The Retardation Spectrum. In an entirely analogous manner, if the Voigt model in Figure 12 is made infinite in extent, it represents a continuous spectrum of retardation times, L, alternatively defined by the continuous analog of equation 47:

$$J(t) = J_g + \int_{-\infty}^{\infty} L\,(1 - e^{-t/\tau})\, d \ln \tau + t/\eta \tag{49}$$

In this case an instantaneous compliance J_g must be added to allow for the possibility of a discrete contribution with $\tau = 0$. (Although J_g may be inaccessible experimentally, its presence must be inferred or else instantaneous deformation would require infinite stress (28).)

The two spectra are of the nature of distribution functions, although they have the dimensions of a modulus (H) and a compliance (L), respectively, rather than the dimensionless character of the usual distribution function. In some treatments of linear viscoelastic behavior, normalized dimensionless distributions are employed:

$$h = H/G_g$$

$$l = L/(J_e - J_g)$$

where G_g is the stress/strain ratio for an instantaneous deformation and, for a viscoelastic solid, J_e is the equilibrium compliance; for a viscoelastic liquid, J_e would be replaced by J_e^0. Although these functions have mathematical convenience (their integrals over $d \ln \tau$ from $-\infty$ to ∞ are unity), they are less practical to apply to experimental data, because the normalizing factors are usually known only with poor precision or else are operationally inaccessible. Hence the non-normalized functions H and L will be used throughout this article.

Actually, H and L are not essential to the treatment of linear viscoelastic behavior. The predictions of molecular theories can, for example, be expressed in the form of directly measurable quantities such as $G(t)$, $G'(\omega)$, and $G''(\omega)$, and compared directly with experiments. Also simultaneous measurements of $G'(\omega)$ and $G''(\omega)$ over a range of frequencies can in principle be interconverted by approximation formulas to check the internal consistency of experiments. However, the easiest and most rapid method of testing the consistency of measured values of G', G'', and $G(t)$ is to convert all of them to H by simple approximation formulas; the simplest way to compare J', J'', and $J(t)$ is through the function L. Moreover, the spectra H and L are useful qualitatively in gaging the distribution of relaxation or retardation mechanisms in different regions of the time scale.

The Boltzmann Superposition Principle; Elastic Recovery

Equations 7 and 8 are two of many possible expressions of the Boltzmann superposition principle (1,29) that the effects of mechanical history are linearly additive, where the stress is described as a function of strain history or rate of strain history or alternatively the strain is described as a function of the history of rate of change of stress. Many early treatises have elaborated on the effects of different sequences of stressing or straining according to this principle, and tested it by a variety of experiments. The procedure may be illustrated by some characteristic sequences of stress history.

We take first a trivial example of just two events in stress history. With the understanding that the deformation is simple shear, the coordinate subscripts for γ and σ will be omitted. If a stress σ_A is applied at $t = 0$ and an additional stress σ_B is applied at $t = t_1$, the total strain at time t is found by applying equation 8 to be

$$\gamma(t) = \sigma_A J(t) + \sigma_B J(t - t_1) \tag{50}$$

namely, the linear superposition of the two strains specified by the creep compliances at their respective elapsed times. This can be generalized for a sequence of finite stress changes σ_i, each at a time point u_i, to give

$$\gamma(t) = \sum_{u_i=-\infty}^{u_i=t} \sigma_i J(t - u_i) \tag{51}$$

A sequence of particular interest is the case where a creep experiment has progressed for some time and the stress is then suddenly removed. The rate of deformation will change sign and the body will gradually return more or less toward its initial state (ie, for shear deformation, toward its initial shape). The course of this reverse deformation is called creep recovery. The results depend in a critical manner on whether the material is a viscoelastic solid or a viscoelastic liquid as distinguished above; ie, whether or not it eventually reaches an equilibrium deformation in creep, characterized by an equilibrium compliance (in shear, J_e). As explained above, attainment of equilibrium corresponds to a Maxwell model (Fig. 11) in which one dashpot is missing—a viscoelastic solid. In bulk compression, an equilibrium compliance no doubt always exists—it is simply the thermodynamic compressibility. For shear and other deformation of polymers, the presence or absence of an equilibrium compliance corresponds approximately to the presence or absence of a crosslinked network in the molecular structure.

In a crosslinked polymer, shear creep followed by creep recovery is essentially as shown schematically in Figure 13. If the stress σ_0 has been applied for a time t_2 long enough for the strain to reach within experimental error its equilibrium value $\gamma(\infty) = \sigma_0 J_e$, then in accordance with equation 51 the course of creep recovery will be given by

$$\gamma_r(t) = \sigma_0[J_e - J(t - t_2)] \tag{52}$$

(noting that removal of stress is equivalent to applying an additional stress $-\sigma_0$). This is just a mirror image of the creep itself in the time axis, displaced vertically by $\gamma(\infty)$ and horizontally by t_2. If, on the other hand, the load is removed at a time t_1 before $\gamma(\infty)$ is reached, the course of recovery will be

$$\gamma_r(t) = \sigma_0[J(t) - J(t - t_1)] \tag{53}$$

which, although not a simple mirror image, is readily susceptible of experimental test by calculating the right side of the equation from a duplicate creep run carried to longer times. The results of numerous such tests, which support the validity of the super-

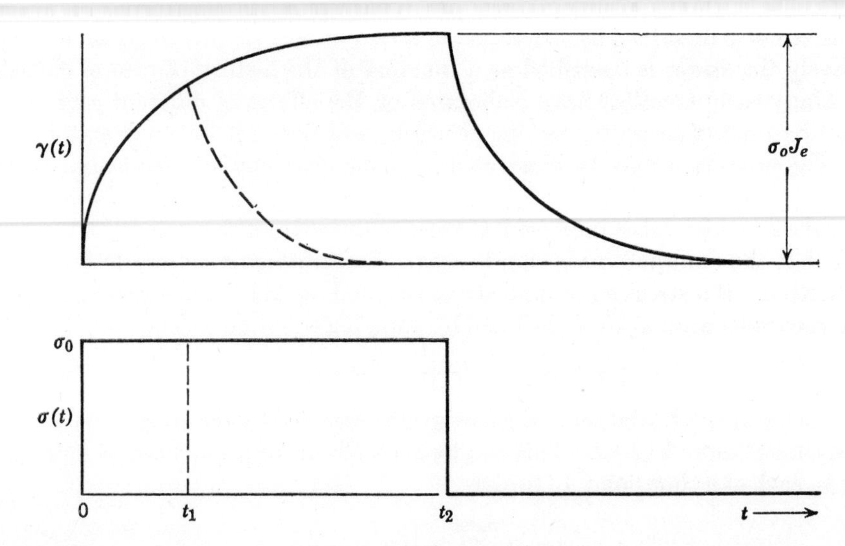

Fig. 13. Shear creep and creep recovery shown schematically for a viscoelastic solid, eg, a crosslinked polymer.

position principle and equations 7 and 8, have been summarized by Leaderman (1). These tests have more often been performed for deformation in simple extension than in shear; then equations 52 and 53 hold with D (defined below) substituted for J and appropriate changes in the stress and strain components.

If the deformations are not kept small, but are carried to the point where the elastic behavior is nonlinear, equations 52 and 53 do not hold. For soft polymeric solids, deviations from linearity appear sooner (ie, at smaller strains) in extension than in shear, because of the geometrical effects of finite deformations. At substantial extensions, the relations between creep and recovery are considerably more complicated than those given above.

As explained above, a viscoelastic liquid, such as an uncrosslinked polymer whose threadlike molecules are not permanently attached to each other, has no equilibrium compliance. Then under constant stress the rate of strain approaches a limiting value and a situation of steady-state flow is eventually attained, governed by a Newtonian viscosity η which is most readily visualized as the sum of all the viscosities in Figure 11. Concomitantly, in this steady state the springs of the models are stretched to equilibrium extensions, representing elastic energy storage which is recoverable after the stress is removed. Thus, even though there is no equilibrium compliance, there is a steady-state compliance which is most readily visualized as the sum of the deformations of the springs in Figure 12 (equipped with a viscosity in series to make it a viscoelastic liquid) divided by the stress.

For such a material, the shear creep followed by creep recovery is shown schematically in Figure 14. The creep is the sum of a deformation approaching a constant value $J_e^0\sigma_0$, analogous to $J_e\sigma_0$ in Figure 13, plus a viscous flow contribution $\sigma_0 t/\eta$. Here J_e^0 is the steady-state compliance, a measure of the elastic deformation during steady flow. After a sufficiently long time (but before the stress is removed at t_2) the creep strain is given by

$$\gamma(t) = \sigma_0(J_e^0 + t/\eta) \tag{54}$$

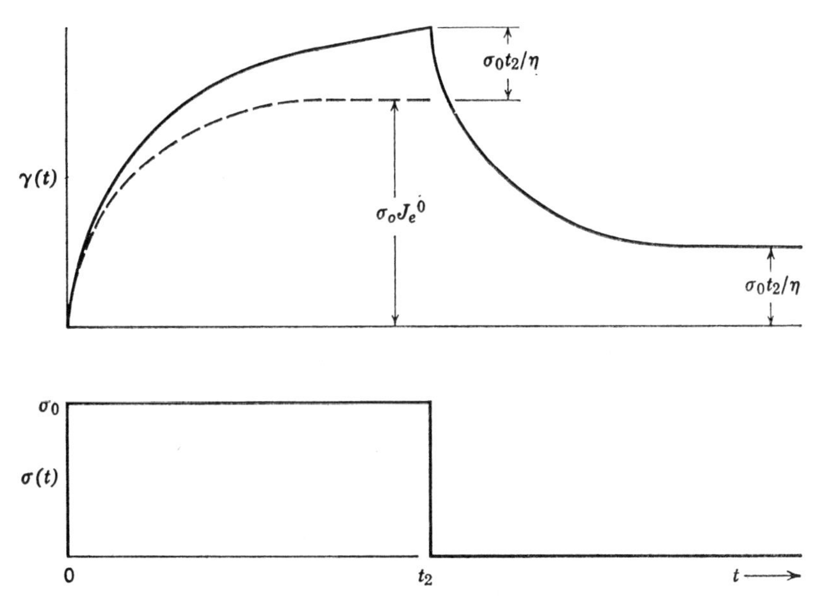

Fig. 14. Shear creep and creep recovery shown schematically for a viscoelastic liquid, eg, an uncrosslinked polymer.

Thus both J_e^0 and η can be obtained from the geometry of a linear plot in this region; whereas the recovery is given by

$$\gamma_r(t) = \sigma_0[J_e^0 + t/\eta - J(t - t_2)] \tag{55}$$

and it approaches a final value of $\sigma_0 t_2/\eta$, providing an alternative determination of η. At stresses sufficiently high to evoke substantial strain rates, these linear equations do not hold, and elastic recovery must be formulated in a more complicated manner.

Equations 50–55 are written for simple shear. Analogous relations hold for simple extension with all J's replaced by D's and the shear viscosity η replaced by the tensile viscosity $\bar{\eta}$, which is defined as $\sigma_{11}/\dot{\epsilon}$ in steady flow.

The sharp dichotomy between viscoelastic solids (ordinarily, crosslinked polymer systems) and viscoelastic liquids (ordinarily, uncrosslinked polymer systems) is apparent in all the time-dependent and frequency-dependent viscoelastic functions which describe their mechanical behavior in small deformations.

Stress–Strain Relations for Other Types of Deformation

In simple shear, the change in shape is not accompanied by any change in volume; this feature facilitates interpretation of the mechanical behavior in molecular terms. There are other deformation geometries which are characterized by a change in volume or combined change in volume and shape. To describe the mechanical behavior under such conditions it is necessary to introduce a more complicated constitutive equation including all components of the stress tensor with two time-dependent functions (30):

$$\sigma_{ij} = \int_{-\infty}^{t} \{G(t - t')\,[\dot{\gamma}_{ij} - \tfrac{1}{3}\,\Sigma\,\dot{\gamma}_{kk}\,\delta_{ij}] + \tfrac{3}{2}\,K(t - t')\,[\tfrac{1}{3}\,\Sigma\,\dot{\gamma}_{kk}\,\delta_{ij}]\}\,dt' \tag{56}$$

where δ is the Kronecker delta. Here $G(t - t')$ is the same shear stress relaxation modulus as in the preceding section; $K(t - t')$ has the physical significance of a vol-

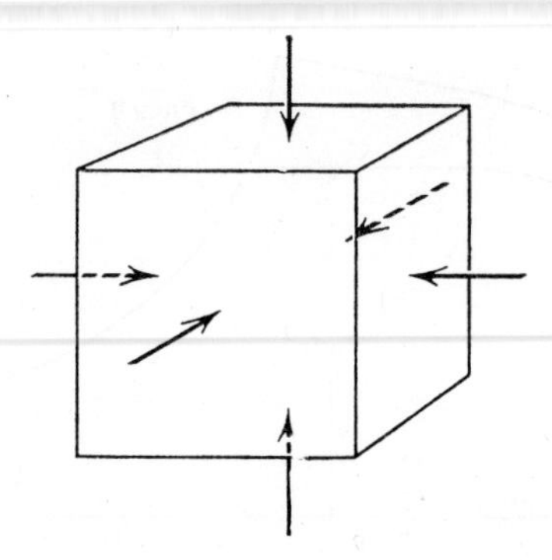

Fig. 15. Geometrical arrangement for bulk compression.

umetric relaxation modulus as will be shown by treatment of a bulk compression experiment, below.

When perceptible changes in volume occur, both $G(t - t')$ and $K(t - t')$ are found to depend strongly on the density or degree of compression or dilatation (30a). Hence linear behavior must be limited to extremely small volume changes.

Bulk Compression or Dilatation. If the dimensions of an isotropic cubical element are increased or decreased uniformly by application of normal forces on all faces (Fig. 15) $\sigma_{11} = \sigma_{22} = \sigma_{23} = -P$, the hydrostatic pressure, and all other stress components are zero; and the three $\dot{\gamma}_{kk}$ are equal. Equation 56 then becomes

$$-P = \tfrac{3}{2} \int_{-\infty}^{t} K(t - t')\dot{\gamma}_{kk}\, dt' \qquad (57)$$

A calculation corresponding to that of equations 9 to 11 will provide the time dependence of pressure following a sudden voluminal deformation. For this purpose, the voluminal strain is conventionally defined as the relative change in volume, $\Delta V/V = 3\,\partial u_1/\partial x_1 = (\tfrac{3}{2})\gamma_{11}$. Positive volume change is dilatation, negative is compression. Then after an initial strain $\Delta V/V$ accomplished in a very small time interval at $t = 0$, we have

$$P(t) = -(\Delta V/V)K(t) \qquad (58)$$

analogous to equation 11 for shear; $K(t)$ is termed the bulk relaxation modulus. Alternatively, if the pressure P is applied suddenly and the volume change is followed as a function of time, the bulk creep experiment is described by the analog of equation 12,

$$(\Delta V/V)\ (t) = -PB(t) \qquad (59)$$

where $B(t)$ is the bulk creep compliance. The complex dynamic modulus $\mathbf{K}^*$ and compliances $\mathbf{B}^*$ can be defined in the same way as those for shear, with $\mathbf{K}^* = K' + iK''$ and $\mathbf{B}^* = B' - iB''$. The storage bulk modulus K' is the component of bulk stress in phase with the volumetric strain divided by the strain, etc. A summary of all moduli and compliances is given at the end of this section. For a solid or liquid in which there is no time-dependent response, or for any material at equilibrium under pressure, $B(t)$ and $\mathbf{B}^*$ become the thermodynamic compressibility $\beta = -1(V)\ (\partial V/\partial P)_T$; and $K(t)$ and $\mathbf{K}^*$ become the equilibrium bulk modulus K_e.

Fig. 16. Geometrical arrangement for simple extension.

Bulk compression would not be expected to involve changes in long-range configuration or contour shape, and in fact the differences between polymers and simple liquids and solids are not so striking in compression as in shear.

The subscript T in the above partial derivative implies an isothermal measurement and indeed all of the experimental examples given here are supposed to be carried out isothermally.

Simple Extension. If an isotropic cubical element is elongated in one direction and the dimensional changes in the two mutually perpendicular directions are equal (Fig. 16), the experiment is termed simple extension; if γ_{11} is positive, γ_{22} and γ_{33} are equal to each other and negative (or possibly zero). The only stress is σ_{11}. This type of deformation results when a rod, strip, or fiber is subjected to a tensile force. Equation 56 for this case becomes

$$\sigma_{11} = \int_{-\infty}^{t} \{G(t - t')\,[\dot{\gamma}_{11} - \tfrac{1}{3}(\dot{\gamma}_{11} + 2\dot{\gamma}_{22})] + \tfrac{3}{2}K(t - t')\,[\tfrac{1}{3}(\dot{\gamma}_{11} + 2\dot{\gamma}_{22})]\}\,dt \quad (60)$$

The calculation of time-dependent stress following a tensile strain imposed in a very brief time interval at $t = 0$ is now somewhat more complicated because both relaxation moduli $G(t)$ and $K(t)$ are involved. The strain γ_{11} is accompanied by equal lateral changes γ_{22} and γ_{33}. An integration corresponding to equations 9–11 gives

$$\sigma_{11}(t) = (\tfrac{2}{3}\,\gamma_{11} - \tfrac{2}{3}\gamma_{22})\,G(t) + \tfrac{3}{2}\,(\tfrac{1}{3}\gamma_{11} + \tfrac{2}{3}\gamma_{22})\,K(t) \quad (61a)$$

$$\sigma_{22}(t) = (-\tfrac{1}{3}\gamma_{11} + \tfrac{1}{3}\gamma_{22})\,G(t) + \tfrac{3}{2}(\tfrac{1}{3}\gamma_{11} + \tfrac{2}{3}\gamma_{22})\,K(t) \quad (61b)$$

By setting $\sigma_{22} = 0$, since there is no lateral stress, and eliminating γ_{22} from equations 61, $\sigma_{11}(t)$ can be expressed in terms of $G(t)$ and $K(t)$. For this purpose, we note that the practical definition of tensile strain is $\epsilon = \partial u_1/\partial x_1 = \tfrac{1}{2}\gamma_{11}$. The final result is

$$\sigma_{11}(t) = \epsilon\left[\frac{9G(t)K(t)}{G(t) + 3K(t)}\right] \quad (62)$$

The quantity in brackets is denoted by $E(t)$, the tensile relaxation modulus. For equilibrium deformation of an elastic solid, $E = 9GK/(G + 3K)$, of course, and this quantity is termed *Young's modulus.*

For creep in extension, the shear and bulk functions combine in a somewhat simpler manner. A sudden tensile stress σ_{11} produces a time-dependent strain

$$\epsilon(t) = \sigma_{11}D(t) \quad (63)$$

where $D(t)$ is the tensile creep function, which is related to shear and bulk creep thus:

$$D(t) = J(t)/3 + B(t)/9 \quad (64)$$

The complex dynamic tensile modulus $\mathbf{E}^* = E' + iE''$ and compliance $\mathbf{D}^* = D' - iD''$ are defined by analogs of equations 16–27.

Tensile experiments are often easy to perform but have the disadvantage that simultaneous changes in both shape and volume make the behavior more difficult to interpret on a molecular basis; moreover, perceptible volume changes may significantly modify the relaxation functions as mentioned above. However, in polymeric systems, in certain broad ranges of time scale, $K(t)$ is often greater than $G(t)$ by two orders of magnitude or more. In this case, equations 62 and 64 become

$$E(t) = 3G(t) \tag{65}$$

$$D(t) = J(t)/3 \tag{66}$$

Then simple extension gives the same information as simple shear, and the results of the two experiments are interconvertible. Physically, this fact arises because the change in volume caused by the extension is negligible in comparison with the change in shape. When $J \gg B$ or $G \ll K$, the material is often loosely but inaccurately called "incompressible"; a "soft elastic solid" might be better. For any viscoelastic material, however, there will in general be some region of time scale where this approximation is inapplicable; G and K will approach each other in magnitude at very short times or very high frequencies.

The relations between G and E can also be written in terms of a dimensionless variable, Poisson's ratio, $\mu = [1 - (1/V)(\partial V/\partial \epsilon)]/2$, corresponding to the degree of lateral contraction accompanying a longitudinal extension. When $J \gg B$, there is no significant volume increase on extension and $\mu = 1/2$. If the volume had increased sufficiently to avoid any lateral contraction of the rod in Figure 16, $(1/V)(\partial V/\partial \epsilon)$ would be unity and $\mu = 0$. The minimum value of μ ordinarily observed is about 0.2 for homogeneous, isotropic materials, but may be smaller for certain heterogeneous materials such as cork or sponge rubber. There are twelve formulas (8a,9) connecting G, K, E, and μ, of which only two will be given here:

$$E = 2G(1 + \mu) \tag{67}$$

$$E = 3K(1 - 2\mu) \tag{68}$$

These show again that, as μ approaches 1/2, E becomes equal to $3G$ and $K \gg E$.

In equations 67 and 68, the moduli can be replaced by the corresponding time-dependent or complex quantities, but then Poisson's ratio must also become either time-dependent or complex.

One-Dimensional Extension in Infinite Medium (Bulk Longitudinal Deformation). An alternative version of extension occurs when the dimensions change in one direction and are *constrained* to be constant in the two mutually perpendicular directions. This is achieved by placing a thin flat sample, whose faces are bonded to rigid members, under tension or compression in the thin direction (Fig. 17). Under these conditions, $\gamma_{22} = \gamma_{33} = 0$. If a sudden tensile strain $\epsilon = \frac{1}{2}\gamma_{11}$ is accomplished and the stress relaxation is followed as a function of time, a calculation analogous to that of the preceding section gives

$$\sigma_{11}(t) = \epsilon[K(t) + (4/3)G(t)] \tag{69}$$

The quantity in brackets is denoted by $M(t)$, and may be called the bulk longitudinal relaxation modulus. For a soft elastic solid with μ only slightly different from 1/2 and $K \gg G$, $M(t)$ is evidently indistinguishable from the bulk modulus. The corresponding complex dynamic modulus $\mathbf{M}^* = M' + iM''$ can be defined in the usual manner. The propagation of a longitudinal elastic wave in a medium all of whose dimensions are large compared with the wavelength is specified by $\mathbf{M}^*$ and the density of the medium $\mathbf{M}^*$ is much more commonly measured experimentally than $M(t)$.

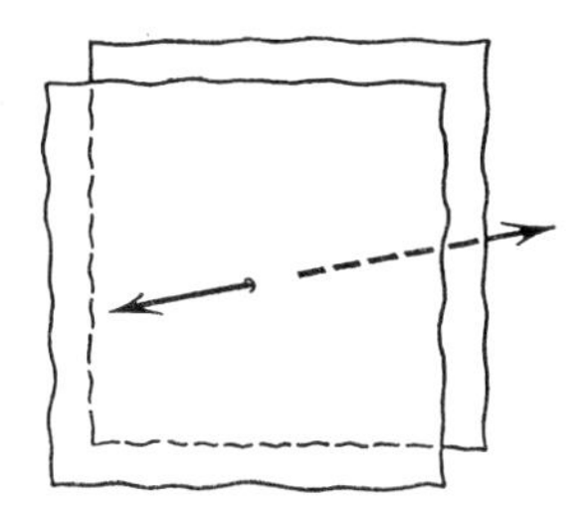

Fig. 17. Geometrical arrangement for bulk longitudinal deformation.

Inhomogeneous Deformations. All the previous examples represent homogeneous deformations in which the strains are the same everywhere in the sample and the macroscopic deformation of the sample is essentially the same as that of any infinitesimal element of it. In many deformation geometries commonly used for experimental measurements, however, such as torsion between coaxial cylinders, torsion of a cylindrical rod, flow through a tube, flexure, etc, the magnitudes of the strains and rates of strain vary from point to point. Application of the equations of continuity and motion and integration over the sample geometry are then necessary to relate external forces and displacements to the viscoelastic functions. For some cases, if the deformations are small, the geometry is no real complication; for example, in any kind of torsion, a sudden angular deformation followed by measurement of the torque as a function of time will provide the shear relaxation modulus $G(t)$. For large deformations, however, different experimental geometries provide different kinds of information (17b).

Finite Strains and Normal Stress Differences

When strains are not infinitesimal, the definitions of equation 3 are inapplicable; there are several alternative ways of defining finite strain. One fortunate simplification is that for most viscoelastic substances which are sufficiently soft to support substantial deformations without breaking, $K \gg G$ and so volume changes can be neglected. (Although, as stated previously, there will in general be some range of time or frequency where $K \gg G$, finite deformations will usually be possible only under conditions where $K(t) \gg G(t)$ and $\mathbf{K}^*(\omega) \gg \mathbf{G}^*(\omega)$.) In this case, a single relaxation modulus suffices for the constitutive equation. If there is a negligible change in volume, the density is constant and the equation of continuity, equation 1, specifies that

$$\sum_{i=1}^{3} \partial v_i/\partial x_i = 0 = \sum_{k=1}^{3} \dot{\gamma}_{kk}$$

Then, for linear behavior, equation 56 simplifies to a very simple expression for infinitesimal deformations of an "incompressible" material:

$$\sigma_{ij} = -P\delta_{ij} + \int_{-\infty}^{t} G(t - t')\dot{\gamma}_{ij}(t')\, dt \qquad (70)$$

The term in the hydrostatic pressure P appears because the normal stresses σ_{ij} are indeterminate if the material undergoes no perceptible compression. This feature does not affect the shear stresses.

For finite deformations, an equation of the same form as equation 70 may be written in coordinates which deform with the material (31) and then transformed back

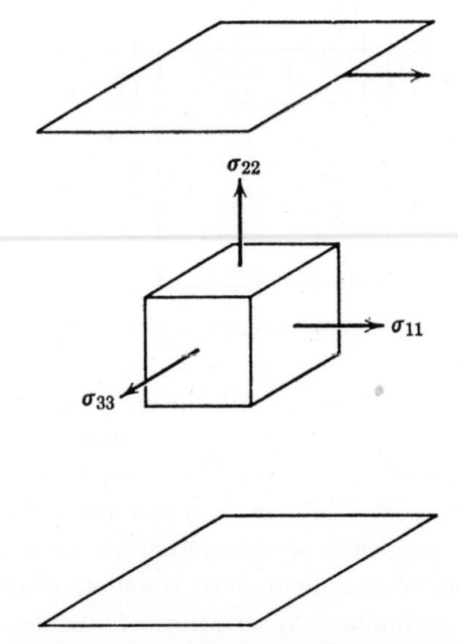

Fig. 18. Normal stresses in steady-state shear flow.

into fixed (laboratory) coordinates; there are various alternative procedures which give equations that have the appropriate invariance properties and reduce to equation 70 for infinitesimal deformations (9,12,13,32). A widely used nonlinear extension of equation 70 is that discussed by Fredrickson (32) which is equivalent to the formulation of the "elastic liquid" by Lodge (13) though the latter is written in terms of γ_{ij} rather than $\dot{\gamma}_{ij}$. It can be represented (16) by the constitutive equation

$$\sigma_{ij} = -P\delta_{ij} + \int_{-\infty}^{t} G(t-t') \sum_{m,n}^{3} \frac{\partial x_i}{\partial x'_m} \frac{\partial x_j}{\partial x'_n} \dot{\gamma}_{mn}(t')/dt' \tag{71}$$

where x_i and x'_1 are the coordinates of a particle of the material at times t and t' respectively. This predicts a number of additional features of behavior.

One important consequence of this more general description is the appearance of normal stresses in simple shearing deformations. Thus, even in steady-state simple shear flow (Fig. 18) where the rate of strain tensor (cf equations 3 and 5) is

$$\dot{\gamma}_{ij} = \begin{pmatrix} 0 & 1 & 0 \\ 1 & 0 & 0 \\ 0 & 0 & 0 \end{pmatrix} \dot{\gamma} \tag{72}$$

the stress tensor has different diagonal components:

$$\sigma_{ij} = \begin{pmatrix} -P + I_2\dot{\gamma}^2 & I_1\dot{\gamma} & 0 \\ I_1\dot{\gamma} & -P & 0 \\ 0 & 0 & -P \end{pmatrix} \tag{73}$$

Here

$$I_1 = \int_0^\infty G(t)\,dt$$

which is simply the steady-flow viscosity η

$$I_2 = 2\int_0^\infty G(t)t\,dt$$

which can be shown to be $2\eta^2 J_e^0$. It should be noted that because of the appearance of the arbitrary hydrostatic pressure P in equation 73, the normal stresses cannot be specified on an absolute basis; the result can only be given in terms of the normal stress differences,

$$\sigma_{11} - \sigma_{22} = \sigma_{11} - \sigma_{33} = 2\eta^2 J_e^0 \dot{\gamma}^2 \tag{74}$$

This result is not sufficiently general, however; it implies that $\sigma_{22} - \sigma_{33} = 0$ (a relation sometimes known as the Weissenberg hypothesis). Experimentally it is observed that $\sigma_{22} - \sigma_{33}$ is not zero though it may be much smaller than $\sigma_{11} - \sigma_{22}$. The more general constitutive relations of the second-order viscoelasticity theory of Coleman and Noll (32) give (33) the result of equation 74 for $\sigma_{11} - \sigma_{22}$ but a different prediction for $\sigma_{11} - \sigma_{33}$. Further generalizations are necessary to describe experimental results satisfactorily. For example, the proportionality of $\sigma_{11} - \sigma_{22}$ to $\dot{\gamma}^2$ implied in equation 74 is observed only at very small values of $\dot{\gamma}$. Moreover, the appearance of nonlinear stress relaxation and non-Newtonian viscosity, as described below, can be predicted only from more complicated constitutive equations, many of which have been proposed (17,34–36).

If steady shear flow of a viscoelastic liquid is abruptly halted, the normal stress differences $\sigma_{11} - \sigma_{22}$ and $\sigma_{22} - \sigma_{33}$ decay gradually with time as does the shear stress (Fig. 5), but following a different relaxation function. Also, for sinusoidally varying shear strains, the normal stresses are sinusoidal functions of time, with oscillatory components at twice the frequency of the shear strain in addition to constant components. Other time-dependent experimental patterns of strain or stress history evoke various characteristic nonlinear phenomena.

A perfectly elastic solid subjected to large shear deformations also exhibits normal stress differences (37). At equilibrium these are constant and strictly speaking outside the scope of viscoelasticity. The simplest extension beyond infinitesimal deformations leads to the relation (13)

$$\sigma_{11} - \sigma_{22} = G_e \gamma_{21}^2 = \sigma_{21}^2 / G_e \tag{75}$$

For viscoelastic solids, time-dependent shearing deformations produce normal stress differences which are time or frequency dependent.

For both viscoelastic liquids and viscoelastic solids, combinations of large shear rates or large static deformations, respectively, with small time-dependent deformations result in a variety of characteristic behavior.

Non-Newtonian Flow and Associated Phenomena

In polymeric liquids, the ratio $\sigma_{21}/\dot{\gamma}_{21}$ in steady shearing flow is not independent of $\dot{\gamma}_{21}$; it falls with increasing $\dot{\gamma}_{21}$, in some cases by several orders of magnitude. This ratio, which for a Newtonian liquid would be the constant viscosity η, is called the non-Newtonian viscosity, denoted here by $\breve{\eta}$.

This feature and several related phenomena can be taken into account by modifying (38) equation 71 to make the function $G(t - t')$ dependent on $\dot{\gamma}(t')$, where

$$\dot{\gamma}(t') = \sum_m \sum_m \dot{\gamma}_{ij}$$

evaluated at t'. It follows that I_1 and I_2 in equation 73 depend on $\dot{\gamma}$; both $\breve{\eta}$ and the ratio $(\sigma_{11} - \sigma_{22})/\dot{\gamma}^2$ (denoted by θ—cf equation 74) diminish with increasing $\dot{\gamma}$. Associated with this behavior are various other nonlinear phenomena including stress relaxation following cessation of steady-state flow, the course of which depends on the magnitude of $\dot{\gamma}$ during flow; stress relaxation following sudden deformation, the course of which depends on the value of γ; and creep recovery, the course of which depends on either γ or $\dot{\gamma}$.

At relatively low values of $\dot{\gamma}$ (again omitting the subscript $_{21}$ for convenience), where deviations from Newtonian flow are slight, it is convenient to formulate the non-Newtonian viscosity thus:

$$\breve{\eta} = \eta[1 - f(\dot{\gamma}\tau_{\breve{\eta}})] \tag{76}$$

in which η is the limiting value of $\breve{\eta}$ at low shear rates. Here $\tau_{\breve{\eta}}$ is a characteristic time the reciprocal of which measures the critical magnitude of $\dot{\gamma}$ for onset of non-Newtonian behavior, and the function f can be expressed as a power series, often in even powers of $\dot{\gamma}\tau_{\breve{\eta}}$. At high shear rates, however, $\breve{\eta}$ is often proportional to $\dot{\gamma}^{-a}$ where a is a fractional exponent. Such power law behavior is observed in ranges of $\dot{\gamma}$ occurring in technological processing of polymeric systems, where $\breve{\eta}$ is a very important quantity. In many cases it is difficult or impossible to make measurements at sufficiently small $\dot{\gamma}$ to obtain a reliable value of η.

In many treatises, the symbol η is used for non-Newtonian viscosity $\sigma_{21}/\dot{\gamma}_{21}$ and η_0 represents the Newtonian viscosity which is approached at vanishing shear rate. In this article, however, reference to Newtonian (vanishing shear rate) viscosity is made so much more frequently than to non-Newtonian viscosity that η is used for the former to avoid incessant use of the subscript 0.

Summary of Moduli and Compliances

The various moduli and compliances which have been introduced for infinitesimal deformations are summarized in Table 1. All moduli have the dimensions of stress (units usually dynes/cm^2); all compliances have the dimensions of reciprocal stress

Table 1. Summary of Moduli and Compliances

Deformation	Simple shear	Bulk compression	Simple extension[a]	Bulk longitudinal
relaxation modulus	$G(t)$	$K(t)$	$E(t)$	$M(t)$
creep compliance	$J(t)$	$B(t)$	$D(t)$	
complex modulus	$\mathbf{G}^*(\omega)$	$\mathbf{K}^*(\omega)$	$\mathbf{E}^*(\omega)$	$\mathbf{M}^*(\omega)$
storage modulus	$G'(\omega)$	$K'(\omega)$	$E'(\omega)$	$M'(\omega)$
loss modulus	$G''(\omega)$	$K''(\omega)$	$E''(\omega)$	$M''(\omega)$
complex compliance	$\mathbf{J}^*(\omega)$	$\mathbf{B}^*(\omega)$	$\mathbf{D}^*(\omega)$	
storage compliance	$J'(\omega)$	$B'(\omega)$	$D'(\omega)$	
loss compliance	$J''(\omega)$	$B''(\omega)$	$D''(\omega)$	
equilibrium modulus	G_e	K_e	E_e	M_e
glasslike modulus	G_g	K_g	E_g	M
equilibrium compliance	J_e	$B_e(=\beta)$	D_e	
glasslike compliance	J_g	B_g	D_g	
steady-state compliance	J_e^0		D_e^0	
steady-flow viscosity[b]	η		$\bar{\eta}$	
dynamic viscosity	$\eta'(\omega)$	$\eta'_v(\omega)$	$\bar{\eta}'(\omega)$	$\eta'_m(\omega)$

[a] For this type of deformation each modulus may be called a Young's modulus (relaxation Young's modulus, storage Young's modulus, etc).

[b] At vanishing shear rate.

(usually $cm^2/dyne$). The quantities in any row of the table can be interrelated by equations of the form of 62, 64, and 69. The symbols follow rather closely the recommendations of a Committee of the Society of Rheology (39).

Illustrations of Viscoelastic Behavior

It seems desirable to familiarize the reader with some concrete examples of the viscoelastic phenomena defined above, to provide an idea of their character as exhibited by various types of polymeric systems. Linear viscoelastic behavior in shear will be illustrated in considerable detail, with a few additional examples of bulk viscoelastic behavior and nonlinear phenomena. The examples are accompanied by some qualitative remarks about molecular interpretation.

Linear Viscoelastic Behavior in Shear or Simple Extension

When the shear creep compliance or the shear relaxation modulus for a particular polymer is plotted against time, or any one of the dynamic functions is plotted against frequency, the most striking feature is the enormous range of magnitudes which the ordinate can assume, changing over several powers of ten. Concomitantly, a still larger range of time or frequency is required on the abscissa scale to encompass these changes. As a result, both coordinates are usually plotted logarithmically. Inspection of such logarithmic graphs reveals a pattern of certain zones of the time (or frequency) scale, where the viscoelastic functions have characteristic shapes: the transition zone from glasslike to rubberlike consistency, the plateau zone, the pseudo-equilibrium zone (in crosslinked polymers), the terminal zone (in uncrosslinked polymers), etc. These regions can be associated qualitatively with different kinds of molecular responses, and appear with different degrees of prominence depending on whether the polymer is of low or high molecular weight, amorphous or crystalline, above or below its glass-transition temperature, and undiluted or mixed with solvent.

Eight polymeric systems have been chosen as examples of these alternatives to illustrate the variety of viscoelastic responses and the gross correlation of behavior in different time and frequency zones with molecular structure. The graphs to be portrayed here represent experimental data from the literature which have been combined by the method of reduced variables (18) to cover as wide a range as possible of the effective time or frequency scale. All the measurements were made on isotropic materials at sufficiently low stresses so that the viscoelastic behavior was linear, and the deformation was usually simple shear though in two cases it was simple extension (in which the shear effects predominate). It was necessary in all cases to calculate some of the viscoelastic functions indirectly from others that were directly measured.

Description of the Polymers Chosen for Illustration. The first four systems involve uncrosslinked polymers; they are viscoelastic liquids in the sense that they do not possess any equilibrium compliance and exhibit viscous flow at sufficiently long times. (The reader is reminded, however, that such a viscoelastic liquid may have a very high viscosity and the superficial appearance of a rubbery solid.)

I. To illustrate a *dilute polymer solution*, in which the viscoelasticity is a relatively minor perturbation of the Newtonian behavior of the solvent: a 2% solution of polystyrene, molecular weight 82,000 with sharp molecular-weight distribution, in Aroclor 1248, a chlorinated diphenyl with viscosity 2.6 poises. Dynamic shear data (40) at low frequencies by Holmes and Ferry and at high frequencies by Lamb and Matheson were combined and reduced to 25°C.

The significance of the reference temperature is that the reduced curves represent the viscoelastic functions as they would have been measured at 25°C over a much wider range of time or frequency scale than the actual experimental measurements provided.

Fairly dilute solutions such as this are of considerable interest because they represent the only type of system to which some molecular theories (40a) are clearly applicable; the polymer molecules are sufficiently separated to move almost independently of each other. The remaining examples described here are all undiluted polymers in which the molecules are extensively intertwined.

II. To illustrate an *amorphous polymer of low molecular weight:* a fractionated poly(vinyl acetate) with molecular weight 10,500. Shear creep data of Ninomiya and Ferry (41) were employed, reduced to 75°C. For a linear molecule as short as these (degree of polymerization about 120), the effects of neighboring molecules on viscoelastic properties can be rather well described in terms of the local frictional forces encountered by a short segment of a moving chain.

III. To illustrate an *amorphous polymer of high molecular weight:* an atactic poly(methyl methacrylate) with moderately broad molecular-weight distribution, weight-average molecular weight 180,000. Shear creep data of Plazek and O'Rourke (42) were employed, reduced to 110°C. This is sharply differentiated from the preceding example by its higher molecular weight. Above a critical molecular weight, which for many polymers is on the order of 20,000, the effect of neighbors on molecular motion can no longer be described solely in terms of local frictional forces; the viscoelastic properties reveal a strong additional coupling to neighbors which acts as though it were localized at a few widely separated points along the molecular chain. This phenomenon, generally known as entanglement coupling, is imperfectly understood, but it clearly prolongs very greatly any molecular rearrangements which are sufficiently long-range to involve regions of a molecule separated from each other by one or more entanglement points. The term "entanglement" has come to be employed in this special sense, and is not currently applied to the short-range intermolecular entwining which must exist in all polymeric systems (other than very dilute solutions) regardless of their molecular length.

IV. To illustrate an *amorphous polymer of high molecular weight with long side groups:* a fractionated poly(*n*-octyl methacrylate) of weight-average molecular weight 3,620,000. Dynamic and shear creep data of Dannhauser, Child, and Ferry (43) and Berge, Saunders, and Ferry (44) were employed, after reduction to a reference temperature of 100°C.

The molecule differs from the preceding in that each monomer unit, with two chain atoms, carries a flexible side ester group which now comprises nearly three-fourths of the monomeric molecular weight. Thus, in a sense, only a small proportion of the total volume is occupied by the chain backbones.

The remaining four examples are viscoelastic solids in the sense that they do not exhibit viscous flow and under a constant stress they eventually reach (or closely approach) an equilibrium deformation.

V. To illustrate an *amorphous polymer of high molecular weight below its glass-transition temperature:* a poly(methyl methacrylate) of high molecular weight. Shear stress relaxation (45) and shear creep (46) data of Iwayanagi were employed, reduced to a reference temperature of −22°C.

The distinguishing feature of this example is that the measurements are all far below the glass-transition temperature (about 100°C) where the chain backbone configurations are largely immobilized. Hence the response to external stress involves primarily very local adjustments somewhat similar to those in the mechanical deformation of an ordinary hard solid (amorphous glucose, for example).

(The glass-transition temperature T_g of any amorphous substance, whether polymeric or not, may be defined as the point where the thermal expansion coefficient α undergoes a discontinuity. Above this temperature, α has the magnitude generally associated with liquids: 6 to 10×10^{-4} deg^{-1}. Decrease in temperature is accompanied by collapse of free volume which is made possible by configurational adjustments. Eventually, the free volume becomes so small that further adjustments are extremely slow or even impossible; then it no longer decreases and the further contraction in total volume with decreasing temperature is much less, so α drops suddenly to between 1 and 3×10^{-4} deg^{-1}.

In polymers, there may be more than one discontinuity in α. The highest is usually associated with the loss of the molecular mobility which permits configurational rearrangements of the chain backbones, and it profoundly alters the viscoelastic behavior; this is "the" glass transition. Others may be associated with the loss of much more specific, local motions, such as the rearrangements of short side groups.)

VI. To illustrate a *lightly crosslinked amorphous polymer:* lightly vulcanized hevea rubber, vulcanized with sulfur and an accelerator to an equilibrium tensile modulus E_e of about 7×10^6 dynes/cm^2. Dynamic data, in simple extension, of Cunningham and Ivey (47) and Payne (48), together with creep data in simple extension of Martin, Roth, and Stiehler (49) were employed, all reduced to a reference temperature of 25°C. Certain minor adjustments in the data are described elsewhere (50).

The molecular structure is a network of highly flexible threadlike strands whose average molecular weight between crosslinks is about 4000. Relatively short-range segmental rearrangements are oblivious of the presence of the linkage points, but of course long-range rearrangements are profoundly affected.

VII. To illustrate a *very lightly crosslinked amorphous polymer:* a styrene–butadiene random copolymer with 23.5% styrene by weight, vulcanized with dicumyl peroxide to an equilibrium shear modulus G_e of about 1.5×10^6 dynes/cm^2. Dynamic shear data and shear creep data of Mancke and Ferry (51) were combined and reduced to 25°C. The feature which differentiates this example from the preceding one is the paucity of crosslinks. The molecular weight between crosslinking points (about 23,000) is sufficiently high to encompass several entanglement loci per network strand, since the molecular weight between coupling entanglements in the uncrosslinked polymer is (52) about 4600. In the rubber of example VI, by contrast, there are about twice as many crosslinks as entanglements. These numbers are subject to some uncertain assumptions, but evidently the network topologies can be expected to be quite different and the viscoelastic functions also show prominent differences.

VIII. To illustrate a *highly crystalline polymer:* a linear polyethylene with a density of 0.965 g/ml at room temperature, corresponding to a high degree of crystallinity. Stress relaxation data of Faucher (53) in simple extension were employed, reduced to a reference temperature of 20°C. (There was no evidence of a change in crystallinity over the temperature range within which experiments were utilized for reduction, viz, −70° to 70°.)

The polymer may be pictured as a matrix of crystalline material with units of various forms such as lamellae and fibrils, containing various kinds of crystal defects and interfaced by regions of greater disorder. Both ordered and disordered regions will contribute to the viscoelastic behavior.

The characteristic linear viscoelastic behavior (in shear except for examples VI and VIII, which are in extension) of these eight representative systems will be reviewed.

The Creep Compliance. Plotted on a linear scale, the creep compliance of a viscoelastic liquid would look like the creep deformation as shown in Figure 14, differing only by a proportionality factor of the stress. In analyzing creep experiments, such linear plots are invariably made to determine the quantities J_e^0 and η. But in presenting the overall aspect of $J(t)$, the ranges of both the magnitude of $J(t)$ and the time scale are so enormous (as mentioned above) that the only way to give a complete representation in a single graph is to make both coordinates logarithmic. This procedure is followed here for depicting all the viscoelastic functions. The units are cgs throughout. In Système International d'Unités (SI) units, a modulus is newtons/meter^2; the conversion factor from dynes/cm^2 is 10^{-1}. A compliance is meters^2/newton, larger by a factor of ten than the value in cm^2/dyne.

In comparing different structural types of polymers, it is the shapes and the magnitudes of the functions which are important; the positions of the curves on the logarithmic time scale are in a sense irrelevant since they depend sharply on the temperature and in any case the reference temperatures for the polymers compared here are not all the same. Purely for clarity in distinguishing the curves, arbitrary shifts along the logarithmic time (or frequency) axis have been made by adding to log t (subtracting from log ω) a constant A with the following values: I, -5; II, -6; III, -7; IV, 0; V, -7; VI, 0; VII, -1; VIII, 2. The same values of A have been used throughout this section. Thus the relative horizontal positions of the curves have no significance; it is their shapes and vertical positions which are to be scrutinized. The vertical positions of curves VI and VIII will not be directly comparable with the others, because they represent deformation in extension instead of shear; the compliances will be lower and the moduli higher by about half an order of magnitude (factor of 2.5 to 3), in accordance with equation 66.

Figure 19 shows the creep compliance thus plotted for the eight typical systems, with the liquids on the left and the solids on the right. The tremendous range of time scale over which response to stress is achieved is immediately apparent.

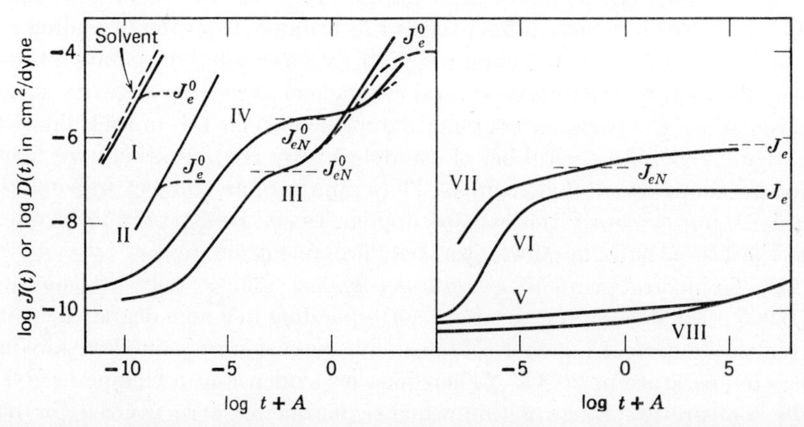

Fig. 19. Creep compliance for eight typical polymer systems: viscoelastic liquids on the left, viscoelastic solids on the right, identified by numbers as described in the text. Deformation is shear, $J(t)$, except for curves VI and VIII, which are for simple extension, $D(t)$. The dashed curves represent the compliance after subtraction of the flow contribution t/η. The solvent for the dilute solution, example I, is also shown as a dashed line.

At short times, $J(t)$ approaches a value on the order of 10^{-10} cm²/dyne, characteristic of a hard glasslike solid. The corresponding region of the time scale is sometimes called the glassy zone. This small compliance corresponds to the absence of any configurational rearrangements of the chain backbones within the interval of the experiment; indeed, curve V shows that below the glass-transition temperature, where the backbone configurations are immobilized, $J(t)$ has this order of magnitude throughout the time scale. Approximately, $J(t)$ may be regarded as possessing a limiting value $J(0)$ at zero time, often written J_g, the subscript standing for glass. This value is often not very well defined operationally, however.

At long times, $J(t)$ for the viscoelastic liquids increases without limit, because it includes a contribution from viscous flow. But if the latter is subtracted, the remainder $J(t) - t/\eta$ approaches a limiting value J_e^0 (cf equation 54). This value has been attained in examples I (10^{-5} cm²/dyne), II (10^{-7} cm²/dyne), and III (10^{-4} cm²/dyne), but not in example IV because the very high molecular weight, together with coupling entanglements, greatly prolongs the time necessary for long-range configurational changes, and at the end of the experiment $J(t) - t/\eta$ is still increasing. In terms of models, J_e^0 is a measure of the energy stored in all the springs (Fig. 11) during steady-state flow; in molecular terms, it measures the average distortion of the polymer coils during flow, when ample time has been allowed for the molecular distribution function to become independent of time.

For the crosslinked rubbers VI and VII, $J(t)$ at long times approaches a limiting value J_e, the equilibrium compliance, which according to the theory of rubberlike elasticity (54) is proportional to the number-average molecular weight between crosslinks in the network. Both J_e^0 and J_e are measures of energy storage but are distinguished since the former refers to steady-state flow and the latter to elastic equilibrium. The other viscoelastic solids, V and VIII, exhibit creep compliance which is still increasing slowly with time at the longest times of observation.

The creep compliance of the dilute solution, I (which could hardly be measured directly but must be calculated from dynamic measurements) is compared with a calculated curve for the solvent. For the latter, a Newtonian liquid, $J(t)$ is simply t/η. The solution compliance is lower than that of the solvent by an amount which is quite small at short times but somewhat larger at long times. The inflection between these regions lies at a point on the time scale related to the relaxation times for configurational rearrangements (40a).

At intermediate times there is for each of the undiluted polymers, except the glasslike (V) and the highly crystalline (VIII), a gradual increase of $J(t)$, which rises by several powers of ten. This reflects the increasing response to external stress by configurational rearrangements, first of the relative positions of chain backbone segments near each other, then farther and farther apart, requiring more and more mutual cooperation and therefore more and more time (cf Fig. 1). In the low-molecular-weight (II) and the moderately crosslinked polymers (VI) this rise occurs in a single stage, usually called the transition from glasslike to rubberlike consistency. At the end of the transition zone, the elapsed time has become long compared with the time required for the slowest rearrangement of a molecule (in the crosslinked polymers, of a strand between two crosslinking points), and the average molecular distortion has approached its maximum. In the uncrosslinked polymers of high molecular weight (III, IV) and the very lightly crosslinked network (VII), there are two stages. The first reflects the relative motion of chain segments between the entanglement coupling loci, and corre-

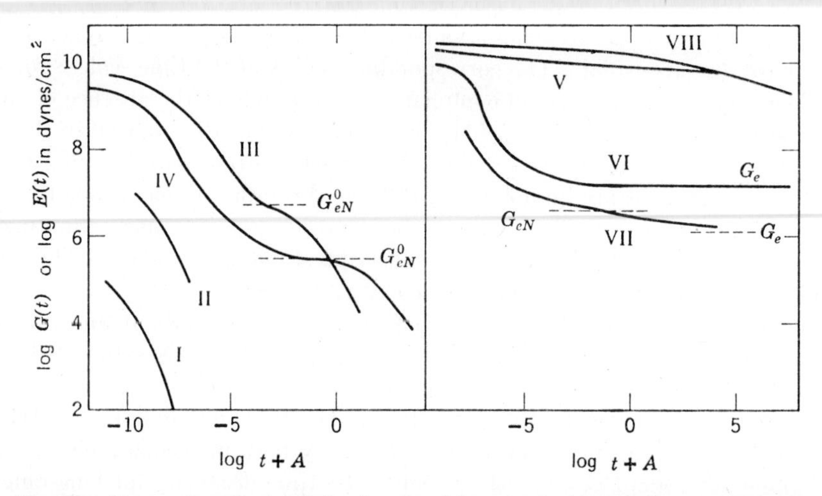

Fig. 20. Stress relaxation modulus for the eight systems identified as in Figure 19.

sponds to the glass–rubber transition. At the end of this stage, $J(t)$ tends to level off at a value denoted by $J_{eN}{}^0$ for the uncrosslinked and J_{eN} for the crosslinked systems. Here the entanglements suppress long-range configurational rearrangements almost as though they were crosslinks. The values of $J_{eN}{}^0$ cannot be determined by inspection, since the curves do not really become flat in this region, but they can be calculated indirectly. They are approximately proportional to the spacings between entanglement loci; the average spacing is evidently greater in example IV than in example III. In the second stage, where $J(t)$ rises from $J_{eN}{}^0$ to $J_e{}^0$ for example III (not achieved in IV), the entanglements slip so that configurational rearrangements of segments separated by entanglements can take place. The corresponding rise from J_{eN} to J_e in example VII is probably attributable to slippage of entanglements on branched structures which are incompletely attached to the network. At a higher degree of crosslinking (VI) such structures do not exist and the second inflection is eliminated.

Stress Relaxation Modulus. The modulus $G(t)$, defined as the stress/strain ratio at constant deformation, is plotted against t with logarithmic scales in Figure 20. In certain regions $G(t)$ is approximately $1/J(t)$, so that the logarithmic plots have roughly the appearance of mirror images of those in Figure 19 reflected in the time axis. The more slowly $J(t)$ changes with time, the more nearly is the reciprocal relation approached. Thus, at short times, $G(t)$ appears to approach a limiting value which (if it exists) is $G_g = 1/J_g$, on the order of 10^{10} dynes/cm^2, and represents the rigidity in the absence of backbone rearrangements. At long times, $G(t)$ for the crosslinked networks VI and VII approaches values which are again nearly constant, written $G_e = 1/J_e$ and representing the equilibrium shear modulus as treated by the theory of rubberlike elasticity (see Elastomers, synthetic).

At long times for the uncrosslinked polymers, however, $G(t)$ falls rapidly and eventually vanishes. In terms of mechanical models, this corresponds to the complete relaxation of all springs in an array such as Figure 11; in molecular terms, it corresponds to the resumption of random average configurations by the macromolecular coils, which have completely freed themselves from the constraints originally imposed on them, even though the external dimensions of the sample remain deformed. (The final residual deformation corresponds to the flow contribution to a creep experiment.)

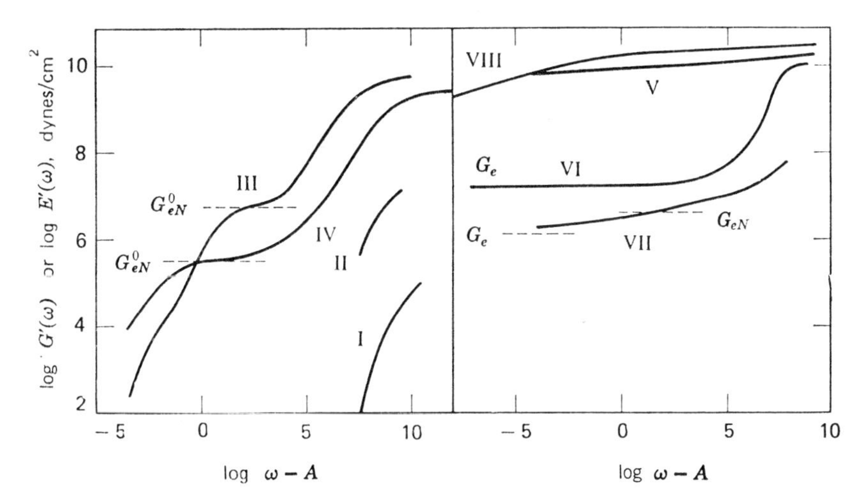

Fig. 21. Storage modulus plotted against frequency, with logarithmic scales, for the eight systems identified as in Figure 19.

The region of time scale in which $G(t)$ falls sharply is often called the terminal zone. If the viscoelastic behavior is represented by a finite mechanical model such as Figure 12, the decay of $G(t)$ must at the end become exponential, proportional to e^{-t/τ_1} where τ_1 is the terminal relaxation time. According to some molecular theories, a well-defined terminal relaxation time should be observed for a polymer with uniform molecular weight, but for a broad molecular-weight distribution the final stage of exponential relaxation may not be attainable.

At intermediate times, the stress gradually falls as the distortion of the chain backbones adjusts itself through Brownian motion, first of segments with respect to closely neighboring segments, then with respect to those farther removed along the backbone contour, and so on (cf Fig. 1). The drop occurs in two stages for the uncrosslinked polymers of high molecular weight III and IV and the very lightly crosslinked network VII, just as the creep compliance rises in either one or two stages, and for the same reasons. Between the transition and terminal zones of II and III, and the transition and slow relaxation zones of VII, $G(t)$ flattens somewhat at a level which again is associated with the average spacing between entanglement coupling points, and is approximately the reciprocal of $J_{eN}{}^0$ or J_{eN}, respectively, viz, $G_{eN}{}^0$ or G_{eN}. From the Boltzmann superposition principle it can be shown that $J(t)G(t) \leq 1$ for all values of t, so each drop in $G(t)$ always occurs at somewhat shorter times than the corresponding rise in $J(t)$.

In the glassy polymer, V, there is very little stress relaxation over many decades of logarithmic time, since no backbone contour changes occur; in the densely crystalline polymer, VIII, there is some relaxation at very long times through whatever mechanism is responsible for the creep which also occurs in this region. Examples of stress relaxation after cessation of steady-state flow (Fig. 5) are not included here because of the limited applicability of this type of experiment. Such stress relaxation is of particular interest in connection with nonlinear phenomena, however, and will be illustrated below.

Storage Modulus. The modulus $G'(\omega)$ is defined as the stress in phase with the strain in a sinusoidal shear deformation divided by the strain; it is a measure of the en-

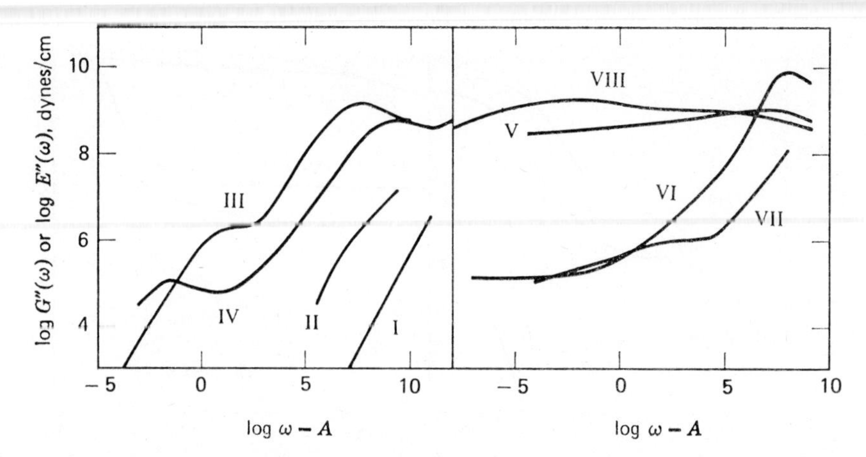

Fig. 22. Loss modulus plotted logarithmically for the eight systems identified as in Figure 19.

ergy stored and recovered per cycle, when different systems are compared at the same strain amplitude. It is plotted against the radian frequency ω with logarithmic scales in Figure 21. Since both $G(t)$ and $G'(\omega)$ are measures of stored elastic energy, and a dynamic measurement at frequency ω is qualitatively equivalent to a transient one at $t = 1/\omega$, these graphs are approximately mirror images of those for the relaxation modulus, reflected in the modulus axis. In particular, when $G(t)$ is changing very slowly, $G(t) \cong G'(1/t)$, so the values G_g and G_e characteristic of high and low frequencies are the same as those characteristic of short and long times, respectively.

At long times for the viscoelastic liquids on the left of the figure, $G'(\omega)$ approaches 0 with decreasing frequency, just as $G(t)$ does with increasing t; macroscopically, this means that the phase angle between stress and strain approaches 90° as the stored energy per cycle of deformation becomes negligible compared with that dissipated as heat. However, the shape of the curve is somewhat different and $G'(1/t) > G(t)$ at all times. At the end of the terminal zone at the left, G' becomes proportional to ω^2 instead of exponentially dependent on t; this relation, seen in the terminal slope of 2 for example I, can be derived readily from model analogies. It is independent of whether the molecular weight is uniform or not. The proportionality constant $G'/|\omega^2|_{\omega \to 0}$, however, is strongly dependent on molecular-weight distribution.

At intermediate times, the behavior is very similar to what has already been described for $G(t)$, except that $G'(1/t)$ always exceeds $G(t)$ to some extent. On a molecular basis, the magnitude of G' depends on what contour rearrangements can take place within the period of the oscillatory deformation.

Loss Modulus. The modulus $G''(\omega)$ is defined as the stress 90° out of phase with the strain divided by the strain; it is a measure of the energy dissipated or lost as heat per cycle of sinusoidal deformation, when different systems are compared at the same strain amplitude. It is plotted with logarithmic scales in Figure 22.

Observation of these curves reveals a feature which can be stated qualitatively as follows. In frequency regions where $G'(\omega)$ changes slowly (undergoes little dispersion), corresponding to very little stress relaxation in the equivalent plot of $G(t)$, the behavior is more nearly perfectly elastic; hence comparatively little energy is dissipated in periodic deformations. Thus, in such regions G'' tends to be considerably less than G'. This effect is prominent in the locations of G'' for the glass (V) and the crystalline polymer (VIII), whose G' curves are relatively flat throughout; G'' is so low that it intersects some of the other curves. Also, the flattening of G' at frequencies below the transition zone for the polymers which exhibit effects of entanglement coupling (III, IV, and VII) is accompanied by a plateau or minimum in G''.

At high frequencies, a mechanical model such as Figure 11 would be expected to approach perfect elastic behavior, as the motion of the dashpots became negligible compared with that of the springs; then G'' should approach zero. On a molecular basis, this would correspond to the absence of any molecular or atomic adjustments capable of dissipating energy within the period of deformation. This situation is not in fact achieved for high polymers or indeed any other solids of simpler structure; however, maxima in G'' appear for the soft polymers III, IV, and VI at rather high frequencies beyond which the losses due to backbone configurational changes diminish. Other maxima, in V and VIII, represent peaks in other dissipative processes.

At very low frequencies, G'' for a viscoelastic liquid should be directly proportional to ω, with a slope of 1 on a logarithmic plot. This is evident in examples I, II, and III. The proportionality constant is the Newtonian steady-flow viscosity η, as shown below. For a simple Newtonian liquid, $G'' = \omega\eta$ over the entire frequency range, and this relation is shown for the solvent of the dilute solution, example I.

For a viscoelastic solid with linear viscoelasticity corresponding to a model with springs and dashpots such as Figures 11 and 12, G'' should also be directly proportional to ω at very low frequencies. Such behavior is not observed for the examples on the right side of Figure 22; experiments have never been carried to sufficiently low frequencies to test this prediction.

Dynamic Viscosity. The dissipative effects of alternating stress can be described just as well by another quantity, the ratio of stress in phase with rate of strain divided by the rate of strain. This has the dimensions of a viscosity, and is the real part η' of a complex viscosity, $\mathbf{n}^* = \eta' - i\eta''$, defined in the same manner as $\mathbf{G}^*$. In sinusoidal deformations, if the strain is $\gamma(t) = \gamma_0 e^{i\omega t}$ the rate of strain $\dot{\gamma}(t)$ is $i\omega\gamma_0 e^{i\omega t} = i\omega\gamma(t)$. Hence $\mathbf{n}^* = \mathbf{G}^*/i\omega$, and the individual components are related by

$$\eta' = G''/\omega \tag{77}$$

$$\eta'' = G'/\omega \tag{78}$$

The in-phase or real component η', often for simplicity called just the dynamic viscosity, is useful especially in discussing uncrosslinked polymers because for these systems at very low frequencies η' approaches η, the ordinary steady-flow viscosity. With increasing frequency, η' falls monotonically as shown in Figure 23, reaching values many orders of magnitude smaller than η.

As is obvious from equation 77, in regions where G'' is flat, η' is inversely proportional to frequency; whereas when G'' rises steeply, on the left side of a maximum, η' may flatten out, as seen particularly in the crosslinked rubber, example VI.

The low-frequency limiting value of η is attained for the viscoelastic liquids I, II, and III. Its magnitude depends of course on temperature, molecular weight, and (for solutions) polymer concentration. For example, IV data do not extend to low enough frequencies to reach the limiting value. For the dilute solution, I, the steady-flow viscosity η is about three times that of the solvent. With increasing frequency, η' falls and approaches a limiting value at high frequencies which is about 1.5 times that of the solvent. The same relations are apparent in the plot of G'' in Figure 22. The fre-

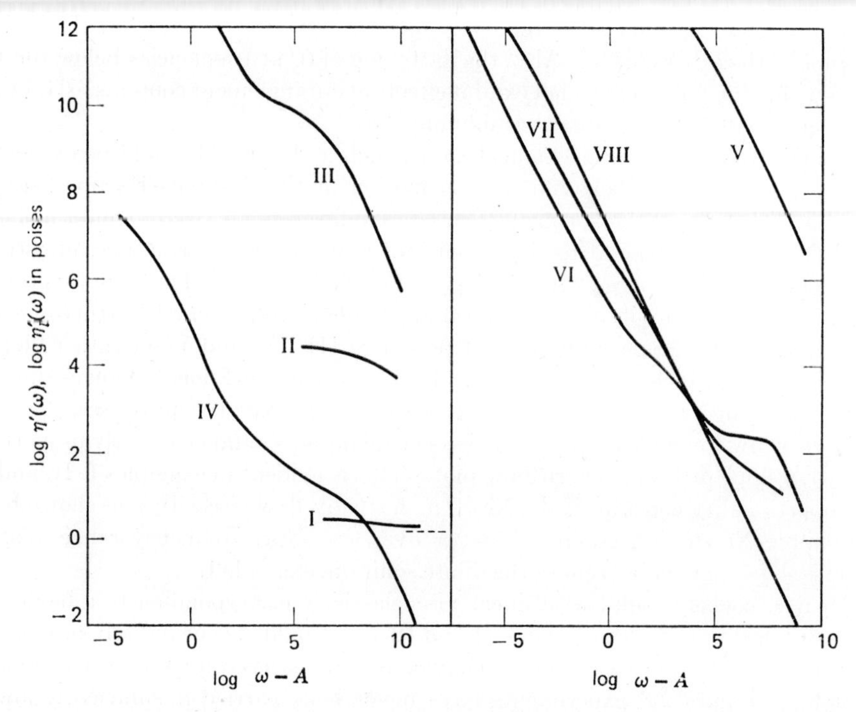

Fig. 23. Real part of the complex dynamic viscosity, plotted logarithmically for the eight systems identified as in Figure 19.

quency region in which the transition occurs is, again, related to configurational relaxation times. For a viscoelastic solid, such as a crosslinked polymer, η is of course infinite, but if the viscoelastic behavior can be represented by a finite mechanical model, η' should approach a finite limiting value at low frequencies; for example, in Figure 11, if one of the viscosities is infinite, there is no finite steady-flow viscosity, but the low-frequency limit of η' is finite and equal to the sum of all the other viscosities. To the author's knowledge, this situation has never been experimentally observed in a crosslinked polymer.

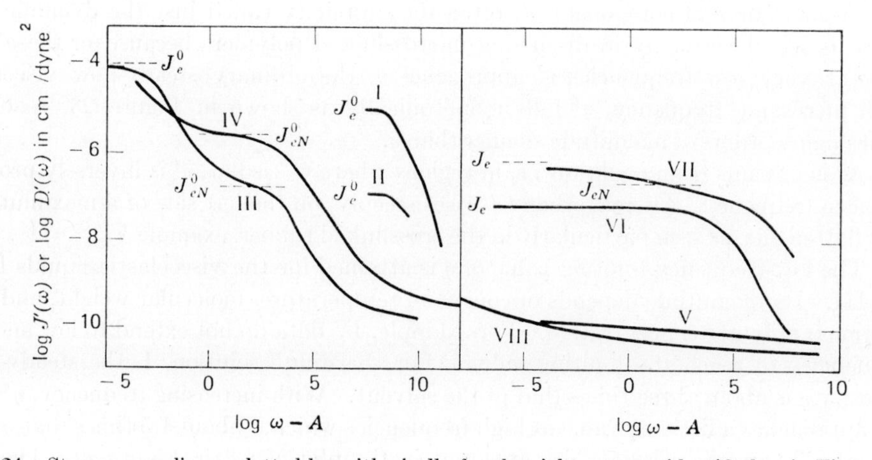

Fig. 24. Storage compliance plotted logarithmically for the eight systems identified as in Figure 19.

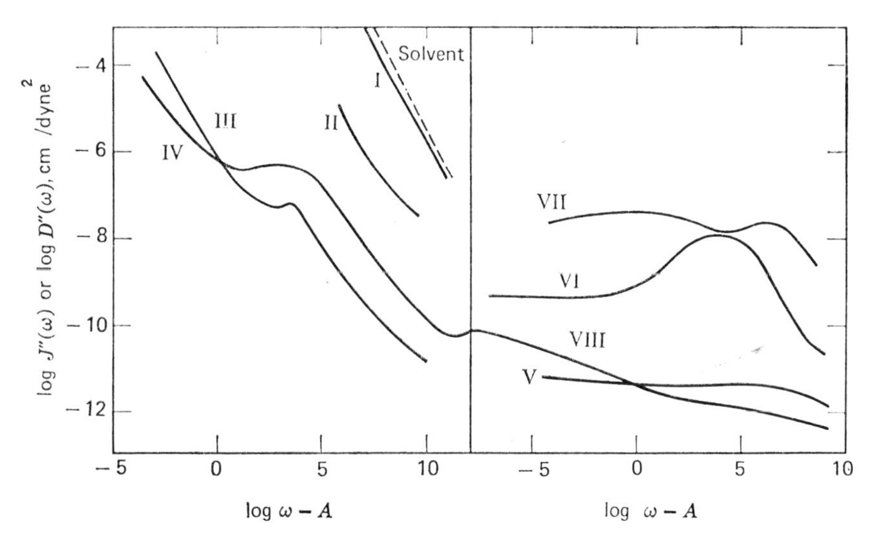

Fig. 25. Loss compliance plotted logarithmically for the eight systems identified as in Figure 19.

An alternative frequency-dependent viscosity, which could also be called the dynamic viscosity, is the absolute value $|\mathbf{n}^*| = (\eta'_2 + \eta''^2)^{1/2} = (G'^2 + G''^2)^{1/2}/\omega$.

Storage Compliance. The compliance $J'(\omega)$ is defined as the strain in a sinusoidal deformation in phase with the stress divided by the stress; it is a measure of the energy stored and recovered per cycle, when different systems are compared at the same stress amplitude. It is plotted with the usual logarithmic scales in Figure 24. For the same reason that $G'(\omega)$ resembles $G(t)$ plotted backwards, $J'(\omega)$ resembles $J(t)$ reflected in the compliance axis. An important distinction, however, appears in those polymers which are viscoelastic liquids: the elastic (recoverable) part of the creep is obtained only after subtracting t/η, and it is the difference $J(t) - t/\eta$ which approaches a limiting value J_e^0; but for J' the phase specification automatically eliminates any flow contribution, so J' itself approaches J_e^0 at low frequencies. This is seen in curves I, II, and III.

In the transition zone and other regions where J' changes rapidly, the shapes of $J'(\omega)$ and $J(t)$ differ such that $J'(1/t) < J(t)$ at all times. From equation 24 it is evident also that $J'(\omega) < 1/G'(\omega)$. The magnitudes of the two compliances and reciprocal moduli fall in the following order: $J'(1/t) < J(t) - t/\eta < 1/G'(1/t) < 1/G(t)$. It is also evident from equations 22 and 23 that $J' < |\mathbf{J}^*| = 1/|\mathbf{G}^*| < 1/G'$. In practice $J(t)$ often lies quite close to $|\mathbf{J}^*|$.

Loss Compliance. The compliance $J''(\omega)$ is defined as the strain 90° out of phase with the stress divided by the stress; it is a measure of the energy dissipated or lost as heat per cycle of sinusoidal deformation, when different systems are compared at the same stress amplitude. It is plotted with logarithmic scales in Figure 25.

For a Newtonian liquid, $J'' = 1/\omega\eta$ over the entire frequency range, as illustrated by the solvent for the dilute solution, example I, with a slope of -1 on the logarithmic scale. For viscoelastic liquids, J'' becomes equal to $1/\omega\eta$ at very low frequencies; this is seen in examples I, II, and III, where slopes of -1 are attained.

A characteristic feature of J'' for viscoelastic solids which are lightly crosslinked networks or polymeric viscoelastic liquids of high molecular weight is a rather broad maximum whose location on the frequency scale corresponds to the low-frequency end

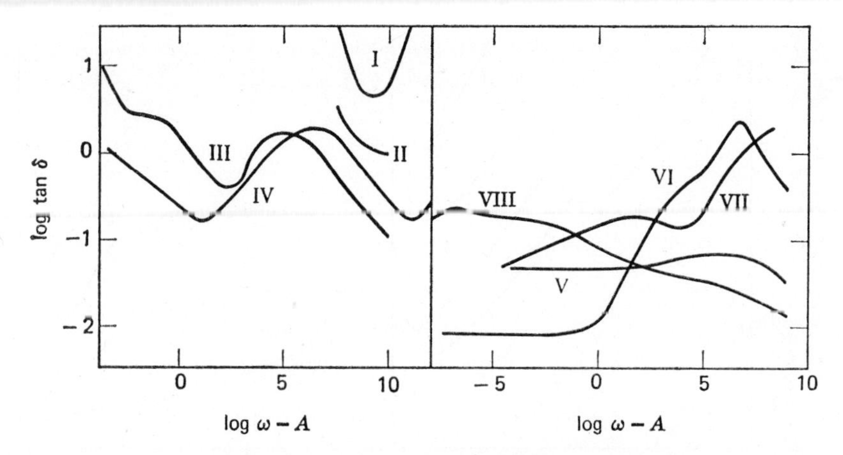

Fig. 26. Loss tangent plotted logarithmically for the eight systems identified as in Figure 19.

of the transition zone as seen in the storage modulus (Fig. 21). This is associated with configurational rearrangements of the strands of a network structure. The maximum for an uncrosslinked polymer of high molecular weight, curves III and IV, is attributable to the entanglement network, in which the entanglements suppress long-range configurational rearrangements almost as though they were crosslinks. In this case, there is a minimum on the low-frequency side of the maximum, and at still lower frequencies, J'' rises to become inversely proportional to ω. For a crosslinked polymer, curve VI of Figure 25, the maximum is attributable to the network netted by the chemical crosslinks together with the entanglements which were present before crosslinking.

For the very lightly crosslinked network, curve VII of Figure 25, there *are two* maxima. The one at higher frequencies corresponds to a network consisting of all strands terminated either by crosslinks or by entanglements (which here considerably outnumber the crosslinks). The other at the left can be attributed to the network which remains after subtraction of those entanglements which are capable of slippage because they are on structures incompletely attached to the chemically crosslinked network, as outlined in the discussion of creep compliance.

Loss Tangent. A useful parameter which is dimensionless and conveys no physical magnitude but is a measure of the ratio of energy lost to energy stored in a cyclic deformation is the loss tangent, $\tan \delta = G''/G' = J''/J'$ (eq. 20). The logarithmic plots in Figure 26 reveal several characteristic levels of $\tan \delta$. First, for the dilute solution (curve I) $\tan \delta$ is very high because both solvent and solute contribute to G'' but only the solute contributes to G'. It goes through a minimum in the frequency range where the transition occurs in η' (Fig. 23) and the other dynamic functions. At low frequencies, $\tan \delta$ is large for all the uncrosslinked polymers (curves I, II, III, IV) and in fact becomes inversely proportional to the frequency. (We recall that in this region G'' is proportional to ω and G' to ω^2; alternatively, J'' is proportional to ω^{-1} and J' approaches a constant; so from either G''/G' or J''/J' it is clear that $\tan \delta$ is proportional to ω^{-1}.) Second, all the amorphous polymers, whether crosslinked or not, have values in the transition zone which are in the neighborhood of $\tan \delta = 1$, ranging perhaps from 0.2 to 3. Third, the glassy and crystalline polymers, V and VIII, have values in the general neighborhood of 0.1; and, finally, the lightly crosslinked polymer attains an extremely small value at low frequencies, on the order of 0.01.

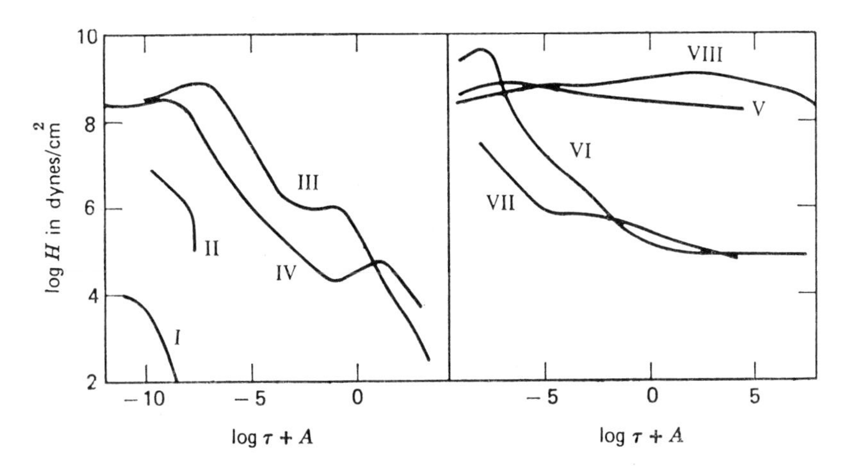

Fig. 27. The relaxation spectrum, plotted with logarithmic scales for the eight typical polymer systems; viscoelastic liquids on the left, viscoelastic solids on the right, identified by numbers as described in the text.

In the transition zone between glasslike and rubberlike consistency, the loss tangent goes through a pronounced maximum for both uncrosslinked polymers of high molecular weight (III and IV) and lightly crosslinked polymers (VI and VII, though the right side of the latter maximum is not encompassed). It is of interest that the maxima in J'' occur to the left of those in tan δ, and the maxima in G'' occur to the right of those in tan δ on the frequency scale; the differences amount to several logarithmic decades. Each of these three functions is a measure of elastic losses or heat dissipation, but it is clear that the frequency region in which the "loss" occurs depends on the choice of function by which the loss is specified.

For the very lightly crosslinked polymer (curve VII), there is a subsidiary maximum at lower frequencies associated with the losses involved in entanglement slippage as discussed in connection with $J(t)$ and J''. In this case, also, the maximum in J'' lies to the left of that in tan δ. Smaller maxima occur in the curves for the glassy and highly crystalline polymers, reflecting other dissipative mechanisms.

The loss tangent determines such macroscopic physical properties as the damping of free vibrations, the attenuation of propagated waves, and the frequency width of a resonance response. It can often be more conveniently measured than any other viscoelastic function, by observations of these phenomena, and is of considerable practical interest. It is less susceptible of direct theoretical interpretation than the other functions, however.

Relaxation Spectrum. Plots of the relaxation spectrum H for the eight polymer types are shown in Figure 27. The constant A has the same significance as before. Their shapes are rather similar to those of G'', reflected in the modulus axis. Their maxima represent concentrations of relaxation processes in certain regions of the logarithmic time scale. At long times, in uncrosslinked viscoelastic liquids, when steady-state flow is reached, H should vanish. It does so for examples I, II, and III; for IV the data do not extend to long enough times. For the viscoeleastic solids on the right, H attains quite low values at long times but gives no evidence of approaching zero; this behavior is associated with the persistence of a small negative slope in stress relaxation (Fig. 20), showing that some degree of relaxation continues apparently

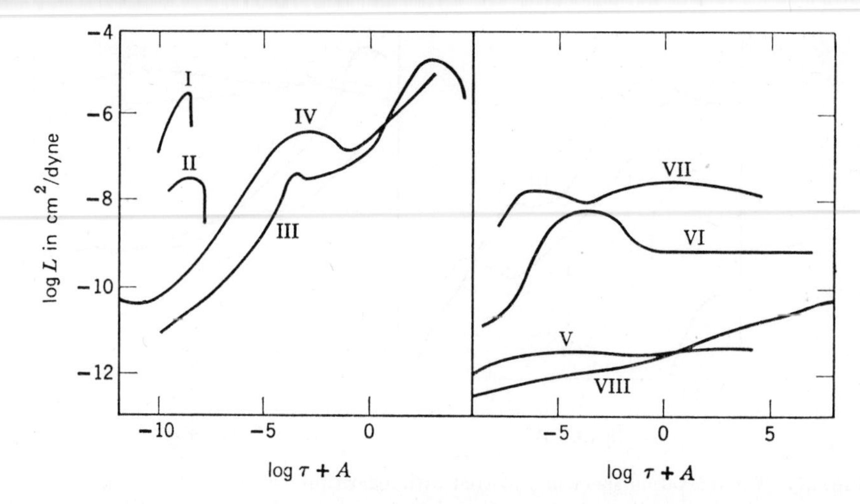

Fig. 28. The retardation spectrum, plotted logarithmically for the eight systems identified as in Figure 27.

indefinitely. At very short times, if the mechanical behavior approaches perfect elasticity, H should also vanish; actually, it remains finite, and in these polymers at a rather high level, since some relaxation processes occur even at the shortest times (dissipative processes at the highest frequencies).

The characteristic zones of the viscoelastic time scale are clearly apparent in H: the glassy zone to the left of the principal maximum, the transition zone where H drops steeply, the terminal zone where it approaches zero, and a region to the right of the transition zone in examples III, IV, and VII were H is relatively flat (the plateau) or passes through a minimum.

Retardation Spectrum. Plots of the retardation spectrum L for the eight polymer types are shown in Figure 28. Their shapes, correspondingly, resemble those of J'' reflected in the compliance axis. Their maxima represent a concentration of retardation processes, measured by their contributions to compliance rather than modulus, in certain regions of the logarithmic time scale; they occur at quite different locations from the maxima in H.

At long times, L, like H, should vanish when an uncrosslinked viscoelastic liquid polymer reaches the state of steady flow. This condition is observed for examples I, II, and III. For example IV, there are compliance mechanisms persisting beyond the longest times for which data are available. The plateau or minimum in the spectrum H corresponds roughly to the maximum in the spectrum L.

As defined above, the spectra H and L refer to deformation in shear. Similar spectra can of course be used for other types of deformation and are specified by modified symbols, eg, H_v for bulk compression and $\overline{H}$ for simple elongation; thus curves VI and VIII in Figure 27 are actually $\overline{H}$. Often in the literature H and $\overline{H}$ are not clearly distinguished.

Linear Viscoelastic Behavior in Bulk (Voluminal) Deformation

As explained above, shear and bulk deformations are essentially different in character, one involving a shape change and the other a volume change; and they are accompanied by quite different molecular processes. There are far fewer experimental data available on bulk viscoelastic properties; however, a much narrower range of behavior among various types of polymeric systems may be expected, since voluminal changes should be dominated by local configurational rearrangements and these are scarcely affected by molecular weight (if sufficiently high), entanglements, or crosslinks in moderate numbers.

As an example of bulk viscoelastic behavior, data for a poly(vinyl acetate) of moderately high molecular weight are shown in Figure 29. Measurements by McKinney and Belcher (55) of the storage and loss bulk compliance B' and B'' at various temperatures and pressures are plotted after reduction to a reference temperature and pressure of 50°C and 1 atm, respectively. The complex bulk compliance is formally analogous to the complex shear compliance, but the two functions present several marked contrasts.

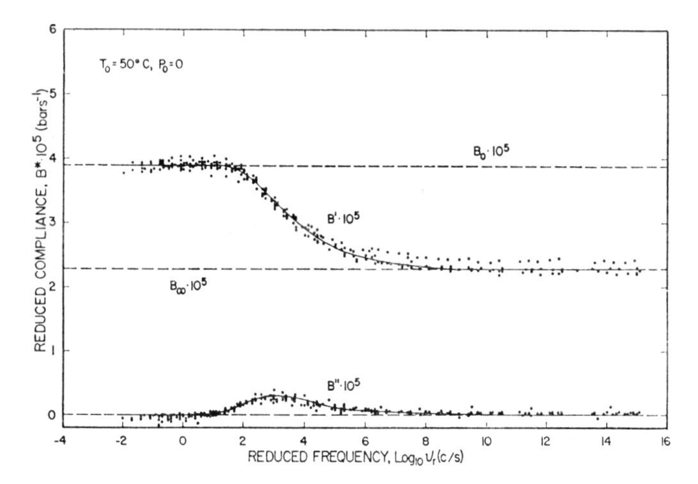

Fig. 29. Storage and loss bulk compliance plotted logarithmically against frequency, for a poly(vinyl acetate) reduced to 50°C and 1 atm, as described in the text (55).

The storage bulk compliance B' falls from a low-frequency limiting value to a high-frequency limiting value, but the change (shown here on a linear scale) is less than a factor of two instead of the many powers of ten displayed by the shear compliance in Figure 24. On a qualitative molecular basis, the low-frequency value reflects volume decreases under pressure due to reduction of atomic and molecular dimensions together with collapse of free volume involving local configurational adjustments which require a finite time. The high-frequency value reflects the volume decrease due to reduction of atomic and molecular dimensions alone. The frequency region in which the transition occurs depends on the relation of the time required for configurational adjustments to the period for cyclic deformation. It is somewhat narrower than the dispersion region for shear viscoelasticity, but it still covers several decades. The maxima in the loss tangents for shear and bulk deformation of this polymer occur at approximately the same frequency when compared at the same temperature, but in general no simple relation between the two can be expected. The loss bulk compliance B'' is zero within experimental error at both low and high frequencies and it passes through a maximum in the region of transition. The maximum loss tangent is on the order of 0.1.

Since the loss tangent is never very large, the storage modulus is nearly the reciprocal of the storage compliance (equation 24 with substitutions) and the loss modulus K'' is nearly B''/B'^2 (analog of equation 27). It is easy to visualize the appearance of these functions. The corresponding transient functions, $B(t)$ and $K(t)$, can be visualized by reflecting mirror images of $B'(\omega)$ and $K'(\omega)$ in the ordinate axis.

Viscoelastic behavior in simple extension or in bulk longitudinal deformation will in general combine the features of shear and bulk viscoelasticity, since the moduli $E(t)$ and $M(t)$ depend on both $G(t)$ and $K(t)$, as shown by equations 62 and 69 (and analogous relations for $\mathbf{E}^*$ and $\mathbf{M}^*$). However, as already pointed out, shear effects predominate in $E(t)$ and $\mathbf{E}^*$, and bulk effects predominate in $M(t)$ and $\mathbf{M}^*$, so the qualitative behavior is evident from the examples already given.

Nonlinear Viscoelastic Phenomena in Shear

When finite strains and/or strain rates are allowed, the variety of behavior is enormously multiplied. As pointed out above, for such situations the simplification $K(t) \gg G(t)$ usually applies, so volume changes can be ignored. However, it is no longer possible to predict behavior in extension from behavior in shear by applying a factor of three, as in linear viscoelasticity (equations 65 and 66); entirely different phenomena may be observed in the two deformational modes. Moreover, different experimental geometries such as torsion and flow through an annulus, which for linear viscoelasticity yield equivalent information concerning the time- or frequency-dependent shear modulus, may now give quite different kinds of information; so a variety of experimental arrangements is desirable not merely for improved accuracy or convenience but to obtain information otherwise inaccessible. Only a few of the more basic nonlinear viscoelastic manifestations are mentioned here.

Normal Stress Differences. The well-known normal stress differences which appear in the shearing deformation of a rubberlike solid in large strains at mechanical equilibrium (37,56) are strictly speaking outside the domain of viscoelasticity. How-

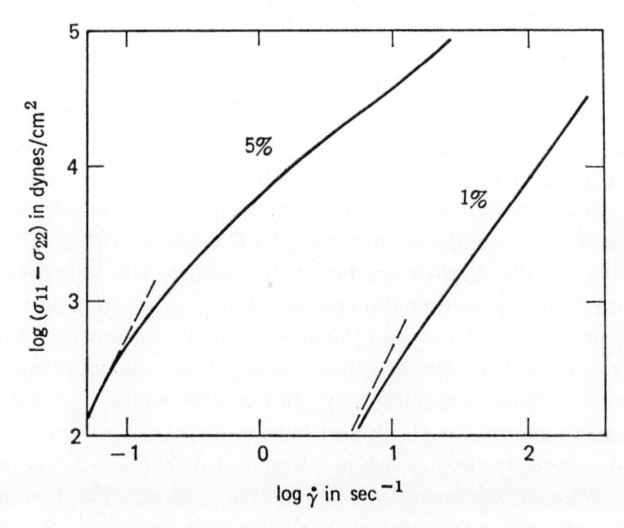

Fig. 30. First normal stress difference plotted logarithmically against shear rate, for two solutions of polystyrene with concentrations indicated, as described in the text (57). Dashed lines have a slope of 2.

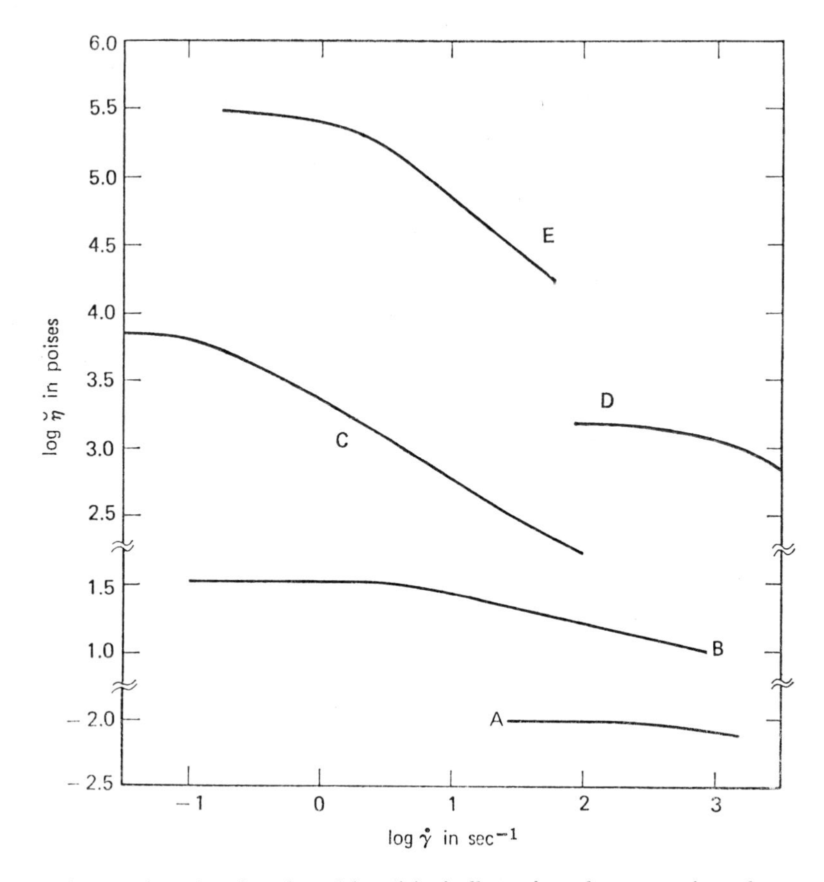

Fig. 31. Non-Newtonian viscosity plotted logarithmically against shear rate, for polymer solutions and undiluted polymers as described in the text.

ever, the normal stresses in time-dependent deformation of viscoelastic liquids, in either steady flow or other time patterns, and in viscoelastic solids for oscillatory or other time-dependent deformations, represent combinations of elastic and viscous effects.

Normal stresses in steady shear flow of moderately dilute polymer solutions are illustrated in Figure 30, which shows measurements by Ashare (57) on solutions of a polystyrene of weight-average molecular weight ($\bar{M}_w$) 1,800,000 and number-average ($\bar{M}_n$) 1,500,000 in a highly viscous solvent (a chlorinated diphenyl with viscosity 3 poises at 25°C). The normal stress difference $\sigma_{11} - \sigma_{22}$ (Fig. 18) is plotted logarithmically against the shear rate $\dot{\gamma}_{21}$. Qualitatively, this quantity represents a tension along streamlines. It increases monotonically with $\dot{\gamma}_{21}$ and at low $\dot{\gamma}_{21}$ appears to approach the proportionality to $\dot{\gamma}_{21}{}^2$ predicted by equation 74 (slope of 2 on logarithmic plot). Thus, $\sigma_{11} - \sigma_{22}$ as a function of $\dot{\gamma}$ bears an analogy to G' as a function of ω for *small* values of the respective arguments. The curve for 5% concentration has an inflection reminiscent of, though far less prominent than, the inflections in G' as a function of ω in Figure 21. It is probably associated with the presence of some degree of entanglement coupling at this concentration, whereas at 1% the molecules do not pervade each others' domains sufficiently for entanglement to occur. (This statement calls attention to the importance of polymer concentration in determining the degree of entanglement, while in the preceding discussion of linear viscoelastic behavior the dependence of entanglement on molecular weight was stressed.)

Non-Newtonian Flow. Examples of non-Newtonian flow in several very different types of polymeric systems are shown in Figure 31, where the non-Newtonian viscosity $\breve{\eta} = \sigma_{21}/\dot{\gamma}_{21}$ is plotted logarithmically against $\dot{\gamma}_{21}$.

First, an extremely dilute solution in which the polymer molecules move essentially independently of each other is represented by curve A, for a polystyrene with molecular weight 6.2×10^6 and sharp molecular-weight distribution at a concentration of 0.05 g/dl in benzene at 30°C: data of Suzuki, Kotaka, and Inagaki (58). From the low-shear rate limiting value of η there is a slight but definite drop in $\breve{\eta}$ with increasing $\dot{\gamma}$, amounting to about 20% in the range investigated. Such dilute solution measurements are often extrapolated to infinite dilution and expressed as a shear-rate dependent intrinsic viscosity, $[\breve{\eta}]$.

Second, data for the somewhat more concentrated polystyrene solutions of Ashare (51) whose normal stress differences are portrayed in Figure 30 are represented by curves B and C. At 1% concentration, where the molecules overlap somewhat but are probably not entangled, there is a moderate drop in $\breve{\eta}$, somewhat more than in the extremely dilute solution; at 5% concentration, where entanglements probably exist, the non-Newtonian effect is much more pronounced and, over about two decades of $\dot{\gamma}$, $\breve{\eta}$ follows rather closely the power-law relation mentioned in the section on non-Newtonian flow. It is of interest that the onset of non-Newtonian flow occurs at a much higher shear rate for 1% than for 5% concentration, just as the normal stresses achieve significant magnitudes at a higher shear rate for 1% than for 5% concentration in Figure 30.

Finally, data for *undiluted* polystyrenes at elevated temperature are represented in curves D and E, from experiments by Stratton (59). The molecular weights (D: $\bar{M}_w$ = 48,000, $\bar{M}_n$ = 45,000; E: $\bar{M}_w$ = 242,000, $\bar{M}_n$ = 236,000) are sufficiently high for entanglements to be present, though for curve D only barely. Qualitatively, curves D and E resemble curves B and C, respectively, though the power law region in curve E has a steeper slope. The positions on the abscissa scale do not have direct significance, because they depend on temperature, concentration, and the solvent viscosity in the case of the solutions.

All these examples are for sharp molecular-weight distribution. The form of the dependence of η on $\dot{\gamma}$ is strongly dependent on the spread of molecular weights; a broad distribution causes a more gradual change of $\breve{\eta}$ at low $\dot{\gamma}$ where the limiting value η is approached.

It may be remarked that the nonlinear behavior of the elongational viscosity, $\breve{\bar{\eta}}$, is entirely different; it increases with increasing $\dot{\gamma}_{11}$ or $\dot{\epsilon}$.

Stress Relaxation after Cessation of Non-Newtonian Flow. The alternative type of stress relaxation experiment portrayed in Figure 5, which follows cessation of steady-state flow, has not been discussed until now. If the shear rate during the steady flow is sufficiently high for non-Newtonian viscosity to be observed, the course of the subsequent relaxation also depends strongly on shear rate. This is illustrated in Figure 32 for a 4% solution of polystyrene, with molecular weight 1.8×10^6 (rather sharp distribution) in a viscous chlorinated diphenyl solvent, by data of Macdonald (60). Here $\sigma_{21}^{ss}(t)$ refers to the shear stress as a function of time in this type of experiment. When the shear rate preceding the relaxation experiment is high, the relaxation is much more rapid. If a suitable constitutive equation were available to describe this behavior accurately, it could be used to predict the course of the relaxation following infinitesimal shear rate, which in this case is not accessible experimentally.

The normal stresses present during the steady-state flow also relax after its cessation, and the course of the first normal stress difference, $(\sigma_{11} - \sigma_{22})^{ss}(t)$, where the

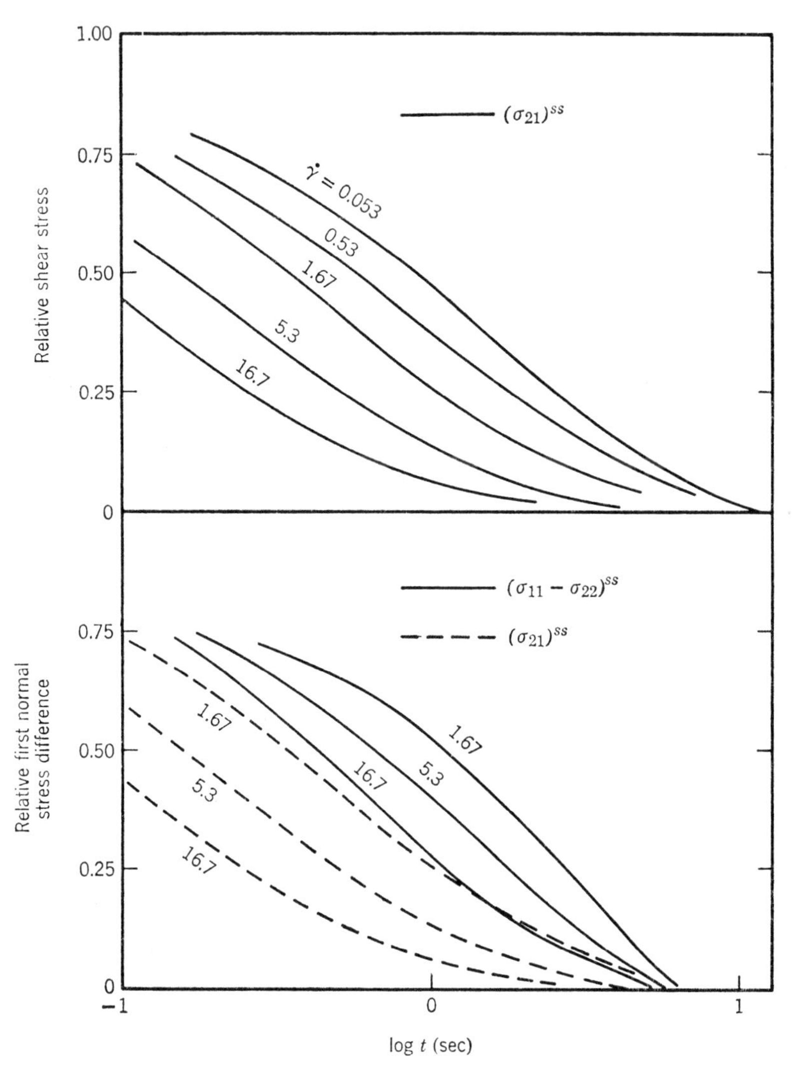

Fig. 32. Relaxation of shear stress and first normal stress difference after cessation of steady-state flow, for a 4% solution of polystyrene with molecular weight of 1.8×10^6 in chlorinated diphenyl as described in the text (60). Numbers refer to the shear rate preceding cessation of flow.

superscript *ss* again refers to this particular type of experiment, is also shown in Figure 32. Here, also, the relaxation is the more rapid, the higher the shear rate which precedes it. However, the normal stress difference relaxes more slowly than the shear stress.

Bibliography

1. H. Leaderman, *Elastic and Creep Properties of Filamentous Materials and Other High Polymers*, The Textile Foundation, Washington, D.C., 1943.
2. W. Weber, *Pogg. Ann.* (2) **4,** 247 (1835).
3. R. Kohlrausch, *Pogg. Ann.* (3) **12,** 393 (1847).
4. L. Boltzmann, *Wied. Ann.* **5,** 430 (1878).
5. C. Zener, *Elasticity and Anelasticity of Metals*, University of Chicago Press, 1948.
6. C. Kittel, *Introduction to Solid-State Physics*, 2nd ed., John Wiley & Sons, Inc., New York, 1956.
7. J. G. Kirkwood, *J. Chem. Phys.* **14,** 180 (1946). S. A. Rice and J. G. Kirkwood, *J. Chem. Phys.* **31,** 901 (1959).
8. J. D. Ferry, *Viscoelastic Properties of Polymers*, 2nd ed., John Wiley & Sons, Inc., New York, 1970.

8a. T. Alfrey, Jr., *Mechanical Behavior of High Polymers*, Interscience Publishers, Inc., New York, 1948.

9. R. B. Bird, W. E. Stewart, and E. N. Lightfoot, *Transport Phenomena*, John Wiley & Sons, Inc., New York, 1960.
10. S. Timoshenko, *Theory of Elasticity*, McGraw-Hill Book Co., New York, 1934.
11. A. J. Staverman and F. Schwarzl, in H. A. Stuart, ed., *Die Physik der Hochpolymeren*, Vol. IV, Chap. I, Springer-Verlag, Berlin, 1956.
12. A. G. Fredrickson, *Principles and Applications of Rheology*, Prentice-Hall, Englewood Cliffs, N.J., 1964.
13. A. S. Lodge, *Elastic Liquids*, Academic Press Inc., New York, 1964.
14. A. E. H. Love, *A Treatise on the Mathematical Theory of Elasticity*, 4th ed., Dover Press, New York, 1944.
15. Y. C. Fung, *Foundations of Solid Mechanics*, Prentice-Hall, Englewood Cliffs, N.J., 1965.
16. R. B. Bird, in R. B. Bird, W. E. Stewart, E. N. Lightfoot, and T. W. Chapman, *Lectures on Transport Phenomena*, American Institute of Chemical Engineering, New York, 1969, Chap. 1.
17. R. B. Bird and P. J. Carreau, *Chem. Eng. Sci.* **23,** 427 (1968).

17a. Ref. 8, Chaps. 3 and 4.

17b. Ref. 8, Chaps. 5–8.

18. Ref. 8, Chap. 11.
19. J. D. Ferry, W. M. Sawyer, and J. N. Ashworth, *J. Polymer Sci.* **2,** 593 (1947).
20. H. Poincaré, *The Foundations of Science*, Science, New York, 1929, p. 181.
21. W. Kuhn, *Helv. Chim. Acta* **30,** 487 (1947).
22. H. F. Olson, *Dynamical Analogies*, Van Nostrand, New York, 1943.
23. G. Kegel, *Kolloid-Z.* **135,** 125 (1954).
24. B. Gross, *J. Polymer Sci.* **20,** 371 (1956).
25. M. Curie, *Radioactivité*, Tome I, Hermann et Cie., Paris, 1935.
26. A. V. Tobolsky, *Properties and Structure of Polymers*, John Wiley & Sons, Inc., New York, 1960, pp. 188–194.
27. R. L. Bergen, Jr., and W. E. Wolstenholme, *SPE J.* **16,** 1235 (1960).
28. B. Gross, *Mathematical Structure of the Theories of Viscoelasticity*, Hermann et Cie., Paris, 1953.
29. L. Boltzmann, *Pogg. Ann. Phys.* **7,** 624 (1876).
30. B. D. Coleman and W. Noll, *Rev. Mod. Phys.* **33,** 239 (1961).

30a. Ref. 8, Chaps. 11 and 18.

31. J. G. Oldroyd, *Proc. Roy. Soc.* (*London*), *Ser. A* **200,** 523 (1950); **245,** 278 (1958).
32. A. G. Fredrickson, *Chem. Eng. Sci.* **17,** 155 (1962).
33. B. D. Coleman and H. Markovitz, *J. Appl. Phys.* **35,** 1 (1964).
34. T. W. Spriggs, J. D. Huppler, and R. B. Bird, *Trans. Soc. Rheol.* **10,** 191 (1966).
35. B. Bernstein, E. A. Kearsley, and L. J. Zapas, *J. Res. Natl. Bur. Stds.* **68B,** 103 (1964).
36. T. W. Spriggs and R. B. Bird, *Ind. Eng. Chem. Fund.* **4,** 182 (1964).
37. R. S. Rivlin, in F. R. Eirich, ed., *Rheology*, Vol. 1. Chap. 10, Academic Press Inc., New York, 1956.
38. T. W. Spriggs, Ph. D. Thesis, University of Wisconsin, 1966, p. 145.
39. H. Leaderman, *Trans. Soc. Rheol.* **1,** 213 (1957).
40. J. D. Ferry L. A. Holmes, J. Lamb, and A. J. Matheson, *J. Phys. Chem.* **70,** 1685 (1966).

40a. Ref. 8, Chap. 9.
41. K. Ninomiya and J. D. Ferry, *J. Phys. Chem.* **67,** 2292 (1963).
42. D. J. Plazek and V. M. O'Rourke, private communication.
43. W. Dannhauser, W. C. Child, Jr., and J. D. Ferry, *J. Colloid Sci.* **13,** 103 (1958).
44. J. W. Berge, P. R. Saunders, and J. D. Ferry, *J. Colloid Sci.* **14,** 135 (1959).
45. S. Iwayanagi, *J. Sci. Res. Inst. Japan* **49,** 4 (1955).
46. K. Sato, H. Nakane, T. Hideshima, and S. Iwayanagi, *J. Phys. Soc. Japan* **9,** 413 (1954).
47. J. R. Cunningham and D. G. Ivey, *J. Appl. Phys.* **27,** 967 (1956).
48. A. R. Payne, in P. Mason and N. Wookey, eds., *Rheology of Elastomers,* Pergamon Press, London, 1958, p. 86.
49. G. M. Martin, F. L. Roth, and R. D. Stiehler, *Trans. Inst. Rubber Ind.* **32,** 189 (1956).
50. J. D. Ferry and K. Ninomiya, in J. T. Bergen, ed., *Viscoelasticity—Phenomenological Aspects,* Academic Press Inc., New York, 1960, p. 55.
51. R. G. Mancke and J. D. Ferry, *Trans. Soc. Rheol.* **12,** 335 (1968).
52. J. F. Sanders, R. H. Valentine, and J. D. Ferry, *J. Polymer Sci* [A-2] **6,** 967 (1968).
53. J. A. Faucher, *Trans. Soc. Rheol.* **3,** 81 (1959).
54. A. J. Staverman, in S. Flugge, ed., *Handbuch der Physik,* Springer-Verlag, Berlin, 1962, Vol. 13, p. 432.
55. J. E. McKinney and H. V. Belcher, *J. Res. Natl. Bur. Stds.* **67A,** 43 (1963).
56. R. S. Rivlin and D. W. Saunders, *Trans. Faraday Soc.* **48,** 200 (1952).
57. E. Ashare, *Trans. Soc. Rheol.* **12,** 535 (1968).
58. H. Suzuki, T. Kotaka, and H. Inagaki, *Rept. Progr. High Polymer Phys. Japan* **10,** 115 (1967).
59. R. A. Stratton, *J. Colloid Interf. Sci.* **22,** 517 (1966).
60. I. F. Macdonald, unpublished experiments.

John D. Ferry
University of Wisconsin

VISCOMETRY

One of the most striking nonequilibrium properties of a long chainlike molecule is its ability, when dissolved at a very low concentration, to alter the flow properties of the mixture. The quantitative measurement of this effect on solution flow at low polymer concentration is termed "dilute-solution viscometry" and is discussed in this article. The wider application of viscometric measurement to concentrated solutions and bulk polymers is covered in MELT VISCOSITY. See also SOLUTION PROPERTIES; MOLECULAR-WEIGHT DETERMINATION; VISCOELASTICITY.

Definitions

The viscosity of a fluid is a measure of its resistance to flow. Induced motion, such as flow, requires a force to initiate the motion and to sustain it against frictional forces. Solids, when placed under a stress beyond their elastic limits, can maintain a strain, whereas fluids must have a sustaining force (stress) to maintain a strain; otherwise, the motion and relaxation of the molecules in the fluid dissipate the stress. When flow exhibits a smooth response to the force in the direction in which it is applied, the flow is said to be laminar; an uneven response characterizes turbulent flow.

Two basic models to define viscosity have evolved, the Newtonian and the Maxwellian models, and it is not unusual to find the nomenclature intermixed. First, let us consider the Newtonian model of two parallel plates, each of area A and separated by distance y, with a fluid between. If the upper plate only is subject to a force moving it along a line parallel to the bottom plate, ie, in direction x, the fluid between will flow in the same direction and will develop a velocity gradient dv/dy across the distance between the plates. The ratio of the force per unit area on the upper plate to the velocity gradient is defined as the viscosity,

$$\eta = (f/A)/(dv/dy) \tag{1}$$

where f is the force, A the area, v the velocity, and y the direction perpendicular to the applied force. The metric unit for viscosity as defined above is the poise, which is defined as 1 dyn-sec/cm^2, and has the fundamental dimensions mass/length × time. The kinematic viscosity, ν, is defined as the quotient of the dynamic viscosity and the density η/ρ, and has the dimensions of (length)^2/time. The metric units of ν are termed stokes.

The velocity gradient dv/dy is more commonly called the *shear rate*, $d\gamma/dt$, where γ is the shear strain, $\Delta x/y$. A shear strain develops because a tangential force (y direction) develops in the fluid as a result of the force applied to the upper plate (x direction). When the shearing force is proportional to the rate of shear and is constant, the fluid is said to be Newtonian. Otherwise stated, the viscosity of a Newtonian fluid is independent of shear rate (1).

In the Maxwellian model, stress (force/unit area) is considered proportional to strain, S; that is,

$$(\text{force/area}) = \epsilon(\text{strain}) = \epsilon S \tag{2}$$

where ϵ is defined as the coefficient or modulus of elasticity (2). If the body is rigid and has no detectable flow, f/A remains equal to ϵS. However, if the body is fluid, f/A does not remain constant, but dissipates through flow. In the simplest case, the rate of disappearance of stress with time can be set proportional to the stress. Hence,

$$d(f/A)/dt = \epsilon(dS/dt) - (f/A)/\tau \tag{3}$$

where τ is a relaxation time of the material. In mobile fluids τ is much smaller than one second. In fluids to which a constant stress is applied, the deformation in this simple case increases linearly with time so that equation 3 becomes

$$dS/dt = (f/A)/\epsilon\tau \tag{4}$$

Table 1. Nomenclature for Viscosity Models

Model	Symbol	Dimensions	Units	Comments
Newtonian				
applied force	f/A	dyn/cm^2		
velocity gradient	dv/dy	(cm/sec)/cm		also called rate of shear $d\gamma/dt$
dynamic viscosity	η	dyn-sec/cm^2	poise	$= (f/A)\ (dv/dy)$ $= (f/A)\ (d\gamma/dt)$
kinematic viscosity	$\nu = \eta/\rho$	cm^2/sec	stokes	
Maxwellian				
applied force (stress)	f/A	dyn/cm^2		
shear strain	$(dx/y) = S$	unitless		in narrow-differential thin section dy (laminar), (dx/dy)
coefficient of elasticity	ϵ	dyn/cm^2		
stress decay	$d(f/A)/dt$	dyn/cm^2-sec		in simple model $= -(f/A)/\tau$
relaxation time	τ	sec		when $d(f/A)/dt = 0$, $d\gamma/dt = (f/A)/\epsilon\tau = (f/A)/\eta$

It follows directly under these assumptions that $dS/dt = d\gamma/dt$ and that $\eta = \epsilon\tau$. It can be shown further that, under the assumption that the stress decays at a rate proportional to its value, this decay follows the first-order rate law, and that the relaxation time τ is directly related to a decrease in the stress to $1/e$ of its original value. These two models, the Newtonian or fluid model, and the Maxwellian or elastic model, and their attendant nomenclature are summarized in Table 1.

Although not all polymer solutions obey these simple relationships, one can, by proper control of the system, ensure that these criteria are met for the solutions under measurement. Deviations from these relationships do exist, however, and are discussed below.

In polymer science, it is not the absolute viscosity of a solvent or a solution that is of particular interest, but the increase in viscosity attributable to the dissolved polymer. Therefore, in the viscometry of polymer solutions, it is some expression of the relative viscosity that is useful. The *relative viscosity* is defined as the quotient of the viscosity of the solution, η_s, and the viscosity of the solvent, η_0,

$$\eta_r = \eta_s/\eta_0 \tag{5}$$

Another measure of the increase in viscosity by a high-molecular-weight solute is the *specific viscosity*,

$$\eta_{sp} = \eta_r - 1 \tag{5a}$$

Two additional quantities useful in viscometry are the *reduced viscosity*, η_{red}, defined as η_{sp}/c, and the *inherent viscosity*, defined as $\ln \eta_r/c$. The relative viscosity of polymer solutions is concentration dependent, so that the most useful viscosity value is the limiting one when either η_{sp}/c or $\ln \eta_r/c$ is extrapolated to zero concentration. This quantity is termed the *intrinsic viscosity* or the limiting viscosity number, $[\eta]$. Since relative viscosity is dimensionless, the units for $[\eta]$ are those of reciprocal concentration. The proposed (IUPAC) concentration unit is g/ml, although the unit g/dl is most commonly used. The definitions and proposed nomenclature are listed in Table 2.

Table 2. Nomenclature of Solution Viscosity (12)

Common name	Proposed IUPAC name	Symbol and defining equation
relative viscosity	viscosity ratio	$\eta_r = \eta_s/\eta_0 \simeq t_s/t_0$
specific viscosity		$\eta_{sp} = \eta_r - 1 = (\eta_s - \eta_0)/\eta_0 \simeq (t_s - t_0)/t_0$
reduced viscosity	viscosity number	$\eta_{red} = \eta_{sp}/c$
inherent viscosity	logarithmic viscosity number	$\eta_{inh} = (\ln \eta_r)/c$
intrinsic viscosity	limiting viscosity number	$[\eta] = (\eta_{sp}/c)_{c \to 0} = [(\ln \eta_r)/c]_{c \to 0}$

Measurement of Viscosity

Many devices, both simple and elaborate, have been invented to measure the viscosity of liquid mixtures or of pure small-molecule fluids. Within the domain of polymer solution viscometry, however, two devices have served as standards. These are the capillary viscometer and the Couette viscometer (3). The first of these has the advantage of simple and inexpensive construction, as well as ease of manipulation. The measurement of a dilute-solution viscosity in a capillary viscometer can be adapted easily to a routine procedure that can yield significant information in a relatively short time and at very low cost.

The Couette viscometer consists of two concentric cylinders with the solution occupying the space between. One cylinder is rotated while the other remains fixed. This device can be modified to provide a wide range of shear rates and is especially useful for measuring the viscosity at very low shear rates. The choice of instrument depends on the accuracy and speed of data collection needed, as well as upon the nature of the polymers being investigated.

Capillary Viscometry. It is impractical to measure the viscosity of a fluid by placing it between flat parallel plates. An alternative method for bringing a fluid under a tangential force and thereby producing a velocity gradient is to cause the liquid to flow under a force through a capillary of uniform bore. The fluid will flow more rapidly in the center than it does at the walls, providing a velocity gradient across the radius of the capillary. The Poiseuille equation (eq. 6) describing capillary flow is derived with the following assumptions: (*a*) the capillary is sufficiently long that end effects can be neglected; (*b*) there are neither tangential nor radial components of the velocity; (*c*) the flow is steady; (*d*) isothermal conditions prevail; (*e*) the velocity at the wall is zero; (*f*) there are no external forces; (*g*) the fluid is incompressible; and (*h*) Newtonian flow.

The force per unit area acting on the fluid through a cylindrical capillary of uniform bore is equal to the product of the differential pressure on the capillary, ΔP, and the cross-sectional area of the capillary of radius R, divided by the area upon which the force is exerted, in this case, the surface area of the walls of the tube of length L. Thus, where dv/dr is the velocity gradient at point r on the radius,

$$-\eta(dv/dr) = \pi r^2 \Delta P/2\pi r L = \Delta P r/2L = f/A \tag{6}$$

This expression for capillary flow, obtained by Poiseuille, shows that the shear rate is directly proportional to r and ΔP and inversely proportional to L and η. The negative sign indicates that the shear stress is opposite to the direction of flow. If the equation is integrated with the boundary condition of zero velocity at the walls ($r = R$), the velocity at the center is given by

$$v(\text{max}) = \Delta P R^2/4\eta L \tag{7}$$

The total volume, V, discharged in t seconds (the efflux time) is

$$V = \int_{r=0}^{r=R} vt\, 2\pi r\, dr = \pi \Delta P R^4 t/8\eta L \tag{8}$$

Rearrangement of equation 8 yields

$$\eta = \pi \Delta P R^4 t/8VL = Kt \tag{9}$$

The constant K is dependent on the geometry of the capillary and the pressure ΔP. If the capillary is topped by a bulb containing a fluid of density ρ and this is allowed to flow under its own weight, the differential pressure is equal to

$$\Delta P = \langle h \rangle \rho g \tag{10}$$

where $\langle h \rangle$ is the average hydrostatic head, and g the gravitational constant. Equation 9 can then be written in simple form as

$$\eta = K'\rho t \tag{11}$$

with K' dependent exclusively on the geometry of the capillary viscometer.

There are two major corrections to this simple equation, one resulting from the problems of end effects in discharge at the end of the capillary, and the other from the fact that dissipation of the force acting on the fluid is not entirely through overcoming viscous drag, but partially through imparting kinetic energy to the molecules. When these two effects are accounted for, equation 11 is transformed into

$$\eta = [\pi R^4 \langle h \rangle g \rho t / 8V(L + nR)] - [mV\rho / 8\pi(L + nR)t] \tag{12}$$

where the quantity $(L + nR)$ (an effective extension of L by nR) arises from the end-effect correction, and the second term arises from the kinetic energy correction. The two new constants n and m have been determined independently for capillary viscometers, and each is close to unity (4). These constants need never be determined in dilute-solution viscometry because ratios or relative measurements are made.

In practice, equation 12 can be reduced to the form

$$\eta = \alpha\rho\left(t - \frac{\beta}{\alpha t}\right) \tag{13}$$

where α and β are constants for a given capillary viscometer determined by a calibration procedure. A detailed description of the calibration method has been published (5).

From equation 13 and the definition of relative viscosity,

$$\eta_r = \eta_s/\eta_0 = [\alpha\rho_s/\alpha\rho_0][(t_s - \beta/\alpha t_s)/(t_0 - \beta/\alpha t_0)] \tag{14}$$

When the ratio of the densities (ρ_s/ρ_0) is close to unity and the efflux time is large compared to $\beta/\alpha t$, then equation 14 can be reduced to the very simple form

$$\eta_r \cong t_s/t_0 \tag{15}$$

With these approximations, the relative viscosity can be obtained by measuring the efflux times of solution (t_s) and solvent (t_0).

In order to determine intrinsic viscosity, $[\eta]$, η_{sp}/c or $\ln \eta_r/c$ must be extrapolated to zero concentration. Empirical expressions for the extrapolation to infinite dilution can be written as follows:

$$\eta_{sp}/c = (\eta_r - 1)/c = [\eta]\{1 + k_1[\eta]c + k'_1([\eta]c)^2 + \cdots\} \tag{16}$$

$$(\ln \eta_r)/c = [\eta]\{1 + k_2[\eta]c + k'_2([\eta]c)^2 - \cdots\} \tag{17}$$

where $k_1 - k_2 = \frac{1}{2}$ and $k'_2 - k'_1 + k_1 = \frac{1}{3}$.

In precision analysis where the data warrants it, these equations can be used in full; however, the usual practice is to truncate these expansions after the second term. The resulting equations are often referred to by the names of scientists originating their use in a particular country:

$$\eta_{sp}/c = [\eta] + k_1[\eta]^2c \quad \text{Huggins equation} \tag{18}$$

$$(\ln \eta_r)/c = [\eta] + k_2[\eta]^2c \quad \text{Kraemer equation} \tag{19}$$

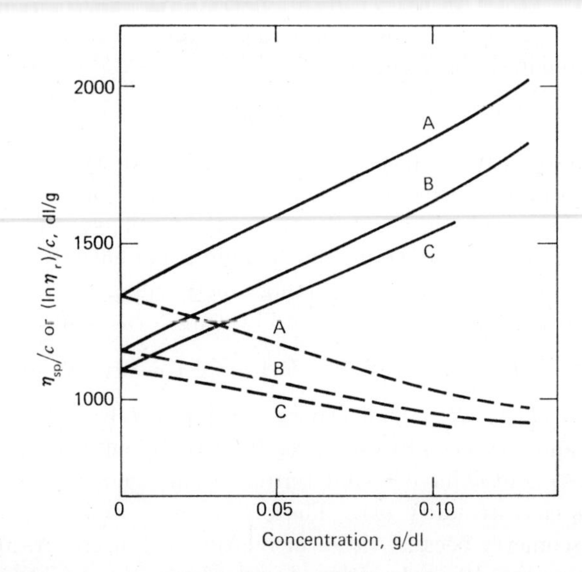

Fig. 1. An example of viscosity–concentration plots for polystyrene ($M_w = 7.14 \times 10^6$) in benzene at 30°C (7). Solid lines, η_{sp}/c; broken lines, $\ln \eta_r/c$. A, data obtained with a Couette viscometer with rotors operating at shear stresses 0.00372, 0.00727, and 0.0764 dyn/cm²; B, with an Ubbelohde viscometer at 8.67 dyn/cm²; C, with an Ubbelohde viscometer at 12.2 dyn/cm². Courtesy *The Journal of Chemical Physics.*

These truncated equations are useful only when a plot of η_{sp}/c or $(\ln \eta_r)/c$ against c is linear. It is not unusual to find curvature in the plots when η_r is >1.8 or <1.2. At the higher end, the curvature results from second-order concentration effects, and at the lower end very often from absorption of polymer onto the capillary walls or inaccurate timing (11). In so-called "good" solvents, meaning those with a very small enthalpy of mixing, k_1 lies close to a value of $\frac{1}{3}$. However, as the Flory temperature, θ, is approached, this parameter can reach values greater than 1.

It is common practice to use both equations 18 and 19 to plot data and extrapolate to a common intercept to obtain $[\eta]$. Such a procedure is reliable, but has been shown to have limitations, especially when k_1 is $\frac{1}{2}$ or larger (6). Figure 1 shows a typical data plot (7).

Equipment. The three basic components for capillary viscometry are the viscometer, the thermostat, and the timer. Capillary viscometers of various types have been reported in the literature. Figure 2 shows some examples. The viscometer most commonly used in polymer chemistry is the Ubbelohde type shown in Figure 2b. This type is called a *suspended-level viscometer;* by the introduction of a side arm just below the end of the capillary, the same external pressure is established on the fluid in the capillary above and below the flowing column. A large lower bulb is often included as a solution reservoir in order to make possible dilutions directly in the viscometer. Ubbelohde viscometers with multiple bulbs topping the capillary are used to provide a viscometer with several shear rates by providing several different hydrostatic heads. Figure 3 shows a Ubbelohde viscometer modified for attachment to a vacuum line so that a controlled atmosphere can be maintained in contact with the fluid. This viscometer has been used to determine the viscosity of polyethylene at high temperatures (>125°C) (10).

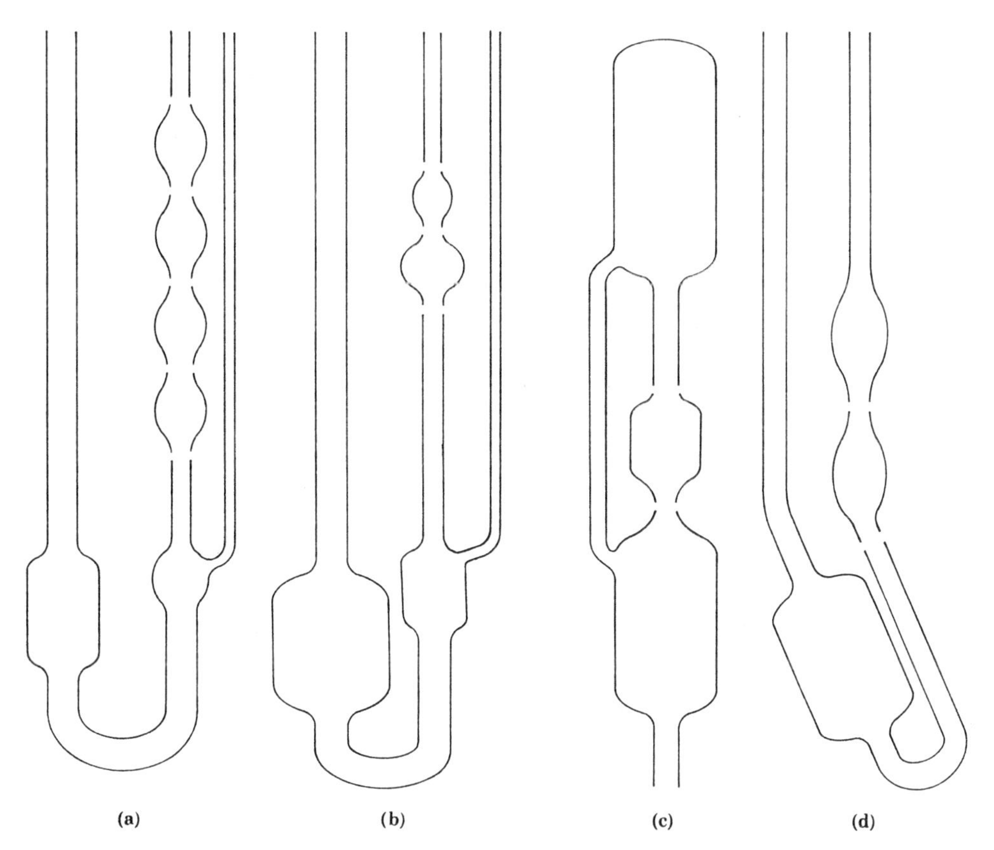

Fig. 2. Viscometer designs. (**a**) Ubbelohde variable-shear-rate viscometer. (**b**) Ubbelohde dilution viscometer. (**c**) Ubbelohde vacuum viscometer. (**d**) Ostwald viscometer.

Capillary viscometers are easily clogged by small particles. It is necessary, therefore, to filter all solutions carefully as they are introduced into the viscometer and to maintain scrupulous cleanliness when rinsing the viscometer. Commercial viscometers are modified by researchers to include a coarse-grade sintered-glass filter in the dilution end of the viscometer so that no fluid can be placed inside without filtration. The solution is either drawn or pushed through the capillary until the reservoir above is filled. After this, the pressure is released and the fluid is allowed to flow freely. The efflux time is measured from the instant the meniscus passes the upper mark on the bulb to the time it passes the lower mark. If the solvent has high volatility, it is best to push the fluid through the capillary by slight pressure on the side opposite the capillary.

When solutions are very dilute and the capillary is very small, it is not unusual to find inconsistencies in viscosity data. It has been shown that many polymers are adsorbed onto the surface of the capillaries, changing their effective radius; in a small capillary this can have a noticeable effect (11). In addition, adsorption changes the concentration of a very dilute solution, thereby introducing further error. Capillaries should be checked regularly for adsorption by running through pure solvents until reproducible efflux times are obtained. Some polymers, when adsorbed, are very difficult to remove by dissolution in a solvent or with cleaning fluid. It is expedient to

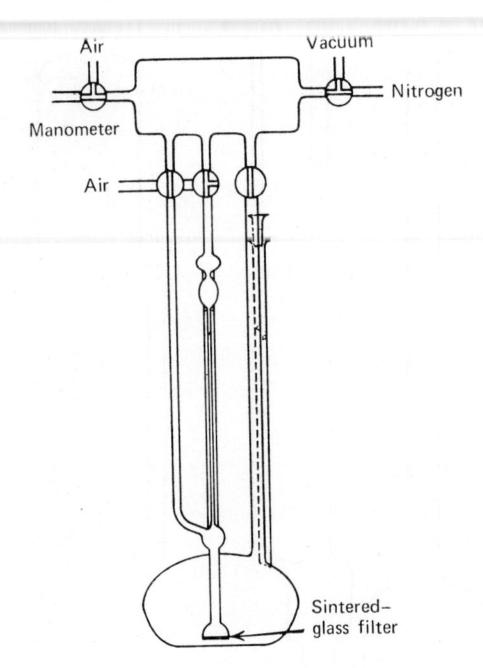

Fig. 3. Ubbelohde viscometer modified for high-temperature studies and for a controlled atmosphere (10).

degrade adsorbed polymer by placing the viscometer in a glass-annealing oven overnight.

For precision work, the viscometer should be placed rigidly and reproducibly vertical in a constant-temperature bath, for which the thermostat should be maintained within ±0.01°C. This is usually attained in most high-quality commercial thermostatic baths. If high-temperature viscometry is desired, the bath must be designed carefully and insulated to prevent thermal gradients from developing.

Manually triggered timing devices should be readable to at least 0.1 sec. Reproducibility of the efflux time to ±0.1 sec can be obtained easily by an experienced operator, when the temperature is maintained within the limits stated above. Within recent years, commercial firms have developed automatic viscometers whose most useful feature is timing to ±0.01 sec or less. The timing circuits are usually triggered by a pair of photoelectric sensors attached to the viscometer which respond as the meniscus passes (15). Sensors that detect a small change in electrical capacitance can also be used as triggers. The major value of the automatic instruments is their precision timing, which reduces operator error.

Couette Viscometry. The all-glass capillary viscometer has proved to be a versatile and inexpensive device for polymer chemists. However, in the past decade, as the study of biological macromolecules progressed, it became apparent that many of these molecules degraded even under the small shear forces found in a capillary viscometer. In addition, some biological and synthetic molecules are so very large that their solutions behave in a non-Newtonian manner when subject to the shear rates experienced in ordinary-sized glass capillaries.

To provide a viscometer with an extremely low shear rate, Zimm and others returned to the earlier concentric cylinder design of Couette (8). In the Zimm viscom-

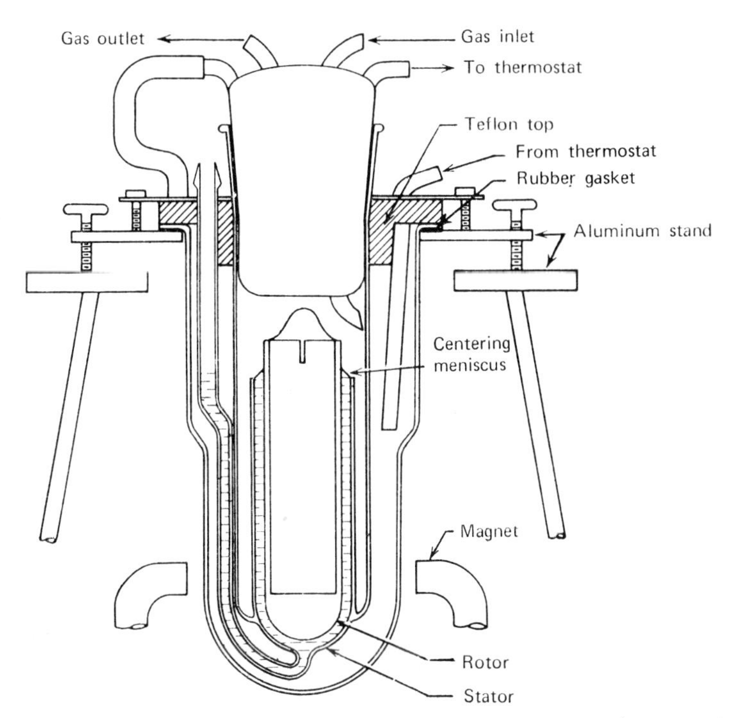

Fig. 4. Schematic drawing of a low-shear-rate, concentric-cylinder Couette viscometer (9). Courtesy *The Journal of Chemical Physics.*

eter, the fluid is placed between two concentric glass cylinders. The outer cylinder (the stator) remains fixed and the inner cylinder (the rotor) rotates. A torque is developed on the rotor through an external rotating magnet which acts on a small pellet of steel embedded in the rotor. The rotor is weighted to float freely in the solution and is held in place by surface forces. A fiducial mark is scribed on the rotor and its time of revolution is recorded. One advantage of the Couette-type viscometer is that the rotational motion (constant stress) can be sustained over long periods of time so that temporal effects can be measured.

There are three stable arrangements of the two cylinders when supported by a meniscus; in the most suitable of these the rotor is suspended slightly above the top of the stator with the meniscus surface forces centering the rotor. When it was found that a constant torque could not be obtained over a wide temperature range using the Zimm glass rotor with a steel pellet, Berry and co-workers substituted an anodized aluminum rotor with an external rotating permanent magnet driven by a synchronous motor (9). These investigators claim this configuration produces a torque essentially independent of temperature. The Berry design, which is patterned after that of Zimm, is shown in Figure 4.

The Couette viscometer is maintained at a constant temperature and a nitrogen atmosphere flows over the exposed surface of the solution.

The period of revolution (seconds per revolution) is measured for the solution (P), the solvent (P_0), and the external magnet (P_m). The relative viscosity η_r is given by equation 20,

$$\eta_r = (P - P_m)/(P_0 - P_m) \tag{20}$$

if the geometry of the system remains constant. The period of revolution for the solution, in terms of instrument constants, is

$$(P - P_m) = (\eta/\rho)[\pi W R_1/T(R_2 - R_1)][(R_1 + R_2)/R_1]^2 [p(R_1,R_2)][1 + \Delta] \quad (21)$$

where R_1 and R_2 are the radii of the rotor and the stator, respectively, W is the weight of the rotor, T is the torque, Δ is a parameter that accounts for end effects, and η is the viscosity of the enclosed fluid of density ρ. When $P = P_m$, $p(R_1,R_2)$ is a function approximately unity related to the ratio of the radii of the rotor and stator. The average shear rate is given by equation 22,

$$(d\gamma/dt) = (\pi/P)[(R_2 + R_1)/(R_2 - R_1)]\, f(R_1,R_2) \quad (22)$$

where the function $f(R_1,R_2)$ is approximately unity, and is related as p is to the ratio of the radii (9). From equation 22 it can be seen that the shear rate is inversely proportional to the period of revolution. Zimm and Crothers report measurements in their Couette viscometers with shear rates as low as 2.5×10^{-2} sec^{-1}. They show also that viscosity measurements made on the same samples in a capillary viscometer and their Couette viscometer give equivalent values for the intrinsic viscosity (7,8).

Interpretation of Results

The increase in viscosity due to a polymeric solute is a function of the volume of the polymer molecule in solution and thus is not a direct or absolute measure of its molecular weight. It has long been known that for a series of linear homologous polymers with similar molecular-weight distributions, the intrinsic viscosity could be related to the absolute measured molecular weights, M, of the samples by the empirical relation,

$$[\eta] = KM^a \quad (23)$$

where K and a are empirical constants determined for a polymer–solvent pair at a single temperature. This equation is often referred to as the *Mark-Houwink equation.* Hundreds of such empirical expressions now exist in the literature (13). Historically, the molecular weight M has been related to either a weight-average or a number-average molecular weight, depending on the manner in which the absolute molecular-weight values were determined in the calibration (23). The constant a is usually between 0.5 and 0.80, although it is occasionally higher if the chain is extended to any appreciable extent from the random-coil conformation. The value of a is 0.5 when the polymer–solvent system is at the Flory, or θ, temperature.

For a polydisperse system, the intrinsic viscosity is related to the molecular weight in the following manner:

$$\eta = K[\sum N_i M_i^{(1+a)}/\sum N_i M_i] = K\bar{M}_v^{\,a} \quad (24)$$

where N_i is the number of polymer molecules of molecular weight M_i. The term in brackets defines the viscosity-average molecular weight, $\bar{M}_v^{\,a}$. This average is never independently determined since the value is dependent on the hydrodynamic volume of the polymer in the solvent system for a particular temperature. When $a = 1$, the viscosity average is equal to the weight average. If a distribution function for a polydisperse system is known, then a true viscosity-molecular average weight may be calculated. The molecular weight values calculated from an empirical expression such as equation 23 using measured values of $[\eta]$ provide a molecular-weight average the same as that used in the calibration. It is the usual practice to determine the constants a and K with a series of narrow-molecular-weight fractions. If intrinsic viscosity measurements are then made on a series of samples with a broad molecular-weight distribution, molecular weights calculated from the calibrated expression (eq. 23) will be in error to an extent depending on the breadth of the distribution. Since the viscosity-average molecular weight approximates the weight-average molecular weight more closely than it does the number-average molecular weight, it is preferable to calibrate equation 23 with weight-average values.

Most recorded data show that K and a in equation 23 hold constant except at low molecular weight (<10,000), the exact values depending on the system (9). Caution is urged in the application of equation 23 in this region without verification by calibrating against absolute values of molecular weight. Very few synthetic polymers have been measured with molecular-weight values larger than 5.0×10^5.

One-Point Determination of $[\eta]$. The principal application of viscometry is the rapid determination of a relative measure of molecular size or weight. In applied viscometry, the usual practice is to restrict measurements of η_{sp} to a single concentration. Provided η_{sp} is obtained at a fixed concentration in a given solvent and temperature, values so obtained are useful in a relative sense to detect possible changes in the molecular weight of polymers prepared by various methods (see also MOLECULAR-WEIGHT DETERMINATION). It is possible to estimate a value for the intrinsic viscosity $[\eta]$ from such a single-point determination. The justification for a single-point extrapolation lies in the fact that the parameters k_1 and k_2 associated with the empirical equations 18 and 19 have constant values over a wide range of concentrations of polymer in many different solvent–polymer systems. Since the intrinsic viscosity is an extrapolated value at infinite dilution, the empirical equations meet the test of linearity well near this region. Under these circumstances, one can eliminate constants between the two empirical equations. One useful equation for a one-point determination of $[\eta]$ is provided by Solomon-Ciuta (37) as

$$\eta_{sp} - \ln \eta_{rel} = c^2[\eta]^2/2 \tag{25}$$

All information in this equation is obtained from a single determination of η_{rel} at one concentration. Solomon has shown that this equation can be improved further in its reliability in determining $[\eta]$ through the following considerations (38):

(*1*) Define two single-point measurable parameters,

$$\Delta = \eta_{sp} - \ln \eta_{rel} \tag{26}$$

$$\pi = (\eta_{sp})(\ln \eta_{rel}) \tag{27}$$

(*2*) Relate the Kraemer constant k_2 to these parameters and estimate k_2 from the data and either of the following equations:

$$k_2 = \frac{1}{\ln \eta_{rel}} - \frac{1}{\sqrt{2\Delta}} \tag{28}$$

$$k_2 = \frac{1}{\ln \eta_{rel}} - \frac{1}{\sqrt{\pi}} \tag{29}$$

(*3*) Calculate $[\eta]$ from the Kraemer equation (eq. 19) using a value of k_2 calculated from *2* above.

Although other one-point methods based on other empirical expressions have been developed, Solomon and Gottesman (38) have shown that estimates based on the Kraemer equation above are accurate over a wide range of experimental variables.

Shear Corrections. When the hydrodynamic volume of a molecule is so large that the polymer tends to elongate under the shearing forces in a capillary, the increase in viscosity due to the polymer is less than that expected in Newtonian behavior. The polymer molecules are sheared into ellipsoidal shapes and align themselves with the flow stream, thus presenting less resistance to flow than they would if they maintained a random conformation. Hence, for polymers of very high molecular weight (ca 10^6) or of very high hydrodynamic volume (extended polymers), corrections to zero shear rate must be made. No simple a priori means of estimating whether shear corrections are needed can be provided. If doubt about this effect exists, viscosity measurements taken at several shear rates provide the best general test.

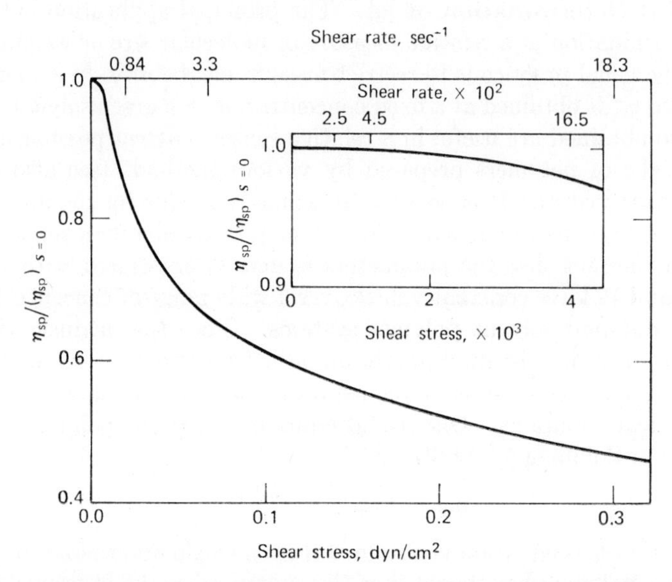

Fig. 5. Shear dependence of the specific viscosity of T2 DNA, at 3.3×10^{-3} g/dl of the sodium salt (8). For this solution, the actual value of $(\eta_{sp})_{S=0}$ was 1.730. Courtesy *Proceedings of the National Academy of Sciences.*

The simplest experimental method for determining the relative viscosity of a sample as a function of shear stress is to use a multiple-bulb Ubbelohde viscometer such as that depicted in Figure 2**a**. Detailed descriptions of capillary shear viscometers and their geometrically related correction factors are available in the literature (14). Shear rates can be calculated by measuring the viscosity of standard fluids and determining the geometrical factors needed for equations 6 and 10. The relative viscosity is then plotted as a function of the shear stress and extrapolated to zero stress. Although this procedure provides a value for η_r at zero shear, the accuracy of the extrapolation is subject to several sources of error: (*a*) the data may be taken in a region where η_r is not a linear function of shear stress; and (*b*) theory indicates that the limiting slope of the curve is zero. Figure 5 shows data for a large DNA polymer at very low shear stress (8). These data clearly indicate the possible error in a linear extrapolation procedure. Where data are needed in the low shear rate region, use of the

Couette viscometer is warranted. Theoretical considerations show that the initial dependence of $[\eta]$ on the shear rate is

$$[\eta]_\gamma/[\eta]_{\gamma=0} = 1 - \Gamma(d\gamma/dt)^2 + \cdots \tag{30}$$

where $[\eta]_\gamma$ is the intrinsic viscosity at shear rate $d\gamma/dt$ and $[\eta]_{\gamma=0}$ is the intrinsie viscosity for $d\gamma/dt = 0$. Experimentally determined values of the constant Γ and theoretical calculations show that this parameter is inversely proportional to temperature, and directly proportional to the square root of chain length (16, 17).

Theory

Theories for the dependence of the intrinsic viscosity on chain length have been developed and refined by many authors. The results can be given in the form

$$[\eta] = \phi(h)(\langle s_0^2\rangle/M)^{3/2}M^{1/2}\alpha^3 \tag{31}$$

where $\langle s_0^2\rangle$ is the mean-square radius of gyration, M is the molecular weight of the chain, α is an expansion factor equal to the ratio of the radius of gyration of the polymer in the solution to what it would be at the θ temperature, and ϕ and h are parameters related to the chain length, the distribution of molecular weights, and the unperturbed dimensions at $T = \theta$. The many theories that relate to this expression are covered in an excellent review article (18). Equation 31 can be reduced to give equation 32.

$$[\eta] = K'M^{1/2}\alpha^3 \tag{32}$$

When the θ temperature is reached, the polymer molecule reaches zero expansion, α reaches 1, and $[\eta]_\theta$ is proportional to $M^{1/2}$. This relationship at $T = \theta$ has been verified many times for many polymer–solvent systems. If one accepts a theoretical numerical value for $\phi(h)$, the value $(\langle s_0^2\rangle/M)^{3/2}$ can be calculated from data taken at $T = \theta$. This ratio is a measure of the reduced size of the polymer chain in the unperturbed state. Values for various polymers show that measured unperturbed dimensions correspond reasonably well with calculations based on statistical models which include information about bond angles, bond lengths, and conformational states taken from data on small molecules or information derived from crystallographic data of polymers. Flory has shown that α varies as M^0 to $M^{0.3}$, depending on the solvent–polymer system (19). His theory, with the approximations made, agrees with the empirical equation 23.

The hydrodynamic theory for the viscosity of polymer molecules has evolved by relating $\phi(h)$ to parameters descriptive of molecular parameters for the chain, and relating the expansion factor α to thermodynamic parameters. Although beautifully developed theories do exist, all introduce parameters related to the models upon which the calculations are based (18). Independent assessment of the various parameters is not possible, so that these theories must be judged on a relative basis. High-precision viscosity data over a wide range of experimental variables have been taken on only a limited number of fully characterized polymer samples. A good example of such data with a concomitant analysis of the theory is given in Ref. 9.

Temperature Dependence of [η]. Excellent viscosity data as a function of temperature for samples of an ionically polymerized polystyrene of narrow molecular-weight distribution are shown in Figure 6 (9). This shows clearly how the temperature dependence changes with molecular weight, with the high-molecular-weight samples showing the greatest change and largest curvature. The solvent Decalin used in this

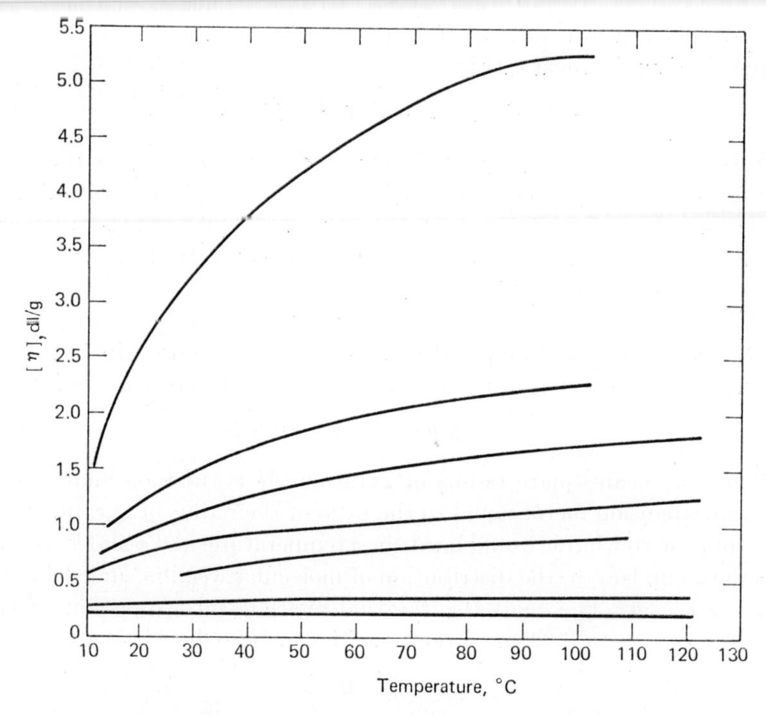

Fig. 6. The intrinsic viscosity $[\eta]$ versus temperature for various nearly monodisperse polystyrene samples of different molecular weight measured in Decalin solutions (9). Courtesy *The Journal of Chemical Physics.*

system is a thermodynamically poor solvent (large ΔH of mixing) for polystyrene. As T increases, there is an increase in $[\eta]$ that represents the expansion of the chain as the polymer–solvent contacts become more favorable (ΔH of mixing decreases). Figure 7 shows that the intrinsic viscosity can go through a maximum when measured over a wide temperature range in a mixture that has both a lower and an upper (θ) critical temperature (20). Polymers dissolved in thermodynamically good solvents have a very low temperature coefficient for $[\eta]$. The temperature dependence of $[\eta]$ lies largely in the dependence of α on T. There is sufficient evidence to show that a term containing a function $C(1 - \theta/T)$ accounts for the temperature effects described above, where C is a temperature-independent parameter. Although various functions relating $(1 - \theta/T)$ to α exist, the term in parentheses shows how the temperature effect increases rapidly near $T = \theta$. At temperatures far removed from θ, the contribution of θ/T compared to one becomes negligible. An article by Berry contains a detailed review of this subject (9).

Unperturbed Dimensions. The unperturbed dimensions of polymer chains can be obtained by light-scattering measurements at $T = \theta$, as well as by determining $[\eta]_{T=\theta}$ and applying equation 32. The unperturbed dimensions or $\langle s_0{}^2\rangle/M$ can be calculated from $K_{T=\theta}$ by taking a calculated value for $\phi(h)$ (18). However, it is also possible to estimate theoretical values of $\langle s_0{}^2\rangle/M$ from viscosity data in thermodynamically good solvents. Many authors have suggested appropriate equations for this procedure; equation 33 suggested by Berry applies over a wide range of values for α

$$([\eta]/M^{1/2})^{1/2} = K_{T=\theta}^{1/2} + 0.42\, K_{T=\theta}^{3/2} B(\langle s_0{}^2\rangle/M)^{-3/2} M/[\eta] \qquad (33)$$

(the expansion factor) (9). The intercept of $K^{1/2}_{T=\theta}$ can be used to estimate the unperturbed dimensions; however, since the parameter B, determined from the slope, is related through α to $K_{T=\theta}$, an iterative procedure must be followed to refine the extrapolation. Procedures are given by Stockmayer and Berry (9,18). Although

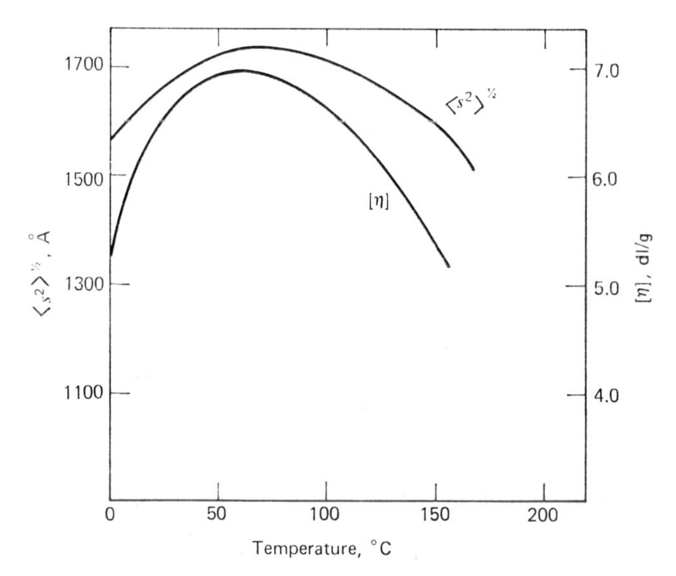

Fig. 7. Polyisobutylene–dibutyl ether mixture: radius of gyration and intrinsic viscosity as functions of temperature (20). Lower critical solution temperature is 200°C. Courtesy *Polymer*.

various authors suggest different functions as the most appropriate to make the iterations and extrapolations, the data in the literature for [η] and concomitant measurements of absolute molecular weight have an error too large to support a single function as the most accurate and the most reliable for all polymer–solvent systems.

Effects of Polymer Structure

Branched Polymers. Any structural change that affects the geometry of the polymer or the polymer–solvent interactions may have marked effects on the hydrodynamic size of the macromolecule in solution. The first such structural effect to consider is branching in homopolymers. Polymers can be branched with short chains and/or long chains, but it is the latter that have a significant effect on [η]. Long-chain branching can derive from a polymerization process involving chain transfer to polymer to form "branch-on-branch" chains, or branches can be introduced at preselected sites on a chain to produce comblike molecules or molecules with a single branch center and several arms radiating from the branch point. The ratio of the intrinsic viscosity of a branched molecule, $[\eta]_B$, to that of a linear molecule, $[\eta]_L$, of a given polymer, each containing the same number of monomer units (ie, the same DP), is always less than one and is dependent on the specific geometry of the chain (21). Viscosity characteristics of carefully prepared and well-characterized cruciform polystyrene molecules are shown in Figure 8 (22). Branched molecules, unless extremely simple in geometry, need several parameters to describe their average structure and their structure distribution (23). Unfortunately, viscosity measurements alone are insufficient to provide information for more than one unknown related to the branch structure.

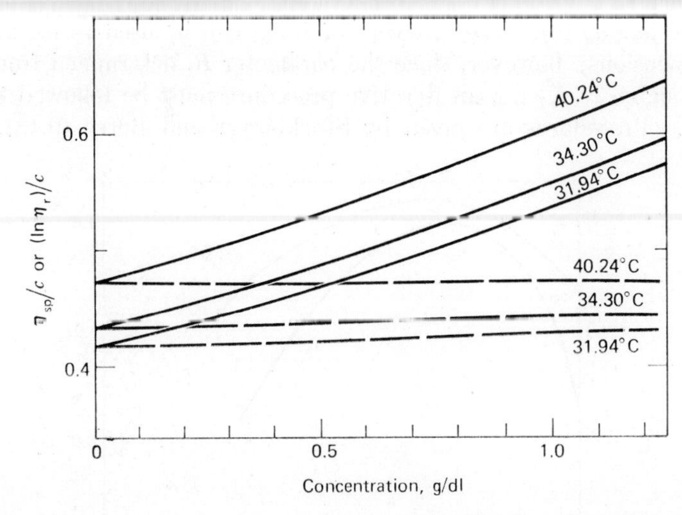

Fig. 8. Variation of η_{sp}/c (solid lines) and $(\ln \eta_r)/c$ (broken lines) with c for star-shaped polystyrene in cyclohexane at various temperatures (22). Courtesy American Chemistry Society.

Application of equation 31, which is based on hydrodynamic theory at $T = \theta$, for branched polymers and linear polymers with equal degrees of polymerization (24) gives

$$[\eta]_B^\theta/[\eta]_L^\theta = \langle s_0^2\rangle_B^{3/2}/\langle s_0^2\rangle_L^{3/2} \tag{34}$$

Statistical calculations of the unperturbed radius of gyration of branched and linear molecules can be used to obtain g (eq. 35), where g is a parameter related exclusively to

$$\langle s_0^2\rangle_B/\langle s_0^2\rangle_L = g \tag{35}$$

random-walk statistics taking into account the number of branches, n, per chain. Combining equations 34 and 35 gives equation 36. When both the radius of gyration

$$[\eta]_B^\theta/[\eta]_L^\theta = g^{3/2} \tag{36}$$

and viscosity measurements have been made, it has been shown experimentally that equations 34 and 36 exaggerate the effect of branching on the viscosity (25). It is suggested that the hydrodynamic radius depends upon the segmental density distribution and therefore should be structure dependent (21). For a cruciform molecule with f arms of equal length, the ratio of the hydrodynamic radius to the linear radius is given by equation 37:

$$h^2 = f/[2 - f + \sqrt{2}(f - 1)]^2 \tag{37}$$

For these same structures

$$g = (3f - 2)/f^2 \tag{38}$$

According to this view, separate structure-dependent viscosity ratios are required for different branch configurations. For a cruciform molecule as just defined, the ratio becomes

$$[\eta]_B^\theta/[\eta]_L^\theta = h^3 g \tag{39}$$

At the limit of low branch densities, the viscosity ratio is proportional to $g^{0.52}$; for high branch densities, the viscosity ratio is proportional to $g^{3/2}$. Zimm and Kilb (26) calculated directly the intrinsic viscosity for star molecules and show that the viscosity ratio is proportional to $g^{1/2}$. Data determined on various types of branched systems, including highly characterized star molecules, favor h^3 or $g^{1/2}$ over $g^{3/2}$ (23). Insufficient data exist to resolve the difference between theory and experiment, but through measurement of better characterized samples and more accurate data, the discrepancies are beginning to be understood (27). Polydisperse branched systems complicate the theoretical picture considerably.

Careful experimentation with well-characterized cruciform molecules shows that the Huggins constant k_1 from equation 18 has a normal value for these branched molecules (22). Equation 23 fails with branched molecules, especially if the samples are polydisperse in both branching and molecular weight.

Copolymers. Given the interrelationships among viscosity, molecular weight, solvent, and temperature, it is not possible to predict the behavior of copolymers when only the mole fraction of the two species in the copolymers is known. This is true even under the best circumstances, for one or more of the following reasons.

(*a*) Copolymers may be heterogeneous in composition as well as in molecular weight; both these heterogeneities affect viscosity.

(*b*) In a three-component system (mers A and B, and solvent C) the AB thermodynamic interactions are not known from homopolymer–solvent data.

(*c*) Copolymers have a variety of structures; this includes both those copolymers prepared in a statistically random fashion as well as those specifically prepared as block or graft copolymers.

Only some relatively simple problems associated with the viscosity of copolymers are mentioned in this section. Copolymers, like branched polymers, will yield detailed knowledge when careful experimentation can be carried out on highly characterized samples. However, good data do exist and provide valuable insights.

Random Copolymers. Statistically random copolymers are those prepared under kinetic processes related directly to the reactivity of the growing chain end during polymerization at a given temperature. This is especially pertinent to copolymers prepared by free-radical polymerization, and is discussed in detail in the article on COPOLYMERIZATION. The effects of compositional heterogeneity on the physical properties of copolymers can be eliminated by making measurements on "azeotropic" copolymers (28). The intrinsic viscosity of azeotropic linear copolymers is larger than that predicted from information on either homopolymer in the same solvent at the same temperature (28). This increase in hydrodynamic volume arises from repulsive interactions between unlike segments in the copolymer chain.

Strictly alternating AB copolymers behave as a simple homopolymer composed of the double mer unit. One exception to this simple outcome is that the unperturbed dimensions for an alternating copolymer in polar and nonpolar (θ) solvents are quite dissimilar (29).

Polymers that can be considered as copolymers of stereoisomers have been shown to have hydrodynamic properties related to their sequential structure (30); however, the effect is less than for true copolymers since the mers have identical chemical structure and the repulsive interactions are less. The Mark-Houwink expressions for highly stereoregular polymers of a given monomer but differing steric sequential order have been shown to differ, especially in poor solvents (30). Detailed theories relating the differences in unperturbed dimensions and their dependence on temperature have been developed for this class of polymer (31).

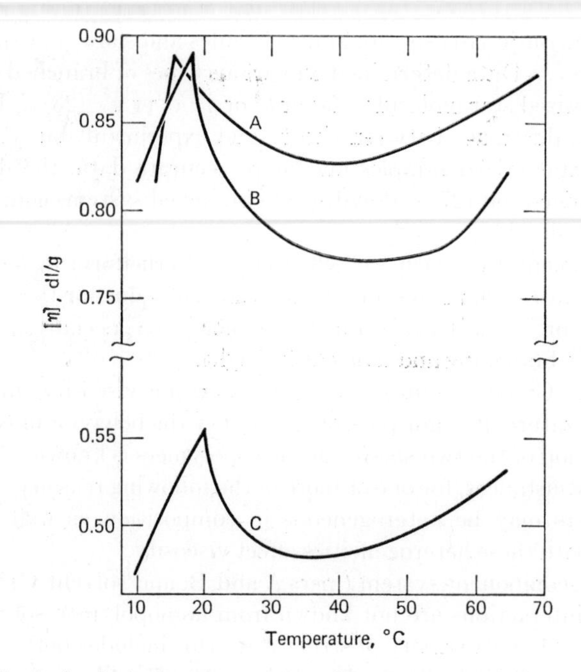

Fig. 9. Variation in the intrinsic viscosity as a function of temperature for a graft copolymer (32). A, in dioxane; B, in benzene; and C, in ethyl acetate.

Block Copolymers. When long sequences of mers exist in a copolymer, the hydrodynamic properties of the system reflect the separate behavior of the individual species rather than an averaged behavior. This immediately suggests dissolving these copolymers in a solvent in which only one of the mers is soluble. Experiments of this type have been performed on well-characterized block and graft copolymers, which have some of the most interesting hydrodynamic properties of any molecular system (32). Dondos has measured $[\eta]$ for a block copolymer system over a wide temperature range in solvents that are nonsolvents for one of the mers. His data are depicted in Figure 9. Rather sharp transitions occur at a particular temperature independent of solvent; in addition, a sharp maximum in $[\eta]$ is followed by a gradual minimum. These data clearly indicate a change in hydrodynamic volume with temperature and imply a molecular folding or coiling–uncoiling phenomenon. A change in $[\eta]$ of this nature, but less sharp, has also been shown to exist in a mixed-solvent system (four components) as the composition of the solvent mixture is changed (33,34). These same graft copolymers when dissolved in solvents thermodynamically good for both mers show the usual behavior of statistically random copolymers.

When random copolymers and block copolymers of equivalent chemical composition are subject to viscometric measurement at their respective (but different) θ temperatures, their unperturbed dimensions are nearly equivalent (35). This fact indicates that solvent–polymer contacts are effective only over very short distances and that under this "ideal" condition, the chains lose their structure identity.

Polyelectrolytes

Polymers that have pendant groups capable of ionizing have remarkable hydrodynamic properties. When pendant groups with similar charge ionize in solution, the

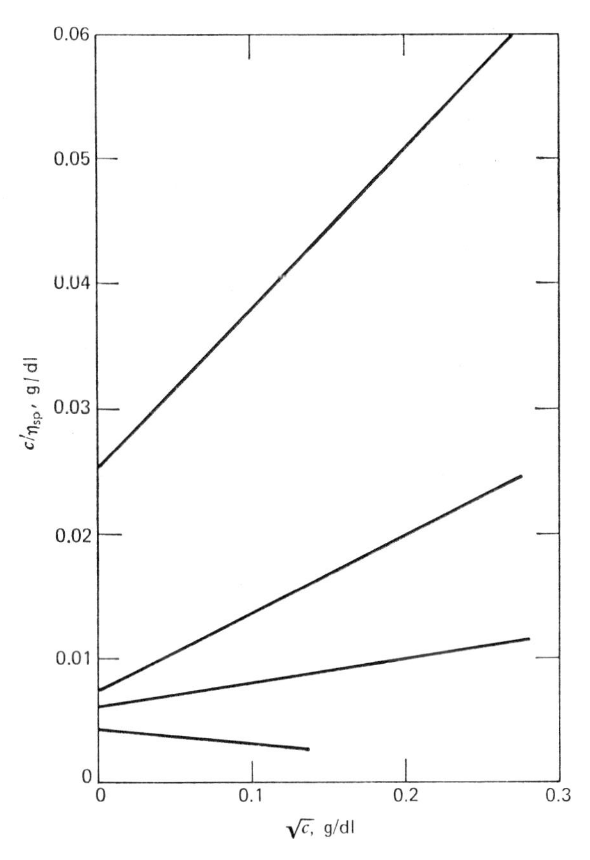

Fig. 10. Viscosity data for various samples of carboxymethylcelluloses with various degrees of substitution (39). The molecular weights of these samples increase from top to bottom. The solvent is pure water. Courtesy *Makromolekulare Chemie.*

chain undergoes an expansion in response to the electrostatic repulsion between the charges spaced along the chain. The hydrodynamic size of a polyion is determined by (*1*) the sign and extent of its charge, (*2*) the distribution of charge, (*3*) the degree of ionization, (*4*) the dielectric constant of the medium, (*5*) the molecular weight of the polyion, and (*6*) the ionic strength of added electrolyte (see POLYELECTROLYTES).

Weak polyelectrolytes may exhibit normal viscosity behavior in organic solvents where the low dielectric constant of the medium suppresses ionization. However, in aqueous solutions or in solvents of high dielectric constant, the ionic equilibrium shifts to produce additional charges on the chain and, just as in small-molecule weak electrolytes, the degree of ionization increases with increasing dilution. When η_{sp}/c is plotted against c, a minimum in the curve is found at low concentrations, and extrapolation to infinite dilution becomes impractical because of the steep upward curvature in this region as the concentration approaches zero. This phenomenon is caused by the polyion expanding rapidly under the influence of increasing ionization at high dilution. Determination of $[\eta]$ by extrapolation can be effected in two ways: (*1*) by plotting (c/η_{sp}) against $\sqrt{c}$ with extrapolation to infinite dilution to give $1/[\eta]$ (this is in accord with an empirical equation suggested by Fuoss of the form $\eta_{sp}/c = A/(1 + B\sqrt{c})$ where A and B are constants); or (*2*) by adding sufficient neutral ionic salts

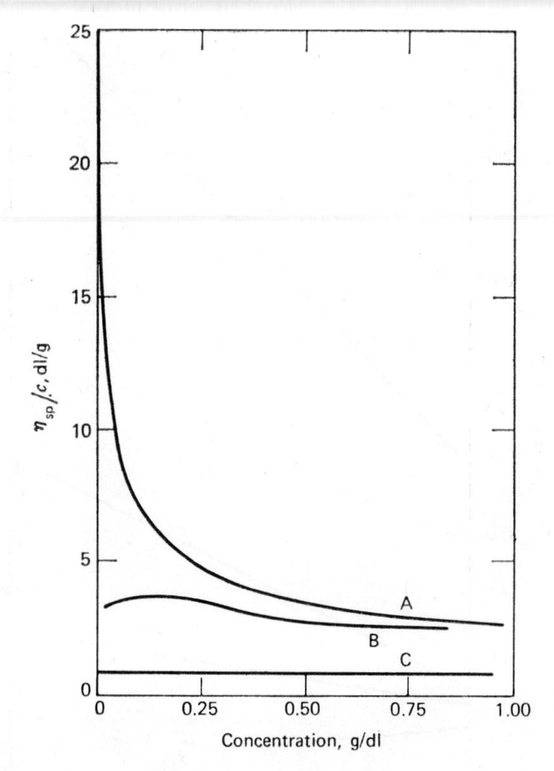

Fig. 11. The viscosity of poly(vinylbutylpyridinium bromide) (39). A, dissolved in water; B, in 0.001 *M* potassium bromide; C, in 0.0335 *M* potassium bromide. Courtesy *Makromolekulare Chemie.*

to suppress the polyion's charge effect. When neutral salts are added, the efflux time of the solvent t_0 is taken for the solvent–neutral salt mixture. Examples of these two extrapolation methods are depicted in Figures 10 and 11. Mark-Houwink constants for weak electrolytes are obtained in the same manner as described for nonionic polymers but it is the usual practice to define the solvent, the polymer, and the concentration of added neutral salt. Since polyions often have very large values for $[\eta]$, it is wise to extrapolate all measurements to zero shear.

Strong polyelectrolytes consist of those polymers whose ionizable groups are fully ionized upon dissolution. The hydrodynamic size of the chain is not drastically affected as the concentration of the polyelectrolyte is changed so that η_{sp}/c versus concentrated c curves usually follow the normal linear extrapolation procedure. However, it is customary to add a small amount of neutral salt to solutions of strong polyelectrolytes in aqueous media to suppress abnormal behavior. Figure 12 depicts some standard viscosity plots for a high-molecular-weight sample of poly(acrylic acid), neutralized to different degrees with sodium hydroxide (40). The number of ionized groups is so large that the small ionic contribution from the weak acid is not noticeable. Figure 13 shows the relationship between the intrinsic viscosity of poly(acrylic acid) and poly-(methacrylic acid) neutralized to their sodium salts from 10 to 100% and the degree of ionization. The curves for a low- and a high-molecular-weight species of poly(sodium acrylate) show a smooth increase with the degree of ionization, whereas poly-(sodium methacrylate) displays little effect from ionization up to 10% neutraliza-

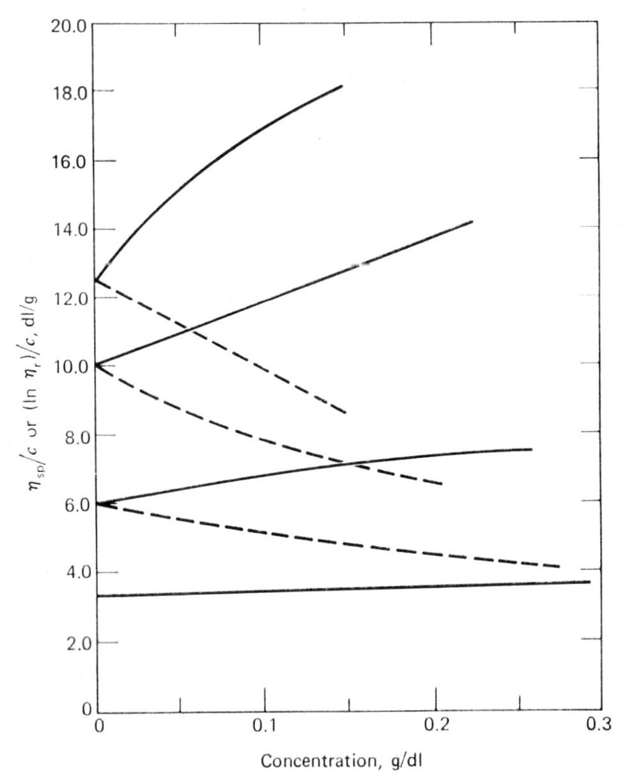

Fig. 12. Selected viscosity–concentration plots for poly(acrylic acid) (mol wt ca 500,000) with various degrees of ionization (40). Solid lines, η_{sp}/c; broken line, $\ln \eta_r/c$. Concentration of sodium bromide, 0.025 *N*; degrees of ionization, 0.103, 0.200, 0.600, and 1.0 from bottom to top (40). Courtesy American Chemistry Society.

tion, but at high degrees of ionization it displays a behavior similar to poly(sodium acrylate). These data were taken in aqueous solutions of a neutral salt. Table 3 shows the Mark-Houwink constants for partially neutralized poly(acrylic acid) as a function of charge density and the ionic strength of the medium. The values of *K* reflect a

Table 3. Dependence of Mark-Houwink Constants *K* and *a* for Poly(acrylic Acid) on Ionic Strength and Charge Density at 25°C (40)[a]

		Concentration of sodium bromide, mole/liter			
Constant	Degree of ionization	0.5	0.1	0.025	0.01
$K \times 10^4$	0.103	23.7	2.07	1.04	0.35
	0.200	9.28	1.61	0.94_2	0.51
	0.400	3.6_9	1.93	1.08_5	1.0
	0.600	3.74	2.35	1.38	1.34
	1.0	5.06	3.12	1.76	1.32
a	0.103	0.384	0.67_9	0.78_5	0.90
	0.200	0.521	0.74_3	0.83_8	0.92
	0.400	0.65_5	0.74_3	0.86_7	
	0.600	0.67_5	0.76_8	0.86_0	0.89
	1.0	0.65_6	0.75_5	0.85_0	0.91

[a] Courtesy American Chemical Society.

change in the unperturbed dimensions of the polyelectrolyte as a function of these two variables and the factor a is related to a change in the expansion of the chain for the same variables. It is to be noted that the value of a for strong polyelectrolytes even in solutions of neutral salts can be larger than 0.8. It can be shown that $[\eta]/M^{1/2}$ plotted against $M^{1/2}$ is a linear function at a given degree of ionization (40). This behavior is in agreement with the Stockmayer-Fixman treatment of polymer solvent interaction already mentioned. When the same data are treated according to equation 27, linear behavior is likewise obtained, provided that the data for $[\eta]$ are extrapolated to zero shear (40). See also ACRYLIC ACID POLYMERS.

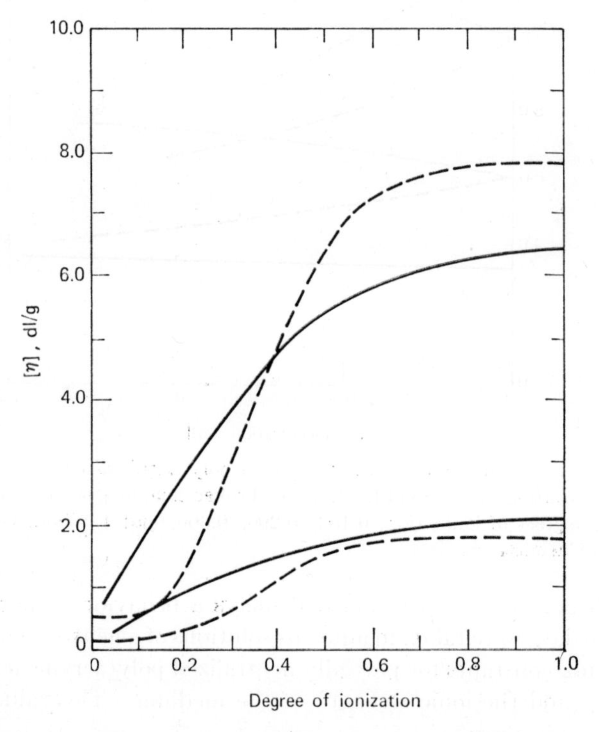

Fig. 13. Typical examples of the relationship between intrinsic viscosity and degree of ionization; concentration of sodium bromide, 0.1 *N*. Solid lines denote the data for two samples of poly(acrylic acid) differing in molecular weight; broken lines denote the data for two samples of poly(methacrylic acid) differing in molecular weight (40). Courtesy American Chemistry Society.

Many polyelectrolytes do not fit conveniently into either of the classes already discussed. For instance, polyampholytes (41) contain both positive and negative charges; some nonionic polymers are found to complex with transition-metal ions and multiple equilibria involving ions are detected (42); copolymers can be formed where the polyions are formed in blocks (43); examples of interactions between simple alkali metal ions and polyethers (44) are shown to have unusual behavior in their hydrodynamic properties. Finally, it should be mentioned that many anomalous hydrodynamic effects have been reported in homopolymers, copolymers, and polyelectrolytes in mixed solvents (45).

Bibliography

1. P. F. Onyon, *Techniques of Polymer Characterization,* Butterworth Scientific Publications, 1959, Chap. 6.
2. E. A. Moelwyn-Hughes, *Physical Chemistry,* 2nd ed., Pergamon Press, 1964, Chap. 16.
3. J. R. Van Wazer et al., *Viscosity and Flow Measurement,* Interscience Publishers, a division of John Wiley & Sons, Inc., New York, 1963, Chap. 4.
4. J. Schurz and H. Pippan, *Monatsh. Chem.* **94,** 859 (1963).
5. *ASTM Designation D445-53T, Tentative Method of Test for Kinematic Viscosity, ASTM Standards,* Part 7, American Society for Testing Materials, Philadelphia, Pa.
6. F. Ibrahim and H. G. Elias, *Makromol. Chem.* **76,** 1 (1964).
7. T. Kotaka, H. Suzuki, and H. Inagaki, *J. Chem. Phys.* **45,** 2777 (1966).
8. B. Zimm and D. M. Crothers, *Proc. Natl. Acad. Sci., U.S.* **48,** 905 (1962).
9. G. C. Berry, *J. Chem. Phys.* **46,** 1338 (1967).
10. A. Nakajima, F. Hamoda, and S. Hayashi, *J. Polymer Sci.* [C] **15,** 285 (1966).
11. C. A. F. Tuijman and J. J. Hermans, *J. Polymer Sci.* **25,** 385 (1957).
12. F. W. Billmeyer, Jr., *Textbook of Polymer Science,* Interscience Publishers, New York, 1962.
13. M. Karata, M. Iwawa, and K. Kamada, in J. Brandrup and E. H. Immergut, eds., *Polymer Handbook,* Interscience Publishers, a division of John Wiley & Sons, Inc., New York, 1966, pp. IV 1–IV 72.
14. H. Van Oene, in D. McIntyre, ed., *Characterization of Macromolecules Structure,* National Academy of Sciences, Washington, D.C., 1968, Chap. 5.
15. P. Gramain and R. Libeyere, *J. Appl. Polymer Sci.* **14,** 383 (1970).
16. E. Passaglia, J. T. Yang, and N. Megemer, *J. Polymer Sci.* **47,** 333 (1960).
17. M. Fixman, *J. Chem. Phys.* **42,** 3831 (1965).
18. M. Kurata and W. H. Stockmayer, *Fortschr. Hochpolymer. Forsch.* **3,** 196 (1963).
19. P. J. Flory, *Principles of Polymer Chemistry,* Cornell University Press, Ithaca, New York, 1953.
20. G. Delmas and D. Patterson, *Polymer* **7,** 513 (1966).
21. B. H. Zimm and W. H. Stockmayer, *J. Chem. Phys.* **17,** 1301 (1949).
22. T. A. Orofino and F. Wenger, *J. Phys. Chem.* **67,** 566 (1963).
23. T. A. Orofino, *Polymer* **2,** 305 (1961); A. R. Shultz, *J. Polymer Sci.* [A] **3,** 4211 (1965); O. Saito, K. Nagasubramanian, and W. W. Graessley, *J. Polymer Sci.* [A-2] **7,** 1937 (1969).
24. B. H. Zimm, *J. Chem. Phys.* **24,** 269 (1956).
25. M. Morton, T. E. Helminiak, S. D. Gadgary, and F. Bueche, *J. Polymer Sci* **57,** 471 (1962).
26. B. H. Zimm and R. W. Kilb, *J. Polymer Sci.* **37,** 19 (1959).
27. W. W. Graessley, in D. McIntyre, ed., *Characterization of Macromolecular Structure,* National Academy of Sciences, Washington, D.C., 1965, pp. 371–388.
28. W. H. Stockmayer, L. D. Moore, M. Fixman, and B. W. Epstein, *J. Polymer Sci.* **16,** 517 (1955).
29. J. J. Hermans and H. A. Ende, *J. Polymer Sci.* [C] **4,** 519 (1964).
30. J. B. Kinsinger and R. A. Wessling, *J. Am. Chem. Soc.* **81,** 2908 (1959); H. Inagaki, T. Miyamoto, and S. Ohta, *J. Phys. Chem.* **70,** 3420 (1966).
31. P. J. Flory, *Statistical Mechanics of Chain Molecules,* Interscience Publishers, a division of John Wiley & Sons, Inc., New York, 1969.
32. A. Dondos, *Makromol. Chem.* **99,** 275 (1966); A. Dondos, *European Polymer J.* **5,** 767 (1969).
33. A. Dondos, P. Rempp, and H. Benoit, *J. Chim. Phys.* **62,** 821 (1965).
34. H. Benoit, *Ber. Bunsengesch.* **70,** 286 (1966).
35. D. Froelich and H. Benoit, *Makromol. Chem.* **92,** 224 (1966).
36. V. Crescenzi, *Fortschr. Hochpolymer. Forsch.* **5,** 358 (1968).
37. O. F. Solomon and I. Z. Ciuta, *Bull. Inst. Polit. Gh. Gh. Dej. Buc.* **30,** 3 (1968).
38. O. F. Solomon and B. S. Gottesman, *Makromol. Chem.* **127,** 153 (1969).
39. R. M. Fuoss, *Disc. Faraday Soc.* **11,** 125 (1951); H. Vink, *Makromol. Chem.* **131,** 133 (1970).
40. I. Noda, T. Tsuge, and M. Nagasawa, *J. Phys. Chem.* **74,** 710 (1970).
41. T. Alfrey, Jr., R. M. Fuoss, H. Morawetz, and H. Pinner, *J. Am. Chem. Soc.* **74,** 438 (1952).
42. M. Mandel and J. C. Leyte, *J. Polymer Sci.* [A] **2,** 2883 (1964).
43. E. E. Magat, I. K. Miller, D. Tanner, and J. Zimmerman, *J. Polymer Sci.* [C] **4,** 615 (1963).
44. F. E. Bailey, Jr., and R. W. Callard, *J. Appl. Polymer Sci.* **1,** 56 (1959).

45. A. Dondos and D. Patterson, *J. Polymer Sci.* [A-2] **7**, 209 (1969); M. J. Blandamer, M. F. Fox, E. Powell, and J. W. Stafford, *Makromol. Chem.* **124**, 222 (1959).

J. B. Kinsinger
Michigan State University

VISCOSE. See CELLOPHANE; RAYON

VISCOSITY-AVERAGE MOLECULAR WEIGHT. See VISCOMETRY

VISCOSITY, MELT. See MELT VISCOSITY

VULCANIZATION

This article discusses reaction mechanisms of sulfur vulcanization of unsaturated elastomers, with emphasis on investigations of simple model compounds. Crosslinking by nonsulfur vulcanization and by other means is discussed in the articles on Chemical crosslinking and on Crosslinking with radiation, under CROSSLINKING. The technology of vulcanization is discussed in RUBBER COMPOUNDING AND PROCESSING; detailed discussion of the various additives used can be found under RUBBER CHEMICALS. Crosslinking of specific elastomers is covered in articles on individual polymers; see ACRYLIC ELASTOMERS; BUTADIENE POLYMERS; BUTYLENE POLYMERS; 2-CHLOROBUTADIENE POLYMERS; the discussion of chlorosulfonated polyethylene under ETHYLENE POLYMERS; ISOPRENE POLYMERS; RUBBER, NATURAL; etc.

The major portion of the elastomeric polymers now in use contain carbon–carbon double bonds; that is, they are unsaturated. Sulfur vulcanization is by far the predominant method employed in the crosslinking of these unsaturated elastomers. The commercial importance of sulfur vulcanization to elastomer technology is illustrated by the fact that ethylene–propylene–diene terpolymers, which are partially unsaturated and thereby sulfur vulcanizable, are in far greater demand than ethylene–propylene copolymers, which are saturated and therefore not sulfur vulcanizable. This is true despite the fact that the copolymers have substantially identical elastomeric properties and are both less expensive and easier to synthesize than the terpolymers. The factors that account for this predominance of sulfur vulcanization over other crosslinking methods are in part: low cost, compatibility with other additives such as reinforcing fillers and extender oils, low toxicity, excellent vulcanizate properties, and a long history of successful use in the rubber industry.

Sulfur vulcanization of an unsaturated elastomer results in monosulfide and polysulfide linkages being formed between polymer chains. If a sufficient number of crosslinks are formed, what once was a moldable material becomes fixed in shape. Although no longer moldable, the crosslinked material is flexible and elastic. If too many crosslinks are formed, however, the unsaturated elastomer is converted to a rigid solid.

A typical sulfur vulcanization system is composed of sulfur, a metal oxide (usually zinc oxide), and one or more organic promoters. Sulfur vulcanization systems will not crosslink saturated hydrocarbon polymers. There are sulfur-containing curing systems utilizing free radicals derived from peroxides, halogenated compounds, or high-energy radiation that will crosslink saturated hydrocarbons. These systems are not usually classified as sulfur vulcanization systems, since their chemistry and manufacturing technology are quite different owing to their ability to attack saturated as well as unsaturated hydrocarbons (see also CROSSLINKING).

The sulfur in sulfur vulcanization systems is the basis of the crosslinks. The sulfur vulcanization promoters, the metal oxide as well as the organic promoters, primarily affect the length of the delay period before the onset of cure, the rate of cure, and, if too long a cure time is chosen, the rate of cure reversion due to crosslink scission. The eventual state of cure or crosslink density is a function of cure time and temperature, cure site concentration, sulfur concentration, and types and concentrations of promoters.

Zinc oxide is almost always chosen as the metal oxide used in sulfur vulcanization because of its effectiveness and lack of toxicity. The role of zinc oxide is complex. Zinc sulfide is formed during sulfur vulcanization rather than hydrogen sulfide, which would form in the absence of zinc oxide. The presence of hydrogen sulfide would result in vulcanizate porosity and scission of polysulfide crosslinks. Many vulcanization promoters react with zinc oxide to form salts of even greater vulcanization activity. Zinc oxide itself is a promoter, speeding the rate of reaction of sulfur with unsaturated cure sites.

Most of the organic sulfur vulcanization promoters can be categorized as either ultraaccelerators, accelerators, or delayed-action accelerators. These three kinds of promoters typically function by interacting with sulfur to increase the rate of vulcanization. Ultraaccelerators are extensively employed in the vulcanization of elastomers containing a low concentration of unsaturated linkages, such as many of the ethylene–propylene–diene terpolymers. Many ultraaccelerators are derivatives of dithiocarbamic acid such as zinc dimethyldithiocarbamate.

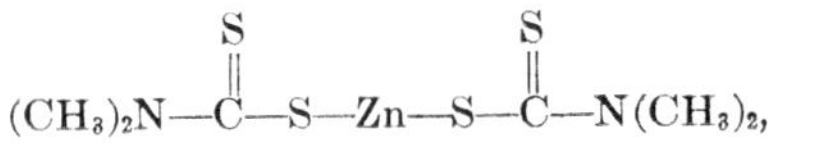

$$(CH_3)_2N-\overset{\overset{\displaystyle S}{\|}}{C}-S-Zn-S-\overset{\overset{\displaystyle S}{\|}}{C}-N(CH_3)_2,$$

zinc dimethyldithiocarbamate

Accelerators and delayed-action accelerators are widely used in elastomers having a high degree of unsaturation, such as polyisoprene and styrene–butadiene rubber. Many accelerators and delayed-action accelerators are derivatives of thiazole, such as zinc benzothiazyl sulfide and *N*-cyclohexyl-2-benzothiazylsulfenamide. There are

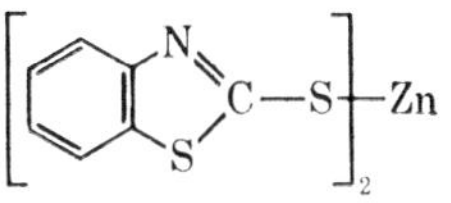

$$\left[\text{C}_6\text{H}_4\text{(N=C–S)}\!-\!S\right]_2 Zn$$

zinc benzothiazyl sulfide

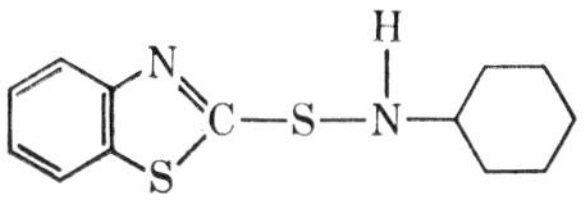

$$\text{C}_6\text{H}_4\text{(N=C–S)}-S-\overset{\overset{\displaystyle H}{|}}{N}-C_6H_{11}$$

N-cyclohexyl-2-benzothiazylsulfenamide

also sulfur-donor accelerators such as 4,4′-dithiomorpholine, which can be used without added sulfur since they donate sulfur atoms for crosslink formation during vulcanization.

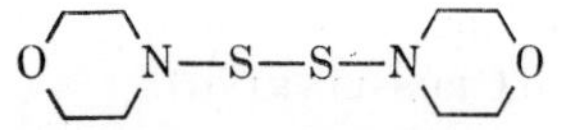

O N—S—S—N O

4, 4′-dithiomorpholine

Another class of sulfur vulcanization promoters acts not on sulfur but rather on the accelerators to increase the effectiveness of the accelerators. Fatty acids and various amines are included in this group, which forms complexes with accelerator zinc salts thereby increasing the solubility of the zinc salts in elastomers.

The nature of the chemical reactions involved in the sulfur vulcanization of unsaturated elastomers has engaged the attention of chemists for many years. Despite the great amount of effort expended and the multitude of experimental data accumulated, the reaction mechanisms involved are not yet completely defined. One reason for this is the great variety of compounds employed in the promotion and modification of sulfur vulcanization reactions. It is not the mechanism of one reaction that is under study but rather of a host of reactions involving combinations of many compounds.

The remainder of this article is devoted to a discussion of the chemical reactions involved in sulfur vulcanization. The reactions of sulfur with unsaturated cure sites in the absence of other additives is considered first. Next zinc oxide, present in almost all sulfur vulcanizations, is examined with regard to current theories concerning its role. Then the most widely utilized accelerator types are discussed class by class in the approximate order of their activity, from the most active to the least.

Sulfur, the crucial component of the vulcanization reaction, is capable of reacting as a free radical, as a cation, or as an anion. Only a mild "push" from an appropriate promoter is needed to convert sulfur to one of these active forms. Olefinic cure sites are readily attacked by sulfur free radicals and by sulfur cations.

In the reaction of sulfur with olefinic cure sites in the usual range of vulcanization temperature ($\sim$150° $\pm$ 30°C) attack is on the double bond and on hydrogens allylic to the double bond. The initial products of these reactions are polysulfides, although small amounts of monosulfide crosslinks are sometimes formed. Further heating gradually transforms the polysulfides to monosulfides. The vulcanization mechanisms discussed in this article are primarily concerned with the reactions leading to the formation of the initial crosslinks. Many of these reactions are also involved in the gradual transformation of the polysulfide crosslinks to monosulfide crosslinks.

Thermally Induced Reactions of Sulfur

Basic to the understanding of sulfur vulcanization chemistry is a determination of the reaction mechanism of the simplest system, the thermally induced reaction of sulfur with olefinic cure sites in the absence of other additives. Although there is not yet complete agreement as to whether these thermally induced reactions proceed by free-radical or by ionic processes (1–5), the bulk of the evidence presently available appears to favor the view that they proceed primarily by a free-radical chain reaction sequence accompanied by secondary polar reactions of the hydrosulfide by-products. Support for the free-radical mechanism comes from kinetic and product studies with model olefins. The more important points of evidence are:

1. Thermally induced sulfur–olefin reactions are autocatalytic, the polysulfide products of the reaction acting as catalysts of the reaction (3). Thermally induced cleavage of simple polysulfide molecules is a homolytic process leading to thiyl (R—S·) and polysulfenyl ($R—S_x·$) free radicals (6–8).

2. Reaction of *cis*- or *trans*-3-methyl-2-pentene with sulfur results in extensive cis–trans interconversion of the starting olefin (9). Thiyl radicals are noted for causing cis–trans interconversion of olefins (10–12).

3. *cis*-1,2-Cyclohexanedithiol is the major product of the reaction of sulfur with cyclohexene after reductive cleavage of the polysulfide linkages (Table 1). When zinc oxide is present, the reaction rate is accelerated and *trans*-1,2-cyclohexanedithiol is the major product. The change in reaction rate and product geometry is attributed to a change in mechanism from a thermally induced free-radical reaction to a zinc oxide-initiated ionic reaction (2).

Table 1. Effect of Zinc Oxide on the Model Reaction of Sulfur with Cyclohexene (2)

Products after $LiAlH_4$ reduction	Product yields, mole %	
	No zinc oxide	Zinc oxide present
cyclohexyl–SH	25	4
cyclohexenyl–SH	23	5
cyclohexane-1,2-(SH)₂ *cis*	45	22
cyclohexane-1,2-(SH)₂ *trans*	1	69
cyclohexyl–S–cyclohexyl	6	0

According to the free-radical reaction mechanism (2), the initial phase of the thermally induced reaction of sulfur with olefins involves the gradual formation of organic polysulfide molecules, which function as initiators of a free-radical chain reaction. This slow initial phase involves the reaction of sulfur radicals (eqs. 1 and 2) or molecules (eqs. 3 and 4) with olefin molecules to form polysulfides. Although sulfur

$$S_8 \rightleftarrows \cdot S_x \cdot \tag{1}$$

$$\cdot S_x \cdot + 2\ \text{olefin} \longrightarrow R{-}S_x{-}R \tag{2}$$

$$\text{—C=C— (olefin)} + S{-}S_x{-}S\ \text{(ring)} \longrightarrow \text{—C(S—)—C(S—)— } (S_x \text{ ring}) \tag{3}$$

$$\text{—C(H)—C=C—} + S{-}S_x{-}S \longrightarrow \text{—C=C—C(}S_xH\text{)—} \xrightarrow{\text{olefin}} R{-}S_x{-}R \tag{4}$$

free radicals are reported to exist in sulfur above 159°C (13–15), it is not known if sufficient sulfur free radicals are present at lower temperatures to allow for the gradual formation of polysulfide molecules as shown in equation 2. All three reactions (eqs. 2–4) may play a part in the formation of the first polysulfide molecules.

Once sufficient polysulfide molecules are present they serve as the primary source of polysulfenyl radicals (eq. 5) and the reaction rate accelerates. This catalytic effect

$$R{-}S_x{-}S_x{-}R \rightleftharpoons 2\,R{-}S_x\cdot \qquad (5)$$

of the polysulfide products is attributed to the greater ease of thermally induced homolytic cleavage of linear organic polysulfide molecules relative to cyclic S_8 sulfur molecules (2, 16).

The bulk of the reaction products are formed by the free-radical chain sequence shown in equations 6–11:

$$\begin{array}{c} -C \\ \| \\ -C \end{array} + R{-}S_x\cdot \rightleftharpoons \begin{array}{l} -C{-}S_x{-}R \\ \;\;| \\ -C\cdot \end{array} \qquad (6)$$

$$(1)$$

$$\underset{(1)}{\begin{array}{l} -C{-}S_x \\ \;\;|\quad\; | \\ -C\cdot\; S_x{-}R \end{array}} \longrightarrow \underset{(cis)}{\begin{array}{l} -C{-}S_x\cdot \\ \;\;| \\ -C{-}S_x{-}R \end{array}} \equiv R{-}S_x\cdot \qquad (7)$$

$$(1) + S_8 \longrightarrow \underset{(cis\ \text{or}\ trans)}{\begin{array}{l} -C{-}S_x{-}R \\ \;\;| \\ -C{-}S_8\cdot \end{array}} \equiv R{-}S_x\cdot \qquad (8)$$

$$(1) + R'{-}S_xH \longrightarrow \begin{array}{l} -C{-}S_x{-}R \\ \;\;| \\ -C{-}H \end{array} + R'{-}S_x\cdot \qquad (9)$$

$$C{=}C{-}C{-}H + R{-}S_x\cdot \rightleftharpoons C{=}C{-}C\cdot + R{-}S_x{-}H \qquad (10)$$

$$C{=}C{-}C\cdot + S_8 \longrightarrow C{=}C{-}C{-}S_8\cdot \equiv R{-}S_x\cdot \qquad (11)$$

in any of the above or subsequent equations where sulfur is shown as a reactant, an organic polysulfide molecule, $R{-}S_x{-}R$, can act as a substitute for the sulfur molecule.

Accompanying the free-radical chain sequence are polar addition reactions (eqs. 12 and 13) to the olefin of the hydrosulfide products of the hydrogen abstraction re-

$$2\,R{-}S_xH \rightleftharpoons H_2S_x + R{-}S_x{-}R \qquad (12)$$

$$\begin{array}{c} -C \\ \| \\ -C \end{array} + H{-}S_x{-}R(H) \longrightarrow \begin{array}{l} -C{-}S_x{-}R(H) \\ \;\;| \\ -C{-}H \end{array} \qquad (13)$$

actions (eq. 10). The addition of hydrogen sulfide to olefins in the presence of sulfur is considered to be a polar reaction as the additions take place in a Markovnikov manner (17,18).

The products shown in Table 1 are typical of the structures obtained from sulfur–olefin reactions after lithium aluminum hydride reduction of the polysulfide linkages. The relative yields of the various structures are a measure of the relative rates of the reactions involved in crosslink formation. The high ratio of *cis*- to *trans*-dithiol (Table 1) indicates that intermediate (**1**) reacts preferably by the intramolecular reaction path of equation 7, which leads only to cis products and only rarely by the intermolecular reaction path of equation 8, which can lead to both cis and trans products. This preference of (**1**) for the intramolecular reaction path in turn implies that the intermolecular hydrogen transfer reaction (eq. 9) involving (**1**) is also of minor importance.

Another reaction of intermediate (**1**) which occurs frequently is reversion to olefin and polysulfenyl radical (eq. 6). The importance of this reaction is evidenced by the previously noted cis–trans interconversion of *cis*- and *trans*-2-methyl-3-pentene during reaction with sulfur (9).

A possibility not previously considered which may serve to reconcile the opposing viewpoints of the proponents of the free-radical and the ionic reaction mechanisms is that the mechanisms of these thermally induced reactions may be controlled by the structure of the olefinic cure sites. The disubstituted polysulfide products of some olefins may be sufficiently unstable at vulcanization temperatures to partially fragment by an ionic process (eq. 14). If such a fragmentation does occur, the cyclic poly-

$$\begin{array}{c} | \\ -\mathrm{C}-\mathrm{S}_x- \\ | \\ -\mathrm{S}_x-\mathrm{C}- \\ | \end{array} \rightleftarrows -\mathrm{S}_x-\overset{\oplus}{\mathrm{S}}\!\!<\!\!\begin{array}{c} | \\ \mathrm{C}- \\ | \\ \mathrm{C}- \\ | \end{array} + -\mathrm{S}_x^{\ominus} \qquad (14)$$

sulfonium ions formed would be expected to initiate ionic chain reactions since these ions are crucial intermediates in the zinc oxide-promoted ionic sulfur–olefin reactions discussed in the next section. It is apparent from the low yield of *trans*-1,2-cyclohexanedithiol (Table 1) that such an ionic chain reaction does not occur with cyclohexene; however, it would be least likely to occur in cyclohexene due to the cis geometry of the dipolysulfide product. Noncyclic olefins with two substituents on at least one end of the double bond such as in polyisoprene would be the most susceptible to ionic cleavage of the disubstituted addition product due to the lability of the tertiary carbon to sulfur linkage and to the neighboring group effect of the adjacent polysulfide substituent.

Zinc Oxide-Promoted Reactions

Zinc oxide has two effects on the reactions of sulfur with olefins, it increases the rates and it changes the product distributions (Table 1). In contrast to the lack of agreement on the mechanism of thermally induced sulfur–olefin reactions, there is general agreement that zinc oxide-promoted reactions proceed by ionic reaction paths (2,19).

Initiation of the reactions is postulated to occur by interaction of sulfur with zinc oxide to form polysulfenium ions, $R{-}S_x^{\oplus}$ (eq. 15). These ions add to olefins to form

$$\text{sulfur} + \mathrm{ZnO} \longrightarrow -\mathrm{S}_x^{\oplus} + -\mathrm{S}_x-\mathrm{ZnO}^{\ominus} \qquad (15)$$

cyclic polysulfonium ions (eq. 16). Subsequent reactions of the cyclic polysulfonium

$$\begin{array}{c} | \\ -C \\ \| \\ -C \\ | \end{array} + -S_x^{\oplus} \longrightarrow \begin{array}{c} | \\ -C \diagdown^{\oplus} \\ | \;\; S-S_x- \\ -C \diagup \\ | \end{array} \quad (2) \tag{16}$$

ions account for most if not all of the reaction products (eqs. 17–22).

$$(2) + S_8 \longrightarrow \begin{array}{c} | \\ -C-S-S_x- \\ | \\ {}^{\oplus}S_8-C- \\ | \end{array} \equiv R-S_x^{\oplus} \quad (trans) \tag{17}$$

$$(2) \rightleftarrows \begin{array}{c} | \\ -C-S-S_x- \\ | \\ -C^{\oplus} \\ | \end{array} \tag{18}$$

$$\begin{array}{c} | \\ -C-S_x-S- \\ | \\ -C^{\oplus} \\ | \end{array} \longrightarrow \begin{array}{c} | \\ -C-S_x \\ | \;\;\; | \\ -C-S- \\ | \;\;\; \oplus \end{array} \rightleftarrows \begin{array}{c} | \\ -C-S_x^{\oplus} \\ | \\ -C-S- \\ | \end{array} \equiv R-S_x^{\oplus} \quad (cis) \tag{19}$$

$$\begin{array}{c} | \\ H-C- \\ | \\ -C \diagdown \\ | \;\; S-S_x- \\ -C \diagup \; \oplus \\ | \end{array} \text{ or } \begin{array}{c} | \\ H-C- \\ | \\ -C^{\oplus} \\ | \\ -C-S-S_x- \\ | \end{array} + B \longrightarrow \begin{array}{c} -C- \\ \| \\ -C \\ | \\ -C-S-S_x- \\ | \end{array} + BH^{\oplus} \tag{20}$$

(B = Lewis base = olefin, polysulfide, sulfur, zinc oxide, etc)

$$\begin{array}{c} | \\ H-C \;\; \oplus \\ | \;\; S-S_x- \\ -C \diagup \\ | \end{array} \text{ or } \begin{array}{c} | \\ H-C-S-S_x- \\ | \\ -C^{\oplus} \\ | \end{array} + B \longrightarrow \begin{array}{c} -C-S-S_x- \\ \| \\ -C- \end{array} + BH^{\oplus} \tag{21}$$

$$\begin{array}{c} | \\ -C^{\oplus} \\ | \\ -C- \\ | \end{array} + S_8 \longrightarrow \begin{array}{c} | \\ -C-S_8^{\oplus} \\ | \\ -C- \\ | \end{array} \tag{22}$$

Reaction of zinc oxide with hydrogen sulfide (eq. 23) suppresses the polar hy-

$$ZnO + H_2S \longrightarrow ZnS + H_2O \tag{23}$$

drosulfide addition reactions (eqs. 12 and 13). An additional reaction involving transfer of a hydride ion to intermediate (2) or to carbonium ion intermediates (eq. 24) has been advocated by some authors (19,20).

$$\begin{array}{c} -\mathrm{C}-\mathrm{H} \\ | \\ -\mathrm{C} \\ \| \\ -\mathrm{C}- \\ | \end{array} + -\overset{|}{\underset{|}{\mathrm{C}}}{}^{\oplus} \longrightarrow -\overset{|}{\underset{|}{\mathrm{C}}}-\mathrm{H} + \begin{array}{c} | \\ -\mathrm{C}^{\oplus} \\ | \\ -\mathrm{C} \\ \| \\ -\mathrm{C}- \end{array} \qquad (24)$$

The primary evidence for a change in mechanism from free-radical to ionic when zinc oxide is added to sulfur–olefin reactions is the change in geometry of the major product from *cis-* to *trans*-1,2-cyclohexanedithiol when cyclohexene is the olefin (Table 1). This change in product geometry is ascribed to the differences in the structures of the crucial intermediates, (**1**) and (**2**), of the free-radical and ionic mechanisms. The free-radical adduct (**1**) has an open structure which favors an intramolecular reaction leading to cis disubstituted products (eq. 7). The polysulfenium ion adduct (**2**) has a cyclic structure which favors an intermolecular ring opening displacement reaction leading to trans disubstituted products (eq. 17).

The products of the zinc oxide-promoted reactions of sulfur with olefins are polysulfide molecules. Since polysulfides are catalysts of the free-radical sulfur–olefin reaction (3), it is reasonable to assume that both ionic and free-radical reaction sequences are proceeding simultaneously in zinc oxide-promoted reactions. Lacking data concerning the relative rates of the ionic and free-radical reactions, their relative contributions are best determined by analyses of product distributions. For the reaction of cyclohexene with sulfur, an estimate of the upper limit of the contribution of the free-radical reaction can be made by comparison of the relative yields of the various products in Table 1. In the thermally induced reaction (no zinc oxide), the yield of *cis*-dithiol is about twice the yield of either cyclohexanethiol or 2-cyclohexene-1-thiol. If this 2:1 ratio is applied to the products of the zinc oxide-promoted reaction, with the arbitrary assumption that all of the cyclohexanethiol and 2-cyclohexene-1-thiol are produced via the free-radical reactions of equations 4, 9, 11, and 13 rather than by the ionic reactions of equations 20 and 22, the expected yield of *cis*-1,2-cyclohexanedithiol from the free-radical reactions in the presence of zinc oxide would be about $2 \times 5 = 10$ mole %. Thus at best in the presence of zinc oxide the thermally induced free-radical reactions can account for approximately only $4 + 5 + 10 = 19$ mole % of all products found.

Effect of Accelerators

Zinc Dialkyldithiocarbamate. Zinc dialkyldithiocarbamate-accelerated vulcanization reactions, which are among the fastest known, are considered to proceed by polar reaction sequences. Differences of opinion exist as to whether the crucial intermediates are ionic molecular fragments (21) or polarized molecules (20,22). The evidence for a nonradical mechanism for reactions accelerated by zinc dialkyldithiocarbamate is primarily based on studies of products of reactions with cyclohexene (21). When zinc dimethyldithiocarbamate is present during the reaction of sulfur with cyclohexene, the reaction rate is increased and more *trans-* than *cis*-1,2-cyclohexanedithiol results (Table 2). This is interpreted to indicate that the accelerator is functioning by an ionic mechanism, because of the similarity of the products to those from the zinc oxide-promoted ionic reaction (Table 1).

The presence of zinc dialkyldithiocarbamates in the reactions of sulfur with model olefins causes a marked increase in the rate of allylic substitution relative to double-bond addition (Tables 1 and 2). This shift in product distribution indicates that zinc

Table 2. Zinc Dimethyldithiocarbamate-Accelerated Reactions of Sulfur with Cyclohexene[a]

Products after $LiAlH_4$ reduction	Product yields, mole %: No zinc oxide	Product yields, mole %: Zinc oxide present
cyclohexyl–SH	10	2
cyclohexenyl–SH	49	65
cyclohexane-1,2-(SH)$_2$ cis	14	8
cyclohexane-1,2-(SH)$_2$ trans	27	25

[a] Ref. 21.

dialkyldithiocarbamates do more than just increase the rate of formation of polysulfenium ions. They must enter into the reactions in a more intimate manner in order to alter the product distribution so markedly.

The first step in zinc dialkyldithiocarbamate-accelerated vulcanization reactions is considered to be reaction of the accelerator with sulfur to form polysulfide homologs of the accelerator (eq. 25). According to the ionic mechanism (21), these polysulfide

$$R_2N{-}\overset{\displaystyle S}{\overset{\|}{C}}{-}S{-}Zn{-}S{-}\overset{\displaystyle S}{\overset{\|}{C}}{-}NR_2 \xrightarrow{S_8} R_2N{-}\overset{\displaystyle S}{\overset{\|}{C}}{-}S_x{-}Zn{-}S_x{-}\overset{\displaystyle S}{\overset{\|}{C}}{-}NR_2 \qquad (25)$$

homologs undergo heterolytic cleavage to form dialkylthiocarbamoyl polysulfenium ions (eq. 26). These ions add to olefins to form cyclic polysulfonium ions (eq. 27).

$$R_2N{-}\overset{\displaystyle S}{\overset{\|}{C}}{-}S_x{-}Zn{-}S_x{-}\overset{\displaystyle S}{\overset{\|}{C}}{-}NR_2 \longrightarrow R_2N{-}\overset{\displaystyle S}{\overset{\|}{C}}{-}S_x^{\oplus} + ZnS + R_2N{-}\overset{\displaystyle S}{\overset{\|}{C}}{-}S_x^{\ominus} \qquad (26)$$

$$\begin{matrix} | \\ -C \\ \| \\ -C \\ | \end{matrix} + R_2N{-}\overset{\displaystyle S}{\overset{\|}{C}}{-}S_x{-}S^{\oplus} \longrightarrow \begin{matrix} | \\ -C \\ | \\ -C \\ | \end{matrix}\!\!>\overset{\oplus}{S}{-}S_x{-}\overset{\displaystyle S}{\overset{\|}{C}}{-}NR_2 \qquad (27)$$

Subsequent reaction of the cyclic polysulfonium ions as outlined for zinc oxide-promoted reactions (eqs. 16–23) serves to account for all of the products of the zinc dialkyldithiocarbamate-accelerated reactions except the increased yields of allylic product.

Two reactions, both dependent upon the presence of the thiocarbamoyl group and either or both of which may be operative, have been proposed (21) to account for the increased yields of allylic products, which are characteristically the predominant products of zinc dialkyldithiocarbamate-accelerated reactions (21,23). The first, internal fragmentation of the cyclic adduct (eq. 28), is an intramolecular reaction analo-

$$\begin{matrix} | \\ -C{-}H \quad S{=}C{-}NR_2 \\ | \qquad\qquad | \\ -C \qquad\qquad S \\ | \;\; \overset{\oplus}{S}{-}S_x \\ -C \\ | \end{matrix} \longrightarrow \begin{matrix} -C- \\ \| \\ -C \\ | \\ -C{-}S{-}S_x^{\oplus} \end{matrix} + R_2N{-}\overset{\displaystyle S}{\overset{\|}{C}}{-}SH \qquad (28)$$

gous to equation 20 proposed for the zinc oxide-promoted sulfur–olefin reaction sequence. The second proposal (eq. 29) involves a simultaneous attack of the dialkyl-

$$\begin{array}{c}|\\ -C-H\quad S{=}C(NR_2)\\ |\qquad\qquad\quad S\\ -C\\ \| \rightarrow {}^{\oplus}S-S_x\\ -C\\ |\end{array} \longrightarrow \begin{array}{c}-C-\\ \|\\ -C\\ |\\ -C-S-S_x^{\oplus}\\ |\end{array} + R_2N-\overset{\overset{S}{\|}}{C}-SH \qquad (29)$$

thiocarbamoyl polysulfenium ion on the double bond and on the allylic hydrogen. In order for this reaction to have a reasonable probability of taking place it is necessary that the positively charged sulfur atom and the thiocarbamoyl sulfur of the dithiocarbamoyl polysulfenium ion normally be in close proximity to each other. Such a geometric arrangement of the two sulfur atoms is considered likely for dialkylthiocarbamoyl trisulfenium ions, owing to resonance stabilization of these ions as shown in equation 30. As a result of their resonance stabilization, dialkylthiocarbamoyl tri-

$$R_2N-C\begin{array}{l}{=}S\quad S^{\oplus}\\ \qquad\quad |\\ -S-S\end{array} \longleftrightarrow R_2N-C\begin{array}{l}-S-S\\ \qquad\quad |\\ {=}S\quad S^{\oplus}\end{array} \longleftrightarrow R_2N-C^{\oplus}\begin{array}{l}-S-S\\ \qquad\quad |\\ -S-S\end{array} \qquad (30)$$

sulfenium ions are expected to be present in the reaction mixtures in greater concentration than are other polysulfenium ions.

When zinc oxide is present in zinc dialkyldithiocarbamate-accelerated reactions, as it usually is during sulfur vulcanization of elastomers, the dialkyldithiocarbamic acid produced reacts to regenerate the original accelerator (eq. 31). In the absence of zinc oxide the unstable acid decomposes to other products.

$$2\,R_2N-\overset{\overset{S}{\|}}{C}-SH + ZnO \longrightarrow (R_2N-\overset{\overset{S}{\|}}{C}-S)_2Zn + H_2O \qquad (31)$$

Thermally induced free-radical reactions and zinc oxide-promoted ionic reactions must proceed simultaneously in the sulfur–olefin reaction mixtures, along with the reactions accelerated by zinc dialkyldithiocarbamate. The products of the first two classes are overshadowed, however, by those from the more rapid zinc dialkyldithiocarbamate-accelerated reaction. In one instance, in a study of the kinetics of the vulcanization of an ethylene–propylene–1,4-hexadiene terpolymer, it was possible to detect the simultaneous occurrence of the zinc oxide-initiated and the zinc dimethyldithiocarbamate-accelerated vulcanization reactions (24).

An alternative mechanism proposed for zinc dialkyldithiocarbamate and other accelerated sulfur–olefin reactions (20,22) involves the interaction of a polarized polysulfide homolog of the accelerator molecule with the allylic carbon and hydrogen of the olefin. This reaction is considered to be a front-side assisted S_N2 displacement of a hydride ion (eq. 32). A shortcoming of this mechanism as it is represented in

$$\begin{array}{c}Zn-S\\ R-S_x-S\quad |\quad S-S_x-R\\ -C-H\\ |\\ C-\\ \|\\ -C-\end{array} \longrightarrow R-S_x-S-\begin{array}{c}|\\ C-\\ |\\ C-\\ \|\\ -C-\end{array} + ZnS + R-S_x-SH \qquad (32)$$

equation 32 is the necessity for the zinc polysulfide molecule and the olefin molecule to assume an unusual geometric configuration with respect to each other, in which one molecule quite closely half surrounds a particular portion of the other molecule. In the absence of strong dipolar, ionic, or other driving forces to cause these molecules to assume this highly structured conformation, it seems unlikely that such a reaction could occur with sufficient frequency to function as the predominant mechanism of accelerated vulcanizations.

Tetraalkylthiuram Monosulfide and Disulfide. Tetraalkylthiuram monosulfides appear to function as accelerators of sulfur olefin reactions through conversion to the corresponding tetraalkylthiuram disulfides or polysulfides (21). Evidence of this is the reaction of tetramethylthiuram monosulfide with sulfur to form the tetramethylthiuram disulfide (25) and the identity of the product distributions from tetramethylthiuram monosulfide- and disulfide-accelerated reactions of sulfur with cyclohexene (Table 3).

Table 3. Tetramethylthiuram Monosulfide- and Disulfide-Accelerated Reactions of Sulfur with Cyclohexene[a]

Products after $LiAlH_4$ reduction	Product yields, mole %		
	Monosulfide with ZnO	Disulfide with ZnO	Disulfide without ZnO
cyclohexanethiol (C₆H₁₁–SH)	2	3	6
cyclohexenethiol (–SH)	83	81	64
1,2-cyclohexanedithiol, *cis*	10	10	27
1,2-cyclohexanedithiol, *trans*	5	6	3

[a] Ref. 21.

Tetralkylthiuram disulfides are postulated to promote free-radical reactions of sulfur with olefins (21). Support for a free-radical mechanism comes from both kinetic and product studies. The following points of evidence are cited:

1. Tetramethylthiuram monosulfide- and disulfide-accelerated reactions of sulfur with cyclohexene result in the formation of more *cis*- than *trans*-1,2-cyclohexanedithiol (Table 3), analogous to the thermally induced free-radical reaction (Table 1).
2. Tetraalkylthiuram disulfides are effective initiators of free-radical polymerization reactions (26–29).
3. As the concentration of tetramethylthiuram monosulfide or disulfide is increased, the rate of vulcanization of an ethylene–propylene–1,4-hexadiene terpolymer containing sulfur and zinc oxide increases to a maximum and then decreases (30). Similar behavior is found for the rate of the free-radical polymerization of methyl methacrylate when the concentration of the tetramethylthiuram disulfide initiator is increased (27,29).

Initiation of the tetraalkylthiuram disulfide-accelerated reactions is postulated to occur by homolytic scission of the accelerator (eq. 33) and interaction of the fragments

$$R_2N{-}\overset{\displaystyle S}{\overset{\|}{C}}{-}S{-}S{-}\overset{\displaystyle S}{\overset{\|}{C}}{-}NR_2 \rightleftarrows 2\,R_2N{-}\overset{\displaystyle S}{\overset{\|}{C}}{-}S\cdot \qquad (33)$$

with sulfur to form mixtures of dialkylthiocarbamoyl polysulfenyl radicals (eq. 34) and tetraalkylthiuram polysulfides (eq. 35). The results of mixed melting point

$$R_2N-\overset{\overset{\displaystyle S}{\|}}{C}-S\cdot \overset{S_8}{\rightleftarrows} R_2N-\overset{\overset{\displaystyle S}{\|}}{C}-S_x\cdot \tag{34}$$

$$2\,R_2N\overset{\overset{\displaystyle S}{\|}}{C}-S_x\cdot \rightleftarrows R_2N-\overset{\overset{\displaystyle S}{\|}}{C}-S_x-S_x-\overset{\overset{\displaystyle S}{\|}}{C}-NR_2 \tag{35}$$

experiments with sulfur and tetramethylthiuram disulfide have been interpreted to indicate that the predominant polysulfenyl free radicals formed when sufficient sulfur is present are the dialkylthiocarbamoyl trisulfenyl free radicals (21). Their preferential formation is attributed to stabilization by resonance forms (eq. 36) analogous to those proposed for dialkylthiocarbamoyl trisulfenium ions (eq. 30).

$$R_2N-C\langle{}^{=S\quad S\cdot}_{S-S}\rangle \longleftrightarrow R_2N-C\langle{}^{S-S}_{=S\quad S\cdot}\rangle \longleftrightarrow R_2N-\dot{C}\langle{}^{S-S}_{S-S}\rangle \tag{36}$$

The dialkylthiocarbamoyl polysulfenyl radicals attack olefins by addition to the double bond (eq. 37) and by abstraction of allylic hydrogens (eq. 38). The resulting

$$\begin{matrix}-C \\ \| \\ -C\end{matrix} + R_2N-\overset{\overset{\displaystyle S}{\|}}{C}-S_x\cdot \rightleftarrows \begin{matrix}-C\cdot \\ | \\ -C-S_x-\overset{\overset{\displaystyle S}{\|}}{C}-NR_2\end{matrix} \tag{37}$$

$$\begin{matrix}-C-H \\ | \\ -C \\ \| \\ -C-\end{matrix} + R_2N-\overset{\overset{\displaystyle S}{\|}}{C}-S_x\cdot \rightleftarrows \begin{matrix}-C\cdot \\ | \\ -C \\ \| \\ -C-\end{matrix} + R_2N-\overset{\overset{\displaystyle S}{\|}}{C}-S_xH \tag{38}$$

carbon free-radical intermediates react as outlined for thermally induced sulfur–olefin reactions (eqs. 7–9,11). These reactions serve to account for all of the products of the tetraalkylthiuram monosulfide- and disulfide-accelerated reactions except for the increased yields of allylic thiols.

Two additional reactions dependent upon the presence of the thiocarbamoyl group have been proposed (21) to account for the increased yields of the allylic compounds that are characteristically the predominant products of thiuram-accelerated reactions. The proposed reactions (eqs. 39 and 40) are free-radical analogs of similar

$$\begin{matrix}-C-H\quad S\\ |\qquad\qquad \searrow C-NR_2 \\ -C\cdot\qquad\quad / \\ |\qquad\qquad S \\ -C-S_x-\end{matrix} \longrightarrow \begin{matrix}-C- \\ \| \\ -C \\ | \\ -C-S_x\cdot\end{matrix} + R_2N-\overset{\overset{\displaystyle S}{\|}}{C}-SH \tag{39}$$

$$\begin{matrix}\qquad\qquad NR_2 \\ -C-H\quad S=C \\ |\qquad\qquad\quad \backslash \\ -C\qquad\qquad S \\ \| \longrightarrow \cdot S-S_x \\ -C\end{matrix} \longrightarrow \begin{matrix}-C- \\ \| \\ -C \\ | \\ -C-S-S_x\cdot\end{matrix} + R_2N-\overset{\overset{\displaystyle S}{\|}}{C}-SH \tag{40}$$

ionic reactions proposed to account for the increased yields of allylic products from zinc dialkyldithiocarbamate-accelerated reactions (eqs. 28 and 29).

As the predominant reaction path of the thiuram-accelerated reactions leads to allylic products (eqs. 38–40), the thiuram accelerators are rapidly converted to dialkyldithiocarbamic acids or their polysulfide homologs. In the presence of zinc oxide these acids are converted to zinc salts (eq. 31). Thus vulcanization reactions that began by being accelerated by tetraalkylthiuram sulfides end up being accelerated by zinc dialkyldithiocarbamates (30,31). This means that in an elastomer being vulcanized with a mixture of sulfur, zinc oxide, and tetraalkylthiuram monosulfide or disulfide there are four crosslinking sequences proceeding: the thermally induced free-radical reaction; the zinc oxide-initiated ionic reaction; the tetraalkylthiuram disulfide-initiated free-radical reaction; and the zinc dialkyldithiocarbamate-initiated ionic reaction.

Tetramethylthiuram monosulfide- and disulfide-accelerated vulcanizations have distinct induction periods (32). The induction period for the monosulfide is longer than that for the disulfide. This can be attributed to the time required to convert the monosulfide to disulfide or polysulfide. Although the delay period of the disulfide can be attributed to the time required for interaction of the accelerator with sulfur and olefin, another significant factor may be the abilities of tetralkylthiuram disulfides and polysulfides and their free-radical fragments to take part in free-radical chain transfer. Although chain-transfer reactions would not retard attack on olefinic cure sites, they can retard crosslinking through combination of free radicals (eq. 5) by tying up active cure site fragments (eq. 41) and they can cause destruction of existing crosslinks (eq. 42).

$$\mathrm{R'{-}S_x\cdot} + \mathrm{R_2N{-}\overset{\displaystyle S}{\overset{\|}{C}}{-}S_x{-}S_x{-}\overset{\displaystyle S}{\overset{\|}{C}}{-}NR_2} \rightleftarrows \mathrm{R'{-}S_x{-}S_x{-}\overset{\displaystyle S}{\overset{\|}{C}}{-}NR_2} + \mathrm{R_2N{-}\overset{\displaystyle S}{\overset{\|}{C}}{-}S_x\cdot} \qquad (41)$$

$$\mathrm{R'{-}S_x{-}S_x{-}R'} + \mathrm{R_2N{-}\overset{\displaystyle S}{\overset{\|}{C}}{-}S_x\cdot} \rightleftarrows \mathrm{R'{-}S_x{-}S_x{-}\overset{\displaystyle S}{\overset{\|}{C}}{-}NR_2} + \mathrm{R'{-}S_x\cdot} \qquad (42)$$

These general factors, ie, the need to convert the accelerators to their polysulfide homologs, the free-radical chain-transfer ability of the disulfides and polysulfides, and the existence of polysulfenyl free radicals terminated by accelerator fragments, seem to be common to many delayed-action accelerators that eventually are converted to zinc salts during vulcanization. The formation of the zinc salts, which are active accelerators, may end the induction periods through the initiation of ionic crosslinking reactions and the termination of the bulk of the free-radical chain-transfer reactions (eq. 41) and crosslinking cleavage (eq. 42).

Mercaptobenzothiazole Derivatives. The derivatives discussed are 2-benzothiazolyl disulfide, zinc benzothiazolyl mercaptide, mercaptobenzothiazole, and 2-benzothiazolylsulfenamides.

2-Benzothiazolyl Disulfide. Vulcanization reactions accelerated by 2-benzothiazolyl disulfide, frequently referred to as MBTS, are generally considered to be free radical in nature (30,33–36). However, the details of the postulated free-radical mechanisms differ. The most likely reaction sequence is one similar to that discussed above for tetraalkylthiuram disulfides and involving homolytic cleavage of the accelerator (eq. 43) followed by interaction of the fragments with sulfur to form poly-

$$\mathrm{BtS{-}SBt} \rightleftarrows 2\ \mathrm{BtS\cdot} \qquad (43)$$

(Bt = 2-benzothiazolyl; structures drawn as benzothiazole rings with N and S in the five-membered ring.)

sulfenyl radicals (eq. 44) and polysulfide homologs of the accelerator (eq. 45). The

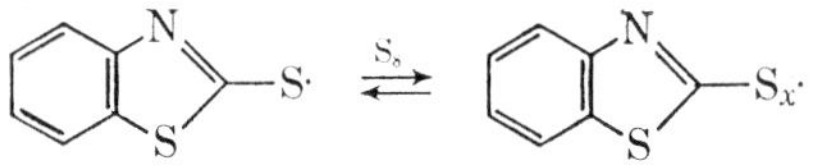

$$\text{C}_7\text{H}_4\text{NS-S}\cdot \underset{}{\overset{S_8}{\rightleftarrows}} \text{C}_7\text{H}_4\text{NS-S}_x\cdot \qquad (44)$$

$$2\ \text{C}_7\text{H}_4\text{NS-S}_x\cdot \rightleftarrows \text{C}_7\text{H}_4\text{NS-S}_x\text{-S}_x\text{-SNC}_7\text{H}_4 \qquad (45)$$

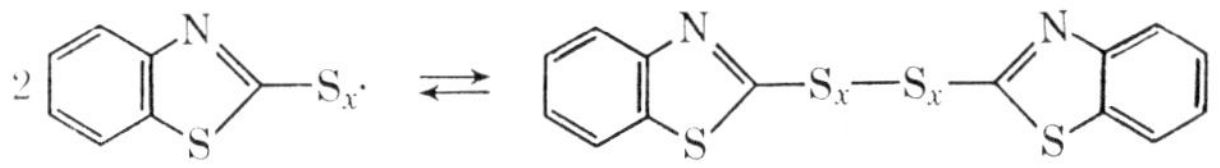

resulting free radicals initiate a free-radical crosslinking reaction sequence similar to those outlined for the thermally induced (eqs. 6–11) and the tetraalkylthiuram disulfide-accelerated (eqs. 6–11,37–40) reactions.

As a consequence of these initial free-radical reactions, the 2-benzothiazolyl disulfide accelerator is converted to mercaptobenzothiazole or one of its polysulfide homologs (30,33–35). The zinc oxide normally present in vulcanization reactions in turn converts these to their zinc salts. As a result sulfur–olefin reactions initially accelerated by 2-benzothiazolyl disulfide end being accelerated by zinc benzothiazolyl mercaptide.

The induction periods characteristic of 2-benzothiazolyl disulfide-accelerated vulcanizations can be attributed to the same factors described for tetraalkylthiuram disulfide-accelerated reactions.

Zinc Benzothiazolyl Mercaptide and Mercaptobenzothiazole. Although considerably less active as accelerators than zinc dialkyldithiocarbamates, mercaptobenzothiazole and its zinc salt are frequently used as primary accelerators in highly unsaturated elastomers and as secondary accelerators in elastomers with a low degree of unsaturation. Mercaptobenzothiazole is commonly employed as an accelerator in the presence of zinc oxide or the zinc salt of a fatty acid. Under these conditions it is converted to zinc benzothiazolyl mercaptide (37,38) which is the actual accelerator of the vulcanization reactions (30).

As a primary accelerator, zinc benzothiazolyl mercaptide is quite responsive to promotion by fatty acids and by amines, both of which are thought to form complexes centered around the zinc atom of the accelerator (22,30,39,40). The promoting effects of these compounds are due at least in part to their increasing the solubility of the zinc benzothiazolyl mercaptide. They may also alter the reaction mechanisms involved in the vulcanization process.

There is at this time no general agreement as to the reaction mechanisms by which zinc benzothiazolyl mercaptide functions as an accelerator. Both ionic (21) and polarized molecule (20) mechanisms have been suggested similar to those proposed for zinc dialkyldithiocarbamates (eqs. 16–32). A free-radical reaction mechanism has also been proposed (38,41). At present it is not possible to decide which if any of the suggested mechanisms are correct. In view of the considerable similarity in the structures and reactions of the two groups of accelerators, the derivatives of mercaptobenzothiazole and the derivatives of dialkyldithiocarbamic acids, it is reasonable to suggest that the two accelerator groups function by similar reaction mechanisms; that is the zinc salts of both groups function by an ionic or polar reaction mechanism and the disulfides of both groups by an initial free-radical mechanism prior to their conversion to zinc salts.

2-Benzothiazolylsulfenamides. Sulfenamides are widely used delayed action accelerators for highly unsaturated elastomers. Although there is general agreement that their cleavage products rather than the sulfenamides themselves are the actual accelerators, the mechanism of their cleavage reactions is in dispute. Both free-radical (36,42) and hydrogen sulfide-promoted (20) cleavage reactions have been proposed. Ionic cleavage reactions have not been ruled out. The eventual end products of the cleavage reactions are salts of mercaptobenzothiazole, which are considered to be the primary accelerators of vulcanization (30).

Hydrogen sulfide-promoted and ionic cleavage reactions would lead directly to salts of mercaptobenzothiazole. In the presence of zinc oxide, zinc benzothiazolyl mercaptide would be formed and would function as the dominant accelerator with assistance from the amine cleavage fractions.

Free-radical cleavage reactions might proceed as shown in equations 46–51 Although initially a free-radical reaction sequence similar to those previously dis-

$$\text{BT}\text{—S—NR}_2 \rightleftharpoons \text{BT}\text{—S}\cdot + \text{R}_2\text{N}\cdot \quad (46)$$

$$\text{BT}\text{—S}\cdot \underset{}{\overset{S_8}{\rightleftharpoons}} \text{BT}\text{—S}_x\cdot \quad (47)$$

$$2\ \text{BT}\text{—S}_x\cdot \rightleftharpoons \text{BT}\text{—S}_x\text{—S}_x\text{—BT} \quad (48)$$

$$\text{R}_2\text{N}\cdot \overset{S_8}{\rightleftharpoons} \text{R}_2\text{N—S}_x\cdot \quad (49)$$

$$2\ \text{R}_2\text{N—S}_x\cdot \rightleftharpoons \text{R}_2\text{N—S}_x\text{—S}_x\text{—NR}_2 \quad (50)$$

$$\text{R}_2\text{N—S}_x\cdot + \text{BT}\text{—S}_x\cdot \rightleftharpoons \text{BT}\text{—S}_x\text{—S}_x\text{—NR}_2 \quad (51)$$

(BT = benzothiazol-2-yl, drawn as structure in the original)

cussed would prevail, the benzothiazolyl fragments would eventually be converted to zinc benzothiazolyl mercaptide, assuming zinc oxide present. As a result amine-assisted zinc benzothiazolyl mercaptide-initiated reactions would eventually dominate the vulcanization process.

Nitrogen Bases. Amines react with sulfur at vulcanization temperatures to form thioamides and hydrogen sulfides (eq. 52) (43). Both the amines (44–46) and

$$(C_2H_5)_2NH \xrightarrow[140°C]{S_8} CH_3\overset{\overset{S}{\|}}{C}NHC_2H_5 + H_2S_x \quad (52)$$

their hydrosulfides (45,47,48) are reported to cleave cyclic S_8 molecules to linear polysulfides. This results in the acceleration of the free-radical sulfur vulcanization process since linear polysulfides are more susceptible to thermally induced homolytic cleavage than cyclic S_8 molecules (2,16). The hydrosulfide products of amine–sulfur reactions can also promote crosslink formation by addition to double bonds to form thiols and monosulfides (43). The thioamide products are postulated to form zinc salts in the presence of zinc oxide, which act as accelerators in a manner similar to zinc benzothiazolyl mercaptide (20).

Aldehyde–amine condensation products are postulated to react with sulfur and act as accelerators in much the same manner as amines (49,50).

Guanidines such as diphenylguanidine react under vulcanization conditions to form nitrogen bases and thioureas all of which, including the original guanidine, can function as accelerators (36,46,51). Thioureas are considered to act as accelerators by forming zinc salts, if zinc oxide is present, which can react in a manner similar to zinc benzothiazolyl mercaptide (20,52).

Thiobisamines. The initial reactions of thiobisamines during vulcanization are believed to be cleavages to active fragments. The mechanisms of the cleavage reactions are uncertain. Free-radical (eqs. 53,54) (36), ionic (eq. 55) (53), and hydro-

$$O(C_4H_8)N{-}S{-}S{-}N(C_4H_8)O \rightleftharpoons 2\,O(C_4H_8)N{-}S\cdot \qquad (53)$$

$$O(C_4H_8)N{-}S{-}S{-}N(C_4H_8)O \rightleftharpoons O(C_4H_8)N{-}S{-}S\cdot + O(C_4H_8)N\cdot \qquad (54)$$

$$(C_6H_5{-}CH_2)_2N{-}S{-}S{-}N(CH_2{-}C_6H_5)_2 \longrightarrow$$

$$(C_6H_5{-}CH_2)_2N{-}S{-}S^{\oplus} + (C_6H_5{-}CH_2)_2N^{\ominus} \qquad (55)$$

sulfide-promoted (53) cleavage reactions have been postulated. All three processes may be operative, even in the same thiobisamine, with the predominating mode of cleavage dependent upon the structure of the thiobisamine and upon the other vulcanization additives present. Free-radical cleavage would initiate a free-radical reaction sequence, ionic cleavage an ionic reaction sequence, and hydrosulfide-promoted cleavage an amine-accelerated reaction. The end products of thiobisamine cleavage reactions, amines, thioamides and amine hydrosulfides, are all promoters of sulfur crosslinking reactions as previously noted.

A major difficulty in the elucidation of the reaction mechanisms of weak accelerators such as amines and thiobisamines is the problem of separating the effects of the accelerator from those of its decomposition products and from those of the thermally induced and zinc oxide-promoted reactions. This problem, which has received little attention in the past, should be given greater consideration in future studies with these and other weak accelerators.

Bibliography

1. E. H. Farmer and F. W. Shipley, *J. Polymer Sci.* **1**, 293 (1946); *J. Chem. Soc.* **1947**, 1519.
2. J. R. Wolfe, Jr., T. L. Pugh, and A. S. Killian, *Rubber Chem. Technol.* **41** (5), 1329 (1968).
3. G. W. Ross, *J. Chem. Soc.* **1958**, 2856.
4. L. Bateman, C. G. Moore, and M. Porter, *J. Chem. Soc.*, 2866.
5. W. A. Pryor, *Mechanisms of Sulfur Reactions*, McGraw-Hill, New York, 1962, p. 96.
6. I. Kende, T. L. Pickering, and A. V. Tobolsky, *J. Am. Chem. Soc.* **87**, 5582 (1965).
7. C. D. Trivette, Jr. and A. Y. Coran, *J. Org. Chem.* **31**, 100 (1966).
8. T. L. Pickering, K. J. Saunders, and A. V. Tobolsky, *J. Am. Chem. Soc.* **89**, 2364 (1967).
9. C. G. Moore and M. Porter, forthcoming publication referred to in Ref. 20, p. 477.
10. R. H. Pallen and C. Sivertz, *Can. J. Chem.* **35**, 723 (1957).
11. C. Walling and W. Helmreich, *J. Am. Chem. Soc.* **81**, 1144 (1959).
12. N. P. Neureiter and F. G. Bordwell, *J. Am. Chem. Soc.*, **82**, 5354 (1960).
13. F. Fairbrother, G. Gee, and G. T. Merrall, *J. Polymer Sci.* **16**, 459 (1955).

14. D. M. Gardner and G. K. Fraenkel, *J. Am. Chem. Soc.* **78,** 3279 (1956).
15. A. V. Tobolsky and A. Eisenberg, *J. Am. Chem. Soc.* **81,** 780 (1959).
16. Ref. 5, p. 9.
17. S. O. Jones and E. E. Reid, *J. Am. Chem. Soc.* **60,** 2452 (1938).
18. R. F. Naylor, *J. Polymer Sci.* **1,** 305 (1946); *J. Chem. Soc.* **1947,** 1532.
19. L. Bateman, R. W. Glazebrook, and C. G. Moore, *J. Appl. Polymer Sci.* **1,** 257 (1959).
20. L. Bateman, C. G. Moore, M. Porter, and B. Saville, "Chemistry of Vulcanization" in L. Bateman, ed., *The Chemistry and Physics of Rubber-Like Substances,* John Wiley & Sons, Inc., New York, 1963.
21. J. R. Wolfe, Jr., *Rubber Chem. Technol.* **41,** 1339 (1968).
22. M. Porter, "The Chemistry of the Sulfur Vulcanization of Natural Rubber" in A. V. Tobolsky, ed., *The Chemistry of Sulfides,* Interscience Publishers, a division of John Wiley & Sons, New York, 1968.
23. E. H. Farmer, J. F. Ford, and J. A. Lyons, *J. Appl. Chem.* **4,** 554 (1954).
24. J. R. Wolfe, Jr., *J. Appl. Polymer Sci.* **12,** 1167 (1968).
25. D. Craig, W. L. Davidson, A. E. Juve, and I. G. Geib, *J. Polymer Sci.* **6,** 1 (1951).
26. R. J. Kern, *J. Am. Chem. Soc.* **77,** 1382 (1955).
27. T. E. Ferington and A. V. Tobolsky, *J. Am. Chem. Soc.,* **77,** 4510 (1955).
28. T. Otsu, *J. Polymer Sci.* **21,** 559 (1956).
29. T. Ferington and A. V. Tobolsky, *J. Am. Chem. Soc.* **80,** 3215 (1958).
30. W. Scheele, *Rubber Chem. Technol.* **34,** 1306 (1961).
31. H. K. Frensdorff, *Rubber Chem. Technol.* **41,** 316 (1968).
32. J R. Wolfe, Jr., *J. Appl. Polymer Sci.* **12,** 1183 (1968).
33. C. G. Moore, *J. Chem. Soc.* **1952,** 4232.
34. J. Tsurugi and H. Fukuda, *Bull. Univ. Osako Pref.* [A] **6,** 145 (1958); *Rubber Chem. Technol.* **31,** 788 (1958).
35. J. Tsurugi and H. Fukuda, *J. Chem. Soc. Japan, Ind. Chem. Sect.* **60,** 362 (1957); *Rubber Chem. Technol.* **31,** 800 (1958).
36. B. A. Dogadkin and V. A. Shershnev, *Rubber Chem. Technol.* **35,** 1 (1962).
37. I. Auerbach, *Ind. Eng. Chem.* **45,** 1526 (1953).
38. J. Tsurugi and H. Fukuda, *J. Chem. Soc. Japan, Ind. Chem. Sect.* **61,** 1377 (1958); *Rubber Chem. Technol.* **33,** 217 (1960).
39. R. H. Campbell and R. W. Wise, *Rubber Chem. Technol.* **37,** 650 (1964).
40. A. Y. Coran, *Rubber Chem. Technol.,* **37,** 679 (1964).
41. J. Tsurugi and H. Fukuda, *J. Chem. Soc. Japan, Ind. Chem. Sect.* **61,** 140 (1958); *Rubber Chem. Technol.* **33,** 211 (1960).
42. H. Fukuda and J. Tsurugi, *J. Chem. Soc. Japan, Ind. Chem. Sect.* **64,** 479, 483 (1961); *Rubber Chem. Technol.* **35,** 484, 492 (1962).
43. C. G. Moore and R. W. Saville, *J. Chem. Soc.* **1954,** 2082, 2089.
44. J. Tsurugi, *Bull. Univ. Osaka Pref.* [A] **5,** 173 (1957); *Rubber Chem. Technol.* **31,** 773 (1958).
45. H. Krebs, *Gummi Asbest.* **8,** 68 (1955); *Rubber Chem. Technol.* **30,** 962 (1957).
46. W. Scheele and M. Cherubim, *Kautschuk Gummi* **10,** WT 185 (1957); *Rubber Chem. Technol.* **31,** 286 (1958).
47. P. D. Bartlett, E. F. Cox, and R. E. Davis, *J. Am. Chem. Soc.* **83,** 103 (1961).
48. P. D. Bartlett, A. K. Colter, R. E. Davis, and W. R. Roderick, *J. Am. Chem. Soc.* **83,** 109 (1961).
49. C. W. Bedford and W. Scott, *Ind. Eng. Chem.* **12,** 31 (1920).
50. D. Craig, L. Schaefgen, and W. P. Tyler, *J. Am. Chem. Soc.* **70,** 1624 (1948).
51. W. Scott and C. W. Bedford, *Ind. Eng. Chem.* **13,** 125 (1921).
52. G. D. Kratz, A. H. Flower, and B. J. Shapiro, *Ind. Eng. Chem.* **13,** 128 (1921).
53. R. W. Saville, *J. Chem. Soc.* **1958,** 2880.

General References

W. Hofmann, *Vulcanization and Vulcanizing Agents,* Palmerton Publishing Co., New York, 1967.

G. Alliger and I. J. Sjothun, ed., *Vulcanization of Elastomers,* Reinhold Publishing Corp., New York, 1964.

D. Craig, "Crosslinking of Elastomers" in "Chemical Reactions of Polymers," in E. M. Fettes, ed., *High Polymers,* Vol. 19, Interscience Publishers, a division of John Wiley & Sons, New York, 1964.

H. L. Fisher, *Chemistry of Natural and Synthetic Rubbers*, Reinhold Publishing Corp., New York, 1957.
L. Bateman,, C. G. Moore, M. Porter, and B. Saville, "Chemistry of Vulcanization," in L. Bateman, ed., *The Chemistry and Physics of Rubber-Like Substances*, John Wiley & Sons, Inc., New York, 1963.

James R. Wolfe, Jr.
E. I. du Pont de Nemours & Co., Inc.

VULCANIZED FIBER

Vulcanized fiber is a dense material of partially regenerated cellulose in which the fibrous structure of the original cellulose is retained to varying degrees, depending upon the grade of fiber. It is available in four main forms: sheets, rolls, rods, and tubes. The rod form (round) is actually machined from strips cut from sheet stock.

Vulcanized fiber is made and used throughout the world. In the United States, industry standards have been set by the National Electrical Manufacturers Association (NEMA) (4). The American Society for Testing and Materials and various industry associations, as well as the government, have also set vulcanized fiber standards. There are three major standard grades, Commercial, Electrical Insulation, and Bone, which account for most of the volume. Table 1 gives the NEMA standards for these grades. Commercial grade is the general-purpose grade that is used for mechanical and electrical applications for which there are no special requirements that dictate selection of one of the special grades. Electrical Insulation grade is a thin (0.004–⅛ in.) grade made particularly to give better electrical properties than Commercial grade. It is also recommended for applications involving difficult forming or bending operations. Bone grade has the highest density obtained from more complete gelatinization of the cellulosic fibers. It possesses maximum hardness, stiffness, and wear resistance as well as excellent turning, threading, and milling characteristics.

History

Until the mid-1800s, animal parchment was used for legal documents and other certificates for which permanence and good aging properties were important. Thus interest was aroused when, in 1846, Poumarède and Figuier (1) noted that a material with many of the properties of animal parchment could be made by dipping filter paper into 66° Baumé sulfuric acid for a short time and immediately washing free of acid. Within a short time commercial products were made by this method and called "vegetable" parchment to differentiate them from the true, natural animal parchment. The first commercial development in England took place in 1853.

Until 1858 sulfuric acid was used as the sole active agent in the parchmentizing process. In 1859 Taylor found that 76% zinc chloride could produce the same general effect as sulfuric acid (2) and patented this process in England. The name "vulcanized fiber" was coined by Taylor for the product resulting from paper treated with sulfuric acid or zinc chloride, pressed together, and washed free of the active chemical. This name, although confusing and misleading since the advent of the rubber industry, has remained in use. (Fibers made from various types of elastomer are discussed under FIBERS, ELASTOMERIC.)

Taylor came to the United States and started vulcanized fiber production about 1873. The first continuous machine was built by the Marshall family in the early

1900s. Through the years many companies were founded to make vulcanized fiber in the United States. Many of these went out of business or merged until at present there are only seven manufacturers in the United States. In 1968 they produced approximately 47 million pounds of vulcanized fiber (3).

Manufacture

Essentially the manufacturing of parchment and vulcanized fiber consists of passing one or more plies of absorbent paper through a strong acid and bringing them together under heat and pressure. After lamination the strong acid is leached from the web to give a material that is dense, stiff, and durable. Parchment making is limited to a maximum thickness of 0.015 in. owing to the strong action of the sulfuric acid, which cannot be leached out quickly enough to prevent cellulose degradation in thicker laminates. However, in vulcanized fiber manufacture the action of zinc chloride is not as severe, and laminates up to 2 in. thick can be made.

Zinc Chloride. Zinc chloride is generally purchased as an aqueous solution of 71% zinc chloride with the concentration measured in terms of specific gravity or in

Table 1. Physical Properties of Vulcanized Fiber Sheets[a] (5)

Property	Typical value	Value for Commercial grade[b]		Value for Bone grade[c]		Value for Electrical Insulation grade[d]	
		Cross-wise	Length-wise	Cross-wise	Length-wise	Cross-wise	Length-wise
tensile strength, psi (min)							
thickness up to 1/8 in.		7,500	13,500	8,000	14,000	7,500	13,500
over 1/8 up to 1/2 in., incl.		7,000	11,000	7,000	11,000		
over 1/2 in.		6,000	7,000				
flexural strength, psi (min)							
thickness 1/16 up to 1/8 in., incl.		13,000	15,000	14,000	16,000	13,000	15,000
over 1/8 up to 1/2 in., incl.		12,000	14,000	13,000	15,000		
over 1/2 up to 1 in., incl.		11,000	13,000				
compressive strength, psi (min), flatwise, all thicknesses		....25,000....		....30,000....			
density, g/cm³ (min)							
thickness under 0.010 in.						....1.00....	
0.010 up to 3/32 in., incl.		.....1.15.....		.....1.30.....		....1.15....	
over 3/32 up to 5/8 in., incl.		.....1.20.....		.....1.30.....		....1.20....	
over 5/8 up to 1 in., incl.		.....1.10.....					
over 1 up to 1 1/4 in., incl.		.....1.05.....					
over 1 1/4 in.		.....1.01.....					
water absorption, change in weight, % (max)		For 2 hr	For 24 hr	For 2 hr	For 24 hr	For 2 hr	For 24 hr
thickness 1/32 in.		60	68	55	63	60	68
1/16 in.		52	66	30	55	52	66
1/8 in.		35	61	20	48	35	61
1/2 in.		13	36	10	25		
1 in.		8	21				

continued

Table 1 (*continued*)

Property	Typical value	Value for Commercial grade[b]	Value for Bone grade[c]	Value for Electrical Insulation grade[d]
dielectric strength, V/mil (min)				
thickness up to 1/16 in., incl.		175	175	
over 1/16 up to 1/8 in., incl.		150	150	
over 1/8 up to 3/8 in., incl.		100	100	
over 3/8 up to 1/2 in., incl.		50	50	
over 1/2 in.		total 25,000 V	total 25,000 V	
0.004–0.005 in.				200
0.005–0.015 in.				300
0.015–0.040 in.				250
0.040 up to 1/8 in. incl.				175
bursting strength, psi (min)				
thickness 0.005 in.				75
0.007 in.				105
0.010 in.		125		150
0.015 in.		185		225
0.020 in.		250		325
tearing strength, g/min				Cross-wise / Length-wise
thickness 0.005 in.				120 / 100
0.007 in.				220 / 190
0.010 in.				300 / 250
0.015 in.				450 / 375
Rockwell hardness (min)		R50	R80	R50
ash content, % (max)				
red	7.0			
gray	1.5			
black	1.5			
silica content, % (max)	0.3			
zinc chloride content, % (max)	0.1			
thermal conductivity, Btu/hr-ft²-(°F/ft)	0.168			
specific heat, Btu/lb-°F	0.403			

[a] Based on NEMA standards and on ASTM Specification D 710.
[b] For general and mechanical use.
[c] Highest density, maximum hardness.
[d] Fish paper.

degrees Baumé. The solution is acidic and in the vulcanized fiber industry the term "acid" is commonly used to mean zinc chloride.

Zinc chloride must meet closely controlled standards of purity, especially in terms of calcium and ammonium content. It has been found that these impurities are detrimental to the strength properties of vulcanized fiber. Metallic impurities, such as iron, preclude good electrical properties in the fiber.

Paper. The paper used in the manufacture of vulcanized fiber is made from either cotton or wood fibers, which are both major sources of cellulose. In the past, most of the paper used has been cotton-base because some fiber properties, especially flexibility and impact strength, exceeded those then attainable with wood pulp paper.

Recent advances in wood pulp technology and developments in fiber making now allow almost any fiber grade to be made from wood cellulose, lessening the dependency of the vulcanized fiber industry on the cotton rag market. Both cotton staple and cotton linters are used successfully to make fiber. However, the longer-fibered staple cotton is more desirable than the shorter-fibered linters when a higher degree of flexibility is desired. See also COTTON; PAPER.

Cotton staple fibers are obtained from old, used rags and from new textile cuttings. These rags are passed through cutters to obtain small uniform pieces, and are then charged to boilers or digesters of 1400–1500 cu ft capacity. During rag charging, water and alkali, usually sodium hydroxide, are added with the rags to wet them evenly. The amount of alkali used varies from 3 to 15% of the rag weight, and the water is about three times the rag weight.

After charging, the boilers are capped, rotation is begun, and live steam is introduced directly into the boiler. Pressure is increased to the desired level, which varies from 40 to 60 psig. Cooking times vary from 6 to 12 hr.

Although one intention of rag cooking is to remove soil, oils, and waxes from the rags, the primary purpose is to soften the rags and degrade or shorten the cellulose chains. After cooking, the moist and hot rags are usually dumped into bins or large areas underneath the boilers. The rags may remain in the bins for several days so that further softening or aging of the rags may take place.

The extent to which the cellulose is degraded is important not only in the papermaking but especially in making vulcanized fiber. The degree of degradation can be determined by cellulose viscosity measurements; there are two or three methods for making these measurements but the most common ones use cupriethylenediamine or cuprammonium hydroxide as the cellulose solvent (see CELLULOSE).

Following aging, the rags are loaded into Hollander-type beaters equipped with washing cylinders. Water is used to distribute the rags in the beater and is added and extracted continuously to wash out loosened soil and the residual liquor from cooking. Washing cylinders are used to extract the wash water from the beater. During washing the beating action breaks down the weave of the rag and a mass of pulp called "half-stock" is eventually obtained. The total time for this process is 3–4 hr, but washing of the rags may be done only for the first half of the total cycle. The half-stock may be bleached in the washer with calcium or sodium hypochlorite to lighten the color of the stock for light-colored fiber. The unreacted bleach is removed by further washing in the Hollander washer.

As many of the rags are old clothing and include buttons, zippers, dirt, etc, there is considerable metal, plastic, and other undesirable material in the half-stock. Therefore, the half-stock coming from the washer beaters is diluted and passed through centrifugal-type cleaners to remove the debris. Other equipment such as vibrating screens is also used. The acceptable fibers are then thickened and pumped to beaters, usually of the Hollander type. The half-stock is beaten for 3–4 hr to the desired fiber length. Dyes and pigments are added at the beater to achieve the desired paper color. Usually no other additives are used. The resultant pulp or stock is further refined by conical-type refiners, such as a Jordan refiner.

If wood pulp or cotton linters are being used as the raw material, they may be added directly to the final Hollander beater or first slurried with water in a pulper and then pumped to the beater. The stock is beaten, colored, and refined in the same manner as the rag stock.

Upon completion of this refining step the pulp or stock is again cleaned by centrifugal cleaners and vibrating screens. The acceptable material is then formed into paper. The type of forming equipment normally used is the Fourdrinier paper machine.

The final product is an absorbent paper which is correctly called a waterleaf grade. In making this grade for vulcanized fiber, all refining and forming is adjusted to give a paper that is absorbent and soft but which has a smooth surface and sufficient strength to be carried through the subsequent fiber-making process. Control of the absorption characteristics of the paper is necessary in order to regulate the zinc chloride treatment. Absorption can be measured by castor oil impregnation and by vertical capillary rise of water. Typical castor oil penetration results are 10–60 seconds to impregnate 97% of the area tested. Capillary rise (Klemm absorption) is usually measured in sixteenths of an inch, typical values being 10–30 sixteenths of an inch rise up a vertical sample for a 5-min test. Paper strength can be determined by burst strength, tensile strength, or tearing strength. Burst strengths range from 20 to 30 psi, and tensile strengths from 16 to 20 psi in the machine direction, 10 to 15 pounds per inch in the cross direction. Tearing strengths range from 24 to 144 g in the machine direction and from 32 to 156 g in the cross direction. Basis weight, the weight per unit area, is reported in terms of a 24 in. × 36 in., 480-sheet or 500-sheet ream. Basis weights range from 30 to 125 lb depending on the desired thickness and properties of the fiber.

Fiber. Basically, vulcanized fiber is made by saturating one or more plies of paper with concentrated zinc chloride; laminating the treated plies; removing the zinc chloride from the composite by leaching; drying the laminate that is completely free from zinc chloride; and calendering the fiber to give the desired surface finish and thickness. Fiber sheet can be made by continuous, semicontinuous, or batch (cut-down) methods. Fiber tube making is a batch operation.

Continuous fiber making begins with a reel or unwind stand that can hold several rolls of paper. The correct number of plies for the desired fiber thickness are fed from the reel stand into a bath of zinc chloride at a concentration of 68–73° Baumé at a temperature of 85–100°F. The plies of paper must not be allowed to contact each other until they are throughly wetted. Separation is accomplished by use of guide bars immediately ahead of the bath and by the tension applied to pull the plies through the bath. Once the plies are wetted they come together under a submerged roll in the bath and are drawn upward through the remainder of the bath and out of the bath to making cylinders.

As the plies of paper are treated with zinc chloride, their surfaces are gelatinized and intimate contact of the plies is begun as they pass under the submerged roll. However, to achieve more complete lamination the ply buildup is passed in an "S" wrap through a pair of vertically mounted cylinders. These cylinders are heated to a surface temperature of 90–150°F so as to continue the action of the zinc chloride; the top making cylinder applies the necessary nip pressure.

The fiber then passes through an aeration section of the continuous machine for 1 to 30 min to continue the zinc chloride–cellulose interaction. This action may continue at ambient conditions or further heat may be applied in aeration by some means such as steam radiators to achieve an air temperature of 100–180°F.

After aeration the fiber enters a series of leaching tanks. The fiber at this point still contains zinc chloride at a concentration near 70%. The leach liquor in the first leach tank has a 10–20% zinc chloride concentration. Therefore, there is a tremendous driving force at this point and leaching takes place rapidly. The succeeding leach tanks contain diminishing concentrations of zinc chloride so as to maintain a transfer force. Leaching of continuous fiber takes less than 24 hr and the fiber leaving the leach system contains less than 0.05% zinc chloride on an oven-dry basis.

Generally the fiber is dried in-line following leaching with steam-heated drum-type driers. After drying, the fiber is calendered to give the desired thickness and smoothness. Following calendering it is either wound into rolls or cut into sheets.

Continuous fiber-making is limited to finished fiber thicknesses of 3⁄32 in. or less, due to practical limitations on leaching capacity. There are several variations of the continuous process used to produce thicknessess of 1⁄32–1⁄4 in., as leaching and/or drying of these thicknesses may take longer than is practical for a continuous machine. These "semi-continuous" machines are also designed to minimize flexing and bending of the wet web. These machines use the same basic zinc chloride saturating and laminating steps as the continuous machines described earlier. However, the leaching and drying steps vary according to the machine and the fiber thickness.

In one system the partially leached fiber is wound loosely onto reels with spacers placed between fiber thicknesses, and the roll of fiber is then processed through a series of leach tanks to complete leaching. Leaching times vary from 2 to 4 weeks depending on thickness. After leaching, the rolls of fiber are unwound and dried on drum driers.

Fiber above 3⁄32-in. thickness may be cut into sheets after saturation with zinc chloride. These sheets are placed vertically in tanks and the leaching solution is pumped intermittently through these tanks in a countercurrent manner. The sheets are dried by hanging in rooms or ovens maintained at 100–230°F. Sheet fiber warps on drying if it is not restrained; it therefore must be pressed flat while hot in a platen press. The fiber is then calendered to give the desired finish and thickness.

Sheets 1⁄4–2 1⁄4-in. thick are made by the "*cut-down*" method. Only one roll of paper is used to make fiber. The paper is unwound under tension and goes through a zinc chloride bath. As it comes out of the zinc chloride bath, it is wrapped around a heated cylinder (95–105°F) having a diameter of 4–4.5 ft. The finished thickness is determined by the number of revolutions of treated paper that are wound on the cylinder. When the desired number is reached, the paper is cut off and the treated build-up is then rolled by a heated cylinder riding on top of the making cylinder to ensure complete and even bonding of the plies. When the desired amount of rolling is accomplished the machine is stopped and a blade or knife is used to cut through the fiber across the face of the roll and parallel to the axis of the cylinder. The machine is then started and the fiber is "peeled" from the cylinder onto a table or platform. This sheet of fiber is usually cut into two pieces. The fiber is allowed to stand at ambient conditions for 2–24 hr. It is then put into leach tanks for a period of 3–12 months depending on thickness. Fiber of this type is dried in ovens. As in making semicontinuous sheet stock, this fiber is pressed while hot to produce flat material and is then calendered to the desired thickness.

Fiber tubes are made by saturating a ply of paper with zinc chloride and wrapping it around a mandrel. The diameter of the mandrel largely determines the inside diameter of the tube, whereas wall thickness is mainly determined by the number of wraps around the mandrel. After making, the tubes are placed in racks and the racks are processed through a series of leach tanks which are filled and refilled with zinc chloride solutions of diminishing concentrations. To facilitate efficient and faster leaching, the mandrels are removed from the tubes after the early stages of leaching.

When the tubes are free from zinc chloride they are removed from the tanks and are oven dried. Small-diameter tubes may be placed on drying mandrels to prevent warping and twisting during oven drying. Oven temperatures range from 150 to 230°F. Following drying, the tubes are removed from the drying mandrels and placed on rolling mandrels. The tubes are rolled in the axial direction to the desired dimensions under pressure. The rolling mandrel determines the finished inside diameter. Wall thicknesses as low as 1/32 in. and as high as 3/8 in. can be made in this process. Inside diameters of 1/8–4 7/8 in. are typical and 5 1/2 in. is maximum for the outside diameter. After rolling, the tubes are ground in centerless grinders to obtain the desired outside diameter.

Seamless can shells are made by a process similar to that for fiber tubes. A zinc chloride saturated web of paper is wound on a mandrel of the desired size; the wall thickness wanted determines the number of plies. The paper web is cut off, the mandrel collapsed, and the can shell removed. The shells are leached in a series of tanks, with a countercurrent pumping of water. The shells, when free of zinc chloride, are dried on steam-heated, expanding cylinders or "cans." The dried "can" size corresponds to the desired inner diameter of the shell. The dried shells are converted into materials-handling cans for industrial and commercial use by adding a rolled top edge or a metal rim, some type of bottom, casters, handles, varnish or lacquer finish, etc, as needed for the intended use.

Process Control. In the production of vulcanized fiber there are several variables that must be controlled to give the desired properties in the final fiber. Whereas paper properties generally vary, the most important property that must be controlled is the cellulose viscosity. This is determined in the papermaking process and is controlled at that point. The viscosity is important in that it can be used to determine the reactivity of the cellulose to zinc chloride. Generally, at lower viscosities the cellulose swells more easily in the zinc chloride solution. The absorptivity of the paper, as noted earlier, is also varied to change the amount of zinc chloride absorbed and to produce variations in the final vulcanized fiber. Other determining factors are zinc chloride concentration and temperature, as well as immersion time; within limits, as concentration and temperature are increased, the greater the action of the zinc chloride solution on the cellulose. Immersion time is determined mainly by machine speed, but also by the distance the paper plies travel through the zinc chloride bath.

The action of the absorbed zinc chloride continues after the paper leaves the acid bath; the making cylinders are heated and temperature is controlled to effect the desired degree of interaction. In aeration of the fiber, temperature and time are important. If further reaction is desired, heat may be applied. Time of aeration can be varied by changing the length of the aeration section of the fiber machine, as well as machine speed.

The reaction is stopped when the fiber enters the leaching process. The leach water temperatures and the decrease in zinc chloride concentration apparently precipitate the dissolved cellulose as a gel. At the beginning of leaching, the temperatures are ambient but near the end of leaching when the fiber is almost free of zinc chloride, elevated temperatures are used. In leaching of fiber it is important that zinc chloride concentrations in the tanks be controlled to utilize the optimum leaching rate as set by the wet-fiber thickness, and to ensure purity of fiber at the end of the leach system. Also of importance is the temperature of the leach solutions, as leaching rates increase with increase in temperature.

As the fiber is leached, 2–3% shrinkage takes place across the width of the fiber. The greatest amount of shrinkage takes place in the drying process. Final width of dried, continuous-made fiber is 80–85% of the original width of the paper. In unrestrained drying of sheet fiber, there is a 20% reduction in width, 12% reduction in length, and 50% reduction in thickness due to shrinkage.

The final product quality is determined by various tests which include internal bond strength, density, tensile strength, impact strength, arc resistance, and dielectric strength. These tests are used primarily to determine the suitability of the material to its end use. However, efforts have been made to determine the molecular structure changes within the cellulosic fibers and to relate them to the physical properties (see the section on Mechanism of Fiber Vulcanization below).

Special Fiber Treatments. Sometimes it is desirable to make specialty grades of fiber having greater flexibility, flame retardancy, or mold resistance. These properties are obtained by the impregnation of the fiber with some combination of reagents (see for example, BIOCIDES; FIRE RETARDANCY; PAPER ADDITIVES AND RESINS). After the fiber is leached free of zinc chloride, the fiber, while still wet, is retained in a leach tank containing the chemical to be used. Although in the case of thicker fiber complete impregnation may not be obtained, the amount of chemical absorbed is sufficient to achieve the desired effect.

Mechanism of Fiber Vulcanization

The actual mechanisms of fiber vulcanization by treating paper with zinc chloride are not well understood. It is known that cellulosic fibers swell or are partially dissolved in zinc chloride solutions of 70% concentration and at temperatures somewhat above 68°F (see CELLULOSE). Centola maintains that the cellulose is transformed from cellulose I to cellulose II as in mercerization (6). According to Lee, the zinc chloride treatment causes the fibers to separate into fibrils, in a manner similar to the fibrillation caused by mechanical action on the fiber in the beating operation of papermaking (see PAPER) (7). The increased surface would afford more bonding area and result in greater bond strength between plies.

When paper plies are saturated with a zinc chloride solution of the proper concentration, the zinc chloride begins to dissolve the surfaces of the cellulose fibers. This forms a zinc chloride–cellulose phase in the interstices between the remaining cellulose fibers containing both dissolved cellulose chains and solvated, freed portions of cellulose chains still bound into a cellulose fiber. The rate of solution is dependent on the concentration of the zinc chloride and the temperature. The extent of solution depends on the relative chain lengths in the cellulose fibers, the ratio of zinc chloride solution to cellulose, and the temperature and time allowed (saturation plus aeration time).

As the saturated web passes through the leaching process, the concentration of the absorbed zinc chloride solution is lowered and the dissolved cellulose precipitates as a gel. In the ideal case, this gel completely fills the voids within the cellulose fiber web.

When the wet fiber web is dried, the amorphous cellulose gel begins to form large areas of intensive hydrogen bonding among the cellulose chains in the gel and to the cellulose fibers forming the structure of the web. As the gel dries, it contracts. This, plus the drying and shrinking of the cellulose fibers themselves, causes the web to shrink and become dense and strong. The extensive, and apparently irreversible, bonding resulting from the cellulose gel gives vulcanized fiber its strength, density, and lack of porosity.

Control of the relative amount of dissolved cellulose, the amount and strength of the residual cellulose fibers, the amount of mechanical and chemical disruption of the gel formation, and the shrinkage allowed during drying are necessary to produce the combination of properties desired for a specific grade of vulcanized fiber.

In cotton and wood fibers, there are regions of crystallinity where the cellulose chains are highly ordered. In the same fibers there are regions of disorder known to be amorphous. Due to the disorder of the amorphous regions, the cellulose chains there are more accessible than in the crystalline regions.

When cellulose fibers were treated with zinc chloride there was a decrease in crystallinity and an increase in accessibility as measured by x-ray diffraction and infrared spectroscopy or by moisture regain (7). With increasing concentrations of zinc chloride, up to 70% concentration, and with increasing temperature of the zinc chloride solution, crystallinity decreased and accessibility increased almost linearly. Generally, the physical properties of the fiber improved with reduction of crystallinity or increase of accessibility. However, above certain zinc chloride solution temperatures, even though these effects were obtained, strength properties decreased. The material was highly reacted, and dense rigid fiber was obtained that was less elastic and quickly reached a failure point when tested.

Although work by Lee (7) showed increasing reaction with increasing temperature, Centola (6) found that high temperatures may prevent complete reaction throughout a mass of cellulosic fibers. When the paper thickness was great, high-temperature (approximately 104°F) zinc chloride caused rapid reaction on the surfaces of the paper which blocked further penetration into the middle of the ply.

Applications

Vulcanized fiber is available in a number of grades suitable to the various applications to which it may be put.

Commercial grade is usually made in red, black, and gray colors. This grade is a mechanical and electrical insulating material that has good forming, punching, and stamping characteristics. It is used for washers, terminal block covers, coil spool heads, insulating plates, switch covers, switch and appliance insulation, and arc barriers.

Electrical Insulation grade is available primarily in a gray color. This grade is primarily used for electrical applications and other applications involving difficult forming operations. Thin insulation grade is sometimes referred to as "fishpaper." It is used for electrical motor slot cell insulation, stator and rotor end laminations, arc shields, metal switch-box liners, field coil insulation, washers, and formed slot wedges.

Bone grade is usually made in a gray color only. This is an electrical and mechanical grade that is tougher and more dense than Commercial grade, thus providing smoother machined surfaces. It is used for cams, terminal blocks, armature slot wedges, bowling alley kick-back plates, and switch-gear arc barriers.

Trunk and case grade is especially adapted for trunks, cases, and formed parts such as athletic articles and welding helmets. It has good flexibility and a plain, smooth surface in addition to the mechanical properties of Commercial grade and is available in a variety of colors.

Abrasive grade is tougher and more resilient than Commercial grades. It has exceptional tear resistance, ply adhesion, resilience, and toughness. It is used as the support for sanding discs on which abrasive grit is resin bonded.

Railroad grade is a tough, high-density material with good mechanical and electrical properties. It is used for railroad block-signal system joint installations, and must withstand the repeated shock and high compressive loads occurring in the track joint.

Flexible (grade) fiber is a special glycerine-treated material that is softer and more pliable than other grades. It is used for gaskets, packings, and similar applications, but is not generally recommended for electrical use.

Bobbin grade is a dense, high-impact material with excellent machining properties and smooth, clean surfaces. It is used for bobbin heads and other textile applications requiring smooth finish.

Pattern grade has good dimensional stability and wear resistance. It is used to make patterns for cutting cloth, leather, etc.

Shuttle grade is a hard, dense material to be glued to wood shuttles, as protection against the wear and repeated pounding received in power looms.

White grade is used for refrigerator hardware shims, suture reels, and file index cards where color and cleanliness are important. It is a slightly harder, more brittle material than other grades; it is not recommended for forming.

Hermetic grade is used to insulate electric motors in sealed refrigeration units. It must have high purity and low content of methanol extractibles since it must be immersed in the refrigerant.

White tag grade is made primarily in white color, with smooth, clean surfaces for lettering and printing. It is used for tags in drycleaning solutions and other rough service, and is more flexible than white-grade fiber.

Laminated Fiber. Laminated fiber is a built-up material, usually of 1/16-in. plies of fiber, adhesive bonded by hot pressing. Laminates are available in thicknesses up to 4 in. In addition to the standard properties of vulcanized fiber, this material has improved dimensional stability and is free from warpage; it also can be produced in much less time than solid vulcanized fiber of comparable thickness. See also LAMINATES.

Special Properties. The electrical equipment industry makes good use of the arc resistance and arc quenching properties of vulcanized fiber in many products. Arc barriers in circuit breakers are usually made from vulcanized fiber. When the breaker opens, an intense electrical arc is formed. This arc jumps the gap between the contact points and is drawn through an "arc chute." In a fraction of a second the gas emitted by the vulcanized fiber quenches the hot arc, protecting the equipment from damage.

Vulcanized fiber is used in making lightning arrestors for outdoor pole-type transformers. A typical arrestor is in tubular form and contains a threaded, spiral fiber insert. When lightning strikes the power line, the overload surge of current is forced through the fiber section to ground and the heat of the arc apparently causes the fiber to release gas, deionizing the air so that the normal line current does not follow to ground.

Fiber Overlay. A thin layer of vulcanized fiber can be bonded to low-grade, wood house siding or to trim lumber to provide a smooth surface The fiber surface holds paint well, and expands and contracts at a uniform rate, eliminating localized stressing of the paint film and therby avoiding fracture and chipping of the paint film.

Novelties. The use of vulcanized fiber to manufacture spoons, tags, combs, vacuum cleaner nozzles, pipe insulation, grommets, flashlight cases, rollers, etc, serves to illustrate the extreme versatility of this material. In addition, special-use grades such as moisture-resistant and fire-resistant vulcanized fiber can be produced. Also, combination of materials can be produced by bonding vulcanized fiber to rubber, laminated plastics, asbestos, and metal.

Economic Aspects

Vulcanized fiber is manufactured from inexpensive, readily available materials. The raw material costs are 20–30% of the total cost of manufacture. Processing costs are relatively high, owing to the length of time necessary to manufacture the product. In the continuous operations, higher manufacturing speeds may be obtained leading to lower costs. However, in the batch manufacture of thick sheets, months of leaching are required before a pure product is obtained. Investment in equipment also affects material costs. Zinc chloride is very corrosive and the making and leaching equipment must withstand attack or be replaced frequently. Space requirements for leaching are quite high and also contribute to the product cost.

Bibliography

1. J. A. Poumarède and L. Figuier, *Compt. Rend.* **23**, 918 (1846).
2. T. Taylor, *Dinglers Polytech. Journal* **155**, 397 (1860).
3. NEMA Summary Report for 1968.
4. NEMA Standards Publication—*Vulcanized Fibre Publication No. VU 1-1963.*
5. *Product Information for Design Engineers*, A105, Spaulding Fibre Company, Inc., p. 17.
6. G. Centola and F. Pancirolli, *L'Industria della Carta* **1** (2), 1–4 (1947).
7. L. T. C. Lee, *Tappi* **47**, 7 (1964).

Kent H. Alverson, Harold E. Parker, and Richard A. Preibisch
Spaulding Fibre Company, Inc.

VULCANIZED OIL. See FACTICE

VULCANIZING AGENTS. See RUBBER CHEMICALS; VULCANIZATION

WALLING-MAYO THEORY. See Copolymerization

WATER-SOLUBLE POLYMERS. See Acrylamide polymers; Acrylic acid polymers; Cellulose ethers; Dextran; 1,2-Epoxide polymers; Polyamines; Polyelectrolytes; Polysaccharides: Proteins; Starch; Vinyl alcohol polymers; Vinyl amide polymers; Vinyl ether polymers

WAXES

Waxes may be either natural or synthetic, and of petroleum, mineral, vegetable, or animal origin. They are generally smooth, glossy, lustrous, and relatively firm solids at room temperature and are fusible when warmed. Originally, the term "wax" referred to beeswax, but now has the broader meaning of all materials that have wax-like properties. Waxes include various types of chemical composition, such as paraffin hydrocarbons, fatty esters, acids, alcohols, and ketones. The utilization of waxes in polymers and of polymer additives in wax is based on the improvement in performance or properties conferred by the components in the blend.

Waxes are discussed in this article in the following four classes: petroleum waxes covering paraffin and microcrystalline wax; natural waxes including plant, insect, and animal waxes; mineral waxes such as montan and ozokerite; and synthetic waxes including polyethylene and Fischer-Tropsch wax made from nonwax raw materials. The subject of waxes is covered in texts by Warth (1), by Bennett (2), and by Guthrie (3), which are quite thorough, although not recent. See also Hydrocarbon Resins.

Petroleum Waxes

Petroleum waxes, paraffin and microcrystalline waxes, essentially are saturated hydrocarbon mixtures obtained by the refining of crude waxes from petroleum. These petroleum waxes comprise by far the largest amount of all the different kinds of wax used in the United States and in the world (4). Paraffin waxes and microcrystalline waxes differ sufficiently from each other in hydrocarbon compositions, physical properties, and the crystal form of the solid so that there are marked differences in their functional properties and uses in polymers and other industrial formulations (see Table 3).

Paraffin Waxes. Paraffin waxes are solid, firm materials that are basically mixtures of saturated straight-chain hydrocarbons obtained from refining waxy distillates derived from paraffinic crude oils. Fully refined paraffin wax is usually obtained by deoiling crude scale wax, which is a soft paraffin wax intermediate containing up to 5% oil. Scale wax is principally made from slack wax, obtained from wax-bearing crude oil distillates, by reducing its oil content by sweating or by solvent deoiling (5). The old pressing and sweating processes and the newer dewaxing and deoiling technology have been described by Nelson (6) and by Tuttle (7). Typical petroleum test properties of most commercial paraffin wax, scale wax, and slack wax fall in the range given in Table 1.

Table 1. Properties of Paraffin Wax, Scale Wax, and Slack Wax

Physical property	ASTM method	Paraffin wax	Scale wax	Slack wax
melting point, °C	D 87	51–58	49–54	42–49
hardness (penetration) at 25°C (77°F)	D 1321	10–20	20–35	50–80
oil content, %	D 721	0.1–0.5	1–5	10–25
Saybolt viscosity at 99°C (210°F), SUs	D 88	38–42	37–38	35–38
tensile strength, lb/in.^{2}	D 1320	160–400	40–200	40–100

Paraffin waxes consist primarily of straight-chain saturated hydrocarbons with only a small amount of branching, such as 2-methyl groups, near the end of the chain (8). The n-alkane content of hydrocarbon waxes can be determined by molecular sieve adsorption (9) or by urea adduction (10). The amount of n-alkanes in paraffin wax usually exceeds 75% and may reach almost 100%. The molecular weights of the hydrocarbons in paraffin wax range from about 280 to 560 (C_{20}–C_{40}), with each specific wax having a range of about eight to fifteen carbon numbers (8).

The ranges of properties representative of several different paraffin waxes are presented in Table 2. Some of the commercial waxes included in the listed property

Table 2. Typical Properties of Paraffin Wax in Different Melting Point Ranges

	A[a]	B[b]	C[c]	D[d]	E[e]
melting point, °C (ASTM D 87)	51–53	54–56	56–58	60–62	67–71
oil content, % (ASTM D 721)	0.1–0.4	0.1–0.3	0.1–0.2	0.1–0.2	0.2–0.5
hardness (penetration) at 25°C (77°F) (ASTM D 1321)	14–20	12–15	10–14	9–14	9–18
hardness (penetration) at 38°C (100°F) (ASTM D 1321)	80–200	50–100	40–60	20–40	13–38
Saybolt color (ASTM D 156)	30	30	30	30	30
Saybolt viscosity at 99°C (210°F), SUs (ASTM D 88)	37–39	37–39	38–40	40–42	45–53
kinematic viscosity at 99°C (210°F), cSt (ASTM D 445)	3.2–3.8	3.2–3.8	3.6–3.9	4.1–4.9	5.7–8.2
flash point, COC,[f] °C (ASTM D 92)	196–210	206–222	212–222	218–238	238–268

[a] Includes Atlantic Wax 151, Atlantic Richfield Co.; Essowax 2530, Humble Oil & Refining Co.; Gulfwax 27, Gulf Oil Corp.; Mobilwax 128/130, Mobil Oil Corp.; Shellwax 100, Shell Oil Co.; Sunoco Wax 3425, Sun Oil Co.

[b] Includes Aristowax 130/134, Union Oil Company of California, distributed by Sonneborn Division, Witco Chemical Corp.; Atlantic Wax 171, Atlantic Richfield Co.; Boron Wax 133/135, Boron Oil Company, Standard Oil Company of Ohio; Essowax 3050, Humble Oil & Refining Co.; Shellwax 120, Shell Oil Co.; Sunoco Wax 3422, Sun Oil Co.

[c] Includes Boron Wax 138/140, Boron Oil Company, Standard Oil Company of Ohio; Eskar Wax R-35, American Oil Co.; Essowax 3250, Humble Oil & Refining Co.; Mobil Wax 138/140, Mobil Oil Corp.; Sinclair Wax 133, Sinclair Oil Corp.; Sunoco Wax 3420, Sun Oil Co.

[d] Includes Atlantic Wax 1115, Atlantic Richfield Co.; Essowax 4030, Humble Oil and Refining Co.; Gulfwax 40, Gulf Oil Corp.; Pacemaker Wax 45, Cities Service Co.; Sinclair Wax 141, Sinclair Oil Corp.; Shellwax 270, Shell Oil Co.

[e] Includes Aristowax 165, Union Oil Company of California, distributed by Sonneborn Division, Witco Chemical Corp.; Essowax 5250, Humble Oil & Refining Co.; Shellwax 300, Shell Oil Co.; Sunoco Wax 5512, Sun Oil Co.

[f] Cleveland open cup.

Table 3. Typical Physical Properties of Petroleum Waxes

	Paraffin wax	Microcrystalline wax
density of solid at 20°C, g/ml	0.88–0.93	0.89–0.94
density of liquid at 100°C, g/ml	0.73–0.77	0.78–0.81
refractive index in solid state, 20°C	1.526–1.535	
refractive index, liquid, 100°C	1.418–1.433	1.435–1.445
specific heat of liquid, cal/g°C	0.50–0.53	0.50–0.55
latent heat of fusion, cal/g	55–61	50–58
flash point, COC[a] (ASTM D 92), °C	194–250	260–304

[a] Cleveland open cup.

range are also given. Paraffin waxes are generally lower melting, have lower molecular weights, and have lower viscosities when liquid than microcrystalline waxes. Paraffin waxes, in the solid state, exist in the form of large, distinct crystals, in contrast to the microscopic crystals of microcrystalline waxes. Physical properties of paraffin wax of an average molecular weight of 400 are listed in Table 3 (1–2). Paraffin wax is soluble in nonpolar organic solvents such as benzene, chloroform, carbon tetrachloride, and naphtha, and insoluble in polar solvents such as water and methanol.

Paraffin waxes show the same general lack of chemical reactivity as the *n*-alkanes that are their principal components. However, paraffin waxes do undergo a number of chemical reactions, including formation of adducts, cracking reactions, and free-radical substitution reactions. The *n*-alkane components in paraffin wax can be reacted with urea to give crystalline clathrated adducts (10–13). Also, normal paraffins can be separated from hydrocarbons by use of molecular sieves, such as Linde's type 5A, Linde Molecular Sieves, Union Carbide Corp. (9). The thermal cracking of paraffin wax is a commercial process for making α-olefins (14).

Paraffin wax can be chlorinated to introduce various percentages of chlorine up to about 70%, which corresponds to an average of twenty-two chlorine atoms per molecule of hydrocarbon (11,15). The highly chlorinated paraffins are very stable and lose very little hydrogen chloride at 150°C. Fluorinated paraffin wax has been made by fluorination of paraffin wax particles coated with sodium fluoride (16).

The liquid-phase oxidation of paraffin wax, as well as the chlorination reaction, has been known and investigated for more than a century (11). It has been used in Germany as a commercial source of fatty acids. The crude oxidized product also contains esters, alcohols, aldehydes, ketones, lactones, hydroxy acids, and unoxidized alkanes.

Microcrystalline Waxes. Microcrystalline waxes are the solid hydrocarbon mixtures refined by deoiling crude petrolatums, which are obtained from the dewaxing of residual lubricating oil stocks and tank-bottom wax. The refining technology has been well described by Nelson (6) and by Warth (1). These waxes are known as microcrystalline because their relatively small crystals give an amorphous appearance to the waxes in the solid state.

Microcrystalline waxes vary considerably in composition and properties, in contrast to paraffin wax. Generally, microcrystalline waxes consist of branched-chain hydrocarbons and alkylcycloaliphatic (naphthenic) hydrocarbons as well as some straight-chain molecules, depending on the particular wax. The molecular weights range from about 450 to 800 (C_{35}–C_{60}). Some physical properties are listed in Table 3.

Table 4. Properties of Some Commercial Microcrystalline Waxes

	Melting point, °C (ASTM D 127)	Hardness (penetration) at 25°C (77°F) (ASTM D 1321)	Viscosity at 99°C (210°F), SUs (ASTM D 88)	Color (ASTM D 1500)
Multiwax W-445[a]	78.3	30	85	L 0.5
Mobilwax 2305[b]	76.6	27	80	L 2.0
Ultraflex[c]	64.0	26	90	1.5
Ceretak[c]	74.4	26	80	1.5
Shellmax 500[d]	61.5	25	79.5	L 2.0
Superflex S-100[e]	64.4	25	87	1.0
Sunoco Wax 5825[f]	66.0	21	72	1.5
Mobilwax Cerese[b]	82.2	16	76	1.5
Shellmax 400[d]	81.1	11	77.5	1.5
Be Square 190/195[c]	87.7	7	85	0.5
Multiwax 195-M[a]	90.5	7	86	1.0
Sunoco Wax 985[f]	89.5	5	84	1.5
Multiwax 200-M[a]	92.2	5	85	1.0
Petrolite C-700[c]	90.5	3	85	0.5

[a] Sonneborn Division, Witco Chemical Corp.
[b] Mobil Oil Corp.
[c] Bareco Division, Petrolite Corp.
[d] Shell Oil Corp.
[e] Quaker State Oil Refining Corp.
[f] Sun Oil Co.

Typical properties of a number of microcrystalline waxes are listed in Table 4 in order of increasing hardness (decreasing penetration).

Microcrystalline wax has moderate chemical reactivity compared to paraffin wax, but no reactions have been utilized other than the air oxidation of the higher-melting, harder microcrystalline waxes (17) to make oxidized microcrystalline waxes that are useful as carnauba wax substitutes in wax emulsions (see Table 5).

Petrolatum, a semisolid, unctuous material, can be considered as a mixture of low-melting microcrystalline wax and about 10–15% mineral oil. It is obtained directly from fractions of mid-continent and Pennsylvania type crude oils after the light fractions have been removed.

Uses. Approximately 60% of all petroleum wax domestically consumed is accounted for by various packaging applications, such as paper wrappers, paperboard containers, some dairy wax, and corrugated paperboard (see also CELLOPHANE; PACKAGING MATERIALS; PAPER). The balance is used for such things as lubricants,

Table 5. Typical Properties of Some Oxidized Microcrystalline Waxes

	Melting point, °C (ASTM D 127)	Color (ASTM D 1500)	Hardness (penetration) at 25°C (77°F) (ASTM D 1321)	Acid no. (ASTM D 1386)	Saponification no. (ASTM D 1387)
Cardis 320[a]	82–85	1–3	5–7	28–30	75–80
Litene OX-88[b]	82–85	1–2	5–7	28–32	50–65
Petronauba C[c]	82–85	1–3	5–7	22–28	50–60

[a] Warwick Wax Division, Western Petrochemical Corp.
[b] Sonneborn Division, Witco Chemical Corp.
[c] Bareco Division, Petrolite Corp.

mold release agents (qv), candles, carbon paper, crayons, polishes (see FLOOR POLISHES), chlorinated waxes, and explosives (18).

Packaging. The packaging industry employs wax and wax-coating formulations generally containing hydrocarbon resins (qv) or other polymers almost exclusively as hot-melt, 100% solids coatings. The use of these coating systems (both paraffin and microcrystalline) in corrugated paper processing has made possible new markets for both wax and corrugated board. The addition of microcrystalline wax to paraffin wax improves its creased moisture-vapor transmission rate, oil resistance, and heat-sealing characteristics. In the production of frozen-food wrappers, it is customary to use 25% microcrystalline and 75% paraffin wax, although up to 50 or 60% microcrystalline wax is sometimes employed.

The combination of small quantities of polymer with paraffin and microcrystalline wax normally will result in improved flexibility, tensile strength, heat sealability, and scuff resistance, and also in higher softening temperatures and chemical resistance (19). The principal resins or polymers used for modifying wax are ethylene–vinyl acetate copolymer, butyl rubber, low-molecular-weight polyethylene, polyisobutylene, ethylcellulose, hydrogenated rosin esters, and polyamides.

Ethylene–vinyl acetate copolymers and ethylene–acrylate copolymers have been used very successfully in modifying wax for hot-melt systems acceptable for coating paperboard. Hot melts generally consist of between 20 and 40% of resin or polymer, eg, ethylene–vinyl acetate copolymer or rosin (qv), between 3 and 20% high-melting microcrystalline wax (90°C), with the balance predominately high-melting paraffin wax (65–70°C). Variations in concentration and ratio of high-melting microcrystalline to high-melting paraffin wax depend, of course, upon the particular application and the end properties resulting from the blend. Wax coatings may also contain other ingredients, such as oils, metallic soaps, plasticizers, and other waxes such as carnauba and montan.

Varnishes and Lacquers. Paraffin has also been added to varnishes and lacquers used in the manufacture of packaging materials such as cellophane (20). Polyisoprene-based lacquers containing wax are excellent for glassine. Wax has also been incorporated in such lacquer bases as ethylcellulose, poly(vinyl chloride), alkyd resins, and other film formers. Very small amounts of wax are highly efficient in producing moisture barriers in these materials. In many cases, the wax provides improved slip, gloss, and resistance to rubbing.

Rubber. Paraffin wax and microcrystalline wax are also used in the manufacture of antichecking waxes for rubber. Automobile tire compositions generally have incorporated a blend of 70–99% petroleum wax and 1–30% ethylene–vinyl acetate copolymer (21). Antichecking waxes are used to inhibit oxidative degradation of rubber better known as ozone cracking.

Chlorinated Paraffins. Approximately 20% of the chlorinated paraffin domestically consumed is used as a flame retardant (by itself or in conjunction with antimony oxide) and secondary plasticizer in various vinyl formulations, such as for vinyl flooring (22). The chlorinated paraffin found most compatible in such hard vinyl formulations usually contains between 40 and 50% chlorine. See also FIRE RETARDANCY; FLOORING MATERIALS. Smaller amounts of chlorinated paraffins are used as flame retardants in synthetic rubber, especially nitrile, butyl, and neoprene.

Oxidized Microcrystalline Wax. In certain applications, it is desirable to have an appreciable saponifiable content in the microcrystalline wax. To obtain this, microcrystalline waxes are partially oxidized. The oxidized microcrystalline waxes thus produced come closer to replacing the more expensive natural waxes such as carnauba and candelilla. Thixotropic compositions used for modifying rheological properties of nonaqueous paint products are based on such modified, oxidized microcrystalline waxes (23) as well as modified, oxidized Fischer-Tropsch waxes. These waxes impart such properties to a paint product as effective pigment suspension, good "anti-sag" properties, increased thixotropic body, and increased viscosity.

Other Uses. Petroleum wax has a variety of protective coating applications outside the packaging field. Strippable coating compositions having a petroleum wax base (paraffin or microcrystalline wax or a blend of paraffin and microcrystalline wax) in conjunction with up to 50% of a resin such as ethylene–vinyl acetate or ethylene–ethyl acrylate copolymers have been prepared (24). The combination of resin and wax improves abrasion resistance, crease resistance at sharp corners, and flexibility. Strippable or temporary coatings have growing usage in the automotive industry (25).

"Mothballing" utilizes another form of temporary protective wax–polymer coating. These "cocoon" coatings are used on large pieces of equipment such as locomotives, railway cranes, and naval vessels. They are based on a petroleum wax–vinyl chloride copolymer that has a high copolymer content and also contains poly(vinylidene chloride), which promotes "cob-webbing." Microcrystalline waxes or blends of paraffin and microcrystalline waxes in conjunction with vinyl chloride copolymers are generally used in this application.

Wax peroxides capable of entering into the vulcanization reactions of rubber have been reported (26). These peroxides apparently form free radicals at elevated temperatures and should be capable of reacting with numerous systems. Olefin polymers such as polyethylene and polypropylene, have been blended with microcrystalline wax (27) and used as extrusion coatings for electrical conductors. These waxes have also been incorporated into polyethylene and polypropylene blow-molded containers by immersion of the container in a molten wax bath. Waterproof cable sheathing has also been prepared from a blend of polyethylene and petrolatum (28). A composition containing 85% petrolatum and 15% polyethylene has been used as a sheathing or waterproof protective outer coating for underground cables, thereby preventing water-induced circuits.

Copolymer compositions of maleic anhydride, citraconic, and itaconic acids, etc, with either microcrystalline or Fischer-Tropsch waxes (of melting point greater than 65°C and a penetration of 0–50 (ASTM D 1321)) are reacted with polyols (29), polyamines, or alcohol amines for use in the preparation of carbon-based inks or polishes.

Petroleum waxes are also used as external lubricants in polymer processing, for mold release agents, mold lubricants, and antiblocking agents. For example, between ½% and 2% of a high-melting paraffin (melting point of about 75°C) can be used in place of stearates as a lubricant in the extrusion of rigid poly(vinyl chloride) compounds. See also RELEASE AGENTS; LUBRICANTS.

Test Methods. Many test methods have been developed by wax manufacturers and consumers. The American Society for Testing and Materials has published these to serve as quality control tests (30).

Melting Point. ASTM D 87-66 is used to measure the melting point of paraffin wax from a temperature–time cooling curve. ASTM D 127-63 is the method used to measure the drop melting point of microcrystalline wax. The melting points are an indication of the molecular weight and purity of the petroleum wax. Generally, the higher-melting waxes (as long as they are unadulterated) are costlier.

Needle Penetration. ASTM D 1321-65 indicates the hardness of the wax; the lower the numerical value, the harder the wax. Generally, harder waxes contain less oil impurity and are more useful in polymer applications.

Oil Content. ASTM D 721-68 gives a direct measurement of the presence of oil and lower melting wax hydrocarbons soluble in methyl ethyl ketone at −32°C. Other than indicating the degree of refinement of the petroleum wax, oil content has no particular significance.

Viscosity. ASTM D 445-65 and ASTM D 88-56 are methods for the empirical measurement of the kinematic viscosity in centistokes and of the Saybolt viscosity in Saybolt Universal sec (SUs). ASTM D 2161-66 covers conversion tables for converting kinematic viscosity into Saybolt viscosity. Viscosity is an indication of the composition of petroleum wax, lower values indicating paraffinic materials and higher values, microcrystalline waxes.

Color. The Saybolt color of petroleum products (ASTM D 56-64) indicates the degree of refinement of paraffin wax. ASTM D 1500-64 is the method for the ASTM color of petroleum products and indicates the degree of decolorization of microcrystalline wax.

Tensile Strength. ASTM D 1320-67 is the method for obtaining the tensile strength of paraffin wax. It is a general guide to the overall refinement of the wax (see Table 1).

Flash Point. ASTM D 92-66 gives the procedure for determining the flash and fire points of petroleum products by the Cleveland open-cup tester. Flash point indicates the degree of refinement and the removal of low-molecular-weight hydrocarbons, cracked material, and solvents.

Acid Number. The amount of free acidity (ASTM D 1386-59) in oxidized petroleum wax or in natural waxes such as carnauba is an indication of the degree of oxidation of an oxidized wax or of the purity and composition of a natural wax.

Saponification Number. ASTM D 1387-59 covers the determination of the saponifiable matter present in oxidized petroleum wax or in natural waxes. It indicates the total amount of acid available either in the ester form or as free acid.

Safety. Petroleum waxes of proper purity have been approved by the U.S. Food and Drug Administration for use in the processing and packaging of foods. Approved petroleum waxes must meet a test for ultraviolet absorbance described in the *Federal Register* (30a). Petroleum waxes with low ultraviolet absorbance are noncarcinogenic (31).

Natural Waxes

Vegetable Waxes. Vegetable waxes are obtained from the coatings on leaves, stems, grasses, fruits, and barks of various plants and trees (32). These waxes are mixtures of esters of fatty acids and high-molecular-weight alcohols and unsaponifiable materials. Properties of a number of the more commercial vegetable waxes are listed in Table 6.

The most important commercial vegetable waxes are carnauba, candelilla, ouricuri, and Japan. These waxes are used to a great extent, either by themselves or in conjunction with the others in the formulation of resin–wax polishes (eg, liquid polishes for wood, floor waxes, shoe pastes). The value of these waxes in polishes lies in the fact

Table 6. Properties of Vegetable Waxes

Wax	Color	Melting point, °C	Specific gravity	Acid no.	Saponification no.	Iodine no.
bayberry	greenish-white	39–49	0.87–0.98	3–4	205–217	2–10
candelilla	brownish to yellow-brown	65–77	0.97–0.99	9–22	35–85	14–37
carnauba	light yellow to brown	78–90	0.97–1.00	1–10	68–88	7–14
esparto	brown	70–78	0.98–0.99	22–33	62–79	8–23
Japan	pale cream	50–56	0.97–0.99	6–20	206–237	4–13
ouricuri	brown	78–84	0.97–1.06	8–24	70–109	6–8
palm	olive-brown	71–86	0.99–1.05	4–16	80–104	3–17
raffia	light brown	82–83	0.83–0.84	5–7	50–52	8–11
sugar cane	yellow to brown	61–82	0.96–1.00	8–28	55–95	5–29

that they produce polishes with very durable luster and hardness. Carnauba is most preferred (33), but candelilla, ouricuri, and Japan wax have been used as substitutes. These waxes are generally used in conjunction with resins such as acrylics, polystyrene, and poly(vinyl chloride). See also FLOOR POLISHES.

Ouricuri wax comes closest to carnauba wax in properties and is lower in price. Its principal use, therefore, is as a substitute for carnauba in products where darker color is acceptable (eg, polishes, carbon paper, and mold-release lubricants). Ouricuri, as well as carnauba wax, has been used to increase the melting point of paraffin waxes in the preparation of hot-melt adhesives and coatings. Candelilla wax has been effectively used with Japan wax and rosin in emulsion recipes for the preparation of material for textile finishing materials (34). (See also TEXTILE RESINS.) Japan wax was at one time imported in large quantities from Japan and used as an activating agent in the vulcanization of rubber. However, in recent years, it has been too expensive for that purpose and has, therefore, been replaced by mixtures of other natural and synthetic waxes with chemical analyses closely approximating those of Japan wax.

Animal Waxes. Beeswax is the most important commercial insect wax; its properties are summarized in Table 7 (35). Other insect waxes include Chinese insect wax and shellac wax. Wool wax (called lanolin when refined) is the major animal wax; its properties are included in Table 7 (35–36).

Table 7. Properties of Beeswax and of Wool Wax

	Beeswax	Wool wax
color	pale yellow to dark brown	brownish-yellow
melting point, °C	62–70	37–43
specific gravity at 15°C	0.96–0.98	0.93–0.95
acid number	7–30	6–22
saponification number	83–104	84–127
iodine number	5–13	15–47

The more important animal waxes include beeswax, spermaceti wax, and wool wax. The majority of these waxes are used in the manufacture of cosmetic creams, lotions and soaps, medicinal ointments and salves, and candles. Little, if any, of these waxes are currently used in the polymer field. Beeswax, which at one time was used in the manufacture of floor polishes (in conjunction with acrylic resins) has been almost entirely replaced by carnauba wax.

Mineral Waxes

Mineral waxes are obtained from fossil remains such as lignite, bitumens, peat, and shale, excluding the waxes derived from petroleum. The properties of montan wax, ozocerite, and ceresin are listed in Table 8 (33).

Table 8. Properties of Some Mineral Waxes

Wax	Hardness	Color	Melting point, °C	Specific gravity	Acid number	Saponification number
montan	very hard	brownish-black	84–90	1.02–1.03	40–50	90–120
ozocerite	fairly hard	white to black	60–90	0.85–0.95	0	0
ceresin	fairly hard	yellow to white	50–80	0.88–0.93	0	0–1

The leading commercial mineral waxes are montan, ozocerite, and to a much lesser extent, ceresin. Montan wax has a major use in the manufacture of carbon paper dopes; this use has increased steadily over a period of years. Montan has been used, to a great extent, in conjunction with various polymers, such as oxidized polyethylene and acrylic resins, in the manufacture of a variety of polish emulsions (37), and as an internal lubricant for poly(vinyl chloride).

Ozocerite has been blended with polystyrene alkylated with a mixture of C_9 olefins to produce a wax composition suitable for the impregnation and coating of fibrous products, such as textiles, leather, wood, and papers (1), for the purpose of improving gloss, adhesiveness, hardness, tensile strength, and elongation characteristics. Mixtures preferably contain 80% or more ozocerite blended with an alkylated polystyrene having a molecular weight of 60,000 to 300,000. Certain hard rubber compounds for electrical purposes contain between 2 and 5% ozocerite, imparting to the rubber system durability, elasticity, and resistance to marked changes in temperature. The name ceresin or ceresin wax is loosely applied to describe refined ozocerite. At present, commercially available ceresin waxes are actually blends of paraffin and compatible materials such as carnauba wax and/or bleached montan wax.

Synthetic Waxes

Synthetic Hydrocarbon Waxes. Hydrocarbon waxes of interest in polymer chemistry and technology include Fischer-Tropsch and polyethylene waxes made from nonwax raw materials.

Fischer-Tropsch Wax. This wax is obtained by the Fischer-Tropsch catalytic hydrogenation of carbon monoxide at high temperature and pressure (11). Fischer-Tropsch wax is white and very hard (penetration of 0–5 at 25°C) with a melting point of 95–110°C. This wax is composed of *n*-paraffins in the molecular-weight range of 600–950 (38).

There is no synthetic Fischer-Tropsch paraffin wax currently produced in the United States. Domestic requirements for this type of wax are met by material imported primarily from South Africa (39) and from West Germany. Synthetic paraffin wax is used to improve the resistance of vulcanized rubber to oxidative degradation.

It also has growing use in a variety of resin–wax polish formulations (40), wherein the synthetic wax tends to complement, rather than replace petroleum-derived products. It is also used, in conjunction with polyamide vehicles, in printing ink formulations, eg, in flexographic inks (see also PRINTING-INK VEHICLES).

Polyethylene Waxes. Polyethylene with molecular weights in the range 2000 to 4000 has the properties of high-molecular-weight hydrocarbon wax. These low-density polyethylenes are generally produced by the high-pressure polymerization of ethylene (41) (see ETHYLENE POLYMERS). The polyethylene produced by this process contains many side branches and has a specific gravity of 0.91–0.96, depending on operating conditions. Melt index is close to 3.5, tensile and yield strength close to 1500 psi, melting point range between 99 and 120°C, and needle penetration at 25°C is 1–10.

A little over 10% of the low-density polyethylene produced in the United States finds use in typical wax applications, such as paper coatings and floor polishes. Non-emulsifiable low-molecular-weight polyethylenes have major applications in paper coatings (42), where they are employed alone as a coating material or, more commonly, as an ingredient of paraffin wax blends. Approximately half of the low-density polyethylene used in paper coating is used to coat paperboard in the manufacture of milk cartons. Other applications include paperboard, paper, and film and foil coatings. Emulsifiable low-molecular-weight polyethylenes, obtained by partial oxidation, are used primarily as ingredients in floor polishes (43). They are usually used in combination with naturally derived waxes and other polymers. Emulsifiable low-molecular-weight polyethylenes are also used as softeners for resin-treated textiles. They tend to give improved tear and tensile strength to durable press fabrics. See also FLOOR POLISHES; PAPER.

Ethylene Telomer Wax. The reaction of ethylene with telogens, such as dialkyl ether or saturated halogenated carboxylic acids using a peroxy catalyst (44) or toluene with an organolithium catalyst (45) at temperatures of 40–125°C in an autoclave leads to wax products. This telomerization process is a potential economical source of high-melting, hard, synthetic waxes for use in polish formulations and hot melts.

Fatty Acids and Amides. Stearic, palmitic, and myristic acids, as well as hydrogenated tallow fatty acids, are waxy materials and serve as internal lubricants and mold-release agents either in the acid form or as metal salts. The fatty acid amides such as stearamide and oleamide are used for lubricant additives; they make good antislip agents and also impart antistatic properties. See also LUBRICANTS; RELEASE AGENTS.

Ethylene bis(stearamide) is produced by condensing ethylenediamine with stearic acid. The bis(stearamide) is a light-colored, waxy solid (sold in a bead or powdered form) with a melting point of 138–144°C and a density of 0.97–0.99 at 25°C (Chemetronwax 100; Advawax 280; Acrawax C). It serves as a lubricant for the extrusion of poly(vinyl chloride), polystyrene, and polyester, and as a antiblocking agent and lubricant for rubber (46).

Miscellaneous Synthetic Waxes. Castorwax (Baker Castor Oil Co.) and Opalwax (Du Pont), commercial forms of fully hydrogenated castor oil, are used in resin adhesive formulations, hot melts, mold lubricants, resin polish formulations, rubber compounding, and textile finishing. Montan wax esters (Hoechst Wax OP) and calcium-saponified montan esters (Hoechst Wax E) are used as internal and external lubricants for poly(vinyl chloride) (47). Poly(oxyethylene) glycols, such as

Union Carbide's Carbowaxes, find some use as rubber mold release agents (48) and as modifiers for alkyd resins (49), methacrylate resins, natural resins (shellac), and polyurethaus. Cyclic hydrocarbons, such as Monsanto's Santowaxes, and chlorinated naphthalenes, such as Koppers' Halowaxes, are used as flame retardants or processing aids for rubber, polystyrene, phenolic resins, and resinous paints. A relatively new α-olefin wax, Gulf's C_{30}^{+}, somewhat resembling low-molecular-weight polyethylene, appears to have some use in resin–wax polish formulations.

Bibliography

1. A. H. Warth, *The Chemistry and Technology of Waxes*, Reinhold Publishing Corp., New York, 1956.
2. H. Bennett, *Industrial Waxes: I. Natural & Synthetic Waxes and II. Compounded Waxes & Technology*, Chemical Publishing Co., Inc., New York, 1963.
3. V. B. Guthrie, *Petroleum Products Handbook*, McGraw-Hill Book Co., Inc., New York, 1960.
4. *Summaries of Trade and Tariff Information, Schedule 4*, Vol. 12, 1968, pp. 121–170.
5. S. Marple, Jr. and L. J. Landry, "Modern Dewaxing Technology," in J. J. McKetta, Jr., ed., *Advances in Petroleum Chemistry and Refining*, Vol. 10, Interscience Publishers, a division of John Wiley & Sons, Inc., New York, 1965, pp. 215–216.
6. P. Zurcher, "Dewaxing," in W. L. Nelson, ed., *Petroleum Refinery Engineering*, 4th ed., McGraw Hill Book Co., Inc., New York, 1958, Chap. 12, pp. 374–394.
7. J. B. Tuttle, "Petroleum Waxes," in A. Standen, ed., *Kirk-Othmer Encyclopedia of Chemical Technology*, Vol. 15, 2nd ed., Interscience Publishers, a division of John Wiley & Sons, Inc., New York, 1968, pp. 92–112.
8. R. H. Hunt and M. J. O'Neal, Jr., "The Composition of Petroleum," in J. J. McKetta, Jr., ed., *Advances in Petroleum Chemistry and Refining*, Vol. 10, Interscience Publishers, a division of John Wiley & Sons, Inc., New York, 1965, pp. 18–20.
9. J. G. O'Connor, F. H. Burow, and M. S. Norris, *Anal. Chem.* **34**, 82 (1962).
10. W. P. Ridenour, I. J. Spilners, and P. R. Templin, *TAPPI* **41**, 257 (1958); A. Briok and C. Kleynjan, *Fette, Seifen, Anstrichmittel* **71**, 1017 (1969).
11. F. Asinger, in H. M. E. Steiner, ed., *Paraffins, Chemistry and Technology* (transl. by B. J. Hazzard), Pergamon Press, New York, 1968 (English ed.).
12. R. L. McLauglin, "Separation of Paraffins by Urea and Thiourea," in B. T. Brooks, S. S. Kurtz, Jr., C. E. Boord, and L. Schmerling, eds., *The Chemistry of Petroleum Hydrocarbons*, Vol. 1, Reinhold Publishing Corp., New York, 1954, Chap. 10, pp. 241–274.
13. L. C. Fetterly, "Organic Adducts," in L. Mandelcorn, ed., *Non-Stoichiometric Compounds*, Academic Press, New York, 1964, Chap. 8, pp. 491–567.
14. A. J. Van Peski (to Shell Development Co.), U.S. Pat. 2,172,228 (1939); California Research Corp., Brit. Pat. 848,385 (1960); W. A. Pardee and F. Chapel (to Gulf Research and Development Co.), U.S. Pat. 2,945,076 (1960); and R. P. Cahn (to Esso Research and Engineering Co.), U.S. Pat. 3,103,485 (1963).
15. F. F. Gardner, *Ind. Eng. Chem.* **25**, 1211 (1933).
16. W. D. Blackley, W. R. Siegart, and H. Chafetz (to Texaco Inc.), U.S. Pat. 3,515,582 (1970).
17. J. Phillips and B. R. Bluestein (to Witco Chemical Corp.), U.S. Pat. 3,224,956 (1965).
18. *Annual Sales of Wax By Broad End-Use Categories In The United States 1963–1968*, Department of Statistics, American Petroleum Institute, Washington, D.C., 1969.
19. "Ethylene–Vinyl Acetate Copolymers: Hot Melts, Adhesives, Boost Demand," *Oil Paint & Drug Reporter*, 40 (Nov. 6, 1967).
20. W. H. Charch (to E. I. du Pont de Nemours & Co., Inc.), U.S. Pat. 2,147,628 (1939).
21. J. S. Boyer (to Sun Oil Co.), U.S. Pat. 3,412,058 (1968).
22. *Oil, Paint & Drug Reporter, Chemical Profiles*, 9 (Aug. 4, 1969).
23. F. M. Frank (to Baker Castor Oil Co.), U.S. Pat. 3,407,160 (1968).
24. R. P. Zmitrovis and E. W. Sanders (to Cities Service Oil Co.), U.S. Pat. 3,489,705 (1970).
25. J. Oliver, *Prod. Finishing* **34**, 82–87 (Sept. 1970).
26. G. G. Rumberger (to Marathon Corp.), U.S. Pat. 2,662,864 (1953).
27. B. R. Bluestein, T. D. Hindman, and C. Bluestein, *TAPPI Special Technical Association Publication, Technical Association of the Pulp and Paper Industry*, 1963, pp. 146–159; R. C. Fox, pp. 160–175; R. Eells and M. O. Brunson, pp. 176–187.
28. M. C. Biskeborn and D. P. Dobbin, "Jelly Blend Waterproofs Cable," *Bell Laboratories Record*, Bell Laboratories, Florham Park, N.J., March 1969, pp. 71–75.

29. Petrolite Corp., Brit. Pat. 1,182,161 (1970).
30. *1969 Book of ASTM Standards*, Parts 17, 18, and 22, American Society for Testing and Materials, Philadelphia, Pa., 1969.
30a. *Amendment to the Federal Register 121.1156*, July 8, 1964, pp. 57–57.4.
31. P. Shubik et al., *Toxicol. Appl. Pharmacol.* **4,** Suppl. (Nov. 1962).
32. W. J. Hackett, *Detergents and Specialties* **6,** 49 (1969).
33. R. E. Sievert and F. W. Rau, *Detergent Age* **3,** 52 (1966).
34. W. Herbig, *Die Öle u. Fette in der Textile Industrie, Monograph Series*, Vol. 3, Stuttgart, Wissenschafliche Verlagegesellschaft m.f.H., 1923.
35. W. J. Hackett, *Detergent and Specialties* **7,** 22 (1970).
36. E. V. Truter, *Wool Wax Chemistry and Technology*, Cleaver-Hume Press Ltd., London, 1956.
37. R. L. Drew, *Soap Chem. Specialties* **41,** 96 (Nov. 1965).
38. J. H. le Roux, *J. Appl. Chem. (London)* **19,** 39, 86, 230 (1969); **20,** 203 (1970).
39. *Soap Chem. Specialties* **32,** 182 (Dec. 1956).
40. C. J. Marsel, *Soap Chem. Specialties* **31,** 131 (Feb. 1955).
41. E. W. Fawcett et al., Brit. Pat. 471,590 (1937); *Chem. Eng.* **75,** 86 (Jan. 15, 1968).
42. *Chem. Eng. News* **42,** 34–35 (Jan. 27, 1964).
43. M. O. Brunson and L. D. Queen, *Soap Chem. Specialties* **41,** 103 (Oct. 1965).
44. W. E. Hanford and J. R. Roland (to E. I. du Pont de Nemours & Co., Inc.), U.S. Pat. 2,457,229 (1948); W. E. Hanford and R. M. Joyce, Jr. (to E. I. du Pont de Nemours & Co., Inc.), U.S. Pat. 2,507,568 (1950).
45. G. G. Eberhardt (to Sun Oil Co.), U.S. Pat. 3,206,519 (1965); A. W. Langer, Jr. (to Esso Research and Eng. Corp.), U.S. Pat. 3,451,988 (1969).
46. S. A. Riethmayer, *Gummi, Asbest, Kunstoffe* **18,** 425–432 (1965).
47. G. Illman, *SPE J.* **23,** 71 (June 1967).
48. C. P. McClelland, *India Rubber World* **125,** 579 (1952).
49. J. P. Dunne (to Columbian Carbon Co.), U.S. Pat. 2,489,763 (1949).

Bernard R. Bluestein and Norman Kudisch
Witco Chemical Corporation

WEAR RESISTANCE. See ABRASION RESISTANCE

WEATHERING

Polymeric materials, especially in the form of paints, plastics, rubbers, and textiles, are frequently exposed to the weather. Paint is often used to protect surfaces from the environment; in 1968, about 850 million gallons of paint, varnish, and lacquer were sold at a cost of over 2.5 billion dollars. Plastics have their largest market in construction where weatherability is a primary design factor; in 1968 about 2.5 billion pounds of plastic were used in building construction alone (see also BUILDING AND CONSTRUCTION APPLICATIONS). Elastomers (synthetic and natural) are commonly seen in automobile tires (qv), gaskets, and similar applications where weather exposure may be expected. Textile fibers and fabrics are exposed in wearing apparel, in automotive interiors, and even as reinforcing for plastics. With the projected growth of such markets and applications, outdoor durability of polymeric materials is destined to become even more critical. However, the state of the art in predicting weatherability lags far behind the need. Exact prediction of the useful lifetime of a given polymer in a specific geographic location is still the dream of both consumers and manufacturers.

Two distinct technical areas are involved in this problem; (a) polymer degradation (see DEGRADATION), and (b) the weather. Each must be defined carefully, then their interactions can be considered in studying the course of deterioration.

The Weather

According to Webster, the *weather* is the "state of the atmosphere with respect to heat or cold, wetness or dryness, calm or storm, clearness or cloudiness." Furthermore, *climate* is defined as "the average course or condition of the weather at a place over a period of years as exhibited by temperature, wind velocity, and precipitation." Both the instantaneous weather and the average climate affect the weatherability of a material. The most significant factors causing weathering of polymeric materials may be listed as: solar radiation, temperature, moisture, wind, and atmospheric constituents or contaminants. Nonclimatological factors that sometimes enter into outdoor deterioration are biological (fungi, bacteria, animals) and physical stress (such as flexing of a plastic structural component). See Microbiological Degradation under DEGRADATION; FRACTURE.

Solar Radiation. There is mounting evidence that the most important weather factor in the deterioration of polymeric materials is solar radiation.

Sunlight is commonly measured in langleys per minute, where one langley equals one calorie per square centimeter of solar energy (1). About 700 meteorological stations all over the world report solar radiation in langleys measured at ground level on a horizontal surface. In addition, commercial stations which expose materials outdoors usually record and report langley data as well as other pertinent weather variables. Radiation intensity depends on the hour of the day; the day and season of the year; the altitude, latitude, and geographical location: and cloud cover, dust, and clarity of atmosphere. The most intense radiation is received on a surface that is at a right angle to the sun; the cosine of the sun's angle to the horizontal equals the ratio of intensity on the horizontal to intensity at a right angle to the sun. It is common practice to expose experimental paints and plastics in Arizona or southern Florida, which are sunny areas where an average summer day would have 700 or 500 langleys on the horizontal, respectively.

Besides intensity, spectral distribution of solar radiation at the earth's surface is variable (2). Of the solar radiation reaching the outer atmosphere of the earth, some is reflected back into outer space from the atmosphere and the tops of clouds, some is absorbed by molecules such as water, carbon dioxide, and ozone, and some is scattered by particles of water and dust. This reduces the solar constant from about 2 cal/sq cm/min above the earth's atmosphere (ie, the power input to a surface normal to the sun's direction, at the earth's mean distance from the sun) to between 0 and 1.5 at the earth's surface. This earth-level insolation will contain much scattered light, which is rich in the ultraviolet and blue portion of the spectrum. The reflected sunlight, or albedo, is rich in infrared, since long-wavelength infrared is re-radiated by the warm earth to the cooler sky. Because of the importance of such variations, it is usually considered desirable to measure solar radiation as near as possible to polymeric specimens being weathered experimentally.

Earth-level insolation extends from a wavelength of about 290 nanometers (millimicrons) in the ultraviolet to beyond 0.01 millimeters (10 microns) in the far infrared. Maximum intensity is in the visible to red region at about 500–1000 nanometers. The highly energetic ultraviolet region (290–400 nm) contains less than 6% of the total radiation on earth (3). Ultraviolet radiation varies much more than the total radiation. It undergoes a cyclic seasonal change with maximum intensity in summertime. Furthermore, the shorter the wavelength the greater the change with season and local conditions and especially with the time of day. These changes in sunlight radiation are reflected in skylight radiation to an even greater extent. While on a clear day the intensity of short-wavelength ultraviolet in skylight may be equal to that in sunlight, when the total radiation is low and there is much scattered radiation, the proportion of short to long wavelength ultraviolet increases.

Because of the variability of natural sunlight it is desirable to have a reference for the spectral distribution of solar radiation at the earth's surface. A reasonable standard which has been proposed is normally incident solar radiation at the mean solar distance (14). At sea level, the intensity and distribution have been calculated for an air mass of 2, ie, for sunlight passing through a daily average of two optical thicknesses of atmosphere before reaching the earth's surface (15). One solar constant under such conditions is 739 watts per square meter. However, variations in ultraviolet radiation are virtually unnoticed in measurements of total solar radiation. An estimate of the amount of ultraviolet radiation below 315 nm in terms of "Coblentz langleys" may be calculated by multiplying langley values by monthly correction factors (5a). Comparison of measurements of ultraviolet and total radiation of direct sunlight in Arizona, Florida, and Washington, D.C., indicates that the ratio of short-wavelength ultraviolet to total radiation may be approximately the same throughout the United States on any given day (4). Recent work has shown that divergent data on strength loss of fibers for winter and summer exposures fall on a single curve when ultraviolet radiation instead of total radiation is used as the measure of exposure (5b). Such results provide evidence for the large contribution of ultraviolet photodegradation to outdoor deterioration of polymeric materials. Because the variations in both sun and sky radiation affect the weathering of polymeric materials, actinic solar radiation should ideally be measured under the same conditions as the polymeric specimens being weathered experimentally. See also DEGRADATION; RADIATION-INDUCED REACTIONS.

Temperature. It is well known that temperature can significantly influence the rate of chemical reactions, including hydrolysis and oxidation, which are factors in weathering.

Solar radiation is the source of the earth's heat. The prime contributor to temperature increase is infrared radiation, which constitutes about 50% of sunlight. Climate similar to almost any other part of the world can be found within the United States (6); normal annual temperature in the United States ranges from about 4°C (40°F) to above 24°C (75°F), yielding an average annual temperature of about 17°C (62°F). An official temperature of air in the shade has been recorded in California of 57°C (134°F). At the other extreme, −55°C (−68°F) has been reported at fairly moderate elevations in Montana. Average temperatures of the world's zones have been given as (7):

torrid zone, ~24°C (75°F)

temperate zone, ~10°C (50°F)

polar zone, ~−7°C (20°F)

at the poles, ~−25°C (−13°F)

The surfaces of materials exposed to temperature extremes can be expected to reach even more extreme surface temperatures. Exposed plastic specimens have been reported (8) to reach 77°C (170°F). Insulated roofing specimens reached 40°C (104°F) above ambient temperature, and cooled as much as 11°C (52°F) below ambient temperature at night (9).

Temperature of the exposed material, not temperature of the surrounding air, is the important factor in deterioration. The effective temperature of the exposed specimen depends on total incident sunlight, ambient temperature and wind, and the material's thermal conductivity, absorptivity, and emissivity. Environmental engineers frequently use a factor known as "sol air temperature," S.A.T., to estimate the temperature of an exposed material (10):

$$\text{S.A.T.} = T_A + \frac{aI}{h}$$

where T_A is ambient air temperature, a is a constant for the material called solar absorptivity, I is total incident solar radiation, and h is a constant for the specific material and climate surrounding, called surface conductance (11). Surface conductance is the time rate of heat exchange by radiation, conduction, and convection of a unit area of a surface with its surroundings. It increases approximately linearly with wind velocity. Solar absorptivity is closely related to color, varying from about 0.9 for black materials such as asphalt to about 0.2 for white poly(vinyl fluoride) film. Thus, a typical bright summer day with an air temperature about 90°F and a breeze of 5 mph could yield a sol-air temperature for a black material of about 120°F. Without a cooling breeze, the calculated surface temperature is 165°F, which is in good agreement with the experimental value of 166°F for black asphalt (9). Temperatures in this range are probably not high enough to cause purely thermal decomposition, but they are probably high enough to increase the rate of hydrolytic, oxidative, and secondary photochemical processes. See DEGRADATION.

Moisture. Water, in the various forms moisture assumes, can promote deterioration by physical means, such as by dissolving or by the consequences of expansion on freezing, and by chemical means such as hydrolysis and catalysis.

All the visible features in the atmosphere, except clouds, that are due to the various forms of water are called hydrometeors (12). Some important forms of hydrometeors that have been distinguished are: snow, rain, sleet, showers, drizzle, fog, mist, haze, hail, dew, and frost. Their relative importance in weathering has not been established.

Rain can be measured by a rain gage, which is generally an open-mouthed container whose catch is funneled to a device that measures it by weight or volume. Essentially the same device is used to measure snow or hail. *Dew* is sometimes measured with a drosometer, which weights the moisture that collects on a given area of a chosen material. Total dew measured on artificial surfaces is approximately equal to that on natural surfaces (16). Time of wetness has been measured principally through a response to moisture by change in (a) weight, (b) length, or (c) electrical resistance of a sensing element.

An invisible moisture factor in the atmosphere is *humidity*. It is known to affect weathering processes but its influence is often subtle. Consequently, its effects on polymers have not received sufficient study for any conclusions to be drawn.

Relative humidity is the most common measure of humidity in the air (17). It is the ratio, usually expressed in percent, of the pressure of water vapor in a gas to saturation pressure of water vapor at the temperature of the gas. Equivalently, it may be considered the ratio of the weight per unit volume of water vapor in a gas mixture and the weight of saturated vapor at the temperature of the gas mixture. Absolute humidity and specific humidity are other terms sometimes used. A useful concept is "dew-point," which is the temperature to which water vapor must be reduced in order to obtain a relative humidity of 100%.

Relative humidity is usually measured with wet-bulb psychrometers or with direct indicators called hygrometers. A psychrometer consists essentially of two similar thermometers with the bulb of one kept wet so that the cooling that results from evaporation makes it register a lower temperature than the dry one; the difference between the readings is a measure of the dryness of the atmosphere. Hygrometers are of various types: mechanical (using length change of absorbing material, such as hair), electrical (using resistance change of a polymer film), or gravimetric (using weight change due to absorbed moisture). Other hygrometers use changes in thermal conductivity, spectral-band intensity, index of refraction, pressure, or volume. A short critical review of the state of this art has recently appeared (18).

Rain is produced by cooling of moisture-laden air. The largest raindrops are about 0.2 in. (5 mm) in diameter, and this size is not exceeded because the drops break up on attaining their terminal velocity of about 18 miles (29 kilometers) per hour in falling. Some record rainfall rates are approximately 1 in. (26 mm) in 1 minute, in California; approximately 46 in. (117 cm) in 1 day, in the Phillipines; and approximately 264 in. (671 cm) in 1 month, in Assam. Obviously, the distribution of rainfall in time, and over the earth, is greatly variable.

Rainfall can best be understood by comprehending the processes leading to its occurrence (13). It is obvious that when there is not much water vapor in the air, as when the temperature is low, little precipitation can occur. High-rainfall regions are in the tropical latitudes and low-rainfall regions are along the high-pressure belts at 30° latitude. As expected, the mid-latitudes are moderately well supplied with rain and snow, and the areas within the polar circles have little precipitation. Of course, these general conditions are modified by the presence of land and water masses, such as mountains, deserts, oceans, and lakes, in a given location.

Snowfall as great as 882 in. has been recorded in mountainous areas of California. Since snow weighs 5–10 lb/cu ft (80–160 kg/cu m), heavy snowfalls mean great weight loads that could severely strain polymeric building materials.

Hailstones, often driven by wind, damage building materials because of the great impact they carry. In the central United States there are frequent storms with large hailstones; sizes up to 3 in. (7.6 cm) in diameter are not uncommon, and there is a record of a 4.5-lb (2-kg) hailstone in Spain. However, the vast majority of hailstones are relatively small. The precise damage attributable to hail is difficult to estimate, although recent laboratory tests with artificial hailstones indicate a significant problem for some roofing products (19).

Freezing of absorbed water in the body of materials can cause severe physical stresses, because ice expands about 10% on solidification. With alternate freezing and thawing, conditions resulting in physical fatigue can be established.

Air Contaminants. Industrial and urban atmospheres can be extremely damaging to polymeric materials because of the presence of airborne chemicals. In the United

States, the five most common primary air pollutants are carbon monoxide, sulfur oxides, hydrocarbons, nitrogen oxides, and particulate matter (20). The major sources of these contaminants are automobiles, industry, electric power plants, space heating, and refuse disposal. Some of these pollutants are more reactive chemically than others, but all contribute to outdoor deterioration of polymeric materials.

Only in the last two decades have there been many reliable data on concentrations of air pollutants. Now there are many measurements available from state and local agencies and the Federal National Air Pollution Control Administration (NAPCA). Recent data from the latter source (21) indicate the following approximate average concentrations of pollutants in various cities in the United States:

particles, 105 μg/cu m
sulfur dioxide, 0.024 ppm
nitrogen dioxide, 0.051 ppm

Average maximum concentrations reported were:

particles, 1254 μg/cu m
sulfur dioxide, 0.420 ppm
nitrogen dioxide, 0.198 ppm

The following paragraphs briefly summarize the nature of the contaminants, their sources, concentrations, chemical interactions, and methods of measurement. A good general reference on methods of measurement is the *ASTM Book of Standards* on water and atmospheric analysis (26).

Sulfurous pollutants in the air include sulfur dioxide, sulfur trioxide, hydrogen sulfide, sulfuric acid, and sulfate salts. About 20% of the sulfur dioxide in the air results directly from emissions. The remaining 80% is from hydrogen sulfide which is later converted to sulfur dioxide. Hydrogen sulfide is eventually oxidized in the air to sulfur, sulfur dioxide, sulfuric acid, and sulfate salts. Of the sulfur dioxide emitted as such more than 80% comes from combustion of sulfur-containing fuels, that is, about 16% of the total in the air at any given time comes from combustion of sulfur-containing fuels. Smelting of nonferrous metals and petroleum refining add the remaining 4% of man-made emissions of sulfur dioxide. The only apparent natural source of sulfur dioxide is volcanic gas, whose contribution is believed to be quite small.

Sulfur trioxide dissolves in water droplets to form sulfuric acid and salts such as ammonium sulfate. Rain and snow absorb sulfur compounds, resulting in more acidic precipitation. In the presence of nitrogen dioxide and hydrocarbons, sulfur dioxide photooxidizes to produce aerosol-containing sulfuric acid. Because of these interactions and the complex environmental sulfur cycle, most measurements of sulfurous pollutants are made of sulfur dioxide.

As indicated above, the average maximum concentration of sulfur dioxide in cities of the United States is about 0.42 ppm; however, instantaneous concentrations have been observed as high as several ppm.

The NAPCA measures sulfation with lead peroxide candles (22). This is an approximate measurement of the activity or dose of sulfur dioxide in the atmosphere at the sampling location. Other recommended measurement techniques (23) are the West-Gaecke and the hydrogen peroxide methods.

Three of the eight possible *oxides of nitrogen* are known to be important constituents of the atmosphere. Nitrous oxide, N_2O, is more plentiful than either of the other two (nitric oxide, NO, and nitrogen dioxide, NO_2). Nitrous oxide is relatively inert; it is not known to be man-made, and has a global concentration of about 0.25 ppm. Data of nitric oxide concentrations are too sketchy to support an estimate of average concentration, but it may be a natural trace constituent of air. Nitrogen dioxide appears to have a natural concentration in the range of a few parts per billion (ppb) or less.

The major sources of man-made nitrogen oxides are combustion processes. Combustion converts the nitrogen in the air to nitric oxide, which in urban air is oxidized slowly by oxygen, and quite rapidly by ozone, to nitrogen dioxide. In cities of the United States the average annual concentrations of nitrogen oxides are less than 0.1 ppm. Man-made sources including automobile emissions, cannot account for even a major part of the nitrogen dioxide content in our air. Lightning and biological processes involving organic nitrogen compounds are among the possible sources, but the question of the primary source remains basically unanswered.

Nitrogen dioxide is a strong absorber of ultraviolet radiation, and triggers smog-producing reactions. It can also combine with water vapor to form nitric acid, which in turn can react to form nitrate salts such as ammonium nitrate.

To measure nitrogen oxides, the Air Pollution Division of the Public Health Service has recommended the colorimetric Saltzman method (23).

Hydrocarbons in the air include methane and higher alkanes, olefins, benzene and other aromatics, terpenes, and soot (mostly carbon). The chief man-made sources of hydrocarbons are in the processing and use of petroleum products, gasoline being the major source. Other sources include incineration, evaporation of industrial solvents, and combustion of coal and wood. In cities in the United States the annual average concentration of the total hydrocarbons is several parts per million.

The unsaturated alkyl and aromatic compounds have been postulated as catalysts in photooxidation of polymers. Soot plays a dual role in outdoor deterioration: first, by simply precipitating on exposed surfaces and thus making them dirty, and, second, by interacting with moisture and other particles to form "acidic soot" described in the following paragraphs.

Particles, both liquid and solid, are extremely complex and are perhaps the most widespread of all the substances that are usually considered pollutants. Particles spring from a range of sources. Those larger than 10 microns in diameter come mainly from mechanical processes such as erosion, grinding, and spraying. Smaller particles between 1 and 10 microns are more numerous in the atmosphere and generally include the largest weight fraction of particles. These come from mechanical processes and also include industrial dusts and ash. Particles of even smaller size, between 0.1 and 1 micron, usually contain more combustion products than larger particles; that is, ammonium sulfate and combustion products begin to predominate in this size range along with aerosols formed in smog. Very little is known of the chemical nature of particles below 0.1 micron in size. The concentration of these very small particles is higher in urban areas and the difference seems to be due largely to combustion. Particles in the range of 0.1 to 10 microns account for most of the mass and a large fraction of the numbers of particles present in urban atmospheres.

Larger particles, such as fly ash and soil, cannot travel very far from their sources due to gravitational settling. Normally in urban air particles remain airborne for only a few days; however, depending on their size, they may remain airborne for several weeks. Another important variable is the height at which a particle is introduced into the air. Data on the concentration of airborne particles at remote points are scanty but atmospheric turbidity measurements indicate that the global level may be rising. Furthermore, the physicochemical behavior of the particles is not well understood because the particles are so diverse and because so little is known of them. The particles grow by condensation, adsorb and absorb vapors and gases, coagulate or disperse, and absorb and scatter light. Particles in the air must inter-react chemically, especially in the very small size range, below 0.1 micron, but such particle–particle reactions have been studied very little.

Even though particle–gas reactions have been studied somewhat more, their chemical reactions are still understood only to a very limited extent. Water vapor is a very important parameter in particle–gas reactions. For example, humidity is known to affect the reaction rate between gaseous ammonia and sulfuric acid mist in forming the salt ammonium sulfate. Rain removes particles from the air but this effect is negligible with particles less than 2 microns in diameter. Particles smaller than 2 microns can occur in rain water if the particles originate in clouds. Thus, we can see that the formation of "acidic rain" or "acidic soot" is quite logical.

Acidic soot is measured by the Air Pollution Division of the Public Health Service by the deterioration of nylon (25). After exposure, nylon hose samples are examined for breaks in the fiber and the number of breaks or defects on the sample is reported as the number of breaks per unit time. These investigators have also developed a soiling index based on measurements of dustfall (26) and wind-blown particulates (27). Particles larger in size than 20 microns settle within a dustfall collector in the same fashion in which they settle upon the surfaces of the earth. Because of their size, these particulates settle relatively close to their origin. Particles between 20 and 100 microns are sampled by the "sticky paper" method and are trapped by the adhesive surface. Microscopic examination of the collected samples gives quantitative estimates as well as qualitative indications of whether the particles are predominantly wind-erosion products, vegetal material, products of incineration, or industrial products.

The fifth common air pollutant, *carbon monoxide*, must not be overlooked. Carbon monoxide appears to be almost exclusively a man-made pollutant. The only significant source known is combustion processes in which carbon is oxidized partially to carbon monoxide instead of fully to carbon dioxide. The automobile is estimated to contribute more than 80% of carbon monoxide emissions with smaller amounts coming from other combustion processes, and still smaller amounts from biological reactions.

Carbon monoxide is essentially chemically inert. It apparently reacts with no other constituents of urban air to a significant degree. NAPCA data show a three-year average of about 7 ppm of carbon monoxide for off-street sites in five major U.S. cities. Instantaneous concentrations of 100 ppm and higher have been found, particularly near heavy auto traffic in restricted areas. The maximum global level of carbon monoxide is probably in the range of 0.1 ppm.

Monitoring of carbon monoxide is commonly done by photometric methods.

Finally, ozone, apart from its role in rapid oxidation of nitric oxide (NO) to nitrogen dioxide (NO_2), can have significant effects on unsaturated polymers. Ozone forms naturally by the irradiation of oxygen with ultraviolet radiation shorter in wavelength than 220 nanometers. Such radiation is present outside the earth's surface. Miscellaneous natural and man-made processes also contribute to the normal concentration of ozone, which is in the range of 0–100 ppm. There is some doubt as to the accuracy of reported concentrations of ozone owing to interference in the measurements by other oxidants. However, the occasionally noted characteristic odor of ozone is detectable even at concentrations of about 0.1 ppm.

The solubility and decomposition of ozone in water lead to suspicion, but difficult proof, of photohydrolytic ozonization. The reaction of ozone that is unquestioned is ozonization of unsaturated hydrocarbons, which leads to cracking and stiffening of polymeric materials such as rubber. This reaction is greatly accelerated by mechanical stress (see ANTIOZONANTS).

Artificial Weathering Devices

Useful outdoor life of many polymeric materials is now measured in decades. However, it is not practical to wait for ten or twenty years for evaluation of weatherability by outdoor exposure. Therefore, industry has searched for accurate rapid evaluation techniques for more than forty years. Recent reviews (29,30) of chemical and spectroscopic techniques for early detection of changes indicate moderate success. The weakness of this approach is in the difficulty of establishing *general* relationships between the properties that change early in deterioration (eg, concentration of carbonyl groups in polyethylene) and the ultimate performance properties (eg, modulus of polyethylene).

Spectroscopic techniques that have been used for early detection of weather-caused changes include infrared absorption, ultraviolet absorption, nuclear magnetic resonance, electron-spin resonance, and fluorescence. Chemical analysis has been conducted on solid oxidation products and on gases evolved during deterioration. Other approaches explored include molecular-weight averages and distributions, optical reflection and transmission, gel content, x-ray diffraction, thermal analysis, electrical properties, ultrasonics, and torsion pendulum analysis.

The major search has been for a device in which very brief exposure would result in the same deterioration produced by normal lengthy outdoor exposure. Acceleration

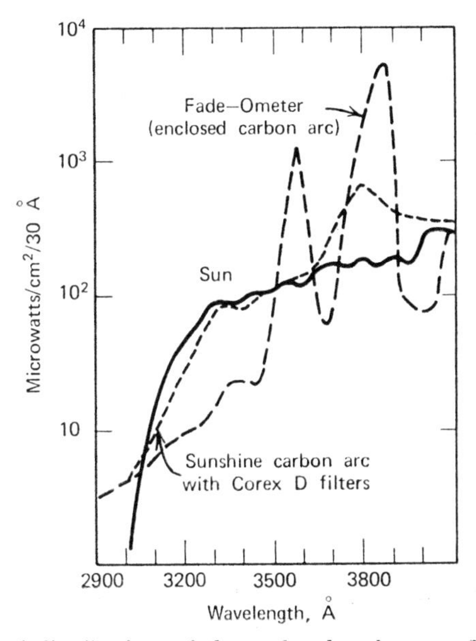

Fig. 1. Ultraviolet spectral distributions of the enclosed carbon arc Fade-Ometer and sunshine carbon arc compared with that of sunlight (3).

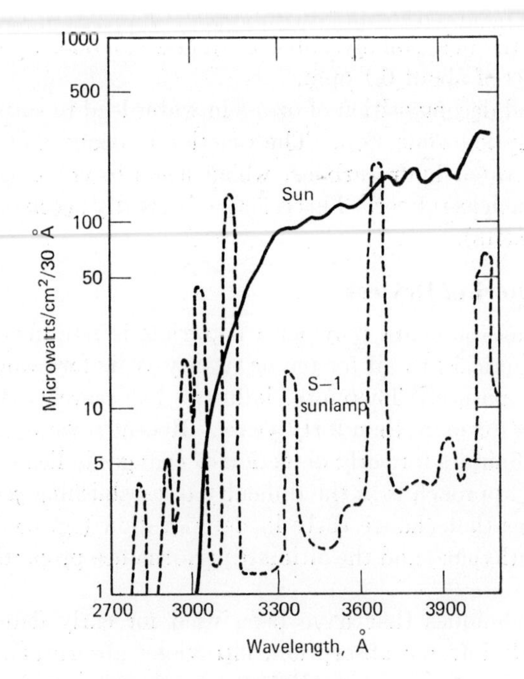

Fig. 2. Spectral-energy distribution of General Electric S-1 400-W mercury vapor sunlamp compared with that of sunlight (3).

factors of a thousand are sought, so that a week's accelerated exposure would accurately predict twenty years of outdoor performance. Commonly achieved acceleration factors are on the order of tenfold. Even tenfold acceleration is not very useful, however, unless the prediction is accurate, ie, unless correlation is achieved between the results of outdoor and accelerated exposures.

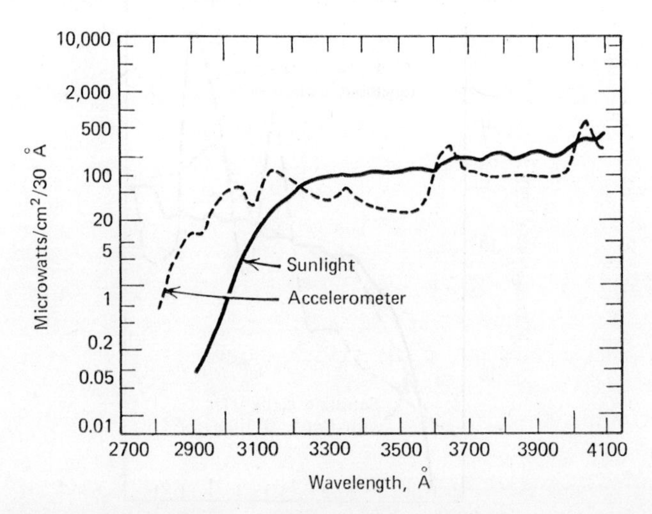

Fig. 3. Spectral-energy distribution of ultraviolet accelerometer compared with that of sunlight. This instrument incorporates a General Electric Type VA-11B quartz high-pressure 1200-W mercury arc and a Corex filter as an outer jacket (3). The accelerometer provides 106 W of ultraviolet, 131 W visible light, and 963 W of infrared radiation.

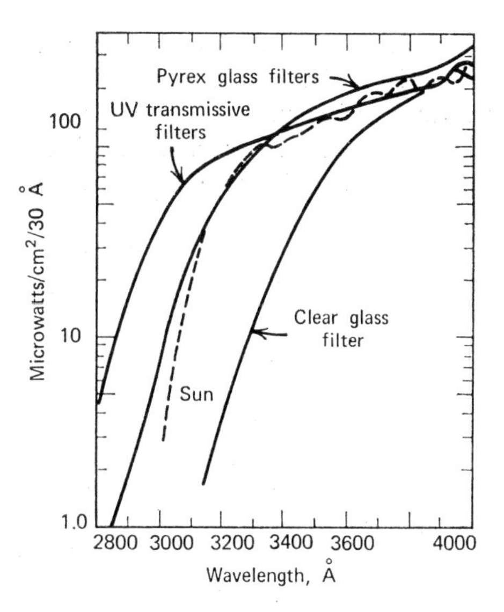

Fig. 4. Spectral-energy distribution of 600-W Osram xenon arc through various types of filters compared with sunlight at the earth's surface (3).

Solar Simulation. The heart of the search for the optimum device lies in simulating the important effects of sunlight on polymeric materials. "Artificial sunlight" has been sought primarily in carbon arcs (Fig. 1), mercury arcs (Figs. 2 and 3), fluorescent lamps, and xenon arcs (Fig. 4). Each of these sources contains actinic radiation, ie, radiation having marked photochemical action. Actinic radiation is found in the green, blue, violet, and, especially, ultraviolet regions of the solar spectrum. The characteristics of sun and sky radiation and of different types of laboratory devices have been reviewed (3,14). The various actinic effects of the accelerated test sources can be related to their specific emission characteristics through information on the activation spectra of the polymeric materials. (Activation spectra quantitatively describe photodegradation as a function of incident wavelength for the source of interest.) Xenon-arc emission has been found to give the closest simulation to actinic solar radiation. Therefore, a xenon arc is the "artificial sunlight" of choice in weathering studies. A reasonable second choice is the sunshine carbon arc. The recently developed combination of fluorescent sunlamps and blacklights (3) offers good simulation of solar radiation in the 290–360 nanometer range only.

It has been shown (31) that exposure of plastics to xenon-arc radiation produces results which correspond to a large degree with those obtained in nature, and acceleration factors up to twenty have been achieved. Higher acceleration factors may be achieved by increasing irradiance on the specimen and/or by increasing temperature. It must be kept in mind, however, that different modes of degradation will dominate under different conditions; thus, only those environmental conditions should be changed that affect the property of interest.

Heat and Moisture. In small closed chambers such as those common in artificial weathering devices the control of temperature and moisture factors is relatively simple. Either constant or cycled temperatures are readily obtained with thermostats. Similarly, hygrometers allow measurement and control of humidity. Rainfall is easily simulated by a fine spray of distilled, deionized water. However, since variation and interaction of these factors is complex, the selected heat and moisture conditions are usually arbitrary and not closely related to the weather in any given location.

"Black-panel" temperature of about 60°C is often selected for xenon-arc exposures (32). A "black-panel" thermometer consists of a thin stainless-steel panel painted with black glossy enamel, to which is fastened a thermocouple or dial-type thermometer.

Common water-spray cycles are totally arbitrary, and are not related to climatic conditions. Humidity selection ranges from ambient to saturation (dew) point.

High-humidity exposure with cycles of light and dark allows for dew formation on the exposed polymer surface. Such conditions have been found to have excellent correlation with exposures in various climates.

Air Contaminants. The effects of gaseous airborne chemicals may be studied in the laboratory either by enclosing the test specimen in a small glass or quartz envelope containing the gas, or by filling the entire test chamber with the gas. In either case, the exposure may be done under dynamic gas-flow, or static, conditions. Measuring and controlling the small gas concentrations typical of our environment present a difficult but possible task. An old example of such a study is the acceleration of cracking of stretched rubber strips by exposure to ozone. In recent studies, polyethylene, poly(vinyl chloride), polystyrene, poly(ethylene terephthalate), and poly(vinyl fluoride) discolored and/or embrittled within three days' exposure to filtered mercury-arc radiation in 13,000 ppm of sulfur dioxide. At lower concentrations of sulfur dioxide (18 ppm SO_2), the deterioration was much slower.

Such results clearly demonstrate degradative effects of air pollutants, but only a few papers have been published on the fundamental processes involved (33).

Equatorial Mounts—Outdoors. Exposing specimens on equatorial mounts which "follow the sun" allows maximum available sunlight to reach the specimens. Mirrors have been placed on such a mount to intensify the irradiation tenfold; forced air is blown over the specimens to prevent overheating. Such devices bear the acronym EMMA: Equatorial Mount with Mirrors for Acceleration (29). When water is sprayed over the specimen to further reduce overheating and allow for hydrolytic effects, the device becomes known as EMMAQUA.

Degradation is accelerated in these exposures, and correlation of the results with normal exposures is often quite good. However, the acceleration factor varies from material to material and from property to property, since radiation intensity only is accelerated and the balance of nature is upset.

Correlation and Prediction

It has been stated (29) that an ideal procedure for analyzing and predicting weatherability would be to: (a) establish the effects of specific weather factors for given properties of given materials by using controlled artificial environments; (b) analyze the weather at a given place in terms of these weather factors; and (c) compute the expected results of exposure of the given material to the given weather by suitable mathematical models.

This basic approach has been independently used by several workers (29,34,35) in the last few years, and has proved to be a vastly improved method of evaluating and predicting weatherability. It has been shown that weathering can be characterized statistically as a "wear-out" process. Wear-out failure is due to the gradual exhaustion of physical or other properties which in some way are directly related to the length of life. Thus, quantitative description of weathering (wear-out) data can be accomplished by an exponential model (36,37). Parameters of the exponential model have physical significance and can be related to weather factors causing the deterioration.

Models for Property Decay. Various forms of exponential models have been proposed and used to describe the change in properties of polymeric materials exposed to weathering.

Changes in concentration of reacting substances in chemical kinetics are described by equations such as $Y = A(1 - e^{-kt})$ where Y is concentration of the reaction product at time t, A is initial concentration of reactant, and k is the rate constant. In a study of accelerated weathering the equation $Y = Ae^{B(t - C)}$ was used, where Y is the value of the property at time t, and A, B, and C are constants found to be dependent on exposure parameters (39). The expression $Y = K - Ae^{-Bt^2}$ was used in paint evaluation studies; Y is red reflectance of paint at time t, and A, B, and K are constants (35). The form of these equations and the assumption that the deterioration of a polymer is a gradual process over time suggested that a general distribution function might represent change in properties as a function of exposure time. Of course, property change is actually a function of the weather, which is different in different locations, whereas exposure time is not different.

One of the most general mathematical forms investigated results from the Weibull distribution. It is frequently used with success as the basis for time-to-failure models for wear-out of electron tubes, relays, and ball bearings. A Weibull-type model has been found useful for describing the decrease of ultimate tensile elongation of plastics caused by weathering (eq. 1). P is the value of the property at time t, and the values of

$$P = b_1 \exp\left[-\left(\frac{t + b_2}{b_3}\right)^{b_4}\right] + b_5 \tag{1}$$

b are parameters determined by nonlinear regression. The five parameters completely describe the degradation curve, and furthermore, have physical significance: b_1 is related to maximum property value ($b_1 + b_5$ = maximum); b_2 is related to pre- or post-aging ($b_2 = 0$, for most cases); b_3 is related to "characteristic life" (time to reach 37% of initial property value); b_4 is related to the shape of the curve ($b_4 \leq 1$ indicates rapid initial decay; $b_4 > 1$ indicates induction period); b_5 is related to the asymptotic value of the property. When both b_2 and b_5 are zero, the model is identical to the Weibull function.

One more parameter for predicting and analyzing weatherability, and the most useful, can be obtained from this model. "Characteristic life," L, is the time that it takes the measured property to decay to 37% of its original value. Parameter b_3 is the characteristic life when b_2 is zero, which is in most cases, and b_5 is zero. For any case in which b_2 is zero, b_3 is a very good approximation of characteristic life for small b_5.

The Weibull-type model was fitted to data from exposures in Arizona, Florida, and Washington, D.C., of polyethylene, poly(ethylene terephthalate), and various types of poly(vinyl chloride). Several methods of model fitting were investigated: analog computers, plotting on probability papers, hazard plotting, and nonlinear least squares analysis. The latter was the best method when a computer program was employed with Marquardt's algorithm (38). Criteria used to judge were:

1. Reality of the parameter values. Since the parameters represent physical reality, the proper range of the parameter values could be judged from the property data.

2. Statistical measures of the fit, such as unexplained variance and standard error of the fit.

3. Standard error of the parameters, including 95% confidence intervals.

Results of the fits were good quantitative confirmation of qualitative knowledge of weathering of these materials. Generally, "characteristic life" of the materials was shortest in Arizona and longest in Washington, D.C. For example, a 60-mil polyethylene film showed the "characteristic life" of its ultimate tensile elongation to be 25 months in Arizona and 167 months in Washington, D.C. Typical 60-mil poly(vinyl chloride) films had values of 1–3 months in Arizona, 6–12 months in Florida, and 9–13 months in Washington, D.C. In many cases, "characteristic life" of the ultimate tensile elongation property was a year or less. Since the plastics were still serviceable for some time after this, it is apparent that ultimate tensile elongation was a good early indication of later physical failure of the materials.

Thus, we have an accurate, compact, and precisely defined description of property change as a function of exposure time. An immediate use of mathematical models could be in conjunction with indoor simulated weathering devices. Much better simulators of outdoor weathering could be designed and constructed by comparing simulator results with mathematical models derived from outdoor exposures. Once the experimental results of accelerated weathering devices are correlated with mathematical models of property changes observed outdoors, models for new materials can be constructed directly from artificial weathering data collected indoors. Thus, in a very short time the outdoor behavior of a new composition could be predicted. This approach would reach its optimum in comparisons between materials of the same general family.

Relation of Time-to-Failure to Weather. In a study described above, deterioration of properties is described as a function of both time and ultraviolet radiation. An alternative approach is to relate the parameters of the Weibull-type model to the weather.

Parameters describing the weather were detailed in the first section of this article. Briefly, it is generally considered that significant weather factors in weathering of polymeric materials are: *solar radiation* (actinic and thermal), *temperature, moisture* (especially humidity and rainfall), *oxygen, air contaminants* (especially O_3, NO_x, SO_x, particulates), and ***wind***. Stress and biological factors are nonclimatological, and are not considered here. See Biological Degradation under DEGRADATION; FRACTURE.

Data on the above variables can be treated in several ways. The problem is to reach optimum balance between a minimum number of variables and an adequate description of weather variation. It is frequently stated in discussions of weatherability that *weather* is not reproducible—from day to day, hour to hour, or year to year. This is certainly true. On the other hand, one can generalize that *climate* in regions of the country is quite stable—on the average, Arizona is hot and dry, Florida is warm and moist, and so on. Thus it is reasonable to describe the climates for exposure in terms of average weather. In the example used here, annual averages were selected for total solar radiation (langleys), ultraviolet radiation ("Coblentz langleys"), relative humidity, inches of rainfall, and sol-air temperature. The constancy of climate is confirmed by the close averages of these factors over several years in Arizona, Florida, and Washington, D.C.

A relationship can be postulated between the fitted parameters from the Weibull-type model (the various b factors in equation 1) and the weather variables. For simplicity, a linear combination of weather variables can be used (equation 2, where the

$$b_i = \sum_j C_{ij} W_j \tag{2}$$

values of C are coefficients of the fit and the values of W are weather variables). For example,

$$b_i = C_i + C_{iL} L + C_{iU} U + C_{iH} H + C_{iR} R + C_{iT} T + E_i \tag{3}$$

where the various different b_i are the parameters fitted to the ultimate tensile elongation data, L is total solar radiation (langleys), U is ultraviolet radiation (Coblentz langleys), H is relative humidity, R is inches of rainfall, and T is sol-air temperature. E_i is the residual or unexplained variation.

The climatological data were used in the five-variable linear model (eq. 3) in their standardized form. In equation 4 W is the standardized weather variable, W' is the

$$W = \frac{W' - \overline{W}}{s} \tag{4}$$

observed weather variable, $\overline{W}$ is the average of the weather variable for the time under study, and s is the standard deviation from the mean. By using such dimensionless weather variables, the solution is unaffected by whether one uses, for example, temperature in °C, °F, or °K.

This approach is analogous to the second stage of Kamal's analysis (39). That is, a mathematical relationship is postulated between the fitted parameters (b's) and the exposure variables.

The simultaneous equations relating the values of b and "characteristic life" to the climatic variables were solved by computerized multiple linear regression. There are five basic methods (40) for selecting the climatic factors and determining the coefficients: all possible regressions, backward elimination, forward selection, stagewise regression, and stepwise regression. The stepwise regression procedure was adopted as optimal since it consists of forward selection of the most significant variables with a "backward glance" to eliminate statistically insignificant variables. The variable added is the one which makes the greatest reduction in the unexplained variance. Variables are automatically removed when they become statistically insignificant, as calculated by the F test. A computer program is available for carrying out this stepwise regression (41). This program computes a sequence of multiple linear regression equations in a stepwise manner. At each step, one variable is added to the regression equation. The variable added is the one which makes the greatest reduction in the error sum of squares. Variables are automatically removed when their F values become too low. The algorithm for the program is analogous to the stepwise regression procedures of Draper and Smith (40).

The results of the multiple linear regressions indicated that each of the values of b in the Weibull-type model has a different relation to the weather variables. Since the b's are mathematically and implicitly related to one another, interpreting their relation to climatic factors is difficult. However, such relations may be useful for synthesizing deterioration curves for climates other than the one in which exposures were actually made.

"Characteristic life" was found to be closely related by the linear model to a small number of climatic variables. Three weather variables, or less, explained most of the variance in "characteristic life" of the tensile elongation of the plastic under consideration. The relation between "characteristic life" and average climate indicates that the most significant factors for this property (a) for polyethylene are temperature and ultraviolet radiation; (b) for poly(ethylene terephthalate) are humidity, temperature, and rain; and (c) for poly(vinyl chloride) are temperature and ultraviolet radiation. Rainfall is a significant, but less important, variable for poly(vinyl chloride). Rainfall and humidity appear to have separate and distinct effects on poly(ethylene terephthalate).

These results are in excellent agreement with knowledge of the effects of these factors on the degradation of these polymers. The relative nonsignificance of moisture factors in the weathering of hydrocarbon and chlorinated hydrocarbon polymers agrees well with experience. Conversely, the significance of humidity and rainfall in degrading the polar poly(ethylene terephthalate) is encouraging. Ultraviolet radiation and temperature are clearly confirmed as highly significant weather factors.

Usefulness of Models. It has been demonstrated that exponential models, especially those of the Weibull type, fit the process of weathering of polymeric materials. The models represent the physical phenomena well. Furthermore, simple relationships describe the significance of climatic variables in determining the values of the parameters in the exponential models. These models are useful for describing: (a) the relation between exposure time and property value, and (b) the relation between time to failure of the material and the weather factors causing the failure. These relations allow quantitative correlation between results of exposure in various climates and between outdoor exposures and simulated weathering tests. Given frequent measurements of a property of a plastic at a small number of exposure sites, behavior of the plastic at other sites can be predicted by such mathematical techniques with a known degree of confidence. The models also allow prediction of failure before it actually occurs. The more measurements are made of the property under study, the more confident the prediction of behavior.

Optimum conditions for operating simulated weathering devices can be determined by comparing laboratory-derived models with those derived from outdoor exposures. Finally, these models may present the basis for development of a unified theory needed to understand the weathering of polymeric materials.

Bibliography

1. F. Daniels, *Direct Use of the Sun's Energy*, Yale University Press, New Haven, Conn., 1964.
2. D. M. Gates, *Sci.* **151**, 523–529 (1966).
3. R. C. Hirt and N. Z. Searle, *Appl. Polymer Symp.* **4**, 61–83 (1967).
4. R. W. Singleton and P. A. C. Cook, *Textile Res. J.* **39**, 43–49 (1969).

5a. W. W. Coblentz, *Bull. Am. Meteorological Soc.* **30**, 204 (1949).

5b. R. W. Singleton, R. K. Kunkel, and B. S. Sprague, *Textile Res. J.* **35**, 228–237 (1965).

6. S. S. Visher, *Climatic Atlas of the United States*, Harvard University Press, Cambridge, 1954.
7. Negretti and Zambra, *Scientific Facts & Data*, Negretti & Zambra, London, 1940.
8. B. L. Garner and P. J. Papillo, *Ind. Eng. Chem., Prod. Res. Develop.* **1**, 249 (1962).
9. W. C. Cullen, *Natl. Bur. Stds. Tech. Note* **231**, 1963.
10. J. L. Threlkeld, *Thermal Environmental Engineering*, Prentice-Hall, Englewood Cliffs, N.J., 1962.
11. *ASHRAE Handbook of Fundamentals*, American Society of Heating, Refrigerating and Air-Conditioning Engineers, New York, 1967.
12. H. R. Byers, *General Meteorology*, McGraw-Hill Book Co., New York, 1944.
13. W. J. Humphreys, *Ways of the Weather*, Jacques Cattell Press, Lancaster, Pa., 1942.
14. J. E. Clark and C. W. Harrison, *Appl. Polymer Symp.* **4**, 97–110 (1967).

15. *Handbook of Geophysics,* Macmillan Book Co., New York, 1961, Chap. 16.
16. A. Wexler, ed., *Humidity and Moisture: Measurement and Control in Science and Industry.* Vol. 2; *Applications,* Reinhold Publishing Co., New York, 1965.
17. A. Wexler and W. G. Brombacher, "Methods of Measuring Humidity and Testing Hygrometers," *Natl. Bur. Stds. Circ.* **512,** 1951.
18. C. F. Quinn, *Test Eng.* **19,** 6–24 (July 1968).
19. S. H. Greenfeld, "Hail Resistance of Roofing Products," *Natl. Bur. Stds., Building Sci. Ser.* **23,** 1969.
20. *Cleaning Our Environment/The Chemical Basis for Action,* American Chemical Society, Washington, D.C., 1969.
21. *Air Quality Data from the National Air Sampling Network and Contributing State and Local Networks, 1964–65,* U.S. Department of Health, Education, and Welfare, Washington, D.C., 1966.
22. N. A. Huey, *J. Air Pollution Control Assoc.* **18,** 610–611 (1968).
23. *Selected Methods for the Measurement of Air Pollutants,* Division of Air Pollution, Public Health Service, Dept. of Health, Education, and Welfare, PHS Publication No. 999-AP-11, May 1965.
24. W. H. Perry and E. C. Tabor, *Arch. Environmental Health* **4,** 254 (1962).
25. L. Greenburg and M. B. Jacobs, *Am. Paint J.* **39,** 64–78 (1955).
26. Part 23, "Water; Atmospheric Analysis," *1970 Book of ASTM Standards,* American Society for Testing and Materials, Philadelphia.
27. C. W. Gruber and G. A. Jutze, *J. Air Pollution Control Assoc.* **7,** 115–117 (1957).
28. L. S. Jaffe, *J. Air Pollution Control Assoc.* **17,** 38 (1967).
29. M. R. Kamal and R. Saxon, *Appl. Polymer Symp.* **4,** 1–28 (1967).
30. V. E. Gray and B. C. Cadoff, *Appl. Polymer Symp.* **4,** 85–95 (1967).
31. V. Schafer, *Appl. Polymer Symp.* **4,** 111–118 (1967).
32. ASTM E 239-70, "Recommended Practice of Operating Light- and Water-Exposure Apparatus (Xenon-Arc Type) for Exposure of Nonmetallic Materials," *1970 Annual Book of ASTM Standards,* American Society for Testing and Materials, Philadelphia.
33. H. H. G. Jellinek and F. Flajsman, *J. Polymer Sci.* [A-1] **8,** 711 (1970).
34. J. E. Clark, unpublished results.
35. W. H. Daiger and W. H. Madson, *J. Paint Technol.* **39,** 399 (1967).
36. W. R. Buckland, *Statistical Assessment of the Life Characteristic,* Hafner Publishing Co., New York, 1964.
37. G. J. Hahn and S. S. Shapiro, *Statistical Models in Engineering,* John Wiley & Sons, Inc., New York, 1967.
38a. D. W. Marquardt, "An Algorithm for Least Squares Estimation of Nonlinear Parameters," *J. Soc. Ind. Appl. Math.* **11,** 431–444 (1963).
38b. D. W. Marquardt and R. M. Stanley, "NLIN 2—Least Squares Estimation of Nonlinear Parameters," A Computer Program in Fortran IV Language; IBM Share Library, Distribution Number 3094-01, Revision of August 1966 (Successor to Distribution No. 1428 and 3094).
39. M. R. Kamal, *Polymer Eng. Sci.* **6,** 333 (1966).
40. N. R. Draper and H. Smith, *Applied Regression Analysis,* John Wiley & Sons, Inc., New York, 1966.
41. W. J. Dixon, Biomedical Computer Programs Section BMD-02R, Health Science Computing Facility, University of California at Los Angeles, 1965.

Joseph E. Clark
National Bureau of Standards

WEIGHT-AVERAGE MOLECULAR WEIGHT. See MOLECULAR-WEIGHT DETERMINATION

WELDING

Welding is one of the most commonly used methods of joining thermoplastic materials. As with metals, welding of thermoplastics is accomplished by application of localized heat sufficient to produce fusion of the areas to be joined. The major difference among the various techniques is in the method of applying heat to the materials. Several different procedures have been developed including hot-gas welding, heated-tool welding, induction welding, friction welding (also known as spin welding), and ultrasonic welding (1). See Bonding under ADHESION AND BONDING; DIELECTRIC HEATING; and ULTRASONIC FABRICATION.

1. S. J. Kaminsky and J. A. Williams, *Handbook for Welding and Fabricating Thermoplastic Materials*, Kamweld Products Co., Norwood, Mass.

WET LAY-UP. See REINFORCED PLASTICS

WETTING. See ADHESION AND BONDING; SURFACE PROPERTIES

WHISKERS. See FIBERS, INORGANIC

WIRE AND CABLE COVERINGS

From the very beginning of the electrical industry, when the first underground lighting cables on the Cornell University campus were insulated with layers of muslin that had been impregnated with tallow, polymers have been used for insulating electrical wires and cables. With the exception of a small amount of minerals (such as asbestos (qv) and mica), a few ceramics, and metallic coverings (such as in the familiar "BX" building wire and the once ubiquitous lead sheath), wire and cable coverings have been largely organic. In recent years, the use of plastics has proliferated so extensively that one major electrical manufacturer now spends more money for wire and cable plastics than for any other raw material except copper.

This article is organized primarily along product lines according to the design properties desired in the final product, with some further subdivision according to method of manufacture. In addition, the discussion differentiates between the insulation and the jacket, or sheathing. Insulation normally refers to the primary dielectric material applied to individual conductors; sheathing is the outer protection which may serve a number of functions, such as providing resistance to abrasion, moisture, and chemical attack, in addition to binding many conductors together into a cohesive unit. See also ELECTRICAL APPLICATIONS; ELECTRICAL PROPERTIES.

Products and Properties

The diversity of products described as "wire and cable" can be subdivided as follows: magnet wire, hook-up wire, building wire and cable, communications cable, and power cable. The following definitions have been extracted from *United States of America Standards Institute: Definitions of Electrical Terms* (see Ref. 1).

Cable: either a stranded conductor with or without insulation and other coverings (single-conductor cable), or a combination of conductors insulated from one another (multiple-conductor cable).

Cable Core: the portion of an insulated cable lying under the protective covering or coverings.

Cable Filler: the material used in multiple-conductor cables to occupy the interstices formed by the assembly of the insulated conductors, thus forming a cable core of the desired shape (usually circular).

Cable Sheath: the protective covering applied to cables.

Jacket: a rubber or synthetic covering, sometimes fabric-reinforced, over the insulation, core, or sheath of a cable.

Semiconducting Jacket: a jacket having a sufficiently low resistance so that its outer surface can be kept at substantially ground potential by a grounded conductor in contact with it at frequent intervals.

Serving of a Cable: a wrapping applied over the core of a cable before the cable is leaded, or over the lead if the cable is armored. Materials commonly used for serving are jute or cotton. The serving is for mechanical protection and not for insulation purposes.

Conductor: a wire or combination of wires not insulated from one another, suitable for carrying electric current.

Insulation of a Cable: that part that is relied upon to insulate the conductor from other conductors or conducting parts or from ground.

Taped Insulation: insulation of helically wound tapes applied over a conductor or over an assembled group of insulated conductors. When successive convolutions of a tape overlie each other for a fraction of the tape width, the taped insulation is "lap wound." This is also called "positive lap wound." When a tape is applied so that there is an open space between successive convolutions, this construction is known as "open butt" or "negative lap wound." When a tape is applied so that the space between successive convolutions is too small to measure with the unaided eye, it is a "closed butt taping."

Strip Process Insulation: insulation consisting of one or more strips of unvulcanized thermosetting material folded around a conductor and vulcanized after application.

Design Properties. Cable and wires are probably utilized in a wider variety of conditions than virtually any other electrical component. Thus the range of properties required extends over almost the entire spectrum available to the polymer scientists. Major design properties influencing the choice of material can be separated into four major areas: thermal, electrical, mechanical, and chemical.

Thermal properties (qv) of cable and wire covering which must be considered typically include softening point, freezing, and melting points, coefficient of thermal expansion, thermal conductivity, specific heat, thermal aging, and low-temperature flexibility. Thermal properties are becoming increasingly important in military (qv), and aerospace uses (qv), and in many other areas where wires and cables are subjected to excessive temperatures during service.

Electrical properties (qv) influencing plastic material selection include insulation resistance, conductivity, dielectric strength, dielectric constant, dielectric loss, capacitance, inductance, power factor, impulse strength, corona level, and surface and volume resistivity.

Chemical properties which must be considered include flammability, chemical stability, stress-cracking resistance, and resistance to acid, alkalies, oils, and solvents. Of all of these characteristics, stress-cracking resistance is perhaps of greatest commercial importance because of the tremendous tonnage of plastics utilized in jacketing communications cable and buried power cable. See also CHEMICALLY RESISTANT POLYMERS; FIRE RETARDANCY; FRACTURE.

Mechanical properties (qv) of plastics which must be considered in cable design include abrasion resistance (qv), flexibility, shear strength, hardness (qv), adhesion (qv), coefficient of friction (see SURFACE PROPERTIES), cold flow, fatigue, and tear resistance. The degree of porosity is vital for many applications. In some cases, no porosity whatsoever can be permitted, whereas some applications of foamed insulation may be up to 80% porous.

Other design characteristics that must be taken into consideration include color, color fading, resistance to degradation by radiation and resistance to weather (see also RADIATION-INDUCED REACTIONS; RADIATION-RESISTANT POLYMERS; WEATHERING). The following is a discussion of the application of these properties to the individual characteristics demanded in each of the major product categories.

Magnet Wire

Magnet wire is manufactured by the immersion coating process which follows essentially the same procedure as candlemaking. The filament (wire) is drawn through a liquid and some of the liquid adheres to the filament. When this dries through the evaporation of volatile constituents, a layer of insulating material is formed on the out-

Table 1. Insulating Materials for Magnet Wire

Material	Important properties	Major use
acrylic resin (copolymer of acrylonitrile and an acrylate together with phenolic resin)	resistance to solvents and refrigerants	hermetically sealed motors where refrigerant contacts the wire
solderable acrylic resin	solders easily at 450°C by melting film	motor windings and as replacement for a variety of other enamels
epoxy resins (high epoxide equivalent with urea–formaldehyde modifying resins)	excellent moisture resistance	oil-filled transformers
epoxy resin with self-bonding overcoat (overcoat is combination of epoxy and poly(vinyl formal) resins)	good windability and bondability	self-supporting coils
poly(hexamethylene adipamide)	excellent windability	general magnet wire where moisture is not a problem
oleoresinous plain enamel (cured varnish with a natural resin and a drying oil)	good high-speed windability	paper-filled coils
polyamide-imide (based on trimellitic anhydride)	toughness, high-temperature resistance, chemical resistance	high performance coils and windings
polyester (primarily synthetic resin based on polyesters of terephthalic acid and polyhydric alcohol with or without a superimposed polyester film; the polyesters may be modified with other resins)	depending on type of modifying resin and outer coating, is resistant to chemical attack; good windability; good abrasion resistance, thermal stability and heat-shock properties	general magnet wire
polyimide	excellent heat resistance (220°C)	hermetically sealed motors
polytetrafluoroethylene	heat resistance	electrical devices intended for high-temperature use
polyurethan	solderability without removal of insulation	high-speed magnet production
poly(vinyl formal)	excellent windability	in Class "A" electrical devices

side of the filament. The process can be repeated to build the thickness to the desired level. In practice, the coated wire is normally passed through an oven to hasten the evaporation of solvents and/or curing of the insulation. Table 1 lists the major insulating materials used in this process (2).

Hook-Up Wire

Hook-up wire has been described by many as "black-box wire." One author, in using almost an entire page to define it (3), stated, " . . . hook-up wires are a collective classification for thin walls and odd-balls, an insulated wire catch-all, the ultimate dust bin of insulation specialists who categorize wire and cables as to types, classes, or applications."

End-use requirements for hook-up wire are so diverse that an enormous range of polymer properties may be considered. For example, back-panel wiring for computers and information processing equipment is usually subjected only to low voltages and room temperatures, but must have high abrasion resistance and excellent uniformity for proper performance in the new automated wiring machines. Semirigid poly (vinyl chloride) is still the most widely used material for this application.

At the other side of the end-use spectrum may be found the polyimide coatings. Polyimide has the advantages of a melting point above 480°C and electrical properties among the most desirable of all insulating materials. Its primary drawback lies in the difficulty in manufacture. The very properties that make it so attractive to the user preclude normal fabrication techniques. Some success has been achieved in the application of thin (less than 1 mil) coatings to magnet wires and over a wire coated with polytetrafluoroethylene or fluorinated ethylene–propylene polymers. Polyimide in tape form may also be applied helically, and heat sealed. To achieve this requires that the polyimide film be laminated with fluorinated ethylene–propylene polymer film. Wire in this form offers substantial weight savings (3).

A table outlining some of the more important materials for hook-up wire may be found under ELECTRICAL PROPERTIES.

Building Wire and Cable

Insulations utilized in building wire and cable have profited substantially from continued development in the plastics industry. Some years ago, natural rubber was extensively used; however, sufficient moisture resistance was difficult to attain, and such cables had limited mechanical strength, abrasion resistance, and high-temperature resistance. Improvements came through the introduction of poly(vinyl chloride) (PVC), styrene–butadiene rubber (SBR), and butyl rubber. A nylon jacket is sometimes applied over a PVC insulation. In some applications chlorosulfonated polyethylene or crosslinked polyethylene have made possible the elimination of the outer protective jacket formerly required. Table 2 lists most of the current constructions (4).

Communications Cable

The development of multiconductor communications cable has proceeded apace with the growth of the communications industry. In spite of the proliferation of microwave towers and satellite transmission, the production of communications cable continues to grow; annual outputs are now measured in many billions of conductor feet.

Most multiconductor communications cable consists of the following five elements:

1. Twisted pair. Two conductors insulated from each other, making a voice circuit.

Table 2. Types of Building Wire and Cable

Trade name	Type letter	Insulation	Outer covering
code	R	code rubber	moisture-resistant, flame-retardant, nonmetallic covering[a]
heat-resistant	RH RHH	heat-resistant rubber	moisture-resistant, flame-retardant, nonmetallic covering[a]
moisture-resistant	RW	moisture-resistant rubber	moisture resistant, flame-retardant, nonmetallic covering[a]
moisture- and heat-resistant	RHW	moisture- and heat-resistant rubber	moisture-resistant, flame-retardant, nonmetallic covering[a]
heat-resistant latex rubber	RUH	90% unmilled, grainless rubber	moisture-resistant flame-retardant, nonmetallic covering
moisture-resistant latex rubber	RUW	90% unmilled grainless rubber	moisture-resistant, flame-retardant, nonmetallic covering
thermoplastic	T	flame-retardant thermoplastic compound	none
moisture-resistant thermoplastic	TW	flame-retardant, moisture-resistant thermoplastic	none
heat-resistant thermoplastic	THHN	flame-retardant, heat-resistant thermoplastic	nylon jacket
moisture- and heat-resistant thermoplastic	THW	flame-retardant, moisture- and heat-resistant thermoplastic	none
moisture- and heat-resistant thermoplastic	THWN	flame-retardant, moisture- and heat-resistant thermoplastic	nylon jacket
thermoplastic and asbestos	TA	thermoplastic and asbestos	flame-retardant, nonmetallic covering
thermoplastic and fibrous braid	TBS	thermoplastic	flame-retardant, nonmetallic covering
synthetic heat resistant	SIS	heat-resistant rubber	none
mineral-insulated metal-sheathed	MI	magnesium oxide	copper
silicone–asbestos	SA	silicone rubber	asbestos or glass
fluorinated ethylene–propylene copolymer	FEP	fluorinated ethylene–propylene copolymer	none
	FEPB	fluorinated ethylene–propylene	glass braid or asbestos braid
varnished cambric	V	varnished cambric	nonmetallic covering or lead sheath
asbestos and varnished cambric	AVA	impregnated asbestos and varnished cambric	AVA-asbestos braid or glass
asbestos and varnished cambric	AVL	impregnated asbestos and varnished cambric	AVL-lead sheath
asbestos and varnished cambric	AVB	impregnated asbestos and varnished cambric	flame-retardant cotton braid (switchboard wiring) flame-retardant cotton braid
asbestos	A	asbestos	without asbestos braid
asbestos	AA	asbestos	with asbestos braid or glass
asbestos	AI	impregnated asbestos	without asbestos braid
asbestos	AIA	impregnated asbestos	with asbestos braid or glass
paper	none	paper	lead sheath

[a] Outer covering is not required over rubber insulations which have been specifically approved for the purpose.

2. Cable core. Several pairs (from 6 to as many as 4000) stranded together.
3. Corewrap. A helically or longitudinally applied tape over the core to provide mechanical and thermal protection during manufacture and dielectric protection when the cable is in service.
4. Shield. Metal to provide electrostatic shielding and mechanical protection.
5. Jacket. The outer protective covering.

The two major electrical design considerations are conductor resistance, and capacitance; the latter is determined by the thickness of insulation, the dielectric constant, and the amount of air present in the cable core. Paper, polyethylene, polypropylene and poly(vinyl chloride) are the four major materials used for insulation.

Thin paper ribbon, wrapped helically around the bare conductor, has been used successfully for eighty years. Successful application of paper ribbon requires that it be preconditioned at about 40–50% relative humidity to provide the necessary pliability required for the wrapping operation. It is maintained in the moisturized condition until immediately prior to the application of the outer jacket, when it is dried at about 135°C.

The greatest quantity of paper cable is pulp insulated. This technique, developed by the Western Electric Company, combines papermaking with the insulation process in a continuous operation simultaneously on as many as 60 wires. Either a modified cylinder type or a Foudrinier paper machine may be used in the following manner:

1. 100% wood pulp is prepared by beating as in normal paper manufacturing operations and fed to the machine in dilute form.
2. Bare conductors are fed through the machine so as to become imbedded in the middle of narrow strips of wet pulp.
3. Excess moisture is squeezed from the paper by press rolls.
4. High-speed polishers rotate around each wire to form a very wet (70% water) cylinder of paper.
5. The wire then travels through a furnace to eliminate most of the remaining water, generating a continuous cylinder of insulation.

As mentioned above, paper insulation, be it ribbon or pulp, requires a significant moisture content in order to survive the rigors of the manufacturing operations. However, optimum electrical properties of low capacitance, low conductance, and high insulation resistance demand extremely dry material. Hence, paper cable cores are vacuum dried prior to the application of the moistureproof outer sheath.

Polyethylene, poly(vinyl chloride), and polypropylene insulations are all applied by extrusion. Minimum facilities required for this process include a wire supply stand, a preheater, an extruder, a cooling trough, and packaging equipment. Many more wiremaking functions are usually integrated with extrusion in modern plants. Some very high production lines now include facilities for wire drawing to final size, annealing, preheating, extruding, cooling, "spark testing" to check insulation integrity, and spooling or barrel packaging. In addition, some facilities are set up to electroplate a tin coating on the wire prior to extrusion, and additional identification in the form of colored stripes, dot-dash coding, or printed legends may be applied in tandem with the operations outlined above. Speeds vary considerably and are greatly dependent upon wire size, extruder size, and type of plastic insulation. Speeds in excess of 4000 ft/min have been achieved in regular production operations.

The most widely used plastic for communication cable insulation is low-density, high-molecular-weight polyethylene. Some compounds have a slight amount of polypropylene added to improve processing characteristics. High-density polyethylene is a harder and stronger material than low-density polyethylene and has a slightly higher dielectric constant. Balancing the advantage of this additional toughness are two slightly detrimental characteristics: (*1*) The insulation resistance decays with time more rapidly than in the case with low-density polyethylene. (*2*) The higher crystalline structure makes high density polyethylene more susceptible to the development of internal stresses caused by poor handling procedures (stretching) during subsequent cablemaking operations, or by unfavorable conditions during extrusion.

Polypropylene has recently been used in major quantities as an additional insulating material. In comparison with either high-density or low-density polyethylene, it displays: a lower dielectric constant, higher tensile strength and tensile yield, higher crystallinity (harder), lower density, the higher softening point, and a higher melt index (a measure of processability). Offsetting these obviously desirable properties are two definite liabilities: a rather high brittleness temperature, of about −18°C, below which the insulation will crack when bent (it is sometimes necessary to splice cables in severe winter weather); and a tendency toward rapid degradation because of a reaction between copper and polypropylene. The latter can be avoided by the addition of stabilizers.

The flame resistance of poly(vinyl chloride) renders this polymer highly desirable for cable applications inside telephone exchanges and other buildings. These products, known as switchboard cable and inside wiring cable, also require positive identification of each separate conductor. The ease with which PVC may be colored by incorporating color concentrates in the compound and by its ready acceptance of surface inks adds to this desirability.

Following insulation, most communication cable conductors are twisted into pairs or quads and stranded into a cable core.

Core Wrap. Core wrap, which is applied either longitudinally and held in place by a filament binding, or by helical wrapping, may serve one or more of three functions: (*1*) To provide mechanical and thermal protection of the core during subsequent manufacturing operations. (*2*) To provide dielectric protection between the electrostatic shield and core. (*3*) To serve as a mechanical binder to hold the core together. Depending on the configuration of the cable core, and the number and severity of subsequent sheathing operations, the core wrap may consist of one of the following: 0.005 in. or 0.007 in. thick polypropylene, 0.005 in., 0.007 in., or 0.010 in., kraft paper tape, 0.003 in. thick polyester film, 0.001 in. polyester film laminated to 0.015 in. rubber, or 0.00142 in. polyester film corrugated to a depth of approximately 0.009 in.

A recent addition to the growing list of core wrap materials is a nonwoven spin-bonded polyester fabric marketed by Du Pont under the trademark "Reemay." The major advantage of this material is in its superior heat resistance. In current practice the thickness varies between 0.010 in. and 0.017 in., depending upon the diameter of the cable core (see NONWOVEN FABRICS).

Shield and Outer Sheath. The shield and outer sheath of communication cables are so far integrated that it is appropriate to discuss them together. Plastic–metal sheath combinations have been used in the United States communication cables since 1947. Depending upon the ultimate application of the cable, as well as the constitution of the inner core, a wide variety of sheath constructions has evolved. These include the following:

Alpeth	*A*luminum–*P*oly*eth*ylene
Stalpeth	*St*eel–*A*luminum–*P*oly*eth*ylene
PAP	*P*olyethylene–*A*luminum–*P*olyethylene
PASP	*P*olyethylene–*A*luminum–*S*teel–*P*olyethylene
ARPAP	*A*luminum–*R*esin–*PAP*
ARPASP	*A*luminum–*R*esin–*PASP*

These acronyms represent a continuing series of efforts to overcome the well-known moisture–vapor permeability of polyethylene as well as to provide additional mechanical and electrical protection. The aluminum, in all of these cases, is 0.008 in. thick and in larger sizes, is transversely corrugated for greater flexibility. Its function is to provide protection against lightning and external electrical "noise." The steel is tin- or solder-coated, about 0.007 in. thick, transversely corrugated, and soldered longitudinally to form a hermetic seal. The resulting steel tube is flooded with an asphaltic corrosion barrier immediately prior to the final extrusion of polyethylene.

Two different specifications have been developed for polyethylene sheathing: as inner jacketing material and for the outer jacket. Thus, the sheath construction for PASP cable sheath, for example, consists of a layer of extruded inner-jacket polyethylene, aluminum, and soldered steel; the assembly is finally covered with an extruded outer jacket of polyethylene compound.

Reliability is a universal requirement for communications equipment. Thus, wire and cable built for outdoor use must serve reliably for at least 20 years. Because continued exposure to many of the common environments of our planet—sunlight, air, oxygen, water, cold, heat, and microorganisms—can cause polymers to degrade and eventually fail, the polyethylene jacket must be specially compounded to survive these hazards for many years (see also DEGRADATION; STABILIZATION).

The most important factors demanding protection are: (a) oxidation from weather exposure, (b) thermal degradation, and (c) environmental stress cracking. As is well known, polyethylene failure in an outdoor environment is primarily due to the absorption of ultraviolet radiation from the sun, and subsequent oxidation. The breakdown mechanism involves the development of carbonyl groups, the formation of which is accelerated by ultraviolet radiation. In the absence of light and at normal ambient temperatures, the reaction between polyethylene and oxygen (from the air) proceeds slowly or not at all. Therefore, finely divided, well-dispersed carbon black is used to minimize the ultraviolet absorption (see ETHYLENE POLYMERS; WEATHERING) (5).

High temperatures, sometimes encountered in some applications of communications cable as well as during processing, can lead to premature failure from oxidation. Hindered thiocresols are the usual antioxidants added to prevent this (see also ANTIOXIDANTS).

Of very great interest to the communications industry is the phenomenon of *environmental stress cracking* (see FRACTURE). When some plastics are subjected to stress in certain environments, a failure can occur. Without this environment, the material often displays no evidence of failure, regardless of the length of time the stress is applied. Such common materials as soaps, wetting agents, detergents, grease pencils, and certain alcohols can cause rapid cracking and catastrophic failure. Preventive measures, in addition to the obvious elimination of the agents, include control of the molecular weight (the lower the molecular weight, the more easily stress cracking may occur) and avoidance of stress in the critical region of the yield point. It has been determined that a material stressed below or much above its yield point in the presence of known stress-cracking agents will not fail; only when the stress is near the yield point will cracking appear.

Low-density polyethylene is used as the inner jacket of PAP and PASP cables, and high-density polyethylene is used on ARPAP and ARPASP. In these applications, the material normally does not contain carbon black, since, of course, it will not be exposed to sunlight. The first layer of aluminum in the ARPAP and ARPASP sheaths is coated on both sides with an ethylene–acrylic acid copolymer adhesive. This resin is the "R" in ARPAP and ARPASP. When this material is wrapped longitudinally and overlapped around the core, the resin heat seals the aluminum when high-density polyethylene is extruded over it at about 500°F. The moisture ingression rate through the resulting seam is reduced by a factor of about 1000 over that experienced through a normal outer layer of polyethylene.

Power Cables

Because of the enormous variety of conditions under which power cables must operate, a correspondingly large variety of materials are employed to meet performance criteria. Rubber compounds have been constantly upgraded since World War II accelerated the substitution of *styrene–butadiene rubber* for the natural rubber previously used. *Butyl rubber* demonstrates excellent resistance to ozone attack, good heat and moisture resistance, as well as high insulation resistance and good dielectric strength (see BUTYLENE POLYMERS). *Neoprene* performs exceptionally well as a jacketing material. It is ozone resistant, extremely resistant to corona effects, has fine mechanical strength and excellent resistance to oil, chemicals, and flame (see also CHLOROBUTADIENE POLYMERS). *Chlorosulfonated polyethylene* possesses a sufficiently good set of properties to be used as an integral insulation and jacket for lower voltage cables. Like neoprene, its power factor and dielectric constant are too high for voltages in excess of 2000 (see also ETHYLENE POLYMERS).

Silicone rubber is used as an insulation in some applications requiring exposure to very high ambient temperatures. It is the only power cable insulation presently rated at 125°C. It is also highly resistant to degradation from radiation, making it desirable for some uses in the nuclear power industry. Another useful feature of silicone rubber is the fact that when it burns, as it does with difficulty, the remaining ash is a nonconductive silica. Thus, it is very useful in some military applications where circuits must remain operative even after fire damage. Because of its comparatively fragile nature when compounded for electrical use, silicone rubber must be protected from abuse. (See also SILICONES.)

Chemically crosslinked polyethylene (the usual activating agent is an organic peroxide) has gained much favor as an insulating material for power cables. It has better electrical properties than any other thermosetting compound. This, coupled with superior mechanical strength, permits thinner insulation walls than were possible with the older rubber compounds at equivalent voltage ratings (see also CROSSLINKING). Among the thermoplastics utilized in power cable, *poly(vinyl chloride)* is the leading material. Its use in the USA has been confined to low-voltage applications (up to 600 V), although in some instances in Europe, poly(vinyl chloride) has been used to insulate cables up to 12,000 volts. Poly(vinyl chloride)'s low cost, its fair degree of mechanical strength, as well as its flame and chemical resistance are such that most building wire and 600-volt circuit wiring utilizes this material almost exclusively. It is also frequently used as a jacket over metallic sheaths to provide corrosion resistance (7).

Impregnated-Paper Power Cables. The insulating material in modern high voltage power cables is exclusively an oil-impregnated paper. The mechanical properties of paper enable it to withstand the rigors of being wrapped onto conductors and cabled in subsequent manufacturing operations. Furthermore, its good electrical properties are considerably enhanced by a thorough impregnation of high-quality insulating oil. The popularity of this combination of oil and paper may be attributed to the following properties: (*1*) The excellent chemical and physical stability of cellulose. (In the absence of oxygen, it is highly resistant to degradation at temperatures of 100°C and higher.) (*2*) Excellent resistance to deterioration from ionic discharge (corona). (*3*) Exceptional oil absorption capacity of the cellulose fibers, which produces a virtually homogeneous dielectric. (*4*) The ease with which the physical properties of paper can be altered to the desired dielectric specifications.

The paper-insulated conductors are placed in a metallic sheath (usually lead or steel pipe) and subsequently filled with a highly refined polybutene oil. The entire assembly is usually protected from corrosion by an outer jacket of poly(vinyl chloride), neoprene, or polyethylene (8).

Bibliography

1. *United States of America Standards Institute: Definitions of Electrical Terms*, ASA C42.35, Group 35, Section 80, the Institute of Electrical and Electronic Engineers, New York, 1957.
2. *Insulation* **15** [8], 322–372 (June/July, 1969).
3. A. H. Lybeck, *Insulation* **12** [5], 60 (May, 1966).
4. *National Electrical Code*, Table 310–2(b), 1965.
5. V. T. Wallder, *Bell Laboratories Record* **46** [5], 150 (May, 1968).
6. J. M. Peacock and R. M. Riley (to Bell Telephone Laboratories), U. S. Pat. 3,534,149 (1969).
7. E. G. Driscoll, *Insulation* **12** [7], 41 (June 1966).
8. G. N. Everest, *Insulation* **12** [9], 49 (Aug. 1966).

Lloyd W. Myers
Western Electric Company